S0-AKY-174

The chemistry of the **quinonoid compounds** Volume 2

Part 2

Edited by

SAUL PATAI

ZVI RAPPOPORT

The Hebrew University, Jerusalem

1988

JOHN WILEY & SONS

CHICHESTER – NEW YORK – BRISBANE – TORONTO – SINGAPORE

An Interscience ® Publication

Library of Congress Cataloging-in-Publication Data:

The Chemistry of the quinonoid compounds.

(Chemistry of functional groups)
'An Interscience publication.'
1. Quinone I. Patai, Saul
II. Rappoport, Zvi
III. Series
QD341.Q4C47 1987 547′.633 86-32494

ISBN 0 471 91285 9 (Part 1)
ISBN 0 471 91914 4 (Part 2)
ISBN 0 471 91916 0 (Set)

British Library Cataloguing in Publication Data:

The chemistry of the quinonoid compounds.—
(The chemistry of functional groups)
Vol. 2
I. Patai, Saul
II. Rappoport, Zvi
547′.636 QD341.Q4

ISBN 0 471 91285 9 (Part 1)
ISBN 0 471 91914 4 (Part 2)
ISBN 0 471 91916 0 (Set)

Typeset by Macmillan India Ltd, Bangalore 25
Printed and bound in Great Britain by Bath Press Ltd, Bath, Avon.

The chemistry of the

quinonoid compounds

Volume 2

Part 2

THE CHEMISTRY OF FUNCTIONAL GROUPS

A series of advanced treatises under the general editorship of Professor Saul Patai

The chemistry of alkenes (2 volumes)
The chemistry of the carbonyl group (2 volumes)
The chemistry of the ether linkage
The chemistry of the amino group
The chemistry of the nitro and nitroso groups (2 parts)
The chemistry of carboxylic acids and esters
The chemistry of the carbon–nitrogen double bond
The chemistry of amides
The chemistry of the cyano group
The chemistry of the hydroxyl group (2 parts)
The chemistry of the azido group
The chemistry of the acyl halides
The chemistry of the carbon–halogen bond (2 parts)
The chemistry of the quinonoid compounds (2 parts)
The chemistry of the thiol group (2 parts)
The chemistry of the hydrazo, azo and azoxy groups (2 parts)
The chemistry of amidines and imidates
The chemistry of cyanates and their thio derivatives (2 parts)
The chemistry of diazonium and diazo groups (2 parts)
The chemistry of the carbon–carbon triple bond (2 parts)
The chemistry of ketenes, allenes and related compounds (2 parts)
The chemistry of the sulphonium group (2 parts)
Supplement A: The chemistry of double-bonded functional groups (2 parts)
Supplement B: The chemistry of acid derivatives (2 parts)
Supplement C: The chemistry of triple-bonded functional groups (2 parts)
Supplement D: The chemistry of halides, pseudo-halides and azides (2 parts)
Supplement E: The chemistry of ethers, crown ethers, hydroxyl groups and their sulphur analogues (2 parts)
Supplement F: The chemistry of amino, nitroso and nitro compounds and their derivatives (2 parts)
The chemistry of the metal–carbon bond (4 volumes)
The chemistry of peroxides
The chemistry of organic selenium and tellurium compounds (2 volumes)
The chemistry of the cyclopropyl group (2 parts)

Contributing Authors

Hans-Dieter Becker	Department of Organic Chemistry, Chalmers University of Technology and University of Gothenburg, S-412 96 Gothenburg, Sweden
S. Berger	Fachbereich Chemie der Universität Marburg, D-355 Marburg, Germany
Jerome A. Berson	Department of Chemistry, Yale University, P.O. Box 6666, New Haven, Connecticut 06511-8118, USA
Peter Boldt	Institut für Organische Chemie der Technischen Universität Braunschweig, D-3300 Braunschweig, FRG
Eric R. Brown	Color Negative Technology Division, Photographic Research Laboratories, Eastman Kodak Company, 1669 Lake Avenue, Rochester, New York 14650, USA
James Q. Chambers	Department of Chemistry, University of Tennessee, Knoxville, TN 37996, USA
Tze-Lock Chan	Department of Chemistry, The Chinese University of Hong Kong, Shatin, New Territories, Hong Kong
M. Catherine Depew	Department of Chemistry, Queen's University, Kingston, Ontario, Canada K7L 3N6
K. Thomas Finley	Department of Chemistry, State University College, Brockport, New York 14420, USA
Colin W. W. Fishwick	Organic Chemistry Department, The University, Leeds LS2 9JT, UK
Karl-Dietrick Gundermann	Institut für Organische Chemie, Technische Universität Clausthal, 3392 Clausthal-Zellerfeld, Leibnizstrasse 6, FRG
P. Hertl	Institut für Organische Chemie der Universität Tübingen, D-7400 Tübingen, Germany
Hiroyuki Inouye	Faculty of Pharmaceutical Sciences, Kyoto University, Sakyo-ku, Kyoto, Japan
Shouji Iwatsuki	Department of Chemical Research for Resources, Faculty of Engineering, Mie University, Tsu, Japan
David W. Jones	Organic Chemistry Department, The University, Leeds LS2 9JT, UK
Marianna Kańska	Department of Chemistry, University of Warsaw, Warsaw, Poland
L. Klasinc	Department of Chemistry, Louisiana State University, Baton Rouge, Louisiana 70803, USA
Eckhard Leistner	Institut für Pharmazeutische Biologie, Rheinische Friedrich-Wilhelms-Universität Bonn, Bonn, FRG

Dieter Lieske	Institut für Organische Chemie, Technische Universität Clausthal, 3392 Clausthal-Zellerfeld, Leibnizstrasse 6, FRG
Tien-Yau Luh	Department of Chemistry, The Chinese University of Hong Kong, Shatin, New Territories, Hong Kong
Kazuhiro Maruyama	Department of Chemistry, Faculty of Science, Kyoto University, Kyoto 606, Japan
S. P. McGlynn	Department of Chemistry, Louisiana State University, Baton Rouge, Louisiana 70803, USA
Richard W. Middleton	The Cancer Research Campaign's Gray Laboratory, Northwood, Middlesex, UK and Brunel University, Uxbridge, Middlesex UB8 3PH, UK
Roland Muller	Institut für Organische Chemie, Universität Tübingen, Auf der Morgenstelle, D-7400 Tübingen, FRG
K. A. Muszkat	Department of Structural Chemistry, The Weizmann Institute of Science, Rehovot, Israel
Yoshinori Naruta	Department of Chemistry, Faculty of Science, Kyoto University, Kyoto 606, Japan
P. Neta	Center for Chemical Physics, National Bureau of Standards, Gaithersburg, Maryland 20899, USA
Atsuhiro Osuka	Department of Chemistry, Faculty of Science, Kyoto University, Kitashirakawa Oiwakecho, Kyoto 606, Japan
John Parrick	Brunel University, Uxbridge, Middlesex UB8 3PH, UK
A. Rieker	Institut für Organische Chemie der Universität Tübingen, D-7400 Tübingen, Germany
John R. Scheffer	University of British Columbia, Vancouver, Canada, V6T 1Y6
Lawrence T. Scott	Department of Chemistry and Center for Advanced Study, College of Arts and Science, University of Nevada-Reno, Reno, Nevada 89557, USA
Anne Skancke	Department of Chemistry, University of Tromsø, P.O.B. 953, N-9001 Tromsø, Norway
Per N. Skancke	Department of Chemistry, University of Tromsø, P.O.B. 953, N-9001 Tromsø, Norway
Howard E. Smith	Department of Chemistry, Vanderbilt University, Nashville, Tennessee 37235, USA
John S. Swenton	Department of Chemistry, The Ohio State University, Columbus, OH 43210, USA
James Trotter	University of British Columbia, Vancouver, Canada, V6T 1Y6
Alan B. Turner	Department of Organic Chemistry, Chalmers University of Technology and University of Gothenburg, S-412 96 Gothenburg, Sweden
Jeffrey K. S. Wan	Department of Chemistry, Queen's University, Kingston, Ontario, Canada K7L 3N6
Henry N. C. Wong	Department of Chemistry, The Chinese University of Hong Kong, Shatin, New Territories, Hong Kong
Klaus-Peter Zeller	Institut für Organische Chemie, Universität Tübingen, Auf der Morgenstelle, D-7400 Tübingen, FRG
Mieczysław Zieliňski	Isotope Laboratory, Faculty of Chemistry, Jagiellonian University, Cracow, Poland

Foreword

The first volume on quinones in 'The Chemistry of Functional Groups' appeared (in two parts) in 1974. In Supplement A (1977) there was no new material on quinones. However, in the decade which has passed since, much new information has accumulated, on quite new subjects as well as regarding rapid and significant developments on subjects which were already included in the main quinone volume.

Hence we decided that it would be timely to publish a second volume on quinones and indeed this has turned out to be a weighty tome, even though we attempted to avoid duplication as far as possible between the two volumes.

Several subjects were intended to be covered but the invited chapters did not materialize. These were updates on quinone methides, on complexes and on rearrangements of quinones as well as a chapter on quinonoid semiconductors and organic metals.

Literature coverage in most chapters is up to 1986.

SAUL PATAI
ZVI RAPPOPORT

Jerusalem
August 1987

The Chemistry of Functional Groups
Preface to the Series

The series 'The Chemistry of Functional Groups' is planned to cover in each volume all aspects of the chemistry of one of the important functional groups in organic chemistry. The emphasis is laid on the functional group trated and on the effects which it exerts on the chemical and physical properties, primarily in the immediate vicinity of the group in question, and secondarily on the behaviour of the whole molecule. For instance, the volume *The Chemistry of the Ether Linkage* deals with reactions in which the C—O—C group is involved, as well as with the effects of the C—O—C group on the reactions of alkyl or aryl groups connected to the ether oxygen. It is the purpose of the volume to give a complete coverage of all properties and reactions of ethers in as far as these depend on the presence of the ether group but the primary subject matter is not the whole molecule, but the C—O—C functional group.

A further restriction in the treatment of the various functional groups in these volumes is that material included in easily and generally available secondary or tertiary sources, such as Chemical Reviews, Quarterly Reviews, Organic Reactions, various 'Advances' and 'Progress' series as well as textbooks (i.e. in books which are usually found in the chemical libraries of universities and research institutes) should not, as a rule, be repeated in detail, unless it is necessary for the balanced treatment of the subject. Therefore each of the authors is asked *not* to give an encyclopaedic coverage of his subject, but to concentrate on the most important recent developments and mainly on material that has not been adequately covered by reviews or other secondary sources by the time of writing of the chapter, and to address himself to a reader who is assumed to be at a fairly advanced post-graduate level.

With these restrictions, it is realized that no plan can be devised for a volume that would give a *complete* coverage of the subject with *no* overlap between chapters, while at the same time preserving the readability of the text. The Editor set himself the goal of attaining *reasonable* coverage with *moderate* overlap, with a minimum of cross-references between the chapters of each volume. In this manner, sufficient freedom is given to each author to produce readable quasi-monographic chapters.

The general plan of each volume includes the following main sections:

(a) An introductory chapter dealing with the general and theoretical aspects of the group.

(b) One or more chapters dealing with the formation of the functional group in question, either from groups present in the molecule, or by introducing the new group directly or indirectly.

(c) Chapters describing the characterization and characteristics of the functional groups, i.e. a chapter dealing with qualitative and quantitative methods of determination including chemical and physical methods, ultraviolet, infrared, nuclear magnetic resonance and mass spectra: a chapter dealing with activating and directive effects exerted by the group and/or a chapter on the basicity, acidity or complex-forming ability of the group if applicable).

(d) Chapters on the reactions, transformations and rearrangements which the functional group can undergo, either alone or in conjunction with other reagents.

(e) Special topics which do not fit any of the above sections, such as photochemistry, radiation chemistry, biochemical formations and reactions. Depending on the nature of each functional group treated, these special topics may include short monographs on related functional groups on which no separate volume is planned (e.g. a chapter on 'Thioketones' is included in the volume *The Chemistry of the Carbonyl Group*, and a chapter on 'Ketenes' is included in the volume *The Chemistry of Alkenes*). In other cases certain compounds, though containing only the functional group of the title, may have special features so as to be best treated in a separate chapter, as e.g. 'Polyethers' in *The Chemistry of the Ether Linkage*, or 'Tetraaminoethylenes' in *The Chemistry of the Amino Group*.

This plan entails that the breadth, depth and thought-provoking nature of each chapter will differ with the views and inclinations of the author and the presentation will necessarily be somewhat uneven. Moreover, a serious problem is caused by authors who deliver their manuscript late or not at all. In order to overcome this problem at least to some extent, it was decided to publish certain volumes in several parts, without giving consideration to the originally planned logical order of the chapters. If after the appearance of the originally planned parts of a volume it is found that either owing to non-delivery of chapters, or to new developments in the subject, sufficient material has accumulated for publication of a supplementary volume, containing material on related functional groups, this will be done as soon as possible.

The overall plan of the volumes in the series 'The Chemistry of Functional Groups' includes the titles listed below:

The Chemistry of Alkenes (*two volumes*)
The Chemistry of the Carbonyl Group (*two volumes*)
The Chemistry of the Ether Linkage
The Chemistry of the Amino Group
The Chemistry of the Nitro and Nitroso Groups (*two parts*)
The Chemistry of Carboxylic Acids and Esters
The Chemistry of the Carbon–Nitrogen Double Bond
The Chemistry of the Cyano Group
The Chemistry of Amides
The Chemistry of the Hydroxyl Group (*two parts*)
The Chemistry of the Azido Group
The Chemistry of Acyl Halides
The Chemistry of the Carbon–Halogen Bond (*two parts*)
The Chemistry of the Quinonoid Compounds (*two parts*)
The Chemistry of the Thiol Group (*two parts*)
The Chemistry of Amidines and Imidates
The Chemistry of the Hydrazo, Azo and Azoxy Groups (*two parts*)
The Chemistry of Cyanates and their Thio Derivatives (*two parts*)
The Chemistry of Diazonium and Diazo Groups (*two parts*)

The Chemistry of the Carbon–Carbon Triple Bond (*two parts*)
Supplement A: The Chemistry of Double-bonded Functional Groups (*two parts*)
The Chemistry of Ketenes, Allenes and Related Compounds (*two parts*)
Supplement B: The Chemistry of Acid Derivatives (*two parts*)
Supplement C: The Chemistry of Triple-Bonded Functional Groups (*two parts*)
Supplement D: The Chemistry of Halides, Pseudo-halides and Azides (*two parts*)
Supplement E: The Chemistry of Ethers, Crown Ethers, Hydroxyl Groups and their Sulphur Analogues (*two parts*)
The Chemistry of the Sulphonium Group (*two parts*)
Supplement F: The Chemistry of Amino, Nitroso and Nitro Groups and their Derivatives (*two parts*)
The Chemistry of the Metal–Carbon Bond (*four volumes*)
The Chemistry of Peroxides
The Chemistry of Organic Se and Te Compounds Vol. 1
The Chemistry of the Cyclopropyl Group (*two parts*)
The Chemistry of Organic Se and Te Compounds Vol. 2

Titles in press

The Chemistry of Sulphones and Sulphoxides
The Chemistry of Organosilicon Compounds
The Chemistry of Enones
Supplement A2: The Chemistry of the Double-Bonded Functional Groups, Volume 2.

Titles in Preparation

The Chemistry of Enols
The Chemistry of Sulphinic Acids, Esters and Derivatives
The Chemistry of Sulphenic Acids and Esters.

Advice or criticism regarding the plan and execution of this series will we welcomed by the Editor.

The publication of this series would never have started, let alone continued, without the support of many persons. First and foremost among these is Dr Arnold Weissberger, whose reassurance and trust encouraged me to tackle this task. The efficient and patient cooperation of several staff-members of the Publisher also rendered me invaluable aid (but unfortunately their code of ethics does not allow me to thank them by name). Many of my friends and colleagues in Israel and overseas helped me in the solution of various major and minor matters, and my thanks are due to all of them, especially to Professor Z. Rappoport. Carrying out such a long-range project would be quite impossible without the non-professional but none the less essential participation and partnership of my wife.

The Hebrew University
Jerusalem, ISRAEL

SAUL PATAI

Contents

The Chemistry of Quinonoid Compounds, Vol. II
Edited by S. Patai and Z. Rappoport
Published by John Wiley & Sons Ltd

CHAPTER **14**

Radiation chemistry of quinonoid compounds

P. NETA
Center for Chemical Physics, National Bureau of Standards, Gaithersburg, Maryland 20899, USA

I. INTRODUCTION

Quinonoid compounds have played a major role in radiation chemistry because of their importance in biological systems and industrial applications and in electron transfer reactions in general. In turn, the techniques of radiation chemistry have proven to be extremely useful for the study of free radical and one-electron transfer reactions of quinones as well as their excited states. Since the publication of the original chapter[1] in this series a large body of data has accumulated on this topic. Various aspects of this information have been partly reviewed[2-6] and recent compilations have summarized the rate constants for reactions of quinonoid compounds with hydrated electrons[7,8], hydrogen atoms[9], hydroxyl radicals[10], perhydroxyl and superoxide[11], organic radicals[12,13], as well as the one-electron redox potentials[14,15].

The majority of the radiation chemical studies on quinones were carried out with solutions at room temperature, in particular aqueous solutions. The effect of radiation on water and other solvents and the manipulation of the primary radicals toward a desired particular reaction have been discussed in detail (see, e.g. Refs 1, 3, 5 and references therein). In this chapter we concentrate on the kinetics and mechanisms of the radiolytic reactions of quinones and on the properties of the intermediate species.

II. REACTIONS OF QUINONES WITH RADICALS

A. Reactions with Solvated Electrons and Reducing Radicals

Quinones and related compounds react with solvated electrons with diffusion-controlled rate constants ($k \sim 10^{10}\ \mathrm{M}^{-1}\mathrm{s}^{-1}$)[7,8] to form the semiquinone radical anions. They react with hydrogen atoms nearly as rapidly ($k \sim 10^{9}\ \mathrm{M}^{-1}\mathrm{s}^{-1}$)[9] to form the same radical, unless the molecule contains an additional site where H atoms may react partially to form a different radical.

$$\text{(1,4-benzoquinone)} + e_s^- \longrightarrow \text{(semiquinone radical anion, } {}^-\mathrm{O}\text{–}C_6H_4\text{–}\dot{\mathrm{O}}) \qquad (1)$$

Quinones are very readily reduced by a wide variety of organic radicals. Typical reducing radicals are α-hydroxy- or α-amino-substituted alkyls, such as those formed from methanol ($\dot{C}H_2OH$), 2-propanol ($(Me)_2\dot{C}OH$), formate ($\dot{C}O_2^-$), or glycine ($^-O_2C\dot{C}HNH_2$), which transfer an electron to quinones very rapidly ($k \sim 10^{9}\ \mathrm{M}^{-1}\mathrm{s}^{-1}$)[12,13], e.g.

$$\text{(1,4-benzoquinone)} + R_2\dot{C}OH \longrightarrow \text{(semiquinone radical anion)} + R_2CO + H^+ \qquad (2)$$

Many other aliphatic and aromatic radicals reduce quinones by a similar mechanism but with lower rate constants. Consequently, quinones have been used to probe the reducing nature of organic radicals and other species as well. These reactions will be discussed in Section V.

B. Reactions with Hydroxyl, Alkyl and Phenyl

Quinones also react with certain radicals via addition to the double bond. Such was suggested to be the mechanism of their reaction with OH radicals, e.g. reaction 3 for benzoquinone with $k_3 = 1.2 \times 10^9\,\text{M}^{-1}\text{s}^{-1}$ [16]. Similar reactions and rate constants have been reported for various substituted anthraquinones[17–19]. These reactions, followed by disproportionation or oxidation of the radical adducts, result in the radiolytic hydroxylation of the quinones.

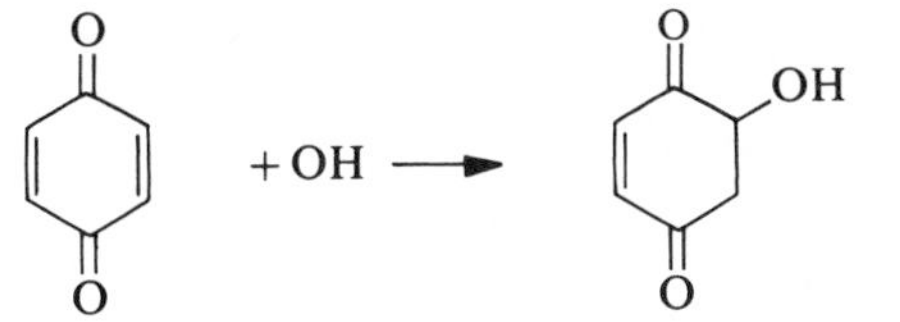

(3)

An addition mechanism was demonstrated also for the reaction of quinones with methyl and phenyl radicals[20], e.g.

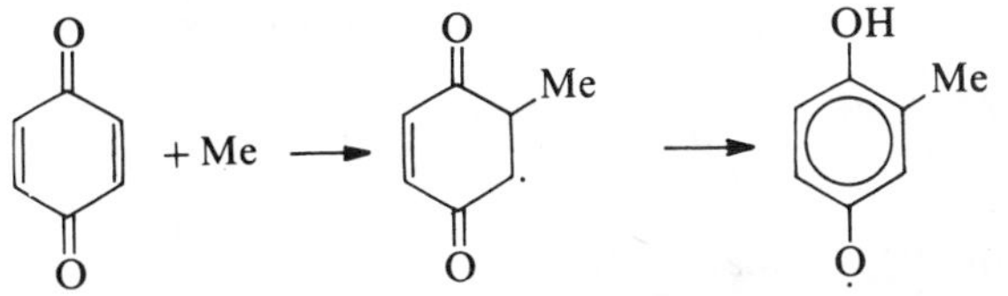

(4)

This reaction, with $k = 4.5 \times 10^7\ \text{M}^{-1}\text{s}^{-1}$ for Me and $1.2 \times 10^9\ \text{M}^{-1}\text{s}^{-1}$ for Ph, was found to be followed by a rapid electron transfer from the adduct radicals to benzoquinone. This results in the quantitative formation of methyl- or phenylbenzoquinone from each $\text{Me}^{\cdot}$ or $\text{Ph}^{\cdot}$ produced in the solution.

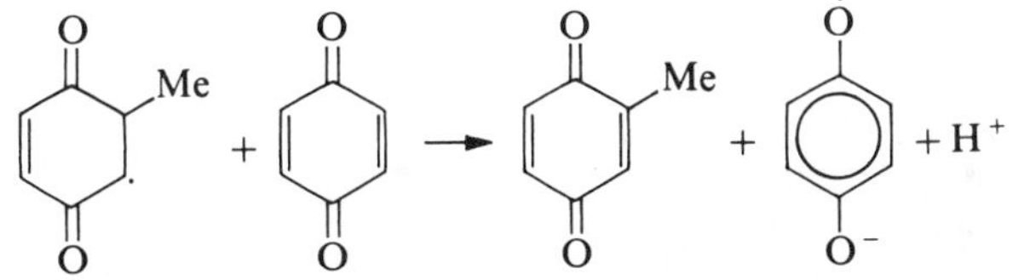

(5)

III. FORMATION OF SEMIQUINONES

Semiquinone radicals are produced by one-electron reduction of quinones as described in Section II. They are also produced by one-electron oxidation of hydroquinones with various radicals and by the reaction of OH radicals with hydroquinones and with certain substituted phenols.

A. One-electron Oxidation of Hydroquinones

Hydroquinones react with a variety of organic and inorganic one-electron oxidizing radicals to yield the semiquinones. The typical oxidants are $X_2^{-\cdot}$ (X = Cl, Br, I, SCN), $N_3^{\cdot}$ and $\dot{C}H_2CHO$. Rate constants for oxidation by $X_2^{-\cdot}$ are moderate in neutral solutions and become very rapid in alkaline solution due to the acid dissociation of the hydroquinone[21].

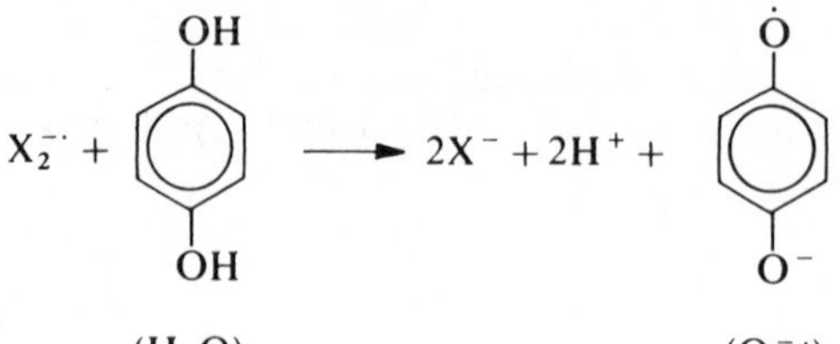

$$X_2^{-\cdot} + H_2Q \longrightarrow 2X^- + 2H^+ + Q^{-\cdot} \quad (6)$$

(H₂Q) (Q⁻˙)

$$X_2^{-\cdot} + HQ^- \rightarrow 2X^- + H^+ + Q^{-\cdot} \quad (7)$$

$$X_2^{-\cdot} + Q^{2-} \rightarrow 2X^- + Q^{-\cdot} \quad (8)$$

The $N_3^{\cdot}$ radical is a more powerful oxidant which reacts with hydroquinones very rapidly ($k \sim 10^9$ $M^{-1}s^{-1}$) even in neutral solutions[22,23]. The $\dot{C}H_2CHO$ radical is a weaker oxidant which is useful for his purpose mainly in alkaline solutions[24,25].

$$\dot{C}H_2CHO + H^+ + Q^{2-} \rightarrow CH_3CHO + Q^{-\cdot} \quad (9)$$

Other radicals which were found to oxidize hydroquinones include $\dot{N}H_2$[26], $\dot{S}O_3^-$ and $SO_5^{-\cdot}$[27], alkoxyl[28] and phenoxyl[25].

B. Reaction of OH Radicals with Hydroquinones

Although the hydroxyl radical is a strong oxidant it tends to react with aromatic compounds by addition to the ring and this mechanism holds for hydroquinones as well. The adduct radicals undergo water elimination to form the semiquinone[16,29–32]. Reaction 11 is a stepwise acid- or base-catalyzed reaction, which in neutral solutions is catalyzed by buffers. The rate of elimination is strongly dependent on the structure of the aromatic compound.

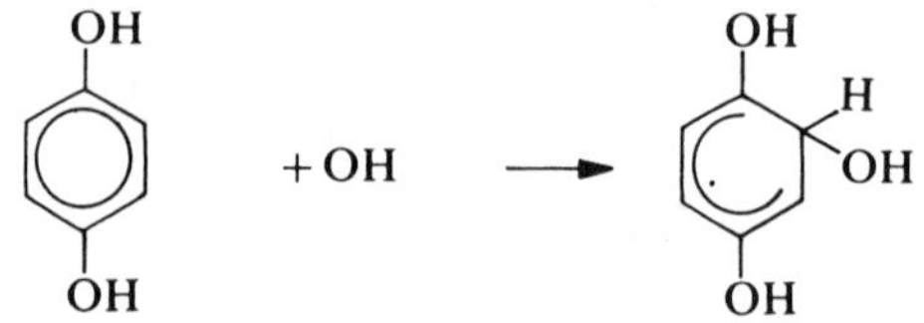

$$H_2Q + OH \longrightarrow \text{HO–C}_6\text{H}_4(\text{H})(\text{OH})^{\cdot} \quad (10)$$

$$\text{HO–C}_6\text{H}_4(\text{H})(\text{OH})^{\cdot} \xrightarrow[-H^+]{-H_2O} Q^{-\cdot} \quad (11)$$

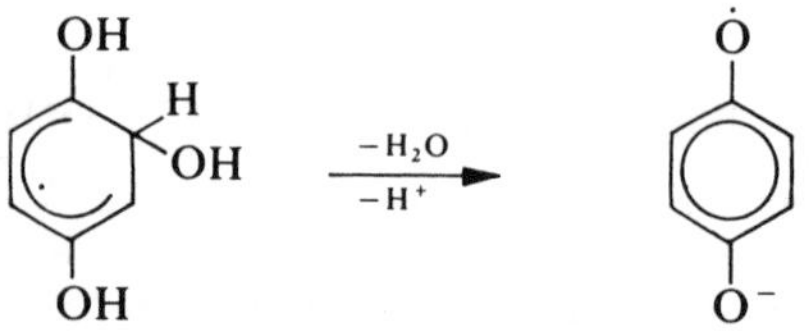

At very high pH, where the OH radical dissociates to $O^{-\cdot}$ ($pK_a = 11.8$), the reaction becomes a direct electron transfer, as suggested for phenoxide ions[33].

$$O^{-\cdot} + Q^{2-} + H^+ \rightarrow OH^- + Q^{-\cdot} \quad (12)$$

C. Reaction of OH Radicals with Other Substituted Phenols

Addition of OH radicals to substituted phenols takes place at various ring carbons, including those bearing the substituent (*ipso* position)

$$p\text{-}XC_6H_4OH + {}^{\bullet}OH \longrightarrow X(OH)C_6H_4OH^{\bullet} + \text{isomers} \qquad (13)$$

Ipso adducts with certain substituents (X = halogen, NH_2, OMe, NO_2) undergo very rapid elimination of HX to form semiquinones[29, 34–37].

$$X(OH)C_6H_4OH^{\bullet} \longrightarrow HX + H^+ + {}^{\bullet}OC_6H_4O^- \qquad (14)$$

IV. PROPERTIES OF SEMIQUINONES

A. Optical Absorption Spectra

Pulse radiolytic studies on the reduction of quinones or the oxidation of hydroquinones permitted the determination of the optical absorption spectra of the semiquinone radicals. These spectra were used, in turn, to characterize the semiquinones, their rates of formation and reactions, their acid–base equilibria and redox potentials. It is important, therefore, to summarize the absorption spectra. The main parameters are given in Table 1.

TABLE 1. Optical absorption spectra of semiquinones

	λ_{max}	ε_{max}	Reference
p-Benzosemiquinones			
Unsubstituted	430	6100	16, 38, 39
	404	~ 5000	
	(316)	(40 000)[a]	
2-Methyl-	430	6200	38
	405	4500	
2,3-Dimethyl-	430	6700	38
	415	5100	
2,5-Dimethyl-	435	6800	38, 39
	415	5000	
2,6-Dimethyl-	430	6100	38
	405	4900	
2,3,5-Trimethyl-	435	6700	38
	410	4300	
Tetramethyl-	440	7600	38, 39
	420	4700	
2-Carboxymethyl-	430	6000	25
	405	5200	
	310	11 000	

TABLE 1. (*Contd.*)

	λ_{max}	ε_{max}	Reference
2,3-Dimethoxy-5-methyl-6-R-(ubiquinone; R = isoprenoid side chain)	445	8600	38
	425	5300	
o-Benzosemiquinones			
Unsubstituted	300		16
4-*t*-Butyl-	313	12 200	31
	(350)	(2400)	
4-(2-Amino-2-carboxyethyl)- (from DOPA)	310	5800	25
4-(2-Amino-1-hydroxyethyl)- (from norepinephrine)	310	5400	25
m-Benzosemiquinones			
Unsubstituted	450	~ 2200	25
	~ 430	~ 2000	
1,4-Naphthosemiquinones			
Unsubstituted	390	12 500	38
	370	7100	
2-Methyl-	390	12 500	38, 39
	370	9500	
2,3-Dimethyl-	400	11 000	38
	380	7300	
2-Hydroxy-	390	6300	39
2-Methyl-3-phytyl-(vitamin K_1)	400	10 200	38
	380	9900	
1,2-Naphthosemiquinones			
Unsubstituted	265	40 000	39
9,10-Anthrasemiquinones			
Unsubstituted	480	7300	39
	395	7800	
1-Sulfonate	500	8000	17, 39
	400	8000	
2-Sulfonate	505	7600	17, 40
	405	8000	
1-Amino-	490	8000	17
	400	6500	
1,4-Diamino-	500	14 000	41
Miscellaneous semiquinones from[b]			
Gallic acid (ox)	337	3500	25
5-Hydroxydopamine (ox)	315	4800	25
Catechin (ox)	315	5800	25
6-Hydroxydopamine (ox)	440	3500	25
	420	2700	
	345	7600	

TABLE 1. (*Contd.*)

	λ_{max}	ε_{max}	Reference
6,7-Dihydroxycoumarin (ox)	530	2400	25
Ellagic acid (ox)	525	6400	25
Quercetin (ox)	520	17 000	25
7-Hydroxycoumarin (ox)	575	2700	25
	525	1750	
Quinalizarin (ox)	720	17 000	25
Naphthazarin (red)[c]	~ 700		42
	~ 400	~ 10 000	
Adriamycin (red)	660	1300	43
	480	12 400	
Mitomycin c (red)	510	5500	43
	360	15 600	43
Daunomycin (red)	700	7200	43
	500	10 500	
Lumazine (red)	~ 450	1000	44
	360	~ 7000	
Lumichrome (red)[c]	450	3900	45
	360	~ 12 000	
Riboflavin (red)	570	5000	46
	510	6000	
	340	11 000	
Flavin-adenine-dinucleotide (red)	~ 550		47
	340		

[a] Determined by non-radiolytic means by S. Fukuzumi, Y. Ono and T. Keii, *Bull. Chem. Soc. Japan*, **46**, 3353 (1973).
[b] Semiquinones produced from the following compounds by oxidation (ox) or by reduction (red).
[c] Spectral parameters given for neutral solutions. Similar values were found for 1-methyl, 3-methyl and dimethyl derivatives. Spectral changes are observed upon protonation of the radical in acid solutions or deprotonation in alkaline solutions (see Ref. 45).

It is seen from Table 1 that *p*-benzosemiquinones exhibit absorption maxima at ~ 430 nm with $\varepsilon \sim$ 6000–7000 $\text{M}^{-1}\,\text{cm}^{-1}$, a second peak at slightly lower wavelengths, and a third very intense peak around 310 nm. In contrast, *o*-benzosemiquinones absorb only at the lower wavelength region, ~ 310 nm. The *m*-benzosemiquinone radical, produced by one-electron oxidation of resorcinol, exhibits a relatively weak absorption at 450 nm ($\varepsilon \sim$ 2200 $\text{M}^{-1}\,\text{cm}^{-1}$). These differences are important in that they permit optical monitoring of reactions, e.g. electron transfer, between different types of semiquinones but not within the same type.

A parallel situation occurs with naphthosemiquinones. The 1,4-isomers absorb at 380–400 nm while the 1,2-isomers only at < 300 nm. The spectra of anthrasemiquinones are red shifted to ~ 500 nm. The miscellaneous semiquinones at the end of Table 1 also conform to the above generalities if their basic structures are not perturbed greatly.

The absorption spectra of the semiquinones given in Table 1 have been determined predominantly in neutral solutions, where all the simple cases exist as the radical anions. The spectra of the simple semiquinones are found to be shifted to lower wavelengths in acid solutions owing to protonation of the O^-. These spectral changes have been utilized for the determination of the acid–base equilibria.

B. Acid–Base Equilibria

The pK_a values for semiquinones are summarized in Table 2.

$$\cdot O-C_6H_4-O^- + H^+ \rightleftharpoons \cdot O-C_6H_4-OH \quad (15)$$

It is seen from Table 2 that most semiquinones have pK_a values around 3–5, i.e. much lower than the values for the same OH group in the parent reduced molecule. Clearly, the

TABLE 2. Acid dissociation constants of semiquinones

	pK_a	Reference
p-Benzosemiquinones		
Unsubstituted	4.0, 4.1	16, 39
2-Methyl-	4.5	38
2,3-Dimethyl-	4.7	38
2,5-Dimethyl-	4.6	39, 48
2,6-Dimethyl-	4.8	38
2,3,5-Trimethyl-	5.0	38
Tetramethyl-	4.9, 5.1	38, 39, 48
Ubisemiquinone	5.9	38
o-Benzosemiquinones		
Unsubstituted	5.0	34
4-*t*-Butyl-	5.2	31
m-Benzosemiquinone		
Unsubstituted	7.1	49
1,4-Naphthosemiquinones		
Unsubstituted	4.1	39, 48
2-Methyl-	4.4–4.7	38, 39, 48
2,3-Dimethyl-	4.3	38
2-Hydroxy-	4.7	39
2-Methyl-3-phytyl-(vitamin K_1)	5.5	38
9,10-Anthrasemiquinones		
Unsubstituted	5.3	39
1-Sulfonate	5.4	17
2-Sulfonate	3.2	17
2,6-Disulfonate	3.2	48
Miscellaneous semiquinones from[a]		
Adriamycin	2.9, 9.2, > 14	42, 50
Lumazine	2.9, ~ 8.6, ~ 12.6	44
1-Methyllumichrome	~ 3.5, 10.5	45, 51
3-Methyllumichrome	~ 3.5, 7.8	45, 51
Dimethyllumichrome	3.5, 10.2	51
Lumichrome	3.5, 8.8, 12.5	51
Riboflavin	2.3, 8.3	46

[a] Radicals from the following compounds undergo several acid-base equilibria with different pK values. For the possible sites of protonation see original references.

unpaired electron causes a considerable increase in the acidity of phenolic OH groups and the magnitude of this effect is dependent on the relative positions. For example, *m*-benzosemiquinone has $pK_a = 7.1$, *o*-benzosemiquinone 5.0 and *p*-benzosemiquinone 4.0. In the *meta* isomer the pK_a of the radical is about 3 units lower than that of resorcinol while in the *para* isomer it is 6 units lower. This large difference is a result of differences in spin density distributions in the radicals. In *m*-benzosemiquinone most of the spin density is on carbons 4 and 6 with very little on the oxygens, while in the *para* isomer ~60% is on the oxygens. From spin density considerations the *ortho* isomer should have a pK_a value similar to that of the *para* isomer. The value observed is, in fact, higher by 1 unit. The reason for this difference must be intramolecular hydrogen bonding in the *ortho*, which inhibits the proton dissociation. This *ortho* effect is general for other semiquinones (cf. 1,4- vs. 1,2-naphthosemiquinones) and other types of radicals as well.

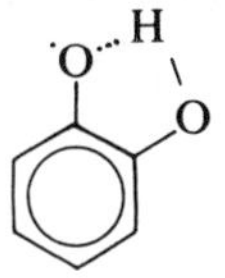

Semiquinones undergo a second protonation step in strong acid.

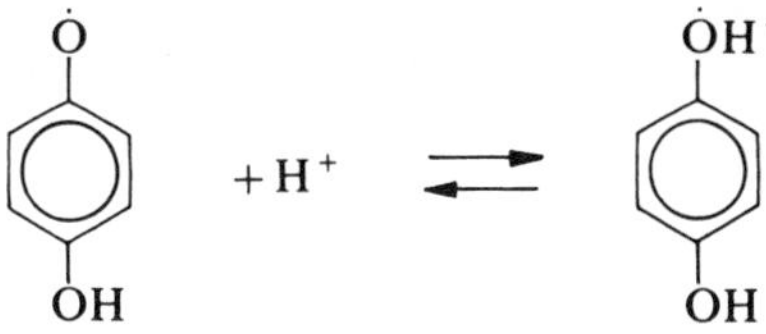

$$\dot{O}\text{-}C_6H_4\text{-}OH + H^+ \rightleftharpoons \dot{O}H^+\text{-}C_6H_4\text{-}OH \qquad (16)$$

These equilibria were monitored by ESR in conjunction with non-radiolytic radical production and pK_a values around -2 were determined[52].

C. ESR Spectroscopy

While optical absorption spectroscopy provided most of the kinetic information and acid–base equilibria for semiquinones, ESR spectroscopy allowed the determination of the second protonation step[52] and gave detailed information on the spin density distribution within the radicals. Although ESR spectroscopy of semiquinones was carried out to some extent in conjunction with radiolytic methods of radical production, most of the ESR information was obtained by other techniques. It is, therefore, beyond the scope of this chapter to present a comprehensive summary of the ESR spectroscopy of semiquinones. Only a few points derived from radiolytic studies will be discussed here.

Steady-state *in situ* radiolysis ESR experiments were carried out with solutions of quinones or hydroquinones and the ESR spectra of semiquinones were recorded. The hyperfine splitting constants and *g*-factors are[29, 34, 53]:

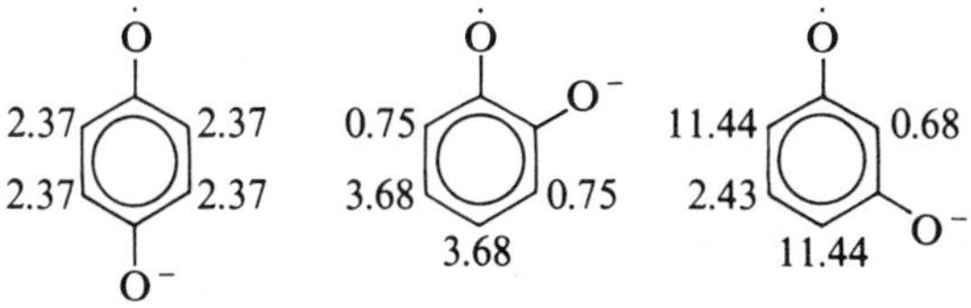

p-benzosemiquinone	*o*-benzosemiquinone	*m*-benzosemiquinone
2.37, 2.37, 2.37, 2.37	0.75, 3.68, 3.68, 0.75	11.44, 0.68, 2.43, 11.44
$g = 2.00456$	2.00455	2.00383

In all cases the two oxygens are equivalent as are all pairs of symmetric carbons. In the case of the *o*- and *p*-semiquinone about 60–65% of the unpaired spin density resides on the oxygens and the rest is distributed over the carbons. The *m*-isomer, on the other hand, has only ~20% of the spin density on the oxygens. This difference is reflected also in the *g*-factors. Protonation of the O^- removes the symmetry and changes the hyperfine splittings considerably. Semiquinones with several substituents have also been studied[29,34] and the ^{13}C hyperfine splitting constants were determined for certain cases[53]. The latter parameters enabled more detailed estimates of spin distributions.

Radiolytic formation of semiquinones was not confined to reduction of quinones or oxidation of hydroquinones. In the ESR study of various phenols[29], reactions with OH radicals were found to yield the corresponding phenoxyl as the main product, but secondary reactions led to the formation of *o*- and *p*-semiquinones as well. With substituents like methoxy or nitro, addition of OH at the *ipso* position was found to be followed by demethoxylation or denitration and production of semiquinone[34,35].

ESR studies of semiquinones are unique in that they permit determination of electron exchange rate constants from line broadening measurements[54]. The values for benzoquinone and its methylated derivatives were found to be in the range of $0.5–2 \times 10^8$ $M^{-1}s^{-1}$.

$$\cdot O\text{-}C_6H_4\text{-}O^- + O{=}C_6H_4{=}O \rightleftharpoons O{=}C_6H_4{=}O + \cdot O\text{-}C_6H_4\text{-}O^- \quad (17)$$

Another study adapted the radiolysis–ESR technique to the measurement of spin-lattice relaxation times for semiquinones[55]. A value of 2 μs was found for *p*-benzosemiquinone anion and a longer T_1 (11.5 μs) for the 2,5-di-*t*-butyl derivative.

Radiolysis of benzoquinone in frozen $CFCl_3$ solutions gave the radical cation and the ESR spectrum indicated that the radical is localized on one oxygen only[56].

D. Raman Spectroscopy

Recent adaptation of time-resolved resonance Raman spectroscopy to pulse radiolysis has led to several studies on semiquinones[57–61]. Raman frequencies have been ascribed to C–O and C–C stretching modes and C–C–C bending modes. For *p*-benzosemiquinones the frequency assigned to the C–O bond was found to be intermediate between those for the corresponding bonds in hydroquinone and benzoquinone. This led to the conclusion that the C–O bond order in the semiquinone is ~1.5[57]. The frequency of the C–O bond increases upon halogenation indicating a higher double bond character[58,60]. On the other hand, deuteration decreases the C–C stretching frequencies with little effect on the C–O frequencies[61].

Raman spectroscopy also yielded information on the excited states of the radicals from examination of the frequencies which are resonance-enhanced. Time-resolved experiments have also been carried out, which allow kinetic measurements on a specific intermediate unmasked by changes in other species.

V. QUINONES AS ELECTRON ACCEPTORS

It has been established in early pulse radiolysis experiments that benzoquinone accepts an electron rapidly from a wide variety of radicals[62]. These radicals include not only typical reducing species (see reaction 2) but also radicals like $O_2^{\bar{\cdot}}$ and NAD. In fact, quinones are such strong electron acceptors that they have been used to detect mild reducing radicals and to distinguish between radicals in a mixture through differences in electron transfer reactions.

2-Methyl-1,4-naphthoquinone has been used to detect reducing radicals in a wide variety of systems[63]. These include radicals produced by hydrogen abstraction from alcohols, sugars, carboxylic acids, amino acids, dipeptides, amines, and amides, by OH addition to aromatic and heterocyclic compounds, and by electron addition to ketones, pyridines, other heterocycles, and metal ions. In most cases reduction of the quinone took place by a certain portion of the radicals and with rate constants of $\sim 10^8$–10^9 $M^{-1}s^{-1}$. Lower rate constants could not be measured because of the competing rapid decay of the radicals under those conditions. Partial yields were rationalized by partial decay of the radical, in competition with electron transfer to the quinone, or by formation of several radicals of which only some are capable of reducing the quinone. By measuring the rate and degree of electron transfer from radicals to several quinones and dyes, an attempt was made to correlate these parameters with the redox potentials of the acceptors[64–67]. The correlation, however, was only qualitative and partially inaccurate.

Because many quinones are reduced by $O_2^{\bar{\cdot}}$ to form the easily detectable semiquinones, they have been used to monitor $O_2^{\bar{\cdot}}$ formation upon reaction of organic radicals with O_2 [68–71]. This was principally done to distinguish between the two possible mechanisms: electron transfer ($R + O_2 \rightarrow R^+ + O_2^{\bar{\cdot}}$) and addition ($\dot{R} + O_2 \rightarrow R\dot{O}_2$). While $O_2^{\bar{\cdot}}$ reduces benzoquinone, peroxyl radicals do not. With $\dot{R}$ bearing an OH or NH_2 group in the alpha position the product of reaction with O_2 was also found to be pH dependent. The initial step is always formation of peroxyl radical (e.g. $(HO)R\dot{O}_2$) but at higher pH values these radicals decompose to give $O_2^{\bar{\cdot}}$.

The reaction of $O_2^{\bar{\cdot}}$ with benzoquinone was used also as a reference to measure rate constants for other reactions of $O_2^{\bar{\cdot}}$ which are not readily monitored[71].

Addition of OH radicals to aromatic and heterocyclic compounds may take place at several positions and thus produce different radicals which may or may not reduce quinones and may react with different rate constants. For example, addition of OH to anisole yields three isomeric radicals, of which the *o*- and *p*- reduce benzoquinone rapidly ($k = 1.2$ and 4.4×10^9 $M^{-1}s^{-1}$, respectively) but the *m*- does not ($k \leqslant 8 \times 10^5$ $M^{-1}s^{-1}$)[72]. These differences helped in the determination of the isomeric distribution of OH adducts. Similar studies were carried out on the OH adducts of phenol and using several quinones as oxidants[73]. Addition of OH to phenol was found to take place 48% at the *ortho* positions, 36% at the *para*, 8% at the *meta* and 8% at the *ipso* position[73].

Addition of OH to pyrimidine bases also yields different radicals which partially reduce quinones[74]. It was established later that 5-OH adducts are the reducing radicals while 6-OH adducts are oxidants[75].

Several studies have utilized quinones to demonstrate the reducing power of unstable metal ions. For example, benzoquinone is reduced very rapidly ($k \sim 3$–5×10^9 $M^{-1}s^{-1}$) by Cd^+, Co^+, Pb^+, Zn^+, moderately rapidly ($k \sim 10^8$ $M^{-1}s^{-1}$) by Ag_3^+, Cr^{2+}, Ni^+ and only very slowly ($k < 10^6$ $M^{-1}s^{-1}$) by Cu^+ [76]. Ru(bipyridyl)$_3^+$ was found to reduce duroquinone extremely rapidly, $k = 4.0 \times 10^9$ $M^{-1}s^{-1}$ [77] and Co(I) complexes with macrocyclic ligands were found to reduce several quinones also with $k \sim 4 \times 10^9$ $M^{-1}s^{-1}$ [78].

The above discussion included representative examples of the utilization of quinones as electron acceptors from various organic and inorganic species. No attempt is made here to cover this topic comprehensively or to tabulate the rate constants. The reader is referred to compilations which give many more rate constants[12,13,79].

Despite the high electron affinity of quinones the semiquinone radicals have been found to donate electrons to nitroxyl radicals[80], superoxide dismutase[81], triphenyltetrazolium[82] and of course to other quinones of higher electron affinity. The latter will be discussed in more detail in the next section.

VI. ONE-ELECTRON REDOX POTENTIALS

Quinones undergo two successive one-electron reduction steps:

$$Q \underset{E^1}{\overset{+e^-}{\rightleftharpoons}} Q^{\cdot -} \underset{E^2}{\overset{+e^-}{\rightleftharpoons}} Q^{2-} \tag{18}$$

The overall two-electron redox potentials have been determined by classical methods and the individual one-electron redox potentials were accessible only under limited conditions where the semiquinone is infinitely stable. Measurement of E^1 and E^2 in aqueous solutions, particularly at neutral pH when $Q^{-\cdot}$ is relatively short-lived, necessitates the use of a fast detection technique. Pulse radiolysis is the most useful method for this purpose. It permits determination of one-electron redox potentials from equilibria such as

$$S^{-\cdot} + Q \rightleftharpoons S + Q^{-\cdot} \tag{19}$$

if E^1 for the $Q/Q^{-\cdot}$ or for the other substrate $S/S^{-\cdot}$ is known, and provided equilibrium 19 is achieved before the radicals decay.

Equilibria between the quinones and oxygen

$$O_2^{-\cdot} + Q \rightleftharpoons O_2 + Q^{-\cdot} \tag{20}$$

have been demonstrated and equilibrium constants derived[38]. $K_{20} = 2.3 \times 10^{-2}$ was determined for duroquinone. E^1 for duroquinone was derived from its two-electron redox potential and its semiquinone formation constant[83,84]. Thus $E^1(O_2/O_2^{-})$ could be calculated from K_{20}[83,84] and further confirmation obtained from experiments with 2,5-dimethylbenzoquinone[85].

This technique was further developed for determination of E^1 for quinones by equilibria with reference quinones, for example naphthoquinone or anthraquinone versus duroquinone[40,86]. Using a set of established E^1 values for several quinones it was possible to determine by pulse radiolysis the one-electron reduction potentials for many other compounds. In particular, redox potentials for nitroaromatic and nitroheterocyclic compounds have been measured[40,87–90] because of the importance of these compounds as radiosensitizers[89]. In addition, several biologically important molecules have been studied by this method (see below).

The values of E^1 for several quinones are summarized in Table 3. A more complete listing is found in Ref. 14. The redox potentials decrease in going from benzoquinone to

TABLE 3. One-electron reduction potentials of quinones at pH 7

Quinone	E^1 $(Q/Q^{-\cdot})$ (mV vs. NHE)	Ref.
1,4-Benzoquinone	+99	84
2-Methylbenzoquinone	+23	84
2,3-Dimethylbenzoquinone	−74	84
2,5-Dimethylbenzoquinone	−67	84
2,3,5-Trimethylbenzoquinone	−165	84
Duroquinone	−235	86
1,4-Naphthoquinone-2-sulfonate	−60	84
2-Methyl-1,4-naphthoquinone	−203	86
2,3-Dimethyl-1,4-naphthoquinone	−240	84
9,10-Anthraquinone-2-sulfonate	−375	40, 87

naphthoquinone to anthraquinone by ~200 mV in each step. Within each series methyl substitution is found to decrease the redox potential substantially. Further generalizations await additional experimental data.

The one-electron reduction potentials have been correlated with rates of electron transfer[54,91] according to the Marcus theory. Table 4 indicates the general trend of increasing rate constant upon increase in driving force (ΔE). Rate constants of $\sim 10^9$ $M^{-1}s^{-1}$ or above may be diffusion-limited. With no driving force the rate constants of self-exchange were found to be $\sim 5 \times 10^7$–2×10^8 $M^{-1}s^{-1}$ for several quinones. With negative driving force, i.e. for the less favorable back reactions, the rate constants are $\sim 10^{-6}$ $M^{-1}s^{-1}$ or less. Quantitative correlations have been discussed before[54,91]. Such correlations are important for indicating the outer sphere nature of the electron transfer, for detecting special circumstances (e.g. geometric constraints) and for predicting unknown rate constants.

TABLE 4. Rate constants of electron transfer from semiquinones to quinones

Semiquinone[a]	Quinone[a]	ΔE^1 (mV,pH 7)	k($M^{-1}s^{-1}$)[b]
$DQ^{-\cdot}$	BQ	334	1.1×10^9
2,6-$DMBQ^{-\cdot}$	BQ	179	1.0×10^9
$DQ^{-\cdot}$	2,5-DMBQ	170	1.1×10^9
2,5-$DMBQ^{-\cdot}$	BQ	164	6.5×10^8
$DQ^{-\cdot}$	2,6-DMBQ	155	9.6×10^8
$AQS^{-\cdot}$	DQ	155	4×10^8
$DQ^{-\cdot}$	DQ	0	2.0×10^8
2,6-$DMBQ^{-\cdot}$	2,6-DMBQ	0	1.7×10^8
$BQ^{-\cdot}$	BQ	0	6.2×10^7
2,5-$DMBQ^{-\cdot}$	2,5-DMBQ	0	5.5×10^7
$DQ^{-\cdot}$	AQS	−155	1.6×10^6
2,6-$DMBQ^{-\cdot}$	DQ	−155	2.0×10^6
$BQ^{-\cdot}$	2,5-DMBQ	−164	9.7×10^5
2,5-$DMBQ^{-\cdot}$	DQ	−170	1.2×10^6
$BQ^{-\cdot}$	2,6-DMBQ	−179	8.3×10^5
$BQ^{-\cdot}$	DQ	−334	1.9×10^3

[a] Abbreviations: BQ, 1,4-benzoquinones; DQ, duroquinone; DMBQ, dimethylbenzoquinone; AQS, 9,10-anthraquinone-2-sulfonate.
[b] From Ref. 54 except values for AQS which are from Ref. 40.

Redox potentials for the second one-electron reduction of quinone, E^2, can be calculated from the overall two-electron potential and E^1 (see e.g. Ref. 84) or can be determined experimentally from equilibria with reference pairs.

$$Q_1^{-\cdot} + Q_2^{2-} \rightleftharpoons Q_1^{2-} + Q_2^{-\cdot} \qquad (21)$$

Such equilibria for semiquinone/hydroquinone pairs have been determined by pulse radiolysis[25] and are partially summarized in Table 5. These values also correspond to the oxidation potentials of the doubly ionized dihydroxybenzenes. It is seen from Table 5 that catechol is oxidized only slightly less readily than hydroquinone while resorcinol differs by ~350 mV. Substitution affects the redox potentials as expected, i.e. the electron-donating substituents methyl, methoxy and hydroxy enhance the oxidation, in that order, while the electron-withdrawing substituents carboxy, sulfonate, vinyl and acetyl exert the opposite effect.

TABLE 5. One electron reduction potentials of semiquinones

Semiquinone	E^2(mV vs. NHE)[a]
1,4-Benzosemiquinones	
Unsubstituted	23
2-Hydroxy-	−110
2-Methoxy-	−85
Tetramethyl-	−54
2-Carboxymethyl-	−50
2-Carboxy-	33
2,5-Disulfonate	116
2-Acetyl	118
1,2-Benzosemiquinones	
Unsubstituted	43
4-Hydroxy-	−110
3-Hydroxy-	−9
4-(2-Amino-2-carboxyethyl)-	18
4-Carboxymethyl-	21
3-Hydroxy-5-(2-aminoethyl)-	42
4-(2-Amino-1-hydroxyethyl)-	44
4-(2-Carboxyvinyl)-	84
3-Carboxy-	118
4-Carboxy-	119
1,3-Benzosemiquinone	385

[a] From Ref. 25 except the value for the 2,5-disulfonate which is from Ref. 27. Determined at 0.5 M KOH.

The redox potentials E^1 and E^2 are pH dependent due to the acid–base equilibria of the semiquinones and the hydroquinones. From known pK_a values, redox potentials at different pHs have been calculated[25,40,84,86].

The rate constants leading to equilibrium 21 (k_{21} and k_{-21}) were also found to be in the range of 10^5–10^9 $M^{-1}s^{-1}$ and to depend on the driving force[25]. These rate constants are strongly dependent upon pH since, in general, the rate of oxidation of QH_2 is much slower than that of QH^- or Q^{2-}. Consequently, equilibrium 21 could not be monitored in neutral solutions.

VII. QUINONES OF BIOLOGICAL IMPORTANCE

Naturally occurring quinones are important in electron transport, in photosynthesis and as vitamins. Other quinones have biological importance as drugs. Radiation chemical studies of these quinones have helped in characterizing the chemical behavior of these compounds, especially their electron transfer rates and redox potentials, and often contributed to the understanding of their biological action.

Quinones have been shown to act as sensitizers of radiation damage to hypoxic cells and thus may serve as drugs to enhance tumor radiotherapy (see e.g. Refs 92–94). Their efficiency as radiosensitizers is related to their ability to oxidize or trap radiation-produced free radicals and thus 'fix' the radiation damage[95–97]. Their efficiency has been correlated with their one-electron reduction potentials determined by pulse radiolysis[93,94]. Although other pharmacological properties are obviously very important, the reduction potential serves as a preliminary test to predict the radiosensitization efficiency of a drug. It should be pointed out, however, that the majority of research on radiosensitizers has concentrated on nitroheterocyclic compounds and not on quinones. The redox potentials for the nitro compounds have been determined in most cases by reference to a quinone using pulse radiolysis[14].

Several studies dealt with anthracycline drugs, such as adriamycin and mitomycin, in an attempt to understand their cancer chemotherapeutic effect from pulse radiolytic observations on their semiquinone radicals[43, 50, 98–101]. To help in the interpretation of those results additional information was obtained on model compounds[42, 102]. In these studies the absorption spectra, acid dissociation constants and redox potentials of the semiquinones have been determined. More importantly, their reactions with oxygen and iron complexes have been monitored. Based on the rates of these reactions, a mechanism has been proposed to account for the enhanced free radical damage caused by these drugs[98, 99, 102]. The main reactions appear to be those of the semiquinones with O_2 to give O_2^- (which disproportionate to yield H_2O_2) and with Fe^{3+} to give Fe^{2+}. This enhances the probability of a Fenton reaction ($Fe^{2+} + H_2O_2 \rightarrow OH$) which forms the damaging OH radicals.

For the drug daunorubicin, reaction of its doubly reduced form with H_2O_2 has been suggested[100, 101]. Furthermore, radiolytic reduction of this drug was found to result in cleavage of the glycosidic bond.

The effect of radiation on another drug, tetracycline, in the solid state was found to result in cleavage of the dimethylamino group[103]. Significant cleavage was detected only at doses much higher than those used for sterilization purposes.

Pulse radiolysis studies on other semiquinones of biological importance have been discussed in previous sections. These include rates of electron transfer to vitamin K and ubiquinone and the spectra and pK_a values of their semiquinones[38]. Other studies dealt with the semiquinones produced by oxidation of catecholamines and related compounds, including their spectra and redox potentials (see e.g. Refs 25, 30). Disproportionation of the semiquinone derived from 3,4-dihydroxyphenylalanine (DOPA) was found to form dopaquinone which cyclizes to yield dopachrome and eventually polymerizes to melanin[23].

Other researchers were concerned with the involvement of plastoquinone and ubiquinone in the reaction centers of photosynthesis. Pulse radiolysis studies[104] on plastoquinone in methanolic solutions showed that the spectrum of the semiquinone is similar to that observed after photoexcitation of Photosystem II of plants, where plastoquinone serves as an electron acceptor. Experiments with ubiquinone found similarity between its semiquinone spectrum and that observed upon photoexcitation of bacterial reaction centers, thus suggesting that ubiquinone may be the electron acceptor in these systems[104]. That quinones are good electron acceptors from excited chlorophyll and porphyrins has been shown by many photochemical studies. Pulse radiolytic experiments demonstrated that quinones accept electrons rapidly ($k \sim 10^8$–10^9 $M^{-1}s^{-1}$) from the radical anions of chlorophyll and porphyrins[105].

VIII. FLAVINS AND RELATED COMPOUNDS

The main biological function of flavins is electron transport. Therefore, they were frequently studied by pulse radiolysis. Rate constants for electron transfer to flavins and from flavin semiquinones to other acceptors, as well as one-electron redox potentials, have been determined. Studies were carried out on riboflavin, FMN (flavinmononucleotide) and FAD[46, 47, 106–111] (flavin adenine dinucleotide) and on simpler model compounds, lumiflavin[112, 113], lumichrome[45, 51, 114] and lumazine[44].

Flavins are reduced by hydrated electrons and by organic radicals such as CO_2^- and $(Me)_2\dot{C}OH$ very rapidly to produce the semiquinones. These radicals exhibit intense absorptions in the 300–600 nm region (see Table 1) which are dependent on pH. From this dependence pK_a values for the semiquinones were determined and found to be in the region of ~3 and ~8 (see Table 2). The acid form of the radical is very stable while the neutral and basic forms are short-lived[46,108]. They decay by disproportionation to form flavin and dihydroflavin.

The one-electron reduction potentials for flavins have been determined from equilibria with several reference quinones and pyridine derivatives. The values at pH 7 for riboflavin, FMN and FAD were $E^1_7(Fl/Fl^{-\cdot}) \sim -310$ mV[47,106,107] and for the lumichromes ~ -520 mV[51]. These values change with pH according to the acid–base equilibria of the radicals (mentioned above) and of the parent flavins. From the known two-electron potential $E^\circ_7(Fl/FlH_2) = -219$ mV the value of the second one-electron reduction potential $E^2_7(Fl^{-\cdot}/FlH_2) \sim -120$ mV was calculated[107].

Flavins are reduced also by α-aminoalkyl radicals, e.g. those derived from glycine and EDTA[113]. $RSSR^{-\cdot}$ radicals also reduce flavins very rapidly[110,111], while RS radicals ($RS + RS^- \rightleftarrows RSSR^-$) oxidize dihydroflavin, as do other typical oxidizing radicals[115].

Electron transfer from the semiquinone of FMN to cytochrome c also was examined by pulse radiolysis and found to have a rate constant of 4×10^7 $M^{-1}s^{-1}$ [109]. This value is about an order of magnitude higher than the rate constant for reduction of cytochrome c by cytochrome reductase.

Rates of electron transfer to flavins were examined by pulse radiolysis also for flavins bound to proteins[116–119]. It is interesting to note that much of the initial reduction occurs on protein sites and subsequently an intramolecular electron transfer leads to formation of the flavin semiquinone. The flavin semiquinone anion is first produced and is then stabilized by accepting a proton from other sites of the protein[119].

IX. EXCITED STATES

The above sections dealt with radiolytic studies of quinones involving free radicals. In certain solvents, however, it is possible to form the excited states of quinones by radiolysis. Thus pulse radiolysis has been utilized also for monitoring spectra and kinetics of quinones in their excited states, predominantly in benzene solutions[120–127].

Triplet–triplet extinction coefficients were determined by pulse radiolysis by monitoring energy transfer processes[120]. This method has an advantage over earlier techniques in that it permits interconnecting a large number of triplets as donor–acceptor pairs[120]. The rate constants for energy transfer were near $\sim 10^{10}$ $M^{-1}s^{-1}$ [120].

The triplet excited states of anthraquinones were also characterized by the energy transfer method[121]. Their lifetimes in benzene and their rate of reaction with O_2 and isopropanol were determined. No interaction was detected between the triplets and their ground states[121].

It is noted, however, that anthraquinones 1,4-disubstituted with amino or hydroxyl groups were suggested to exist in benzene solutions in an associated form even at very low concentrations[124]. This suggestion was based on the observation that triplet energy transfer from biphenyl to the anthraquinone appears to remove more than one ground state molecule. No aggregation was indicated for the triplets of these quinones.

Pulse radiolysis permitted the first observation of the triplet excited state of ubiquinone[122]. Various derivatives of ubiquinones were subsequently studied[123]. The results with these derivatives led to the conclusion that the low triplet energy and quantum yield of triplet ubiquinone is due to the methoxy groups on the ring and not the isoprenoid side chain[123]. These results further suggested that the ubisemiquinone observed in bacterial photosynthesis is most likely formed by electron transfer from excited chlorophyll rather than via triplet ubiquinone[123].

X. MISCELLANEOUS TOPICS

Quinones, being efficient traps for electrons, have been used frequently to probe various effects on the reactivity of solvated electrons and on electron transfer reactions[128–136]. In a study on the kinetics of electron attachment to benzoquinone in non-polar solvents an unusual dependence on solvent and temperature has been noted[130]. In pentane and similar solvents the reaction was found to be fast and with a positive activation energy but in solvents like neopentane and tetramethysilane the reaction is much slower and has a negative activation energy. These results have been rationalized by suggesting that an electron reacting with benzoquinone leads to an excited benzoquinone anion. In pentane this product relaxes rapidly to a stable semiquinone anion but in tetramethylsilane the excited anion detaches its electron rapidly because the energy level of the electron in this liquid is much lower[130].

Quinones have been used to study the effect of exothermicity on rates of electron transfer reactions in order to test electron transfer theories. By measuring the rates in rigid organic glasses the problems of diffusion control and reactant complexation were avoided. The rates of electron transfer were slow for reactions with low driving force and became faster upon increasing exothermicity. However, at very high exothermicities the rates were found to decrease, as predicted by the theories[135]. Further confirmation of this 'inverted region' was obtained from measurements of intramolecular electron transfer rates[136]. A series of electron acceptors, including several quinones, were attached to one end of a rigid molecular spacer (androstane skeleton) which was bound to biphenyl at its other end. Rates of intramolecular electron transfer from the biphenyl anion to the acceptor were measured by pulse radiolysis. They were found to increase from $10^6\,s^{-1}$ to $> 10^9\,s^{-1}$ upon increasing ΔG° from 0.05 to 1.2 eV. Further increase in ΔG° to 2.4 eV was accompanied by a gradual decrease in rate constant to $< 10^8\,s^{-1}$ [136].

Bianthrone, because of steric interactions, assumes two distinct conformations. The anion radicals of bianthrones were found to undergo conformational changes. The rate constants for these processes were also monitored by pulse radiolysis[137]. A rate of $7 \times 10^4\,s^{-1}$ was found for bianthrone anion but only $1.1 \times 10^3\,s^{-1}$ for the 1,1′-dimethyl derivative, in which the conformational change is sterically hindered.

Several studies examined the radiation-induced bleaching of anthraquinone dyes in solution[138–140] and in the solid state[141]. Bleaching was found to result from the reactions of electrons and other reducing radicals as well as by addition of OH to the ring. The effects of O_2 and H_2O_2 concentrations were examined. The interest in these systems was primarily for their application in radiation dosimetry and no detailed mechanistic studies were carried out.

XI. ACKNOWLEDGEMENTS

I wish to thank Dr R. E. Huie for his comments on this manuscript and the Office of Basic Energy Sciences of the US Department of Energy for financial support.

XII. REFERENCES

1. J. H. Fendler and E. J. Fendler, in *The Chemistry of the Quinonoid Compounds* (Ed. S. Patai), Wiley, London, 1974, p. 539.
2. I. V. Khudyakov and V. A. Kuz'min, *Russ. Chem. Rev.*, **44**, 801 (1975); **47**, 22 (1978).
3. P. Neta, *Adv. Phys. Org. Chem.*, **12**, 223 (1976).
4. R. L. Willson, *Chem. Ind.* (*London*), 183 (1977).
5. A. J. Swallow, *Progr. React. Kinet.*, **9**, 195 (1978).

6. P. Neta, *J. Chem. Ed.*, **58**, 110 (1981).
7. M. Anbar, M. Bambenek and A. B. Ross, *Natl. Stand. Ref. Data Ser., Natl. Bur. Stand.*, Report No. 43 (1973).
8. A. B. Ross, *Natl. Stand. Ref. Data Ser., Natl. Bur. Stand.*, Report No. 43, Supplement (1975).
9. M. Anbar, Farhataziz and A. B. Ross, *Natl. Stand. Ref. Data. Ser., Natl. Bur. Stand.*, Report No. 51 (1975).
10. Farhataziz and A. B. Ross, *Natl. Stand. Ref. Data Ser., Natl. Bur. Stand.*, Report No. 59 (1977).
11. B. H. J. Bielski, D. E. Cabelli, R. L. Arudi and A. B. Ross, *J. Phys. Chem. Ref. Data*, **14**, 1041 (1985).
12. A. B. Ross and P. Neta, *Natl. Stand. Ref. Data Ser., Natl. Bur. Stand.*, Report No. 70 (1982).
13. K. D. Asmus and M. Bonifacic, in *Landolt-Bornstein Numerical Data and Functional Relationships in Science and Technology*, New Series, Vol. 13, Part b, 1984.
14. P. Wardman, *J. Phys. Chem. Ref. Data*, to be published.
15. S. Steenken, in *Landolt-Bornstein Numerical Data and Functional Relationships in Science and Technology*, New Series, Vol. 13, Part e, ch. 10, 1985.
16. G. E. Adams and B. D. Michael, *Trans. Faraday Soc.*, **63**, 1171 (1967).
17. B. E. Hulme, E. J. Land and G. O. Phillips, *J. Chem. Soc. Faraday Trans. 1*, **68**, 1992 (1972).
18. C. E. Burchill, D. M. Smith and J. L. Charlton, *Can. J. Chem.*, **54**, 505 (1976).
19. K. P. Clark and H. I. Stonehill, *J. Chem. Soc. Faraday Trans. 1*, **73**, 722 (1977).
20. D. Veltwisch and K. D. Asmus, *J. Chem. Soc. Perkin Trans. 2*, 1147 (1982).
21. A. B. Ross and P. Neta, *Natl. Stand. Ref. Data Ser., Natl. Bur. Stand.*, Report No. 65 (1979).
22. Z. B. Alfassi and R. H. Schuler, *J. Phys. Chem.*, **89**, 3359 (1985).
23. M. R. Chedekel, E. J. Land, A. Thompson and T. G. Truscott, *J. Chem. Soc. Chem. Commun.*, 1170 (1984); A. Thompson, E. J. Land, M. R. Chedekel, K. V. Subbarao and T. G. Truscott, *Biochim. Biophys. Acta*, **843**, 49 (1985).
24. S. Steenken, *J. Phys. Chem.*, **83**, 595 (1979).
25. S. Steenken and P. Neta, *J. Phys. Chem.*, **86**, 3661 (1982).
26. P. Neta, P. Maruthamuthu, P. M. Carton and R. W. Fessenden, *J. Phys. Chem.*, **82**, 1875 (1978).
27. R. E. Huie and P. Neta, *J. Phys. Chem.*, **89**, 3918 (1985).
28. W. Bors, D. Tait, C. Michel, M. Saran and M. Erben-Russ, *Isr. J. Chem.*, **24**, 17 (1984).
29. P. Neta and R. W. Fessenden, *J. Phys. Chem.*, **78**, 523 (1974).
30. W. Bors, M. Saran, C. Michel, E. Lengfelder, C. Fuchs and R. Spottl, *Int. J. Radiat. Biol.*, **28**, 353 (1975).
31. H. W. Richter, *J. Phys. Chem.*, **83**, 1123 (1979).
32. M. Gohn and N. Getoff, *J. Chem. Soc. Faraday Trans. 1*, **73**, 1207 (1977).
33. P. Neta and R. H. Schuler, *Radiat. Res.*, **64**, 233 (1975).
34. S. Steenken and P. O'Neill, *J. Phys. Chem.*, **81**, 505 (1977).
35. P. O'Neill, S. Steenken, H. van der Linde and D. Schulte-Frohlinde, *Radiat. Phys. Chem.*, **12**, 13 (1978).
36. R. H. Schuler, *Radiat. Res.*, **69**, 417 (1977).
37. K. Bhatia and R. H. Schuler, *J. Phys. Chem.*, **77**, 1356 (1973).
38. K. B. Patel and R. L. Willson, *J. Chem. Soc. Faraday Trans. 1*, **69**, 814 (1973).
39. P. S. Rao and E. Hayon, *J. Phys. Chem.*, **77**, 2274 (1973).
40. D. Meisel and P. Neta, *J. Am. Chem. Soc.*, **97**, 5198 (1975).
41. E. McAlpine, R. S. Sinclair, T. G. Truscott and E. J. Land, *J. Chem. Soc. Faraday Trans. 1*, **74**, 597 (1978).
42. E. J. Land, T. Mukherjee, A. J. Swallow and J. M. Bruce, *J. Chem. Soc. Faraday Trans. 1*, **79**, 391 and 405 (1983).
43. B. A. Svingen and G. Powis, *Arch. Biochem. Biophys.*, **209**, 119 (1981).
44. P. N. Moorthy and E. Hayon, *J. Phys. Chem.*, **79**, 1059 (1975).
45. P. F. Heelis, B. J. Parsons, G. O. Phillips, E. J. Land and A. J. Swallow, *J. Phys. Chem.*, **86**, 5169 (1982).
46. E. J. Land and A. J. Swallow, *Biochemistry*, **8**, 2117 (1969).
47. R. F. Anderson, *Ber. Bunsenges. Phys. Chem.*, **80**, 969 (1976).
48. R. L. Willson, *Chem. Commun.* 1249 (1971).
49. P. O'Neill, S. Steenken and D. Schulte-Frohlinde, unpublished result.
50. E. J. Land, T. Mukherjee, A. J. Swallow and J. M. Bruce, *Arch. Biochem. Biophys.*, **225**, 116, (1983).

51. P. F. Heelis, B. J. Parsons, G. O. Phillips, E. J. Land and A. J. Swallow, *J. Chem. Soc. Faraday Trans. 1*, **81**, 1225 (1985).
52. W. T. Dixon and D. Murphy, *J. Chem. Soc. Faraday Trans. 2*, **72**, 1221 (1976).
53. K. M. Madden, H. J. McManus and R. H. Schuler, *J. Phys. Chem.*, **86**, 2926 (1982).
54. D. Meisel and R. W. Fessenden, *J. Am. Chem. Soc.*, **98**, 7505 (1976).
55. R. W. Fessenden, J. P. Hornak and B. Venkataraman, *J. Chem. Phys.*, **74**, 3694 (1981).
56. H. Chandra and M. C. R. Symons, *J. Chem. Soc. Chem. Commun.*, 29 (1983).
57. G. N. R. Tripathi, *J. Chem. Phys.*, **74**, 6044 (1981).
58. G. N. R. Tripathi and R. H. Schuler, *J. Chem. Phys.*, **76**, 2139 (1982).
59. G. N. R. Tripathi, *Am. Chem. Soc. Symp. Ser.*, **236**, 171 (1983).
60. G. N. R. Tripathi and R. H. Schuler, *J. Phys. Chem.*, **87**, 3101 (1983).
61. R. H. Schuler, G. N. R. Tripathi, M. F. Prebenda and D. M. Chipman, *J. Phys. Chem.*, **87**, 5357 (1983).
62. R. L. Willson, *Trans. Faraday Soc.*, **67**, 3020 (1971).
63. P. S. Rao and E. Hayon, *Biochim. Biophys. Acta*, **292**, 516 (1973).
64. P. S. Rao and E. Hayon, *J. Phys. Chem.*, **77**, 2753 (1973).
65. P. S. Rao and E. Hayon, *Nature* (*Lond.*), **243**, 344 (1973).
66. P. S. Rao and E. Hayon, *J. Am. Chem. Soc.*, **96**, 1287, 1295 (1974).
67. P. S. Rao and E. Hayon, *J. Phys. Chem.*, **79**, 397 (1975).
68. M. Simic and E. Hayon, *Biochem. Biophys. Res. Commun.*, **50**, 364 (1973).
69. E. Hayon and M. Simic, *J. Am. Chem. Soc.*, **95**, 6681 (1973).
70. S. Abramovitch and J. Rabani, *J. Phys. Chem.*, **80**, 1562 (1976).
71. C. L. Greenstock and G. W. Ruddock, *Int. J. Radiat. Phys. Chem.*, **8**, 367 (1976).
72. S. Steenken and N. V. Raghavan, *J. Phys. Chem.*, **83**, 3101 (1979).
73. N. V. Raghavan and S. Steenken, *J. Am. Chem. Soc.*, **102**, 3495 (1980).
74. E. Hayon and M. Simic, *J. Am. Chem. Soc.*, **95**, 1029 (1973).
75. S. Fujita and S. Steenken, *J. Am. Chem. Soc.*, **103**, 2540 (1981).
76. R. M. Sellers and M. G. Simic, *J. Am. Chem. Soc.*, **98**, 6145 (1976).
77. D. Meisel, M. S. Matheson, W. A. Mulac and J. Rabani, *J. Phys. Chem.*, **81**, 1449 (1977).
78. A. M. Tait, M. Z. Hoffman and E. Hayon, *J. Am. Chem. Soc.*, **98**, 86 (1976).
79. G. V. Buxton and R. M. Sellers, *Natl. Stand. Ref. Data Ser., Natl. Bur. Stand.*, Report No. 62 (1978).
80. P. O'Neill and T. C. Jenkins, *J. Chem. Soc. Faraday Trans. 1*, **75**, 1912 (1979).
81. P. Wardman, in *Radiation Biology and Chemistry: Research Developments* (Eds H. E. Edwards, S. Navaratnam, B. J. Parsons and G. O. Phillips), Elsevier, New York, 1979, p. 189.
82. Z. K. Kriminskaya, K. M. Dyumaev, G. V. Fomin and A. K. Pikaev, *High Energy Chem.*, **17**, 231 (1983).
83. P. M. Wood, *FEBS Lett.*, **44**, 22 (1974).
84. Y. A. Ilan, G. Czapski and D. Meisel, *Biochem. Biophys. Acta*, **430**, 209 (1976).
85. Y. A. Ilan, D. Meisel and G. Czapski, *Isr. J. Chem.*, **12**, 891 (1974).
86. D. Meisel and G. Czapski, *J. Phys. Chem.*, **79**, 1503 (1975).
87. P. Wardman and E. D. Clarke, *J. Chem. Soc. Faraday Trans. 1*, **72**, 1377 (1976).
88. P. Neta, M. G. Simic and M. Z. Hoffman, *J. Phys. Chem.*, **80**, 2018 (1976).
89. G. E. Adams, I. Ahmed, E. D. Clarke, P. O'Neill, J. Parrick, I. J. Stratford, R. G. Wallace, P. Wardman and M. E. Watts, *Int. J. Radiat. Biol.*, **38**, 613 (1980).
90. L. Sjoberg, T. E. Erikson, I. Mustea and L. Revesz, *Radiochem. Radioanal. Lett.*, **29**, 19 (1977).
91. D. Meisel, *Chem. Phys. Lett.*, **34**, 263 (1975).
92. G. E. Adams and M. S. Cooke, *Int. J. Radiat. Biol.*, **15**, 457 (1969).
93. G. A. Infante, P. Gonzalez, D. Cruz, J. Correa, J. A. Myers, M. F. Ahmad, W. L. Whitter, A. Santos and P. Neta, *Radiat. Res.*, **92**, 296 (1982).
94. G. A. Infante, P. Guzman, R. Alvarez, A. Figueroa, J. N. Correa, J. A. Myers, L. J. Lanier, T. M. Williams, S. Burgos, J. Vera and P. Neta, *Radiat. Res.*, **98**, 234 (1984).
95. G. E. Adams, C. L. Greenstock, J. J. van Hemmen and R. L. Willson, *Radiat. Res.*, **49**, 85 (1972).
96. C. L. Greenstock, J. D. Chapman, J. A. Raleigh, E. Shierman and A. P. Reuvers, *Radiat. Res.*, **59**, 556 (1974).
97. S. Nishimoto, H. Ide, T. Wada and T. Kagiya, *Int. J. Radiat. Biol.*, **44**, 585 (1983).
98. J. Butler, B. M. Hoey and A. J. Swallow, *FEBS Lett.*, **182**, 95 (1985).

99. E. J. Land, R. Mukherjee, A. J. Swallow and J. M. Bruce, *Br. J. Cancer*, **51**, 515 (1985).
100. C. Houee-Levin, M. Gardes-Albert and C. Ferradini, *FEBS Lett.*, **173**, 27 (1984).
101. C. Houee-Levin, M. Gardes-Albert, C. Ferradini, M. Faraggi and M. Klapper, *FEBS Lett.*, **179**, 46 (1985).
102. H. C. Sutton and D. F. Sangster, *J. Chem. Soc. Faraday Trans. 1*, **78**, 695 (1982).
103. J. Kuduk-Jaworska and B. Jezowska-Trzebiatowska, *Pol. J. Chem.*, **54**, 973 (1980).
104. R. Bensasson and E. J. Land, *Biochem. Biophys. Acta*, **325**, 175 (1973).
105. P. Neta, A. Scherz and H. Levanon, *J. Am. Chem. Soc.*, **101**, 3624 (1979).
106. D. Meisel and P. Neta, *J. Phys. Chem.*, **79**, 2459 (1975).
107. R. F. Anderson, *Biochim. Biophys. Acta*, **722**, 158 (1983).
108. R. F. Anderson, *Biochim. Biophys. Acta*, **723**, 78 (1983).
109. C. Salet, E. J. Land and R. Santus, *Photochem. Photobiol.*, **33**, 753 (1981).
110. R. Ahmad and D. A. Armstrong, *Can. J. Chem.*, **62**, 171 (1984).
111. P. S. Surdhar, R. Ahmad and D. A. Armstrong, *Can. J. Chem.*, **62**, 580 (1984).
112. R. Ahmad, Z. Wu and D. A. Armstrong, *Biochemistry*, **22**, 1806 (1983).
113. P. S. Surdhar, D. E. Bader and D. A. Armstrong, *Can. J. Chem.*, **63**, 1357 (1985).
114. R. Ahmad and D. A. Armstrong, *Int. J. Radiat. Biol.*, **45**, 607 (1984).
115. P. S. Surdhar and D. A. Armstrong, *J. Phys. Chem.*, **89**, 5514 (1985).
116. A. J. Elliot, P. L. Munk, K. J. Stevenson and D. A. Armstrong, *Biochemistry*, **19**, 4945 (1980).
117. K. Kobayashi, K. Hirota, H. Ohara, K. Hayashi, R. Miura and T. Yamano, *Biochemistry*, **22**, 2239 (1983).
118. M. Faraggi and M. H. Klapper, *J. Biol. Chem.*, **254**, 8139 (1979).
119. M. Faraggi, J. P. Steiner and M. H. Klapper, *Biochemistry*, **24**, 3273 (1985).
120. R. Bensasson and E. J. Land, *Trans. Faraday Soc.*, **67**, 1904 (1971).
121. B. E. Hulme, E. J. Land and G. O. Phillips, *J. Chem. Soc. Faraday Trans. 1*, **68**, 2003 (1972).
122. R. Bensasson, C. Chachaty, E. J. Land and C. Salet, *Photochem. Photobiol.*, **16**, 27 (1972).
123. E. Amouyal, R. Bensasson, and E. J. Land, *Photochem. Photobiol.*, **20**, 415 (1974).
124. E. J. Land, E. McAlpine, R. S. Sinclair and T. G. Truscott, *J. Chem. Soc. Faraday Trans. 1*, **72**, 2091 (1976).
125. F. Wilkinson and A. Garner, *J. Chem. Soc. Faraday Trans. 2*, **73**, 222 (1977).
126. A. Garner and F. Wilkinson, *Chem. Phys. Lett.*, **45**, 432 (1977).
127. N. H. Jensen, R. Wilbrandt, P. Pagsberg, A. H. Sillesen and K. B. Hansen, *J. Photochem.*, **9**, 227 (1978).
128. J. Teply and I. Janovsky, *Chem. Phys. Lett.*, **17**, 373 (1972).
129. B. H. Milosavljevic and O. I. Micic, *J. Phys. Chem.*, **82**, 1359 (1978).
130. R. A. Holroyd, *J. Phys. Chem.*, **86**, 3541 (1982).
131. A. Kira and M. Imamura, *J. Phys. Chem.*, **82**, 1966 (1978).
132. A. Kira, Y. Nosaka, M. Imamura and T. Ichikawa, *J. Phys. Chem.*, **86**, 1866 (1982).
133. J. V. Beitz and J. R. Miller, *J. Chem. Phys.*, **71**, 4579 (1979).
134. R. K. Huddleston and J. R. Miller, *J. Phys. Chem.*, **87**, 4867 (1983).
135. J. R. Miller, J. V. Beitz and R. K. Huddleston, *J. Am. Chem. Soc.*, **106**, 5057 (1984).
136. J. R. Miller, L. T. Calcaterra and G. L. Closs, *J. Am. Chem. Soc.*, **106**, 3047 (1984).
137. P. Neta and D. H. Evans, *J. Am. Chem. Soc.*, **103**, 7041 (1981).
138. N. Suzuki and H. Hotta, *Bull. Chem. Soc. Japan*, **50**, 1441 (1977).
139. S. Hashimoto, T. Miyata, N. Suzuki and W. Kawakami, *Radiat. Phys. Chem.*, **13**, 107 (1979).
140. N. B. El-Assy, A. Alian, F. A. Rahim and H. Roushdy, *Int. J. Appl. Radiat. Isot.*, **33**, 433 (1982).
141. M. M. Abou Sekkina and S. S. Assar, *Int. J. Appl. Radiat. Isot.*, **32**, 847 (1981).

The Chemistry of Quinonoid Compounds, Vol. II
Edited by S. Patai and Z. Rappoport

CHAPTER 15

Chemistry of quinone bis- and monoketals

JOHN S. SWENTON
Department of Chemistry, The Ohio State University, Columbus, OH 43210, USA

I. INTRODUCTION

Quinone and quinone-related natural products have occupied a pivotal position in organic chemistry throughout the years; thus, methods for the preparation of quinones and the chemistry of the quinone moiety have been well-studied[1]. The simple quinone unit is quite reactive toward nucleophiles and is easily reduced via electron transfer to form the corresponding radical anion. Thus, in the synthesis of functionalized quinones, the usual approach is to perform many of the synthetic operations on the corresponding hydroquinone ether. The quinone unit is then introduced at one of the later stages of the synthesis by oxidation of the functionalized hydroquinone ether. Aside from the high reactivity of the quinone linkage, 1,4 additions to unsymmetrical quinones often afford mixtures of regioisomeric products. This complicates synthetic processes initiated by addition at the β position of the quinone, one example being the preparation of certain indoles via the reaction of quinones with enamines—the Nenitzescu reaction[2].

In the past ten years two types of protected quinone derivatives—*p*-quinone bisketals (**1**) and quinone monoketals (**2**)—have become readily available, leading to increased use of the compounds in synthesis[3]. The quinone bisketal and monoketal serve as valuable

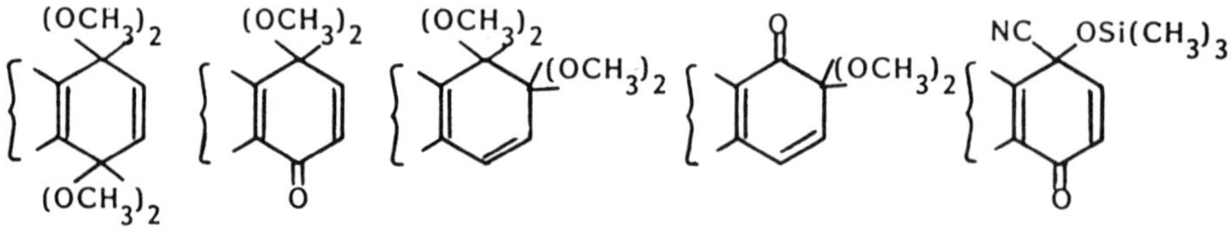

$(OCH_3)_2$ $(OCH_3)_2$ $(OCH_3)_2$ $(OCH_3)_2$ $(OCH_3)_2$ NC $OSi(CH_3)_3$

$(OCH_3)_2$ O O O

1 2 3 4 5

protected quinone derivatives since they possess the quinone oxidation state; yet, they often circumvent reactivity and regiochemical problems encountered in reactions of the quinone entity itself. This review will discuss first the preparation and chemistry of quinone bisketals, followed by an analogous treatment of the preparation and chemistry of quinone monoketals. Sufficient literature has appeared in the past several years to establish the chemical utility of these compounds. A comparison of this chemistry with that of the analogous quinones will be made where possible. Although less chemistry of the *o*-quinone analogs (**3**) and (**4**) has been published, a brief survey of this chemistry is also given. Excluded from this detailed coverage is the chemistry of trimethylsilyl cyanide-blocked quinone derivatives[4], e.g. **5**, although this chemistry will be noted wherever a direct comparison can be made with reactions reported for quinone monoketals.

Several conventions will be observed throughout the chapter. Chemical Abstracts and IUPAC rules no longer use the term ketal, favoring the term acetal for all such structural units; nevertheless, the term bisketal will be used when describing a molecule such as **1**. Furthermore, although bisketals of both *p*- and *o*-quinones are known, much more chemistry has been published on the former series of compounds. Thus, the *p*-quinone bisketals will be referred to simply as quinone bisketals while similar derivatives of *o*-quinones will explicitly be called *o*-quinone bisketals. Finally, the symbol Ⓔ used in equations denotes electrochemical oxidation.

II. PREPARATION OF QUINONE BISKETALS

A. Anodic Oxidation of 1,4-Dimethoxyaromatic Systems

Although reaction of a quinone with alcohol would be the most direct route to quinone bisketals, this method is usually unsatisfactory. Presumably, 1,4 addition of the alcohol to the quinone system followed by irreversible aromatization dominates the reversible 1,2 addition required for ketal formation. The one non-electrochemical[5] preparation of the quinone bisethylene glycol ketal **9** involved bromination of the 1,4-cyclohexanedione

$$\mathbf{6} \xrightarrow[\text{Et}_2\text{O},\ 77\%]{\text{Br}_2} \mathbf{7} + \mathbf{8} \xrightarrow[\text{DMSO},\ 86\%]{t\text{-BuOK}} \mathbf{9} \qquad (1)$$

bisethylene glycol ketal **6** followed by dehydrobromination of **7** and **8** (equation 1). The yields for this reaction were good; however, the generality of the method as a route to substituted benzoquinone bisketals rests on the availability of the requisite 1,4-cyclohexanedione. As will be discussed below, the high yields of quinone bisketals from anodic oxidation of readily available 1,4-dimethoxybenzene derivatives make the above chemistry of historical interest only. Furthermore, this bromination/dehydrobromination sequence does not appear to be directly applicable to the preparation of naphthoquinone bisketals.

1. Anodic oxidations in a single cell

A route that has proved to be the most general for the preparation of quinone bisketals was first reported by Weinberg and Belleau in 1963[6]. Thus, anodic oxidation of 1,4-dimethoxybenzene (**10**) at constant current in a single cell (anode and cathode not separated) using 1% methanolic potassium hydroxide as both solvent and electrolyte afforded the benzoquinone dimethyl ketal **11a** in 88% yield (equation 2). Apparently, this

$$\mathbf{10} \xrightarrow[\text{KOH/CH}_3\text{OH},\ 88\%]{Ⓔ} \mathbf{11a} \qquad (2)$$

unique route was inspired by anodic oxidation studies of furans in the early 1950s by Clauson-Kaas and coworkers[7a–d] who showed that a wide variety of furans (e.g the parent **12**) could be electrochemically oxidized to the bisketals of structure **13** (equation 3). This transformation was initially conducted with ammonium bromide[7a] as electrolyte in methanol and involved electrogenerated bromine as the reactive species. Subsequent work demonstrated that electrochemical oxidation performed on furans with non-halogen electrolytes[7b–d] also afforded products analogous to **13** in good yield.

$$\mathbf{12} \xrightarrow[CH_3OH,\ H_2SO_4]{(E)} \mathbf{13}\ \text{(2,5-dimethoxy-2,5-dihydrofuran: } H, OCH_3;\ H_3CO, H) \qquad (3)$$

This anodic oxidation provided a unique and efficient route to a totally protected quinone derivative in one step from 1,4-dimethoxybenzene; however, the reaction attracted little attention in the intervening years. Several papers of a mechanistic nature appeared in 1973–5[8], but little synthetic use was made of the chemistry. Shortly thereafter, however, the potential utility of the reaction for the preparation of functionalized quinones was recognized[9]. This electrochemical oxidation was originally envisioned as a method for the preparation of functionalized protected quinone derivatives which subsequently could be transformed into anthraquinone natural products[10].

Of particular concern was the generality of the anodic oxidation to produce quinone bisketals. The concern was that oxidation of substituents on the benzenoid system would compete with the desired ring oxidation[11a–d]. In fact, under conditions similar to those employed for the anodic oxidation to form **11a**, oxidation of benzylic methylene groups, benzylic secondary and tertiary alcohols, dimethyl acetals, aldehydes, amides and conjugated esters had been reported[11d]. However, as illustrated in Table 1, this particular oxidation occurs in a single cell without potential control for a wide range of substituted 1,4-dimethoxybenzene and 1,4-dimethoxynaphthalene[11,12] systems as well as for some heterocyclic compounds[13,14]. Furthermore, some highly functionalized 1,4-dimethoxy-aromatic systems were oxidized to the respective quinone bisketals in good yields[15–17]. Some representative examples are given in Table 2.

The majority of these reactions were performed in 1 % methanolic potassium hydroxide; the presence of the water in the potassium hydroxide and adventitious water in the methanol had little effect on the reaction. However, the efficiency of the anodic oxidation decreased with increasing concentration of potassium hydroxide (Figure 1)[18]. For cases wherein base hydrolysis of a functional group is a problem (e.g. methyl esters), sodium methoxide in anhydrous methanol can be employed as the electrolyte and solvent system. Certain easily oxidized functional groups in an unprotected form such as amines, aldehydes and primary and secondary alcohols undergo competing oxidation and complicate the product mixture. However, amino groups not directly attached to the ring can be protected from oxidation by conversion to their trifluoroacetates[19]. Likewise, aldehydes and alcohols can be converted to their corresponding acetals and ethers. The latter linkages are stable to these electrochemical oxidation conditions.

2. *Anodic oxidations in a divided cell*

The single-cell oxidations described above are conducted with the compound in contact with both the anode and the cathode. This arrangement is acceptable for anodic oxidations in which the substrate is not easily reduced at the cathode. However, for compounds which

TABLE 1. Anodic oxidation in single cell[11d]

(Scheme: 1,4-dimethoxybenzene bearing R^1, R^2, R^3 → (Ⓔ, CH_3OH, KOH) → quinone bisketal with two $(OCH_3)_2$ groups)

R^1	R^2	R^3	Yield (%)
Br	H	H	78
Br	H	Br	58
CH_3	H	H	80
$Si(CH_3)_3$	H	H	93
CH_3	CH_3	CH_3	63
$CH(OCH_3)CH_3$	H	H	92
1,3-dioxolan-2-yl	H	H	88
$CH(OH)CH_3$	H	H	50[a]
$(CH_2)_3OH$	H	H	48[a]
$CH_2CH{=}CH_2$	H	H	81
SCH_3	H	H	54[b]

(Scheme: 1,4-dimethoxynaphthalene bearing R^1, R^2 → (Ⓔ, CH_3OH, KOH) → naphthoquinone bisketal with two $(OCH_3)_2$ groups)

R^1	R^2	Yield (%)
H	H	74
CH_3	H	75
OCH_3	H	83
CH_3	OCH_3	82
CH_3	$Si(CH_3)_3$	80

[a] Other products characterized from the reaction mixture.
[b] Yield of quinone monoketal from direct hydrolysis of the reaction mixture.

TABLE 2. Some representative anodic oxidations of functionalized 1,4-dimethoxyaromatic and heterocyclic systems in CH_3OH/KOH

Starting aromatic compound	Product (Yield)
	(98%)[15]
	(93%)[16]
	(97%)[16]
	(55%)[17]

Table 2. (continued)

Starting aromatic compound	Product (Yield)
4,7-dimethoxy-5-(methylthio)-6-bromo-2-(trimethylsilyl)benzothiophene (H_3CS, Br, OCH_3, OCH_3, $Si(CH_3)_3$)	bisketal with $(OCH_3)_2$, $(OCH_3)_2$, H_3CS, Br, $Si(CH_3)_3$ (85%)[13]
4,7-dimethoxy-1-(phenylsulfonyl)indole (H_3CO, H_3CO, N–SO_2Ph)	bisketal with $(OCH_3)_2$, $(OCH_3)_2$, N–H† (76%)[14a]
4,7-dimethoxybenzofuran (OCH_3, OCH_3, O)	bisketal with $(OCH_3)_2$, $(OCH_3)_2$, O (74%)[14b]
ethyl 5,8-dimethoxy-6-methoxy-7-methyl-2,3-dihydro-1H-pyrrolo[1,2-a]indole-9-carboxylate (OCH_3, H_3CO, H_3C, OCH_3, $COCH_2CH_3$ with C=O, N)	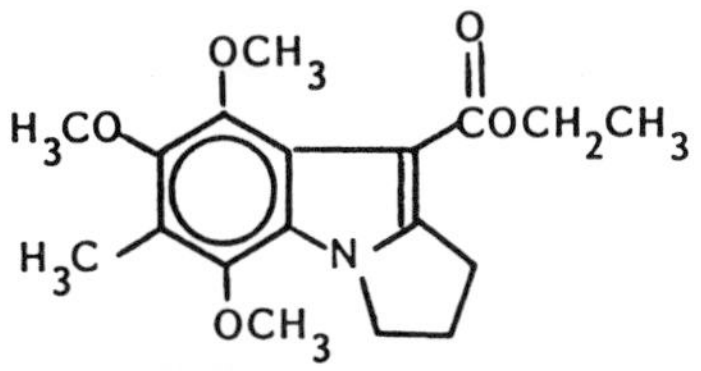 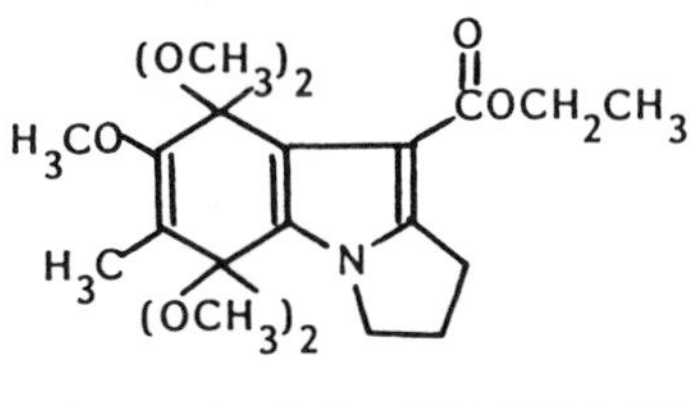 bisketal with $(OCH_3)_2$, $(OCH_3)_2$, H_3CO, H_3C, $COCH_2CH_3$, N (50%)[14a]

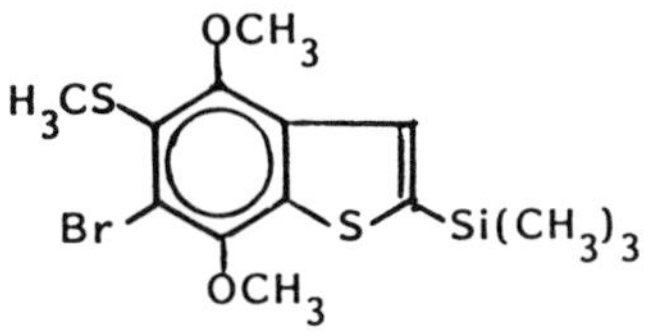

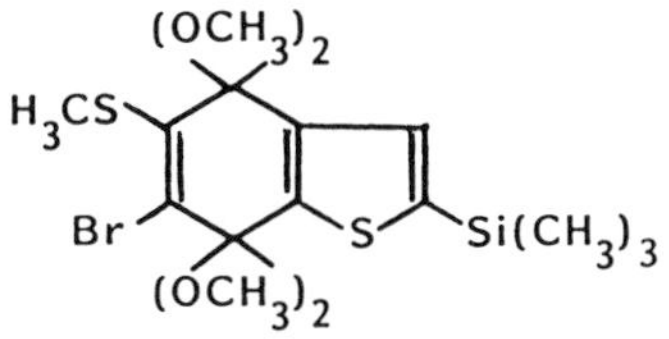

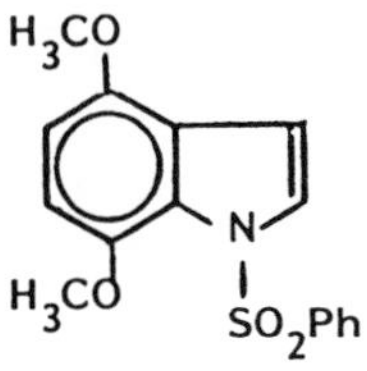

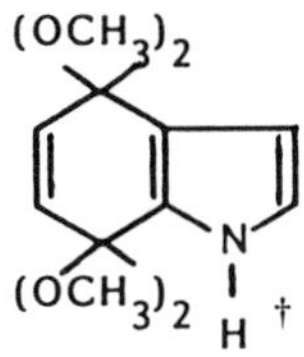

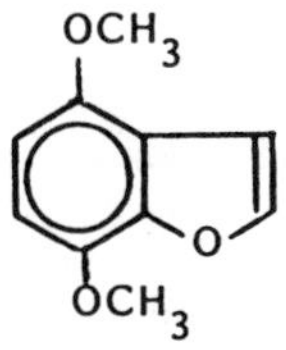

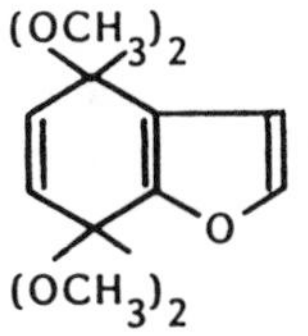

† The phenylsulfonyl group is lost either during the reaction or in the workup.

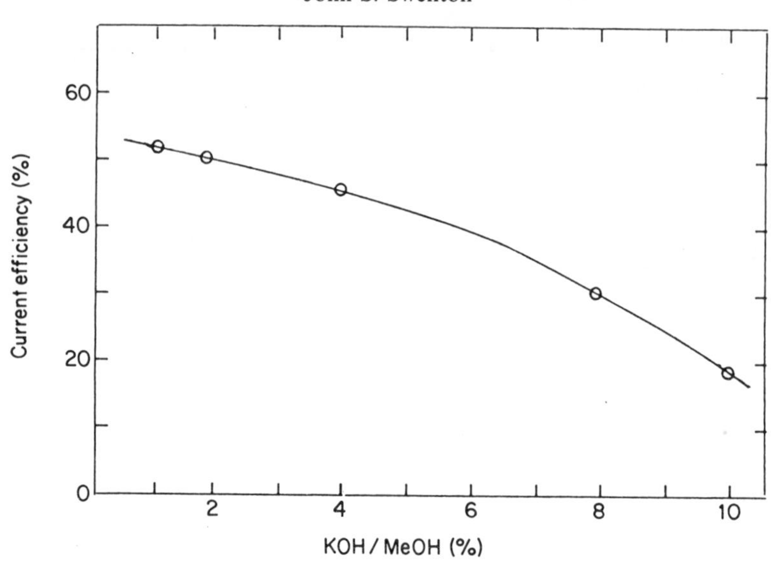

FIGURE 1. Current efficiency for anodic oxidation of 1-methoxynaphthalene as a function of base concentration

have reducible functional groups, this cell design can lead to undesired reduction processes of the substrate. In these cases successful oxidation of the compounds to quinone bisketals can be performed by conducting the electrochemical experiment in a cell which allows current to flow between the electrodes but keeps the substrate from coming into contact with the cathode. A simple method of conducting divided-cell oxidations is to employ an H-type cell with the anode and cathode compartments separated by a glass frit. More elaborate cell designs have the anode and cathode separated by a variety of membranes. Simple alkyl, chloro and bromo groups (in the benzenoid series only), non-conjugated double bonds, acetals and ethers are stable to the aforementioned anodic oxidation conditions in a single cell[11d]. However, other substituents undergo a competing reduction at the cathode under these conditions. Table 3 lists a series of anodic oxidations which, while giving complex mixtures of products in a single-cell system, usually gave acceptable yields of quinone bisketals in divided-cell electrolyses[11d]. In some cases the lower yields for anodic oxidations in divided cells may be attributed to diffusion of the compound into the cathode chamber during the course of the electrolysis and do not reflect any inherent inefficiency in the electrolysis reaction itself.

3. Mechanistic and experimental aspects of the anodic oxidation of other 1,4-oxygenated aromatics to quinone bisketals

Although anodic oxidation studies have been performed mainly on 1,4-dimethoxybenzene systems, it would seem that other simple 1,4-dialkoxybenzenes should also undergo smooth oxidation to the corresponding quinone bisketals[20]. Anodic oxidation of 1,4-diethoxybenzene **14** in 1% methanolic potassium hydroxide afforded in high yield an approximate 1:1 mixture of stereoisomeric bisketals **15** (equation 4) as determined by ^{13}C-NMR analysis at 125 MHz in these laboratories. Furthermore, quinone bisketals were formed in good yield from anodic oxidation of the compounds shown in Table 4. For the

TABLE 3. Anodic oxidations in a divided cell[11d]

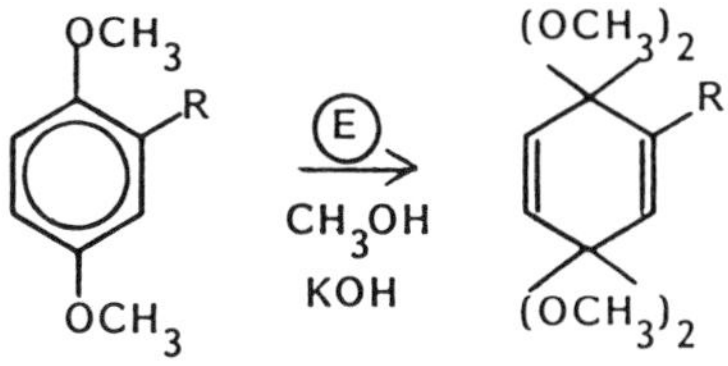

R	Yield (%)
$NHCOCH_3$	17
$(CH_2)_2CO_2CH_3$	61
$(CH_2)_2CON(CH_3)_2$	68
$(CH_2)_2CONH_2$	50
$CH{=}CHCO_2CH_3$	46
CHO	26, 59[a]

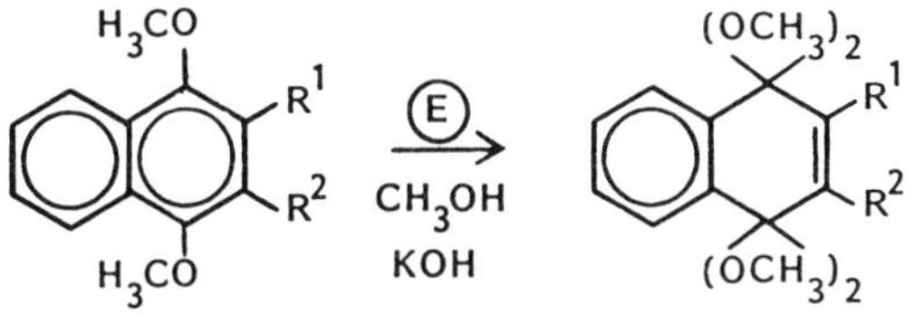

R^1	R^2	Yield (%)
Br	H	84
Br	CH_3	85
Br	Br	50

[a] Reaction at a vitreous carbon electrode.
The product is the bis(ketal) ester ($R = CO_2CH_3$).

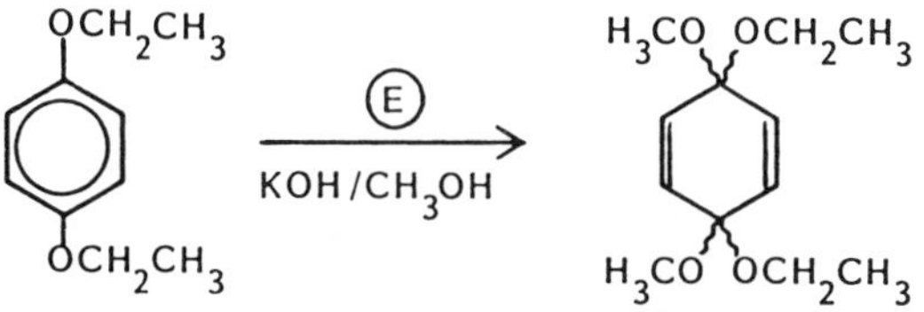

14 → 15 (78%) (KOH/CH_3OH, anodic oxidation) (4)

TABLE 4. Anodic oxidations to form ethylene glycol ketals of benzo- and naphthoquinone in CH_3OH/KOH

Entry	Reactant	Product(s) (Yield)	
	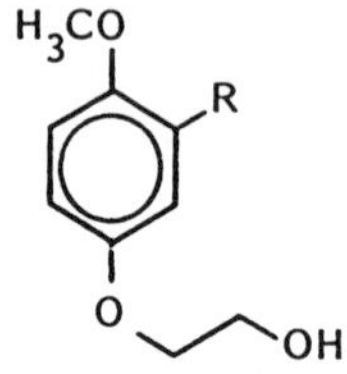	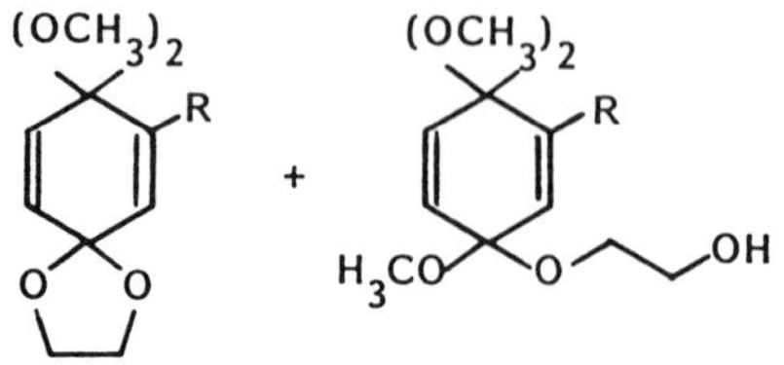	
1	R = H	(66%)	(15%)[21]
2	R = CH_3	(83%)	(7%)[21]
3	R = Br	(81%)	(13%)[21]
4	R = OCH_3	(86%)	—
5	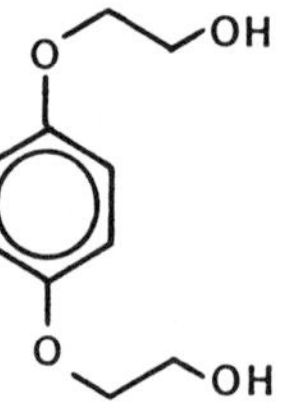	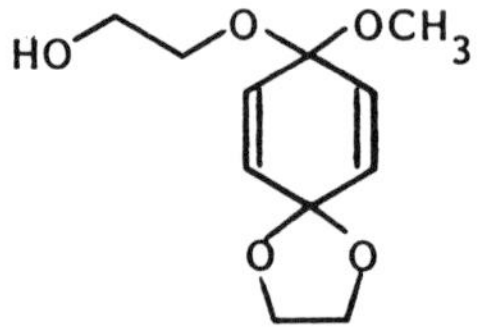	(93%)[8c]
6	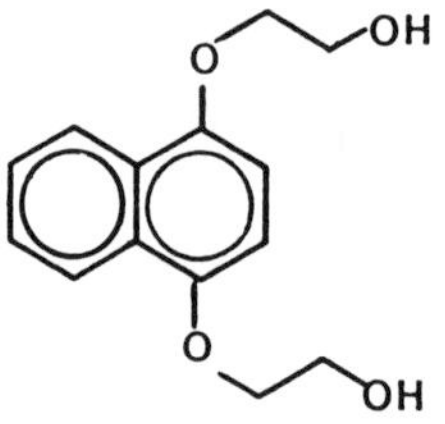	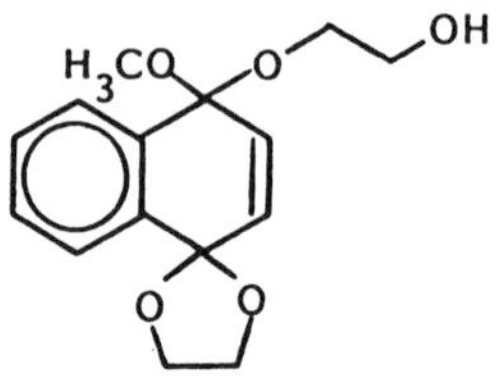	(85%)[21]

reactions cited in Table 4, anodic oxidation produced the mixed methanol–ethylene glycol ketal—a result of mechanistic significance[8c, 18]. Although not extensively studied, the nature of the alcohol component in the bisketal formation appears to have a bearing on the yield and current efficiency of the reaction, judging from the results of anodic oxidation of **14**. Electrolysis of **14** in ethanolic potassium hydroxide did form the analogous bisketal **16** (equation 5) but in lower isolated yield (63%) and current efficiency (21%) than the

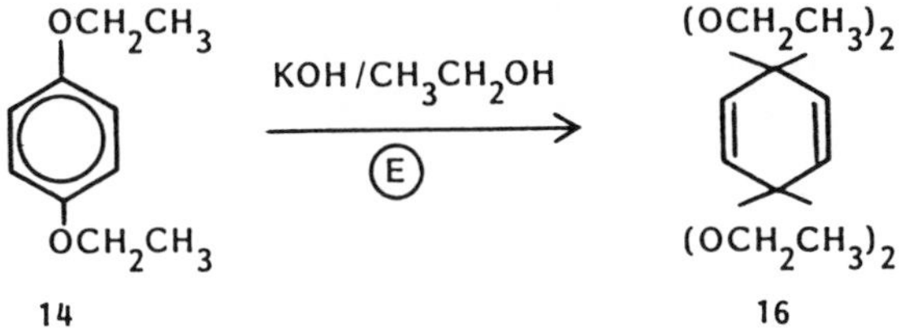

(5)

14 16

analogous reaction in methanol[11d]. Furthermore, the crude reaction mixture appeared much more complex as compared to analogous anodic oxidations done in methanol. This limitation on the reaction is not serious since the bisketals from other alcohols may often be prepared by performing exchange reactions on the readily available tetramethyl bisketal (*vide infra*).

An important, yet unanswered, mechanistic question is the detailed steps in the anodic oxidation of 1,4-dimethoxybenzenes to quinone bisketals. Several mechanisms have been offered for the process, but only the most recent proposal[18, 22] will be illustrated for 1,4-dimethoxybenzene (Scheme 1). The key feature of this mechanism is that two electrochemical intermediates are generated at the anode. The first is the radical cation of the

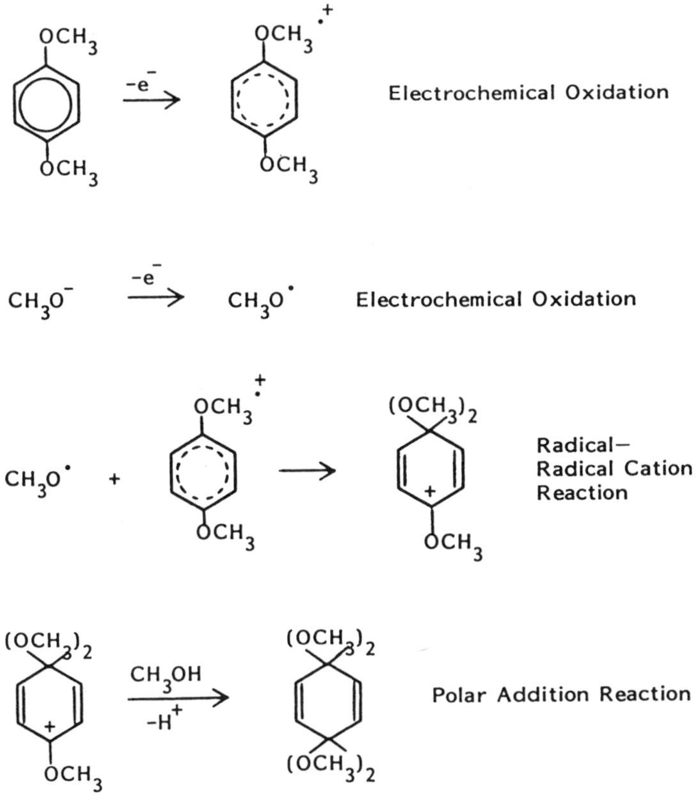

Electrochemical Oxidation

$CH_3O^- \xrightarrow{-e^-} CH_3O^\bullet$ Electrochemical Oxidation

Radical–Radical Cation Reaction

Polar Addition Reaction

SCHEME 1. EEC_rC_p mechanism for anodic oxidation of 1,4-dimethoxybenzene[18]

aromatic substrate, and the second is the methoxyl radical. Reaction of these two transients on the electrode surface affords a cationic intermediate which then reacts with methanol to produce the observed quinone bisketal. These steps were termed the EEC_rC_p mechanism[18], denoting that two electrochemical steps (EE) were followed by a radical combination step (C_r) and a polar-addition step (C_p). Although the detailed arguments supporting this mechanistic proposal will not be covered in this review, interested readers are referred to the original papers[18, 22] for further details.

In summary, anodic oxidations of 1,4-dimethoxybenzene and naphthalene systems are usually high-yield routes to their respective quinone bisketals. The current efficiencies of the oxidations are in the 20–80% range, most simple systems having current efficiencies of greater than 50%. Experimentally, the reactions require only a source of direct current and electrodes. Only in some highly functionalized systems has controlled potential electrolysis markedly improved the yield of the reaction. Platinum anodes and cathodes were used in the majority of the studies because of their convenience: however, inexpensive vitreous carbon anodes gave comparable—in some cases better—yields than platinum anodes[11d].

The reactions are usually conducted at 0°C but have been performed at temperatures as high as the boiling point of the solvent and as low as −25°C. The progress of the reaction is conveniently followed by UV spectroscopy since the product absorbs light at a considerably shorter wavelength than the starting material, although thin-layer chromatography can also be used if a UV spectrometer is not available. In fact, for many of the systems studied, an isosbestic point is observed as the reaction progresses: at the completion of the reaction, the optical density of the starting material at its absorption maximum is less than 5% of its initial value. Although it is more efficient to oxidize a homogeneous solution, compounds only partially soluble in methanol can be successfully electrolyzed as a slurry and/or with some tetrahydrofuran added as cosolvent. This works acceptably provided that the product is soluble in the reaction media so it does not precipitate onto the electrode.

The bisketals of quinones are usually quite stable in the absence of acid. However, for extended storage the bisketals should be handled with and stored in glassware which has been washed with dilute ammonium hydroxide. Magnesium sulfate should never be used as a drying agent for solutions of bisketals since hydrolysis to the monoketal will occur for some systems.

B. Quinone Bisketals from Anodic Oxidation of Other Aromatic Compounds

The anodic oxidation of 1,4-dialkoxyaromatic derivatives (hydroquinone ethers) is a very convenient method for preparation of quinone bisketals in the laboratory. However, this method requires the respective hydroquinone or the quinone as starting material. If the starting hydroquinone ethers are prepared from the quinone itself, the three-step reaction sequence—quinone reduction/alkylation/anodic oxidation—does not change the oxidation state of the compound and is a circuitous route to quinone bisketals. Furthermore, the availability of the hydroquinone ether determines the convenience of the electrochemical oxidation. Often, more readily available starting materials for the preparation of quinone bisketals are aromatic compounds below the hydroquinone oxidation state. Direct conversion of such systems to quinone bisketals also couples the oxidation step with the formation of the bisketal and involves efficient use of oxidation equivalents.

The anodic oxidation of benzene and simple benzene derivatives has been studied by the Hoechst group[23]. Although anodic oxidation of benzene does not appear to proceed satisfactorily in methanolic potassium hydroxide, this oxidation can be conducted in methanol at a platinum anode using a variety of other electrolytes. Most of the reported anodic oxidations were conducted only to partial conversion, the current efficiency for the formation of quinone bisketals being about 40%. This reaction has been conducted successfully for benzene (**17a**), anisole (**17b**), *o*-chloroanisole (**17c**), and *m*-methoxytoluene (**17d**) (equation 6).

$$\mathbf{17} \xrightarrow[CH_3OH]{(E),\ KF} \mathbf{11} \qquad (6)$$

	R^1	R^2	R^3
a,	H	H	H
b,	OCH_3	H	H
c,	OCH_3	Cl	H
d,	OCH_3	H	CH_3

Oxidation potentials for benzene and its simple alkyl derivatives are apparently too high for their efficient oxidation to quinone bisketals in methanolic potassium hydroxide. Even anodic oxidations of anisole, 2,6-di-*i*-propylanisole (**18**) and 2,6-di-*t*-butylanisole (**19**) give only low yields (20–40%) of their respective quinone bisketals in addition to other oxidation products[22]. However, anodic oxidation of 1,3-dimethoxybenzene (**20**) in methanolic potassium hydroxide gives the bisketal (**11e**) in 66% yield (equation 7)[6].

18 19

$$\mathbf{20} \xrightarrow[CH_3OH]{(E),\ KOH} \mathbf{11e} \qquad (7)$$

Naphthalene (**21**) is more easily oxidized than benzene: thus, conversion of a methoxylated naphthalene[11c, e, 12, 18] to a 1,4-naphthoquinone bisketal should be more facile than the analogous reaction of anisole. Anodic oxidation of naphthalene itself in methanolic potassium hydroxide heated to reflux gave the bisketal **22** in 42% isolated yield (22% current efficiency, equation 8)[18]. At the time the work was performed, the investigators were unaware of the Hoechst work[23]; presumably, the efficiency and yield of the reaction could be markedly improved by conducting the reaction in methanol with potassium fluoride as the electrolyte.

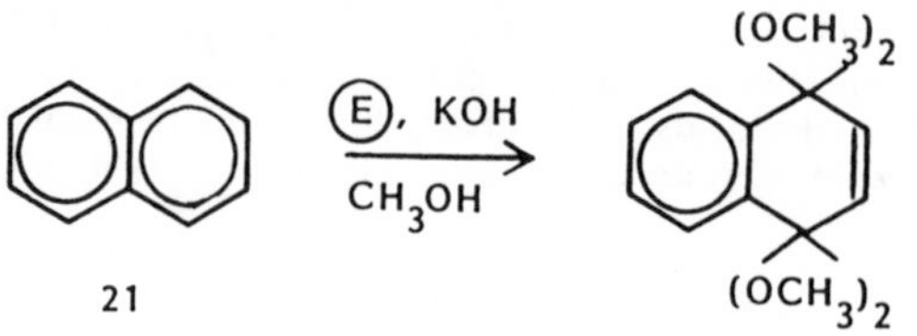

(8)

An extensive investigation of the products and their mechanisms of formation for anodic oxidation of 1- and 2-methoxynaphthalene and 1,2-, 1,3-, 1,4-, 1,5-, 1,6-, 1,7-, 2,3-, 2,6- and 2,7-dimethoxynaphthalene in methanolic potassium hydroxide has been reported[18]. Table 5 summarizes the reactions relevant to the present discussion. Interested readers are referred to the original article[18] for other electrochemical processes and the mechanistic discussion of these reactions; however, several general points concerning the synthetic utility of the reactions are noteworthy. The ratio of two- and four-electron oxidation products was dependent upon the particular compound being oxidized and the temperature of the oxidation. The origin of the temperature dependence of the products from anodic oxidation of 1-methoxynaphthalene is outlined in Scheme 2. Thus, the major process is two-electron oxidation and 1,4 addition of methanol to give 1,1,4-trimethoxy-1,4-dihydronaphthalene. At low temperature this product is moderately stable toward 1,4 elimination of methanol, but elimination occurs during workup of the reaction mixture to give 1,4-dimethoxynaphthalene. When the reaction is conducted in methanolic potassium hydroxide heated to reflux, an *in situ* 1,4 elimination of methanol occurs, followed by further oxidation to produce the bisketal.

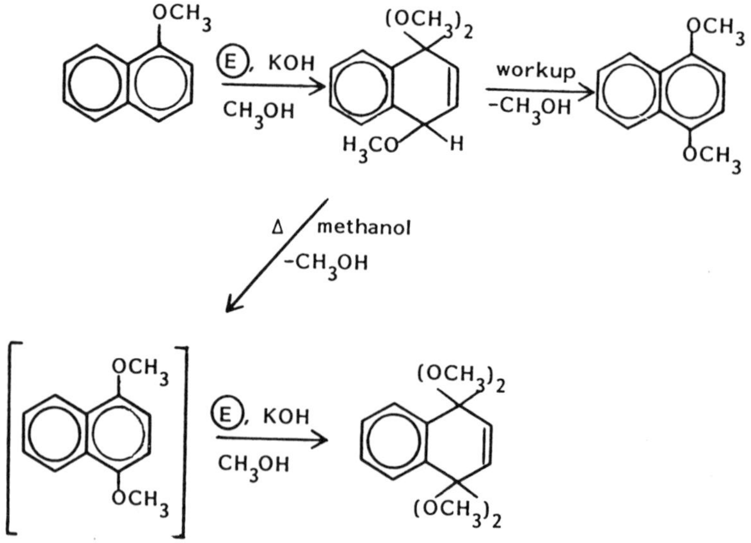

SCHEME 2. Rationalization of the temperature dependence of the anodic oxidation of 1-methoxynaphthalene

Due to the temperature-dependent nature of the electrochemical oxidation, this chemistry can also serve as a valuable method for the preparation of methoxylated

TABLE 5. Anodic oxidation products of selected methoxylated naphthalenes[18]

Reactant (Temperature)	Products (Yields)		
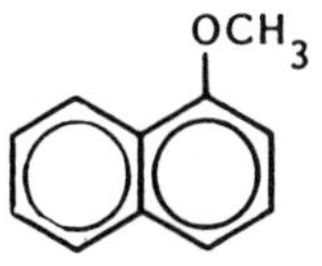	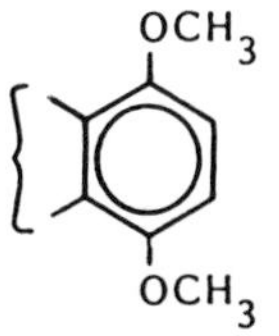 +	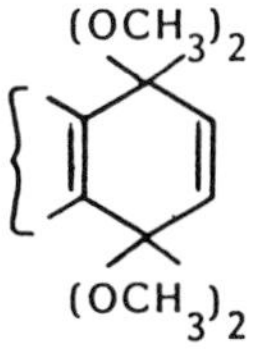 +	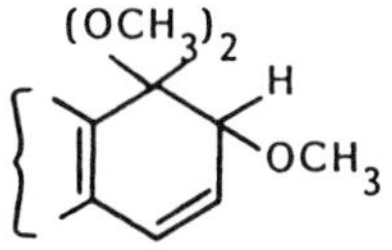
(0 °C)	(66%)	(21%)	(8%)
(reflux)	--	(83%)	(9%)
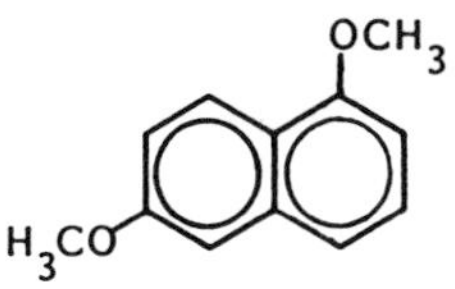 (reflux)	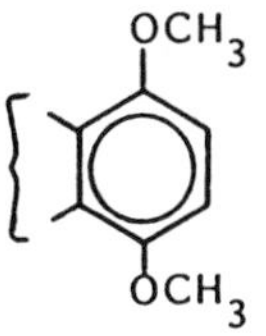 +	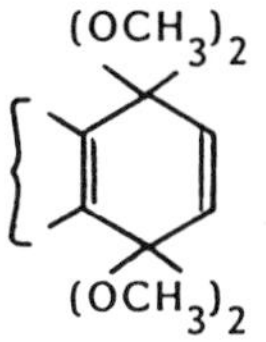 +	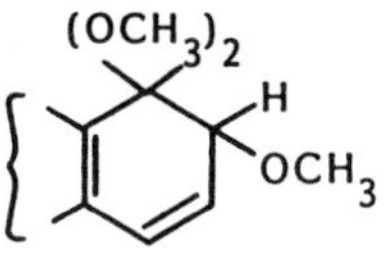
	(10%)	(45%)	(14%)
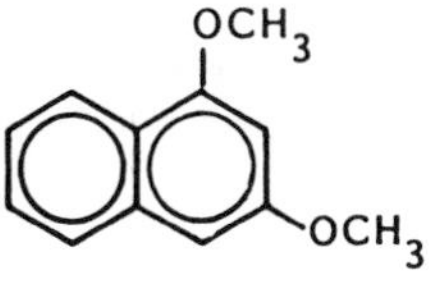 (reflux)	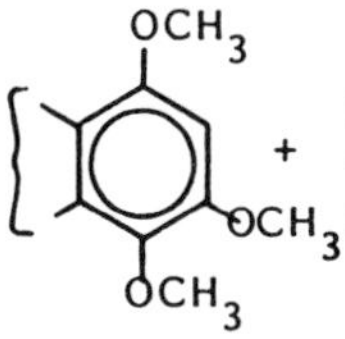 +	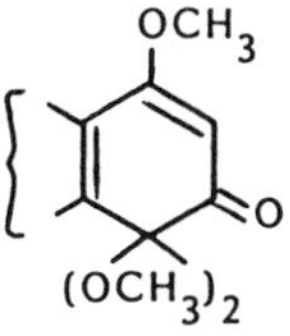 +	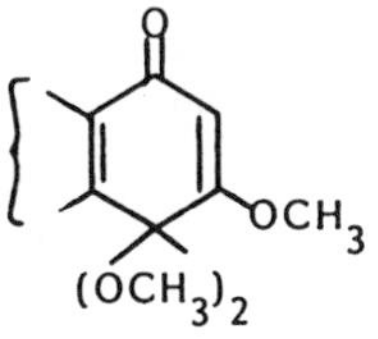
	(14%)	(59%)	(20%)
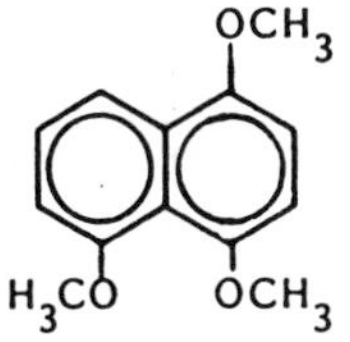	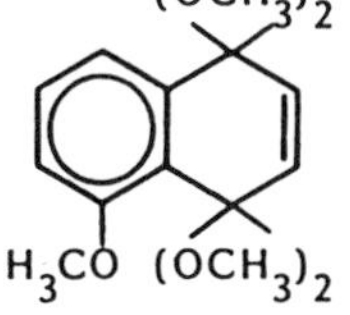		
(25 °C)	(> 80%)		

naphthalene derivatives. This is illustrated by the preparation of 1,4,5-trimethoxynaphthalene (**25**)—a valuable intermediate for production of juglone methyl ether (5-methoxy-1,4-naphthoquinone)—by anodic oxidation of readily available 1,5-dimethoxynaphthalene (**23**) (equation 9)[12,16].

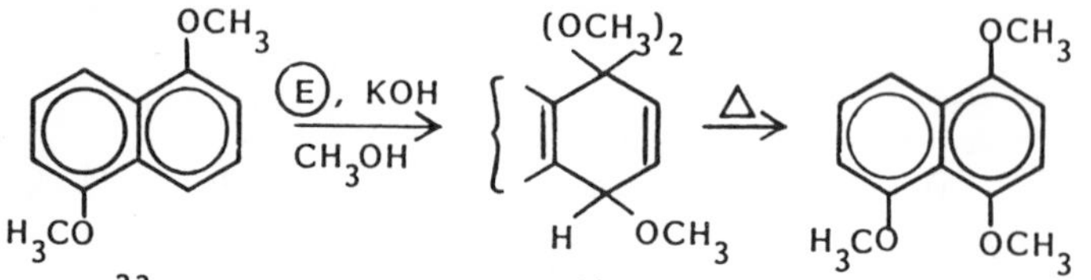

(9)

Anodic oxidation can be the most advantageous method for conversion of methoxylated aromatic systems to the quinone oxidation state. The preparation of benzo[*b*]thiophene quinones is illustrative of this point. Conversion of the phenols **26a,b** to their respective quinones (**27**) by classical chemical methods was effected in overall

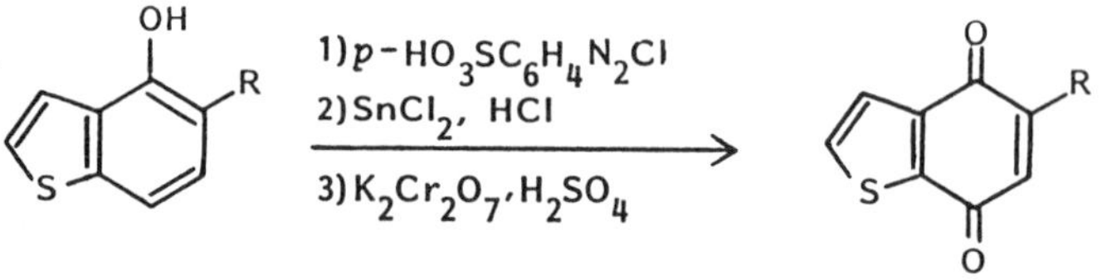

(10)

yields of 44%[24a] and 15%[24b] (equation 10). Anodic oxidation of the corresponding methoxyl derivatives **28** afforded the quinone bisketals **29a–d** in yields of 76–85% (equation 11)[13]. These compounds can be hydrolyzed with aqueous acid to the corresponding quinones in excellent yields. Thus, this four-electron oxidation procedure

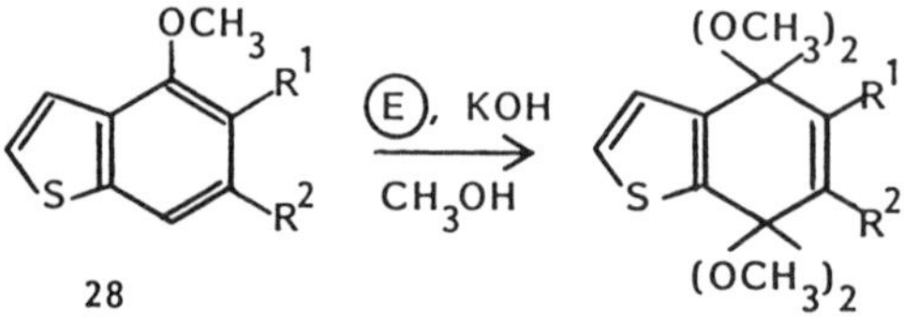

(11)

	R^1	R^2	%
a,	H	H	78
b,	CH_3	H	85
c,	Br	H	79
d,	H	Br	76

more than complements the classical method for preparation of benzo[*b*]thiophene quinones and quinone bisketals.

III. REACTIONS OF QUINONE BISKETALS

A. General Considerations

Many quinone bisketals are easily hydrolyzed by adventitious acid and moisture in the air to quinone monoketals—the source of acid catalysis is most often found on the surface of glassware. As noted earlier, all glassware which is to come in contact with these molecules should be rinsed with dilute ammonium hydroxide and dried in an acid-free environment. Furthermore, chromatography is best conducted on neutral alumina or silica gel which has been washed with ca. 5% methanolic ammonia and then dried *in vacuo* at ca. 100°C overnight. Ordinary silica gel often causes hydrolysis of quinone bisketals to quinone monoketals. With this acid lability in mind, quinone bisketals can be handled and reacted in an otherwise normal fashion. The bisketals are reasonably stable thermally since benzoquinone bisketal **11a** can be isolated by vacuum distillation (b.p. 85–89 °C/0.3 mm). Electrophilic additions to the double bond of quinone bisketals have not been studied extensively. However, hydrogenation[25] of benzoquinone bisketal **11a** to the bisketals of 1,4-

$$\mathbf{30} \xleftarrow[\text{Pd/C},\ H_2,\ 92\%]{CH_3OH/Et_3N} \mathbf{11a} \xrightarrow[\text{Pd/C},\ 2H_2,\ 90\%]{CH_3OH} \mathbf{31} \qquad (12)$$

cyclohexanedione (**31**) and 1,4-cyclohex-2-ene-1,4-dione **30** has been reported (equation 12). Interestingly, reaction of the Simmons–Smith reagent or dimethylsulfoxonium methylide and **11** led to **10**, **32** and **33** (equation 13)[26].

$$\mathbf{11a} \xrightarrow[CH_2{=}S(O)(CH_3)_2]{Zn(Cu),\ CH_2I_2\ \text{or}} \underset{\text{major}}{\mathbf{32}} + \underset{\text{minor}}{\mathbf{10}} + \underset{\text{minor}}{\mathbf{33}} \qquad (13)$$

B. Exchange Reactions with Quinone Bisketals

As was noted previously, anodic oxidation of 1,4-diethoxybenzene in ethanolic potassium hydroxide was not a clean reaction. Although the anodic oxidations of

$$\mathbf{11a} + ROH \xrightarrow[-CH_3OH]{H^+} \mathbf{34} \qquad (14)$$

34a, R = Et (91%)
b, R = *n*-Bu (90%)
c, R = $PhCH_2$ (96%)

hydroquinone ethers in solvents other than methanol and ethanol have not been extensively studied, many of the bisketals that can be potentially derived from these types of reactions may be prepared from the readily available tetramethoxy compound, **11a**. Thus, the acid-catalyzed reaction of **11a** with an alcohol as solvent at room temperature gave good yields of the bisketals **34a–c** (equation 14)[27]. This chemistry suggests that the cation **35** reacts under kinetic control at the 1-position with oxygen nucleophiles since heating **11a** in methanol with an acid catalyst affords 1,2,4-trimethoxybenzene, presumably via **36** (equation 15).

34 $\xrightleftharpoons{H^+, -CH_3OH}$ 35 $\xrightarrow{CH_3OH, -H^+}$ 36 (15)

OCH_3, $(OCH_3)_2$, OCH_3, H, $(OCH_3)_2$

C. Metalated Quinone Bisketals

One of the early uses of quinone bisketals in synthesis was as a metalated quinone equivalent[9]. Brominated benzoquinone and naphthoquinone bisketals, readily available from anodic oxidation of corresponding 1,4-dimethoxy bromoaromatic derivatives, undergo rapid metal–halogen exchange reactions at −78 °C. The lithio derivatives react with selected electrophiles to give good yields of the corresponding functionalized quinone bisketals and thence the functionalized quinone by acid hydrolysis. Table 6 shows the results from reaction of the organolithium derived from the parent benzoquinone bisketal. Several general points are noteworthy.

(1) The reagent reacts with difficultly enolizable ketones to give adducts in good yields while protonation of the bisketal is a problem with substrates that are more readily enolized.

(2) This hindered reagent reacts cleanly with aryl esters to give ketones.

(3) Hydrolysis of the adducts from aldehydes and ketones affords the parent quinones in high yields.

(4) Low yields of adducts are obtained from reactions with simple alkyl and allylic halides.

The problem of competing enolization observed in the reaction of the lithiated quinone bisketal with aldehydes and ketones might have been less serious if the recently developed organocerium derivatives[28] had been employed for the reaction. The functionalized lithiated quinone bisketal (**38**) was employed in one of the first regiospecific routes to anthracyclinones (e.g. **39**, equation 16)[29].

37 + 38 $\xrightarrow[2)\ H_3O^+]{1)\ -70\ °C}$ 39 (56%)

H_3CO; Li, $(OCH_3)_2$, CH_3, OLi, $(OCH_3)_2$, OCH_3; OCH_3, CCH_3, OH, H_3CO, OCH_3, OCH_3

(16)

TABLE 6. Functionalization chemistry of the 2-lithio derivative of benzoquinone bisketal[9a,b]

Reactant	Yield (%)	Yield (%)	R	R'
cyclohexanone	80	--	$(CH_2)_4$	
cycloheptanone	40	--	$(CH_2)_5$	
benzaldehyde	68	--	Ph	H
benzophenone	72	--	Ph	Ph
methyl benzoate	--	78	Ph	
N-benzoylpiperidine	--	68	Ph	
dimethyl phthalate	--	70*	$2\text{-}CO_2CH_3(C_6H_4)$	
dimethyl 3-methoxyphthalate	--	70	$3\text{-}OCH_3\text{-}2\text{-}CO_2CH_3(C_6H_3)$	

No/low yield of products with: CH_3I, CH_3CH_2I, CH_3CHO, CH_3COCH_3, $PhCH_2Br$, $PhCH_2Cl$, CH_3COCl

* Product was mixture of ester and pseudo ester.

Although introduction of allylic substituents by reaction of the lithiated quinone bisketal with allylic halides was unsuccessful, the reaction proceeded satisfactorily when the corresponding cuprate of the bisketal was employed[30]. Functionalization of quinone bisketals via the cuprate was a key aspect in the synthesis of **40a**, menaquinone-2 (**40b**), phylloquinone (**40c**), cympol (**41a**) and cympol methyl ether (**41b**)[30].

D. Bisketals as Synthons for Nucleophilic Substitution Reactions of 1,4-Dimethoxyaromatic Systems

Nucleophilic substitution reactions of quinone bisketals give rise to products in the hydroquinone oxidation state, and this has been exploited by the Hoechst group for the preparation of a variety of 1,4-dimethoxybenzene derivatives[31]. Table 7 illustrates the range of nucleophilic reagents which afforded substitution products under acid-catalyzed conditions with benzoquinone bisketals. Although analogous substitution reactions have not been studied extensively for substituted quinone bisketals, the regiochemical outcome of such reactions should show good selectivity based on the regiochemistry of the

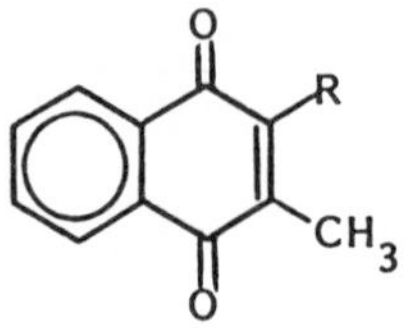

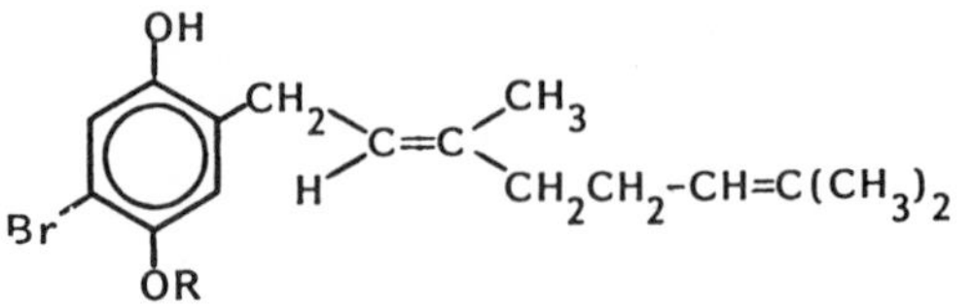

40 a, R = $-CH_2-CH=C(CH_3)_2$

b, R = $-H_2C(H)C=C(CH_3)CH_2CH_2CH=C(CH_3)_2$

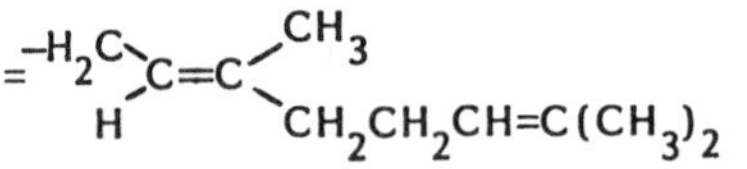

c, R = $-H_2C(H)C=C(CH_3)[CH_2CH_2CH(CH_3)]_3CH_3$

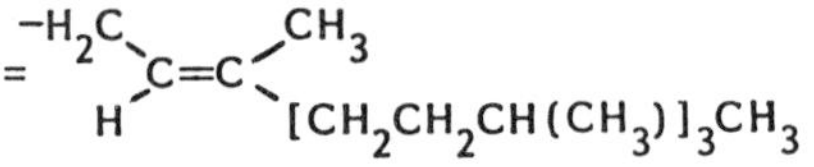

41 a, R = H

b, R = CH_3

monohydrolysis of the quinone bisketals (*vide infra*). Much useful aromatic substitution chemistry could result from further studies in this area.

IV. PREPARATION OF QUINONE MONOKETALS

A. Chemical Oxidation of Phenols

Monoprotected quinones have been known for many years and have been prepared by a variety of methods. Compounds in which one of the quinone carbonyl groups is masked as a dihalide derivative (see Table 8) are well characterized. Although these compounds might permit selective reaction at the unprotected carbonyl group of the quinone, their preparation is not general and often proceeds in low yield. More recently, the trimethylsilyl cyanide blocked quinone derivatives, e.g. **5**, have been used in synthesis[4]; however, in general, they appear to be less versatile than the quinone monoketals described below, and their chemistry will not be presented.

There is an extensive body of literature on the formation of quinone monoketals via oxidation of functionalized phenols with a variety of reagents: ferric chloride[35], potassium hexacyanoferrate (III)[35, 36], ceric ammonium sulfate[37], tetrachlorobenzoquinone[37], *N*-bromosuccinimide[37], manganese dioxide[38], dichlorodicyanobenzoquinone[38], silver oxide[38], copper (II) pyridine complex and oxygen[39], periodic acid[40], thallium (III) nitrate[41] and mercuric oxide with iodine[42]. In some cases, the yields of quinone monoketals are low, and in other cases they are not reported. Furthermore, many of these reactions are unique to a particular molecule because they possess structural features which lead to intramolecular formation of the quinone monoketal. Thus, an extensive discussion of this area is beyond the scope of the present review. However, Table 9 shows some of the unique quinone monoketals obtained from these studies, and readers are referred to the original papers for further examples and details.

Although a number of simple quinone monoketals were known in the literature, there was not a reasonably general method for their preparation until the work of McKillop, Taylor and coworkers[41]. These investigators found that thallium trinitrate, in either methanol or a mixture of methanol and trimethyl orthoformate, smoothly converted a number of *p*-alkoxyphenols to quinone monoketals in good yields. Table 10 lists a representative number of oxidations successfully performed using thallium trinitrate. Aside from the toxicity of thallium, which is not a small consideration for the large-scale preparation of monoketals, this is a good procedure for the preparation of these

TABLE 7. Benzoquinone bisketals in the synthesis[31] of substituted 1,4-dimethoxybenzenes

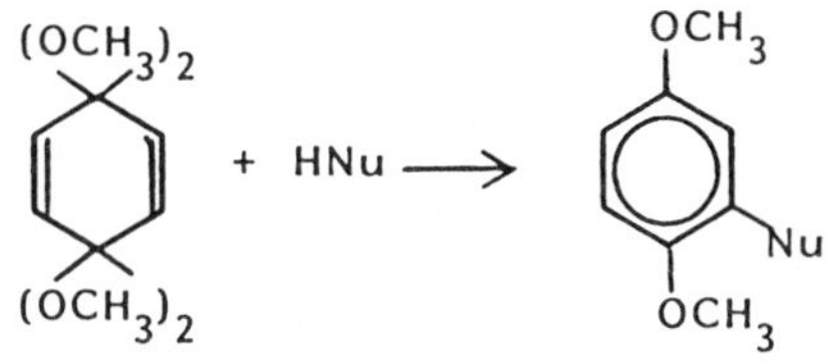

$(OCH_3)_2$ / $(OCH_3)_2$ + HNu ⟶ OCH_3 / Nu / OCH_3

H-Nu	Yield (%)	H-Nu	Yield (%)
CH_3OH	90	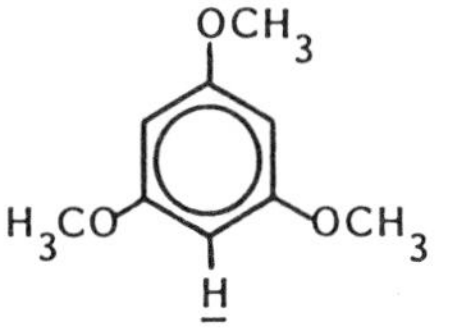 OCH_3, H_3CO, OCH_3, H	50
H-Cl	93		
$CH_3C(=O)OH$	92	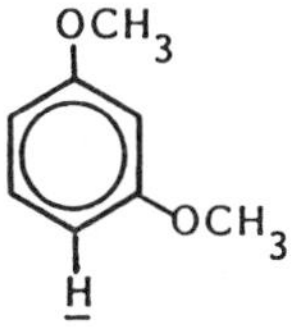 OCH_3, OCH_3, H	20
HN_3	87[a]		
C_2H_5SH	69	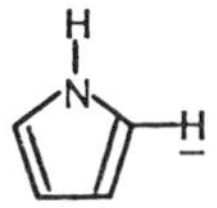 H–N, H	51
$HSCH_2CO_2Et$	61		
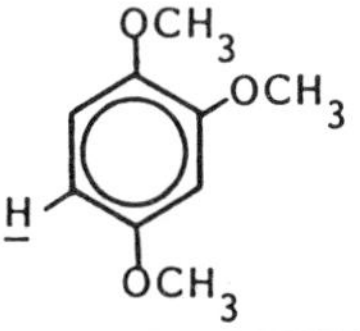 OCH_3, OCH_3, H, OCH_3	61	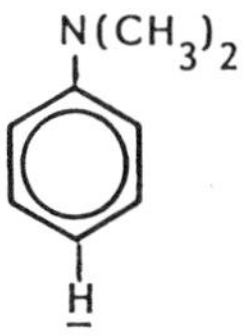 $N(CH_3)_2$, H	48

[a] Characterized as the amine after reduction.

TABLE 8. Some quinone derivatives in which one carbonyl group is masked as a dihalide

Compound	Product(s)
Cl_2, HOAc/pyridine	(60%)[32]
$FClO_3$, pyridine	(1:1 mixture, good yield)[33]
Cl_2, $CHCl_3$	(30%) + (26%)[34]; HCl slow; CH_3OH, room temp.

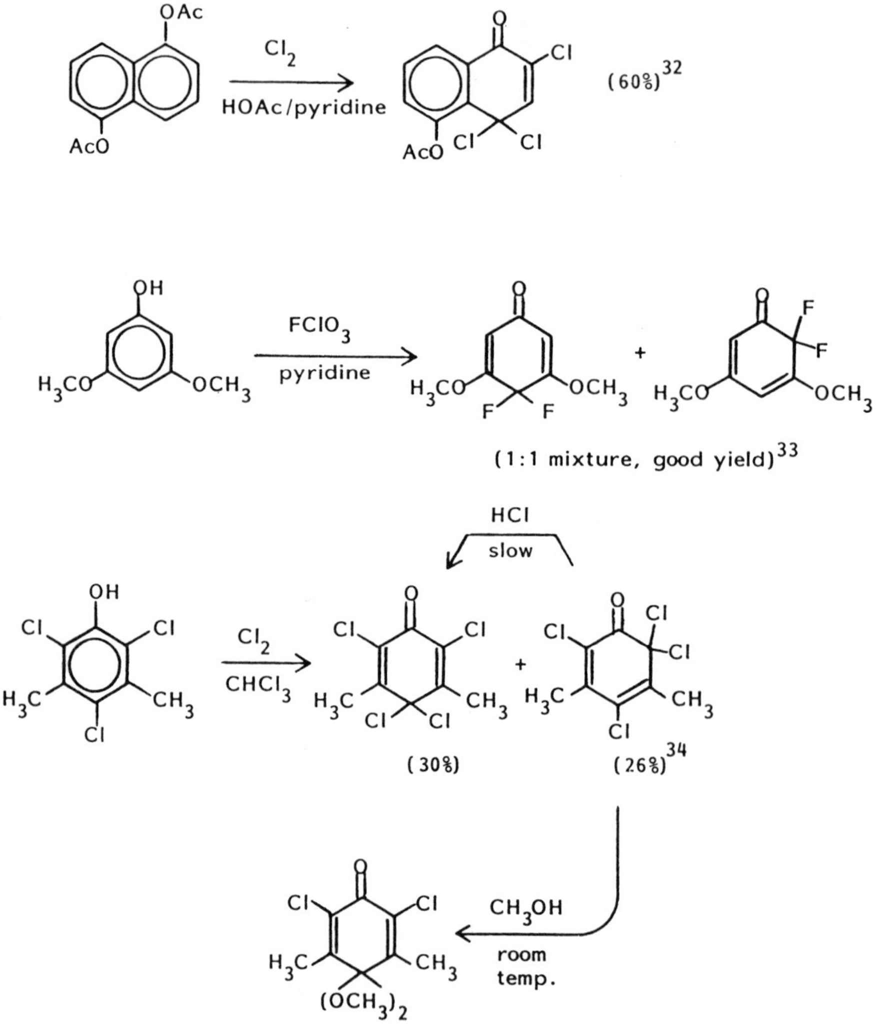

synthetically interesting quinone derivatives. The availability of quinone monoketals via the thallium trinitrate oxidation undoubtedly played a role in the subsequent increase in the use of these compounds in synthesis.

Further work has refined the preparation of quinone monoketals using thallium trinitrate as an oxidant by using suspended potassium bicarbonate in the reaction media—

TABLE 9. Selected examples of quinone monoketals and an aminal obtained via phenol oxidations

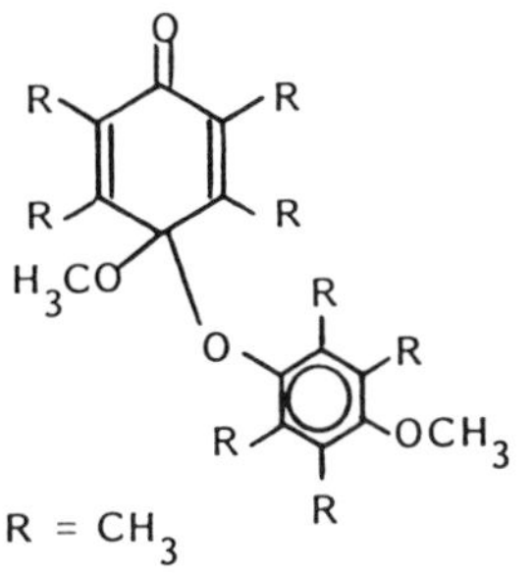

R = CH_3 (ref 35)

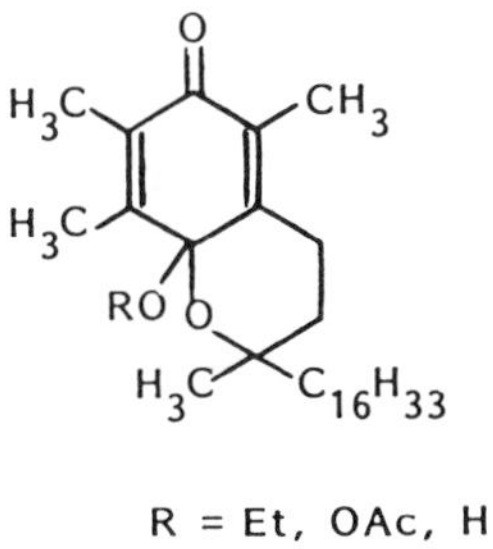

R = Et, OAc, H (ref 36)

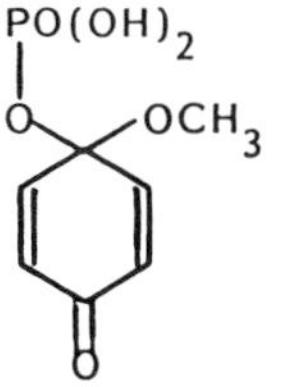

(ref 37b)

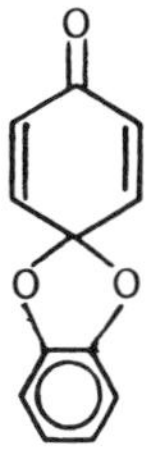

(ref 38c)

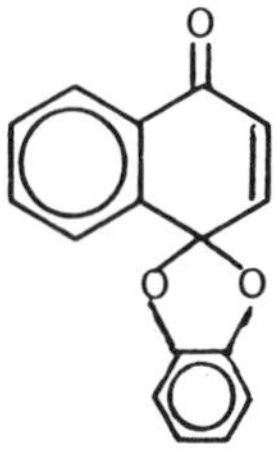

(ref 38d)

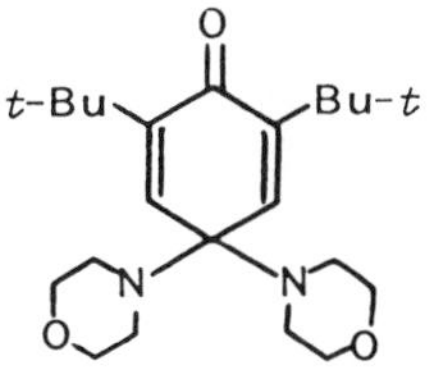

(ref 39)

presumably to neutralize acid generated in the reaction[43]. This synthetic expedient improved the yield of acid-sensitive quinone monoketals. In addition, a comparison was made of the efficiency of the three most common oxidizing agents: dichlorodicyanoquinone (DDQ), ferric chloride and thallium trinitrate[43]. The results, summarized in Table 11, illustrate the complementary nature of these oxidizing agents in the conversion of phenols to quinone monoketals. In many cases, DDQ and ferric chloride gave yields of quinone monoketals comparable to those of the toxic thallium trinitrate.

These investigations have made available a variety of quinone monoketals for use in synthesis subject to the availability of the requisite *p*-alkoxyphenol. Although many quinone monoketals are best prepared—especially on a large scale—by hydrolysis of quinone bisketals, for some compounds, chemical oxidation may still be the method of choice.

B. Electrochemical Oxidation of Phenols

As is apparent from Section IV.A, quinone monoketals are conveniently available by chemical oxidation of *p*-alkoxyphenols. However, in some cases relatively expensive and toxic oxidizing agents are required. Ronlan and coworkers have extensively studied the anodic oxidation of phenols in methanol as a function of anode material (platinum vs. carbon), anode potential, concentration, supporting electrolyte and temperature[47].

TABLE 10. Anodic oxidation of *p*-alkoxyphenols with thallic(III) nitrate

Phenol				Monoketal
R^1	R^2	R^3	R^4	Yield (%)[41]
H	H	H	H	97
CH_3	H	H	H	89
CH_3	H	H	CH_3	87
t-Bu	H	H	*t*-Bu	96
H	OCH_3	OCH_3	H	95
H	OCH_3	OCH_3	$C(O)CH_3$	92
Cl	H	H	H	97
Br	H	H	H	91

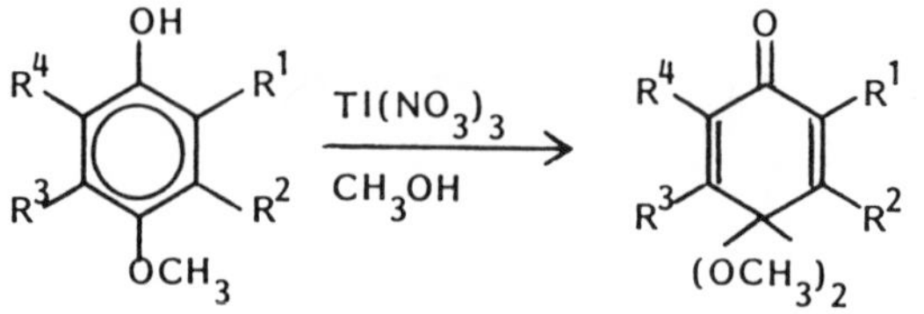

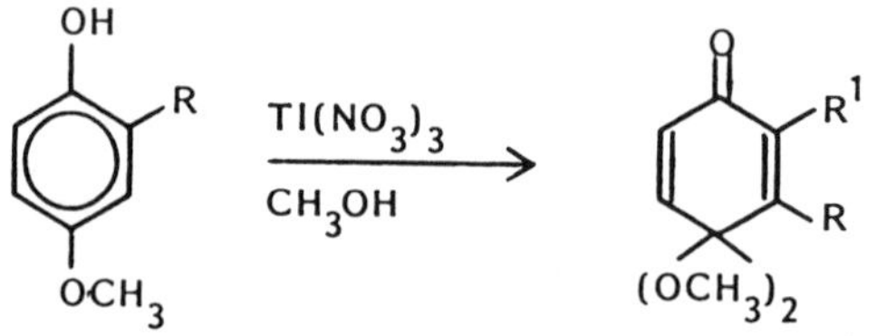

$Tl(NO_3)_3$, CH_3OH

R = CH=O → R = H, R^1 = $CH(OCH_3)_2$

R = CO_2CH_3 → R = OCH_3, R^1 = CO_2CH_3 (90)[44]

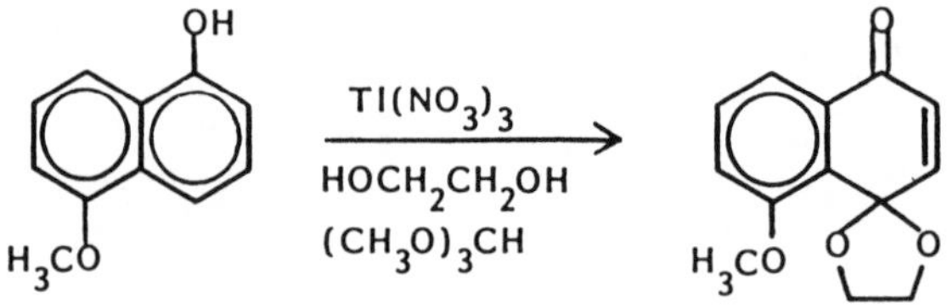

$Tl(NO_3)_3$, $HOCH_2CH_2OH$, $(CH_3O)_3CH$ (33)[45]

$Tl(NO_3)_3$, CH_3OH (efficient)[46]

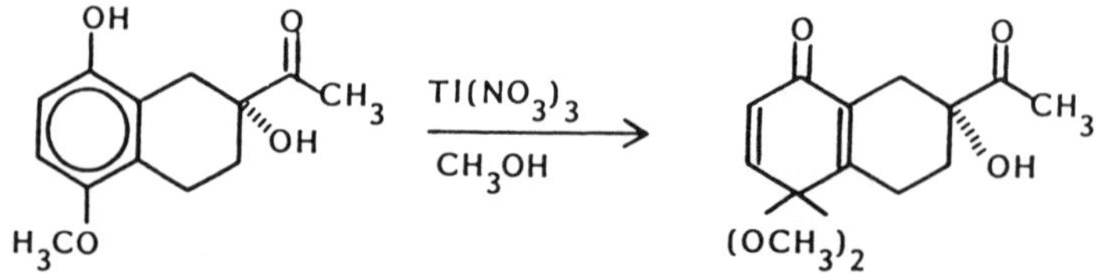

TABLE 11. Comparison of various oxidizing agents for oxidation of *p*-alkoxyphenols to quinone monoketals in methanol[43]

Phenol	Quinone monoketal	R^1	R^2	Yield (%)		
				DDQ	$FeCl_3$	$Tl(NO_3)_3$
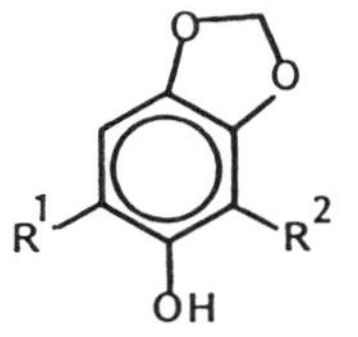	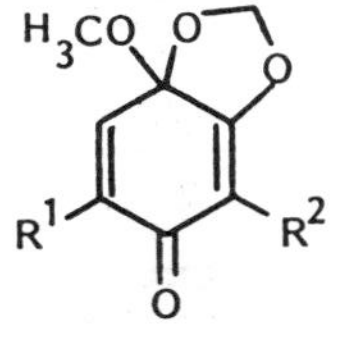	H	H	< 5	< 5	83
		$CH_2CH{=}CH_2$	H	88	84	78
		H	$CH_2CH{=}CH_2$	31	0	21
		H	OCH_3	57	0	68
		OCH_3	H	50	50	70
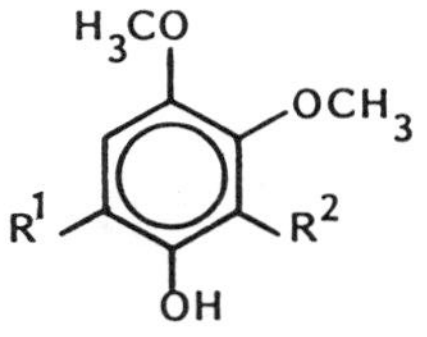	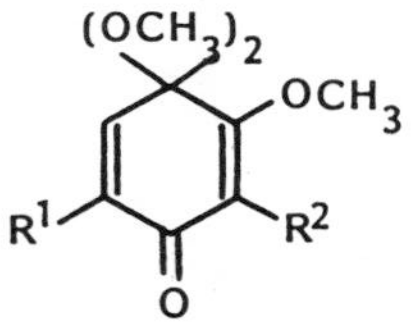	H	H	< 5	< 5	66
		$CH_3CH_2CH_2$	H	85	88	80
		H	$CH_2CH{=}CH_2$	75	0	0
		OCH_3	H	< 5	78	0

Although phenol **42** underwent anodic oxidation in methanol to form the quinone monoketal **43** in 46% yield, the authors noted it was quite difficult to separate the product from the oxidation mixture. However, for the more hindered 2,6-di-*t*-butylphenol (**44**) the corresponding quinone monoketal **45** was obtained in 77% yield from a 10-gram electrolysis (equation 17).

$$\text{42, R = H; 44, R} = t\text{-Bu} \xrightarrow[\text{LiClO}_4,\ \text{lutidine}]{\text{(E), carbon}} \text{43 (46\%); 45 (77\%)}, (\text{OCH}_3)_2 \quad (17)$$

These workers also published a procedure for the preparation of the parent quinone monoketal **43** by oxidation of *p*-methoxyphenol in methanol using lithium perchlorate as supporting electrolyte[25b]. Although this electrochemical oxidation is an excellent method[48] most investigators have preferred to use chemical oxidants for the conversion of *p*-alkoxyphenol to quinone monoketals. However, this electrochemical oxidation was the method of choice for conversion of functionalized *p*-alkoxyphenols to their respective quinone monoketals (Table 12)[49].

TABLE 12. Anodic oxidation of *p*-methoxyphenols to quinone monoketals[49]

Phenol		Quinone Monoketal	R	Yield (%)
OH, R, OCH_3	(E), $LiClO_4$, CH_3OH	O, R, $(OCH_3)_2$	CH_3CH_2	(82)
			t-$BuCH_2$	(83)
			H	(97)
R, OH, OCH_3	(E), $LiClO_4$, CH_3OH	R, O, $(OCH_3)_2$	CH_3	(82)
			CH_3CH_2	(85)
			t-Bu	(82)

One variant of the above phenol oxidation is the electrolysis of the corresponding silyl ether of the phenol[50]. Thus, anodic oxidation of the silyl ether of *p*-methoxyphenol in methanol–acetonitrile using lithium perchlorate as the electrolyte in the presence of lithium carbonate affords benzoquinone monoketal **43** in 99% yield[50].

Another electrochemical method to furnish quinone monoketal systems is the oxidation of phenol ethers in a non-nucleophilic solvent—although the yields in these cases are usually not high. Thus, while anodic oxidation of **46** in acetonitrile/methanol gave **47b**[51a,b], anodic oxidation of **46** in the absence of methanol in the presence of Et_3N afforded the quinone monoketal-like compounds (**47a**) in modest yield (equation 18) in addition to other products[52].

$(OCH_3)_2$ — Ⓔ, CH_3OH, CH_3CN ← OCH_3 … $C-CO_2H$ → Ⓔ C, CH_3CN, Et_3N (18)

47b (90%) 46 47a

R	R	Yield (%)
CH_3	CH_3	(41)
CH_3	H	(22)
$(CH_2)_5$		(40)

In summary, the preparation of quinone monoketals via electrochemical oxidation of *p*-methoxyphenols is an excellent alternative to the chemical oxidations discussed above. The reactions have been conducted at constant current on a moderate scale (ca. 0.2 mol) using simple apparatus and afford good yields of product. While the scope of these anodic oxidations has not been extensively studied, this method of phenol oxidation allows preparation of quinone monoketals without the use of toxic and expensive oxidizing agents.

C. Quinone Monoketals via Hydrolysis of Quinone Bisketals

1. *Background and general comments*

Provided that the required *p*-methoxyphenol is available, the oxidation methods discussed above serve as excellent preparations of the corresponding quinone monoketal. However, for compounds in which the ring is substituted, the 1,4-dialkoxy ether (typically the dimethoxy ether) is often a more readily available starting material since it is prepared by reduction of the quinone and alkylation of the resulting hydroquinone. Since quinone bisketals can be prepared in high yields by anodic oxidation of 1,4-dimethoxy benzene and naphthalene derivatives, hydrolysis of one of the two ketal groups would afford quinone monoketals via a route complementary to that discussed above. Furthermore, some quinone bisketals are available by direct oxidation of aromatic compounds having oxidation states lower than the hydroquinone: this would then serve as a more efficient preparation of quinone monoketals.

The monohydrolysis of the bis(ethylene glycol) ketal of benzoquinone was one of the first routes employed for preparation of quinone monoketals[5]. Even though a kinetic study had reported that the acid-catalyzed hydrolysis of the second ketal of quinone bisketal (**11a**) was 300 times slower than the first ketal[53], apparently some difficulty was encountered in taking advantage of this difference in rate preparatively[26]. Thus, hydrolysis of **11a** under the conditions[8a] reported to give the monoketal **43** (equation 19) gave only

$(OCH_3)_2$ $\xrightarrow{H_3O^+}$ (19)

$(OCH_3)_2$ $(OCH_3)_2$

11a 43

benzoquinone when repeated by a second group[26]; however, the **11a** to **43** conversion could be conducted by hydrolysis with warm water. The contradictory results reported in these studies probably arose from the presence of adventitious acid catalyzing hydrolysis of the quinone monoketal to the quinone during the reaction.

The hydrolysis of quinone bisketals to quinone monoketals is usually a routine preparative procedure when attention is given to the items noted below[54]. It is advantageous to use a weakly acidic media and to mix a cooled solution of the quinone bisketal in acetone with the cooled acid solution. This is advisable even in cases wherein the actual hydrolysis is conducted at room temperature. Most of the quinone bisketal hydrolyses have been conducted in 1–8% aqueous acetic acid–acetone at temperatures from −20 °C to 35 °C, depending upon the particular compound. Ordinarily, the quinone monoketals obtained from standard workup of the reaction mixture are suitable for most preparative purposes. If chromatographic separation of regioisomeric quinone monoketals is required, neutral alumina or silica gel are suitable adsorbents. However, the silica gel should be washed with ca. 5% methanolic ammonia and dried overnight under vacuum above 100 °C for quinone monoketals which are especially labile toward hydrolysis to the quinone. Most quinone monoketals are indefinitely stable when stored in base-washed glassware in the absence of light at 0 °C, although this low temperature is probably not critical to their stability.

2. *Stereoelectronic considerations for the regiochemistry of bisketal hydrolysis*

A major synthetic advantage of quinone monoketals over quinones is the regiochemistry inherent in nucleophilic additions at the carbon β to the carbonyl group. Thus, the regiochemical outcome of the monohydrolysis of quinone bisketals is of major concern. Tables 13 and 14 illustrate some major points concerning the regioselectivity of quinone bisketal hydrolysis as a function of the substituents on the vinylic carbons. Several general comments about the regiochemistry are noteworthy. First, in both benzene and naphthalene derivatives, the major monoketal arises from hydrolysis of the ketal having the smaller adjacent vinyl substituent. In monosubstituted naphthalene derivatives, virtually one regioisomer is formed since one of the vinylic substituents is hydrogen. In benzoquinone bisketals monosubstituted with an acetamido, bromo, methoxy, benzoyl, or thiomethyl group the regioselectivity of the hydrolysis is very high, the alternate hydrolysis product not being characterized. A simple methyl substituent in benzoquinone bisketal leads to a 64:11 mixture of regioisomers while a trimethylsilyl group gives nearly a 1:1 mixture of the regioisomeric monoketals.

The effect of substituents on the regiochemistry of quinone bisketal hydrolysis undoubtedly results from both the steric and electronic effects of the substituents.

TABLE 13. Monohydrolysis of benzoquinone bisketals[54]

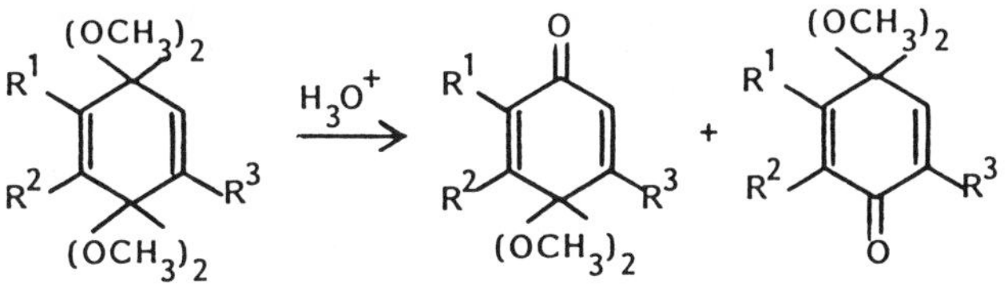

R^1	R^2	R^3	Yield (%)	Yield (%)
H	H	Br	88	3
H	H	CH_3	64 [a]	11 [a]
H	H	$Si(CH_3)_3$	29 [a]	38 [a]
H	H	$CH(CH_3)(OCH_3)$	58 [a]	19 [a]
H	H	$NHCOCH_3$	79	[b]
CH_3	CH_3	CH_3	90	[b]
H	H	OCH_3	66 [a]	[b,c]
H	H	SCH_3	60 [a]	[b]
H	H	COPh	42	[b]

[a] Overall yields of purified monoketals, based on the aromatic precursor to the bisketal.
[b] Alternate hydrolysis product not seen.
[c] Other products observed.

Extensive studies of ketal and orthoester hydrolysis and on the breakdown of tetrahedral intermediates have been reported [55,56] and a discussion of these principles as applied to the regiochemistry of quinone bisketal hydrolysis is instructive. A major point of the previous mechanistic work is that in the transition state for breakage of the carbon–oxygen bond of a protonated ketal, the remaining oxygen should orient one of its lone pairs of electrons in such a way as to stabilize the resultant positive charge. Thus, as the oxonium ion is formed, the starred atoms pictured in Scheme 3 must be coplanar with the C–R bonds to achieve the best overlap of one pair of the oxygens' non-bonding electrons with the cationic center. Any steric effects encountered in achieving this planar intermediate should raise the activation energy for ketal hydrolysis.

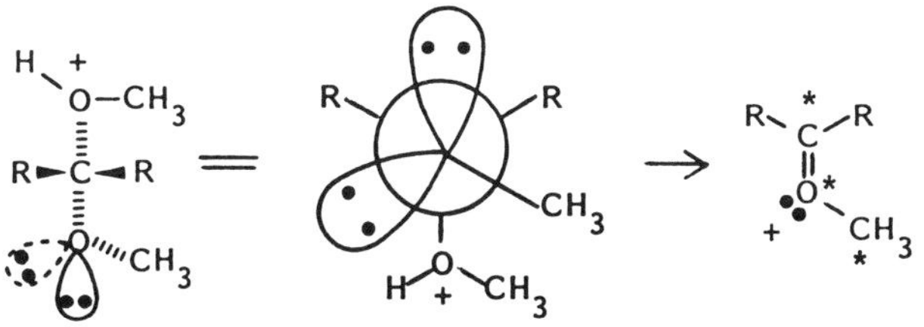

SCHEME 3. Stereochemical outcome of ketal hydrolysis

TABLE 14. Monohydrolysis of naphthoquinone bisketals[54]

R^1	R^2	Yield (%)	Yield (%)
H	H	93 [a]	
CH_3	H	90 [a]	
Br	H	85 [a]	
SCH_3	H	56 [a,b]	
CH_3	$Si(CH_3)_3$	57 [b,c]	
OCH_3	H	27	64
OCH_3	CH_3	36 [d]	24
SCH_3	CH_3	58	
Br	CH_3	94 [a]	

[a] Alternate hydrolysis product not seen.
[b] Yield is based on aromatic precursor.
[c] Other isomer not isolated; NMR of Raney Ni reduction products indicated a ratio of ca. 9:1.
[d] Isomeric monoketal formed in 19% yield.

A brief discussion of the higher regioselectivity of ketal hydrolysis in similarly substituted naphthoquinone versus benzoquinone systems illustrates the above discussion. In monosubstituted naphthalenes only one of the four possible intermediates is relatively free of steric interaction as the transition state is approached. Two of the other intermediates for ketal hydrolysis have methoxy–peri hydrogen interactions while the third has a methoxy–R interaction (Scheme 4). This rationale would account for the higher regioselectivity of bisketal hydrolysis observed for naphthoquinone bisketals.

no steric hindrance to planarity

steric hindrance to planarity

SCHEME 4. Intermediates for hydrolysis of naphthoquinone bisketals

In addition to steric considerations, the effect of the vinyl substituent in stabilizing or destabilizing a carbonium ion center will also influence the direction of ketal hydrolysis. Thus, vinylic substituents such as methyl and methoxy would be expected to stabilize positive centers because of allylic resonance and this would favor hydrolysis of the more distant ketal function. In disubstituted systems such as those listed at the bottom of Table 14, neither ketal can achieve the ideal planar geometry discussed above, and the effect of these substituents on carbonium ion stability may be a primary factor in the regiochemical outcome of the hydrolysis. In summary, as long as steric and electronic effects are complementary, the monohydrolysis of quinone bisketals will be highly regioselective. Where the effects are opposed, the regioselectivity is less predictable. The importance of steric as well as inductive effects in bisketals having more remote functionalities is discussed below.

The generalizations on regiochemistry of bisketal hydrolysis derived from the results of Tables 13 and 14 can be used to predict the regiochemical outcome of the monohydrolysis of bisketals of more synthetic interest. For example, hydrolysis of the bisketal **48** afforded a mixture of **49** and the quinone **50**[30]. The quinone decomposed during alumina chromatography, facilitating the separation of **49** and **50**, thus giving **49** in 73% yield. Reduction with zinc–copper couple afforded pure cymopol monomethyl ether (**41b**, equation 20). A number of tetralin bisketals possessing vinylic bromine substituents have been hydrolyzed to quinone monoketals with high regioselectivity[16,29].

$(OCH_3)_2$ R Br $(OCH_3)_2$ 48 $\xrightarrow{H_3O^+}$ O R Br $(OCH_3)_2$ 49 (73%) + O R Br O 50

$\xrightarrow[\text{THF, } H_2O,\ \Delta]{\text{Zn/Cu, HOAc}}$ OH R Br OCH_3 41b (20)

R = CH_3 CH_3 CH_3

The convenience of preparing substituted quinone monoketals by oxidation of *p*-methoxyphenols is dependent upon availability of the phenol, while the monoketal obtained from bisketal hydrolysis is determined by substituents on the ring. A study has been made using *p*-methoxyphenols as the common intermediates to two alternate quinone monoketals, **51** and **54** (Scheme 5)[21]. In many reactions, **51** and **54** would effectively serve as regioisomeric quinone monoketals. The basis of the strategy for obtaining **51** and **54** from the same *p*-methoxyphenol was the expected slower rate of

	R	Yields (%) from 53: 54	Yields (%) from 53: 51
a	H	(94)	--
b	CH_3	(93)	--
c	Br	(78)	--[a]
d	OCH_3	--[a]	(68)

SCHEME 5. Strategy and results for preparation of alternate monoketals from a common *p*-methoxyphenol. [a]The alternate monoketal was detected but not isolated

hydrolysis of the ethylene glycol ketal vs. a dimethyl ketal in compounds such as **53**. The electrolysis of mixed ethylene glycol–methanol ethers of substituted hydroquinones proceeds in good to excellent yield (Table 4) to afford the quinone bisketals, **53**. This approach is successful when the substituents on the quinone bisketal are bromo and methyl, since the monoketals of type **54** were obtained in good yield. However, the results with **53d** were disappointing since the influence of a methoxyl group overrides the lower hydrolytic reactivity of the ethylene glycol ketal, affording the same monoketal as that obtained directly from phenol oxidation; thus, the method has limited utility. A general route to both quinone monoketal regioisomers of a given system is yet to be developed.

3. Inductive effects of allylic substituents on the regioselectivity of bisketal hydrolysis

Finally, an interesting effect of an allylic oxygen functionality on the regiochemistry of hydrolysis of bisketal **55b** was noted in connection with synthetic studies of anthracyclinones[15,57]. Although **55a** shows no selectivity in its monohydrolysis, giving a 1:1 mixture of **56a** and **57a**, bisketal **55b** shows a synthetically useful preference for **56b** (equation 21). The products and kinetics for a series of bisketals with different allylic groups was

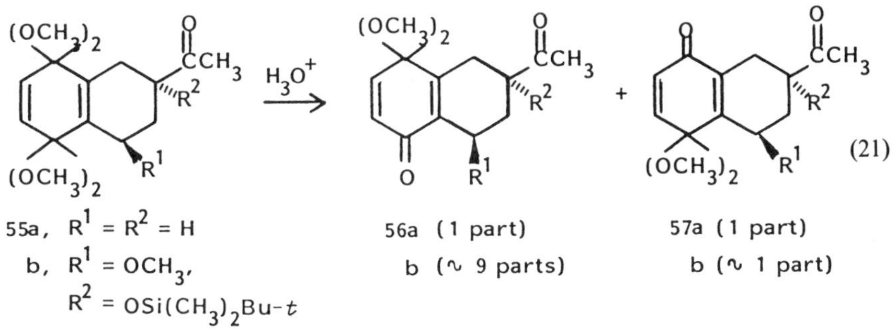

(21)

55a, $R^1 = R^2 = H$
b, $R^1 = OCH_3$, $R^2 = OSi(CH_3)_2Bu\text{-}t$

56a (1 part)
b (~ 9 parts)

57a (1 part)
b (~ 1 part)

studied[58] and the results are shown in Table 15. The kinetic data illustrate the interplay of steric and inductive effects on the regiochemistry of quinone bisketal hydrolysis. While the ether- (entries 3, 4) and fluoro- (entry 5) substituted systems give the same major regioisomer as the alkyl-substituted systems (entries 6 and 7), they do so for different reasons. The selective formation of monoketal **59** from the ether- (entries 3, 4) and the fluoro- (entry 5) substituted systems is not due to an increased rate of formation of **59**, but rather to a decreased rate of formation of **60**. For the alkyl-substituted systems (entries 6 and 7), the overall rate of bisketal hydrolysis increases, and the selective formation of monoketals of type **59** are due largely to the increased rate of formation of **59**. Thus, the similar regioselectivities observed for entries 2–5 and entries 6 and 7 are due to entirely different reasons. Furthermore, the vinyl-substituted systems (entries 8 and 9) favor the formation of regioisomer **63**, not because its formation is increased by the presence of the double bond but largely due to the retarding effect of the double bond on the rate of formation of **62**.

The accelerated rate of hydrolysis for the alkyl-substituted systems (entries 6 and 7) is best attributed to relief of steric interaction in the ionization step. For the bisketals having allylic oxygen and fluoro substituents, the rate of formation of the intermediate leading to

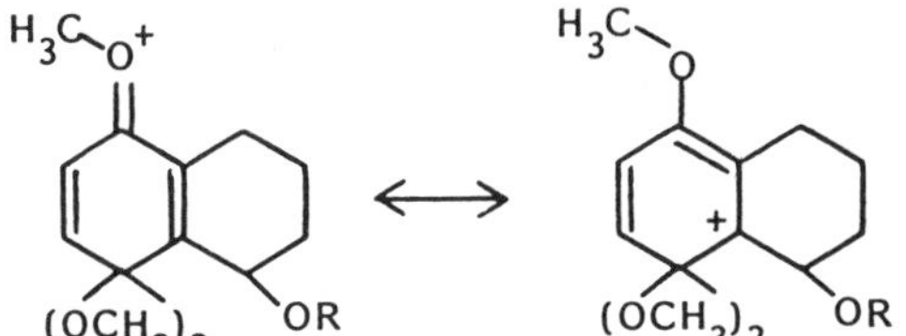

64

64 would be decreased by the inductive effect of the substituent on the allylic cation being formed. Likewise, the decreased rate of formation of monoketal **62** for the triene systems (entries 8 and 9) must be due partly to the inductive effect of the added sp^2 centers. For the intermediate leading to formation of **63**, the inductive effect of the sp^2 center is counterbalanced by the benefit of allylic resonance leading to little change in overall rates. However, for the intermediate leading to the formation of **62**, only the inductive effect of the sp^2 center is operative.

As can be ascertained from the above discussion the regiochemistry of hydrolysis of a given bisketal is an interplay of steric, stereoelectronic and inductive effects. While the individual importance of these effects changes with the compound of interest, sufficient information is available to allow one to make a reasonable prediction as to the regiochemical outcome of the hydrolysis of a functionalized quinone bisketal.

TABLE 15. Effect of allylic substituents on rates and regioselectivity of bisketal hydrolysis[58]

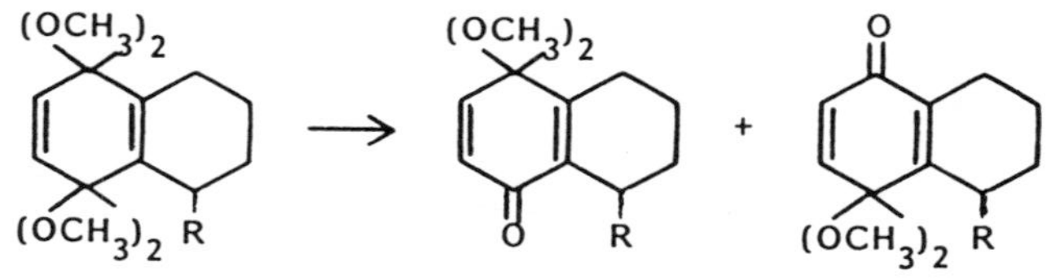

58 → 59 + 60

Entry	R	58 Relative Rate[a]	59 Relative Rate[b]	60 Relative Rate[b]	Ratio 59:60
1	H	7.7	3.85	3.85	1:1
2	OH	2.9	2.6	0.3	8.6:1
3	OCH_3	4.6	3.9	0.7	5.6:1
4	$OSi(CH_3)_2Bu$-*t*	3.6	2.6	1.0	2.6:1
5	F	1.0	0.9	0.08	11.2:1
6	CH_3	10.0	6.0	4.0	1.5:1
7	Bu-*t*	20.1	17.5	2.6	6.7:1

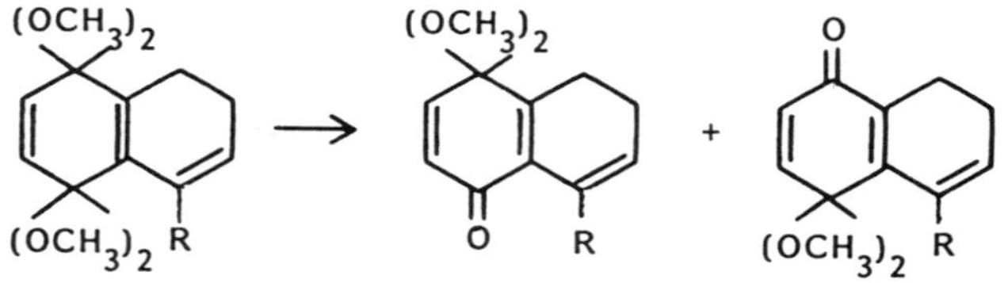

61 → 62 + 63

Entry	R	61	62	63	
8	H	5.1[c]	0.8	4.3	1:5.3
9	CH_3	2.9[c]	0.3	2.6	1:8.6

[a] For disappearance of **58**.
[b] For formation of product.
[c] These relative rate data are on the scale as for entries 1–7.

V. REACTIONS OF QUINONE MONOKETALS

A. 1,2 Additions to Quinone Monoketals

The 1,2 addition of Grignard and organolithium reagents to the carbonyl group of quinone monoketals comprises a synthetically useful approach to a variety of functionalized, protected *p*-quinol derivatives. Since the resultant *p*-quinol ketals can often be hydrolyzed under mild acidic conditions to the respective *p*-quinol, it is interesting to briefly compare analogous additions to quinone monoketals and quinones (Scheme 6). In

1) RM, 2) H_3O^+ ; HO, R ; 1) RM, 2) H_3O^+ ; $(OCH_3)_2$

[M = Li or MgX]

SCHEME 6. Routes to *p*-quinols from quinones and quinone monoketals

addition, the trimethylsilyl cyanide derivaties of quinones[4a–f] react with Grignard and organolithium reagents to give *p*-quinol derivatives after deblocking[3d,4]. Thus, additions to both quinone monoketals and the trimethylsilyl cyanide derivatives of quinones serve as viable routes to *p*-quinol derivatives.

The early literature reports variable results for reactions of Grignard reagents with simple quinones. In a 1952 paper[59] it is stated, 'In general, the reactions between quinones and Grignard reagents do not give good yields'. However, direct addition of organometallic reagents to quinones to produce mono- and bisaddition products can be synthetically useful for appropriate substrates and reaction conditions. Thus, reaction of quinones such as **65** with organolithium and Grignard reagents gives the functionalized quinones **67** in 50–60% overall yields after direct hydrolysis of the bisaddition adducts **66**

OEt, EtO ; 2RLi ; LiO, R, OEt, EtO, LiO, R ; H^+ ; R, R

65 66 67 (22)

R = Ph (62%)
R = CH_3 (52%)

(equation 22)[60]. It is also possible to selectively perform monoadditions of organolithium reagents to quinones, especially if the reaction is conducted at low temperature (−78 to −120°C)[62]. The regioselectivity of these monoadditions is influenced not only by the substituents on the quinone but also by the organometallic reagent (Grignard vs. alkyllithium), the solvent and the presence of tetramethylethylenediamine (TMEDA) in the reaction[62,63]. Thus, reaction of simple alkyllithium and Grignard reagents with quinones at low temperatures constitutes a synthetic route to *p*-quinols with some regiochemical control (Table 16). However, these reactions can be complicated by electron transfer processes and bisadditions of the organometallic reagents to the quinone.

The formation of *p*-quinols by reaction of quinone monoketals with alkyllithium reagents followed by acid hydrolysis has the advantage of affording only one regioisomeric product (Table 17). However, as would be expected from the extensive studies of organometallic additions to *p*-quinol derivatives[64], reduction of the quinone monoketal to

TABLE 16. Selected monoadditions of organometallic reagents to quinones

t-Bu, Bu-t; ArMgBr, room temperature, ref 61 → t-Bu, Bu-t, HO, Ar; 80%

H_3C, CH_3; ref 63 → H_3C, CH_3, R, OH + HO, R, H_3C, CH_3

Conditions	R	Yield (%)	Yield (%)
CH_3MgBr, −78 °C	CH_3	60	10
CH_3Li, TMEDA, −107 °C	CH_3	10	87
n-BuLi, −78 °C	*n*-Bu	60	15
n-BuLi, TMEDA, −107 °C	*n*-Bu	12	66

OCH_3; ref 63 → H_3C, OH, OCH_3

Conditions	
CH_3Li, TMEDA, −107 °C	80%
CH_3MgBr, −78 °C	complex mixture

TABLE 17. Selected examples of *p*-quinol ketals and *p*-quinols prepared from quinone monoketals

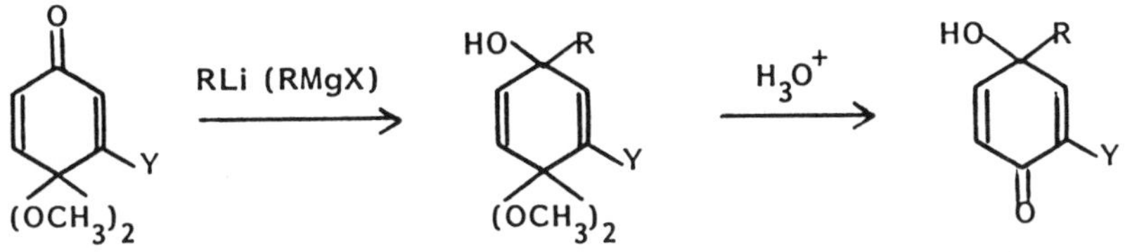

O, Y, $(OCH_3)_2$ —RLi (RMgX)→ HO, R, Y, $(OCH_3)_2$ —H_3O^+→ HO, R, Y, O

RLi(RMgX)	Y	Yield (%)	Yield (%)	Ref
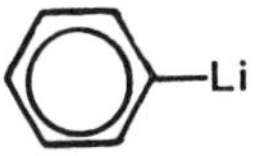 —Li (phenyl)	H	85	81	14a
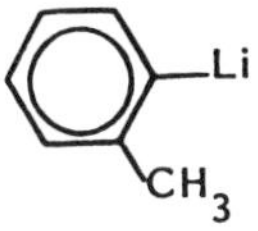 —Li, CH_3	H	84	77	14a
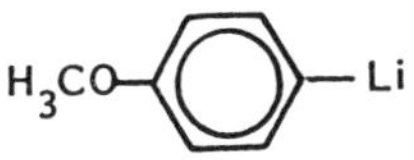 H_3CO— —Li	H	90	75	14a
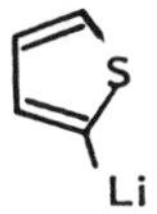 S, Li	H	79	--	14a
CH_3MgBr	Br	84	--	14a
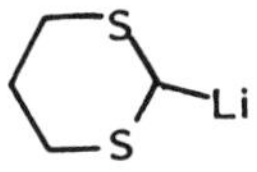 S, S, Li	OCH_3	90	--	3b
$LiCH_2CO_2CH_3$	H	91	54	3b
$PhSO_2CH_2Li$	OCH_3	86	--	3b

the respective phenol by electron transfer[65] from the organometallic reagent can complicate some of these reactions. In general, primary (methyl, *n*-butyl, etc.) and aryl organometallic reagents react with quinone monoketals at low temperature to afford excellent yields of the corresponding *p*-quinol ketals (Table 17). The 1,2-addition process is favored for a given system if the quinone monoketal is reacted with the alkyllithium reagent at −78°C. However, even under these conditions, reduction of the quinone monoketal to the respective phenol can compete with the 1,2 addition for the case of secondary organolithium and Grignard reagents. While the reaction of tertiary and allylic organometallic reagents with quinone monoketals has not been extensively studied, the reaction of *t*-butyl and allyl Grignard reagents with monoketal **51c** afforded major products that were derived from reduction (**68**) and rearrangement (**69, 70**, equation 23)[54].

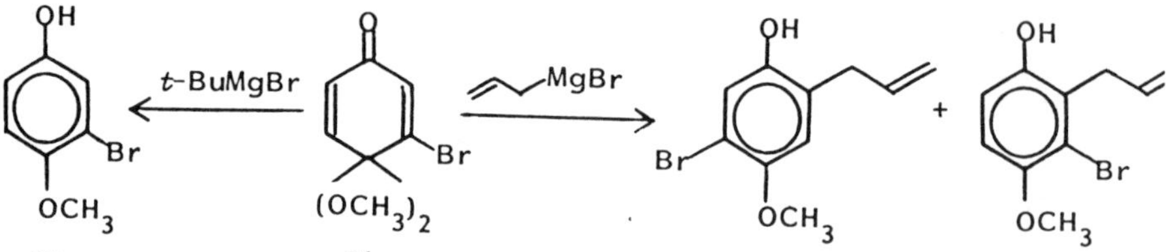

68 ← *t*-BuMgBr — 51c — allyl-MgBr → 69 (44%) + 70 (11%)

(23)

While there are only a limited number of examples, Peterson[4c] (**43** → **71**) and Wittig[3c,66] (**43** → **72**) reactions of quinone monoketals produce the corresponding protected quinone methides in good yields (equation 24). These compounds have proven to be valuable intermediates in the synthesis of several natural products[3,4] and have recently been employed in the synthesis of aryltetralin lignans[66].

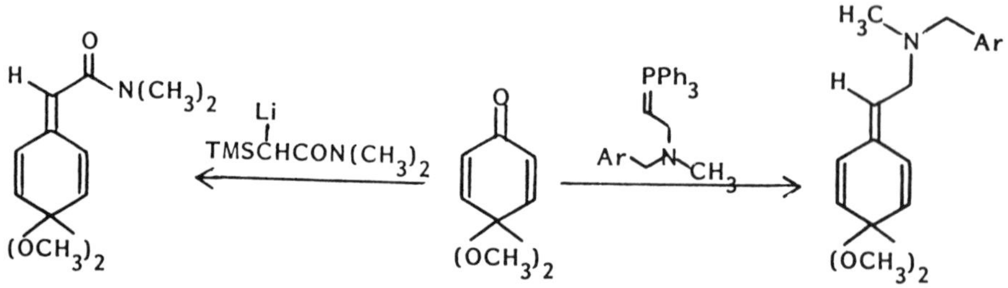

71 (77%) ← TMSCH(Li)CON(CH$_3$)$_2$ — 43 — Ph$_3$P=CHCH$_2$N(CH$_3$)CH$_2$Ar → 72 (47%)

(24)

With the limitation of reduction as a competing reaction, the 1,2 addition of organometallic reagents to quinone monoketals is a useful approach to certain *p*-quinol ketals and *p*-quinols. The major advantage of additions to quinone monoketals relative to quinones is the well-defined regiochemical outcome of the reaction and the formation of the *p*-quinol in a protected form. The *p*-quinol ketals formed in these reactions can not only be hydrolyzed to *p*-quinols but can also be useful synthetic intermediates in their own right. A recent example is the acid-catalyzed rearrangements of the aryl-substituted *p*-quinol ethers to substituted phenanthrenes (Table 18). A similar cyclization (compound not shown) produced a substituted fluorene derivative.

TABLE 18. Acid-catalyzed cyclization of *p*-quinol ketals to substituted phenanthrenes[4b]

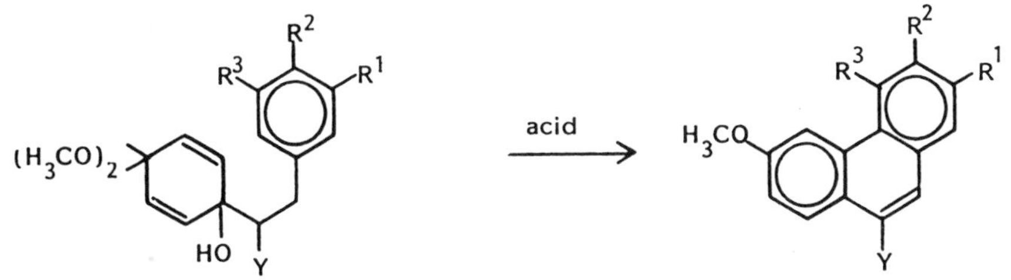

Y	R^1	R^2	R^3	Acid	Yield (%)
CO_2CH_3	OCH_3	OCH_3	OCH_3	P_2O_5/CH_3SO_3H	77
CO_2CH_3	OCH_3	OCH_3	H	$SnCl_4$	74
SO_2Ph	OCH_3	H	H	$BF_3 \cdot Et_2O$	62
CO_2CH_3	H	H	H	$BF_3 \cdot Et_2O$	64

B. Simple Michael Additions to Quinone Monoketals

The high reactivity of simple quinones in 1,4-addition reactions is well-documented. Although the replacement of one carbonyl group of the quinone with a ketal linkage would afford some steric hindrance toward nucleophilic attack at the β position of the monoketal, the inductive effect of the ketal linkage would not markedly alter the electron-deficient character at this position. Thus, as might be expected, quinone monoketals are efficient Michael acceptors and undergo a wide variety of transformations initiated by 1,4 addition of nucleophiles at the β carbon. In contrast to the corresponding quinones, Michael additions to quinone monoketals are regiospecific and the resultant adducts are much less likely to undergo aromatization under the reaction conditions and subsequent adventitious oxidation or reduction. Although virtually unexplored until recently, conjugate additions to quinone monoketals and the subsequent elaboration of the adducts provides a facile method for the preparation of interesting synthetic intermediates. This section will highlight some of the simple 1,4-addition reactions of quinone monoketals. The following sections will present reactions of quinone monoketals initiated by 1,4 addition but having ring formation as the final result.

Some early examples of 1,4 additions of oxygen, nitrogen and sulfur nucleophiles to the parent quinone monoketal **43** are outlined in Scheme 7[25a,48]. Especially interesting is the novel caged compound formed from reaction of **43** with sodium sulfide. Likewise, the additions of soft carbon nucleophiles such as diethyl malonate and ethyl cyanopropionate to quinone monoketals (Table 19) afford good yields of 1,4-addition products[67]. These compounds may be aromatized in a second step by heating with acid to afford the functionalized hydroquinone ethers in high yield (Table 19). This sequence provides an excellent two-step procedure for effecting overall nucleophilic substitution on a quinone, regiospecifically producing a functionalized hydroquinone monoether.

Although of reasonable generality, the 1,4 additions to quinone monoketals are not without limitation, especially when the addition involves the transfer of simple alkyl

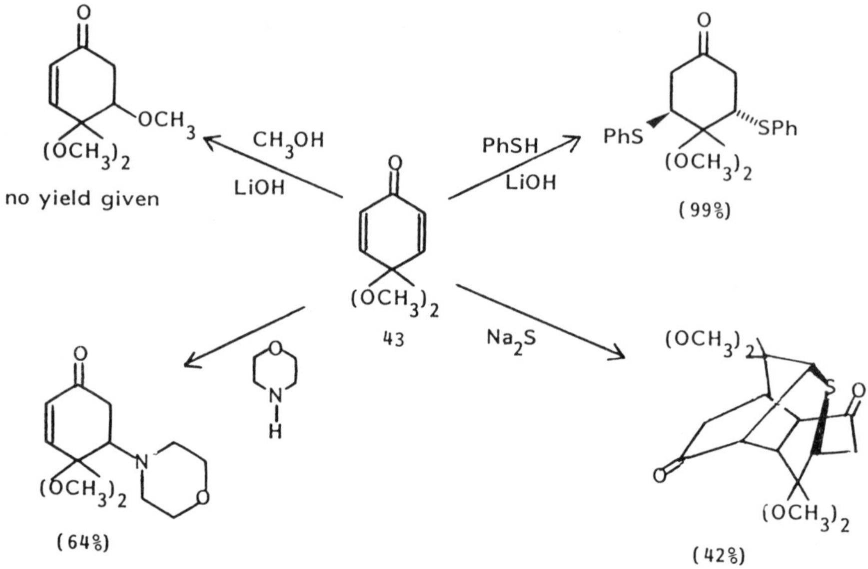

SCHEME 7. Selected examples of Michael addition reactions of nitrogen, oxygen and sulfur nucleophiles to quinone monoketals[48]

groups. Attempted conjugate addition of lithium dimethyl cuprate to quinone monoketals results in reduction of the quinone monoketal to the corresponding phenol, presumably via an electron transfer process as illustrated by the formation of **73** from **43** (equation 25)[25a, 65]. Michael addition of simple carbon nucleophiles except cyanide ion) to

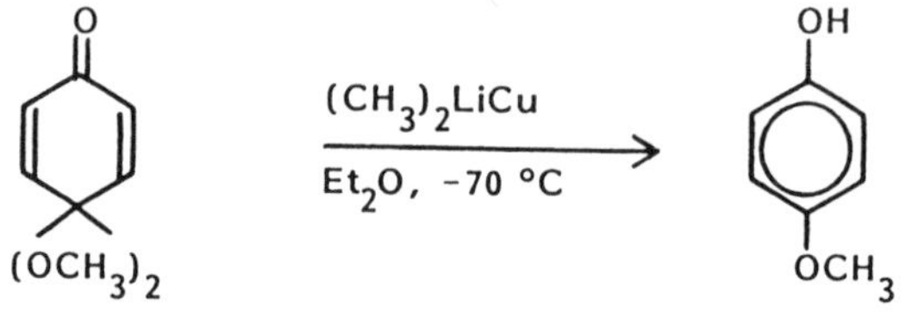

$$\mathbf{43} \xrightarrow[Et_2O,\ -70\ ^{\circ}C]{(CH_3)_2LiCu} \mathbf{73} \qquad (25)$$

quinone monoketals is often not a trivial matter, judging from some reports in the literature. In general, quinone monoketals react with 2-lithio-2-alkyl-1,3-dithianes, lithio trimethylsilyl cyanide derivatives of aldehydes and 1-lithio-1-[(methoxymethyl)oxy] butane to give low yields of 1,4-addition products, a mixture of 1,2-addition products and reduction processes usually being observed[68]. However, in one instance, the use of HMPA led to a useful addition of a dithiane anion to the naphthoquinone monoketal **74** to form **75** (equation 26)[68]. The anions from nitromethane and nitroethane have been reported to afford 1,4-addition products in good yield with certain naphthoquinone monoketals, but the scope of the reaction was not studied[68].

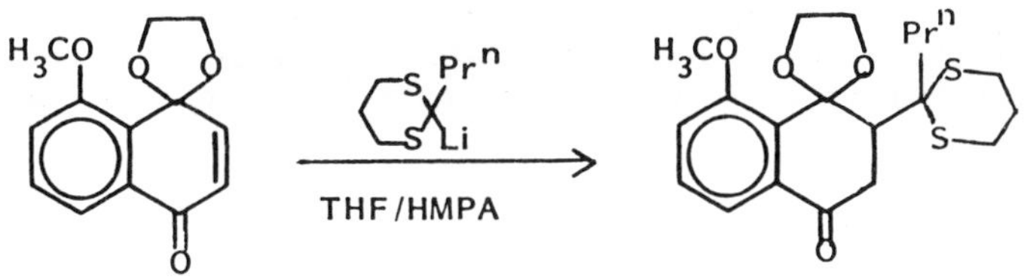

H_3CO ... THF/HMPA ... (26)

74 → 75 (86%)

TABLE 19. Diethyl malonate and ethyl cyanopropionate additions to quinone monoketals using sodium ethoxide in ethanol

Monoketal		1,4-Adduct	Hydroquinone Ether
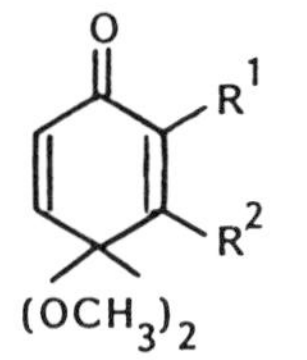 $(OCH_3)_2$		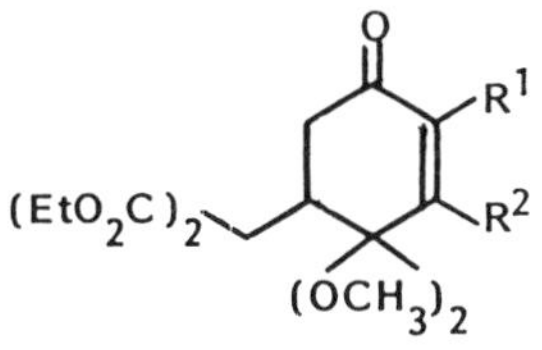 $(EtO_2C)_2$, $(OCH_3)_2$	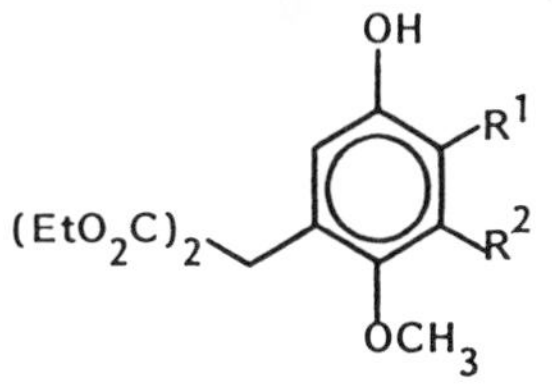 OH, $(EtO_2C)_2$, OCH_3
R^1	R^2	Yield (%)	Yield (%)
H	H	91	88
OCH_3	H	83	63
OCH_3 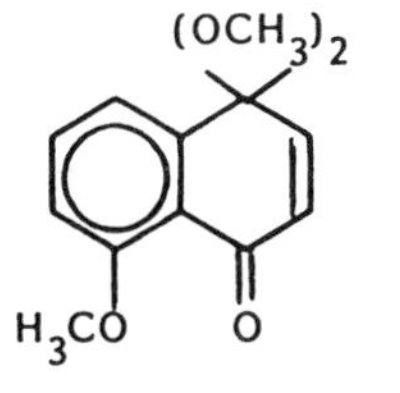	CH_3	70 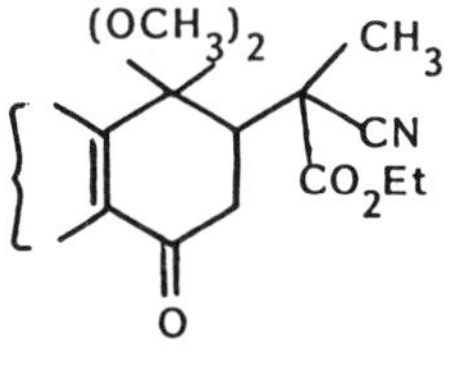	86 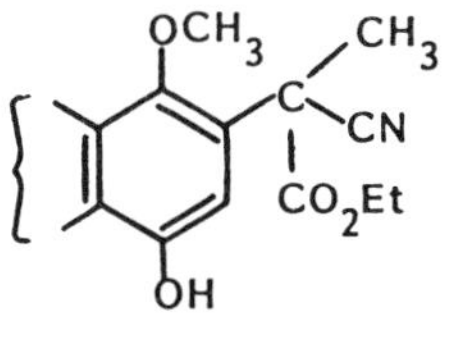
$(OCH_3)_2$, H_3CO, O		$(OCH_3)_2$, CH_3, CN, CO_2Et, O	OCH_3, CH_3, C, CN, CO_2Et, OH (84% overall)

Somewhat more successful 1,4 additions to selected naphthoquinone monoketals were found with acyl nickel complexes of simple alkyllithium compounds formed from the alkyllithium and nickel tetracarbonyl[68]. This reagent reacted with **76** to afford **77** in good yield (equation 27); however, reduction was the major reaction pathway when benzoquinone monoketals were used as reactants. Thus, this chemistry does not serve as a general method for introduction of acyl groups to the β position of quinone monoketals.

1) n-PrLi, $Ni(CO)_4$; 2) allyl iodide

76 → 77 (81%) (27)

A special type of formal 1,4 addition of acyl groups to quinone monoketals involves reaction of the bis-silyl ether **78** with both benzo- and naphthoquinone monoketals to give 1,4-addition products[69]. For the case of the naphthoquinone monoketal, the reaction afforded **79** via air oxidation of the initially formed hydronaphthoquinone. The benzoquinone monoketal reacted with **78** to give after acidic workup a mixture of **80a** and **80b** (equation 28). In both systems the keto acid moiety was in equilibrium with its

$OSi(CH_3)_3$ / $OSi(CH_3)_3$

78

1) naphthoquinone monoketal $(OCH_3)_2$; 2) H^+ → 79 (51%) (HO_2C)

1) benzoquinone monoketal $(OCH_3)_2$; 2) H^+ → RO, HO, HO_2C

80a, R = H (49%)

b, R = CH_3 (15%) (28)

respective cyclic lactol form. Although these products formally derive from a 1,4-addition reaction, they may be formed via an initial Diels–Alder reaction followed by a reverse aldol-type cleavage. Interestingly, **81** reacts with benzoquinone **82** in a rapid redox process to form **83** and **84** with no 1,4-addition product reported (equation 29).

C. Annelations of Quinone Monoketals via Nucleophilic β,β' Addition

The previous section illustrated the high reactivity of quinone monoketals in Michael addition reactions. Nucleophilic addition at the reactive β position is also the initial step in

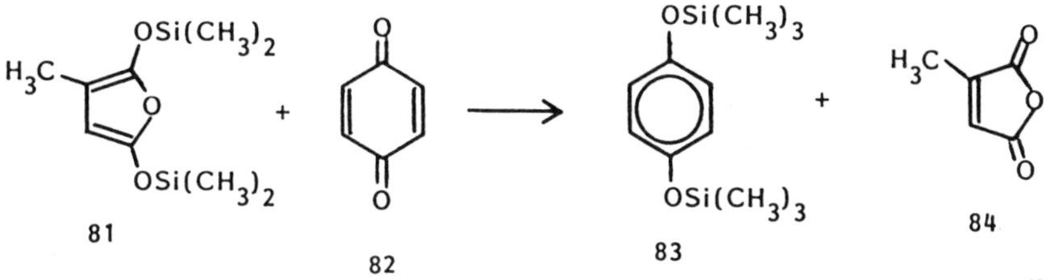

81 + 82 → 83 + 84

(29)

a number of efficient annelation reactions of quinone monoketals. These reactions have been grouped into two categories: β,β' additions, wherein a bimolecular nucleophilic addition to the β carbon of the quinone monoketal is followed by a second intramolecular addition to produce a bicyclo [3.3.1] ring system; and β, α-additions, which involve initial addition to the β carbon, followed by intramolecular reaction of the nucleophilic center generated at the α position with an electrophilic center.

A number of different active methylene compounds have been reacted with quinone monoketals to give β,β'-annelation products in good to excellent yields (Table 20). In reactions with unsymmetrically substituted quinone monoketals, the product is formally

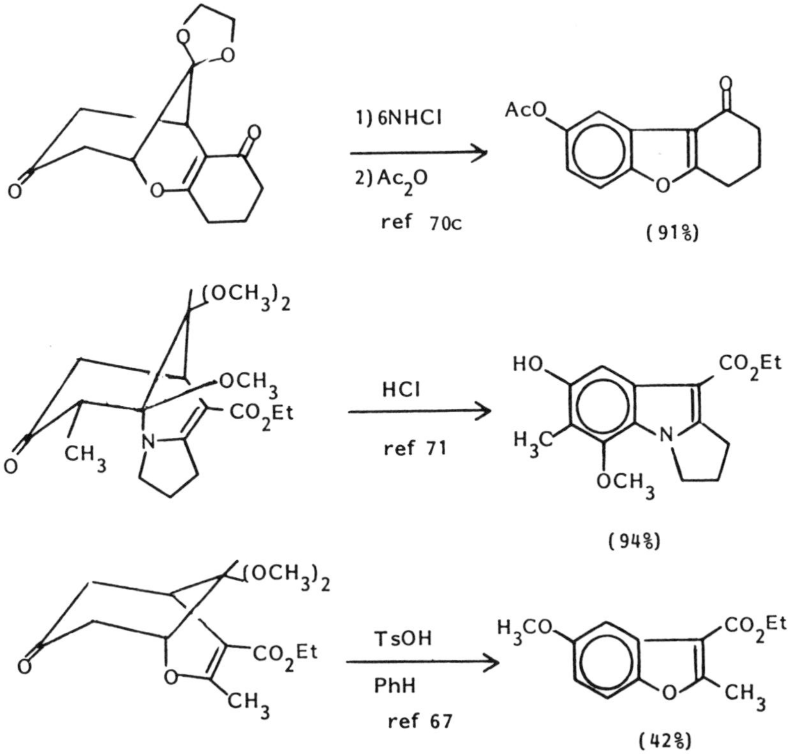

1) 6NHCl, 2) Ac_2O, ref 70c → (91%)

HCl, ref 71 → (94%)

TsOH, PhH, ref 67 → (42%)

SCHEME 8. Transformations of β,β'-annelation products of quinone monoketals to benzofurans and indoles

TABLE 20. Selected annelations derived from β,β' additions to quinone monoketals

Entry	Monoketal	Anion Precursor Conditions	Product
1	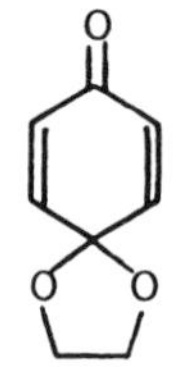 O; O O	$CH_3O_2CCH_2\overset{O}{\overset{\Vert}{C}}CH_2CO_2CH_3$ NaOEt (catalytic) ref 70a	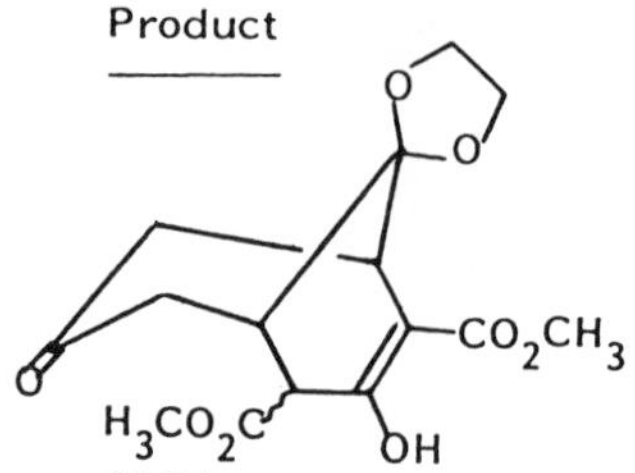 O O; O; CO_2CH_3; H_3CO_2C; OH (86%)
2	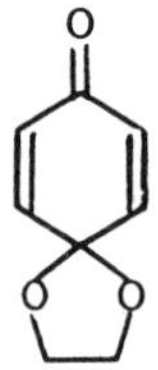 O; O O	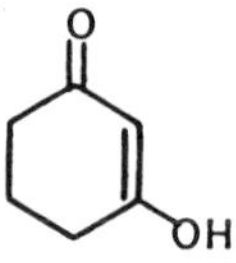 O; OH *t*-BuOK, *t*-BuOH ref 70c	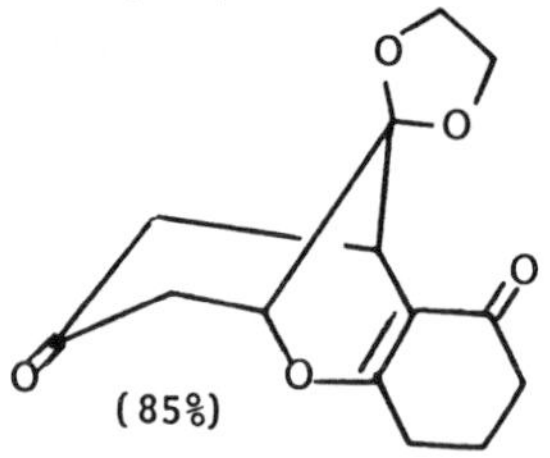 O O; O; O; O (85%)
3	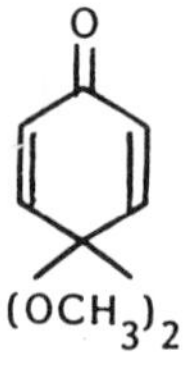 O; $(OCH_3)_2$	$CH_3\overset{O}{\overset{\Vert}{C}}CH_2\overset{O}{\overset{\Vert}{C}}OEt$ NaOEt, EtOH ref 67	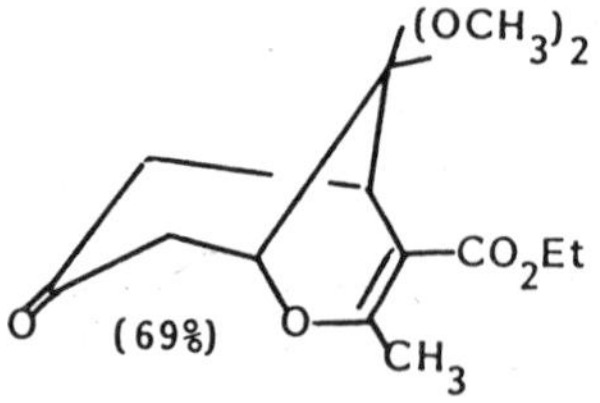 $(OCH_3)_2$; CO_2Et; O; O; CH_3 (69%)
4	 O; $(OCH_3)_2$	$CH_3\overset{O}{\overset{\Vert}{C}}CH_2\overset{O}{\overset{\Vert}{C}}OEt$ NaH, THF ref 67	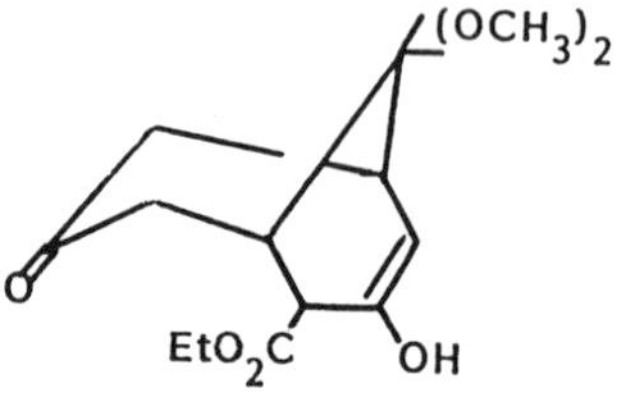 $(OCH_3)_2$; O; EtO_2C; OH (66%)
5	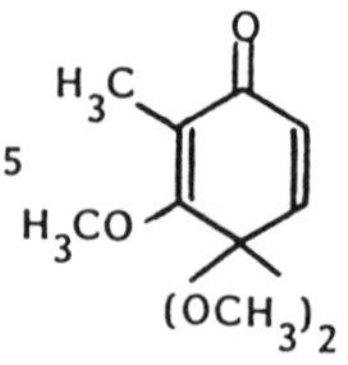 O; H_3C; H_3CO; $(OCH_3)_2$	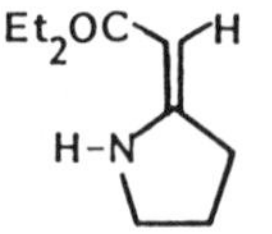 Et_2OC, H; H-N 1) NaH 2) H_2O ref 71	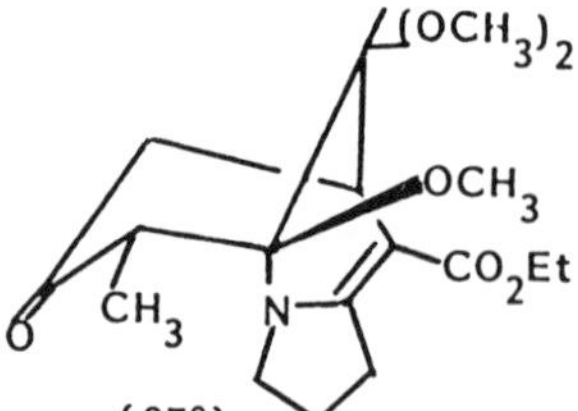 $(OCH_3)_2$; OCH_3; CO_2Et; O; CH_3; N (87%)

derived from initial addition at the less hindered, more reactive β' position (entry 5), followed by cyclization at the more substituted β position. An interesting dependence on reaction conditions was noted for the reactions of quinone monoketals and ethyl acetoacetate (entries 3 and 4). Thus, with sodium ethoxide in ethanol, the second step involves bond formation between the oxygen of the ambident anion and the β' carbon while the sodium enolate in tetrahydrofuran gives the product derived from bonding of the carbon to the β' position.

These β,β'-annelation products can often be converted to functionalized benzofuran and indole ring systems in high yield as illustrated in Scheme 8. These heterocyclic products are formally derived from acid-catalyzed ionization of the ketal followed by 1,2 migration of the heteroatom and subsequent aromatization. Thus, the products from β,β' addition of 1,3-cyclohexanedione and ethyl acetoacetate with quinone monoketals under acidic conditions afford benzofurans, whereas products from the anions of enaminoesters and monoketals give indoles. This latter reaction is only one of several examples and constitutes a regiospecific variant of the Nenitzescu indole synthesis. However, conversion to indoles apparently is limited to adducts derived from quinone monoketals having a β-methoxy group.

D. Annelations of Quinone Monoketals via Nucleophilic β Addition, Followed by Electrophilic α Functionalization

Annelation reactions of quinone monoketals analogous to those of conjugated ketones should be especially facile as long as the nucleophilic species is a poor reducing agent. Fortunately, with many of the common annelating reagents, reduction is seldom observed, and high yields of annelation products are formed. The cyclopropanation of **43** and **54a** by dimethylsulfoxonium methylide was one of the first examples of this process and was a key step in the synthesis of bishomoquinone **85** (equation 30)[5,26]. A similar annelation

1) $(CH_3)_3SO^+, I^-$, NaH; 2) H_3O^+ (30)

(OR)$_2$

43, R = CH_3
54a, $(OR)_2$ = OCH_2CH_2O

85 (56% overall from 54a)

product of **54d** was employed as a key intermediate in a synthesis of (±)-deacetamido-isocolchicine (Scheme 9) and (±)-colchicine[72].

The recent interest in anthracyclinone chemistry has brought about a resurgence in the chemistry of anthraquinones[73]. Thus, an attractive convergent and regiospecific approach to tetracyclic quinone systems would involve β,α annelation of 1,4-dipole equivalents to quinone monoketals (Scheme 10). Initially, this strategy was explored with the anions of homophthalic esters, i.e. **86** → **88** (equation 31)[57b,c]. Subsequently, the anion of the cyanophthalides[15,57b,74] **89** and sulfone[57a,b,75] **90** were used to produce the corresponding anthraquinone **91** directly, with **89** generally affording significantly better yields in this annelation (equation 32). The lower yields of anthraquinone product when using **90** in the reaction was attributed to competing 1,4 addition of the phenylsulfinate anion, X = $PhSO_2$, released in the final step of the annelation (i.e. **92** → **93**), with the unreacted

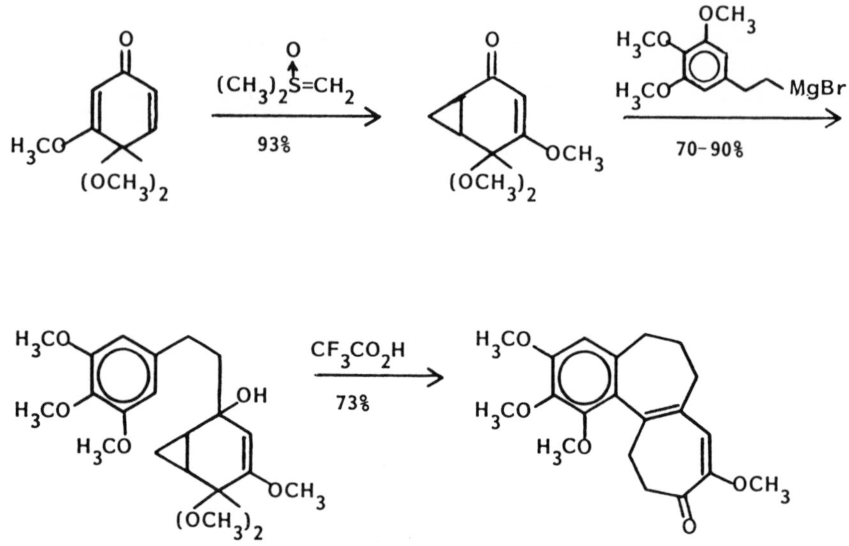

SCHEME 9. Key synthetic steps in the synthesis of racemic deacetamidoisocolchicine

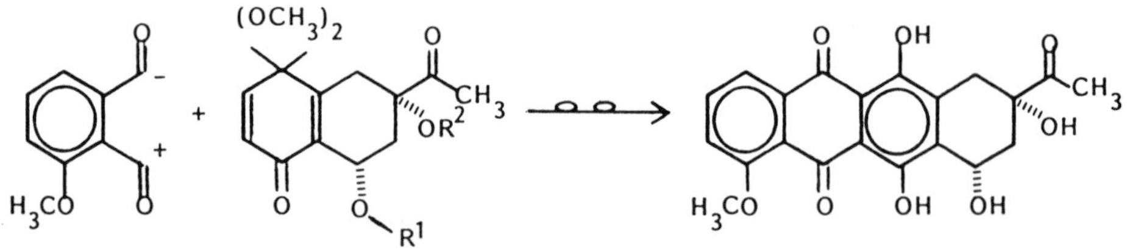

SCHEME 10. Synthetic strategy to daunomycinone using a β,α annelation of a quinone monoketal

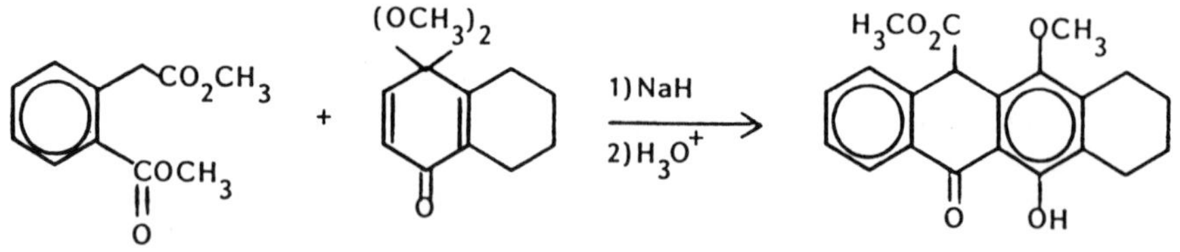

86 87 88 (60%)

(31)

quinone monoketal[57b]. This serves again to emphasize the Michael reactivity of quinone monoketals with soft nucleophiles.

Using the highly functionalized quinone monoketal **94** and the cyanophthalides **89a–e**,

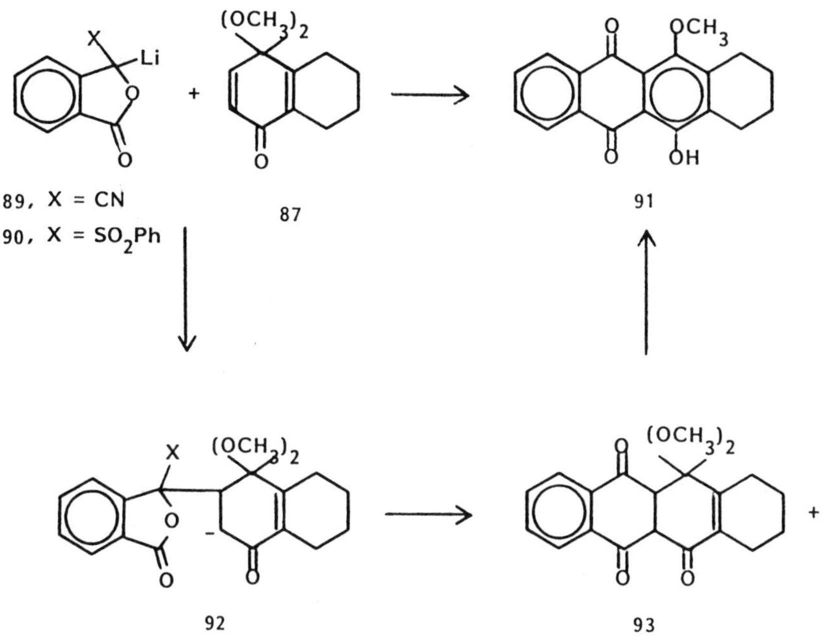

89, X = CN
90, X = SO_2Ph

87

91

(32)

92

93

a regiospecific route to fully functionalized anthracyclinones **95** was achieved (equation 33)[15, 57, 76]. Over 30 g of anthracyclinone analog, **95a**[15], as well as a variety of other

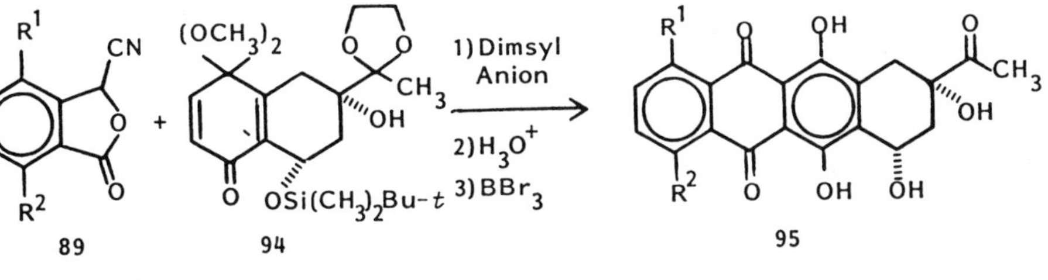

1) Dimsyl Anion
2) H_3O^+
3) BBr_3

89 94 95

R^1, R^2 (overall yield): a, H,H (84%); b, H,OCH_3 (42%); c, H,F (70%); d, F,H (48%); e, F,F (52%)

(33)

analogs, **95b–e**, were prepared, using this chemistry. This particular annelation of quinone monoketals is an excellent regiospecific procedure for the preparation of anthraquinones having acid-sensitive and thermally labile substituents and is now a preferred method for preparation of anthracyclinone analogs[74, 75]. Interestingly, reaction of **89a** with the quinone derived from **94** gave very low yields of annelation product. A number of variants of the β, α annelation such as the reaction of **96** with a monoketal to give **97** (equation 34) are possible[77]. Undoubtedly, many other annelations of this type will be investigated in the coming years.

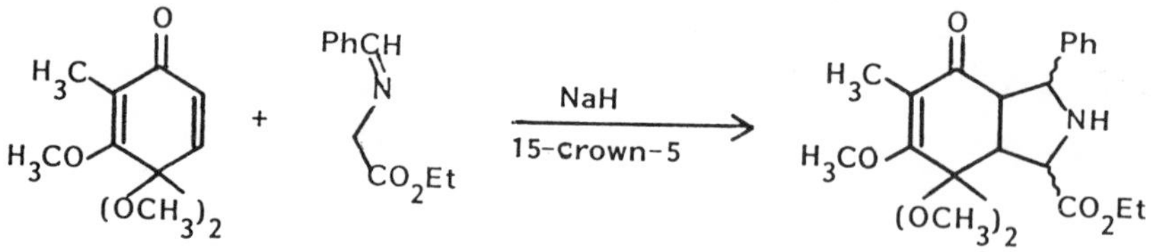

(34)

96

97 (45%)

NaH

15-Crown-5

E. Reactions of Quinone Monoketals with Derivatives of Ammonia

In principle, ammonia derivatives have two major pathways available for reaction with quinone monoketals. The 1,4 addition of secondary amines to the parent quinone monoketal was mentioned in Section V.B. Addition to a carbonyl group followed by loss of water to form an imine is commonplace in carbonyl chemistry, and products derived from this mode of addition would be especially favored if subsequent chemistry rendered the addition irreversible. The reaction of quinone monoketals with several ammonia derivatives was examined as a method for conversion of the carbonyl group of the quinone monoketal to other functionalities. Scheme 11 illustrates the use of this chemistry for the conversion of quinone monoketals to the corresponding nitroso, amino, phenylazo and hydrogen compounds[78, 79]. The condensation of the NH_2 group of the reagent with the carbonyl group of the monoketal is undoubtedly the initial step in all of these reactions.

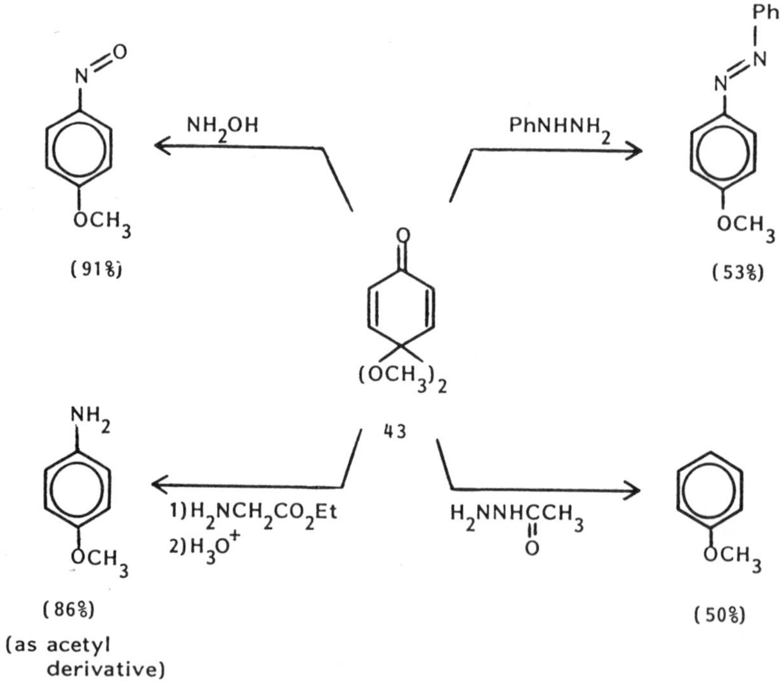

NH_2OH (91%)

$PhNHNH_2$ (53%)

43

1) $H_2NCH_2CO_2Et$
2) H_3O^+

(86%)
(as acetyl derivative)

$H_2NNHCOCH_3$ (50%)

SCHEME 11. Reaction of quinone monoketal **43** with ammonia derivatives

Since the quinone monoketal is often derived from the corresponding *p*-methoxyphenol, this constitutes a method for replacement of the phenolic hydroxyl group by NO, N = N–Ph, NH_2 and H.

The quinone imine unit has been of long-standing interest in chemistry[80, 81], but protected forms of these molecules appear not to have been reported. The intramolecular condensation of an amino group with a quinone monoketal carbonyl group serves as a route to the ketals of quinone imines. Two variants of this chemistry are known. Anodic oxidation of **98a–d** followed by direct hydrolysis of the reaction mixture affords in one operation the quinone imine ketals **101a–d** via intermediates **99** and **100** (equation 35)[81, 82]. However, the success of this method depends on the selective hydrolysis of the ketal linkage adjacent to the side chain bearing the amino group of **99** to give the intermediate **100**. It was not possible to prepare the unsubstituted quinone imine ketal using this approach.

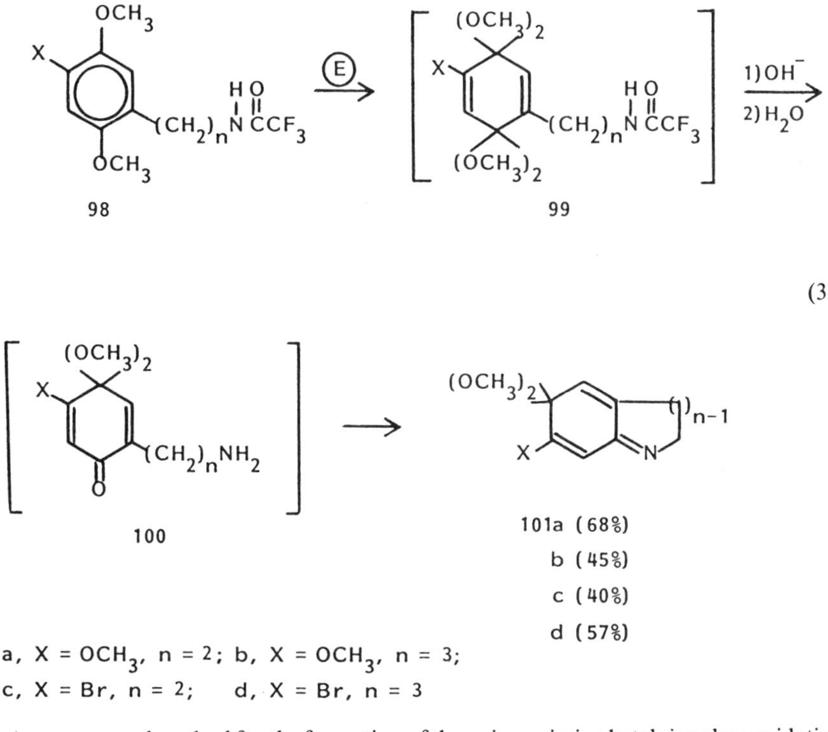

98 → (E) → [99] → 1) OH^-, 2) H_2O → (35)

[100] → 101a (68%) b (45%) c (40%) d (57%)

a, X = OCH_3, n = 2; b, X = OCH_3, n = 3;
c, X = Br, n = 2; d, X = Br, n = 3

A more general method for the formation of the quinone imine ketals involves oxidation of the respective *p*-methoxyphenol followed by *in situ* hydrolysis of the trifluoroacetyl group and intramolecular condensation (Table 21)[81]. The protection of the amine from anodic oxidation via its trifluoroacetamide derivative is especially convenient since this amide linkage is cleaved under mild basic conditions to generate the reactive amino moiety which then undergoes intramolecular condensation with the carbonyl group.

F. Acid-catalyzed Cycloaddition Reactions of Quinone Monoketals

Neolignans[83] are a group of secondary plant metabolites structurally characterized by the presence of two arylpropanoid units. Several members of this class of natural products

TABLE 21. Quinone imine ketals from phenol derivatives

Compound		Imine
H_3CO, OH, R, R', $NHCCF_3$ (C=O)	1) (E) 2) KOH	$(OCH_3)_2$, R, R', N
R = R' = H		(82%)
R = OH, R' = H		(91%)
R = OH, R' = CH_3		(81%)
H_3CO, OH, R, R', H–$NCCF_3$ (C=O)	1) (E) 2) KOH	$(OCH_3)_2$, R, R', N
R = OH, R' = H		(91%)
R = OH, R' = CH_3		(89%)

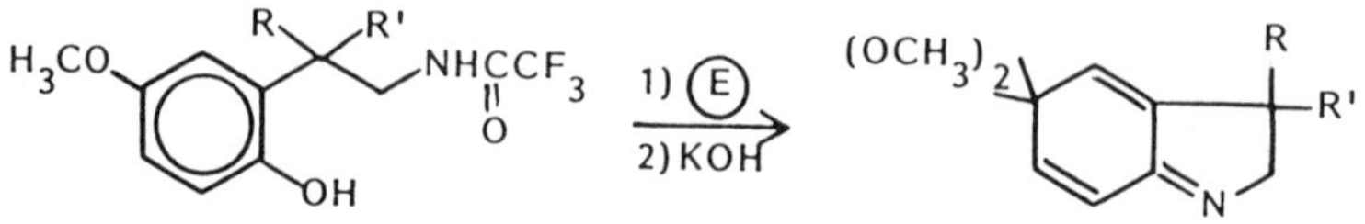

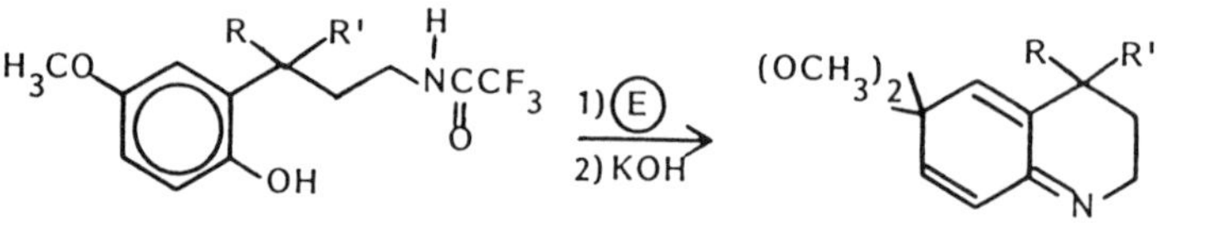

have been synthesized by an interesting cycloaddition reaction of a 3-alkoxy-substituted quinone monoketal and isosafrole[84, 85]. The chemistry involved is conveniently rationalized by a cycloaddition of a positively charged 1,3-dipole intermediate, **104**, formed by acid-catalyzed loss of methanol from the quinone monoketal **103**, and an olefin as outlined in Scheme 12. The cations **105** and **108** produced from this initial cycloaddition can undergo hydrolysis, or rearrangement and hydrolysis, to afford the observed products **106** and **107**. In some cases, the vinyl ether linkage of **107** undergoes hydrolysis under the workup to give the enol.

The products from this type of cycloaddition are markedly dependent upon the acid catalyst and the reaction media as illustrated by the reactions of **109** and **110** below[84]. This reaction in acetonitrile catalyzed by 2,4,6-trinitrobenzenesulfonic acid gives a mixture of **111** and **112**. However, if the same reaction was performed in methylene chloride at $-78^\circ C$ using trifluoromethanesulfonic acid as catalyst, **113** was formed in addition to **112**. Finally, when using triethyloxonium hexachloroantimonate as catalyst in methylene chloride, the spiro dienone **114** was the major product (equation 36). The products from the first set of conditions can be rationalized by assuming an initial cycloaddition to produce intermediates analogous to **105** and **108** which undergo hydrolysis to afford **111** and **112**. When the reactions are conducted in methylene chloride, a non-nucleophilic medium, these initially formed cations undergo rearrangement and cyclization to the allyl linkage, affording **113** and **114**.

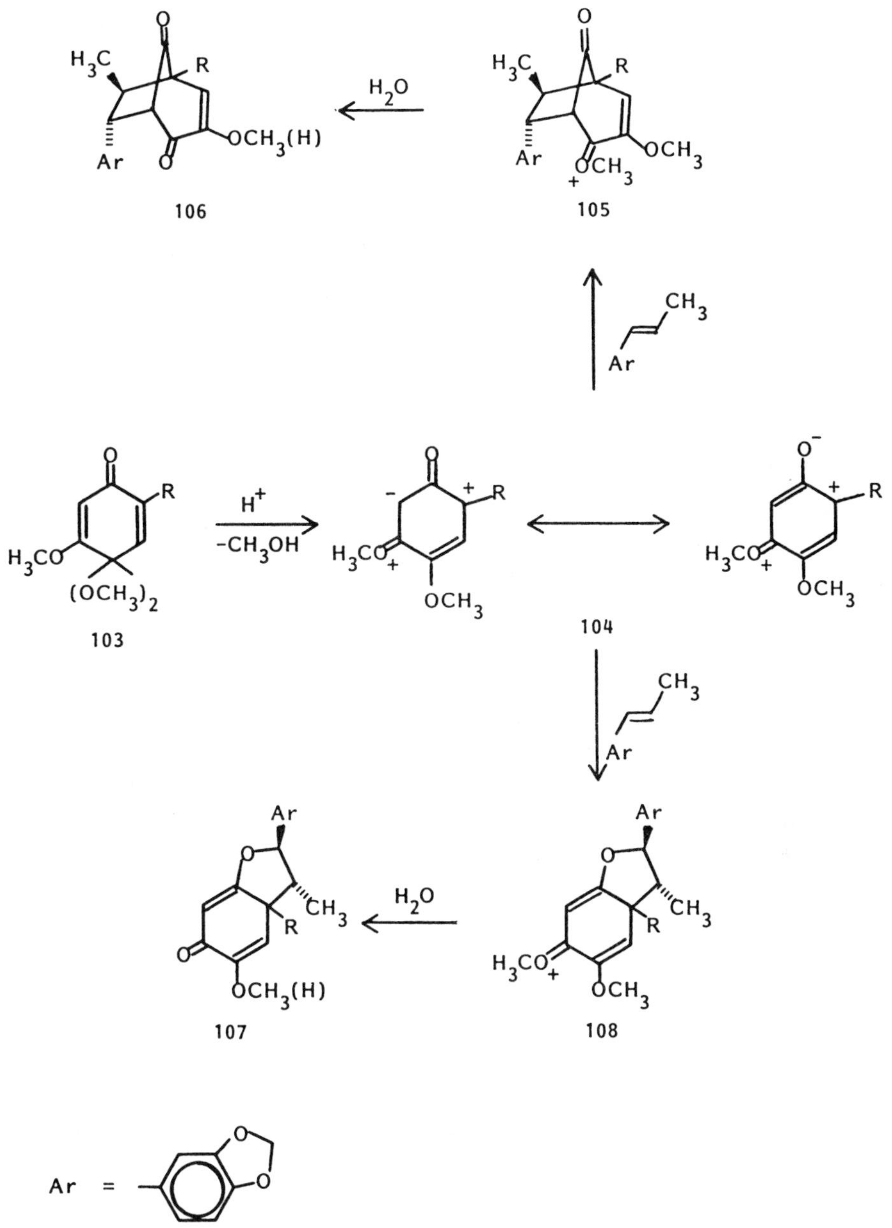

H_3C R O Ar O $OCH_3(H)$

106

H_2O

H_3C R O Ar $\overset{+}{O}CH_3$ OCH_3

105

CH_3 Ar

O R H_3CO $(OCH_3)_2$

103

H^+

$-CH_3OH$

O − + R H_3CO + OCH_3

104

O^- + R H_3CO + OCH_3

CH_3 Ar

Ar O CH_3 R O $OCH_3(H)$

107

H_2O

Ar O CH_3 R H_3CO + OCH_3

108

Ar =

SCHEME 12. Generalized reaction scheme for acid-catalyzed cycloaddition of quinone monoketals and olefins

(36)

110 + 109

2,4,6-$(O_2N)_3C_6H_2COOH$, 0 °C → 111 (20%) + 112 (27%)

CF_3SO_3H, −78 °C → 113 (18%) + 112 (33%)

$Et_3O^+SbCl_6^-$ → 114 (47%)

Ar =

The overall yields of these cycloadditions are not high, and the reactions may be specific for 3-oxygenated quinone monoketals. However, the availability of the starting materials and the one-step formation of a rather complex ring system with good control of the stereochemistry makes this type of reaction an attractive, albeit limited, synthetic method. Variants of this cycloaddition have been used as key steps in the synthesis of Gymnomitrol[86] and Megaphone[87] and in a route to tropolones[88]. The reaction of **115** with **116** to give **117**, a key intermediate in the synthesis of Gymnomitrol, illustrates a common competing reaction when quinone monoketals are reacted with Lewis acid even under mild conditions: reduction of the quinone monoketal to the respective phenol (e.g. **116** → **118**, equation 37). Presumably, the methoxy moiety of the quinone monoketal, **116**, is the reducing agent and is oxidized to formaldehyde in the course of the reaction.

115 + 116 —(1) H^+; 2) $NaBH_4$)→ 117 (10%) + 118 (30–45%) (37)

An especially interesting variant of this chemistry is the direct electrochemical oxidation of **119** in the presence of isosafrole (**110**) to produce **120** in 81% yield (equation 38)[89]. It

110 + 119 —(Ⓔ, CH_3OH, HOAc)→ 120 (81%) (38)

is proposed that the electrochemical oxidation of the phenol forms a cation analogous to **104** which then undergoes the cycloaddition reaction. In general, this electrochemical procedure would offer an attractive alternative to using the quinone monoketal as the positively charged 1,3-dipole precursor in the cycloaddition reaction.

G. Diels–Alder Reactions of Quinone Monoketals

While the thermal stability of quinone monoketals has not been extensively investigated, they appear to be moderately stable when heated in a non-nucleophilic solvent. The parent quinone monoketal ($R^2 = R^3 = H$ in Scheme 13) undergoes thermolysis at 180°C in tetrahydrofuran solution (sealed tube) to produce *p*-methoxyphenol and presumably formaldehyde, or a product derived from formaldehyde[27]. The half-lives of **43** at 165°C and 190°C are about 7 and 1 hours, respectively. Thus, the thermal stability of **43** and presumably of other dimethyl ketals of this type is sufficient to perform many bimolecular thermal cycloaddition reactions at temperatures below 165°C. In addition, the ethylene glycol quinone monoketal, **54a**, is more thermally stable than **43**, further extending the temperature range for thermal cycloaddition reactions of **54a**[27].

Thus far, Diels–Alder reactions of the parent quinone monoketal **43** have been studied with 1-[20a,b] and 2-substituted[90] dienes, 1-substituted isobenzofurans[91] and cyclobutadiene[92] (Scheme 13). The reaction of **43** with 1-methoxybutadiene is fast and highly regioselective while reactions with 2-substituted dienes proceed more slowly (ca. 130–140°C for 140–200 hours) and produce a mixture of regioisomers. The reactions of **43** with benzocyclobutenol[20a] and isobenzofurans[91] are also highly regioselective and allow a facile entry into linear polycyclic ring systems. This latter reaction was investigated for a number of 1-substituted isobenzofurans [R^1 = H, Me, $Si(Me)_3$, CH_2OH, CO_2Me] with quinone monoketals (R^2, R^3 = H, H; H, Me; H, OMe; OMe, H). The reaction proceeded with high regioselectivity for all of the compounds studied. However, whereas 1-methyl isobenzofurans and **43** reacted at room temperature to afford a quantitative yield of the Diels–Alder product, isobenzofurans having electron-withdrawing groups at position 1 (e.g. CO_2Me) required higher temperatures and gave lower yields for the reaction.

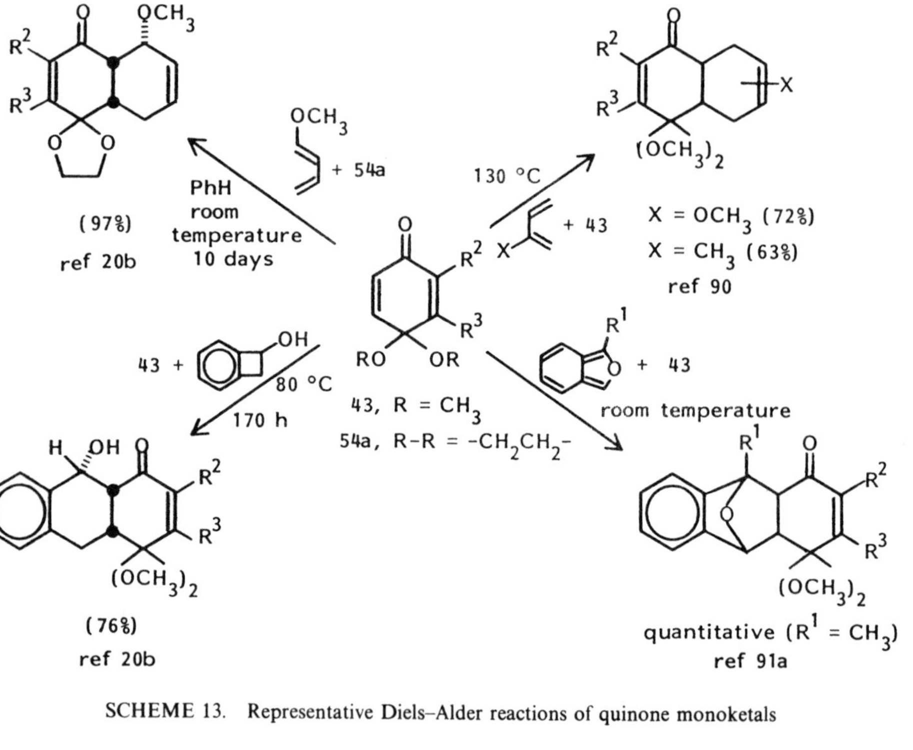

SCHEME 13. Representative Diels–Alder reactions of quinone monoketals

The major advantages of conducting Diels–Alder reactions with the quinone monoketal relative to the quinone is the higher regioselectivity[91] of the addition and the formation of a product in which the two quinone carbonyl groups are differentiated chemically. In some cases, this compensates for the somewhat lower reactivity of the quinone monoketal as compared to the corresponding quinone in the Diels–Alder reaction.

H. Concluding Remarks on Quinone Monoketal Chemistry

The previous sections have dealt with quinone monoketal chemistry that could be reasonably placed into certain categories. This section illustrates those reactions of quinone monoketals which are not included in the prior discussions. The previous sections have been concerned with using quinone monoketals as quinone equivalents. However, the generality of the anodic oxidation of 1,4-dimethoxyaromatic systems to quinone bisketals and the often facile hydrolysis of these compounds to monoketals and thence to quinones, has prompted the use of this anodic oxidation/hydrolysis sequence as a method of oxidation of 1,2,4-trimethoxy-3-methylbenzene to 2-methoxy-3-methylbenzoquinone[93]. Thus, in addition to the use of anodic oxidation/hydrolysis as a route to quinone monoketals, the convenience of the method makes it a practical route to the analogous quinones.

An interesting route to amino-substituted quinone monoketals involves the regiospecific attack of methoxide ion on 4-alkoxy-*o*-quinone imines[94]. Since the imines are available from manganese dioxide oxidation of the corresponding phenol, e.g. **121**, this comprises another method for conversion of phenols to quinone monoketals. However, the yields for the methoxide addition reaction to **122** to form **123** (equation 39) are less than 20%.

OH, $NHSO_2Ar$, H_3C, OR — MnO_2 → O, $N{-}SO_2Ar$, H_3C, OR — CH_3OH, NaOH → O, $NHSO_2Ar$, H_3C, H_3CO, OR

121 122 123

$R = CH_3, (CH_2)_{15}CH_3$ (39)

Finally, the photochemistry of four quinone monoketals has been studied, and the results are given in Scheme 14[95]. All four quinone monoketals afford different products in poor-to-moderate yields. Thus, it is not possible to present any general discussion of the photochemistry of quinone monoketals.

VI. *o*-BENZO- AND NAPHTHOQUINONE BISKETALS AND MONOKETALS

A. *o*-Benzo- and Naphthoquinone Bisketals

The preceding pages illustrate the diverse range of chemistry of *p*-quinone bisketals and monoketals; however, much less information is available on their *o*-quinone analogs. There are relatively few electrochemical preparations of *o*-quinone bisketals. The parent *o*-quinone bisketal **125** was first reported in 1963 as one of the products [together with **11e** (R^2 = MeO, R^3 = H), **126** and **127**] from anodic oxidation of 1,2-dimethoxybenzene **124** in methanolic potassium hydroxide (equation 40)[6]. The interesting bis-*o*-ester **127** formed in 10% yield from the oxidation of **124** is apparently a secondary product since it is formed in 77% yield from the oxidation of **125**. Presumably, the low yield of the material discouraged the study of its chemistry. The anodic oxidation of the veratrole analog **128**

O
t-Bu
Bu-t
$(OCH_3)_2$
hν
PhH
O
t-Bu
Bu-t
(38%)
+
O
t-Bu
OCH_3
t-Bu
OCH_3
(28%)

O
t-Bu
Bu-t
$(OCH_3)_2$
hν
PhH
O
H
t-Bu
Bu-t
H
CO_2CH_3
(60%)

O
$(OCH_3)_2$
hν
PhH
O
OCH_3
H_3CO
major product
(unstable)
+
O
CO_2CH_3
+
O
CO_2CH_3
(10–15%)

O
O O
hν
polymeric material

SCHEME 14. Photochemistry of quinone monoketals

afforded the analogous *o*-quinone bisketal **129** (equation 41), and the *trans*-stereochemistry of the addition product was established by detailed ^{1}H-NMR spectroscopic studies[96]. However, it was noted in 1977 that anodic oxidation of **124** on a 38-g scale at −30°C afforded **125** in 54% yield after distillation—making the compound now readily available[97].

The simple *o*-naphthoquinone bisketals are formed from anodic oxidation of 2-methoxy- and 1,2-dimethoxynaphthalene derivatives[18]. In most cases, the compounds were not characterized but instead hydrolyzed directly to *o*-naphthoquinone monoketals. Thus, anodic oxidation of 1,2-, 2,6- and 2,7-dimethoxynaphthalene followed by acid

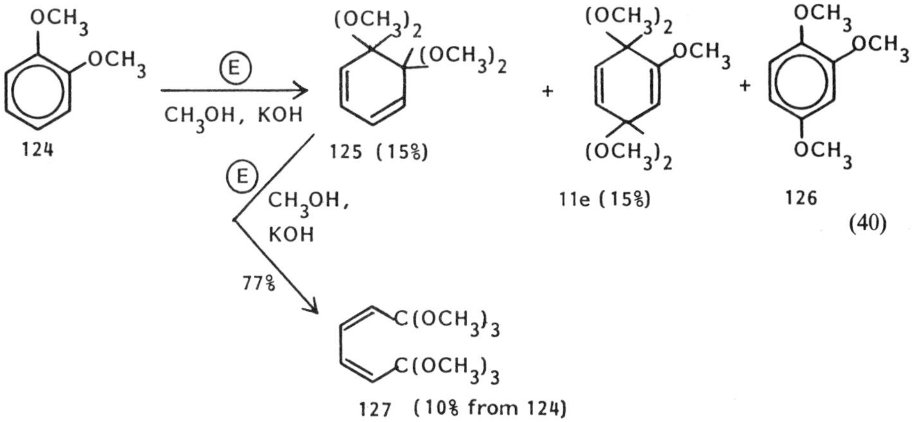

(40)

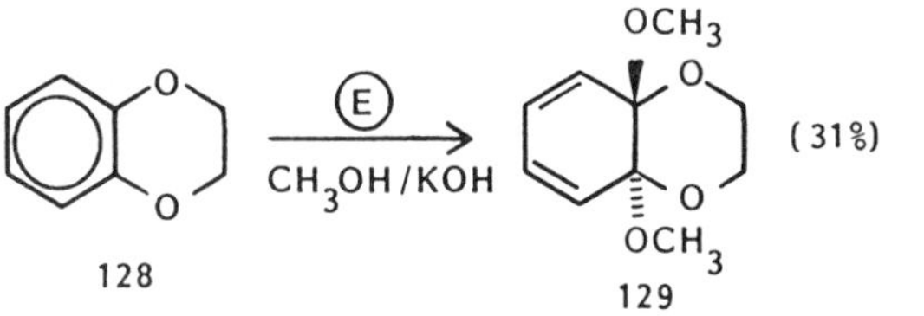

(41)

hydrolysis of the crude reaction mixture afforded the *o*-naphthoquinone monoketals shown in Table 22. In contrast to the *o*-quinone monoketals of the benzenoid series, these compounds are quite stable (*vide infra*) since one double bond of the diene unit is part of an aromatic ring.

The chemistry of *o*-benzoquinone bisketals has not been extensively studied. Reaction of **125** under the Simmons–Smith conditions afforded a mixture of **130** and **131** (equation 42)[97]. This contrasts with similar unsuccessful efforts to cyclopropanate quinone bisketals

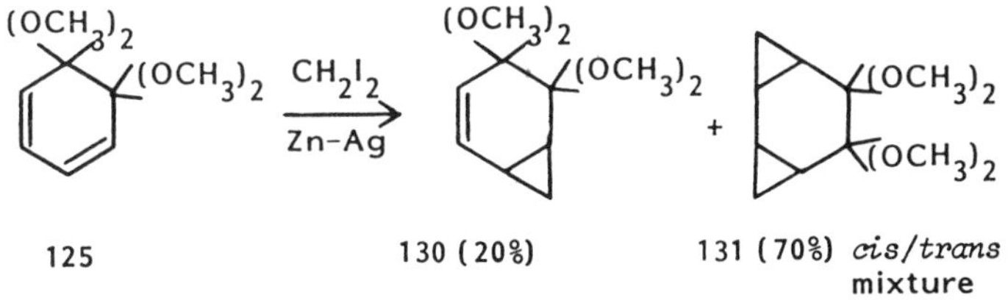

(42)

discussed earlier. However, the reaction of organolithium reagents with **125** has been studied in detail, and this serves as a useful route to substituted 1,2-dimethoxybenzenes (veratroles). The lithium amides shown in equation 43 react regiospecifically with **125** to afford initially the adduct **132** which is subsequently aromatized to 3-substituted veratrole derivatives **133**[98a]. This mode of addition contrasts with the 4-substituted products formed from reaction of amines with *o*-quinones. Alkyllithium reagents also react with **125** to give mixtures of 3- and 4-substituted veratrole derivatives[98b], the ratio of which is

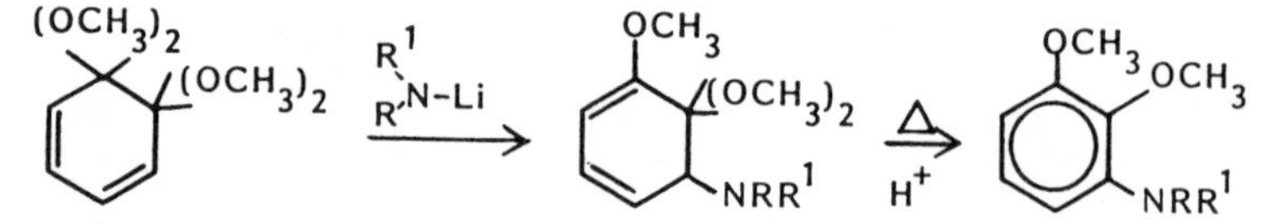

(43)

RR^1NLi (% Yield): CH_3NHLi (66); $PhCH_2NHLi$ (69); $CH_2{=}CH{-}CH_2NHLi$ (84); $p{-}CH_3C_6H_4NHLi$ (64); NH_2Li (0)

TABLE 22. Preparation *o*-naphthoquinone bis- and monoketals

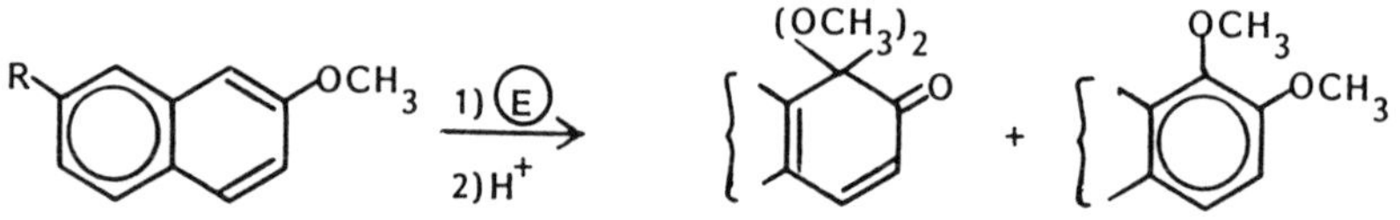

R = H	(36%)	(26%)
R = OCH_3	(53%)	(10%)

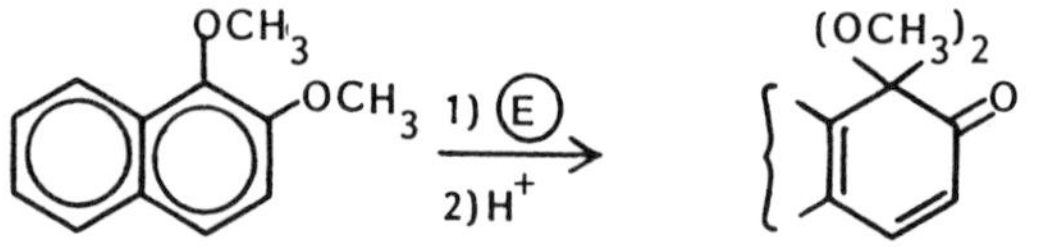

(70%)

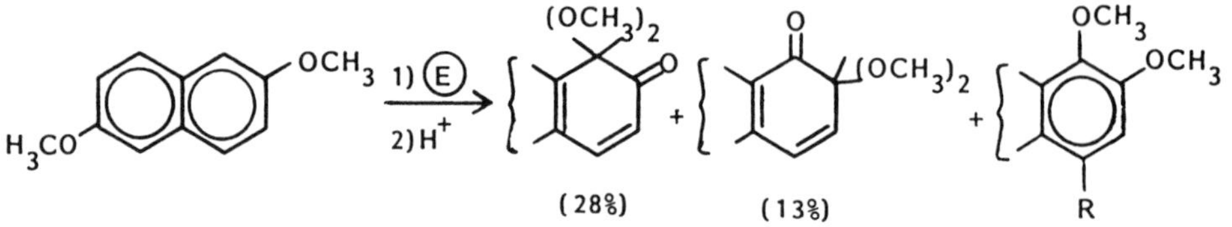

(28%) (13%)

R = H (23%)

R = OCH_3 (16%)

dependent upon the particular alkyllithium reagent (Table 23). A priori, it appears that the functionalization chemistry of quinone bisketals depicted in Table 7 should be applicable to the preparation of functionalized 1,2-dimethoxybenzenes.

B. *o*-Benzo- and Naphthoquinone Monoketals

Monoprotected derivatives of *o*-benzo- and naphthoquinones have been known for many years. Naphthoquinone bromo- and chlorophenoxyhydrins such as **134** were

TABLE 23. Reaction of *o*-quinone bisketal (**125**) with organolithiums

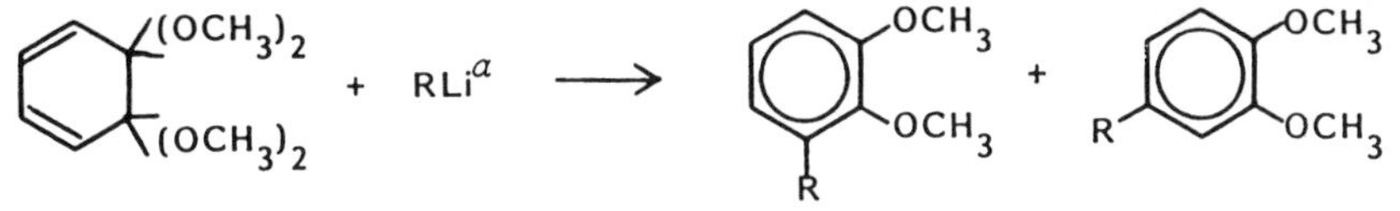

R	Temperature (°C)	Yield (%)	Yield (%)
CH_3	-78	67	9
n-$C_{15}H_{31}$	room temp.	62	13
n-$C_{17}H_{35}$	room temp.	64	11
$CH_3CH{=}CH$	room temp	32	0
C_6H_5	-78	83	0
$CH_2{=}CHCH_2$	-78[b]	13	46
$(CH_3)_2CH$	-78	26	14
$(CH_3)_3C$	-78	33	41

[a] 2.0–2.2 equivalents of RLi were used in ether.
[b] THF was used as solvent.

reported as far back as 1919[99]. Diacetates of *o*-benzoquinones such as **135** were prepared in poor yields from lead tetraacetate oxidation of the corresponding phenol[100]. The *o*-benzo- and naphthoquinone monoketal moieties have been obtained in both the benzene and naphthalene series from oxidation of *o*-alkoxy phenols[39,40] and naphthols[38d].

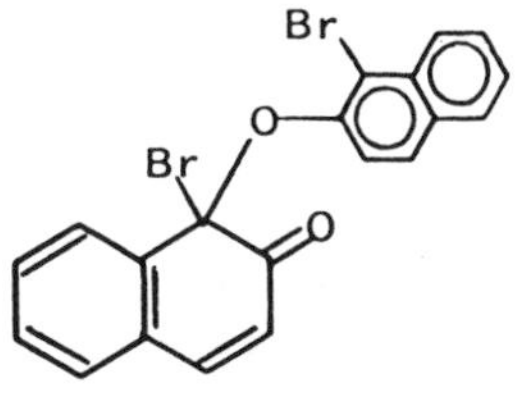

134

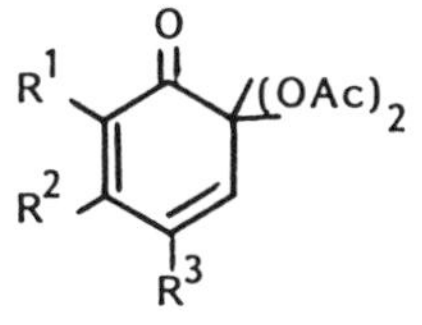

135, $R^1 = R^2 = H$, $R^3 = CH_3$
$R^1 = R^3 = CH_3$, $R^2 = H$
$R^1 = H$, $R^2 = R^3 = CH_3$

Relatively little chemistry has been reported for these ketal derivatives although the Diels–Alder reaction of an analog of an *o*-quinone monoketal served as a key step in the synthesis of Ryanodol[101].

The dearth of chemistry for simple benzenoid systems is undoubtedly related to their facile dimerization **137** → **138** (equation 44). Andersson and coworkers[40] found in their

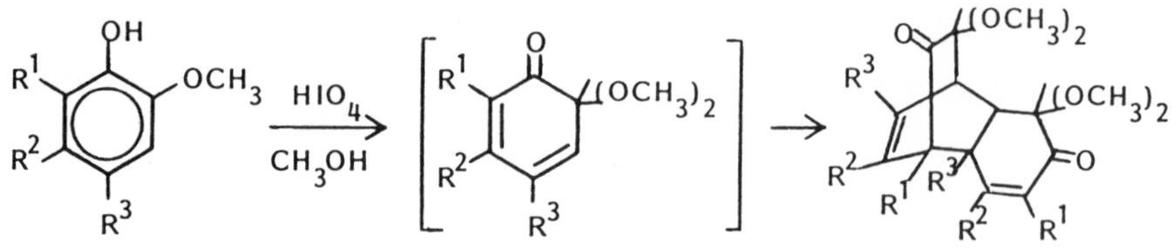

136 137 138 (44)

R^1	R^2	R^3	Yield (%)
H	H	H	23
CH_3	H	H	34
H	H	CH_3	12

studies of *o*-quinone monoketals and other 2,4-cyclohexadienone derivatives that only *o*-quinone monoketals having substituents at the 5-position were stable as monomers (e.g. 5,6,6-trimethoxy-2,4-cyclohexadienones). However, in one case the dimerization of an electrochemically generated *o*-quinone monoketal was shown to be advantageous in a novel but low-yielding synthesis of astone, **141**, a neolignan natural product, from **139** via **140** (equation 45)[102]. Interestingly, the related 6,6-diacetoxy-2,4-cyclohexadienones are

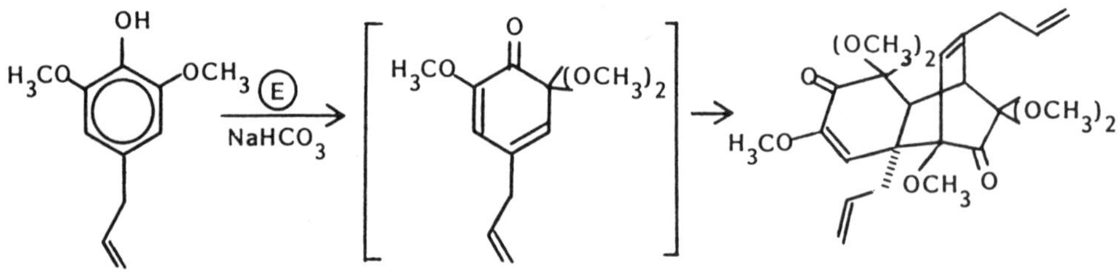

141 (3.3%)

139 140 (45)

less reactive toward dimerization and can be isolated as monomers[100]. Although anodic oxidation of naphthalene derivatives followed by acid hydrolysis is a useful route to *o*-naphthoquinone monoketals (see Table 22), their chemistry has not been extensively

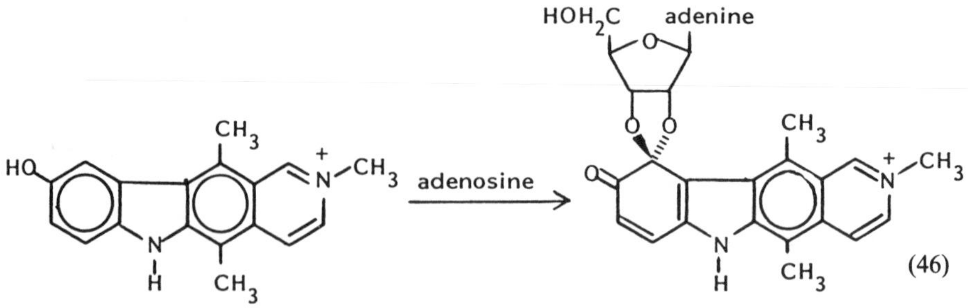

(46)

142 143

studied[18]. Finally, an *o*-quinone monoketal fused to a furan ring was postulated as an intermediate in a thallium(III) oxidation of an *o*-methoxyphenol[103].

Recently, an especially interesting *o*-quinone monoketal (**143**) has been reported as a major product from the biological oxidation of 9-hydroxyellipticinium acetate (**142**, equation 46)[104]. The product is of obvious importance in developing an understanding of the mode of biological activity of 9-hydroxyellipticinium acetate, and undoubtedly the chemistry of these special types of *o*-quinone monoketals will be the subject of future publications.

VII. ACKNOWLEDGEMENTS

The National Science Foundation is thanked for generously supporting our research efforts in this area. Research efforts emanating in this laboratory would not have been possible without a group of capable and dedicated researchers whose names are given in the reference section. Special thanks go to Mr Mike Capparelli and Mr Gary Morrow who critically aided in evaluating this material and to the other students who proofread the manuscript.

VIII. REFERENCES

1. S. Patai (Ed.), *Chemistry of Functional Groups, The Chemistry of the Quinonoid Compounds*, Parts 1 and 2, John Wiley and Sons, New York, 1974.
2. See G. R. Allen, *Org. React.*, **20**, 337 (1968).
3. (a) J. S. Swenton, *Acc. Chem. Res.*, **16**, 74 (1983); (b) P. M. Koelsch and S. P. Tanis, *Kodak Lab. Chem. Bull.*, **52**, 1 (1980); (c) D. A. Evans, D. J. Hart, P. M. Koelsch and P. A. Cain, *Pure Appl. Chem.*, **51**, 1285 (1979); (d) S. Fijita, *Yuki Gosei Kagaku Kyokaishi*, 307 (1981).
4. (a) D. A. Evans, J. M. Hoffman and L. K. Truesdale, *J. Am. Chem. Soc.*, **95**, 5822 (1973); (b) D. A. Evans, P. A. Cain and R. Y. Wong, *J. Am. Chem. Soc.*, **99**, 7083 (1977); (c) D. J. Hart, P. A. Cain and D. A. Evans, *J. Am. Chem. Soc.*, **100**, 1548 (1978); (d) K. A. Parker and J. R. Andrade, *J. Org. Chem.*, **44**, 3964 (1979); (e) A. J. Guildford and R. W. Turner, *Tetrahedron Lett.*, **22**, 4835 (1981); (f) D. A. Evans and R. Y. Wong, *J. Org. Chem.*, **42**, 350 (1977).
5. J. E. Heller, A. S. Dreiding, B. R. O'Connor, H. E. Simmons, G. L. Buchanan, R. A. Raphael and R. Taylor, *Helv. Chim. Acta*, **56**, 272 (1973).
6. N. L. Weinberg and B. Belleau, *J. Am. Chem. Soc.*, **85**, 2525 (1963).
7. (a) N. Clauson-Kaas, F. Limborg and K. Glens, *Acta Chem. Scand.*, **6**, 531 (1952); (b) N. Clauson-Kaas and Z. Tyle, *Acta Chem. Scand.*, **6**, 962 (1952); (c) N. Clauson-Kaas and F. Limborg, *Acta Chem. Scand.*, **6**, 551 (1952); (d) P. Nedenskov, N. Elming, J. T. Nielsen and N. Clauson-Kaas, *Acta Chem. Scand.*, **9**, 17 (1955).
8. (a) N. L. Weinberg and B. Belleau, *Tetrahedron*, **29**, 279 (1973); (b) N. L. Weinberg, D. H. Marr and C. N. Wu, *J. Am. Chem. Soc.*, **97**, 1499 (1975); (c) P. Margaretha and P. Tissot, *Helv. Chim. Acta*, **58**, 933 (1975).
9. (a) M. J. Manning, P. W. Raynolds and J. S. Swenton, *J. Am. Chem. Soc.*, **98**, 5008 (1976); (b) J. S. Swenton, D. K. Jackson, M. J. Manning and P. W. Raynolds, *J. Am. Chem. Soc.*, **100**, 6182 (1978).
10. (a) P. W. Raynolds, M. J. Manning and J. S. Swenton, *Tetrahedron Lett.*, 2383 (1977); (b) J. S. Swenton and P. W. Raynolds, *J. Am. Chem. Soc.*, **100**, 6188 (1978).
11. (a) M. J. Manning, D. R. Henton and J. S. Swenton, *Tetrahedron Lett.*, 1679 (1977); (b) D. R. Henton, B. L. Chenard and J. S. Swenton, *J. Chem. Soc., Chem. Commun.*, 326 (1979); (c) M. G. Dolson, D. K. Jackson and J. S. Swenton, *J. Chem. Soc., Chem. Commun.*, 327 (1979); (d) D. R. Henton, R. L. McCreery and J. S. Swenton, *J. Org. Chem.*, **45**, 369 (1980); (e) See also G. Bockmair and H. P. Fritz, *Electrochim. Acta*, **21**, 1099 (1976); L. Eberson and B. Helgee, *Chem. Sci.*, **5**, 47 (1974).
12. D. K. Jackson and J. S. Swenton, *Synth. Commun.*, **7**, 333 (1977).
13. (a) B. L. Chenard and J. S. Swenton, *J. Chem. Soc., Chem. Commun.*, 1172 (1979); (b) B. L. Chenard, J. R. McConnell and J. S. Swenton, *J. Org. Chem.*, **48**, 4312 (1983).

14. (a) Unpublished results of R. DeSchepper; (b) Unpublished results of W.-B. Shu.
15. J. S. Swenton, J. N. Freskos, G. W. Morrow and A. D. Sercel, *Tetrahedron*, **40**, 4625 (1984).
16. J. S. Swenton, D. K. Anderson, C. E. Coburn and A. P. Haag, *Tetrahedron*, **40**, 4633 (1984).
17. W. Bornatsch and E. Vogel, *Angew. Chem.*, **14**, 420 (1975).
18. M. G. Dolson and J. S. Swenton, *J. Am. Chem. Soc.*, **103**, 2361 (1981).
19. C. Chen, C. Shih and J. S. Swenton, *Tetrahedron Lett.*, **27**, 1891 (1986).
20. (a) M. C. Carreño, F. Fariña, A. Galań and J. S. Garcia Ruano, *J. Chem. Res.*, *Synop.*, 296 (1979); (b) M. C. Carreño, F. Fariña, A. Galań and J. S. Garcia Ruano, *J. Chem. Res.*, *Miniprint*, 3443 (1979); (c) F. Fariña, A. Galań and J. L. Barcia, *An. Quin.*, **74**, 954 (1978).
21. M. G. Dolson and J. S. Swenton, *J. Org. Chem.*, **46**, 177 (1981).
22. A. Nilsson, U. Palmquist, T. Pettersson and A. Ronlán, *J. Chem. Soc.*, *Perkin Trans. 1*, 708 (1978).
23. R. Pistorius and H. Mallanuer, U.S. Patent 4,046,652.
24. (a) L. F. Fieser and R. G. Kennelly, *J. Am. Chem. Soc.*, **57**, 1611 (1935); (b) R. Kitchen and R. B. Sandin, *J. Am. Chem. Soc.*, **67**, 1645 (1945).
25. (a) A. Nilsson and A. Ronlán, *Tetrahedron Lett.*, 1107 (1975); (b) E. Konz and R. Pistorius, *Synth.*, 603 (1979).
26. G. L. Buchanan, R. A. Raphael and R. Taylor, *J. Chem. Soc.*, *Perkin Trans. 1*, 373 (1973).
27. Unpublished results of M. Capparelli.
28. (a) T. Imamoto, T. Kusumoto, Y. Tawarayama, Y. Suguira, T. Mita, Y. Hatanka and M. Yokoyama, *J. Org. Chem.*, **49**, 3904 (1984); (b) T. Imamoto, Y. Suguira and N. Takiyama, *Tetrahedron Lett.* **25**, 4233 (1984); (c) T. Imamoto, N. Takiyama, and K. Nakamura, *Tetrahedron Lett.*, **26**, 4763 (1985).
29. (a) D. K. Jackson, L. Narasimhan and J. S. Swenton, *J. Am. Chem. Soc.*, **101**, 3989 (1979); (b) J. S. Swenton, D. K. Anderson and D. K. Jackson, *J. Org. Chem.*, **46**, 4825 (1981).
30. (a) P. W. Raynolds, M. J. Manning and J. S. Swenton, *J. Chem. Soc.*, *Chem. Commun.*, 499 (1977); (b) B. L. Chenard, M. J. Manning, P. W. Raynolds and J. S. Swenton, *J. Org. Chem.*, **45**, 378 (1980).
31. B.-T. Groebel, E. Konz, H. Millauer and R. Pistorius, *Synth.*, 605 (1979).
32. R. H. Thomson, *J. Org. Chem.*, **13**, 371 (1948).
33. D. Taub, *Chem. Ind. (Lond.)*, 558 (1962).
34. S. Kumamoto and T. Kato, *Kogyo Kagaku Zasshi*, **60**, 1325 (1957); *Chem. Abs.*, **53**, 16997e (1959).
35. C. Martius and H. Eilingsfeld, *Ann. Chem.*, **607**, 159 (1967).
36. V. D. Parker, *J. Am. Chem. Soc.*, **91**, 5380 (1969).
37. (a) W. Durckheimer and L. A. Cohen, *J. Am. Chem. Soc.*, **86**, 4388 (1964); (b) W. Durckheimer and L. A. Cohen, *Biochem.*, **3**, 1948 (1964).
38. (a) T. R. Kasturi and T. Arunachalam, *Can. J. Chem.*, **46**, 3625 (1968); (b) T. R. Kasturi, T. Arunachalam and G. Subrahmanyam, *J. Chem. Soc. (C)*, 1257 (1970); (c) I. G. C. Coutts, M. R. Hamblin and S. E. Welsby, *J. Chem. Soc.*, *Perkin Trans 1*, 493 (1981); (d) I. G. C. Coutts, D. J. Humphreys and K. Schofield, *J. Chem. Soc. (C)*, 1982 (1969).
39. D. G. Hewitt, *J. Chem. Soc., (C)*, 2967 (1971).
40. (a) G. Andersson and P. Berntsson, *Acta Chem. Scand.*, *Ser. B.*, **29**, 948 (1975); (b) G. Andersson, *Acta Chem. Scand.*, *Ser. B.*, **30**, 64 (1976); (c) G. Andersson, *Acta Chem. Scand.*, *Ser. B.*, **30**, 403 (1976).
41. A. McKillop, D. H. Perry, M. Edwards, S. Antus, L. Farkas, M. Nogradi and E. C. Taylor, *J. Org. Chem.*, **41**, 282 (1976).
42. (a) A. Goosen and C. W. McCleland, *J. Chem. Soc.*, *Perkin Trans 1*, 646 (1978); (b) A. Goosen and C. W. McCleland, *J. Chem. Soc.*, *Chem. Commun.*, 655 (1975).
43. G. Buchi, P.-S. Chu, A. Hoppmann, C.-P. Mak and A. Pearce, *J. Org. Chem.*, **43**, 3983 (1978).
44. T. W. Hart and F. Scheinmann, *Tetrahedron Lett.*, **21**, 2295 (1980).
45. (a) D. J. Crouse and D. M. S. Wheeler, *Tetrahedron Lett.*, **50**, 4797 (1979); (b) D. J. Crouse, M. M. Wheeler, M. Goemann, P. S. Tobin, S. K. Basu and D. M. S. Wheeler, *J. Org. Chem.*, **46**, 1814 (1981).
46. R. N. Warrener, P. S. Gee and R. A. Russell, *J. Chem. Soc.*, *Chem. Commun.*, 1100 (1981).
47. A. Nilsson, U. Palmquist, T. Pettersson and A. Ronlán, *J. Chem. Soc.*, *Perkin Trans 1*, 696 (1978).
48. C. H. Foster and D. A. Payne, *J. Am. Chem. Soc.*, **100**, 2834 (1978).

49. C-P. Chen and J. S. Swenton, *J. Chem. Soc., Chem. Commun.*, 1291 (1985). See also Y. Shizuri, K. Nakamura, S. Yamamura, S. Ohba, H. Yamashita and Y. Saito, *Tetrahedron Lett.*, **27**, 727 (1986).
50. R. F. Stewart and L. L. Miller, *J. Am. Chem. Soc.*, **102**, 4999 (1980).
51. (a) E. J. Corey, S. Barcza and G. Klotman, *J. Am. Chem. Soc.*, **91**, 4782 (1969); (b) A. Ronlán and V. Parker, *J. Chem. Soc. C*, 3214 (1971).
52. H. G. Thomas and H.-W. Schwager, *Tetrahedron Lett.*, **25**, 4471 (1984).
53. R. Chaturvedi, J. Adams and E. H. Cordes, *J. Org. Chem.*, **33**, 1652 (1968).
54. D. R. Henton, K. Anderson, M. J. Manning and J. S. Swenton, *J. Org. Chem.*, **45**, 3422 (1980).
55. (a) P. Deslongchamps, *Tetrahedron*, **31**, 2463 (1975); (b) P. Deslongchamps, *Stereoelectronic Effects in Organic Chemistry*, Pergamon Press, New York, 1983.
56. (a) For a general discussion of ketal and orthoester hydrolysis, see: E. H. Cordes and H. G. Bull, *Chem. Rev.*, **74**, 581 (1974); T. Fife, *Acc. Chem. Res.*, **5**, 264 (1972); (b) A. J. Kirby and R. J. Martin, *J. Chem. Soc., Chem. Commun.*, 803 (1978).
57. (a) M. G. Dolson, B. L. Chenard and J. S. Swenton, *J. Am. Chem. Soc.*, **103**, 5263 (1981); (b) B. L. Chenard, M. G. Dolson, A. D. Sercel and J. S. Swenton, *J. Org. Chem.*, **49**, 318 (1984); (c) B. L. Chenard, D. K. Anderson and J. S. Swenton, *J. Chem. Soc., Chem. Commun.*, 932 (1980).
58. C-P. Chen and J. S. Swenton, *J. Org. Chem.*, **50**, 4569 (1985).
59. H. M. Crawford, M. Lumpkin and M. McDonald, *J. Am. Chem. Soc.*, **74**, 4087 (1952).
60. (a) H. W. Moore, Y. L. Sing and R. S. Sidhu, *J. Org. Chem.*, **42**, 3320 (1977); (b) H. W. Moore, Y. L. Sing and R. S. Sidhu, *J. Org. Chem.*, **45**, 5057 (1980).
61. A. Rieker and G. Henes, *Tetrahedron Lett.*, 3775 (1968) and references therein.
62. A. Fischer and G. N. Henderson, *Tetrahedron Lett.*, **24**, 131 (1983).
63. D. Liotta, M. Saindane and C. Barnum, *J. Org. Chem.*, **46**, 3369 (1981).
64. (a) F. Wessely, L. Holzer, and H. Vilcsek, *Monatsch. Chem.*, **83**, 1252 (1952); (b) F. Wessely, L. Holzer and H. Viscsek, *Monatsch. Chem.*, **84**, 655 (1953); (c) O. Polansky, E. Schinzel and F. Wessely, *Monatsch. Chem.*, **87**, 24 (1956) and papers cited therein; (d) For recent mechanistic studies see: B. Miller and J. G. Haggerty, *J. Chem. Soc., Chem. Commun.*, 1617 (1984).
65. (a) For a mechanistic study of additions of an organolithium reagent to a quinone monoketal see: D. Liotta, M. Saindane, and L. Waykole, *J. Am. Chem. Soc.*, **105**, 2922 (1983); (b) S. Hunig and G. Wehner, *Chem. Ber.*, **113**, 324 (1980).
66. (a) A. Pelter, R. S. Ward and R. R. Rao, *Tetrahedron Lett.*, **24**, 621 (1983); (b) A. Pelter, R. S. Ward and R. R. Rao, *Tetrahedron*, **41**, 2933 (1985).
67. K. A. Parker and S-K. Kang, *J. Org. Chem.*, **45**, 1218 (1980).
68. (a) M. F. Semmelhack, L. Keller, T. Sato and E. Speiss, *J. Org. Chem.*, **47**, 4382 (1982); (b) M. F. Semmelhack, L. Keller, T. Sato, E. Speiss and W. Wulff, *J. Org. Chem.*, **50**, 5566 (1985).
69. P. Brownbridge and T-H. Chan, *Tetrahedron Lett.*, **21**, 3431 (1980).
70. (a) I. A. McDonald and A. S. Dreiding, *Helv. Chim. Acta*, **56**, 2523 (1973); (b) H. Stetter and J. Lennartz, *Justus Liebigs Ann. Chem.*, 1807 (1977); (c) R. O. Duthaler and U. H.-U. Wegmann, *Helv. Chim. Acta*, **67**, 1755 (1984).
71. R. M. Coates and P. A. MacManus, *J. Org. Chem.*, **47**, 4822 (1982).
72. (a) D. A. Evans, D. J. Hart and P. M. Koelsch, *J. Am. Chem. Soc.*, **100**, 4593 (1978); (b) D. A. Evans, S. P. Tanis and D. J. Hart, *J. Am. Chem. Soc.*, **103**, 5813 (1981).
73. For leading references to this work, see: *Tetrahedron Symposium-In-Print Number 17*, **40**, 4539–4789 (1984).
74. B. A. Keay and R. Rodrigo, *Can J. Chem.*, **61**, 637 (1983).
75. (a) R. A. Russell and R. N. Warrener, *J. Chem. Soc., Chem. Commun.*, 108 (1981); (b) R. A. Russell, A. S. Krauss and R. N. Warrener, *Tetrahedron Lett.*, **25**, 1517 (1984); (c) R. A. Russell, R. W. Irvine and A. S. Krauss, *Tetrahedron Lett.*, **25**, 5817 (1984); (d) R. A. Russell, P. S. Gee, R. W. Irvine and R. N. Warrener, *Aust. J. Chem.*, **37**, 1709 (1984); (e) A. M. Becker, R. W. Irvine, A. S. McCormick, R. A. Russell and R. N. Warrener, *Tetrahedron Lett.*, 3431 (1986).
76. G. W. Morrow and J. S. Swenton, *J. Org. Chem.*, accepted for publication.
77. K. A. Parker, I. D. Cohen and R. E. Babine, *Tetrahedron Lett.*, **25**, 3543 (1984).
78. E. C. Taylor, G. E. Jagdmann and A. McKillop, *J. Org. Chem.*, **43**, 4385 (1978).
79. For a similar example with *p*-quinols see: E. Hecker and R. Lattrell, *Justus Liebigs Ann. Chem.*, **662**, 48 (1963).

80. K. T. Finley and L. K. J. Tong, in *The Chemistry of the Carbon–Nitrogen Double Bond* (Ed. S. Patai), Interscience, New York, 1970, pp. 633–729.
81. C-P. Chen, C. Shih and J. S. Swenton, *Tetrahedron Lett.*, **27**, 1891 (1986).
82. C. Shih, PhD thesis, The Ohio State University, 1982.
83. O. R. Gottlieb, *Phytochem.*, **11**, 1537 (1972).
84. G. Buchi and C-P. Mak, *J. Am. Chem. Soc.*, **99**, 8073 (1977).
85. G. Buchi and P-S. Chu, *J. Org. Chem.*, **43**, 3717 (1978).
86. G. Buchi and P-S. Chu, *J. Am. Chem. Soc.*, **101**, 6767 (1979).
87. G. Buchi and P-S. Chu, *J. Am. Chem. Soc.*, **103**, 2718 (1981).
88. C-P. Mak and G. Buchi, *J. Org. Chem.*, **46**, 1 (1981).
89. (a) Y. Shizuri and S. Yamamura, *Tetrahedron Lett.*, **24**, 5011 (1983); (b) Y. Shizuri, K. Nakamura and S. Yamamura, *J. Chem. Soc., Chem. Commun.*, 530 (1985).
90. M. Carmen Carreño, F. Fariña, A. Galan and J. L. Garcia Ruano, *J. Chem. Res. (S)*, 370 (1981); *J. Chem. Res. (M)*, 4310 (1981).
91. (a) R. N. Warrener and B. C. Hammer, *J. Chem. Soc., Chem. Commun.*, 942 (1981); (b) R. A. Russell, D. A. C. Evans and R. N. Warrener, *Aust. J. Chem.*, **37**, 1699 (1984).
92. W. G. Dauben and A. F. Cunningham Jr, *J. Org. Chem.*, **48**, 2842 (1983).
93. (a) K. A. Parker, I. D. Cohen, A. Padwa and W. Dent, *Tetrahedron Lett.*, **25**, 4917 (1984); (b) A. Padwa, Y-Y. Chen, W. Dent and H. Nimmesgern, *J. Org. Chem.*, **50**, 4006 (1985).
94. (a) S. Fujita, *J. Chem. Soc., Chem. Commun.*, 425 (1981); (b) S. Fujita, *J. Org. Chem.*, **48**, 177 (1983).
95. (a) D. G. Hewitt and R. F. Taylor, *J. Chem. Soc., Chem. Commun.*, 493 (1972); (b) P. Margaretha, *Helv. Chim. Acta*, **59**, 661 (1976).
96. R. R. Fraser and C. Reves-Zamora, *Can. J. Chem.*, **45**, 929 (1967).
97. M. Engelhard and W. Luttke, *Chem. Ber.*, **110**, 3759 (1977).
98. (a) Y. Kikuchi, Y. Hasegawa, and M. Matsumoto, *J. Chem. Soc., Chem. Commun.*, 878 (1982); (b) Y. Kikuchi, Y. Hasegawa, and M. Matsumoto, *Tetrahedron Lett.*, **23**, 2199 (1982).
99. R. Pummerer, *Chem. Ber.*, **52**, 1403 (1919).
100. F. Wessely and F. Sinwel, *Monatsh. Chem.*, **81**, 1055 (1950).
101. P. Deslongchamps, *Pure Appl. Chem.*, **49**, 1329 (1977).
102. M. Iguchi, A. Nishiyama, Y. Terada and S. Yamamura, *Tetrahedron Lett.*, 4511 (1977).
103. R. B. Gammill, *Tetrahedron Lett.*, **26**, 1385 (1985).
104. (a) J-B. LePecq, Nguyen-Dat-Xuong, C. Gosse and C. Paoletti, *Proc. Natl. Acad. Sci.*, **71**, 5078 (1974); (b) G. Meunier, B. Meunier, C. Auclair, J. Bernadou and C. Paoletti, *Tetrahedron Lett.*, **24**, 365 (1983); (c) G. Pratviel, J. Bernadou and B. Meunier, *J. Chem. Soc., Chem. Commun.*, 60 (1985); (d) V. K. Kansal, S. Funakoshi, P. Mangeney, P. Potier, B. Gillet, E. Guittet and J. Y. Lallemand, *Tetrahedron Lett.*, **25**, 2351 (1984); (e) J. Bernadou, B. Meunier, G. Meunier, C. Auclair and C. Paoletti, *Proc. Natl. Acad. Sci.*, **81**, 1297 (1984); (f) V. K. Dansal, R. Sundaramoorth, B. C. Das and P. Potier, *Tetrahedron Lett.*, **26**, 4933 (1985).

The Chemistry of Quinonoid Compounds, Vol. II
Edited by S. Patai and Z. Rappoport

CHAPTER **16**

Quinhydrones and semiquinones

M. CATHERINE DEPEW and JEFFREY K. S. WAN
Department of Chemistry, Queen's University,
Kingston, Ontario, Canada K7L 3N6

ABBREVIATIONS

AQ	anthraquinone
AZQ	3,6-diaziridinyl-2,5-bis(carboethoxyamino)-1,4-benzoquinone
BQ	1,4-benzoquinone
BQH_2	1,4-benzohydroquinone
CQ	camphorquinone
CIDMP	chemically induced dynamic magnetic polarization
CIDEP	chemically induced dynamic electron polarization
CIDNP	chemically induced dynamic nuclear polarization
DIOP	2,3-*o*-isopropylidene-2,3-dihydroxy-1,4-bis-(diphenylphosphino)butane
2,6-DTBQ	2,6-di-*t*-butyl-1,4-benzoquinone
3,5-DTBQ	3,5-di-*t*-butyl-1,2-benzoquinone
DQ	1,4-duroquinone (2,3,5,6-tetramethyl-1,4-benzoquinone)
DQH_2	1,4-durohydroquinone
E/A	emissive/absorptive
FQ	2,3,5,6-tetrafluoro-1,4-benzoquinone
FQH_2	2,3,5,6-tetrafluoro-1,4-benzohydroquinone
FT-IR	Fourier transform-infrared spectroscopy
HFS	hyperfine splitting
ISC	intersystem crossing
$MBPh_4$	alkali metal tetraphenylborate
NQ	1,4-naphthoquinone
NQH_2	1,4-naphthohydroquinone
PBN	phenyl-*N*-*t*-butyl-nitrone
P–Q	porphyrin–quinone
RPM	radical pair mechanism
TM	triplet mechanism
T_1	spin lattice relaxation time
TEA	triethylamine
TFA	trifluoroacetic acid
SDS	sodium dodecyl sulphate
SOD	superoxide dismutase

I. INTRODUCTION

Historically and today semiquinones as a class of organic radicals continue to play a major role in chemistry and biological chemistry. In the early development of ESR applications to free radical chemistry, the apparent ease of preparation and generally interesting structural aspects of many semiquinones have provided important models for the spectroscopists to advance a better understanding of the nature and correlation of ESR parameters with the structure of free radicals. The parents of these semiquinones usually contain two reactive carbonyl groups structurally integrated into an aromatic ring system which are attractively amenable to photochemical and photobiological investigations. Indeed the basic understanding of the primary photochemical processes of simple *para*-quinones has greatly enhanced the development of the photoexcited triplet mechanism in the CIDEP (chemically induced dynamic electron polarization) and CIDNP (chemically induced dynamic nuclear polarization) phenomena, as the earlier, critical studies employed exclusively the semiquinone radicals in photochemical systems. Today a systematic study, combining both ESR and time-resolved CIDMP techniques, on quinone reactions can yield rather detailed information not normally obtained from conventional magnetic resonance experiments.

It is not our intention to present here a review of the techniques and theories of ESR and CIDMP phenomena since many past and current reviews are readily available in literature. Rather, we shall attempt to focus our attention on a number of aspects directly concerned with semiquinones and to emphasize the contributions to semiquinone chemistry advanced by the modern experimental techniques, especially in ESR and time-resolved CIDEP methods. As well, we shall touch upon some other important areas such as photosynthesis in which semiquinones play significant roles. Some of these related fields are clearly outside our expertise and no effort will be claimed to give a complete list of references to the enormous amount of work published in these fields. Although the original edition of this book contained excellent discussions concerning the quinhydrones, we shall begin in this chapter by considering some further work concerning this interesting class of complexes.

II. QUINHYDRONES

Quinhydrones are 1:1 complexes formed between quinones (Q) and hydroquinones (QH_2); their existence has been known for many years. Although quinhydrones are largely dissociated in solution, in the solid state X-ray structures have shown alternating Q and QH_2 molecules held together in chains by H-bonding in one dimension and π-bonding in a second. The parent or prototype is of course the complex of 1,4-benzoquinone (BQ) and its corresponding hydroquinone (BQH_2) and is usually referred to as quinhydrone; it exists as a conventional donor–acceptor complex in solution and crystallizes in both a monoclinic[1] and a triclinic form[2]. The crystal structures and properties of quinhydrones have been discussed in some detail in the first edition of this series[3, 4]. This information will therefore be referred to only very briefly and most of the focus will be on the literature published in this field since 1974.

Unsymmetrically substituted quinhydrones readily undergo redox reactions in solution; this also occurs in the solid state but at a much lower rate. For example, the quinhydrones formed from deuterium-labelled 1,4-benzoquinone and hydroquinone (**1**), and ^{14}C-duroquinone (DQ) and durohydroquinone (DQH_2) (**2**) were found not to exchange at room temperature and although **1** and **2** did exchange slowly[5] at 107–120 °C, the phenyl-substituted quinhydrones **3** and **4** did not[6]. Indeed these compounds could not be made to

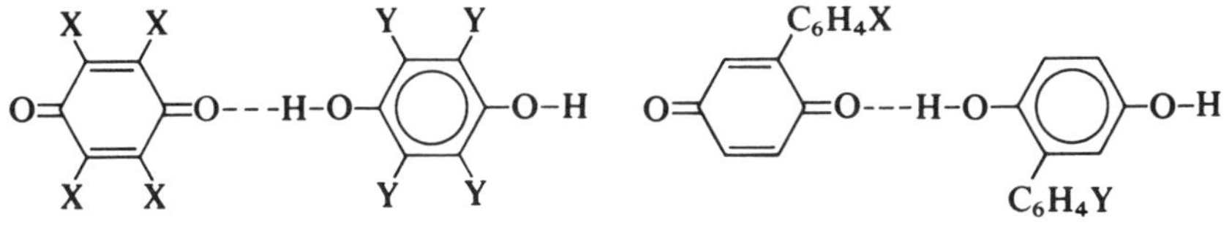

(**1**) X = D, Y = H
(**2**) X = $^{14}CH_3$, Y = CH_3

(**3**) X = H, Y = Cl
(**4**) X = Cl, Y = H

isomerize under any known conditions in the solid state. A number of unsymmetrically substituted quinhydrones have been formed both by crystallizing from solution and by grinding together in a mortar and pestle the desired Q/QH_2 pair[7]. While grinding the initially yellow solids together the authors noted a rapid darkening in colour as the π complex formed. The product quinhydrones whether synthesized by either method have been shown to have the identical spectral and X-ray powder patterns.

This solid state method has enabled the synthesis of several quinhydrones which would be inaccessible by crystallization from solution due to rapid redox hydrogen exchange. For example, BQ and 2-methylhydroquinone as well as 2-phenylquinone and naphthohydroquinone form quinhydrone products which have distinctly different X-ray powder patterns from those of the isomeric complex[7]. The reactions forming these 'unstable' quinhydrones were readily followed by differential scanning calorimetry and by FT-IR. The formation of a quinhydrone complex has previously been shown by Slifkin and Walmsley[8] to result in a shift of the carbonyl resonance of the starting quinone to a lower frequency. Formation of both the 'unstable' quinhydrone and its stable isomer resulted in this expected carbonyl shift; however, the spectra of the two redox isomers were quite different. Thus no hydrogen transfer resulted from the initial synthesis (grinding together) or from preparation of the paraffin mull; however, standing for longer periods of time in the paraffin suspension did result in some conversion of the unstable to the more stable redox isomer.

The formation of quinhydrone complexes is influenced by the donor properties of the hydroquinone, the acceptor strength of the quinone, and also steric factors. For example, chloranil and tetrachlorohydroquinone do not form a quinhydrone presumably because of the weak donor capabilities of the hydroquinone[6]. These factors also affect the redox behaviour of unsymmetrically substituted quinhydrones in solution. Slow exchange rates are favoured by a close balance of the redox potentials of the two component pairs. Substitution of the starting materials also reduces the exchange rate and can in fact stop formation of the desired complex. For example, quinhydrone complexes were not formed from the following Q/QH_2 pairs in DMSO: 1,4-naphthoquinone/durohydroquinone, chloranil/hydroquinone, 2,5-diphenyl-1,4-benzoquinone/2,5-di-*t*-butylhydroquinone. In addition the strong H-bonding solvent DMSO retards exchange compared with benzene.

Although reaction of BQ with 2,5-dimethylhydroquinone gave the expected quinhydrone, reaction of 2,5-dimethylquinone with BQH_2 produced a 1:2 complex[7,9]. Similar behaviour was noted for 2-methylquinone and BQH_2 which formed a complex in a ratio ranging from 1:1.5 to 1:2, while the corresponding redox partners BQ and 2-methylhydroquinone gave the quinhydrone as previously mentioned. These non-equimolar products were formed whether the reaction was solid state or by crystallization of the product from solution and manipulation of the molar ratios of the starting compounds did not affect the outcome. Apparently the products of these reactions are controlled by the energetics of crystal packing rather than the stoichiometry of the starting materials. In addition 2,5-dimethylquinone and 2,5-dimethylhydroquinone formed a 2:1 complex on crystallization and not the quinhydrone[10]. The crystal structure of this 2:1 complex has been determined to consist of basic triplet structural units of one hydroquinone molecule forming hydrogen bonded bridges to two quinone molecules. The quinone therefore has one carbonyl group hydrogen bonded to a neighbouring hydroquinone and the other carbonyl group directed towards an aromatic C–H of an adjacent triplet. The FT-IR spectrum of this complex shows two carbonyl stretching frequencies at 1628 cm^{-1} and 1663 cm^{-1} consistent with this structure. This unusual behaviour has so far only been noted for quinones having one or two methyl substituents; in contrast DQ and DQH_2 form a 1:1 quinhydrone complex[10].

Although quinhydrones are largely dissociated in solution their ^{13}C-NMR spectra have been studied in the solid state to search for evidence of the effects of complexation and charge transfer stacking in the crystal[11]. The quinone carbonyl and the hydroquinone hydroxylic carbons are the atoms most likely to be perturbed by these effects. Both the monoclinic and triclinic polymorphs of quinhydrone have relatively simple spectra with single resonances for the carbonyl carbon of the quinone and also for the phenolic carbon of the hydroquinone component of the complex. The resonances occurred at the same chemical shift position in both the mono- and triclinic complexes. The authors found upfield shifts (relative to the uncomplexed quinone) of 3–4 ppm for the quinone carbonyl resonances of BQ complexed with BQH_2 and also in the 1:2 complex of 2,5-dimethyl-1,4-benzoquinone with BQH_2. This effect is similar in magnitude but opposite in direction to that noted due to hydrogen bonding between the phenolic hydroquinone H and the quinone carbonyl O in, for example, the quinhydrone formed from DQ and DQH_2. This latter quinhydrone is of interest because the hydroxyl stretch is at a higher frequency (3495 cm^{-1}) than expected suggesting substantial differences in hydrogen bonding than that found in less substituted quinhydrones[10]. Apparently the crystal structure geometry is distorted by the steric requirements of the *ortho* methyl groups. The authors have noted that the differences in chemical shifts between the complexed quinhydrones and the uncomplexed starting materials are substantial enough to make solid state ^{13}C-NMR a useful technique for their differentiation although it is difficult to establish the source of the hydrogen bonding or crystallographic effects.

This technique has also been used to follow the solid state redox reaction of the 1:1 complex of BQ and 2,5-dimethyl-1,4-hydroquinone complex to form the 1:2 complex of 2,5-dimethyl-1,4-benzoquinone and BQH_2 shown in equation 1. Monitoring this reaction

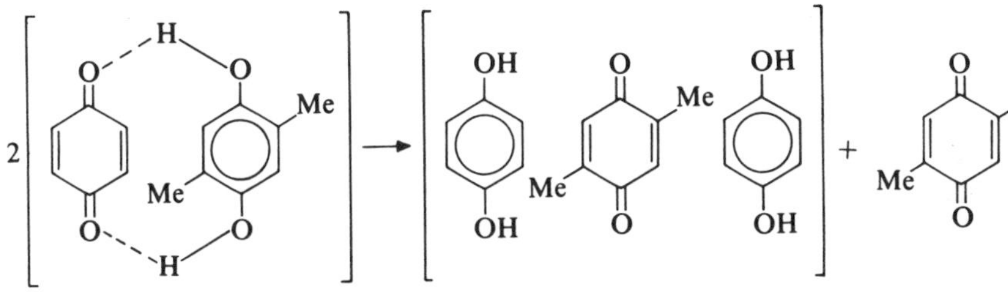

(1)

in nujol mulls by FT-IR was complicated by redox exchange in this medium; it was hoped that solid state NMR would circumvent this problem. By mixing all the possible combinations of the starting quinones, hydroquinones and quinhydrones involved it was found that exchange of one hydroquinone for another was rapid relative to the redox hydrogen transfer between quinone and hydroquinone. In the solid state the redox reaction was complete after 8 h at 85 °C or 30 min at 115 °C.

The charge transfer energies of several simple substituted BQ/BQH_2 complexes in solution and in the solid phase[12, 13] have been studied. Correlations were made between the E_{CT} and the activation energy for the redox transformations (equation 2).

$$Q' + QH_2 \rightleftharpoons Q'H_2 + Q \qquad (2)$$

Curtin and Paul[14] have discussed an interesting centre of symmetry noted in the X-ray crystal structure of the phenyl and *p*-chlorophenyl quinhydrones **5** and **6**. The centre of symmetry lies midway between an oxygen atom of a hydroquinone and the oxygen atom of

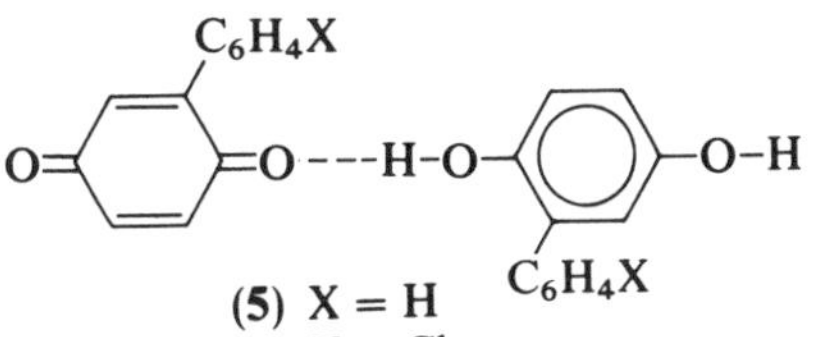

(5) X = H
(6) X = Cl

the adjacent hydrogen bonded quinone; this suggests a structure (**7**) having the hydrogen equidistant from the two oxygens in the quinhydrone. However, the chemical non-equivalence of the quinhydrones prepared from the isomeric phenylquinone and hydroquinone pairs **3** and **4** has been demonstrated as well as their lack of interconversion in the solid phase. Also since it is known that in quinhydrones the hydrogen is generally unsymmetrically bonded as shown[15] (**8**), they explained the anomalous X-ray result as due to the formation of large ordered polar regions in the crystal which when averaged over a large volume give the appearance of being centrosymmetric to the X-ray diffractometer.

(**7**)

(**8**)

III. SEMIQUINONES

One-electron oxidation of hydroquinones or reduction of quinones results in the formation of semiquinones; they can exist as radical anions ($Q^{-\cdot}$) or neutral radicals ($Q\dot{H}$). In addition, in this chapter we will discuss the formation of the hydroquinone cation radical ($QH_2^{+\cdot}$), and the quinone cation radical ($Q^{+\cdot}$). The structures of these radicals are represented below. These radicals are often the intermediates in redox reactions and they will be our primary focus in this chapter.

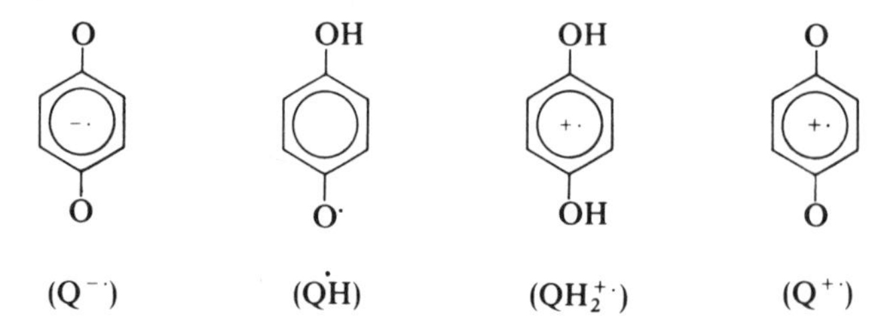

($Q^{-\cdot}$) ($Q\dot{H}$) ($QH_2^{+\cdot}$) ($Q^{+\cdot}$)

Semiquinones can be generated by reduction of quinones in a variety of ways, several of the more common of these being:

(1) radiolysis of aqueous solutions of quinones

$$e_{aq}^{-} + Q \rightarrow Q^{-\cdot}$$

(2) electron abstraction from oxygen species or other inorganic ions

$$Q + OH^{-} \rightarrow Q^{-\cdot} + OH^{\cdot}$$

(3) quenching of photoexcited molecules such as triplet chlorophyll

$$Chl^{*} + Q \rightarrow Chl^{+\cdot} + Q^{-\cdot}$$

(4) electron transfer

$$Q + RH \rightarrow Q^{-\cdot} + RH^{+\cdot}$$

(5) hydrogen abstraction

$$Q + RH \rightarrow QH^{\cdot} + R^{\cdot}$$

Corresponding oxidation of hydroquinones can occur by reactions such as:

(1) photoionization

$$QH_2 \rightarrow QH^{\cdot} + e^-_{aq} + H^+$$

(2) oxidation by excited state molecules such as dyes (D)

$$D^* + QH_2 \rightarrow DH^{\cdot} + QH^{\cdot}$$

(3) oxidation by radicals such as hydroxyl radical

$$OH^{\cdot} + QH_2 \rightarrow QH_2^{\cdot}OH \rightarrow QH^{\cdot} + H_2O$$

Clearly the interconversion of the neutral and anionic semiquinone radicals shown in equation 3 may occur in any of these systems with the equilibrium being strongly influenced by the medium.

$$QH^{\cdot} \rightleftharpoons Q^{-\cdot} + H^+ \tag{3}$$

Recently a comprehensive tabulation of ESR data up to the end of 1984 concerning the radicals derived from the quinones has been published[16]. Obviously this chapter cannot hope to discuss all the radical species documented and will not attempt to do so. Some general trends in the ESR parameters for certain families of quinones will be described, applications of LCAO-MO and INDO calculations for the interpretation of the spectra, and ESR and related techniques such as CIDEP and CIDNP amenable to the study of radical reaction intermediates, rates of formation and decay, and mechanisms will also be briefly described. In addition, although other chapters in this volume are concerned with the spectral, redox and acid–base properties of the quinones we will briefly document some of the corresponding information for the radical intermediates.

Subsequent sections deal with radical intermediates in some model systems of substitution and addition reactions of quinones, and with quinone metal and organometal radical complexes. A brief summary of the literature describing the importance of semiquinones in antibiotics, and in biologically important systems such as micelles, vitamins, photosynthesis and respiration will be given in Section VI.

A. ESR Spectral Parameters of Semiquinones

It has been shown that a principle of additivity reasonably describes the changes in the hyperfine coupling parameters that result from substitution of the aromatic ring protons of the simplest quinone, 1,4-benzoquinone. The principle holds for alkyl and halogen substituents and also for several bulkier chemical functions[16]. This principle can be expressed as shown in equation 4

$$a_i^{j,k} = a_i^j + a_i^k + a_i^o \tag{4}$$

where a_i^o is the coupling constant of the proton at position i in the unsubstituted radical, a_i^j is the coupling constant at position i observed when R^1 is introduced at position j and $a_i^{j,k}$ is the value of the coupling constant observed at position i when R^1 and R^2 are introduced at positions j and k respectively.

Pedersen presents a table of additivity parameters for several substituents to 1,4-benzoquinone[16]. Limitations of this method for calculation of hyperfine coupling constants appear to be that the additivity principle applies only to sets of spectra obtained under identical experimental conditions; also any conformational changes will obviously perturb the values of the coupling constants. Although used for the qualitative prediction of coupling constants for a number of simple substituted quinones[16–19] both *o*- and *p*-, a deviation from expected values was noted for the methoxyl-substituted quinones. The breakdown occurs when two methoxyl groups are adjacent to one another; in this instance the usually observed methoxyl hydrogen splitting of ~ 0.8–1.0 G is reduced to zero[20]. This indicates a breakdown of hyperconjugation between the two adjacent methoxyls due to steric effects. This phenomenon is not observed for the corresponding methyl-substituted quinones (Table 1). One of the biologically important methoxyl-containing quinones, ubiquinone (**9**), has an ESR spectrum showing no observable methoxyl hydrogen hyperfine couplings[21].

TABLE 1. Hyperfine coupling constants for some methyl[b]- and methoxyl[b, c]-substituted semiquinone radicals

Semiquinone	Hyperfine coupling (a_H G)				
	a_2	a_3	a_4	a_5	a_6
1,4-BQ	2.33	2.33		2.33	2.33
2-OMe-BQ	0.86[a]	3.69		1.99	0.52
2,3-$(OMe)_2$-BQ	0.00[a]	0.00[a]		2.65	2.65
2,5-$(OMe)_2$-BQ	1.01[a]	1.01		1.01[a]	1.01
2,6-$(OMe)_2$-BQ	0.78[a]	1.45		1.45	0.78[a]
2,3,5-$(OMe)_3$-BQ	0.00[a]	0.00[a]		0.84[a]	0.67
2,3,5,6-$(OMe)_4$-BQ	0.00[a]	0.00[a]		0.00[a]	0.00[a]
Ubiquinone	1.02[a]	2.04[a]		0.00[a]	0.00[a]
2-MeBQ	2.10[a]	1.70		2.55	2.37
2,3-Me_2BQ	1.67[a]	1.67[a]		2.54	2.54
2,5-Me_2BQ	2.36[a]	1.70		2.36[a]	1.70
2,6-Me_2BQ	2.12[a]	1.87		1.87	2.12[a]
2,3,5-Me_3BQ	2.33[a]	1.91[a]		1.76[a]	1.94
Duroquinone	1.90[a]	1.90[a]		1.90[a]	1.90[a]
1,2-BQ					
3,4-$(OMe)_2$-BQ		0.00[a]	0.44[a]	3.92	1.70
3,4-$(OMe)_2$,-6-Me-BQ		0.00[a]	0.53[a]	2.95	1.85[a]
3,6-$(OMe)_2$,-4-Me-BQ		0.00[a]	5.61	1.70	0.61[a]
4,5-$(OMe)_2$-BQ		0.32	1.10[a]	1.10[a]	0.32
3,4,5-$(OMe)_3$-BQ		0.00[a]	0.10[a]	1.20[a]	0.60
3,4,6-$(OMe)_3$-BQ		0.00[a]	0.90[a]	1.10	0.68[a]

[a] Splittings at substituted positions are for 3 equivalent Hs of the Me or OMe group.
[b] Ref. 20.
[c] Ref. 22.

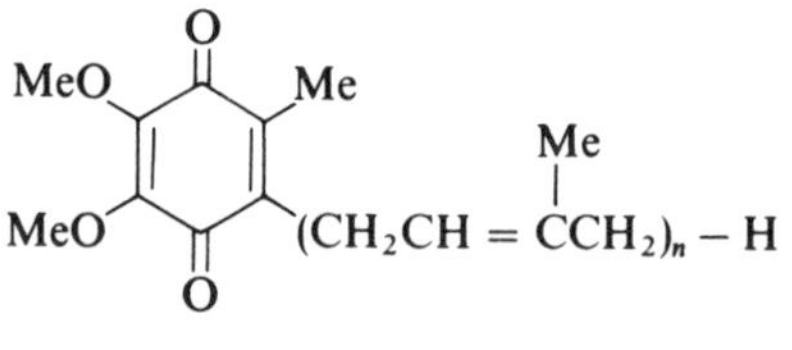

MeO, MeO, Me, $(CH_2CH=C(Me)CH_2)_n-H$ $n = 1–12$

(**9**)

The methoxyl-substituted quinones shown in Table 1 also do not exhibit a linear shift in electrochemical midpoint potential with increasing substitution (Hammett substituent relationship). Gascoyne and Szent-Gyorgyi[20] suggest $\Delta E = -101$ mV for a freely mobile methoxyl group and $\Delta E = -41$ mV for a methoxyl group with a methoxyl neighbour as appropriate values for estimation of the midpoint potentials for this family of quinones. Several *o*-quinones having methoxyl substituents show similar effects in their ESR spectra[22, 23] with sterically crowded OMe groups having no observable hydrogen couplings although adjacent unhindered OMe groups show measurable splittings (Table 1). This is quite noticeable for the *o*-quinones 4,5-dimethoxy-1,2-benzoquinone and 3,4,5-trimethoxy-1,2-benzoquinone in which the coupling constant decreases almost to zero from 1.0 G with the addition of the adjacent substituent.

The ESR spectra of several quinones having fused heterocyclic ring systems have been assigned recently by Clay and Murphy[17]. Simplified McLachlan SCF calculations were inadequate to verify the assignments but by comparisons with related species and by noting smooth variations in the splittings with substitution they were able to explain the spectra. Very little alteration of spin density results from heterocyclic substitution as can be seen by examination of the splittings of the hydrogens at positions 5 and 6 in compounds **10**–**16**.

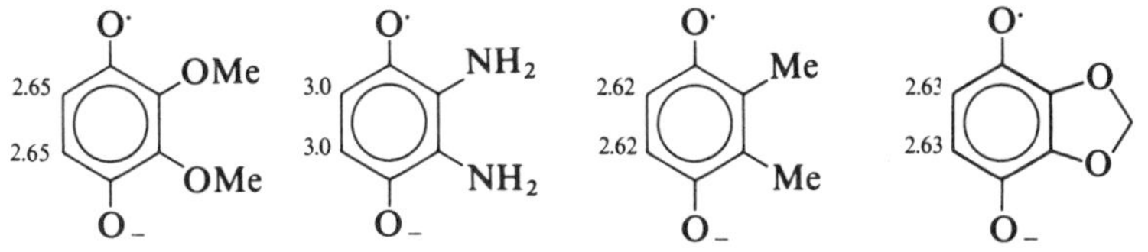

(**10**) (**11**) (**12**) (**13**)

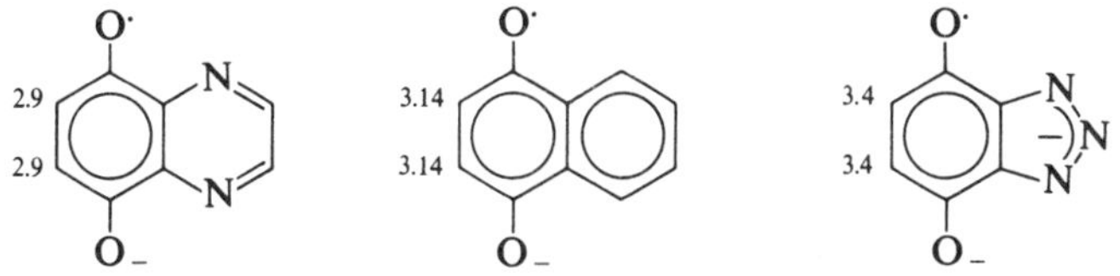

(**14**) (**15**) (**16**)

These authors also assigned the ESR spectra of adrenochrome as well as benzo[1,2-c:4,5-c′]dipyrazole-4,8-(1*H*,5*H*)-quinone (**17**) and 3-ethoxycarbonylnaphthindiazole-4,9-semiquinone (**18**) by comparisons with simpler model systems.

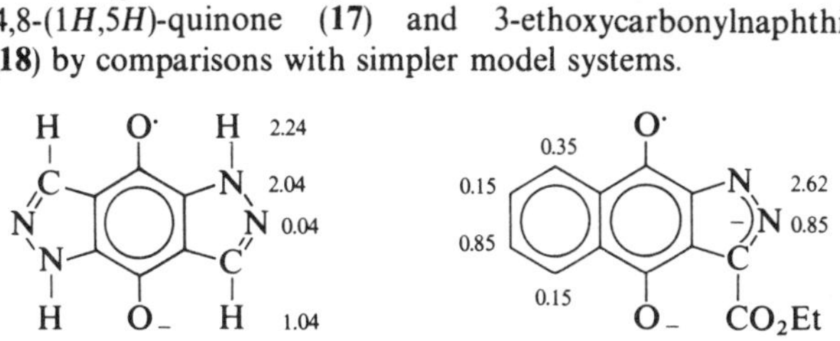

(**17**) (**18**)

The ESR spectral parameters of the 2-amino- and 2,5-diamino-1,4-naphthosemiquinone anions have also recently been described[24].

A similarity noted for ratios of the splittings for the *o*-semiquinones and their radical cations (comparable to that observed for phenoxyl radicals and phenol radical cations) shows the correspondence of the spin densities in these compounds[22, 25, 26]. In each pair the magnitude for the cations exceeds that of the anions by about 20%. Using this observation the radical cation from 1,2-methylene-dioxynaphthalene (**20**), a potent synergist for carbamate insecticides, has been assigned by comparison with 1,2-

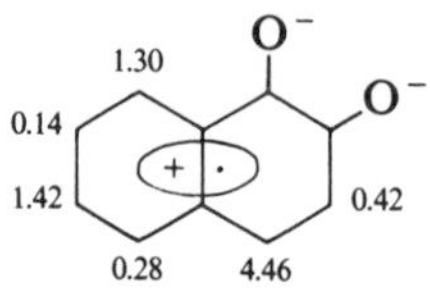

(19) 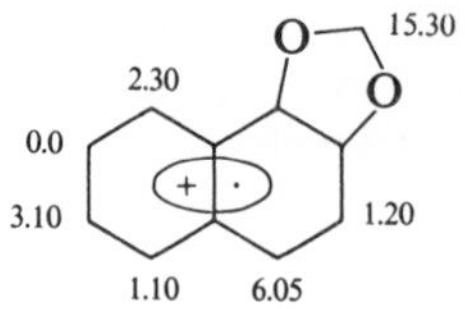 **(20)**

naphthosemiquinone (**19**). The ESR spectrum of the radical anion from fumigatin (**21**) has also been assigned using this procedure.

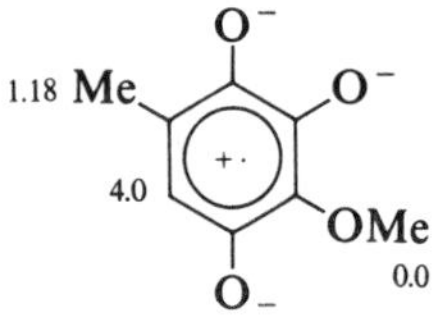

(21)

Many of the hyperfine coupling constants observed in semiquinone ESR spectra have been discussed using molecular orbital theory with the McLachlan SCF refinement[27] in order to explain or predict the magnitudes, signs and trends [26, 28, 29]. For example, Das and coworkers[21] used such a treatment to show that the long alkyl side chains in the semiquinone anions of vitamins E, K_1 or ubiquinone perturbed the spin density in the aromatic ring to about the same extent as did a methyl substituent. For these same radicals an alternating linewidth effect was attributed to hindered rotation about the bond connecting the carbon atom of the alkyl side chain to the aromatic ring.

The observed hyperfine coupling from the hydrogen bonded to an aromatic ring carbon a^{H} can be related to the spin density on the atom to which the hydrogen is bonded (ρ_{C}^{π}) using McConnell's equation (5),

$$a^{\mathrm{H}} = Q_{\mathrm{CH}}^{\mathrm{H}} \rho_{\mathrm{C}}^{\pi} \tag{5}$$

where $Q_{\mathrm{CH}}^{\mathrm{H}}$ is considered to be constant for similar systems. The ^{13}C hyperfine coupling a^{C} from a carbon atom bonded to three other atoms X_1, X_2, X_3 can be related to the spin densities also[30] as shown in equation 6.

$$a^{\mathrm{C}} = \left(S^{\mathrm{C}} + \sum_{j=1}^{3} Q_{\mathrm{CX}_j}^{\mathrm{C}} \right) \rho_{\mathrm{C}}^{\pi} + \sum_{j=1}^{3} Q_{\mathrm{X}_j\mathrm{C}}^{\mathrm{C}} \rho_{\mathrm{X}_j}^{\pi} \tag{6}$$

Karplus and Fraenkel calculated the values $S^{\mathrm{C}} = -12.7$ G, $Q_{\mathrm{CC'}}^{\mathrm{C}} = 14.4$ G, $Q_{\mathrm{C'C}}^{\mathrm{C}} = -13.9$ G and $Q_{\mathrm{CH}}^{\mathrm{C}} = 19.5$ G so that for the $\mathrm{C}_2'\mathrm{CH}$ fragment

$$a^{\mathrm{C}} = 35.6\, \rho_{\mathrm{C}}^{\pi} - 13.9 \sum_{\mathrm{C'}} \rho_{\mathrm{C}}^{\pi} \tag{7}$$

and for $C_3'\mathrm{C}$

$$a^{\mathrm{C}} = 30.5\, \rho_{\mathrm{C}}^{\pi} - 13.9 \sum_{\mathrm{C'}} \rho_{\mathrm{C}}^{\pi} \tag{8}$$

and these equations have been tested and found reliable in several systems[31]. (A more complete discussion of the derivation of these parameters can be found in Refs 31 and 159, pp. 511–524.) There is, however, some disagreement in the literature concerning the appropriate parameters to relate the ^{13}C and ^{17}O hyperfine splittings to the spin densities in the carbonyl-containing fragment C'_2CO of the semiquinone radicals. Several authors[32–34] use the value $Q^{H}_{CH} = -27$ G although Luz and coworkers suggested quite different Q parameters[35]. Prabhananda has reconsidered the systems of Das and Fraenkel[32]: (1) 2,5-dioxo-1,4-benzosemiquinone in KOH/H_2O, (2) $BQ^{-\cdot}$ in $DMSO/H_2O$, (3) $BQ^{-\cdot}$ in $EtOH/H_2O$ and also (4) $BQ^{-\cdot}$ in H_2O and obtained the values $Q^{H}_{CH} = -26.2$ G, $Q^{C}_{CO} = 18.66$ G and $Q^{C}_{OC} = -30.64$ G. This author also introduced the parameter $Q^{H}_{OH} = 6.0$ G to acknowledge the effects of hydrogen bonding in aqueous and alcohol solutions, ($Q^{H}_{OH} = 0$ in aprotic solvents). Using these parameters he also predicted spin densities in some phenoxyl radicals and the benzophenone ketyl radical[36].

Strauss and Fraenkel[37] have estimated $Q^{C}_{CC''} = 30$ G for the C'_2CC'' fragment when C″ is sp^3 hybridized and has zero spin density. Fessenden[38] predicted that $Q^{C}_{CC''}$ is unlikely to be much greater than 19.5 G (the value of Q^{C}_{CH}). However, Prabhananda's calculations of $Q^{C}_{CC''}$ of 16.9 G for 2,6-di-*t*-butylphenol and 22.2 G for durosemiquinone are more consistent with Fessenden's prediction[36].

Pedersen in a study of naphthoquinones (NQ) and anthraquinones (AQ) has noted that α-OH groups give rise to an observable proton hyperfine splitting even in relatively basic solutions[39]. This is a consequence of hydrogen bonding to the adjacent carbonyl group which disfavours dissociation. The assignments of splittings in anthrasemiquinones having β-OH groups are complicated by the small size of these splittings. However, β-OH groups in a number of substituted anthrasemiquinone radicals were found to dissociate at pH $\simeq$ 12; following dissociation the resulting O^- substituents cause a greater perturbation in the spin density distribution than does OH. Thus comparisons of splittings in spectra obtained below and above the pH of dissociation can frequently aid in the assignments of splittings in these relatively complex spectra. In addition it has been noted that replacing OH by OMe in these positions causes very little alteration of the splitting pattern[16, 39].

Considerable controversy has surrounded the calculations of spin density distribution and the signs of the hyperfine coupling constants for the 1,2-benzosemiquinone radicals and their derivatives. Experimental values measured under a variety of conditions are given[16]; for example Felix and Sealy[40] have observed $a_{H_{3,6}} = 0.75$ G and $a_{H_{4,5}} = 3.67$ G in aqueous solutions. Theoretical calculations, however, have generally been in poor agreement with these results. For example, Huckel calculations in which the coulomb integral was varied to give $|\rho_4| > |\rho_3|$ suggest that the values are both positive[41]. McLachlan calculations suggest that ρ_3 is negative[42]. The INDO method has predicted the opposite order of spin densities[43], i.e. $|\rho_3| > |\rho_4|$; although by using the molecular geometry optimization refinement described by Shinagawa[44, 45] the correct relative magnitudes of spin density are predicted. However, this method has been criticized for requiring an extremely distorted radical structure[46]. Spanget-Larsen has suggested addition of effective solvent field parameters to correct the ordering predicted by INDO calculations[46, 47]. Finally Kuwata and Shimizu[48] have described an open shell calculation which reproduces the experimental splittings and predicts both ρ_3 and ρ_4 are positive. This is in agreement with the observations and calculations described by Felix and Sealy using both proton and ^{13}C-ESR measurements and Karplus–Fraenkel theory relating the ^{13}C splittings to the aromatic spin densities[40].

The INDO method of Shinagawa has also been applied to 1,3-semiquinone radicals[49]. The relative magnitudes of the spin densities at positions two and five were found to change when the radicals were generated by alkaline or acidic oxidation.

The regioselective coupling of acyl and alkyl radicals with 1,2-naphthosemiquinone anion has been explained in terms of the estimated spin densities in the radical. Radicals such as benzyl and diphenylmethyl preferentially couple at C(4) while phenacetyl attacks at C(3) of the semiquinone[50].

The ESR spectra of substituted 1,2-benzosemiquinones from molecules such as L-dopa (**22**) and tyrosine exhibit the effects of magnetic inequivalence of the diastereotopic methylene protons in the side chain. Although slightly different hyperfine coupling constants are measured for these hydrogens some selective broadening of the lines can occur due to restricted rotation about the $ArCH_2$–$CHNH_3^+CO_2^-$ bond[51,52].

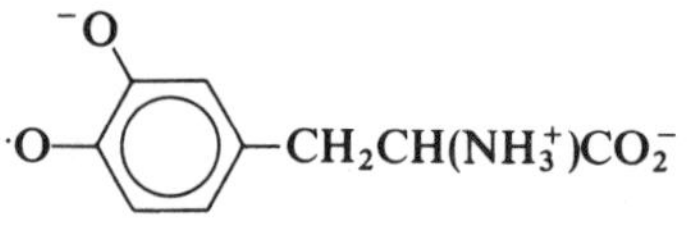

$^-$O

·O–(ring)–$CH_2CH(NH_3^+)CO_2^-$

(**22**)

Another characteristic parameter obtainable from ESR spectra is the g factor. For semiquinones the g factor is usually considerably higher than values for aromatic hydrocarbons; the appreciable spin density on oxygen is the main reason for the deviation from near free spin value. For several semiquinones g factors were found to be linearly dependent on $\Sigma\rho_{\mathrm{O}}^{\pi}$ in agreement with equation 9 and[53] where ρ_{O}^{π} is the spin density on the oxygen. Γ_0 depends on the spin-orbit coupling on oxygen as well as energy differences involving the unpaired electron π orbital, non-bonding orbitals on the oxygen and the C–O bonding orbitals. It was determined that $\Gamma_0 = 0.008$ and $g_0 = 2.0025$; thus, as expected when the spin density on the oxygens is zero the g factor

$$g = g_0 + \Sigma\Gamma_0\rho_{\mathrm{O}}^{\pi} \tag{9}$$

approaches that of the aromatic hydrocarbons[36]. This linear relationship also applies to the hydroquinone cation radicals of BQ, NQ, or 9,10-AQ in acid solvents[54]. The authors again found it was necessary to use an additional parameter $Q_{\mathrm{OH}} \simeq 50$ G to account for the observed ^{17}O hyperfine splitting which is affected by hydrogen bonding between the hydroxyl group and the solvent. This study estimates spin density on oxygen lower than that of Sullivan and coworkers[55].

Q-band ESR spectra of the tetrahalogenated semiquinones have been studied to determine the origin of the effects of the halogens on the g tensor. It was noted[56] that g_x and g_y (the components in the plane of the aromatic ring) were increased with respect to the unsubstituted parent $BQ^{-\cdot}$ following the series Cl, Br, I while the perpendicular component g_z was decreased. The authors suggest that the halogen d orbitals are not predominantly responsible for the noted g anisotropy but that the effects are due to σ–π interactions changing the spin density on halogens. This is consistent with the observed order of magnitude of the effect on g_z, i.e. tetrachloro < tetrabromo < tetraiodosemiquinone. The contributions to g_x and g_y are positive and are attributed to spin density delocalized in the π^* orbital. Further perturbation of the g_z component for tetraiodosemiquinone on changing the solvent from DMSO to ethanol was attributed to changes in the energy separation between the orbital of the unpaired electron and the lone-pair orbitals on the halogens.

B. Solvent Effects in ESR Spectra of Semiquinones

Changes in solvation of radicals in solution are frequently observed by changes in the magnitude and even sign of hyperfine splitting constants. The radicals derived from quinones are solvated in protic solvents in such a way as to perturb these values substantially due to the formation of hydrogen bonds. As well, aprotic solvents may interact with the quinone radical anions via ion–dipole and ion–induced-dipole effects.

Although it was originally thought that hydrogen bonding to a C=O function resulted in lowering of the ground state energy with little effect on the excited state[57, 58] it has been suggested that the excited state is also involved and that geometry changes for both the ground and excited states are affected differently as a result of hydrogen bonding. This is reflected in a blue shift of 0.13–0.25 eV in the carbonyl n-π^* transition in H-bonding solvents. This shift has been interpreted by Beecham and coworkers[57] to arise from a differential lengthening of the CO bond in the excited state with respect to the ground state. This lengthening (of ~ 0.002 Å) resulted in a redistribution of intensity in the vibrational subbands which are resolved in CD spectra. The use of additional parameters in MO calculations to account for changes in the spin density at oxygen for semiquinone radicals in hydrogen bonding solvents was mentioned previously (Section II.A).

Simple semiquinones have been found to have $a_{^{13}C}$ and $a_{^{17}O}$ correlated in a linear fashion to the Kosower Z value (or less well to the solvent dielectric constant ε) for solvents such as water, ethanol and DMSO; however, the correlation is not good for sterically hindered semiquinone radicals[59] such as 2,6-dimethyl-1,4-$BQ^{-\cdot}$. Steric effects were also noted by Gough[60] for the neutral semiquinone radicals of BQ and DQ. He concluded that the hydroxylic proton lies in the plane of the ring and that its splitting varies linearly with solvent polarity. For $DQH^{\cdot}$ the steric hindrance of the Me groups caused a solvent dependence of a_H reflecting the effective molecular size of the surrounding solvent molecules which would be expected to distort the hydroxylic bond out of the plane of the ring on hydrogen bonding.

In several mixed protic–aprotic solvent systems such as H_2O/DMSO and H_2O/HMPA there appeared to be competition between the solute and the aprotic solvent for hydrogen bonds. This complicates the measurements of equilibrium constants for solvation in the mixed solvent systems and it is suggested that the equilibrium being measured by these ESR studies is between H-bonded and non-H-bonded radicals[59]. Previously it had been determined that the dianion radicals from the trihydroxybenzenes **23** and **24** could not

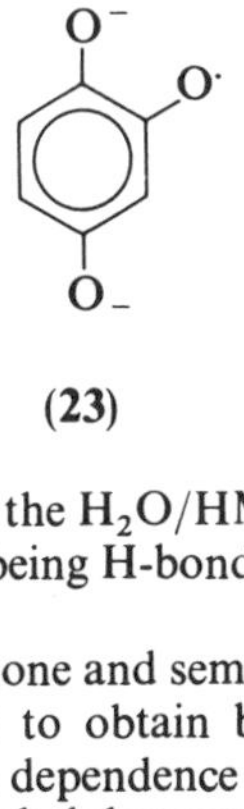

(23) 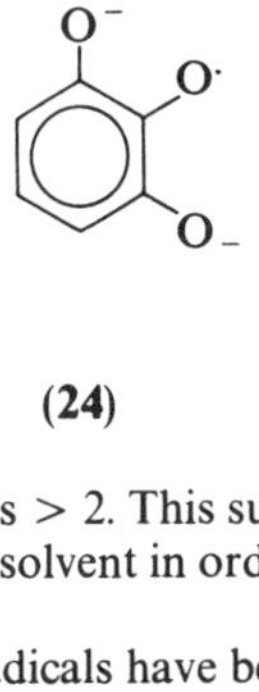 **(24)**

persist in solution unless the H_2O/HMPA ratio was > 2. This suggested that water must be sufficiently free from being H-bonded to aprotic solvent in order to solvate the dianion radical[61].

A number of semiquinone and semidione type radicals have been investigated by Loth and Graf[62, 63] in order to obtain both structural and kinetic information from the temperature and solvent dependence of their ESR spectra. For example, the tautomeric radicals (2-hydroxy-4-methylphenoxyl and 2-hydroxy-5-methylphenoxyl) derived from 4-methylcatechol were observed and their splittings compared with INDO calculations. In general the formation of intramolecular hydrogen bonds in these and related radicals is disturbed and sometimes prevented by steric hindrance, internal rotation or intra-molecular proton exchange.

An unusual solvent effect was observed in the study of $BQ^{-\cdot}$ in frozen DMSO and DMSO–EtOH solvents[64]. Even at 50 K below the freezing points of the solutions the ESR spectrum of $BQ^{-\cdot}$ remained isotropic; the authors suggest that liquid-like pockets exist in the solvent in which the quinone can tumble rapidly. The dominant contributions to spin lattice relaxation in this medium are apparently spin rotation in origin while the dominant linewidth effect comes from anisotropic g and hyperfine tensor modulation.

Dynamic processes such as H^+ or e^- exchange can also have dramatic effects on the ESR spectra of quinone radicals. The effects usually involve line-broadening of the resonances due to the species undergoing the exchange processes. The study of the neutral radicals of the *o*-, *m*-, and *p*-quinones in acid solution has been complicated by this problem; poorly resolved and unanalysable spectra are frequently obtained (Ref 65 and references therein). Dixon and Murphy[65] analysed the spectra of the neutral radicals from hydroquinone, catechol, resorcinol and phloroglucinol and estimated rate constants for protonation of the neutral radicals to be 1.1, 1.0, 4.1 and 4.4 ($\times 10^9\ \mathrm{dm^3\,mol^{-1}\,s^{-1}}$) respectively. Degenerate electron exchange has been cited by Hore and McLauchlan[66] to have significant effects on the time-resolved ESR spectrum of radical anions such as duroquinone (equation 10).

$$DQ^{-\cdot} + DQ \rightleftharpoons DQ + DQ^{-\cdot} \tag{10}$$

Since time-resolved spectra of systems exhibiting CIDEP provide early (0.5 μs to ~ 200 μs) information about the behaviour of the radicals generated, fast processes such as electron transfer would be expected to be a factor in their appearance and analysis. The authors suggest that degenerate electron exchange is an important factor affecting both longitudinal and transverse relaxation in such systems. They discuss methods of dealing with exchange processes in the calculations of T_1 and T_2 and the factors affecting recording and analysis of the radical time profiles[66,67].

Meisel and coworkers[68,69] measured degenerate electron exchange rates for several *p*-quinone radical anions in acetone/isopropanol from the effects of concentration on the linewidth of their ESR spectra. The values they obtained are given in Table 2. Equation 11 was used to describe the linewidth effect of electron transfer between pairs of non-equivalent quinones in the slow exchange region.

$$\Delta H = \Delta H_o + \frac{k_{ex}[Q][1 - g_i/\sum_i g_i]}{\sqrt{3}\pi(2.83 \times 10^6)} \tag{11}$$

TABLE 2. Rates of degenerate electron exchange and electron transfer in benzosemiquinones

Quinone	Electron exchange k_{ex} ($M^{-1}\,s^{-1}$)	E_7 (V vs. NHE)
Benzoquinone	6.2×10^7	0.99
2,5-Dimethylbenzoquinone	5.5×10^7	−0.65
2,6-Dimethylbenzoquinone	1.7×10^8	−0.80
Duroquinone	2.0×10^8	−0.247

Reaction	Electron transfer k_f ($M^{-1}\,s^{-1}$)[a]	k_b ($M^{-1}\,s^{-1}$)[a]	K_{eq}^b
$2,5\text{-}DMBQ^- + BQ$	$6.5 \pm 0.3 \times 10^8$	9.7×10^5	6.7×10^2
$2,6\text{-}DMBQ^- + BQ$	$1.0 \pm 0.1 \times 10^9$	8.3×10^5	1.2×10^3
$DQ^- + BQ$	$1.1 \pm 0.05 \times 10^9$	1.9×10^3	5.7×10^5
$DQ^- + 2,5\text{-}DMBQ$	$1.0 \pm 0.1 \times 10^9$	1.2×10^6	8.5×10^2
$DQ^- + 2,6\text{-}DMBQ$	$9.6 \pm 1.0 \times 10^8$	2.0×10^6	4.7×10^2

[a] k_f and k_b are the rate constants for the forward and back reactions as shown in equation 12.
[b] K_{eq} = the equilibrium constant for equation 12.

In equation 11 g_i is the statistical line intensity of the ith line[70]. Since one-electron redox potentials for the quinone/semiquinone couples they studied were available from pulse radiolysis the electron transfer rates for equation 12 could be predicted using Marcus theory from the self-exchange reaction rates. The experimentally measured and calculated rates were in good agreement, some experimental values are shown in Table 2.

$$Q_1^{-\cdot} + Q_2 \underset{k_b}{\overset{k_f}{\rightleftharpoons}} Q_1 + Q_2^{-\cdot} \tag{12}$$

C. Photoreduction and CIDMP of the Quinones

The photochemistry of simple quinones has been quite well understood for several decades; thus these compounds have been extensively used as model compounds for the investigation of phenomena such as chemically induced dynamic electron (CIDEP) and nuclear polarization (CIDNP). Early experiments were largely directed towards understanding the mechanisms by which these processes occurred and the development and testing of the theory describing them. Indeed the study of quinone photoreductions was largely responsible for the development of the Triplet Mechanism (TM) of CIDEP. Now, however, CIDEP and CIDNP experiments can be applied to probe the nature of more complex quinone reactions especially those initiated by photolysis or radiolysis. With time-resolved CIDEP studies much information may be obtained including determination of reaction mechanisms, identification of transient radical intermediates often undetectable by other experiments, measurement of radical spin lattice relaxation times, precursor triplet dynamic properties, radical ion-pair interactions in the solid state, and often relative reaction rates by the employment of appropriate experimental conditions. On the other hand, CIDNP provides complimentary information on the products formed from the radical reactions whether these are in or escaped from the primary radical cage.

Absorption of a photon of light by a quinone results in excitation to the first excited singlet state S_1 followed by rapid intersystem crossing to the triplet state T_1 from which the subsequent reactions occur. In the presence of a magnetic field semiquinone radicals produced by irradiation of quinones in the presence of reducing agents show CIDEP arising from both triplet (TM) and radical pair (RPM) mechanisms. Briefly the TM results from anisotropic intersystem crossing (ISC) into the triplet sublevels resulting in a non-Boltzmann population (i.e. polarization) which can be transferred to subsequently produced radicals if the reactions forming them are faster than the triplet state spin lattice relaxation. Since triplet T_1s are of the order of nanoseconds in liquids this requires near diffusion-controlled reaction rates. Secondarily polarized radicals can also arise if the initially formed polarized radicals react faster than their spin lattice relaxation (not an unlikely occurrence since radical T_1s are of the order of microseconds). In the radical-pair mechanism the polarization arises from the magnetic interactions of encountering pairs of radicals which may involve either the initial pair (correlated pair) formed in the primary photochemical reactions, or pairs of radicals encountered at random (random pair).

The coexistence of the two mechanisms of CIDEP in quinone photolyses is now well established[71–78]. In general, the intersystem crossing process led to a strongly spin-polarized quinone triplet state and its polarization was transferred during the chemical reactions to the primary radicals such as the semiquinone radicals and to some secondary radicals derived from the intermediate semiquinones. In viscous solvents and/or using high intensity light sources such as an excimer laser, the semiquinone radical concentrations would be increased while their separation rates (diffusion) would be reduced. These factors combined to enhance the magnitude of the RPM polarization. Thus, Pedersen and coworkers[76] have estimated that less than 20% of the polarization in the photolysis of BQ in ethylene glycol is due to RPM. The contributions due to the RPM can also be enhanced for semiquinone radicals such as durosemiquinone which have larger hyperfine couplings[77,79]. Several authors have discussed the separation of initial polarization and secondary polarization in quinone systems and assessed the agreement with theoretical predictions[71,72,76]. The effects of solvent viscosity and heavy atom perturbations on the TM enhancement factor and the rates of ISC have been described[73].

In time-resolved experiments CIDEP studies provide a simple method to estimate the transient radical spin lattice relaxation times. The radical spin lattice relaxation offers a sensitive probe of local radical environments. The large initial polarizations of radicals created by the TM permit calculation of effective T_1s from extrapolation or fitting of their exponential decay curves. Advancements in experimental techniques have involved gating the microwave power after the laser flash[80] and the use of rapidly modulated light sources coupled with phase-sensitive ESR detection[81]. These techniques are hampered somewhat by the requirement of multiple parameter fitting (to account for among other things microwave power levels and inhomogeneous contributions to linewidths)[66,82].

Earlier experiments using the direct detection ESR methods after laser flash photolysis estimated that T_1s for semiquinone radicals from a series of benzoquinones, naphthoquinones and anthraquinones depended both on the extent of the aromatic system and on the effective molecular size of the semiquinone radical[83,84]. In addition a simplified 'rotating rod' primitive model has been developed to describe the effect on T_1 of restricting rotation in one direction by addition of a long polymer chain to the radical[85]. The polymer radicals were treated as freely rotating about the rod axis but motion was frozen about the transverse axis. The model neglects complex vibrational and crankshaft motions of the polymeric chain; it predicts that T_1 values will be relatively insensitive to chain length but will be sensitive to values of individual second rank g tensor components in such hindered rotation situations. Ullman[86] and Monnerie and coworkers[87] have also proposed models for the study of the dynamics of polymers in solution; in the latter case the ESR line shape analysis can also probe effects on the transverse relaxation time T_2. The T_1 values for some quinone–organometallic radical adducts have recently been studied; the formation of these species will be described later (Section IV). For the uranyl-phenanthrenequinone radical ion $(UO_2PQ)^{+\cdot}$ the value of T_1 was found to be considerably shorter than for the parent $PQ^{-\cdot}$ semiquinone (i.e. $< 0.8\ \mu s$ and $4\ \mu s$ respectively). This significant reduction in relaxation time is due to the extremely large spin-orbit coupling of the uranyl atom which provides an additional relaxation mechanism[88,89]. This large spin-orbit coupling is also evidenced by the low g factor of the radical ion complex (1.9940 ± 0.0001). T_1 values of benzoquinone–uranyl radical complexes having bulky *t*-butyl substituents on the quinone showed little or no effect of radical size on the magnitude of T_1. Organotin–quinone radical adducts also show little variation in T_1 compared with the values of the parent semiquinones[90]; apparently the organotin moiety does not appreciably contribute to the quinone system.

As predicted by the TM theory no hyperfine dependence for T_1 or T_2 has been noted[71,72,83]. Fessenden and coworkers[91] have obtained values of T_1 and T_2 for benzoquinone radicals using a saturation recovery method applied to steady state radical concentrations. Although somewhat hampered by transient mutations due to the high microwave powers used in the experiments the values obtained were in good agreement with those described previously. The spin lattice relaxation times are extremely sensitive to changes in viscosity, H-bonding and temperature and correlations can only be made reliably for a series of compounds of similar structure measured under identical experimental conditions.

Bartels and coworkers have recently described a dynamic polarization recovery method for T_1 measurements that is particularly successful for small, very reactive transient radicals which can be generated in high concentrations by laser photolysis or pulse radiolysis[92]. The method is valid even in the presence of CIDEP or fast chemical decay. They conclude that spin rotation is the dominant relaxation mechanism in the series of small radicals studied.

CIDEP effects have also been utilized in order to obtain information about the precursor triplet properties of the quinones. The time-resolved ESR spectra for the triplet quinone molecules can frequently be obtained in glassy matrices at 77 K[93–96]. Values for the zero field splitting parameters D and E for several quinones in various matrices were obtained as shown in Table 3. The values obtained in these glassy media are believed to be more representative of the triplet states in liquids than are those obtained in crystalline matrices[96]. It was found necessary to postulate a distribution of D and E values in glassy matrices for the quinones since the zero field splitting parameters are sensitive to environment and the guest molecules are probably trapped in sites which are not homogenous. Therefore, the time-resolved spectra can be more accurately simulated if Gaussian distributions of D and E values are used[95,96]. For these 1,4-quinones the authors find the centre of the distribution of D values to be larger than 0.30 cm^{-1}. The value of D for the 9,10-anthraquinone triplet state was slightly lower in non-polar than in polar solvents. In non-polar solvents the triplet states of 1,4-BQ and 1,4-NQ reacted too quickly for adequate triplet state time-resolved spectra to be obtained. The values of $|D'|$ estimated by Murai and coworkers[96] are larger than those estimated earlier for 1,4-BQ (700 G[76] as a lower limit, and 3000 G[71]) but these earlier values were indirectly measured and have wide uncertainties.

TABLE 3. Zero field splitting parameters for some 1,4-quinones obtained in glassy matrices at 77 K

Quinone	Solvent	$D(cm^{-1})$	$E(cm^{-1})$	Reference
1,4-BQ	1,4-dibromobenzene	−0.1767	0.0026	291, 292
1,4-BQ	EPA[b]	−0.330	0.019	96
1,4-BQ	1,4-BQ-d_4	−0.0684	0.0038	292
1,4-NQ	EPA[b]	−0.330	0.019	96
9,10-AQ	EPA[b]	−0.351	0.005	95
9,10-AQ	PM[a]	−0.318	0.005	96
9,10-AQ	*n*-Octane site 1	−0.2894	0.041	293
	n-Octane site 2	−0.309	0.007	293

[a] PM = isopentane : methylcyclohexane (1 : 4 v/v).
[b] EPA = diethyl ether : isopentane : ethanol (5 : 5 : 2 v/v).

The negative sign of D indicates that the Z-spin sublevel is exclusively populated by intersystem crossing in these quinones. Conservation of triplet spin polarization in energy transfers between a triplet molecule and ground state acceptor has been recently demonstrated in glassy matrices[97,98] and the direction of polarization, absorptive or emissive, correlated to the signs of the zero field splitting parameters for triplet states of biacetyl[99] and benzophenone[100]. Triplet T_1s in solution for duroquinone have been estimated from CIDEP experiments to range from 2.7 ns in methanol to 17 ns in cyclohexanol[73] at 260 K.

A technique called CIDEP-enhanced-ENDOR in which the pumped NMR transitions in a transient radical are observed by changes in the CIDEP intensities in the ESR spectrum has been used to study 1,4-BQ. The ENDOR enhancement of the ESR signal was ~ 10%, somewhat larger than obtained in continuous wave (CW) ENDOR[101,102].

The coexistence of the TM and RPM in CIDEP is now well accepted; however, for CIDNP the origin of the polarization is generally attributed to the RPM which can briefly be illustrated as in Scheme 1. Several excellent reviews of this technique and its theory and applications exist[103–106].

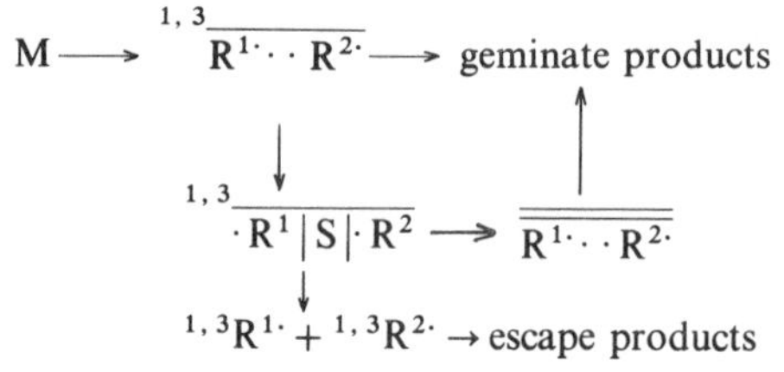

$$
\begin{array}{ccc}
M \longrightarrow & {}^{1,3}\overline{R^{1\cdot}\cdot\cdot R^{2\cdot}} \longrightarrow & \text{geminate products} \\
 & \downarrow & \uparrow \\
 & {}^{1,3}\overline{\cdot R^{1}|S|\cdot R^{2}} \longrightarrow & \overline{\overline{R^{1\cdot}\cdot\cdot R^{2\cdot}}} \\
 & \downarrow & \\
 & {}^{1,3}R^{1\cdot} + {}^{1,3}R^{2\cdot} \rightarrow & \text{escape products}
\end{array}
$$

SCHEME 1

A singlet or triplet state excited molecule can dissociate to form a corresponding singlet or triplet radical pair $\overline{R^{1}\cdot\cdot R^{2}}$ which can then give geminate recombination or disproportionation products. Alternatively it can diffuse apart, subsequently to re-encounter, or the radicals can form escape products. The spin sorting that occurs during the diffusion of the radicals and the nuclear spin dependence of the reactions is responsible for the nuclear polarizations in the geminate products and the equally large polarization of opposite sign in the escape products. There has also been described a relatively little known CIDNP mechanism called the triplet Overhauser mechanism which has been invoked in certain quinone photolyses. This mechanism involves large initial electron polarization of the radicals by the TM followed by a key electron-nuclear cross-relaxation step (Overhauser effect) prior to formation of the diamagnetic products exhibiting the abnormal nuclear polarizations. The theory has been described in the literature[107]. This mechanism was initially proposed by Vyas and Wan[108] to account for observed polarization in tetrafluoro-1,4-benzoquinone (FQ) in chloroform or with FQH_2 in benzene.

Roth and coworkers[109] recently re-examined the same systems and studied the magnetic field dependence and quencher concentration dependence of the reactions. For photolysis of FQ and the corresponding tetrafluorohydroquinone in benzene at magnetic fields below 100 G the experimental observations are consistent with RPM involving S–T_1 mixing[107, 109]. At higher fields two independent contributions to the polarization are operating. One is assigned to be due to a biradical adduct (**25**) between triplet quinone and solvent benzene. Minor cross-combination products between semiquinone radical and solvent have also been observed in CIDNP studies of fluoranil with dioxane and chloranil

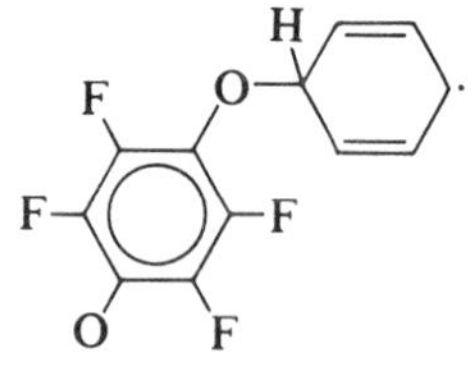

(25)

with 3,5-di-*t*-butylphenol[108, 110]. The second contribution to the polarization which increases monotonically with magnetic field is attributed to the triplet Overhauser mechanism.

CIDNP studies of benzoquinone photolysis with BQH_2 in $CDCl_3$ noted that the sign of the polarization changed from enhanced absorptive to emissive as the hydroquinone concentration was increased[107]. A similar effect was observed in the electron transfer quenching of trifluoroacetophenone. Competing RPM and triplet Overhauser polarizations with the triplet mechanism 'exposed' at the higher hydroquinone concentration can account for these results. Similarly simultaneous operation of TM and RPM CIDNP has been observed by Kuznets and coworkers[111] for the photolysis of BQ and diphenylamine in hexafluorobenzene.

The study of BQ photolysis in isopropanol by combined CIDNP and CIDEP concluded that the formation of hydroquinone and the enol $CH_2 = C(Me)OH$ were the main in-cage products while acetone was formed largely from scavenged reactions[106].

A study of the effects of pumping the ESR transitions of radical pairs formed in the photolysis of quinones and observing the changes in the CIDNP spectra has been reported[112].

Magnetic field effects on reaction rates and products are of considerable interest in micelles; quinone photolyses in micelles will be discussed briefly in Section VI. In solution, application of a magnetic field of 3360 G was found to increase the yield of escaping alkoxy radicals from photolysis of BQ in isopropanol[113].

Most triplet state quinones are strong electron and hydrogen acceptors and it is frequently difficult to determine whether the primary photochemical event in their photoreduction involves direct electron transfer or hydrogen abstraction or both. The question was first raised by Bridge and Porter[114] and attempts to provide an unequivocal answer have employed primarily ns and ps laser flash photolysis and ESR/CIDEP experiments. Although H-abstraction is generally accepted as the primary event in solvents such as alcohols there have been dissenting opinions.

The reactions following photolysis of a quinone in hydrogen donating solvents can be written as shown below:

$$Q \xrightarrow{h\nu} {}^1Q \xrightarrow{\text{ISC}} {}^3Q^* \tag{13}$$

$$^3Q^* + RCH_2OH \rightarrow QH^{\cdot *} + R\dot{C}HOH^{\cdot *} \tag{14}$$

$$Q + R\dot{C}HOH \rightarrow QH^{\cdot} + RCHO \tag{15}$$

$$2QH^{\cdot} \rightarrow QH_2 + Q \tag{16}$$

$$QH^{\cdot} \rightleftharpoons Q^{-} + H^{+} \tag{17}$$

$$^3Q + QH_2 \rightarrow 2QH^{\cdot} \tag{18}$$

If oxygen is present several other steps are possible; however, ESR studies frequently are carried out in deoxygenated solutions.

Kambara and Yoshida[115, 116] have proposed an anionic mechanism to account for their ESR observations for the photoreduction of 1,4-benzoquinone in alcohols. The mechanism describing the one-electron transfer from ethanol to excited state quinone is shown in reactions 19–22 below.

$$^3BQ + EtOH \longrightarrow BQ^{-\cdot} + EtOH^{+\cdot} \tag{19}$$

$$EtOH^{+\cdot} + EtOH \longrightarrow EtO^{\cdot} + EtOH_2^{+} \tag{20}$$

$$2BQ^{-\cdot} \longrightarrow BQ + BQ^{2-} \tag{21}$$

$$BQ^{2-} + 2H^{+} \longrightarrow BQH_2 \tag{22}$$

The authors observed a radical having hyperfine coupling constants of 14.6 G (N) and 2.6 G (H) when the quinone was photolysed in the presence of phenyl-*N*-*t*-butyl-nitrone in ethanol. This major component was attributed to the trapped ethoxyl radical which can be produced as shown in equation 20. They also note the presence of a minor component due to the trapping of the hydroxyethyl radical ($a_N = 15.3$ G and $a_H = 3.7$ G) and although this may arise by hydrogen transfer from the triplet quinone it is suggested that it is produced from the ethoxyl radical as shown in equation 23.

$$CH_3CH_2O^{\cdot} + C_2H_5OH \rightarrow C_2H_5OH + CH_3\dot{C}HOH \quad (23)$$

Although they could observe both $BQ^{-\cdot}$ and $BQH^{\cdot}$ in the experiments the neutral radical was noted only when high concentrations of the starting quinone were employed. In addition the authors saw no increase in the concentration of $BQH^{\cdot}$ when acetic acid was added to the medium suggesting that equilibrium 24 was not operating. Instead they propose that the neutral radical was generated by reaction between the quinone and the product hydroquinone as shown in equation 18; a large increase in the signal due to BQH was noted when additional hydroquinone was added. A trapped alkoxyl radical was also detected by McLauchlan and Sealy[117] in the spin-trap ESR study of several quinones and solvent alcohols.

$$BQ^{-\cdot} + H^{+} \rightleftharpoons BQH^{\cdot} \quad (24)$$

A similar one-electron transfer mechanism was invoked by Scheerer and Gratzel[118] in a study to correlate the net rate constants for electron transfer and free energy changes involved in the reactions. They measured the yields of $DQ^{-\cdot}$ produced when ^3DQ was photolysed with a variety of electron-donating quenchers. Using their reaction conditions (detection by fast conductance measurements and aqueous ethanol 2:1 v/v as solvent) it is quite possible that electron transfer was favoured or that if any neutral radical were produced it would be undetected.

Wong[119] investigated the photolysis of 1,4-naphthoquinone in isopropanol and 2-butanol and concluded on the basis of the CIDEP behaviour of the system that the experimental results are due to a hydrogen abstraction mechanism even in the presence of pyridine which stabilized the anion radical ($NQ^{-\cdot}$) after deprotonation of the initially formed $NQH^{\cdot}$. In isopropanol the neutral $NQH^{\cdot}$ was strongly emissively polarized consistent with TM CIDEP; the $NQ^{-\cdot}$ observed was not polarized indicating that deprotonation was slower than the T_1 for $NQH^{\cdot}$. In the presence of triethylamine no $NQH^{\cdot}$ was observed and $NQ^{-\cdot}$ was strongly emissively polarized indicating either photoreduction by a different mechanism or that in the presence of this strong base the rates of H-abstraction and deprotonation are dramatically increased.

Kobashi and coworkers[120, 121] have studied the photoreduction of chloranil by dioxane and tetrachlorohydroquinone and determined the relative efficiencies for H-atom abstraction to be 0.13 and 0.58 respectively. The mechanism of H-abstraction in the reaction of triplet chloranil with durene[122] was proposed by these authors to be a simultaneous competition of two mechanisms. One is a two-step process of electron transfer followed by proton transfer via a triplet ion pair, and the other is rapid H-atom transfer in the non-relaxed encounter complex.

Kanemoto and coworkers detected only the neutral durosemiquinone radical in a CIDEP study of the photoreduction of duroquinone and phenol in alcohols; no $DQ^{-\cdot}$ was observed suggesting direct H-abstraction[73]. In a later publication[123] both the neutral and anionic semiquinones, $NQH^{\cdot}$ and $NQ^{-\cdot}$ were observed when naphthoquinone was photolysed with phenol in ethanol or isopropanol. TM CIDEP indicated the initial formation of the neutral radical in these solvents (by H-abstraction) followed by deprotonation to the anion. No polarization was detected for the anion indicating that spin lattice relaxation had returned the system to equilibrium before deprotonation. In isopropanol the authors did not observe any deprotonation to the anion.

The CIDEP behaviour of triplet quinones has been employed by Wan and Elliot[79] to estimate the relative triplet quenching (H-abstraction) rate constants when these quinones are phororeduced in the presence of a variety of hydrogen donors such as alcohols and phenols. They noted a much lower quenching rate for DQ than for BQ or its methyl-substituted derivatives, or AQ when alcohols were the donors. In addition 2-methylphenol,

phenol and pentachlorophenol were found to have quenching reaction rates approximately the same as triethylamine towards ^{3}DQ; durohydroquinone, however, reacts approximately one order of magnitude faster than the phenols. Of the phenols, pentachlorophenol is the most effective H-donor. The phenols are of course much more reactive as H-donors than isopropanol.

Recent improvements in the traditional[93] experimental set-ups for conducting CIDEP studies include the use of commercial boxcar integrators[100, 124] and similar devices[125]. With these modifications it is possible to obtain polarization spectra by sweeping the ESR spectrum at various time intervals (0.5–200 μs) after an initiating laser pulse. Thus it is possible to observe the evolution of the time profile of the CIDEP signal.

In addition, Wan and coworkers have described a method for the simultaneous recording and display of both the CW ESR spectrum and the CIDEP spectrum for a given experiment. This involves the installation of an external narrow bandwidth microwave amplifier between the ESR cavity and the crystal detector and division of the signal prior to detection[126]. This experimental set-up has proven extremely useful for the study of CIDEP in a number of quinone photoreductions including benzoquinone, 2,5-dimethylbenzoquinone, menadione, naphthoquinone and vitamin K_1. As an example the CW ESR and CIDEP spectra simultaneously observed for benzoquinone at time intervals 0.5 and 1.5 μs after the laser flash are shown (Figure 1). It is immediately apparent that the polarization spectrum of the neutral radical (BQH$^{\cdot}$) is much more intensely emissively polarized at 0.5 μs than at the longer delay time. In addition signals due to the anion radical $BQ^{-\cdot}$ are seen to appear in the second spectrum consistent with its formation by deprotonation from the neutral radical. When BQ was photolysed in acetic acid/isopropanol 1:1 the polarized neutral radical was the only species observed, the equilibrium being shifted away from deprotonation to the anion. Frequently the neutral semiquinone radical is observable only in the time-resolved CIDEP spectra and not in the CW ESR spectra. In such cases, its intermediacy in reaction mechanisms can only be surmised unless appropriate CIDEP experiments are performed. For example, menadione and vitamin K_1 exhibit well documented CW ESR spectra and strongly emissively polarized time-resolved spectra due to their respective semiquinone anions when photolysed in basic alcoholic solvents. Photolysis in a variety of polar and non-polar solvents even with added phenol resulted in no neutral radical detectable by CW ESR. However, strong emissively polarized spectra were observed under these conditions and attributed to the neutral radicals of these quinones (**26**). Although not completely resolved the quartet structure observable in the

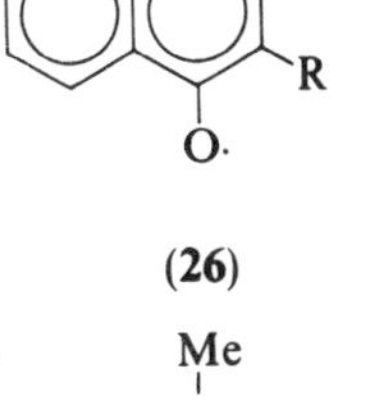

OH, Me, R, O·

(**26**)

menadione: R = H

vitamin K_1:R = $CH_2CH{=}C(Me){-}[(CH_2)_3{-}CH(Me){-}]_3{-}Me$

vitamin K_1 neutral radical spectrum had $a_H^{Me} = 8.0$ G consistent with the value reported in the literature for this species[127]. The CW and polarization spectra for the vitamin K_1 and menadione radical anions and the polarization spectra of their corresponding neutral

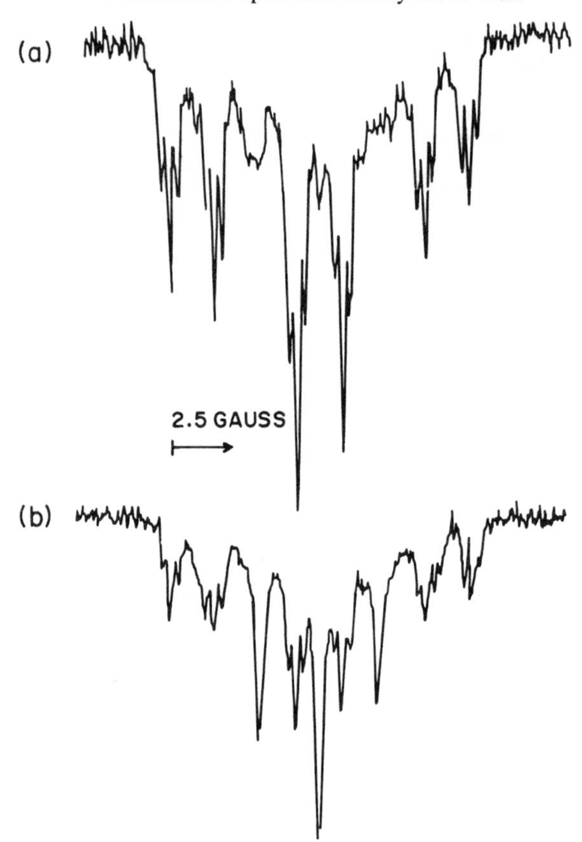

FIGURE 1. Time-resolved CIDEP ESR spectra of benzosemiquinone radicals (a) 0.5 μs and (b) 1.5 μs after the laser flash

radicals are shown (Figure 2). The lack of resolution in the anion radical polarization spectra can be attributed to line broadening as a result of the dynamic equilibrium of deprotonation to the anion after the initial formation of the neutral radical. The fast time resolution (500 ns) of the CIDEP experiment can detect this effect while the CW ESR spectrum is unaffected.

Another attempt to shed some light on the initial event problem in quinone photoreductions was made by Lazarev and coworkers[128]. They photolysed a single crystal of 3,6-di-*t*-butylpyrocatechol (**27**) doped with 10^{-2} M 3,6-di-*t*-butyl-*o*-benzoquinone (**28**) at 77 K. They observed both ion-radical ($Q^{-\cdot} + AH^{+\cdot}$) and neutral ($QH + A^{\cdot}$) radical pairs and although the former were short-lived they appeared not to be converted to the neutral pairs but rather to recombine to form initial reactants. Thus the authors suggest that the processes of H-abstraction and electron transfer compete in this system.

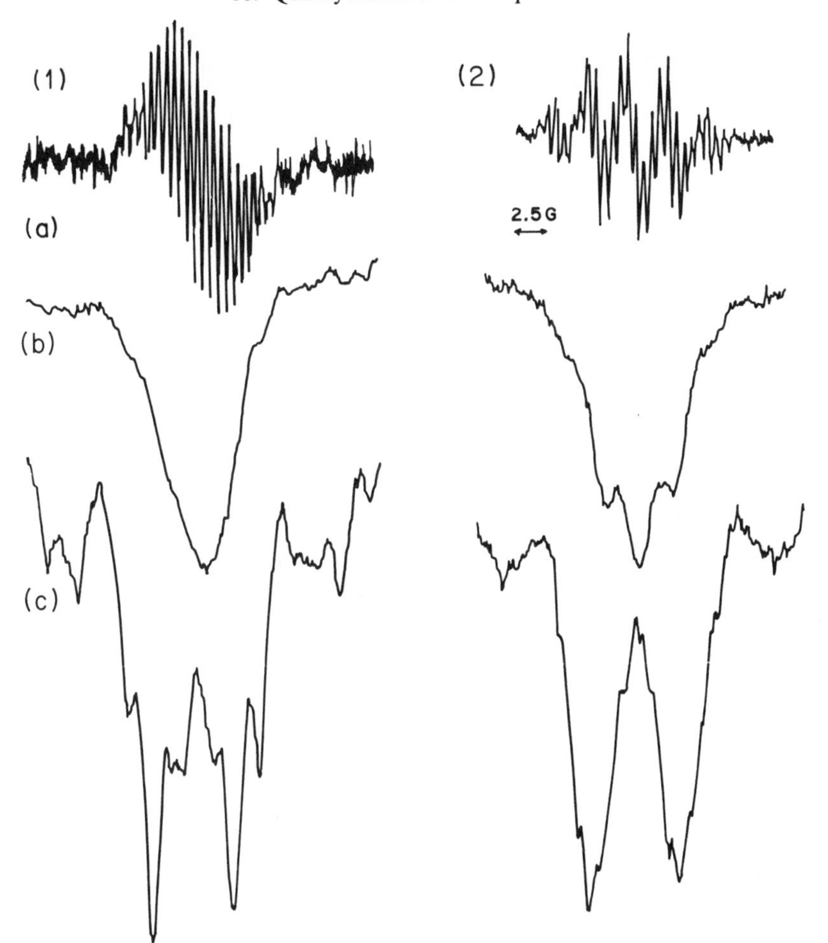

FIGURE 2. ESR spectra of menadione (1) and vitamin K_1 (2) semiquinone radicals at 25°C: (a) CW ESR spectra of the semiquinone anions in basic ethanol solution; (b) time-resolved CIDEP spectra for the semiquinone anions in basic ethanol solution taken 1.0 μs after the laser flash; (c) time-resolved CIDEP spectra in isopropanol/toluene 3:7 v/v containing 10^{-2} M phenol taken 0.5 μs after the initiating laser flash

OH OH O O

(27) **(28)**

In the study of the photochemical behaviour of ketones and quinones it has been suggested that the mechanism of photoreduction is dependent on the nπ* or $\pi\pi$* character of the lowest triplet state (T_1). For example, Porter and Suppan[129], and Formosinho[130]

have noted that the $\pi\pi^*$ states of substituted ketones have approximately 10^{-2} to 10^{-4} times lower predilection for H-abstraction than do the $n\pi^*$ states. Kemp and Porter[131] have suggested that changing solvent from ethanol to water switches the lowest triplet state from $n\pi^*$ to $\pi\pi^*$; this accounts for their lack of observation of BQH˙ when BQ was photolysed in water. Increasing methyl substitution in the quinone is suggested to stabilize the $\pi\pi^*$ state with respect to the $n\pi^*$ (for example, the lowest triplet states of duroquinone and benzoquinone have $\pi\pi^*$ and $n\pi^*$ character respectively).

Effects of substitution on the relative excited state energies are also noted for the halogenoanthraquinones. For the α-halogenoanthraquinones the lowest triplet states are mixed $n\pi^*$–$\pi\pi^*$ or $\pi\pi^*$ in character while the β-substituted isomers have $n\pi^*$ character. $n\pi^*$ triplet anthraquinones react by abstraction of a hydrogen atom from solvent to give the ketyl radical[132]. Inoue and coworkers[133] have proposed that as a result of their $\pi\pi^*$ triplet character the photoreduction of the α-halogenoanthraquinones occurs by direct electron transfer from ethanol; Hamanoue and colleagues[134, 135] do not agree. Their pico- and nanosecond laser photolyses of the chloroanthraquinones showed no indication of electron transfer; however, the greater $\pi\pi^*$ triplet state character of, for example, 1,8-dichloroanthraquinone with respect to 1-chloroanthraquinone is reflected in their H-abstraction rates in ethanol which are $1.6 \times 10^4\ s^{-1}$ and $2.3 \times 10^6\ s^{-1}$ respectively. Several other haloanthraquinones including 1,5-dichloro, 1-bromo, 1,5-dibromo and 1,8-dibromo, all of which have triplet states of mixed $n\pi^*$–$\pi\pi^*$ or $\pi\pi^*$ character, apparently react by H-abstraction from ethanol. Thus it appears that lowest triplet $n\pi^*$ character favours H-abstraction in the photoreduction of quinones in alcohol solvents; the switching of the lowest triplet state to $\pi\pi^*$ in nature, although it will decrease the rate of H-abstraction, may not always cause a switch to another mechanism such as electron transfer.

Photolysis of ketones and quinones is enhanced dramatically by the addition of amines; in general, the mechanism is believed to result from transfer of an electron from the amine to the triplet carbonyl compound forming an exciplex or ion pair followed by proton transfer[136, 137]. Roth deduced the presence of both neutral and anionic radical intermediates in the CIDNP study of quinone photoreductions in the presence of triethylamine (TEA)[138]. For BQ in acetonitrile the CIDNP observations on the product diethylvinylamine which must be formed via the neutral aminoalkyl radical could be explained only by considering the contribution from the radical ion pair. It was therefore concluded that net H-abstraction is a two-step process in such a system.

Hamanoue and coworkers[139] noted a dramatic increase in the quantum yields of photoreduction of AQ, 1-chloro-AQ, and 1-bromo-AQ in the presence of Et_3N. They proposed that the initially formed exciplex between lowest triplet AQ and Et_3N changes into a contact ion pair ($AQ^{-\cdot} + TEA^{+\cdot}$) and then following proton transfer into AQH˙ and the triethylamine radical. The contact ion pair was much more stable in ethanol than in toluene as would be expected. The authors also suggest[140] that the electron transfer mechanism from Et_3N to triplet quinones and dissociation of the exciplex depend on the nature of the solvent. For example Chen and Wan have remarked that a flash photolysis study of 2,6-di-*t*-butylbenzoquinone by triethylamine in benzene appears to give only the neutral radical; no anion radical was detected[141].

CIDEP experiments have also been an asset in the study of photoreductions of quinones having substituents such as *t*-butyl or isopropyl which can undergo intramolecular H-abstraction and rearrangements. The *t*-butyl quinones have been shown to form substituted 1,3-benzodioxole derivatives (**29**) when photolysed in benzene or acetic acid and hydroquinones such as **30** with rearranged side chains when photolysed in isopropanol[142]. Farid has suggested a biradical intermediate resulting from internal H-abstraction from the *t*-butyl side chain which can then react in a number of ways depending on experimental conditions including formation of a spirocyclopropyl ketone[143, 144].

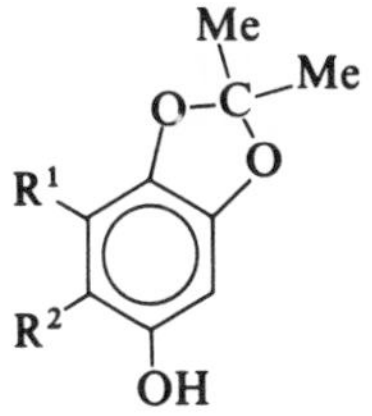

(29) 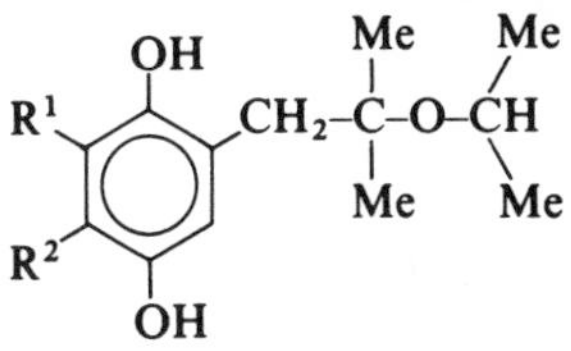 (30)

Wan and coworkers[142] have detected two primary polarized radicals **31** and **32** which arise from the same excited triplet quinone when 2-*t*-butyl-1,4-benzoquinone and a phenol were photolysed in isopropanol. CIDEP studies monitoring the dependence of the initial polarization on phenol concentration indicated that although the radicals were formed in the ratio 2:1 the rate of formation of **32** was five to seven times faster than that of the major component **31**. An additional secondary benzohydrofuran radical **33** was formed when toluene was used as solvent. It is interesting to note that although two radicals corresponding to structures **31** and **32** are noted for the photolysis of certain quinones such as 2-methylbenzoquinone[145] this is not usually the case. Indeed 2,6-di-*t*-butyl and 2,5-di-*t*-butyl-benzoquinone form only the radical corresponding to **31**.

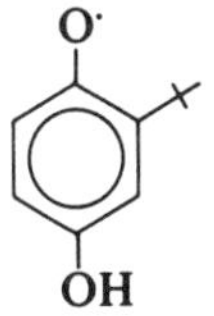

(31) 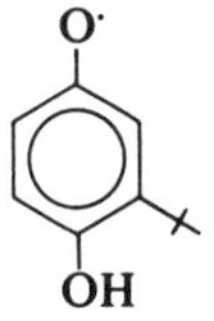 (32) 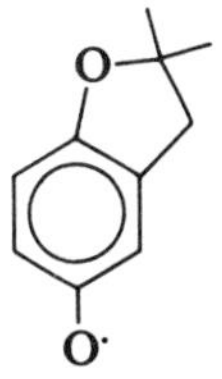 (33)

A biradical intermediate was proposed in the photolysis of 2-isopropoxy-1,4-naphthoquinone to account for the observed formation of the rearranged radical (**34**)

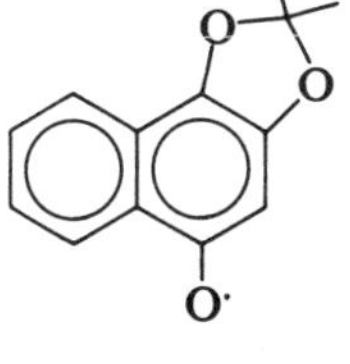

(34)

observed[146]. This compound and other substituted quinones are of commercial interest as photoredox agents in non-silver imaging processes and have been used for the development of novel high gain photothermographic imaging processes[147]. The ethereal oxygen at the 2-position in such compounds has been shown to participate efficiently in formation and stabilization of organometallic radical adducts, thus the 2-isopropoxy-substituted *p*-quinones can behave as 1,2-*ortho*-substituted quinones (see Section IV).

D. Quinone Cation Radicals

ESR spectra for a considerable number of *o*-, *m*- and *p*-hydroquinone cation radicals have been recorded as well as those from a few substituted 1,4-naphthohydroquinones, and 9,10-anthrahydroquinones (Ref. 16 and references therein). In general the radicals are formed in strong acid solutions by oxidation of the corresponding hydroquinone. For several of the cations such as those from 1,4-hydroquinone and durohydroquinone both *cis* and *trans* rotamers can be observed at reduced temperatures and their splittings measured.

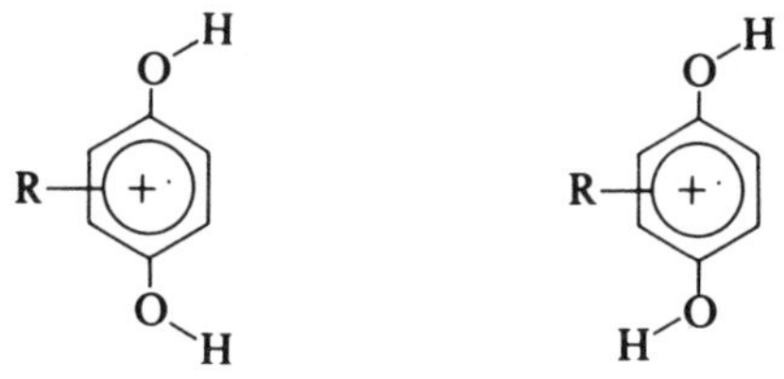

cis *trans*

		a_1^{OH}	a_4^{OH}	a_2	a_3	a_5	a_6
R = H	*cis*	3.294	3.294	2.147	2.147	2.356	2.356
	trans	3.294	3.294	2.456	2.055	2.456	2.055
R = Me	*cis*	2.887	2.887	2.171	2.171	1.931	1.931
	trans	2.887	2.887	2.785	1.389	2.785	1.389

The ESR spectra of these hydroquinone cation radicals show a relatively large temperature dependence for the hydroxyl proton splitting; this observation can therefore be used to estimate potential barriers to rotation about the carbon–oxygen bond. For the hydroquinone cation radicals from benzoquinone, duroquinone, naphthoquinone, 2,3-dimethyl-1,4-naphthoquinone, and 9,10-anthraquinone these were 10 ± 2, 6.6 ± 1.5, 5.9 ± 1.5, 4.6 ± 1.3 and 3.2 ± 4.0 kcal mol^{-1} respectively[148].

Although the hydroquinone cation radicals are reasonably well documented, the one-electron oxidized quinone radical, i.e. the quinone cation radical $Q^{+\cdot}$, has rarely been reported.

A novel stable radical cation (**35**) has been generated from tetrakis(dimethylamino)-*p*-benzoquinone[149]. The radical exhibits a dark purple colour and is stabilized by the strong donor effect of the dimethylamino function which can delocalize the positive charge. The cation was generated at a platinum anode in DMF/0.1 M TBAP (tetrabutylammonium perchlorate) solution at room temperature and gives a 19 line ESR spectrum having g factor 2.0032 and $a_H \simeq a_N = 3.32$ G for 24 protons and four nitrogens. The corresponding

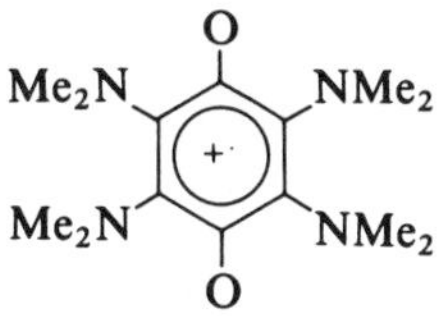

(35)

radical anion was generated in tetrahydrofuran with Na likely as the triple ion complex $[Na^+Q^{-\cdot}Na^+]^{+\cdot}$. This radical exhibited an ESR spectrum having $g = 2.0046$, $a_H = 0.112$ G, and $a_{Na} = 1.43$ G (from ENDOR spectra).

Few other quinone radical cation species have appeared in the literature. A radical cation of benzoquinone was claimed to have been generated at 77 K but the spectrum was reassigned to unidentified impurity[150].

Wan and coworkers have studied the formation of several quinone radical cations in trifluoroacetic acid (TFA)[151–153] including those from benzoquinone, menadione, vitamin K_1, 2,5-dimethyl- and 2,5-diphenylbenzoquinone. Duroquinone and chloranil do not form cations in this solvent. TFA has been suggested to stabilize the cations primarily by interactions with the trifluoromethyl group[154–156]. In TFA benzoquinone gives a five line ESR spectrum ($g = 2.0040$, $a_H = 2.22$ G) thermally which increases in intensity when irradiated[151]. The stability of this cation radical in TFA has permitted its use in charge transfer studies forming known S- and N-containing cation radicals such as *N,N,N′,N′*-tetramethyl-*p*-phenylenediamine, diphenylamine, thianthrene, phenothiazene and thianaphthene as shown in equation 25.

$$BQ^{+\cdot *} + S \rightarrow BQ + S^{+\cdot *} \tag{25}$$

Time-resolved CIDEP spectra show that the emissive TM polarization of the quinone cation is transferred to the newly formed S-containing cation[157].

A possible mechanism for formation of $BQ^{+\cdot}$ which is consistent with the CW and time-resolved CIDEP ESR observations is shown in equations 26–30. Initial protonation of benzoquinone in the acidic solvent may occur at either the carbonyl oxygen or at the quinonoid ring. In the photolysis of cyclopentadiene in TFA Davies and coworkers[158] have proposed an intermediate carbenium ion which subsequently loses a hydrogen atom to form the cyclopentadienyl radical cation. Both in the thermal and photochemical production of $BQ^{+\cdot}$ the oxidation of the quinone (reactions 27 and 29 respectively) may involve either H-atom transfer or charge transfer between BQH^+ and BQ. The product cation radical is stabilized by solvation with TFA but the neutral radical $B\dot{Q}H$ rapidly decays in this medium. Although no hydroquinone cation is observed with benzoquinone a small amount of the deuterated hydroquinone cation is observed when BQ-d_4 is irradiated in TFA. This species may arise as shown in reaction 30. Observations supporting this mechanism involving initial protonation of the quinone are:

$$BQ + CF_3COOH \rightarrow BQH^+ + CF_3COO^- \tag{26}$$

$$BQ + BQH^+ \rightarrow BQ^{+\cdot} + BQH^\cdot \tag{27}$$

$$BQ \xrightarrow{h\nu} {}^3BQ^* \tag{28}$$

$$^3BQ^* + BQH^+ \rightarrow BQ^{+\cdot *} + BQH^{\cdot *} \tag{29}$$

$$BQH^{\cdot *} + H^+ \rightarrow BQH_2^{+\cdot} \tag{30}$$

(1) No reaction was detectable when trifluoroacetic anhydride was used as solvent; addition of one drop of water, however, resulted in immediate formation of $BQ^{+\cdot}$.

(2) When formic acid is used as the solvent, the $BQ^{+\cdot}$ formed decays at least three times faster than in TFA. Thus formic acid is acidic enough for protonation of the quinone but lacks the stabilizing trifluoromethyl moiety.

(3) No $BQ^{+\cdot}$ is formed when acetic acid is employed as solvent. Only $B\dot{Q}H$ is observed in this less acidic medium, likely as a result of H-abstraction from the solvent by the photoexcited quinone.

Although the ESR parameters of $BQ^{+\cdot}$ and $BQ^{-\cdot}$ show an initially surprising similarity the much greater linewidth in the CIDEP spectra of the $BQ^{+\cdot}$ species demonstrates the differing relaxation times for the two radicals. INDO calculations on the possible different radical species derived from benzoquinone show very little difference in the expected ring proton hyperfine couplings[151]. This observation is consistent with the reported small difference in hydrogen coupling constants for $C_6H_6^{+\cdot}$ compared with $C_6H_6^{-\cdot}$. The values of a_H are 4.44 and 3.82 respectively, i.e. a difference of only 15%[159]. Since the quinone radicals have proportionally less spin density in the aromatic ring the hyperfine coupling constants may well vary only slightly between the possible paramagnetic species. The total behaviour of quinones in acidic solvents such as TFA and formic acid is not yet completely understood. Hopefully investigations of the formation, reactions and properties of such cation radical species will be stimulated by these few reports.

IV. METAL AND ORGANOMETAL SEMIQUINONE COMPLEXES

A. Introduction

In the last decade the use of *o*- and *p*-quinones to trap reactive transient organometallic radicals has attracted increasing attention. The use of a variety of nitroxides as spin traps is well known, but the formation of quinone spin adducts with organometals is of interest for a variety of reasons. Primarily of course, the increased stability which results from the quinone–metal association increases the radical lifetime and thus facilitates identification of either semiquinone or metal radical structure. Often additional information is provided by the observation of metal hyperfine splittings or HFS constants associated with other ligands attached to the metal centre or changes in g factors in the radical adducts. In extreme examples, such as the spin adducts resulting from quinone trapping of rhenium carbonyl radicals, the resultant radical complex is so persistent that it can be isolated and its properties, spectral and physical, extensively investigated. This has included a rare example of an emission–fluorescence characterization of a radical, the 3,5-di-*t*-butyl-1,4-benzoquinone–$Re(CO)_4$ adduct. The persistence of the group VIIB metal carbonyl–quinone radical adducts and the lability of the CO ligand makes these radicals ideal substrates for ligand-exchange studies[160–162]. Ligand exchange using optically active ligands has resulted in the formation of stable optically active quinone–organometal radical adducts whose spectral and optical properties have been characterized. The stereoselective nature of subsequent reactions of such optically active adducts has been demonstrated[163, 164].

Quinones and semiquinones are often associated with metal ions or organometals in biological systems and for the most part their interactions are not well understood. In a quinone–organometal radical complex both the metal and the quinone have redox potentials in easily accessible and biologically useful ranges. Metal–quinone electron transfer reactions have been found to occur in mitochondria and in bacterial photosynthesis[165–167]. The catecholate ligands are well known for their affinity for ferric ion; it has been estimated[168] that enterochelin, an iron sequestering agent found in enteric bacteria which uses three catecholate groups to encapsulate the ferric ion, has a complex formation constant of 10^{52}.

The literature on quinone–metal radical adducts is not comprehensively covered in this section. Our focus is on the intermediate quinone–metal ion pairs and quinone–organometal complexes and not on product complexes which frequently form complex solvates. The use of CIDEP and ESR techniques to elucidate some of the reaction mechanisms involved in the radical adduct formation will be briefly discussed. Free radical intermediates in photochemical reactions of carbonyl compounds (including quinones) with organometals[169] and transition metal *o*-quinone complexes[170] have been reviewed recently.

B. Radical Ion Pairs

Radical ion pairs can be generated electrochemically but are usually formed by alkali metal reduction in ethereal solvents. Depending on the stability of the ion pair, this process can also lead to the formation of triple ions[171].

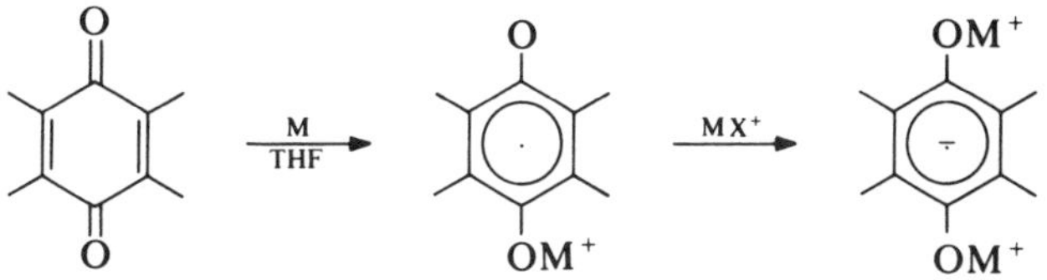

$$Q \xrightarrow[\text{THF}]{\text{M}} Q^{\cdot-}M^{+} \xrightarrow{\text{MX}^{+}} M^{+}Q^{2-}M^{+}$$

Crown ethers have been used to facilitate the observation of ion pairs in non-polar solvents when the quinones were reduced thermally by alkali metals[172, 173]. In addition the use of the appropriate crown ether and alkali metal alkoxide in the photolysis of 2,6-di-*t*-butylbenzoquinone in benzene resulted in the observation of the $Na^{+}Q^{-\cdot}$ contact ion pair[141] for which the Na HFS could be seen.

A photochemical method for the generation of radical ion pairs involving irradiation of a quinone with alkali tetraphenylborates in ethereal solvents has been described[174,175]. This technique is applicable also to other carbonyl containing compounds. In general, it was discovered that for *o*-quinones and sterically hindered *p*-quinones the photolysis results in the formation of radical ion pairs; however, triple ions result with unhindered *p*-quinones.

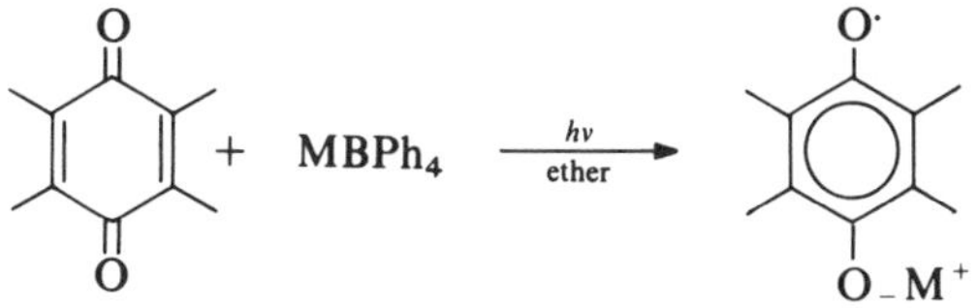

$$Q + \text{MBPh}_4 \xrightarrow[\text{ether}]{h\nu} {}^{\cdot}\text{O}-\text{C}_6\text{Me}_4-\text{O}^{-}\text{M}^{+} \qquad (31)$$

M = Li; K, Na

In solvents such as HMPA alkali metal reductions of quinones yield the free radical anions; however, addition of metal salts such as perchlorates or iodides to the anion radical solutions can result in the formation of the ion pairs which exist in rapid (on the ESR time-scale) equilibrium with the free ions. For 2,6-di-*t*-butylbenzoquinone (2,6-DTBQ) in HMPA the ion pair with K^{+} was formed only at the unhindered quinone oxygen while Na^{+} and Li^{+} could interact at either oxygen[176, 177]. The rapid equilibrium between the ion-paired and free semiquinone results in the observation of time-averaged spectra from the species involved. When the alkali metal counterion has two alternative sites for ion pairing it is sometimes possible to observe cation migration between the two sites. For example, when the perchlorate salts of Mg^{2+}, Ca^{2+} and Ba^{2+} were added to HMPA solutions of 2,6-DTBQ the equilibria between the ion pairs and free ions were sufficiently slowed down that both species could be observed simultaneously in the ESR spectra. For Ca^{2+} different ESR spectra for the hindered and unhindered radical-ion pairs could be detected. The hydrogen HFS constants for a series of metal ion pairs with DTBQ are shown in Table 4[178].

TABLE 4. Coupling constants for 2,6-di-*t*-butylbenzoquinone (2,6-DTBQ) radical ion pairs

Ion pair	a_H (G)
2,6-DTBQ$^{-\cdot}$ (free)	2.39
O–(2,6-DTBQ$^{\cdot-}$)–O,K^+	2.10
Mg^{2+},O–(2,6-DTBQ$^{\cdot-}$)–O	0.86
Ca^{2+},O–(2,6-DTBQ$^{\cdot-}$)–O	1.12
O–(2,6-DTBQ$^{\cdot-}$)–O, Ca^{2+}	1.46
–(2,6-DTBQ$^{\cdot-}$)–O, Ba^{2+}	1.42

A novel method for the formation of *o*-quinone metal ion chelates was described by Felix and Sealy[40]. These complexes were formed by photolysis of catechols in the presence of diamagnetic metal ions from groups IIA, IIB, IIIA and IIIB in aqueous solution at neutral pH. The quinone–metal ion pairs were characterized by lower *g* factors and reduced spin densities at the carbonyl oxygens. The changes in quinone hyperfine couplings were correlated to the ratio of charge to ionic radius for the metal ion. The group IIA metals were associated more weakly with the quinone anions as evidenced by the lack of observed HFS constants for the major metal magnetic isotopes. These couplings were observed with the IIB, IIIA and IIIB metals. The metal ion pairs of the substituted *o*-quinone L-dopa (**36**) were also studied by this method[52]. The zwitterionic form of this *o*-quinone is observed at

M^{n+} ··· O,O-semiquinone ring with $CH_2CH(NH_3^+)CO_2^-$ substituent

$n = 1, 2$

(**36**)

neutral pH; this species deprotonates at pH 9. The complexity of the ESR spectrum of this radical is partially due to the non-equivalent chiral methylene hydrogens in the amino acid side chain. Formation of the metal ion pairs especially those involving Cd^{2+} or Zn^{2+} considerably increases the stability of the dopa semiquinone radical as well as the radicals of dopamine and adrenaline[179].

A number of *o*-quinones and ene-diols such as vitamin C form organothallium complexes which have been suggested to be radical-ion pairs[180–182].

Ion-exchange reactions are frequently used to convert less stable radical ion pairs to more stable ones (equation 32). Calcium ions in blood can be replaced by potassium ions by an exchange reaction with potassium phenanthrasemiquinone (equation 33)[169].

$$K^+PQ^{-\cdot} \xrightarrow{Na^+} Na^+PQ^{-\cdot} + K^+ \qquad (32)$$

$$K^+PQ^{-\cdot} \xrightarrow[H_2O/THF]{Ca^{2+} \text{ in blood}} Ca^{2+}PQ^{-\cdot} + K^+ \qquad (33)$$

Uranyl–quinone ion pairs have been generated by photolysis in THF of the quinone and uranyl salts (equation 34). Using CIDEP techniques as well as CW ESR observations a mechanism for the reaction has been suggested[88, 89]. This involves transfer of energy from the polarized excited phenanthroquinone triplet to the uranyl ion followed by water splitting and then formation of the quinone–organometal ion pair (reactions 35–38). The $[UO_2HPQ]^{2+\cdot}$ which is also formed in the system is believed to result from a secondary process and may form as shown in equations 39–41.

$$PQ + UO_2(NO_3)_2 \cdot nH_2O \xrightarrow[THF]{h\nu} UO_2^{2+}PQ^{-\cdot} + UO_2H^{3+}PQ^{-\cdot} \qquad (34)$$

$$PQ \xrightarrow[ISC]{h\nu} {}^3PQ^* \qquad (35)$$

$${}^3PQ^* + UO_2^{2+} \cdot nH_2O \rightarrow {}^3(UO_2^{2+} \cdot nH_2O)^* + PQ \qquad (36)$$

$${}^3(UO_2^{2+} \cdot nH_2O)^* \rightarrow UO_2^{+\cdot *} + H^+ + \dot{O}H \qquad (37)$$

$$UO_2^{+\cdot *} + PQ \rightarrow [UO_2PQ]^{+\cdot *} \qquad (38)$$

$$[UO_2PQ]^{+\cdot} \xrightarrow{h\nu} UO_2^{2+} + PQ^{-\cdot} \qquad (39)$$

$$PQ^{-\cdot} + H^+ \rightarrow HP\dot{Q} \qquad (40)$$

$$HP\dot{Q} + UO_2^{2+} \rightarrow [UO_2HP\dot{Q}]^{2+} \qquad (41)$$

C. Organometal–Quinone Radical Adducts

There are several methods by which quinone–organometal radical adducts can be formed. For example, ion-exchange reactions were mentioned in the previous section. Metal–quinone ion pairs which are generated by methods such as alkali metal reduction can be exchanged for organometals to form stable heavy metal–quinone radical adducts[183] (equation 42).

$$K^{+}PQ^{-\cdot} + MeHgCl \xrightarrow{THF} MeHgP\dot{Q} + KCl \qquad (42)$$

Oxidation of catechols by organometallic hydroxides can also result in the organometal–quinone complex (**37**). This method was extensively used by Stegmann and coworkers in the formation of organothallium complexes of quinones[180, 181, 184].

S_N2 reactions by excited quinones have been described[185, 186] (equation 43).

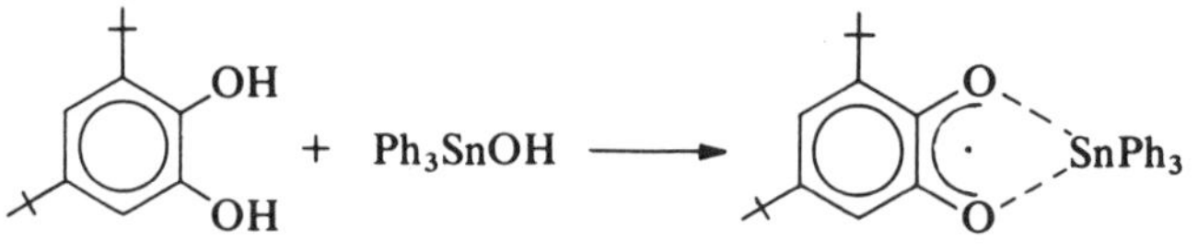

$$\text{3,5-di-}t\text{-butylcatechol} + Ph_3SnOH \longrightarrow \text{(semiquinone)}\cdot\cdot\cdot SnPh_3$$

(**37**)

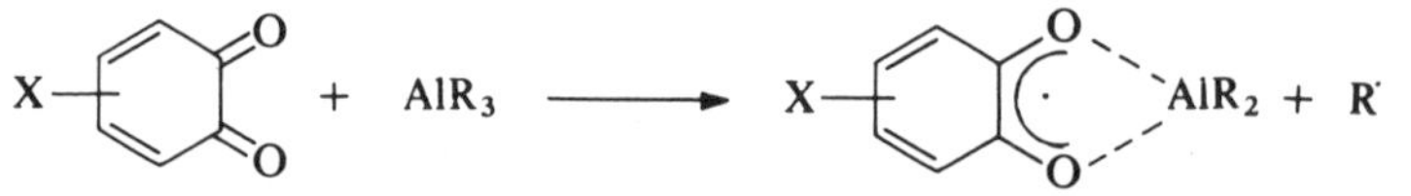

$$X\text{-}o\text{-quinone} + AlR_3 \longrightarrow X\text{-(semiquinone)}\cdot\cdot\cdot AlR_2 + R^{\cdot} \qquad (43)$$

Most commonly, however, quinone–organometal radical adducts are formed as a result of radical addition reactions. Often the initial organometal radicals can be generated by thermal or photochemical homolysis of species containing a M–M′ bond (equations 44 and 45), followed by addition of the radicals to quinones. For example photolysis of $Me_3SnMn(CO)_5$ resulted in addition of $\dot{M}n(CO)_4$ where two adjacent oxygens were available for chelation while $Me_3\dot{S}n$ added to one carbonyl oxygen only[187].

$$R_xMM'R_y \longrightarrow R_xM^{\cdot} + M'R_y^{\cdot} \qquad (44)$$

$$Q + R_xM^{\cdot} \longrightarrow Q^{\cdot}MR_x \qquad (45)$$

Organometals can react with a variety of compounds to produce paramagnetic intermediates by a charge transfer mechanism which may be initiated thermally, or photochemically by irradiation of the charge transfer band or via a triplet exciplex. CIDEP experiments were consistent with the charge transfer mechanism involving quenching of excited quinone triplet by organometals such as R_3SnX to form a primary radical ion pair of the quinone and organometal[90, 188]. Organotin compounds are effective electron donors and reactions with phenanthrenequinone[189], substituted *o*-quinones[190, 191] and *p*-quinones[192, 193] supported this mechanism. In addition direct spectroscopic evidence was obtained when the quinones in host organotin crystals were photolysed at 77 K. The ESR spectrum of the triplet state of the primary radical ion pair was observed consistent with the following mechanism (equations 46–48)[194].

$$Q + Ph_3SnX \xrightarrow{h\nu} {}^3[Q^{-\cdot} \ldots Ph_3SnX^{+\cdot}] \rightarrow {}^{77K}[Q^{-\cdot} + Ph_3SnX^{+\cdot}]\text{pair} \qquad (46)$$

$$[Q^{-\cdot} + Ph_3SnX^{+\cdot}] \rightarrow Q^{-\cdot} + Ph_3Sn^{+} + \dot{X} \qquad (47)$$

$$Q^{-\cdot} + Ph_3Sn^{+} \rightarrow Ph_3Sn^{\cdot}\ Q \qquad (48)$$

1. *p-Quinones*

Spin adducts of organosilyl, -germyl and -stannyl radicals with several *p*-quinones have been described[192, 195–197]. Relatively stable adducts have been obtained with the hindered 2,6-DTBQ and trialkyl C, Si, Ge, Sn and Pb radicals as well as with the diphenylphosphinyl radical ($\dot{P}Ph_2$) and the phenylthiyl radical (PhS)[192]. Some controversy concerning radical addition to such quinones still persists due to the presence of two distinct reactive sites, the carbon–carbon double bond and the carbonyl group. Isomeric radical adducts in the addition of silyl and germyl radicals to unsaturated carbonyl containing compounds such as maleic anhydride have been observed[198] with addition to C=C predominating at low temperatures and addition only to C=O oxygen being observed at room temperature. (Certain organometals such as the organotin and lead adducts prefer to add only to the carbonyl oxygen.)

The addition of trialkylsilyl radicals to 3,6-dimethylthieno[3,2-b]thiophen-2,5-dione was felt to occur faster at the carbon–carbon double bond although addition of this radical to the carbonyl function was also observed. This prompted a thorough CIDEP investigation of the addition of trialkylsilyl radicals to substituted *p*-benzoquinones[124]. Addition to both the ring and the carbonyl group of 2,6-DTBQ was observed giving adducts **38** and **39**. The oxygen adduct was confirmed to be the more thermally stable species.

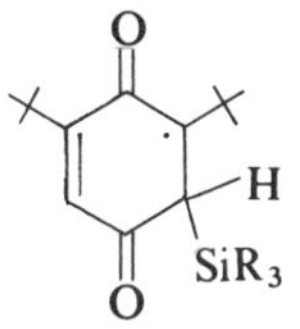

O H SiR$_3$ O   O· OSiR$_3$

(38) 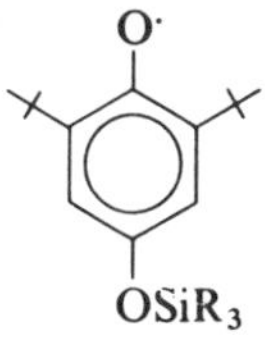   **(39)**

The effect of differences in the organometal character on the site of addition to such *p*-quinones was evident when 2,6-DTBQ was photolysed with $Me_3Sn–Mn(CO)_5$. As mentioned earlier in this section the two primary radicals $Me_3\dot{S}n$ and $\dot{M}n(CO)_5$ showed very different behaviours; the trialkyltin radical adding to the carbonyl group and the manganese radical at the ring carbon.

The dual reactivity of the quinones is also exhibited in reaction with the methyl radical which has been shown to form a ring adduct[199] and the *t*-butyl radical which formed a stable phenoxyl radical by addition to the C=O group[192]. A recent study of the reaction of several phosphorus containing radicals with quinones demonstrated that for 2,6-DTBQ, $\dot{P}(O)Ph_2$ and $\dot{P}(O)(OEt)_2$ added to both C=C and C=O functions giving ESR spectra for both ring and oxygen adducts. However $\dot{P}Ph_2$ and $\dot{P}(S)Et_2$ gave only observable oxygen adducts. With duroquinone the only adducts which were definitely identified were the ring adducts of DQ with $O\dot{P}Ph_2$ and $O\dot{P}(OEt)_2$. The heterocyclic sulphur containing quinone benzo[1,2-b;5,4-b′]dithiophen-4,8-dione (**40**) was observed to form oxygen adducts with all the above mentioned phosphorus containing radicals; both isomeric forms (**41a** and **b**) were observed except in the case of $S\dot{P}Et_2$ which appeared to form only the isomer corresponding to **41b**[200].

Some adducts of 2,6-DTBQ and several phosphine oxides have been described[201]. These authors noted only oxygen adducts and suggested that rotation about the phosphorus–oxygen bond is strongly hindered.

The reaction of quinones with nitrosoarenes gives the 2,3-disubstituted dinitrone product[202] rather than the expected 2,5-disubstituted product. The mechanism of the

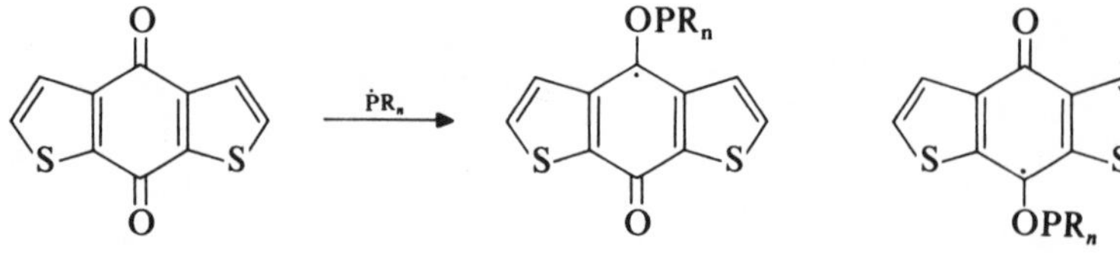

(40) (41a) (41b)

reaction is not known but was suggested to be a radical process involving electron transfer and to involve either trapping of the PhN–O radical by the quinone or addition of the nitrosobenzene to the semiquinone anion.

Spin adducts of metals and *p*-quinones can in certain instances be stabilized by the presence of a heteroatom in a position available for participation in chelation of the metal centred radical. The trialkyllead adduct **42** was described when 2-methoxy-*p*-benzoquinone was photolysed with Pb_2R_6[169]. The 2-isopropoxy-1,4-naphthoquinones **43a** and **b** efficiently formed organometal adducts **44** with Re, Mn, V and Mo radicals. Comparison of the ESR parameters of these organometallic radical adducts with those of

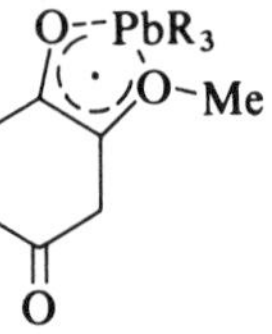

(42) 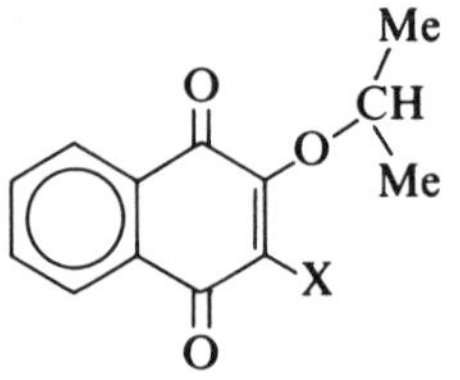 (43a) X = H (43b) X = Cl (44)

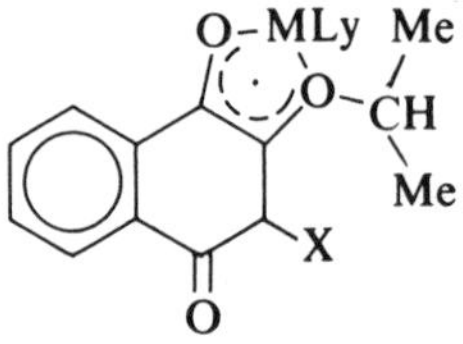

the corresponding 1,2-naphthoquinone metal adducts supported the participation of the ethereal oxygen in the chelation[146]. The 2-isopropoxyl–naphthoquinone complexes with $\dot{R}e(CO)_5$ were unusual in that both 1:1 and 1:2 rhenium–quinone adducts could be observed simultaneously, relative concentrations of the two complexes could be varied by changing the ratio of metal to quinone.

2. *o*-Quinones

o-Quinones are able to interact with a great variety of organometallic radicals by chelation with the adjacent carbonyl oxygens. Examples of adducts from metals of all the main groups, transition metals[170] and the actinides have been recorded in the literature (Ref. 169 and references therein). The *o*-quinones can act as bidentate ligands towards organometals (**45**) or as monodentate ligands forming unsymmetric adducts **46** and **47** which may or may not be in rapid dynamic equilibrium. Determination of the structure of

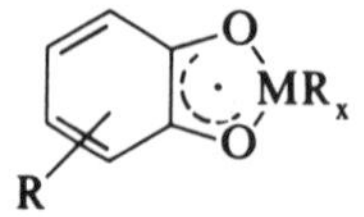

(45) 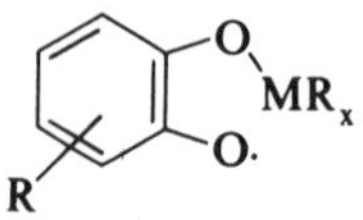 (46) 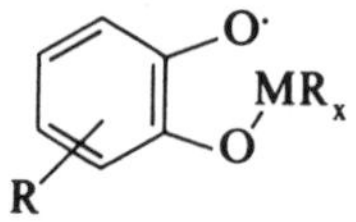 (47)

the radical adducts is usually based on ESR parameters, especially linewidth alternation studies and analysis of quinone and metal ligand splitting patterns. Factors affecting the relative stability of the mono- or bidentate adducts and the rate of cation migration between the two oxygens in the monodentate structures include the nature of the metal and its other ligands as well as the steric and electronic properties of the quinone.

For a series of tin radicals α-dicarbonyl compounds such as biacetyl have been shown to chelate $\dot{Sn}Cl_3$ and $\dot{Sn}BuCl_2$ through both oxygens; however, $\dot{Sn}Bu_2Cl$ and $\dot{Sn}Bu_3$ coordinate through one oxygen only, the adducts being non-fluxional and fluxional respectively in the latter two cases[195]. Similarly the activation parameters for intramolecular cation migration were found to increase with increasing chlorine substitution in alkyltin–quinone radical adducts[191]. Activation energy for intramolecular migration was found to decrease in the order $SiR_3 > GeR_3 > SnR_3$ for such radical adducts.

The structure of adducts of aluminium salts with 3,6-di-*t*-butyl-*o*-quinone were influenced by solvent; non-coordinating solvents such as toluene favoured symmetrical coordination of the metal between the quinone oxygens but more strongly coordinating solvents such as ether, or THF converted the structure to a monodentate radical as shown

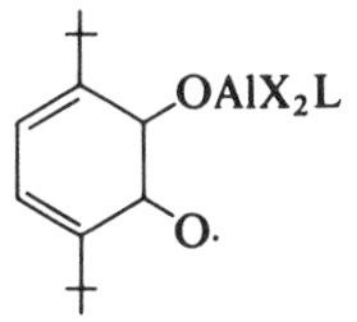

$OAlX_2L$

O·

(48)

below (**48**) where L is the coordinating solvent molecule. Depending on the nature of the coordinating solvent the aluminium moiety may be fluxional or non-fluxional on the ESR time-scale with strongly coordinating solvents increasing the rate of intramolecular migration[186]. These observations are consistent with the postulation of two rapidly equilibrating tautomers for $PQ^{\dot{-}}AlCl_2$ in THF[203].

Complex tautomeric behaviour of tris-quinone metal radical complexes has also been observed[204–206].

Complexes of *o*-quinones with organomercury complexes have been studied quite extensively. Since heavy metal cations such as HgR^+ readily displace alkali metal cations in $M^+Q^{-\cdot}$ complexes this method was regarded as an important procedure for the detection and removal of trace amounts of toxic organomercurials[183, 207]. Interactions of organomercury and mercuric salts with several *o*-quinones and also with vitamin C which has three oxygen functions available for complexation with metal centres have been reported[208]; ^{199}Hg splittings were observed for the *o*-quinone and vitamin C adducts. Many of the quinone–organomercury radicals exhibit very low *g* factors; similar dramatic *g*-shifts were noted in the uranyl–quinone radical complexes. Therefore, a simple theoretical model describing the relative contributions to the *g*-shift from spin-orbit coupling of the metal and from d-p_π-bonding has been discussed[208]. The mercury salts were also found to interact with vitamin C; however, the observed *g*-shifts were small and no metal hyperfine splittings were detected. These observations were consistent with the formation of mercury–vitamin C ion pairs similar to that proposed for VTC-Ph_2TlOH[182].

A charge transfer mechanism has been suggested for the reactions of organomercury with *o*-quinones (equations 49–51); however, thermal or photochemical decomposition of the organomercury is another possible route for initiation of the adduct formation which cannot be eliminated (equations 52, 53)[208].

$$\mathrm{Q + HgR_2 \rightarrow Q^{-\cdot} + R_2Hg^{+\cdot}} \quad (49)$$

$$\mathrm{R_2Hg^{+\cdot} \rightarrow RHg^+ + R^\cdot} \quad (50)$$

$$\mathrm{Q^{-\cdot} + HgR^+ \rightarrow Q\dot{-}HgR} \quad (51)$$

$$\mathrm{R_2Hg \rightarrow RHg^\cdot + R^\cdot} \quad (52)$$

$$\mathrm{RHg^\cdot + Q \rightarrow RHg\dot{-}Q} \quad (53)$$

No evidence for intramolecular migration of mercury between the quinone carbonyl groups has been found[208, 209] consistent with a structure having Hg coordinated equally to the two functional groups. For the quinone–mercury radical adducts of compound **40** the spin density at the metal atom was approximately one order of magnitude lower than that for the paramagnetic complexes of this compound with manganese or rhenium carbonyls or with As, Sb or Bi biphenyls[209].

In *o*-quinones charge delocalization onto the quinone–metal chelate ring is not as extensive as in their sulphur-containing analogues such as the 1,2-benzodithiolates (Ref. 170 and references therein). Dithiocarbonyl compounds are well known as superior spin traps compared to the dicarbonyls[210].

Considerable work in our laboratory has been concerned with the reactions of group VIIB organometals (especially the carbonyls) with *o*-quinones. The extreme persistence of the $\mathrm{Re(CO)_4\dot{-}Q}$ radical adducts and the lability of the carbonyl ligands makes them ideal subjects for ligand-exchange studies. In addition development of ESR–HPLC techniques for isolation of the radicals enabled their complete spectral characterization even including emission–fluorescence studies. Formation of the adducts results from trapping of the primary $\mathrm{\dot{R}e(CO)_5}$ radicals formed by photolysis of the Re–Re bond (equations 54, 55). The metal atom chelates symmetrically to all the *o*-quinones studied to form persistent 1:1 Q:M adducts. The sole exception to this noted so far is the previously mentioned 2-isopropoxyl-1,4-naphthoquinone. However, different ratios of quinone to metal have been observed for other group members. $\mathrm{Cr_2(CO)_6}$ forms $\mathrm{Cr(CO)_2PQ_2^\cdot}$ as the major component whether the reaction with phenanthrenequinone is carried out in the dark or under irradiation. $\mathrm{Cr(CO)_4PQ^\cdot}$ is produced as a minor component but only by photolysis[161].

$$\mathrm{Re_2(CO)_{10} \xrightarrow{h\nu} 2\dot{R}e(CO)_5} \quad (54)$$

$$\mathrm{\dot{R}e(CO)_5 + Q \longrightarrow Q\dot{-}Re(CO)_4 + CO} \quad (55)$$

For the $\mathrm{Q\dot{-}Re(CO)_4}$ radical adducts the order of stability is 3,5-di-*t*-butyl-1,2-benzoquinone (3,5-DTBQ), tetrabromo-1,2-quinone, 1,2-naphthoquinone ~ acenaphthoquinone ~ phenanthrenequinone > camphorquinone ≫ tetrachloro-1,2-quinone. Once the initial quinone–rhenium complex is formed substitution of one or more of the CO ligands by a variety of P, N, As, Sb containing species is possible. For the most part only one CO is substituted even when bidentate substituents are employed; however, for the phosphites disubstituted adducts were also formed. This can be observed in Figure 3 in which the initial $\mathrm{Re(CO)_4\dot{-}3,5\text{-}DTBQ}$ second derivative ESR spectrum (3a) changes on addition of one $\mathrm{P(OMe)_3}$ substituent (3b) and then further splitting results on substitution of the second $\mathrm{P(OMe)_3}$ group (3c). The g factors and hydrogen HFS constants can be seen to decrease on substitution while the values of a_{Re} increase consistent with withdrawal of the electron density from the quinone ring towards the phosphorus ligands.

During reaction of these adducts with a series of N-containing substituent ligands it was discovered that effective substitution required that the ligand not be sterically hindered at the donor atom and that a possibility for delocalization of electron density into, for example, an aromatic π system or empty d orbitals must exist. The substitution mechanism is not proven but appears to be dissociative rather than associative[161].

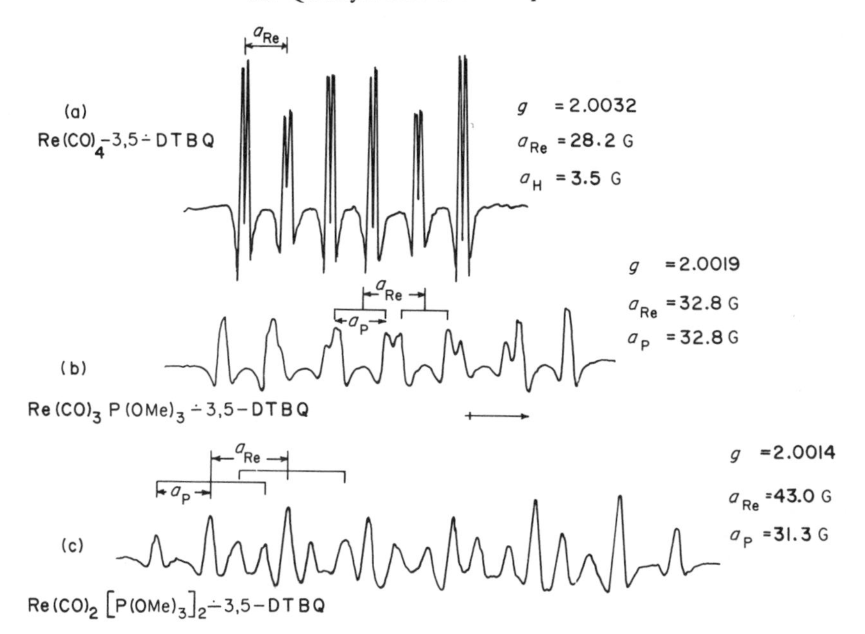

FIGURE 3. Second derivative ESR spectra in toluene at 25°C for: (a) $Re(CO)_4\dot{-}3,5$-DTBQ; (b) $Re(CO)_3P(OMe)_3\dot{-}3,5$-DTBQ; (c) $Re(CO)_2[P(OMe)_3]_2\dot{-}3,5$-DTBQ

Similar ligand substitution reactions occur for the other group VIIB metal carbonyl–phenanthrenequinone radical adducts. However, differences in reactivity are apparent; PPh_3 substitutes two equatorial CO groups in $Mo(CO)_4\dot{-}PQ$ while in $Cr(CO)_2\dot{-}PQ_2$ only one axial carbon monoxide is replaced.

Substitution of $Re(CO)_4\dot{-}3,5$-DTBQ with the optically active phosphorus-containing ligands (+) or (−)-DIOP results in the formation of an optically active metal–quinone radical (**49**). The isolation by HPLC and full spectral characterization of these optically active intermediates is unique[161]. The word unique was used in this description since to our knowledge no other *optically active radical* has been isolated, purified and characterized. Usually the reactivity of such radicals precludes their determination other than in reaction mixtures containing other starting materials and products. The optical rotation values were obtained for both the (+) and (−) DIOP complexes.

CO CO Re CO (+)-DIOP

(49)

H CH_2PPh_2 Me O Me O H CH_2PPh_2

(+)-DIOP

In addition a rare example of the emission–fluorescence characterization with quantitative fluorescence quantum yield and lifetime measurement for a radical in solution at room temperature was obtained in the study of the $Re(CO)_4\dot{-}3,5$-DTBQ radical adduct[211]. The radical was discovered to have λ^{abs}_{max} 230 and 500 nm in cyclohexane and $\lambda^{em}_{max} = 320$ nm.

The fluorescence yield and lifetimes were found to be 0.06 ± 0.01 (compared with standard anthracene) and 10.3 ± 0.1 ns respectively[211]. Another quinone radical whose fluorescence behaviour has been recently examined is that of camphorquinone[212] (see Section V.B).

V. PHYSICAL CHEMISTRY OF THE SEMIQUINONES

Much of the data concerned with the physical properties of the semiquinones has relied on the formation of these radicals by pulse radiolysis or flash photolysis. Generation and stabilization of the radical anions in basic aqueous media is fairly simple but accurate spectral data for the less stable neutral radicals is more difficult. The spectral, acid–base and redox properties of the hydroquinone and quinone cation radicals have not yet been systematically studied.

Swallow[213, 214] has recently reviewed the physical data relevant to the semiquinones.

A. Acid–Base Properties

The equilibrium reaction 56 between the neutral and anionic semiquinone radicals has been studied for a variety of quinones in aqueous solution and pK_a values obtained. Frequently the data were obtained from pH dependence of optical density measurements at a wavelength where the extinction coefficients of the anion and neutral radicals are quite different (usually at λ_{max} for $Q^{-\cdot}$) after the radicals have been generated by pulse radiolysis, or from conductivity measurements. The pK_a values for several *o*- and *p*-semiquinones are collected in Table 5. In general, the pK_a values for the semiquinones are approximately 5 pH units lower than those for the corresponding quinones.

$$QH^{\cdot} \rightleftharpoons Q^{-\cdot} + H^{+} \qquad (56)$$

Often the measurements are made in solvents of differing composition which makes the absolute numbers difficult to compare. However, for the series of simple substituted BQs the electron-donating capacity of the methyl group increases the pK_a value by 0.25 units per substituent. This trend is not followed as well in the 1,4-naphthoquinone series. The pK_a value for 2,3-dimethyl-1,4-naphthoquinone is lower than that for the 2-methyl derivative, menadione.

The pK_a values for the semiquinones from vitamin K_1 and ubiquinone appear to be considerably higher than the analogous 2,3-dimethylnaphthoquinones; however, the higher percentage of alcohol in the solvent required to dissolve these compounds is likely the causative factor[213]. This effect is also noted for durosemiquinone which exhibits a pK_a value of 6.0 ± 0.1 in aqueous isopropanol/acetone 7:1 v/v and 5.0 ± 0.1 in aqueous isopropanol/acetone 1:1 v/v. Indeed determination of ubisemiquinone pK_a in methanol gives a value of 6.45 ± 0.15, even higher than in isopropanol/acetone 7:1 (5.90). Swallow suggests that if ubisemiquinone and vitamin K_1 semiquinone could be prepared in aqueous solutions the pK_a values expected would be 4.9 and 4.5 respectively. The small differences in pK_a values for the semiquinones of DQ, vitamin K_1, and ubiquinone indicate that this property is relatively insensitive to the nature of the long alkyl side chain. This is not unexpected since these alkyl substituents have been shown to perturb the aromatic spin density in the semiquinones by an amount similar to the methyl group[21].

The anthraquinone-sulphonates studied show a dramatic decrease in pK_a value when the substituent is in the 2-position but little or no effect for substitution at positions 1 or 6.

The *o*-semiquinones have pK_a values very similar to the *p*-semiquinones; the only value observed for a *m*-semiquinone generated from resorcinol was somewhat higher (7.0). The semiquinone radicals generated from epinephrine, adrenalone and camphorquinone are somewhat more acidic than those of the other quinones; the reasons for this are not readily apparent. Complications in the acid–base properties due to participation of the amino acid

TABLE 5. Acid–base properties of semiquinones

Compound	pK_a	Reference
1,4-Benzoquinone	3.90	23
	4.10	222, 225, 294
2-Methylbenzoquinone	4.45	225
2,3-Dimethylbenzoquinone	4.65	225
2,5-Dimethylbenzoquinone	4.60	225, 294, 295
2,6-Dimethylbenzoquinone	4.75	225
2,3,5-Trimethylbenzoquinone	4.95	225
Duroquinone	5.10	295
	5.00	294
	6.00	225
Diphenoquinone	3.20	295
1,4-Naphthoquinone	4.10	294, 295
2-Methylnaphthoquinone	4.40	225
	4.50	295
2,3-Dimethylnaphthoquinone	4.25	225
2-Hydroxynaphthoquinone	4.70	295
Vitamin K_1	5.50	225
Ubiquinone	5.90	225
	6.45	296
Anthraquinone	5.30	295
Anthraquinone-1-sulphonate	5.40	214
Anthraquinone-2-sulphonate	3.25	214
Anthraquinone-2,6-disulphonate	3.20	214
1,2-Benzoquinone	5.00	23, 222
3-Methoxy-1,2-benzoquinone	5.00	23
4-Methyl-1,2-benzoquinone	4.50	213
4-*t*-Butyl-1,2-benzoquinone	5.20	213, 297
Resorcinol	7.10	23
	7.00	222
1,2-Naphthoquinone	4.80	295
Epinephrine	3.70	295
Adrenalone	3.60	295
Camphorquinone	4.10	214

side chains in the first two may have an effect, and in camphorquinone of course, the usual planar aromatic geometry of the quinones is absent.

Further protonation of the neutral semiquinone is possible in strong acid solutions (equation 57) but has not been extensively studied. Land and Porter measured a pK_a value of -1.1 for the ionization of the durohydroquinone cation radical in 50% aqueous ethanol[215].

$$QH^{\cdot} + H^+ \rightleftharpoons QH_2^{+\cdot} \qquad (57)$$

B. Optical Spectra

Absorption maxima and extinction coefficients have been tabulated quite extensively in the literature from pulse radiolysis and flash photolysis experiments. Although the spectra for both neutral and anionic radicals can be obtained from either technique, accurate determination of extinction coefficients is more readily obtained using pulse radiolysis where standard radiation dosimetry can estimate the radical yields to within about 5% accuracy. The radical anion is the more stable species in alkaline aqueous solutions, alcohol or in organic solutes like DMF. As demonstrated in Table 6 the values of λ_{max} for the

TABLE 6. Optical absorption data for semiquinones

Compound	$\dot{Q}H$		$Q^{-\cdot}$		Reference
	λ_{max}(nm)	ε($M^{-1}cm^{-1}$)	λ_{max}(nm)	ε($M^{-1}s^{-1}$)	
Benzoquinone	415	4700	425	6900	213
	410	5500	435	7100	298
2-Methylbenzoquinone	405	4500	430	6200	225
2,3-Dimethylbenzoquinone	415	5100	430	6700	225
2,5-Dimethylbenzoquinone	415	5000	435	7000	225
	415	3600	440	6800	295
2,6-Dimethylbenzoquinone	405	4900	430	6100	225
2,3,5-Trimethylbenzoquinone	410	4300	435	6700	225
Duroquinone	420	4700	440	7600	225
	425	4000	445	7100	295
1,4-Naphthoquinone	370	7300	390	13 000	225, 295
2-Methylnaphthoquinone	370	9500	390	12 500	225, 295
2,3-Dimethylnaphthoquinone	380	7300	400	11 000	225
2-Hydroxynaphthoquinone	370	5900	390	6300	295
Vitamin K_1	380	9900	400	10 200	225
Ubiquinone	425	5300	445	8600	225
	420	3000	445	8000	296
9,10-Anthraquinone	375	11 000	395	7800	213, 295
			480	7300	
Anthraquinone-1-sulphonate	385	12 000	400	8000	213, 295
			500	8000	
Anthraquinone-2-sulphonate	390	12 500	400	8000	213
			500	8000	
1,2-Naphthoquinone	< 260	16 000	265	40 000	295
4-*t*-Butyl-1,2-benzoquinone	290	7700	313	12 200	297
	390	1850			
Epinephrine	< 260	13 000	265	3300	295
Adrenalone	280	10 000	290	17 000	295

different solvents and methods are in good agreement although as would be expected there is greater variation in the estimates for ε.

The semiquinone radical anions all have λ_{max} at approximately 10–30 nm longer wavelength (red shifted) than the neutral species; the similarity between the two values is not unexpected since protonation causes little perturbation in the orbitals involved[213]. In general it has been observed that basic forms of free radicals absorb at lower energies than the neutral forms. The extinction coefficients are considerably higher in most cases for the anions than for the neutral radicals.

As is readily observed from Table 6 neither λ_{max} nor ε show much variation with substitution for the series of methyl-substituted benzosemiquinones or naphthosemiquinones. The anthrasemiquinone anions exhibit two bands in their absorption spectra in contrast to the single band for the anions of the benzo- or naphthosemiquinones.

Although the values obtained show fairly good agreement there are some differing observations made by flash photolysis experiments for durosemiquinone and ubiquinone. The neutral durosemiquinone radical has been estimated to have $\varepsilon_{420_{nm}}$ of 5500 ± 500 in ethanol and $\varepsilon_{407.5} = 8850\ M^{-1}cm^{-1}$ in cyclohexane[216]. The radicals of ubisemiquinones-0* and 10*, vitamin K_1 and plastoquinone-9 all exhibit very similar absorption

* The semiquinones of structure **9**, having $n = 0$, 10, respectively.

spectra[217]. Although these radicals can be prepared in alcoholic media, estimates of their values of ε and λ_{max} in aqueous solution were obtained by extrapolation of the behaviour of durosemiquinone from alcohol to aqueous solvent. For example, although ubiquinone has $\varepsilon_{275-280} = 7400\,\mathrm{M}^{-1}\,\mathrm{cm}^{-1}$ for $\mathrm{U\dot{Q}H}$ and $\varepsilon_{320} = 10\,700\,\mathrm{M}^{-1}\,\mathrm{cm}^{-1}$ for $\mathrm{UQ}^{-\cdot}$ in methanol, the predicted values for these neutral and anionic radicals in aqueous solution were $\varepsilon_{425} = 5300\,\mathrm{M}^{-1}\,\mathrm{cm}^{-1}$ and $\varepsilon_{445} = 8600\,\mathrm{M}^{-1}\,\mathrm{cm}^{-1}$ respectively. Similarly it was estimated that the neutral semiquinone of vitamin K_1 would have $\varepsilon_{380} = 9900\,\mathrm{M}^{-1}\,\mathrm{cm}^{-1}$ while the anion would have $\varepsilon_{400} = 10\,200\,\mathrm{M}^{-1}\,\mathrm{cm}^{-1}$ if they could be prepared in aqueous solution[213]. Values of λ_{max} for the neutral and anionic semiquinones of a related quinone, α-tocopherylquinone (**60**) are reported as 415 and 435 nm in ethanol respectively[127]. Although no extinction coefficients were reported the similarity of these spectra to those of the durosemiquinone radicals again is consistent with the small perturbation of the aromatic ring orbitals due to the long alkyl substituent.

Land and coworkers[218, 219] have suggested that in alkaline methanol solutions of 2-$\mathrm{NH_2}$-9,10-AQ substituted at the 4-position with NHMe, $\mathrm{NH_2}$ or OH, pulse radiolysis results in more than one quinone being lost for each reducing radical introduced into the system. The radical ions from these anthraquinones had extinction coefficients approximately double the values for the monosubstituted anthrasemiquinone anions. They interpret these observations to indicate that the quinone molecules may be aggregated in solution.

The absorption and emission–fluorescence characteristics of camphoroquinone and the corresponding camphorsemiquinone radical anion ($\mathrm{CQ}^{-\cdot}$) generated thermally with Na in ethanol have been determined. The parent quinone has $\lambda_{max} = 468$ nm while the $\mathrm{CQ}^{-\cdot}$ species has $\lambda_{max} = 324$ nm. At high concentrations the radical appears to dimerize likely as the species $(\mathrm{Na^+\,Q^{-\cdot}})_2$ and exhibits a strong absorption band at $\lambda_{max} = 266$ nm. The fluorescence wavelength bands for the three species CQ, $\mathrm{Q}^{-\cdot}$ and $(\mathrm{Na^+\,Q^{-\cdot}})_2$ are 459 nm, 390 nm and 290 nm respectively[212].

C. Redox Properties

Quinones are reduced to hydroquinones in two one-electron steps which can be described by the redox potentials; $E(\mathrm{Q/Q^{-\cdot}})$ for reduction to the semiquinone and $E(\mathrm{Q^{-\cdot}/Q^{2-}})$ for reduction to the hydroquinone. These quantities can be related by

$$E(\mathrm{Q^{-\cdot}/Q^{2-}}) = 2E(\mathrm{Q/Q^{2-}}) - E(\mathrm{Q/Q^{-\cdot}}) \tag{58}$$

where $E(\mathrm{Q/Q^{2-}})$ is the two-electron reduction potential of the quinone.

Values of $E(\mathrm{Q/Q^{-\cdot}})$ in aprotic solvents such as DMF have been extensively studied and are well described in the first edition of this series[220] and also in Ref. 221, and since a chapter on electrochemistry of quinones (Chapter 12) is included in this edition we will not discuss these data here. However, the redox chemistry of semiquinones in aqueous systems is of substantial biological significance and we will therefore collate some of the information applicable to aqueous systems in this section. Measurements of redox potentials in aqueous solutions are complicated by dependence on the pH and by the relative instability of the semiquinone radicals except in basic media. When only a single ionization of the semiquinone is involved the value of $E(\mathrm{Q/Q^{-\cdot}})$ depends on pH according to

$$E(\mathrm{Q/Q^{-\cdot}}) = E(\mathrm{Q/Q^{-\cdot}})_7 + 59\log\left(\frac{k_i + [\mathrm{H^+}]}{k_i + 10^{-7}}\right) \tag{59}$$

where k_i is the ionization constant. This has been used to calculate redox potentials at pH 7 for a variety of quinones measured at higher pHs. Swallow[213] has collected and discussed values from several authors[68, 69, 222–224]. Polarographic half-wave potentials have been found to be good approximations for equilibrium redox midpoint potentials since the electrochemical reduction of the quinones is usually reversible. The half-wave potentials have been qualitatively correlated with substituent and positional constants in Hammett equations[220].

From Table 7 it can be observed that methyl substitution in the *p*-benzoquinones decreases $E(Q/Q^{-\cdot})$ by about 85 mV. For the methoxy-substituted quinones a similar additive effect is not followed as was discussed (Section III.A). The values suggested to account for the observed redox changes in the methoxy-substituted compounds were $\Delta E = -101$ mV for a freely mobile methoxy group and $\Delta E = -41$ mV for a sterically hindered methoxy group[20]. A substituent effect of this magnitude is consistent with the standard midpoint potential for ubiquinone reported by Patel and Willson[225]. Swallow has estimated values of $E(Q/Q^{-\cdot})$ at pH 7 to be -230 ± 20 mV for ubiquinone and -130 ± 20 mV for plastoquinone[213]. These quinones can only be studied in aqueous media containing substantial concentrations of organic solutes.

TABLE 7. Reduction potentials of quinones in aqueous solutions

Compound	$E(Q/Q^{-\cdot})_7$	mV vs. NHE $E(Q/Q^{2-})_7$	$E(Q^{-\cdot}/Q^{2-})_7$
1,2-Benzoquinone	+210	+370	+530
1,4-Benzoquinone	+99	+286	+473
2-Methylbenzoquinone	+23	+230	+437
2,3-Dimethylbenzoquinone	−74	+175	+424
2,5-Dimethylbenzoquinone	−66	+180	+426
2,6-Dimethylbenzoquinone	−80	—	—
2,3,5-Trimethylbenzoquinone	−165	+114	+393
Duroquinone	−240	+57	+354
1,4-Naphthoquinone	—	+70	—
2-Methylnaphthoquinone	−203	−5	+193
2,3-Dimethylnaphthoquinone	−240	−70	—
Vitamin K_1^a	−170	−60	+220
Ubiquinone[a]	—	+90	—
1,4-Naphthoquinone-2-sulphonate	−60	+120	+300
9,10-Anthraquinone-2-sulphonate	−380	−228	−76
Anthraquinone-1,5-disulphonate	—	−170	—

[a] Values were measured in a solution containing isopropanol (5 M) and acetone (2 M). Otherwise solutions contained less than 1 M added solutes.

Steenken and Neta[222, 226] have measured redox potentials for a number of biologically important phenols, catechols and hydroquinone derivatives. From measured values of $E(Q^{-\cdot}/Q^{2-})$ at pH 13.5 for these systems they have calculated the values at pH 7 (see Table 8). They noted that hydroquinone and catechol have similar potentials while resorcinol is much less readily oxidized, having a potential 350 mV more positive. Conversion of O^- to OMe, giving 4-methoxyphenol, for example, also raises the redox potential. Several biologically important catechols such as norepinephrine, L-dopa, 5-hydroxydopamine and L-epicatechin have redox potentials similar to catechol; adrenalone is, however, deactivated by a carbonyl group.

Relative electron affinities (EA) for a variety of quinones have been obtained by measuring gas-phase equilibrium constants for the electron transfer reactions using pulsed ion cyclotron resonance[227]. Similar additive substituent effects (i.e. of -1.7 kcal mol^{-1} for each methyl group) are noted for the methyl-1,4-benzoquinones. However, chloro

TABLE 8. One-electron redox potentials

Compound	$E(Q^{-\cdot}/Q^{2-})$ pH 7	mV vs. NHE pH 13.5
Resorcinol	810	385
4-Methoxyphenol	600	400
Catechol	530	43
HTCC[a]	480	192
Hydroquinone	459	23
Ascorbate	300	15
Durohydroquinone	—	−54
2-Methoxyhydroquinone	—	−85

[a] HTCC = 2,2,5,7,8-pentamethylchroman-6-ol.

substituents do not affect EA linearly; an increase of 6.2 kcal mol^{-1} is noted for the introduction of the first chloro group while only 5.25 kcal mol^{-1} is observed when a second *o*- or *p*-substituent is introduced.

$$Q^{-} + R \rightarrow R^{-} + Q \qquad (60)$$

Comparisons of gas-phase electron affinities with those solution values calculated from polarographic half-wave potentials ($E_{1/2}$) and charge transfer spectroscopic measurements (E_{CT}) for quinones, nitrobenzenes and benzophenones showed reasonably good agreement[228]. The solution values were higher for low electron affinity compounds and lower for those of high EA. This likely reflects the changing solvation energy of the anion with changing substituent charge delocalization.

The half-wave oxidation and reduction potentials of tetrakis(dimethylamino)-*p*-benzoquinone were measured in DMF[149]. The strongly electron-donating substituents shifted the redox potentials substantially as shown below. This reflects the much lower ionization energy of the amino-containing quinone which allowed facile formation of its stable radical cation.

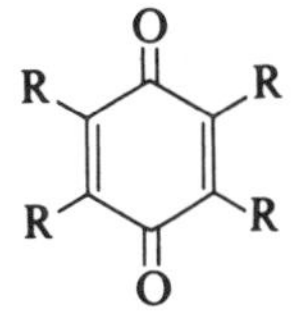

	$E^{ox}_{1/2}$	$E^{red}_{1/2}$
R=H	+2.85 V	−0.40 V
R=NMe$_2$	+0.25 V	−1.07 V

VI. OTHER SIGNIFICANT CHEMICAL AND BIOLOGICAL ASPECTS OF SEMIQUINONES

A. Semiquinones in Micelles

Micelles are dynamic associations of large numbers of surfactant molecules. They have been extensively studied both because their properties mimic many of those of natural biological membranes and also as effective cages for chemical reactions. The effects of external magnetic fields on primary photochemical processes have been studied using laser flash spectroscopy and time-resolved ESR spectroscopy. A recent review on magnetic field effects in micelles is available[229]. In homogeneous fluids these magnetic field effects are rather small due to the rapidity with which initially generated radicals escape from the primary solvent cage, however, the increased lifetime of the radical pair in the micelle greatly enhances the magnitude of the effects.

Generally micelles such as those consisting of sodium dodecyl sulphate (SDS) act as good H-donors in reactions with photoexcited carbonyl compounds such as benzophenone and the quinones. The reactions of a quinone in a micelle can be written as follows (equations 61–65).

$$Q \xrightarrow{h\nu} {}^{1}Q^{*} \xrightarrow{ISC} {}^{3}Q^{*} \quad (61)$$

$${}^{3}Q + RH \rightarrow {}^{3}(QH^{\cdot\cdot}R) \quad (62)$$

$${}^{3}(QH^{\cdot\cdot}R) \rightarrow {}^{1}(QH^{\cdot\cdot}R) \quad (63)$$

$${}^{1,3}(QH^{\cdot\cdot}R) \rightarrow \dot{Q}H + R^{\cdot} \quad (64)$$

$${}^{1}(QH^{\cdot\cdot}R) \rightarrow \text{products} \quad (65)$$

$$\dot{Q}H \rightleftharpoons Q^{-\cdot} + H^{+} \quad (66)$$

Laser irradiation of the quinone promotes it to its first excited singlet state from whence it rapidly intersystem crosses to the excited triplet state. Abstraction of a hydrogen from a detergent molecule by the triplet quinone forms the triplet radical pair (reaction 62). The mixing of the singlet and triplet spin states of the radical pair is influenced by the external magnetic field; the intersystem crossing rate may also be perturbed. The radicals escaping from the initial pair are assumed not to react unless they are in the singlet state. Some dissociation of the neutral semiquinone to the anion is to be expected since the pK_a values for the semiquinones are fairly acidic (~ 4–5). Disproportionation of two semiquinone radicals to form quinone and hydroquinone is disfavoured relative to the possibility of this reaction in solution, since the semiquinone radicals are isolated in the micellar matrix.

In general for the quinones the radical pair mechanism provides a reasonable explanation for the magnetic field effects in the micelles at low magnetic fields. Thus it has been found that for the photolysis of quinones in micelles the effect of increasing the magnetic field is to decrease the recombination rate and thus to increase the yield of escaping radicals[230–233]. These effects can be quite dramatic; Sakaguchi and Hayashi have noted[230] that for NQ (or 2-methyl-NQ) the increase in yield of escaping radicals at 1.34 $\times 10^4$ G compared to no external magnetic field is enhanced by a factor of 2.4 (3.1). The magnetic field effects for these quinones at fields below 0.1×10^4 G have been found to be much greater in magnitude than for benzophenone derivatives[230].

It was necessary, however, to invoke an alternative mechanism involving the relaxation of the odd electrons in the radical pairs to explain the magnetic field effects at higher fields[234, 235]. This mechanism has also been applied in other systems[236].

For NQ and menadione the magnetic field effects were observed to behave as follows:

(1) At zero applied magnetic field the radical pairs decayed exponentially at a rate of $4 \times 10^6\ s^{-1}$.

(2) At intermediate fields a biexponential decay was noted with one rate constant decreasing with increasing magnetic field. This decrease saturated at an applied field of 1.34×10^4 G at which point the radical pair decay rate was $6 \times 10^5\ s^{-1}$.

(3) The yields of escaping radicals increased with increasing magnetic field strengths[230].

The effects of the external magnetic field on ISC rates were described for the anthraquinones[231] and benzoquinones[232]; the triplet–singlet ISC rates decreased with increasing magnetic field.

It is difficult to determine the nature of the alkyl radical R generated by hydrogen abstraction from the detergent. In a CIDEP study it was noted that the time-resolved ESR spectrum of $\dot{R}$ could be analysed in terms of four hydrogens, indicating that abstraction had occurred somewhere other than the terminal regions of the SDS chain. In this study the fast hydrogen abstraction by ^{3}NQ (much faster than that of ^{3}BP) resulted in both the NQ̇H and Ṙ being observed initially in the emissive mode consistent with the TM of CIDEP polarization. Subsequent E/A RPM polarization was superimposed as the radical pairs evolved[237–239]. An attempt to identify the nature of the alkyl radical Ṙ was made by using the spin trap phenyl-*t*-butylnitrone (PBN). The resulting T-N(R)Ȯ adduct showed a magnetic field dependence; also the concentration of the adduct which is due to escaped Ṙ increased with increasing magnetic field[240].

The influence of paramagnetic ions, the lanthanides, on the course of photoreactions of quinones in micelles has also been investigated. Addition of these paramagnetic species is, in essence, the application of an internal magnetic field. Sakaguchi and Hayashi[241] have shown that the decay of the NQH˙˙R) pairs depends on the number of unpaired electron spins associated with the micelle. This effect is observed only in the presence of an external magnetic field. The magnitude of the effect decreased in the order $Gd^{3+} > Sm^{3+} > Dy^{3+} > Nd^{3+} > Ho^{3+} > Er^{3+} > La^{3+} > Lu^{3+}$. The increased decay rate is presumed to be effected by increased relaxation of the radical pair electron spins due to tumbling of the paramagnetic ions around the micelle. A similar phenomenon was noted for benzyl radical pairs in the photolysis of dibenzyl ketone when lanthanide ions were added[242].

B. Antitumour Antibiotics

Several of the anthraquinone antibiotics are effective antitumour agents. Some of the best known are adriamycin (**50**), daunomycin (**51**), carminomycin (**52**) and aclacinomycin A (**53**). Although the structures of these compounds are quite similar their biological

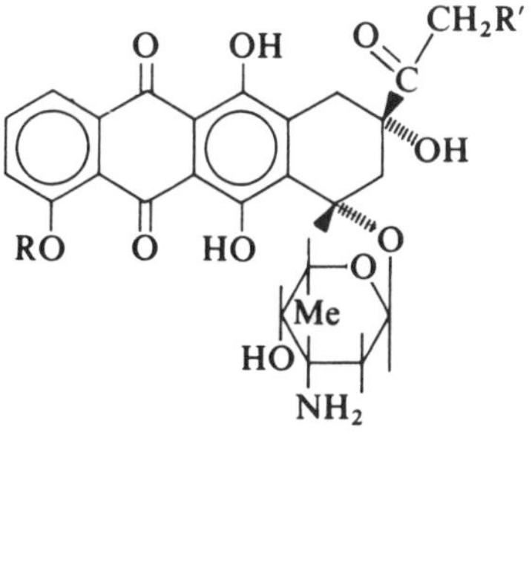

(50) R = Me, R′ = OH
(51) R = Me, R′ = H
(52) R = R′ = H

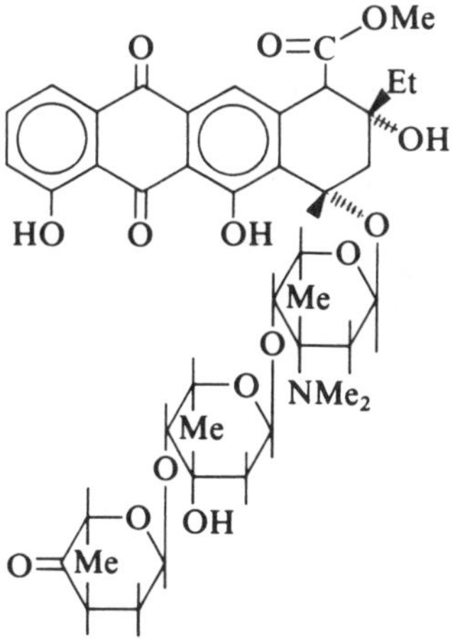

(53)

activities are diverse. Their pharmacological modes of action appear to be associated with suppression of nucleic acid and to some extent protein biosynthesis. This results in chromosomal defects in dividing cells. ESR studies of a spin-labelled derivative of adriamycin show that it forms a complex with DNA by intercalation between the purine and pyrimidine bases (Ref. 243 and references therein).

An alternative view of the molecular mechanisms of action of these compounds involves the formation of free radical intermediates. The possibility of either oxidation of the hydroquinone moiety or reduction of the quinone may provide some selectivity in their biological effects (i.e. antitumour activity relative to toxicity). The radicals formed by reduction of the anthracycline antibiotics have been studied by ESR[243–245] and indeed at physiological pH spontaneous formation of semiquinone radicals has been observed. A variety of enzymes are also known to catalyse the formation of such semiquinone free radicals. Other quinone-containing antibiotics such as mitomycin C, carboquinone and bruneomycin are polyfunctional and may act as alkylating or cross-linking agents towards DNA in addition to their capacity to form semiquinone radicals. Indeed activation of the anthracyclines by reduction may foster covalent attachment to biomolecules. Reduction of mitomycin C in DMSO with sodium tetrahydroborate gives a 36 line ESR spectrum of a stable semiquinone radical (**54**) having $a_{NH_2} = 2.3$ G, $a_{N\,ring} = 1.7$ G, $a_H^{Me} = 1.0$ G, $a_H^{CH_2} = 0.65$ G and $g = 2.0046$[246]; the semiquinone radical is also generated in microsomes. The alkylating agent AZQ (**55**) can also be reduced to semiquinone radicals in microsomes[247].

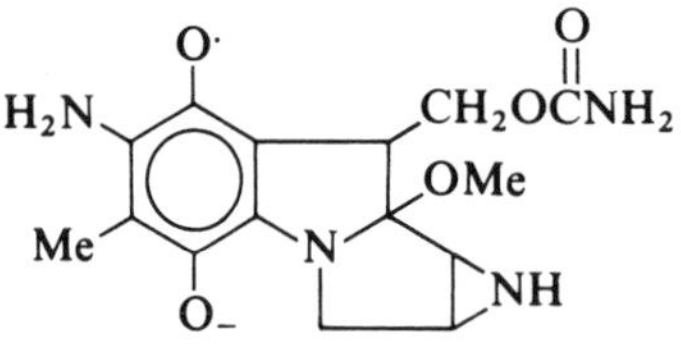

(**54**) 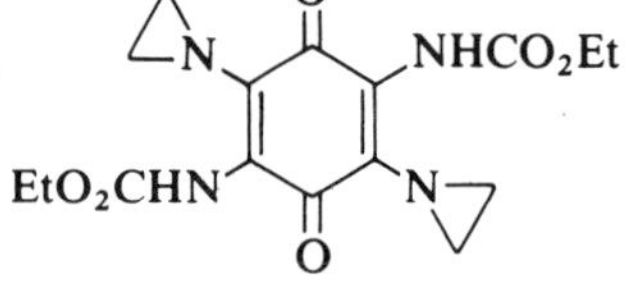 (**55**)

Several quinones which are model compounds for the anthracycline antibiotics have been studied; their semiquinone radicals are stable in basic media in the absence of oxygen and give well resolved ESR spectra[244, 248].

There is considerable evidence that reactive oxygen species: superoxide, peroxide and hydroxyl radical are generated in vivo as a result of the metabolism or pharmacological activity of the anthracycline antibiotics, possibly as shown in reactions 67–70. Superoxide ion has been demonstrated by spin trapping in microsomal incubations of these drugs[249].

$$Q^{-\cdot} + O_2 \rightarrow Q + O_2^{-\cdot} \qquad (67)$$

$$2O_2^{-\cdot} + 2H^+ \xrightarrow{SOD} H_2O_2 + O_2 \qquad (68)$$

$$2H_2O_2 \xrightarrow{catalase} 2H_2O + O_2 \qquad (69)$$

$$Q^{-\cdot} + H_2O_2 \rightarrow Q + OH^{-\cdot} + OH^{\cdot} \qquad (70)$$

Superoxide and peroxide have been established as intermediates in the action of anthracyclines on DNA by their selective removal by the enzymes superoxide dismutase (SOD) and catalase respectively[250]. The cardiotoxicity of these antibiotics may be related to the reduced concentration of these cell protective enzymes as well as glutathione peroxidase in the heart relative to other organs[251, 252].

A scheme for intracellular enzymatic reductive activation of the anthraquinone antibiotics and subsequent generation of active oxygen species is shown below[243, 249].

C. Photosynthesis

Quinones also play important roles in photosynthesis and respiratory electron transport chains; their activities in these systems involve the intermediacy of semiquinone radicals. Although ubiquinone and the plastoquinones are the biologically important quinones

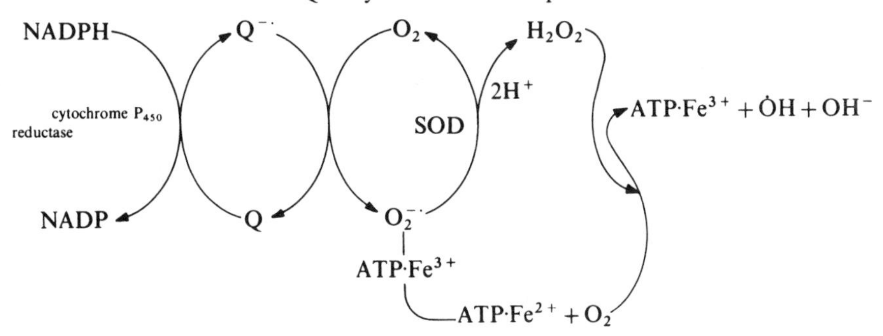

involved, chemical investigation of such biological systems has relied heavily on the use of model compounds in which simpler quinones are frequently incorporated. The literature on the chemical nature of both bacterial and plant photosynthesis is vast and complex and will not be discussed in detail here. Recent reviews are available[253–255]. Contributions of CIDMP to determination of the nature of the primary radical pairs involved in photosynthesis and the routes of subsequent electron transfer to other components of the reaction centre also have been discussed (Refs 254, 256–264, and references therein).

The primary event in photosynthesis involves oxidation of the lowest excited singlet state of a chlorophyll electron donor by a nearby electron acceptor. The distance between the two is restricted by a surrounding reaction centre protein. The initial reaction steps can be written as:

$$\mathrm{PIX} \rightarrow \mathrm{P^{*}IX} \rightarrow \mathrm{P^{+}I^{-}X} \rightarrow \mathrm{P^{+}IX^{-}} \rightarrow \ldots\ldots \tag{71}$$

where X is the first stable electron acceptor and P* is the first excited singlet state of the primary donor P. The primary photoreactants arise from an excited singlet state[265]. The lifetime of the initial radical pairs is approximately 200 ps but this can be lengthened to $\sim$ 15 ns by prereducing or removing the ubiquinone or menaquinone component X. The initial radical pair then decays by recombination of charges and the yields and reaction rates of the possible recombination processes can be affected by external applied magnetic fields (Ref. 266 and references therein).

Many model systems have been synthesized to study the intramolecular electron transfer processes between donors and acceptors fixed at certain distances and orientations relative to one another. Most often the donors are chlorophyll or porphyrin derivatives and the acceptors quinones (Refs 267–271 and references therein). For example, the photoinitiated electron transfer from tetratolylporphyrin to benzoquinone in a model compound linking these two moieties by a diester bridge was found to be most efficient when the bridge contained three methylene groups. A folded conformation minimizing the donor–acceptor separation was proposed for this structure[270]. A comparison of electron transfer rates in mesophenyloctaalkylporphyrins coupled to quinone via one or two bicyclo[2.2.2]octyl spacers (**56**) has been reported[272]. The effect of the second spacer in **56** is to decrease the electron transfer rate by a factor of 500–1600 depending on the solvent. In general electron transfer rates have been found to decrease with increasing solvent polarity in such porphyrin–quinone (P–Q) models.

Wasielewski and coworkers[270] have correlated rate constants for the forward electron transfer and recombination reactions with the free energies for the reactions using rigidly coupled P–Q complexes. The use of such complexes with constrained geometries can eliminate the problem of contributions from a variety of conformational isomers and provide better distance control between the donor and acceptor sites in the intramolecular electron transfer reactions.

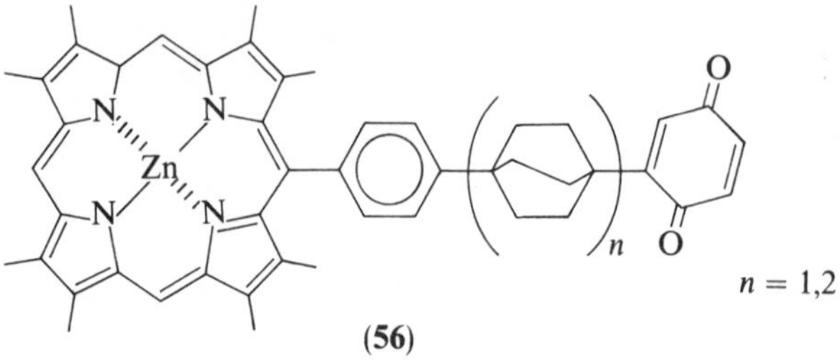

(56)

Further charge separation and sequential charge transfer can be studied in model compounds such as **57** in which a caroteneoid portion is linked to the porphyrin–quinone moiety[273]. Irradiation of this compound results in rapid electron transfer to $C–P^+–Q^-$ and then formation of $C^+–P–Q^-$, the charge separated intermediate. These reactions are extremely fast, occurring in less than 100 ps. The insertion of the neutral porphyrin molecule prolongs the lifetime to several μs and increases the quantum yield of the charged intermediate $C^+–P–Q^-$.

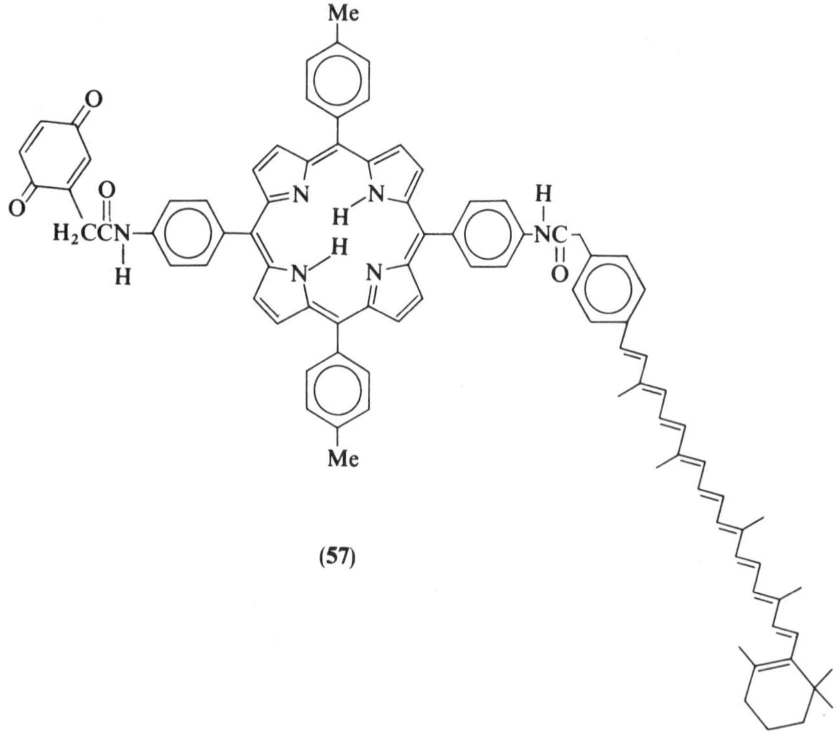

(57)

D. Interactions of Quinones with Active Oxygen Species

The interactions of quinones with oxygen and various reactive oxygen species such as $\dot{O}H$, $O_2^{-\cdot}$ and H_2O_2 may have considerable biological significance in a variety of systems. The toxicity of quinones has been suggested to be mediated by the formation of their semiquinone radicals and the subsequent production of reactive oxygen species[274, 275].

For example, menadione, which causes both DNA and cell membrane damage, is activated in hepatocytes to the semiquinone radical which can then be reoxidized to the quinone producing superoxide, peroxide and hydroxyl radical[276]. The neurotoxicity of 6-aminodopamine and 6-hydroxydopamine have been correlated with the reactions of their respective semiquinone or semiquinone–imine radicals[277] which produce activated oxygen species.

In addition to being produced by reactions of quinones with oxygen in biological systems, hydroxide ion, superoxide ion and singlet oxygen also can react with quinones and hydroquinones in a variety of ways. Superoxide ion has been suggested to react with catechols and hydroquinones first by protonation to form HO_2 followed by subsequent reactions which may involve charge transfer to form the semiquinone radical[278]. Pulse radiolysis of aqueous hydroquinones has been suggested to involve formation of the trihydroxycyclohexadienyl radical ($H_2\dot{Q}$–OH) which decays unimolecularly to the semiquinone radical ($H\dot{Q}$)[279]. An alternative mechanism for disappearance of H_2Q–OH involving a peroxy intermediate such as **58** has also been proposed to occur[279,280]. Hydroperoxides have also been identified as intermediates in reactions of vitamin E and vitamin K_1, especially in reactions with $^1O_2$[281–284]. It has been suggested that

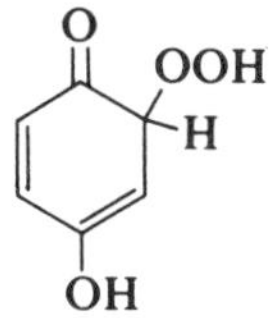

O OOH H OH

(58)

hydroperoxides of vitamin K_1 may be involved in the carboxylation of glutamic acid residues of blood proteins. Wilson and Tharp have shown that hydroperoxides of vitamin K_1 models can be readily formed from molecular oxygen and that subsequent ionic decomposition of these peroxides leads to active acylating species[285].

Reaction of hydroxide anion radical ($OH^{-\cdot}$) with anthraquinones in aprotic media has been shown to result in significant yields of the radicals ($AQ^{-\cdot}$) and the dioxygen species (H_2O_2, $^1O_2$ and $O_2^{-\cdot}$) through the initial formation of the addition complex [$AQ(OH)^{-\cdot}$] (**59**) which can react with additional AQ to give the semiquinone and the radical [$A\dot{Q}(OH)$][286]. Reactions such as these may be important in biomembranes.

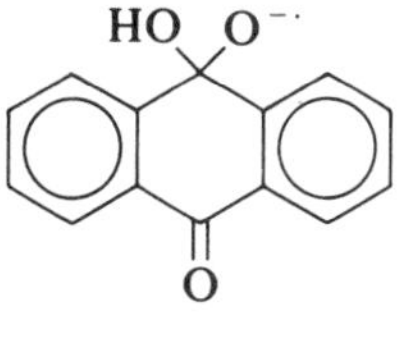

HO O$^{-\cdot}$ O

(59)

An ESR study of oxidation of a variety of catechols and naphthalenediols with H_2O_2 has shown that the oxidation occurs to give preferential introduction of the new oxygen centre *ortho* rather than *para* to the initial hydroxyl groups[287].

Although not itself a quinone vitamin E is known to react with quinones as well as a variety of biologically significant antioxidants such as vitamin C and glutathione. Although it functions primarily as an antioxidant, protecting lipids from peroxidation by scavenging peroxyl and alkoxyl radicals, $^1O_2$, and superoxide anion radical, vitamin E also has a variety of less well established roles. Its oxygen scavenging properties may be involved in its ability to protect cardiac tissue from the toxicity of the antitumour quinones adriamycin and daunomycin[275]. Vitamin E can react with $^1O_2$ either by reaction to form hydroperoxydienones which subsequently hydrolyse to stable products, or by a physical quenching process (Ref. 288 and references therein). Gorman and coworkers have recently suggested that reversible formation of an exciplex intermediate is involved in this quenching[289]. Photolysis of vitamin E in the presence of $^1O_2$ has been shown to produce, among other species, the α-tocopherylquinone **60**[281]. This quinone may also have

O

O

$C_{16}H_{33}$

(60)

considerable biological significance since it has been involved in the reactions of vitamin E with hepatotoxins such as CCl_4, and in the protection of lung tissue from airborne pollutants (Ref. 288 and references therein). The ESR parameters for the semiquinone radical derived from **60** have been described[21].

Photolysis of vitamin E in the presence of quinones such as benzoquinone results in its rapid oxidation to the chromanoxyl radical E . Time-resolved CIDEP experiments show total emissive polarization by the TM for this radical indicating that the oxidant is the triplet quinone and that the reaction is very fast[288, 290]. However, quinones such as menadione and vitamin K_1 which have lower redox potentials do not photooxidize vitamin E in alcohol solutions. Steenken has suggested a value of 0.48 V for the redox potential of vitamin E, similar to that measured for the model compound 2,2,5,7,8-pentamethylchroman-6-ol[226]. The presence of a variety of different quinones in lipophilic environments in which vitamin E is biologically active suggests that further investigation of their interactions might be productive.

VII. ACKNOWLEDGEMENTS

The authors are grateful to the Natural Sciences and Engineering Research Council of Canada for financial support. It is a pleasure to acknowledge the many delightful and enlightening discussions with many of our past and present colleagues of this laboratory.

VIII. REFERENCES

1. T. Sakurai, *Acta Crystallogr.*, **B24**, 403 (1968).
2. T. Sakurai, *Acta Crystallogr.*, **19**, 320 (1965).
3. R. Foster and M. I. Foreman, in *The Chemistry of the Quinonoid Compounds* (Ed. S. Patai), John Wiley and Sons, Chichester, 1974, p. 257.
4. J. Bernstein, M. D. Cohen and L. Leiserowitz, in *The Chemistry of the Quinonoid Compounds* (Ed. S. Patai), John Wiley and Sons, Chichester, 1974, p. 37.
5. A. I. Brodskii, I. P. Gragerov and L. V. Pisarzhevskii, *Dokl. Akad. Nauk SSSR*, **79**, 277 (1951).
6. G. R. Desiraju, D. Y. Curtin and I. C. Paul, *J. Org. Chem.*, **42**, 4071 (1977).
7. A. O. Patil, D. Y. Curtin and I. C. Paul, *J. Am. Chem. Soc.*, **106**, 348 (1984).
8. M. A. Slifkin and R. H. Walmsley, *Spectrochim. Acta*, **26A**, 1237 (1970).

9. A. O. Patil, D. Y. Curtin and I. C. Paul, *J. Am. Chem. Soc.*, **106**, 4010 (1984).
10. A. O. Patil, S. R. Wilson, D. Y. Curtin and I. C. Paul, *J. Chem. Soc. Perkin Trans. 2*, 1107 (1984).
11. J. R. Scheffer, Y. F. Wong, A. O. Patil, D. Y. Curtin and I. C. Paul, *J. Am. Chem. Soc.*, **107**, 4898 (1985).
12. K. K. Kalnin'sh and V. V. Shchukareva, *Bull. Acad. Sci. USSR Div. Chem. Sci.*, **34**, 90 (1985).
13. K. K. Kalnin'sh, *J. Chem. Soc. Faraday Trans. 2*, **78**, 327 (1982).
14. D. Y. Curtin and I. C. Paul, *Chem. Revs*, **81**, 525 (1981).
15. G. R. Desiraju, D. Y. Curtin and I. C. Paul, *Mol. Cryst. Liq. Cryst.*, **52**, 259 (1979).
16. J. A. Pedersen, *CRC handbook of EPR Spectra from Quinones and Quinols*, CRC Press Inc., Florida, USA, 1985, 382 pages.
17. J. D. R. Clay and D. Murphy, *J. Chem. Soc. Perkin Trans. 2*, 1781 (1984).
18. P. Ashworth, *Mol. Phys.*, **30**, 313 (1975).
19. B. Venkataraman, B. G. Segal and G. K. Fraenkel, *J. Chem. Phys.*, **30**, 1006 (1959).
20. P. R. C. Gascoyne and A. Szent-Gyorgyi, *Int. J. Quant. Chem.: Quant. Biol. Symp. II*, 217 (1984).
21. M. R. Das, H. D. Connor, D. S. Leniart and J. H. Freed, *J. Am. Chem. Soc.*, **92**, 2258 (1970).
22. D. M. Holton and D. Murphy, *J. Chem. Soc. Perkin Trans. 2*, 1757 (1980).
23. S. Steenken and P. O'Neill, *J. Phys. Chem.*, **81**, 505 (1977).
24. C. Sieiro, A. Sanchez and J. Castaner, *Spectrochimica Acta*, **41A**, 987 (1985).
25. W. T. Dixon and D. Murphy, *J. Chem. Soc. Perkin Trans. 2*, 1823 (1976).
26. D. M. Holton and D. Murphy, *J. Chem. Soc. Faraday Trans. 2*, **75**, 1185 (1979).
27. A. D. McLachlan, *Mol. Phys.*, **3**, 233 (1960).
28. W. T. Dixon, P. M. Kok and D. Murphy, *J. Chem. Soc. Faraday Trans. 2*, **73**, 709 (1977).
29. W. T. Dixon, M. Moghimi and D. Murphy, *J. Chem. Soc. Faraday Trans. 2*, **70**, 1713 (1974).
30. M. Karplus and G. K. Fraenkel, *J. Chem. Phys.*, **35**, 1312 (1961).
31. J. R. Bolton, in *Radical Ions* (Eds E. T. Kaiser and L. Kevan), Interscience, N.Y., 1968, p. 1.
32. M. R. Das and G. K. Fraenkel, *J. Chem. Phys.*, **42**, 1350 (1965).
33. P. D. Sullivan and J. R. Bolton, *J. Am. Chem. Soc.*, **90**, 5366 (1968).
34. C. A. Claxton and D. McWilliams, *Trans. Faraday Soc.*, **64**, 2593 (1968).
35. M. Broze and Z. Luz, *J. Chem. Phys.*, **51**, 749 (1969).
36. B. S. Prabhananda, *J. Chem. Phys.*, **79**, 5752 (1983).
37. H. L. Strauss and G. K. Fraenkel, *J. Chem. Phys.*, **35**, 1738 (1961).
38. R. W. Fessenden, *J. Phys. Chem.*, **71**, 74 (1967).
39. J. A. Pedersen, *J. Mag. Res.*, **60**, 136 (1984).
40. C. C. Felix and R. C. Sealy, *J. Am. Chem. Soc.*, **104**, 1555 (1982).
41. D. R. Eaton, *Inorg. Chem.*, **3**, 1268 (1964).
42. G. Vincow, *J. Chem. Phys.*, **38**, 917 (1963).
43. J. A. Pople, D. L. Beveridge and P. A. Dobosh, *J. Am. Chem. Soc.*, **90**, 4201 (1968).
44. Y. Shinagawa and Y. Shinagawa, *J. Am. Chem. Soc.*, **100**, 67 (1978).
45. Y. Shinagawa, Y. Shinagawa, N. U. Yesaka and K. Fukuda, *Int. J. Quant. Chem.*, **18**, 357 (1980).
46. J. Spanget-Larsen, *Int. J. Quant. Chem.*, **18**, 365 (1980).
47. J. A. Pedersen and J. Spanget-Larsen, *Chem. Phys. Lett.*, **35**, 41 (1975).
48. K. Kuwata and Y. Shimizu, *Bull. Chem. Soc. Jpn*, **42**, 864 (1969).
49. Y. Shinagawa, S. Koyama and Y. Shinagawa, *Int. J. Quant. Chem.*, **23**, 477 (1983).
50. A. Takuwa, O. Soga and K. Maruyama, *J. Chem. Soc. Perkin Trans. 2*, 409 (1985).
51. R. C. Sealy, L. Harman, P. R. West and R. P. Mason, *J. Am. Chem. Soc.*, **107**, 3401 (1985).
52. C. C. Felix and R. C. Sealy, *J. Am. Chem. Soc.*, **103**, 2831 (1981).
53. A. J. Stone, *Mol. Phys.*, **6**, 509 (1963).
54. C. C. Felix and B. S. Prabhananda, *J. Chem. Phys.*, **80**, 3078 (1984).
55. P. D. Sullivan, J. R. Bolton and W. E. Geiger, Jr, *J. Am. Chem. Soc.*, **92**, 4176 (1970).
56. B. S. Prabhananda, C. C. Felix, J. S. Hyde and A. Walvekar, *J. Chem. Phys.*, **83**, 6121 (1985).
57. A. F. Beecham, A. C. Hurley and C. H. I. Johnson, *Aust. J. Chem.*, **33**, 699 (1980).
58. P. R. Taylor, *J. Am. Chem. Soc.*, **104**, 5248 (1982).
59. D. M. Holton and D. Murphy, *J. Chem. Soc. Faraday Trans. 1*, **78**, 1223 (1982).
60. T. Gough, *Trans. Faraday Soc.*, **62**, 2321 (1966).
61. W. T. Dixon, D. M. Holton and D. Murphy, *J. Chem. Soc. Faraday Trans. 2*, **74**, 521 (1977).
62. K. Loth and F. Graf, *Helv. Chim. Acta*, **64**, 1910 (1981).

63. K. Loth, F. Graf and H. H. Gunthard, *Chem. Phys. Lett.*, **45**, 191 (1977).
64. G. Krishnamoorthy and B. S. Prabhananda, *J. Chem. Phys.*, **76**, 108 (1982).
65. W. T. Dixon and D. Murphy, *J. Chem. Soc. Faraday Trans. 2*, **72**, 135 (1976).
66. P. J. Hore and K. A. McLauchlan, *Mol. Phys.*, **42**, 1009 (1981).
67. P. J. Hore, K. A. McLauchlan, S. Frydkjaer and L. T. Muus, *Chem. Phys. Lett.*, **77**, 127 (1981).
68. D. Meisel and G. Czapski, *J. Phys. Chem.*, **79**, 1503 (1975).
69. D. Meisel and R. W. Fessenden *J. Am. Chem. Soc.*, **98**, 7505 (1976).
70. R. L. Ward and S. I. Weissman, *J. Am. Chem. Soc.*, **79**, 2086 (1957).
71. S. Frydkjaer and L. T. Muus, *Chem. Phys.*, **51**, 335 (1980).
72. L. T. Muus, S. Frydkjaer and K. Bondrup Nielsen, *Chem. Phys.*, **30**, 163 (1978).
73. A. Kanemoto, S. Niizuma, S. Konishi and H. Kokubun, *Bull. Chem. Soc. Jpn*, **56**, 46 (1983).
74. H. M. Vyas, S. K. Wong, B. B. Adeleke and J. K. S. Wan, *J. Am. Chem. Soc.*, **97**, 1385 (1975).
75. K. Y. Choo and J. K. S. Wan, *J. Am. Chem. Soc.*, **97**, 7127 (1975).
76. J. B. Pedersen, C. E. M. Hansen, H. Parbo and L. T. Muus, *J. Chem. Phys.*, **63**, 2398 (1975).
77. A. D. Trifunac and J. R. Norris, *Chem. Phys. Lett.*, **59**, 140 (1978).
78. B. B. Adeleke and J. K. S. Wan, *J. Chem. Soc. Faraday Trans. 1*, **72**, 1799 (1976).
79. A. J. Elliot and J. K. S. Wan, *J. Phys. Chem.*, **82**, 444 (1978).
80. K. A. McLauchlan and G. R. Sealy, *Mol. Phys.*, **52**, 783 (1984).
81. R. Baer and H. Paul, *Chem. Phys.*, **87**, 73 (1984).
82. P. J. Hore and K. A. McLauchlan, *Mol. Phys.*, **42**, 533 (1981).
83. J. W. M. DeBoer, T. Y. C. Chan Chung and J. K. S. Wan, *Can. J. Chem.*, **57**, 2971 (1979).
84. D. A. Hutchinson, J. Russell, J. W. M. DeBoer and J. K. S. Wan, *Chem. Phys.*, **53**, 149 (1980).
85. D. A. Hutchinson, M. C. Depew, K. E. Russell and J. K. S. Wan, *Macromolecules*, **15**, 602 (1982).
86. R. Ullman, *Macromolecules*, **14**, 746 (1981).
87. C. Friedrich, F. Laupretre, C. Noel and L. Monnerie, *Macromolecules*, **14**, 1119 (1981).
88. D. A. Hutchinson, K. S. Chen, J. Russell and J. K. S. Wan, *J. Chem. Phys.*, **73**, 1862 (1980).
89. J. W. M. DeBoer, K. S. Chen, Y. C. Chan Chung and J. K. S. Wan, *J. Am. Chem. Soc.*, **101**, 5425 (1979).
90. K. S. Chen, J. K. S. Wan and J. K. Kochi, *J. Phys. Chem.*, **85**, 1726 (1981).
91. R. W. Fessenden, J. P. Hornak and B. Venkataraman, *J. Chem. Phys.*, **74**, 3694 (1981).
92. D. M. Bartels, R. G. Lawler and A. D. Trifunac, *J. Chem. Phys.*, **83**, 2686 (1985).
93. S. S. Kim and S. I. Weissman, *J. Mag. Res.*, **24**, 167 (1976).
94. S. S. Kim and S. I. Weissman, *Rev. Chem. Intermed.*, **3**, 107 (1979).
95. H. Murai, T. Hayashi and Y. J. I'Haya, *Chem. Phys. Lett.*, **106**, 139 (1984).
96. H. Murai, M. Minami, T. Hayashi and Y. J. I'Haya, *Chem. Phys.*, **93**, 333 (1985).
97. D. Weir and J. K. S. Wan, *J. Am. Chem. Soc.*, **106**, 427 (1984).
98. K. Akiyama, S. Tero-Kubota, Y. Ikegami and T. Ikenoue, *J. Am. Chem. Soc.*, **106**, 8322 (1984).
99. H. Murai, T. Imamura and K. Obi, *J. Phys. Chem.*, **86**, 3279 (1982).
100. H. Murai, T. Imamura and K. Obi, *Chem. Phys. Lett.*, **87**, 295 (1982).
101. R. Z. Sagdeev, W. Mohl and K. Mobius, *J. Phys. Chem.*, **87**, 3183 (1983).
102. F. Lendzian, P. Jaegermann and K. Mobius, *Chem. Phys. Lett.*, **120**, 195 (1985).
103. F. J. Adrian, *Rev. Chem. Intermed.*, **1**, 3 (1979).
104. G. L. Closs, R. J. Miller and O. D. Redwine, *Acc. Chem. Res.*, **18**, 196 (1985).
105. B. Brocklehurst, *Int. Rev. Phys. Chem.*, **4**, 279 (1985).
106. J. K. S. Wan, in *Advances in Photochemistry*, Vol. 12 (Eds J. N. Pitts, Jr, G. S. Hammond and K. Gollnick), J. Wiley and Sons, Inc., N.Y., 1980, p. 283.
107. F. J. Adrian, H. M. Vyas and J. K. S. Wan, *J. Chem. Phys.*, **65**, 1454 (1976).
108. H. M. Vyas and J. K. S. Wan, *Can. J. Chem.*, **54**, 979 (1976).
109. R. S. Hutton, H. D. Roth, M. L. M. Schilling, A. M. Trozzolo and T. M. Leslie, *J. Am. Chem. Soc.*, **104**, 5878 (1982).
110. B. B. Adeleke and J. K. S. Wan, *Spectrosc. Lett.*, **10**, 871 (1977).
111. V. M. Kuznets, D. N. Shigorin, A. L. Buchachenko and A. Z. Yankelevich, *Dokl. Akad. Nauk. SSSR*, **253**, 585 (1980).
112. E. G. Bagryanskaya, Yu. A. Grishin, R. Z. Sagdeev, T. V. Leshina, N. E. Polyakov and Yu. N. Molin, *Chem. Phys. Lett.*, **117**, 220 (1985).
113. I. V. Khudyakov, A. I. Prokof'ev, L. A. Margulis and V. A. Kuzmin, *Chem. Phys. Lett.*, **104**, 409 (1984).

114. N. K. Bridge and G. Porter, *Proc. R. Soc. Lond. Ser. A*, **244**, 276 (1958).
115. Y. Kambara and H. Yoshida, *Bull. Chem. Soc. Jpn*, **50**, 1367 (1977).
116. S. Noda, T. Doba, T. Mizuta, M. Miura and H. Yoshida, *J. Chem. Soc. Perkin Trans. 2*, 61 (1980).
117. K. A. McLauchlan and R. C. Sealy, *J. Chem. Soc. Chem. Commun.*, 115 (1976).
118. R. Scheerer and M. Gratzel, *J. Am. Chem. Soc.*, **99**, 865 (1977).
119. S. K. Wong, *J. Am. Chem. Soc.*, **100**, 5488 (1978).
120. H. Kobashi, Y. Tomioka and T. Morita, *Bull. Chem. Jpn*, **52**, 1568 (1979).
121. H. Kobashi, T. Nagumo and T. Morita, *Chem. Phys. Lett.*, **57**, 369 (1978).
122. H. Kobashi, M. Funabashi, T. Kondo, T. Morita, T. Okada and N. Mataga, *Bull. Chem. Soc. Jpn*, **57**, 3557 (1984).
123. S. Niizuma, N. Sato, H. Kawata, O. Murakami, A. Kanemoto and H. Kokubun, *Bull. Chem. Soc. Jpn*, **58**, 2769 (1985).
124. M. T. Craw, A. Alberti, M. C. Depew and J. K. S. Wan, *Bull. Chem. Soc. Jpn*, **58**, 3675 (1985).
125. S. Basu, K. A. McLauchlan and G. R. Sealy, *Chem. Phys. Lett.*, **88**, 84 (1982).
126. E. Kam, M. T. Craw, M. C. Depew and J. K. S. Wan, *J. Mag. Reson.*, **67**, 556 (1986).
127. G. Leary and G. Porter, *J. Chem. Soc. A*, 2273 (1970).
128. G. G. Lazarev, Ya. S. Lebedev, A. I. Prokof'ev and R. R. Rakhimov, *Chem. Phys. Lett.*, **95**, 262 (1983).
129. G. Porter and P. Suppan, *Trans. Faraday Soc.*, **61**, 1664 (1965).
130. S. J. Formosinho, *J. Chem. Soc. Faraday Trans. 2*, **74**, 1978 (1978).
131. D. Kemp and G. Porter, *Proc. R. Soc. Lond. A.*, **326**, 117 (1971).
132. K. Tickle and F. Wilkinson, *Trans. Faraday Soc.*, **61**, 1981 (1965).
133. H. Inoue, K. Ikeda, H. Mihara, M. Hida, N. Nakashima and K. Yoshihara, *Chem. Phys. Lett.*, **95**, 60 (1983).
134. K. Hamanoue, K. Yokoyama, Y. Kajiwara, K. Nakajima, T. Nakayama and H. Teranishi, *Chem. Phys. Lett.*, **110**, 25 (1984).
135. K. Hamanoue, K. Nakajima, Y. Kajiwara, T. Nakayama and H. Teranishi, *Chem. Phys. Lett.*, **110**, 178 (1984).
136. S. G. Cohen, A. Parola and G. H. Parsons, Jr, *Chem. Revs*, **73**, 141 (1973).
137. E. Amouyal and R. Bensasson, *J. Chem. Soc. Faraday Trans. 1*, **73**, 1561 (1977).
138. H. D. Roth, in *Chemically Induced Magnetic Polarization* (Eds L. T. Muus, P. W. Atkins, K. A. McLauchlan and J. P. Pedersen), D. Reidel Publishing Co., Dordrecht, Holland, 1977, pp. 39–76.
139. K. Hamanoue, K. Yokoyama, Y. Kajiwara, M. Kimoto, T. Nakayama and H. Teranishi, *Chem. Phys. Lett.*, **113**, 207 (1985).
140. K. Hamanoue, T. Nakayama, K. Sugiura, H. Teranishi, M. Washio, S. Tagawa and Y. Tabata, *Chem. Phys. Lett.*, **118**, 503 (1985).
141. K. S. Chen and J. K. S. Wan, *Spectrosc. Lett.*, **12**, 647 (1979).
142. T. Foster, A. J. Elliot, B. B. Adeleke and J. K. S. Wan, *Can. J. Chem.*, **56**, 869 (1978).
143. S. Farid, *J. Chem. Soc. Chem. Commun.*, 303 (1970).
144. S. Farid, *J. Chem. Soc. Chem. Commun.*, 73 (1971).
145. H. M. Vyas, B. B. Adeleke and J. K. S. Wan, *Spectrosc. Lett.*, **9**, 663 (1976).
146. M. C. Depew, B. B. Adeleke, A. Rutter and J. K. S. Wan, *Can. J. Chem.*, **63**, 2281 (1985).
147. T. DoMinh, *Abstr. 65th Can. Chem. Conf.*, **PH**, 22 (1982); U.S. Patent 4308341 (1982).
148. P. D. Sullivan, *J. Phys. Chem.*, **75**, 2195 (1970).
149. H. Bock, P. Hanel, W. Kaim and U. Lechner-Knoblauch, *Tetrahedron Lett.*, **26**, 5115 (1985).
150. H. Chandra and M. C. R. Symons, *J. Chem. Soc. Chem. Commun.*, 29 (1983); 1044 (1984).
151. M. T. Craw, M. C. Depew and J. K. S. Wan, *Can. J. Chem.*, **64**, 1414 (1986).
152. M. T. Craw, M. C. Depew and J. K. S. Wan, *Phosphorus and Sulfur*, **25**, 369 (1985).
153. M. T. Craw, M. C. Depew and J. K. S. Wan, *J. Mag. Res.*, **65**, 339 (1985).
154. J. Dannenberg, *Angew. Chem. Int. Ed. Engl.*, **14**, 641 (1975).
155. U. Svanholm and V. D. Parker, *Tetrahedron Lett.*, 471 (1972).
156. M. C. Depew, L. Zhongli and J. K. S. Wan, *Spectrosc. Lett.*, **16**, 451 (1983).
157. M. C. Depew, L. Zhongli and J. K. S. Wan, *J. Am. Chem. Soc.*, **105**, 2480 (1983).
158. J. L. Courtneidge, A. G. Davies and S. N. Yazdi, *J. Chem. Soc. Chem. Commun.*, 570 (1984).
159. W. Gordy, in *Theory and Application of ESR* (Ed. W. West), J. Wiley and Sons, N.Y., 1980, pp. 516–522.
160. K. A. M. Creber and J. K. S. Wan, *Chem. Phys. Lett.*, **81**, 453 (1981).

161. K. A. M. Creber, T.-I. Ho, M. C. Depew, D. J. Weir and J. K. S. Wan, *Can. J. Chem.*, **60**, 1504 (1982).
162. K. A. M. Creber and J. K. S. Wan, *Can. J. Chem.*, **61**, 1017 (1983).
163. K. A. M. Creber and J. K. S. Wan, *Tran. Metl. Chem.*, **8**, 253 (1983).
164. T.-I. Ho, K. A. M. Creber and J. K. S. Wan, *J. Am. Chem. Soc.*, **103**, 6524 (1981).
165. E. K. Runge, *Bioorg. Khim.*, **3**, 787 (1977).
166. C. A. Wraight, *Biochim. Biophys. Acta*, **459**, 525 (1977).
167. C. A. Wraight, *FEBS Lett.*, **93**, 283 (1978).
168. C. J. Carrano and K. N. Raymond, *Acc. Chem. Res.*, **12**, 183 (1979).
169. K. A. M. Creber, K. S. Chen and J. K. S. Wan, *Rev. Chem. Intermed.*, **5**, 37 (1984).
170. C. G. Pierpont and R. M. Buchanan, *Coord. Chem. Rev.*, **38**, 45 (1981).
171. T. E. Gough and P. R. Hindle, *Can. J. Chem.*, **47**, 3393 (1969).
172. B. Kaempf, S. Raynal, A. Collet, F. Schue, S. Boileau and J. Lehn, *Angew. Chem. Int. Ed. Engl.*, **13**, 611 (1974).
173. M. A. Komarynsky and S. I. Weissman, *J. Am. Chem. Soc.*, **97**, 1589 (1975).
174. K. S. Chen and J. K. S. Wan, *J. Am. Chem. Soc.*, **100**, 6051 (1978).
175. K. S. Chen and J. K. S. Wan, *Chem. Phys. Lett.*, **57**, 285 (1978).
176. G. R. Stevensen and A. Alegria, *J. Phys. Chem.*, **78**, 1771 (1974).
177. A. Alegria, R. Concepción and G. R. Stevenson, *J. Phys. Chem.*, **79**, 361 (1975).
178. L. Echegoyen, I. Nieves and G. R. Stevenson, *J. Phys. Chem.*, **86**, 1611 (1982).
179. D. Plancherel and A. Von Zelewsky, *Helv. Chim. Acta*, **65**, 1929 (1982).
180. H. B. Stegmann, H. U. Bergler and K. Scheffler, *Angew. Chem. Int. Ed. Engl.*, **20**, 389 (1981).
181. H. B. Stegmann, K. B. Ulmschneider and K. Scheffler, *J. Organomet. Chem.*, **101**, 145 (1975).
182. H. B. Stegmann, K. Scheffler and P. Schuler, *Angew. Chem. Int. Ed. Engl.*, **17**, 365 (1978).
183. K. S. Chen, R. T. Smith and J. K. S. Wan, *Can. J. Chem.*, **56**, 2503 (1978).
184. H. B. Stegmann, W. Uber and K. Scheffler, *Tetrahedron Lett.*, 2697 (1977).
185. A. G. Davies, D. Griller, B. P. Roberts and J. C. Scaiano, *Chem. Commun.*, 196 (1971).
186. A. G. Davies, Z. Florjańczyk, E. Lusztyk and J. Lusztyk, *J. Organomet. Chem.*, **229**, 215 (1982).
187. T. Foster, K. S. Chen and J. K. S. Wan, *J. Organomet. Chem.*, **184**, 113 (1980).
188. J. K. Kochi, K. S. Chen and J. K. S. Wan, *Chem. Phys. Lett.*, **73**, 557 (1980).
189. K. Mochida, J. K. Kochi, K. S. Chen and J. K. S. Wan, *J. Am. Chem. Soc.*, **100**, 2927 (1978).
190. G. A. Razuvaev, V. A. Tsarjapkin, L. V. Gorbunova, V. K. Cherkasov, G. A. Abakumov and E. S. Klimov, *J. Organomet. Chem.*, **174**, 47 (1979).
191. A. Alberti, A. Hudson and G. F. Pedulli, *J. Chem. Soc. Faraday Trans. 2*, **76**, 948 (1980).
192. K. S. Chen, T. Foster and J. K. S. Wan, *J. Chem. Soc. Perkin Trans. 2*, 1288 (1979).
193. A. Alberti and A. Hudson, *J. Chem. Soc. Perkin Trans. 2*, 1098 (1978).
194. S. Emori, D. Weir and J. K. S. Wan, *Chem. Phys. Lett.*, **84**, 512 (1981).
195. P. J. Barker, A. G. Davies, J. A. A. Hawari and M. W. Tse, *J. Chem. Soc. Perkin Trans. 2*, 1488 (1980).
196. A. Alberti and A. Hudson, *Chem. Phys. Lett.*, **48**, 331 (1977).
197. A. Alberti and A. Hudson, *J. Organomet. Chem.*, **164**, 219 (1979).
198. A. Alberti, A. Hudson and G. F. Pedulli, *Tetrahedron*, **38**, 3749 (1982).
199. V. A. Roginskii and V. A. Belyakov, *Dokl. Akad. Nauk. SSSR*, **237**, 1404 (1977).
200. A. Alberti, A. Hudson, G. F. Pedulli, W. G. McGimpsey and J. K. S. Wan, *Can. J. Chem.*, **63**, 917 (1985).
201. A. I. Prokof'ev, N. P. Provotorova, N. A. Kardanov, N. N. Bubnov, S. P. Solodovnikov, N. N. Godovikov and M. I. Kabachnikov, *Bull. Acad. Sci. USSR Div. Chem. Sci.*, **31**, 1534 (1982).
202. A. R. Forrester and R. H. Thomson, *Z. Naturforsc. Teilb.*, **40**, 1515 (1985).
203. P. E. Barker, A. Hudson and R. A. Jackson, *J. Organomet. Chem.*, **208**, C1 (1981).
204. A. I. Prokof'ev, T. I. Prokof'eva, N. N. Bubnov, S. P. Solodovnikov, I. S. Belostotskaya, V. V. Ershov and M. I. Kabachnik, *Dokl. Akad. Nauk. USSR*, **234**, 276 (1977).
205. A. I. Prokof'ev, N. N. Bubnov, S. P. Solodovnikov and M. I. Kabachnik, *Dokl. Akad. Nauk. SSSR*, **245**, 178 (1979).
206. A. I. Prokof'ev, T. I. Prokof'eva, I. S. Belostotskaya, N. N. Bubnov, S. P. Solodovnikov, V. V. Ershov and M. I. Kabachnik, *Dokl. Akad. Nauk. USSR*, **246**, 244 (1979).
207. K. S. Chen, R. Smith, M. A. Singer and J. K. S. Wan, *Bull. Environ. Contam. Toxicol.*, **20**, 443 (1978).
208. D. Weir, D. A. Hutchinson, J. Russell and J. K. S. Wan, *Can. J. Chem.*, **60**, 703 (1982).

209. A. Alberti, F. P. Colonna, X. Li and M. C. Depew, *J. Organomet. Chem.*, **292**, 335 (1985).
210. B. B. Adeleke, K. S. Chen and J. K. S. Wan, *J. Organomet. Chem.*, **208**, 317 (1981).
211. W. G. McGimpsey and J. K. S. Wan, *J. Photochem.*, **22**, 87 (1983).
212. S. Sur, T.-I. Ho and J. K. S. Wan, unpublished results,.
213. A. J. Swallow, in *Function of Quinones in Energy Conserving Systems* (Ed. B. L. Trumpower), Academic Press Inc., N.Y., 1982, pp. 59–73.
214. A. J. Swallow, *Prog. React. Kinet.*, **9**, 195 (1978).
215. E. J. Land and G. Porter, *Proc. Chem. Soc. Lond.*, 84 (1960).
216. E. Amouyal and R. Bensasson, *J. Chem. Soc. Faraday Trans. 1*, 1274 (1976).
217. R. Bensasson and E. J. Land, *Biochim. Biophys. Acta*, **325**, 175 (1973).
218. E. McAlpine, R. S. Sinclair, T. G. Truscott and E. J. Land, *J. Chem. Soc. Faraday Trans. 1*, **74**, 597 (1978).
219. E. J. Land, E. McAlpine, R. S. Sinclair and T. G. Truscott, *J. Chem. Soc. Faraday Trans. 1*, **72**, 2091 (1976).
220. J. Q. Chambers, in *The Chemistry of the Quinonoid Compounds*, Part II (Ed. S. Patai), Wiley, N.Y., 1974, p. 737.
221. R. C. Prince, M. R. Gunner and P. L. Dutton, in *Functions of Quinones in Energy Conserving Systems* (Ed. B. L. Trumpower), Academic Press, N.Y., 1982, pp. 29–33.
222. S. Steenken and P. Neta, *J. Phys. Chem.*, **83**, 1134 (1979).
223. Y. A. Ilan, G. Czapski and D. Meisel, *Biochim. Biophys. Acta*, **430**, 209 (1976).
224. P. Wardman and E. D. Clarke, *J. Chem. Soc., Faraday Trans. 1*, **72**, 1377 (1976).
225. K. B. Patel and R. L. Willson, *J. Chem. Soc. Faraday Trans. 1*, **69**, 814 (1973).
226. S. Steenken and P. Neta, *J. Phys. Chem.*, **86**, 3661 (1982).
227. E. K. Fukuda and R. T. McIver, Jr, *J. Am. Chem. Soc.*, **107**, 2291 (1985).
228. E. P. Grimsrud, G. Caldwell, S. Chowdhury and P. Kebarle, *J. Am. Chem. Soc.*, **107**, 4627 (1985).
229. N. J. Turro, *Proc. Natl. Acad. Sci. USA,* **80**, 609 (1983).
230. Y. Sakaguchi and H. Hayashi, *J. Phys. Chem.*, **88**, 1437 (1984).
231. Y. Tanimoto, H. Udagawa and M. Itoh, *J. Phys. Chem.*, **87**, 724 (1983).
232. Y. Tanimoto, H. Udagawa, Y. Katsuda and M. Itoh, *J. Phys. Chem.*, **87**, 3976 (1983).
233. Y. Tanimoto and M. Itoh, *Chem. Phys. Lett.*, **83**, 626 (1981).
234. H. Hayashi and S. Nagakura, *Bull. Chem. Soc. Jpn,* **57**, 322 (1984).
235. Y. Sakaguchi and H. Hayashi, *Chem. Phys. Lett.*, **87**, 539 (1982).
236. T. Ulrich and U. E. Steiner, *Chem. Phys. Lett.*, **112**, 365 (1984).
237. Y. Sakaguchi, H. Hayashi, H. Murai and Y. J. I'Haya, *Chem. Phys. Lett.*, **110**, 275 (1984).
238. K. A. McLauchlan and D. G. Stevens, *Chem. Phys. Lett.*, **115**, 108 (1985).
239. H. Hayashi, Y. Sakaguchi, H. Murai and Y. J. I'Haya, *Chem. Phys. Lett.*, **115**, 111 (1985).
240. M. Okazaki, S. Sakata, R. Konaka and T. Shiga, *J. Am. Chem. Soc.*, **107**, 7214 (1985).
241. Y. Sakaguchi and H. Hayashi, *Chem. Phys. Lett.*, **106**, 420 (1984).
242. N. J. Turro, X. Lei, I. R. Gould and M. B. Zimmt, *Chem. Phys. Lett.*, **120**, 397 (1985).
243. N. M. Emanuel, G. N. Bogdanov and V. S. Orlov, *Russ. Chem. Rev.*, **53**, 1121 (1984).
244. J. W. Lown and H.-H. Chen, *Can. J. Chem.*, **59**, 3212 (1981).
245. J. W. Lown and H.-H. Chen, *Can. J. Chem.*, **59**, 390 (1981).
246. J. W. Lown, S.-K. Kim and H.-H. Chen, *Can. J. Biochem.*, **56**, 1042 (1978).
247. P. L. Gutierris, R. D. Friedman and N. R. Bachur, *Cancer Treat. Repts*, **66**, 340 (1982).
248. N. J. F. Dodd and T. Mukherjee, *Biochem. Pharm.*, **33**, 379 (1984).
249. B. Kalyanaraman, E. Perez-Reyes and R. P. Mason, *Biochim. Biophys. Acta*, **630**, 119 (1980).
250. B. K. Sinha, *Chem. Biol. Interact.*, **30**, 67 (1980).
251. T. H. Doroshow, G. Y. Locker and C. E. Myers, *J. Clin. Invest.*, **65**, 128 (1980).
252. J. Goodman and P. Hochstein, *Biochem. Biophys. Res. Commun.*, **77**, 797 (1977).
253. R. K. Clayton, *Photosynthesis: Physical Mechanisms and Chemical Patterns*, Cambridge University Press, Cambridge, 1980.
254. R. E. Blankenship, *Acc. Chem. Res.*, **14**, 163 (1981).
255. R. E. Blankenship and W. W. Parson, *Ann. Rev. Biochem.*, **47**, 635 (1978).
256. P. Hore, in *Primary Photo-Processes in Biology and Medicine* (Eds R. V. Bensasson, G. Jori, E. J. Land and T. G. Truscott), Plenum Press, N.Y., 1985, pp. 131–145.
257. A. R. McIntosh and J. R. Bolton, *Rev. Chem. Intermed.*, **1**, 121 (1979).
258. J. Tang and J. R. Norris, *Chem. Phys. Lett.*, **94**, 77 (1983).

259. S. G. Boxer, C. E. D. Chidsey and M. G. Roelofs, *J. Am. Chem. Soc.*, **104**, 1452 (1982).
260. S. G. Boxer, C. E. D. Chidsey and M. G. Roelofs, *J. Am. Chem. Soc.*, **104**, 2674 (1982).
261. M. C. Thurnauer and J. R. Norris, *Chem. Phys. Lett.*, **76**, 557 (1980).
262. M. E. Michel-Beyerle, H. Scheer, H. Seidlitz, D. Tempus and R. Haberkorn, *FEBS Lett.*, **100**, 9 (1979).
263. M. K. Bowman, D. E. Budil, G. L. Closs, A. G. Kostka, C. A. Wraight and J. R. Norris, *Proc. Natl. Acad. Sci. USA*, **78**, 3305 (1981).
264. A. J. Hoff and P. J. Hore, *Chem. Phys. Lett.*, **108**, 104 (1984).
265. A. J. Hoff, P. Gast and J. C. Romijn, *FEBS Letts.*, **73**, 185 (1977).
266. K. W. Moehl, E. J. Lous and A. J. Hoff, *Chem. Phys. Lett.*, **121**, 22 (1985).
267. Y. Degani and I. Willner, *J. Phys. Chem.*, **89**, 5685 (1985).
268. Y. Sakata, S. Nishitani, S. Nishimizu, A. Misumi, A. R. McIntosh, J. R. Bolton, Y. Kanda, A. Karen, T. Okada and N. Mataga, *Tetrahedron Lett.*, **26**, 5207 (1985).
269. J. A. Schmidt, A. Siemiarczuk, A. C. Weedon and J. R. Bolton, *J. Am. Chem. Soc.*, **107**, 6112 (1985).
270. M. R. Wasielewski, M. P. Niemczyk, W. A. Svec and E. B. Pewitt, *J. Am. Chem. Soc.*, **107**, 1080 (1985).
271. J. H. Fendler, *J. Phys. Chem.*, **89**, 2730 (1985).
272. B. A. Leland, A. D. Joran, P. M. Felker, J. J. Hopfield, A. H. Zewail and P. B. Dervan, *J. Phys. Chem.*, **89**, 5571 (1985).
273. S. Nishitani, N. Karata, Y. Sakata, S. Misumi, A. Karen, T. Okada and N. Matuga, *J. Am. Chem. Soc.*, **105**, 7771 (1983).
274. C. Lind, P. Hochstein and L. Ernster, *Arch. Biochem. Biophys.*, **216**, 178 (1982).
275. B. N. Ames, *Science, N.Y.*, **221**, 1256 (1983).
276. H. Morrison, B. Jernstrom, M. Nordenskjold, H. Thor and S. Orrenius, *Biochem. Pharmacol.*, **33**, 1763 (1984).
277. E. Perez-Reyes and R. P. Mason, *Mol. Pharm.*, **18**, 594 (1980).
278. D. T. Sawyer, M. J. Gibian, M. M. Morrison and E. Y. Seo, *J. Am. Chem. Soc.*, **100**, 627 (1978).
279. A. A. Al-Suhybani and G. Hughes, *Z. Physik. Chem. Neue Folge*, **141**, 229 (1984).
280. J. A. Pedersen, *J. Chem. Soc. Perkin Trans. 2*, 424 (1973).
281. G. W. Grams, K. Eskins and G. Inglett, *J. Am. Chem. Soc.*, **94**, 866 (1972).
282. L. A. Paquette, F. Bellamy, M. C. Bohm and R. Gleiter, *J. Org. Chem.*, **45**, 4913 (1980).
283. R. W. Wilson, T. F. Walsh and R. Whittle, *J. Am. Chem. Soc.*, **104**, 4162 (1982).
284. R. L. Clough, B. G. Yee and C. S. Foote, *J. Am. Chem. Soc.*, **101**, 683 (1979).
285. R. M. Wilson and G. Tharp, *J. Am. Chem. Soc.*, **107**, 4100 (1985).
286. J. L. Roberts, Jr, H. Sugimoto, W. C. Barrette, Jr and D. T. Sawyer, *J. Am. Chem. Soc.*, **107**, 4556 (1985).
287. P. Ashworth and W. T. Dixon, *J. Chem. Soc. Perkin Trans. 2*, 739 (1974).
288. M. T. Craw and M. C. Depew, *Rev. Chem. Intermed.*, **6**, 1 (1985).
289. A. A. Gorman, I. R. Gould, I. Hamblett and M. C. Standen, *J. Am. Chem. Soc.*, **106**, 6956 (1984).
290. J. K. S. Wan, *J. Photochem.*, **17**, 517 (1981).
291. H. Veenvliet and D. A. Wiersma, *Chem. Phys.*, **8**, 432 (1985).
292. S. J. Sheng and M. A. El-Sayed, *Chem. Phys. Lett.*, **34**, 216 (1975).
293. T. Murao and T. Azumi, *J. Chem. Phys.*, **70**, 4460 (1979).
294. R. L. Willson, *Trans. Faraday Soc.*, **67**, 3020 (1971).
295. P. S. Rao and E. Hayon, *J. Phys. Chem.*, **77**, 2274 (1973).
296. E. J. Land and A. J. Swallow, *J. Biol. Chem.*, **245**, 1890 (1970).
297. H. W. Richter, *J. Phys. Chem.*, **83**, 1123 (1979).
298. G. E. Adams and B. D. Michael, *Trans. Faraday Soc.*, **63**, 1171 (1967).

The Chemistry of Quinonoid Compounds, Vol. II
Edited by S. Patai and Z. Rappoport

CHAPTER **17**

Heterocyclic quinones

RICHARD W. MIDDLETON
The Cancer Research Campaign's Gray Laboratory, Northwood, Middlesex, UK and Brunel University, Uxbridge, Middlesex UB8 3PH, UK
JOHN PARRICK
Brunel University, Uxbridge, Middlesex UB8 3PH, UK

I. INTRODUCTION

This survey of the chemistry of heterocyclic quinones gives consideration to those compounds which have a quinonoid nucleus fused to a heterocycle. Thus, compounds such as **1** and **2** are included whereas **3** is not within the scope of this chapter except as an intermediate from which a heterocyclic quinone might be formed. A subgroup of

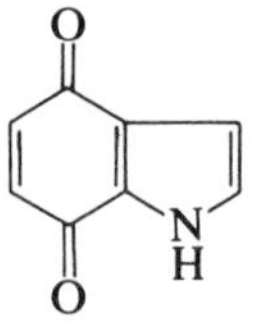

(**1**) 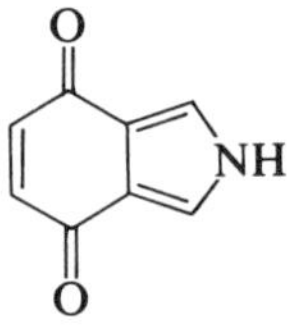 (**2**) 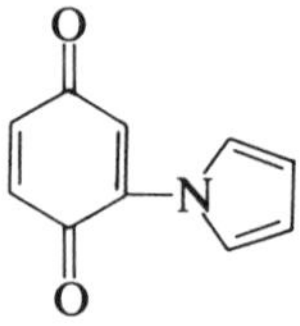 (**3**)

compounds in which the heterocycle is fused to a quinonoid nucleus and includes an atom of the heterocycle as part of the quinonoid system, e.g. **4**, have also been excluded from

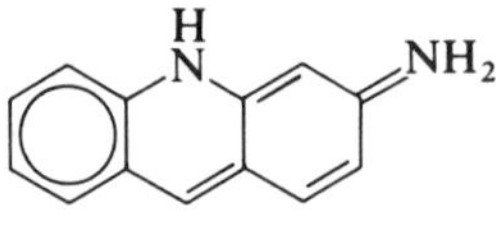

(**4**)

consideration, except where they occur as intermediates or as the reaction products of a heterocyclic quinone as defined above. Compounds having a 1,2- or 1,4-dicarbonyl system in an unsaturated heterocycle as part of their structure have also been excluded from consideration since they do not display properties especially recognizable as quinonoid. On the other hand consideration is given to the chemistry of some representatives of quinone imines and quinone methides (see Section IV) and diazaquinones (see Section V).

Emphasis has been placed in this survey on recent developments in the field of heterocyclic quinone chemistry, and there have been many. The information which follows has been distilled from consultation of in excess of 450 references, the vast majority of which are later than 1970. Much of the stimulation for work in this area has come from the discovery of naturally occurring heterocyclic quinones having potentially important biological properties. In one case, that of isoindole, the first representative of the heterocycle found in nature proved to be a quinone derivative. The occurrence of these compounds has encouraged synthetic work designed to produce the heterocyclic quinone system in an efficient manner in order to effect a total synthesis of the naturally occurring quinone or to produce a series of analogues of the bioactive compound for subsequent structure-activity analysis. In those cases where some understanding of the biochemical mechanism of action of a quinone has been established there has often been a surge of synthetic activity. This is well exemplified in the case of the antitumour antibiotic, mitomycin, and the idea of bioreductive activation of certain quinones (see Section III.D). The indole, quinoline, benzofuran and benzopyran quinones have been the most extensively investigated while very little work has been done on, for instance, cinnoline, phthalazine or benzoxazole quinones. In general, about half the recent work on heterocyclic quinones, as judged by the number of references, has been devoted to those containing one or more nitrogen atoms.

This survey of the preparations and properties of heterocyclic quinones is illustrative rather than comprehensive and is arranged so that the usefulness of reagents and reactions are considered in sequence rather than the alternative arrangement of giving consideration to each type of heterocyclic quinone system in sequence. We have endeavoured to take illustrative examples from a wide range of heterocyclic quinone types and to provide recent references so that we hope the reader will be able to find a useful lead reference to a particular quinonoid system of interest. The discussion here is mainly concerned with recent work and this does mean that earlier work may be neglected. A small number of reviews of earlier work is available and of particular interest at this point are the more widely based reviews of Baxter and Davis[1] who concentrate their attention on synthetic routes but also make mention of some applications of heterocyclic quinones and give references to earlier reviews; of Horspool[2] who describes the synthesis and thermal reactions of *ortho*-quinones including some heterocyclic quinones, though the emphasis is on carbocyclic systems; and of Sartori[3] who describes the heterocyclic quinones obtainable from 2,3-dichloro-1,4-naphthoquinone. It is worth mentioning here that some of the work quoted in Sartori's review has been shown recently[4] to contain erroneous structural assignments. In addition, monographs on particular ring systems may contain mention of the quinonoid derivatives.

II. PREPARATIVE ROUTES

A. Oxidation of Benzoheteroarenes

1. General

It is probably true to say that most preparations of heterocyclic quinones involve the oxidation of the suitably substituted carbocycle of a benzoheteroarene. The reagent of choice depends upon the ease of oxidation and stability of the particular heterocycle under investigation, but the classical reagents such as dichromate, Fremy's salt (potassium nitrosodisulphonate), and nitric acid still find wide application. The more recently introduced combination of silver(II) oxide and nitric acid and ceric ammonium nitrate (CAN) are finding increasing applications. Interestingly, the use of the supported reagent silver carbonate does not seem to have been reported in the heterocyclic quinone field though it is an efficient reagent for the formation of carbocyclic quinones[5] and has the virtue of ease of use. The uses of the frequently employed oxidizing agents are described briefly.

2. Oxidants and their uses

a. Dichromate This vigorous and strongly acidic reagent has obvious limitations on its usefulness but it has been found to be effective for the oxidation of 5,8-dihydroxy-[6] and 5,8-diamino-quinolines[7], in the latter case even in the presence of a 6-methoxyl group. Similar success was achieved with 5,7-diaminothiazole[8] but, on the other hand, strong nitric acid was found to be more useful than dichromate for the oxidation of a 8-hydroxy-5-methoxycoumarin[9].

b. Nitric acid or nitrous acid Nitric acid has been used extensively to oxidize coumarins (here it seems to be the reagent of choice[9]), quinolines, isoquinolines, benzimidazoles, benzothiazoles and benzothiophenes to *p*-quinones. The reagent can also produce *o*-quinones from the corresponding dimethoxy compounds[10]. Nitration of the benzene system is a possible complicating factor in these reactions when several electron-donating substituents are present. The oxidation of the quinoline ester **5** to **6** proceeds smoothly at room temperature in the presence of ceric ammonium nitrate, but at 0°C in the presence of nitric acid the hydroxynitroquinolone quinone **9** is the major product from the acid **7** together with **8** and a small quantity of a monodemethylated product[11]. However, 4,7-

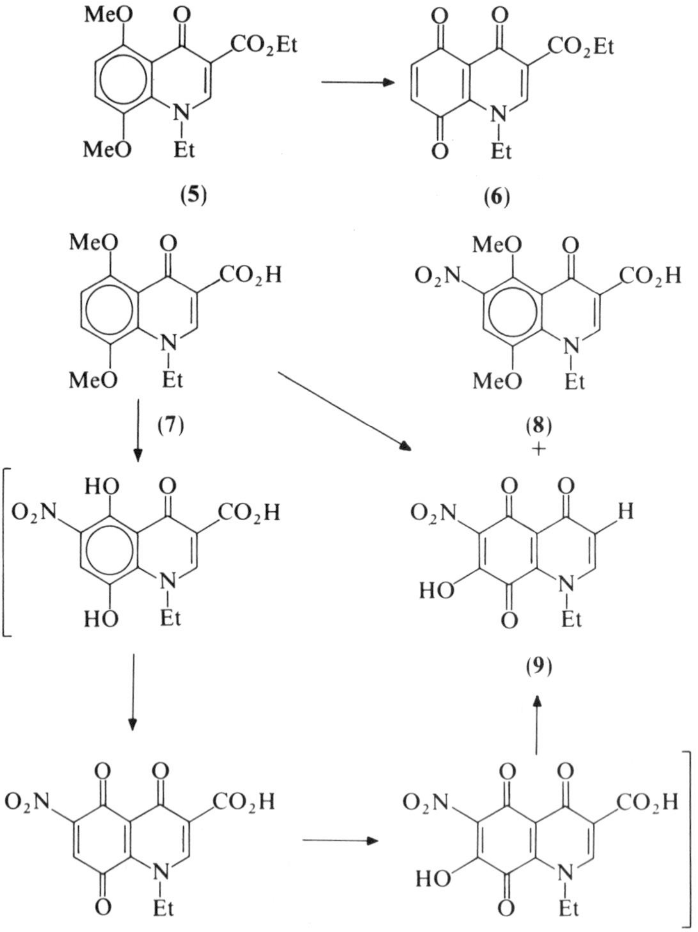

MeO O CO_2Et N MeO Et

(**5**)

O O CO_2Et N O Et

(**6**)

MeO O CO_2H N MeO Et

(**7**)

MeO O O_2N CO_2H N MeO Et

(**8**)

+

HO O O_2N CO_2H N HO Et

O O O_2N H HO N O Et

(**9**)

O O O_2N CO_2H N O Et

O O O_2N CO_2H HO N O Et

dihydroxyindoles carrying an electron-withdrawing 3-substituent have been oxidized to the corresponding quinone with strong nitric acid[12].

Nitrous acid has been used to obtain the mitosene derivative **11** by oxidative demethylation of the 4,5,7-trimethoxyindole derivative **10**. The *o*-quinone **12** was also produced in low yield[13]. The 2,5,6-benzimidazolintrione **13** was first prepared by the nitrous acid oxidation of 5,6-dimethoxy-2-benzimidazolinone[14].

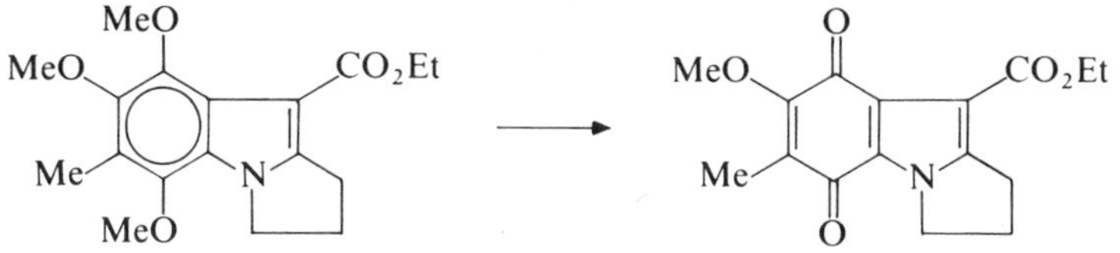

(**10**) (**11**) 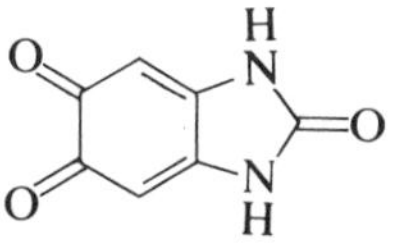 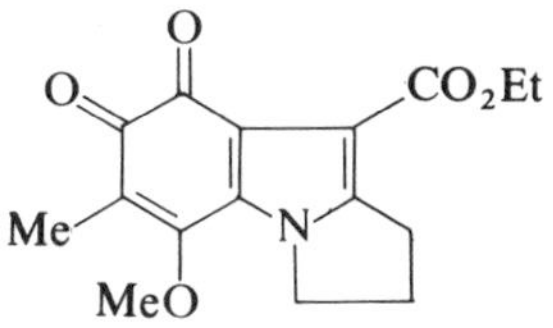 + (**13**) (**12**)

c. Dioxygen (air) Air has been used to produce heterocyclic quinones from suitable precursors fairly infrequently but both **14** and **15** have been oxidized to *p*-quinones by the action of air and, in each case, the amino group *ortho* to a hydroxyl group was replaced by a hydroxyl group either in the oxidation process or the workup procedure. However, the subtle variation in structure between **14** and **15** caused major changes in the conditions

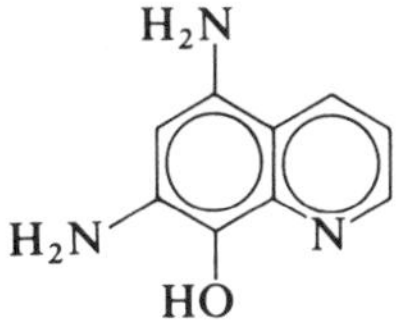

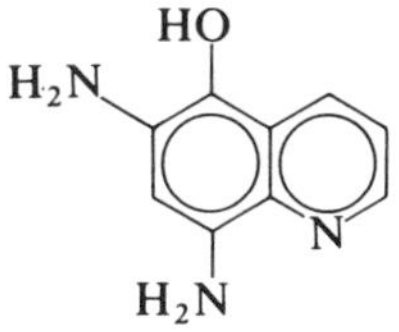

(**14**) (**15**)

necessary for a successful oxidation: **14** required basic conditions whereas **15** yielded the quinone smoothly only under acidic conditions[15]. Copper(II) compounds were used by Russian workers[16] to promote the oxidation of 6-hydroxybenzothiazole to the corresponding 6,7-quinone and, in part of their extensive investigation of heterocyclic quinone imines, copper(II) acetate was used[17] in the oxidation of **16** to **17**. It is thought that 5,6-

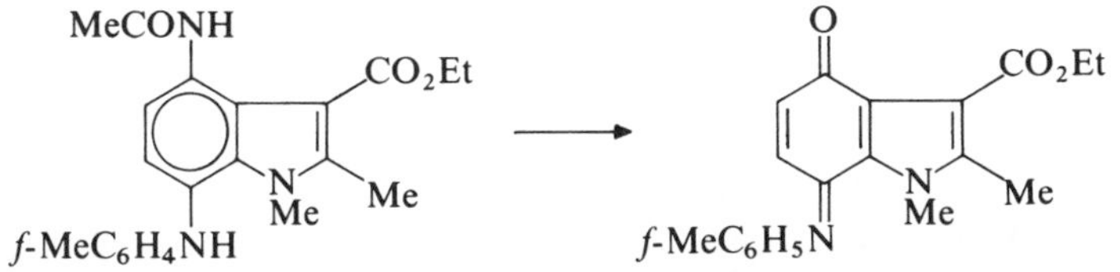

(**16**) (**17**)

dihydroxyindoles are immediate precursors of the skin pigment, melanin, and this idea has stimulated a number of investigations of the autoxidation and oxidation of these compounds. This work has been reviewed[18].

The tetrahydroquinoline-2,5-dione (**18**) has been oxidized with oxygen in the presence of potassium *tert*-butoxide to the hydroxyquinolinetrione (**19**); a similar reaction occurs in the naphthalene series[19].

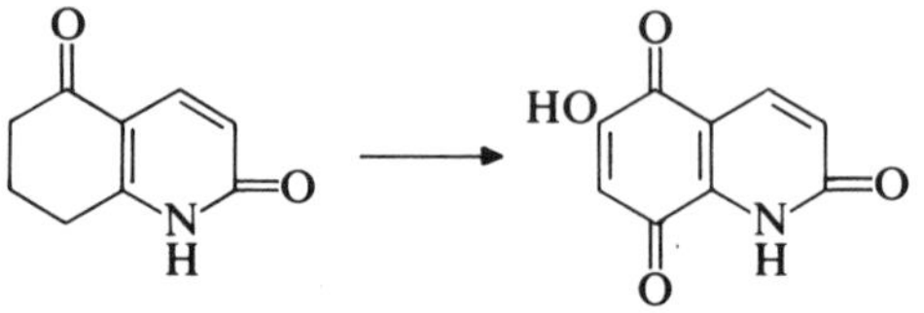

(**18**) (**19**)

d. Fremy's salt (potassium nitrosodisulphonate) Probably the most widely used oxidizing agent in this field. For instance, 4-,5- and 7-hydroxyindoles and 4-aminoindoles have been converted into the corresponding 4,7-quinones. In the case of 7-hydroxy-2,3-diphenylindole, both the 4,7- and 6,7-quinone were formed but the *N*-methyl derivative gave only the 6,7-quinone[20]. The efficiency of the oxidation of amines may be markedly affected by the pH of the reaction mixture. The oxidation of 5-amino-6-methoxyquinazoline[21] or 6-methoxy-8-hydroxyquinolines[22] gave the corresponding 5,8-quinones. Interesting possibilities for the formation of isomeric quinones arise when the quinone precursor contains both 1,4- and 1,2-related oxidizable substituents in the carbocycle. There seems to have been no systematic study of the effects of either the presence of various groups or of the choice of oxidizing agents on the type of quinone formed in any heterocyclic system.

e. Silver oxides Silver(I) oxide has been used to convert 1,4-dihydroxy compounds into the *p*-quinones: for instance, the conversion of **20** to **21**[22]. The introduction of nitric acid

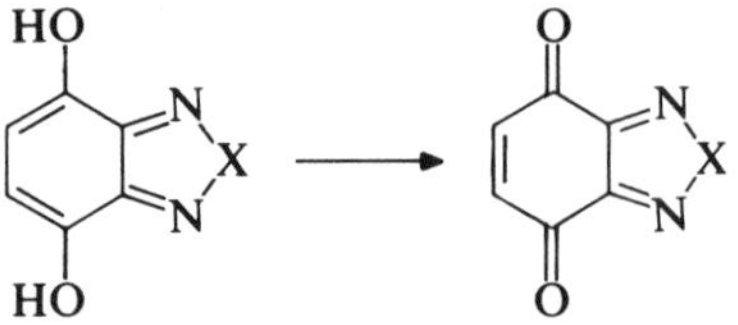

X = S or CH=CH

(**20**) (**21**)

and silver(II) oxide as a combination of reagents causing ether splitting and oxidation in a one-pot process was an important development[23]. The method has found increasing application in the area of heterocyclic chemistry and is illustrated here by the penultimate step in the synthesis[24] of kalafungin (**22**), a member of a growing number of pyranonaphthoquinone antibiotics which can be considered as derivatives of juglone. Kalafungin is the first example that has been mentioned here of a heterocyclic quinone where the heteroatom is β to the quinonoid nucleus and the heterocyclic moiety is not an aromatic nucleus. A similar oxidation step was also used in the synthesis of (+)-deoxygriseusin B, the characterization of which led to the revision of an earlier structure for (−)-griseusin B. The revised structure is **23**[25].

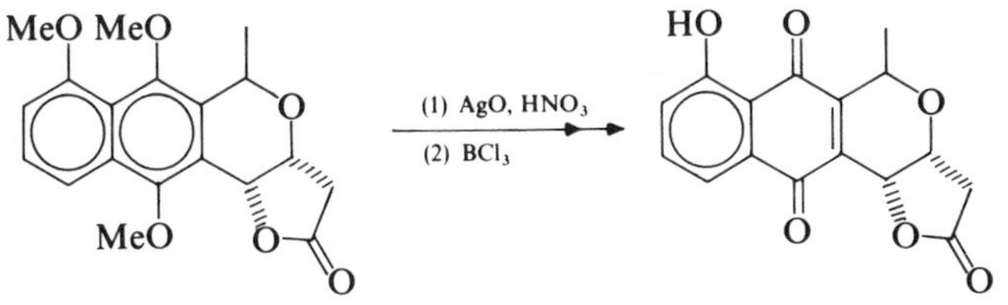

(22)

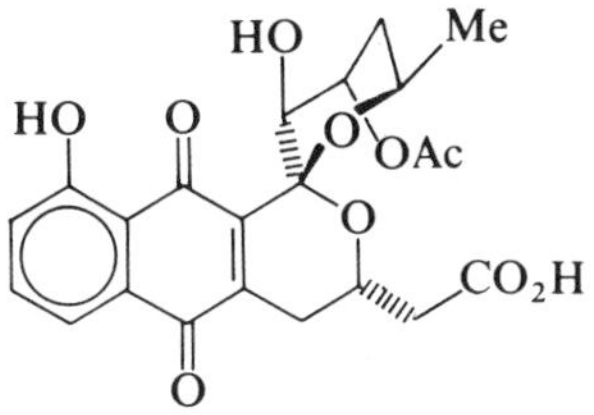

(23)

f. Ceric ammonium nitrate (*CAN*) This is another reagent which is useful for the direct conversion of 1,4-dimethoxyarenes into their corresponding quinones. The reagent has been employed extensively in heterocyclic quinone synthesis in the naphthopyran systems similar to those just described. Investigations of synthetic routes to the reduction products **25**, **27** and **29** from the β-pyranonaphthoquinone aphid pigments protoaphin *fb* (**24**), protoaphin *sl* (**26**) and deoxyprotoaphin (**28**) have shown that silver(II) oxide is a more useful oxidant than CAN when the naphthalene nucleus carries one or more methoxyl groups in addition to those oxidized in quinone formation[26]. In other cases where the nucleus is more stable to oxidation, for instance quinoline and benzimidazole, CAN has proved to be useful for the conversion of 1,4-dihydroxy derivatives to the corresponding quinones when other reagents have proved less effective[27]. The reagent is useful in the indole series provided that steps are taken to delocalize the nitrogen lone pair of electrons

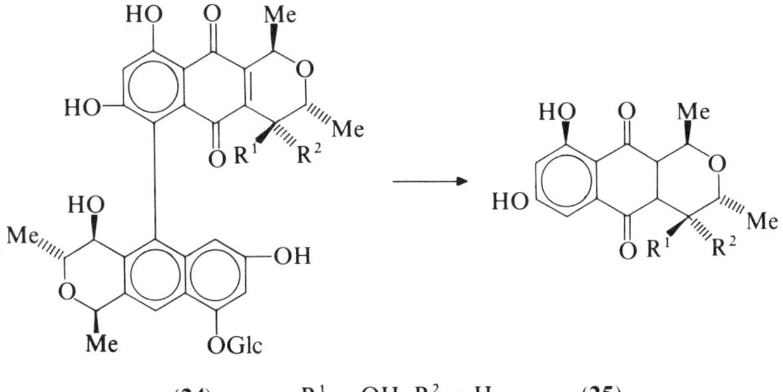

(**24**)	R^1 = OH, R^2 = H	(**25**)
(**26**)	R^1 = H R^2 = OH	(**27**)
(**28**)	R^1 = R^2 = H	(**29**)

other than into the aromatic system prior to the attempt at the oxidation step. In this way the aldehyde **30**, where the carbonyl function is removing electrons from the pyrrole nucleus, was successfully oxidized to the quinone **31** in high yield[28]. A possible

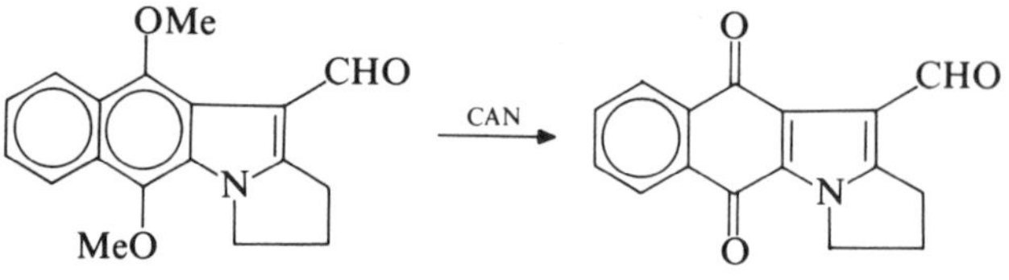

OMe CHO CAN O CHO N N MeO O

(**30**) (**31**)

complicating factor when the 1,4-dimethoxyarene nucleus has an unsubstituted position is the formation of a nitro derivative rather than, or in addition to, quinone formation. The efficiency of the oxidation with CAN may be increased by the addition of pyridine-2,6-dicarboxylic acid *N*-oxide to the reaction mixture and this procedure has been used in the preparation of a range of quinoline- and isoquinoline-5,8-quinones from the corresponding dimethyl ethers[29].

CAN will also oxidize monomethoxy compounds having an unsubstituted *vic* position to the *o*-quinone in certain favourable cases. An interesting example here is the formation of the *o*-quinone **35** as an intermediate[30] in the synthesis of the unusual coenzyme **37**, which is used by bacteria such as *Methylophilus methylotrophus* in the oxidation of methanol. The *o*-quinone (**35**) formed smoothly from the monomethoxy compound (**32**), but hydrolysis of the three ester functions could not be brought about to produce the coenzyme, methoxatin (**37**), in one step. It was necessary to form the intermediate ketal (**36**) in order to allow efficient hydrolysis of the ester functions under basic conditions and then this product yielded **37** upon acidification. Interestingly, the isomeric monomethoxy compound **33** failed to react with CAN[31] and the dimethoxy compound **34** was conveniently oxidized with silver(II) oxide and nitric acid[32].

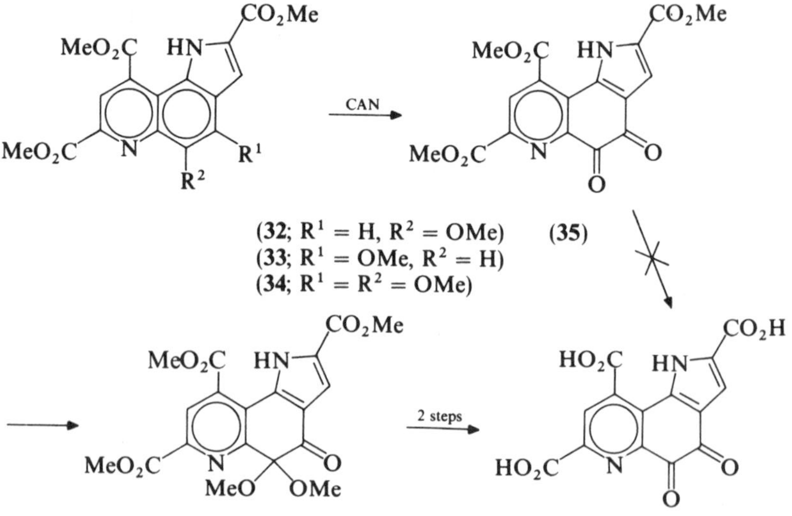

CO_2Me MeO_2C HN CAN CO_2Me MeO_2C HN MeO_2C N R^1 R^2 MeO_2C N O O

(**32**; R^1 = H, R^2 = OMe) (**35**)
(**33**; R^1 = OMe, R^2 = H)
(**34**; R^1 = R^2 = OMe)

CO_2Me MeO_2C HN 2 steps CO_2H HO_2C HN MeO_2C N O MeO OMe HO_2C N O O

(**36**) (**37**)

g. Periodate Sodium metaperiodate has been used in the quinoline series for the oxidation of 6-hydroxyquinolines to quinoline-5,6-quinones[33] and for the formation of 5,8-quinones from highly substituted quinolines. Much of this work has been directed towards the synthesis of the antitumour antibiotic, streptonigrin (**38**), and related compounds. The 8-hydroxy-5-methoxyquinoline (**39**) was oxidized by periodate to the ketal **41** which yielded the quinone **42** on acidification[10]. In the case of the corresponding amine **40**, oxidation with either periodate or CAN yielded both the *p*-quinone **42** and the *o*-quinone **43**[10]. Reports of attempts to use periodate in the formation of heterocyclic quinones are limited in number and types of heterocycle, and it does seem that its potential usefulness in this type of reaction may have been neglected.

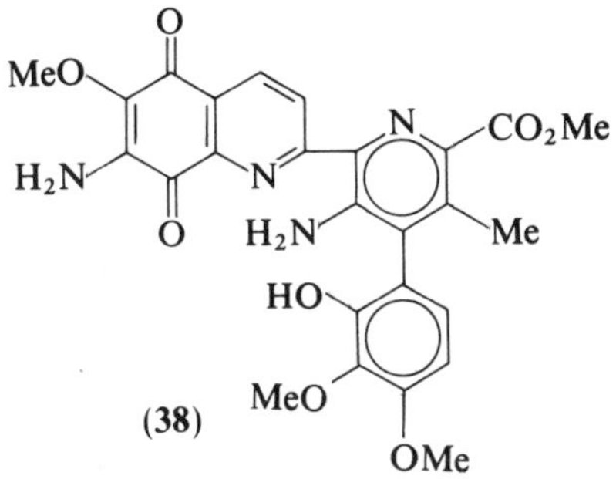

MeO, H_2N, O, O, N, H_2N, N, CO_2Me, Me, HO, MeO, OMe

(**38**)

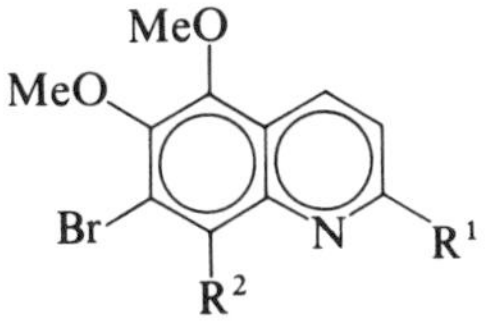

MeO, MeO, Br, N, R^1, R^2

(**39**; R^2 = OH)
(**40**; R^2 = NH_2)

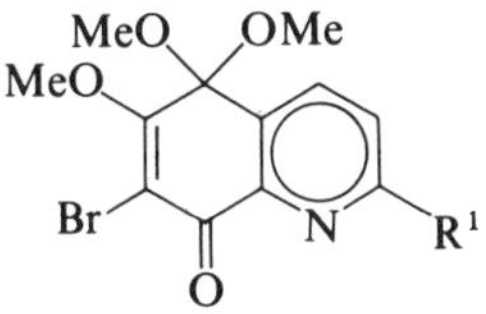

MeO, OMe, MeO, Br, N, R^1, O

(**41**)

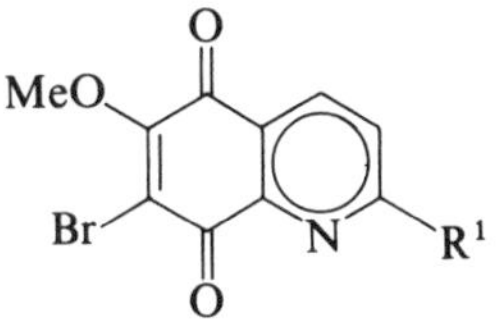

O, MeO, Br, N, R^1, O

(**42**)

O, O, Br, N, R^1, H_2N

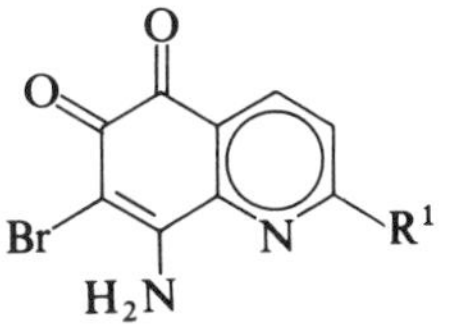

(**43**)

R^1 = substituted 2-pyridyl

h. Hydroxide or alkoxide ions The action of alkali upon certain nitrohetero aromatic compounds may lead to quinones. The formation of the quinone monoxime (**47**) by the action of sodium methoxide in methanol on 4-nitrobenzothiadiazole (**44**) has been explained in terms of the initial formation of the Meisenheimer complex (**45**) and the intermediacy of the oximinoketal (**46**)[34]. In a related reaction[35], the action of hydroxide ion upon 5-nitro-1,10-phenanthroline (**48**) yielded the oximino-*o*-quinone (**50**) by attack of hydroxide ion at the 6-position, formation of the tautomeric arylhydroxylamine **49** and elimination of a molecule of water.

O_2N — $\overset{MeO^-}{\rightleftharpoons}$ — $[NO_2^-$, H, OMe$]$ — $\overset{MeO^-}{\rightleftharpoons}$ — $[^-O$, OH, N, OMe$]$

(**44**) (**45**)

$\xrightarrow{-HO^-}$ $[$NO, OMe$]$ $\longrightarrow$ $[NO^-$, MeO, OMe$]$

(**46**)

$\xrightarrow{H^+,\ H_2O}$ HON, O

(**47**)

NO_2 — $\xrightarrow{HO^-}$ — ^-O, O^-, N^+, H, HO — $\rightleftharpoons$ — ^-O, OH, N, H, O

(**48**) (**49**)

$\xrightarrow{-H_2O}$ NO^-, O $\xrightarrow{H^+}$ NOH, O

(**50**)

i. Miscellaneous methods Oxidizing agents which have been used rarely for the oxidation of hydroxyquinolines to the corresponding quinones include lead tetraacetate[36] and benzoyl *tert*-butyl nitroxide[31]; the latter was employed to provide the *o*-quinone, methoxatin (**37**), from a monohydroxy precursor. Oxidation of 8-hydroxy-5,6-dimethoxyquinoline with bromine yielded, perhaps surprisingly, the non-brominated 6-methoxyquinoline 5,8-quinone as the initial isolable product[10]. The same workers have used nitronium tetrafluoroborate to obtain a 5,6-quinone from a 2-substituted 8-acetamido-5,6-dimethoxyquinoline. Nitrosation at the *o*-position to a hydroxyl group has provided the mono-oxime of an *o*-quinone[37].

An excellent method for the formation of quinones from 4-methoxybenzothiophenes has been shown to be electrochemical anodic oxidation at a platinum electrode and at 65°C[38]. Under these conditions the bis-ketal **51** was formed in high yield and this was readily hydrolysed to the quinone **52** under acidic conditions. Similarly, 5- or 6-substituted derivatives of **51** were obtained from the corresponding substituted 4-methoxythiophenes

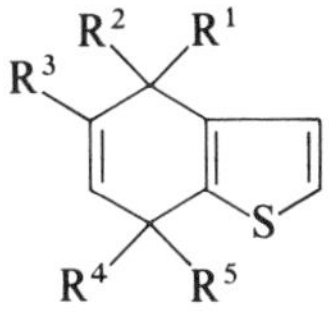

(**51**; $R^1 = R^2 = R^4 = R^5 = OMe$, $R^3 = H$)
(**52**; $R^1R^2 = O$, $R^3 = H$, $R^4R^5 = O$)
(**53**; $R^1 = R^2 = R^4 = R^5 = OMe$, $R^3 \neq H$)
(**54**; $R^1 = R^2 = OMe$, $R^3 \neq H$, $R^4R^5 = O$)

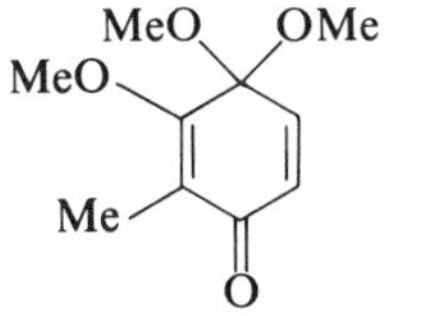

(**55**)

and yielded the quinones upon hydrolysis. Mild hydrolysis of the 5-substituted bis-ketals **53** produced the mono-ketal **54**, whereas under the same mild conditions hydrolysis of **51** afforded the two isomeric mono-ketals. Parker and coworkers[39] have recently utilized Swenton's electrochemical oxidation sequence to obtain an intermediate **55** in the synthesis of an isoindole quinone.

B. Cyclization Reactions Forming the Quinone System

1. Friedel–Crafts reaction

This reaction, which has been extensively applied to the preparation of anthraquinones, may be used for the formation of tricyclic and more extended heterocyclic quinones in those cases where the heterocycles are stable to the conditions of the reaction and are susceptible to electrophilic attack. In a tricyclic product the middle ring is quinonoid and one terminal ring is heterocyclic. The other terminal ring may be benzenoid. When these factors are considered, it is not surprising that the main applications of this method have been in the synthesis of benzothiophene, quinoline and isoquinoline quinones, but it is perhaps less immediately apparent that the method is useful for the preparation of some substituted isobenzofurans. For instance, *o*-(2-thienoyl)benzoic acid (**56**) yields **57** under Friedel–Crafts conditions. Similar type of reaction with thiophene 3,4-dicarboxylic acid chlorides[40] and the appropriate arene yield the isobenzothiophene quinones, **58**, **59** and **60**. Similarly, toluenes and xylenes can be diacylated with 2,5-dimethyl- and 2,5-diphenylfuran 3,4-dicarboxylic acids to give good yields of the quinones **61**[41]. It is necessary to consider the possibility that rearrangement of substituent groups may have

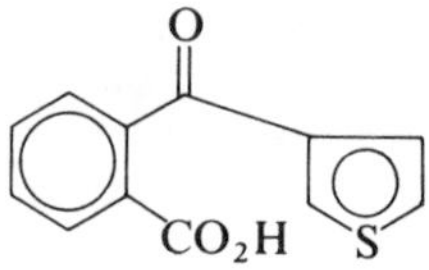

(56) 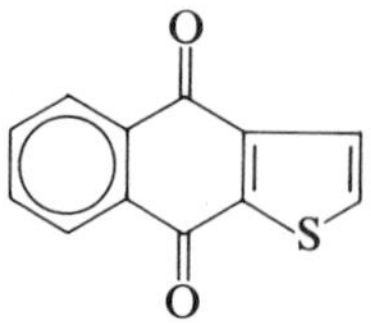 (57)

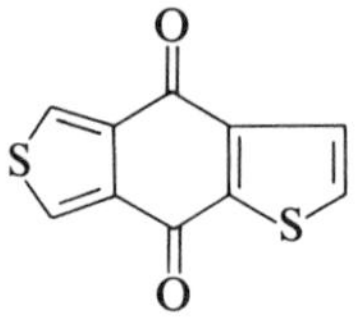

(58) 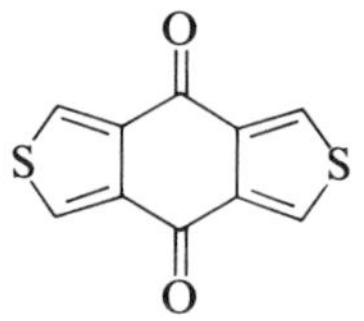 (59) 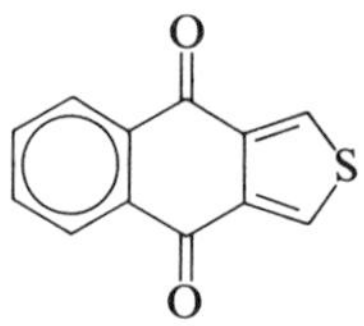 (60)

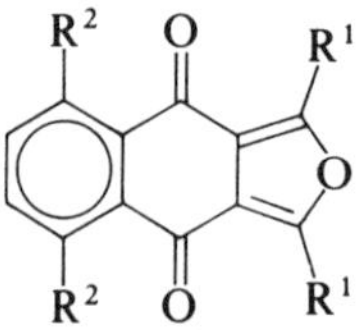

(61) 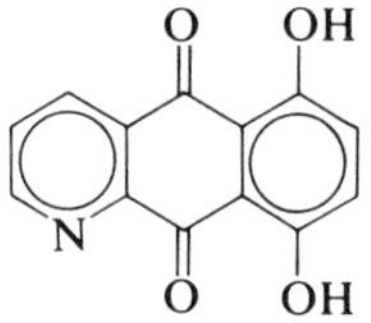 (62) 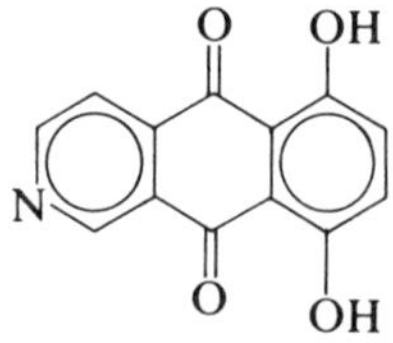 (63)

occurred prior to the cyclization process in reactions of this type[42]. Dihydroxy-1- and -2-azaanthraquinones (**62** and **63**) have been obtained by the action of quinolinic anhydride or cinchomeronic anhydride on 1,4-dimethoxybenzene[43]. However, attempts to use this approach for the synthesis of the pigments phomazarin (**64**) and isophomazarin (**65**) and the antibiotic bostrycoidin (**66**) were not successful, but this failure led to the development of a useful novel cyclization procedure[43] which is described next.

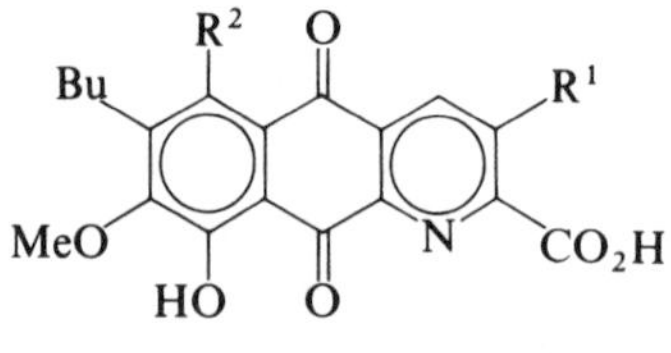

(**64**; R^1 = OH, R^2 = H)
(**65**; R^1 = H, R^2 = OH)

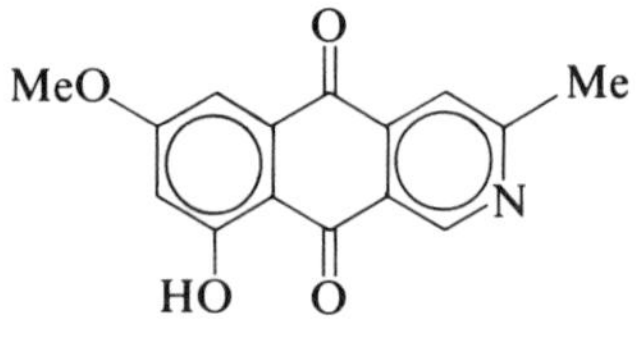

(66)

2. Free radical cyclization

Minisci and coworkers have shown that radicals generated from aldehydes by the action of iron(III) sulphate and *t*-butyl hydroperoxide behave as nucleophiles producing 2- and

4-acylated pyridines from protonated pyridines[44]. In this way it is possible to obtain 2- and 4-benzoylation of pyridine-3-carbonitrile and its derivatives[43]. Thus, 3,5-dimethoxybenzaldehyde and 3-cyanopyridine gave a mixture of **67**, **68** and **69**.

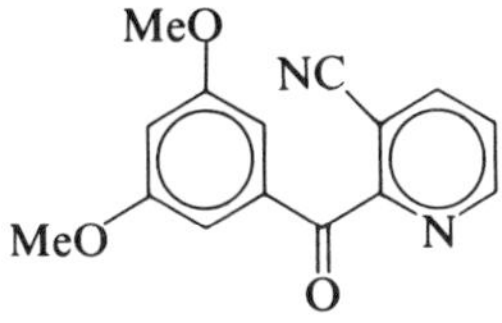

MeO NC MeO N O (67)

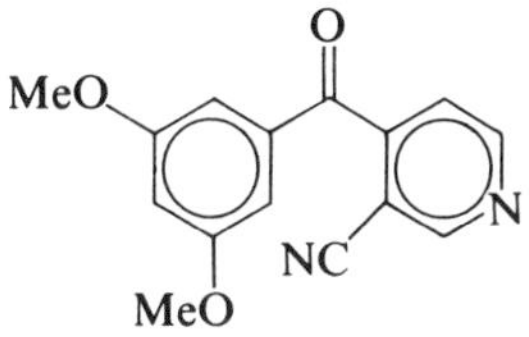

O MeO N NC MeO (68)

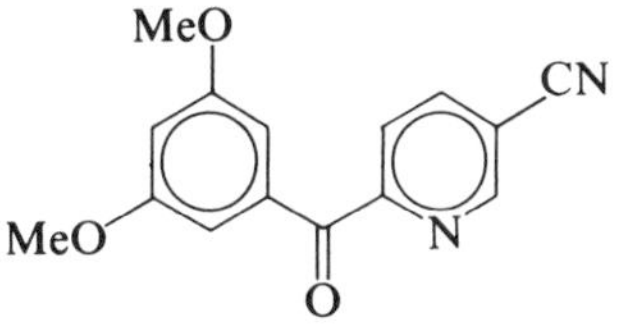

MeO CN MeO N O (69)

Cyclization of the cyano compounds **67** and **68** to the quinones **70** and **71** was achieved by the action of hydrogen chloride followed by ammonia under Houben–Hoesch reaction conditions. Bostrycoidin has been obtained using this approach to the heterocyclic quinone nucleus.

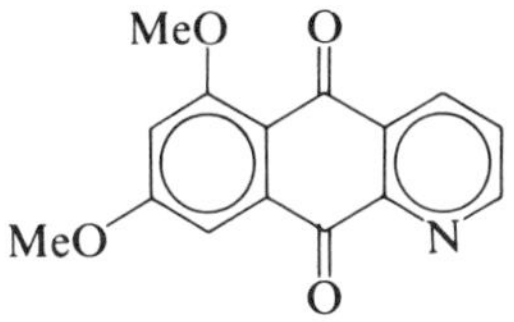

MeO O MeO N O (70)

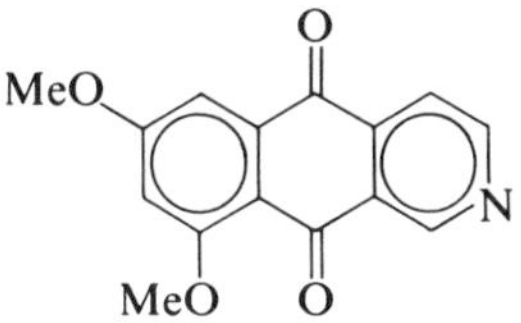

O MeO N MeO O (71)

A reaction which is said to go by a radical mechanism has conveniently yielded examples of those quinones with a heteroatom in a five-membered ring at the position β to the quinonoid nucleus[45]. Attempts to apply the method to other systems do not appear to have been reported. The process requires an *o*-bis(bromoacetyl)heterocycle, e.g. **72**, which undergoes cyclization in the presence of a zinc–copper couple in aprotic medium to the

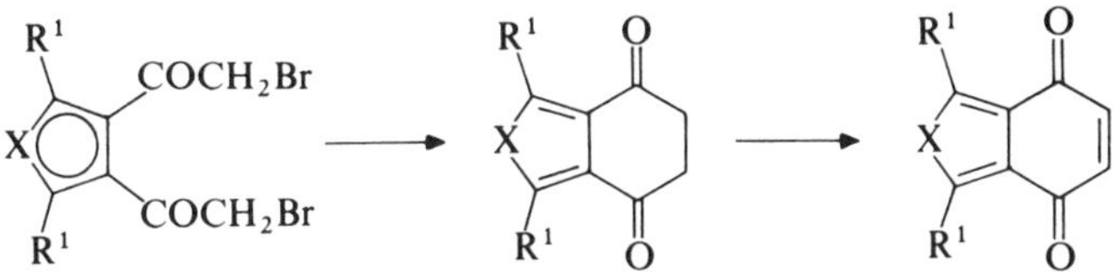

R^1 $COCH_2Br$ X $COCH_2Br$ R^1 (72)

R^1 O X R^1 O (73)

R^1 O X R^1 O (74)

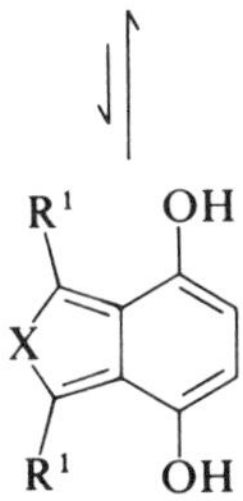

X = MeN, O or S

R^1 OH X R^1 OH (75)

dihydroquinone, e.g. **73**. Oxidation to the quinone, e.g. **74**, occurs readily as might be expected since **73** is the tautomer of the quinol **75**. The tautomeric equilibrium appears to lie heavily towards the keto form **73**. It is suggested[45] that reductive cyclization occurs by a biradical mechanism since the corresponding diacetylpyrrole was obtained when **72** (X = MeN, R^1 = Me) was reduced in a protic medium.

3. Organometallic reagents

Surprisingly little work has been done on the quinones of tricyclic systems derived from indole. The carbazole quinones **76** and **77** are known but there appears to be no reference to quinones from the tricyclic carboline systems. In contrast several routes have been

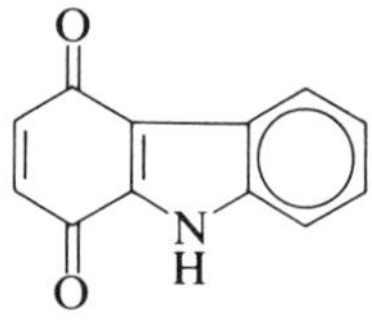

(**76**) 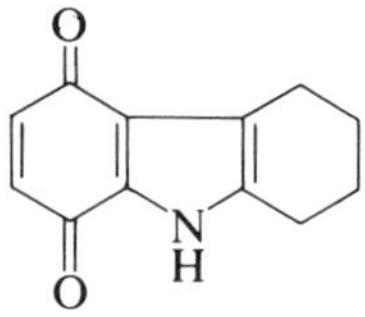 (**77**)

explored for the synthesis of tetracyclic systems derived from **78**. Such compounds have been used as intermediates in the syntheses of the antitumour alkaloid, ellipticine (**79**), and analogues. Convenient routes to the quinone **78** often utilize organometallic reagents.

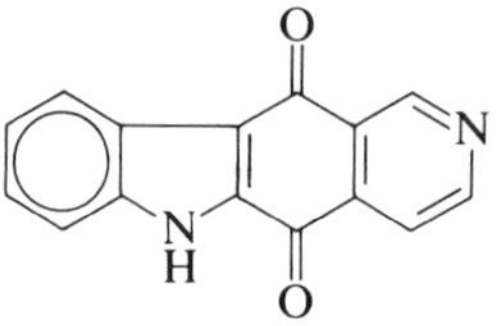

(**78**) 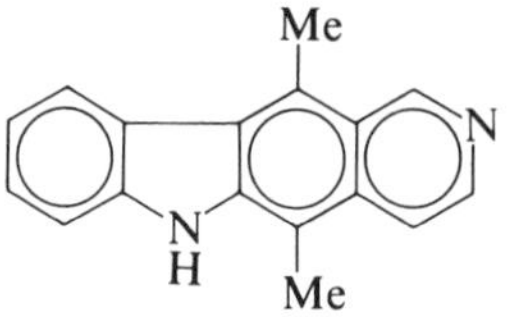 (**79**)

Excellent, high yield syntheses of **78** and the three isomeric compounds with the pyridyl nitrogen in the 1-, 3- or 4-positions have been developed by Joule's group[46]. Their method uses the ready 2-lithiation of 1-benzenesulphonylindole and the reaction of the lithiated species with pyridine lactones. For instance, the lithiated indole **80** undergoes reaction with the lactone **81** to provide the ketoalcohol **82**. Oxidation of the alcohol to the aldehyde, protection of this function as the acetal, followed by solvolysis of the *N*-protecting group in alkali yielded **83**, and this was cyclized to **78** in greater than 90% yield upon deprotection and oxidation. Later[47] it was found that the alcohol **82** takes part in a nucleophilic cyclization on to the 3-position of the indole nucleus in the presence of alcoholic base to give the oxepine **84**. This, in an excess of base and in the presence of oxygen yielded **78** via the oxirane **85**.

Saulnier and Gribble[48] have used the 3-iodoindole (**86**) to obtain lithiation in the 3-position, and subsequent reaction of the 3-lithioindole with cinchomeronic anhydride gave the 4-pyridyl ketone **87** which was converted to the ester **88**. Cyclization to the quinone **89** was achieved by utilizing the acidity of the 2-H on the 1-benzenesulphonylindole nucleus.

Watanabe and Sniekus[49] have used what they term 'tandem directed metalation reactions' in their syntheses of derivatives of ellipticine and other heterocyclic quinones. They used the established property of an *N*,*N*-diethylcarboxamide group to direct

Li N SO_2Ph (80) + O O N (81) → $HOCH_2$ N N O SO_2Ph (82)

→ O O N N H O (83) → (78)

(82) → H O N H N O (84) → [O N N H O^-] (85)

→ [OH N N H O] → (78)

I N SO_2Ph (86) —(1) LiBu; (2) N O O O→ O CO_2H N N SO_2Ph (87)

→ O CO_2Et N N SO_2Ph (88) → O N N H O (89)

lithiation *ortho* to itself in order to obtain the pyridyl lithium **90**. This on reaction with 1-methylindole-3-carboxaldehyde yielded the lithium derivative of the alcohol **91**. Upon reaction of this with more butyllithium, lithiation at the 2-position of the indole nucleus is achieved because the lithium of the attacking organometallic reagent is coordinated to the alkoxide oxygen. The newly formed organolithium reagent then undergoes intramolecular cyclization with the dialkylamide group with elimination of lithium diethylamide. Workup of this one-pot reaction with water in the presence of air yielded the quinone **94** as a

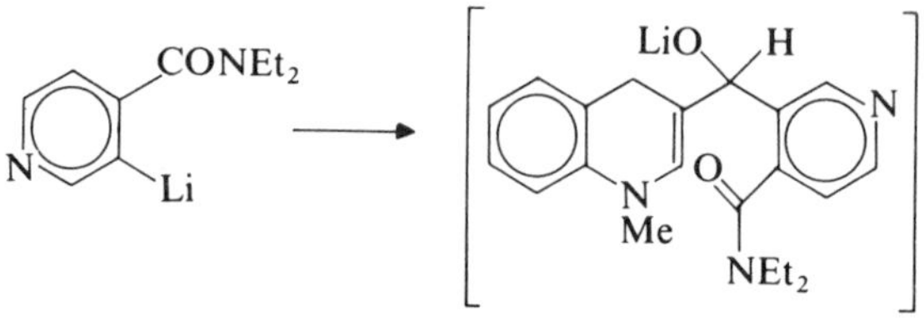

CONEt$_2$ N Li LiO H N N Me O NEt$_2$

(90) (91)

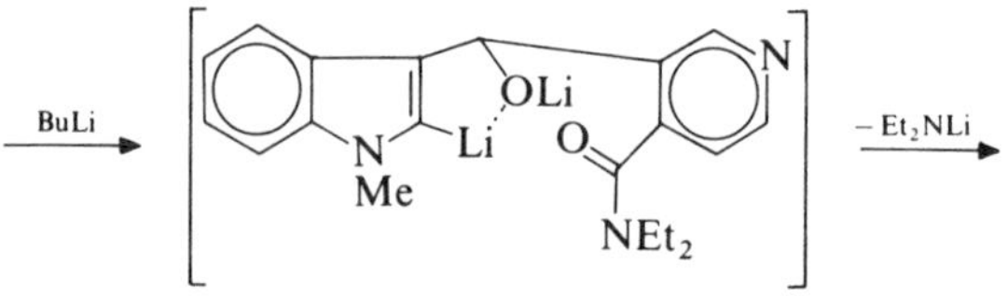

BuLi N Me OLi Li O NEt$_2$ N $-$Et$_2$NLi

(92)

H OLi N N Me O i H$_2$O ii O$_2$ O N N Me O

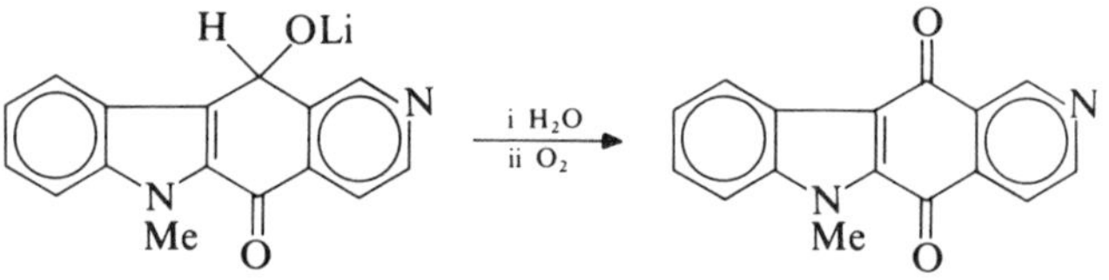

(93) (94)

product of oxidation. A similar reaction sequence but starting from *N*,*N*-diethylbenzamide to form the initial lithium aryl and subsequent reaction of this with 4-chloro-1-methyl-5-azaindole 3-carboxaldehyde gave an intermediate which was subjected to tandem directed metalation with butyllithium to give 5*H*-pyrido[4,3-b]benzo(f)indolo-6,11-quinone (**95**), a benzocarboline quinone, on oxidative workup of the reaction mixture[50].

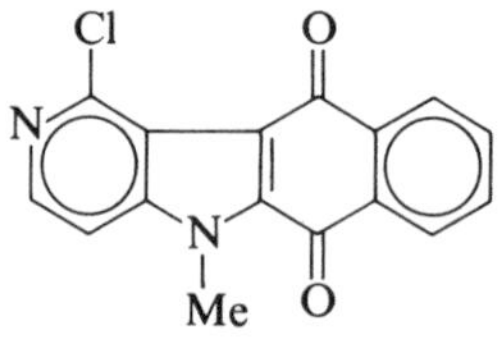

Cl O N N Me O

(95)

In a similar way the lithiated amide **96** upon reaction with thiophene 3-carboxaldehyde yielded **97**, which on tandem directed metalation of the heterocycle with butyllithium and subsequent oxidative hydrolysis gave the thiophene quinone **98**[49], though in only moderate yield. However, the process is essentially a multi-step but one-pot reaction from

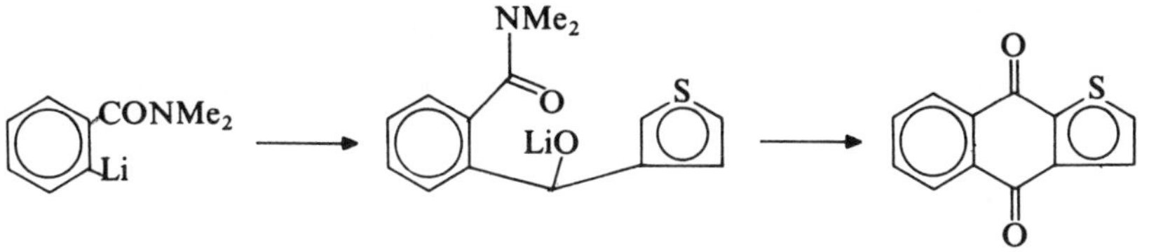

(96) (97) (98)

an aromatic (or heteroaromatic) dialkylamide and appears to be a versatile and convenient route to a range of heterocyclic quinones. Examples of further applications of the method are the preparations of **99**, **101**, **102**[49] and **100**[51].

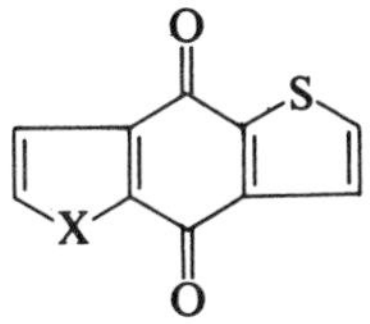

(**99**; X = O)
(**100**; X = S)
(**101**; X = – CH=N–)

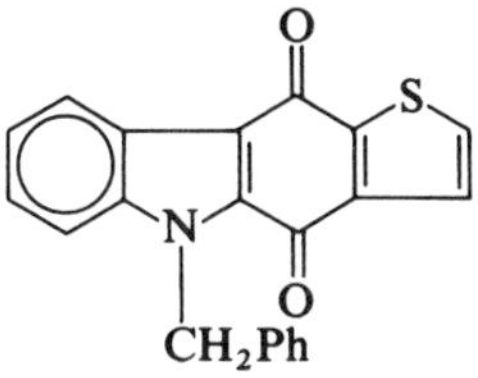

(**102**)

An interesting novel approach to the benzopyran system[52] utilizes the reaction of an alkyne with an α,β-unsaturated chromium carbene complex, e.g. **103**, for the controlled construction of the aromatic nucleus under neutral conditions. Oxidation of the arene–chromium carbonyl complex **104**, without prior isolation, produces demetalation and oxidation to give the quinone **105**.

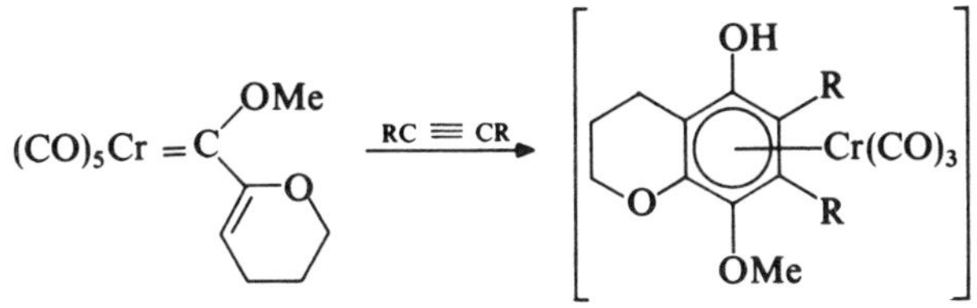

(**103**) (**104**)

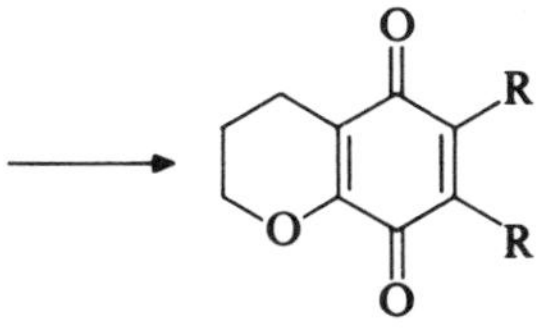

(**105**)

4. *Diyne route*

As part of an extensive investigation of diyne chemistry, Mueller has developed a useful route to certain heterocyclic quinones analogous to isoindole through the intermediacy of an organorhodium quinone. For example, the diyne **106** is cyclized in the presence of tris(triphenylphosphine)rhodium(I) chloride to the quinone **107**. The transition metal can be replaced by the action of nitrosobenzene to give the isoindole quinone **108**[53]. The two steps from **106** to **108** may be conducted as a one-pot reaction process[54]. By replacement of

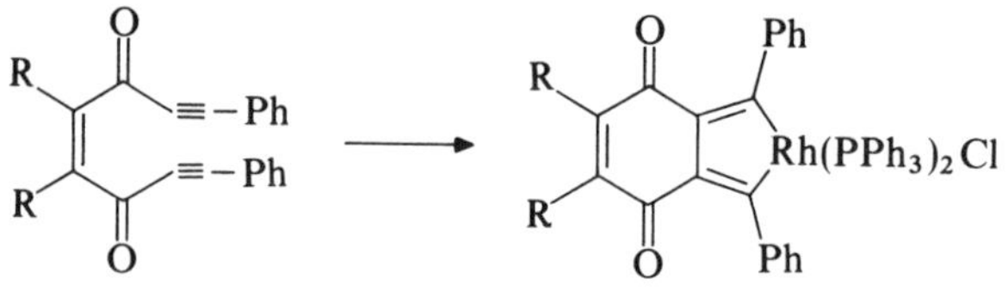

(**106**) (**107**)

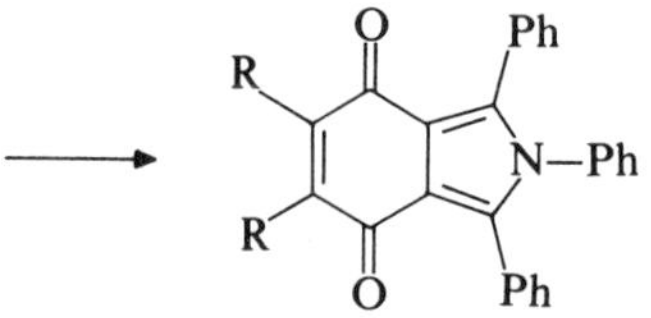

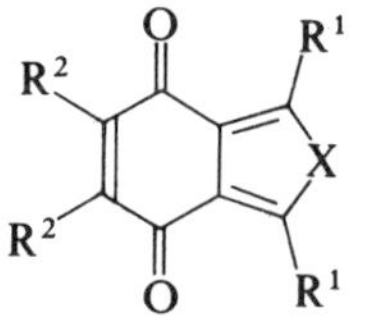

(**108**) (**109**; X = O)
(**110**; X = S)
(**111**; X = Se)

the nitrosobenzene by other reagents it is possible to obtain other heterocyclic quinones: *m*-chloroperbenzoic acid[54], sulphur[55] and selenium[54] separately yield compounds of types **109**, **110** and **111**, respectively. If the bisketoacetylenic system is in the form of vicinal substituents on a heterocycle, it is possible to use this general approach in order to obtain other types of heterocyclic quinone. So, for example, the vicinal dialdehyde **112**[53] yields the diyne **113** and hence the organorhodium quinone **114**, which on treatment with certain acetylenes yields the tricyclic triazole quinone **115**.

5. *Intramolecular benzoin condensation*

Three of the five possible thiophene analogues of phenanthraquinone have been obtained by an intramolecular benzoin condensation of diformylbithienyls followed by aerial oxidation of the product[56]. In this way, 2,2′-diformyl-3,3′-bithienyl (**116**) yielded **117**, and **118** was obtained similarly. The method is applicable to the isomeric compounds having the sulphur atom in the β-position: thus **119** yielded **120**. This approach has not been exploited extensively outside the field of thiophene chemistry.

6. *Intramolecular nucleophilic displacement*

This method has been used to provide the novel *p*-quinone derivative which occurs in part A of the antitumour antibiotic known as CC-1065 (**121**). It seems likely that the already considerable interest in this type of quinone derivative will increase and it is worthy of note here that the heterocyclic quinone antibiotic mycorrhizin A (**122**) also contains a

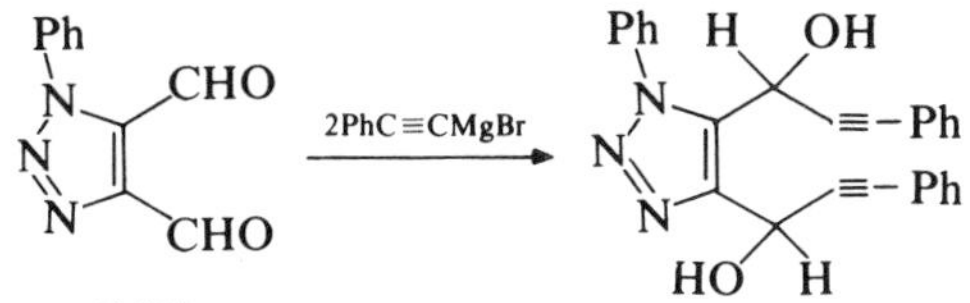

(112)

$2PhC{\equiv}CMgBr$

MnO_2

(113)

$Rh(PPh_3)_3Cl$

(114)

$RC{\equiv}CR$

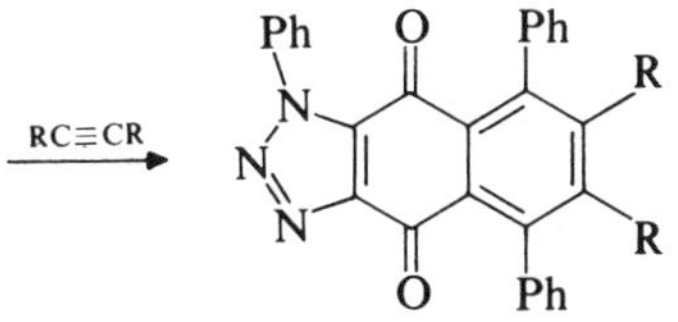

(115)

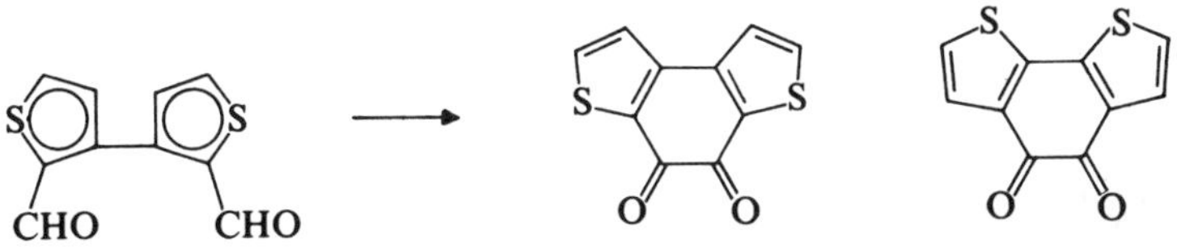

(116)

(117)

(118)

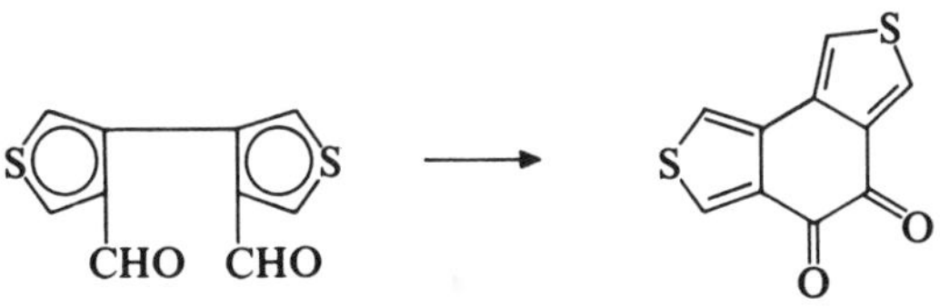

(119)

(120)

cyclopropane ring but not as part of the quinonoid system. Construction of the cyclopropane system at position 6 of the developing cyclohexa-1,4-dien-3-one system has been achieved by treatment of the tricyclic phenol **123** with base[57] or from **124** by use of an intramolecular Mitsunobo reaction with diethyl azodicarboxylate and triphenylphosphine[58].

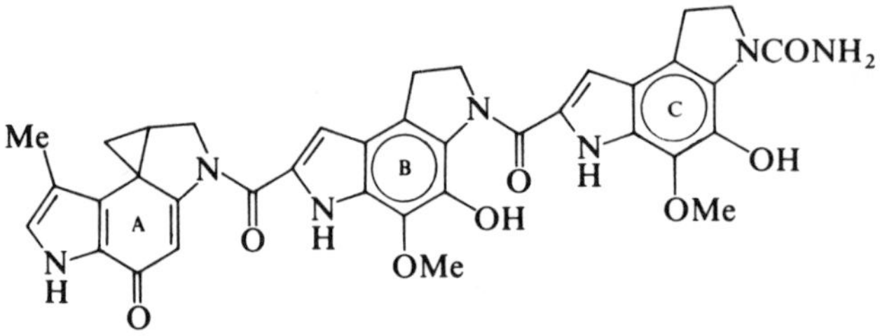

(121)

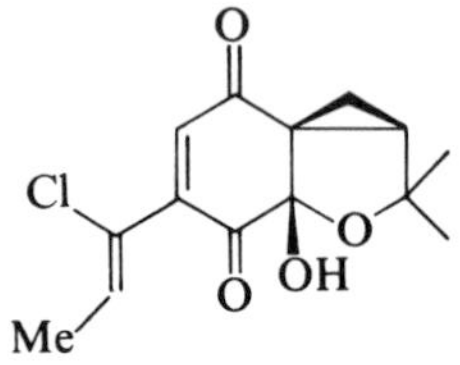

(122)

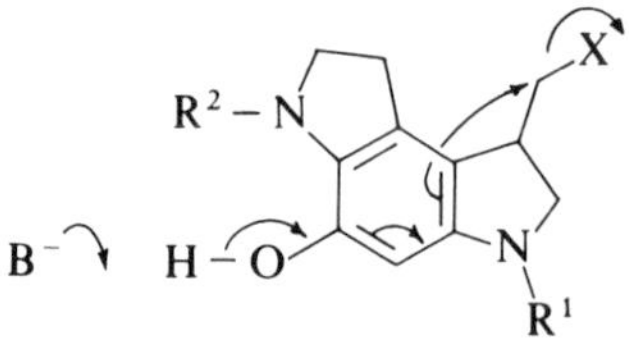

(123; $R^1 = SO_2Me$, $R^2 = H$, X = Br)
(124; $R^1 = COMe$, $R^2 = SO_2Ph$, X = OH)

C. Methods which start from a Quinone

1. Cyclization of substituted benzo- and naphtho-quinones

Several naturally occurring 2,3-dihydronaphtho[1,2-*b*]furan 4,5-quinones have been isolated and found to be derivatives of dunnione (**125**). A recent synthesis of (±)-trypethelone (**126**) has involved the use of a substituted naphthoquinone and a rearrangement reaction[59]. The 2-methoxynaphthoquinone (**127**) was hydrolysed under basic conditions and the diol acetylated to give **128**. This, on treatment with silver(I) oxide and isoprenyl bromide in HMPT gave the allyl ether **129**. The substrate **128** is the vinylogue of an anhydride and is probably converted by the basic oxide to the silver salt

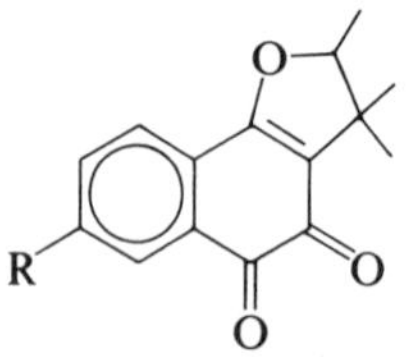

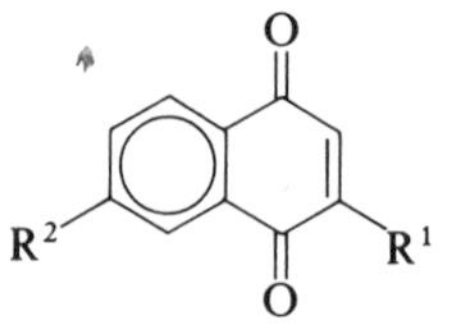

(**125**; R = H)
(**126**; R = OH)

(**127**; $R^1 = OMe$, $R^2 = OH$)
(**128**; $R^1 = R^2 = OAc$)
(**129**; $R^1 = OCH_2CH{=}CMe_2$, $R^2 = OAc$)

and then alkylated in the usual way. Refluxing **129** in ethanol was sufficient to cause two consecutive thermal rearrangements to give the acetates of trypethelone (**126**) and of the expected by-product β-isotrypethelone (**130**) which yielded **126** and **130** after acidification of the thermolysis product.

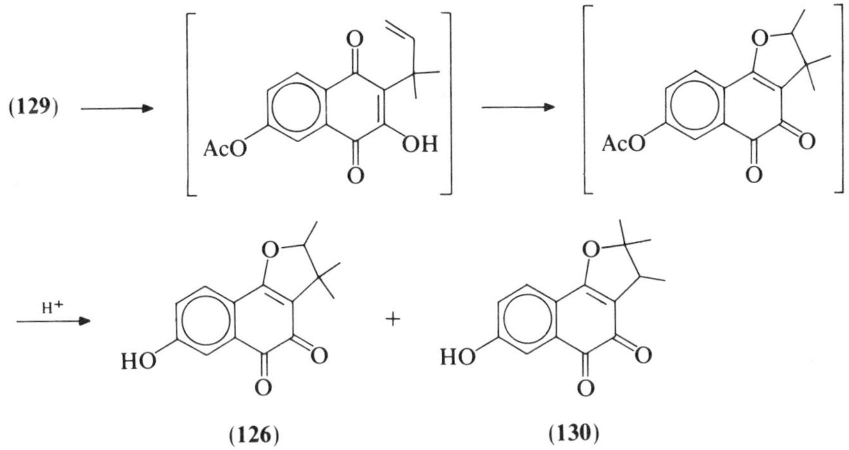

(129) →

H+ →

(126) (130)

Schäfer and Falkner[60] have described the annelation of the trisubstituted benzoquinone **131** with *N'*-phenylbenzamidine to give mainly the trisubstituted quinoline 5,8-quinone **133** with the benzoxazole **135**, the benzimidazole **134** and the trisubstituted quinoline 5,6-quinone **132** as by-products. The proposed reaction scheme requires initial displacement of the 2-anilino group from the quinone **131** by the benzamidine followed by a variety of intramolecular cyclizations and intermolecular substitution reactions.

In a search for a simple route to the tricyclic mitosene nucleus **138** Rapoport has discovered a novel metal catalysed cyclization on to a quinone, although a similar copper(I) bromide catalysed cyclization on to arenes is known. Rapoport's route[13] involves the reaction of the dibromobenzoquinone **136** with the unsaturated amino ester **137** in the presence of copper(II) bromide to give a separable mixture of the mitosene derivative **138** and the linear isomer **139**.

2,3-Dichloro-1,4-naphthoquinone has been used extensively as a starting material in the synthesis of tricyclic and higher members of various types of heterocyclic quinones. The chemistry of quinones containing five-membered heterocycles has been reviewed[3]. Many of these compounds contain bridgehead nitrogen atoms; the syntheses are simple but the yields are often poor. As mentioned earlier, some of the data given in the review[3] has been corrected in the light of later investigations[4]. An example is the reported reaction of *o*-aminophenol with 2,3-dichloro-1,4-naphthoquinone to give the phenoxazine **140**. More recent work[61] has shown that a mixture is formed and among the products are **141** and **142** but **140** is not found (the methoxyl group in **141** is derived from the methanol solvent). The naphthoquinone analogues **143** of the more common benzoheterocyclic quinones **144** are often obtainable from dichloronaphthoquinone.

Kishi's group used a substituted benzoquinone in their approaches to the mitomycin nucleus. For instance, their 19-step total synthesis of deiminomitomycin A[62] from 2,4-dimethoxy-3-methylphenol included the hydrogenolysis of the highly protected **145** followed by treatment of the quinol with oxygen to yield the heterocyclic quinone **147** in reasonable yield, presumably by intramolecular addition of the amine to the quinone **146** despite the requirement to form an eight-membered ring. Conversion of the alcohol to a

PhC:(NH)NHPh

(**131**)

(**132**)

(**133**)

(**134**)

(**135**)

Cu(II)

(**136**) (**137**) (**138**) + (**139**)

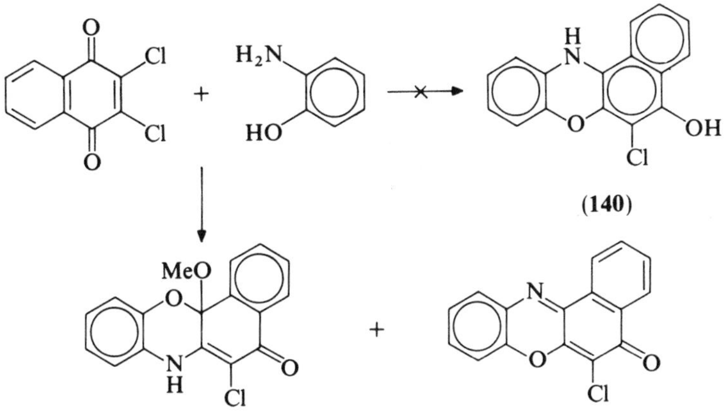

(140)

(141) (142)

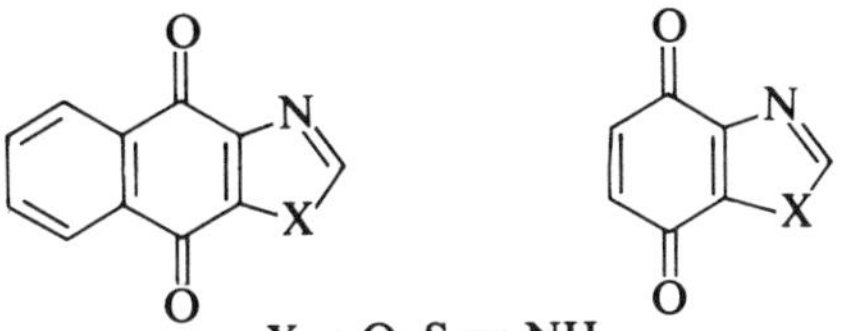

X = O, S or NH

(143) (144)

phenyl carbonate and of the ketal to a hemithioketal gave an intermediate (**148**) which underwent the crucial transannular cyclization in the presence of mercuric chloride and triethylamine to give a (1:1) mixture of the geometrical isomers of the tricyclic system, from

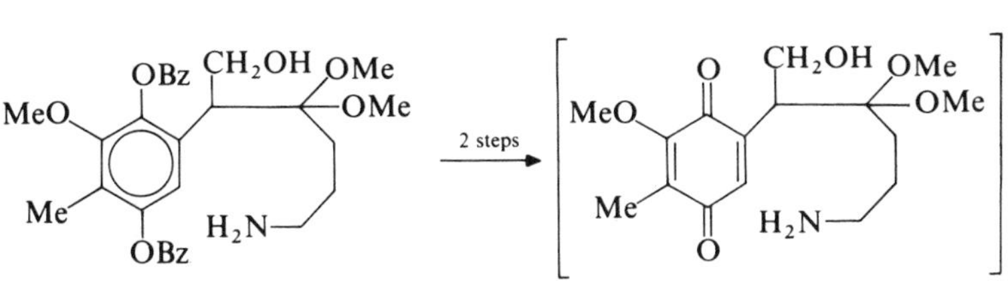

(**145**; Bz = CH_2Ph) (**146**)

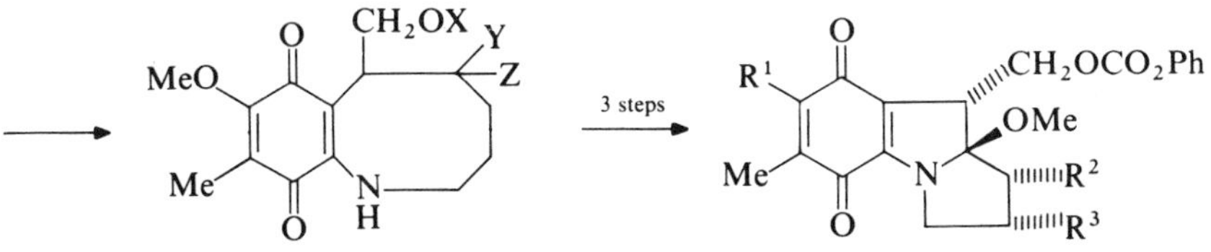

(**147**; X = H, Y = Z = OMe)
(**148**; X = CO_2Ph, Y = OMe, Z = SMe)

(**149**; R^1 = OMe, R^2 = R^3 = H)
(**150**; R^1 = NH_2, R^2,R^3 = NMe)

which the *trans* isomer **149** was isolated by thin-layer chromatography. A similar approach to the indoloquinone system was used in the same group's total synthesis of the related natural product, porfiromycin[63] (**150**).

2. Cycloaddition reactions of quinones

Dipolar cycloaddition to benzoquinone occurs readily and has been exploited in a number of ways to produce heterocyclic quinones. Recently, a number of novel aziridinyl-substituted 1(2*H*)-indazole-4,7-quinones have been synthesized this way and shown to have antitumour activity[64]. Diazomethane adds to the unsubstituted double bond of 2,3-dichloro-1,4-benzoquinone and the intermediate is oxidized by an excess of the benzoquinone to give the dichloroindazole quinone (**151**). The 5-chlorine atom is replaceable by aziridine to afford **152**.

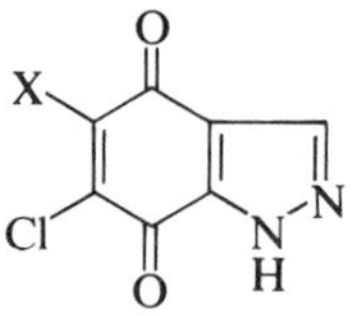

(**151**; X = Cl)
(**152**; X = $-N\!\!<$ (aziridinyl))

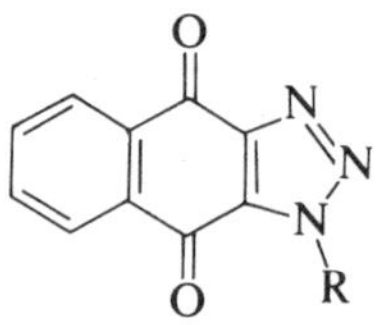

(**153**; R = $CH_2C_6H_4OMe$-*p*)
(**154**; R = H)

Approaches to benzo- and naphtho-1,2,3-triazoles have included 1,3-dipolar addition of azides to 1,4-quinones. For example, the use of 4-methoxybenzyl azide with 1,4-naphthoquinone[65] produced a 1-substituted triazole **153** from which the unsubstituted quinone **154** was obtained by treatment with trifluoroacetic acid. In a similar way aliphatic[64] and aromatic nitrile oxides[66] add to substituted benzoquinones to give the isoxazoline addition products **155** which are readily converted into the isoxazoloquinones **156**. The site- and regioselectivity of the reaction has been examined in some detail[67].

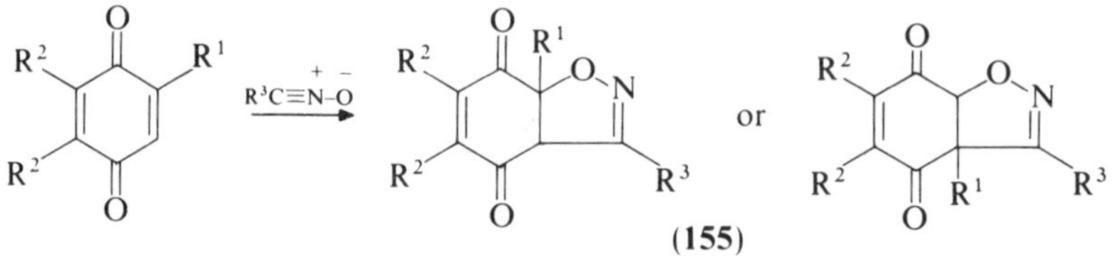

$R^3C{\equiv}\overset{+}{N}{-}\overset{-}{O}$

(**155**)

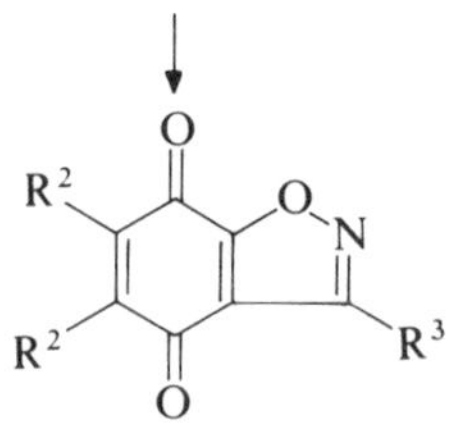

(**156**)

Preparative routes to isoindole quinones rely heavily upon dipolar cycloaddition reactions. 1,3-Diphenylisoindole-4,7-quinone (**158**) is conveniently obtained by photolysis of a solution of benzoquinone and 2,3-diphenyl-2*H*-azirine in the presence of oxygen[68]. The process is thought to involve the intermediate **157**. Other cycloaddition reactions leading to isoindoles have utilized the mesoionic oxazolium oxides **159** which are readily

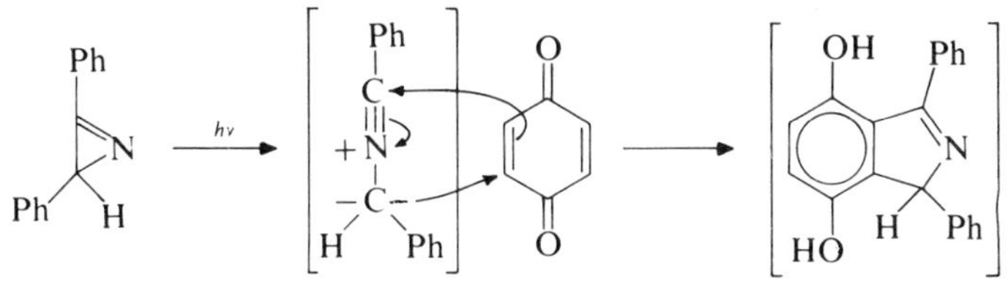

(**157**)

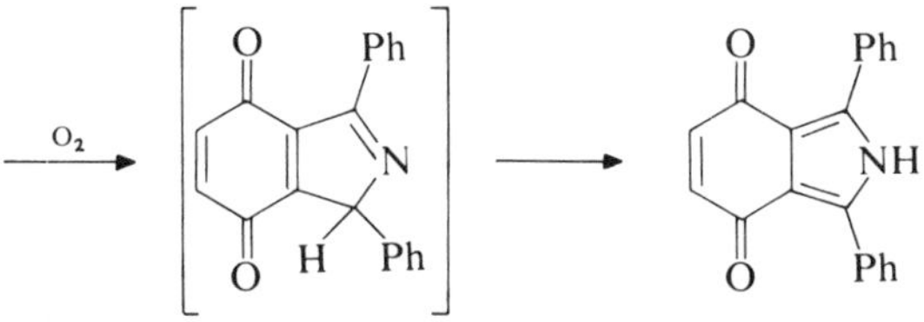

(**158**)

generated from *N*-acylglycine derivatives[69]. Addition to the quinone occurs across the 2,4-positions of the mesoionic system to give intermediates such as **160**, which on loss of carbon dioxide yield the isoindole **161**. This heterocyclic quinone may undergo further reaction to give the tricyclic **162**. Similarly, the reaction of the glycine derivative **163** gives the oxazolium oxide **164**, which yields the partially reduced indolizine quinone **165** upon reaction with benzoquinone[70]. Isobenzothiophene quinones such as **167** are obtained from cycloaddition reactions of benzoquinone with 1,3-dithiolylium-4-enolates (**166**) in a one-pot reaction[71].

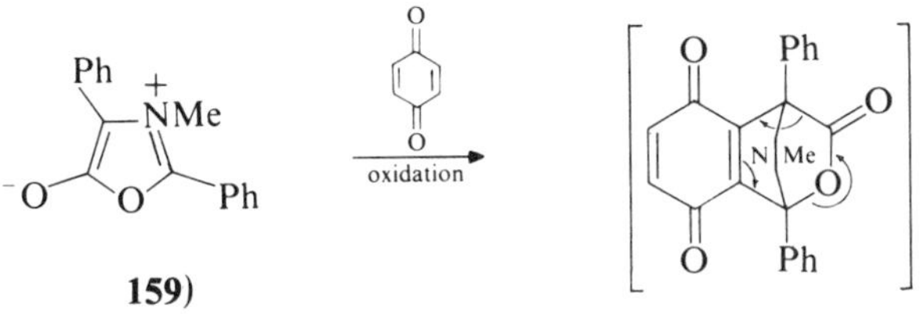

159)

(**160**)

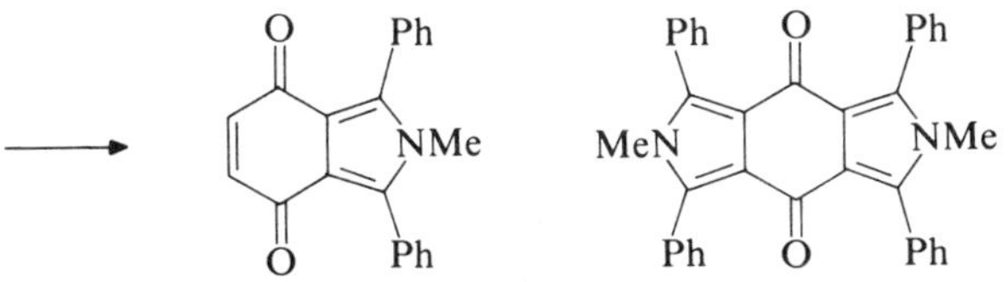

(**161**)

(**162**)

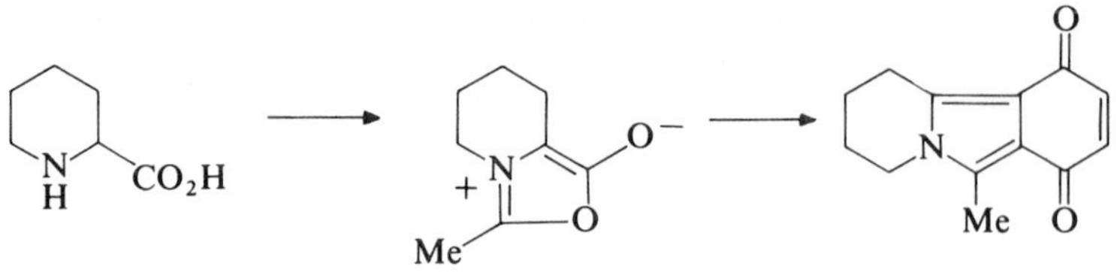

(163) (164) (165)

(166) (167)

The first naturally occurring isoindole to be isolated was obtained from the sponge *Reniera* and was identified as the quinone **171**. A synthesis of this and other isoindole quinones[39] uses non-stabilized azomethine ylides, e.g. **169**, formed from cyanomethyl trimethylsilylmethylamines, e.g. **168**, by the action of silver fluoride. The reaction of the ylide with 2-methoxy-3-methylbenzoquinone (**170**) gave **171** in one step.

$$NCCH_2NMeCH_2SiMe_3 \xrightarrow{AgF} CH_2=\overset{+}{N}Me-\overset{-}{C}H_2 \leftrightarrow \overset{+}{C}H_2-NMe-\overset{-}{C}H_2$$

(**168**) (**169**)

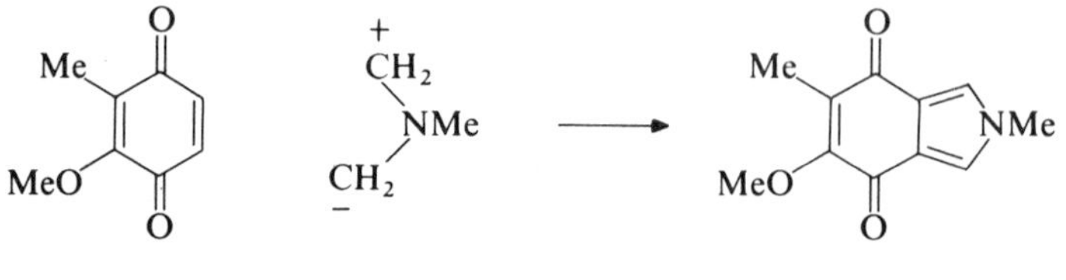

(170) (169) (171)

3. Retro-Diels–Alder reactions

The parent isoindole-4,7-quinone (**175**) has been synthesized by subjecting **174** to a retro-Diels–Alder reaction in the presence of the ethyne scavenger, 3,6-di-(2-pyridyl)-*s*-

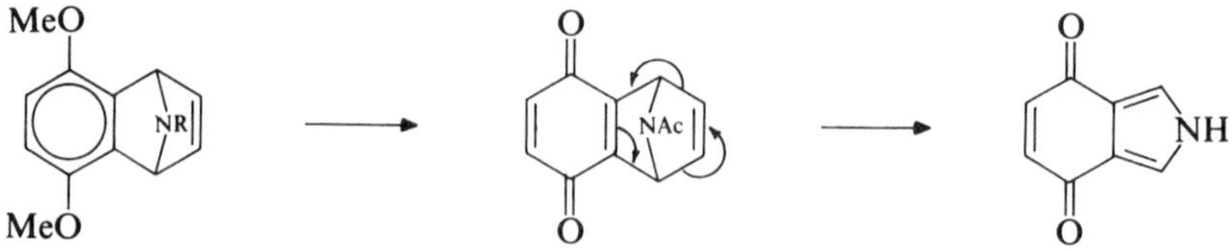

(**172**; R = CO_2Et)
(**173**; R = Ac)

(**174**) (**175**)

tetrazine[72]. The heterocyclic quinone **174** was obtained by the action of 3,6-dimethoxybenzyne (from the appropriate dimethoxyanthranilic acid or 1-aminobenzotriazole) on *N*-ethoxycarbonylpyrrole to give **172**, conversion of this to the *N*-acetyl compound **173**, and subsequent oxidation to the quinone **174**. An analogous but simpler process from furan leads to **176** and thence to the parent isobenzofuran quinone (**177**) which is obtained as a

(**176**) (**177**)

stable yellow solid in high yield[72]. An alternative and somewhat more complex route[73] to **177** starts with the reaction of 1,4-benzoquinone and 3,4-dimethoxyfuran to give the *endo*- and *exo*-Diels–Alder adducts **178** and **179**, respectively. The adducts add a molecule of chlorine in a stereospecific *cis* manner from the *exo* side to give **180**. Solvolysis with methanol then leads to the bis-dimethoxy derivative **181** which, on oxidation yields the

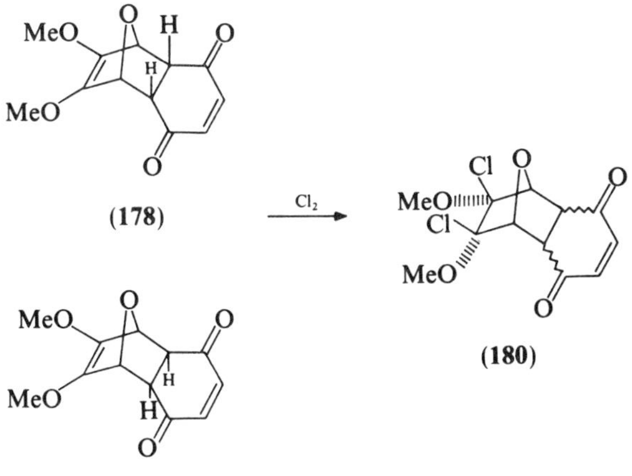

(**178**) $\xrightarrow{Cl_2}$ (**180**)

(**179**)

(**180**) $\xrightarrow{MeOH}$ (**181**) $\longrightarrow$ (**182**)

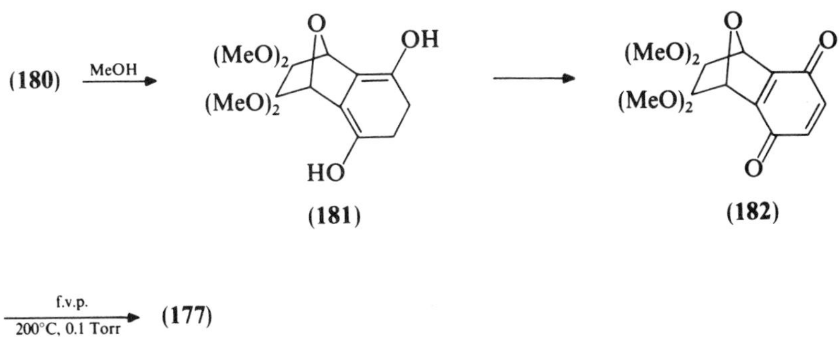

$\xrightarrow[200°C,\ 0.1\ Torr]{f.v.p.}$ (**177**)

heterocyclic quinone **182**. Flash vacuum pyrolysis (f.v.p.) of this at relatively low temperature gives isobenzofuran quinone by a retro-Diels–Alder reaction.

4. Photochemical reactions

Photolysis of 2-acetyl-3-(2-furyl)-1,4-benzoquinones (**183**) is suggested[74] to yield a diradical (**184**) through an $n \rightarrow \pi^*$ excitation of the acetyl group. Intramolecular rearrangement then gives the isobenzofuran products **185** in useful yields.

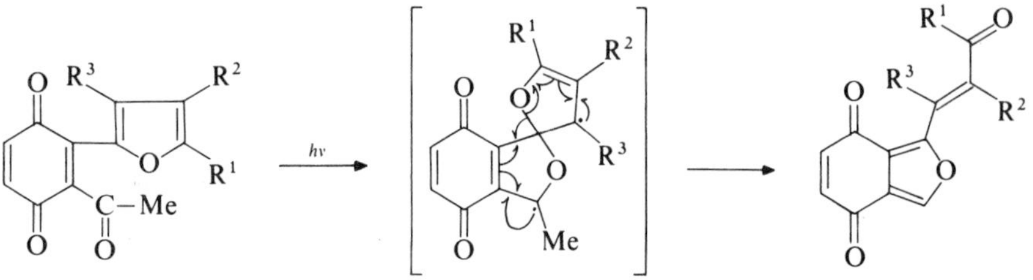

(**183**) (**184**) (**185**)

Sensitized photo-oxygenation with singlet oxygen of vinyl arenes in the presence of base and a vinyl ether has been shown to give benzopyran quinones[75]. It is suggested that the substituted vinyl arene, e.g. **186**, undergoes two attacks by singlet oxygen to yield the diperoxide **187**, this then decomposes in the presence of base to give the quinone methide which is trapped to yield the product **188**.

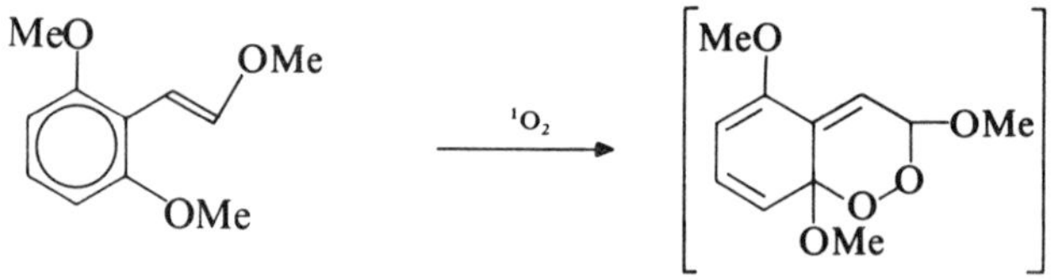

(**186**)

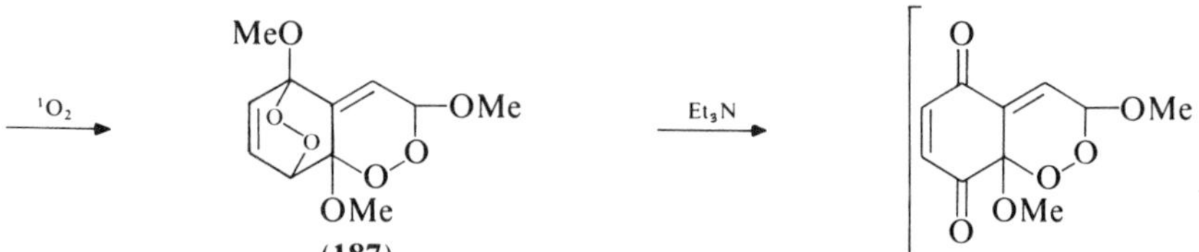

(**187**)

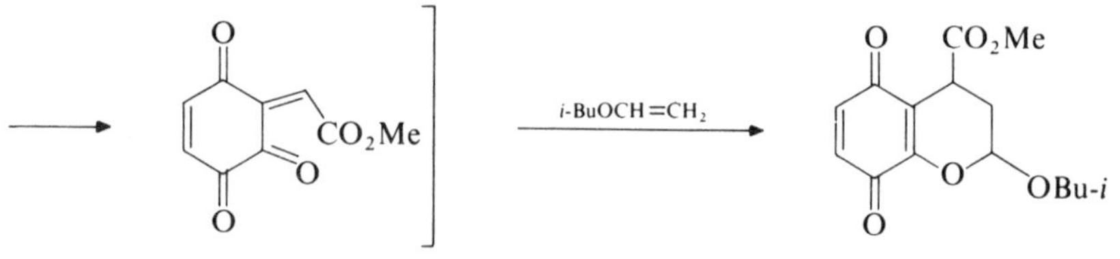

(**188**)

III. PROPERTIES OF HETEROCYCLIC QUINONES

A. General Considerations

Your reviewers have been impressed by the general lack of systematic investigations of the reactivity of heterocyclic quinones and their derivatives. There are very few cases where data are available from comparative studies of the reactivity of a series of heterocyclic quinones and one or more carbocyclic analogues. Therefore, quantitative data showing the effects of the presence of a particular heteroatom and of its position relative to the quinonoid nucleus are not available. Recently there have been some moves to rectify this situation, at least in a semiquantitative sense, with the publication of results from studies of the addition of dienes to heterocyclic quinones. An understanding of the factors which control the regioselectivity of these reactions has been gained for the relatively few heterocycles studied this far.

When we turn to substitution reactions the detailed knowledge is very sparse. What is known has been acquired almost entirely from the frustrations and successes of work primarily directed towards the synthesis of naturally occurring or potentially biologically active quinones. Because of these origins the information tends to be fragmentary and there are even less data of a comparative or quantitative nature than in the case of the addition reactions.

There has been, and there continues to be much interest in the potential use of heterocyclic quinones in medicinal chemistry. Some time ago the major interest was in the field of anti-malarials, now the activity is centred more on the development of heterocyclic quinones for use in cancer chemotherapy or as adjuvants in the radiotherapy of cancer. The finding of anticancer activity in compounds such as mitomycin, CC-1065 and streptonigrin, together with their novel structures, has stimulated much research activity. Studies of mitomycin and related compounds have produced interesting theories to explain their biological activity, and these ideas suggest that other heterocyclic quinones may have useful anticancer activity. The recent discovery that methylotrophic bacteria utilize a dehydrogenase enzyme which requires a coenzyme, methoxatin (**37**), having a novel heterocyclic quinone structure adds another facet to the importance of heterocyclic quinones.

The industrial use of heterocyclic quinones has been small, presumably because the requirements can be satisfied with the cheaper carbocyclic quinones. The chemical properties, biological and medicinal interests and uses, and (briefly) industrial applications of heterocyclic quinones are discussed in more detail in the remainder of this chapter.

B. Addition Reactions

1. Addition of 1,3-dienes

Reactions of the Diels–Alder type present a useful method for the annelation of heterocyclic *p*-quinones. Clearly the reaction of an unsymmetrical heterocyclic quinone and/or an unsymmetrical diene may lead to more than one product. Studies of factors influencing regioselectivity in this type of reaction with quinoline- and isoquinoline-5,8-quinones have been completed recently. The use of other heterocyclic quinones has been studied much less in these reactions.

The electron-rich diene, *trans*, *trans*-1,4-bis(ethoxycarbonylamino)-1,3-butadiene adds to quinoline-5,8-quinone (conveniently[76] obtained from 8-hydroxyquinoline) to give the addition product **189** which is oxidized[77] in air to the aromatic system **190**. Similar reactions occur with isoquinoline and quinazoline quinones. The quinone 6,7-diester **191** undergoes addition of cyclopentadiene to give the adduct **192** which cannot become fully aromatic[36].

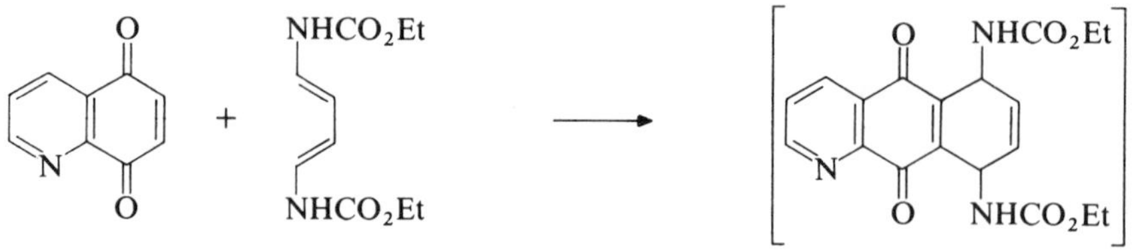

(189)

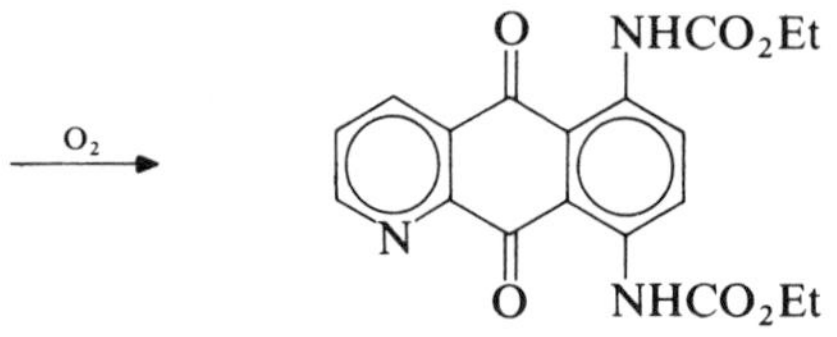

(190)

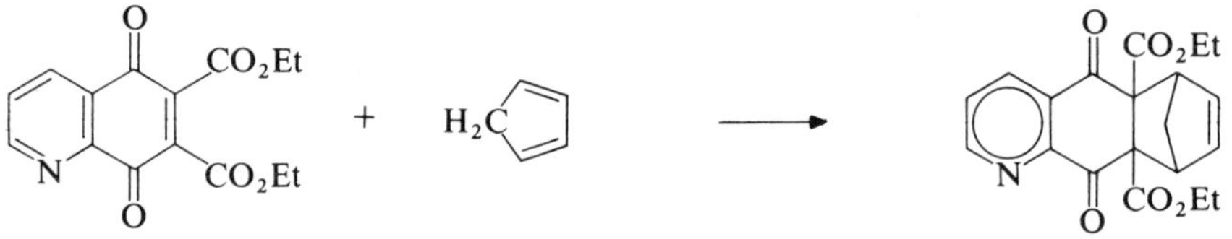

(191)

(192)

Recent investigations with unsymmetrical dienes have produced results which allow some discussion of the factors which influence regioselectivity in these reactions. The addition of 1-methoxycyclohexa-1,3-diene to quinoline-5,8-quinone yielded compounds which exist mainly in the dihydroxy forms **193** and **194**. Pyrolysis of the mixture gave a product from which the major component **195** was readily separated[78]. These results taken together with studies of similar reactions of isoquinoline-5,8-quinone[78] indicate that the position of substitution of the carbonyl group on the pyridine nucleus determines the site of attack and can produce what is, for preparative purposes, a regiospecific reaction. In the reaction shown, the 6-position of the quinoline quinone is the one affected most by the carbonyl group bonded to the electron-withdrawing 2-position of the pyridine nucleus, while the 7-position is affected by the 5-carbonyl group which is bonded to the much less electron-demanding 3-position of the pyridine ring. Hence, overwhelmingly, the preferred site for nucleophilic attack in the cycloaddition reaction is the 6-position. This is in line with the findings for the substitution reactions of these quinones and seems to indicate that the reactions are probably not concerted and certainly not synchronous. In agreement is the finding that methacrolein *N,N*-dimethylhydrazone (**196**) adds to quinoline-5,8-quinone in a regioselective way to give the diazaanthraquinone **197**[78].

2. *Addition of olefins and isocyanides*

Interest in the aza- and diaza-anthraquinones is stimulated by the occurrence of the fungal pigments phomazarin (**198**) and isophomazarin (**199**)[79] and the antibiotic diazaquinomycin (**200**)[80]. In reactions of quinoline-5,8-quinone with two molar equivalents of 1,1-dimethoxyethene, the product is a mixture of isomeric 1-azaanthraquinones,

O O N OMe → O H O H N OMe + O H O H N OMe

HO HO N OMe

(193)

Δ

O O N OMe

(195)

HO OMe HO N

(194)

Δ

O OMe O N

N NMe$_2$ ⟷ N + NMe$_2$ −

(196)

O O N → H H O N NMe$_2$ O N

Δ

O O N N

(197)

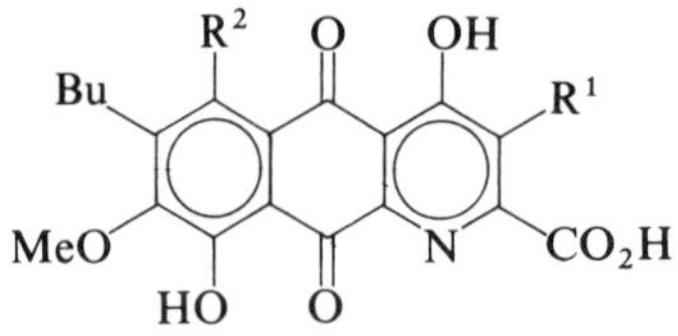

(**198**; $R^1 = OH, R^2 = H$)
(**199**; $R^1 = H, R^2 = OH$)

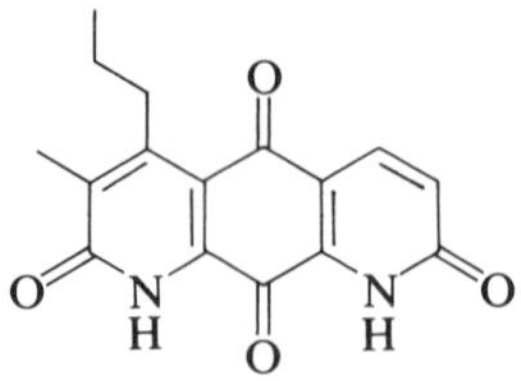

(**200**)

in a ratio of 4:3. In this type of reaction, whose mechanism is uncertain, the quinoline quinone shows much less regioselectivity than the corresponding isoquinoline quinone[76].

Isoindole quinones **201**, which can be formed from benzoquinone and an aryl isocyanide, undergo further reaction with the aryl isocyanide to give the isomeric, dark blue quinones **202** and **203**. A mechanism for the reaction has been proposed[81].

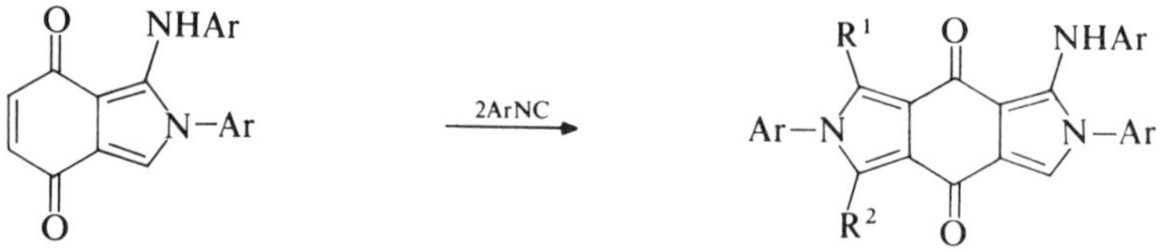

2ArNC

(**201**)

(**202**) $R^1 = NHAr, R^2 = H$
(**203**) $R^1 = H, R^2 = NHAr$

3. *Miscellaneous addition reactions*

A group of compounds which may be considered as the products of addition to benzofuran 4,7-quinone are produced by the fungus *Gilmaniella humicola* Barron. Two of these, mycorrhizin A (**204**) and chloromycorrhizin A (**205**) are strongly inhibitory to the root rot fungus *Fomes annosus* (Fr.) Cke. In a recent approach to this nucleus, R. F. C.

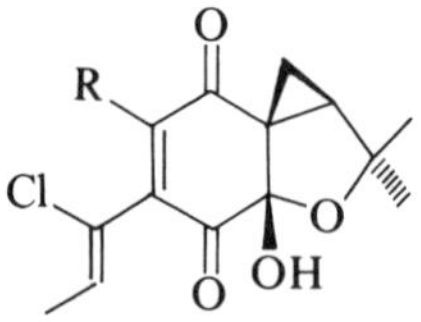

(**204**; R = H)
(**205**; R = Cl)

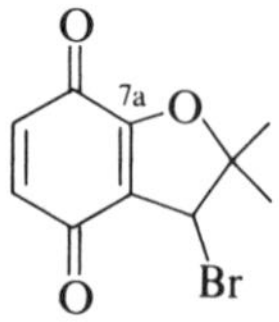

(**206**)

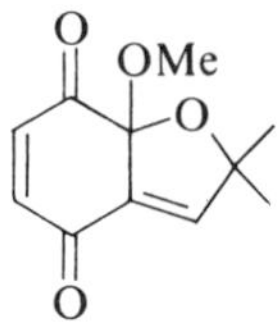

(**207**)

Brown's group have found an unusual addition of methanol to 3-methyl-2,2-dimethylbenzofuran-4,7-quinone (**206**) in the presence of silver oxide to give **207**[82]. It is suggested that the process involves kinetically preferred addition of methanol at the 7a-position of the allylic cation formed on loss of bromide ion from **206**.

The products of the formal addition of oxygen to either the carbon–carbon bond of a heterocyclic quinone to give an epoxide or to an azine nitrogen atom in such a compound to give an *N*-oxide are almost unknown. A start has been made recently to uncover their chemistry in the quinoline- and isoquinoline-5,8-quinone series[83]. The epoxides such as **208** are formed readily using conditions similar to those used in order to form naphthoquinone epoxide. The quinone *N*-oxide **211** was obtained by *N*-oxidation of 5-acetamido-8-hydroxyquinoline with hydrogen peroxide in the presence of sodium tungstate to give **209**. Acidic hydrolysis of the acetyl group and oxidation of the 5-amino-8-hydroxyquinoline 1-oxide (**210**) gave **211**.

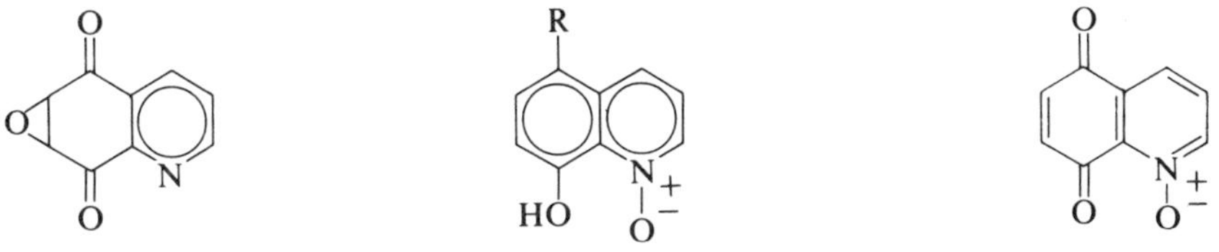

(**208**)

(**209**; R = NHCOMe)
(**210**; R = NH_2)

(**211**)

Addition of hydrogen chloride to the quinone **212** affords the chloroquinol **213**[27]. Probably the most commonly used addition reaction in this series is the hydrogenation of quinones to quinols but there appears to be nothing remarkable about this reaction in the heterocyclic series.

HCl

(**212**) (**213**)

C. Substitution Reactions

1. Replacement of a hydrogen atom

We have already seen (Section III.B.1) that the 6-position in quinoline-5,8-quinones is the most susceptible to nucleophilic attack. Aniline derivatives react with quinoline quinones to give products such as **214** when the reaction is catalysed by cerium(III) chloride[84]. Presumably the cerium(III) ion coordinates with the pyridyl nitrogen atom so promoting attack by the nucleophile. Replacement of a hydrogen atom in the 7-position is possible even when there is an electron-releasing substituent in the 6-position of a quinoline 5,8-quinone. For instance, a 7-alkylthio substituent may be introduced into 6-

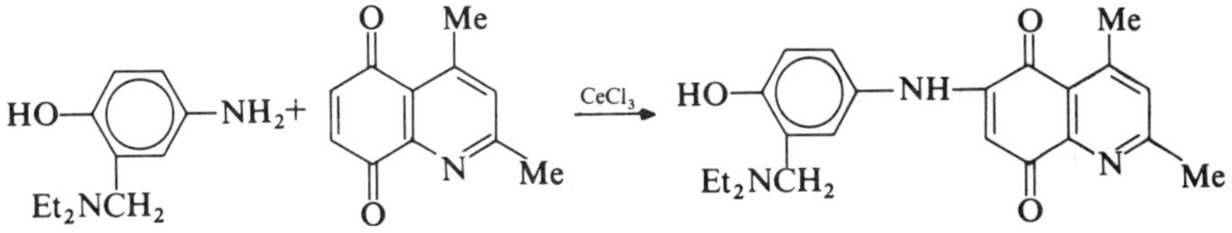

$CeCl_3$

(**214**)

hydroxyquinoline-5,8-quinone and the same hydroxyquinone undergoes a modified Mannich reaction to give a 7-alkylaminomethyl derivative[85]. The presence of a 6-hydroxyl substituent on the quinoline-5,8-quinone[86] or the benzothiazole-4,7-quinone[8] nucleus allows the introduction of an alkyl group into the 7- or 5-position, respectively, by the generation of alkyl radicals from diacyl peroxides. Much of this type of work was done in the 1960s and 70s in the search for an effective antimalarial based on heterocyclic quinones. Other important work involving *vic*-alkylhydroxyquinones is to be found in the α- and β-lapachone series of benzopyran quinones, but the alkylation reactions themselves are on naphthoquinones. Little seems to have been done with heterocyclic quinones in this area of substitution chemistry.

2. *Replacement of groups or atoms other than hydrogen*

The displacement of halogen or methoxyl have been the most used reactions. The 5-chlorine atom is readily replaced by amines from 5,6-dichloroindazole-4,7-quinone[64,87] whereas 6,7-dichloroquinoline-5,8-quinone gives the 6,7-bis-imidazole derivative[87]. Regiospecific substitution in the 6-position is often achievable[88] in this series by nucleophilic displacement of a 6-methoxyl group even when a 7-bromine is present. At first sight, the azide ion appears to be exceptional as a nucleophile in this reaction because the action of azide ion on **215** is to give the bromine displaced product **218**. However, it is

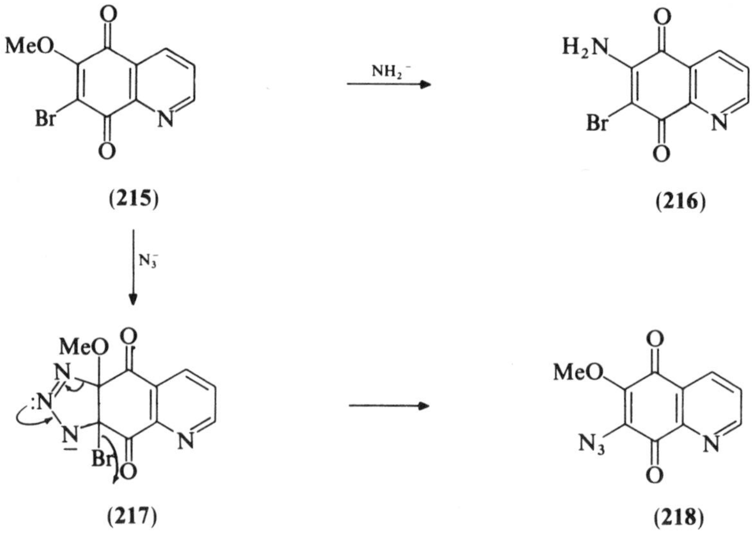

MeO Br N O O NH_2^- H_2N Br N O O

(215) **(216)**

N_3^-

MeO N :N N Br O O N MeO N_3 N O O

(217) **(218)**

thought that the reaction goes by initial nucleophilic addition at the 6-position followed by cyclization to give the triazole **217**. Subsequent displacement of bromine in a ring-opening process gives **218**. Reduction of **218** gives the corresponding amine, and this procedure complements the action of amide ion on **215** to give the aminobromoquinone **216**. In this connection it is interesting to note that treatment of a 7-methoxyquinone of the type **219** with azide ion in DMF gives the corresponding 6-amino-7-methoxy compound (**220**) directly[88]. As would be expected, the 6-methoxyl group of the quinazoline **221** is readily replaceable by a secondary amine. Vigorous treatment of **221** with aziridine gives **223**, presumably via the intermediate **222**[21].

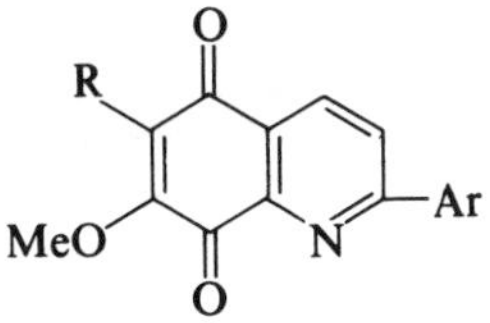

(**219**; R = H)
(**220**; R = NH_2)

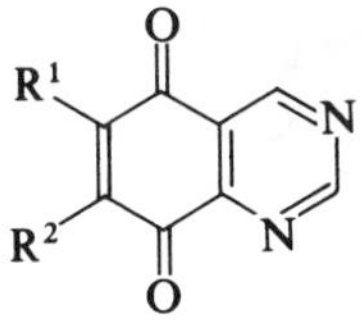

(**221**; R^1 = OMe, R^2 = H)
(**222**; R^1 = 1-aziridinyl, R^2 = H)
(**223**; R^1 = R^2 = 1-aziridinyl)

The reaction of 5-bromo-6-methylbenzothiophen-4,7-quinone (**224**) with methylthiolate ion gave the expected 5-substituted derivative **225** as the major product but the minor component of the reaction mixture was shown to be **226**. The mechanism suggested for the formation of **226** involves an intermediate anion with some *o*-quinone methide character[89].

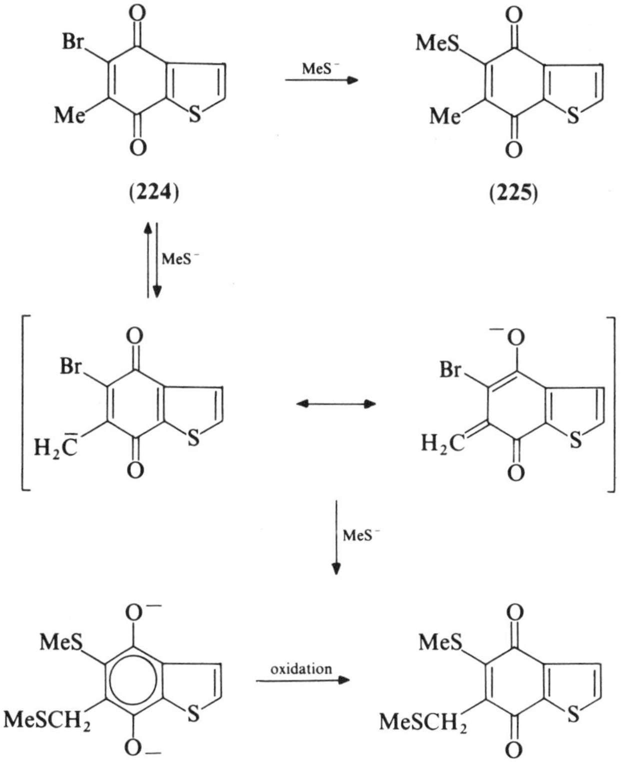

(**224**)

(**225**)

(**226**)

Nucleophilic displacement of nitro groups from the carbocyclic and non-quinonoid ring of tricyclic heterocyclic quinones had been found to occur in both the systems represented by the triazole quinone **227**[90] and the thiadiazole quinone **228**[22]. Thermal replacement of a methoxyl group from the related tricyclic system **229** by primary amines was successful only under vigorous conditions. However, on irradiation with light the reaction proceeded smoothly to give mainly the monosubstitution product **230** in the presence of methylamine, together with a smaller proportion of **231**[91].

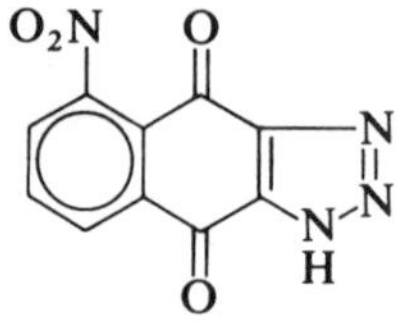

(**227**) 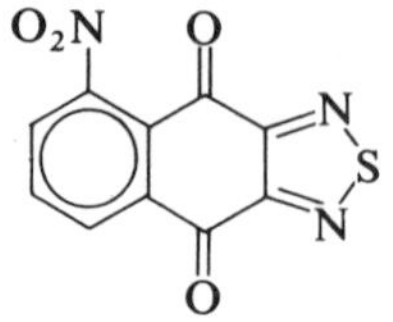 (**228**)

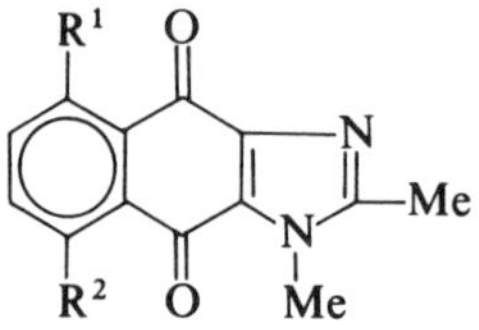

(**229**; $R^1 = R^2 = OMe$)
(**230**; $R^1 = NHMe$, $R^2 = OMe$)
(**231**; $R^1 = OMe$, $R^2 = NHMe$)

D. Reactions at the Carbonyl Groups

Probably the most common reaction of heterocyclic quinones is their reduction, either when they are used in a one molar excess in order that the quinone may oxidize the initially formed quinol in an addition reaction or when the quinone is reduced to give a quinol product. The latter type of reaction appears to proceed smoothly with no especially

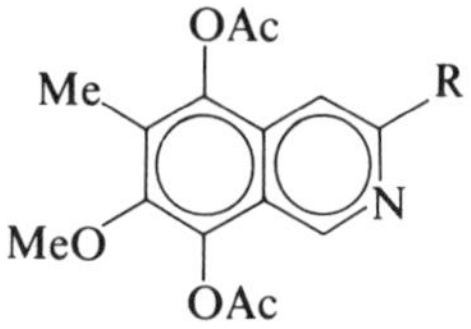

(**232**; R = H)
(**233**; R = 1-morpholino)

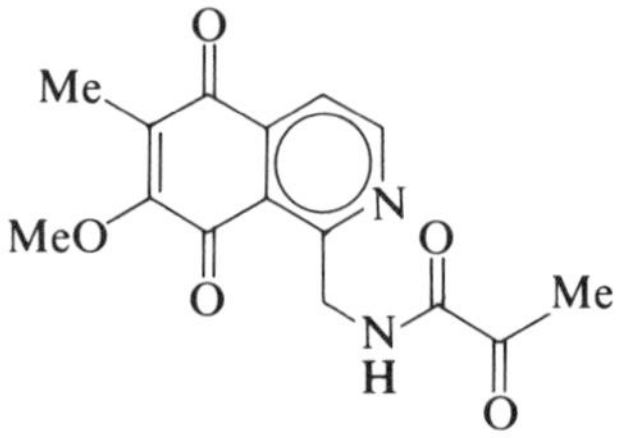

(**234**)

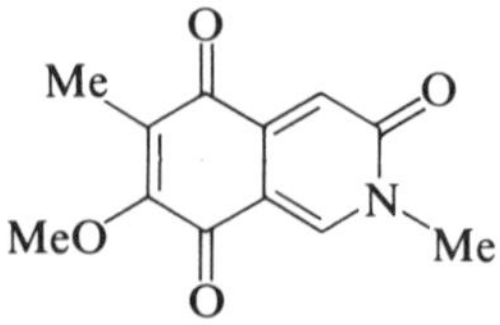

(**235**) 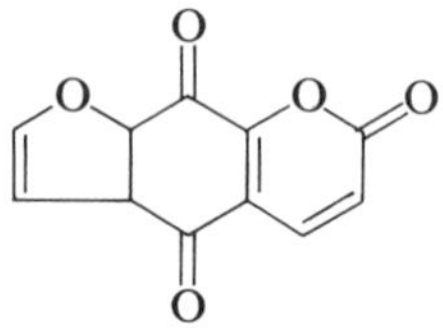 (**236**)

noteworthy features. Of course the quinol may present manipulation problems and it is sometimes more convenient to form a derivative of the hydroxyl groups *in situ*. An example of this procedure is the formation of **232** and **233** by reduction with zinc dust and acetic acid in the presence of acetic anhydride of the corresponding quinones. These quinol derivatives were used as intermediates in synthesis yielding the antibiotics mimocin (**234**)[92] and mimosamycin (**235**)[93], first isolated from the culture medium of *Streptomyces lavendulae*. The reduction of a quinone and the formation of derivatives of the quinol by the action of di- and trimethyl esters of phosphorous acid upon the quinone, psoralen (**236**) have been investigated[94]. The chromone quinone (**237**) yielded **238** with trimethyl phosphite[94].

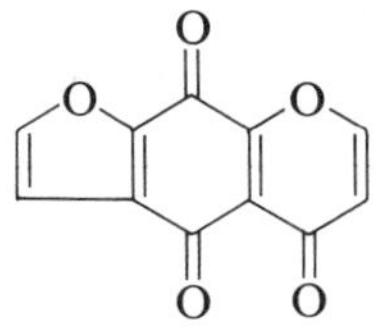

(**237**) 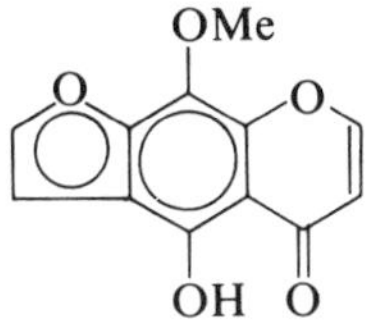 (**238**)

Thiele acetylation has been used in the heterocyclic series of quinones in order to introduce a 2-acetoxy group into a 1,4-quinone. Perhaps a more unusual use of the reaction is shown in the conversion of the *o*-quinone **240**, which is obtained from the Nenitzescu reaction derived 5-hydroxyindole **239**, into the triacetyl derivative of the *p*-quinol **241** and then into the *p*-quinone **242**[95].

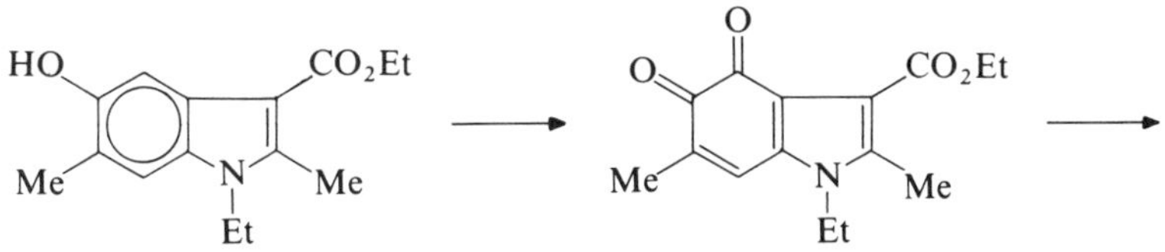

(**239**) (**240**)

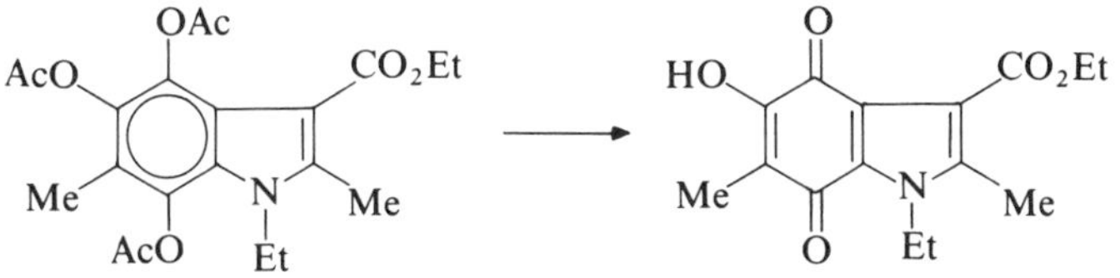

(**241**) (**242**)

Intramolecular cyclization of 3-alkyl-2-hydroxy-1,4-quinones, where the alkyl group carries a reactive function (usually hydroxyl or olefin) in the γ-position, is an extensively used route (the Hooker reaction) to pyran-containing quinones. There are many examples where the starting quinone is a 1,4-naphthoquinone and the product is a naphthopyran quinone but there seem to be very few examples of the reaction where the initial quinone is heterocyclic. An important and increasingly numerous group of naturally occurring naphthopyran quinones with the heteroatom in the β-position is being identified. Eleutherin (**243**) and isoeleutherin (**244**) are the simplest representatives of the group and

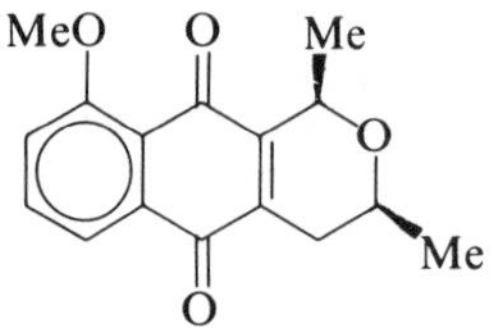

(243) 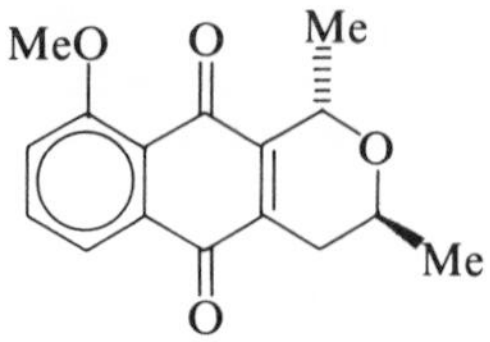 (244)

illustrate that the compounds may be considered to be derivatives of juglone. The aerial oxidation of nanaomycin A (**245**) to a mixture of nanaomycin D (**248**) and its enantiomer, kalafungin[96] has been explained by postulating the quinone methide **246** and the quinol lactone **247** as intermediates[97].

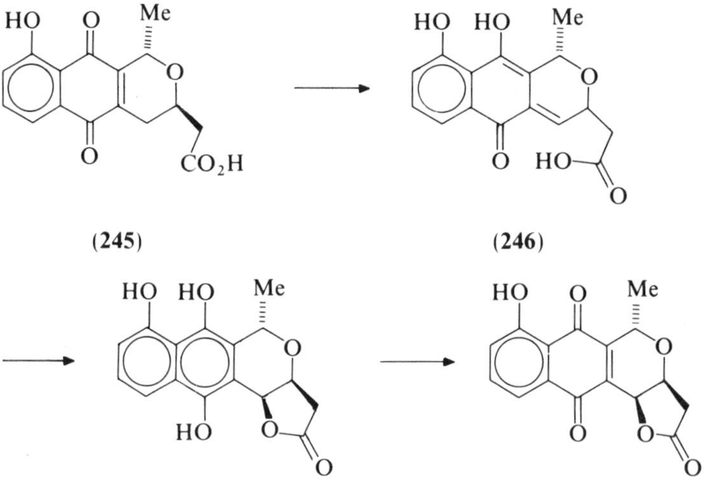

(245) (246)

(247) (248)

Extensive studies have been made of the mono- and dioximes of 2,1,3-benzothiadiazolediones and 2,1,3-benzofurazandiones[98] and configurations assigned to geometrical isomers on the basis of NMR evidence. 5-Hydroxy-4-nitroso-1,3,2-benzofurazan (**249**) is the minor component of an equilibrium mixture with the quinone monoxime **250**. A similar relationship exists between the 7-hydroxy-4-nitroso-1,3,2-benzofurazan and the *p*-quinone monoxime **251**[99]. In their nitroso form these compounds can take part in the

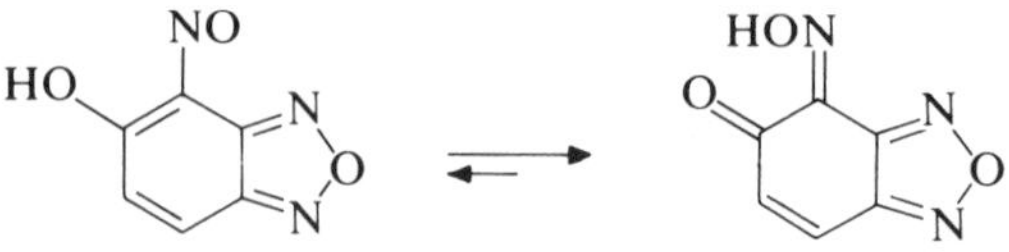

(249) (250) 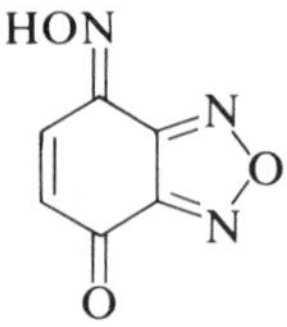 (251)

Boulton–Katritzky rearrangement, **252** to **253**, so that in organic solvents the major form of **249** is **250**, while in aqueous solvents the *p*-quinone derivative **251** is the major isomer[99]. Similar conversions from *o*- and *p*-quinones occur with 1,3,2-benzofurazan-4,5-quinone dioximes[100], presumably through the 5-hydroxyamino intermediate in this case. Some tricyclic 1,2,3-thiadiazole quinone derivatives form complexes with tetrathianaphthacene having high electrical conductivity[101].

(**252**) (**253**)

Studies have been made of the interaction of thiophene quinones with metals and of the autoxidation of coumarin quinones. Electron spin resonance has been used to investigate the nature of the species formed when dithienobenzoquinones react with organometallic derivatives of group IVB[102] and VB elements[103] and of ions formed by reduction of these quinones with alkali metals[104]. ESR studies of the photoreactions of a range of quinones, including a dithienobenzoquinone, with phosphorus derivatives show that some derived phosphorus radicals add to both carbon–carbon and carbon–oxygen double bonds whilst others are regioselective and add to the latter only[105]. The autoxidation of hydroxycoumarins in aqueous alkali has been studied by ESR spectroscopy. 6,7-Dihydroxycoumarins undergo normal autoxidation through semiquinone radicals in dilute alkali to give the corresponding *o*-quinone. In stronger alkali the pyrone ring is cleaved to produce hydroxylated cinnamic acid semiquinones[106]. The inhibition of succinate oxidase activity by oxidized 6,7- and 7,8-dihydroxycoumarins is thought to be due to the activity of the semiquinone radicals formed in the oxidation[107].

The coenzyme, methoxatin (**254**), isolated from the methanol dehydrogenases of methylotrophic bacteria, has been mentioned earlier (Section II.A.2.f). It has been found to serve as a covalently bound coenzyme for bovine serum amine oxidase and it may be required as a cofactor for other mammalian enzymes. Indeed, methoxatin may be a vitamin. Methoxatin has been synthesized (see Section II.A.2.f) but it is not a readily available compound. Recently, the more accessible 7,9-didecarboxymethoxatin (**255**) has been prepared and shown to have electrochemical properties indistinguishable from those of methoxatin. Comparisons have been made of both the redox and amine oxidizing abilities of a series of related tricyclic quinones. The monocarboxylic acid **255** oxidizes primary but not secondary amines and, under certain conditions in the presence of oxygen

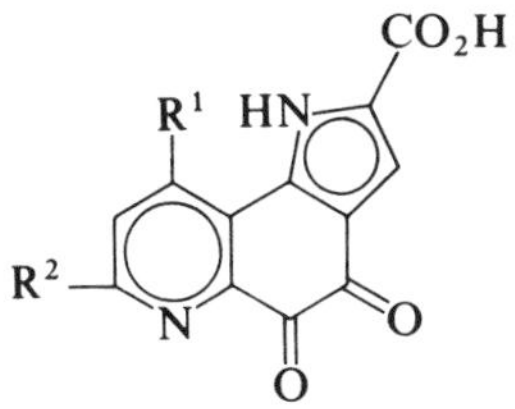

(**254**; $R^1 = R^2 = CO_2H$)
(**255**; $R^1 = R^2 = H$)

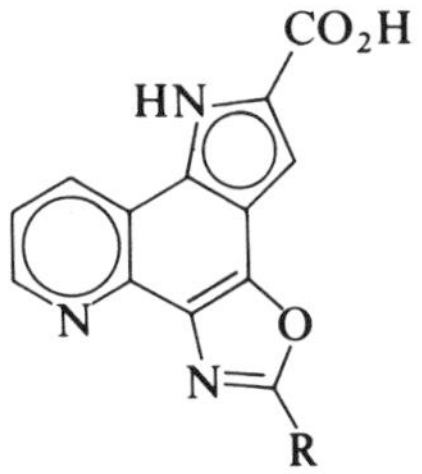

(**256**; R = Ph)
(**257**; R = H)

and primary amines, **255** is converted into oxazoles; for example **256** from benzylamine and **257** from glycine, presumably by decarboxylation following oxazole formation[108] in the latter case.

The naturally occurring antibiotic, mitomycin C (**258**) is an antineoplastic agent but, unfortunately, is highly toxic. The anticancer action of mitomycin is known to require an initial reduction step; perhaps in the first place to the semiquinone radical (**259**) and then to the quinol (**260**). This compound is now activated to form the vinylogue of a quinone methide (**261**) which is attacked by a nucleophile (DNA in the cell) to give **262** and this may subsequently cross-link by displacement of the carbamate to yield **263**. The mechanism of the biological activity of the mitomycins is being explored extensively[109].

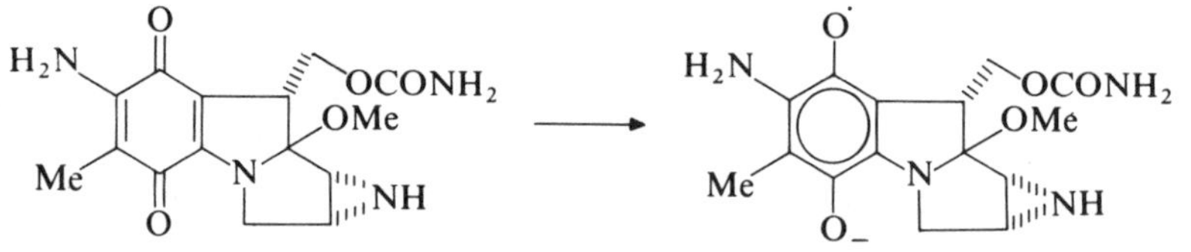

H_2N $OCONH_2$ OMe Me NH O O N

(**258**)

H_2N $OCONH_2$ OMe Me NH O˙ O_- N

(**259**)

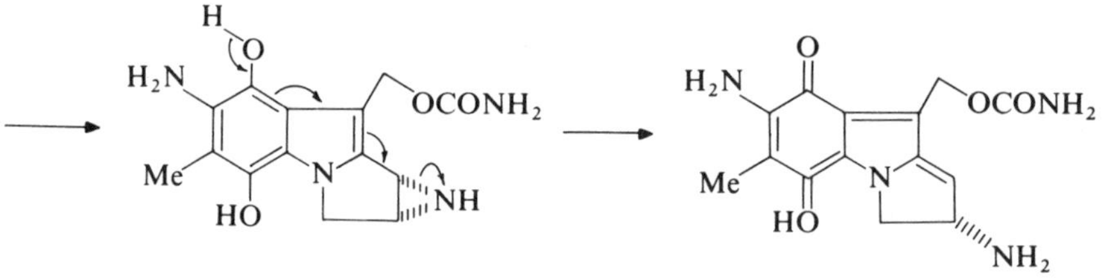

H O H_2N $OCONH_2$ Me N NH HO

(**260**)

O H_2N $OCONH_2$ Me N HO NH_2

(**261**)

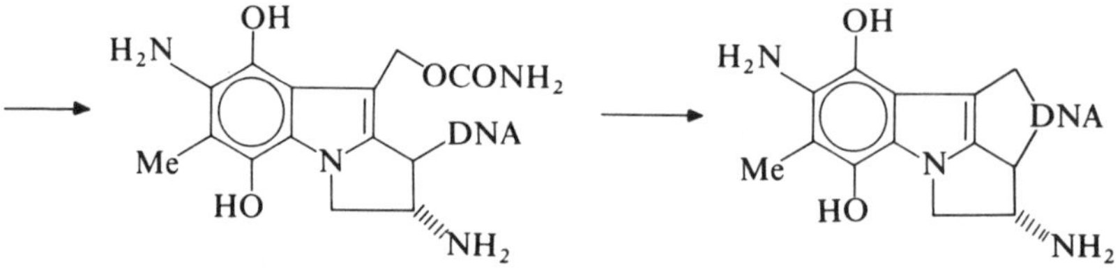

OH H_2N $OCONH_2$ DNA Me N HO NH_2

(**262**)

OH H_2N DNA Me N HO NH_2

(**263**)

The proposed mode of action of mitomycin described above is a fairly complex example of an antitumour agent taking part in a process that has become known as bioreductive alkylation[110]. Support for this mechanism occurring in chemical systems has been obtained with mitomycin C[111, 112]. It has been suggested that the principle of bioreductive activation to yield a highly reactive quinone methide may be usefully exploited in compounds which would be expected to show selective toxicity to oxygen-deficient (hypoxic) cells in solid tumours. These cells are known to be resistant to killing by X-rays and may also be resistant to chemotherapeutic agents. In a simple example, the principle of bioreductive activation and subsequent alkylation can be illustrated by the conversion of **264** through to **265**. The idea of quinone methide formation in biological systems has been

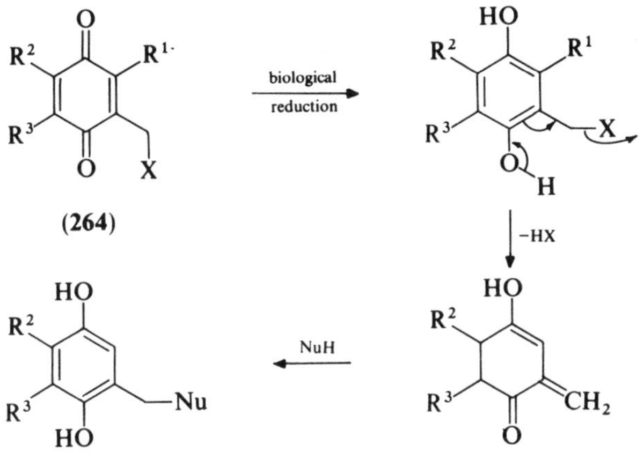

biological reduction

−HX

NuH

(264)

(265)

extended to include routes involving reductive phenolic deoxygenation and quinone–quinone methide equilibration. Nearly 200 naturally occurring quinones, many of them heterocyclic, have been shown to possess the structural features necessary for quinone methide formation by one of the three mechanisms[113]. In most cases the medicinal applications of these quinones have not been evaluated.

IV. QUINONE METHIDES

Heterocyclic quinone methides are a little investigated group of compounds, and this is particularly so for methides derived from those heterocyclic quinones which fall within the definition and limitations set in the introduction to this chapter. However, quinone methides which can be considered to be derived from compounds having a 1,2-dicarbonyl system as part of the heterocyclic moiety are well recognized and are the subject of considerable interest. The most extensively investigated example is indole-2,3-quinodimethane (**266**), a derivative of isatin (**267**) which is not usually considered to be a quinone and certainly was not recognized as such in the previous part of the chapter. Perhaps some justification for considering here such compounds as **266** is necessary.

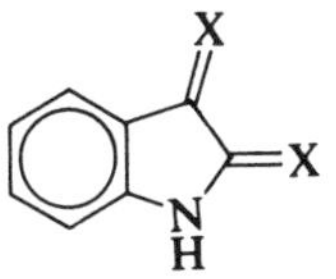

(**266**; X = CH_2)
(**267**; X = O)

In a discussion of carbocyclic non-benzenoid quinones, Turney[114] quotes a definition due to Trost of non-benzenoid quinones as any dicarbonyl species whose two-electron reduction product would generate a non-benzenoid aromatic system. If the scope of the definition is widened to include heterocyclic systems, then isatin falls into the category of a quinone, since two electrons are used in order to reduce the dicarbonyl system. In this way, **266** can be seen as a methide derived from a quinone. Probably the most important

(267) $+6H^+ + 6e \longrightarrow$ [indole] $+ 2H_2O$

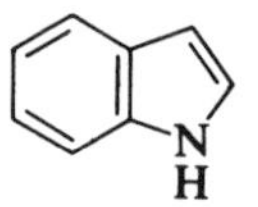

justification for a brief discussion of these compounds in this chapter is that they are useful synthetic intermediates whose importance is likely to increase.

The major interest in heterocyclic quinone methides is as synthetic intermediates which can either be trapped in intermolecular reactions with dienophiles or can undergo intramolecular reactions with a suitably disposed unsaturated system within the same molecule. The ease with which the heterocyclic quinone methides can be generated is crucial in the development of their use, and several routes have been devised. For example, the 2,3-bisbromomethylindole **268** yields **270** on treatment with sodium iodide[115] and the quaternary ammonium salt **269** gives **271** upon treatment with fluoride ion[116]. The

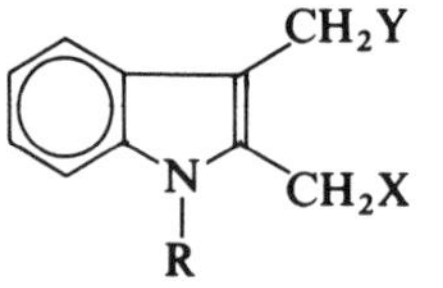

(**268**; R = COPh, X = Y = Br)
(**269**; R = Me, X = $SiMe_3$, Y = $\overset{+}{N}Me_3\bar{I}$)

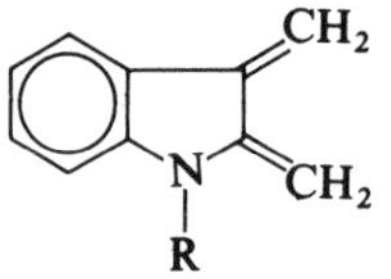

(**270**; R = COPh)
(**271**; R = Me)

reactive methides such as **270** or **271** are readily trapped with dimethyl acetylenedicarboxylate to yield carbazoles or with *N*-phenylmaleimide or *p*-benzoquinone to give tetracyclic compounds. Similar intermolecular Diels–Alder reactions have been used to trap methides generated from tricyclic systems of the type **272**, where X is the heteroatom O, S or Se or the group NR^1. All these methods require that the indole nitrogen atom be protected. However, mild thermolysis of the readily available and unprotected pyrano[3,4-6] indole-3-ones (**273**) in the presence of acetylenes yield carbazoles **274**[117].

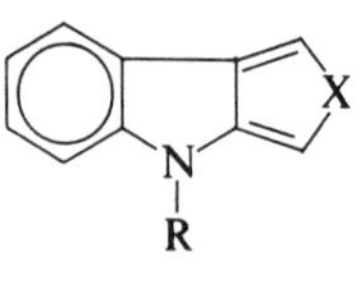

$R^2C{\equiv}CR^2$

(272) 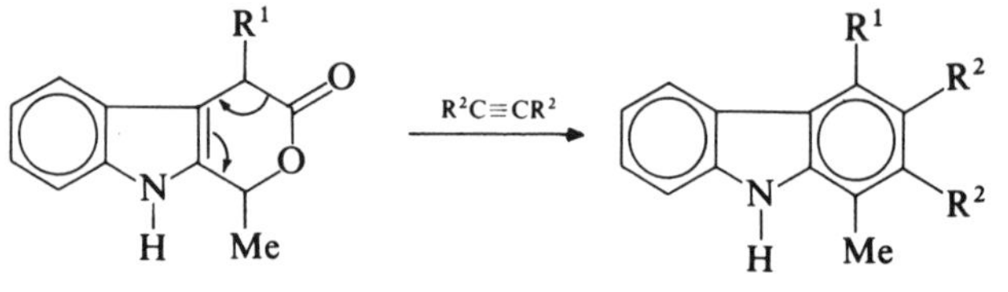 (273) (274)

The concept of forming and trapping a heterocyclic quinodimethane has been taken one stage further by the inclusion of a dienophilic group within the quinodimethane structure at such a position that trapping by an intramolecular Diels–Alder cyclization occurs. This elegant approach to certain indole alkaloids has been exploited particularly by Magnus and coworkers[118]. The use of the intramolecular Diels–Alder process is illustrated by the route from **275** to the heterocyclic **277** via the quinodimethane intermediate **276**.

Some discussion of additional aspects of quinone methide chemistry is to be found in Section III.D and quinone imines are mentioned in Section II.A.2.C and occur in a reaction scheme in Section II.C.1.

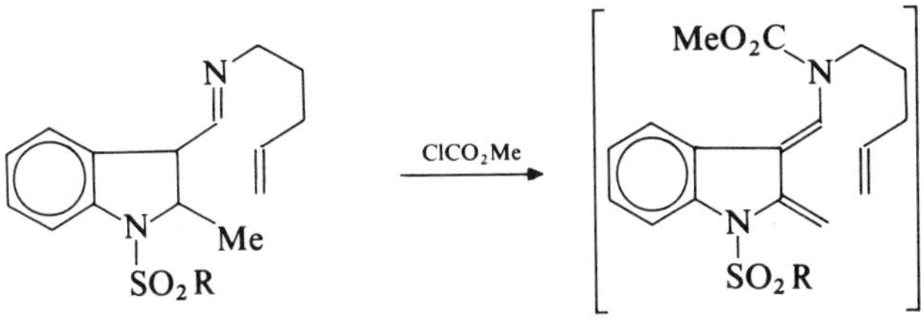

(275) (276)

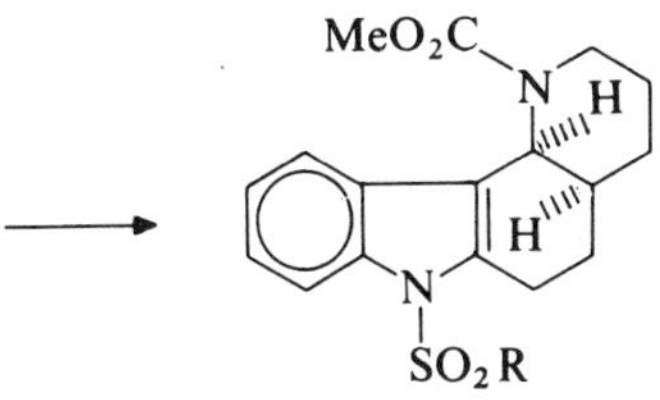

(277)

R = C_6H_4OMe-*p*

V. DIAZAQUINONES

One of the few simple heteroaromatic systems not mentioned this far is phthalazine. This is because phthalazine quinones of the types so far discussed do not appear to have been recorded. However, 6-hydroxypyridazin-3(2*H*)-one (maleic hydrazide) (**278**), 4-hydroxy-1(2*H*)-phthalazinone and other heterocycles containing the cyclic hydrazide system are potentially capable of undergoing oxidation to the reactive diazaquinone: for example, the formation of **279** from **278**. The usual oxidizing agents are *t*-butyl

(278) (279)

hypochlorite or lead tetraacetate[119,120] but chlorine and nickel peroxide have been used. The first of these reagents has been used in the formation of a number of diazaquinones from the cyclic hydrazides derived from a range of heterocyclic 1,2-dicarboxylic acids[121].

Diazaquinones are generated *in situ* in the reaction mixture but some can be isolated at low temperature[121]. They readily undergo addition reactions and, indeed, are among the most powerful dienophiles. As expected then, diazaquinones undergo [4 + 2] cycloaddition reactions of the Diels–Alder type with a wide range of dienes to form cyclic systems having two bridgehead nitrogen atoms. Thus, maleic hydrazide (**278**) reacts with 2,3-disubstituted butadienes in the presence of an oxidizing agent to give products of the type **280**. Similarly, 4-hydroxy-1(2*H*)-phthalazinone (**281**) yields **283** and **284** with cyclohexa-1,3-diene (**285**) and 1-methylpyridin-2(1*H*)-one (**286**) via the diazaquinone **282**[122].

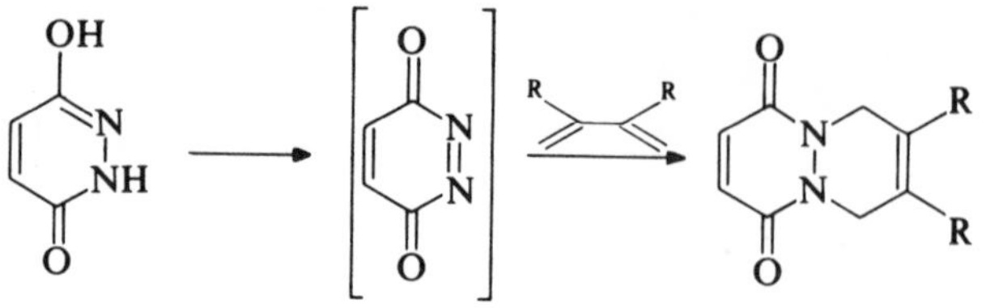

OH N NH O → O N N O → R R → O N N O R R

(280)

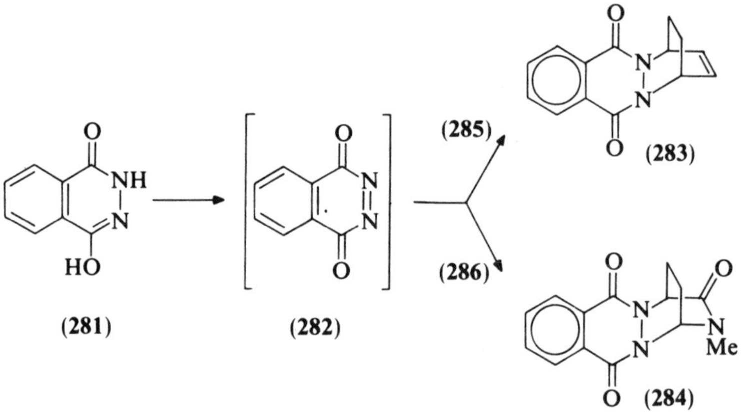

O NH N HO → O N N O → (285) O N N O (283)

(281) (282) (286) O N N O N Me (284)

Diazaquinones also undergo [2 + 2] cycloaddition reactions with olefins. For instance, diazaquinone **279** reacts with styrene to give the diazacyclobutane **287** which produces the ring-opened product **288** on treatment with water or *t*-butanol. A similar reaction occurs with enamines. For instance, the enamine **289** reacts with phthalazinedione **282** to give **290**[123].

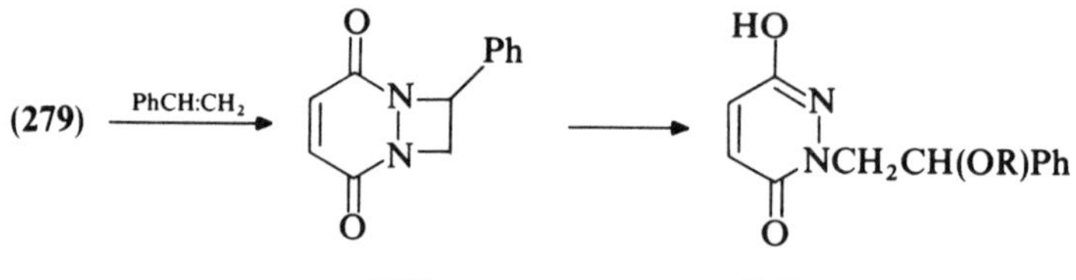

(279) PhCH:CH$_2$ → O N N O Ph → HO N $NCH_2CH(OR)Ph$ O

(287) (288)

CO_2Et Me O NH_2 → OH N N CO_2Et O HN O

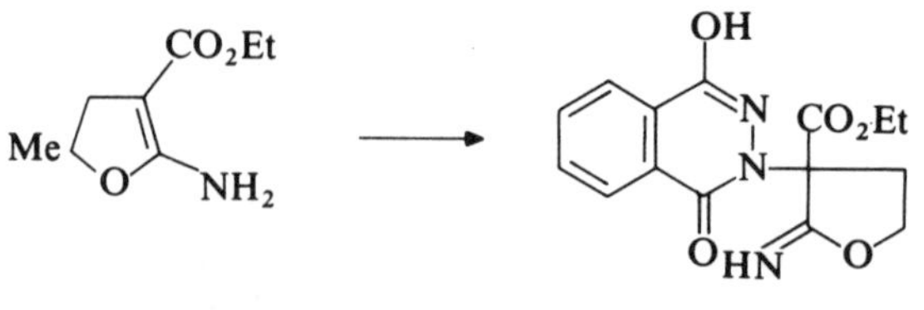

(289) (290)

The mechanistic interpretations of the chemiluminescence of luminol (**291**) under oxidizing conditions have included the formation of the intermediate diazaquinone **292** at an early stage in the reaction process[121, 124]. Evidence in support of this includes the fact that the presence of cyclopentadiene in the reaction mixture prevents chemiluminescence as does the presence of 2- or 3-substituents on hydroxyphthalazinone nucleus. Recent studies of the reaction kinetics[125] lend weight to earlier suggestions that a carbon-centred peroxide (**293**) is formed from the diazaquinone (**292**). The peroxide **293** decomposes

(**291**) (**292**) (**293**)

through several steps to the aminophthalate ion with the emission of light. The mechanistic details of the chemiluminescent processes under both protic and aprotic conditions have yet to be firmly established.

VI. REFERENCES

1. I. Baxter and B. A. Davis, *Chem. Soc. Quart. Rev.*, **25**, 239 (1971).
2. W. M. Horspool, *Chem. Soc. Quart. Rev.*, **23**, 204 (1969).
3. M. F. Sartori, *Chem. Rev.*, **63**, 279 (1963).
4. N. L. Agarwal and W. Schafer, *J. Org. Chem.*, **45**, 5144 (1980).
5. A. McKillop and D. W. Young, *Synthesis*, 422 and 481 (1979).
6. C. Temple, J. D. Rose and J. A. Montgomery, *J. Med. Chem.*, **17**, 615 (1974).
7. Y.-P. Wan, T. H. Porter and K. Folkers, *J. Heterocyclic Chem.*, **11**, 519 (1974).
8. M. D. Friedman, P. L. Stotter, T. H. Porter and K. Folkers, *J. Med. Chem.*, **16**, 1314 (1973).
9. S. K. Mukerjee, T. Saroja and T. R. Seshadri, *Tetrahedron*, **24**, 6527 (1968).
10. K. V. Rao and H.-S. Kuo, *J. Heterocyclic Chem.*, **16**, 1241 (1979).
11. H. Link, K. Bernauer and G. Englert, *Helv. Chim. Acta*, **65**, 2645 (1982).
12. G. Malesani, G. Chiareletto, M. G. Ferlin and S. Masiero, *J. Heterocyclic Chem.*, **18**, 613 (1981).
13. J. R. Luly and H. Rapoport, *J. Am. Chem. Soc.*, **105**, 2859 (1983).
14. A. V. El'tsov and L. S. Efros, *J. Gen. Chem. U.S.S.R.*, **30**, 3319 (1960).
15. W. C. Fleming and G. R. Pettit, *J. Org. Chem.*, **36**, 3490 (1971).
16. A. V. Luk'yanov, V. G. Voronin and Yu. S. Tsizin, *Khim. Geterosikl. Soedin.*, 196 (1971); *Chem. Abstr.*, **75**, 35925f (1971).
17. G. N. Kurilo, N. L. Rostova, A. A. Cherkasova, K. F. Turchin, L. M. Alekseeva and A. N. Grinev, *Khim. Geterosikl. Soedin.*, 1374 (1980); *Chem. Abstr.*, **94**, 121229 (1981).
18. T. F. Spande, in *Indoles*, Part 3 (Ed. W. J. Houlihan), J. Wiley Interscience, New York, 1979, p. 143.
19. G. R. Pettit, W. C. Fleming and K. D. Paul, *J. Org. Chem.*, **33**, 1089 (1968).
20. J. Mott and W. A. Remers, *J. Med. Chem.*, **21**, 493 (1978).
21. J. Renault, S. G. Renault, M. Baron, P. Mailliet, C. Paoletti, S. Cros and E. Voisin, *J. Med. Chem.*, **26**, 1715 (1983).
22. J. D. Warren, V. J. Lee and R. B. Angier, *J. Heterocyclic Chem.*, **16**, 1617 (1979).
23. C. D. Snyder, W. E. Bondinell and H. Rapoport, *J. Org. Chem.*, **36**, 3951 (1971).
24. G. Kraus, H. Cho, S. Crowley, B. Roth and H. Sugimoto, *J. Org. Chem.*, **48**, 3439 (1983).
25. T. Kometani, Y. Takendi and E. Yoshu, *J. Org. Chem.*, **47**, 4725 (1982).

26. R. G. Giles, I. R. Green, V. I. Hugo and P. R. K. Mitchell, *J. Chem. Soc., Perkin Trans. 1*, 2383 (1984).
27. I. Baxter and W. R. Phillips, *J. Chem. Soc., Perkin Trans. 1*, 2374 (1973).
28. S. N. Falling and H. Rapoport, *J. Org. Chem.*, **45**, 1260 (1980).
29. A. Kubo, Y. Kitahara, S. Nakahara and R. Numata, *Chem. Pharm. Bull.*, **31**, 341 (1983).
30. E. J. Corey and A. Tramontano, *J. Am. Chem. Soc.*, **103**, 5599 (1981).
31. A. R. Mackenzie, C. J. Moody and C. W. Rees, *J. Chem. Soc., Chem. Commun.*, 1372 (1983).
32. J. A. Gainor and S. M. Weinreb, *J. Org. Chem.*, **46**, 4319 (1981).
33. K. V. Rao, *J. Heterocyclic. Chem.*, **14**, 653 (1977).
34. L. Di Nunno and S. Florio, *Tetrahedron*, **33**, 855 (1977).
35. R. D. Gillard, R. P. Houghton and J. N. Tucker, *J. Chem. Soc., Dalton Trans.*, 2102 (1980).
36. G. Jones and R. K. Jones, *J. Chem. Soc., Perkin Trans. 1*, 26 (1973).
37. A. V. El'tsov, V-Yu. Kukushkin and L. M. Bykova, *Zh. Obshch. Khim.*, **51**, 2116 (1981), *Chem. Abstr.*, **96**, 104128w (1982).
38. B. L. Chenard, J. R. McConnell and J. S. Swenton, *J. Org. Chem.*, **48**, 4312 (1983).
39. K. A. Parker, I. D. Cohen, A. Padwa and W. Dent, *Tetrahedron Lett.*, **25**, 4917 (1984).
40. D. W. H. MacDowell and J. C. Wisowaty, *J. Org. Chem.*, **37**, 1712 (1972).
41. D. V. Nightingale and H. L. Needles, *J. Heterocyclic Chem.*, **1**, 74 (1964).
42. M. S. Newman and K. G. Ihrman, *J. Am. Chem. Soc.*, **80**, 3652 (1958).
43. D. W. Cameron, K. R. Deutscher, G. I. Feutrill and D. E. Hunt, *Aust. J. Chem.*, **35**, 1451 (1982).
44. F. Minisci and O. Porta, in *Advances in Heterocyclic Chemistry* (Eds A. R. Katritzky and A. J. Boulton), Academic Press, New York and London, 1974, Vol. 16, pp. 123–180.
45. E. Ghera, Y. Gaoni and D. H. Perry, *J. Chem. Soc., Chem. Commun.*, 1034 (1974).
46. D. A. Taylor, M. M. Boradarani, S. J. Martinez and J. A. Joule, *J. Chem. Res. (S)*, 387 (1979); (*M*) 4801 (1979).
47. M. G. Beal, W. B. Ashcroft, M. M. Cooper and J. A. Joule, *J. Chem. Soc., Perkin Trans. 1*, 435 (1982).
48. M. G. Saulnier and G. W. Gribble, *J. Org. Chem.*, **48**, 2690 (1983).
49. M. Watanabe and V. Sniekus, *J. Am. Chem. Soc.*, **102**, 1457 (1980).
50. C. Robaut, C. Rivalle, M. Rautureau, J. M. Lhoste, E. Bisagni and J.-C. Chermann, *Tetrahedron*, **41**, 1945 (1985).
51. D. W. Slocum and P. L. Gierer, *J. Org. Chem.*, **41**, 3668 (1976).
52. W. D. Wulff, K. S. Chen and P. C. Tang, *J. Org. Chem.*, **49**, 2293 (1984).
53. E. Mueller and W. Winter, *Justus Liebigs Ann. Chem.*, 1876 (1974).
54. J. Hambrecht and E. Mueller, *Justus Liebigs Ann. Chem.*, 387 (1977).
55. E. Mueller, E. Luppold and W. Winter, *Chem. Ber.*, **108**, 237 (1975).
56. H. Wynberg and H. J. M. Synnige, *Rec. Trav. Chim. Pays-Bas*, **88**, 1244 (1969).
57. W. Wierenga, *J. Am. Chem. Soc.*, **103**, 5298 (1981).
58. P. Magnus and T. Gallagher, *J. Chem. Soc., Chem. Commun.*, 389 (1984).
59. V. Guay and P. Brassard, *J. Org. Chem.*, **49**, 1853 (1984).
60. W. Schäfer and C. Falkner, *Justus Liebigs Ann. Chem.*, 1445 (1977).
61. N. L. Agarwal and W. Schäfer, *J. Org. Chem.*, **45**, 2155 (1980).
62. F. Nakatsubo, A. J. Cocuzza, D. E. Keeley and Y. Kishi, *J. Am. Chem. Soc.*, **99**, 4835 (1977).
63. F. Nakatsubo, F. Fukuyama, A. J. Cocuzza and Y. Kishi, *J. Am. Chem. Soc.*, **99**, 8116 (1977).
64. G. A. Conway, L. F. Loeffler and I. H. Hall, *J. Med. Chem.*, **26**, 876 (1983).
65. D. R. Buckle, H. Smith, B. A. Spicer and J. M. Tedder, *J. Med. Chem.*, **26**, 714 (1983).
66. S. Shiraishi, B. S. Holla and K. Imamura, *Bull. Chem. Soc. Jpn*, **56**, 3457 (1983).
67. T. Hayakawa, K. Araki and S. Shiraishi, *Bull. Chem. Soc. Jpn*, **57**, 1643 and 2216 (1984).
68. P. Gilgen, B. Jackson, H. J. Hansen, H. Heimgarten and H. Schmid, *Helv. Chim. Acta*, **57**, 2634 (1974).
69. J. A. Myers, L. D. Moore, W. L. Whitter, S. L. Council, R. M. Waldo, J. L. Lanier and B. U. Omoji, *J. Org. Chem.*, **45**, 1202 (1980).
70. T. Uchida, S. Tsubokawa, K. Harihara and K. Matisumoto, *J. Heterocyclic Chem.*, **15**, 1303 (1978).
71. H. Gotthardt, C. M. Weisshuhn and B. Christl, *Chem. Ber.*, **111**, 3037 (1978).
72. G. M. L. Cragg, G. R. F. Giles and G. H. P. Ross, *J. Chem. Soc., Perkin Trans. 1*, 1339 (1975).
73. A. A. Hofmann, I. Wyrsch-Walraf, P. X. Iten and C. H. Eugster, *Helv. Chim. Acta*, **62**, 2211 (1979).

74. G. Weissberger and C. H. Eugster, *Helv. Chim. Acta*, **49**, 1806 (1966).
75. M. Matsumoto, K. Kuroda and Y. Suzuki, *Tetrahedron Lett.*, **22**, 3253 (1981).
76. D. W. Cameron, K. R. Deutscher and G. I. Feutrill, *Aust. J. Chem.*, **35**, 1439 (1982).
77. K. T. Potts and D. Bhattacharjee, *Synthesis*, 31 (1983).
78. K. T. Potts, D. Bhattacharjee and E. B. Walsh, *J. Chem. Soc., Chem. Commun.*, 114 (1984).
79. D. W. Cameron, K. R. Deutscher, G. I. Feutrill and D. E. Hunt, *Aust. J. Chem.*, **35**, 1451 (1982).
80. S. Omura, A. Nakagawa, H. Aoyama, K. Hinotozawa and H. Sano, *Tetrahedron Lett.*, **24**, 3643 (1983).
81. W. Ott, V. Formacek and M. Hubert, *Justus Liebigs Ann. Chem.*, 1003 (1984).
82. R. F. C. Brown, G. D. Fallon, B. M. Gateheruse, C. M. Jones and I. D. Rae, *Aust. J. Chem.*, **35**, 1665 (1982).
83. J. Mtochowski, K. Kloc and J. Piatkowska, *Heterocycles*, **19**, 1889 (1982).
84. C. Temple, J. D. Rose and J. A. Montgomery, *J. Med. Chem.*, **17**, 972 (1974).
85. T. H. Porter, F. S. Skelton and K. Folkers, *J. Med. Chem.*, **15**, 34 (1972).
86. T. H. Porter, F. S. Skelton and K. Folkers, *J. Med. Chem.*, **14**, 1029 (1971).
87. W. Gauss, H. Heitzer and S. Petersen, *Justus Liebigs Ann. Chem.*, **764**, 131 (1972).
88. T. K. Liao, W. H. Nyberg and C. C. Cheng, *J. Heterocyclic Chem.*, **13**, 1063 (1976).
89. R. H. Thomson and R. D. Worthington, *J. Chem. Soc., Perkin Trans. 1*, 282 (1980).
90. J. M. Tedder and D. R. Buckle, *J. Chem. Res. (S)*, 12 (1983).
91. G. Green-Buckley and J. Griffiths, *J. Chem. Soc., Perkin Trans. 1*, 702 (1979).
92. K. Matsuo, M. Okumura and K. Tanaka, *Chem. Pharm. Bull. Jpn*, **30**, 4170 (1982).
93. H. Fukumi, H. Kurihara and H. Mishima, *Chem. Pharm. Bull. Jpn*, **26**, 2175 (1978).
94. M. R. Mahran, W. B. Abdon and T. S. Hafez, *Egypt J. Chem.*, **24**, 401 (1982); *Chem. Abstr.*, **99**, 38285d (1983).
95. G. R. Allen Jr and M. J. Weiss, *J. Med. Chem.*, **10**, 1 (1967).
96. S. Omura, H. Tanaka, Y. Okada and H. Marumo, *J. Chem. Soc., Chem. Commun.*, 320 (1976).
97. T. Li and R. Ellison, *J. Am. Chem. Soc.*, **100**, 6263 (1978).
98. A. S. Angeloni, D. Dal Monte, S. Pollicino, E. Sandri and G. Scapini, *Tetrahedron*, **30**, 3839 (1974).
99. A. S. Angeloni, D. Dal Monte, E. Sandri and G. Scapini, *Tetrahedron*, **30**, 3849 (1974).
100. A. S. Angeloni, V. Cere, D. Dal Monte and E. Sandri, *Tetrahedron*, **28**, 303 (1972).
101. Y. Yamashita, T. Suzuki, T. Mukai and G. Saito, *J. Chem. Soc., Chem. Commun.*, 1044 (1985).
102. A. Alberti, A. Hudson, G. F. Pedulli and P. Zanirato, *J. Organomet. Chem.*, **198**, 145 (1980).
103. A. Alberti, A. Hudson, A. Maccioni, G. Podda and G. F. Pedulli, *J. Chem. Soc., Perkin Trans. 2*, 1274 (1981).
104. G. F. Pedulli, A. Alberti, L. Testaferri and M. Tiecco, *J. Chem. Soc., Perkin Trans. 2*, 1701 (1974).
105. A. Alberti, A. Hudson, G. F. Pedulli, W. G. McGimpsey and J. K. S. Wan, *Can. J. Chem.*, **63**, 917 (1985).
106. P. Ashworth, *Tetrahedron Lett.*, 1435 (1975).
107. P. Zboril, *Coll. Czech. Chem. Commun.*, **45**, 641 (1980).
108. P. R. Sleath, J. B. Noar, G. A. Eberlein and T. C. Bruce, *J. Am. Chem. Soc.*, **107**, 3328 (1985).
109. S. K. Carter and S. T. Crooke (Eds), *Mitomycin C: Current Status and New Developments*, Academic Press, New York, 1979.
110. A. J. Lin, L. A. Cosby, C. W. Shansky and A. C. Sartorelli, *J. Med. Chem.*, **15**, 1247 (1972).
111. P. J. Keller, J. F. Kozlowski and U. Hornemann, *J. Am. Chem. Soc.*, **101**, 7121 (1979).
112. Y. Hashimoto, K. Shudo and T. Okamoto, *Chem. Pharm. Bull.*, **28**, 1961 (1980).
113. H. W. Moore, K. F. West, K. Srinivasacher and R. Czerniak, in *Structure–Activity Relationships of Anti–tumour Agents* (Eds D. N. Reinhoudt, T. A. Connors, H. M. Pinedo and K. W. van de Poll), Martinus Nijhoff Publishers, The Hague, 1983, pp. 93–110.
114. T. A. Turney, in *The Chemistry of Quinonoid Compounds*, Part 2 (Ed. S. Patai), John Wiley Interscience, London and New York, 1974, p. 857.
115. B. Saroja and P. C. Srinivasan, *Tetrahedron Lett.*, **25**, 5429 (1984).
116. E. R. Marinelli, *Tetrahedron Lett.*, **23**, 2745 (1982).
117. C. J. Moody, *J. Chem. Soc., Perkin Trans. 1*, 2505 (1985).
118. C. Exon, T. Gallagher and P. Magnus, *J. Am. Chem. Soc.*, **105**, 4739 (1983).
119. T. J. Kealy, *J. Am. Chem. Soc.*, **84**, 966 (1962).
120. R. A. Clement, *J. Org. Chem.*, **27**, 1115 (1962).

121. D. B. Paul, *Aust. J. Chem.*, **37**, 1001 (1984).
122. V. V. Kane, H. Werblood and S. D. Levine, *J. Heterocyclic Chem.*, **13**, 673 (1976).
123. H. Warnhoff and K. Wald, *Chem. Ber.*, **110**, 1716 (1977).
124. M. M. Rauhaut, A. M. Semsel and B. G. Roberts, *J. Org. Chem.*, **31**, 2431 (1966).
125. T. E. Ericksen, J. Lind and G. Merenyi, *J. Chem. Soc., Faraday Trans. 1*, **77**, 2137 (1981).

The Chemistry of Quinonoid Compounds. Vol. II
Edited by S. Patai and Z. Rappoport

CHAPTER **18**

Polymerization and polymers of quinonoid compounds

SHOUJI IWATSUKI
Department of Chemical Research for Resources,
Faculty of Engineering, Mie University,
Tsu, Japan

I. INTRODUCTION

Benzoquinone is a stable, yellow solid at room temperature and benzoquinone methides in general are far more reactive to dimerization and polymerization unless stabilized by appropriate substituents. Thus 1,4-benzoquinone dimethide (QM)* is so reactive that it gives poly(*p*-xylylene) spontaneously at room temperature, and 7,7,8,8-tetracyano-1,4-

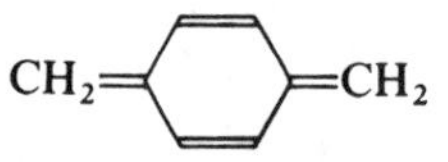

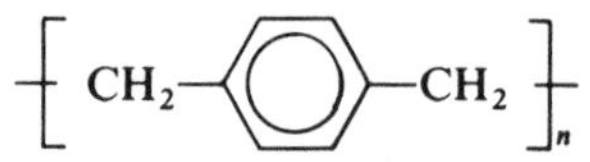

1,4-benzoquinone dimethide,
p-quinodimethane, *p*-xylylene
(QM)

poly-(*p*-xylylene)
poly-QM

quinodimethane (TCNQ) and 7,7,8,8-tetrakis(ethoxycarbonyl)-1,4-quinodimethane (TECQ) stabilized by four strongly electron-withdrawing substituents are stable indefinitely and scarcely or sparingly homopolymerizable. However, these electron-

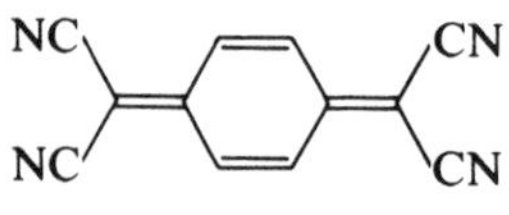

TCNQ

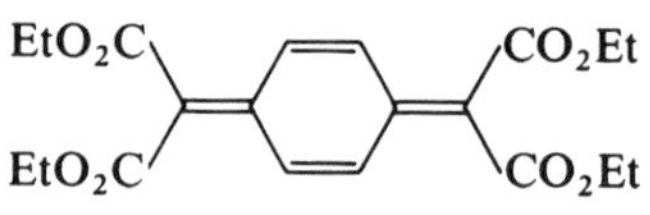

TECQ

accepting derivatives are copolymerizable readily and alternatingly with an electron-donating monomer such as styrene.

Unexpectedly, 7,8-bis(alkoxycarbonyl)-7,8-dicyanoquinodimethane (ACQ) is considered to occupy an intermediate position in physical and chemical properties between those of TCNQ and TECQ, is homopolymerizable via free radical and anionic initiation, the latter of which gives the high polymer with molecular weight above some millions, and is

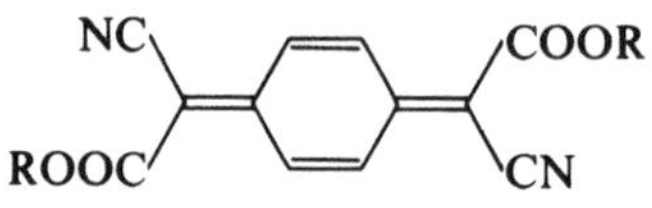

$R = CH_3$ (MCQ)
$R = C_2H_5$ (ECQ)
$R = C_4H_9$ (BCQ)

ACQ

copolymerizable in random fashion with styrene. This unexpected polymerization behavior is believed to be useful both for the chemistry of quinonoid compounds and for polymer chemistry. In this chapter, polymerization and polymers of quinonoid compounds are described with emphasis on the usual polymerization behavior since the comprehensive review of Errede and Szwarc[1] mainly for unsubstituted quinonoid compounds and others[2] are already published.

*The IUPAC nomenclature for *p*-xylylene (*p*-quinodimethane, 1,4-benzoquinone dimethide) is 1,4-dimethylene-2,5-cyclohexadiene. However, since in this review reference is frequently made to the original papers, the trivial names are retained.

II. FLASH PYROLYSIS OF *p*-XYLENE

In 1947 Szwarc prepared a white polymeric material by a rapid flow (flash) pyrolysis of *p*-xylene under reduced pressure[3]. Since *p*-xylylene diiodide was detected among the pyrolysis products with iodine gas[4] he proposed the formation of *p*-xylylene (*p*-quinodimethane) (QM) in this pyrolysis [3, 5]. He claimed the polymeric material to be poly-*p*-xylylene (poly-QM)[3] and proposed a mechanism for its formation[4] which involves thermal cleavage of carbon–hydrogen bonds of *p*-xylene to yield *p*-xylyl radicals, which in turn collide with each other to give *p*-xylene and QM through disproportionation. QM condenses and polymerizes to produce poly-QM, a high melting point substance which is inert to organic and inorganic reagents. Subsequently this material has been extensively studied on account of its unusual chemical and physical properties[6–11]. It has also attracted the interest of many quantum chemists[12–15]. Coulson and coworkers[12] calculated the energy difference of the QM molecule between the singlet ground state and the triplet excited state of QM to be 8–9 kcal mol^{-1}. The corresponding value for ethylene was determined by Evans to be 82 kcal mol^{-1} [16]. This unusually low energy difference is responsible for the very high reactivity of the QM molecule. In a series of studies on QM derivatives[17–27], Errede and Landrum prepared a 0.12 molar solution of the QM monomer. Rapid flow pyrolysis of *p*-xylene was carried out under reduced pressure of 4 mmHg at 1000°C and the pyrolysis product was condensed at −78°C to obtain solutions of monomeric QM up to 0.12 molar concentration[17]. In addition to QM, toluene, styrene, *p*-ethylstyrene, 1,2-di-(*p*-tolyl)ethane, a diarylmethane, anthracence and 4,4-dimethylstilbene were produced as by-products[21]. Even when kept at −78°C the solution of the pyrolysis product polymerizes very slowly. When an aliquot is drawn up into a warm pipet and allowed to flow back into the solution, the polymerization rate is markedly increased as shown in Figure 1, presumably due to formation of diradicals with *n*-mers. The

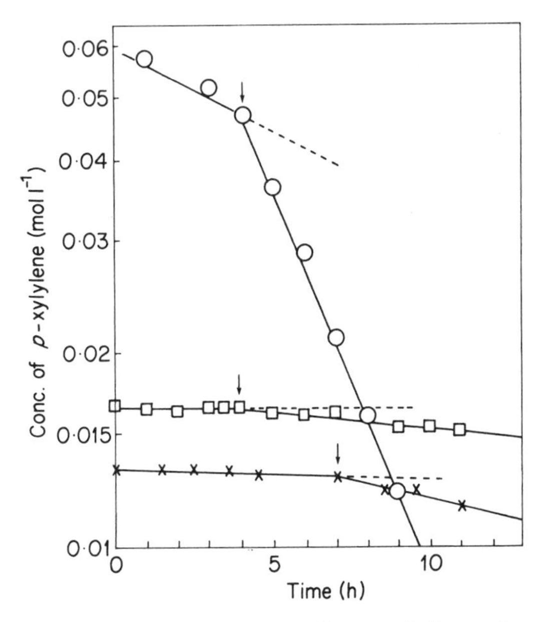

FIGURE 1. Polymerization of *p*-xylylene at −78 °C: arrow indicates time when solution was disturbed by contact with object at room temperature. *Reproduced with permission from Ref.* 22. *Copyright* (1960) *American Chemical Society*

polymerization of QM takes place by successive addition of QM monomer until all the monomer is consumed or the polymeric free radical ends are entrapped in the polymer mesh. A linear polymer with molecular weight above 2×10^5 (obtained by measuring the number of radioactive iodine end groups incorporated into the polymer when the polymerization mixture is quenched with radioactive iodine after the polymerization has proceeded to 95% completion) is obtained[22]. Errede and coworkers have found that solutions which polymerize with apparent first-order rate constants of $9 \pm 1 \times 10^{-6}\ s^{-1}$ could be reproduced fairly consistently if the solution of the pyrolysis product was filtered through a bed of crystalline *p*-xylene using an apparatus that was prechilled to −78°C. Such solutions were used to determine the rate of polymerization at various temperatures above −78°C. The rates were found to obey a first-order law with respect to monomer as shown in Figure 2. The Arrhenius plot of the apparent first-order rate constants was linear and from its slope the activation energy for the polymerization was calculated to be 8.7 kcal mol^{-1}. The first-order plot of polymerization at −78°C appears to be linear for the first 10 h but the deviation from the first-order kinetics becomes appreciable at longer reaction times as shown in Figure 3, corresponding to the slow but steady decrease in apparent rate constants. A plot of the reciprocal of the apparent rate constants against time is linear (Figure 4), indicating that the disappearance of the polymerization active sites is second-order with respect to the sites[20]. This treatment gives the ratio of apparent rate

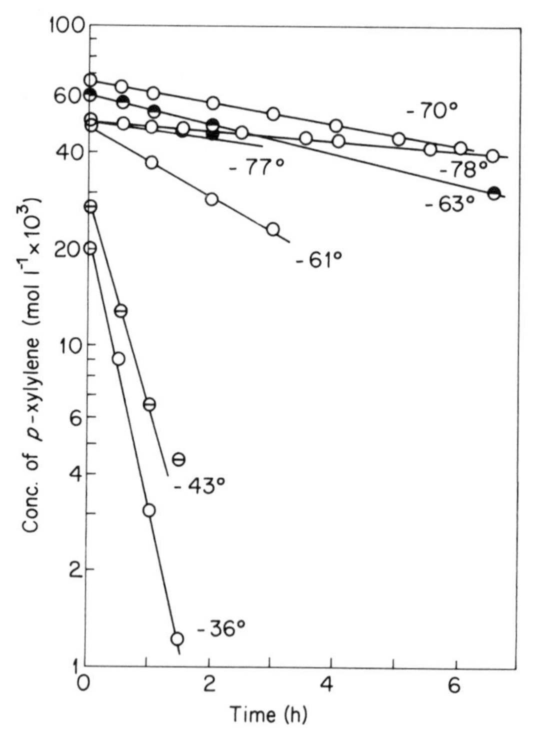

FIGURE 2. Polymerization of *p*-xylylene solution having as apparent rate constant $(k_P) = 9 \pm 1 \times 10^{-6}$ at −78 °C as a function of temperature. *Reproduced with permission from Ref.* 22. *Copyright* (1960) *American Chemical Society*

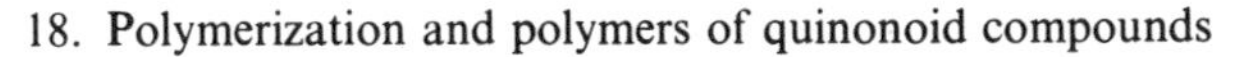

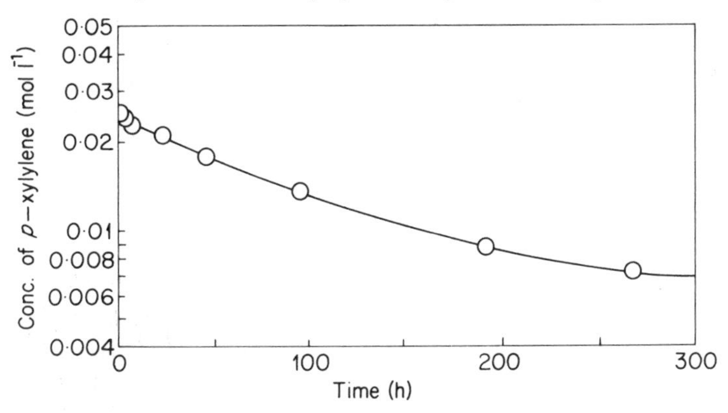

FIGURE 3. Polymerization of *p*-xylylene at $-78\,^{\circ}$C. *Reproduced with permission from Ref.* 22. *Copyright* (1960) *American Chemical Society*

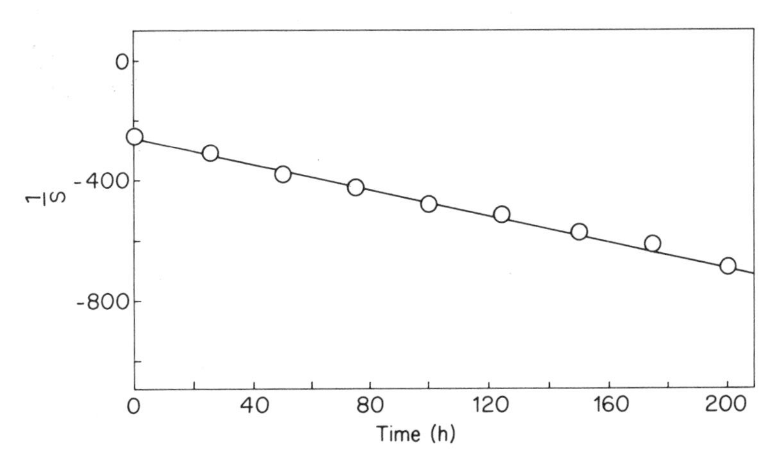

FIGURE 4. Reciprocal of the apparent rate constant (k_P) of *p*-xylylene polymerization at $-78\,^{\circ}$C as a function of time. $1/S - 1/S_0 = \frac{4.6\,K_t}{k}$ where $S = \frac{k_P}{2.3}$, then $\frac{2K}{k} = 0.904$. K is the specific rate constant for coupling of one insoluble free radical end-group with another. *Reproduced with permission from Ref.* 22. *Copyright* (1960) *American Chemical Society*

constant of disappearance of the polymerization active site to that of the polymerization to be 0.45, in sharp contrast to the conventional free radical vinyl polymerization in which termination is about $10^4 \sim 10^6$ times faster than propagation. Interestingly, one characteristics of QM polymerization is that the propagation, a radical addition reaction between a very stable polymer radical and a very reactive monomer, takes place with a rate similar to that of the termination, which is a radical coupling reaction between very stable polymer radicals.

QM does not copolymerize with conventional olefinic monomers at $-78\,^{\circ}$C in the usual way that both monomers are mixed, but the homopolymer of QM is obtained[19]. However, when a solution of QM at $-78°C$ is added to a solution of a conventional monomer maintained at about 100°C, a copolymer can be produced[19]. On the other hand, QM is copolymerizable in the usual way with pseudomonomers such as sulfur dioxide[19], nitroso compounds[19] and phosphorous trichloride[25]. When oxygen or air is bubbled through a solution of QM, QM is copolymerized with oxygen to yield poly-QM peroxide with an oxygen content ranging from 1 to 23% (molar ratios of QM to oxygen from 31:1 up to 1:1)[18]. The polymerization of QM is not influenced by conventional chain transfer agents such as carbon tetrachloride, chloroform, *p*-cumene, nitrobenzene and hydroquinone[19]. When a three-fold excess of thiophenol, a highly reactive chain transfer agent, is added to a solution of QM, a telomer with a 21:1 ratio of QM to thiophenol units is obtained, i.e. the chain transfer reaction takes place with difficulty[19].

When a solution of QM is heated at a temperature higher than $-78°C$, in addition to the insoluble high molecular weight polymer, some soluble low molecular weight products such as a cyclo-trimer, cyclo-tetramer, 1,4-bis(2′-*p*-tolylethyl)benzene and oligomers are obtained[22]. Furthermore, when a solution of QM at $-78°C$ is added dropwise to a hot inert solvent such as toluene at 100°C a cyclo-dimer is obtained in good yield[22]. The reaction scheme proposed by Errede and coworkers is shown in Scheme 1[22].

(a) Isothermal polymerization at low temperature

$$n\mathrm{CH_2{=}C_6H_4{=}CH_2} \xrightarrow[\text{surface}]{\text{warm}} \cdot(\mathrm{CH_2{-}C_6H_4{-}CH_2})_n\cdot \xrightarrow{\mathrm{M}} \cdot\mathrm{CH_2{-}C_6H_4{-}CH_2M_mCH_2{-}C_6H_4{-}CH_2}\cdot$$

$$\mathrm{CH_2{-}C_6H_4{-}CH_2M_mCH_2{-}C_6H_4{-}CH_2}\cdot + \cdot\mathrm{CH_2{-}C_6H_4{-}CH_2M_nCH_2{-}C_6H_4{-}CH_2}\cdot \longrightarrow \cdot\mathrm{CH_2{-}C_6H_4{-}CH_2M_{m+n+2}CH_2{-}C_6H_4{-}CH_2}\cdot$$

(b) Non-isothermal polymerization

$$2\mathrm{CH_2{=}C_6H_4{=}CH_2} \rightleftharpoons \cdot\mathrm{CH_2{-}C_6H_4{-}CH_2{-}CH_2{-}C_6H_4{-}CH_2}\cdot \longrightarrow \text{cyclo-dimer } (\mathrm{CH_2{-}CH_2} \text{ bridged } \mathrm{C_6H_4} \text{ rings, } \mathrm{CH_2{-}CH_2})$$

$$\downarrow \mathrm{M}$$

cyclo-trimer $\longleftarrow$ $\cdot\mathrm{CH_2}$ … $\cdot\mathrm{CH_2}$ (open tetramer diradical) $\xrightarrow{\mathrm{M}}$ $\cdot[\mathrm{CH_2{-}C_6H_4{-}CH_2}]_4\cdot$ → cyclization; $\xrightarrow{\mathrm{M}}$ propagation

M: QM monomer

SCHEME 1

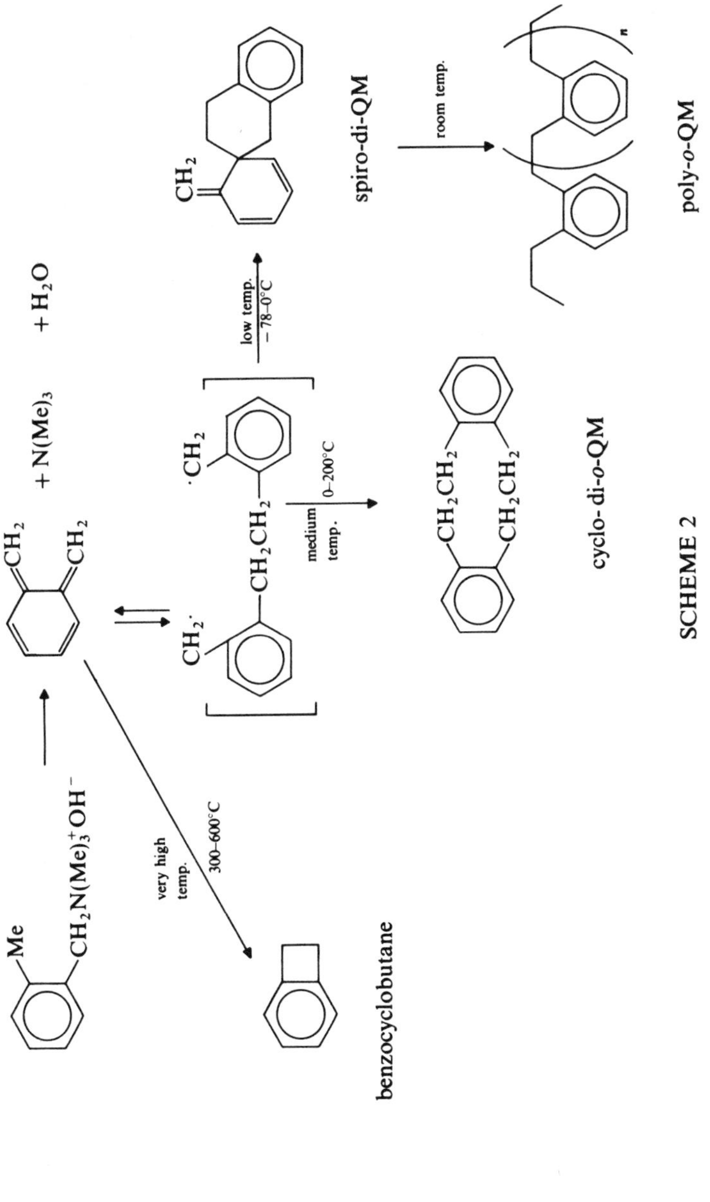

Me

$CH_2N(Me)_3^+OH^-$

CH_2 CH_2 + $N(Me)_3$ + H_2O

very high temp. 300–600°C

benzocyclobutane

$\cdot CH_2$ CH_2CH_2 $\cdot CH_2$

low temp. −78–0°C

medium temp. 0–200°C

CH_2

spiro-di-QM

room temp.

CH_2CH_2 CH_2CH_2

cyclo-di-*o*-QM

n

poly-*o*-QM

SCHEME 2

Furthermore, Errede successfully prepared *o*-xylylene(*o*-quinodimethane) (*o*-QM) by Hofmann degradation of *o*-methylbenzyltrimethylammonium hydroxide at low pressure using a modified flow process[24]. Bis(*o*-methylbenzyl)ether and *o*-methylbenzyl alcohol were formed as by-products. When *o*-QM is warmed from − 78°C to room temperature, spiro-(5,5)-2,3-benz-6-methyleneundeca-7,9-diene(spiro-di-QM) is obtained in a good yield (Scheme 2). When *o*-QM is heated at temperatures from 0 to 200°C, cyclo-di-*o*-QM is obtained (Scheme 2). Furthermore, at 300–600°C, benzocyclobutane is predominantly formed (Scheme 2). Apparently, the formation of the spiro compound is favored at temperatures lower than 0°C. Spiro-di-QM can be preserved without any change at − 15°C. It solidifies slowly when cooled to − 20°C and remelts at − 5 to 0°C. When it is warmed to room temperature, it polymerizes very slowly to give high molecular weight poly-*o*-QM with an intrinsic solution viscosity of about 0.6 dl g^{-1} and glass temperature of 9°C[24]. Spiro-di-QM is copolymerized with conventional olefinic monomers such as styrene, acrylonitrile, methyl methacrylate, vinylidene fluoride and 1,3-butadiene[24]. In its polymerization an effective chain transfer with conventional reagents such as mercaptans and carbon tetrahalides[24] takes place in contrast to QM polymerization. When an acid catalyst is added to a concentrated solution of spiro-di-QM in hexane, poly-*o*-QM is obtained (Scheme 3). When the catalyst is added to a dilute solution of spiro-di-QM, 1-

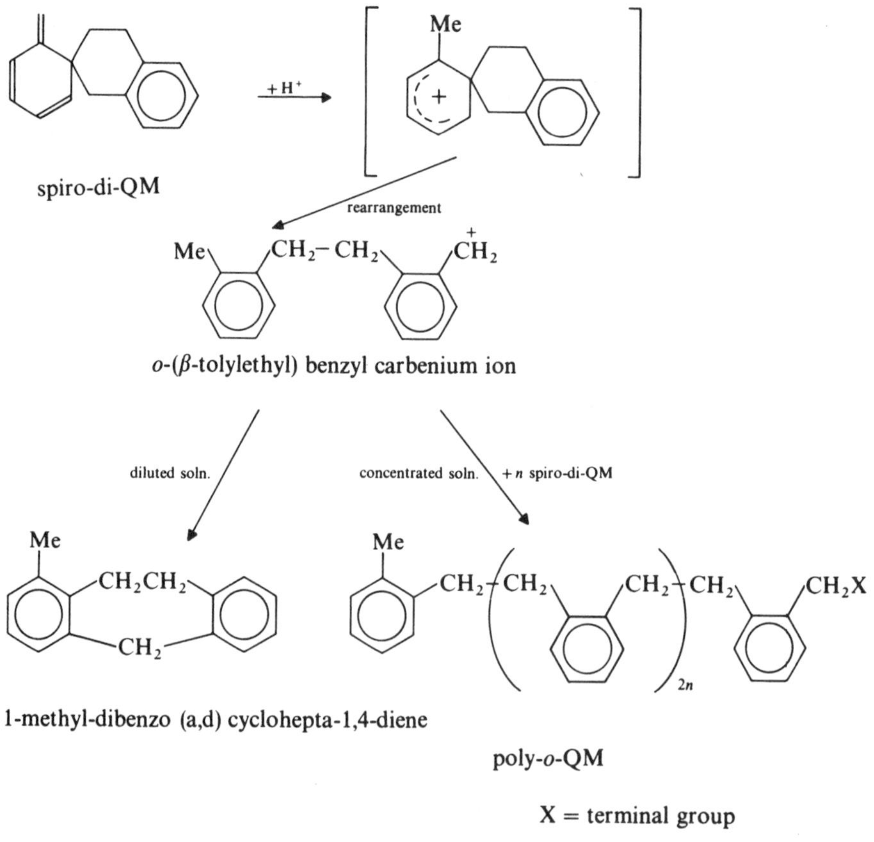

SCHEME 3

methyl-dibenzo (a, d) cyclohepta-1,4-diene is produced in good yield via intramolecular aromatic substitution of *o*-(β-tolylethyl)benzyl carbenium ion, formed in turn by rearrangement of the carbenium ion formed by addition of proton to the spiro-dimer[27] (Scheme 3). Spiro-di-QM is copolymerized with formaldehyde in the presence of an acid catalyst to give the corresponding polymeric ether[27].

III. HOFMANN DEGRADATION AND OTHER METHODS

It was pointed out that the flash pyrolysis method of *p*-xylene has several limitations[28], i.e. (1) At most 25% yields of QM are obtained at the extreme pyrolysis temperature of 1150°C[6]; (2) the polymers obtained are loosely cross-linked[7, 8] and (3) the vapor-deposited polymeric products formed by this method are contaminated with 10–20% of low molecular weight by-products[6, 8].

Fawcett[29] found that degradation of *p*-xylyltrimethylammonium hydroxide can take place at temperature as low as 100°C, and the immediate and concurrent polymerization of the monomer affords linear, soluble poly-QM in high yield. This method was successfully

$$\text{Me}-C_6H_4-CH_2\overset{+}{N}(Me)_3OH^- \xrightarrow{100\,^\circ C} (CH_2{=}C_6H_4{=}CH_2) \longrightarrow -\!\!\left[CH_2-C_6H_4-CH_2\right]\!\!-_n$$

QM

poly-QM

applied to 5-methyl-2-furfuryltrimethylammonium hydroxide and 5-methyl-2-thienyltrimethylammonium hydroxide to obtain 2,5-dimethylene-2,5-dihydrofuran and -thiophene, respectively[30]. Both monomers are very reactive and they either polymerize readily, or in the presence of polymerization inhibitors form a heterocyclophane, crystalline cyclic dimer[30]. The furan monomer can be isolated at -78°C[30]. This method is

$$\text{Me}-C_4H_2X-CH_2\overset{+}{N}(Me)_3OH \longrightarrow CH_2{=}C_4H_2X{=}CH_2$$

X = O or S

$$\longrightarrow -\!\!\left[CH_2-C_4H_2X-CH_2\right]\!\!-_n$$

$\xrightarrow{\text{with free-radical inhibitor}}$ [2.2](2,5)heterocyclophane (cyclic dimer, X)

widely applicable to the synthesis of other QM polymers substituted with groups which are insensitive to the strongly basic medium; e.g. poly-2,5-dimethoxy-QM, which can be hydrolyzed to poly-2,5-dihydroxy-QM of redox properties[31]. In a modified process, *p*-

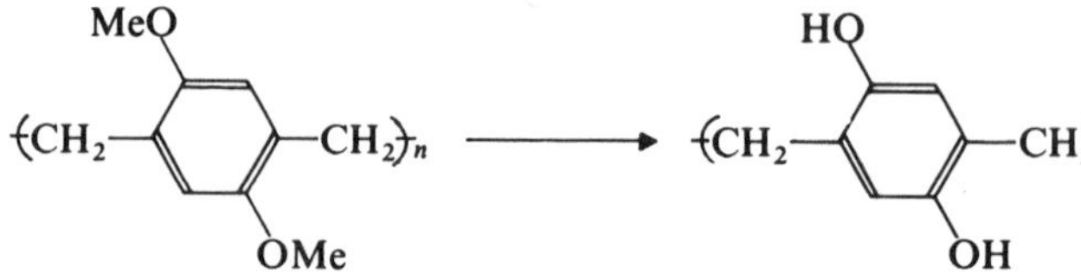

[(trimethylsilyl)methyl] benzyl trimethyl ammonium iodide can decompose with tetrabutylammonium fluoride in acetonitrile at room temperature to give poly-QM (51% yield) and cyclo-di-QM (6% yield), or at refluxing temperature to give 50% cyclo-di-QM[32].

Amorphous, low molecular weight (~ 3000) poly-*p*-xylylene (poly-QM) was synthesized by the Wurtz–Fitting reaction of 7,8-dibromo-*p*-xylene with sodium or magnesium metal[32]. This reaction was improved in various ways to give satisfactory yield of the crystalline polymer[9, 34]. Thus, different dehalogenation agents such as reduced iron and cobalt powder suspended in water[35], Urushibara Nickel[35], naphthalene–alkali complex[36] and tin(II) chloride[37] were used. Electrolytic reduction was also applied[38]. When strong bases such as sodium amide[39, 40], potassium *t*-butoxide[41] and sodium methoxide[42] are used, 7,8-dihalo-*p*-xylenes are converted to a copolymer of xylylidene and halo-QM with the following structure:

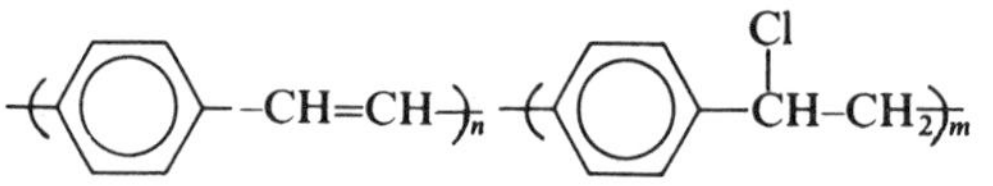

The reaction of 7-chloro-*p*-xylene with potassium *t*-butoxide in *p*-xylene in the presence of stable *N*-oxy biradicals such as 2,2,6,6-tetramethylpiperidinoxy-4-spiro-2′-(1′,3′-dioxane)-5′-spiro-5″-(1″,3″-dioxane)-2″-spiro-4″-(2‴,2‴,6‴,6‴-tetramethyl-piperidinoxy) gives the

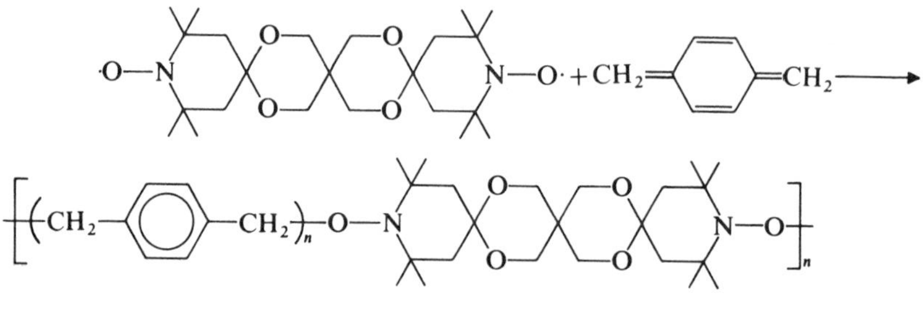

corresponding copolymer[43] indicating the formation of QM as a reaction intermediate. The Friedel–Crafts reaction of benzene with 1,2-dichloroethane affords non-linear polymer since it is insoluble in any solvent and does not melt[44, 45]. The same reaction of (2-chloroethyl)benzene likewise yields only an intractable polymer[45]. Di-azo-*p*-tolylmethane may undergo a cationic rearrangement to poly-QM[46].

IV. VAPOR-COATING PROCESS: GORHAM'S STUDY

Gorham[28, 47, 48] had developed the vapor-coating process in which [2,2]-paracyclophane, cyclo-di-xylylene(cyclo-di-QM), is pyrolyzed in vacuum at 600°C and the pyrolyzed gas is

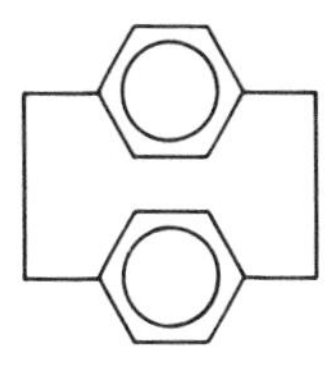

cyclo-di-QM

condensed on glass or metal surface to yield a tough, transparent polymeric film. This pyrolysis of cyclo-di-QM under milder and more readily controlled conditions than described by Szwarc and Errede results in almost quantitative preparation of the polymer containing less than 1% carbon tetrachloride extractable material, most of which is unreacted dimer. The polymer film obtained is readily soluble in hot chlorinated biphenyls and benzyl benzoate, indicating that it is free from cross-linking. This process has much greater advantages than the Szwarc–Errede direct pyrolysis of *p*-xylene. Due to the milder pyrolysis temperature the vapor-coating process may be applied to the preparation of a variety of substituted QM polymers.

Cyclo-di-QM was identified first by Brown and Farthing[11,49,50]. In the Szwarc's pyrolysis of *p*-xylene under reduced pressure at 680–850°C, the products obtained in 10–20% yield were mainly polymeric materials which were found to contain a small portion of low molecular weight compounds soluble in chloroform. Thus, 4,4′-dimethyldibenzyl was detected in the chloroform extract[9] together with traces of acetone insoluble compound[9] which was identified as [2,2]-paracyclophane by X-ray diffraction measurement[50]. Independently, Cram and Steinberg prepared a cyclic dimer in a poor yield by Wurtz–Fitting reaction of 4,4′-dibromomethyl dibenzyl[51]. Due to its distorted structure

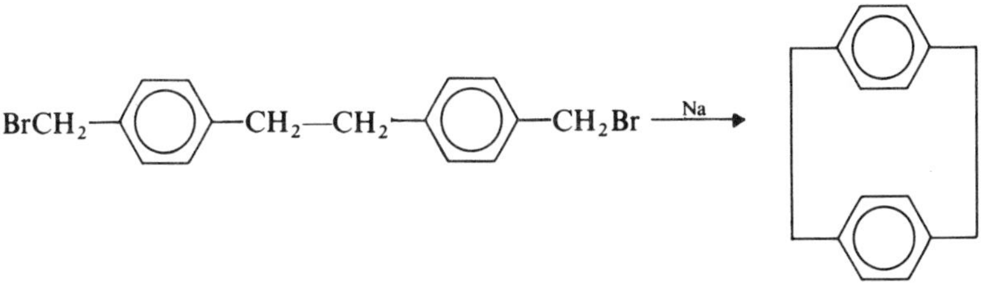

$BrCH_2$—C₆H₄—CH_2—CH_2—C₆H₄—CH_2Br $\xrightarrow{Na}$ [2,2]-paracyclophane

this dimer is sterically hindered and is generally referred to as cyclophane. It has been widely studied[52,53]. Errede and coworkers selectively prepared cyclo-di-QM by adding dropwise a 0.1 M QM solution in hexane maintained at −78°C to toluene heated at 100°C[22]. Pollart had developed a solvent quenching technique for the synthesis of cyclo-di-QM[54,55]. The condensation of QM vapor directly into an organic solvent at a temperature of 50–200°C results in the formation of cyclo-di-QM in a yield higher than 90%[54]. The rapid pyrolysis of a mixture of steam and *p*-xylene at 850–900°C followed by condensation of the vapor in an organic solvent such as *p*-xylene at 50°C gives cyclo-di-QM in 8–10% yield with only 0.1% polymeric material[55].

Gorham prepared about 30 types of substituted paracyclophanes including the dichloro, dibromo, dicyano, dimethyl, diethyl and tetrachloro derivatives[28] for the preparation of polymers of the respective substituted QMs.

The various substituted QM monomers condense and polymerize on the surface at temperatures lower than the threshold condensation temperature which is related to the molecular weight and volatility of the respective monomer. The threshold condensation temperature is defined as the highest temperature of the surface on which the QM monomers condense and polymerize at an appreciable rate. At normal pressure (about 0.1 mmHg) the threshold temperatures are 30°C for QM, 60°C for 2-methyl-QM, 90°C for 2-ethyl- and 2-chloro-QM, and 130°C for 2-cyano-, 2-bromo-, and dichloro-QM.

The mechanism of the vapor-coating process of unsymmetrically substituted cyclo-di-QM has been studied[28]. The pyrolysis gas from mono-acetyl-cyclo-di-QM is initially led through a glass tube maintained at 90°C and subsequently through another glass tube kept at 20°C. The polymer deposited at 90°C has been identified as poly-acetyl-QM on the basis of its elemental analysis and IR spectrum. The second polymer obtained at 20°C has been characterized as poly-QM by its IR spectrum and by properties such as its melting point of 400°C and the insolubility in any organic solvent below 250°C. These results reasonably

2-acetyl-[2,2]-paracyclophane → (600 °C) acetyl-QM + QM

acetyl-QM → (90 °C) poly-acetyl-QM

QM → (30 °C) Poly-QM

SCHEME 4

suggest that acetyl-cyclo-di-QM is cleaved to two species, QM and acetyl-QM, (Scheme 4) instead of to a ring-opened biradical product such as

$$\cdot CH_2-C_6H_4-CH_2-CH_2-C_6H_3(COMe)-CH_2\cdot$$

The subsequent fractional polymerizations which then take place depend upon the threshold condensation temperatures of these fragments.

Immediately after its preparation by the vapor-coating process, the poly-QM was found to be paramagnetic (radical concentration of 5–10×10^{-4} mol g^{-1})[28]. When the polymer is annealed at an elevated temperature, the ESR signal disappears. It was therefore concluded that this polymerization takes place via a free radical mechanism similar to the scheme of Errede[22]. The polymer films deposited at room temperature in the vapor-coating process are always formed in the metastable α-polymorph, which transforms to the stable β-modification upon heating to 220°C or higher[47]. The irreversible transformation was originally observed by Brown and Farthing[11]. Niegish found that single crystals of the poly-QM are prepared by heating a 0.05% (w/v) α-chloronaphthalene solution to 238°C, followed by slow cooling to 208°C[56]. These single crystals give two entirely different morphological structures. One is the pseudorectangular crystal due to the α-polymorph of monoclinic structure with parameters $a = 5.92$ Å, $b = 10.64$ Å, c (chain axis) $= 6.55$ Å, and $\beta = 134.7°$ and the other is the hexagonal crystal due to the β-hexagonal polymorph with parameters $a = 20.52$ Å and c (chain axis) $= 6.58$ Å[57]. Niegish stated that the α-polymorph is metastable because of irreversible α to β transition. However Wunderlich and coworkers mentioned that α-polymorph is stable[58]. They studied the crystallization during the polymerization of poly-QM in the vapor-coating process at temperatures ranging from −196 to 200°C and made the following qualitative observations[58, 59]. The monomer is weakly absorbed on the surface, and at temperatures of 26 to −17°C, its concentration is high enough and the mobility is sufficient to initiate polymerization which can be fast if the monomer is available. Before the polymerization is terminated, successive crystallization starts to take place. Since the glass temperature of the polymer is at 80°C, crystallization from the bulk polymer should not be possible in this temperature range. The necessary mobility of the molecular chains is maintained since the chains are on the surface and are swollen within the monomer. Under these conditions, folded-chain crystals of the stable α-polymorph oriented epitactically with chain axis in the crystal surface are grown with a rate determined by temperature (preferential orientation of the chains on the surface is observed). A steady-state separation between polymerization and crystallization sites might be established. At a low temperature, the polymerization rate would be determined by the higher monomer concentration while the crystallization rate would slow down due to decrease in chain mobility, resulting in the further separation between both sites. At higher temperatures, the polymerization rate decreases due to decrease in monomer concentration in the surface. At temperatures lower than −78°C, the surface concentration increases to such an extent that the monomer is immobilized due to surface condensation, yielding a change in the mechanism. The monomer either crystallizes first in the solid state and polymerizes and crystallizes then in the solid state or it polymerizes and immediately crystallizes to the polymer crystal, which is irregularly folded and unoriented with metastable β-polymorph. In either case, polymerization and crystallization are simultaneous. At temperatures higher than the glass temperature, the polymer is produced in a two-stage process where initially a small amount of the polymer (seed material) is deposited at a temperature lower than the threshold condensation temperature of 30°C and the second-stage polymerization is carried out at high temperatures. The crystallization rate is thought to increase rapidly with temperature while the polymerization rate decreases due to the lower monomer concentration on the surface, probably leading to an approach of the crystallization site to the polymerization site (i.e. simultaneous polymerization and crystallization). The crystals obtained are of the β-polymorph as at the lower temperature. Beach[60] developed a reaction model for the vapor-coating process of QM in which the temperature, pressure and rate of growth variables are correlated with the molecular rate constants, diffusional mass transport and molecular weights. The parameters obtained by numerical analysis in his model are: a propagation rate constant $k_p = 6.11 \times 10^3\ cm^3 g^{-1} s^{-1}$ at 20°C from Errede's data[22], an initiation rate constant $k_i = 6.34 \times 10^2\ cm^6\ g^{-2} s^{-1}$ at 20°C and a diffusivity between monomer and polymer of $D = 1.13 \times 10^{-9}\ cm^2\ s^{-1}$. The initiation is the step of a formation of a trimer diradical with activation energy of 24.8 kcal mol^{-1}. The rates of consumption of the monomer by initiation at −30 and 20°C are calculated to be 100 000 times and 320 times, respectively, slower than by propagation.

The vapor-coating process was developed by Union Carbide Corporation which commercially manufactures the unusual polymers under the trade name Parylene; Parylene N refers to unsubstituted poly-QM and Parylene C to poly-2-chloro-QM.

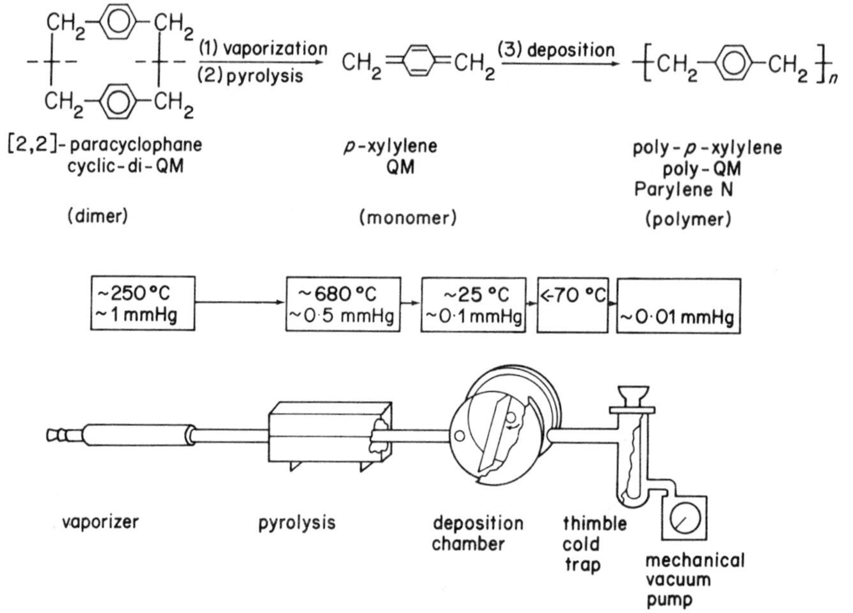

[2,2]-paracyclophane cyclic-di-QM (dimer) —(1) vaporization (2) pyrolysis→ $CH_2{=}C_6H_4{=}CH_2$ p-xylylene QM (monomer) —(3) deposition→ $\lbrack CH_2{-}C_6H_4{-}CH_2 \rbrack_n$ poly-p-xylylene poly-QM Parylene N (polymer)

~250 °C ~1 mmHg → ~680 °C ~0·5 mmHg → ~25 °C ~0·1 mmHg → <-70 °C → ~0·01 mmHg

vaporizer, pyrolysis, deposition chamber, thimble cold trap, mechanical vacuum pump

FIGURE 5. Schematic representation of the vapor-coating process, using [2,2]-paracyclophane as an example

The process involves three steps as outlined in Figure 5 for Parylene N. The film deposited on the surface in this process is free of pinhole and can be adjusted to a thickness of several submicrons to several millimeters. The physical and electrical properties of these polymers are compiled in Table 1. They exhibit high moduli at room temperature and tensile moduli above 300 000 psi. Their glass transition temperatures are in the range of 60–90°C and their melting points are as high as 290–400°C. These polymers have low gas-permeability characteristics especially of Parylene C. Parlylene N is a dielectric exhibiting a very low dissipation factor, a high dielectric strength and a dielectric constant invariable with frequency. It is used as a dielectric of a plastic-film capacitor. Parylene C additionally exhibits a very low permeability to moisture and other corrosive gases and is especially useful for the coating of critical electric assemblies.

7,7,8,8-Tetrafluoro-QM polymer was prepared by a similar process starting from the corresponding paracyclophane derivative[61,62]. This polymer also exhibits physical and electrical properties similar to those of Parylene polymers (Table 1). It is extremely resistant to sunlight even after exposure for 3600 h while Parylene N is changed to a brittle material after exposure for 535 h[61].

V. POLYMERIZATION OF HALO-*p*-XYLYLENES

A. 7,7,8,8-Tetrachloro-*p*-xylylene

7,7,7,8,8,8-Hexachloro-*p*-xylene is dechlorinated on a copper mesh under reduced pressure (0.1–1.0 mmHg) at 300–600°C to 7,7,8,8-tetrachloro-*p*-xylylene (TCX) in yields up to 90%[63]. The pyrolysis product is absorbed in toluene maintained at −78°C and the

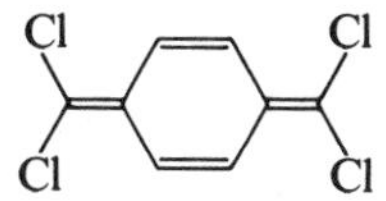

TCX

resulting yellow colored suspension is cooled down to give yellow solid. Repeated recrystallization from tetrahydrofuran under nitrogen at temperatures ranging from −60 to 0°C gives yellow needles. When these crystals are kept at room temperature, their yellow color gradually fades and poly-TCX is formed. For example, a tetrahydrofuran solution of 0.0244 M TCX polymerizes at 20°C for 30 min up to a conversion of 50%[63]. When the gaseous pyrolysis product is condensed on the surface maintained at a temperature above 120°C, a transparent film of poly-TCX is formed[64]. The freshly prepared film still contains monomeric TCX and exhibits a strong ESR signal, indicating the presence of free radicals. Polymerization is completed by annealing of the film at 190°C for 30 min[64]. When the pyrolysis gas is deposited on the surface below 90°C, crystalline monomeric TCX is formed and then gradually polymerizes[64].

At 100°C, both crystals and a large amount of a transparent film are obtained simultaneously[64]. Poly-TCX displays a tensile modulus of 480 000 psi, a tensile strength of 8000 psi, a softing range of 280–290°C, a dielectric constant of 2.81, and a dissipation factor 2.6×10^{-4} at a frequency range of 60 cycles to 100 kilocycles[64]. TCX may be recrystallized and kept at temperatures below −10°C without polymerization. It is therefore clearly more stable than QM but is much more reactive than conventional olefinic monomers.

TABLE 1. Physical and electrical properties of poly-QMs

	Poly-*p*-xylylene (Parylene N)[28]	Polychloro-*p*-xylylene (Parylene C)[28]	Poly-7,7,8,8-tetrafluoro-*p*-xylylene[61]
Tensile properties (at room temperature)			
Tensile strength (psi)	6800	10 600	6200
Tensile modulus (psi)	350 000	460 000	360 000
Elongation at break (psi)	10–15	220	100
Tensile modulus at 200°C (psi)	25 000	25 000	—
Thermal properties			
Crystalline melting point (°C)	400	290	
Glass transition temperature (°C)	80	80	90
Permeability at 77°F (cm^3 (STP) mil/100 in^2 24 h)			
H_2	250	200	
CO_2	225	21	
O_2	30	8	
N_2	9	1	
H_2O (g mil/100 in^2 24 h)	6.0	0.6	
Electrical properties (1–3 mil film)			
Dielectric constant (1 kc/s^{-1})	2.65	3.2	2.36
Dissipation factor (1 kc/s^{-1})	0.0002	0.04	0.0008
Dielectric strength (v/mil^{-1})	7000	5000	5250

B. 2,5,7,7,8,8-Hexachloro-*p*-xylylene

2,5,7,7,8,8-Hexachloro-*p*-xylylene (HCX) is prepared by passing 2,5,7,7,7,8,8,8-octachloro-*p*-xylene vapor over a copper mesh at 500°C at reduced pressure[65]. Pure yellow

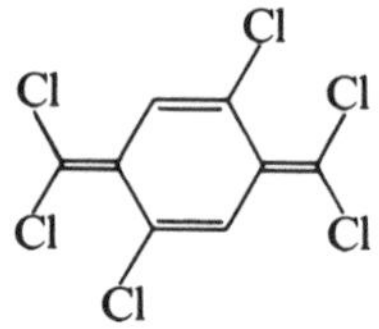

HCX

crystalline HCX is obtained by the same recrystallization method used for the isolation of TCX. HCX can be kept without any change below 0°C for a long time, but at room temperature it changes gradually to a white powder which has been identified as poly-HCX on the basis of its elemental analysis. The change of HCX crystals to its amorphous polymer on standing was examined by X-ray diffraction in order to follow the rate of the solid state polymerization of HCX. The height (or intensity) of the X-ray diffraction profile, corresponding to the crystalline portion (monomer) in the probe, decreased linearly with time, indicating the zero reaction order of the spontaneous solid state polymerization. After 66 h the X-ray diffraction of the monomer disappeared completely. The zero reaction order can be explained in terms of the effective monomer concentration in the solid state. Thus, when an active site migrates into a crystal in the course of polymerization, the number of HCX monomers around the active site, which are susceptible to polymerization, is considered to be constant. The polymer chain already formed does not influence this number because it always exists just in the near of the active site and is excluded from the HCX crystal. In addition, the formation of the active sites is regarded to be independent of the number of HCX monomers in the volume unit of the crystal. The rate of spontaneous polymerization of HCX in benzene follows first-order kinetics with respect to monomer concentration with apparent first-order rate constants of $5.56 \times 10^{-5}\,s^{-1}$ at 30°C and $13.33 \times 10^{-5}\,s^{-1}$ at 50°C. The linear Arrhenius plot gives an apparent activation energy of 8.2 kcal mol^{-1} for the polymerization of HCX. The temperature at which the apparent first-order constant for the polymerization of QM in toluene is $5.56 \times 10^{-5}\,s^{-1}$ is calculated to be −68°C by using an apparent first-order rate constant of $9 \times 10^{-6}\,s^{-1}$ at −78°C and the apparent activation energy of the polymerization of 8.7 kcal mol^{-1} [22]. Since the corresponding temperature for HCX is 30°C, QM is much more reactive than HCX, i.e. HCX is a much more stable monomer than QM.

C. 2-Cyano-7,7,8,8-tetrachloro-*p*-xylylene

2-Cyano-7,7,8,8-tetrachloro-*p*-xylylene (CTCX) is prepared similarly by gas-phase dechlorination of 2,5-bis(trichloromethyl)benzonitrile on a copper mesh[66]. CTCX also

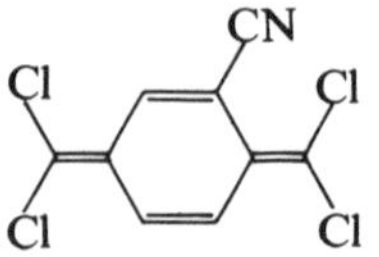

CTCX

readily undergoes spontaneous polymerization with approximately first-order kinetics with respect to monomer at monomer concentrations higher than $2\text{–}3 \times 10^{-3}$ mol l^{-1} whereas it is second order below this monomer concentration. The apparent first-order rate constant at an initial monomer concentration of 0.01 mol l^{-1} is 1.3×10^{-4} s^{-1} at 30°C and the apparent activation energy is 8.8 kcal mol^{-1}. The apparent rate constant is 5.6 $\times 10^{-5}$ s^{-1} at 15°C. Consequently, the tendency of CTCX for homopolymerization is higher than that of HCX (30°C) and much lower than that of QM (−68°C).

D. Copolymerizations with Styrene

In contrast to QM, the three monomers TCX, HCX and CTCX have been found to undergo spontaneous copolymerization with various vinyl monomers such as styrene (St), isoprene, vinyl acetate, acrylonitrile and methyl methacrylate[60–62].

For the copolymerization systems of TCX–St, HCX–St, CTCX–St (Figure 6), CTCX–HCX (Figure 7) and CTCX–TCX (Figure 8) the following monomer reactivity ratios have been obtained; r_1 (TCX) = 85 and r_2(St) = 0 at 22°C[65], r_1(HCX) = 3 ± 0.8 and r_2(St) = 0.02 ± 0.05 at 50°C[63], r_1(CTCX) = 12 ± 6 and r_2(St) = 0.03 ± 0.02 at 20°C[64], r_1(CTCX) = 0.8 and r_2(HCX) = 0.95 at 20°C and r_1(CTCX) = 0.25 and r_2(TCX) = 1.7 at 20°C[66]. The relative reactivities of TCX, HCX and CTCX toward the active site of the polymer chain with a terminal St unit have been estimated from comparison of the reciprocals of r_2(St) of the TCX–St, HCX–St and CTCX–St systems: TCX($1/r_2 = 1/0$) > HCX(1/0.02) ≧ CTCX(1/0.03). The monomer reactivity ratios are thereby assumed to remain essentially unchanged in the temperature range of the polymerization (20–50°C). A comparison of the reciprocals of r_1 (CTCX) of the TCX–CTCX and HCX–CTCX systems gives another order of the relative reactivities of the monomers toward the active site of the polymer chain with a terminal CTCX unit: TCX (1/0.25) > HCX (1/0.8) ⩾ CTCX (1/1). Both reactivity orders are in good agreement. In addition, the relative reactivities have been

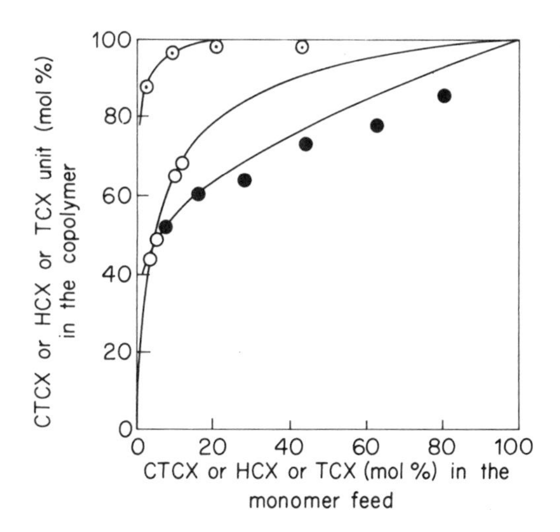

FIGURE 6. Composition of the copolymerization of CTCX with styrene (St) (O), copolymerization of HCX with St (●), and the copolymerization of TCX with St (O). The lines are obtained by a theoretical equation using the monomer reactivity ratios; r_1 (CTCX) = 12 and r_2 (St) = 0.03 for the CTCX–St system, r_1 (HCX) = 3.0 and r_2 (St) = 0.02 at 50°C for the HCX–St system and r_1 (TCX) = 85 and r_2 (St) = 0 at 22°C for the TCX–St system

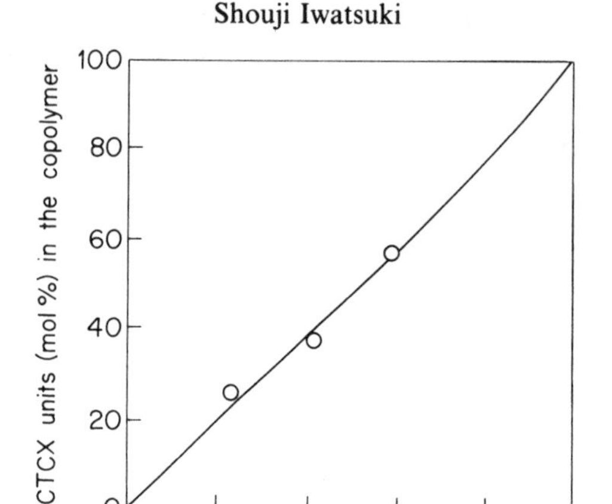

FIGURE 7. Composition of the copolymerization of CTCX with HCX. (O) refers to experimental value, and solid line is calculated from the theoretical equation using r_1 (CTCX) = 0.8 and r_2 (HCX) = 0.95

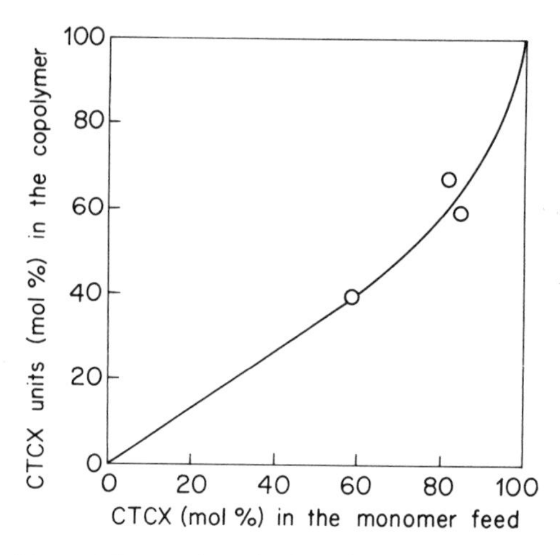

FIGURE 8. Composition of the copolymerization of CTCX with TCX. (O) refers to experimental value and solid line is calculated from the theoretical equation using r_1 (CTCX) = 0.25 and r_2 (TCX) = 1.7

compared with some parameters estimated by quantum chemical calculations such as π-electron density, frontier electron density, and the free valence at the exocyclic carbon (Table 2). CTCX has different values at the two exocyclic carbons because of its unsymmetric structure. The results are in agreement with the relationship of Kooyman and Farenhorst[68] between the relative reactivity of the trichloromethyl radical toward the aromatic hydrocarbons (such as benzene, naphthalene, anthracene, pyrene, etc.) and the highest free valence index of the hydrocarbon and also with the relationship of Hush[14]

TABLE 2. π-Electron density, frontier density and free valence at the exocyclic carbon of CTCX, HCX and TCX[a]

		π-Electron density			
		Ground state	Singlet excited state	Frontier density	Free valence
CTCX[b]	C(7)	0.9642	1.0169	0.4048	0.412
	C(8)	0.9556	1.0675	0.4598	0.421
HCX		0.9691	1.0676	0.4469	0.424
TCX		0.9869	1.0716	0.4620	0.440

[a] Calculated by the ASMO–SCF method.

[b] (structure: Cl, Cl, C(7), C(8), Cl, Cl, CN)

between the polymerizability of QM compounds and the free valence at their corresponding exocyclic atom.

Another interesting phenomenon has been found in the copolymerization of HCX with St starting with a high St monomer feed such as 92.3 mol%. A change of the content of the

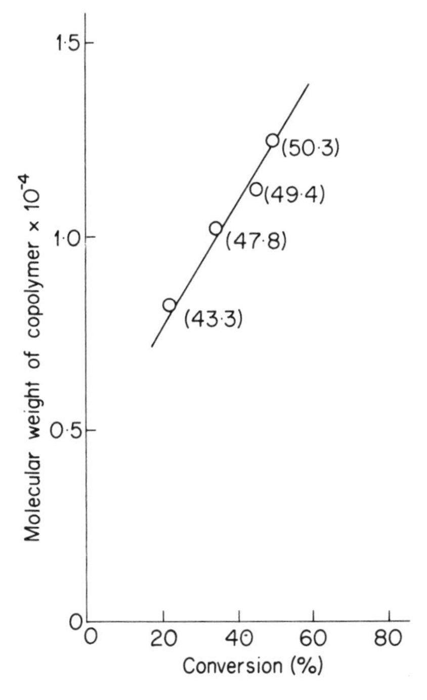

FIGURE 9. Relationship between the molecular weight of the copolymer and conversion in the copolymerization of HCX with St starting with high St monomer concentration (92.3 mol%). Figures in parentheses refer to the content of the St unit in mol% in the copolymer obtained

St unit in the copolymers with conversion can be observed similarly as in common vinyl copolymerization but its magnitude is very small (see Figure 9). The molecular weight of the obtained copolymers increases significantly with conversion as shown in Figure 9. This implies that this polymerization partially proceeds by a stepwise addition mechanism, i.e. it exhibits a somewhat 'living character' which is in sharp contrast to a conventional free radical vinyl polymerization. Therefore, it may be presumed that the active site of a polymer molecule with a terminal HCX unit reacts not only with St and HCX monomers via the radical addition mechanism but also with HCX via radical coupling.

Perchloro-*p*-xylylene prepared by Ballester and coworkers[69] exhibits no tendency for polymerization and is very stable even at elevated temperatures.

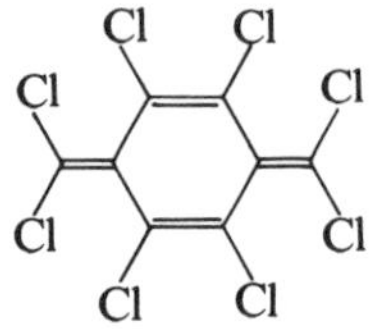

Perchloro-*p*-xylylene

VI. POLYMERIZATION OF 7,8-BIS(ALKOXYCARBONYL)-7,8-DICYANOQUINODIMETHANE

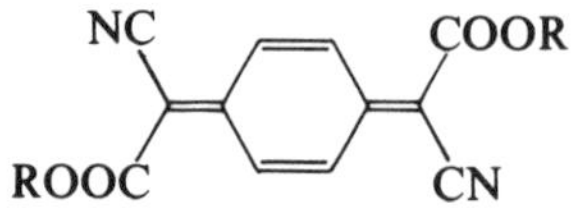

ACQ

R = Me (MCQ)
R = Et (ECQ)
R = Bu (BCQ)

Preparation and polymerization of 7,8-bis(alkoxycarbonyl)-7,8-dicyanoquinodimethane (ACQ) with ethoxy (ECQ) and methoxy groups (MCQ) as the alkoxy group has been briefly reported independently by Hall and coworkers[70] and Iwatsuki and coworkers[71] in 1982. On the basis of the chemical structure, ACQ was expected to occupy an intermediate position in physical and chemical properties between 7,7,8,8-tetracyanoquinodimethane (TCNQ)[72,73] and 7,7,8,8-tetrakis(alkoxycarbonyl)quinodimethanes with methoxy (TMCQ)[72,74] and ethoxy (TECQ)[75] groups as the alkoxy group. On the other hand, ACQ was also expected to have properties different from those of TCNQ, TMCQ and TECQ, in which the substituents at the 7 and 8 positions are identical, because ACQ has two different substituents at the 7 and 8 positions.

The polymerization behavior of 7,8-bis(butoxycarbonyl)-7,8-dicyanoquinodimethane (BCQ) was investigated in detail as a representative of ACQ[76]. Some physical properties of its high polymer were also studied because BCQ could be prepared more easily than other ACQ monomers and because its polymer is more soluble in many conventional organic solvents[76]. The electron-accepting character of BCQ was measured by means of the charge transfer complex method by using an empirical relationship[77] between the electron affinities of the acceptors and the energy of the intermolecular charge transfer absorption

(ν_{CT}) of the corresponding complexes with a given donor such as hexamethylbenzene (HMB). The solvent used was benzene. *p*-Chloranil (PCA) which was reported to have an electron affinity of 2.48 eV[78] was employed as a reference acceptor compound. Table 3 gives the intermolecular charge transfer absorption (λ_{max}^{CT}) and the electron affinity of BCQ as well as of TMCQ, TECQ, MCQ, ECQ, TCNQ, PCA and maleic anhydride (MAnh). It is apparent that ACQ is intermediate in electron-accepting character between TMCQ and TCNQ, as expected from their chemical structures.

TABLE 3. Electron-accepting ability of ACQ and other related quinodimethanes, PCA and maleic anhydride (MAnh)

	MAnh	TMCQ,	TECQ	ACQ			PCA	TCNQ
				Me	Et	Bu		
Reduction potential[a]/V		−0.83		−0.65				−0.20
CT band[b] HMB (nm)		< 440–550[c]		500	497	496	516	594
EA (eV)				2.16[d]	2.10[d]	2.07[d]	2.48[e]	2.87[e]
CT band[f] DMA(nm)	422		457					
EA (eV)	1.33[e]		2.27[g]					

[a] Measured by cyclovoltammetry. In MeCN Ag/Ag$^+$.
[b] λ_{max} of CT band with hexamethylbenzene (HMB).
[c] CT bands are overlapped seriously with absorption of TMCQ and TECQ.
[d] Calculated by using the equation; $h\nu = \text{Ip} - \text{EA} + \text{C}$, and 2.48 eV for *p*-chloranil[78].
[e] Cited from literature[78].
[f] λ_{max} of CT band with dimethylaniline (DMA).
[g] Calculated by using the equation $h\nu = \text{Ip} - \text{EA} + \text{C}$, and 1.33 eV for MAnh[78].

TABLE 4. Homopolymerizations[a] of BCQ initiated by various catalysts at 0°C

Run no.	Catalyst, 1	[BCQ]/[1]	Solvent	ml	Time h	$\bar{M}n/10^4$ [e]
1	Et_3N	103	$CHCl_3$	5	1.0	9.5
2	Et_3N	103	$(CH_2Cl)_2$	5	1.0	35.0
3	Et_3N	100	Toluene	10	1.0	21.5
4	Proton sponge[b]	102	$CHCl_3$	5	1.0	10.6
5	Pyrrolidine	108	$CHCl_3$	5	1.25	5.6
6	Pyridine	93	$CHCl_3$	5	0.5	No polymer
7	Ph_3P	101	$CHCl_3$	5	1.0	No polymer
8	TPP–Al[c]	59	CH_2Cl_2	5	56	10.1
9	BuLi	106	Toluene	10	0.7	103.0
10	EtMgBr	6	Toluene	10	0.5	1.9
11	AIBN[d]	8	$CHCl_3$	5	5.5	4.4
12	$BF_3 \cdot Et_2O$	101	CH_2Cl_2	5	1.0	No polymer

[a] [BCQ] = 28 mM for runs 1, 2, 4–8, 11 and 12 and [BCQ] = 14 mM for runs 3, 9 and 10.
[b] Proton sponge: 1,8-bis(dimethylamino)naphthalene.
[c] TPP–Al: 5,10,15,20-tetraphenylporphine–Et_2AlCl. Reaction at room temperature.
[d] Azobisisobutyronitrile to which one drop of AcOH was added.
[e] Number-average molecular weight, $\bar{M}n$, determined by GPC, THF eluent.

Table 4 summarizes the results of the polymerizations of BCQ with various anionic, cationic and free radical initiators. It is obvious that BCQ is homopolymerizable with anionic and free radical initiators but not polymerizable with cationic ones. On the other hand, TCNQ[72, 73] and TECQ[75] were reported to be not homopolymerizable with any initiators, though they react as powerful electron-accepting monomers.

The fact that BCQ carries two different kinds of substituents (cyano and butoxycarbonyl) at the 7 and 8 positions whereas TCNQ and TECQ have identical substituents at the 7 and 8 positions, is considered to be responsible for the high homopolymerizability of BCQ.

The copolymerization of BCQ with styrene was found to be really in random fashion as shown by the shape of the copolymerization composition curve (Figure 10) and the fine straight lines for its Kelen–Tüdős and Fineman–Ross plots. The monomer reactivity ratios were obtained as r_1(BCQ) $= 0.9 \pm 0.3$ and r_2(St) $= 0.02 \pm 0.02$ at 60°C. TCNQ and TMCQ were reported to copolymerize with styrene alternatingly and spontaneously even though TMCQ is a weaker acceptor monomer than BCQ. The random behavior of the copolymerization of BCQ with styrene was thought to be closely associated with the homopolymerizability of BCQ, due to the lack of identity of both substituents either at the 7 or at the 8 position. Monomer reactivity ratios of the copolymerization of BCQ with styrene allow one to calculate the monomer reactivity parameters such as Alfrey–Price's Q and e values of BCQ as 9.9 and $+1.20$, respectively. When these values are compared with values of maleic anhydride (MAh) ($Q = 0.23$, $e = 2.25$)[79] and ethyl α-cyanoacrylate ($Q = 2.14$, $e = 0.78$)[79], it is found that BCQ has the higher electron affinity but the lower e value than MAnh. On the other hand, the e value of BCQ seems to be in good agreement with that of ethyl α-cyanoacrylate, which has a similar substitution pattern at the terminal carbons.

When BCQ was dissolved in some conventional basic solvents, the yellow color of the solution changed rapidly to colorless, indicating a rapid conversion of the BCQ monomer to its polymer. Table 5 summarizes the results of the polymerization of BCQ in five basic solvents. As shown in Table 4, most amine compounds except pyridine are capable of initiating the polymerization of BCQ in chloroform or toluene. The basicity of the basic

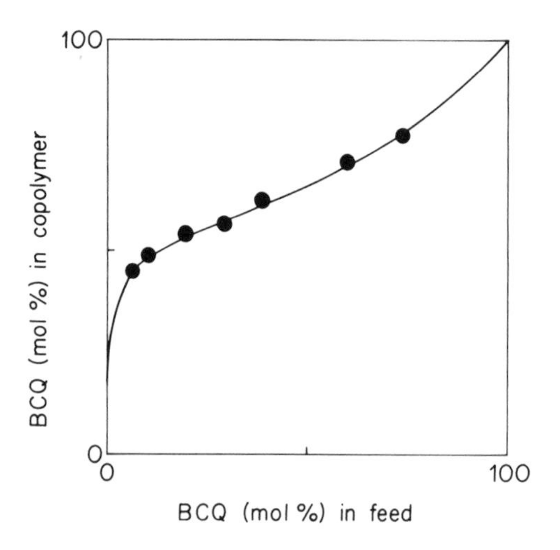

FIGURE 10. Copolymerization composition diagrams of BCQ with styrene

TABLE 5. Spontaneous homopolymerizations[a] of BCQ in various solvents

Solvent	$\Delta\nu_{OD}$[b]	$\bar{M}n/10^4$[c]	$\bar{M}w/\bar{M}n$[c]
Acetonitrile	49	410	1.72
Acetone	64	455	1.94
THF	93	86	1.80
DMF	107	340	1.89
DMSO	141	360	1.76

[a] BCQ 5.5 mg; solvent 1 ml; temp., room temp.
[b] OD stretching frequency shift of MeOD-base hydrogen bonding.
[c] Number-average and weight-average molecular weight, ($\bar{M}n$) and ($\bar{M}w$), determined by GPC using THF as eluent.

compounds could be expressed in terms of the strength of the hydrogen bond formed between a proton-donating alcohol such as methanol-d and the basic compounds[80]. In this respect, pyridine is a weaker base ($\nu_{OH} = 168\ cm^{-1}$) than triethylamine ($\nu_{OH} = 238\ cm^{-1}$)[81] and indeed pyridine is incapable of initiating the polymerization, but triethylamine is capable. Thus basicity is apparently related to the capability of initiating the polymerization of BCQ. However, even solvents such as acetonitrile and acetone, which are much less basic than pyridine, are capable of initiating the polymerization of BCQ, suggesting that additional factors other than basicity also exert an influence on the initiation of the anionic polymerization of BCQ. The relationships of the ratio of the concentration of the monomer to the initiator vs. the molecular weight of the polymers ($\bar{M}n$) in polymerizations with various amounts of butyllithium and triethylamine as initiator at an almost fixed concentration of BCQ as the monomer in toluene are shown in Figure 11.

In almost all the experimental runs with butyllithium the yellow color of the reaction systems disappeared within 0.5 h, suggesting that a complete conversion to the polymer is reached rapidly. However, when the concentration of butylithium was very small especially for [BCQ]/[BuLi] > 100, the polymerization was found to become slow and the reproducibility of the experiments to become poor. It is likely that the ratio of the concentration of the monomer to that of the initiator is linearly related to the molecular weight of the polymer produced. The results of the polymerization with additional monomer additions are shown in Figure 12. Apparently, after all the monomer additions, the polymers exhibit one fairly sharp peak in their gel permeation chromatography, even though their peak widths become a little bit broader with each monomer addition. The peak positions move to the higher molecular weight side with each monomer addition, indicating that the polymeric species produced after the monomer has completely polymerized is still able to react with added monomer molecules and to grow to a polymeric species with a higher molecular weight. Hence, the polymerization of BCQ with butyllithium is a living-like type process.

The polymerization with triethylamine was also found to give a polymer with high molecular weight. In the concentrated range of triethylamine, e.g. for [BCQ]/[triethylamine] < 5, the polymerization reaches completion within 0.5 h and the molecular weight of the polymers obtained increases with a decrease in the concentration of triethylamine in the monomer feed, as expected. However, in the more dilute range of triethylamine, i.e. for [monomer]/[initiator] > 6, the polymerization cannot be completed within an hour and molecular weight of the polymer obtained decreased with a decrease of the triethylamine concentration in the monomer feed. In spite of the low concentration of

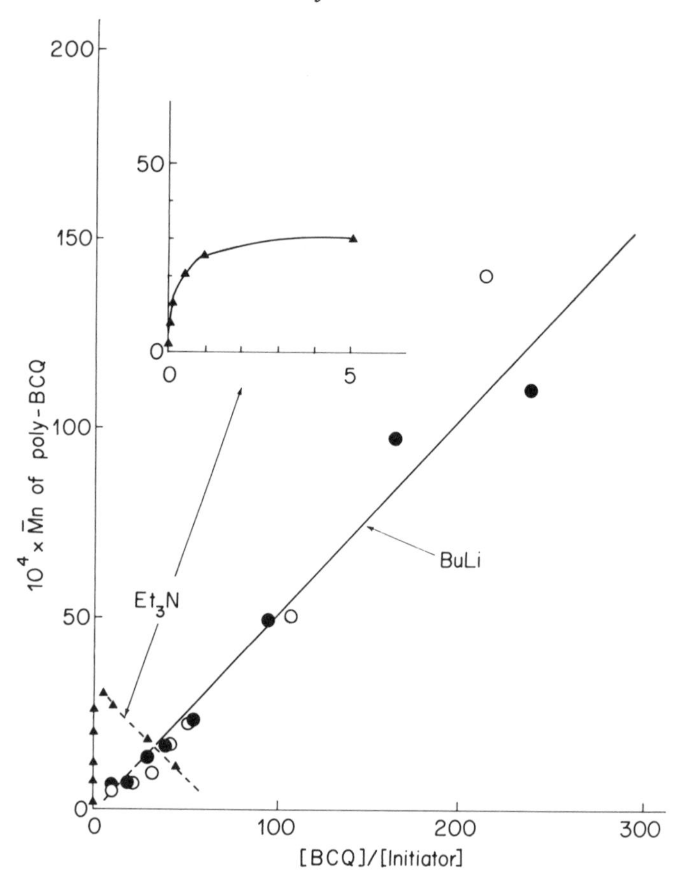

FIGURE 11. Relationships of the molecular weight of the polymer vs. the concentration ratio of monomer to initiator in polymerizations of BCQ with butyllithium ((○ and (●) correspond to duplicated series of runs) and triethylamine (▲) at 0 °C in toluene (monomer concentration, 10 mM: solution volume, 10 ml)

triethylamine, a polymer with a lower molecular weight is produced, contrary to the general behavior in a living type of anionic polymerization. The calculated values of the initiator efficiency were found to increase extensively with the ratio of monomer to initiator. At the moment no satisfactory mechanism has been given for the polymerization with triethylamine, especially for the decrease in the molecular weight of the polymer with a decrease in initiator concentration in the monomer feed.

When pyrrolidine was employed as an initiator (i.e. $[BCQ]/[I] = ca.\ 10^{-3}$) a 1:1 adduct was formed, while triethylamine gave a polymer with high molecular weight of the order of 10^4 under the same experimental condition of high initiator concentration. Therefore, it is conceivable that an electron transfer reaction takes place between pyrrolidine and BCQ to give a zwitterion with positive and negative charges on the pyrrolidinium and α-cyano-α-(butoxycarbonyl)benzyl moieties, respectively. The latter moiety is considered to be able to add a BCQ monomer. When excess pyrrolidine is present, it is conceivable that the

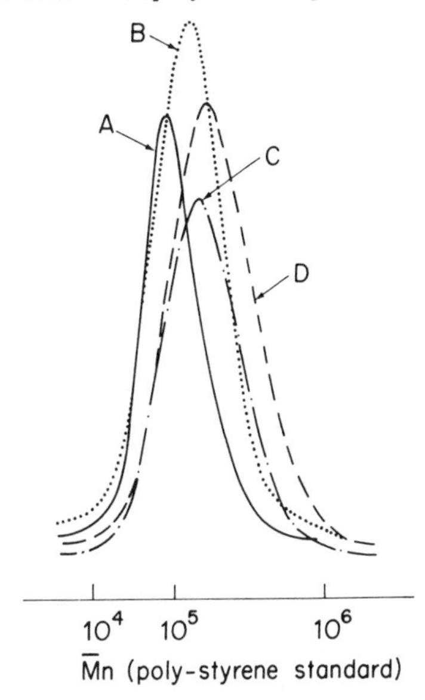

FIGURE 12. Gel permeation chromatogram of poly-BCQs with three additional monomer additions. Initiator butyllithium; polymerization temperature, 0 °C; solvent, toluene; concentration of the monomer, 8.13 mM; volume of the monomer solution at each addition, 7 ml. (A) The initial polymerization at [BCQ]/[BuLi] = 21(——). Molecular weight ($\bar{M}n$) of the polymer = 9.1×10^4 and the index of $\bar{M}w/\bar{M}n$ = 1.59. (B) After the first additional monomer addition (. . . .). $\bar{M}n$ of the polymer = 14.0×10^4 and $\bar{M}w/\bar{M}n$ = 1.76. (C) After the second addition (—·—). $\bar{M}n$ of the polymer = 18.0×10^4 and $\bar{M}w/\bar{M}n$ = 1.79. (D) After the third addition (— — —). $\bar{M}n$ of the polymer = 20.0 $\times 10^4$ and $\bar{M}w/\bar{M}n$ = 1.83

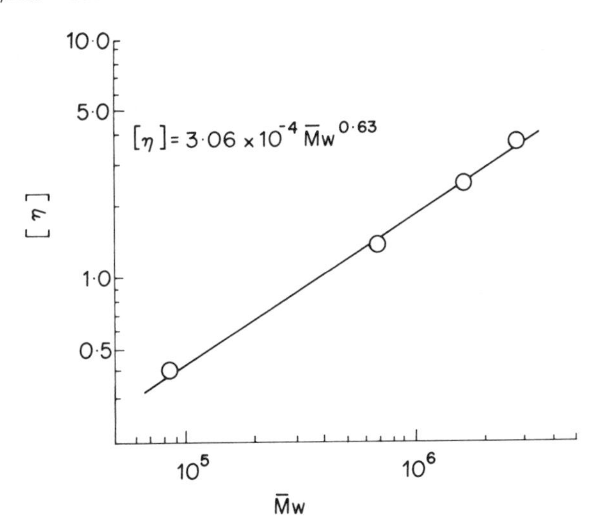

FIGURE 13. Relationship of solution viscosity[a] vs. molecular weight[b] for poly-BCQ. [a] Limiting viscosity number, using Ubbelohde viscometer in tetrahydrofuran, at 25 °C. [b] Light scattering measurement

concentration of the zwitterions will be sufficiently high and an intermolecular proton transfer from a pyrrolidinium moiety to the anionic moiety will give the 1:1 adduct before reaction with another BCQ monomer. In the case of triethylamine, a similar formation of a

BCQ + pyrrolidine (N–H) ⟶ [BCQ^- + pyrrolidinium (N^+–H)]

⟶ (H)(+N-pyrrolidinium)–C(CN)(COOBu)–C_6H_4–C^-(CN)(COOBu)

[BCQ]/[I] = 10^{-3} ⟶ (pyrrolidin-1-yl)–C(CN)(COOBu)–C_6H_4–C(CN)(COOBu)–H

[BCQ]/[I] > 1 ⟶ poly-BCQ

zwitterion also takes place, but since the quaternary ammonium ion has no hydrogen for the proton transfer reaction, the zwitterion is able to add the BCQ monomer in preference to termination reactions to give a polymer with a molecular weight of ca. 10^4. Moreover, an efficient electron transfer reaction of triethylamine to BCQ (the first step of an initiation reaction) was assumed to be important only at relatively high concentration of triethylamine molecules, due to very low initiator efficiency and the slow polymerization rate at the low initiator concentration.

The Mark–Houwink equation for poly-BCQ in tetrahydrofuran has been obtained. The polymer obtained with butyllithium was fractionated by means of fractional gel permeation chromatography to give four fractions for which the limiting viscosity number and the weight-average molecular weight were obtained from solution viscosity and light scattering measurements, respectively. The results are shown in Figure 13, where the α-index was found to be 0.63, indicating that this polymer chain is fairly flexible in solution even though it was expected to be stiff due to the presence of phenylene and tetra-substituted ethylene groups in the backbone chain.

The solubility of poly-ACQ (poly-BCQ, poly-MCQ and poly-ECQ) is summarized in Table 6. The glass transition temperature (Tg) for poly-BCQ, poly-MCQ and poly-ECQ

TABLE 6. Solubility of poly-ACQ

Poly-ACQ	Soluble	Swell	Insoluble
Poly-BCQ Poly-ECQ	(Benzene, acetone, THF $CHCl_3$, DMSO, DMF)		(MeOH, hexane isopropyl ether (IPE))
Poly-MCQ	conc. H_2SO_4	DMSO, DMF	(Benzene, acetone, THF, $CHCl_3$, MeOH, hexane, IPE)

was measured by differential scanning calorimetry. The results are summarized in Table 7, together with data[82] for poly(alkyl methacrylate), poly(alkyl α-cyanoacrylate), poly(*p*-alkylstyrene), and poly-QM for comparison. Tgs for poly-ACQs at 108°C are independent of the alkoxy group, whereas Tgs for polymers of vinyl compounds such as poly(alkyl methacrylate), poly(alkyl α-cyanoacrylate), and poly (*p*-alkyl-styrene) vary significantly, depending upon the alkyl group. The difference in Tg dependence on substituent was thought to be attributable to the difference in the backbone chain structure. Poly-ACQs exhibit higher Tg by ca. 30°C than poly-QM. The difference was considered to arise from the fact that poly-ACQs carry many highly polar substituents, such as cyano and alkoxycarbonyl groups.

VII. POLYMERIZATION BEHAVIOR OF QUINODIMETHANES AS ACCEPTOR MONOMERS

A. Introduction

p-Benzoquinone(BQ) which displays electron-accepting properties is a well-known inhibitor[83,84] and retarder[85] in free radical polymerization; it undergoes copolymerization with styrene despite its very low susceptibility to copolymerization[85,86]. *p*-Chloranil (PCA), which is a much stronger electron acceptor than BQ, undergoes alternating copolymerization with styrene in the presence of free radical initiators[87,88]. 2,3-Dichloro-5,6-dicyano-*p*-benzoquinone (DDQ), an even stronger electron acceptor than PCA, is also

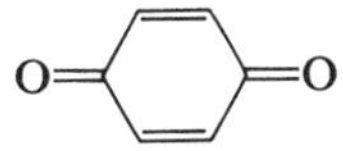

BQ 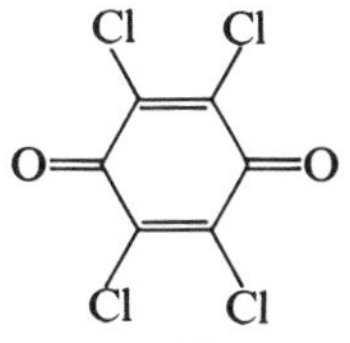 PCA 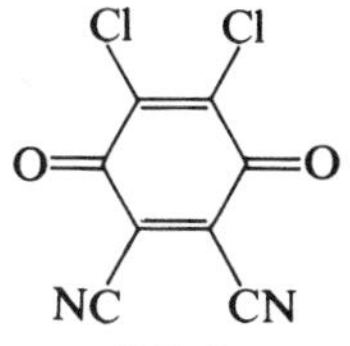 DDQ

alternatingly copolymerized with styrene even in the absence of a free radical initiator[89,90]. The relative reactivity of these benzoquinones as acceptor monomers toward the polymer radical with a terminal styrene unit is closely related to their electron-accepting character[90,91].

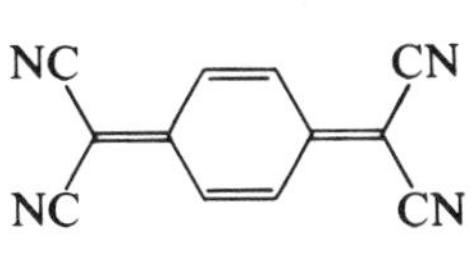

TCNQ 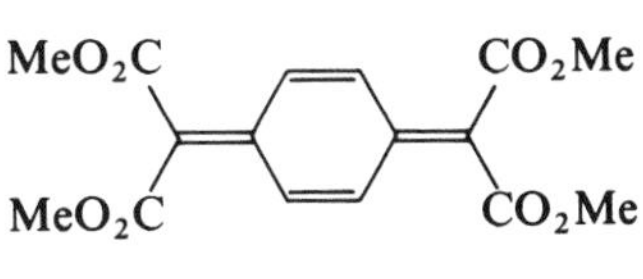 TMCQ

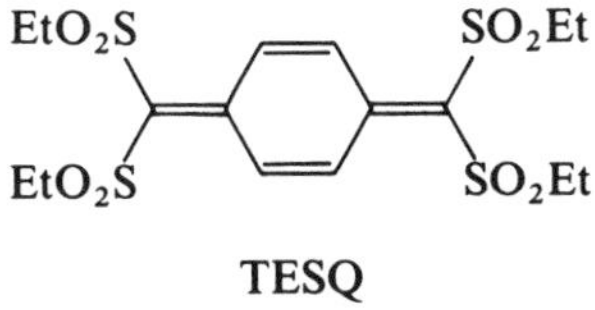

TESQ

TABLE 7. Glass transition temperature (Tg) of the poly-ACQ, poly-alkyl methacrylates, poly-(alkyl α-cyanoacrylates), poly (*p*-alkyl styrenes) and poly-*p*-xylylene

R	$-[C(CN)(COOR)-C_6H_4-C(CN)(COOR)]-$ [a]	$-[CH_2-C(Me)(COOR)]-$ [b]	$-[CH_2-C(CN)(COOR)]-$ [b]	$-[CH_2-CH(C_6H_4R)]-$ [b]	$-[CH_2-C_6H_4-CH_2]-$ [b]
-H				100	
-Me	108	105	170	93	
-Et	108	65	175	< 78	∼ 80
-Bu	108	20	(140)	6	

[a] Obtained by DSC.
[b] Ref. 82.

Chemists at Du Pont described the preparation of a series of new compounds with electron-accepting properties such as 7,7,8,8-tetracyanoquinodimethane (TCNQ)[92], 7,7,8,8-tetrakis(methoxycarbonyl)quinodimethane (TMCQ)[92], 7,7,8,8-tetrakis(ethylsulfonyl)quinodimethane (TESQ)[93] and 11,11,12,12-tetracyanonaphtho-2,6-quinodimethane(TNAP)[94] in the early 1960s.

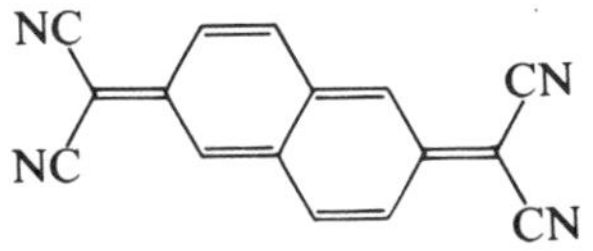

TNAP 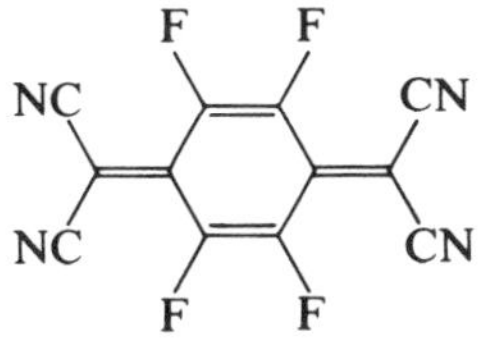 $TCNQF_4$ 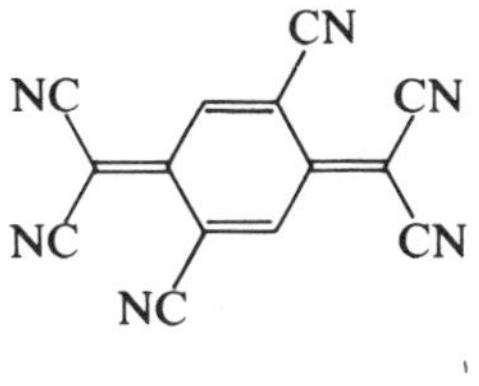 $TCNQ(CN)_2$

In addition, 2,3,5,6-tetrafluoro-7,7,8,8-tetracyanoquinodimethane ($TCNQF_4$)[95] and 2,5,7,7,8,8-hexacyanoquinodimethane ($TCNQ(CN)_2$)[95], which display stronger electron-accepting properties than TCNQ, were prepared in 1975[95]. These compounds have been extensively studied due to their powerful electron-accepting character, in connection with their charge transfer complexes with high electric conductivity referred to as organic metal[96]. Whereas TCNQ was reported to initiate a cationic polymerization of alkyl vinyl ethers[97, 98], polymerization of these quinodimethanes as acceptor monomers had not been studied in detail until the spontaneous alternating copolymerization of TCNQ with styrene was reported in 1978[73].

B. TCNQ–Styrene System

When TCNQ is mixed with a styrene solution in acetonitrile, a dark red color attributed to the formation of a charge transfer complex between TCNQ and styrene develops instantaneously. On standing at room temperature for a day, TCNQ dissolves slowly in acetonitrile and reacts with styrene at the interface of the TCNQ crystals, thereby producing a gelatinous shell of swollen pink-colored copolymer[73]. The copolymer is insoluble in conventional organic solvents such as benzene and chloroform, and it swells in aprotic polar solvents such as *N,N*-dimethylformamide and dimethyl sulfoxide at room temperature and eventually dissolves on prolonged heating at higher temperatures (e.g. 80 °C). Elemental analysis and NMR data reveal that the copolymer is a truely alternating copolymer. Its ^{1}H-NMR spectrum contains only two kinds of peaks; an aromatic peak of δ7.0–7.5 ppm and a peak at δ3–3.5 ppm assigned to the methine and methylene protons of the styrene units which are much more deshielded than the corresponding protons of homopolystyrene whose peaks generally appear between δ1.0 and 3.0 ppm[99]. This deshielding is presumed to arise primarily from the powerful electron-withdrawing effect of the neighboring dicyanomethylene groups when the styrene unit directly links two TCNQ units on both its sides.

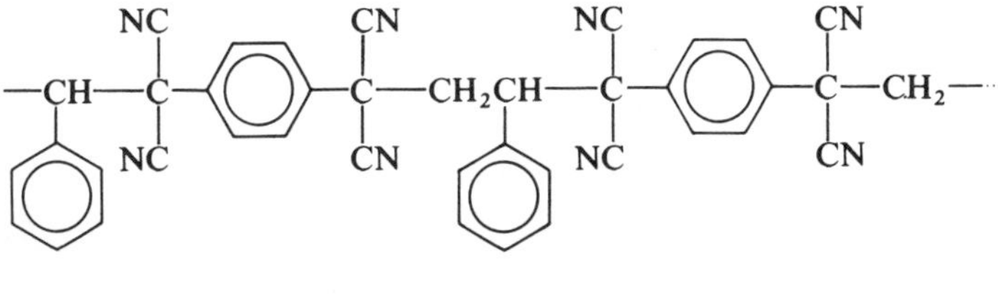

C. Systems of Styrene with $TCNQF_4$, TNAP and $TCNQ(CN)_2$

$TCNQF_4$ displays considerably stronger electron-accepting properties[100] and is better soluble in organic solvents than TCNQ and is therefore conveniently used for kinetic studies. The electron-accepting character of TNAP is intermediate between that of TCNQ and $TCNQF_4$, and $TCNQ(CN)_2$ displays the strongest electron-accepting properties[100]. The addition of styrene to a solution of TCNQ, TNAP, or $TCNQF_4$ in acetonitrile causes a deepening of the color of the respective acceptor solution, due to the formation of the colored charge transfer complexes. TCNQ–styrene (St), TNAP–St and $TCNQF_4$–St systems absorb light in the range of 450–580 nm[73], 530–639 nm[101] and 500–750 nm[102] respectively. The charge transfer transition absorption of the $TCNQ(CN)_2$–St system cannot be measured because the absorbance of the mixture decreases so rapidly that it disappears completely within a minute, probably due to a very rapid polymerization[101]. $TCNQF_4$, TNAP and $TCNQ(CN)_2$ undergo alternating copolymerization with styrene without any initiator, similarly to TCNQ[101,102]. Kinetic studies of the spontaneous alternating copolymerization of $TCNQF_4$–St system revealed that the copolymerization follows the three-halves order with respect to each of $TCNQF_4$ and styrene[102]. The copolymerizations of the TCNQ–St[102] and TNAP–St[101] systems were found to obey the same kinetics. On the other hand, the copolymerization of the $TCNQ(CN)_2$–St system follows first-order kinetics with respect to both $TCNQ(CN)_2$ and styrene[101]. A three-halves order kinetics has previously been found for the spontaneous alternating copolymerizations of the systems *p*-dioxene–maleic anhydride and 1,2-dimethoxy-ethylene–maleic anhydride[103] and a similar multi-step copolymerization reaction scheme[103] was suggested in those cases: (*a*) The donor and the acceptor monomers form a charge transfer complex and an intramolecular first-order reaction of the complex gives the propagating radical species, (*b*) the radical species adds to the complex to give an alternating copolymer, and (*c*) termination takes place between the propagating polymer radicals which are assumed to be in stationary state (Scheme 5). The first-order kinetics observed in the copolymerization of the $TCNQ(CN)_2$–St system was also found for the TCNQ–methyl methacrylate (MMA) system as mentioned below. The rate constants, overall activation energies and the half-life times (for the acceptor monomer under a given monomer concentration) of the TCNQ–St, TNAP–St, $TCNQF_4$–St and $TCNQ(CN)_2$–St systems, are compiled in Table 8. These systems have similar overall activation energies of copolymerization. The $TCNQ(CN)_2$–St system copolymerizes about 1000 times as rapidly as the TCNQ–St system. The rates of copolymerizations are closely related to the electron-accepting ability of the acceptor monomer as measured by its electron affinity (EA) (see Tables 8 and 9).

When acceptor monomers with a low positive e value of the Alfrey–Price Q–e scheme such as methyl methacrylate (MMA) ($e = 0.4$)[79], methyl acrylate (MA) ($e = 0.6$)[79] and acrylonitrile (AN) ($e = 1.2$)[79] are used as comonomers in the copolymerization with $TCNQF_4$, it has been found[102] that MMA is alternatingly and spontaneously copolymerized, MA undergoes alternating copolymerization only by means of a radical initiator, and AN is not susceptible to copolymerization. It is noteworthy that MMA and MA with positive e value undergo alternating copolymerization as donor monomers instead of acceptor monomers with $TCNQF_4$ which is a very strong electron-acceptor monomer. This alternating tendency in the $TCNQF_4$–MMA and $TCNQF_4$–MA systems cannot be explained in terms of the Alfrey–Price Q–e scheme because all monomers of these systems have positive e values and repulsive forces instead of attractive forces would be expected. It has therefore been proposed that the large difference in the polar character between $TCNQF_4$ and the alternatingly copolymerizable comonomers, which causes a charge transfer interaction, is one of the primary factors responsible for their alternating tendency. In addition, TCNQ, which is a weaker acceptor monomer than $TCNQF_4$, also

$$(a)\quad TCNQF_4 + St \underset{}{\overset{K_{CT}}{\rightleftharpoons}} \text{CT Complex} \xrightarrow{k_i} 2\ R^{\cdot}$$

$$\text{Rate of initiation: } = \frac{d[R\cdot]}{dt} = 2k_i K_{CT}[TCNQF_4][St]$$

$$\text{when } K_{CT} \ll 1$$

$$(b)\quad P\cdot_{n-1} + n\text{CT Complex} \xrightarrow{k_p}$$

R–(CH₂–CH(C₆H₅)–C(CN)₂–C₆F₄–C(CN)₂)ₙ₋₁–CH₂–CH(C₆H₅)–C(CN)₂–C₆F₄–C·(CN)₂

propagating polymer radical $[P_n{}^{\cdot}]$

$$\text{Rate of polymerization: } \frac{-d[TCNQF_4]}{dt} = \frac{-d[St]}{dt} = k_p K_{CT}[P_n{}^{\cdot}][TCNQF_4][St]$$

$$(c)\quad 2\ P_n\cdot \xrightarrow{k_t} \text{dead Polymer}$$

$$\text{Rate of termination: } -\frac{d[P_n\cdot]}{dt} = 2k_t[P_n\cdot]^2$$

At stationary state of propagating polymeric radical

$$R_i = R_t$$

$$2k_i K_{CT}[TCNQF_4][St] = 2k_t[P_n\cdot]^2$$

$$[P_n\cdot] = \sqrt{\frac{k_i K_{CT}}{k_t}[TCNQF_4][St]}$$

then

$$R_p = k_p \sqrt{\frac{k_i}{k_t}} K_{CT}{}^{3/2}[TCNQF_4]^{3/2}[St]^{3/2}$$

SCHEME 5

TABLE 8. Rate constants (k_p), overall activation energies (E_a), and half-life times ($T_{1/2}$) of the copolymerization of the TCNQ–St, TNAP–St, $TCNQF_4$–St, and $TCNQ(CN)_2$–St systems

Systems	Kinetic order in the complex	$10^3 k_p/l^{1/2}\ mol^{-1/2}\ s^{-1}$	At T°C	E_a (kcal mol^{-1})	$T_{1/2}$ at 34.5°C (min)
TCNQ–St	1.5	2.75	34.5	17.3	1000[a]
TNAP–St	1.5	21.4	34.5	16.3	151
$TCNQF_4$–St	1.5	52.9	34.5	16.6	4.0
$TCNQ(CN)_2$–St	1.0	1.05	10	16.7	1.2

[a] Calculated from the rate constant.

TABLE 9. Modes of polymerization in acetonitrile

Vinyloxy compound (e value)[77]		TCNO(CN)$_2$	TCNQF$_4$[102]	DDQ[90]	TNAP[101]	TCNQ[105]	TESQ[106]	TMCQ[106]
	Electron affinity (eV)		3.22[78]	3.00[78]		2.88[78]		
	Reduction potential(V)	0.65[100]	0.53[100]	0.51[107]	0.21[94] 0.20[100]	0.17[100] −0.2[117]	0.092[108]	−0.83[117]
VAc (−0.22)		Adduct	Alternating copolymer	Alternating copolymer	Adduct	Alternating copolymer	Adduct	Alternating copolymer
PhVE (−1.21)		Adduct	Alternating copolymer	Alternating copolymer	Adduct	Alternating copolymer	Adduct	Alternating copolymer
CEVE (−1.41)		Homopolymer	Homopolymer	Homopolymer	Homopolymer	Alternating copolymer	Homopolymer	Alternating copolymer
n-BVE (−1.20)		Homopolymer	Homopolymer	Homopolymer	Homopolymer	Homopolymer	Homopolymer	Alternating copolymer
i-BVE (−1.77)		Homopolymer	Homopolymer	Homopolymer	Homopolymer	Homopolymer	Homopolymer	Alternating copolymer

copolymerizes alternatingly and spontaneously with MMA whereas MA is not copolymerizable with TCNQ[104]. The rate of spontaneous alternating copolymerization between TCNQ and MMA is about one thousandth as slow as that between $TCNQF_4$ and MMA[104]. Moreover, the slow rate of the copolymerization obeys first-order kinetics with respect to the TCNQ monomer concentration[104].

D. Modes of Polymerization of Vinyloxy Monomers with Electron-accepting Quinodimethanes

As vinyloxy monomers *n*-butyl vinyl ether (*n*-BVE), isobutyl vinyl ether (*i*-BVE), 2-chloroethyl vinyl ether (CEVE), phenyl vinyl ether (PhVE), and vinyl acetate (VAc) have been used. The electron-donating character of these compounds may be arranged in the above order by means of Taft and Hammett substituent constants of the vinyloxy and vinyl groups. Stille and coworkers[97, 98] reported that TCNQ initiates the cationic homopolymerization of alkyl vinyl ethers in acetonitrile in line with its powerful electron-accepting character which causes an electron transfer. In the polymerization of TCNQ with each of the five monomers in acetonitrile, it has been found[105] that *n*-BVE and *i*-BVE homopolymerize whereas CEVE, PhVE and VAc copolymerize in an alternating fashion with TCNQ. The two modes of polymerization are consistently correlated with the electron-donating character of the vinyloxy monomers[105]. Moreover, when other electron-accepting quinodimethane derivatives such as $TCNQ(CN)_2$[101], $TCNQF_4$[102], DDQ[90], TNAP[101], TESQ[106] and TMCQ[74] are used, the modes of polymerization indicated in Table 9 are observed. These results, except for the case of TESQ, suggest that the modes of polymerization are also correlated with the electron-accepting character of these monomers. It is concluded, therefore, that the difference in polar character between the donor and acceptor monomers is responsible for an electron transfer reaction and strongly affects the determination of the mode of polymerization.

Furthermore, it has been found[109] that the mode of polymerization of TCNQ with CEVE depends upon the solvent used. An alternating copolymer is obtained in acetonitrile, whereas in ethylene carbonate a homopolymer of CEVE results. Low molecular weight products composed of TCNQ and CEVE units are obtained when dimethyl sulfoxide (Me_2SO) and *N*-methylformamide are employed as solvents. This solvent effects on the mode of polymerization may be ascribed to the polarity and basicity of the solvent. The dielectric constants of acetonitrile, Me_2SO, ethylene carbonate and *N*-methylformamide are 37.5, 46.68, 89.6, and 182.4, respectively[110]. The frequency shifts, $\Delta\nu_{OH}$, of phenol with acetonitrile, ethylene carbonate and Me_2SO are 155, 159 and 350 cm^{-1}, respectively[111]. Frequency shifts, $\Delta\nu_{OH}$, of *p*-fluorophenol with *N*-methylformamide and Me_2SO are 271 and 367 cm^{-1}, respectively[112]. It is therefore obvious that the basicity of these solvents increases in the following order: acetonitrile $\leqslant$ ethylene carbonate $<$ N-methylformamide $<$ Me_2SO. From the difference in the dielectric constants the following order of reactivity of the electron transfer reaction between TCNQ and CEVE may be deduced: *N*-methylformamide $>$ ethylene carbonate $>$ Me_2SO $>$ acetonitrile. Since *N*-methylformamide is more basic than ethylene carbonate, the cationic end of the radical cation species formed by the electron transfer reaction can be more tightly solvated in it by solvent molecules, probably leading to a deactivation of the cationic end[113]. Another free radical end may add monomer molecules to give low molecular weight products composed of both monomer units. Likewise, the formation of low molecular weight products in Me_2SO may also be attributed to its high basicity. The drastic difference in the mode of polymerization between acetonitrile and ethylene carbonate conceivably arises primarily from the difference in polarity of these solvents. The more polar ethylene carbonate permits the electron transfer reaction between TCNQ and CEVE to occur, while the less polar acetonitrile does not. Because both solvents have a low basicity of similar magnitude, the cationic polymer end is only weakly solvated and does not inhibit the cationic polymerization.

E. Amphoteric Behavior of TMCQ and TECQ in Alternating Copolymerization

7,7,8,8-Tetrakis(methoxycarbonyl)- and tetrakis(ethoxycarbonyl)-quinodimethane, TMCQ[72,74] and TECQ[75], are quinodimethane derivatives with an electron-withdrawing functional group similar to that of TCNQ. A study of the charge transfer absorption bands between TMCQ or TECQ and conventional donor compounds revealed that TMCQ and TECQ display electron-accepting properties which are much weaker than those of TCNQ[75]. TECQ has been found to exhibit only a slightly weaker electron-accepting nature than TMCQ. Moreover, it has been found that TMCQ[74] and TECQ[75] behave as electron donors when they meet with TCNQ which has very strong electron-accepting properties. A comparison of the absorption bands reveals that TECQ[74] displays only a slightly more electron-donating character toward TCNQ than TMCQ. Consequently, it is

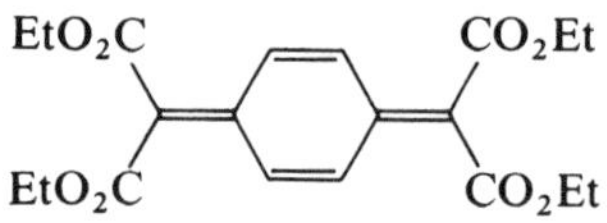

EtO_2C CO_2Et

EtO_2C CO_2Et

TECQ

concluded[75] that TMCQ and TECQ exhibit an amphoteric polar character in the formation of charge transfer complexes. Their amphoteric polar nature may be explained consistently in terms of a π-electron density scheme. Styrene has the highest π-electron density followed by TECQ which has only a slightly higher density than TMCQ, and TCNQ clearly has the lowest density. The sufficiently large difference in the π-electron density therefore gives rise to the formation of charge transfer complexes between styrene and TECQ or TMCQ as well as between TECQ or TMCQ and TCNQ.

TMCQ and TECQ copolymerize alternatingly and spontaneously as acceptor monomers with conventional electron-donating comonomers such as styrene, *i*-BVE, *n*-BVE, CEVE, PhVE and VAc. TMCQ and TECQ are also copolymerized alternatingly and spontaneously as donor monomers with the acceptor monomer TCNQ. This amphoteric behavior in alternating copolymerization was found first in the terpolymerization[74] of styrene, TMCQ and TCNQ, where the terpolymers obtained contain 50 mol % of TCNQ units regardless of the monomer feed ratio. This indicates that TMCQ and styrene copolymerize as donor monomers with TCNQ, contrary to the expectation that TMCQ would copolymerize as an acceptor monomer. The compositional relationships of the terpolymerization of styrene, TMCQ and TCNQ, as well as of styrene, TECQ and TCNQ are shown in Figures 14 and 15, respectively, where open and closed circles refer to the monomer feed and the terpolymer composition, respectively. The terpolymerization composition relationships can be illustrated by the composition diagrams of the binary copolymerization between TMCQ and styrene and between TECQ and styrene (St) (Figure 16), because the content of the TCNQ unit is always constant (50 mol %) in any run. According to the mechanism involving the complex formation[114] in the alternating copolymerization, the apparent monomer reactivity ratios of the complexes are calculated as follows: $r_1(K_1/K_2)$ (TMCQ–TCNQ complex) $= 7 \pm 3$ and $r_2(K_2/K_1)$ (St–TCNQ complex) $= 0.7 \pm 0.3$ for the St–TMCQ–TCNQ system and $r_1(K_1/K_2)$ (TECQ–TCNQ complex) $= 15 \pm 10$ and $r_2(K_2/K_1)$ (St–TCNQ complex) $= 0.5 \pm 0.3$ for the St–TECQ–TCNQ system. The relative reactivities of the TMCQ–TCNQ and

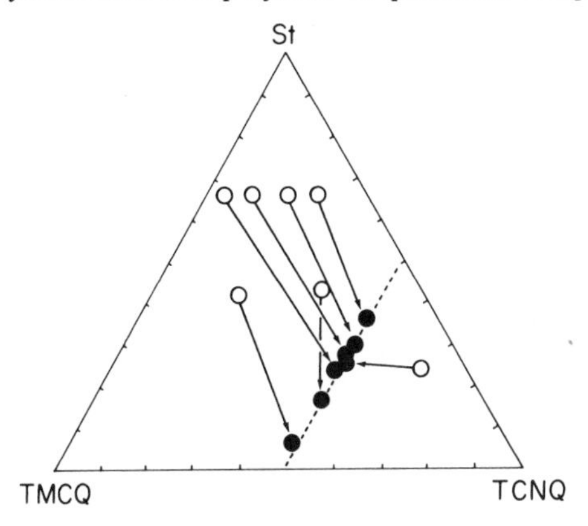

FIGURE 14. Triangular diagram of the composition of the terpolymer of TCNQ, TMCQ and styrene (St): (○), feed composition; (●) terpolymer composition. Arrows denote change in the composition from the feed to the terpolymer obtained

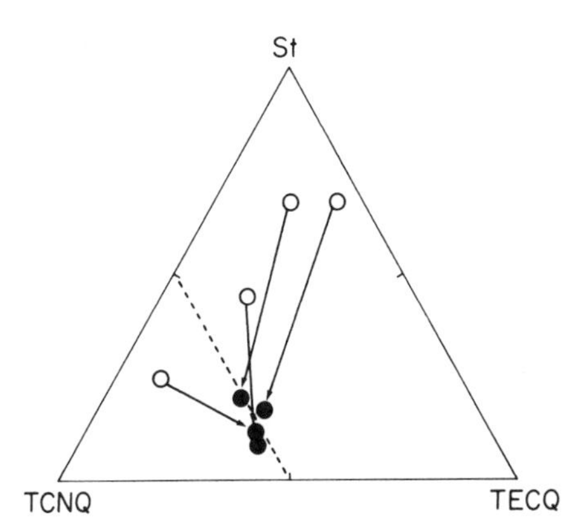

FIGURE 15. Triangular diagram of the composition of the terpolymer of TECQ, TCNQ and styrene (St): (○), feed composition; (●), terpolymer composition. Arrows denote change in the composition from the feed to the terpolymer obtained

TECQ–TCNQ complexes toward the polymer radical with a terminal St–TCNQ complex unit are as follows: St–TCNQ complex (1) < TMCQ–TCNQ complex (1.4) < TECQ–TCNQ complex (2.0). Thus, the TECQ–TCNQ complex is more reactive than the TMCQ–TCNQ complex. The reactivity of these complexes coincides with the electron-donating efficiency of the donor with respect to TCNQ, TECQ being a better electron donor than TMCQ.

The terpolymerization of TECQ–TMCQ–St[75] using monomer feed mol ratios of TECQ/TMCQ/St = 14.2/14.9/70.9 and 11.2/36.4/52.4 at 60°C gave the terpolymers (conversion of 11.5 and 7.6%) with the mol ratios TECQ/TMCQ/St = 22.6/27.4/50.0 and 10.2/39.8/50.0, respectively. From the difference of the ratios of the TECQ to TMCQ content in monomer feed (TECQ/TMCQ = 0.49/0.51 and 0.24/0.76) and in the terpolymers (TECQ/TMCQ = 0.45/0.55 and 0.20/0.80, respectively) it may be assumed that

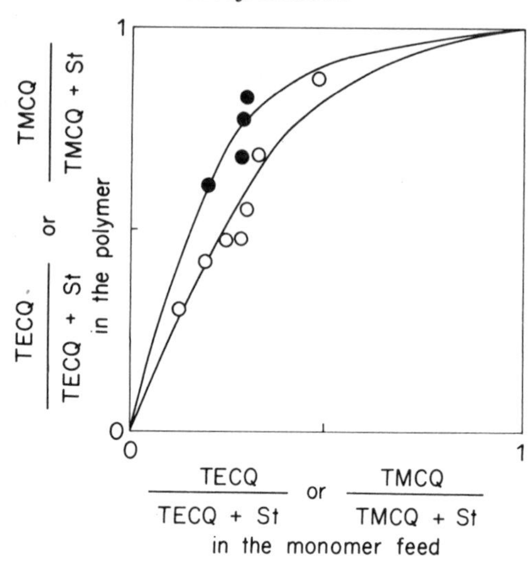

FIGURE 16. Diagram of the composition of the terpolymerizations of TECQ, TCNQ and styrene (St), and of TMCQ, TCNQ and St as binary copolymerizations between TECQ and St and between TMCQ and St, respectively. The lines are calculated using $r_1(K_1/K_2) = 15 \pm 10$ and $r_2(K_2/K_1) = 0.5 \pm 0.3$ for the terpolymerization of the TECQ–TCNQ–St system (●) and $r_1(K_1/K_2) = 7 \pm 3$ and $r_2(K_2/K_1) = 0.7 \pm 0.3$ for the terpolymerization of the TMCQ–TCNQ–St system (O)

TMCQ is somewhat more reactive than TECQ in the alternating copolymerization with styrene, i.e. the TMCQ–St complex is more reactive than the TECQ–St complex. Consequently, the reactivity order of the acceptor monomers TMCQ and TECQ in their alternating copolymerization with styrene is in good agreement with their electron-accepting character in their charge transfer complex formation with styrene. However, it is difficult to ascribe the small difference between TMCQ and TECQ to an inductive substituent effect reflected in the Hammett constant[115] and to a steric substituent effect reflected in the Taft steric parameter[116] between methyl and ethyl groups since differences are very small.

When TMCQ is heated above 175 °C or exposed to light, it polymerizes even though the product appears dimeric or trimeric[92]. Recently, Hall and Bently[117] reported that TMCQ polymerizes with free radical and anionic initiators to give homopolymer with a melting point of about 300 °C and an intrinsic solution viscosity of $0.91\,\mathrm{dl\,g^{-1}}$ (as a polycarboxonium salt). Thus, TMCQ readily undergoes homopolymerization. However, TECQ cannot be homopolymerized by means of azobisisobutyronitrile (AIBN), *n*-butyllithium and boron trifluoride etherate[75]. Only when TECQ is kept in the crystalline state at room temperature for a month, a white powder insoluble in benzene with molecular weight of 2600 ($\overline{\mathrm{DP}} = 6.6$), is formed in poor yield[75]. Therefore, it is obvious that TECQ exhibits only a very slight tendency to homopolymerize, in contrast to TMCQ. This difference in homopolymerizability cannot be attributed to a steric effect due to the very small difference in the Taft steric parameters between methyl ($E_s = 0.00$) and ethyl ($E_s = -0.07$)[116]. Presumably, the specific structure of the tetrakis(alkoxycarbonyl)quinodimethane may amplify significantly the small difference between methyl and ethyl groups, and leads to the different amphoteric character of the monomers observed in the charge transfer complexation and in the alternating copolymerization and the difference in the homopolymerizability of TMCQ and TECQ.

F. Polymerization Behavior of TESQ

Since the ethylsulfonyl group ($\sigma_p = 0.68$) exhibits the same electron-withdrawing power as cyano group ($\sigma_p = 0.66$) judged by the Hammett substituent constant[115], TESQ was expected to display similar polymerization behavior as an acceptor monomer to TCNQ. However, it has been found[106] that in the charge transfer transition TESQ (EA = 1.17 eV) exhibits a much lower electron affinity than TCNQ (EA = 2.84 eV)[78]. TESQ and styrene have been subjected to spontaneous alternating copolymerization in nitromethane, but when *p*-dioxane or dichloromethane is used instead of nitromethane, an alternating copolymer is not obtained, and the content of the styrene unit is higher than 50 mol%. From this it may be assumed that a cationic polymerization of styrene takes place simultaneously. This solvent effect cannot be explained in terms of its polarity and basicity[104]. Indeed, 1-phenylethanol and TESQ may readily undergo dehydration and polymerization to polystyrene, suggesting that TESQ and its hydrogenation product may initiate simultaneously both cationic polymerization of styrene and the alternating copolymerization of styrene with TESQ. Moreover, when the TESQ fraction in monomer feed is above 40 mol%, no copolymer but only the 1:1 adduct in high yield is obtained, in contrast to the copolymerization of TCNQ and styrene.

TESQ initiates the cationic oligomerization of *i*-BVE, *n*-BVE, CEVE and PhVE, but a reaction of TESQ with VAc has not been observed. Consequently, TESQ is considered to be more acidic (as proton acid) than TCNQ from the observed modes of polymerization of a series of those vinyloxy monomers whereas TESQ exhibits lower electron affinity than TCNQ. Concerning these differences between TESQ and TCNQ, it should be taken into account that in the π conjugation between the substituents and the quinodimethane part, the 3p orbital of sulfur participates for the ethylsulfonyl group, and the 2p orbital of carbon for the cyano group. Price and Oae[118] suggested that the 2p–3p π bond is less stable than the 2p–2p π bond. According to the theory of hard and soft acids and bases it may be assumed that TESQ is a much harder acid than TCNQ.

G. Polymerization Behavior of QBS

Quinone diimine, prepared as a colorless crystalline compound by Willstätter and Mayer[119], is expected to show an intermediate behavior between *p*-benzoquinone and quinodimethane from the relationship of Coppinger and Bauer[15] between the stability of hetero *p*-benzoquinones and the electronegativity of their exocyclic atoms, carbon (2.50), nitrogen (3.07) and oxygen (3.50)[120]. Unsubstituted *p*-quinone diimine is very susceptible to light and acid, especially in solution, and may undergo reactions such as hydrolysis and polymerization[121]. Adams and Nagarkatti[121] reported that *p*-quinone diimines carrying electron-withdrawing substituents such as acyl, alkylsulfonyl and arylsulfonyl groups at exocyclic nitrogen atoms become less susceptible to hydrolysis.

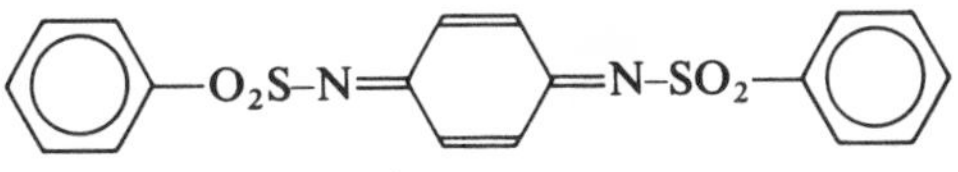

$C_6H_5-O_2S-N=C_6H_4=N-SO_2-C_6H_5$

QBS

The electron-accepting character of *p*-quinone bis(benzenesulfonimide) (QBS) was examined. The charge transfer transition between QBS and hexamethylbenzene (HMB) appears in benzene at room temperature at 495 nm as shown in Figure 17[122]. The electron affinity (EA) of QBS could be estimated as 2.17 eV on the basis of the value of 2.48 eV[78] for the EA of *p*-chloranil (PCA) and the charge transfer transition at 510 nm between PCA

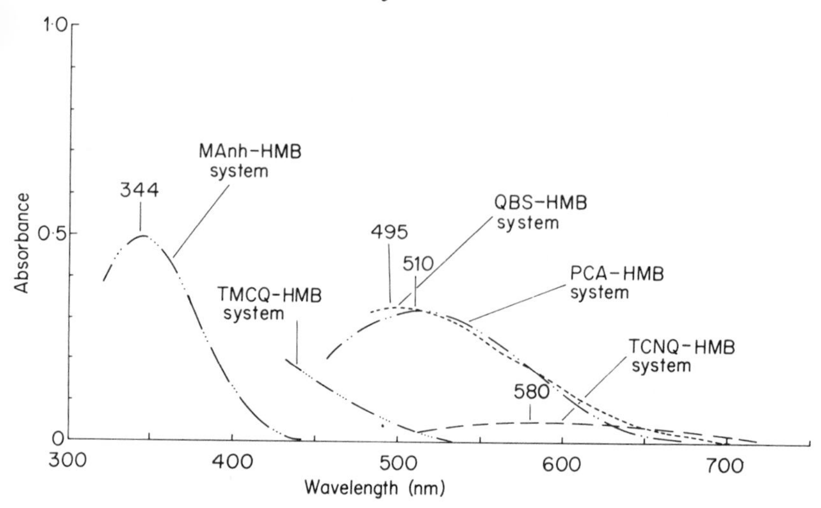

FIGURE 17. UV–VIS difference spectra between HMB and various acceptor compounds in benzene at room temperature. Concentrations of solutions employed are [TCNQ] = 1.69×10^{-4} mol l^{-1} and [HMB] = 0.102 mol l^{-1} for the TCNQ–HMB system, [PCA] = 1.06×10^{-2} mol l^{-1} and [HMB] = 1.0×10^{-2} mol l^{-1} for the PCA–HMB system, [QBS] = 1.0×10 mol l^{-1} and [HMB] = 0.1 mol l^{-1} for the QBS–HMB system, [TMCQ] = 7.11×10^{-3} mol l^{-1} and [HMB] = 9.95×10^{-2} mol l^{-1} for the TMCQ–HMB system, and [MAnh] = 1.05×10^{-2} mol l^{-1} and [HMB] = 6.10×10^{-2} mol l^{-1} for the MAnh–HMB system, respectively

and HMB, as shown in Figure 17. It was concluded, therefore, that QBS is intermediate in electron-accepting character between PCA and maleic anhydride (MAnh) (EA = 1.33 eV)[78]. Electron-accepting character was compared among TCNQ, PCA, QBS and MAnh by means of their charge transfer transition maxima with HMB, appearing at 580, 510, 495 and 344 nm, respectively, as shown additionally in Figure 17. The maximum for the TMCQ–HMB system was not observed because it overlaps seriously with the absorption of TMCQ alone. However, it certainly exists in the wavelength range below 450 nm, implying that TMCQ is a weaker electron acceptor than QBS. Comparison between TMCQ and MAnh could not be carried out by this charge transfer transition method. Consequently, the following order of the electron-accepting character of the acceptor compounds was found: TCNQ > PCA > QBS > TMCQ, MAnh.

The copolymerization of QBS with styrene and acenaphthylene were attempted at 60 °C in benzene without initiator for 168 h and 23.2 h, respectively[122]. In both cases no polymeric material could be obtained, and only the starting materials were recovered quantitatively. When a free radical initiator such as azobisisobutyronitrile (AIBN) was added, copolymers were obtained[122]. Figure 18 shows the composition diagrams of the homogeneous copolymerizations of QBS with styrene and acenaphthylene. The copolymers obtained as a white powder and their elemental analysis showed almost fixed amounts of carbon, hydrogen and nitrogen regardless of monomer feed ratio. The analysis is in good agreement with the calculated values for the copolymers composed of equimolar amounts of each component monomer. The molecular weight of the copolymers obtained for the QBS–styrene and QBS–acenaphthylene systems were measured by vapor pressure osmometry in chloroform to be about 7000 and 4000–7000, respectively, corresponding to degrees of polymerization of about 15 and 8–13, respectively, based upon an alternating

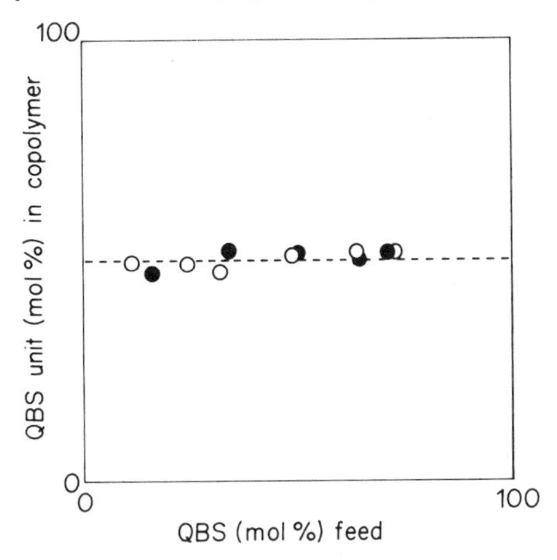

FIGURE 18. Composition diagram of the copolymerizations of QBS with styrene (St) (O) and with acenaphthylene (●)

structure of donor and acceptor units[122]. These molecular weights are very low, in contrast to those of the alternating copolymers of styrene with TCNQ ($[\eta] = 0.407$ dl g^{-1} in *N*,*N*-dimethylformamide–0.1 wt % LiCl at 30 °C)[73], TMCQ ($\eta_{sp}/C = 0.1$–0.4 dl g^{-1} in benzene at 30 °C)[74] and PCA ($\eta_{sp}/C = 0.19$ dl g^{-1} in benzene, molecular weight 15 900)[88]. In the IR spectra, the QBS monomer shows absorption at 1550 cm^{-1} due to the stretching vibration of the imide group but the QBS–styrene copolymer does not. In the ^{1}H-NMR spectrum of this copolymer, the methine and methylene protons appear in the $\delta 5.2$ and $\delta 4.0$ regions, respectively, being much more deshielded than the corresponding hydrogens of homopolystyrene, which generally appear in the $\delta 1$–2 ppm region[99]. Presumably, the deshielding arises from the powerful electron withdrawal by the neighboring benzenesulfonamide group when the styrene monomer unit is sandwiched between QBS monomer units in the copolymer. In addition, these methine and methylene protons in the copolymer appear as very broad signals, presumably due both to the influence of the quadrupole moment of the neighboring nitrogen nuclei and to the decrease in the flexibility of the main chain. It can be concluded therefore that QBS can copolymerize (co-oligomerize) in an alternating fashion with styrene and acenaphthylene when a free radical initiator is used and QBS reacts at exocyclic nitrogen sites[122].

The copolymerization between QBS and vinyl monomers with small positive *e* values such as methyl methacrylate (MMA), methyl acrylate (MA) and acrylonitrile (AN) were attempted with AIBN in benzene at 60 °C for 48 h. In no case was polymeric material obtained, and unreacted QBS was recovered almost quantitatively, similarly to the reactions of PCA with those monomers[88]. Since QBS has a much lower electron-accepting character than $TCNQF_4$ and TCNQ, the gap in π-electron density between QBS and MMA was considered to be too small to enable formation of a charge transfer complex between them and consequent further alternating copolymerization.

The copolymerizations of QBS with *n*-BVE, *i*-BVE, CEVE, PhVE and VAc gave the reaction products as white powders except in the case of VAc, in which no reaction took place and the starting materials were recovered quantitatively[122]. It is conceivable that the gap in π-electron density between VAc, which is the weakest donor monomer among the five vinyloxy monomers[105], and QBS, the weak acceptor monomer, is not sufficient to enable charge transfer complex formation and alternating copolymerization. It is evident from the composition data of the copolymers obtained that the copolymers are composed of equimolar amounts of QBS and donor comonomers, indicating the alternating copolymer structures. It can be pointed out that QBS, the weak acceptor monomer, cannot initiate the cationic polymerization of the vinyloxy compounds in benzene similarly to TECQ[75], TMCQ[74] and MAnh, whereas the stronger acceptor monomers, TCNQ and $TCNQF_4$, can initiate the cationic polymerization of a strong donor monomer such as *n*-BVE and *i*-BVE[102,105]. The results correspond well to the low electron-accepting character of QBS.

Terpolymerizations of the QBS–MAnh–styrene (St), QBS–PCA–St, QBS–TMCQ–St and QBS–TCNQ–St systems were carried out at 60 °C for a quantitative comparison in the polymerizability of the five acceptor monomers. The terpolymers of all systems were obtained as white powders and were always composed of about 50 mol% of the styrene monomer unit regardless of the monomer feed ratio, and thus the sum of the QBS and other acceptor monomer (MAnh, PCA, TMCQ, or TCNQ) unit was about 50 mol%. Consequently, the terpolymerization composition relationships of the QBS–MAnh–St, QBS–PCA–St, QBS–TMCQ–St and QBS–TCNQ–St systems can be illustrated by their composition diagrams of binary copolymerizations between QBS and MAnh, between QBS and PCA, between QBS and TMCQ, and between QBS and TCNQ, shown in Figures 19–22, respectively. According to the complex mechanism treatment[114] the modified monomer reactivity ratios of the complexes were calculated to be $r_1\ (K_1/K_2) = 30 \pm 20$ and $r_2\ (K_2/K_1) = 0.1 \pm 0.1$ for the QBS–MAnh–St system (C_1 is QBS–St complex and C_2 is MAnh–St complex), $r_1\ (K_1/K_2) = 15 \pm 10$ and $r_2\ (K_2/K_1) = 0.2 \pm 0.2$ for the QBS–PCA–St system (C_1 is QBS–St complex and C_2 is PCA–St complex), $r_1\ (K_1/K_2) = 1.18 \pm 0.1$ and $r_2\ (K_2/K_1) = 0.15 \pm 0.05$ for the QBS–TMCQ–St system (C_1 is QBS–St complex and C_2 is TMCQ–St complex), and $r_1\ (K_1/K_2) = 0.01 \pm 0.01$ and $r_2\ (K_2/K_1) = 45 \pm 10$ for the QBS–TCNQ–St system (C_1 is QBS–St complex and C_2 is TCNQ–St complex), respectively. K_1 and K_2 refer to equilibrium constants for formation

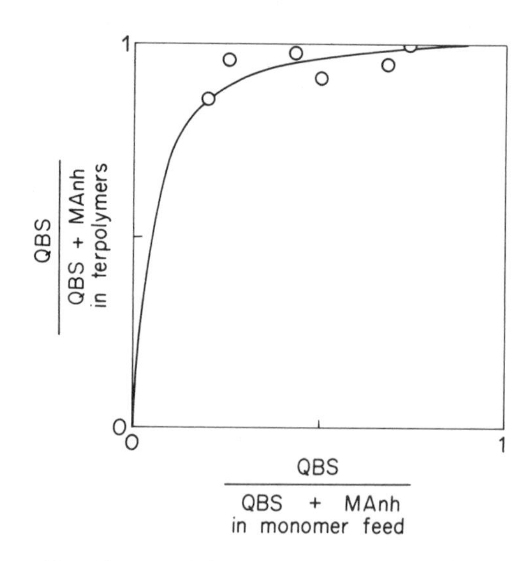

FIGURE 19. Composition diagram of the terpolymerization of QBS, MAnh and St as binary copolymerization between QBS and MAnh. The line is calculated by using $r_1\,(K_1/K_2)$ (QBS–St complex) = 30 and $r_2\,(K_2/K_1)$ (MAnh–St complex) = 0.1

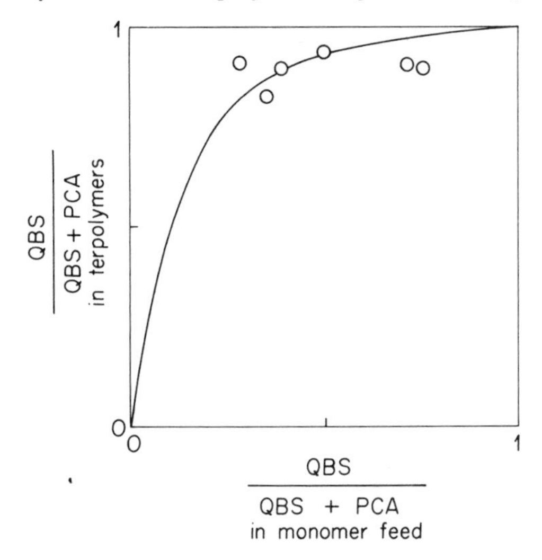

FIGURE 20. Composition diagram of the terpolymerization of QBS, PCA and St as binary copolymerization between QBS and PCA. The line is calculated by using $r_1(K_1/K_2)$ (QBS–St complex) = 15 and $r_2(K_2/K_1)$ (PCA–St complex) = 0.2

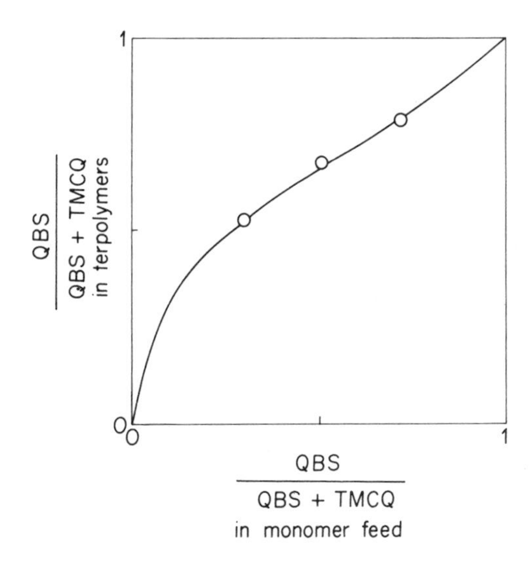

FIGURE 21. Composition diagram of the terpolymerization of QBS, TMCQ and St as binary copolymerization between QBS and TMCQ. The line is calculated by using $r_1(K_1/K_2)$ (QBS–St complex) = 1.18 and $r_2(K_2/K_1)$ (TMCQ–St complex) = 0.15

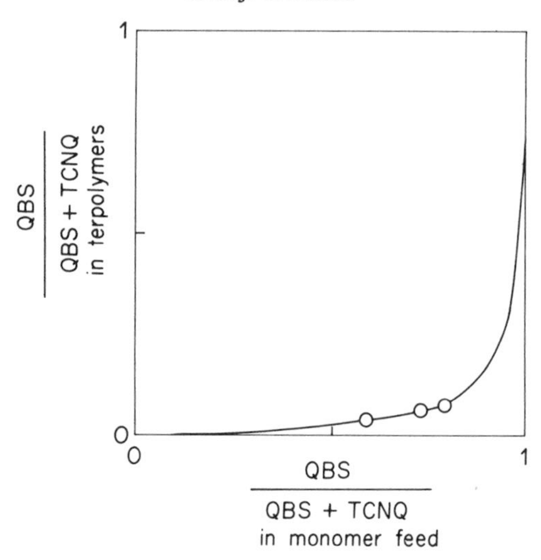

FIGURE 22. Composition diagram of the terpolymerization of QBS, TCNQ and St as binary copolymerization between QBS and TCNQ. The line is calculated by using $r_1(K_1/K_2)$ (QBS–St complex) = 0.01 and $r_2(K_2/K_1)$ TCNQ–St complex) = 45

of complex 1 (C_1) and complex 2 (C_2), respectively. The reciprocals of the modified monomer reactivity ratios were used as a measure of the relative reactivity of the complexes toward the polymer radical with a given terminal complex unit. The relative reactivity order of the MAnh–St, PCA–St, TMCQ–St and TCNQ–St complexes toward the polymer radical with the QBS–St complex unit was obtained as MAnh–St complex (1/30) < PCA–St complex (1/5) < TMCQ–St complex (1/1.18) < QBS–St complex (1) < TCNQ–St complex (1/0.01). Previously, it was pointed out[101,102] from the alternating copolymerizations of those electron-accepting quinodimethane derivatives with styrene that reactivity of their styrene complexes is related intimately to the electron-accepting character of the quinodimethanes. EA values of MAnh, QBS, PCA and TCNQ were reported to be 1.33, 2.17, 2.48 and 2.88 eV, respectively[78]. Although the EA value of TMCQ has not yet been reported, it is likely from the charge transfer complexation profile as shown in Figure 17 that TMCQ is a weaker electron acceptor than QBS. Thus, the order of electron-accepting character for the acceptor monomers is assumed to be MAnh, TMCQ < QBS < PCA < TCNQ and is in good agreement with the reactivity order of their styrene complexes except for PCA. The PCA–St complex is regarded as much less reactive than expected from the electron-accepting character of PCA. Coppinger and Bauer[15] pointed out on the basis of experimental data on the π–π electron transition of Hückel molecular orbital calculation that the stability of hetero *p*-benzoquinone compounds is related well to the energy difference between the quinonoid ground state and the benzenoid transition state. Increase in electronegativity of the exocyclic atom results in a decrease in the highest occupied bonding energy level and an increase in the lowest unoccupied antibonding level, leading to an increase in energy difference between ground and transition states and a large stability of the compound. Consequently, the exocyclic atom of hetero *p*-benzoquinones affects not only the stability of their compounds, that is, the reactivity, but also their electron-accepting character. In the case of the PCA–St system, it may be presumed that the exocyclic electronegative oxygen atom of PCA affects its

7,7-Dicyanoquinonemethide

stability more effectively than its electron-accepting character. Furthermore, 7,7-dicyanoquinonemethide[123] was found to be alternatingly copolymerizable with styrene[124].

VIII. REFERENCES

1. L. A. Errede and M. Szwarc, *Q. Rev. (Lond.)*, **12**, 301 (1958).
2. (a) Y. Minoura, in *Polymerization of p-Xylylene in New Polymerization Reactions* (Ed. T. Saegusa), Kagaku-Dojin Co., Kyoto, 1971 (in Japanese), p. 7; (b) S. Iwatsuki, *Kobunshi*, **23**, 135 (1974) (in Japanese); (c) S. Iwatsuki, *Adv. Polym. Sci.*, **58**, 94 (1984).
3. M. Szwarc, *Nature*, **160**, 403 (1947).
4. M. Szwarc, *Disc. Faraday Soc.*, **2**, 46 (1947).
5. M. Szwarc, *J. Chem. Phys.*, **16**, 128 (1948): *J. Polym. Sci.*, **6**, 319 (1951).
6. R. S. Corey, H. C. Haas, M. W. Kane and D. I. Livingston, *J. Polym. Sci.*, **13**, 137 (1954).
7. M. H. Kaufman, H. F. Mark and R. B. Mesrobian, *J. Polym. Sci.*, **13**, 3 (1954).
8. L. A. Auspos, L. A. R. Hall, J. K. Hubbard, W. Kirk Jr, J. R. Schaefgen and S. B. Speck, *J. Polym. Sci.*, **15**, 9, 19 (1955).
9. J. R. Schaefgen, *J. Polym. Sci.*, **15**, 203 (1955).
10. A. C. Farthing, *J. Chem. Soc.*, 3261 (1953).
11. C. J. Brown and A. C. Farthing, *J. Chem. Soc.*, 3270 (1953).
12. C. A. Coulson, D. P. Craig, A. Maccoll and A. Pullman, *Disc. Faraday Soc.*, **2**, 36 (1947).
13. A. I. Namiot, M. E. Dyatkin and Y. K. Syrkin, *Compt. Rend. Acad. Sci. USSR*, **48**, 267 (1945); *Chem. Abstr.*, **40**, 4927 (1946).
14. N. S. Hush, *J. Polym. Sci.*, **11**, 289 (1953).
15. G. M. Coppinger and R. H. Bauer, *J. Phys. Chem.*, **67**, 2846 (1963).
16. D. F. Evans, *J. Chem. Soc.*, 2753 (1959).
17. L. A. Errede and B. F. Landrum, *J. Am. Chem. Soc.*, **79**, 4952 (1957).
18. L. A. Errede and S. L. Hopwood Jr, *J. Am. Chem. Soc.*, **79**, 6507 (1957).
19. L. A. Errede and J. M. Hoyt, *J. Am. Chem. Soc.*, **82**, 436 (1960).
20. L. A. Errede and J. P. Cassidy, *J. Org. Chem.*, **24**, 1890 (1959).
21. L. A. Errede and J. P. Cassidy, *J. Am. Chem. Soc.*, **82**, 3653 (1960).
22. L. A. Errede, R. S. Gregorian and J. M. Hoyt, *J. Am. Chem. Soc.*, **82**, 5218 (1960).
23. L. A. Errede, J. M. Hoyt and R. S. Gregorian, *J. Am. Chem. Soc.*, **82**, 5224 (1960).
24. L. A. Errede, *J. Am. Chem. Soc.*, **83**, 949 (1961).
25. L. A. Errede and W. A. Pearson, *J. Am. Chem. Soc.*, **83**, 954 (1961).
26. L. A. Errede, *J. Polym. Sci.*, **49**, 253 (1961).
27. L. A. Errede, *J. Am. Chem. Soc.*, **83**, 959 (1961).
28. W. F. Gorham, *J. Polym. Sci.*, **4**, 3027 (1966).
29. F. S. Fawcett, US. Pat. 2757146 (1956).
30. N. E. Winberg, F. S. Fawcett, W. E. Mochel and C. W. Theobald, *J. Am. Chem. Soc.*, **82**, 1428 (1960).
31. L. D. Taylor and H. S. Kolesinski, *J. Polym. Sci.*, **B1**, 117 (1963).
32. Y. Ito, S. Miyata, M. Nakatsuka and T. Saegusa, *J. Org. Chem.*, **46**, 1043 (1981).
33. R. A. Jacobson, *J. Am. Chem. Soc.*, **54**, 1513 (1932).
34. A. A. Vanscheidt, E. P. Mel'Nikova, M. G. Krakovyak, L. A. Kukhareva and G. A. Gladkovkii, *J. Polym. Sci.*, **52**, 179 (1961).
35. K. Sisido and N. Kusano, *J. Polym. Sci.*, **A1**, 2101 (1963).
36. Y. Minoura, O. Shiina and S. Okabe, *Kogyo Kagaku Zasshi*, **70**, 1243, 1247 (1967).
37. H. E. Lunk and E. A. Youngman, *J. Polym. Sci.*, **A3**, 2983 (1965).
38. H. G. Gilch, *J. Polym. Sci., A-1*, **4**, 1351 (1966).

39. D. F. Hoeg, D. I. Luck and E. P. Goldberg, *J. Polym. Sci.*, **B2**, 697 (1964).
40. W. R. Dunnavant and R. A. Markle, *J. Polym. Sci., A-1*, **3**, 3649 (1965).
41. H. G. Gilch and W. L. Wheelwright, *J. Polym. Sci., A-1*, **4**, 1337 (1966).
42. R. H. Wade, *J. Polym. Sci.*, **B5**, 567 (1967).
43. T. Fujita, T. Yoshioka and N. Soma, *J. Polym. Sci. Polym. Lett. Ed.*, **16**, 1515 (1978); *J. Polym. Sci. Polym. Chem. Ed.*, **18**, 3253 (1980).
44. V. V. Korshak, G. S. Kelenikov and A. V. Kharchevikova, *Dokl. Akad. Nauk. USSR*, **56**, 169 (1947).
45. K. Shishido and S. Kato, *Kogyo Kagaku Zasshi*, **43**, 565, 952 (1940).
46. V. V. Korshak, V. A. Sergeev, V. K. Shifikov and P. S. Burenko, *Vysokomol. Soed.*, **5**, 1957 (1963): *Chem. Abstr.*, **60**, 5643 (1964).
47. W. F. Gorham, *ACS Polym. Prep.*, 73 (1965).
48. W. F. Gorham, Br. Pat. 883937-883941 (1961); Ger. Pat. 1085673 (1960); *Chem. Abstr.*, **55**, 22920 (1961).
49. C. J. Brown and A. C. Farthing, *Nature*, **164**, 915 (1949).
50. C. J. Brown, *J. Chem. Soc.*, 3265 (1953).
51. D. J. Cram and H. Steinberg, *J. Am. Chem. Soc.*, **73**, 5691 (1951).
52. D. J. Cram and J. M. Cram, *Acc. Chem. Res.*, **4**, 204 (1971).
53. F. Vögtle and P. Neumann, *Angew. Chem. Int. Ed.*, **11**, 73 (1972).
54. D. F. Pollart, *Am. Chem. Soc. Div. Pertol. Chem., Prep.*, **10**, 175 (1965); *Chem. Abstr.*, **67**, 21512 (1967).
55. D. F. Pollart, US Pat. 3149175, Ger. Pat. 1155444 (1963); *Chem. Abstr.*, **60**, 2817 (1964), US Pat. 3247274 (1966); *Chem. Abstr.*, **65**, 8816 (1966).
56. W. D. Niegish, *Bull. Am. Phys. Soc.*, **11**, 248 (1966).
57. W. D. Niegish, *J. Polym. Sci. Polym. Lett. Ed.*, **4**, 531 (1966).
58. S. Kubo and B. Wunderlich, *J. Polym. Sci. Polym. Phys. Ed.*, **10**, 1949 (1972).
59. G. Triber, K. Boehlke, A. Weitz and B. Wunderlich, *J. Polym. Sci. Polym. Phys. Ed.*, **11**, 1111 (1973).
60. W. F. Beach, *Macromolecules*, **11**, 72 (1978).
61. S. W. Chow, W. E. Loeb and C. E. White, *J. Polym. Sci.*, **13**, 2325 (1969).
62. S. W. Chow, L. A. Pilato and Wheelwright, *J. Org. Chem.*, **35**, 20 (1970).
63. H. G. Glich, *Angew. Chem.*, **77**, 592 (1965).
64. H. G. Glich, *J. Polym. Sci., A-1*, **4**, 438 (1966).
65. S. Iwatsuki and H. Kamiya, *Macromolecules*, **7**, 732 (1974).
66. S. Iwatsuki and K. Inoue, *Macromolecules*, **10**, 58 (1977).
67. S. Iwatsuki and T. Kokubo, unpublished results.
68. E. C. Kooyman and E. Farenhorst, *Trans. Faraday Soc.*, **49**, 58 (1953).
69. M. Ballester, J. Castaner and J. Riera, *J. Am. Chem. Soc.*, **88**, 957 (1966).
70. H. K. Hall Jr, R. J. Cramer and J. E. Mulvaney, *Polym. Bull. (Berl.)*, 165 (1982).
71. S. Iwatsuki, T. Itoh, K. Nishihara and H. Furuhashi, *Chem. Lett.*, 517 (1982).
72. D. S. Acker and W. R. Hertler, *J. Am. Chem. Soc.*, **84**, 3370 (1962).
73. S. Iwatsuki, T. Itoh and K. Horiuchi, *Macromolecules*, **11**, 497 (1978).
74. S. Iwatsuki and T. Itoh, *Macromolecules*, **13**, 983 (1980).
75. S. Iwatsuki, T. Itoh and I. Yokotani, *Macromolecules*, **16**, 1817 (1983).
76. S. Iwatsuki, T. Itoh, T. Iwai and H. Sawada, *Macromolecules*, **18**, 2726 (1985).
77. R. Foster, *Organic Charge-transfer Complexes*, Academic Press, London, 1969, Chapter 3.
78. E. C. M. Chen and W. E. Wentworth, *J. Chem. Phys.*, **63**, 3183 (1975).
79. L. J. Young, in *Polymer Handbook* (Eds J. Brandrup and E. H. Immergut), Wiley-Interscience, New York, 1975, Vol. II, p. 387.
80. W. Gordy and S. C. Stanford, *J. Chem. Phys.*, **7**, 93 (1939); **8**, 170 (1940); **10**, 204, 215 (1941).
81. T. Kagiya, Y. Shimada and T. Inoue, *Bull. Chem. Soc. Jpn*, **41**, 767 (1968).
82. W. A. Lee and R. A. Rutherford, in *Polymer Handbook* (Eds J. Brandrup and E. H. Immergut), Wiley-Interscience, New York, 1975, Vol. III, p. 139.
83. J. C. Bevington, N. A. Ghanem and H. W. Melville, *Trans. Faraday Soc.*, **51**, 946 (1955).
84. F. Tüdös, *J. Polym. Sci.*, **30**, 343 (1958).
85. J. C. Bevington, N. A. Ghanem and H. W. Melville, *J. Chem. Soc.*, 2822 (1955).
86. C. H. Bamford, W. C. Barb, A. D. Jenkins and P. F. Onyon, *The Kinetics of Vinyl Polymerization by Radical Mechanism*, Butterworth, London, 1958, pp. 188–247.

87. J. W. Breitenbach, *Can. J. Res.*, **28B**, 507 (1950).
88. C. F. Hauser and N. L. Zutty, *J. Polym. Sci., A-1*, **8**, 1385 (1970).
89. C. F. Hauser and N. L. Zutty, *Macromolecules*, **4**, 478 (1971).
90. S. Iwatsuki and T. Itoh, *J. Polym. Sci., Polym. Chem. Ed.*, **18**, 2971 (1980).
91. S. Iwatsuki and T. Itoh, *Makromol. Chem.*, **182**, 2161 (1981).
92. D. S. Acker and W. R. Hertler, *J. Am. Chem. Soc.*, **84**, 3370 (1962).
93. W. R. Hertler and R. E. Benson, *J. Am. Chem. Soc.*, **84**, 3475 (1962).
94. J. Diekmann, W. R. Hertler and R. E. Benson, *J. Org. Chem.*, **28**, 2719 (1963).
95. R. C. Wheland and E. L. Martin, *J. Org. Chem.*, **40**, 3101 (1975).
96. For instance see J. B. Torrance, *Acc. Chem. Res.*, **12**, 79 (1979).
97. S. Aoki and J. K. Stille, *Macromolecules*, **3**, 473 (1970).
98. R. F. Tarvin, S. Aoki and J. K. Stille, *Macromolecules*, **5**, 663 (1972).
99. F. A. Bovey, G. V. D. Tiers and G. Filipovich, *J. Polym. Sci.*, **38**, 73 (1959).
100. R. C. Wheland and J. L. Gillson, *J. Am. Chem. Soc.*, **98**, 3916 (1976).
101. S. Iwatsuki, T. Itoh, H. Saito and J. Okada, *Macromolecules*, **16**, 1571 (1983).
102. S. Iwatsuki and T. Itoh, *Macromolecules*, **15**, 347 (1982).
103. T. Kokubo, S. Iwatsuki and Y. Yamashita, *Makromol. Chem.*, **123**, 256 (1969).
104. S. Iwatsuki and T. Itoh, *Macromolecules*, **16**, 332 (1983).
105. S. Iwatsuki and T. Itoh, *Macromolecules*, **12**, 208 (1979).
106. S. Iwatsuki, T. Itoh and Y. Shimizu, *Macromolecules*, **16**, 532 (1983).
107. M. E. Peover, *J. Chem. Soc.*, 4540 (1962).
108. L. R. Melby, R. J. Harder, W. R. Hertler, W. Mahler, R. E. Benson and W. E. Mochel, *J. Am. Chem. Soc.*, **84**, 3374 (1962).
109. S. Iwatsuki, T. Itoh and S. Sadaike, *Macromolecules*, **14**, 1608 (1981).
110. J. A. Riddick and W. Bunger, *Organic Solvents*, 3rd edn, Wiley, New York, 1979, p. 536.
111. C. A. L. Filgueiras and J. E. Huheey, *J. Org. Chem.*, **41**, 79 (1970).
112. E. M. Arnett, L. Joris, E. Mitchell, T. S. S. R. Murty, T. M. Gorrie and P. v. R. Schleyer, *J. Am. Chem. Soc.*, **92**, 2365 (1970).
113. Y. Shirota and H. Mikawa, *J. Macromol. Sci. Rev. Macromol. Chem.*, **C16(2)**, 129 (1977–1978).
114. S. Iwatsuki and Y. Yamashita, *Makromol. Chem.*, **104**, 263 (1967).
115. A. J. Gordon and R. D. Rord, *The Chemist's Companion*, Wiley, New York, 1972, p. 146.
116. R. W. Taft Jr, in *Steric Effects in Organic Chemistry* (Ed. M. S. Newman), Wiley, New York, 1968, Chapter 13.
117. H. K. Hall Jr and J. H. Bentley, *Polym. Bull.*, **3**, 203 (1980).
118. C. C. Price and S. Oae, *Sulfur Bonding*, Ronald Press Co., New York, 1962, p. 5.
119. R. Willstätter and E. Mayer, *Ber. Dtsch. Chem. Ges.*, **37**, 1494 (1904).
120. A. J. Gordon and R. D. Rord, *The Chemist's Companion*, Wiley, New York, 1972, p. 82.
121. R. Adams and A. S. Nagarkatti, *J. Am. Chem. Soc.*, **72**, 4601 (1950).
122. S. Iwatsuki and T. Itoh, *Macromolecules*, **17**, 1425 (1984).
123. J. A. Hyatt, *J. Org. Chem.*, **48**, 129 (1983).
124. S. Iwatsuki and T. Itoh, *Macromolecules*, **20**, (1987).

The Chemistry of Quinonoid Compounds, Vol. II
Edited by S. Patai and Z. Rappoport

CHAPTER 19

Isotopically labelled quinones

MIECZYSŁAW ZIELIŃSKI
Isotope Laboratory, Faculty of Chemistry, Jagiellonian University, Cracow, Poland
MARIANNA KAŃSKA
Department of Chemistry, University of Warsaw, Warsaw, Poland

I. SYNTHESES OF LABELLED QUINONES AND RELATED COMPOUNDS

Numerous isotopically labelled quinonoid compounds have been obtained in the course of isotopic biochemical studies reviewed in Section III. In the present section the recently elaborated synthetic procedures used for the preparation of labelled quinones or related compounds of practical importance are briefly described.

A. Syntheses of Labelled Benzoquinones and Related Compounds

1. *Syntheses of ^{125}I- and ^{123}I-labelled quinones*

a. Synthesis of 2,3-^{125}*diiodo*-5-t-*butyl*-1,4-*benzoquinone*

The ^{125}I-labelled benzoquinone **1**, a powerful inhibitor of photosynthetic electron transfer, was prepared with 49% overall radiochemical yield with specific radioactivity of

$$\text{2,3-dibromo-5-}t\text{-Bu-1,4-benzoquinone} \xrightarrow[100\,^{\circ}\mathrm{C},\ 5\ \mathrm{h}]{\mathrm{Na^{125}I/Et_2CO}} \text{2,3-}^{125}\mathrm{I}_2\text{-5-}t\text{-Bu-1,4-benzoquinone} \ + \tag{1}$$

$$\text{2-bromo-3-}^{125}\mathrm{I}\text{-5-}t\text{-Bu-1,4-benzoquinone} \xrightarrow{\mathrm{Na^{125}I}} \text{2,3-}^{125}\mathrm{I}_2\text{-5-}t\text{-Bu-1,4-benzoquinone}\ \mathbf{(1)}$$

49 mCi $mmol^{-1}$ based on $Na^{125}I$ using the exchange reaction (1) between 2,3-dibromo-5-*t*-butyl-1,4-benzoquinone and $Na^{125}I$[1].

b. Synthesis of ^{123}*I-labelled 4-iodo-2,5-dimethoxyphenylisopropylamine* (**2**)

Labelled with short half-life iodine-123 ($t_{1/2} = 13$ h) the title compound which has clinical potential for the imaging of normal brain tissue and in the study of mental disorder, has been prepared according to a rapid synthetic scheme (equation 2)[2]. It comprises the protection of the amine against oxidation with a phthalimido group to give **3** followed by a direct iodination of the ring of **3** with iodine monochloride. The amide (**4**) was quickly hydrolysed with hydrazine in butanol without isolation. Compound **2** was obtained as its hydrochloride with 10% radioiodine incorporation efficiency.

MeO, OMe, CH_2–CH(Me)–N(phthalimido) (**3**) $\xrightarrow{^{123}ICl}$ MeO, OMe, ^{123}I, CH_2–CH(Me)–N(phthalimido) (**4**)

$\xrightarrow[BuOH]{H_2NNH_2}$ MeO, OMe, ^{123}I, CH_2–CH(Me)–NH_2 (**2**) (2)

2. Syntheses of ^{11}C-, ^{14}C- and ^{13}C-labelled quinones and related compounds

a. Synthesis of ^{11}*C-labelled coenzyme* Q_{10}.

Coenzyme ^{3}H-Q_{10} and ^{14}C-Q_{10} function as one component of the electron transfer sequence in mitochondrial membrane, act as an antioxidant toward the superoxidative reaction *in vivo* and are used as a therapeutic agent for myocardial ischaemia. The coenzyme has been also labelled with short-lived carbon-11 ($t_{1/2} = 20.34$ min), by *O*-methylation of 3-demethyl Co-Q_{10} (**5**) with ^{11}MeI, synthesized in furan from $^{11}CO_2$ by the automated cyclotron synthesis within approximately 25 min (equation 3). The specific

$$^{14}N(p,\alpha)^{11}CO_2 \xrightarrow[-20\,^{\circ}C]{LiAlH_4/THF} \xrightarrow[70\,^{\circ}C]{THF} \xrightarrow[-20\,^{\circ}C]{H_2O} {}^{11}MeOH \xrightarrow[100\,^{\circ}C]{HI} {}^{11}MeI \quad (3)$$

radioactivity of the ^{11}C-coenzyme-Q_{10} (**6**), obtained within 35–50 min according to equation 4, was 4–5 mCi μmol^{-1}, radiochemical purity of 95%. Good radiochemical yields (6–16% based on trapping ^{11}MeI) were achieved by using Ag_2O as a base. In the presence of NaOH and other bases a chromene derivative was mainly obtained[3].

b. 4-Nitrocatechol-[UL-^{14}*C]* (**7**).

Uniformly ring-labelled compound (**7**) with specific radioactivity of 1.25 mCi mg^{-1} or 1.76 mCi $mmol^{-1}$, with purity higher than 99%, was synthesized in 49% yield by

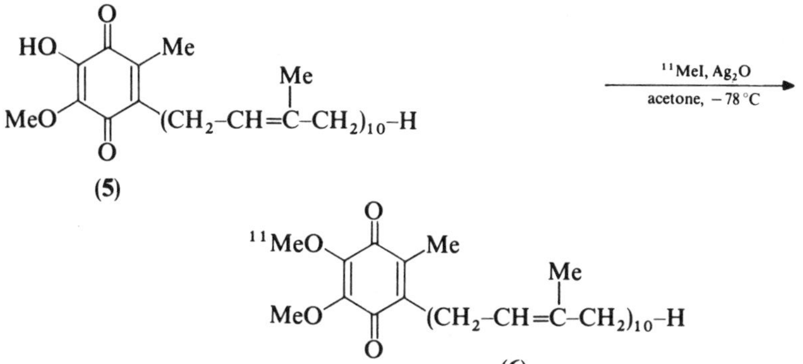

$$\text{(5)}\ \xrightarrow[\text{acetone, }-78\,^\circ C]{^{11}MeI,\ Ag_2O}\ \text{(6)} \qquad (4)$$

(5): 2-HO-3-MeO-5-Me-6-$(CH_2\text{-}CH{=}C(Me)\text{-}CH_2)_{10}\text{-}H$-1,4-benzoquinone

(6): 2-^{11}MeO-3-MeO-5-Me-6-$(CH_2\text{-}CH{=}C(Me)\text{-}CH_2)_{10}\text{-}H$-1,4-benzoquinone

irradiation of 4-nitrophenol-[UL-^{14}C] (0.57 mmol, 1.0 mCi) in aqueous solution of hydrogen peroxide for 1 hour, at 45–50 °C[4].

$$4\text{-}NO_2C_6H_3(OH)_2$$

(7)

c. Preparation of specifically ^{14}C- and ^{3}H-labelled shikimic acids

Tritium([2-T]) and carbon-14 ([1-^{14}C], [6-^{14}C], [7-^{14}C], [2,3,4,5-^{14}C])-labelled D-shikimates (**8**), which are used as precursors in biosynthetic investigations of numerous

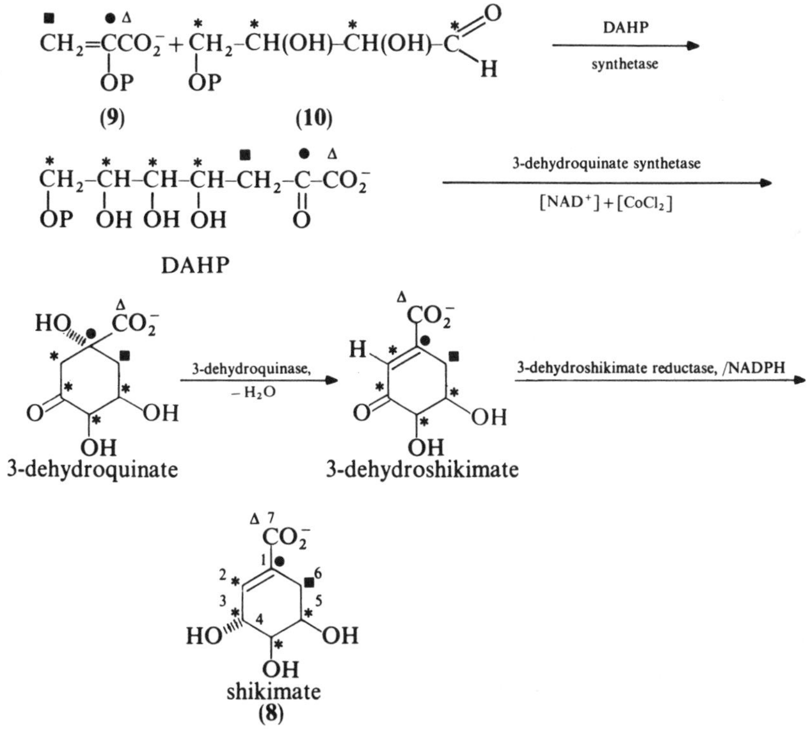

$$CH_2{=}C(OP)CO_2^- \ \textbf{(9)} + CH_2(OP)\text{-}CH(OH)\text{-}CH(OH)\text{-}CHO\ \textbf{(10)} \xrightarrow[\text{synthetase}]{\text{DAHP}}$$

$$CH_2(OP)\text{-}CH(OH)\text{-}CH(OH)\text{-}CH(OH)\text{-}CH_2\text{-}C(=O)\text{-}CO_2^-\ \text{(DAHP)} \xrightarrow[[NAD^+]+[CoCl_2]]{\text{3-dehydroquinate synthetase}}$$

$$\text{3-dehydroquinate} \xrightarrow[-H_2O]{\text{3-dehydroquinase,}} \text{3-dehydroshikimate} \xrightarrow{\text{3-dehydroshikimate reductase, /NADPH}} \text{shikimate}\ \textbf{(8)} \qquad (5)$$

DAHP

3-dehydroquinate

3-dehydroshikimate

shikimate
(**8**)

quinonoid compounds are usually synthesized according to Scharf and Zenk[5] as exemplified in equation 5. The precursor [^{14}C] phosphoenolpyruvate (**9**) is synthesized from pyruvate-1,2 or 3-^{14}C (ca. 25 mCi mmol^{-1}) and ATP using pyruvate synthase (equation 6). The yield of purified **9** was 78–85% of the labelled pyruvate. The precursor

$$\overset{*}{C}H_3\text{–}\overset{*}{C}O\text{–}COO^- + ATP + H_2O \xrightleftharpoons{30\,°C,\ 15\ min} \overset{*}{C}H_2{=}\overset{*}{C}(OPO_3H_2)\text{–}COO^- + AMP + \text{phosphate} \quad (6)$$

(**9**)

erythrose-4-phosphate [1,2,3,4-^{14}C] and 4-T (**10**) was prepared from glucose-[U-^{14}C] or glucose-[6-T] by enzymatic phosphorylation to labelled glucose-6-phosphate and subsequent treatment with lead tetraacetate. Condensation of **9**-[^{14}C] with unlabelled **10** or of labelled **10** with unlabelled **9** to yield shikimic acids labelled with ^{14}C or T has been achieved in the presence of cell-free extract of the *E. coli* mutant 83-24 which lacks shikimate kinase. An 86% maximal conversion of **9**-[1-^{14}C] to shikimate carboxyl-[^{14}C] was reached by incubating the reactants for 2.5 hours at 37 °C.

d. Synthesis of ^{13}C-labelled vitamin E

To elucidate the interactions[6] between vitamin E and lipids in biomembrane by measurement of ^{13}C-relaxation time on vitamin E in biomembrane, [4′a-^{13}C] all-rac-α-tocopherol (**11**) was synthesized by coupling a solution of 2-([4-methyl-^{13}C]-5-bromo-4-methylpent-1-yl)-6-methoxymethoxy-2,5,7,8-tetramethylchroman (**12**) in a mixture of THF and HMPA with 3,7-dimethyl-1-(thiazolin-2-yl)thio-2,6-octadiene (**13**) in THF in the presence of *n*-butyllithium in hexane (equation 7). The total yield of the labelled α-tocopherol based on [^{13}C]methyl iodide was 59%[6].

e. Synthesis of [7α-^{14}C]-methoxycephalosporin, antibacterial agent, CS-1170

The quinonoidal compound **16** has been utilized in the synthesis of the ^{14}C-labelled 7-methoxy Schiff base (**17**) in 49% yield based on ^{14}MeOH, and of the key intermediate (**18**)

(**12**) R = CH$_2$OMe + (**13**) $\xrightarrow{n\text{-BuLi}}$ (**14**)

$$\xrightarrow[\text{THF}]{\text{AcOH/Zn}} (\mathbf{15}) \xrightarrow[(2)\ 10\%\ \text{HCl–MeOH, 30 min, RT}]{(1)\ 50\ \text{atm}\ H_2/PtO_2,\ \text{EtOAc, RT}} (\mathbf{11}) \quad (7)$$

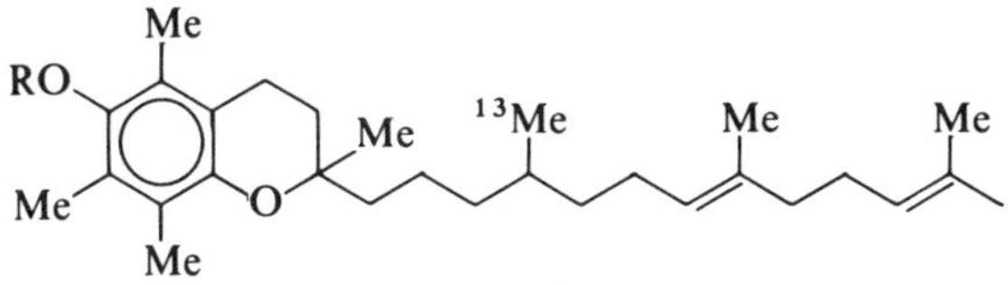

(**15**)

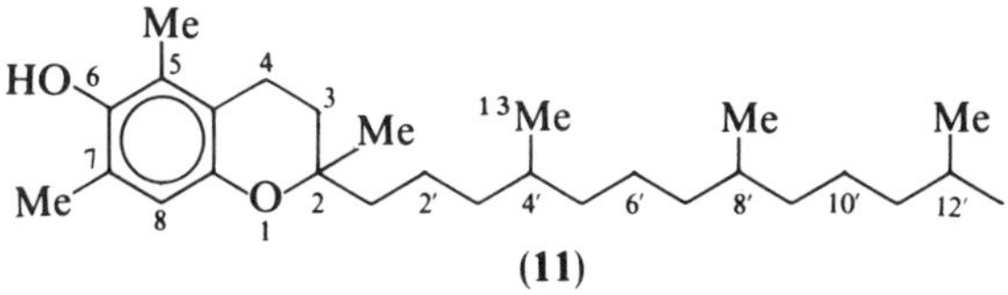

(**11**)

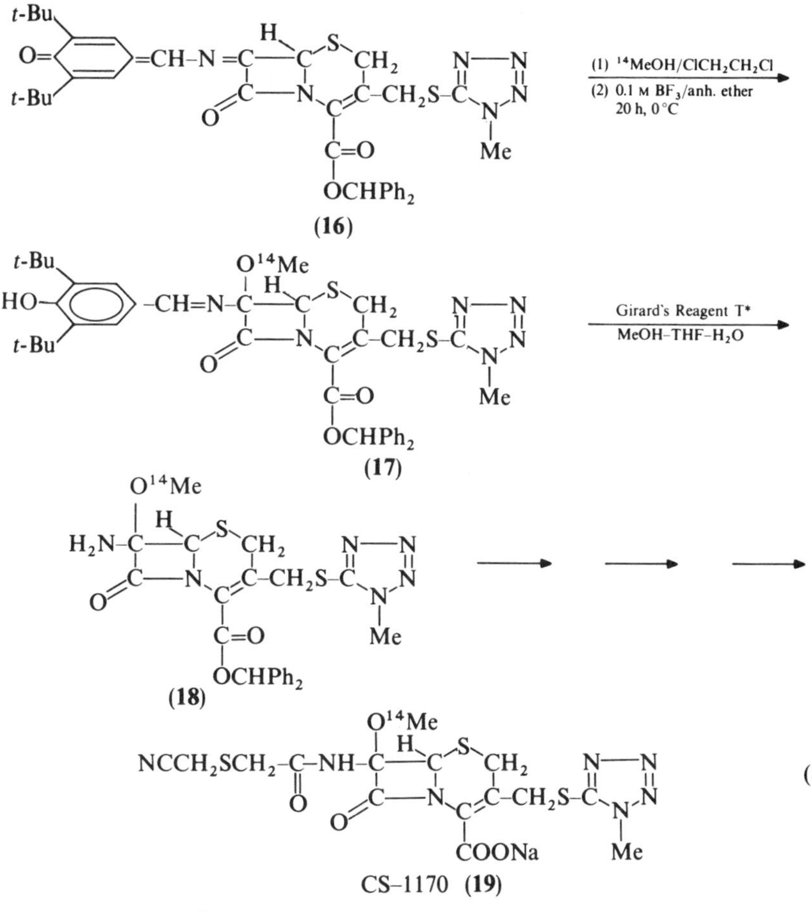

(1) $^{14}MeOH/ClCH_2CH_2Cl$

(2) 0.1 M BF_3/anh. ether 20 h, 0 °C

(**16**)

Girard's Reagent T*

$MeOH{-}THF{-}H_2O$

(**17**)

(**18**)

CS-1170 (**19**) (8)

$[Me_3\overset{+}{N}CH_2CONHNH_2]Cl^-$

*Girard's Reagent T (2-hydrazino-*N*,*N*,*N*-trimethyl-2-oxoethanaminium chloride) in the synthesis of the new semisynthetic cephamycin derivative CS-1170 (**19**), containing labelled 7α-methoxy-^{14}C group[7] (equation 8). The specific activity of the diluted CS-1170 drug was 6.25 mCi mmol^{-1} and the radiochemical purity was 96.7%.

f. Synthesis of 1-*methoxy* [^{14}*C*] *colchicine*

Colchicine (**20**), a poisonous alkaloid, was used for many centuries in treating rheumatism and gout. It was labelled with a non-metabolizable 1-methoxy ^{14}C group by selective demethylation of **20** and condensing 1-demethylcolchicine (**22**) with ^{14}MeI. The specific radioactivity of the purified product (**20***), measured by mass spectrometry, was 57 mCi mmol^{-1} with a radiochemical purity of 99%[8]. Metabolic studies[9,10] have shown that in the course of biotransformations '*in vivo*' colchicine is transformed to products demethylated at the 2 and the 3 positions and partly at the 10 position. Consequently in order to study the metabolism of **20** in rats the reaction scheme of equation 9 has been chosen for the synthesis of labelled alkaloid.

MeO, MeO, MeO (positions 4, 1), 7 —NHCOMe, 10, O, MeO **(20)** —$SnCl_4$ / MeCOCl→ MeO, MeO, O–C(=O)–Me **(21)** —K_2CO_3→

MeO, MeO, OH **(22)** —^{14}MeI / DMF→ MeO, MeO, ^{14}MeO **(20*)** (9)

g. Synthesis of quercetin-[4-^{14}*C*]

Radiolabelled quercetin-4-^{14}C (**23**) has been synthesized[11] in order to study its physiological mode of action upon insect larvae (growth inhibition) and its mutagenicity and to clarify the question if the ingestion of flavonoids is a source of human cancer. Since all the hydrogens of the quercetin nucleus are potentially exchangeable[12] the ^{14}C was incorporated into the flavone skeleton. The synthesis starts with benzyloxyacetonitrile-^{14}C (**24**), which is prepared from benzyl chloromethyl ether and $H^{14}CN$ (equation 10)

$$K^{14}CN + PhCH_2OCH_2Cl \xrightarrow[\text{benzene}]{\text{18-crown}} PhCH_2OCH_2{}^{14}CN \quad (10)$$

(24)

utilizing phase transfer with 18-crown-6 ether. Condensation of **24** with phluoroglucinol gave the ketone (**25**). Esterification of **25** with veratroyl chloride, subsequent rearrangement with triethylamine and base hydrolysis of the intermediate flavone ester yielded **26**. Simultaneous debenzylation and demethylation affected at high temperature using pyridine hydrochloride provided quercetin-4-^{14}C in 75% yield and specific activity of 2.07 mCi mmol^{-1} (equation 11).

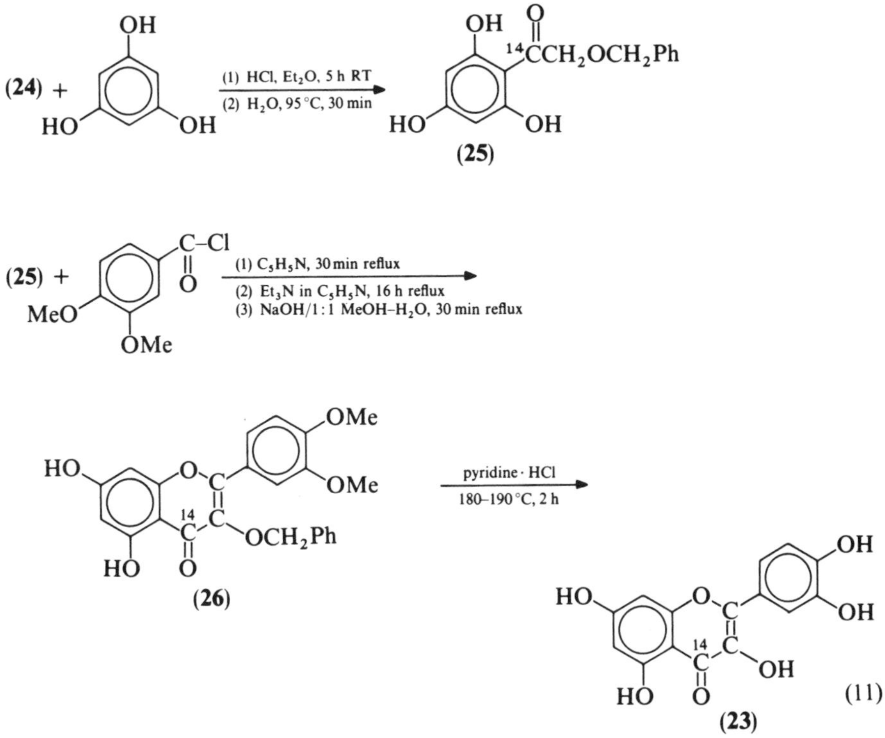

(24) + phloroglucinol $\xrightarrow[\text{(2) } H_2O,\ 95\,^\circ C,\ 30\ \text{min}]{\text{(1) HCl, } Et_2O,\ 5\ \text{h RT}}$ (25)

(25) + 3,4-dimethoxybenzoyl chloride $\xrightarrow{}$ (26)

(1) C_5H_5N, 30 min reflux
(2) Et_3N in C_5H_5N, 16 h reflux
(3) NaOH/1:1 $MeOH-H_2O$, 30 min reflux

(26) $\xrightarrow[180\text{–}190\,^\circ C,\ 2\ \text{h}]{\text{pyridine}\cdot \text{HCl}}$ (23) (11)

h. Synthesis of geraniol-7-^{14}C

Labelled geraniol (**27**) has been synthesized in five steps from acetone-2-^{14}C in 46% yield, 99% purity and 0.26 mCi $mmol^{-1}$ specific radioactivity. It was used to test the proposed mechanism of biotransformation of β-thujaplicin (**28**) in the *Thuja plicata* tree involving ring enlargement of the cyclohexano-terpene (**29**) which is derived from geraniol (**27**)[13] (equation 12).

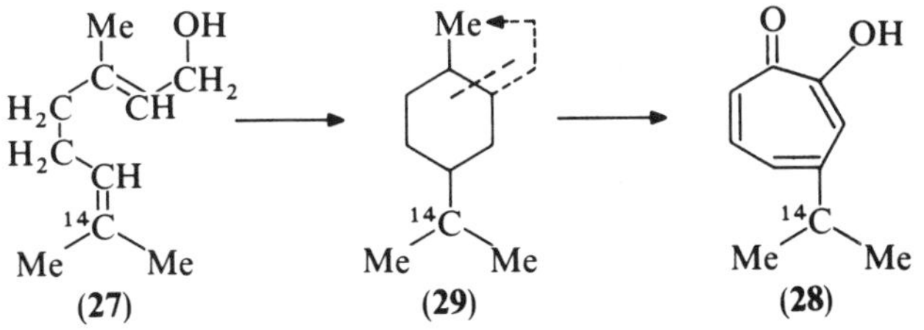

(27) ⟶ (29) ⟶ (28) (12)

3. Synthesis of tritium- and deuterium-labelled quinones and their derivatives

a. Synthesis of [^3H]conduritol C cis-*epoxide*

A considerable kinetic tritium isotope effect was found[14] in the course of the reduction of 5,6-dibromo-2-cyclohexene-1,4-dione (**30**) with tritiated [^3H] $NaBH_4$. The product,

5,6-dibromo-2-cyclohexene-1,4-diol (**31**), had a 2.74×10^{10} cpm $mmol^{-1}$ specific activity and served as the precursor to [^{3}H]conduritol C *cis*-epoxide (**32**), the enzyme inhibitor for β-galactosidase from *E. coli* (equation 13). A satisfactory yield of the dibromodiol (**31**) is

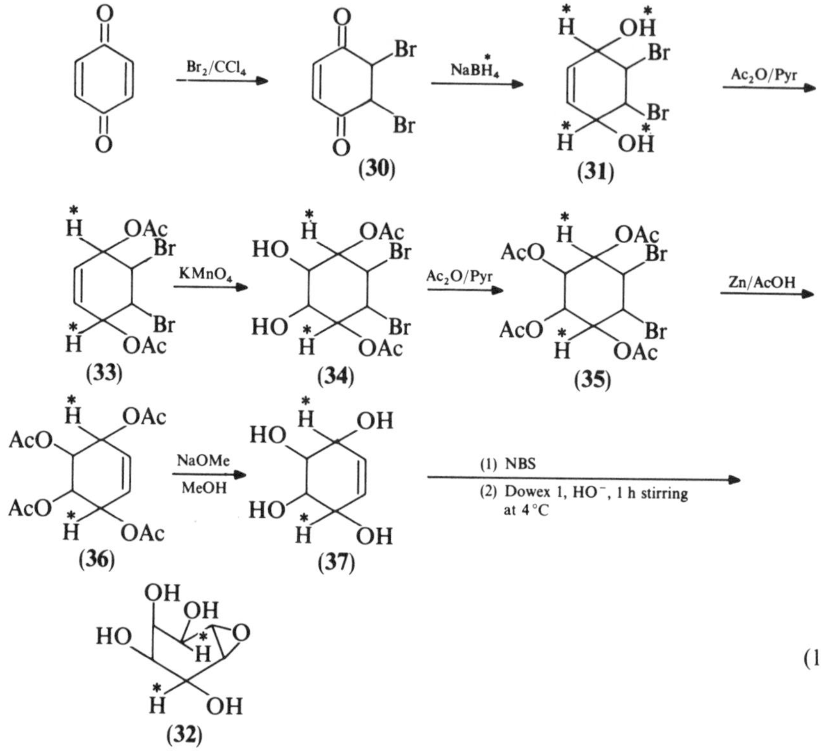

(13)

obtained when equimolar or higher amounts of the tritiated reductant is used. The reaction has to be carried out in the two-phase ether/water system in order to avoid aromatization to bromohydroquinone. Reacting equimolar amounts of **30** and [^{3}H]$NaBH_4$ and using excess of *p*-acetamidobenzaldehyde to trap the unreacted hydride gave a diol with specific radioactivity of 0.45 times that of the original borohydride whereas the specific activity of the unreacted hydride was 2.5 times higher. Thus, the specific activity

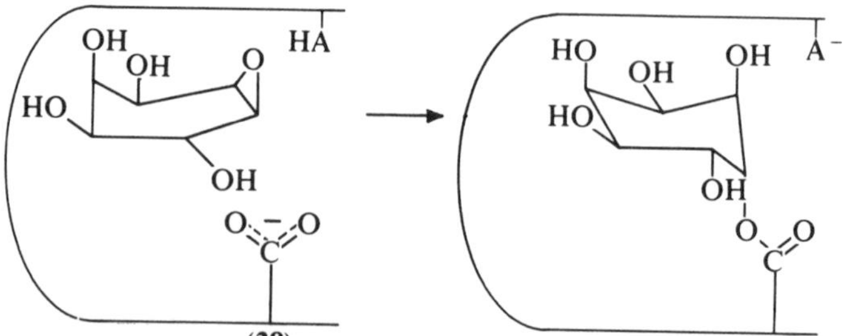

(**38**)

of the reduction product has to be determined experimentally during the synthesis by a liquid scintillation counting technique. By using tritium-labelled **32** it has been shown that the inhibition reaction of β-galactosidase with **32** is due to formation of an ester bond between **32** and a carboxylate group at the active site of **38**[14].

b. Synthesis of tritium-labelled catechol

1,2-Dihydroxybenzene (catechol) is the most abundant phenol in cigarette smoke condensate[15]. It shows biological activity, including carcinogenic activity with benzo[*a*]pyrene. Labelled [U-^{3}H] catechol was synthesized in a one-step catalytic reduction of 1 mmol of tetrabromocatechol in a low pressure hydrogenation apparatus with 1 Ci of tritium gas, followed after 30 minutes by addition of hydrogen gas[15] (equation 14). Purification gave 68% of catechol [U-^{3}H] (**39**), with 2.5 mCi mmol^{-1} specific activity.

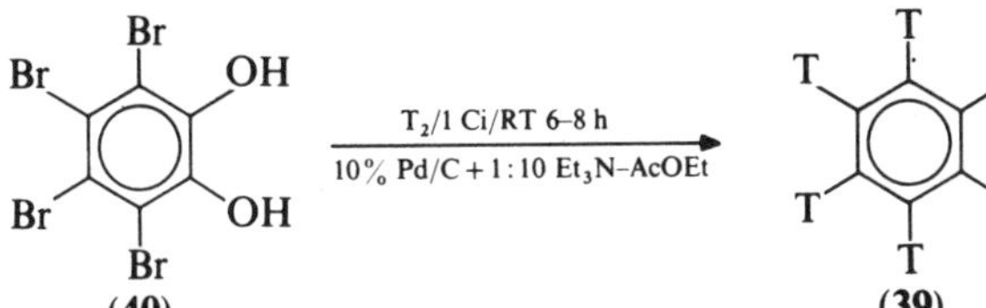

(**40**) $\xrightarrow[10\%\ Pd/C + 1:10\ Et_3N\text{-}AcOEt]{T_2/1\ Ci/RT\ 6\text{-}8\ h}$ (**39**) (14)

c. 2,5,6-*Trideuteriohomovanillic acid,* 2,5,6-*trideuteriovanillactic acid and* 2,5,6-*trideuterio*-3,4-*dihydroxyphenylacetic acid synthesis*

HVA-d_3 (**41**), VLA-d_3 (**42**) and DOPAC-d_3 (**43**) were synthesized from the corresponding unlabelled analogues by dissolving each of them in 9% DCl/D_2O and heating for 6 h at 80 °C in a sealed tube. A quantitative replacement of the exchangeable 2,5,6-hydrogens was achieved under these conditions as judged by mass fragmentography and gas chromatography[16]. These compounds were used for the preparation of internal standards for quantitative mass-fragmentographic analysis of biological materials.

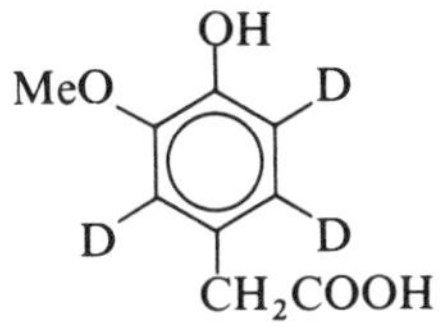

(**41**) 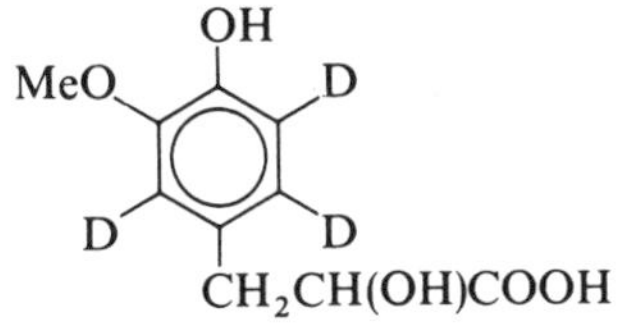 (**42**) 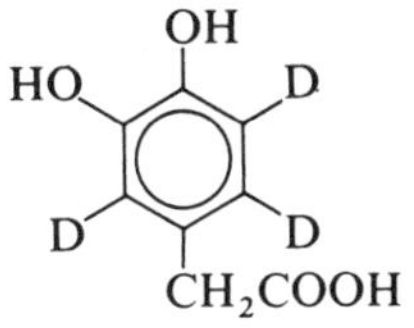 (**43**)

d. Synthesis of methyl-labelled catechol

Monomethylated catechols, i.e. 2-methoxy-d_3-phenol (**44**), 4-hydroxy-3-methoxy-d_3-mandelic acid (**45**), 4-hydroxy-3-methoxy-d_3-phenylacetic acid (**46**) and 4-hydroxy-3-methoxy-d_3-phenylethylene glycol (**47**), were prepared according to the general method of equation 15 (and 15a)[17].

e. Isotopic synthesis by nuclear deuterium exchange in methoxybenzenes

The 6,8-d_2-catechin/epicatechin 5,7,3′,4′-tetramethyl ether (**48**) was obtained when methylated procyanidin, isolated from *Vitis vinifera*, was cleaved with 0.05 N DCl/D_2O, 20% $MeCOOD/D_2O$ or D_2O/dioxan mixtures. The molecular ion of **48** in the mass spectrum showed the presence of two deuterium atoms and the ions derived from the A

OH, OH $\xrightarrow[\text{6 h reflux}]{CD_3I,\ K_2CO_3/Me_2CO}$ OCD_3, OH (**44**) $\xrightarrow[\text{70 °C, 3 h stirring}]{\text{80\% glyoxalic acid}/H_2O\text{, ascorbic acid, pH 9.5}}$

CH(OH)COOH, HO, OCD_3 (**45**)

(1) esterification: 3% HCl/EtOH 25 °C, 30 min
(2) acetylation: 1:1 Pyr-Ac_2O, 25 °C, 2 h
(3) reduction: $NaBH_4$/aq. NaOH

CH_2COOH, HO, OCD_3 (**46**, 54% yield) (15)

(**45**) (1) reduction: B_2H_6/BH_3–THF, 0 °C, 1 h (2) warming to RT, 30–40 h → $CH(OH)CH_2OH$, HO, OCD_3 (**47**, 21% yield from **45**) (15a)

OMe, OMe, D, MeO, O, D, OH, OMe (**48**)

ring of **48** have *m*/*e* values higher by two units[18] than those from the unlabelled compound. This result prompted an NMR investigation of the deuterium exchange between the ring hydrogens of methoxybenzenes or the C(6) and C(8) hydrogens of methylated flavonoids with 3:1 D_2O/dioxan mixtures at 95 °C in Pyrex ampoules. After 16 hours of heating the following % exchange were found: methoxybenzene, 0%; 1,2-dimethoxybenzene, 0%; 1,3-dimethoxybenzene, 36.2%; 1,2,3-trimethoxybenzene, 21.5%; 1,2,4-trimethoxybenzene, 16.1%; 1,3,5-trimethoxybenzene, 100%; catechin 5,7,3′,4′-tetramethyl ether, 100%; 5,7,3′,4′-tetramethoxyflavan, 100%; dihydroquercetin 5,7,3′,4′-tetramethyl ether, 0%; 5,7,3′,4′-tetramethoxy-2,3-*trans*-flavan-3,4-*cis*-diol, 0%. These data are considered to be consistent with an electrophilic aromatic substitution mechanism involving two consecutive slow proton transfers steps, each involving addition of a proton to give an intermediate phenonium ion followed by proton abstraction to yield the exchanged methoxybenzene. In the absence of an acidic catalyst, the exchange is assisted by the Pyrex glass surface. No exchange was found in soda glass NMR tubes.

f. Synthesis of deuterium-labelled rutin

Rutin, a flavonyl glycoside, used in the treatment of capillary bleeding, has been selectively[19] labelled with deuterium in the stable 2′,5′,6′-positions of the catechol ring to give **49** by a two-step hydrogen–deuterium exchange mechanism under mild alkaline

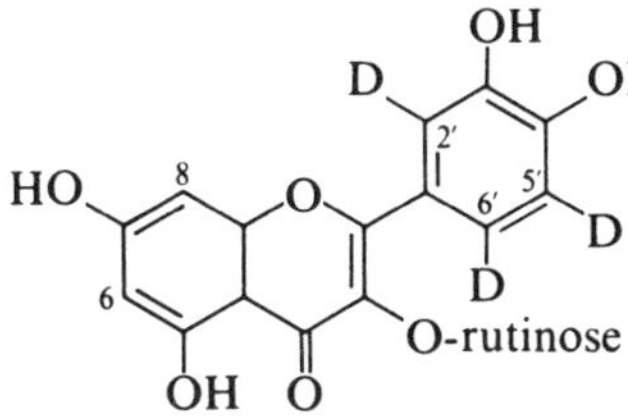

(49)

conditions[19]. In the first step rutin-2′,5′,6′,6,8-d_5 was obtained by heating a mixture of sodium hydroxide and rutin in D_2O at 95 °C during 8 hours under nitrogen. In the second step the labile deuterium atoms in the 6,8 positions of the resorcinol ring were replaced with hydrogen by stirring the solution of sodium hydroxide and rutin-2′,5′,6′,6,8-d_5 in water for 1 h at 25 °C, acidification with 10% acetic acid and repetition of the procedure. No loss of deuterium at any position was observed when the rutin-2′,5′,6′-d_3, which is useful for the metabolic study in man, was heated at 60 °C for 2 hours in methanol or water.

g. Synthesis of deuterium and oxygen-18 labelled norepinephrine

2-Dibenzylamino-1-(3,4-dihydroxyphenyl)ethanone-2,2-d_2 hydrochloride (**50**) was obtained[20] by D/H exchange of the protons α to the keto moiety of 2-dibenzylamino-1-(3,4-dihydroxyphenyl)ethanone-2,2-H_2 (at 80 °C for 4 days) with DCl/D_2O-dioxane. The post-exchange solution was lyophilized to dryness, the fresh solvent added and the exchange reaction repeated for an additional 7 days. Mass spectral analysis showed 87.5 atom % d_2, 11.8 atom % d_1 and 0.7 atom % d_0.

4-(2-Amino-1-hydroxyethyl-1,2,2-d_3)benzene-1,2-diol, NEd_3 (**51**), was synthesized by reduction the deuteriated precursor (**50**) with $D_2/Pd/C$ for 4 hours (equation 16). The

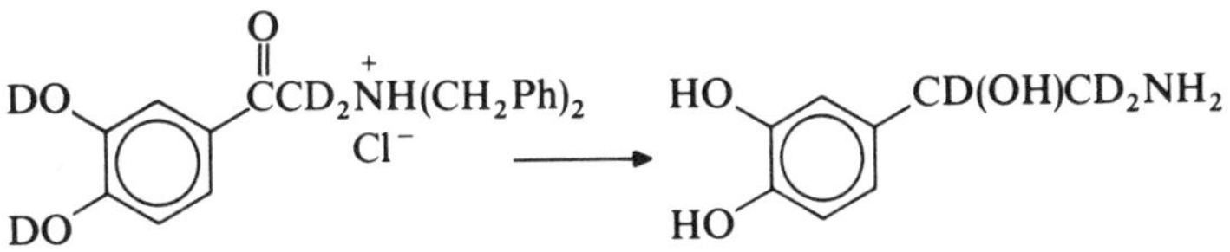

(50) → **(51)** (16)

labile deuterium atoms in the free base of norepinephrine (neurotransmitter) were washed out by back exchange with aqueous ammonia solution. The protecting benzyl groups are removed in the course of reduction. 2-Amino-1-(3,4-dihydroxyphenyl)ethanone-^{18}O hydrochloride (**52**), useful for *in vivo* biochemical studies due to an 'unmeasurable' isotope effect, was prepared according to equation 17.

$$3,4\text{-}(HO)_2C_6H_3\text{-}C(=O)\text{-}CH_2\overset{+}{N}H_3\ Cl^- \xrightleftharpoons[\text{RT, mild conditions, 24 h, isotopic equilibrium}]{H_2{}^{18}O\,(20\text{ atom \% of }{}^{18}O)\text{-MeOH-MeCN }(0.5:8:1,\ v/v/v),\text{ trace of anhyd. HCl}} 3,4\text{-}(HO)_2C_6H_3\text{-}C(={}^{18}O)\text{-}CH_2\overset{+}{N}H_3\ Cl^-\ \textbf{(52)} \quad (17)$$

h. Synthesis of deuterium-enriched erythro-*α-methylnorepinephrine and norepinephrine*

In the course of biomedical studies of (*S*)-α-methyldopa (**53**), the antihypertensive agent[21], *erythro*-α-methylnorepinephrine (**59**), enriched with six or seven deuterium atoms, has been needed as a mass spectrometric stable internal standard and it was synthesized according to equation 18. 1-(3,4-Dimethoxyphenyl)-1-propanone-3,3,3-d_3

$$3,4\text{-}(HO)_2C_6H_3\text{-}CH_2\text{-}C(NH_2)(Me)\text{-}COOH\quad \textbf{(53)}$$

$$3,4\text{-}(MeO)_2C_6H_3\text{-}C(=O)\text{-}Me \xrightarrow[(2)\ CD_3I/\text{anhyd. THF}]{(1)\ LiN(Pr\text{-}i)_2/\text{anhyd. THF}} 3,4\text{-}(MeO)_2C_6H_3\text{-}C(=O)\text{-}CH_2CD_3\ \textbf{(54)} + 3,4\text{-}(MeO)_2C_6H_3\text{-}C(=O)\text{-}CH(CD_3)_2\ \textbf{(55)}$$

$$\textbf{(54)} \longrightarrow 3,4\text{-}(MeO)_2C_6H_3\text{-}C(=O)\text{-}CH(Br)\text{-}CD_3\ \textbf{(56)} \longrightarrow 3,4\text{-}(MeO)_2C_6H_3\text{-}C(=O)\text{-}CH(N(CH_2Ph)_2)\text{-}CD_3\ \textbf{(57)} \xrightarrow[120\,^{\circ}C,\ 20h,\ N_2]{48\%\ DBr/D_2O}$$

$$3,4\text{-}(HO)_2C_6D_3\text{-}C(=O)\text{-}CD(N(CH_2Ph)_2)\text{-}CD_3\ \textbf{(58)} \xrightarrow[(2)\ H_2,\ 2\text{ atm},\ 10\%\ Pd/C,\ 17\text{ h}]{(1)\ 25\%\ NH_3/Et_2O} 3,4\text{-}(HO)_2C_6D_3\text{-}CH(OH)\text{-}CD(NH_2\cdot HCl)\text{-}CD_3\ \textbf{(59)} \quad (18)$$

(**54**) was obtained by trideuteriomethylation of the lithio derivative of 3,4-dimethoxy-acetophenone. Bromination of the propanone-d_3 with phenyltrimethylammonium tribromide in tetrahydrofuran yielded 2-bromo-1-(3,4-dimethoxyphenyl)-1-propanone-3,3,3-d_3 (**56**), which when treated with dibenzylamine gave 2-dibenzylamino-1-(3,4-dimethoxyphenyl)-1-propanone-3,3,3-d_3 (**57**). In the course of the cleavage of methyl ether groups with deuterium bromide three to four additional deuterium atoms were introduced in the resulting 2-dibenzylamino-1-(3,4-dihydroxyphenyl-2,5,6-d_2,d_3)-1-propanone-3,3,3-d_3 (**58**). Catalytic hydrogenation of **58** gave the desired *erythro*-2-amino-1-(3,4-dihydroxyphenyl-2,5,6-d_2,d_3)-1-propanol-2,3,3,3-d_4 hydrochloride (**59**) enriched with seven deuterium atoms. In a similar reaction sequence deuteriochloride (**60**) and 2-amino-d_2-1-(3,4-dihydroxy-d_2-phenyl-2,5,6-d_2,d_3)-1-ethanol-1,2,2,0-d_4 deuteriochloride were prepared as the *tris*-perfluoropropionyl derivatives (**61**) (equation 19)[21].

MeO, MeO–C₆H₃–COMe → MeO, MeO–C₆H₃–$COCH_2Br$ → MeO, MeO–C₆H₃–$COCH_2N(CH_2Ph)_2$

$\xrightarrow[D_2O]{DBr}$ (**60**) R'O, R'O–C₆D₃–$COCD_2N(CH_2Ph)_2$ $\xrightarrow[Pd/C]{D_2}$ (**61**) R'O, R'O–C₆D₃–$CD(OR'')CD_2NHR'$ (19)

$R' = COC_2F_5$
$R'' = C_2H_5$

i. Synthesis of deuteriated methylhydroquinone derivatives and DOM-d_6

In the course of synthesis of deuterium-labelled internal standards for quantitative determination of organic compounds, especially drugs, pesticides and food additives in complex mixtures at low levels, by selected ion monitoring[22] several deuterium-labelled derivatives of methylhydroquinone have been obtained[22]. These include: 2,5-di(methoxy-d_3) toluene (**62**), from reaction of methylhydroquinone with dimethyl-d_6 sulphate (DMS-d_6) under nitrogen; 2,5-di(methoxy-d_3)-4-methylbenzaldehyde (**63**), by treating an ice-

Methylhydroquinone (OH, OH, Me) $\xrightarrow[1\ h\ reflux]{DMS\text{-}d_6,\ NaOH}$ (**62**) (OCD_3, OCD_3, Me) $\xrightarrow{HCN,\ AlCl_3}$ (**63**) (OCD_3, OCD_3, Me, C(=O)–H)

$\xrightarrow{EtNO_2}$ (**64**) (OCD_3, OCD_3, Me, $CH{=}C(Me){-}NO_2$) $\xrightarrow{LiAlH_4}$ (**65**) (OCD_3, OCD_3, Me, $CH_2{-}CH(Me){-}NH_2$) 67% (20)

cooled solution of **62** in benzene with dry hydrogen cyanide, followed by addition of aluminium chloride and gaseous hydrogen chloride; 2-nitro-1-[2,5-di(methoxy-d_3)-4-methylphenyl]-1-propene (**64**), by reacting a solution of **63** in glacial acetic acid with nitroethane and ammonium acetate; 2,5-di(methoxy-d_3)-4-methylamphetamine (DOM-6) (**65**), by addition of ether solution of nitropropene (**64**) to solution of lithium aluminium hydride in anhydrous ether and subsequent careful addition of water and 30% Rochelle salt solution (equation 20). The free base (**65**) obtained was converted to the hydrochloride with hydrogen chloride–ether complex. The product had the following isotopic composition: d_6, 96.1%; d_5, 3.1%; d_4, 0.6%; d_1 and d_0 < 0.1%.

B. Synthesis of ^{14}C-, ^{35}S- and ^{3}H-labelled Complex Anthraquinones and Labelled Drugs

1. Synthesis of [^{14}C]anthralin

The chemically stable 1,8-dimethoxy-[10-^{14}C]anthraquinone (**66**) was found[23] to be a suitable precursor of the unstable 1,8-dihydroxy-[10-^{14}C]-9-anthrone, anthralin (**67**), extensively used in the topical treatment of psoriasis. Quinone **66** can be stored indefinitely at low temperatures as such and converted efficiently to **67** when required (equation 21).

OMe OMe OMe OMe

Br ^{14}COOH

(1) Mg, THF, 40 °C
(2) $^{14}CO_2$, −78 °C

(**68**) (**69**)

OMe O OMe

(1) 96% H_2SO_4, 1 h stir
(2) $K_2Cr_2O_7$ in H_2O, 0°C
AcOH, 75 min reflux

$AlCl_3$ in CH_2Cl_2, 16 h

(**66**)

OH O OH OH O OH

12N HCl, AcOH
Sn, 1 h reflux

(**70**) (**67**) (21)

Isotopic carbon-14 was introduced to the anthraquinone **66** by carboxylation of the Grignard reagent derived from 3-bromo-2-(2-methoxybenzyl)anisole (**68**), using inexpensive $^{14}CO_2$. Cyclization/oxidation of the resultant 3-methoxy-2-(2-methoxybenzyl) (carbonyl-^{14}C)benzoic acid (**69**) gave the labelled **66**, which is easily transformed in 80% yield to **67** by demethylation with aluminium chloride in dichloromethane to form 1,8-dihydroxy-[10-^{14}C]anthraquinone (**70**), followed by reduction with powdered tin and hydrochloric acid. The final product (**67**) was obtained with specific activity of 57 mCi $mmol^{-1}$. The bromo compound (**68**) was obtained from inactive 3-methoxy-2-(2-methoxybenzyl)benzoic acid by a Curtius/Sandmeyer reaction sequence[23].

2. *Synthesis of endocrocin and endocrocin-9-one*

Endocrocin-9-one (**73**) and endocrocin (**74b**) are important intermediates in the biosynthesis of emodin and related anthraquinones. They have been prepared[24] by treatment of the dicarboxylic acid (**72**), ^{14}C-labelled at both carboxyl groups, with polyphosphoric acid or anhydrous HF, oxidation of the resulting **73** with H_2O_2 in 1 N NaOH and *O*-demethylation of the formed endocrocin 6,8-dimethyl ether (**74a**) with BBr_3 in refluxing CH_2Cl_2 followed by chromatography. The ^{14}C-labelled Mühlemann's dicarboxylic acid (**72**) has been synthesized by condensation of the diketone (**71**) with dimethyl[1,5-$^{14}C_2$]acetonedicarboxylate (equation 22).

(**71**) → (**72**) —anhyd.HF→ (**73**) —H_2O_2, 1N NaOH→ (**74a**) —BBr_3, CH_2Cl_2 reflux→ (**74b**) (22)

3. *Synthesis of [14-^{14}C]adriamycin and daunorubicin*

a. [14-^{14}C]Adriamycin·HCl (**75**)

This important anticancer antibiotic has been prepared from unlabelled adriamycin using ^{14}C-diazald as the source of the label, followed by reaction scheme of equation 23[25] which does not require protection of the phenolic hydroxyl groups. *N*-TFA-adriamycin (**76**), in tetrahydrofuran reacted with aqueous periodic acid at room temperature to give the carboxylic acid (**77**) in 82% yield. Stirring a suspension of the isolated unpurified key intermediate (**77**) in chloroform with triethylamine and isobutyl chloroformate in the cold afforded the unstable mixed anhydride (**78**). Addition of **78** to etheral solution of ^{14}C-diazomethane afforded, after workup, the diazoketone (**79**) in 25% yield. The methyl ester (**80**) was the major 20% by-product in this step. The intermediate **79** was converted with excess of hydrogen bromide in 95% yield to bromoketone (**81**). Treatment of **81** in THF with aqueous potassium carbonate yielded *N*-trifluoroacetyl-adriamycin[14-^{14}C] (**76**^{x}). Protection of the 14-hydroxyl group of *N*-TFA-adriamycin in 56% yield was achieved with *p*-anisyldiphenylchloromethane in pyridine[26]. [14-^{14}C]Adriamycin (**75**) was obtained from **82** by removal of the *N*-TFA group with sodium hydroxide solution and successive removal of the *p*-anisyldiphenylmethyl group by 80% acetic acid. The formed **75** was converted to its hydrochloride with methanolic HCl in 45% yield. The resulting red [14-^{14}C]-(**75**) was identical in all respects with adriamycin · HCl.

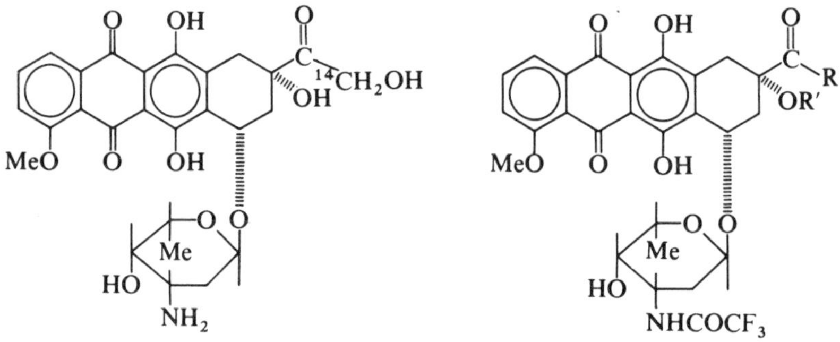

(**75**)

(**76**) R = CH_2OH, R′ = H
(**77**) R = OH, R′ = H
(**78**) R = CO_2Bu-*i*, R′ = CO_2Bu-*i*
(**79**) R = $^{14}CHN_2$, R′ = CO_2Bu-*i*
(**80**) R = OMe, R′ = CO_2Bu-*i*
(**81**) R = $^{14}CH_2Br$, R′ = CO_2Bu-*i*
(**82**) R = $^{14}CH_2OCPh_2C_6H_4OMe$-*p*, R′ = H

(**76**) → (**77**) → (**78**) → (**79**) → (**81**) → (**76**^{x}) → (**82**) → (**75**) (23)

b. Synthesis of daunorubicin-[14-^{14}C] *and adriamycin-*[14-^{14}C]

The clinically useful antineoplastic agents, anthracycline antibiotics adriamycin (**75**) and daunorubicin (**83**), were also labelled with ^{14}C at the C(14) position by using the reaction scheme of equation 24[27]. Treatment of adriamycinone (**85**) in THF with 15:1 molar excess of $^{14}CH_3MgI$ and periodate oxidative cleavage of the glycol (**86**) affords daunomycinone [14-^{14}C] (**87**). Condensation of **87** with the protected 1-chlorodaunosamine (**91**), prepared from *N*-trifluoroacetyl-1,4-di-*O*-*p*-nitrobenzoyldaunosamine with HCl, followed by deacylation of the resulting α-glycoside with aqueous sodium hydroxide and addition of HCl, afforded daunorubicin-[14-^{14}C] · HCl (**83**), with 6.9 mCi mmol^{-1} activity. Bromination of daunomycinone-[14-^{14}C] (**87**) and hydrolysis gave **88** and adriamycinone-[14-^{14}C] (**89**), respectively. The 14-OH group of **89** was protected by coupling with *p*-anisylchlorodiphenylmethane to give 14-*O* (*p*-anisyldiphenylmethyl)-adriamycinone [14-^{14}C] (**90**). The latter was condensed with **91** to produce the α-glycoside. Basic deacetylation gave crude 14-*O*(*p*-anisyldiphenylmethyl) adriamycin [14-^{14}C] (**84**). Deprotection of **84** with 80% acetic acid afforded adriamycin-[14-^{14}C] which with HCl gave adriamycin · HCl (**75**), in 40% yield (specific activity 6.5 mCi mmol^{-1}).

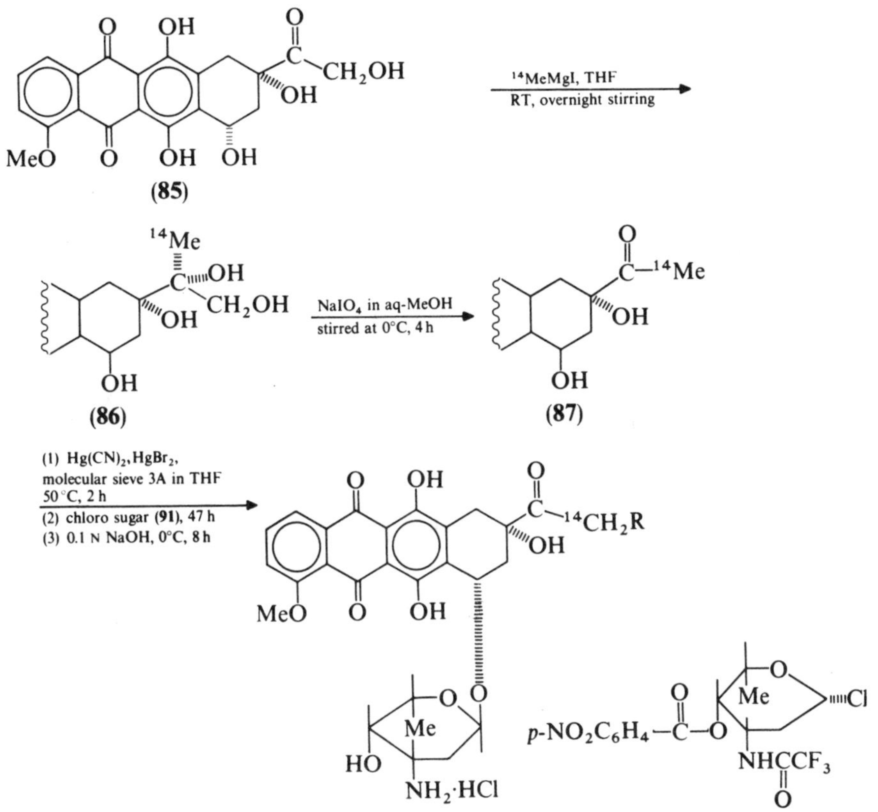

(85)

^{14}MeMgI, THF
RT, overnight stirring

(86)

$NaIO_4$ in aq-MeOH
stirred at 0°C, 4 h

(87)

(1) $Hg(CN)_2$, $HgBr_2$, molecular sieve 3A in THF 50°C, 2 h
(2) chloro sugar (**91**), 47 h
(3) 0.1 N NaOH, 0°C, 8 h

(**83**) R = H
(**75**) R = OH
(**84**) R = $OCPh_2C_6H_4OMe$-*p*

(**91**)

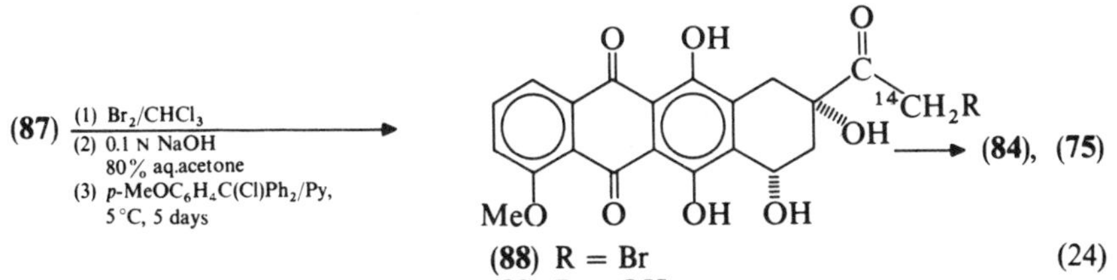

(**87**) (1) $Br_2/CHCl_3$
(2) 0.1 N NaOH 80% aq.acetone
(3) *p*-$MeOC_6H_4C(Cl)Ph_2$/Py, 5°C, 5 days

→ (**84**), (**75**)

(**88**) R = Br
(**89**) R = OH
(**90**) R = $OCPh_2C_6H_4OMe$-*p*

(24)

4. *Synthesis of ^{14}C- or ^{35}S-labelled 2,3-dicyano-1,4-dithia-9,10-anthraquinone*

The ^{14}C- and ^{35}S-labelled title compounds **92**, **93** and **94** which exhibit pesticidal and fungicidal properties and are used in the protection of agricultural and fruit production, have been prepared on a milligram scale in closed all-glass apparatus useful for the

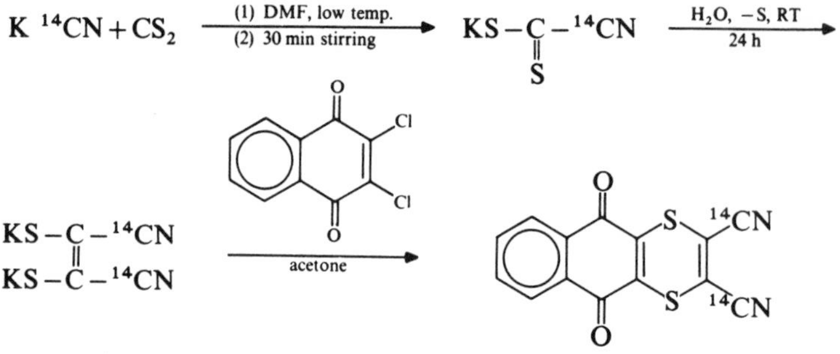

$$K^{14}CN + CS_2 \xrightarrow[\text{(2) 30 min stirring}]{\text{(1) DMF, low temp.}} KS-\underset{\underset{S}{\|}}{C}-{}^{14}CN \xrightarrow[\text{24 h}]{H_2O,\ -S,\ RT}$$

$$\begin{matrix} KS-C-{}^{14}CN \\ \| \\ KS-C-{}^{14}CN \end{matrix} \xrightarrow[\text{acetone}]{\text{2,3-dichloro-1,4-naphthoquinone}} \textbf{(92)} \quad (25)$$

(92)

production of labelled compounds[28] with high specific radioactivity. Dithianones having 88 Ci of $^{14}C\ mol^{-1}$ and 44 Ci of $^{35}S\ mol^{-1}$ were obtained on a microscale according to reaction scheme 25. Dithianon-^{35}S (**93**), was also prepared according to equation 25. The

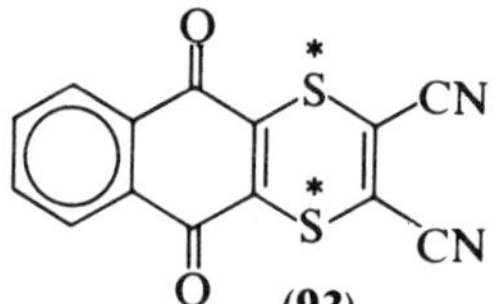

(93)

required ^{35}S-labelled dipotassium 1,2-dicyanoethenedithiolate was obtained by the exchange with elemental sulphur under reduced pressure (equation 26). Dithianon (**93**)

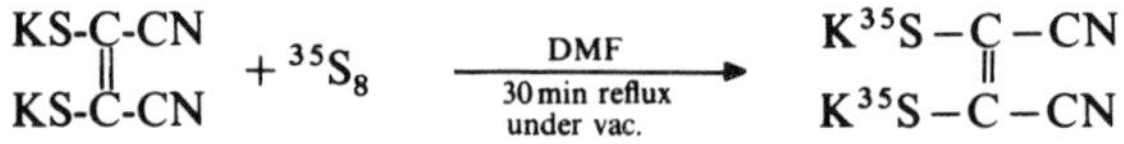

$$\begin{matrix} KS\text{-}C\text{-}CN \\ \| \\ KS\text{-}C\text{-}CN \end{matrix} + {}^{35}S_8 \xrightarrow[\text{30 min reflux under vac.}]{\text{DMF}} \begin{matrix} K^{35}S-C-CN \\ \| \\ K^{35}S-C-CN \end{matrix} \quad (26)$$

was also synthesized by a direct rapid exchange of the pure dithianon with elemental sulphur-^{35}S in dimethylformamide. A dithianon doubly labelled with both ^{14}C and ^{35}S (**94**) was prepared similarly by the direct exchange of dithianon-^{14}C with ^{35}S (equation 27), as well as according to equation 28. The use of the exchange scheme (27) for simultaneous labelling with ^{14}C and ^{35}S is recommended since the potassium 1,2-dicyanoethenedi-

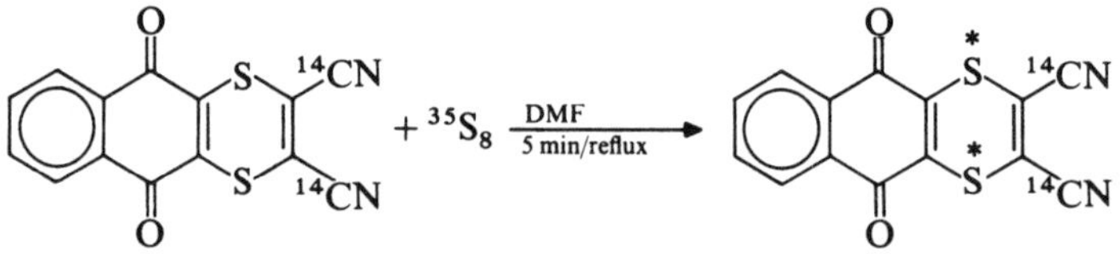

$$\textbf{(92)} + {}^{35}S_8 \xrightarrow[\text{5 min/reflux}]{\text{DMF}} \textbf{(94)} \quad (27)$$

(94)

$$\begin{matrix} KS-C-{}^{14}CN \\ \| \\ KS-C-{}^{14}CN \end{matrix} \xrightarrow[\text{30 min at 155°C}]{^{35}S_8 \text{ in DMF, scaled under vac.}} \begin{matrix} K\overset{*}{S}-C-{}^{14}CN \\ \| \\ K\overset{*}{S}-C-{}^{14}CN \end{matrix}$$

$$\xrightarrow[\text{acetone}]{\text{2,3-dichloro-1,4-naphthoquinone}} \textbf{(94)} \quad (28)$$

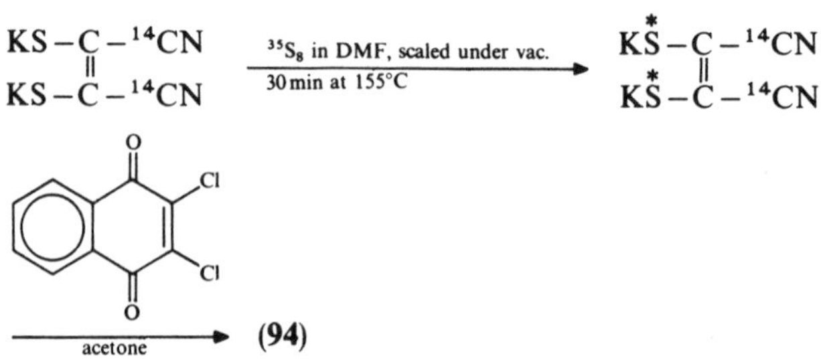

thiolate undergoes a rapid thermal decomposition, and the final radiochemical and chemical yields of the labelled substance obtained according to scheme 28 are rather low (radiochem. yield 16.3%, chem. yield 13.2%, radiochem. purity 97.2%). The lower specific activity of ^{35}S can be improved by the use of higher activities of elemental ^{35}S in the exchange scheme (27).

5. *Synthesis of ^{14}C- and ^{3}H-labelled islandicin, skyrin, emodin, emodinanthrone, secalon acid and moniliformin*

The ^{14}C-labelled islandicin (**95**) and skyrin (**96**) were obtained in significant yield[29] by feeding the synthetically ^{14}C-labelled diketonaphthol (**97**) to the surface cultures of the mould fungus *Penicillium islandicum* (equation 29). The chemical synthesis of the ^{14}C

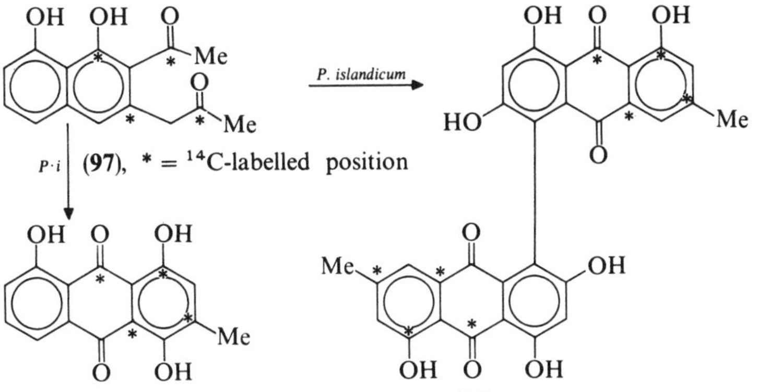

(**97**), * = ^{14}C-labelled position

P. islandicum

P. i

(**95**) (**96**)

(29)

intermediate (**97**) was carried out following reaction scheme 30[30–32]. The diethyl 3-pyrrolidinoglutarate was obtained in 82% yield by reacting pyrrolidine with diethyl

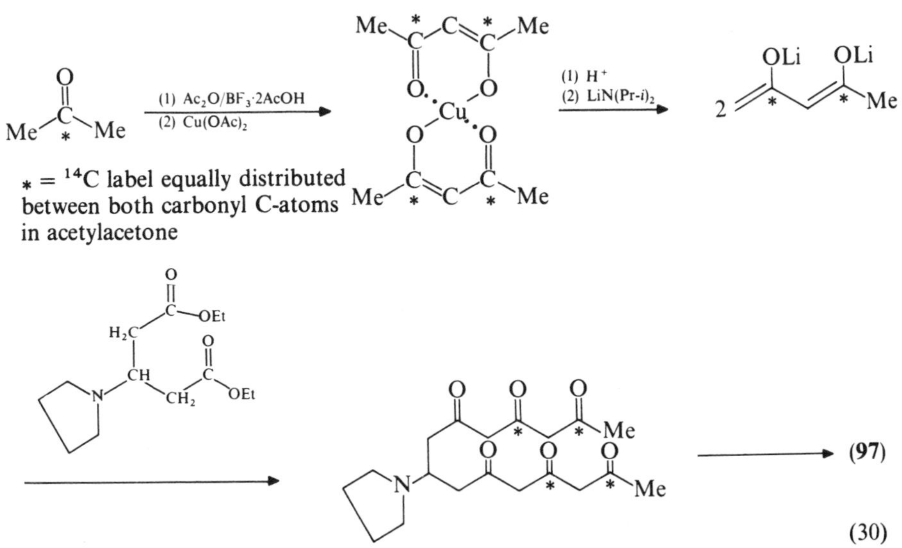

(1) $Ac_2O/BF_3{\cdot}2AcOH$
(2) $Cu(OAc)_2$

(1) H^+
(2) $LiN(Pr\text{-}i)_2$

* = ^{14}C label equally distributed between both carbonyl C-atoms in acetylacetone

→ (**97**)

(30)

glutaconate. By these labelling experiments the participation of the bicyclic intermediate (**97**) in the general polyketide route of anthraquinone biosynthesis by microorganism has been established.

In the course of biochemical studies of the ergochrome synthesis by 8-day-old *Penicillium oxalicum* (ATCC 10476) the 3-^{14}C-labelled emodin (**98**) and the intermediate [11-^{14}C]emodinanthrone were obtained by chemical methods. **98** was synthesized with a rather low (8 %) yield by the Friedel–Crafts reaction of the anhydride of 3,5-dimethoxyphthalic acid with [3-^{14}C]-*m*-cresol. The latter was obtained in a six-step synthesis from [1-^{14}C]benzoic acid with 31 % overall yield (equation 31). **99** was obtained in a 45 % yield

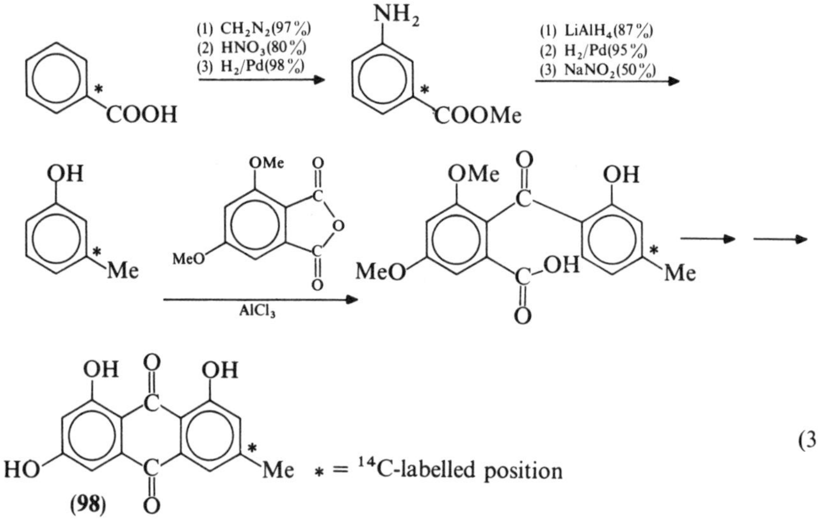

(1) CH_2N_2(97 %) (2) HNO_3(80 %) (3) H_2/Pd(98 %)

(1) $LiAlH_4$(87 %) (2) H_2/Pd(95 %) (3) $NaNO_2$(50 %)

$AlCl_3$

(**98**) * = ^{14}C-labelled position (31)

by condensing the 3-methoxy-5-(methyl-^{14}C)-phthalic anhydride (**100**), with 1,3-dihydroxybenzene (equation 32). The labelled anhydride (**100**) was synthesized according

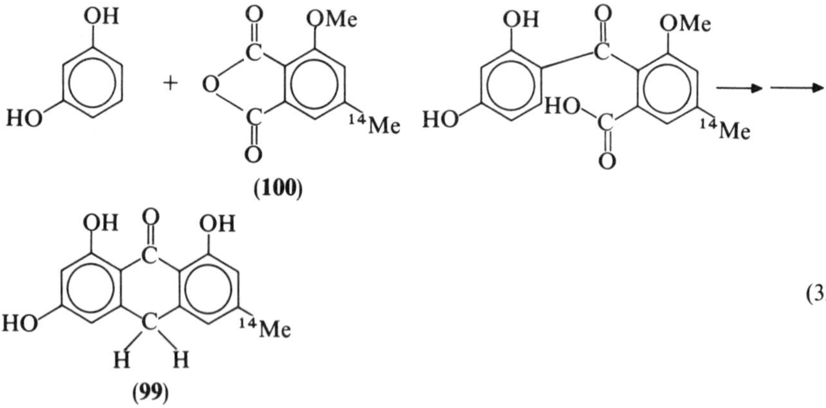

(**100**)

(**99**) (32)

to reaction scheme 33. When a mixture of both ^{14}C-labelled compounds **98** and **99** was added to *Penicillium oxalicum* growing medium it was incorporated effectively into the

(1) n-BuLi, $^{14}CO_2$(66%)
(2) $LiAlH_4$(75%)
(3) H_2/Pd(71%)

Birch reduction

200 °C

$-C_2H_4$ 74%

(100)

MeO-C(O)-C≡C-C(O)-OMe

(33)

^{14}C-labelled acid (**101**). ^{14}C- and tritium-labelled acid (**101**) was also obtained by growing the *Penicillium oxalicum* culture (ATCC-10476) in the presence of biosynthetically

chemical degradation

28 CO_2 + ^{14}Me-$\overset{*}{C}$OOH → $\overset{*}{C}O_2$

$^{14}MeNH_2$ —$KMnO_4$→ $^{14}CO_2$

(34)

(**101**)

uniformly labelled [U-^{3}H]emodin and one of the [U-^{14}C]anthraquinones: [U-^{14}C]emodin, [U-^{14}C]chrysophanol (**102**), [U-^{14}C]islandicin or [U-^{14}C]catenarin (**103**).

[U-^{3}H]-(**98**) → (**102**) ← (**103**)

(**102**) → (**101**)

(35)

The radioactivities of the product (**101**), isolated and purified up to a constant specific radioactivity, were measured and the ratio $^3H/^{14}C$ was established by using a scintillation counter. The cumulative data concerning the incorporation of radioactive anthraquinones into **101** by *P. oxalicum* indicate that chrysophanol (**102**) incorporated into **101** 3.6 times better than emodin (**98**) and 11.8 more effectively than islandicin (**95**). The conclusion was also made that emodin and islandicin are incorporated into **101** through the intermediate

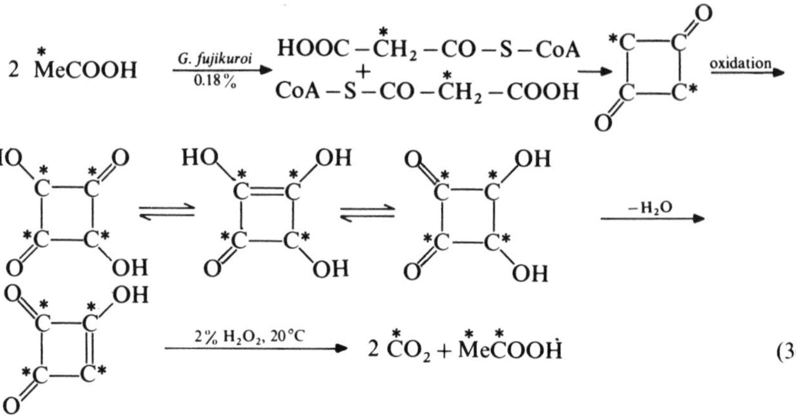

$$2\ \overset{*}{Me}COOH \xrightarrow[0.18\%]{G.\ fujikuroi} \begin{matrix} HOOC-\overset{*}{C}H_2-CO-S-CoA \\ + \\ CoA-S-CO-\overset{*}{C}H_2-COOH \end{matrix} \longrightarrow \xrightarrow{oxidation} \rightleftharpoons \rightleftharpoons \xrightarrow{-H_2O}$$

$$\xrightarrow{2\%\ H_2O_2,\ 20\,^{\circ}C} 2\ \overset{*}{C}O_2 + \overset{*}{Me}\overset{*}{C}OOH \qquad (36)$$

(104)

chrysophanol (**102**)[33, 34]. In Franck's review[35] on the synthesis, structure and applications of mycotoxine derivatives, the biosynthesis of [U-^{14}C]moniliformins (**104**), from [1-^{14}C]- and [2-^{14}C]acetate by *Gibberella fujikuroi* has been outlined (equation 36). All carbon atoms in **104** had the same specific radioactivity[35, 36].

II. ISOTOPIC CHEMICAL STUDIES WITH QUINONES

A. Mass Spectrometric Gas Phase and Liquid Phase Reactions with ^{18}O- and ^{2}H-labelled Quinones

1. Gas phase reactions with positively charged [^{18}O]anthraquinone, [^{18}O]hydroxyanthraquinone and [^{18}O]hydroxyfluorenones

[9,10-bis-^{18}O]Anthraquinone (**105**) (obtained by concentrated HCl-catalysed exchange of parent anthraquinone with $H_2{}^{18}O$) and [9,10-bis-^{18}O]-1-hydroxyanthraquinone (**106**), [9-^{18}O]-4-hydroxy-9-fluorenone (**107**), [9-^{18}O]-1-hydroxy-9-fluorenone (**108**) and [9-^{18}O]-2-hydroxy-9-fluorenone (**109**), prepared by a similar method and containing 40%, 8%, 60%, 55% and 60% of ^{18}O respectively were used for investigation of the gas phase reactions of positively charged ions[37]. Mass spectrometric studies of the gas phase

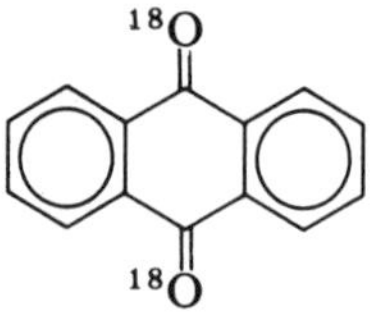

(105) 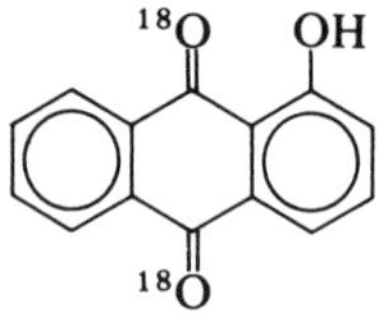 (106) 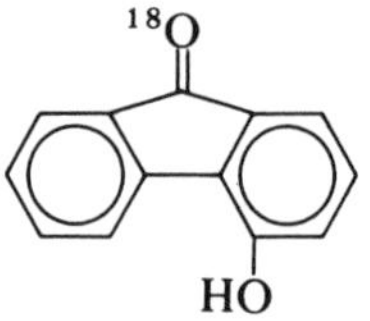 (107)

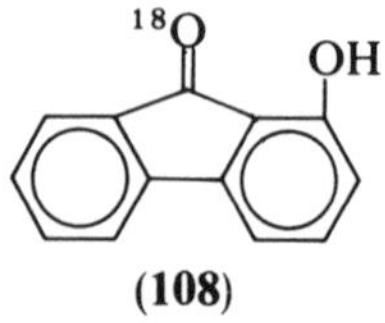

(108) 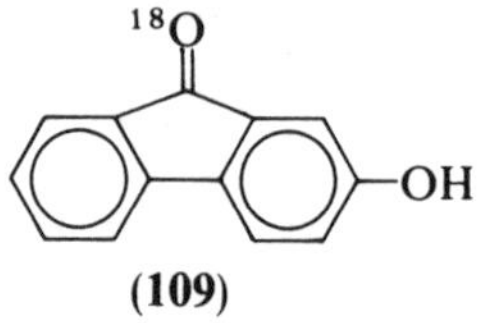 (109)

decomposition of the positive ions derived from **105** confirmed the suggestion that the principal reaction of positively charged quinones in the gas phase is the loss of neutral carbon monoxide from the molecular ion. This is followed by ejection of the second carbon monoxide molecule from the resulting $[M\text{–}CO]^{+\cdot}$ ion (equation 37). A more detailed schematic representation of this consecutive loss of two carbon monoxide molecules from the molecular ion of labelled anthraquinone is given in equation 38. By

$$[M]^{+\cdot} \rightarrow [M-CO]^{+\cdot} \rightarrow [M-CO-CO]^{+\cdot} \qquad (37)$$

m/z 212 → $\rightarrow C\equiv{}^{18}O +$ *m/z* 182

→ $C\equiv{}^{18}O+$ *m/z* 152 (38)

using [9,10-di-^{18}O]-1-hydroxy-9,10-anthraquinone (**106**) it was possible to distinguish between the loss of carbon monoxide from the two different functional groups of the molecular ion of 1-hydroxyanthraquinone whose schematic decomposing is given in equation 39.

$$M^{+\cdot} \rightarrow [M\text{–}CO]^{+\cdot} \rightarrow [M\text{–}CO\text{–}CO]^{+\cdot} \rightarrow [M\text{–}CO\text{–}CO\text{–}CO]^{+\cdot} \qquad (39)$$

It also enabled the estimation of the relative probability of the different decomposition channels shown in equation 40. Thus it has been shown that the *m/z* 228 ion loses $C^{18}O$ and $C^{16}O$ in an approximate ratio of 4 : 1. Consequently the loss of carbon monoxide from the carbonyl position is preferred over its loss from the hydroxyl position. Moreover the loss of the *m/z* 28 fragment is entirely due to $C^{16}O$ loss and not to C_2H_4 loss. It has also been found that the peaks corresponding to reactions ($228^+ \rightarrow 170^+$)/($228^+ \rightarrow 168^+$) are observed in a ratio of 1 : 1.6. This implies that the consecutive loss of two $C^{18}O$ molecules, i.e. the formation of the hydroxybiphenylene structure, is favoured over the loss of carbon monoxide from the hydroxyl position in the second step. The simultaneous loss of *m/z* 56[$C^{16}O + C_2H_4$] from the molecular ion appeared to be a minor process. The *m/z* 200 ions generate the 172^+ and 170^+ ions in a 1 : 5 ratio. In the case of the *m/z* 198 ions the decomposition route depends largely upon the isomer studied. The [9-^{18}O]-1-hydroxy-9-fluorenone ion loses virtually only $C^{16}O$ from the hydroxyl group position, while [9-^{18}O]-4-hydroxy-9-fluorenone prefers to lose $C^{18}O$ from the central ring, as judged by the $C^{18}O/C^{16}O$ ratio of ca. 10 : 1, and probably to form hydroxybiphenylene. In the case of molecular ion of [9-^{18}O]-2-hydroxy-9-fluorenone, $C^{18}O$ and $C^{16}O$ isotopic molecules are lost in a ratio of 2 : 3. It has also been found that the ratio of 4-hydroxyfluorenone to 1-hydroxyfluorenone is 2 : 1 when ^{18}O-enriched 1-hydroxyanthraquinone loses $C^{18}O$ in the ion source of the mass spectrometer. This means that the carbonyl group adjacent to the hydroxyl group is lost preferentially, probably because among the hydroxyfluorenones **107** has a greater stability than **108**.

$-C_2H_4$ → $[C_{11}H_4{}^{18}O_2]^{+\cdot}$ $m/z = 172$

$m/z = 200$ — $-C^{18}O$

$-C^{16}O$

$m/z = 228$ — $-C^{18}O$ → $m/z = 198$ — $-C^{16}O$ → $m/z = 170$

$-C^{18}O$ — $m/z = 168$

$-C^{18}O$ — $-C^{18}O$ — $-C^{16}O$

$m/z = 198$ — $m/z = 170$

(40)

2. Negative-ion mass spectrometric studies with labelled quinones

Isotopically labelled substituted naphthoquinones and anthraquinones have been used in negative-ion mass spectrometry[38–42], especially of esters of the type R^1–COO–R^2, where R^1 or R^2 is a quinone residue. The basic fragmentation of ethers of the type **110** proceeds by a loss of an alkyl radical with formation of the resonance-stabilized form (**111**) (equation 41). Electron impact studies with 1-d_3-methoxyanthraquinone and the

$-R^{\cdot}$

(**110**) (**111**) $m/z = 223$ (41)

observed elimination of the $CD_3O^{\cdot}$ group established that the '$M - RO^{\cdot}$' process is the minor fragmentation path in the mass spectrometry of complex ethers[39]. By using deuterium-labelled anthraquinone esters (**112**), prepared from alizarin-2-acetate and [$^2H_6$]acetic anhydride, and naphthoquinone ester (**113**) obtained by treatment of 2-hydroxynaphthoquinone with [$^2H_6$]acetic anhydride it has been demonstrated[38] that **113**

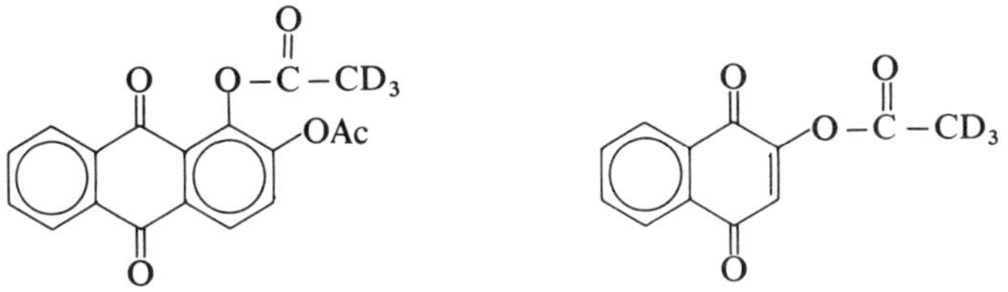

(**112**) (**113**)

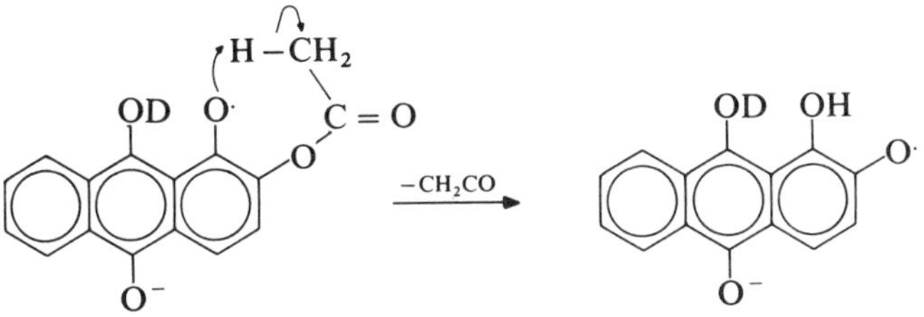

$$\xrightarrow{-CH_2CO} \qquad (42)$$

(**114**) (**115**)

specifically eliminates ketene CD_2CO, while **112** undergoes two fragmentation pathways, M–CD_2CO–CH_2CO and M–CD_2CO–MeCO·. Elimination of ketene from the 1-position of **112** produces **114** which decomposes by loss of a second ketene molecule to form **115** (equation 42). Thus it has been shown that the presence of an adjacent phenoxide radical or anion to the acetoxy group is sufficient prerequisite for elimination of ketene in the acetoxyanthraquinone system.

a. *Deuterium isotope effect study of the mechanism of negative-ion reactions in the gas phase*

Expecting that hydrogen loss or transfer takes place in the rate-determining step of the unimolecular elimination of ketene from negatively charged quinone acetate molecular ions, deuterium-labelled anthraquinone 1-acetate and 1-propionate and deuterium-labelled analogous 2- and 8-substituted 1,4-naphthoquinones have been prepared from the corresponding phenols with anhydrides. The negative-ion mass spectra of the esters were investigated and the deuterium isotope effects were determined for the quinone derivatives **116–133**[43]. The estimated ratios of the unimolecular rate constants for hydrogen atom transfer (i.e. formation of M–CHDCO) and for deuterium transfer (i.e. formation of M–CH_2CO) for quinones singly labelled with deuterium at the tertiary or secondary carbon are listed in Table 1. (Data corresponding to MeCO· and to CD_3CO· analogues were probably utilized in the course of evaluations of these isotope effects.) The data shown in Table 1 indicate that the hydrogen transfer is taking place in the rate-determining step of the reaction. The k_H/k_D values are relatively low and imply rather the unsymmetrical 'product-like' transition state characteristic for endothermic radical reactions. The tunnelling contributions are therefore ignored. The authors ascribed the slightly lower k_H/k_D values for 2-acetoxy-1,4-naphthoquinone (**121**) than for 8-acetoxy-1,4-naphthoquinone (**125**) to stabilization by delocalization of the radical centre at position 1 in compounds **121** and **123**, with a consequent effective stabilization of both

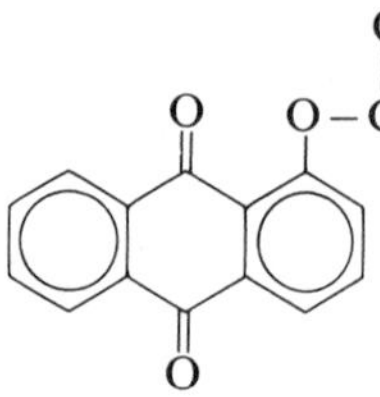

(**116**) R = Me
(**117**) R = CH_2D
(**118**) R = Et
(**119**) R = CHDMe

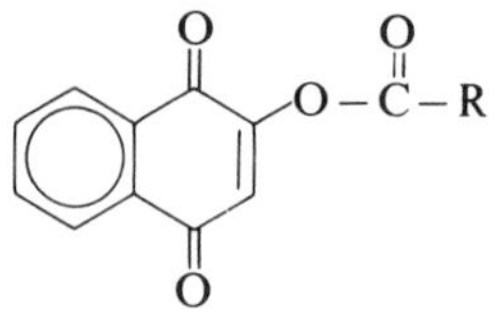

(**120**) R = Me
(**121**) R = CH_2D
(**122**) R = Et
(**123**) R = CHDMe

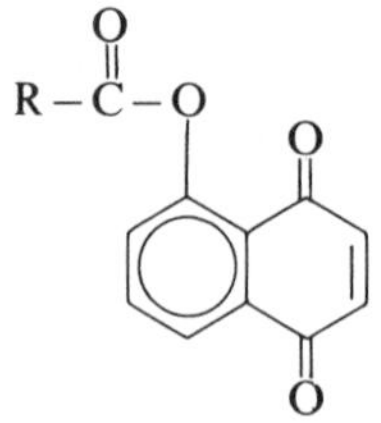

(**124**) R = Me
(**125**) R = CH_2D
(**126**) R = Et
(**127**) R = CHDMe

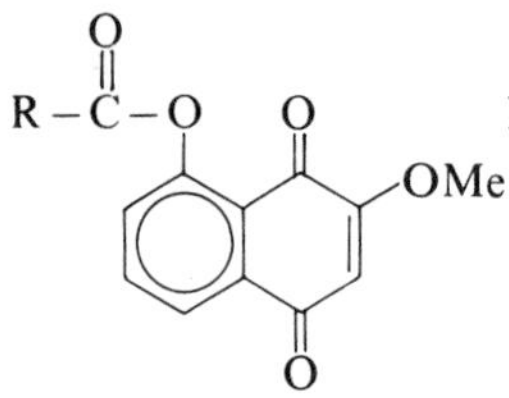

(**128**) R = Me
(**129**) R = CH_2D

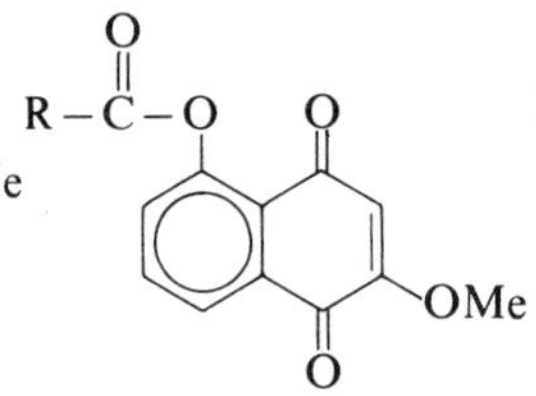

(**130**) R = Me
(**131**) R = CH_2D

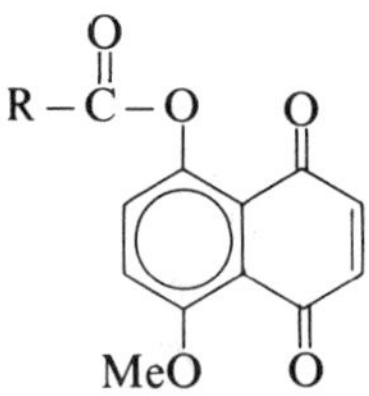

(**132**) R = Me
(**133**) R = CH_2D

TABLE 1. Kinetic isotope effect in the gas phase elimination of CH_2=C=O or MeCH=C=O from 1-acetates and 1-propionates of substituted quinones.

	$k_{\text{H transfer}}/k_{\text{D transfer}}$	
Compound	Ion source	First field free region
117	1.7	1.8
119	1.8	1.9
125	2.3	2.45
127	2.35	2.5
121	1.45	1.5
123	1.4	1.55
129	1.45	1.5
131	2.1	2.25
133	1.55	1.65

reactant and product with respect to the transition state, thus increasing asymmetry of the transition state. In contrast the acyloxy group of **125** and **127** does not stabilize the radical centre and the isotope effect is higher (cf. structures **134** and **135**).

The proposal that these rearrangements are endothermic radical reactions with 'unsymmetrical product-like' transition state was corroborated by studying the methoxy derivatives. The 2-methoxy group of **129** and the 5-methoxy group in **133** decreased the isotope effect by stabilizing the radical centre in reactant and product while the 3-methoxy

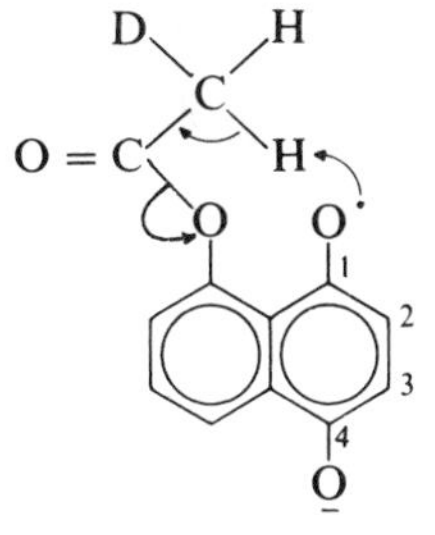

(134) 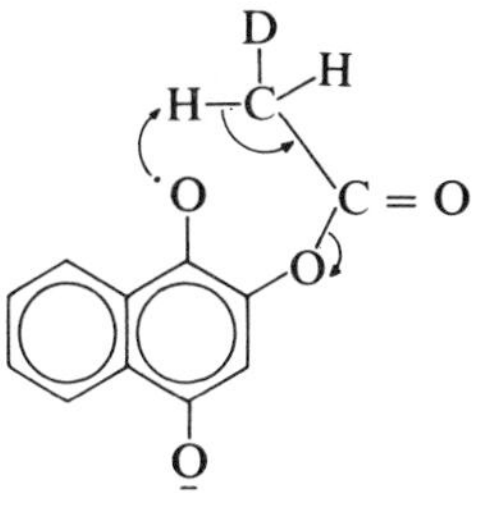 (135)

group of **131** which stabilizes only the radical centre at position 4 did not lower the isotope effect. It is also suggested that in rearrangements in which $k_H/k_D > 2.1$ the possibility that the rate-determining step involves a proton transfer cannot be completely excluded. (One can distinguish between a reactant-like and a product-like hydrogen transfer in the ketene elimination reactions studying the ^{13}C and ^{14}C kinetic isotope effects in the course of the carbon–hydrogen bond rupture.)

3. Mechanism of C–C bond cleavage of cyclic 1,2-diketones with alkaline hydrogen [^{18}O]peroxide

Oxygen-18 labelling was also applied[44] to investigate the mechanism of the carbon–carbon bond cleavage of cyclic 1,2-diketones with alkaline hydrogen peroxide. ^{18}O tracer study of the reaction of 3,5-di-*t*-butyl-1,2-benzoquinone (**136**) and 9,10-phenanthrenequinone (**137**) indicated that the carbon–carbon cleavage reaction proceeds

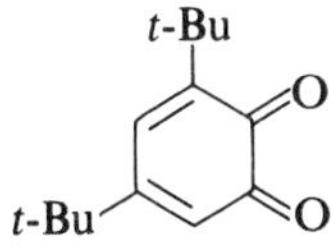

(136) 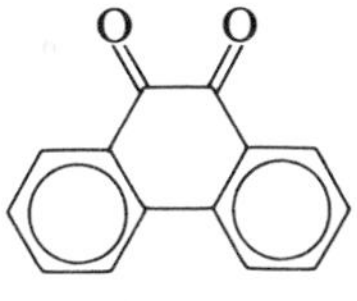 (137) 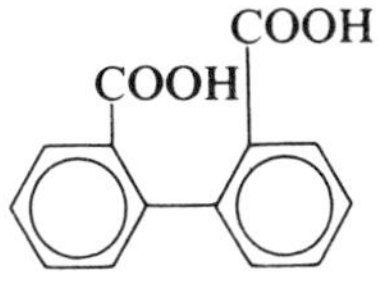 (138)

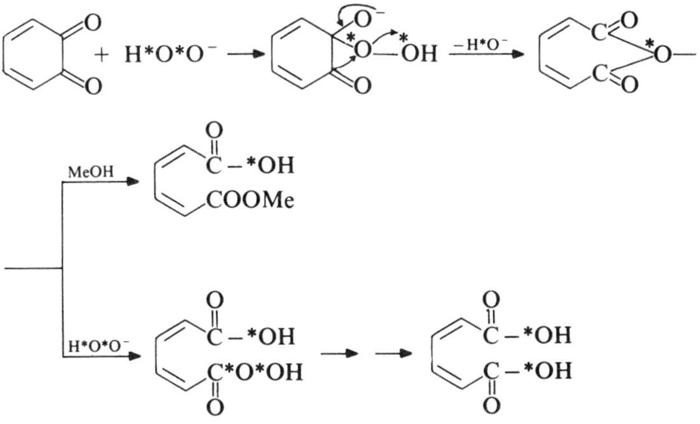

(43)

$* = {}^{18}O$ label

via an acyclic Baeyer–Villiger type mechanism (equation 43). Mass spectral determination of the ^{18}O% excess in the molecular fragments of the oxidation products eliminates the rather attractive cyclic dioxetane mechanism in the oxidation with $H_2^*O_2$ (equation 44), as

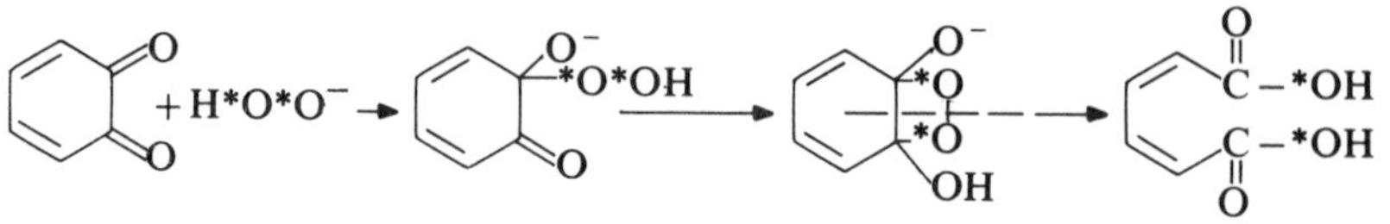

$$\text{o-quinone} + H^*O^*O^- \rightarrow \text{[}O^-, {}^*O^*OH\text{ adduct]} \longrightarrow \text{[dioxetane: }{}^*O, {}^*O, O^-, OH\text{]} \dashrightarrow C(=O){-}^*OH,\ C(=O){-}^*OH \qquad (44)$$

well as a mechanism involving an intermediate peroxide. Thus in the reaction of phenanthrenequinone with $H_2^*O_2$/NaOH in 1:1 THF/MeOH the diphenic acid (**138**) and its monoester (**139**) are obtained, each in about 34% and 25% isolated yields respectively. One oxygen atom in the monomethyl diphenate (**139**) should be derived from $H_2^*O_2$ since mass spectral cleavage of the acid group of **139** results in loss of the excess ^{18}O (equation 45), while cleavage of the COOMe group yields [^{18}O]carboxyl labelled acid[44]. These

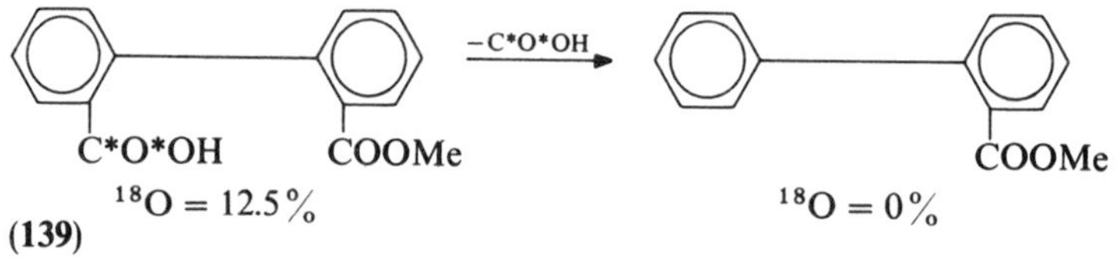

$$\underset{(\mathbf{139})\quad {}^{18}O = 12.5\%}{\text{biphenyl}(C^*O^*OH)(COOMe)} \xrightarrow{-C^*O^*OH} \underset{{}^{18}O = 0\%}{\text{biphenyl}(COOMe)} \qquad (45)$$

experimental observations are interpreted as strongly supporting the acyclic type mechanism of equation 43. However, it needs further theoretical and isotope effect investigation.

B. Spectroscopic, Radiation and Chemical Investigations of Labelled Quinones

1. Spectroscopic studies of labelled quinones

a. ^{13}C Nuclear magnetic resonance studies with quinones

Carbon-13 nuclear magnetic resonance spectra of hydroxymethoxyanthraquinones, acetoxymethoxyanthraquinones and naturally occurring anthraquinone analogues were measured for isotopically enriched compounds in deuteriated $CDCl_3$ and $(CD_3)_2SO$ solvents[45]. The structures of averufin (**140**), tri-*O*-methylaverufin (**141**) and tri-*O*-acetylaverufin (**142**) have been elucidated by using labelled averufin obtained from the

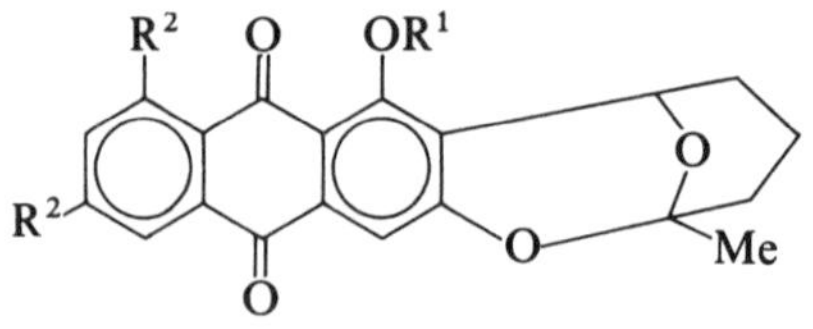

(**140**) $R^1 = H, R^2 = OH$
(**141**) $R^2 = OMe; R^1 = Me$
(**142**) $R^2 = OCOMe; R^1 = COMe$

^{13}C(1)-enriched acetate[46,47] by *Aspergillus versicolor*. The precursor–product relationship between the acetate and averufin has been demonstrated by using [^{13}C]acetate[48a]. Compound **143** was proposed as the precursor of averufin, which in turn has been suggested to be the intermediate in the biosynthetic production of aflatoxin (**144**) B_1, a potent hepatocarcinogen, by *Aspergillus flavus* (equation (46)[48b,49]. Deuteriation effects

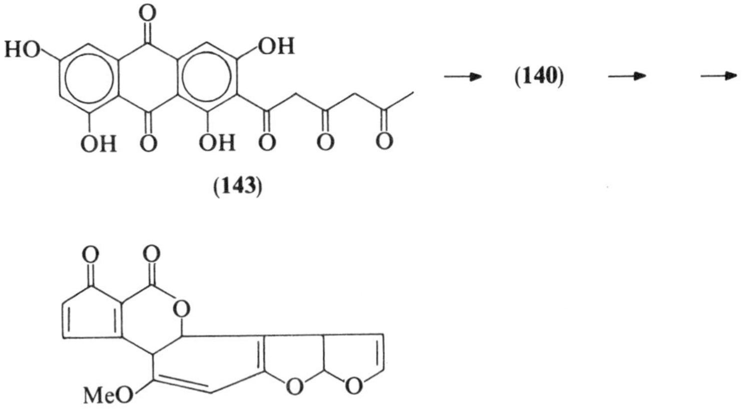

HO OH OH O OH O O O → (**140**) → →

(**143**)

O O O MeO O O (46)

(**144**)

were used to determine the ^{13}C chemical shift[50,51] δ_C for 1,4-naphthoquinone (**145**), vitamin K_3(**146**), juglone (**147**), naphthazarin (**148**) and their methyl ethers and acetates **149** and **150**[51].

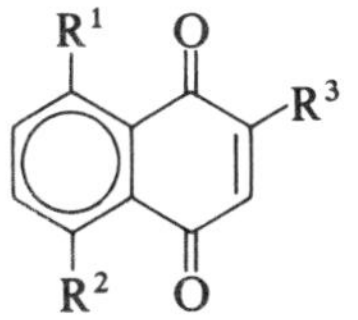

R^1 O R^3 R^2 O

(**145**) $R^1 = R^2 = R^3 = H$
(**146**) $R^1 = R^2 = H$, $R^3 = Me$
(**147**) $R^1 = OH$, $R^2 = R^3 = H$
(**148**) $R^1 = R^2 = OH$, $R^3 = H$
(**149**) $R^1 = OMe$, $R^2 = OH$, $R^3 = H$
(**150**) $R^1 = OAc$, $R^2 = OH$, $R^3 = H$

b. EPR study of hydroxyanthrasemiquinones

Deuterium-labelling experiments had to be carried out in the course of determinations of the hydroxyl proton coupling constants of six α-hydroxylated, two β-hydroxylated and three α,β-dihydroxylated anthrasemiquinones by EPR[52]. The semiquinones studied were prepared by reduction of quinones with sodium dithionite in alkaline solvent (pH ca. 12) composed of D_2O, EtOD and NaOD. Any hydroxyl proton doublet was then replaced by a 1:1:1 triplet and the splitting was reduced to about one-sixth of the proton splitting. The deuterium triplet splitting often becomes smaller than the experimental line width, leading to a considerable simplification of the spectrum. This has been found for instance in the EPR spectra of 1,2,5,8-tetrahydroxyanthrasemiquinone taken in deuteriated solvents.

Biological semiquinone anions and neutral semiquinone anions and neutral semiquinone radicals generated by irradiation with a 250 Hg/Xe lamp in quartz ESR tubes and immobilized in a solvent matrix frozen in liquid nitrogen were reviewed by Hales and Case[53].

c. Fluorescence studies with 1,5-*dihydroxyanthraquinone*

The effect of deuteration on the absorbance and fluorescence spectra of 1,5-dihydroxyanthraquinone (**151**) has also been investigated. Deuterium substitution of the

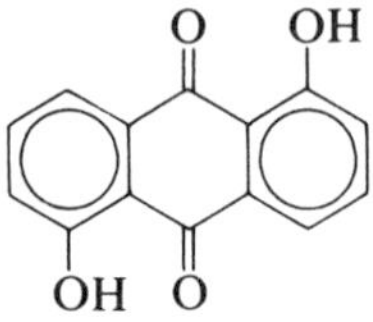

(**151**)

hydroxy protons almost[54] totally eliminates the structure in the room temperature absorbance, and it also regularizes the room temperature fluorescence profile. The deuterium replacement has a very minor effect on the S_1–S_2 oscillator strength but it shifts the transition by 3–4 nm (ca. 150 cm^{-1}) to the blue. At wavelengths shorter than 375 nm the absorbance spectra of non-deuteriated and deuteriated **151** are virtually indistinguishable since the π–π^* transitions of the anthraquinone framework have only little charge-transfer character. Deuterium substitution increases the fluorescence quantum yield by nearly four-fold and diminishes the short wavelength component of the emission. Changes at $\lambda > 575$ nm are comparatively minor. The marked fluorescence intensity increase in the 560–575 nm region is interpreted as the symptom of the O–H stretching vibration activity in this wavelength region (ω_{OH} = ca. 3000 cm^{-1} in the IR spectrum, $\omega_{O\text{-}D} = 2300\,cm^{-1}$)[55]. The intensity of the low-temperature short wavelength fluorescence (SWF) is greatly reduced when the hydroxy protons are deuteriated (the beginning—the 'origin'—of the very weak bands is shifted 86 ± 3 cm^{-1} to the blue from its normal isotopic species counterparts). The long wavelength fluorescence (LWS) commencing around 560 nm is much stronger, devoid of any sharp structure and is qualitatively the same at room temperature as at 10 K and intrinsically broader vibronically than SWF. Also in hexane solvent a very substantial effect of isotopic substitution on the LWF/SWF intensity ratio (vibronic intensity distribution) was found, while the changes of the fundamentals are small and within 10 cm^{-1} of the 339, 377 and 400 cm^{-1} values. The intensity ratio of SWF to LWF always drops drastically when the hydroxy protons are isotopically replaced. The frequencies of a number of fundamentals identified in the fluorescence and excitation spectra of **151** are not particularly sensitive to deuterium substitution of the hydroxy protons because of their mostly skeletal character. However, the decoupling of modes is not complete since some significant intensity effects are caused by deuteriation. The observations listed above have been interpreted as indicating that in the photoexcited state a single proton transfer is taking place with very small potential energy barrier creating a 1,10-quinone stable form in the S_1 state. A similar excited-state proton transfer (ESPT) was found for other α-hydroxyanthraquinones in contrast to quinizarin, 1,2,4-trihydroxyanthraquinone, 1,2,5,8- and 1,4,5,8-tetrahydroxyanthraquinones where the 1,4-substitution pattern, which stabilizes the system against excited-state proton transfer, is operating.

d. Isotope effect study of the optical absorption–emission by p-*benzoquinone*

Deuterium-labelled *p*-benzoquinones BQ-d[56], BQ-2,6-d_2, BQ-d_4 and d_3-Me-toluquinone were used to investigate spectroscopically the lowest $n\pi^*$ triplet state of BQ-h_4 at 1.8 K. Vibrational analysis of the 537, 537 nm absorptions of all the isotopically substituted *p*-benzoquinones and vibrational analysis of the phosphorescence spectrum of BQ-h_4 in a BQ-d_4 crystal permitted the location at $18\,609 \pm 1\ cm^{-1}$ of the unobserved origin of the B_{1g} ($n\pi^*$) triplet state of BQ-h_4 monomer as a guest in the BQ-d_4 crystal. The electronic origin of the emitting state is directly observed in the asymmetrically substituted isotopic quinones. The origin of the phosphorescence spectrum of BQ-2,6-d_2 as a guest (1 mol %) in BQ-d_4 at 1.8 K was found to be at $18\,627\ cm^{-1}$. It has been shown that isotope effects observed in the singlet–triplet absorption in isotopic mixed crystals of BQ-h_4 in BQ-d_4 at 1.8 K are due to hydrogen (deuterium) bonding effects that shift the mainly centred on oxygen electronic excitation of BQ-h_4. It was also concluded that the observed isotope effect on the vibronic structure in the phosphorescence spectra of the *p*-benzoquinones is at least partly due to an isotope-dependent excited state geometry[56]. Absorption studies of the BQ-h_4 in BQ-d_4 isotopic mixed crystals revealed the existence of the so-called 'cluster state' absorption caused by formation of 'translationally inequivalent dimer' (**152**), which does not have inversion symmetry. Molecule No. 1 in this 'dimer' absorbs $4.0\ cm^{-1}$ to lower energy than molecule no. 2. In the trimer, formed by the BQ molecules numbered 1, 2 and 3, the inversion symmetry is preserved.

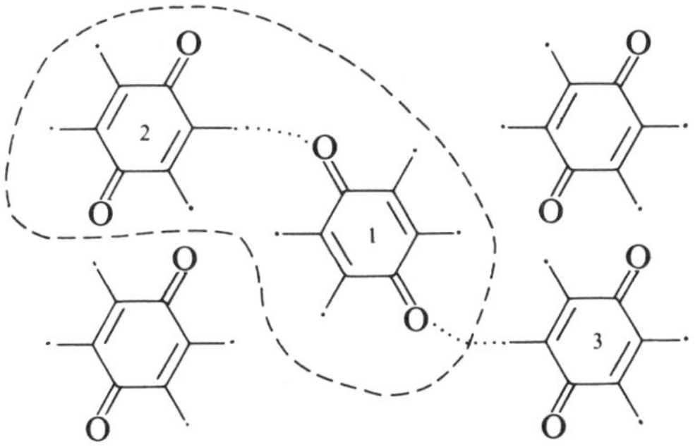

(**152**)

No. 1 and no. 2 (encircled with dashed lines) BQ-h_4 molecules form the translationally inequivalent dimer in the BQ-d_4 host crystal. Dotted lines indicate the hydrogen bonds responsible for the cluster formation.

The splitting observed in the absorption spectra of pure BQ-d and BQ-2,6-d_2 crystals at 1.8 K was ascribed to the existence of 'chemical disordering', implying that the excitation energy of a particular molecule depends upon the positions of isotopic hydrogens of the neighbouring molecules. (The observed deuterium isotope effects are partly due to change in short range static forces upon deuteriation.) The EPR spectrum of BQ-h_4 in BQ-d_4 host crystal was also taken and the large isotope effect on the ZFS parameters of the lowest triplet state of BQ-h_4 found was interpreted as an intramolecular phenomenon caused by isotope-dependent spin-orbit coupling effects.

e. Deuterium isotope effects on the quenching of the triplet state with substituted phenols

In the course of the detailed kinetic studies of the mechanism of the quenching of the triplet state of 2,6-diphenyl-*p*-benzoquinone (**153**) and anthanthrone (**154**)[57a] by sub-

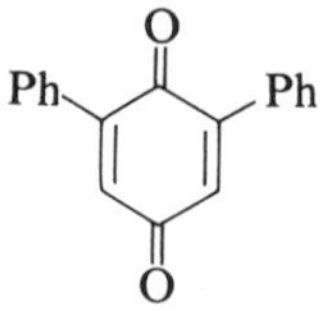

(153)

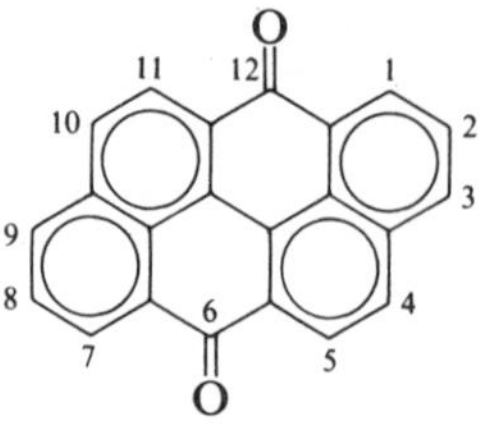

(154)

stituted phenols the effect of deuteriation of the OH group of phenols (obtained by exchange with D_2O or MeOD), on the rate constants of the quenching of the triplet state of **154** with phenols in benzene was also examined[57b]. The deuterium isotope effect, $k_T(H)/k_T(D)$, expressed as a ratio of rate constants of quenching of the ketone triplet state with undeuterated and with deuterated phenolic group depends to a large extent on the nature and the position of the substituent in the aromatic ring. For example, for quenching of anthanthrone triplet state with 2,6-di-*t*-butyl-4-methylphenol $k_T(H)/k_T(D) = 1.5$. For quenching of **154** with *p*-$PhC_6H_4OH(D)$ and with 2,4-$(NO_2)_2C_6H_3OH(D)$ no deuterium isotope effect is observed. In the quenching reaction with 3-nitro- and 2,6-di-*t*-butyl-4-nitrophenol $k_T(H)/k_T(D)$ equals 2.0 ± 0.1 but $k_T(H)/k_T(D) = 9.5$ in the quenching reaction of anthanthrone triplet state with 2,3-$Cl_2C_6H_3OH(D)$ in benzene. These isotope effects were interpreted as indicating that in quenching reactions in which $k_T(H)/k_T(D) = 1.0$, formation of a hydrogen bonded complex is the rate-limiting step while in the pair of isotopic reactions characterized by $k_T(H)/k_T(D) = 9.5$ the homolytic one-step hydrogen transfer in the hydrogen bonded complex is rate limiting. Introduction of a nitro group into the phenolic ring diminishes the deuterium isotope effect partly due to some steric hindrances in the formation of the hydrogen bonded complex. In the case of quenching of the triplet state of **154** with very acidic nitro-substituted phenols, the deuterium isotope effect is very small or it nearly disappears since it is determined by the hydrogen/proton equilibrium isotope effect in the reversible transfer expressed by equation 47:

$$\mathrm{s}\left(\rangle\mathrm{C{=}O} \ldots \mathrm{PhOH}\right) \rightleftharpoons \mathrm{s}\left(\rangle\mathrm{C{-}OH^+} \ldots \mathrm{PhO^-}\right) \rightarrow \begin{cases} \rangle\dot{\mathrm{C}}\mathrm{{-}OH + PhO^{\cdot}} \\ \rangle\mathrm{C{=}O + PhOH} \end{cases} \quad (47)$$

The authors of this work[57b] also suggested the general conclusion that the deuterium isotope effect on the quenching rate constant of triplet states of quinones with phenols increases on increasing the acidity of the substituted phenols and then diminishes at further increase of the acidity of hydrogen donor, $XC_6H_4OH(D)$. The dependence of the polarization ratio 'P' on the rate constant k_q of the triplet chemical reaction with hydrogen donor SH, appears in equation 48, where T_1 (triplet) is the spin-lattice relaxation time of the triplet and P_0 is the polarization ratio at infinite concentration of the hydrogen donor $[SH]_\infty$.

$$P = P_0\{k_q[SH]T_{1(\mathrm{triplet})}/(1 + k_q[SH]\,T_{1(\mathrm{triplet})})\} \quad (48)$$

A preliminary experiment concerning the photolysis of 1,4-naphthoquinone in isopropanol and ($^2H_8$) isopropanol, supports the dominant role of the phototriplet mechanism of the chemically induced dynamic electron polarization[58a]. The polarization ratios measured at $-20°C$ for the same hyperfine line of the seminaphthoquinone radical were: 0.39 in MeCH(OH)Me and 0.15 in $CD_3CD(OD)CD_3$. In a separate series of experiments using 2,6-di-*t*-butylphenol-OH and 2,6-di-*t*-butylphenol-OD as donors the polarization ratios of the phenoxy radical were measured in the presence of 2-methylbenzoquinone. Since the same hyperfine line of the identical phenoxy radical was monitored, the same methylbenzoquinone triplet was involved, the concentrations were also kept the same, the variation of the polarization ratio could be attributed to the kinetic isotope effect. The estimated experimental k_H/k_D value for this system is 1.6 ± 0.1[58a].

2. *Photochemical and free radical studies of labelled quinones*

Magnetic isotope effects in the radical pairs and diradical pathways of photolysis and thermolysis of organic compounds have been reviewed recently by Turro and Kraeutler[58b].

a. Deuterium isotope effect on a radical pairs disappearance

In the photochemical reduction of the frozen solutions of 3,6-di-*t*-butyl-1,2-benzoquinone (**155**) and 2,4,6-tri-*t*-butylphenol (**156**) in vaseline and in methylcyclohexane two types of radical pairs are formed. They have different distances between unpaired electrons[59], which are 5.15 ± 0.02 Å and 6.25 ± 0.02 Å, respectively. The rates of disappearance of the radical pairs formed by photolysis of frozen glassy solutions of **155** with deuteriated and non-deuteriated **156** have been measured and activation energies, E,

(**155**)

(**156**)

for the radical pair loss were estimated for the deuteriated and non-deuteriated compounds. The constants E_1, E_2 (in kcal mol^{-1}) and k_0 (in s^{-1}) in equation 49 relate the relative concentrations

$$n_T(t)/n_0 = \frac{E_2}{E_2 - E_1} - \frac{RT}{E_2 - E_1}\ln(k_0) - \frac{RT}{E_2 - E_1}\ln(t) \tag{49}$$

of the radical pairs, $n_T(t)/n_0$, with the logarithm of time, ln (t), and the temperature 'T': $E_1 = 6.7 \pm 1$; $E_2 = 10 \pm 1$; $\lg(k_0) = 8 \pm 2$ were for non-deuteriated phenol, and $E_1 = 11 \pm 1$; $E_2 = 17 \pm 1$; $\lg(k_0) = 13 \pm 2$ for the deuteriated phenol. In the case of non-deuteriated phenol **156**, $\Delta E/\Delta R = 16.5 \pm 6$ kcal mol^{-1} Å^{-1}, where ΔR is the increase of the distance between the hydrogen atom traps (i.e. the unpaired electrons).

b. Deuterium isotope effect on the photochemical reactions of quinones with water

In the photolysis of an aqueous solution of *p*-benzoquinone, besides the isolated stable product hydroquinone, the transient formation of benzene-1,2,4-triol and 2-hydroxy-1,4-benzoquinone was noticed[60,61]. The photolysis rate of *p*-benzoquinone in D_2O was slightly smaller than in H_2O[60] giving a solvent isotope effect k_{H_2O}/k_{D_2O} of 1.15. This value indicates that direct abstraction of hydrogen from water is not the rate-determining step of the photolysis in water and suggests that an electrophilic attack of the excited *p*-benzoquinone on water is the slow process. Addition of *p*-nitroso-*N*,*N*-dimethylaniline (NDA), an effective hydroxyl radical scavenger, to the aqueous solution of *p*-benzoquinone did not change the absorption peak of the quinone. No direct photolysis of NDA occurred under the conditions of the irradiation. Consequently the radical

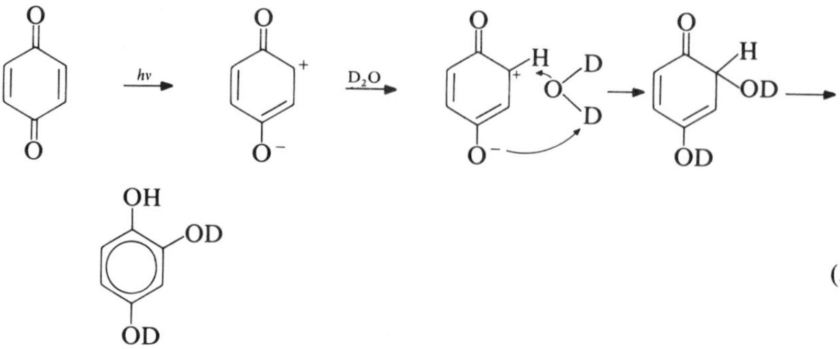

(50)

mechanism for the primary process of the photochemical reaction of *p*-benzoquinone in water has been rejected and equation 50, involving a polar intermediate, has been proposed instead[60]. However, this suggestion should be corroborated by comparative kinetic studies of the photolysis of *p*-benzoquinone-d$_4$ in D_2O.

c. Photoreduction of sodium 1,2-*naphthoquinone*-4-*sulphonate in* $H_2O(D_2O)$ *solutions*

A one-electron transfer from hydroxide ion to photoexcited sodium 1,2-naphthoquinone-4-sulphonate, NQ (**157**), was investigated by irradiating **157** in aerobic H_2O and D_2O[62]. Sodium 1,2-dihydroxynaphthalene-4-sulphonate, NQH_2 (**158**), having a

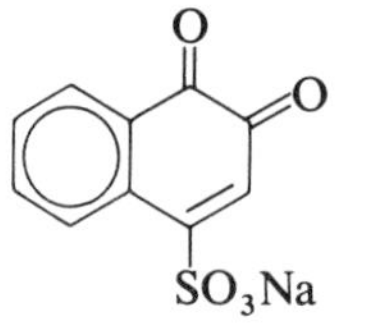

(**157**) 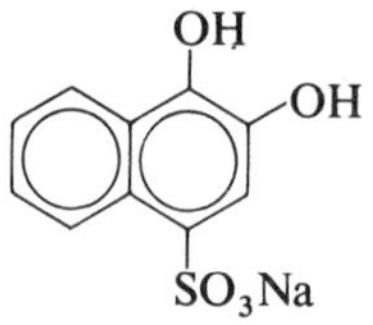 (**158**)

blue fluorescence emission maximum at 470 nm, was produced with 50% yield. The quantum yield at 365 nm of 0.118 ± 0.007 in the aerobic conditions was constant in the 5.0–6.8 pH region. Under nitrogen atmosphere the efficiency of the photoreaction of **157** was ca. 2.2-fold higher than in air. These facts indicate that the photoreaction of **157** in water proceeds via the triplet state. The quantum yield for the disappearance of **157** was reduced by a factor of ca. 2.5 on addition of moderate amounts of hydroquinone as a hydroxyl radical scavenger. The solvent isotope effect, k_{H_2O}/k_{D_2O}, on the initial rate of **158** formation is 1.8, indicating that the reaction involves protonation of the radical anion. On the basis of the above data the multi-step photoreduction scheme of **157** in water (equation 51) has been proposed, where $NQH^{\cdot}$ is the semiquinone radical of NQ.

$$NQ + h\nu \rightarrow NQ^{*1}$$
$$NQ^{*1} \rightarrow NQ^{*3}$$
$$NQ^{*3} + OH^- \rightarrow NQ^{-\cdot} + {}^{\cdot}OH$$
$$NQ^{-\cdot} + H^+ \rightarrow NQH^{\cdot}$$
$$2NQH^{\cdot} \rightarrow NQH_2 + NQ$$
$$NQ + {}^{\cdot}OH \rightarrow NQ(OH^{\cdot}) \qquad (51)$$

The pH independence and the deuterium isotope effect indicate that the electron transfer to photoexcited NQ is fast while the protonation of $NQ^{-\cdot}$ is a relatively slow process and is partly rate determining.

d. Deuterium isotope effect on the homolytic alkylation of benzoquinone

A kinetic isotope effect, $k_H/k_D = 1.9 \pm 0.1$, was found for the competitive phenoxymethylation of 1:1 benzoquinone-h_4 and benzoquinone-d_4 in water at 65 °C. The result was ascribed to the reversibility of the first step in reaction scheme 52 where the k_1, k_{-1} and

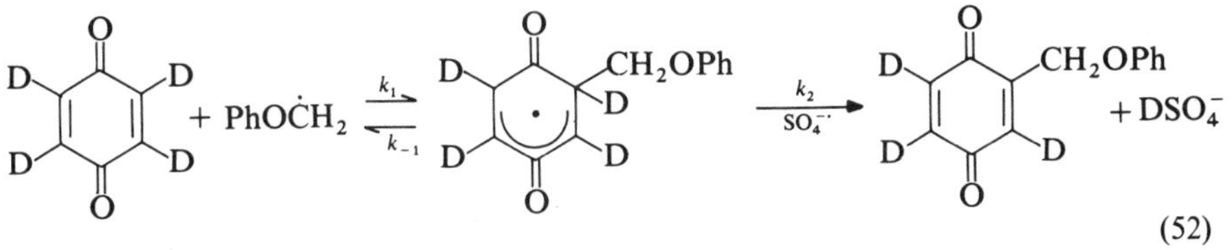

+ $Ph\dot{O}CH_2$ $\underset{k_{-1}}{\overset{k_1}{\rightleftharpoons}}$ $\xrightarrow[SO_4^{-\cdot}]{k_2}$ CH_2OPh + DSO_4^-

(52)

k_2 are of comparable magnitude[63]. The phenoxymethyl radical was generated[64] by a radical route decarboxylation of phenoxyacetic acid with a silver nitrate/ammonium peroxydisulphate couple (equation 53).

$$RCOOH \xrightarrow{Ag^+/S_2O_8^{2-}} R^{\cdot} + CO_2 \qquad (53)$$

3. Deuterium isotope effect on the antioxidant activity of vitamin E and on a two-electron reduction with daunomycinone hydroquinone

a. Chain-breaking activity of vitamin E

In the course of studies of the antioxidant activity of vitamin E component (α-, β-, γ- and δ-tocopherols (**159**)) and related phenols *in vitro*[65–70] the effect of deuteriation of the

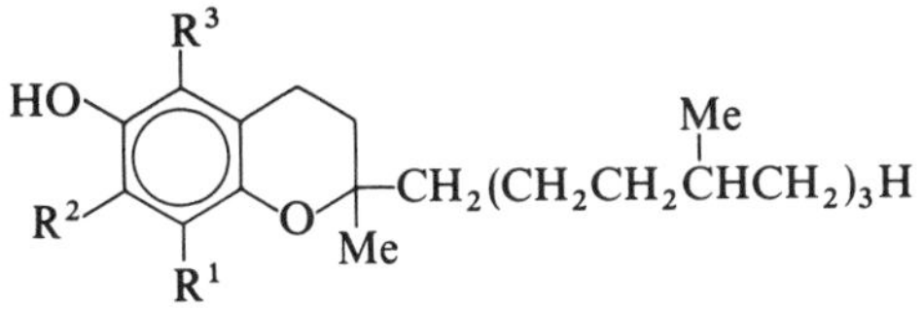

HO, R^1, R^2, R^3, Me, $CH_2(CH_2CH_2\overset{Me}{C}HCH_2)_3H$

(159)

α-T: $R^1 = R^2 = R^3 = Me$
β-T: $R^1 = R^3 = Me, R^2 = H$
γ-T: $R^1 = R^2 = Me, R^3 = H$
δ-T: $R^1 = Me, R^2 = R^3 = H$

phenolic hydrogen of α- and γ-tocopherols has been investigated and compared with related deuterated phenolic chain-breaking antioxidants[66]. The estimated k_H/k_D ratios for reaction 54 of the peroxyl radical with the phenolic group are collected in Table 2.

$$ROO^{\cdot} + ArOH(D) \xrightarrow[k_H/k_D]{} ROOH(D) + ArO^{\cdot} \qquad (54)$$

The deuteriation reduces the antioxidant activity of the tocopherols and related phenols such as **160** and **161** since deuteriated tocopherols react more slowly with the peroxyl radicals. The substantial kinetic deuterium isotope effects indicate that hydrogen atom abstraction in reaction **54** is the rate-controlling step. The above data imply also that the main function of vitamin E (α-tocopherol) *in vivo* is an antioxidant action. The role of the

TABLE 2. Deuterium kinetic isotope effect for inhibition of autooxidation of styrene with oxygen by selected phenolic antioxidants at 30 °C[66]

Antioxidant	k_H/k_D
α-Tocopherol	4.0 ± 0.5[a]
γ-Tocopherol	9.1
2,2,5,7,8-Pentamethyl-6-hydroxychroman(**161**)	5.1 ± 0.5[a]
2,3,5,6-Tetramethyl-4-methoxyphenol (**160**)	10.6 ± 3.7[a]
2,6-Di-*t*-butyl-4-methoxyphenol	9.4
2,6-Di-*t*-butyl-4-methylphenol	6.8
1-Naphthol	4.3

[a] Average of two or more separate measurements.

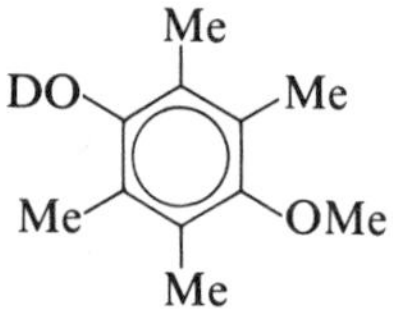

(**160**) 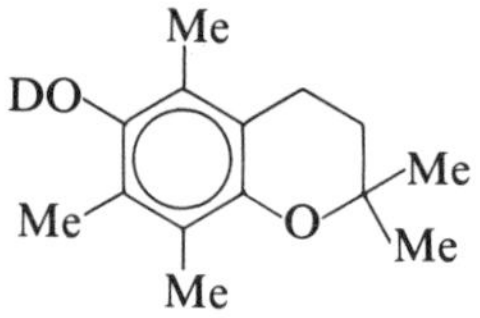 (**161**)

phytyl side chain attached to the hydroxychroman moiety is to increase its solubility in biomembranes and to allow its penetration into monolayers of phospholipid molecules. The chroman fused ring system maintains the *p*-type lone pair of the etheral oxygen nearly perpendicular to the aromatic plane, therefore stabilizing the phenoxyl radical ArO˙ (equation 54).

b. Reduction with 7-deoxydaunomycinone hydroquinone

7-Deoxydaunomycinone (**162**), a redox catalyst bound to DNA, which probably leads to cell death, is the product of a reductive glycoside cleavage of the antileukaemia drug daunomycin. This has been demonstrated by using *dl*-bi-(3,5,5-trimethyl-2-oxomorpholin-3-yl) (**163**) as the agent for reducing this drug[71]. The anaerobic solution of **163** disproportionates in the presence of **162** to **164** and **165**. The kinetics of this reaction

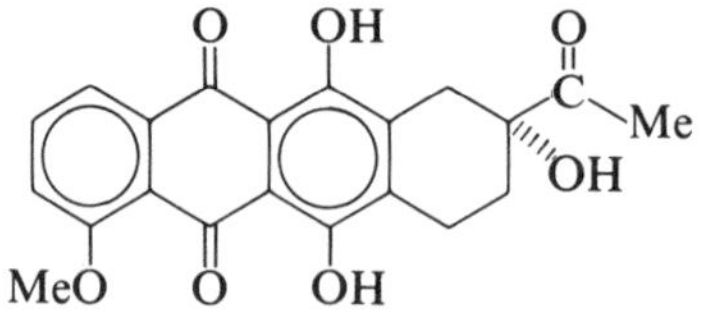

(**162**)

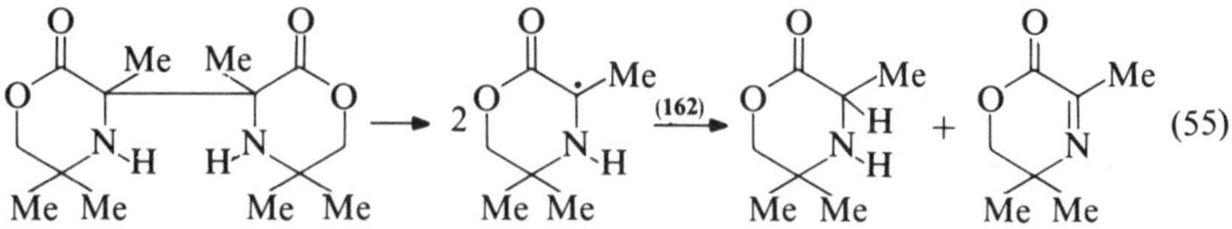

(**163**) → 2 (**166**) —(162)→ (**165**) + (**164**) (55)

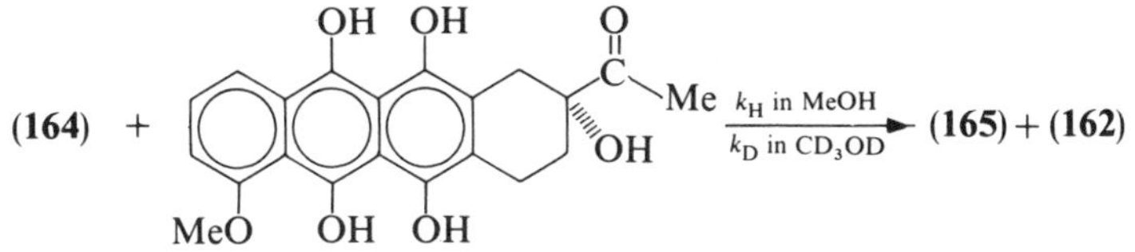

$$(\mathbf{164}) + (\mathbf{167}) \xrightarrow[k_D \text{ in } CD_3OD]{k_H \text{ in MeOH}} (\mathbf{165}) + (\mathbf{162}) \quad (56)$$

(167)

were followed spectrophotometrically in MeOH and in CD_3OD. In the absence of **162** the disproportionation either does not take place or is very slow in both solvents. For example, in CD_3OD no disproportionation of **163** after 135 h at 35 °C was observed. By monitoring the disproportionation process in the presence of **162** the reaction mechanism was established and the second order rate constants for the reduction of **164** by **167** were measured (equation 56). In MeOH at 25 °C $k_H = 2.06\ \text{M}^{-1}\,\text{s}^{-1}$, and in CD_3OD $k_D = 0.69\ \text{M}^{-1}\,\text{s}^{-1}$. The magnitude of the isotope effect, $k_H/k_D = 3.0$, indicates that the bond to hydrogen is broken in the transition state. The isotope effect of a deuteriated solvent[72, 73] on the bond homolysis of **163** and on the disproportionation of **166** to **164** and **165** in the absence of catalyst is small, being 1.10 ± 0.09 at 80 °C[72].

4. *Addition of dithiophosphates to* p-*quinones*

The mechanism of addition of the effective pesticides phosphorus dithioacids and silyl dithiophosphates to *p*-benzoquinone has been investigated[74] by using the deuteriated substrates. The deuteriated dithioacid (**168, D**) reacted with *p*-benzoquinone and with *p*-benzoquinone-d_4 about two times faster than the dithioacid (**168, H**). The k_H/k_D values

$(EtO)_2P(=S)$—SD (**168, D**) $(EtO)_2P(=S)$—SH (**168, H**) $(EtO)_2P(=S)$—S—$SiMe_3$ (**171**)

were in the range of 0.42–0.53 in *n*-heptane, benzene, 1,4-dioxane and acetonitrile. This indicates that the structure of the transition state in the process of proton migration from dithioacid to *p*-benzoquinone resembles the structure of the intermediate (**169**). The isotopic hydrogen atom is more strongly covalently bound in the transition state than in the reactant (**168**). The effect of deuteriation of the quinone on the reaction rate depends on the nature of the solvent. In heptane and benzene the deuteriated and non-deuteriated *p*-benzoquinone reacted at the same rate, implying that the ring-proton migration takes place after the rate-limiting step of the reaction. In 1,4-dioxane and in acetonitrile at 20°C, $k_H/k_D = 2.03$ and 3.58, respectively, implying that in these non-basic solvents the conversion of the intermediate **169** to product **170** is the controlling step of the reaction. A similar behaviour was found in the reaction between *O,O*-diethyl-*S*-trimethylsilyl dithiophosphate (**171**) and *p*-benzoquinone-d_4. In heptane and benzene $k_H/k_D = 1$, and in 1,4-dioxane and in acetonitrile $k_H/k_D = 1.5$ and 2.13, respectively. These results suggest that the addition of dithiophosphates to *p*-benzoquinone takes place according to equation 57. The conversion of the n–π complex (**172**) into intermediate **169** is the rate-controlling step in weakly basic solvents. In nucleophilic solvents the dienone–phenol rearrangement in the intermediate (**169**) becomes rate limiting[74].

5. *Selective reduction of anthraquinone with deuterium*

Anthraquinone can be converted to 9,9,10,10-tetradeuterio-9-10-dihydroanthracene (**173**) in the presence of excess of D_2, carbon monoxide and catalytic amounts of $Co_2(CO)_8$

$$(R^2O)_2P(=S)\text{-}SR^1 + \text{1,4-benzoquinone} \longrightarrow [(R^2O)_2P(=S)\text{-}SR^1 \cdots \text{quinone}]\ (\mathbf{172}) \longrightarrow$$

$$[(R^2O)_2P(=S)\text{-}S\text{-}C_6H_4(=O)(OR^1)]\ (\mathbf{169}) \longrightarrow (R^2O)_2P(=O)\text{-}S\text{-}C_6H_3(OH)(OR^1)\ (\mathbf{170}) \quad (57)$$

$R^1 = H(D)$ or $-SiMe_3$

$$\text{anthraquinone} \xrightarrow[\text{benzene, } Co_2(CO)_8,\ 250^\circ C]{D_2/800\text{ psi; CO/1100 psi}} \text{9,9,10,10-tetradeuterio-9,10-dihydroanthracene}\ (\mathbf{173}) \quad (58)$$

(equation 58). After 6 hours 66% of isotopic equilibrium was attained. In dioxane either D_2 or D_2O can be used as the isotopic source[75]. The selective nature of the reduction-exchange reaction can be used as a synthetic route to **173**. The specific selectivity for positions 9 and 10 in anthracene was confirmed by ^{1}H-NMR and no exchange of ring protons 1–8 was found. The product obtained from the anthraquinone reduction was used as the hydrogen source in the oxo reaction of 1-octene with CO and $Co_2(CO)_8$ and gave C_9 aldehydes randomly substituted with deuterium[75].

C. Isotope Effects in Hydrogen Transfer Reactions to Quinones

Deuterium and tritium isotope effects in hydrogen and proton transfer processes are the subject of continuous theoretical and experimental investigations. Special monographs, chapters in monographs and reviews on this topic have been published[76–90]. Several groups investigate the proton transfer processes to different organic bases[90–100]. We therefore present below only the recent isotopic results concerning hydrogen transfers (or migration) to quinones, resulting in oxidation or dehydrogenation of organic and inorganic molecules.

1. Deuterium migration in the 2,5-dihydroxy-1,4-benzoquinone

In the course of studies on diotropic proton migration between the *ortho* oxygen atoms in a series of 2,5-dihydroxy-1,4-benzoquinones (**174** → **175**) a kinetic deuterium isotope effect $(k_H/k_D)_{-50^\circ C} = 1.96$, was found[101]. It was interpreted as indicating that the probability of simultaneous synchronous proton migration in the exchange reaction of equation 59 is rather low. It is also suggested that the estimated value of the kinetic isotope effect coupled with the high value of the activation enthalpy for deuterium transfer in equation 59 ($\Delta H^{\neq} = 8.3 \pm 0.2$ kcal mol^{-1}, $\Delta S^{\neq} = -8.2 \pm 0.8$ e.u., $\Delta G^{\neq}_{298}$

(174) X = H, Cl, OH (175)

(59)

= 10.7 kcal mol^{-1}), results from the formation of energetically unfavourable zwitterionic intermediates generated in a consecutive intramolecular 1,4-hydrogen transfer processes, and formation of the structure **176** in the case of intermolecular mechanism[101]. The low k_H/k_D value may also be caused by the product-like transition state. The relative contribution of intermolecular and intramolecular mechanisms to the observed deuterium transfer process depends on the temperature of the reaction.

(176)

(59a)

(177)

(60)

2. Dehydrogenation of acenaphthene by quinones

Deuterium isotope effects, k_H/k_D, of 3.49 and 4.14 have been found[102] in the dehydrogenation of a mixture of 1,1,2,2-tetradeuterioacenaphthene and acenaphthene with 2,3-dichloro-5,6-dicyanobenzoquinone (DDQ) and tetrachloro-*o*-benzoquinone (TOQ). These values are considered to be inconsistent with simultaneous cleavage of both C–H bonds in the transition state, but to indicate a considerable cleavage of a single C–H bond in the transition state and to be consistent with a hydride abstraction mechanism. The stepwise carbenium ion nature of this reaction was verified by demonstrating the lack of both intermolecular deuterium scrambling in the intermediate carbenium ions and of 1,2-hydride shifts. Dehydrogenation of *cis*-1,2-dideuterioacenaphthene (**177**) proceeded with 77.7% and 62.9% *cis* elimination using DDQ and TOQ as oxidant respectively. These and other kinetic data corroborate an hypothesis involving an initial ion pair formation, which then collapses to products by *cis* elimination or dissociates into ions (equation 60).

3. Hydrogen isotope effects in the aromatization of 1,4-dihydrobenzene and 1,4-dihydronaphthalene with DDQ and chloranil

There is general agreement that the oxidation of hydroaromatic compounds with quinones proceeds either by direct hydride transfer (equation 61)

$$DH + A \underset{}{\overset{\text{fast}}{\rightleftharpoons}} [DH\ldots A] \xrightarrow{\text{slow}} D^+ + AH^- \qquad (61)$$

or through a sequential electron and hydrogen transfer process (equation 62)

$$DH + A \overset{\text{fast}}{\rightleftharpoons} [DH\ldots A] \overset{\text{fast}}{\rightleftharpoons} [DH^{+\cdot}\ldots A^{-\cdot}] \xrightarrow{\text{slow}} D^+ + AH^- \qquad (62)$$

and that the hydrogen transfer within the charge-transfer complexes [DH...A] is the slow step in the aromatization process. However, Hashish and Hoodless[103] did not find an isotope effect in the aromatization of partially tritiated 1,4-dihydronaphthalene, and therefore proposed a mechanism in which a rate-determining step of oxidation is associated with a slow electron transfer in the charge-transfer complex (equation 63).

$$DH + A \overset{\text{fast}}{\rightleftharpoons} [DH\ldots A] \overset{\text{slow}}{\rightleftharpoons} [DH^{+\cdot}\ldots A^{-\cdot}] \xrightarrow{\text{fast}} D^+ + AH^- \qquad (63)$$

This scheme was incompatible with the rather substantial primary isotope effect measured, for instance[104], for oxidation of 1,4-cyclohexadiene with DDQ. Hence the deuterium isotope effect for the aromatization of 1,4-dihydrobenzene-d_6 and 1,4-dihydrobenzene-d_8 with DDQ in benzene and for the aromatization of 1,4-dihydronaphthalene-d_{10} with DDQ in dichloroethane or with chloranil in 1,2-dichloroethane were reinvestigated[105]. The results collected in Table 3 clearly indicate that the oxidation of both hydroaromatic compounds involves hydrogen transfer in the rate-determining step, but the kinetic data cannot differentiate between scheme 61 and scheme 62. The deuterium isotope effects of similar magnitude, $k_H/k_D = 4.0$ and 6.9, observed in the oxidation of tropilidene-d_8 and 1,2,3-triphenyl-3-deuteriocyclopropene with DDQ in glacial acetic acid serve as models of reactions showing normal or high isotope effects where the carbon–hydrogen bond is broken in the rate-limiting step[106]. Kinetic tritium isotope effects were not observed in the aromatization of partially tritiated 1,4-dihydronaphthalene due to the low sensitivity of the method which is based on the determinations of small rate differences between partially tritiated or deuteriated molecules having several equivalent reaction sites and the unlabelled species. Assuming that the substitution of hydrogen by tritium eliminates completely one C–T reaction site in the labelled 1,4-dihydronaphthalene, the maximum

TABLE 3. Deuterium isotope effects in the aromatization of 1,4-dihydrobenzene and 1,4-dihydronaphthalene with DDQ and chloranil[104, 105]

Compound	Oxidant	Conditions	$10^2 k(\mathrm{M}^{-1}\mathrm{s}^{-1})$	k_H/k_D
1,4-Dihydrobenzene-h_8	DDQ	Benzene, 25°C	2.85	—
1,4-Dihydrobenzene-d_6	DDQ	Benzene, 25°C	1.67	9.8
1,4-Dihydrobenzene-d_8	DDQ	Benzene, 25°C	0.31	9.2
1,4-Dihydronaphthalene-h_{10}	DDQ	1,2-Dichloroethane, 25°C	124.5	—
1,4-Dihydronaphthalene-d_{10}	DDQ	1,2-Dichloroethane, 25°C	12.57	9.9
1,4-Dihydronaphthalene-h_{10}	Chloranil	1,2-Dichloroethane, 120°C	1.235	—
1,4-Dihydronaphthalene-d_{10}	Chloranil	1,2-Dichloroethane, 120°C	0.255	4.84
1,4-Dihydronaphthalene-d_{10}	Chloranil	1,2-Dichloroethane, 25°C		8.0

allowed ratio of aromatization rates of singly tritium-labelled 1,4-dihydronaphthalene and 1,4-dihydronaphthalene-h_{10} will be equal to ca. 1.4 (or to 1.33 neglecting the secondary tritium isotope effect). Hence at low conversion of the labelled substrate, the increase of its specific radioactivity will be very small and might escape detection.

4. *Deuterium isotope effects in the dehydrogenation of alcohols by quinones*

a. *Oxidation of benzyl-α-d alcohol by DDQ*

Kwart and George[107] have determined the isotope partitioning ratio, ipr = PhCHO/PhCDO, in the oxidation of **178** by 2,3-dichloro-5,6-dicyanobenzoquinone (DDQ) (equation 64). The relative amounts of PhCHO and PhCDO in the post-reaction

$$\underset{(\mathbf{178})}{\mathrm{PhC\overset{D}{H}OH}} + \mathrm{DDQ} \begin{cases} \xrightarrow{k_{H'}} \mathrm{PhCDO} + \mathrm{DDQ(H_2)} \\ \xrightarrow{k_D} \mathrm{PhCHO} + \mathrm{DDQ\ (HD)} \end{cases} \tag{64}$$

mixture were estimated by means of the mass spectrometric technique[108] and were found to be 3.94 and temperature independent in the 343–463 K interval. The temperature dependence of the k_H/k_D ratios is usually utilized for assessing the structure of the activated complexes in hydrogen transfer processes. The 'ipr' corresponding to the $k_{H'}/k_D$ ratio can be written as the quotient of the temperature-dependent k_H/k_D and $k_H/k_{H'}$ ratios, or as a quotient of the $k_{H'}/k_{D'}$ and $k_D/k_{D'}$ ratios derived from equations 65 and 66.

$$\mathrm{PhCH_2OH + DDQ} \xrightarrow{2k_H} \mathrm{PhCHO + DDQ\ H_2} \tag{65}$$

$$\mathrm{PhCD_2OH + DDQ} \xrightarrow{2k_{D'}} \mathrm{PhCDO + DDQ\ HD} \tag{66}$$

The primary deuterium isotope effects, k_H/k_D, calculated within the harmonic one bond or symmetric transition complex approximation by using a value of $\omega_{C\text{-}H} = 2895\ \mathrm{cm}^{-1}$ are:

Temp. (K)	343.16	363.16	383.16	403.16	423.16	443.16	463.16
k_H/k_D	5.222	4.768	4.395	4.084	3.822	3.599	3.406

Theoretical calculations of the secondary kinetic isotope effect, $(k_H/k_{H'})$, require the use of at least the four-centre transition state model. Their temperature dependence might be unlike that of the primary k_H/k_D ratio. Consequently the experimentally determined ipr is

(67)

temperature independent. (The temperature dependences of the 'heavy atom' ^{13}C isotope effects in the carbon–carbon bond rupture are discussed in Ref. 109.) Kwart and George suggested that the oxidation of benzyl-α-d alcohol proceeds according to equation 67 via a bent transition state. The observed temperature independence of the $k_{H'}/k_D$ ratio was ascribed to an angular non-linear transfer of the hydrogen in the rate-determining step. A linear symmetric transition state should give a temperature-dependent deuterium isotope effect. Unfortunately the $k_H/k_{D'}$ and $2k_H/(k_{H'}+k_D)$ ratios have not been determined over 120°C temperature range studied. Hence a detailed discussion concerning the structure of the transition state in the benzyl alcohol oxidation with DDQ should be postponed until more experimental and theoretical data pertaining to this reaction are accumulated. The results obtained so far indicate only that the intramolecular hydrogen transfer taking place within adduct **179** is the rate-determining step of the process.

b. Dehydrogenation of deuteriated 1-*phenyl*-1-*propanols*

Deuterium isotope effects in the dehydrogenation of alcohols by quinones have also been investigated by Ohki *et al.*[110]. The initial oxidation rates of PhCH(OH)Et (**180**), PhCH(OD)Et, PhCD(OH)Et and PhCD(OD)Et by 2,3-dichloro-5,6-dicyano-*p*-benzoquinone (DDQ) at 60 °C give relative rates of 8.9, 9.1, 1.0 and 1, respectively. The large primary isotope effect of the α-hydrogen and the undetectable isotope effect of a deuterium atom of the OH group imply that the C_α–H bond cleavage is taking place in the rate-determining step while the rupture of the O–H bond is of secondary importance. Additional studies of solvent and substituent effects, as well as the effect of additives on the yield of the propiophenone have suggested that the hydrogen transfer proceeds via formation of a complex which precedes the rate-determining step, and that the complex and/or the transition state of the rate-determining step are solvated and considerably polarized. A free radical mechanism which is inconsistent with the large negative activation entropy was excluded. The most probable mechanism and the corresponding rate equations are given in equations 68–70,

$$\mathrm{DDQ} + \mathrm{A} \overset{K}{\rightleftharpoons} \text{complex} \overset{k}{\rightarrow} \text{products} \quad (68)$$

$$\text{Rate} = k_{obs}[\mathrm{DDQ}][\mathrm{A}] = k[\text{Complex}] = kK[\mathrm{DDQ}][\mathrm{A}] \quad (69)$$

complex of equation 68.

(70)

where [A] is the concentration of 1-phenyl-1-propanol, K is the equilibrium constant between the reactants and the charge-transfer complex, k is the rate constant of the rate-limiting step and k_{obs} is the observed second order rate constant. The mechanistic scheme (equation 70) is also supported by the fairly large negative ρ value of -2.7 indicating the formation of a positively charged transition state. An electron-deficient carbon centre in the transition state has been found similarly in the oxidation of substituted benzyl alcohols and α,α-dideuteriobenzyl alcohol by chloramine-T in acid solution[111]. The suggested scheme does not rule out the possibility of solvent participation in a subsequent rapid proton transfer step.

c. Oxidation of allyl alcohols with DDQ

The deuterated secondary allyl alcohol, 3α-deuterio-3β-hydroxy-Δ^4 steroid, **181**, underwent oxidation with DDQ at 27 °C in *t*-butyl alcohol at a five-fold slower rate than the 3α-hydrogen compound. Consequently the 3-C–H bond is cleaved in the rate-determining step, and the reaction proceeds via a slow hydride transfer followed by a rapid proton loss (equation 71)[112].

(**181**)

R = 0.65 atoms H
= 0.35 atoms D

(71)

5. Deuterium isotope effect study of the dehydrogenation of alcohols with 7,7,8,8-tetracyanoquinodimethane (TCNQ)

TCNQ (**182**), which is structurally similar to quinones readily dehydrogenates benzyl-type alcohols and hydroaromatic compounds while being reduced to *p*-benzenedimalononitrile (**183**) (equation 72). The reaction proceeds via the intermediacy of a

$$\text{(NC)}_2\text{C}{=}\text{C}_6\text{H}_4{=}\text{C(CN)}_2 \; (\mathbf{182}) + 2\text{H} \longrightarrow \text{(NC)}_2\text{CH}{-}\text{C}_6\text{H}_4{-}\text{CH(CN)}_2 \; (\mathbf{183}) \quad (72)$$

carbenium ion when 1,2-dihydrobenzenes are used as the hydrogen donors. Indeed when 1,2-dihydro-1,1-dimethylnaphthalene was used as the hydrogen donor, 1,2-dimethylnaphthalene was obtained, i.e. the oxidation was accompanied by methyl group migration. The dehydrogenation of hydrogen donors 'HD' by TCNQ was suggested to occur via the intermediacy of a charge-transfer (CT) complex (equations 73 and 74).

$$\text{TCNQ} + \text{HD} \overset{K}{\rightleftharpoons} [\text{CT complex}] \overset{k}{\rightarrow} \text{products} \quad (73)$$

$$\text{rate} = k_{obs}[\text{TCNQ}]\cdot[\text{HD}] = k[\text{CT complex}] = kK[\text{TCNQ}]\cdot[\text{HD}] \quad (74)$$

The rate-determining step and the nature of the hydrogen transfer step in the dehydrogenation by TCNQ were investigated by studying the deuterium isotope effects[113] in the dehydrogenation of isotopic 1-phenylpropanols to propiophenones. The relative reaction rates of PhCH(OH)Et, PhCH(OD)Et, PhCD(OH)Et and PhCD(OD)Et with TCNQ at 140 °C in dioxane were: 4.0, 4.0, 1.0 and 1, respectively. These data indicate that the C–H bond rupture is of primary importance in the rate-determining step whereas the cleavage of the O–H bond is of secondary importance in this hydrogen transfer thermal process. The experimental k_H/k_D value of 4.0 is close to the calculated value of 3.95 at 140°C, neglecting tunnelling and taking $\omega_{C\text{-}H} = 2985\ cm^{-1}$ for the C–H frequency. The deuterium kinetic isotope effect combined with solvent and substituent effects on the reaction rate suggest a similar two-step ionic mechanism for the thermal dehydrogenation of alcohols by TCNQ or tetracyanoethylene (TCNE) and of 1,2-dihydronaphthalene by quinones. The dehydrogenation of 1-arylpropanols by TCNQ involves a rate-limiting hydride transfer with carbenium ion formation, followed by proton loss and ketone formation in subsequent rapid steps (equation 75)[113].

Et, Ph–C–H, O–H; NC CN C, C NC CN; slow; Ph Et C+ OH; + ; H NC–C–CN, C NC CN; fast; Ph, Et C=O + (**183**)

(75)

6. *Deuterium isotope effects in the oxidation of* N*-methylacridan by quinones*

Primary and secondary isotope effects in the oxidation of *N*-methylacridan (**184**) to *N*-methylacridinium ion (**185**) have been determined[114–116a] by studying spectrophotometrically the kinetics of the reaction of (**184**)-9,9-h_2, (**184**)-9,9-hd and (**184**)-9,9-d_2 with *p*-benzoquinone (BQ), TCNQ, *p*-chloranil (CA), TCNE and 2,3-dicyano-1,4-benzoquinone (DCBQ) in acetonitrile or in acetonitrile–water mixture at 25°C (equation 76). Tracer

(D)H H(D), N, Me (**184**); π acceptors in acetonitrile; H(D), N+, Me (**185**) + AH^- (76)

studies have shown that in the course of the reaction hydrogen is transferred to an oxygen of the oxidants and no intermediate of type **186** is formed. The hydroquinone which was isolated from the post-reaction mixture did not contain excess of deuterium. The product

O= =O, H H

(**186**)

(**185**) formed in the oxidation of (**184**)-9,9-d_2 by BQ in 90% acetonitrile did not contain ^1H in the 9 position and no exchange of the 9-hydrogens with the solvent during the oxidation was found. However, indirect kinetic evidence and direct NMR monitoring of solutions containing isotopically labelled **184** and **185** showed scrambling of the deuterium between unreacted **184** and **185** during the oxidation which lowers the observed ipr ratio. When the BQ to **184** ratio increased from 1:1 to 1000:1 the measured ipr in 90% acetonitrile increased from 2.9 to the upper limit of 9.4. This deuterium exchange caused visible discrepancies between the ipr and p/s values (equation 78) only in the case of relatively slow oxidation with BQ in 75% and 90% acetonitrile. The oxidation rate constants for the isotopomeric *N*-methylacridans, the derived primary and secondary deuterium isotope effects, including isotope partitioning ratio 'ipr = $k_{H'}/k_D$', determined by mass spectral analysis of the isolated *N*-methylacridinium chloride, are presented in Table 4. The notations used in Table 4 are defined by equations (77 a–c) and (78).

$$(\mathbf{184})\text{-9,9-h}_2 \xrightarrow{2k_H} (\mathbf{185})\text{-h} \qquad k_{HH} = 2k_H \tag{77a}$$

$$(\mathbf{184})\text{-9,9-hd} \begin{cases} \xrightarrow{k_{H'}} (\mathbf{185})\text{-d} \\ \xrightarrow{k_D} (\mathbf{185})\text{-h} \end{cases} \qquad k_{HD} = (k_{H'} + k_D) \tag{77b}$$

$$(\mathbf{184})\text{-9,9-d}_2 \xrightarrow{2k_{D'}} (\mathbf{185})\text{-d} \qquad k_{DD} = 2k_{D'} \tag{77c}$$

$$p = k_H/k_D = k_{H'}/k_{D'} \qquad k_{HD}/k_{DD} = (p+s)/2$$

$$s = k_H/k_{H'} = k_D/k_{D'} \qquad \text{ipr} = k_{H'}/k_D = \frac{p}{s} = \frac{[\mathbf{185\text{-}d}]}{[\mathbf{185\text{-}h}]} \tag{78}$$

$$k_{HH}/k_{DD} = p \cdot s$$

The kinetically determined p/s values are given in Table 4 for comparison. It is seen that with the slow reacting acceptor BQ it was practically impossible to suppress completely the scrambling in 90% and 100% acetonitrile even at a 1000:1 ratio of BQ to **184**. The data in the third column of Table 4 indicate that the experimental k_{HH}/k_{HD} ratios are not very sensitive to the oxidizing ability of the π acceptors. The lack of useful correlations between the k_{HH}/k_{HD} values and the redox potentials of the oxidant is probably due to the low accuracy in the determinations of the k_{HD} values. The k_{HH}/k_{DD} ratios are much more useful in that respect and are comparable with the calculated primary kinetic isotope effects given in the seventh column. They suggest a contribution of quantum mechanical tunnelling to the k_H/k_D isotope effect in the oxidation with benzoquinone but kinetic studies at other temperatures should be carried out before drawing further conclusions concerning the energetic profiles in these oxidations. The '*p*' values for reaction with TCNQ and CA are quite close to the theoretical isotope effect of 6.7 at 25 °C for a single carbon–hydrogen bond rupture, $k_{C\text{-}H}/k_{C\text{-}D}$, neglecting tunnelling. The last two values of 4.75 and 5.36 found for oxidations with TCNE and DCBQ in acetonitrile are lower than the classical value of 7 expected for hydrogen atom transfer with a symmetrical transition state structure. Data given in the sixth column show quite large scatter, caused probably by experimental uncertainties. Nevertheless the average $k_H/k_{H'}$ ratio of 1.097 ± 0.060 is very useful in interpreting these isotopic kinetic experiments in which only k_{HH}/k_{HD} and especially k_{HH}/k_{HT} ratios have been measured, since determination of k_{HH}/k_{TT} ratios require working with 100% pure R_2CT_2. A quite good agreement between ipr values based on product analysis and the p/s ratios derived from kinetic measurements is noticeable. All the deuterium isotope effect data presented in Table 4 call for a mechanism in which the carbon–hydrogen bond is broken in the rate-determining step. Spectroscopic and chemical

TABLE 4. Rates and deuterium kinetic isotope effects[a] in the oxidation of isotopic *N*-methylacridan molecules with various acceptors in acetonitrile (AN) at 25 °C[115]

Acceptor and solvent	k_{HH}($M^{-1}s^{-1}$)	k_{HH}/k_{HD}	k_{HH}/k_{DD}	k_{HD}/k_{DD}	$k_H/k_{H'}$	k_H/k_D	$k_{H'}/k_D$	ipr
BQ in 60% AN[b]	1.42×10^{-2}	1.919	9.595	5.00	1.075	8.925	8.302	8.14 ± 1.4
BQ in 75% AN[b]	5.51×10^{-3}	1.900	11.479	6.042	1.039	11.044	10.626	9.1 ± 1.9
BQ in 90% AN[b]	1.86×10^{-3}	1.875	12.00	6.40	1.019	11.781	11.567	9.4 ± 2
BQ in AN	5.79×10^{-4}	1.956	14.088	7.202	1.055	13.349	12.648	—
TCNQ in AN	2.32×10^{-1}	1.966	7.227	3.676	1.169	6.183	5.290	—
TCNQ in AN[b]	2.33×10^{-1}	2.009	7.373	3.671	1.201	6.141	5.115	—
CA in AN	1.17×10	1.980	8.731	4.410	1.136	7.685	6.763	6.2 ± 0.7
TCNE in AN	1.01×10^{2}	1781	5.206	2.923	1.096	4.750	4.334	4.5 ± 1.4
DCBQ in AN	8.67×10^{3}	1.806	5.819	3.221	1.086	5.357	4.931	4.7 ± 0.1
		1.910 ± 0.078			1.097 ± 0.060			

[a] The notations are those given in equations 77–78.
[b] Containing 0.01 M AcOH.

evidence indicates that the slow hydride transfer in the oxidation is preceded by a reversible formation of a charge-transfer (CT) complex between the hydride donor DA and the hydride acceptor A. Some fractionation of isotopic hydrogen in the CT complex formation in the case of (**184**) -9,9-hd is to be expected. The complexation constants K differ for the different isotopomeric methylacridans. It is proposed that the transition state of the oxidation has a structure intermediate between the donor–acceptor face-to-face orientation (**187**) and the products of electron (**188**) or hydride (**189**) transfer. The magnitude of deuterium primary isotope effect and the solvent effect favour a one-step hydride transfer mechanism although they do not exclude a sequential electron–hydrogen atom transfer mechanism.

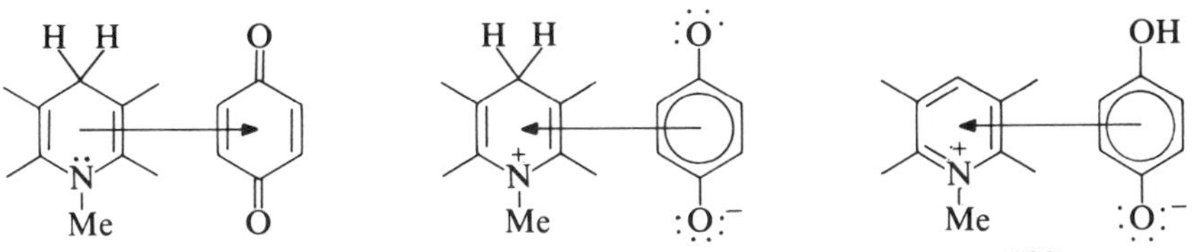

(**187**) (**188**) (**189**)

In connection with the problem of hydride transfer discussed above it is noteworthy that good agreement was recently found[116b] between the kinetic deuterium isotope effect k_H/k_D of 4.4–4.5 in the oxidation of 1-benzyl-1,4-dihydronicotinamide (BNAH, BNAH-4,4-d_2 and BNAH-4-d_1 systems) with 10-methyl- and 10-methyl-9-phenylacridinium ions (equation 79) and the mass spectrometrically determined isotope partitioning ratio in the $AcPh^+$/BNAH system of 4.5 in acetonitrile and 4.7 in 9:1 v/v acetonitrile–water.

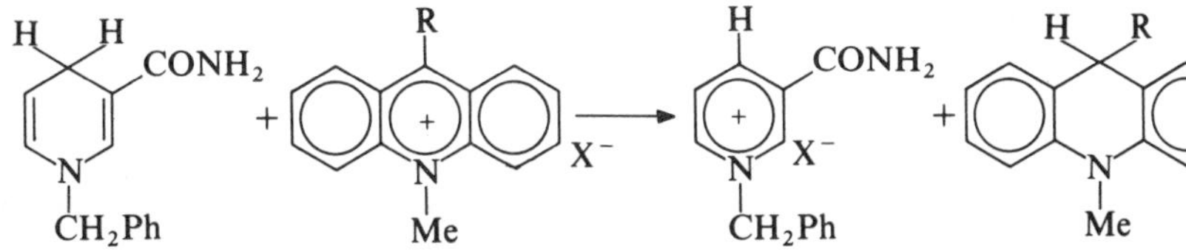

(**190**) (BNAH) + (AcR^+) X^- ⟶ (BNA^+) X^- + (AcRH) (79)

7. Hydride transfer from 1-benzyl-1,4-dihydronicotinamide to p-*benzoquinone derivatives*

In the course of studies of the mechanism of the irreversible hydride transfer from 1-benzyl-1,4-dihydronicotinamide (**190**) (an NADH model compound) to a series of *p*-benzoquinones 'Q', the primary kinetic isotope effects, k_H/k_D have been determined at 298 K in acetonitrile and correlated with redox potentials of quinones $E^\circ(Q/Q^{-\cdot})$. The results are listed in Table 5. The rate constants k_{HH} of BNAH (**190**) and the rate constants k_{DD} of BNAH-4,4-d_2 were used for calculation of k_H/k_D ratios with $\pm 5\%$ error. It has been assumed that the secondary α-deuterium isotope effects are unity, since a secondary α-deuterium kinetic isotope effect value of 1.0 ± 0.1 has been deduced from the rate constants k_{HH} of BNAH, k_{HD} of BNAH-4-d_1 and k_{DD} of BNAH-4,4-d_2 in their reactions with *p*-chloranil, *p*-bromanil and 2,6-dichloro-*p*-benzoquinone. The majority of the k_H/k_D values (of entries 3–11 in Table 5) are located in the range 5.2–6.2. The maximum value of 6.2 for the hydride-transfer reaction with *p*-benzoquinone is again quite close to the theoretical

TABLE 5. Redox potentials, $E^\circ(Q/Q^{-\cdot})$ of Q, rate constants k and primary deuterium kinetic isotope effects, k_H/k_D, for the hydride transfer from BNAH to *p*-benzoquinone (*p*-BQ) derivatives in MeCN at 298 °K[117]

p-Benzoquinone	$E^\circ(Q/Q^{-\cdot})$	$k(M^{-1}s^{-1})$	k_H/k_D
2,3-Dichloro-5,6-dicyano-*p*-BQ	0.51	8.4×10^6	1.5
2,3-Dicyano-*p*-BQ	0.28	7.2×10^5	2.6
p-Chloranil	0.01	1.0×10^3	5.3
p-Bromanil	0	7.3×10^2	5.2
2,6-Dichloro-*p*-BQ	−0.18	7.5×10	5.6
2,5-Dichloro-*p*-BQ	−0.18	5.0×10	5.5
Chloro-*p*-BQ	−0.34	7.6	6.1
p-BQ	−0.50	1.3×10^{-2}	6.2
Methyl-*p*-BQ	−0.58	2.3×10^{-3}	5.9
2,6-Dimethyl-*p*-BQ	−0.67	8.4×10^{-5}	5.6
Trimethyl-*p*-BQ	−0.75	1.3×10^{-5}	5.6
Tetramethyl-*p*-BQ	−0.84	Very slow	—

value of 6.7 calculated for C–H/C–D bond cleavage, neglecting tunnelling. A plot of the k_H/k_D ratios versus the redox potentials of Q, $E^\circ(Q/Q^{-\cdot})$ gives a Bell-shaped dependence with a Westheimer maximum. (There is a linear correlation between pK_a of semiquinone radicals and the redox potential $E^\circ(Q/Q^{-\cdot})$. It would be interesting to confirm the low k_H/k_D values of 1.5 and 2.6 obtained in the fast reactions with 2,3-dichloro-5,6-dicyano-*p*-benzoquinone and 2,3-dicyano-*p*-benzoquinone by determining the ipr ratios in both cases using BNAH-4-d_1. Tritium isotope effects would also be useful as an additional test of the Bell-shape dependence obtained in this hydride transfer reaction for the first time. The authors[117] have shown that transient CT complexes are formed in the course of the hydride transfer reaction studied. Their absorption spectra, with maxima in the 670–735 nm range, were similar to the reflectance spectra of the 1:1 complexes of BNAH with *p*-chloranil, *p*-bromanil and 2,6-dichloro-*p*-benzoquinone isolated from benzene or toluene solutions under nitrogen atmosphere. It is proposed that the reaction proceeds according to the 'sequential electron–proton–electron transfer mechanism'. The radical ion pair $[BNAH^{+\cdot} \ldots Q^{-\cdot}]$ is formed inside the CT complex and thereby no free radical species are involved in the process (equation 80).

$$\begin{aligned} &BNAH + Q \overset{K_{CT}}{\rightleftharpoons} [BNAH \ldots Q] \underset{k_{-1}}{\overset{k_1}{\rightleftharpoons}} [BNAH^{+\cdot} \ldots Q^{-\cdot}] \overset{k_H}{\rightarrow} \\ &[BNA^{\cdot} \ldots QH^{\cdot}] \overset{fast}{\rightarrow} BNA^{+} + QH^{-} \\ &QH^{-} + Q \overset{fast}{\rightarrow} QH^{\cdot} + Q^{-\cdot} \\ &2QH^{\cdot} \overset{fast}{\rightarrow} QH_2 + Q \end{aligned} \tag{80}$$

Reaction series 80 was used to correlate the k and k_H/k_D values with E° by the Marcus theory. This multistep mechanism is practically equivalent to a one-step hydride transfer from the CT complex since the formation constant of the 'radical ion pair' (k_1/k_{-1}) is much smaller than the formation constant of the encounter complex K_{CT} and it is impossible to detect the former by physical or chemical methods.

8. Oxidation of phenylhydrazines by quinones

Deuterium isotope effects have been used to investigate the mechanism of oxidation of phenylhydrazines by quinones[118–121]. The deuteriated phenylhydrazine, $PhNDND_2$, obtained by three-fold exchange of hydrazine with D_2O, contained > 95% deuterium in the hydrazine group. The $k_{(p\text{-}NO_2C_6H_4NHNH_2)}/k_{(p\text{-}NO_2C_6H_4NDND_2)}$ ratio for the oxidation of *p*-nitrophenylhydrazine with 1,4-benzoquinone in acetonitrile was 2.9 at 40°C. The $k_{(PhNHNH_2)}/k_{(PhNDND_2)}$ ratio for oxidation of phenylhydrazine with duroquinone in MeCN was 3.0 at 74°C, but in the reaction of hydrazine with 2,6-di-*t*-butyl-1,4-benzoquinone, 1,4-benzoquinone and 2,3-dichloro-5,6-dicyano-1,4-benzoquinone the k_H/k_D ratios were much smaller, being 1.85 at 60°C, 1.30 at 29°C and 1.10 at 20°C, respectively. These effects were interpreted as primary isotope effects, corresponding to a mechanism which involves initially charge transfer complexation followed by a rate-determining hydrogen transfer with cleavage of the N–H bond (equation 81). A sandwich-type structure (**191**) in which the

$$\text{Ar-NHNH}_2 + \text{(X}_2\text{Y}_2\text{-1,4-benzoquinone)} \underset{k_{-1}}{\overset{k_1}{\rightleftharpoons}} [\text{CT complex}] \xrightarrow{k_2} [\text{Ar-}\dot{\text{N}}(\text{NH}_2)\ldots\text{HO-C}_6\text{X}_2\text{Y}_2\text{-O}^{\bullet}] \rightarrow [\text{Ar-}\dot{\text{N}}\text{-NH}_2 \cdots \dot{\text{O}}\text{-C}_6\text{X}_2\text{Y}_2\text{-OH}] \longrightarrow$$

$$\text{ArN}=\text{NH} + p\text{-HOC}_6\text{X}_2\text{Y}_2\text{OH};$$

$$\text{ArN}=\text{NH} + \text{O}=\text{C}_6\text{X}_2\text{Y}_2=\text{O} \rightleftharpoons [\text{ArN}=\text{NH}\ldots\text{O}=\text{C}_6\text{X}_2\text{Y}_2=\text{O}] \longrightarrow \tag{81}$$

$$[\text{ArN}=\dot{\text{N}}\ldots\dot{\text{O}}\text{C}_6\text{X}_2\text{Y}_2\text{OH}] \rightarrow \text{N}_2 + \dot{\text{A}}\text{r}\ldots\dot{\text{O}}\text{C}_6\text{X}_2\text{Y}_2\text{OH} \xrightarrow{\text{ArNHNH}_2} 2\text{ArH} + \text{N}_2 + p\text{-HOC}_6\text{X}_2\text{Y}_2\text{OH}$$

π-electron ring orbitals of the electron donor hydrazine interact with the ring π electrons of the electron acceptor quinone was proposed for the CT complex. Application of reaction

Z, NH₂, N, H, X, X, O, O, Y, Y

(**191**)

sequence 81 for the reaction with duroquinone gave a calculated $k_{2(H)}/k_{2(D)}$ of 4.4 when the isotope effect on the equilibrium k_1/k_{-1} was neglected. It would be interesting to support sequence 81 by studying the ^{15}N kinetic isotope effect corresponding to the nitrogen–hydrogen bond rupture.

9. *Hydrogen transfer from metal hydrides to quinones*

In completing the discussion of hydrogen transfer reactions to quinones we should mention also that in the spontaneous reduction of tetrachloro-, chloro- and the unsubstituted *p*-benzoquinone and TCNE with metal hydrides of the type HMR_3 (M = Sn, Ge, Si and R = Me, Et, *n*-Bu, *i*-Pr and Ph; equation 82), there is negligible ($\pm 10\%$, if any) deuterium isotope effect, defined[86] as $[(k_H/k_D)\text{-}1]\%$. In the reduction of TCNE with *n*-Bu_3SnH/D[122] $k_H/k_D = 1.0 \pm 0.1$ in cyclohexane, 1.3 ± 0.1 in methylene chloride, 1.4 ± 0.1 in ether and 1.5 ± 0.1 in acetonitrile. These very small isotope effects might represent

$$HMR_3 + O{=}C_6H_4{=}O \longrightarrow HO{-}C_6H_4{-}OMR_3 \qquad (82)$$

hydride transfers from the metal hydrides to TCNE proceeding either via very early or very late transition states. However, the slightly different from unity k_H/k_D ratios do not depend on the reduction potential E° in the range -0.8 to $+0.5$ volts of the various acceptors studied, contrary to the hydride transfer from *N*-methylacridan. In the oxidation of *N*-methylacridan both the second order rate constants and k_H/k_D values depend largely on the solvent polarity. In contrast in the reactions of *n*-Bu_3SnH only the second order rate constants increase by a factor of 10^3 by changing the solvent from cyclohexane to acetonitrile whereas the k_H/k_D ratio does not increase significantly. These differences are interpreted[122] as ruling out a radical chain and a hydride transfer reduction mechanisms with metal hydrides and as supporting the charge transfer mechanism in which the hydrogen transfer occurs after the rate-limiting electron transfer within the CT complex. The found k_H/k_D values of the order of 1.1–1.5, if real, are therefore secondary effects related to the electron transfer step (cf. Section II.C.11).

10. *The effect of pressure on the kinetic isotope effects*

A large primary kinetic deuterium isotope effect ($k_H/k_D = 12.3$ at 25 °C) was observed by Lewis and coworkers[123,124] in the reaction of the colourless leucocrystal violet (LCV) with tetrachloroquinone (CA) in acetonitrile, which gives a cationic purple dye (CV^+) (equation 83). This reaction has been reinvestigated recently by Isaacs and coworkers[125] and by Nishimura and Motoyama[126] at pressures ranging from 0 to 2 kbar.

$$\underset{(LCV)}{Ar_3CH(D)} + \text{chloranil } (C_6Cl_4O_2) \longrightarrow CV^+ + \underset{(192)}{{}^-O{-}C_6Cl_4{-}OH(D)} \qquad (83)$$

$Ar = p\text{-}Me_2NC_6H_4$

The results obtained by Isaacs and confirmed by Nishimura[126] are presented in Tables 6 and 7. They show that the deuterium isotope effect diminishes with increasing the pressure

TABLE 6. Deuterium isotope effect in the reaction of leucocrystal violet with chloranil[125]

Pressure bar	T(°) 21	29	40	29
	k_H/k_D in			
		Acetonitrile		Isobutyronitrile
0	12.3	11.2	9.1	11.5
500	11.6	9.5	8.6	10.3
1000	10.8	8.5	8.2	9.0
1500	10.1	8.2	7.8	8.2
2000	9.3	8.0		8.0

TABLE 7. Activation parameters for the reaction of leucocrystal violet H, D with chloranil in acetonitrile[125]

	LCV (H)	LCV (D)
E_A (kcal mol^{-1})		
1 bar	8.7 ± 0.5	10.3 ± 0.5
490	8.5	11.0
985	8.3	10.6
log A		
1 bar	7.2	7.3
490	7.3	8.1
985	7.3	8.0
$\Delta V^{\ddagger}_{0}$ (cm^3 mol^{-1})^a in MeCN	-25 ± 2	-35 ± 2
$\Delta V^{\ddagger}_{0}$ (cm^3 mol^{-1})^a in *i*-PrCN	-22	-29

^{a} Calculated from: $-RT\partial \ln k/\partial p = (V_{reagents} - V^{\ddagger}) = \Delta V^{\ddagger}$.

from a k_H/k_D value of 11–12 at atmospheric pressure to a value near 8 at 2 kbar. The precision of measurements was sufficient to observe a higher activation energy for the deuterium reaction but insufficient to estimate the effect of pressure upon the barrier dimensions. The very large primary deuterium isotope effect at low pressure, which is characteristic for hydrogen transfer processes occurring within highly hindered cage-like structures, is considered as a manifestation of a quantum mechanical tunnelling. In such cases the solvation of the hydrogen is poor and the effective masses of the hydrogen isotopes are close to their atomic masses. The pressure effect upon the k_H/k_D values is ascribed to increased solvation of the hydride ion and consequently an increase in its effective mass along the reaction coordinate with pressure which reduces the contribution of the quantum tunnelling. This view is strengthened by the pressure effect studies in the bulkier isobutyronitrile solvent, where the decrease of the deuterium isotope effect with pressure increase was found to be slightly less steep than in acetonitrile. In both solvents at a pressure of 2 kbar the value of k_H/k_D is around 8, approaching the value characteristic for processes caused by zero point energy differences between C–H and C–D bonds. Nishimura and Motoyama[126] proposed the two alternative reactions (equations 84 and 85) for the oxidation of LCV with chloranil (CA) which involve the participation of intermediate outer (X_1) and inner (X_2) charge transfer complexes and a polarized partial bond formation of the type $C^{\delta+}\cdots H\cdots{}^{\delta-}O$ in the transition state in equation 85.

$$\mathrm{LCV + CA \rightleftharpoons (X_1) \rightleftharpoons (X_2)}$$

$$\mathrm{X_1 \text{ or } X_2 \xrightarrow{\text{slow}} CV^+ + (\mathbf{192})} \tag{84}$$

$$\mathrm{LCV + CA \rightleftharpoons (X_1) \rightleftharpoons (X_2)}$$

$$\mathrm{\searrow^{\text{slow}}\ CV^+ + (\mathbf{192})} \tag{85}$$

The complexes X_1 and X_2 might have structures **193** and **194**, respectively. The authors are inclined to conclude that the observed large kinetic pressure effect is more consistent with reaction 85 for the oxidation of LCV with CA, but the detailed reaction mechanism

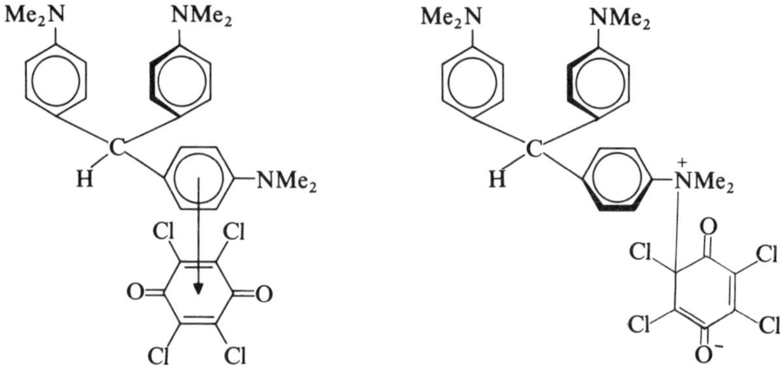

(193) **(194)**

requires further investigations. The effect of pressure on tunnelling was also investigated in the reaction of 2,4,6-trinitrotoluene with 1,8-diazabicyclo[5.4.0]undec-7-ene in aprotic solvents[127, 128a]. It is suggested[128a] that under a moderate pressure of a few kilobars the reduction of the 'free volume' of the solvent and the increase of the number of solvent molecules in the proximity of the reaction centre is the main pressure effect.

11. Concluding remarks

The experimental results presented in Section II.C demonstrate clearly that the hydrogen transfer step in the oxidation of organic compounds with quinones is the rate-determining one. The nature of the organic processes presented in this section was not always fully clarified. Many oxidations with quinones have been carried out at one temperature only and additional time-consuming kinetic determinations are needed to establish without ambiguity the degree of tunnelling and the structures of the transition states in the hydrogen transfer processes. Further studies of the pressure effect upon the primary kinetic deuterium isotope effect will be extremely helpful in that respect also. Heavy atom kinetic isotope effect studies of the hydrogen transfer process should provide an additional insight into the skeletal changes upon activation of the reacting molecules. A unique departure from the common mechanism of reduction of quinones with organic reductants was found with metal reductants. However, the physicochemical state of the metal hydride molecules in the solvent used and the nature of the 'spontaneous reduction' should be clarified before the decisive rejection of the hydrogen transfer step in the reduction process as the rate-determining, and formulating an electron transfer process within the CT complex as the slowest one in the metal hydride reduction. The absolute magnitude of the deuterium isotope effect cannot be used for testing the heterolytic nature of the transition state. Kinetic isotope effects depend on the magnitudes of the force constants at the reaction centre (hydrogen) both in the precursors and in the transition states. The force constants corresponding to the C–H and N–H bonds differ from the force constants corresponding to the metal–hydrogen bonds in the metal hydrides used in studies of isotope effects in the reduction. The low values of deuterium isotope effects found in the latter reactions do not imply that electron transfer is the rate-determining step. Other possibilities related to the formation of new bonds with the central metal atom should also be investigated by using heavy atom isotope effects[128b].

III. BIOCHEMICAL SYNTHESES AND USES OF LABELLED QUINONES

Isotopically labelled quinonoid compounds have continued to be a subject of very active biochemical investigations and applications during the last 10–15 years. Several detailed monographs and reviews covering this field have already appeared. Mitchell[129] reviewed the radioactive drugs used in cancer 'radiochemotherapy' and 'immunoradiotherapy' including the applications of 2-methyl-1,4-naphthoquinol(bisdisodium phosphate), a radioactive drug of high molar tritium specific activity (abbreviated as T-MNDP-Synkavit), developed in Cambridge, England and used with some success in the treatment of inoperable cases of carcinoma. Biosyntheses of quinones, of the vitamin K and other natural naphthoquinones were reviewed by Bentley[130, 131]. Galimov[133] covered in his monograph the various problems of isotopic fractionations in 'nature' with a particular emphasis on the distribution of ^{13}C and ^{18}O isotopes in natural biological systems and in biochemical experiments conducted *in vitro*. He also dealt with the practical uses of isotope effects in unravelling the problem of the genesis of organic matter in nature and in the exploration of geological resources. In this section we review the biochemical studies in which labelled benzoquinones, naphthoquinones and anthraquinones have been isolated and identified by using radiochromatographic techniques. This is preceded by an introductory mathematical treatment of the kinetic isotope method developed by Neiman and Gal[133].

A. A General Treatment of the Applications of Isotopic Tracers in Biochemical Studies

Isotopes are extremely helpful in the elucidation of complex chemical and biochemical reactions, both at the preliminary reconnaissance stage of investigation characteristic for the present biochemical studies with quinones and in advanced physical studies of biochemical processes. In this section a brief presentation of the general relationships concerning intermediate formation and disappearance, derived for simple chemical sequences of reactions and applied chiefly in oxidation and decomposition studies[133a] is given.

1. Determination of the formation and consumption rates of an intermediate

Let us consider a sequence of consecutive chemical transformations comprising the intermediate 'X' to be investigated (equation 86).

$$A \rightarrow B \rightarrow C \xrightarrow{\omega_1} \overset{\alpha}{X} \xrightarrow{\omega_2} D \rightarrow E \tag{86}$$

We further assume that spectroscopic and analytical methods detect the occurrence of the intermediate in the reacting system and enable determination of its concentration changes with time. The observed change of the concentration of X is the difference between its unknown rate of formation ω_1 and its unknown rate of consumption ω_2 (equation 87).

$$d[X]/dt = \omega_1 - \omega_2 \tag{87}$$

By introducing into the reacting system the labelled intermediate X and by observing the simultaneous changes of [X] and of the specific radioactivity α of X (defined by the quotient $\alpha = I/[X]$, where I is total radioactivity of the intermediate X, and [X] is total concentration of the intermediate), we obtain a second equation which enables to find two unknowns ω_1 and ω_2 of equation 87. A closer look at equation (86) shows that the rate of change of the specific activity of X* depends on the rate of formation ω_1 and on the concentration term [X] according to equation 88,

$$\frac{d\alpha}{dt} = -\frac{\alpha\omega_1}{[X]} \tag{88}$$

if small kinetic isotope effects for elements heavier than the hydrogen are neglected (Hevesy approximation). Equation 88 gives directly equation 89 for ω_1.

$$\omega_1 = -[X]\frac{d\ln(\alpha)}{dt} \tag{89}$$

Substitution of equation 89 into equation 87 leads to equation 90 for the rate of consumption of the intermediate:

$$\omega_2 = -[X]\frac{d\ln(\alpha)}{dt} - \frac{d[X]}{dt} \tag{90}$$

The rates of formation and consumption of an intermediate Y can also be determined by labelling the precursor X of the intermediate. The reaction sequence in this case is given in equation 91.

$$A \rightarrow B \rightarrow C \xrightarrow{\omega_1} \overset{\alpha}{X} \xrightarrow{\omega_2} \overset{\beta}{Y} \xrightarrow{\omega_3} D \rightarrow \tag{91}$$

The change of the specific radioactivity α of X is given as before by equation 88. The total activity, I_y, of the intermediate Y is a product of its specific radioactivity, β, and its concentration [Y] (equation 92).

$$I_y = \beta[Y] \tag{92}$$

Differentiation of equation 92 gives equation 93.

$$\frac{dI_y}{dt} = \beta\frac{d[Y]}{dt} + [Y]\frac{d\beta}{dt} \tag{93}$$

The left side of equation 93 is also the difference between the rate of formation of the labelled intermediate Y*, ω_2^*, and the rate, ω_3^*, of its disappearance (equation 94).

$$\frac{dI_y}{dt} = \omega_2^* - \omega_3^* = \alpha\omega_2 - \beta\omega_3 \tag{94}$$

Elimination of $d(I_y)/dt$ from equations 93 and 94 and replacement of $d[Y]/dt$ by $d[Y]/dt = \omega_2 - \omega_3$, leads to equation 95.

$$\frac{d\beta}{dt} = \frac{(\alpha - \beta)\omega_2}{[Y]} \tag{95}$$

At the beginning $\beta = \alpha$ and $(d\beta/dt)_{t=0} = 0$, but since α is decreasing with time, β is also decreasing, $d\beta/dt < 0$ and $\beta > \alpha$.

When the reaction between the intermediate Y and its precursor X is reversible (equation 96):

$$A \rightarrow B \xrightarrow{\omega_1} \overset{\alpha}{X} \underset{\omega_{-2}}{\overset{\omega_2}{\rightleftharpoons}} \overset{\beta}{Y} \xrightarrow{\omega_3} C \rightarrow \ldots \tag{96}$$

the observed changes of α, β, [X] and [Y] with time give the four differential equations 97–100 which allow evaluation of the four unknowns ω_1, ω_2, ω_{-2}, ω_3.

$$\frac{d\alpha}{dt} = -\frac{\alpha\omega_1}{[X]} + \frac{(\beta-\alpha)\omega_{-2}}{[X]} = \frac{-\alpha(\omega_1+\omega_{-2})+\beta\omega_{-2}}{[X]} \tag{97}$$

$$\frac{d\beta}{dt} = \frac{(\alpha-\beta)\omega_2}{[Y]} \tag{98}$$

$$\frac{d[X]}{dt} = \omega_1 + \omega_{-2} - \omega_2 \tag{99}$$

$$\frac{d[Y]}{dt} = \omega_2 - \omega_{-2} - \omega_3 \tag{100}$$

Many biochemical reactions include several reversible reaction steps and the strict solution of the kinetic equations is much more complicated than the simple case described above.

2. *Determination of the reaction sequence*

In order to show that intermediate Y is formed directly from the precursor X, one introduces into the reacting system a small amount of unlabelled Y together with the labelled X, and the changes of [X], [Y], α and β in the course of the reaction investigated are determined. The answer obtained is positive when the specific activities of X and of Y are changing with time as shown in Figure 1.

When Y is formed partly from the labelled precursor X and partly from one or more inactive precursors (K_i, K_1, K_2, K_3, etc.) (equation 101) then

$$A \rightarrow B \xrightarrow{\omega_1} \overset{\alpha}{X} \xrightarrow{\omega_{2a}} \overset{\beta}{Y} \xrightarrow{\omega_3} D \rightarrow \qquad K_i \xrightarrow{\omega_{2bi}} Y \tag{101}$$

the curve of β against time falls always below the curve of α against time as shown in Figure 2. By using equations 88 and 95 we obtain equation 102.

$$\frac{d\beta}{dt} = \frac{(\alpha-\beta)\omega_{2a}}{Y} - \beta\frac{\omega_{2b}}{Y} = \frac{\left[\alpha - \beta\left(1+\frac{\omega_{2b}}{\omega_{2a}}\right)\right]\omega_{2a}}{Y} \tag{102}$$

The condition for a maximum, $d\beta/dt = 0$, is given in equation 103,

$$\left(\frac{\alpha}{\beta}\right) = \frac{\omega_{2a}+\omega_{2b}}{\omega_{2a}} = \frac{\overline{AB}}{\overline{AC}} \tag{103}$$

which enables the evaluation of the respective rate ratios. If more than one unlabelled precursor K generates intermediate Y then the ω_{2b} term in equation 103 is replaced by the sum of the formation rates of Y, $\sum_i \omega_{2b_i}$.

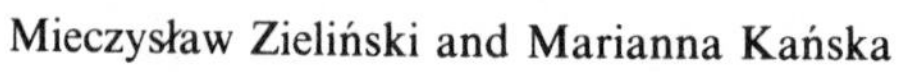

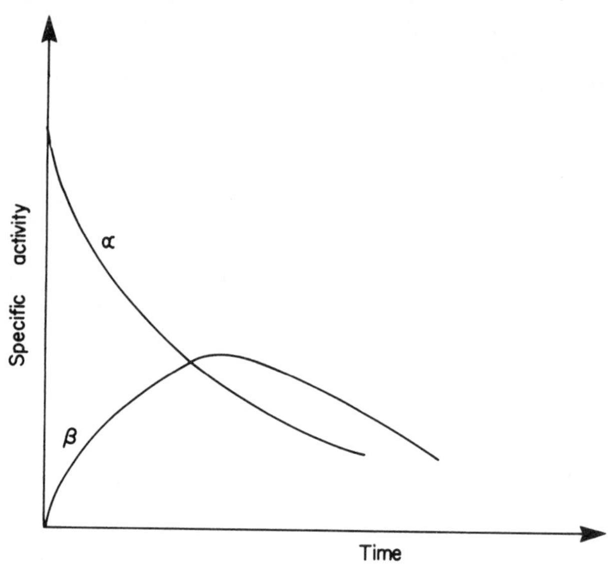

FIGURE 1. Time dependence of the specific activities of the precursor X(α) and of the intermediate Y(β)

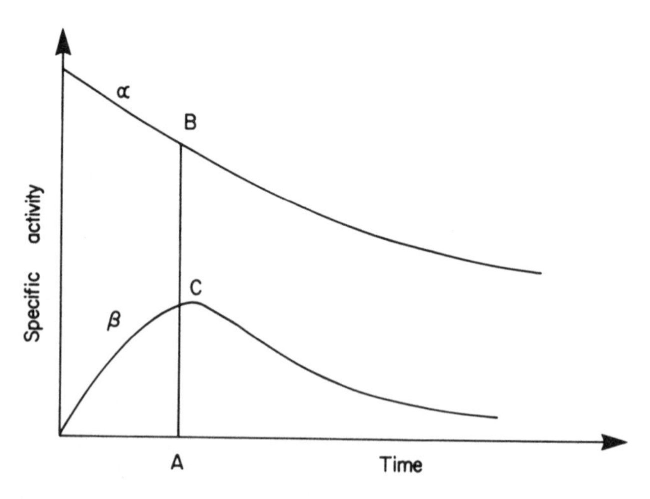

FIGURE 2. Time dependences of α and β when Y is formed also from an inactive precursor

In the case of a reaction sequence represented by equation 104,

$$A \rightarrow B \xrightarrow{\omega_1} X \xrightarrow{\omega_2} Y \xrightarrow{\omega_3} C \rightarrow \qquad B \rightarrow Z \xrightarrow{\omega_4} Y \quad (104)$$

the introduction of the labelled compound B into the reacting system and subsequent isolation of the intermediates [X], [Y], [Z] and measurements of their respective specific radioactivities α, β and γ allows the calculation of the ratio of the rates $\omega_2/\omega_4 = (\beta-\gamma)/(\alpha-\beta)$ which results directly from the condition for a maximum of the $\beta(t)$ time dependence (equation 105).

$$\frac{d\beta}{dt} = \frac{(\alpha-\beta)\omega_2 + (\gamma-\beta)\omega_4}{[Y]} = 0 \quad (105)$$

Usually a given intermediate participates in several different biochemical reactions taking place in the living organism investigated, and the direct use of simple relations 89–105 is of rather limited value. The technical and analytical difficulties also do not allow one to follow exactly the kinetics of the formation of the given compound and its intermediates in most cases. Finally, we will recall that the concentration of the intermediate $[X_f]$, defined by equation 106, can be calculated by graphical integration of equation 107 obtained in turn by substitution of equation 89 into 106.

$$[X_f] = \int_0^t \omega_1 dt \quad (106)$$

$$[X_f] = -\int_{\alpha_0}^{\alpha_t} [X]\, d(\ln\alpha) = \int_{\alpha_t}^{\alpha_0} [X]\, d(\ln\alpha) = 2.303 \int_{\alpha_t}^{\alpha_0} [X]\, d(\log\alpha) \quad (107)$$

The difference between $[X_f]$ and [X] taken at any given time interval gives the 'consumed amount' of intermediate [X].

B. Biosyntheses and Uses of Labelled Benzoquinone Derivatives

1. Biosynthesis of plastoquinone-9 and tocopherols by ethiolated maize shoots

^{14}C-Isotopic tracer studies have established[134] that ^{14}C-tyrosine (**195**), [^{14}C]-*p*-hydroxyphenylpyruvic acid (**196**) and [α-^{14}C] and [U-^{14}C]homogentisic acid (**197**) are utilized by maize shoots for the biosynthesis of ^{14}C-labelled γ-tocopherol (**200**), β-tocopherol (**201**), α-tocopherol (**202**) and α-tocopherolquinone (**203**) (equation 108) and plastoquinone-9 (**204**) (equation 109). It is suggested that in the course of radioactive tocopherols production D-tyrosine is first converted into *p*-hydroxyphenylpyruvic acid[135] which in turn is degraded to homogentisic acid before incorporation into tocopherolquinone (**203a**–**203d**) (equation 108) and plastoquinone (**204**) (equation 109). The sequence of biochemical transformations of equation 108 is supported by several observations:

(1) [^{14}C]homogentisic acid is incorporated into tocopherols and into plasto- and α-tocopherolquinones more effectively than *p*-hydroxy[^{14}C]phenylpyruvic acid (**196**);

OH OH OH
$^{14}CH_2$ $^{14}CH_2$ $^{14}CH_2COOH$
CHNH$_2$ C = O OH
COOH COOH
(**195**) (**196**) (**197**)

PPO … H
phytyl pyrophosphate

(*) = ring ^{14}C label

2-methyl-6-phythyl benzo-hydroquinone (**198**)

δ-tocopherol (**199**)

γ-tocopherol (**200**)

β-tocopherol (**201**)

α-tocopherol (**202**)

(**203**)

a: $R^1 = R^2 = H, R^3 = Me$
b: $R^1 = R^3 = H, R^2 = Me$
c: $R^1 = H, R^2 = R^3 = Me$
d: $R^1 = R^2 = R^3 = Me$

(108)

(2) in isotope competition experiments, addition of *p*-hydroxyphenylpyruvic acid of natural isotopic composition to the maize shoots culture markedly decreased the incorporation of radioactivity from DL-[β-^{14}C]tyrosine into plastoquinone; addition of homogentisic acid of natural isotopic composition into the culture diminishes significantly

the incorporation of radioactivity from *p*-hydroxy-[U-^{14}C]phenylpyruvic acid into plastoquinone-9, tocopherols and α-tocopherolquinone;

(3) *p*-hydroxyphenylacetic acid, as well as ^{14}C-labelled phenylacetic acid which could be considered as an intermediate in the conversion of *p*-hydroxyphenylpyruvic acid into homogentisic acid as well as 2-methylquinol-4-β-D-glucoside (homoarbutin) are not involved in the biosynthesis of labelled plastoquinone-9 (**204**), tocopherol and tocopherolquinone.

The chemical degradation of labelled isoprenoid quinones and chromanols showed that nuclear carbon atoms and the side-chain carbon atom adjacent to the nucleus of D-tyrosine, *p*-hydroxyphenylpyruvic acid and homogentisic acid are incorporated as a C_6–C_1 unit to the *p*-benzoquinone rings and as nuclear methyl groups of plastoquinone-9 (equation 109) and α-tocopherolquinone and as the aromatic nuclei and nuclear methyl groups of **200** and **203**. Ethiolated dark-grown 7-day-old maize shoots, excised and

(**197**) + PPO–[nonaprenyl chain]$_8$H → 2-methyl-6-nonaprenyl hydroquinone → plastoquinol → (**204**) (109)

nonaprenyl pyrophosphate

2-methyl-6-nonaprenyl hydroquinone

plastoquinol

(**204**)

exposed with continuous illumination to D-[G-^{14}C]shikimate for periods up to 24 h produced ^{14}C-labelled dehydrophylloquinone, α-dehydrotocopherolquinone, α-dehydrotocopherol and γ-dehydrotocopherol[136]. It is suggested that the unsaturation occurs beyond the C(4′) position of quinones and beyond the C(1′) position in the case of chromanols. Dehydro compounds are accumulated in substantial amounts in ethiolated tissues, because these tissues are unable to carry out the complete reduction of all the molecules of geranyl–geranyl-substituted intermediates, and a mixture of dehydro and normal labelled products is formed.

2. *Syntheses of* ^{14}*C- and tritium-labelled polyprenyl quinones and tocopherols by* Calendula officinalis

In the course of ^{14}C isotopic studies of the mechanism of the biosynthesis of polyprenyl quinones and tocopherols[137,138] by *C. officinalis*, using $^{14}CO_2$, [1-^{14}C]acetate, [2-^{14}C]mevalonate and [U-^{14}C]tyrosine as labelled precursors, the dynamics of the incorporation of tritium-labelled [2,3-^{3}H]phytol into tocopherols **199**–**202** and tocopherolquinones **203a–d**, produced in different cellular subfractions of *C. officinalis* leaves[139], has also been investigated. By isolating the individual tritium-labelled tocopherols and tocopherolquinones from various cellular subfractions after 1, 2, 4 and 6 hours of feeding with [2,3-^{3}H]phytol protoplasts (obtained from 4 week old *C. officinalis* leaves) it has been shown[139] that condensation of homogentisic acid with phytyl chain and formation of 2-MeTQ (**203a**) and 3-MeTQ (**203b**) takes place both in chloroplast and in microsomal fractions. Subsequent methylation and cyclization leads to the formation of 2,3-MeTQ (**203c**), or 8-MeT (**199**) and 7,8-MeT (**200**), which are methylated to 5,7,8-MeT (**202**) (equation 110).

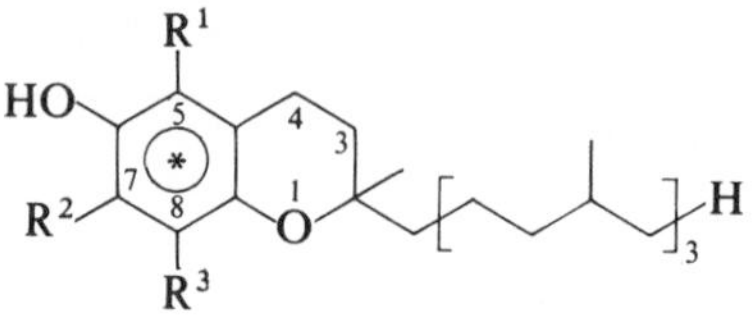

8-MeT (**199**) $R^1 = R^2 = H, R^3 = Me$
7-MeT (**199a**) $R^1 = R^3 = H, R^2 = Me$
7,8-MeT (**200**) $R^1 = H, R^2 = R^3 = Me$
5,7,8-MeT (**202**) $R^1 = R^2 = R^3 = Me$

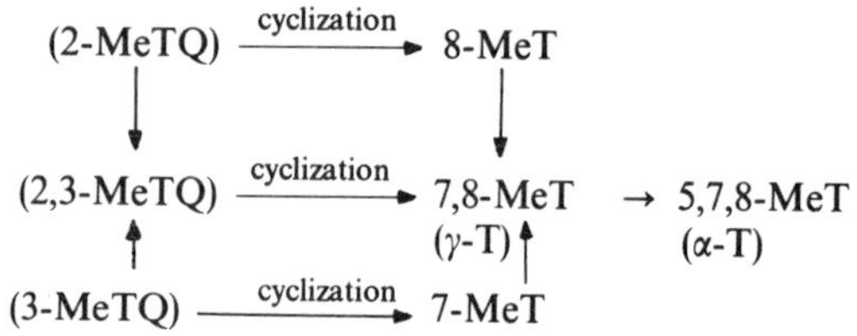

(2-MeTQ) —cyclization→ 8-MeT (110)
(2,3-MeTQ) —cyclization→ 7,8-MeT (γ-T) → 5,7,8-MeT (α-T)
(3-MeTQ) —cyclization→ 7-MeT

biosynthesis of tocopherols by *C. officinalis*

It is suggested that α-TQ (**203d**) does not participate in the biosynthesis of α-T (**202**) but results from the oxidation of α-T (**202**). A conclusion was also made that tocopherols and TQ (**203**) are transported to the mitochondria of plants from their place of synthesis. The biosynthesis of chloroplastidic terpenoid quinones and chromanols has been reviewed recently by Pennock[140].

Using [^{14}C]-4-hydroxybenzoic acid it has been also shown[141–143a] that *Escherichia coli* forms ubiquinone (**204**) which is synthesized by prenylation of 4-hydroxybenzoic acid with an all-*trans* polyisoprenoid side chain which consists predominantly of eight units (equation 111).

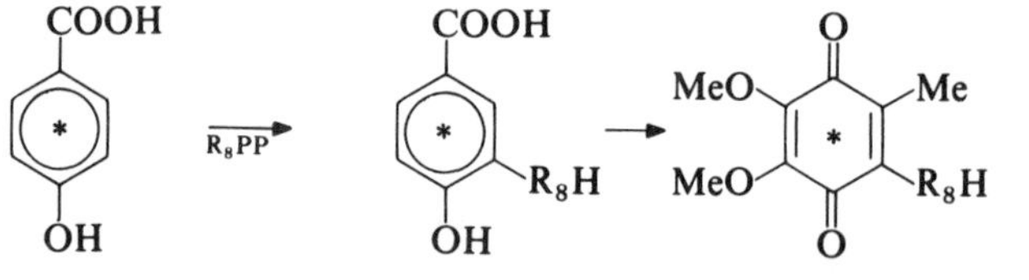

$$\text{4-HOC}_6\text{H}_4\text{COOH} \xrightarrow{R_8PP} \text{3-}R_8H\text{-4-HOC}_6\text{H}_3\text{COOH} \longrightarrow (\mathbf{204}) \qquad (111)$$

R_8PP-octaprenyl pyrophosphate

R_8-octaprenyl side chain

3. *Biosynthesis of [^{14}C]mavioquinone*

Carbon-14 labelled mavioquinone (**205**) has been biosynthesized[143b] by *Mycobacterium avium* in a growth medium containing [^{14}Me]-L-methionine, or [3-^{14}C]propionate and to a lesser extent from [1-^{14}C] or [2-^{14}C]acetate. Shikimic acid, a common precursor of the

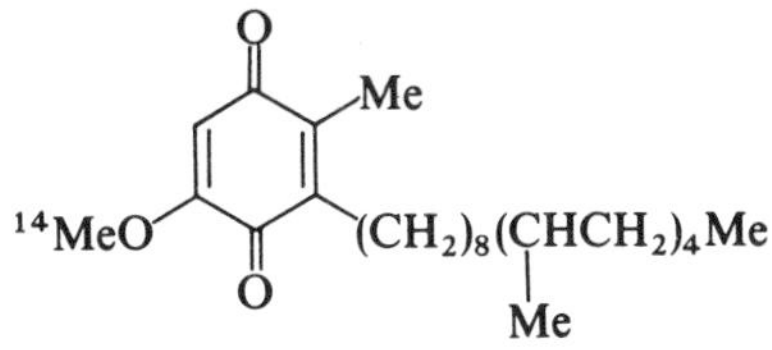

(205)

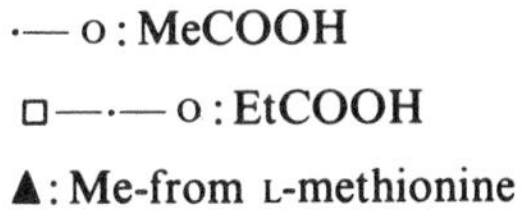

·— o : MeCOOH

□—·— o : EtCOOH

▲: Me-from L-methionine

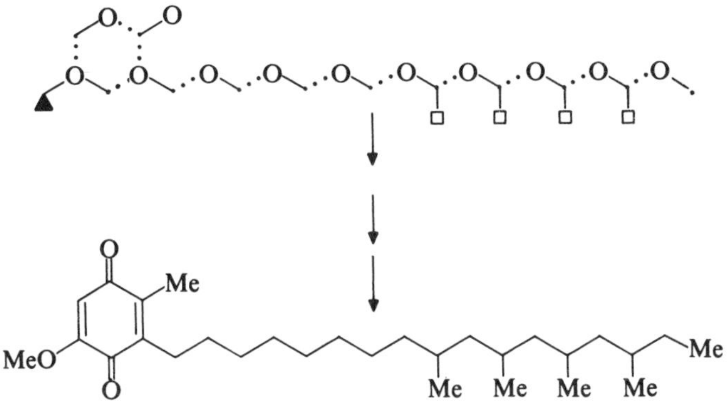

(205a) (112)

quinone ring of mena- and ubiquinones, and *p*-hydroxybenzaldehyde, a precursor of ubiquinones, were not incorporated into **205a**. The degradation experiments have shown that more than 75% of the label from [^{14}Me]-L-methionine is localized in the *O*-methyl group of mavioquinone, the rest being randomized in the side chain and in the ring. This specific incorporation of the Me group of methionine into an *O*-methyl group was also demonstrated by introducing [C^2H_3]-L-methionine into the growth medium. At least 75% of the ^{14}C from [^{14}C]propionic acid was found in the branched part of the side chain. The absence of incorporation of the methyl group of methionine in the ring C-methyl group and the failure to incorporate aromatic precursors into the ring carbon atoms suggest an acetate–methylmalonate condensation as the pathway for the biosynthesis of mavioquinone (equation 112).

C. Biosyntheses and Uses of Labelled Naphthoquinone Derivatives

1. Menaquinone biosynthesis by Escherichia coli *extracts*

The radiochemical evidence for the involvement of 1,4-dihydroxy-2-naphthoic acid in the menaquinones (**210**) biosynthesis by cell-free extracts of *E. coli* was presented by Bryant and Bentley[144,147]. The addition of [2,3-$^{14}C_2$]-O-succinoylbenzoic acid (OSB, **206**), as the disodium salt to incubation mixtures of *E. coli* essentially cell-free extracts, containing naphthoate synthetase (coenzyme A having an approximate molecular weight of 45000) and ATP resulted in the formation of radioactive 1,4-dihydroxy-2-naphthoic acid (**208**). Supplementary addition of the side-chain precursor, farnesyl pyrophosphate, to the extract stimulated conversion of [2,3-^{14}C]OSB into menaquinones. The mechanism

(206) OSB, R = CoA → (207) → (208) $\xrightarrow{R_8PP}$ (209) DMK-n → (210) MK-n (113)

R_8H = prenyl = $[\ldots]_{n=8}$H

n is number of isoprene units, equal for instance to 8 or 9

and the intermediates involved in the transformation sequences OSB → → → MK (**210**) are therefore as given in equation 113.

The Claisen condensation (i.e. elimination of water between a carboxyl and methylene group) which gives the ketotetralone (**207**) requires the formation of a CoA derivative at least of one carboxyl group. The biochemical mechanism of the conversion of 1,4-dihydroxy-2-naphthoic acid (**208**) to DMK (demethylmenaquinone (**209**)) and to menaquinone (MK, **210**) was investigated by Bentley[147] and Shineberg and Young[145]. Despite of the overall similarities, the biosynthetic paths of ubiquinone and menaquinones are different. Incubation of cell extract from *E. coli* with [^{14}C]-1,4-dihydroxy-2-naphthoic acid and solanesylpyrophosphate in the presence of Mg^{2+} and Triton-X-100 resulted in the formation of radioactive demethylmenaquinone-9 as well as radioactive menaquinone-8, by a concerted replacement of the carboxyl of 1,4-dihydroxy-2-naphthoic acid with a prenyl side chain. The isotopic tracer experimental evidence eliminates symmetrical intermediates such as naphthoquinone or its quinol[146]. Men A^- mutants of *E. coli*, which are defective in the prenylation reaction, accumulate 1,4-dihydroxy-2-naphthoate rather than 1,4-naphthoquinone[143]. However, this strain had normal levels of 4-hydroxybenzoate octaprenyltransferase, the enzyme catalysing the analogous prenylation reaction in ubiquinone biosynthesis. This indicates that the two octaprenyltransferases are quite different. In the case of ubiquinone formation the substrate for prenylation is also the aromatic carboxylic acid, but 4-hydroxybenzoate is prenylated without decarboxylation. Another enzyme catalyses the decarboxylation in the next step and 2-prenyl phenol is formed.

2. *Biosynthesis of vitamin K_2*

The role which shikimic acid plays in the biosynthesis of vitamin K_2 (MK-8), ubiquinone and in the aromatic synthesis in general in *E. coli* was investigated by Cox and Gibson[148]. In the preliminary experiments the *E. coli* cells were grown in the presence of

[G-^{14}C]shikimate during 6 h, harvested, extracted and the extracts were chromatographed on silica-gel thin-layer plates. The vitamin K and the ubiquinone appeared as yellow bands. Scanning of the plates for radioactivity showed that shikimate incorporates into both vitamin K and ubiquinone to about equal degrees. Following a kinetic investigation of the likely intermediates in the biosynthesis of vitamin K and ubiquinone, several suspected intermediates of natural isotopic composition were added in excess to the culture containing already labelled shikimate. Addition of unlabelled 4-hydroxybenzoate resulted in an almost disappearance of the radioactive peak corresponding to ubiquinone but did not affect the incorporation of radioactivity of shikimate into vitamin K. Addition of 3,4-dihydroxybenzaldehyde and adrenaline affected the incorporation of [^{14}C]shikimate into vitamin K since the peak of radioactivity corresponding to vitamin K on thin-layer plates largely disappeared. Compounds tested which do not affect the incorporation of [^{14}C]shikimate into either vitamin K or ubiquinone are catechol, 4-hydroxyphenylpyruvate, 2,3-dihydroxybenzoate, phenylpyruvate and menadione.

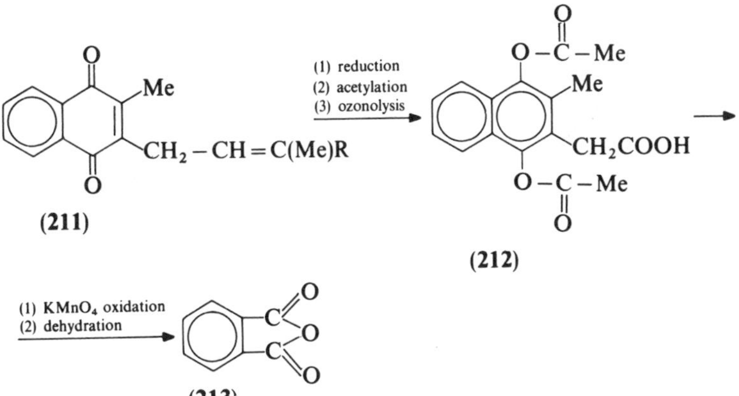

(211) Me; $CH_2-CH=C(Me)R$ —(1) reduction (2) acetylation (3) ozonolysis→ (212) $O-C(=O)-Me$, Me, CH_2COOH, $O-C(=O)-Me$ —(1) $KMnO_4$ oxidation (2) dehydration→ (213) (114)

Stepwise chemical degradation of vitamin K_2 (**211**), synthesized by *E. coli* from [^{14}C]shikimate showed that the benzene ring and carbonyl groups of the naphthoquinone arise from shikimic acid (equation 114). Specific ^{14}C activities of **211**, of 1,4-diacetoxy-2-methylnaphthalene-3-acetic acid (**212**) and of the phthalic anhydride (**213**) were nearly the same. Consequently the ^{14}C radioactivity of the original vitamin K was retained in the phthalic anhydride. When cells of *E. coli* were grown in a glucose–citrate–mineral salts medium to which [1,2-$^{14}C_2$]acetate was added, the extracted and purified vitamin K was

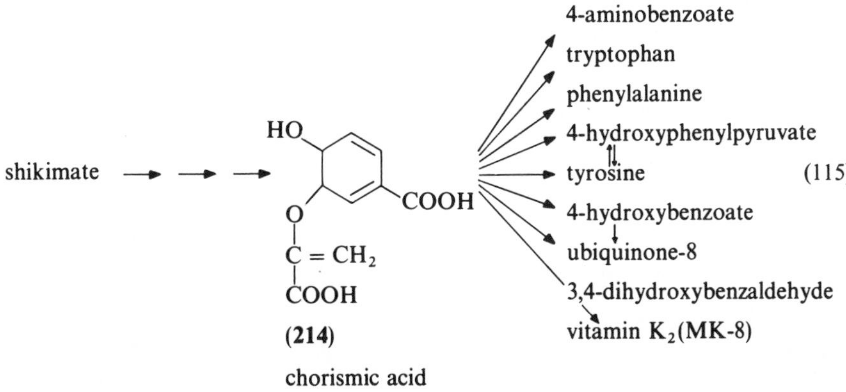

shikimate → → → (214) chorismic acid [HO, COOH, $O-C(=CH_2)-COOH$] → 4-aminobenzoate; tryptophan; phenylalanine; 4-hydroxyphenylpyruvate ⇅ tyrosine; 4-hydroxybenzoate → ubiquinone-8; 3,4-dihydroxybenzaldehyde → vitamin K_2(MK-8) (115)

found to be radioactive. The degradation experiments showed that the specific radioactivities of compounds **211**, **212** and **213** are in 1.4:0.68:0.1 ratio.

The ^{14}C of the [Me-^{14}C]methionine fed to the culture also incorporates into vitamin K_2. After chemical degradation the specific activities of **211** and **212** were about equal. The question of whether quinone ring synthesis is direct from acetate or indirect through the shikimate intermediate is still open for discussion and requires further experimental investigation. Cox and Gibson[148a] suggested equation 115 as a probable pathway for the biosynthesis of aromatic compounds in *E. coli*.

Esmon, Sadowski and Suttie[148b] discovered that vitamin K plays a certain role in the incorporation of $H^{14}CO_3^-$ into plasma proteins in the course of synthesis of prothrombin (one of the four blood-clotting zymogens) by liver extract from vitamin K deficient rats (upon the addition of vitamin K).

3. [^{14}C]*Juglone (5-hydroxy-1,4-naphthoquinone) biosynthesis*

Müller and Leistner[149] have shown that 1,4-naphthoquinone, one of the more than 120 naphthoquinones isolated from plants, is a natural intermediate involved in the biosynthesis of juglone in *Juglans regia*. Besides ^{14}C-labelled juglone (**215**), symmetrically ^{14}C-labelled 1,4-naphthoquinone (**216**) has been isolated from a young *Juglans regia* plant which was kept in contact with an aqueous solution of 4-(2′-carboxyphenyl)-4-oxobutyrate-[2-^{14}C] (**217**), for 25 h (equation 116). Its intermediacy was confirmed also

COOH, *C—COOH, C=O → [X] → (218) lawsone

(217)

*C = ^{14}C-labelled position

[X] ↓ (216) → (215)

(116)

by growing the *Juglans regia* (and *Impatiens balsamina*) plants in aqueous solution of [1,4-^{14}C]-1,4-dihydroxy-2-naphthoic acid which was expected to be the intermediate [X] shown in equation 116. The naphthoquinone from *Impatiens* was inactive whereas 1,4-naphthoquinone from *Juglans* was radioactive. In a more careful radiobioexperiment[150, 151] with *Juglans* plant, infused with ^{14}C-labelled *o*-succinoylbenzoic acid (**217**) and tritium-labelled [6-^{3}H]-D-glucose, the ^{14}C-labelled 4-oxo-α-tetralone (**221**), β-hydrojuglone (**225**) and glucosides **224** and **227** were also extracted. ^{14}C-Labelled glycosides such as 4-hydroxy-1-naphthalenyl-β-D-glucopyranoside as well as 1,4-naphthoquinone, juglone and 4,8-dihydroxy-1-naphthalenyl-β-D-glucopyranoside have been additionally isolated from fruits of different *Juglans* species[151]. A glycosyltransferase, catalysing the glycosylation of hydroquinone has been isolated from leaves of *J. regia* and callus culture of *J. major*. All the accumulated data are presented by the general equation 117 which shows the biosynthetic relationships between naphthalene derivatives in *Juglans*.

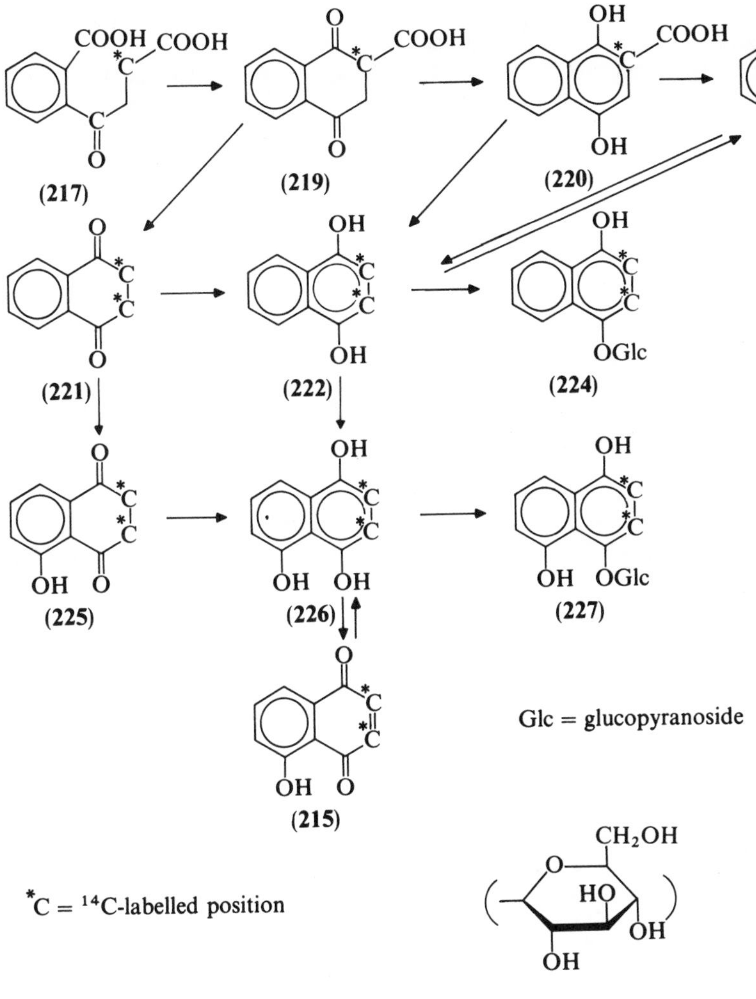

(217) (219) (220) (223)

(221) (222) (224)

(225) (226) (227) (117)

(215)

Glc = glucopyranoside

*C = ^{14}C-labelled position

4. *Biosynthesis of [^{14}C]alkannin*

Alkannin (5,8-dihydroxy-2-((*S*)-1′-hydroxy-4′-methylpent-3-enyl)-1,4-naphthoquinone) (**228**), a coloured dyestuff found in the Boraginaceae, was ^{14}C-labelled heavily during

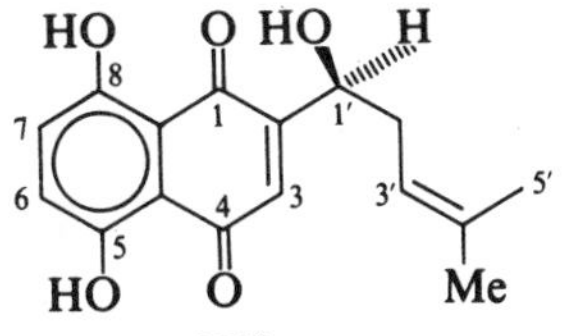

(**228**)
alkannin

feeding *Plagiobothrys arizonicus* with phenylalanine-, or cinnamic acid-ring-[1-^{14}C], *p*-hydroxybenzoic acid-ring-[1,2,6-^{14}C], mevalonic acid-[2-^{14}C], and doubly labelled mevalonic acid-[5-^{14}C] and ^{3}H[152]. No significant incorporation of shikimic acid-[7-^{14}C] and tyrosine-[U-^{14}C] into the pigment was found. The distribution of carbon-14 within the labelled molecule of alkannin as well as data concerning the incorporation of supposed ^{14}C-labelled precursors of this 1,4-naphthoquinone suggest that alkannin is produced by a new biosynthetic sequence which involves the prenylation of *p*-hydroxybenzoic acid with two molecules of prenylpyrophosphate in succession, or with geranylpyrophosphate, subsequent decarboxylation and finally ring-closure of diprenylhydroquinone.

5. *Biosynthesis of [^{14}C]plumbagin*

In the course of study of the mechanism of the biosynthesis of 5-hydroxy-2-methyl-1,4-naphthoquinone (plumbagin, 2-methyl-juglone, **229**) in the genus *Drosera* and in the genus *Plumbago* it has been revealed that young shoots of *Plumbago europaea* L. do not incorporate shikimate-[7-^{14}C], L-[^{14}Me]methionine, DL-tyrosine-[β-^{14}C], DL-phenylalanine[ring-1-^{14}C] or DL-mevalonic acid-[5-^{14}C] into plumbago to a significant extent after 24 hours. However, acetate-[1-^{14}C], [2-^{14}C] and malonate-[2-^{14}C] labelled this naphthoquinone heavily. These findings suggested that **229** is formed via a polyacetate–malonate pathway. This conclusion was confirmed by extensive chemical degradation of plumbagin obtained by labelling with acetate-[1-^{14}C] or acetate-[2-^{14}C] feeding[153]. 7-Methyljuglone (**230**), co-occurring in *Drosera* plants is also formed by the

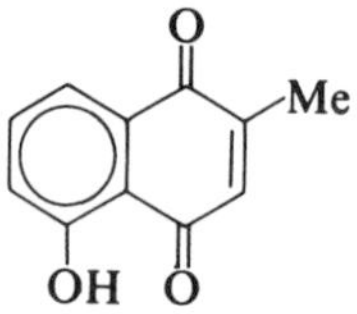

(229) 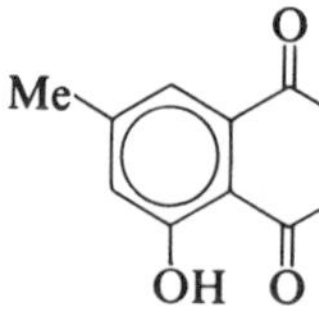 **(230)**

polyacetate–malonate route. In subsequent radiochemical isotopic studies of the biosynthetic routes leading to plumbagin in *Drosera*[154] it has been established that ^{14}C-labelled 2-methyljuglone is produced very efficiently in the young unrolling leaves of *Drosophyllum lusitanicum* from DL-tyrosine-[β-^{14}C] (**195**) via the acetate–polymalonate pathway (equation 118). Tyrosine is broken down to acetate through the homogentisate **197**, which is the central intermediate in this pathway.

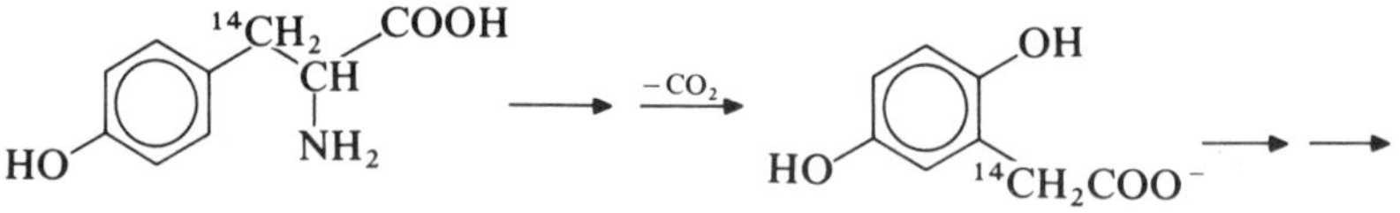

(195) → $\xrightarrow{-CO_2}$ (197) → → ^{14}MeCOO$^-$ → → labelled plumbagin (118)

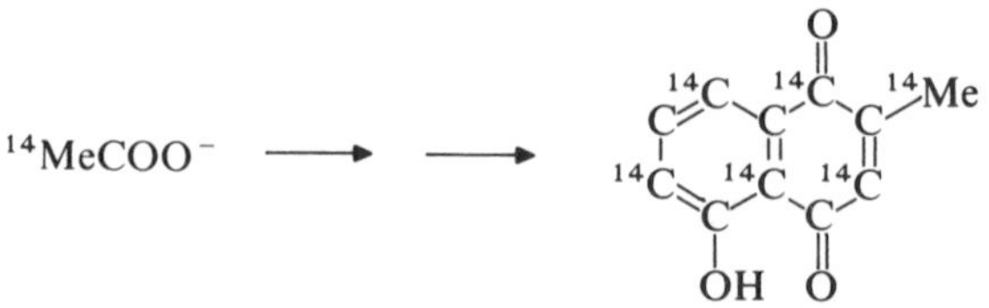

Evidence for reaction 118 has been obtained by carrying out an extensive chemical degradation of the isolated ^{14}C-labelled plumbagin by Kuhn–Roth oxidation, alkaline hydrogen peroxide oxidation, etc. The established distribution pattern of the radioactivity within the plumbagin molecule, obtained after feeding DL-tyrosine-[β-^{14}C] to *Drosophyllum* leaves, almost coincides with the expected (16.7% or 0%) percentage if the acetate–polymalonate pathway is operative. In accordance with equation 118, acetate-[1-^{14}C] and [2-^{14}C] as well as malonate-[2-^{14}C], propionate-[2-^{14}C] and acetogenic amino acids were found to be excellent precursors of plumbagin. Carbon-14 of acetate-[1-^{14}C] is located in positions 2, 4, 5, 7 and 9 in agreement with the theoretically expected yield (20% in each of these positions). Carbon-14 of acetate-[2-^{14}C] gave the same labelling pattern as found after tyrosine-[β-^{14}C] feeding. Incorporation of other possible ^{14}C-labelled precursors and especially incorporation of homogentisic acid-[2-^{14}C] was about equal to that of tyrosine. The highest (ca. 30%) incorporation was found on feeding with intermediates which are formed after the oxidative cleavage of the aromatic ring of homogentisic acid, carried out by an iron-containing enzyme. In the presence of the chelating agent α,α'-bipyridyl which is an inhibitor of the homogentisate oxygenase, the pool of free homogentisic acid increased and the incorporation of the label into plumbagin was depressed. Thus it has been established that carnivorous plants and their cell suspension cultures are able to degrade phenolic compounds[155], like a number of animals and microorganisms, to acetate, both under sterile and non-sterile conditions. The acetate is subsequently incorporated into plumbagin and 7-methyljuglone probably via an intermediate having a hexaacetyl chain.

6. *The role of [^{14}C]-o-succinoylbenzoic acid (OSB) in the biosynthesis of ^{14}C-labelled naphthoquinones*

Dansette and Azerad[156] synthesized [^{14}C]-*o*-succinoylbenzoic acid (**231**) by condensing [^{14}C]phthalic anhydride with succinic acid (equation 119). The OSB formed was administered to *Mycobacterium phlei*, *E. coli* K_{12}, *Aerobacter aerogens* 62-1 and to young shoots of *Impatiens balsamina* or *Juglans regia*.

^{14}C-phthalic anhydride + HOOC–CH_2–CH_2–COOH $\xrightarrow[-CO_2]{KoAc\cdot}$ [intermediate, $(CH_2)_2$] $\xrightarrow{HO^-}$ ^{14}COOH / C=O / $(CH_2)_2$ / COOH (119)

(**231**)

Degradative oxidation by $KMnO_4$ to phthalic acid of isolated MK-9 [II-H_2] (5′6′-dihydromenaquinone-*n*: abbreviation MK-9 [II-H_2]—the Roman numeral refers to the second isoprene unit out from the nucleus, $n = 9$ is number of isoprene units; **210**), menaquinone-8, demethylmenaquinone-8, lawsone, juglone and pseudopurpurine showed that the label is exclusively localized in C(1) and C(4) of the naphthoquinone ring. It has thus been concluded that OSB is a true intermediate in the biosynthesis of the naphthoquinone ring and a general scheme for its production has been proposed. According to Campbell[157] the first step in the OSB production is condensation, with dehydration of chorismic acid with the succinoyl–semialdehyde thiamine–pyrophosphate

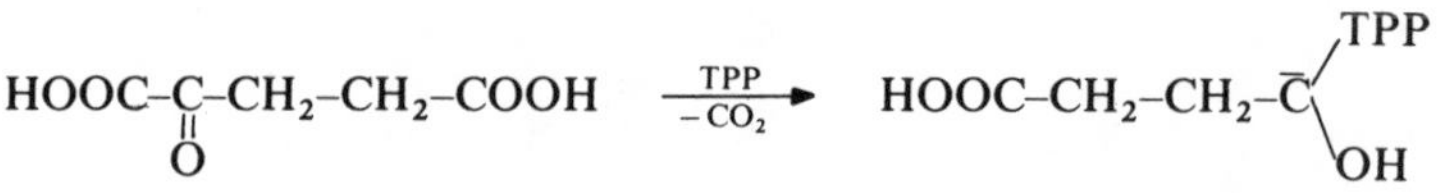

$$\mathrm{HOOC{-}\underset{\underset{\displaystyle O}{\|}}{C}{-}CH_2{-}CH_2{-}COOH \xrightarrow[-CO_2]{TPP} HOOC{-}CH_2{-}CH_2{-}\bar{C}\langle^{TPP}_{OH}} \qquad (120)$$

(232)

(TPP) carbanion (**232**) (equation 120). Grotzinger and Campbell[158] have also shown that **231** is an obligatory intermediate in the lawsone biosynthesis. When [U-^{14}C]glutamate was fed in aqueous solution to fresh *Impatiens balsamina* cuttings for 5 h, glycosidically bound and non-bound forms of ^{14}C-labelled lawsone were isolated from the organic extract of cuttings. The sensitive GLC–MS–proportional counter scanning method used in this study enabled the identification of 3-(2′-carboxyphenyl)-3-oxopropionate (**233**) as a second product of [U-^{14}C]glutamate metabolism[158].

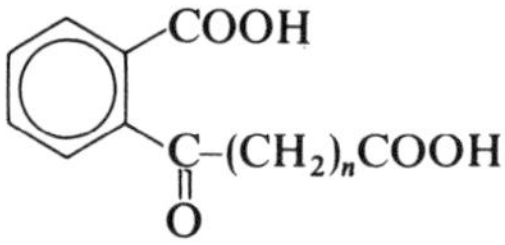

(233) $n = 1$ **(234)**

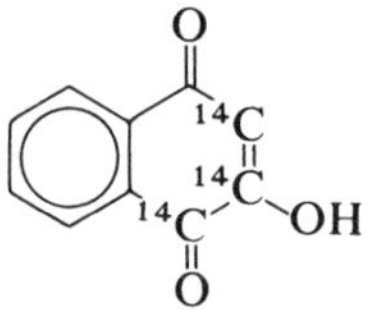

Impatient balsamina plants produced ^{14}C-labelled lawsone (**234**) also from ^{14}C-labelled alanine (**235**) and aspartate[159] (**236**). Robins and coworkers have shown[160] that C(2), C(3)

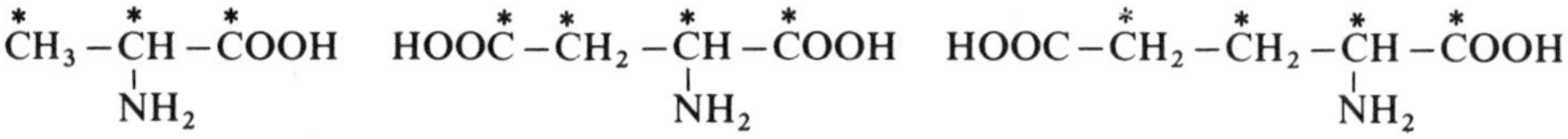

$$\mathrm{\overset{*}{C}H_3-\overset{*}{C}H(NH_2)-\overset{*}{C}OOH} \qquad \mathrm{HOO\overset{*}{C}-\overset{*}{C}H_2-\overset{*}{C}H(NH_2)-\overset{*}{C}OOH} \qquad \mathrm{HOOC-\overset{*}{C}H_2-\overset{*}{C}H_2-\overset{*}{C}H(NH_2)-\overset{*}{C}OOH}$$

(235)   **(236)**   **(237)**

and C(4) atoms of the naphthalene nucleus of menaquinone of *E. coli* and *M. phlei* are derived from [U-^{14}C]glutamate (**237**). Azerard and coworkers[161] obtained 'radioactive dihydromenaquinone-9' and proved that cell-free extracts of *M. phlei* can catalyse the reaction shown in equation 121.

$$\text{DMK-9} \longrightarrow \underset{\mathbf{(210a)}}{\text{MK-9}} \longrightarrow \underset{\mathbf{(238a)}}{\text{MK-9(II-H}_2)} \qquad (121)$$

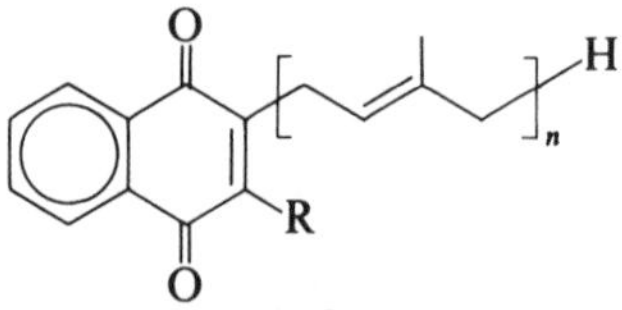

(210a)

R = H, DMK-9 = demethylmenaquinone-*n*
R = Me, MK-9 = menaquinone-*n*

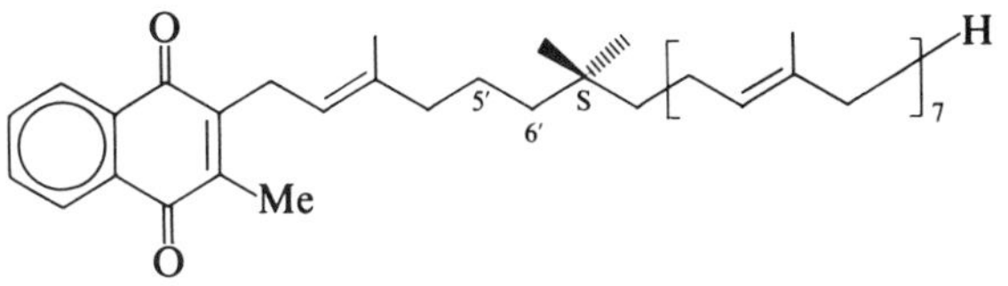

(**238a**)
MK-9(II-H_2)

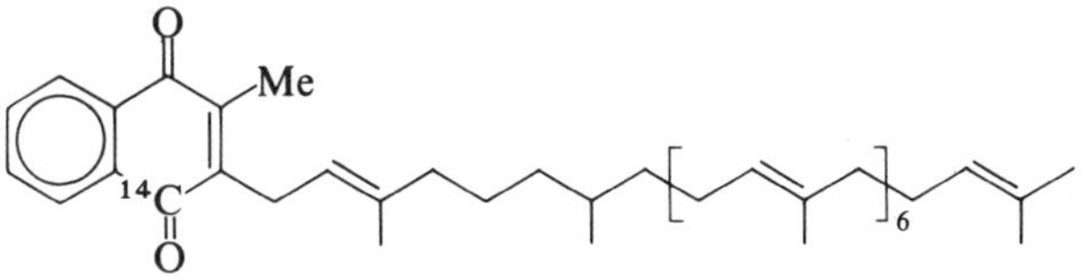

(**238b**)
MK-9(II-H_2)

The mechanism of production of [4-^{14}C]-MK-9 (II-H_2; **238b**) in *Mycobacterium phlei* from [7-^{14}C]shikimic acid, synthesized in turn from (3*R*, 5*R*)-3,4,5-triacetoxy-cyclohexanone and $H^{14}CN$ was investigated by Baldwin and coworkers[162] (equation 122).

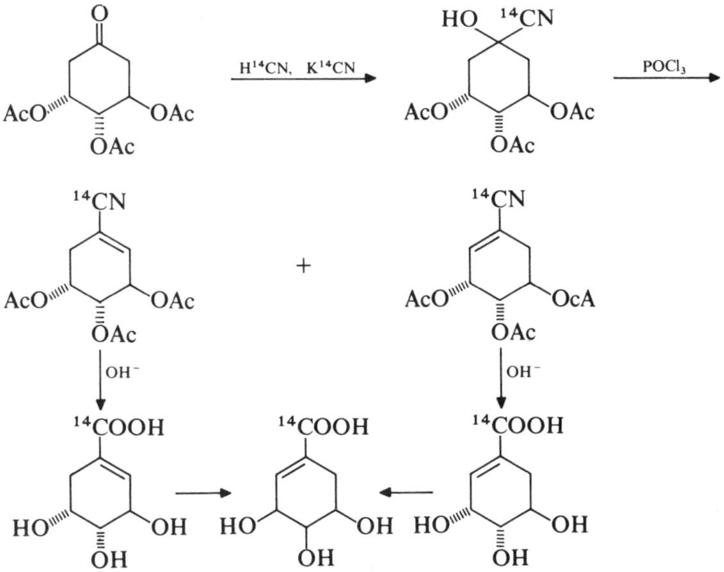

$H^{14}CN$, $K^{14}CN$; $POCl_3$; OH^- ; OH^-

(122)

The following ^{14}C-labelled naphthoquinone congeners were obtained[163,164] in the course of biosynthetic experiments in which 4-([carboxy-^{14}C]-2-carboxyphenyl)-4-oxobutanoic acid (**231**) was administered to callus tissues of *Catalpa ovata* grown for 2 weeks and incubated in the dark for 5 days: catalpalactone (**239**), α-lapachone (**240**), catalponone (**241**), menaquinone-1 (**242**), 3-hydroxydehydro-iso-α-lapachone (**243**), 1-hydroxy-2-methylanthraquinone (**244**) and lapachol (**245**). The above synthetic results and the

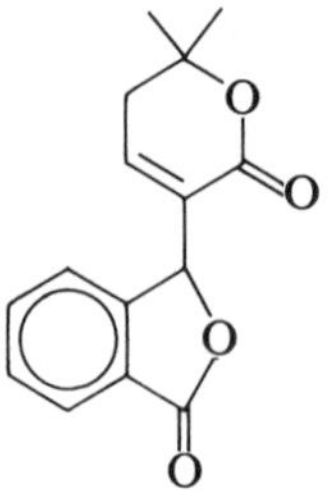

(239) 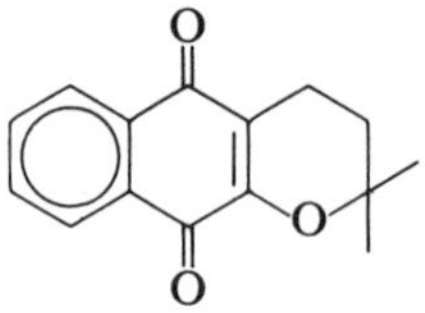 (240) 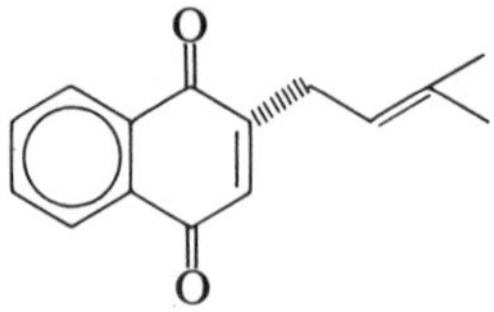 (241)

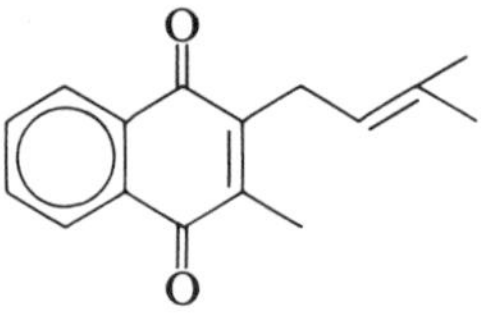

(242) 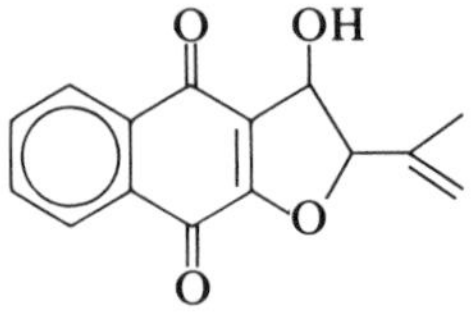 (243)

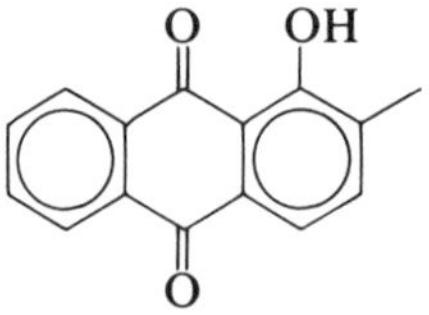

(244) 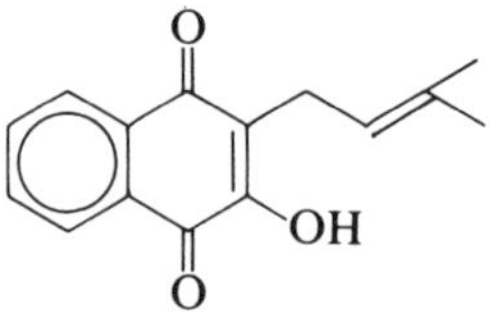 (245)

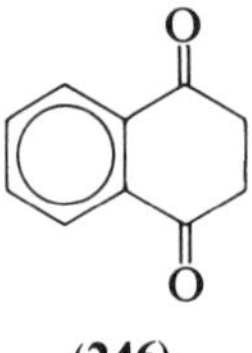

(246) 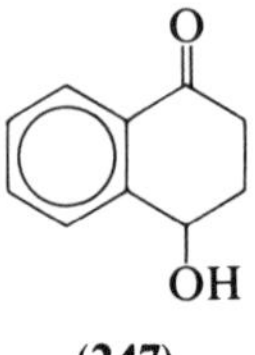 (247)

observation that **231** was incorporated into **246** in 3.75 times higher ratio than into **247** demonstrates that the naphthoquinones are biosynthesized stereospecifically in the callus tissues mainly according to the biosynthetic pathway of equation 123. It involves the intermediate compounds: 2-carboxy-4-oxo-1-tetralone (COT) and 2-carboxy-2-prenyl-4-oxo-1-tetralone (prenyl COT) while the route shown in equation 123a, involving 2-carboxy-4-hydroxy-1-tetralone (CHT) → 2-carboxy-4-hydroxy-2-prenyl-1-tetralone (prenyl CHT) → catalponol (**248**), is a subsidiary one. These conclusions were confirmed by administration of 4-([carboxy-^{14}C]-2-carboxyphenyl)-4-hydroxy[4-^{3}H] butanoic acid, obtained by reduction of [carboxy-^{14}C]-(**231**) with NaB^3H_4, to the callus cultures. Feeding of the doubly labelled compound to the wood of this plant was also performed in order to investigate the biosynthetic route in the wood. It has been demonstrated that COT is biosynthesized both in callus tissues and in wood by direct cyclization of **231** but not via

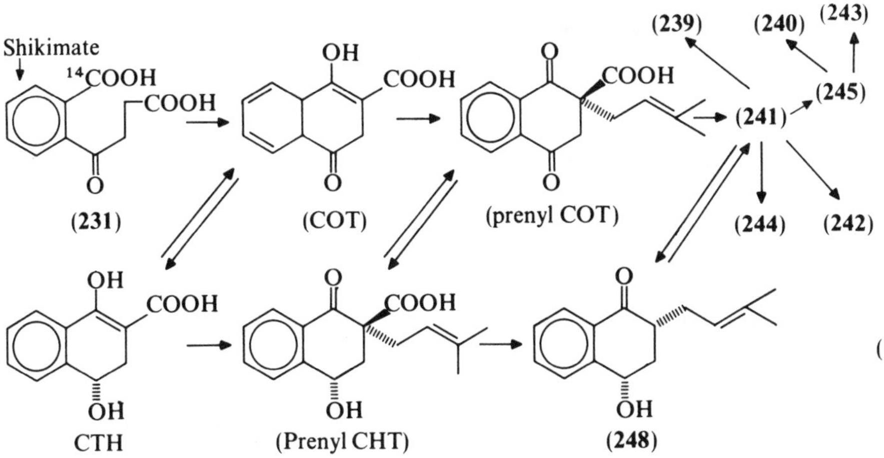

(123)

(123a)

CHT. The observed formation of menaquinone-1 (**242**) from **241** suggests that vitamin K (especially plant origin phylloquinone) is similarly formed through the same prenylation mechanism shown above. All labelled intermediates and products were isolated from the growth medium as the stable methyl esters by isotopic dilution method.

Three unusually prenylated ^{13}C- and ^{2}H-labelled naphthoquinones, **249**, **250** and **251**, of *Streptocarpus dunnii* were biosynthesized by feeding *S. dunnii* cell suspension cultures with ^{13}C- and ^{2}H-labelled precursors, [2-carboxy-^{13}C]-4-(2-carboxyphenyl)-4-oxo-butanoic acid (**252**) and [7-^{2}H]-lawsone (**253**), respectively[165a,b]. In the ^{13}C biosynthesis,

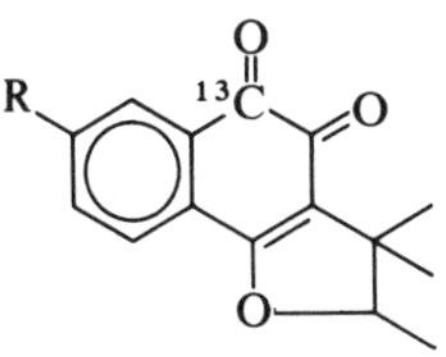

(**249**)

a: R = H
b: R = D

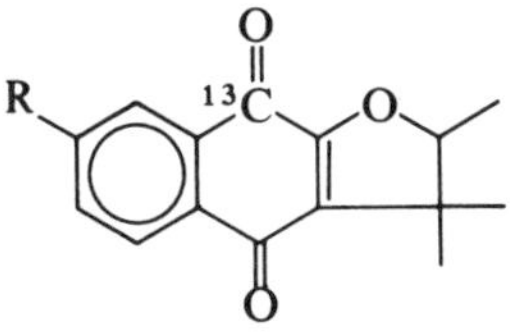

(**250**)

a: R = H
b: R = D

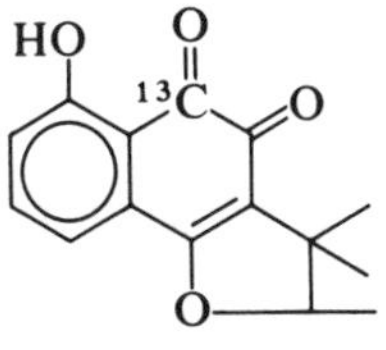

(**251**)

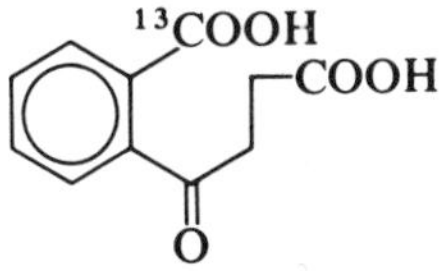

(**252**)

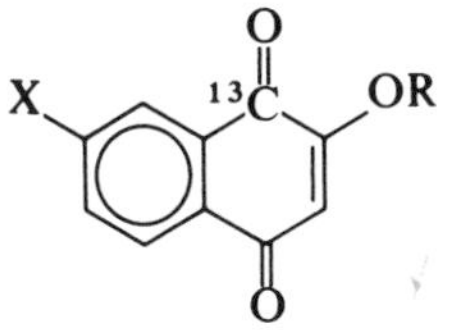

(**253**)

a: R = H, X = H
b: R = H, X = D

(**258**)

a: R= [prenyl], X = H
b: R= [prenyl], X = D

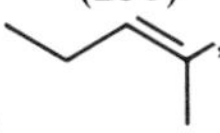

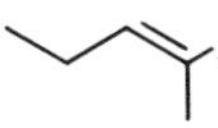

besides the dunniones **249a** and **250a**, the anthraquinones **254** and **255** labelled with ^{13}C in the 10-position were also obtained. Thus, it has been concluded that the naphthoquinones

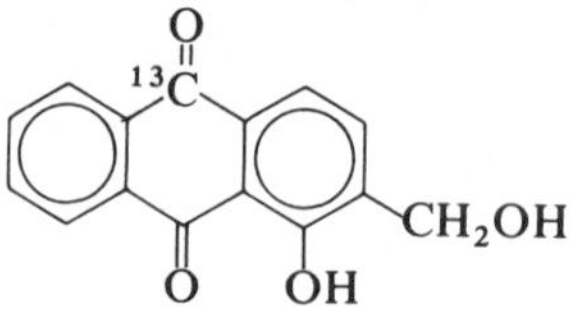

(**254**) 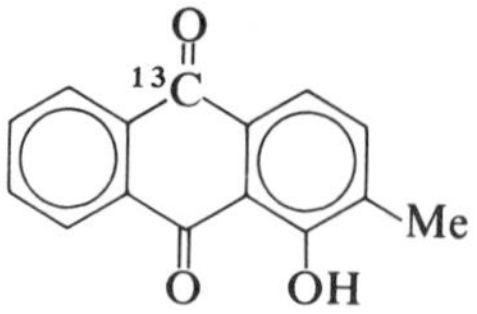 (**255**)

labelled with ^{13}C in the 1-position are biosynthesized from **252** via intermediates **256** and **253a**.

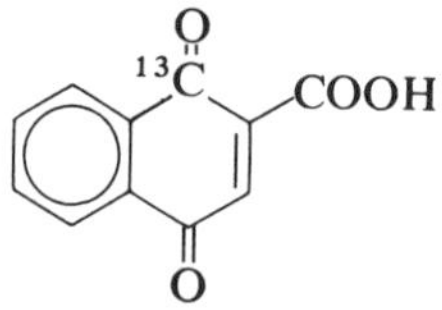

(**256**) 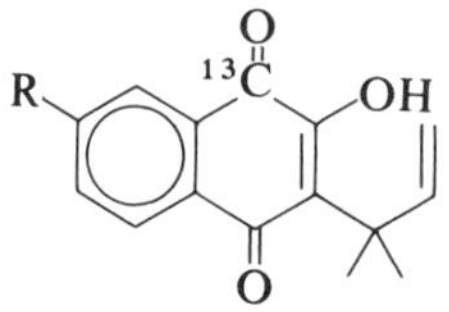 (**257**)

a: R = H
b: R = D

The lawsone **253a** was later prenylated at C(2) and the ether **258a** formed transformed via Claisen-type rearrangement to naphthoquinone (**257a**), which in turn was converted to labelled dunnione (**249a**), α-dunnione (**250a**) and 8-hydroxydunnione (**251**). Anthraquinones **254** and **255** are formed from **256** via the prenylation product **259** or the naphthoic acid (**261**) via the intermediate tectoquinone (**260**) (equation 124). The intermediacy of lawsone (**253**) in the naphthoquinone biosynthesis was established by feeding *S. dunnii* cultures with [7-^{2}H]-lawsone. Naphthoquinones **249b**, **250b** and **251** containing deuterium on C(7) were separated by silica gel chromatography.

No label was found in anthraquinones **254**, **255** and **260** derived from this culture. By adding the [7-^{2}H]-ether (**258b**) to the *S. dunnii* cell suspension cultures and isolating the intermediate **257b** and the deuterium-labelled products **249b**, **250b** and **251** from the post-reaction medium, the authors excluded the possibility of formation of the ether (**262**)

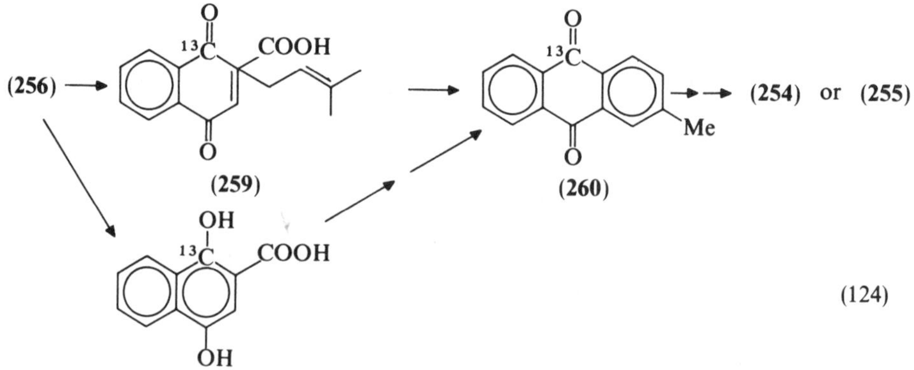

(**256**) → (**259**) → (**260**) →→ (**254**) or (**255**)
(**256**) → (**261**) → (**260**)

(124)

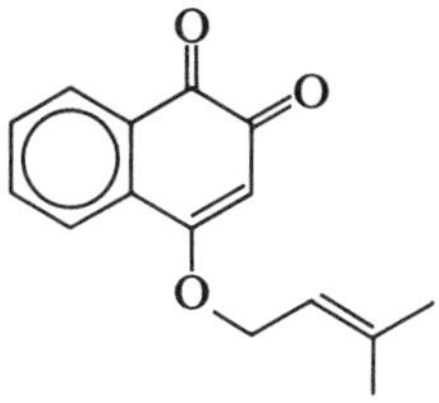

(262)

directly from **253** by prenylation, and its participation in the dunnione (**249**)-type naphthoquinone production[165].

Mass spectrometric evidence for 2-succinoylbenzoic acid and 1,4-dihydroxy-2-naphthoic acid as the menaquinone precursors and intermediates in menaquinone biosynthesis have been presented by Young[166]. The biosynthetic pathway leading to naphthoquinones, benzoquinones and anthraquinones has been reviewed by Leistner[167]. The structure of 1-hydroxy-2-(hydroxymethyl)anthraquinone (**254**), isolated from the leaves and roots of *Streptocarpus dunnii* has also been established by biosynthesizing labelled **254** using as potential precursors shikimic acid-[7-^{14}C], *o*-succinoylbenzoic acid-[2,3-^{14}C] and DL-mevalonic acid-[2-^{14}C][168a]. It has been shown that the OSB pathway is operative in the biosynthesis of this anthraquinone.

7. ^{18}O-*Biooxidation studies with naphthoquinones*

Lapachol (**245**) (possessing antitumour, antibiotic and antimalarial activities) and dichloroallyl lawsone (**245a**), incubated with several fungi and streptomycetes, undergo oxidative ring fission[168b] (equation 124a).

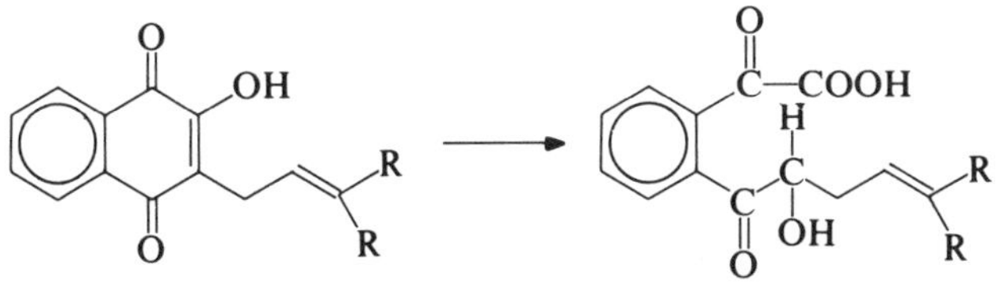

(124a)

(**245**) R = Me (**245b**) R = Me
(**245a**) R = Cl (**245c**) R = Cl

The compound **245c** was prepared also from dichloroallyl lawsone by treatment with H_2O_2 under alkaline conditions. Material obtained in chemical synthesis had the same infra red and mass spectra as the metabolite (**245c**) obtained in the biodegradation experiment. The mechanism of this process has been investigated by growing *Penicillium notatum* in $^{18}O_2$ atmosphere and subsequent mass spectral analysis of the oxygen-18 labelled ketol derivative (**245b**), obtained in the biodegradation of **245**. The location of heavy oxygen atom in the metabolite structure (**245b**) was accomplished by the use of the well defined mass spectral fragmentation pattern of the ketol metabolite (**245b**) prepared synthetically[168c] and metabolite derived from microbiological oxygenation of lapachol (**245**) by *P. notatum*. Analysis of mass spectral fragmentation pattern of **245b** showed that the ^{18}O label is located in the side chain hydroxyl group. No ^{18}O excess was found in the COOH moiety of **245b**. The conclusion was therefore made that molecular oxygen is introduced into lapachol via the monooxygenase pathway (124b), very likely by the initial formation of an epoxide (**245d**). In the acid or base-catalysed opening of the ring the oxygen atom is finally located as the secondary alcohol (**245b, c**).

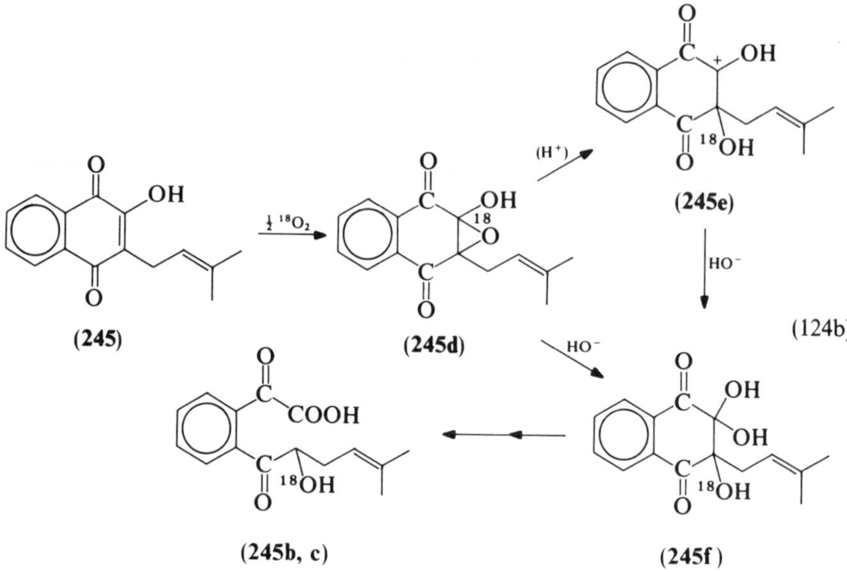

(124b)

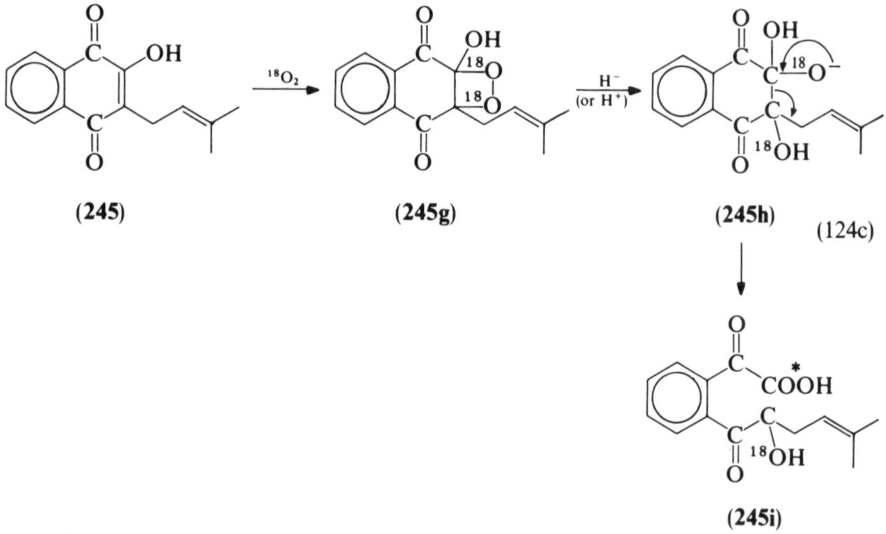

(245) (245d) (245e) (245f) (245b, c)

(124c)

(245) (245g) (245h) (245i)

The second dioxygenase pathway (124c) of the metabolic conversion of **245** into the ketol derivative (**245i**) implies that two ^{18}O atoms, derived from molecular oxygen, are incorporated into a cyclic endoperoxy intermediate (**245g**) which, undergoing 'hydride or acid-catalysed opening', yields (**245i**) having excess oxygen-18 both in hydroxyl and in carboxylic acid groups. The lack of the noticeable excess of heavy oxygen in the carboxyl group forced the authors[168b] to reject the reaction sequence (124c). A special room temperature test experiment showed that oxygen-18 exchange (^{18}O scrambling) between carboxyl group of the metabolite (**245i**) and water during 24 hours of incubation at 27°C and subsequent work up procedure in acidic conditions did not take place. The possibility of the carbon–carbon double bond cleavage and location of the heavy oxygen as the secondary alcohol (**245b**) according to acyclic type mechanism, analogous to scheme 43, is not taken into account in the biochemical mechanistic considerations.

D. Biosyntheses and Uses of Labelled Anthraquinones

1. Biosynthesis of [^{14}C] and [^{13}C]islandicin

Carbon-14 labelled islandicin (**263**), a red pigment isolated from the mould *P. islandicum*, has been biosynthesized by Gatenbeck[169] from a polyketide chain precursor. Incorporation of $Me^{14}COONa$ into the *P. islandicum* culture resulted in the labelling pattern shown in **263**, which was confirmed subsequently[170] by incorporation of $Me^{13}COONa$ into the growing organism. Using 90% enriched ^{13}C doubly labelled acetate, $^{13}Me^{13}COONa$, it has been shown that islandicin is biosynthesized from the precursor **264**. The 2–2.5% incorporation level of ^{13}C was determined by adding $^{14}MeCOONa$ to [^{13}C]acetate. Prior to the ^{13}C-NMR resonance study[171] of the islandicin biosynthesized from $^{13}MeCOONa$, the isolated labelled product was converted with Ac_2O/py to the triacetate.

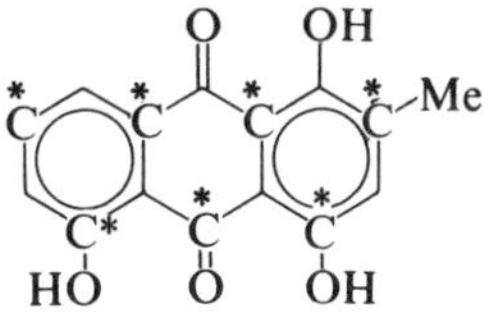

C* = ^{14}C or ^{13}C

(**263**)

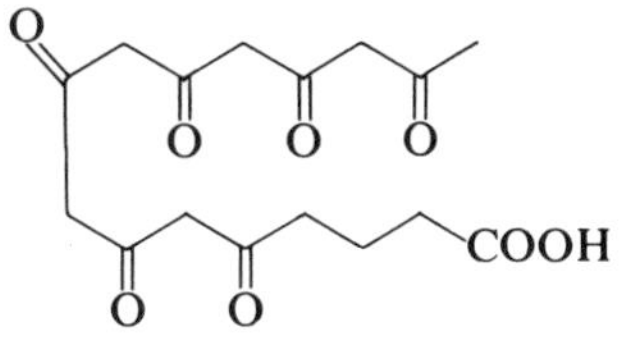

(**264**)

The antibiotic cerulenin, (2*S*),(3*R*) 2,3-epoxy-4-oxo-7,10-dodecadienoylamide (**265**), isolated from the culture filtrate of the fungus *Cephalosporium caerulens*[172] is known as a specific inhibitor of fatty acid biosynthesis[173]. When added[174] to a culture in which

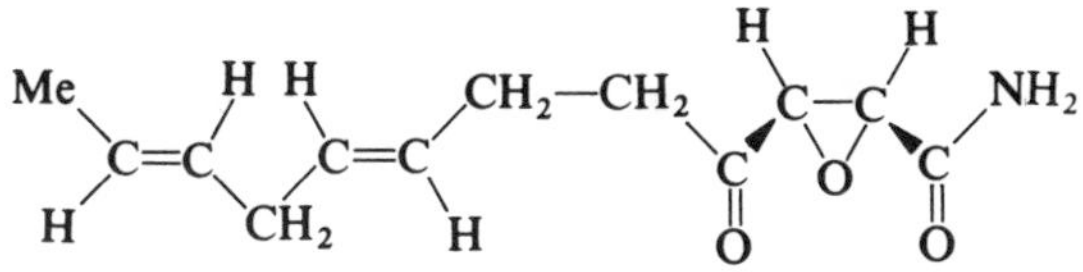

(**265**) cerulenin

vegetative mycelia of *Cortinarius orichalceus* were grown, it caused a simultaneous drastic decrease of the incorporation of the [1-^{14}C]acetate into both the mycelial triacylglycerols and the anthraquinone pigments emodin-6,8-dimethyl ether and anthraquinone physcion (produced also by the *C. orichalceus*). Cerulenin acts specifically on the polyketide formation. Hence the drastic decrease of the radioactivity incorporated into emodin-6,8-dimethyl ether in the presence of cerulenin confirms the suggestion that the biosynthesis of anthraquinones in higher fungi takes place via a condensation reaction of acetyl-CoA and malonyl-CoA, formation of an intermediate polyketo acid which undergoes intramolecular condensation, and cyclization.

2. Biosynthesis of chrysophanol and emodin

Two different modes of incorporation of [^{14}C]acetate into structurally different types of anthraquinones were established[175] by biosynthesis of chrysophanol (**267**) and emodin

(**268**) in *Rhamnus frangula* and *Rumex alpinus* from [1-^{14}C]acetate and [2-^{14}C]acetate. Radioactivity from [^{14}C]acetate enters only into ring C of alizarin (**266**), while the radioactivity from [^{14}C]acetate was found to be equally distributed in chrysophanol. The outlined differences were explained by suggesting that ring C of alizarin is derived from mevalonic acid which itself originates from acetate, whereas the chrysophanol is synthesized from an anthraquinone precursor in the course of cyclization of the polyketide unit derived from labelled acetate by linear combination.

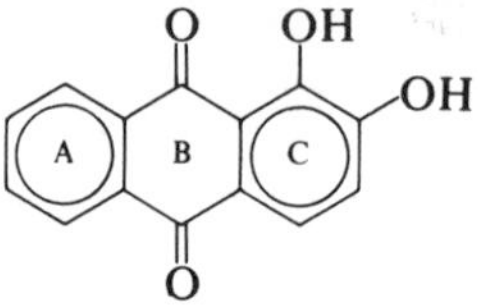

(**266**)

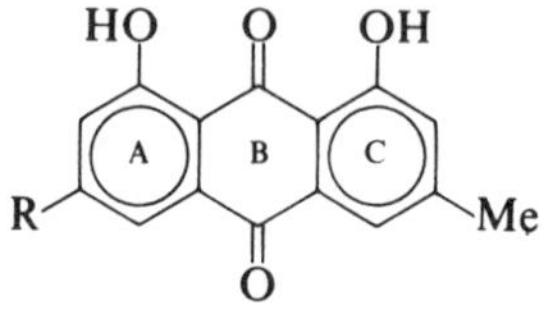

(**267**) R = H

(**268**) R = OH

[^{14}C]Shikimate and [^{14}C]mevalonate did not incorporate into **267** or **268**. Carbon-14 labelled **267**, **268** and chryzasin **269** were also biosynthesized by *Rumex obtusifolius* from [1-^{14}C] and [2-^{14}C]acetates[176].

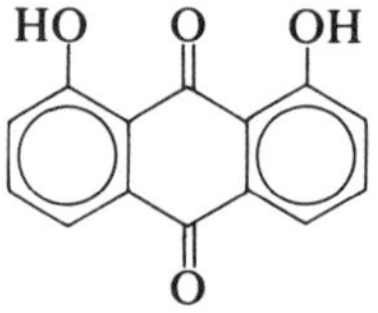

(**269**)

The isolated radioactive anthraquinones were degraded to phthalic and hydroxyphthalic acids. Comparison of the radioactivity present in the isolated acids and in the yellow quinones suggested that the acetate malonate route is used for producing anthraquinones in these higher plants. The metabolism of anthraquinones in the developing inflorescences was investigated by feeding radioactive anthraquinones, obtained from plants which had been grown in $^{14}CO_2$, to 20 young inflorescences. The results indicate that anthraquinones are rapidly utilized in the early stages of fruit formation. Their rate of conversion during the next period is slower though continues to be steady, similar to what has been found in developing fruits of *Cassia acutifolia* L.[177]. Singly and doubly labelled [^{13}C]acetates were also effectively[178] incorporated into griseofulvin (**270**), an important antifungal antibiotic, by a mutant strain of *Penicillium patulum*, by simple folding of a single heptaketide chain (equation 125). By feeding[179] the ammonium salt of

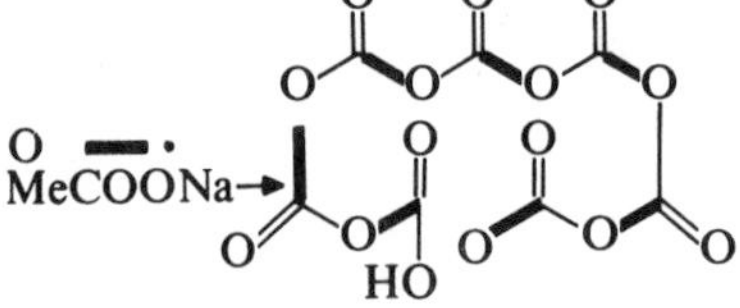

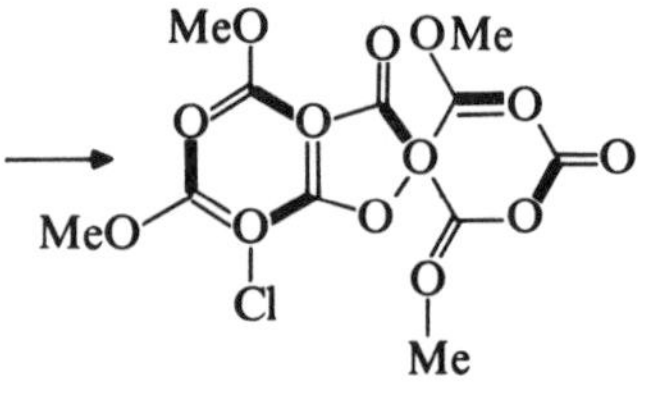

MeCOONa → (**270**) (125)

endocrocin (**271**) ^{14}C-labelled at C(9) and in the carboxyl group[180] to young mushrooms *Dermocybe sanguinea* it has been shown that **271** is the precursor only of the carboxylic acids **272** and **273** (equation 126). In a second experiment the labelled precursor [2,4-

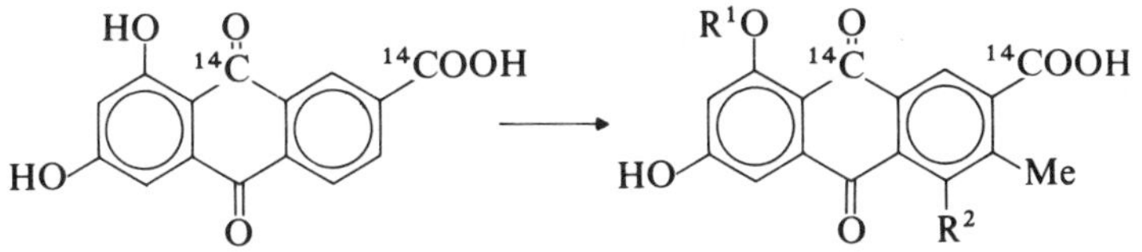

(271) → (126)

(272) $R^1 = R^2 = H$
(273) $R^1 = Me$, $R^2 = H$

$^3H_2$]emodin-6-mono-β-D-glucoside (**274**) was administered to young *D. semisanguinea* (Bull. exFr). Only the tritium-labelled neutral pigments **276** and **275** were obtained (equation 127). It has therefore been suggested that anthraquinones are biosynthesized

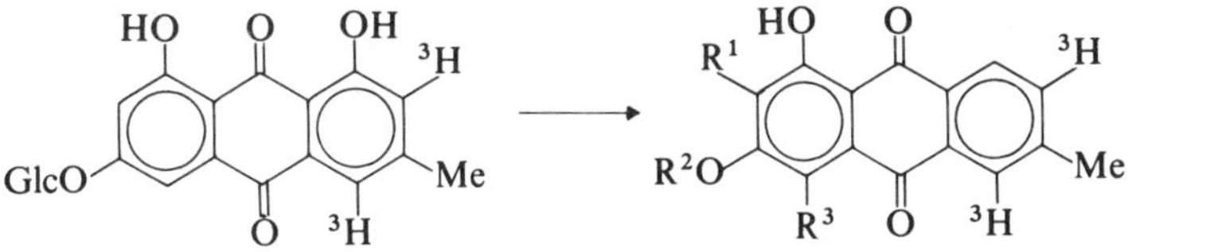

(274) → (127)

(275) $R^1 = OH$, $R^2 = Me$, $R^3 = H$
(276) $R^1 = R^3 = OH$, $R^2 = Me$

according to equation 128, which involves either endocrocin-9-anthrone or a compound in which the ring carrying the carboxyl group is not yet aromatic, such as the common precursor [X] of anthraquinones[179–181]

Polyacetate → [X] → (274) → (275) → (276)
Polyacetate → [X] → (271) → (272) → (273) (128)

Emodin-6-mono-β-D-glucoside (**274**) labelled with tritium in the non-labile 2 and 4 positions has been obtained by tritium exchange reaction between emodin (**268**) and 0.7 N KOT water solution during 24 h at 110 °C. Tritium from the hydroxyl groups and from relatively labile 5- and 7-T positions of the emodin molecule has been washed out by heating hydroxyl-^3H and [2,4,5,7-$^3H_4$]tritium-labelled emodin with 0.7 N KOH during 3 h at 110 °C and converting the product, labelled in [2,4-$^3H_2$] positions, into the monoglucoside[181]. In the preliminary deuterium exchange experiment it has been found that at 100 °C the aromatic protons of emodin are completely exchanged with 0.7 N KOD in the following decreasing order: 7-H (during 15 min) > 5-H (ca. 60 min) > 2-H (ca. 12 h) > 4-H (ca. 24 h)[181].

Efficient biosynthetic incorporation of ^{14}C-labelled emodinanthrone (**277**) as well as of [^{14}C]emodin into both dimeric skyrins (**278**, **279**) and (+) rugulosin (**280**), dimeric anthraquinonoids of *Penicillium brunneum* and *P. islandicum*, has been investigated by Sankawa and coworkers[182].

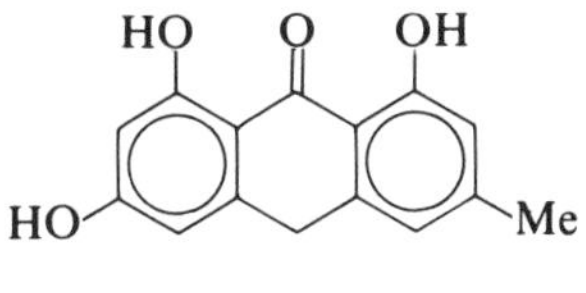

(277)

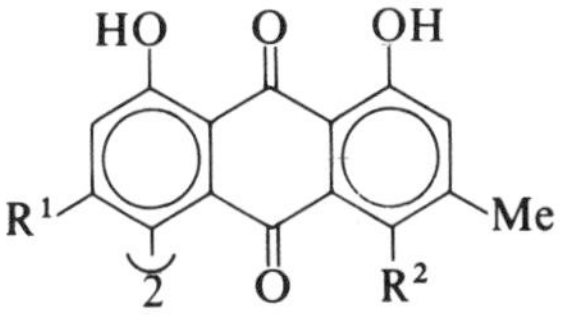

(278) $R^1 = OH$, $R^2 = H$
(279) $R^1 = H$, $R^2 = OH$

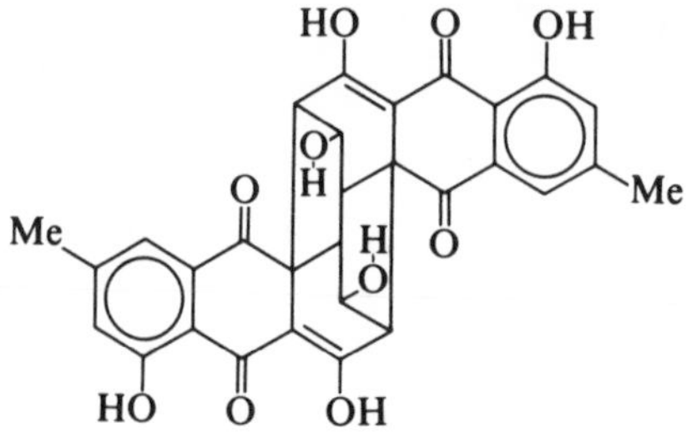

(280)

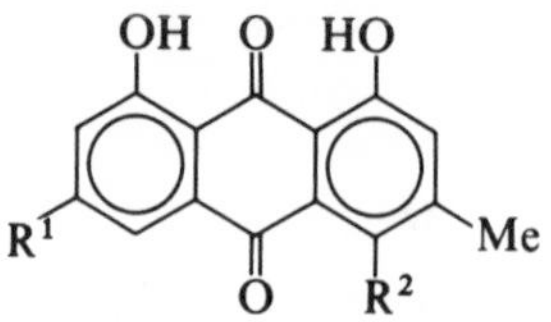

(281) $R^1 = R^2 = OH$
(263) $R^1 = H, R^2 = OH$
(267) $R^1 = R^2 = H$

No incorporation of [^{14}C]catenarin (**281**) into islandicin (**263**) was found and it has been suggested that the hydroxylation must take place after partial hydrogenation of the ring. Catenarin (**281**) produced biosynthetically does not undergo biohydrogenation.

3. Biosynthesis of ^{14}C-labelled alizarin

The mechanism of biosynthetic incorporation of shikimic acid into alizarin (**266**) in *Rubia tinctorum* has been investigated[183] by using carboxyl-^{14}C-labelled shikimic acid which is known to be incorporated as an intact C_7 unit into vitamin K, juglone, lawsone and alizarin. In the case of 5-hydroxy-1,4-naphthoquinone (juglone, **215**) the carboxyl group of shikimic acid is incorporated into both keto-C-atoms to an equal extent[184]. In the case of alizarin biosynthesis in *Rubia tinctorum* a non-symmetrical incorporation of [^{14}C]carboxyl group of shikimic acid was found and the biochemical sequence given in equation 129 was suggested. The phthalic acid obtained in the degradation of [^{14}C]alizarin

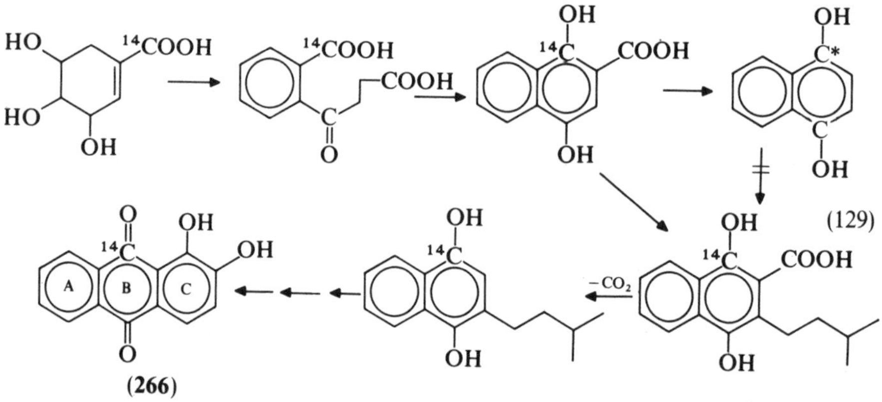

(129)

(266)

had the same specific radioactivity as the starting material, the benzoic acid obtained contained about 50% of the activity of alizarin, whereas veratric acid, whose carboxyl group corresponds to the keto-C(10) atom of alizarin was almost inactive. A symmetric compound like 1,4-naphthoquinol was thus excluded as an intermediate in the biosynthesis of [^{14}C]alizarin. After [1,2-^{14}C]shikimate feeding, the radioactivity in alizarin was confined only to ring A[185]. In a subsequent ^{14}C tracer experiment[186] the synthesis of alizarin was taking place in the cut off *Rubia tinctorum* roots fed with carboxyl-[^{14}C]-D-shikimic acid and [5-^{14}C]mevalonic acid. The labelled alizarin degraded according to equation 130. The results of degradation showed that:

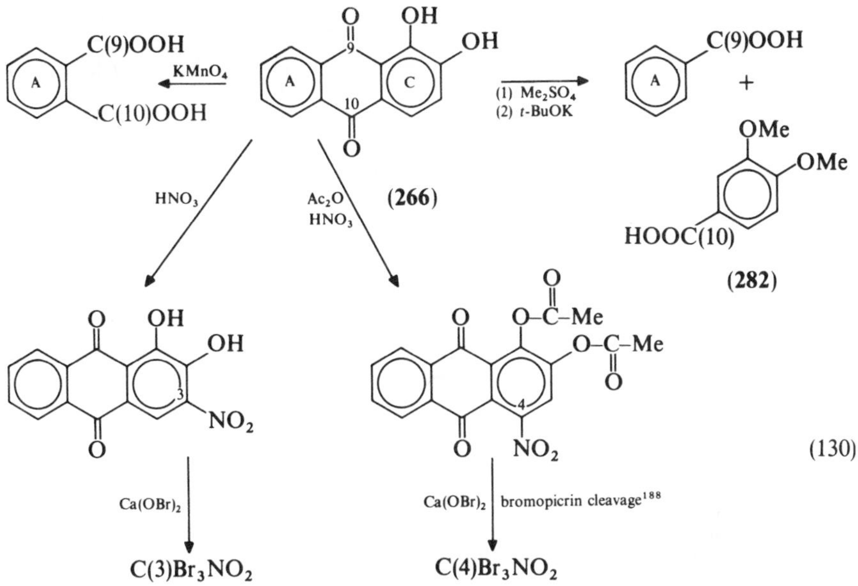

(130)

(1) After [7-^{14}C]shikimate feeding the carboxyl group of shikimic acid is exclusively incorporated non-symmetrically into the C(9) atom of alizarin. Veratric acid (**282**), the carboxyl group of which is derived from the C(10) atom of alizarin contains no radioactivity, while benzoic acid, formed by the degradation procedure of Davies and Hodge[187], contains all the radioactivity. Phthalic acid contained all the radioactivity corresponding to the keto-C-atoms of alizarin.

(2) After [5-^{14}C]mevalonic acid feeding, the activity from mevalonic acid enters specifically into position 4 of alizarin[188]. The C(3) atom did not contain any radioactivity. This suggests that ring C(1) to C(4) carbons are derived from mevalonic acid by way of γ,γ-dimethylallylpyrophosphate.

Further biochemical experiments with [2-^{14}C]glutamate indicated that its carbon-14 incorporates specifically into C(10) of alizarin. Therefore the general scheme of equation

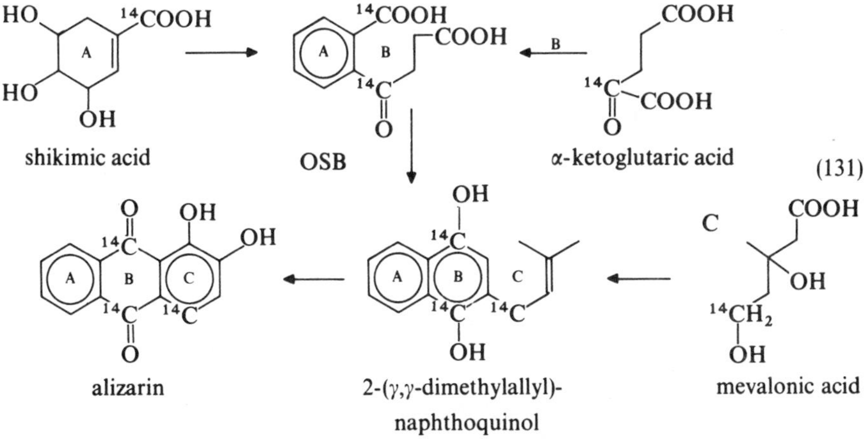

(131)

131 for radioactivity flow from different precursors to alizarin has been proposed. It does not include symmetrical intermediates like 1,4-naphthoquinone or 1,4-naphthoquinol.

Incorporation of ^{14}C-labelled compounds into morindone and alizarin anthraquinone skeleton by intact plants and cell suspension cultures of *Morinda citrifolia* has also been investigated[189]. In the case of anthraquinones produced by cell-suspension cultures, 1-methoxy-3-hydroxyanthraquinone-2-carboxaldehyde ('damnacanthal') was the most abundant, whereas alizarin and morindone (**283**) are produced in minor quantities.

(283) **(284)**

Morindone is the most abundant anthraquinone in the intact 1-year-old plants. The main degradation product of morindone was veratric acid derived from ring A plus C(9). Alizarin was split into benzoic acid and veratric acid. The incorporation of [1-^{14}C]acetic acid, [2-^{14}C]-DL-mevalonic acid and [U-^{14}C]-D-glucose into both intact plants and plant suspension cultures of *Morinda citrifolia* was negligible. No incorporation was obtained with mevalonic acid. Highest incorporation of [7-^{14}C]-D-shikimic acid and *o*-(succinoyl-4-^{14}C) benzoic acid (OSB) carboxyl-^{14}C was found. ^{14}C Tracer degradation studies showed that the radioactivity from [7-^{14}C]shikimic acid and OSB carboxyl-^{14}C incorporates specifically into the C(9) position of alizarin and morindone. It is suggested that ring C of anthraquinones, biosynthesized in *Morindona citrifolia*, is derived from mevalonic acid and the hydroxy groups are introduced at a later stage of the biosynthetic pathway. This is similar to what has been found for the incorporation of OSB into juglone[156]. Purpurin carboxylic acid (**284**) is derived from shikimic acid, glutamic acid and mevalonic acid by way of OSB in a similar way to the alizarin biosynthesis of equation 131.

IV. ACKNOWLEDGEMENTS

M. K. thanks professor A. Fry for generous permisson to use the modern facilities of the University library during her summer research tenure at the University of Arkansas. M. Z. thanks the Jagiellonian University of Cracow for providing him with living quarters in the University House of Creative Work in Zakopane during the summer time where many pages of this chapter were written. The biomedical aspects of the numerous compounds mentioned in the chapter especially in relation to plant pharmacology and physiology were discussed as before with the wife of one of us, mgr. pharm. Halina Papiernik-Zielinska. Dr R. Kanski's help in the course of retyping the revised manuscript is also acknowledged. The work was supported by the Faculty of Chemistry of Jagiellonian University.

V. REFERENCES

1. W. Oettmeir, *J. Labelled Compounds and Radiopharm.*, **15**, 581 (1978).
2. G. Braun, A. T. Shulgin and T. Sargent III, *Second International Symposium on Radiopharmaceutical Chemistry, St. Catherine's College Oxford, 3rd–7th July, 1978*, pp. 44–45.
3. T. Takahashi, T. Ido, R. Iwata, K. Ishiwata, K. Hamamura and K. Kogure, *J. Labelled Compounds and Radiopharm.*, **22**, 565 (1985).
4. F. A. Norris and G. G. Still, *J. Labelled Compounds*, **9**, 661 (1973).
5. K. H. Scharf and M. H. Zenk, *J. Labelled Compounds*, **7**, 525 (1971).
6. S. Urano, R. Muto and M. Matsuo, *J. Labelled Compounds and Radiopharm.*, **22**, 775 (1985).

7. H. Nakao, K. Fujimoto and H. Yanagisawa, *J. Labelled Compounds and Radiopharm.*, **15**, 381 (1978).
8. R. Pontikis, N. H. Nam, H. Hoellinger and L. Pichat, *J. Labelled Compounds and Radiopharm.*, **20**, 549 (1983).
9. M. Schonhartig, G. Mende, G. Siebert and Z. Hoppe-Seyler's, *Physiol. Chem.*, **355**, 1391 (1974).
10. A. Hunter and C. Klaassen, *J. Pharmacol. Exp. Ther.*, **192**, 605 (1975).
11. C. A. Elliger, *J. Labelled Compounds and Radiopharm.*, **20**, 515 (1983).
12. E. S. Hand and R. M. Horowitz, *J. Am. Chem. Soc.*, **86**, 2084 (1964).
13. H. D. Durst and E. Leete, *J. Labelled Compounds*, **7**, 52 (1971).
14. M. Herrchen and G. Legler, *Eur. J. Biochem.*, **138**, 527 (1984).
15. K. K. Hwang, B. Bhooshan, R. E. Kouri and C. J. Henry, *J. Labelled Compounds and Radiopharm.*, **19**, 35 (1982).
16. F. A. J. Muskiet, H. J. Jeuring, C. G. Thomasson, J. van der Meulen and B. G. Wolthers, *J. Labelled Compounds and Radiopharm.*, **14**, 49 (1978).
17. S. P. Markey, K. Powers, D. Dubinsky and I. J. Kopin, *J. Labelled Compounds and Radiopharm.*, **17**, 103 (1980).
18. G. F. Kolar, *J. Labelled Compounds*, **7**, 409 (1971).
19. K. Hiraoka and T. Miyamoto, *J. Labelled Compounds and Radiopharm.*, **28**, 613 (1981).
20. R. C. Murphy, *J. Labelled Compounds*, **11**, 341 (1975).
21. A. Kalir, C. Freed, K. L. Melmon and N. Castagnoli, Jr, *J. Labelled Compounds and Radiopharm.*, **13**, 41 (1977).
22. A. F. Fentiman and R. L. Foltz, *J. Labelled Compounds and Radiopharm.*, **12**, 69 (1976).
23. C. Brown, J. Eustache, J. P. Frideling and B. Shroot, *J. Labelled Compounds and Radiopharm.*, **21**, 973 (1984).
24. W. Steglich and W. Reininger, *Chem. Commun.*, 178 (1970).
25. B. R. Vishnuvajjala, T. Kataoka, F. D. Cazer, D. T. Witiak and L. Malspeis, *J. Labelled Compounds and Radiopharm.*, **14**, 77 (1978).
26. T. H. Smith, A. N. Fujiwara, D. W. Henry and W. W. Lee, *J. Am. Chem. Soc.*, **98**, 1970 (1976).
27. C. R. Chen, M. T. Fong, A. N. Fujiwara, D. W. Henry, M. A. Leaffer, W. W. Lee and T. H. Smith, *J. Labelled Compounds and Radiopharm.*, **14**, 111 (1978).
28. J. Šeda, M. Fuad and R. Tykva, *J. Labelled Compounds and Radiopharm.*, **14**, 673 (1978).
29. B. Franck and A. Stange, *Liebigs Ann. Chem.*, 2106 (1981).
30. B. Franck and H. P. Gehrken, *Angew. Chem.*, **92**, 484 (1980).
31. T. M. Harris, C. M. Harris and K. B. Hindley, *Fortschr. Chem. Org. Naturst.*, **31**, 217 (1974).
32. T. M. Harris, C. M. Harris, A. D. Webb, P. I. Wittek and T. P. Murray, *J. Am. Chem. Soc.*, **98**, 6065 (1976).
33. B. Franck, H. Backhaus and M. Rolf, *Tetrahedron Lett.*, **21**, 1185 (1980).
34. B. Franck, G. Bringmann and G. Flohr, *Angew. Chem.*, **92**, 483 (1980).
35. B. Franck, *Angew. Chem.*, **96**, 462 (1984).
36. G. Breipohl, Dissertation, University of Münster, 1982.
37. C. J. Proctor, B. Kralj, E. A. Larka, C. J. Porter, A. Maquestian and J. H. Beynon, *Org. Mass Spectrom.*, **16**, 312 (1981).
38. A. C. Ho, J. H. Bowie and A. Fry, *J. Chem. Soc. B*, 530 (1971).
39. J. H. Bowie and A. C. Ho, *Aust. J. Chem.*, **24**, 1093 (1971).
40. J. H. Bowie and A. C. Ho, *Aust. J. Chem.*, **26**, 2009 (1973).
41. J. C. Wilson and J. H. Bowie, *Aust. J. Chem.*, **28**, 1993 (1975).
42. T. McAllister, *J. Chem. Soc. Chem. Commun.*, 245 (1972).
43. J. C. Wilson, J. A. Benbow, J. H. Bowie and R. H. Prager, *J. Chem. Soc. Perkin Trans. 2*, 498 (1978).
44. Y. Sawaki and C. S. Foote, *J. Am. Chem. Soc.*, **105**, 5035 (1983).
45. Y. Berger, A. Castonguay and P. Brassard, *Org. Magnetic Res.*, **14**, 103 (1980).
46. Y. Berger and J. Jadot, *Bull. Soc. Chim. Belg.*, **85**, 271 (1976).
47. C. P. Gorst-Allman, K. G. R. Pachler, P. S. Steyn, P. L. Wessels and D. B. Scott, *J. Chem. Soc. Perkin Trans. 1*, 2181 (1977).
48. (a) D. L. Fitzell, D. P. H. Hsieh, R. C. Yao and G. N. la Mar, *J. Agric. Food Chem.*, **23**, 442 (1975); (b) K. G. R. Pachler, P. S. Steyn, R. Vleggaar, P. L. Wessels and D. B. Scott, *J. Chem. Soc. Perkin Trans. 1*, 1182 (1976).
49. M. T. Lin and D. P. H. Hsieh, *J. Am. Chem. Soc.*, **95**, 1668 (1973).

50. F. W. Wehrli, *J. Chem. Soc. Chem. Commun.*, 663 (1975).
51. M. Kobayashi, Y. Terui, K. Tori and N. Tsuji, *Tetrahedron Lett.*, 619 (1976).
52. J. A. Pedersen, *J. Magnetic Res.*, **43**, 373 (1981).
53. B. J. Hales and E. E. Case, *Biochim. Biophys. Acta*, **637**, 291 (1981).
54. M. H. Van Benthem and G. D. Gillispie, *J. Phys. Chem.*, **88**, 2954 (1984).
55. D. Hadzi and N. Sheppard, *Trans. Faraday Soc.*, **50**, 911 (1954).
56. H. Veenvliet and A. Wiersma, *Chem. Phys.*, **8**, 432 (1975).
57. (a) I. A. S. Edwards and H. P. Stadler, *Acta Cryst.*, **B27**, 946 (1971); (b) T. A. Kokrashvili, P. P. Levin and V. A. Kuzmin, *Izv. Akad. Nauk SSSR, Ser. Khim.*, 765 (1984).
58. (a) B. B. Adeleke and J. K. S. Wan, *J. Chem. Soc. Faraday Trans. 1*, 1799 (1976); (b) N. J. Turro and B. Kraeutler, in *Isotopic Effects: Recent Developments in Theory and Experiment* (Eds E. Buncel and C. C. Lee), Elsevier, Amsterdam, 1984, Chapter 3, pp. 107–160.
59. G. G. Lazarev, Ya. S. Lebedev and M. V. Serdobov, *Izv. Akad. Nauk SSSR, Ser. Khim.*, 2358 (1976).
60. M. Shirai, T. Awatsuji and M. Tanaka, *Bull. Chem. Soc. Jpn*, **48**, 1329 (1975).
61. S. Hashimoto, K. Kano and H. Okamoto, *Bull. Chem. Soc. Jpn*, **45**, 966 (1972).
62. K. Kano and T. Matsuo, *Tetrahedron Lett.*, 4323 (1974).
63. M. Jacobsen, S. C. Sharma and K. Torssell, *Acta Chem. Scand.*, **B33**, 499 (1979).
64. N. Jacobsen, *Org. Synth.*, **56**, 68 (1977).
65. G. W. Barton, Y. Le Page, E. J. Gabe and K. U. Ingold, *J. Am. Chem. Soc.*, **102**, 7791 (1980).
66. G. W. Burton and K. U. Ingold, *J. Am. Chem. Soc.*, **103**, 6472 (1981).
67. J. A. Howard and E. Furimsky, *Can. J. Chem.*, **51**, 3738 (1973).
68. J. A. Howard and K. U. Ingold, *Can. J. Chem.*, **40**, 1851 (1962).
69. J. A. Howard and K. U. Ingold, *Can. J. Chem.*, **41**, 1744 (1963).
70. J. A. Howard and K. U. Ingold, *Can. J. Chem.*, **41**, 2800 (1963).
71. D. L. Kleyer and T. H. Koch, *J. Am. Chem. Soc.*, **105**, 5911 (1983).
72. D. L. Kleyer and T. H. Koch, *J. Am. Chem. Soc.*, **105**, 2504 (1983).
73. J. M. Burns, D. L. Wharry and T. H. Koch, *J. Am. Chem. Soc.*, **103**, 849 (1981).
74. G. A. Kutyrev, A. A. Kutyrev, R. A. Cherkasov and A. N. Pudovik, *Phosphorus and Sulfur*, **13**, 135 (1982).
75. T. A. Weil, S. Friedman and I. Wender, *J. Org. Chem.*, **39**, 48 (1974).
76. L. Melander, *Isotope Effects on Reaction Rates*, Ronald Press, New York, 1960.
77. C. J. Collins and N. S. Bowman (Eds), *Isotope Effects in Chemical Reactions*, Van Nostrand Reinhold Co., New York, 1970.
78. E. Caldin and V. Gold (Eds), *Proton-Transfer Reactions*, Chapman and Hall, London, New York, 1975.
79. E. S. Lewis, Chapter 10 in Ref. 78, pp. 317–338.
80. E. S. Lewis and M. M. Butler, *J. Am. Chem. Soc.*, **98**, 225 (1976).
81. E. S. Lewis and K. Ogno, *J. Am. Chem. Soc.*, **98**, 2260 (1976).
82. R. P. Bell, *Chem. Soc. Rev.*, **3**, 513 (1975).
83. K. T. Leffek, in *Isotopes in Hydrogen Transfer Processes* (Eds. E. Buncel and C. C. Lee), Elsevier, Amsterdam, 1976, Chapter 3, pp. 89–125.
84. R. A. More O'Ferrall, *J. Chem. Soc. B*, 785 (1970).
85. A. Jarczewski and P. Pruszyński, *Wiadomości Chemiczne*, **32**, 693 (1978).
86. M. Zieliński, *Isotope Effects in Chemistry*, Polish Sci. Publishers, Warsaw, 1979.
87. L. Melander and W. H. Saunders, Jr, *Reaction Rates of Isotopic Molecules*, John Wiley, New York, 1980.
88. R. P. Bell, *The Tunnel Effect in Chemistry*, Chapman and Hall, New York, 1980.
89. N. S. Isaacs, Ref. 58b, Chapter 2, pp. 67–105.
90. L. B. Sims and D. E. Lewis, in Ref. 58b, Chapter 4, pp. 161–259.
91. A. Jarczewski, P. Pruszyński and K. T. Leffek, *Can. J. Chem.*, **53**, 1176 (1975).
92. A. Jarczewski, P. Pruszyński and K. T. Leffek, *J. Chem. Soc. Perkin Trans. 2*, 814 (1977).
93. A. Jarczewski, P. Pruszyński and K. T. Leffek, *Can. J. Chem.*, **57**, 669 (1979).
94. A. Jarczewski, G. Schroeder and K. T. Leffek, *J. Chem. Soc. Perkin Trans. 2*, 866 (1979).
95. K. T. Leffek and P. Pruszyński, *Can. J. Chem.*, **59**, 3034 (1981).
96. K. T. Leffek and P. Pruszyński, *Can. J. Chem.*, **60**, 1692 (1982).
97. A. Jarczewski, P. Pruszyński and K. T. Leffek, *Can. J. Chem.*, **61**, 2029 (1983).
98. A. Jarczewski, P. Pruszyński, M. Kazi and K. T. Leffek, *Can. J. Chem.*, **62**, 954 (1984).

99. A. Jarczewski, G. Schroeder, W. Galezowski, K. T. Leffek and U. Maciejewska, *Can. J. Chem.*, **63**, 576 (1985).
100. G. Schroeder and A. Jarczewski, *Fast Reactions In Solutions Discussion Group Meeting, 15–17 September 1985*, University of York, England, Abstract p. 20.
101. V. A. Bren, V. A. Chernoivanov, L. E. Konstantinovski, V. I. Minkin, L. E. Nivorozhkin and Yu. A. Zhdanov, *Dokl. Akad. Nauk SSSR*, **251**, 1129 (1980).
102. B. M. Trost, *J. Am. Chem. Soc.*, **89**, 1847 (1967).
103. Z. M. Hashish and I. M. Hoodless, *Can. J. Chem.*, **54**, 2261 (1976).
104. P. Müller, *Helv. Chim. Acta.*, **56**, 1243 (1973).
105. P. Müller and D. Joly, *Tetrahedron Lett.*, **21**, 3033 (1980).
106. P. Müller and J. Roček, *J. Am. Chem. Soc.*, **94**, 2719 (1972).
107. H. Kwart and T. J. George, *J. Org. Chem.*, **44**, 162 (1979).
108. H. Kwart and J. J. Stanulonis, *J. Am. Chem. Soc.*, **98**, 4009 (1976).
109. M. Zieliński, in *The Chemistry of Carboxylic Acids and Esters* (Ed. S. Patai), Interscience, London, 1969, Chapter 10.
110. A. Ohki, T. Nishiguchi and K. Fukuzumi, *Tetrahedron*, **35**, 1737 (1979).
111. K. K. Banerji, *Bull. Chem. Soc. Jpn*, **50**, 1616 (1977).
112. S. H. Burstein and H. J. Ringold, *J. Am. Chem. Soc.*, **86**, 4952 (1964).
113. A. Ohki, T. Nishiguchi and K. Fukuzumi, *J. Org. Chem.*, **44**, 766 (1979).
114. A. K. Colter, G. Saito, F. J. Sharom and A. P. Hong, *J. Am. Chem. Soc.*, **98**, 7833 (1976).
115. A. K. Colter, G. Saito and F. J. Sharom, *Can. J. Chem.*, **55**, 2741 (1977).
116. (a) G. Saito and A. K. Colter, *Tetrahedron Lett.*, 3325 (1977); (b) A. van Laar, H. J. van Ramesdonk and J. W. Verhoeven, *Recl. Trav. Chim. Pays-Bas*, **102**, 157 (1983).
117. S. Fukuzumi, N. N. Nishizawa and T. Tanaka, *J. Org. Chem.*, **49**, 3571 (1984).
118. T. G. Sterleva, L. A. Kiprianova, A. F. Levit and I. P. Gragerov, *Zh. Org. Khim.*, **12**, 2034 (1976).
119. L. A. Rykova, L. A. Kiprianova and I. P. Gragerov, *Teor. Eksp. Khim.*, **16**, 124 (1980).
120. L. A. Rykova, L. A. Kiprianova and I. P. Gragerov, *Teor. Eksp. Khim.*, **16**, 825 (1980).
121. I. P. Gragerov, A. F. Levit, L. A. Kiprianova, T. G. Sterleva and L. A. Rykova, *Teor. Eksp. Khim.*, **17**, 595 (1981).
122. R. J. Klinger, K. Mochida and J. K. Kochi, *J. Am. Chem. Soc.*, **101**, 6626 (1979).
123. C. D. Ritchie, W. F. Sager and E. S. Lewis, *J. Am. Chem. Soc.*, **84**, 2349 (1962).
124. E. S. Lewis and J. K. Robinson, *J. Am. Chem. Soc.*, **90**, 4337 (1968).
125. N. S. Isaacs, K. Javaid and E. Rannala, *J. Chem. Soc., Perkin Trans. 2*, 709 (1978).
126. N. Nishimura and T. Motoyama, *Bull. Chem. Soc. Jpn*, **57**, 1 (1984).
127. N. Sugimoto, M. Sasaki and J. Osugi, *J. Am. Chem. Soc.*, **105**, 7676 (1983).
128. (a) N. Sugimoto, M. Sasaki and J. Osugi, *J. Chem. Soc. Perkin Trans. 2*, 655 (1984). (b) A. Fry, in *Isotope Effects in Chemical Reactions* (Eds C. J. Collins and N. S. Bowman), Van Nostrand Reinhold Co., New York, 1970, pp. 364–411.
129. J. S. Mitchell, in *Radiotracer Techniques and Applications*, Vol. 2 (Eds E. A. Evans and M. Muramatsu), Marcel Dekker Inc., New York and Basel, 1977, pp. 1081–1110.
130. R. Bentley, *Biosynthesis*, **3**, 181 (1975).
131. R. Bentley, *Pure Appl. Chem.*, **41**, 47 (1975).
132. E. M. Galimov, *The Nature of the Biological Fractionation of Isotopes*, Edition "Science", Moscow, 1981, in Russian.
133. (a) M. B. Neiman and D. Gal, *Kinetic Isotope Method and its Applications*, Akademiai Kiado, Budapest 1971; (b) M. B. Neiman, *Nukleonika*, **1**, 147 (1956); (c) N. M. Emanuel and D. Gal, *Oxidation of Ethylbenzene*, Edition "Science", Moscow, 1984, pp. 41–43 and references cited therein, in Russian.
134. G. R. Whistance and D. R. Threlfall, *Biochem. J.*, **117**, 593 (1970).
135. G. R. Whistance and D. R. Threlfall, *Biochem. J.*, **109**, 577 (1968).
136. D. R. Threlfall and G. R. Whistance, *Phytochem.*, **16**, 1903 (1977).
137. W. Janiszowska, W. Michalski and Z. Kasprzyk, *Phytochemistry*, **15**, 125 (1976).
138. W. Janiszowska and R. Jasińska, *Acta Biochim. Polon.*, **29**, 37 (1982).
139. W. Janiszowska, poster presented during Annual Meeting of the Polish Biochem. Soc., Cracow, September 1985.
140. J. F. Pennock, *Biochem. Soc. Trans.*, **11**, 504 (1983).
141. J. A. Hamilton and G. B. Cox, *Biochem. J.*, **123**, 435 (1971).
142. I. G. Young, P. Stoobant, C. G. Macdonald and F. Gibson, *J. Bacteriol.*, **114**, 42 (1973).

143. (a) I. G. Young, *Biochemistry*, **14**, 399 (1975); (b) F. Scherrer and R. Azerad, *J. Chem. Soc. Chem. Commun.*, 128 (1976).
144. R. W. Bryant, Jr and R. Bentley, *Biochemistry*, **15**, 4792 (1976).
145. B. Shineberg and I. G. Young, *Biochemistry*, **15**, 2754 (1976).
146. R. M. Baldwin, C. D. Snyder and H. Rapoport, *Biochemistry*, **13**, 1523 (1974).
147. R. Bentley, *Pure Appl. Chem.*, **41**, 47 (1975).
148. (a) G. B. Cox and F. Gibson, *Biochem J.*, **100**, 1 (1966); (b) C. T. Esmon, J. A. Sadowski and J. W. Suttie, *J. Biol. Chem.*, **250**, 4744 (1975).
149. W. U. Müller and E. Leistner, *Phytochem.*, **15**, 407 (1976).
150. W. U. Müller and E. Leistner, *Phytochem.*, **17**, 1735 (1978).
151. W. U. Müller and E. Leistner, *Phytochem.*, **17**, 1742 (1978).
152. H. V. Schmid and M. H. Zenk, *Tetrahedron Lett.*, 4151 (1971).
153. R. Durand and M. H. Zenk, *Tetrahedron Lett.*, 3009 (1971).
154. R. Durand and M. H. Zenk, *Phytochem.*, **13**, 1483 (1974).
155. M. N. Zaprometov, *Ber. Akad. Wiss. UDSSR*, **125**, 1359 (1959).
156. P. Dansette and R. Azerad, *Biochem. Biophys. Res. Commun.*, **40**, 1090 (1970).
157. I. M. Campbell, *Tetrahedron Lett.*, 4777 (1969).
158. E. Grotzinger and I. M. Campbell, *Phytochem.*, **13**, 923 (1974).
159. I. M. Campbell, *Tetrahedron Lett.*, 4777 (1969).
160. D. J. Robins, I. M. Campbell and R. Bentley, *Biochem. Biophys. Res. Commun.*, **39**, 1081 (1970).
161. R. Azerad, R. Bleiler-Hill, F. Catala, O. Samuel and E. Lederer, *Biochem. Biophys. Res. Commun.*, **27**, 253 (1967).
162. R. M. Baldwin, C. D. Snyder and H. Rapoport, *Biochemistry*, **13**, 1523 (1974).
163. K. Inoue, S. Ueda, Y. Shiobara and H. Inouye, *Tetrahedron Lett.*, **21**, 621 (1980).
164. K. Inoue, S. Ueda, Y. Shiobara, I. Kimura and H. Inouye, *J. Chem. Soc. Perkin Trans. 1*, 1246 (1981).
165. (a) K. Inoue, S. Ueda, H. Nayeshiro and H. Inouye, *J. Chem. Soc. Chem. Commun.*, **993** (1982); (b) H. Inouye, S. Ueda, K. Inoue, H. Nayeshiro and N. Moritome, Proc. 5th Intl. Cong. Plant Tissue and Cell Culture; *Plant Tissue Culture*, 1982, pp. 375–376.
166. I. G. Young, *Biochemistry*, **14**, 399 (1975).
167. E. Leistner, in *Pigment Plants*, 2nd Ed. (Ed. F. C. Czygan), Akad-Verlag, Berlin (G.D.R.), 1981, pp. 352–369.
168. (a) J. Stöckigt, U. Srocka and M. H. Zenk, *Phytochem.*, **12**, 2389 (1973); (b) S. L. Otten and J. P. Rosazza, *J. Biol. Chem.*, **258**, 1610 (1983); (c) S. L. Otten and J. P. Rosazza, *Appl. Environ. Microbiol.*, **35**, 554 (1978).
169. S. Gatenbeck, *Acta Chem. Scand.*, **14**, 296 (1960).
170. R. C. Paulick, M. L. Casey, D. F. Hillenbrand and H. W. Whitlock, Jr, *J. Am. Chem. Soc.*, **97**, 5303 (1975).
171. M. Tanabe, T. Hamasaki and H. Seto, *Chem. Commun.*, 1539 (1970).
172. Y. Iwai, J. Away and T. Hata, *J. Ferment. Technol.*, **51**, 575 (1973).
173. S. Omura, *Bacteriol. Rev.*, **40**, 681 (1976).
174. G. J. A. Gstraunthaler, *Biochim. Biophys. Acta*, **750**, 424 (1983).
175. E. Leistner, *Phytochem.*, **10**, 3015 (1971).
176. J. W. Fairbairn and F. J. Muhtadi, *Phytochem.*, **11**, 215 (1972).
177. J. W. Fairbairn and A. B. Shrestha, *Phytochem.*, **6**, 1203 (1967).
178. T. J. Simpson and J. S. E. Holker, *Phytochem.*, **16**, 229 (1977).
179. W. Steglich, R. Arnold, W. Lösel, and W. Reininger, *J. Chem. Soc. Chem. Commun.*, 102 (1972).
180. W. Steglich and W. Reininger, *J. Chem. Soc. Chem. Commun.*, 178 (1970).
181. W. Lösel, Diplom. Thesis, Technical University, Munchen, 1966.
182. U. Sankawa, Y. Ebizuka and S. Shibata, *Tetrahedron Lett.*, 2125 (1973).
183. E. Leistner and M. H. Zenk, *Tetrahedron Lett.*, 1677 (1971).
184. E. Leistner and M. H. Zenk, *Z. Naturforsch.*, **23b**, 259 (1968).
185. E. Leistner and M. H. Zenk, *Tetrahedron Lett.*, 1395 (1967).
186. E. Leistner, *Phytochem.*, **12**, 337 (1973).
187. D. G. Davies and P. Hodge, *J. Chem. Soc. C*, 3158 (1971).
188. R. Baddiley, G. Ehrensvärd, E. Klein, L. Reio and E. Saluste, *J. Biol. Chem.*, **183**, 777 (1950).
189. E. Leistner, *Phytochem.*, **12**, 1669 (1973).

The Chemistry of Quinonoid Compounds, Vol. II
Edited by S. Patai and Z. Rappoport

CHAPTER **20**

The solid state photochemistry of tetrahydronaphthoquinones: crystal structure–reactivity relationships

JOHN R. SCHEFFER and JAMES TROTTER
Department of Chemistry, University of British Columbia, Vancouver, Canada, V6T 1Y6

I. INTRODUCTION

The photochemistry of quinones, including their photochemical behavior in the solid state, was reviewed in Part 1 of *The Chemistry of Quinonoid Compounds*[1]. The present chapter summarizes the photochemistry of a class of compounds that is closely related to the *p*-quinones, namely the tetrahydro-1,4-naphthoquinones. A great majority of the work described has been carried out in our laboratory over the past 12 years, and we include some results that have not been published previously. The article is concerned primarily with the crystal structures of compounds containing the 2-en-1,4-dione (ene-dione) chromophore and the photochemical reactions that these molecules undergo in the crystalline phase. The solution phase photochemistry of these systems, which has also been investigated in detail in our laboratory, will be discussed in this review only to compare and contrast it with the solid state results.

The study of chemical reactions in the solid state and the interpretation of the results in terms of the conformation and packing arrangement of the molecules that make up the crystal lattice have provided unparalleled insights into the mechanistic features of both bimolecular and unimolecular processes[2]. This arises from two properties that are unique to the crystalline phase: (1) determination of the crystal structure by X-ray diffraction methods provides a detailed, three-dimensional view of the reaction ensemble immediately prior to reaction, and (2) because of the relatively strong forces that hold crystals together, atomic and molecular motions are restricted, and chemical reactions in crystals tend to be least motion in character such that the transition states, intermediates and products closely resemble the reactants in shape and volume. As a result, the chemical reactivity of a given molecule in its crystalline phase frequently differs from its chemistry in liquid phases or the gas phase, and this provides a further impetus for carrying out chemical studies in solids.

The first investigation of ene-dione photochemistry in the solid state was carried out by Diels and Alder in 1929[3]. They found that exposure of crystals of the *p*-benzoquinone/1,3-butadiene adduct **1** (Scheme 1) to sunlight yielded a high-melting isomer or polymer. In 1964, Cookson and coworkers[4] briefly reinvestigated this reaction and concluded that intermolecular [2+2] photocycloaddition had occurred across the ene-dione double bond to give a dimer. The extreme insolubility of the dimer prevented its complete characterization or the determination of its stereochemistry. In the same paper[4], Cookson and colleagues reported that photolysis of **1** in solution 'gave only tars'. In the early 1970s, we became interested in the photochemistry of **1** and its derivatives, and it is this work with which the majority of this article is concerned.

II. INITIAL STUDIES

In agreement with the findings of Cookson and coworkers, we too observed extensive tar formation when **1** was photolyzed in benzene or *t*-butanol[5]. Careful examination of the crude reaction mixture, however, revealed that two rearrangement products, subsequently shown to be diketones **1c** and **1d**, were present in a combined yield of approximately 10% (throughout this review, the letters **a**, **b**, **c** etc. are used to refer to photoproduct structural types. Scheme 1 thus indicates that irradiation of ene-dione **1** gives no type **b** product). The structure of photoproproduct **1d** rests on a crystal structure determination[6]. Also in agreement with Cookson's report, we found that irradiation of crystals of **1** gave dimeric material[7]. The dimer was assigned the structure and stereochemistry **1a** on the basis of an X-ray crystal structure analysis[8] and the crystal and molecular structure of ene-dione **1** was also determined[9].

The results outlined in Scheme 1 raise two major questions: (1) why does ene-dione **1** react differently in solution and the solid state, and (2) of the many possible stereoisomeric dimers, why is only **1a** formed in the solid state? The answer to both questions is found in

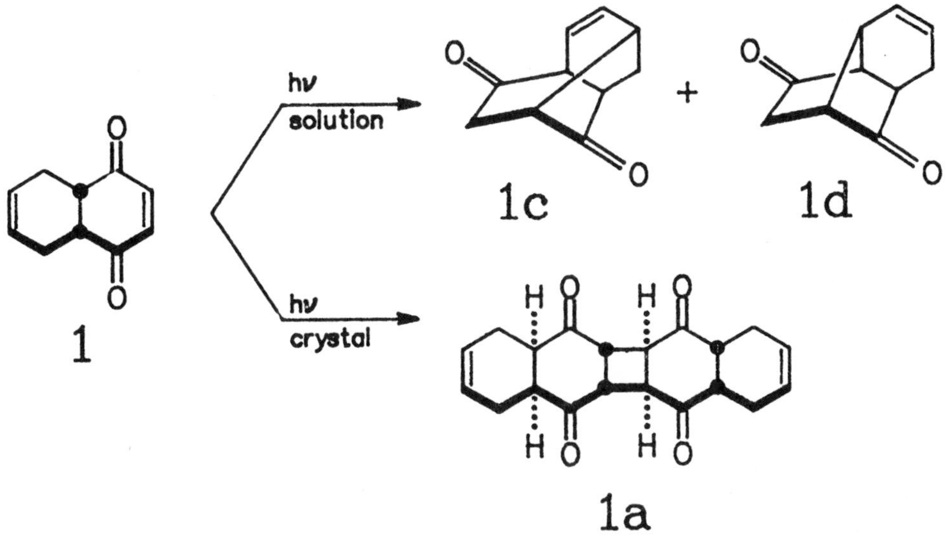

SCHEME 1

the crystal structure of ene-dione **1**, in which two neighboring molecules are arranged as shown schematically below; the ene-dione double bonds of neighboring molecules are oriented in an arrangement very favorable for [2+2] photocycloaddition; the double bonds approach each other in a parallel, top-to-bottom fashion with a center-to-center distance of 3.76 Å. Most significantly, the incipient dimer pair has a stereochemical relationship that is identical to that found in the photoproduct **1a**. Thus the dimer structure and stereochemistry are governed by the positions of the molecules in the crystal lattice (topochemical control), and this accounts for the stereoselectivity of the process. In solution, however, no fixed orientation exists between neighboring molecules, and they undergo unimolecular photorearrangement rather than dimerization. A discussion of the mechanism of the rearrangement is deferred to a later section of the review.

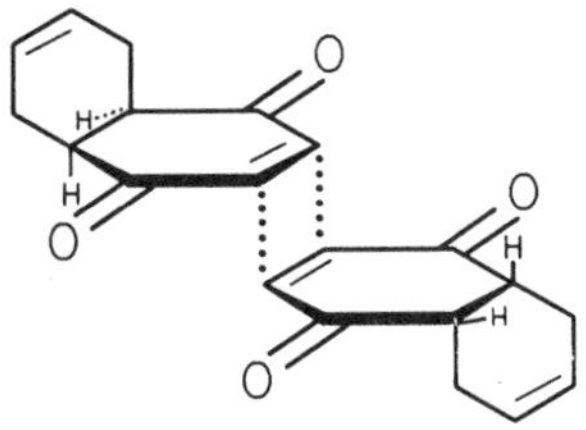

The photochemistry of ene-dione **1** is very similar in principle to the classic case of the cinnamic acids studied by Schmidt[2e]. In solution, the cinnamic acids undergo unimolecular photochemistry (*cis, trans* photoisomerization), whereas in the solid state they take part in [2+2] photodimerization owing to a favorable orientation between the double bonds of adjacent molecules. Based on the study of several examples, Schmidt suggested that a parallel, top-to-bottom approach with a center-to-center distance between double bonds of < 4.1 Å is required for efficient solid state photodimerization;

ene-dione **1** thus obeys Schmidt's rule, and this rule has been verified in a number of other instances as well[10].

An interesting aspect of the crystal structure of ene-dione **1** is that there are two independent molecules in the asymmetric unit as shown below. The two molecules have

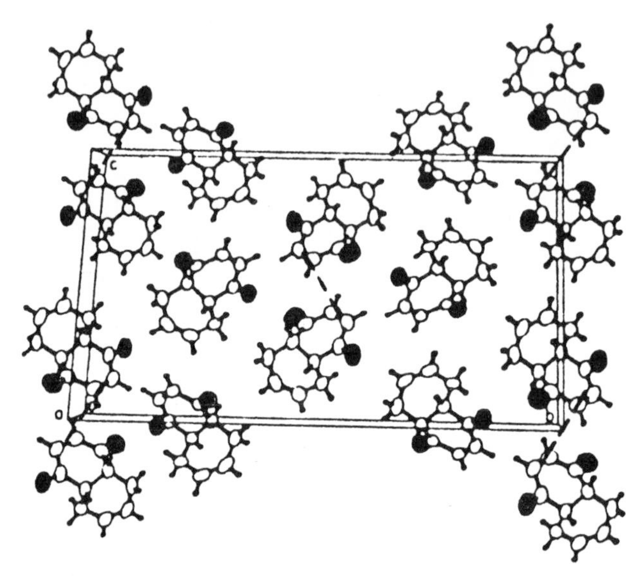

shapes (conformations) which differ only slightly; however, they differ dramatically in packing arrangement. One (discussed above) forms pairs about centers of symmetry, with 3.76 Å between double bonds; the other molecule has a nearest neighbor, non-parallel double bond separation of 5.27 Å, and is unsuitable for dimer formation. Because the two molecules are present in equal amounts in the crystal, the maximum chemical yield of dimer **1a** is predicted to be 50%. The observation of yields considerably above 50% was taken as an indication that some dimer may be forming at dislocation sites and defects that are developed in the crystal during the later stages of reaction[7].

In 1964, Cookson and coworkers[4] also reported that irradiation of crystals of the ene-diones **2** and **3** (Scheme 2), affords high yields of the corresponding *intramolecular* [2 + 2]

hν
solution & crystal

2 4

hν
solution & crystal

3 5

SCHEME 2

cycloadducts **4** and **5**. The same photoproducts were formed in solution, and no photorearrangement or intermolecular [2 + 2] photocycloaddition was observed. In order to investigate the reasons for these reactivity differences, Greenhough and Trotter[11] determined the crystal and molecular structure of ene-dione **3**. This showed that there are no close, parallel double bond contacts between lattice neighbors, hence no intermolecular [2 + 2] photocycloaddition in the solid state. This study also showed that the conformations of ene-diones **3** and **1** are quite different. These conformations are contrasted in the figure below. The carbon–carbon double bonds in **3** are parallel, with a center-to-center distance of 3.53 Å. This conformation clearly favors intramolecular photocycloaddition, as is observed. On the other hand, the conformer present in crystals of ene-dione **1** is not conducive to internal cycloaddition.

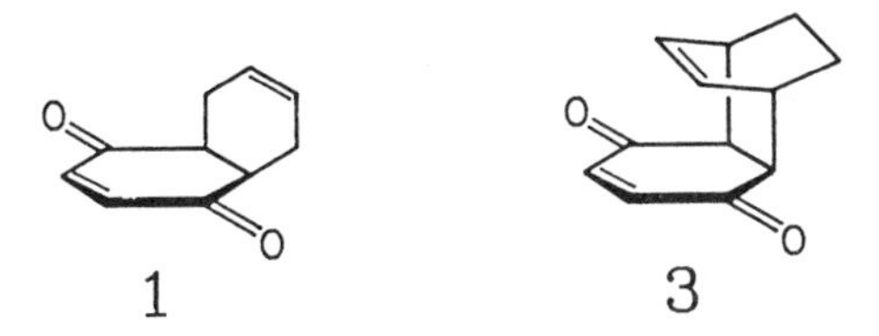

The conformations of ene-diones **1** and **3** in the solid state provide insight into the dynamic situation present in solution. The X-ray crystal structure study of **1** indicates that the molecule exists in a 'twist' conformation with staggered bridgehead hydrogen atoms (torsion angle of 60° about the C(4a)–C(8a) bond):

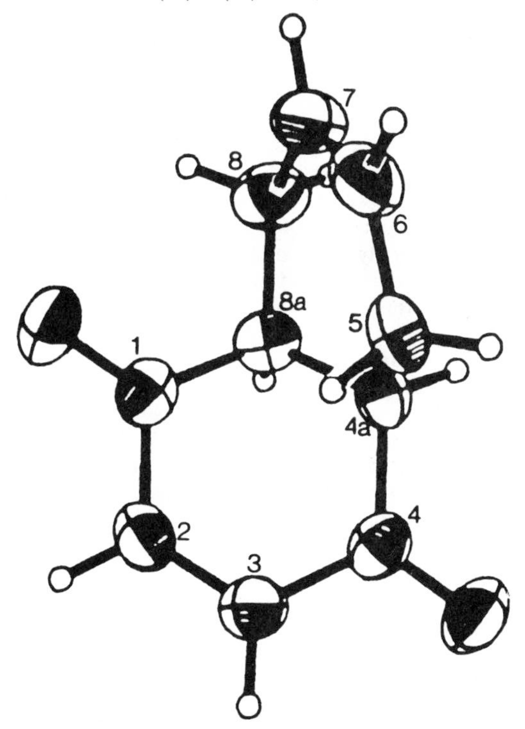

Both six-membered rings have half-chair conformations with approximately planar π-bond systems; bond torsion angles are 0–59°, mean 29° in the cyclohexene ring, and 2–52°, mean 26°, in the ene-dione ring, so that the latter ring is slightly flatter. As illustrated in Scheme 3, two such 'twist' conformations (**A** and **B**) are possible for ene-dione **1**. Conformers **A** and **B** have equal energy (they are enantiomeric), and both are present in the racemic crystals of **1** and are related by a half-chair to half-chair conformational inversion of the cyclohexene ring. The equilibrium between **A** and **B** passes through the eclipsed conformer **C**. **C** can be either *endo* (shown) or *exo* (not shown); only *endo*-**C** is capable of intramolecular [2+2] photocycloaddition. The Diels–Alder adducts **2** and **3**, which are *endo*, are forced into conformation **C** by the one- and two-carbon bridges across positions C(5) and C(8), and this accounts for their photochemistry (internal [2+2] cycloaddition), both in solution and the solid state.

A C B

A'

(Drawn as enantiomer of B)

SCHEME 3

We have carried out variable temperature NMR studies on a number of ene-diones that give quantitative information on the energy barriers separating conformers **A** and **B**[12]. Because **A** and **B** lack symmetry, nominally equivalent atoms [e.g. C(1) and C(4)] are magnetically non-equivalent. However, if the equilibrium between conformers **A** and **B** is rapid on the NMR time-scale, an average plane of symmetry is established, and the spectrum is simplified accordingly. We found[12] that the equilibrium between conformers **A** and **B** for ene-dione **1** could not be frozen out at temperatures as low as −90°C. However, with the introduction of methyl groups at the bridgehead atoms C(4a) and C(8a), the coalescence temperature was in the −60 to −70°C range, and the free energy of activation for the conformational equilibrium was determined to be 8.7–9.2 kcal mol^{-1}. The *solid state* magic angle spinning ^{13}C-NMR spectra of **1** and substituted derivatives of **1** are also most informative[12, 13]. For instance, the number of independent molecules in the asymmetric unit can be established quickly by noting the signal multiplicity of nominally equivalent carbon atoms. In the case of **1** (two independent molecules per asymmetric unit), there are four carbonyl carbon signals, two for each independent molecule; ene-diones with only one molecule in the asymmetric unit show carbonyl carbon doublets.

III. COMPOUNDS STUDIED: SYNTHESIS OF STARTING MATERIALS

In this chapter we shall discuss the X-ray crystallography and solid state photochemistry of 17 separate ene-diones, **1** and **6–21** (Scheme 4). All were prepared through relatively straightforward Diels–Alder chemistry. Compounds **6**, **7**, **8**, **9**, **10**, **12**, **14** and **17** were known when this work began; the others (except for **21**) were prepared for the first time in our laboratory. The unusual ene-dione **21** was prepared, photolyzed and its crystal structure determined by Weisz and coworkers[14]. Ene-diones **15**, **16**, **19** and **20** required the

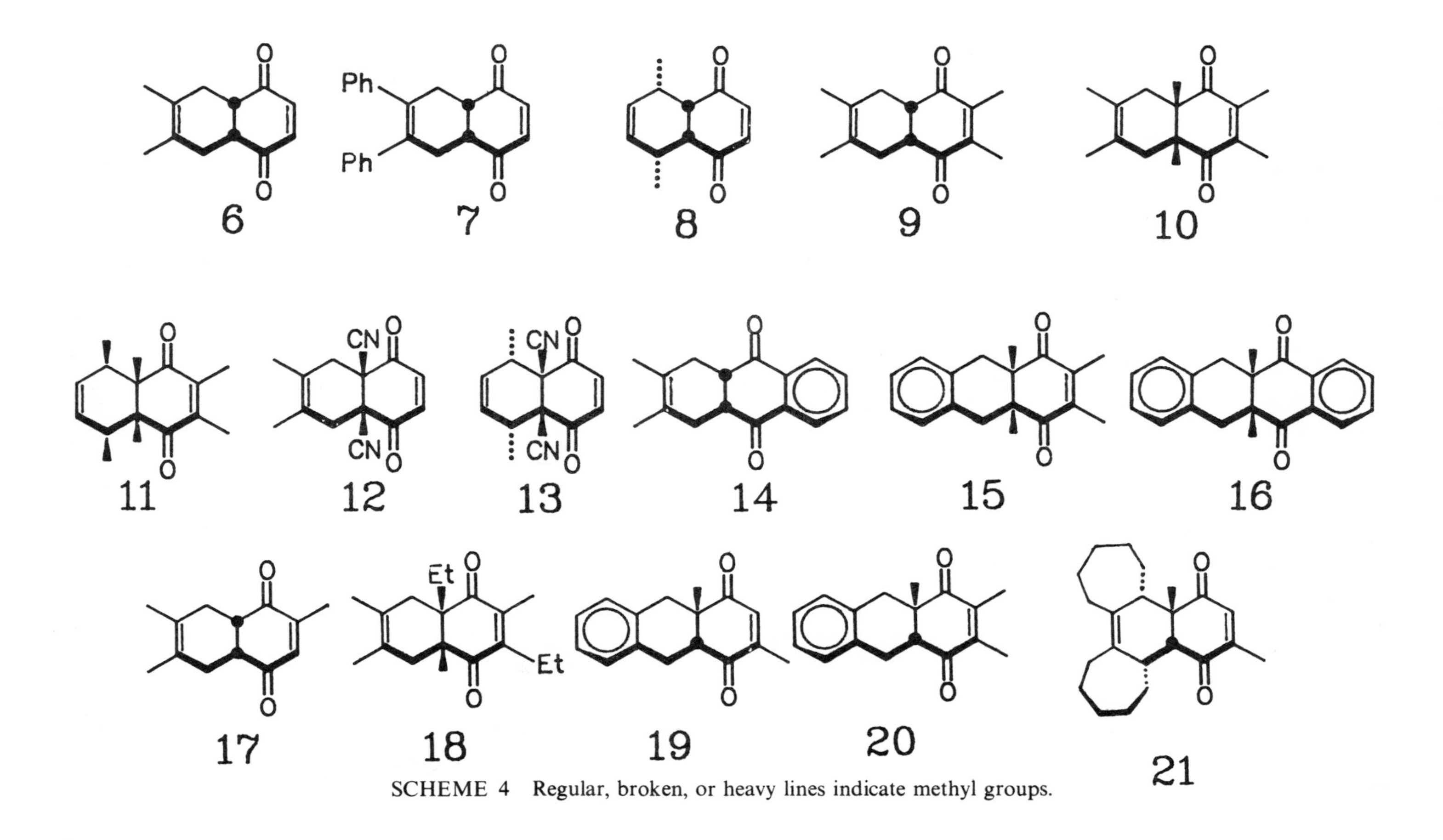

SCHEME 4 Regular, broken, or heavy lines indicate methyl groups.

use of *o*-quinodimethane as the Diels–Alder diene[15]. The general philosophy behind the choice of ene-diones **6–21** was to determine what effect the presence of substituents has on the crystal structures and photochemistry of the basic tetrahydronaphthoquinone system **1**.

IV. CRYSTALLOGRAPHIC STUDIES

The crystal structures of ene-diones **1** and **7–21** have been determined by X-ray diffraction methods (Table 1, crystals of **6** were not suitable for crystallographic studies)[14, 16]. The striking result of the X-ray crystallographic investigations is that *all the compounds studied crystallize in twist conformations*, similar to that described above for **1**.

TABLE 1. Crystal data for compounds **1** and **7–21**

Ene-dione	Space group	**a** α	**b** β	**c** (Å) γ (°)	Z	R
1	$P2_1/c$	5.266	24.267 114.50	14.506	8	0.062
7	$C2/c$	27.092	6.527 120.56	22.112	8	0.053
8	$P2_1/c$	7.189	22.241 106.51	6.843	4	0.048
9	$P2_1/c$	5.245	29.452 106.44	8.278	4	0.072
10	$P2_1/c$	7.312	11.540 92.26	16.674	4	0.088
11	$C2/c$	24.930	7.795 101.13	14.472	8	0.070
12	$P2_1/c$	8.717	12.464 117.87	12.783	4	0.055
13	Pbcn	15.915	11.525	13.157	8	0.046
14	$Pna2_1$	15.643	5.160	15.568	4	0.040
15	Cc	6.877	22.377 101.68	9.972	4	0.037
16	$P\bar{1}$	9.907 80.54	12.094 85.03	13.049 89.73	4	0.044
17	$P2_1/n$	19.191	5.278 90.70	22.330	8	0.094
18	$P2_1/n$	21.557	7.932 95.79	9.525	4	0.039
19	$P\bar{1}$	6.834 85.46	9.817 73.48	10.071 76.71	2	0.041
20	$P2_1/a$	9.833	8.065 102.69	17.377	4	0.052
21	$P2_1/n$	18.389	15.398 90.22	6.370	4	0.061

Particularly noteworthy are the crystalline phase conformations of the asymmetrically substituted ene-diones **17–21**. In contrast to the symmetrically substituted ene-diones (**1** and **6–16**), in which conformers **A** and **B** (see Scheme 3) have equal energy and are present in equal amounts in the solid state (none of the compounds discussed in this review crystallizes in chiral space groups), the asymmetrically substituted systems have conformers **A** and **B** that are *not* isoenergetic. For example, the two possible twist conformations for ene-dione **19** are shown below. The main difference between them in terms of their relative conformational energies is the environment of the methyl group at C(8a); in conformer **A** it is pseudoequatorial with respect to the cyclohexene ring, and in conformer **B**, it is pseudoaxial. The crystal structure of **19** reveals the presence of only conformer **A**. From this we conclude that the cyclohexene ring controls the conformation of this molecule. This is reasonable in view of the fact that the ene-dione ring is more nearly planar than the cyclohexene ring and thus interacts sterically to a lesser extent with the methyl substituent. The vinyl methyl group, which has no significant steric interactions with the rest of the molecule, appears to play no role in determining conformation. These arguments are strengthened by the crystal conformations found for ene-diones **19** and **21**. Both adopt conformations in which the bridgehead methyl group is pseudoequatorial with respect to the cyclohexene ring rather than the ene-dione ring. In the case of ene-dione **18**, the exclusive solid state conformation is the one that places the slightly bulkier ethyl group pseudoequatorial to the cyclohexene ring. The situation in the case of ene-dione **17** is unusual. This molecule is asymmetrically substituted, but contains no substituents that can be pseudoaxial or pseudoequatorial. As a result, the molecule crystallizes with *both* conformers **A** and **B** (below) present in equal amounts, presumably because the C(2) methyl group contributes very little to the molecular conformational energy, and the conformers are isoenergetic.

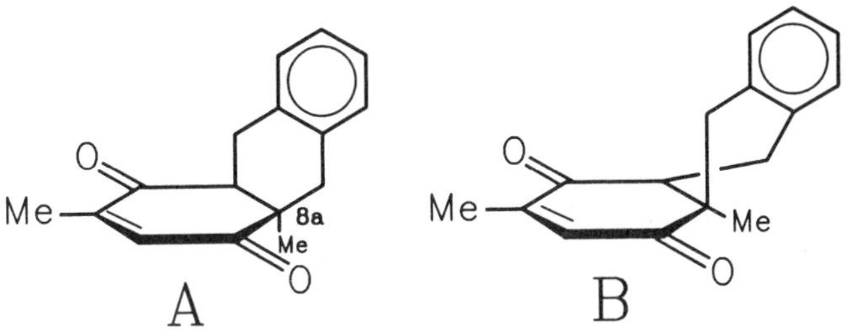

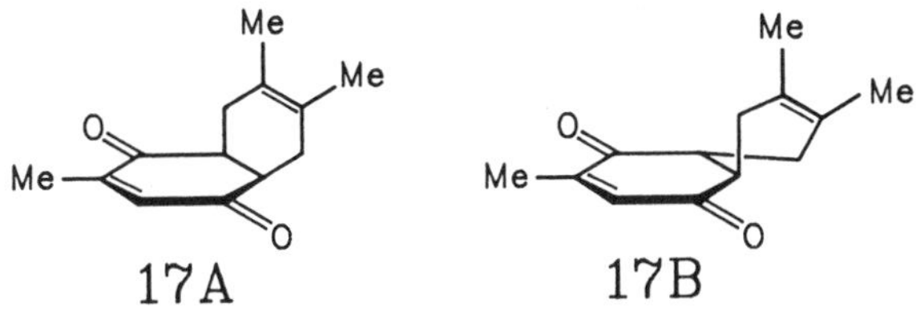

An exception to the rule is ene-dione **20**. This material crystallizes in a conformation that places the bridgehead methyl group *pseudoaxial* with respect to the cyclohexene ring. We believe this to be the less stable conformer because in solution reaction occurs predominantly from the conformation in which the methyl group is pseudoequatorial to the cyclohexene ring (*vide infra*). As yet, we have not identified the crystal packing factors that favor the solid state conformation.

A general conclusion that emerges from the crystallographic studies is that for ene-diones, intrinsic conformational energy rather than crystal packing energetics usually, but not always, determines the conformations that are adopted in the solid state. With the exception of ene-dione **20**, the crystal conformations are those predicted by the rules of classical conformational analysis to be of lowest energy. This provides clear evidence for the idea that organic molecules generally crystallize in a single conformation that corresponds to the minimum energy conformation in solution[17].

Turning now to the crystal packing arrangements in ene-diones **6**–**21**, it was found that only compound **8** packs in an arrangement suitable for intermolecular [2+2] photo-dimerization. The photochemistry of **8** will be discussed in a later section, but we may note at this point that it *does* undergo dimerization in the solid state. A second type of photodimerization reaction is also possible. This is intermolecular hydrogen atom abstraction by a ketone oxygen atom followed by carbon–carbon bond formation between the radicals thus generated. This type of process has been demonstrated by Lahav and coworkers in crystalline inclusion complexes formed between deoxycholic acid and various aliphatic and aromatic ketones[18]. As far as we are aware, however, this reaction has never been observed in a homomolecular crystalline material. It was of considerable interest, therefore, that we encountered evidence for this process as a minor component of the solid state reactivity of ene-diones **17** and **20**. This is discussed in more detail in the sections that follow.

V. PHOTOCHEMICAL RESULTS: SOLID STATE STRUCTURE–REACTIVITY CORRELATIONS

The photochemistry (both solution and solid state) of each of the sixteen ene-diones whose structures are shown in Scheme 4 will be outlined in the sections that follow. In cases where two or more molecules exhibit similar behavior, they will be discussed together. Insofar as is possible, the differences between the solid state and solution results, as well as the mechanisms operating in the solid state, will be explained with the aid of the relevant X-ray crystal structure data.

A. Intermolecular [2+2] Photocycloaddition

Ene-diones **6** and **8** belong to this category[5a, 7]. Irradiation of crystals of ene-dione **6** leads to dimer **6a**; when **6** is photolyzed in solution, mixtures of the rearrangement products **6b**, **6c** and **6d** are obtained. Happily, the overall yield is ca. 90%, much higher than in the case of ene-dione **1**. Photolysis of ene-dione **8** in the solid state also affords dimeric material (**8a**), while in solution it affords the novel rearrangement product **8e**. These reactions are summarized in Scheme 5. The structure and stereochemistry of dimers **6a** and **8a** were determined by X-ray crystallography[19].

As was the case with ene-dione **1** discussed earlier, the X-ray crystal structure of ene-dione **8**[16] shows that it packs favorably for intermolecular [2+2] photocycloaddition. In accord with Schmidt's rule[2e], the distance between double bonds of centrosymmetrically related incipient dimer pairs is 4.04 Å. The dimer structure and stereochemistry correspond exactly to the packing of the incipient dimer pair in the crystal, showing that the reaction is topochemically controlled. One major difference between the packing in ene-diones **1** and **8** is that in **8**, there is only one molecule in the asymmetric unit. Thus, in principle, quantitative dimer yields are possible. High yields were, in fact observed[7]. Turning to ene-dione **6**, it will be recalled that crystals of this material were not suitable for crystallography. The examples of ene-diones **1** and **8**, however, leave little doubt that the

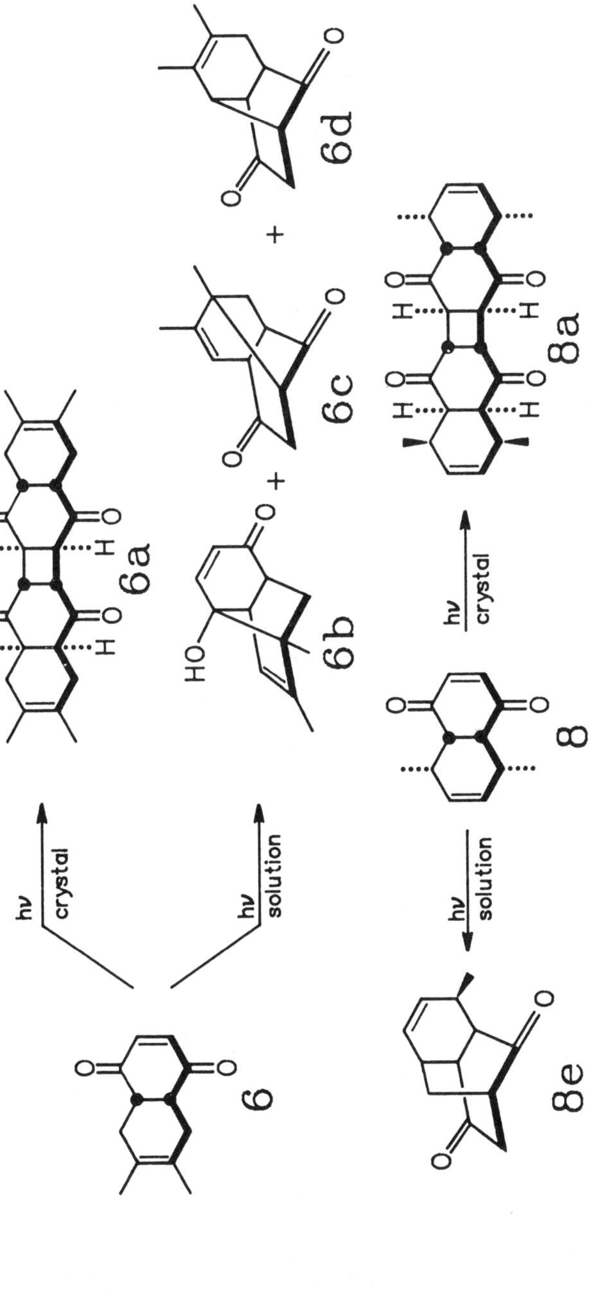

SCHEME 5

solid state photochemistry of **6** is also controlled by way of a crystal packing arrangement that favors dimerization.

Because ene-diones have similar conformations in solution and the solid state, we may use the crystal structure-derived conformations to help us understand the intriguing solution phase photorearrangement mechanisms. Ene-dione **6** undoubtedly adopts the preferred twist conformation common to all the ene-diones that lack bridged cyclohexene rings. This conformation is ideally suited for the first step of the photorearrangement process, namely, five-membered transition state (β) hydrogen atom abstraction of H(8) by O(1) (Scheme 6, R = Me). Abstraction leads to the bis-allylic biradical **6BR** which can close directly to photoproduct **6b**, or in solution, isomerize to **6BR**′. This can close either to **6c** or isomerize to **6BR**‴; closure of this latter species gives photoproduct **6d**. Photoproduct **6d** can also be formed from the *exo*-eclipsed biradical **6BR**″. This scheme also accounts for the products formed when ene-dione **1** is photolyzed in solution (see Scheme 1). Deuterium

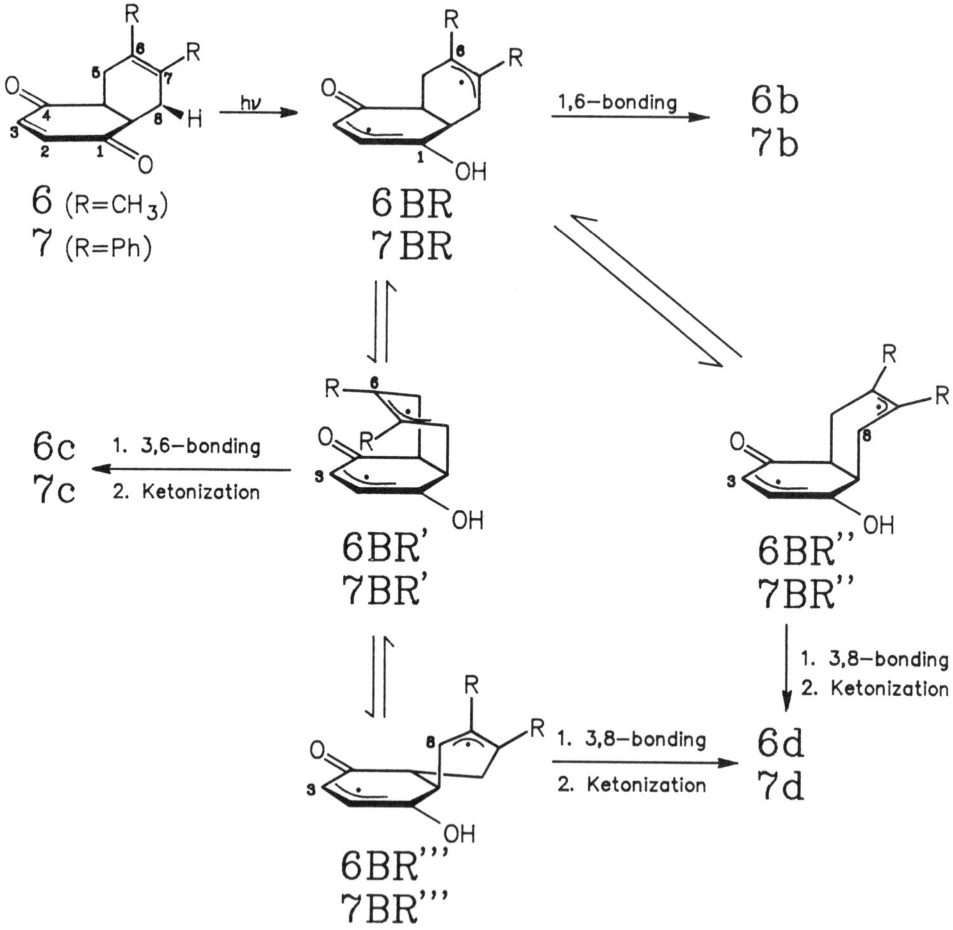

SCHEME 6

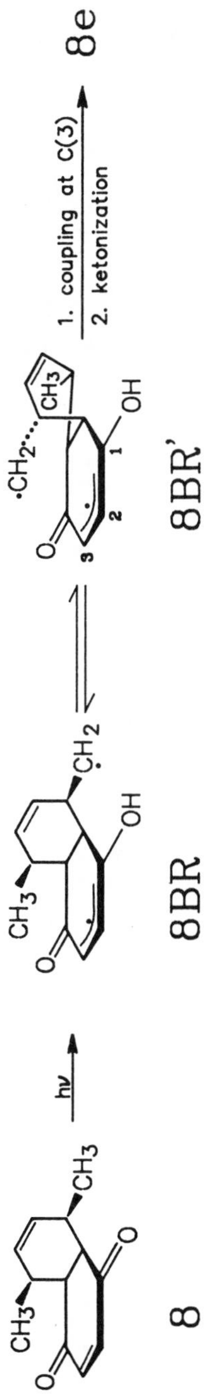

SCHEME 7

labeling and other mechanistic studies that establish these pathways have been published[5a]. As we shall see, for ene-diones other than **1**, **6** and **8**, abstraction also occurs in the solid state, and from the X-ray crystal structure data, we may obtain the distance and angular parameters for the process.

The conformation of ene-dione **8** in the solid state (and the favored conformer in solution) is the same as that of ene-dione **6**, except that the abstractable β-hydrogen atoms of **6** are replaced by methyl groups. Thus β-hydrogen atom abstraction is impossible in **8**, and instead, abstraction of a methyl γ-hydrogen atom is observed in solution (Scheme 7). This leads to biradical **8BR**, which, in solution, undergoes conformational isomerization to biradical **8BR**′ and then closes to yield the enol form of the observed photoproduct **8e**[5a].

B. Enone-alcohol Formation

Ene-diones **7**, **12** and **14** belong in this category and behave similarly upon irradiation in the solid state; each gives the corresponding enone-alcohol type photoproduct **7b**, **12b** and **14b** (Scheme 8)[5, 7]. These products, which are analogous to photoproduct **6b** produced in the photolysis of ene-dione **6** (see Scheme 5), were the only products formed; no dimers of any kind could be detected. Enone-alcohols **12b** and **14b** were also the sole products formed when ene-diones **12** and **14** were photolyzed in solution; ene-dione 7, however,

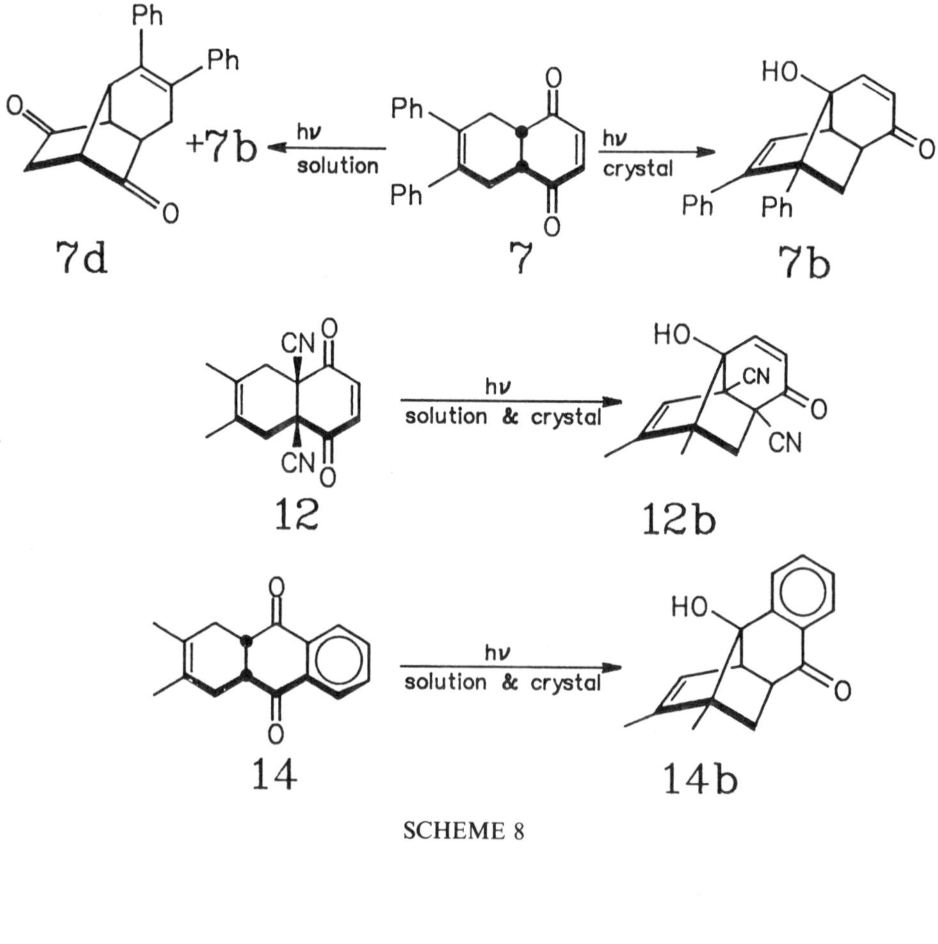

7d + 7b ← hν, solution — 7 — hν, crystal → 7b

12 — hν, solution & crystal → 12b

14 — hν, solution & crystal → 14b

SCHEME 8

gave a mixture of enone-alcohol **7b** and a new photoproduct, **7d**, when irradiated in solution. Diketone **7d** is analogous to **6d**.

These results demonstrate that unimolecular, hydrogen atom abstraction-initiated photochemistry *is* observed in the solid state provided that the crystal packing does not favor [2+2] photocycloaddition. We note also that in the case of ene-dione 7, fewer rearrangement products are formed in the crystal than in solution. This is explained readily with the aid of the mechanism outlined in Scheme 6 (R = Ph). Due to the restraints of the crystal lattice, the initially formed biradical **7BR** cannot undergo conformational isomerization to any of the other biradical intermediates. As a result, only photoproduct **7b**, which has the same basic conformation as that of ene-dione **7** and intermediate **7BR**, is formed in the solid state. No such restrictions exist in solution, however, with the result that both **7b** and **7d** are formed in this medium.

C. Cyclobutanone Formation

As outlined in Scheme 9, irradiation of crystals of ene-diones **10**, **11**, and **15** leads to a new type of photoproduct that contains a four-membered ring ketone moiety, hence the term cyclobutanone (products **10f**, **11f** and **15f**, respectively)[5,7,20]. In the case of ene-diones **10** and **11**, the cyclobutanone photoproducts are accompanied by lesser amounts of the corresponding enone-alcohols **10b** and **11b**; cyclobutanone **15f** was the sole product from the solid state photolysis of **15**. Very similar product mixtures were obtained in solution as in the crystal except for ene-dione **15**; solution photolysis of **15** gave **15d** and **15f**.

What is the mechanism of the formation of the cyclobutanone-containing photoproducts **10f**, **11f** and **15f**? As in the other solid state unimolecular photorearrangements, the conformation of the starting material in the crystal, determined by X-ray crystallography, provides the answer. Scheme 10 depicts the conformation of ene-dione **10** in the solid state; it exists in the now-familiar twist conformation. In this conformation, the *endo* allylic hydrogen atom on C(5) lies almost directly over the ene-dione carbon–carbon double bond. Transfer of this hydrogen (six-membered transition state) to C(2) of the double bond affords biradical **10BR**, and closure of this biradical by C(3) to C(5) bonding leads to the observed photoproduct **10f**. Like the solid state rearrangements of ene-diones **7**, **12** and **14**, the biradical and the final product have the same basic conformation as the starting material. Exactly the same mechanism and structure–reactivity correlation can be applied to the formation of photoproducts **11f** and **15f**.

With regard to the formation of enone-alcohols **10b** and **11b** from ene-diones **10** and **11**, respectively, the same mechanism as discussed earlier (Scheme 6) is operative. We note that no enone-alochol is formed in the case of ene-dione **15**, either in the solid state or in solution. This is due to the presence of the aromatic ring. Enone-alcohol formation requires C(1) · · · C(6) bonding in biradical **6BR**, a process which would disrupt aromaticity in the case of ene-dione **15**. Instead, the biradical analogous to **6BR** derived from **15** undergoes conformational isomerization in solution and gives diketone **15d**, the analogue of **6d**. This process does not disrupt aromaticity. Diketone **15d** is not formed in the solid state because the required conformational changes are not permitted by the crystal lattice.

An interesting point is that enone-alcohol formation and cyclobutanone formation involve the initial abstraction of *different allylic hydrogen atoms*. Abstraction of *endo*-H(8) by oxygen gives enone-alcohol and abstraction of *endo*-H(5) by carbon leads to cyclobutanone. The question thus arises, why do ene-diones **10**, **11** and **15** on the one hand, but not ene-diones **7**, **12** and **14** on the other, give rise to cyclobutanone-type photoproducts? The answer to this question comes from photophysical studies carried out in solution [5b,20]. Quenching and sensitization studies on ene-diones **10** and **15** show that

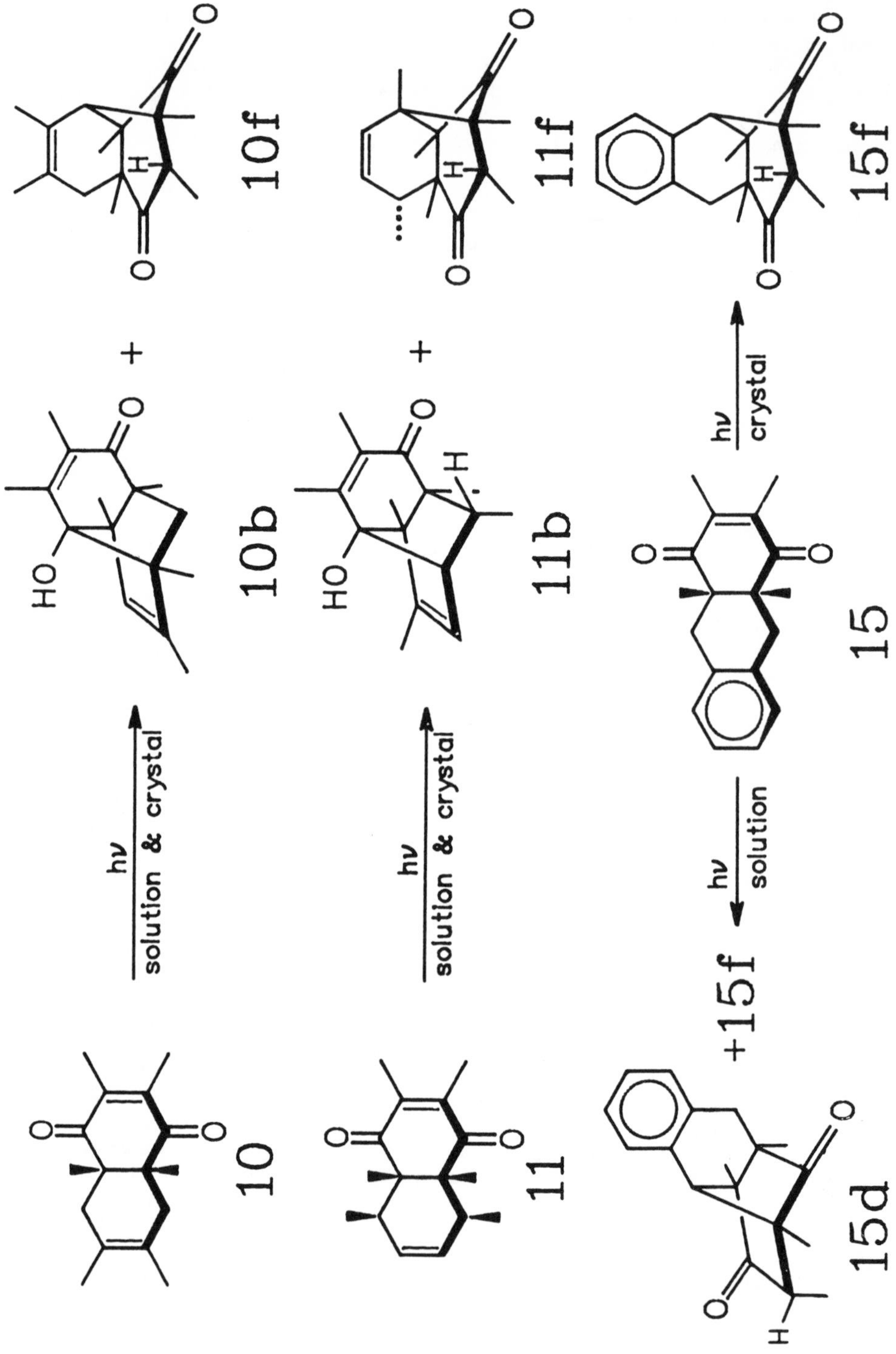

10 —hν, solution & crystal→ 10b + 10f

11 —hν, solution & crystal→ 11b + 11f

15d +15f ←hν, solution— 15 —hν, crystal→ 15f

HO

H

O

SCHEME 9

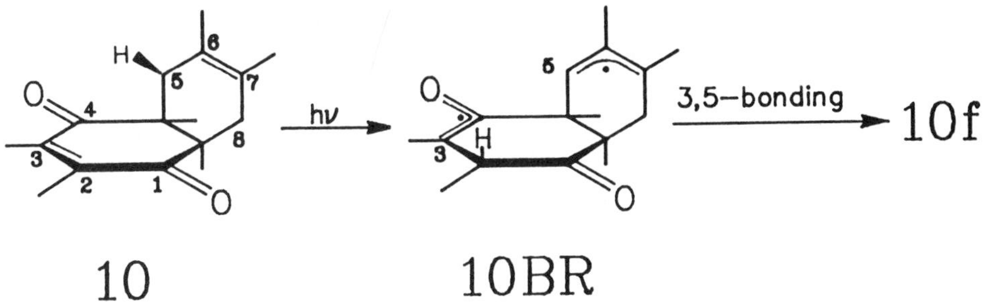

SCHEME 10

the cyclobutanone-type photoproducts are triplet-derived and that the enone-alcohol photoproducts as well as diketone **15d** come from singlet excited states. We assume that this holds in the solid state as well. It seems likely that the singlet excited state is n, π^* in nature and that the triplet is π, π^*. This is consistent with the $(\pi, \pi^*)^3$ being formed faster from the $(n, \pi^*)^1$ and having a lower energy relative to the $(n, \pi^*)^3$ as the result of ene-dione double bond methyl substitution[21,22]. The mechanistic picture is thus clear: those ene-diones with methyl groups on the C(2)–C(3) double bond (**10**, **11** and **15**) undergo relatively rapid intersystem crossing to $(\pi, \pi^*)^3$ excited states that react to give cyclobutanones. In these cases, intersystem crossing is still sufficiently slow that some reaction (enone-alcohol or diketone **15d** formation) is observed from the initially formed $(n, \pi^*)^1$ excited state. In the absence of methyl substituents at C(2) and C(3), intersystem crossing is too slow to compete with reaction from the $(n, \pi^*)^1$ state and only products derived from the singlet manifold are produced. These excited state–reactivity correlations are entirely consistent with what is known concerning the photochemical hydrogen atom abstraction reactions of α,β-unsaturated ketones, namely that π, π^* excited states favor abstraction by the β-carbon atom, and that n, π^* excited states lead to hydrogen atom abstraction by oxygen[23]. The predilection of ketone n, π^* excited states to engage in hydrogen atom abstraction through the use of the oxygen n-orbital is well established[24].

It is interesting to ask why the $(n, \pi^*)^1$ and $(\pi, \pi^*)^3$ excited states of ene-diones **10**, **11** and **15** react differently from one another. One contributing factor may be the preference for triplet excited states to give biradical intermediates that have a greater separation between the radical centers than the biradical intermediates formed from singlet excited states. As Michl has pointed out[25], this stems from the zwitterionic character of singlet biradicals in which radical separation has to overcome coulombic attractive forces. Comparing biradicals **10BR** (Scheme 10) and **6BR** (Scheme 6), it can be seen that the radical centers in the latter (which is a singlet-derived) can approach each other more closely (two-bond separation) than they can in the triplet-derived biradical **10BR** (minimum three-bond separation). From studies in solution on the temperature dependence of the product in the photolysis of ene-dione **15**[20], a singlet/triplet activation energy difference $[E_a(\text{triplet}) - E_a(\text{singlet})]$ of 4.5 kcal mol^{-1} was determined. We assume that the rate-determining step in each case is hydrogen atom abstraction; the activation energy difference may be attributed to the fact that S_1 has a higher excitation energy than T_1, as well as to the greater resonance stabilization of biradical **6BR** as compared to biradical **10BR**.

D. Oxetane Formation

Upon irradiation in either the solid state or solution, ene-dione **13** gave only the intramolecular oxetane **13g** (Scheme 11)[5b,7]. The structure of the oxetane was established by X-ray crystallography[26].

SCHEME 11

The crystal structure of ene-dione **13** shows that it has essentially the same conformation as ene-dione **8** discussed earlier (Scheme 7). The two differ only in the presence of two cyano groups at carbon atoms 4a and 8a, yet their solid state and solution phase photochemistry are completely different. In the solid state, ene-dione **13** does not photodimerize (as did **8**) because the packing precludes it. There are no abstractable allylic hydrogen atoms, therefore photoproducts of the **b**, **c**, **d**, or **f** type are not formed. Abstraction of a γ-hydrogen atom from a methyl group is possible, but the conformational isomerization of the resulting biradical required to form a stable photoproduct (see Scheme 7) is impossible in the solid state. The conformational isomerization of **13BR** is probably slower than that of **8BR** *in solution* as well, due to the bridgehead cyano groups. This may account for the fact that no photoproduct analogous to **8e** is formed in solution. As a last resort, the excited state of ene-dione **13** dissipates its energy by undergoing intramolecular oxetane formation. The fact that this reaction is geometrically feasible, but unobserved, for all of the ene-diones except those with aromatic rings fused in the C(6)/C(7) position, indicates that it is associated with an intrinsically low overall rate constant.

E. Compounds that React in Solution but not in the Solid State

Compounds **9** and **16** fall in this category[5a, 27]. In solution, ene-dione **9** affords products **9b** and **9d** upon irradiation. Photolysis of ene-dione **16** gives a photoproduct type not observed previously, namely the cyclopropanol **16h**. Neither ene-dione reacts when photolyzed in the solid state. These results are summarized in Scheme 12.

The photochemistry of ene-dione **9** is unusual in comparison with its close relative, ene-dione **10**. We expected that **9**, like **10**, would lead to substantial amounts of the corresponding, triplet-derived cyclobutanone (type **f**) photoproduct, both in solution and the solid state. This expectation was based on the fact that both **9** and **10** have methyl substituents on the C(2)–C(3) double bond, a feature that was suggested earlier to be important in bringing about cyclobutanone formation through facilitation of intersystem crossing to a low-lying $(\pi, \pi^*)^3$ excited state. On reflection, it is perhaps not too surprising that ene-dione **9** gives only singlet-type photoproducts. The balance between reaction from the singlet excited state and intersystem crossing is evidently a very delicate one for ene-diones **10**, **11** and **15**, and it is not unreasonable to suggest that the structural difference between ene-diones **9** and **10** (C(4a)/C(8a) methyl group substitution) could tip the scales one way or the other.

The solution phase photochemistry of ene-dione **16** is unique. None of the other ene-diones studied gave cyclopropanol-type photoproducts. Why does **16**? The answer is that

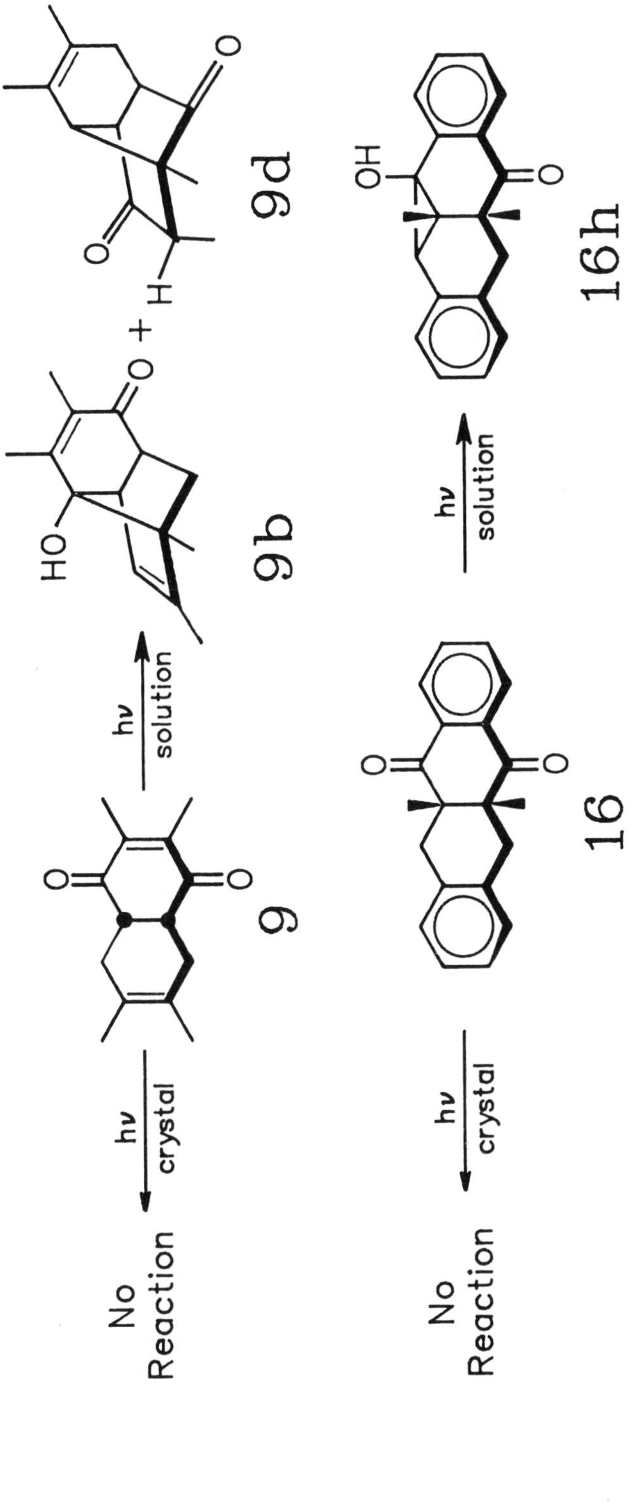

No Reaction ← hν, crystal — 9 — hν, solution → 9b + 9d

No Reaction ← hν, crystal — 16 — hν, solution → 16h

SCHEME 12

formation of any of the other photoproduct types (**a** through **g**) would disrupt aromaticity. Photoproduct **16h** is the only product of biradical collapse that does not. Cyclopropanol **16h** is a stable, crystalline material that does, however, undergo an intriguing secondary rearrangement of its own upon photolysis[27].

The lack of solid state photochemistry in the case of ene-dione **9** is a mystery that tells us there are solid state effects on chemical reactivity that are yet to be discovered. The conformation and packing arrangement of ene-dione **9** appear to be little different from those ene-diones that react unimolecularly with alacrity in the solid state. On the other hand, there are two clear reasons for the unreactivity of ene-dione **16** in the solid state. The first can be appreciated by reference to Scheme 6. Cyclopropanol formation requires C(1) · · · C(8) bonding in one of the intermediate biradicals, but can only occur from either **6BR**″ or **6BR**‴. As we have seen before, the conformational motions required to form these biradical intermediates are not permitted in the solid state. A more detailed understanding of the lack of cyclopropanol formation in the solid state emerges from an inspection of the packing diagram of ene-dione **16** (below). Crystals of ene-dione **16** contain two independent molecules with slightly different conformations in the asymmetric unit, and these pack in stacks consisting of alternating conformers. A methyl group of a lower molecule projects directly into the space between the rings of an upper molecule. Cyclopropanol formation, however, requires that these two rings move considerably closer together, and this is prevented by the presence of the methyl group. Our calculations[27] show that were cyclopropanol formation to occur in the solid state, the methyl-carbon to aromatic-center distances would decrease from 4.1 and 4.8 Å to approximately 2.6 and 3.0 Å, an impossible steric situation in view of the fact that the sum of the van der Waals radii of a methyl group and an aromatic ring is 3.7 Å[28].

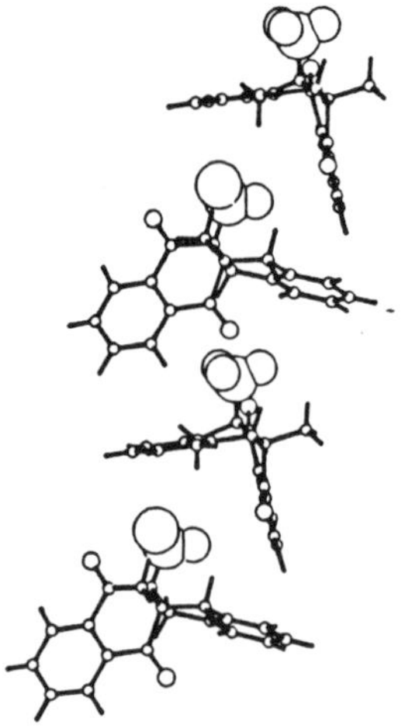

F. Conformation–Specific Solid State Photochemistry

The asymmetrically substituted ene-diones **17–21** belong to this category[29, 30]. As discussed earlier, the two possible twist conformations of these ene-diones are diastereomeric. In solution, both conformers are present in amounts that are of the same order of magnitude because the energy difference between conformers is relatively small. To a first approximation, the free energy difference between conformers **A** and **B** that have a single methyl group at C(4a) or C(8a) is the axial/equatorial free energy difference for a C(4) methyl group in cyclohexene ca. 1 kcal mol^{-1})[31] minus the axial/equatorial free energy difference for a C(5) methyl group in cyclohex-2-ene-1,4-dione (value not known, assumed to be ca. 0.5 kcal mol^{-1}). This leads to an **A**/**B** free energy difference of ca. 0.5 kcal mol^{-1}, which corresponds at 25 °C to a mixture containing ca. 70% of the more stable conformer. With the exception of ene-dione **17**, however, only one of the two twist conformers is present in the solid state. Thus, in contrast to the liquid phase, the crystalline phase provides a medium for studying the chemistry of a single, pure conformer. We have termed this 'conformation-specific' chemistry[29].

We consider first the ene-dione **18**. As outlined in Scheme 13, irradiation in solution leads to a mixture of four photoproducts, two of the enone-alcohol type (**18b** and **18b'**) and two of the cyclobutanone type (**18f** and **18f'**). In contrast, photolysis of crystals of ene-dione **18** gives a single photoproduct, enone-alcohol **18b**. The explanation of these results is straightforward: the conformers **18A** and **18B** are present in roughly equal amounts in solution, and photolysis gives two products from each conformer. Note that the photoreactivity of ene-dione **18** is exactly the same as that of its close relative, **10**. In the crystal, only conformer **18A** is present and irradiation leads to enone-alcohol **18b**.

Why is cyclobutanone **18f** not formed in the solid state? We feel this is due to a unique solid state steric effect. The first step of cyclobutanone formation is transfer of an allylic hydrogen atom from C(5) to C(2). As illustrated below, this causes pyramidalization of C(2) with a concomitant downward movement of the methyl group attached to it. The X-ray crystal structure of ene-dione **18** shows that the ethyl groups at C(3) and C(8a) would sterically impede this movement (dotted lines). This steric effect can be avoided in solution because the ethyl groups can rotate out of the way. In ene-diones **10**, **11** and **15**, this effect is

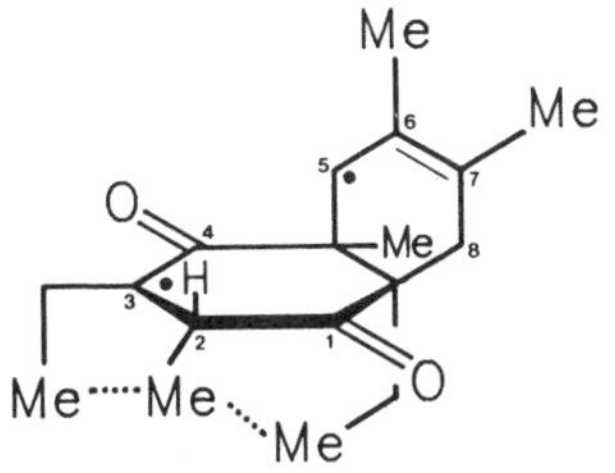

Biradical from 18A

absent because there are no ethyl groups, and cyclobutanone photoproducts *are* formed in the solid state. Computer simulation of the pyramidalization motions verifies this effect in **18**, and we have termed it 'steric compression control'[29, 32].

Ene-dione **19** gives the same cyclobutanone type photoproduct (**19f**) in solution as it does in the solid state (Scheme 14); enone-alcohol type photoproducts are precluded in this case by the presence of the aromatic ring. The crystal structure shows that only conformer **19A** is present in the solid state. This explains the exclusive formation of cyclobutanone **19f** in this medium, but why is the other cyclobutanone isomer (**19f'**) not observed in solution? There are two reasons for this. First of all, conformer **19A** is the major conformer present in solution, and second, the biradical **19BR**, which gives rise to the observed photoproduct, has a tertiary center and is therefore more stable than its secondary isomer, **19BR'**.

Ene-dione **20** undergoes photolysis in acetonitrile to give two isomeric cyclobutanones, **20f** and **20f'** (Scheme 14), in a ratio of 57 to 43; in the solid state, only cyclobutanone **20f'** is formed. These unusual results stem from the fact, discussed earlier, that ene-dione **20**

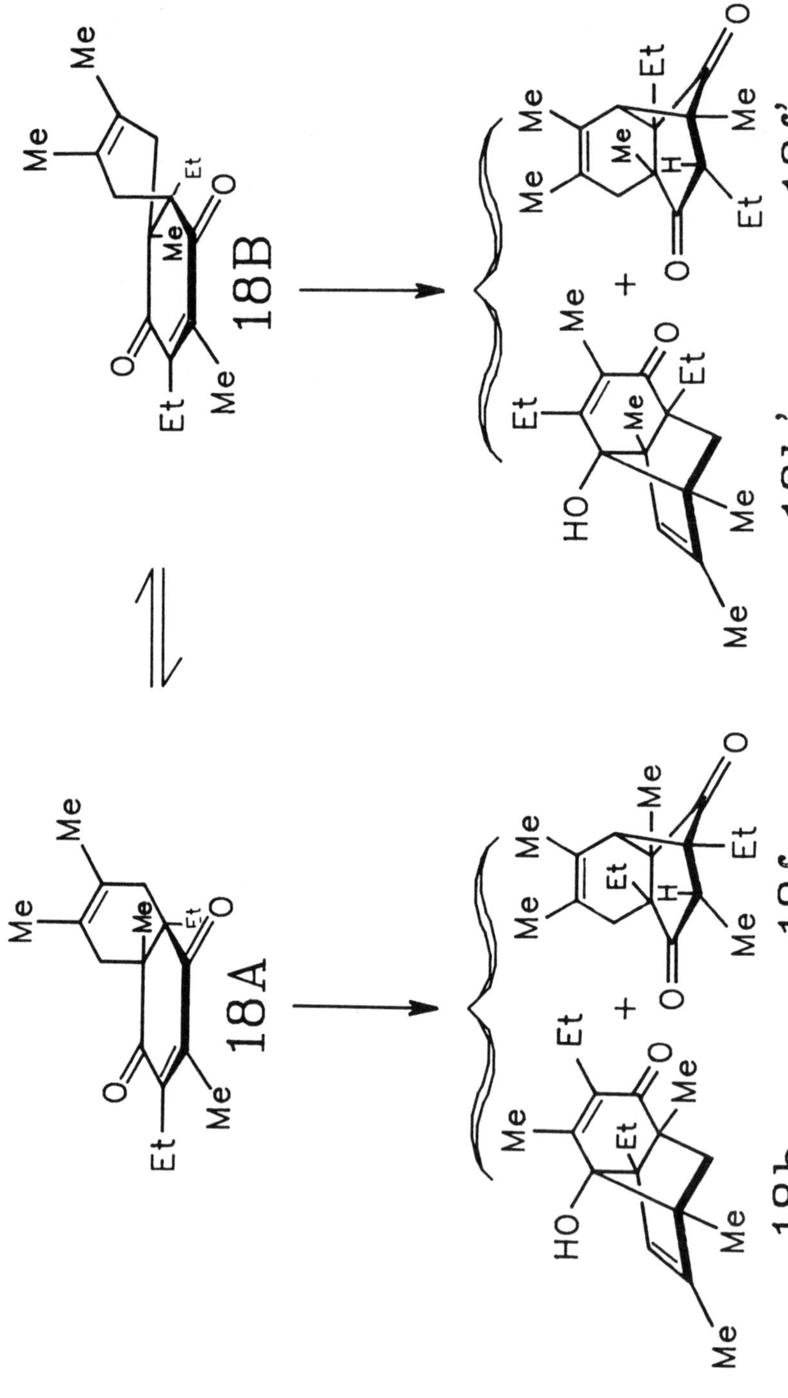

SCHEME 13

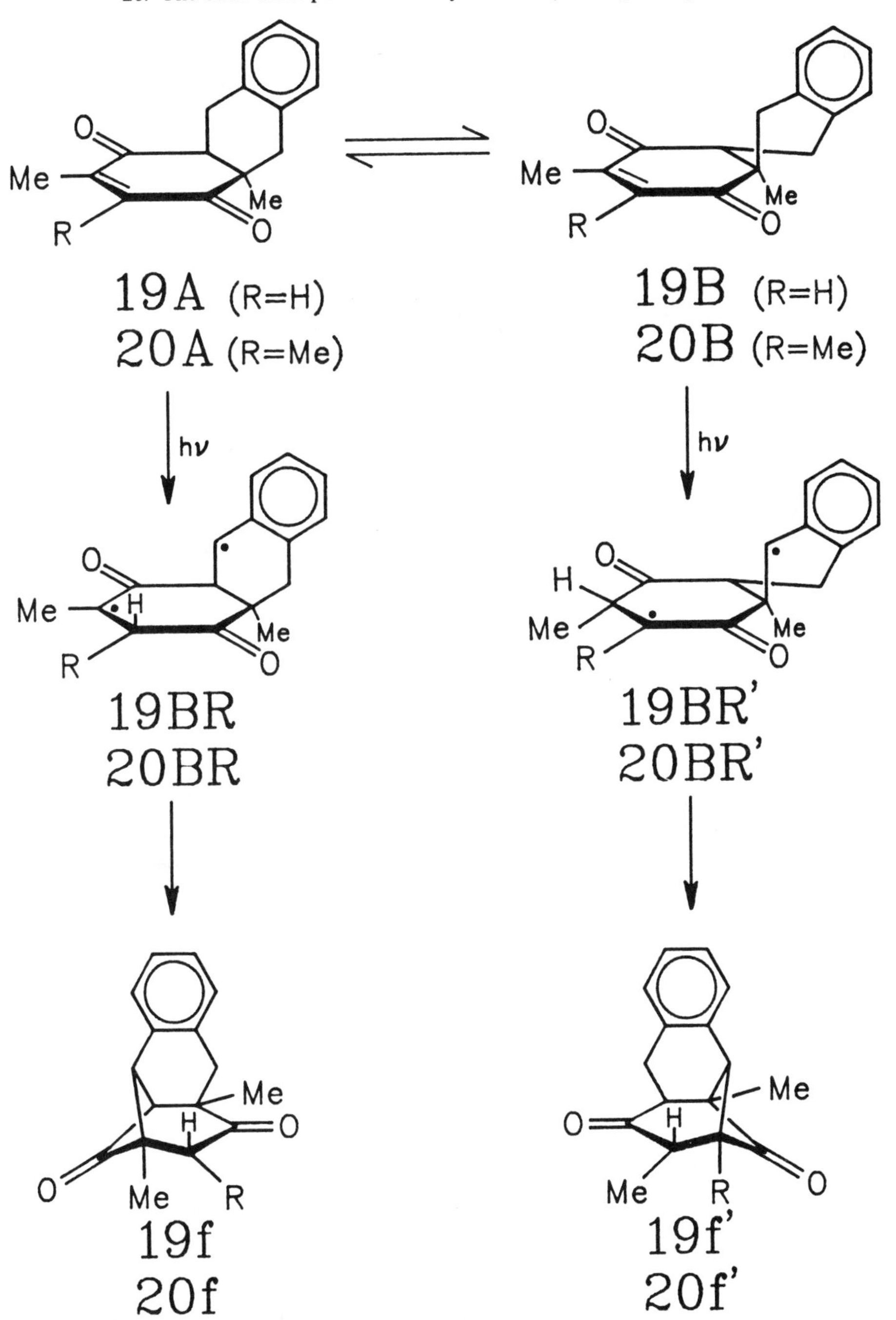
O
Me
R
Me
O
19A (R=H)
20A (R=Me)
O
Me
R
Me
O
19B (R=H)
20B (R=Me)
hν
hν
O
Me
H
R
Me
O
19BR
20BR
O
H
Me
R
Me
O
19BR'
20BR'
Me
H
O
O
Me
R
19f
20f
Me
O
H
Me
R
O
19f'
20f'

SCHEME 14

crystallizes in its *less stable* conformation, **20B**. In solution, conformer **20A** predominates and leads to the major photoproduct **20f**. In contrast to biradicals **19BR** and **19BR′**, intermediates **20BR** and **20BR′** have comparable radical stability and because **20B** is present in solution, some photoproduct **20f′** is formed in this case.

Just as we were beginning to feel that the solid state photochemistry of ene-diones could not provide any further surprises, Weisz and coworkers[14a] reported the synthesis and photochemical behavior of ene-dione **21**. As outlined in Scheme 15, irradiation of this material in ethyl acetate affords a mixture of the intramolecular [2+2] cycloaddition product **21h** and the unusual rearrangement product **21i**; photolysis of crystals of ene-dione **21** gives only **21i**.

Formation of **21i** involves seven-membered transition state abstraction of a δ-hydrogen atom by oxygen atom O(1) from the lower energy twist conformer **21A**, followed by C(1) · · · C(6) bonding of the biradical so produced. This biradical closure step is identical to the topochemically allowed, second step of enone-alcohol (type **b**) photoproduct formation (see Scheme 6). Interestingly, the solid state reaction fails for ene-diones that contain one or two five-membered or six-membered rings fused to the cyclohexane moiety. This was ascribed to unfavorable O · · · H abstraction distances in the solid state[14]. The details of this aspect of the work are deferred to the section on the geometric requirements for hydrogen atom abstraction.

Finally, we briefly discuss the photochemistry of ene-dione **17**. Because, as mentioned earlier, both twist conformers of **17** are present in equal amounts in the crystal, no conformation-specific, solid state chemistry is observed, and irradiation of this material gives a mixture of photoproducts that is essentially the same in the solid state as in solution[5b]. Interestingly, the packing diagram of ene-dione **17** shows that it should be capable of intermolecular [2+2] photocycloaddition in the solid state. The ene-dione double bonds of neighboring molecules in the lattice are situated above one another in a head-to-tail, parallel arrangement; the center-to-center distance between double bonds is 3.9 Å, less than Schmidt's upper limit of 4.1 Å. A second notable feature of the packing in ene-dione **17** is that there is a close (2.5 Å) contact between a carbonyl oxygen atom of one molecule and a hydrogen atom on the C(2) methyl group of a neighboring molecule. As we shall see, this distance is within the upper limit for hydrogen abstraction. Thus a second mode of dimerization is possible for ene-dione **17**, namely, intermolecular hydrogen atom abstraction (photoreduction) followed by coupling of the radicals so produced. However, only a very small dimer peak was found in the mass spectrum of the crude solid state reaction mixture. The mass spectrum also exhibited a peak at 2M-18. This is consistent with the dimer being of the photoreduction type (loss of water from the tertiary alcohol produced by radical coupling). The results with ene-dione **17** show that a crystal packing arrangement that favors [2+2] photocycloaddition does not always cause this process to win out over unimolecular photochemistry.

The crystal packing of ene-dione **20** shows that potentially, it too is capable of photoreductive dimerization in the solid state, but that [2+2] photodimerization is not possible; the O · · · H contact between lattice neighbors in this case is 2.5 Å. The mass spectrum of the crude reaction mixture from solid state photolysis of ene-dione **20** again shows weak 2M and 2M-18 peaks, consistent with photoreductive dimerization; as yet we have been unable to isolate and characterize either photoreductive dimer.

VI. QUANTITATIVE STRUCTURE–REACTIVITY CORRELATIONS: THE GEOMETRIC PARAMETERS ASSOCIATED WITH HYDROGEN ATOM ABSTRACTION

The photochemical studies described above have uncovered two general types of hydrogen atom abstraction processes that occur in the solid state. The first is β or five-membered

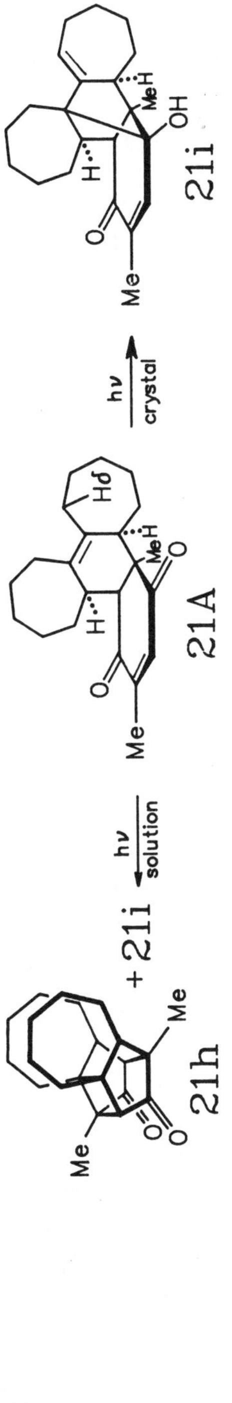

SCHEME 15

transition state abstraction of an allylic or benzylic hydrogen atom by an ene-dione oxygen atom, and the second is γ or six-membered transition state allylic or benzylic hydrogen atom abstraction by one of the central carbon atoms of the ene-dione chromophore. Because these processes occur in crystals, where atomic and molecular motions are severely restricted relative to liquid media, the shape of the molecule undergoing reaction is likely to be very similar to that determined by X-ray crystallography, and detailed structure–reactivity correlations are possible. The question of what geometry changes are induced in ene-diones by electronic excitation will be dealt with later.

The structure–reactivity correlations will be discussed in terms of three parameters that characterize the geometric relationship between the abstracting oxygen or carbon atom and the hydrogen atom being abstracted. These are d, the oxygen $\cdots$ hydrogen or carbon $\cdots$ hydrogen abstraction distance, τ, the angle formed between the O $\cdots$ H or C $\cdots$ H vector and its projection on the mean plane of the carbonyl group or the ene-dione central double bond, and Δ, the C=O $\cdots$ H or C=C $\cdots$ H angle. These parameters are represented schematically in Figure 1.

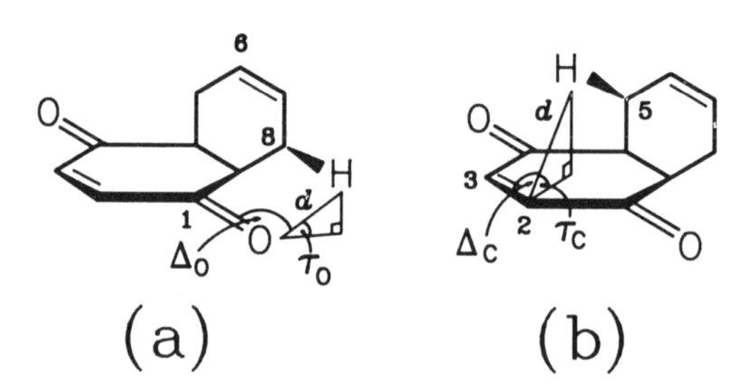

FIGURE 1. Definition of geometric parameters for hydrogen abstraction: d = abstraction distance; τ_0 = angle formed between O(1) $\cdots$ H(8) vector and its projection on mean plane of C(1)=O group; Δ_0 = C(1)=O(1) $\cdots$ H(8) angle; τ_c = angle formed between C(2) $\cdots$ H(5) vector and its projection on mean plane of C(2)=C(3) double bond; Δ_c = C(3)=C(2) $\cdots$ H(5) angle

The values of d, τ, and Δ for each of the compounds that undergo intramolecular hydrogen atom abstraction in the solid state are collected together in Table 2, which also lists a fourth parameter, the distance between the carbon atoms that become bonded to one another in the final photoproduct, and the types of photochemical reaction observed. The parameters for hydrogen atoms are less accurately determined by X-ray crystallography than for the heavier carbon and oxygen atoms, and fewer significant figures are quoted in the interatomic distances involving hydrogen atoms.

The data in Table 2 demonstrate that for ene-diones, intramolecular hydrogen atom abstraction occurs over distances that are less than or equal to the sum of the van der Waals radii of the atoms involved, 2.7 Å for O $\cdots$ H and 2.9 Å for C $\cdots$ H[28]. These distances are considerably longer than those previously considered feasible. The often-quoted limit of 1.8 Å for six-membered transition state hydrogen atom abstraction by oxygen stems from Djerassi's work on the mass spectrometric McLafferty rearrangement of steroidal ketones[33]. Evidently, excited state processes have less stringent distance requirements.

The suggestion that hydrogen atom abstraction occurs over distances less than or equal to the sum of van der Waals radii of the atoms involved has been substantiated by other studies from our laboratory[34]. It is also consistent with the studies on ene-dione **21** and related compounds[14]. These authors found, via X-ray crystallography, that the distance between the reactive δ-hydrogen atom and its nearest carbonyl oxygen atom varied with C/D ring size as shown below. Only ene-dione **21**, with an abstraction distance of 2.7 Å, the

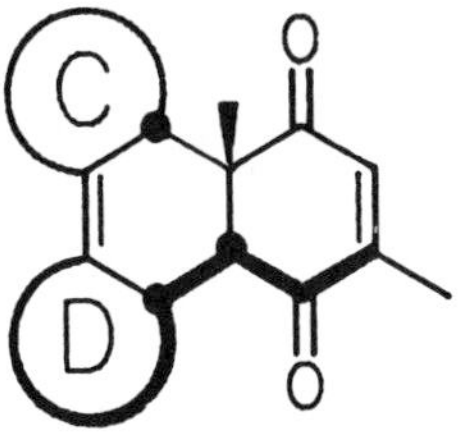

Ring Size		δH ···· O Distance (Å)
C	D	
5	5	3.2
6	5	4.1
6	6	4.8
6	7	4.0
7	7	2.7

van der Waals radii sum, was photochemically reactive in the solid state.

With regard to the angular relationship between the abstracted and the abstracting atoms, we may expect that the most favorable arrangement will be defined by the spatial characteristics of the atomic orbital to which the hydrogen becomes bonded. In the case of hydrogen atom abstraction by oxygen, which occurs in ene-diones through the $(n, \pi^*)^1$ excited state, there is little doubt that the atomic orbital involved in abstraction is the non-bonding 2p orbital on oxygen[24]. This orbital lies in the plane of the carbonyl group and forms an angle of 90° with the C=O axis[35]. Thus the optimum values of τ_0 and Δ_0 for this process should be 0° and 90° respectively. The data in Table 2 reveal that ene-diones crystallize in conformations that, in terms of angle, *are nearly perfect for β-hydrogen abstraction by oxygen.* The average τ_0 angle is 4°, and the average value of Δ_0 is 84°. In the case of hydrogen atom abstraction by sp^2 hybridized carbon, it is the 2p AO that is involved. The preferred geometry of approach of the hydrogen atom here is 90° to the C=C axis and along the long axis of the AO rather than in the nodal plane. This gives optimum values of 90° for both τ_c and Δ_c respectively. The data in Table 2 show that ene-diones are less ideally arranged for carbon abstraction than for oxygen abstraction; the average τ_c value is 50°, and the average value of τ_c is 74°. This may be a third factor that contributes to the higher activation energy observed for this process in solution compared to abstraction of a β-hydrogen atom by oxygen.

The fourth geometric parameter listed in Table 2 is the distance between the carbon atoms that become bonded to one another in the final tricyclic photoproduct. The process of biradical closure is equal in importance to hydrogen atom abstraction to the success of the solid state reaction; if closure is topochemically disallowed, i.e. requires motions that are incompatible with the restraints of the crystal lattice, reverse hydrogen transfer leads to regeneration of starting material, and no net chemistry is observed. The carbon ··· carbon distances listed in Tables 1 and 2 range from 3.13 to 3.52 Å, with an average of 3.33 Å. Of course these distances refer to the ground state, not to the biradical intermediate, but these species certainly have similar gross conformations, and the fact that the distances are all close to the sum of the van der Waals radii for two carbon atoms (3.40 Å) is a strong

TABLE 2. Geometric parameters for intramolecular hydrogen abstraction

Ene-dione	d(Å)	τ_0(°)	Δ_0(°)	$C(1)\cdots C(6)$ (Å)	Reaction type[a] Solution	Solid
β-Hydrogen abstraction by oxygen and enone-alcohol formation, reaction type[a] 2 (refer to Figure 1a for definition of parameters)						
1[b]	2.5	4	80	3.43	2	1
6	—	—	—	—	2	1
7	2.5	3	81	3.51	2	2
8	2.4(γ)	15	101	3.4[c]	4	1
9	2.4	3	82	3.49	2	NR[d]
10	2.5	0	85	3.35	2+3	2+3
11	2.3	1	86	3.33	2+3	2+3
12	2.6	8	84	3.38	2	2
13		no abstractable hydrogen			5	5
14	2.6	5	81	3.46	2	2
16[b]	2.5	4	81	(2.51)[e]	2	NR[d]
17[b]	2.5	6	85	3.52	2	2
18	2.4	5	83	3.35	2+3	2
γ-Hydrogen abstraction by carbon and cyclobutanone formation, reaction type[a] 3 (refer to Figure 1b for definition of parameters)		τ_c(°)	Δ_c(°)	$C(3)\cdots C(5)$ (Å)		
10	2.8	52	73	3.17	2+3	2+3
11	2.7	50	74	3.17	2+3	2+3
15	2.8	51	74	3.13	2+3	3
19	2.9	47	75	3.26	3	3
20	2.9	50	74	3.20	3	3

[a] Reaction types: 1, dimerization; 2, β-H abstraction by O; 3, γ-H abstraction by C; 4 = γ-H abstraction by O; 5 = oxetane.
[b] Averages for two molecules in the asymmetric unit.
[c] Estimated after conformational inversion.
[d] NR = no reaction.
[e] C(1)···C(8) bond formation.

indication of a favorable arrangement for bonding. Another indication that the twist conformation is favorable for enone-alcohol and cyclobutanone formation is the alignment of the orbitals involved. The top lobes of the p orbitals at C(1) and C(6) (biradical collapse to enone-alcohol) point directly at one another (biradical **6BR**, Scheme 6), and the same is true of the orbitals at C(3) and C(5) in the biradical **10BR** (Scheme 10) leading to cyclobutanone-type photoproduct.

We turn now to a discussion of the important question of whether the *ground state* structural parameters can be correlated with *excited state* chemical reactivity. We believe they can. Based on spectroscopic and theoretical studies, there is general agreement that the $(n, \pi^*)^1$ states of α,β-unsaturated ketones have geometries that are essentially identical to their ground state geometries except for slight increases (< 0.1 Å) in the C=O and C=C bond lengths[36]. Assuming that enones and ene-diones have similar excited state characteristics, this indicates that the ground state structural data collected in Table 2 for the process of five-membered transition hydrogen atom abstraction by ene-dione oxygen also apply very closely to the reactive $(n, \pi^*)^1$ excited state. The situation in the case of the ene-dione $(\pi, \pi^*)^3$ excited state is less certain. Using enones again as a model, it was suggested initially that the $(\pi, \pi^*)^3$ excited states of cyclohexenones are twisted about the C=C bond, the torsion angle and consequent triplet energy varying as the structural constraints of the system[37]. More recent work, however, has disclosed a much more complex situation[38] and at present there is no clear consensus as to the number, degree of twisting and the reactivity of the triplet states generated when cyclohexenones are photolyzed. Therefore, we prefer to leave the question of the shape of the reactive excited state responsible for γ-hydrogen atom abstraction by ene-dione carbon unresolved until further work is carried out. If this *is* a twisted species, we note that twisting may actually *facilitate* the process of abstraction by carbon by tilting the abstracting p orbital at C(2) directly toward the γ-hydrogen being abstracted.

VII. EXPERIMENTAL TECHNIQUES IN SOLID STATE PHOTOCHEMISTRY

A useful preliminary test to determine whether or not, as well as how, a given ene-dione will react in the solid state is to irradiate KBr pellets of the material and monitor the reaction by infrared spectroscopy. Prior to 1980, the pure crystal irradiations were carried out in a special variable temperature, evacuable apparatus consisting of an outer jacket with a Pyrex window and an inner, gold-plated reaction surface which is in contact with a reservoir containing liquid coolant[7]. The crystals were grown on the gold plated reaction surface of the inner vessel by slow evaporation of solutions of the ene-dione to be photolyzed. The last traces of solvent were removed by pumping on the sample, and the irradiations (450 W Hanovia lamp) were conducted in the absence of oxygen by maintaining the vacuum or by the introduction of nitrogen. The reaction temperature was chosen to be well below the eutectic temperature of the reaction mixture as determined by differential scanning calorimetry. The wavelength used was > 340 nm (Corning 7380 glass filter), which excites the low intensity ($\varepsilon < 150$), near-visible absorption band common to all the compounds studied. Conversions were kept low (< 30 %) so as to avoid destruction of the parent lattice and possible resulting loss of topochemical control, although in most instances, higher conversions could be achieved without noticeable loss of specificity or change in product ratio.

In recent years, we have turned to nitrogen lasers (337 nm) as the light source of choice in crystal irradiations. The collimated, monochromatic beams of lasers can be focused easily on small, carefully grown single crystals of the compound being studied, giving reasonable conversions in relatively short times without much sample heating. For low temperature photolyses, a specially designed, windowed Teflon cell is used. The temperature is controlled by passing vaporized liquid nitrogen through the cell; the temperature can be varied by adjusting the liquid nitrogen boil-off rate. Both solid state and solution samples are photolyzed using these techniques, generally to < 5 % conversion. In some cases the conversions are varied and the product ratios are extrapolated to 0 % conversion. We are well aware of the possibility that crystal defects, phase changes and changes in crystal packing can affect profoundly reactions occurring in the solid state. We recrystallize our samples routinely from different solvents and sublime them when possible to determine whether they exhibit polymorphism. To date, no ene-dione polymorphs have been identified. In addition to single crystal photolyses, parallel irradiations are conducted on polycrystalline samples to determine if crystal defects play a role in the reactions being studied; no effect of this type has been observed. The possibility that the reaction under investigation may be of the single crystal to single crystal type is also monitored carefully; so far, none has been found.

VIII. SUMMARY

The solid state photochemistry of *cis*-4a,5,8,8a-tetrahydro-1,4-naphthoquinone (ene-dione **1**) and 18 of its substituted derivatives has been reviewed and the results discussed with reference to the X-ray crystal structure data for 17 of the 19 compounds. All but one of the 17 ene-diones whose crystal structures were determined crystallize in a common 'twist' conformation in which the cyclohexene moiety exists in a half-chair form *cis*-fused to a more-nearly planar cyclo-hex-2-ene-1,4-dione ring. The exception, ene-dione **3**, is prevented from adopting this conformation by a two-carbon bridge across positions C(5) and C(8). This leads to an eclipsed conformation in which the ene-dione double bond and the cyclohexene double bond are parallel and 3.53 Å apart; as a result, irradiation of crystals of ene-dione **3** affords quantitative yields of the intramolecular [2 + 2] cage product **5**. In the twist conformation, however, the ene-dione and cyclohexene double bonds are non-parallel and much further apart. Accordingly, intramolecular [2 + 2] cycloaddition is not observed when crystals of these materials are photolyzed.

Half-chair to half-chair cyclohexene ring inversion leads to two possible twist conformations for each ene-dione. Ene-diones with methyl groups attached to the central C(4a)–C(8a) bond have free energies of activation for inversion of ca. 9 kcal mol^{-1}. For symmetrically substituted systems, the twist conformations are isoenergetic and enantiomeric. In all cases of this type, however, the crystals are not chiral, as both enantiomers are present in equal amounts. Asymmetrically substituted ene-diones have twist conformations that are diastereomeric. In four of the five cases of this type studied, only one diastereomer (both enantiomers) was present in the crystal. In solution, both diastereomers are present in rapid equilibrium. The solid state medium thus permits the study of the reactivity of a single, pure conformer, a circumstance we term 'conformation-specific chemistry'. Three times out of four, it was found that the diastereomer present in the crystal is the one in which the perturbing substituent is pseudoequatorial with respect to the cyclohexene ring rather than to the ene-dione ring. The reverse was found for ene-dione **20**, indicating that the diastereomers have similar conformational energies (ΔG ca. 0.5 kcal mol^{-1}). In one instance (ene-dione **17**), where the perturbing substituent is on the ene-dione double bond, both diastereomers are present in equal amounts in the solid state.

Despite the fact that 16 of the ene-diones whose crystal structures were determined crystallize in a twist conformation, identical solid state photoreactivities were not observed. Eleven took part in intramolecular hydrogen abstraction processes, two underwent topochemically controlled, bimolecular [2 + 2] ene-dione double bond photocycloaddition, one gave rise to internal oxetane formation, and two were photochemically inert. The X-ray data showed that bimolecular [2 + 2] photoaddition occurred only when the crystal packing was such that adjacent molecules were oriented so that the reacting double bonds were parallel with center-to-center distances of less than 4.1 Å. Crystal packing of this type does not guarantee dimerization, however. Ene-dione **17**, with parallel ene-dione double bonds 3.9 Å apart, gives intramolecular hydrogen abstraction-derived products when irradiated in the solid state.

Three types of intramolecular hydrogen abstraction reactions were observed in the solid state: (1) abstraction by carbonyl oxygen through a five-membered transition state, (2) abstraction by C=C carbon via a six-membered transition state, and (3) in one instance only (ene-dione **21**), abstraction by carbonyl oxygen through a seven-membered transition state. All three processes are facile in the solid state owing to favorable geometric and distance factors. Also facilitating abstraction is the fact that in each case the hydrogen atom being abstracted is either allylic or benzylic. The structural parameters associated with these abstractions are summarized and compared to values previously estimated in the literature. This reveals that abstraction can take place over distances considerably greater than heretofore considered favorable and the data suggest that, with favorable geometry, these distances can be at least as great as the sum of the van der Waals radii of the abstracting and abstracted atoms (2.7 Å for C=O···H and 2.9 Å for C=C···H). Particularly noteworthy is the finding that the five-membered transition state abstraction process is favored by a nearly perfect angular relationship between the n orbital of the abstracting oxygen atom and the hydrogen atom being abstracted. Not only does the hydrogen lie within a few degrees of the mean plane of the carbonyl group and the n orbital, but the C=O · · · H angle is very close to the ideal of 90°. The excited states giving rise to the hydrogen abstraction processes are tentatively identified as $(n, \pi^*)^1$ for abstraction by oxygen and $(\pi, \pi^*)^3$ for abstraction by carbon, and the general validity of formulating ground state structure–excited state reactivity relationships for these states is discussed. Equally important to the observation of the internal solid state hydrogen transfer reactions is the fact that the biradical intermediates produced by abstraction can collapse directly to stable products without the necessity for conformational isomerization. This is predicated on the reasonable assumption that the biradical has the same basic conformation as its ground state precursor, and the biradical coupling distances as estimated from the ene-dione structural data are all < 3.5 Å, of the order of the sum of the van der Waals radii for two carbon atoms (3.4 Å).

Intramolecular photochemical oxetane formation is observed only for the one substrate that has no neighbors in position to afford [2 + 2] photodimers and for which internal hydrogen abstraction is made geometrically impossible by replacement of the normally abstractable allylic hydrogen atoms by methyl groups. The fact that internal oxetane formation is geometrically feasible but undetected for all the other substrates studied indicates that it is the least favored of the solid state photoreactions. Finally, two of the ene-diones proved to be photochemically unreactive in the solid state. In the case of compound **9**, the reasons for this behavior are not known, but for ene-dione **16**, the photo-inertness is due to the fact that the biradical produced by hydrogen abstraction has no carbon–carbon bonding possibilities that do not either interrupt aromaticity or require a half-chair to half-chair conformational ring inversion, a motion that is impossible in the solid state. This type of reasoning also serves to explain the solid state/solution reactivity differences observed for many of the ene-diones studied. Irradiation in solution leads to biradical intermediates analogous to those formed in the solid state but which have much greater conformational freedom. Products are thus formed in solution that are not possible in the crystal, and in some instances, these become the major photoproducts.

IX. ACKNOWLEDGEMENT

A great deal of credit for the work reported in this review is due to a talented group of coworkers at UBC. The X-ray crystallography was carried out by Dr Simon Phillips, Dr Trevor Greenhough, Dr Sara Ariel, Stephen Evans, Fred Wireko and Christine Hwang. The photochemistry was studied by Alice Dzakpasu, Syed Askari, Jack Jay and Peter Northcote. Funding was provided by the Natural Sciences and Engineering Research Council of Canada and the Petroleum Research Fund, whom we thank.

X. REFERENCES

1. J. M. Bruce, in *The Chemistry of the Quinonoid Compounds* (Ed. S. Patai), John Wiley and Sons, New York, 1974, Part 1, Ch. 9, p. 465.
2. The following is a partial list of review articles on the subject of chemical studies in organic crystals. (a) M. Hasegawa, *Chem. Rev.*, **83**, 507 (1983); (b) J. M. Thomas, S. E. Morsi and J. P. Desvergne, *Adv. Phys. Org. Chem.*, **15**, 63 (1977); (c) L. Addadi, S. Ariel, M. Lahav, L. Leiserowitz, R. Popovitz-Biro and C. P. Tang, in *Chemical Physics of Solids and their Surfaces* (Eds M. W. Roberts and J. M. Thomas), The Royal Society of Chemistry, London, 1980, Specialist Periodical Reports, Vol. 8, Ch. 7; (d) A. Gavezzotti and M. Simonetta, *Chem. Rev.*, **82**, 1 (1982); (e) G. M. J. Schmidt, *Pure Appl. Chem.*, **27**, 647 (1971); (f) M. D. Cohen, *Angew. Chem., Int. Ed. Engl.*, **14**, 386 (1975); (g) J. R. Scheffer, *Acc. Chem. Res.*, **13**, 283 (1980); (h) M. D. Cohen and B. S. Green, *Chem. Br.*, **9**, 490 (1973); (i) I. C. Paul and D. Y. Curtin, *Acc. Chem. Res.*, **6**, 217 (1973); (j) J. M. Thomas, *Chem. Br.*, **13**, 175 (1977); (k) J. M. McBride, *Acc. Chem. Res.*, **16**, 304 (1983); (l) J. M. Thomas, *Pure Appl. Chem.*, **51**, 1065 (1979); (m) M. Lahav, B. S. Green and D. Rabinovich, *Acc. Chem. Res.*, **12**, 191 (1979); (n) S. R. Byrn, *The Solid State Chemistry of Drugs*, Academic Press, New York, 1982; (o)

G. M. J. Schmidt, in *Solid State Photochemistry* (Ed. D. Ginsburg), Verlag Chemie, New York, 1976; (p) J. Trotter, *Acta Cryst.*, **B39**, 373 (1983); (q) D. Y. Curtin and I. C. Paul, *Science*, **187**, 19 (1975).

3. O. Diels and K. Alder, *Chem. Ber.*, **62**, 2362 (1929).
4. R. C. Cookson, E. Crundwell, R. R. Hill and J. Hudec, *J. Chem. Soc.*, 3062 (1964).
5. (a) J. R. Scheffer, K. S. Bhandari, R. E. Gayler and R. A. Wostradowski, *J. Am. Chem. Soc.*, **97**, 2178 (1975); (b) J. R. Scheffer, B. M. Jennings and J. P. Louwerens, *J. Am. Chem. Soc.*, **98**, 7040 (1976).
6. C. S. Gibbons and J. Trotter, *J. Chem. Soc., Perkin Trans. 2*, 737 (1972).
7. J. R. Scheffer and A. A. Dzakpasu, *J. Am. Chem. Soc.*, **100**, 2163 (1978).
8. S. E. V. Phillips and J. Trotter, *Acta Cryst.* **B33**, 991 (1977).
9. S. E. V. Phillips and J. Trotter, *Acta Cryst.*, **B33**, 996 (1977).
10. K. Gnanaguru, N. Ramasubbu, K. Venkatesan and V. Ramamurthy, *J. Org. Chem.*, **50**, 2337 (1985).
11. T. J. Greenhough and J. Trotter, *Acta Cryst.*, **B36**, 2840 (1980).
12. S. Ariel, J. R. Scheffer, J. Trotter and Y-F. Wong, *Tetrahedron Lett.*, **24**, 4555 (1983).
13. C. A. McDowell, A. Naito, J. R. Scheffer and Y-F. Wong, *Tetrahedron Lett.*, **22**, 4779 (1981).
14. (a) A. Weisz, M. Kaftory, I. Vidavsky and A. Mandelbaum, *J. Chem. Soc., Chem. Commun.*, 18 (1984); (b) M. Kaftory and A. Weisz, *Acta Cryst.*, **C40**, 456 (1984).
15. S. Askari, S. Lee, R. R. Perkins and J. R. Scheffer, *Can. J. Chem.*, **63**, 3526 (1985).
16. **7**: S. E. V. Phillips and J. Trotter, *Acta Cryst.*, **B32**, 3098 (1976); **8**: S. E. V. Phillips and J. Trotter, *Acta Cryst.*, **B33**, 984 (1977); **9**: S. E. V. Phillips and J. Trotter, *Acta Cryst.*, **B32**, 3095 (1976); **10**: S. E. V. Phillips and J. Trotter, *Acta Cryst.*, **B32**, 3088 (1976); **11**: S. E. V. Phillips and J. Trotter, *Acta Cryst.*, **B32**, 3091 (1976); **12**: S. E. V. Phillips and J. Trotter, *Acta Cryst.*, **B32**, 3101 (1976); **12b**: S. E. V. Phillips and J. Trotter, *Acta Cryst.*, **B33**, 1599 (1977); **13**: S. E. V. Phillips and J. Trotter, *Acta Cryst.*, **B32**, 3104 (1976); **14**: S. E. V. Phillips and J. Trotter, *Acta Cryst.*, **B33**, 1605 (1977).
17. J. D. Dunitz, *X-Ray Analysis and the Structure of Organic Molecules*, Cornell University Press, Ithaca, N.Y., 1979, pp. 312–318.
18. (a) R. Popovitz-Biro, C. P. Tang, H. C. Chang, M. Lahav and L. Leiserowitz, *J. Am. Chem. Soc.*, **107**, 4043 (1985); (b) C. P. Tang, H. C. Chang, R. Popovitz-Biro, F. Frolow, M. Lahav, L. Leiserowitz and R. K. McMullan, *J. Am. Chem. Soc.*, **107**, 4058 (1985).
19. Dimer **6a**: S. E. V. Phillips and J. Trotter, *Acta Cryst.*, **B33**, 991 (1977); dimer **8a**: S. E. V. Phillips and J. Trotter, *Acta Cryst.*, **B33**, 984 (1977).
20. S. H. Askari, J. R. Scheffer, J. Trotter and F. Wireko, *J. Am. Chem. Soc.*, in press.
21. C. M. O'Donnell and T. S. Spencer, *J. Chem. Ed.*, **49**, 822 (1972).
22. (a) J. A. Barltrop and D. Giles, *J. Chem. Soc. (C)*, 105 (1969); (b) R. L. Cargill, W. A. Bundy, D. M. Pond, A. B. Sears, J. Saltiel and J. Winterle, *Mol. Photochem.*, **3**, 123 (1971).
23. C. B. Chan and D. I. Schuster, *J. Am. Chem. Soc.*, **104**, 2928 (1982).
24. N. J. Turro, *Modern Molecular Photochemistry*, Benjamin/Cummings, Menlo Park, California, 1978, Ch. 10.
25. J. Michl, *Mol. Photochem.*, **4**, 243 and 257 (1972).
26. S. E. V. Phillips and J. Trotter, *Acta Cryst.*, **B33**, 1602 (1977).
27. S. Ariel, S. H. Askari, J. R. Scheffer and J. Trotter, *Tetrahedron Lett.*, **27**, 783 (1986).
28. A. Bondi, *J. Phys. Chem.*, **68**, 441 (1964).
29. S. Ariel, S. Evans, C. Hwang, J. Jay, J. R. Scheffer, J. Trotter and Y-F. Wong, *Tetrahedron Lett.*, **26**, 965 (1985).
30. S. H. Askari and J. R. Scheffer, unpublished results.
31. B. Rickborn and S-Y. Lwo, *J. Org. Chem.*, **30**, 2212 (1965).
32. S. Ariel, S. Askari, J. R. Scheffer, J. Trotter and L. Walsh, *J. Am. Chem. Soc.*, **106**, 5726 (1984).
33. C. Djerassi, *Pure Appl. Chem.*, **9**, 159 (1964).
34. (a) W. K. Appel, Z. Q. Jiang, J. R. Scheffer and L. Walsh, *J. Am. Chem. Soc.*, **105**, 5354 (1983); (b) S. Ariel, V. Ramamurthy, J. R. Scheffer and J. Trotter, *J. Am. Chem. Soc.*, **105**, 6959 (1983).
35. H. E. Zimmerman, *Adv. Photochem.*, **1**, 183 (1963).
36. (a) R. R. Birge, W. C. Pringle and P. Leermakers, *J. Am. Chem. Soc.*, **93**, 6715 (1971); (b) R. R. Birge and P. A. Leermakers, *J. Am. Chem. Soc.*, **93**, 6726 (1971).
37. R. Bonneau, *J. Am. Chem. Soc.*, **102**, 3816 (1980).
38. (a) D. I. Schuster, R. Bonneau, D. A. Dunn, J. M. Rao and J. Joussot-Dubien, *J. Am. Chem. Soc.*, **106**, 2706 (1984); (b) N. J. Pienta, *J. Am. Chem. Soc.*, **106**, 2704 (1984).

The Chemistry of Quinonoid Compounds, Vol. II
Edited by S. Patai and Z. Rappoport

CHAPTER **21**

Quinonediimines, monoimines and related compounds

ERIC R. BROWN
Colour Negative Technology Division, Photographic Research Laboratories, Eastman Kodak Company, B82 RL, Rochester, New York 14650, USA

I. INTRODUCTION

The chapter on quinonediimines and related compounds, written by Finley and Tong[1] in the earlier volume of this series on the chemistry of functional groups, surveyed the literature up to the late 1960s. Since that time many areas of the chemistry of quinonediimines have been extended in great detail. Quinonediimines and monoimines are important in many commercial applications, such as color photography and hair-coloring chemistry. Pertinent reviews in these two areas have been published[2,3]. Pharmaceutical chemistry, especially drug metabolism, is another important area of quinonediimine reactivity which has been studied since 1970; one example of this work is given later.

A major barrier to the quantitative study of quinonediimines is the elusive nature of the species. As Willstätter[4] pointed out, these compounds are unstable to light, water and acids. As a result, much recent work has relied on the oxidation of the precursor phenylenediamines and aminophenols to prepare the species of interest. These oxidations have been carried out using rapid mixing techniques with such chemical oxidants as ferricyanide[5], iodine[6], manganate[7], Mn(III)[8], permanganate in mild acid or alkali[6], persulfate, catalyzed peroxide, *N*-chloramine[9] and silver ion[4]. In photographic applications, exposed silver chloride or bromide is the oxidant[2]. Hair-coloring applications use hydrogen peroxide in mild alkali[3] as the oxidant. Other studies have made use of sophisticated electrochemical methods at inert electrodes to oxidize the precursor, phenylenediamine or aminophenol, to form the corresponding diimine or monoimine. The following chemical reaction coupled to the electrochemical step, perturbs the electrochemical response and forms the basis for studying the rate and mechanism of such reactions[10,11].

Regardless of the mode of oxidation, quinonediimines and quinonemonoimines are produced by two distinct one-electron oxidation steps[2] from the parent phenylenediamine or aminophenol, resulting in a radical intermediate semiquinone of varying stability. For some oxidants, the second oxidation step may be faster than the first, precluding any significant formation of radical when excess oxidant is used.

NH_2 … R^1–N–R^2 $\underset{}{\overset{-e^-}{\rightleftharpoons}}$ NH_2 … (+) … R^1–N–R^2 $\overset{-e^-,\ -H^+}{\rightleftharpoons}$ NH … R^1–N^+–R^2

The stability of the semiquinone, S, depends on the nature of R^1 and R^2, and on whether *p*-quinoneimines or *o*-quinoneimines are formed. Semiquinone species of *o*-diimines and *o*-monoimines have been reported as transient species in such rapid oxidation techniques as pulse radiolysis and flow electron spin resonance (ESR) methods.

p-Semiquinones are much more stable, even for monoimines, when R^1 and R^2 are alkyl groups. These stable, often highly colored, intermediates confound mechanistic studies of diimines when oxidative formation is neither rapid nor complete[5,6].

Stable, isolable quinoneimines are formed when strongly electron-withdrawing groups are attached to nitrogen. Imines such as **1–4** have been studied extensively by Adams and Reifschneider[12], and more recently by Fujita[13].

Reviews of the chemistry of many *o*-quinonediimides have appeared recently by Heine and coworkers[14] and by Friedrichsen and Böttcher[15], and will not be repeated here.

The major goal of this chapter is to discuss the kinetics and mechanism of nucleophilic reactions of quinonemonoimines and quinonediimines after formation in a predominantly aqueous solution. The major reactions include deamination (attack by water or hydroxide), coupling reactions by carbanions to form dyes, sulfonation (addition of sulfite

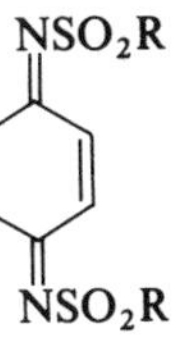

NSO$_2$R NSO$_2$R (1)   O NSO$_2$R (2)   NSO$_2$R NSO$_2$R (3)   O NSO$_2$R (4)

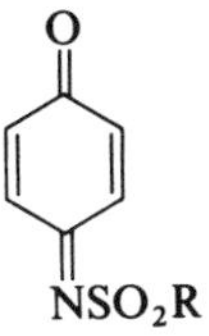

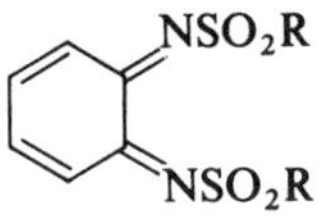

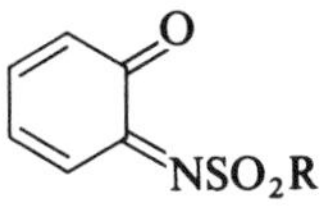

and arylsulfinic acid), and intramolecular nucleophilic attack (self-coupling). Extensions of these same reactions to quinoneimides and diimides are also included. The role of redox equilibration, the semiquinone Michaelis equilibrium, is discussed briefly with regard to condensation reactions involving the quinoneimine with its precursor, phenylenediamine or aminophenol.

II. NUCLEOPHILIC REACTIONS OF QUINONEIMINES

Addition and substitution reactions with quinonediimines or monoimines can occur at several sites on the molecule. In general, there are four reaction sites; the carbon atom attached to the unsubstituted nitrogen, the carbon atom attached to the substituted nitrogen, and the two ring positions, as illustrated for *N,N*-dialkylquinonediimine (**5**). The actual site of attack depends on several factors including the nucleophile and the structure of **5**.

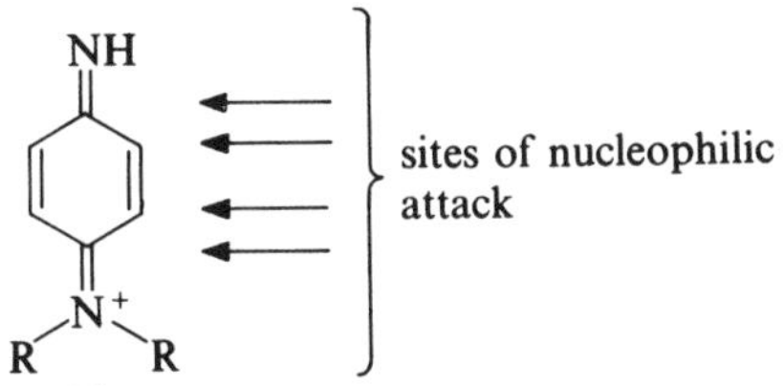

NH N^+ R R sites of nucleophilic attack

(5)

Because of the competitive nature of nucleophilic attack, the product distribution will be determined by the site(s) that reacts fastest. The kinetics at any given site are determined by the electron density at that site, the stability of the products formed and steric considerations of both the electrophilic site and the approaching nucleophile.

The general reaction scheme for attack by the nucleophile, X^-, consists of two steps: (1) reversible formation of an intermediate and (2) irreversible formation of products.

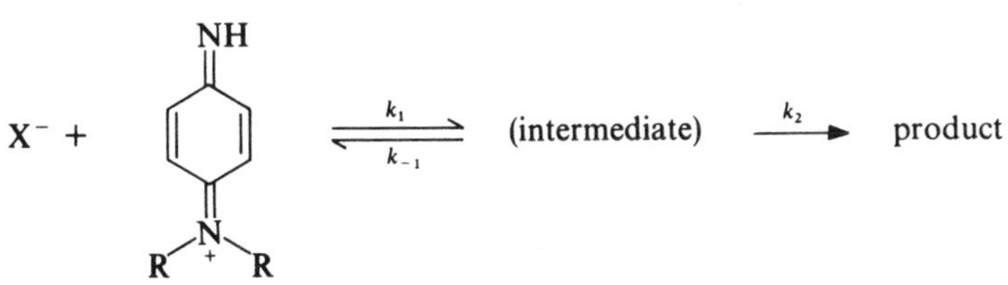

$$X^- + \text{(NH / } N^+R_2 \text{ quinonediimine)} \underset{k_{-1}}{\overset{k_1}{\rightleftharpoons}} \text{(intermediate)} \xrightarrow{k_2} \text{product}$$

The reaction kinetics can be quite simple when very little intermediate accumulates and the steady-state assumption is used. When $[X^-]$ is in large excess or otherwise buffered, the observed first-order rate constant for loss of diimine or formation of product is given by equation 1.

$$k_{obs} = \frac{k_1 k_2[X^-]}{k_{-1} + k_2} \quad (1)$$

If $k_2 > k_{-1}$, equation 1 becomes $k_{obs} = k_1[X^-]$ and the reaction is first order in $[X^-]$ as well as in diimine. Such is the case for attack by OH^- (deamination), as described later. When $k_{-1} > k_2$, equation 1 becomes $k_{obs} = Kk_2[X^-]$ where $K = k_1/k_{-1}$, the equilibrium constant for formation of intermediate.

Another extreme case occurs when the intermediate rapidly forms and maintains its equilibrium value. The kinetics of product formation now are determined only by k_2. The observed rate constant is still first order, but no longer depends on the concentration of X^-. By varying $[X^-]$, it is sometimes possible to obtain values for both k_1 and k_2 as described later for some dye-forming reactions.

$$X^- + H^+ + \text{(6)} \underset{k_{-1}}{\overset{k_1}{\rightleftharpoons}} \text{(intermediate)} \xrightarrow{k_2} \text{product}$$

NH
NH
(6)

If the quinonediimine is neutral, as illustrated by **6**, the formation of an intermediate requires a proton, presumably on nitrogen. This is the case for deamination of quinonediimine in acid solution, where the attacking nucleophile, water, reacts with protonated diimine. Water is too poor a nucleophile to react at a kinetically significant rate with the neutral diimine except when it is the only species present.

The semiquinone species, which is involved in oxidative formation of quinonediimine, is not reactive in the nucleophilic reactions discussed here[16]. The major complication resulting from the presence of semiquinone is that of spectral interference in the observation of product formation or diimine loss. In addition, the diimine concentration as a function of time will be complicated by the redox reaction forming the species, instead of being controlled only by the attack of the nucleophile. Unambiguous analysis of the kinetics to determine k_{obs} in equation 1 is difficult under these conditions. The presence of semiquinone, particularly in acid solutions, further complicates kinetic studies, not because the oxidation of parent phenylenediamine is slow, but because it is incomplete.

III. DEAMINATION STUDIES

The nucleophilic attack of water and hydroxide is an important reaction of quinonediimines and monoimines, but the reaction often prevents isolation of the species under study. Tong[17] and, later, Corbett[18] used ferricyanide to oxidize *p*-phenylenediamines to the corresponding diimine in neutral or alkaline solution. Nickel, Jaenicke and their coworkers[5,6,19,20] observed that, in acid solution, ferricyanide is not a sufficiently strong oxidant to completely form the diimine of most *N*-substituted compounds. They recommend using I_2[6], permanganate[19], Ce(IV)[20], or persulfate catalyzed by Cu(II)[20] to form the diimine or monoimine of interest rapidly and irreversibly. Electrochemical oxidation has also been used to form monoimines and diimines[11,21–24]. These oxidation techniques in rapid flow or stopped-flow equipment, particularly the work of Nickel and Jaenicke[5,6,19,20,25], have led to a good understanding of the deamination (hydrolysis) mechanism for a large number of *para*-quinonediimines and monoimines.

The overall path of diimine deamination is given by Scheme 1[19], giving *p*-benzoquinone as the product in both acidic and alkaline solution.

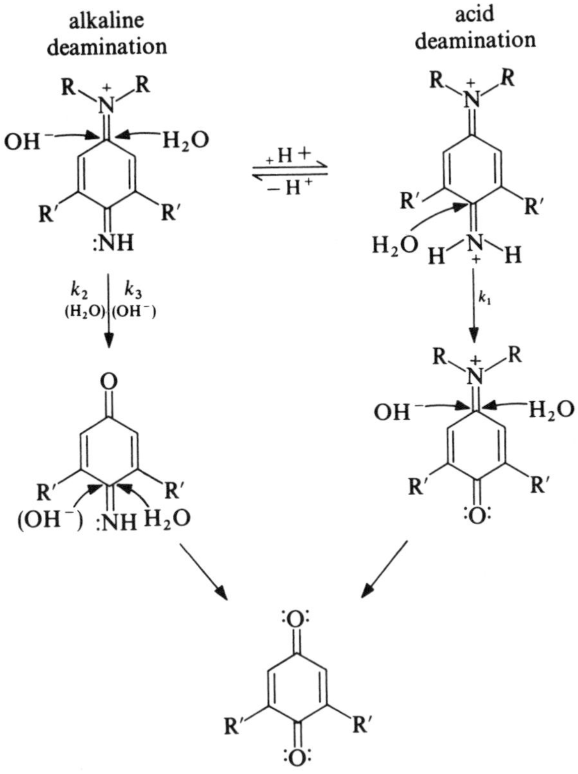

alkaline deamination

acid deamination

SCHEME 1. Overall path of deamination of diimines to quinone[19]

A. Unsubstituted Quinoneimines

For the unsubstituted quinonediimine (**6**), the kinetics of deamination have been reported by both Tong[17] and Corbett[18] over an extended pH range. The pH dependence of $\log k_{obs}$, obtained by following either the loss of diimine or the formation of monoimine spectrally in the ultraviolet, is shown in Figure 1. This dependence is the result of superposition of three parallel pathways for the attack of water on the three forms of the

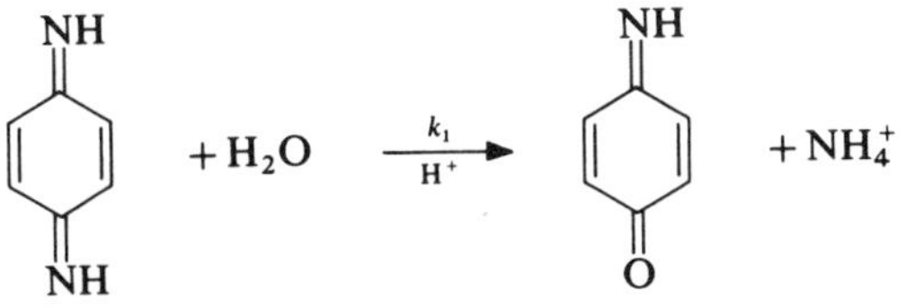

$$\text{D (6)} + H_2O \xrightarrow[H^+]{k_1} \text{(7)} + NH_4^+ \qquad (2a)$$

NH
$+H_2O \xrightarrow{k_2}$ (7) $+NH_4^+$ (2b)
$^+NH_2$
DH^+

$^+NH_2$
$+H_2O \xrightarrow{k_3}$ (7) $+NH_4^+ + H^+$ (2c)
$^+NH_2$
DH_2^{2+}

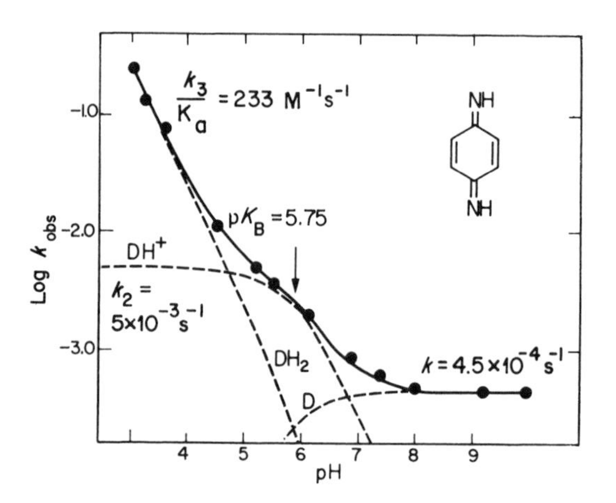

FIGURE 1. The effect of pH on the hydrolysis rate of *p*-benzoquinonediimine at 30 °C, showing the contribution of the hydrolysis of the individual species (broken lines) to the overall rate (full line) calculated, from equations 3 and 4. The points are experimental measurements[18]

diimine (equations 2a–c)[18]. Notice that, although a proton is required in equation 2a to balance the reaction, it is not required in the rate-limiting step, which is discussed in more detail later. The pH-dependent expression for the pseudo first-order rate constant, k_{obs}, in Figure 1 (solid line) is given by equation 3, where the values of k_1 through k_3 are given in Figure 1. These values include the water activity term ($\sim$ 55 M) so they are first-order rate constants. D_T is the total concentration of diimine. The relative concentration

$$k_{obs} = \frac{k_3\,\{DH_2{}^{2+}] + k_2[DH^+] + k_1[D]}{D_T} \tag{3}$$

of each form is given by equation 4, where K_a and K_b are the ionization constants of the doubly protonated and singly protonated species, respectively.

$$\frac{[DH_2^{2+}]}{D_T} = \frac{[H^+]^2/K_aK_b}{1+[H^+]^2/K_b+[H^+]^2/K_aK_b} \quad (4a)$$

$$\frac{[DH^+]}{D_T} = \frac{[H^+]/K_b}{1+[H^+]/K_b+[H^+]^2/K_aK_b} \quad (4b)$$

$$\frac{[D]}{D_T} = \frac{1}{1+[H^+]/K_b+[H^+]^2/K_aK_b} \quad (4c)$$

The ionization equilibria are $pK_a = 1.5$ and $pK_b = 5.75$, shown by the arrow pointing downward in Figure 1. The diimine is most stable in alkaline solution. In acid solution below pH 1, k_{obs} goes through a maximum[26] not shown in Figure 1, indicating the possibility of a more complex mechanism.

As evidence for the rate-limiting addition of water to the diimine in an alkaline solution, Tong and Glesmann[27] have measured the ionic strength dependence of the diimine hydrolysis. The rate is independent of μ up to 0.062 M, indicating no charge on the reactants in the rate-limiting step.

The monoimine (7) in equation 2 also undergoes deamination, to form *p*-benzoquinone (**8**). The pH dependence of $\log k_{obs}$ is given in Figure 2 from data taken from Tong[17] and Corbett[18]. Again, the pH dependence can be accounted for by superposition of parallel reactions of the neutral and protonated monoimine (equations 5a and 5b). The expression

M(7) $+ H_2O \xrightarrow[H^+]{k_1}$ (**8**) $+ NH_4^+$ (5a)

MH$^+$ $+ H_2O \xrightarrow{k_2}$ (**8**) $+ NH_4^+$ (5b)

for k_{obs} is given by equation 6, which is the sum of the rate constants for each of the two steps above;

$$k_{obs} = \frac{k_1[M]+k_2[MH^+]}{M_T} \quad (6)$$

where

$$\frac{[M]}{M_T} = \frac{1}{1+[H^+]/K_a} \quad (7a)$$

and

$$\frac{[MH^+]}{M_T} = \frac{[H^+]/K_a}{1+[H^+]/K_a} \quad (7b)$$

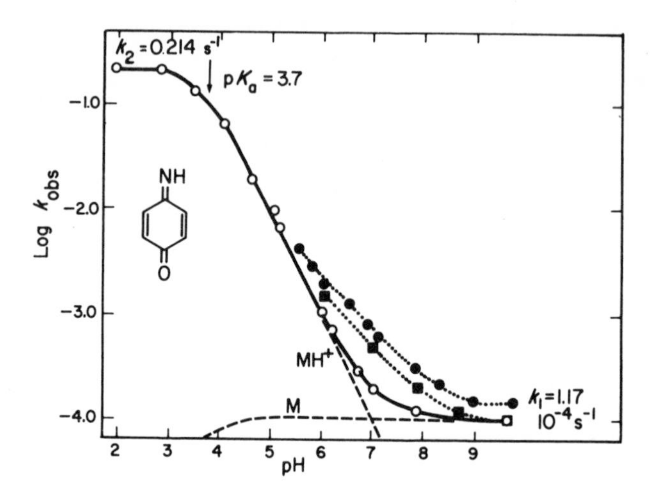

FIGURE 2. The effect of pH on the hydrolysis rate of *p*-benzoquinonemonoimine. ○ Acetate buffer at 30 °C[18]; ● phosphate buffer at 30 °C[18]; and ■ phosphate buffer at 25 °C[17]. The solid line is calculated from equations 6 and 7 using the rate constant values shown above

TABLE 1. Summary of deamination kinetics[17, 18]

Structure	pK_a	k_3 (s^{-1})	pK_b	k_2 (s^{-1})	k_1 (s^{-1})
NH NH	1.5	7.77×10^{3a}	5.75	0.005	4.5×10^{-4} 3.3×10^{-4b}
NH O	—	—	3.7	0.214	1.17×10^{-4}
NH Me O	—	—	—	—	1.0×10^{-4b}

[a] Maximum value of k_3, assuming pK_a = 1.5[18].
[b] Author's analysis of Tong's results[17].

The solid line in Figure 2 is calculated from equations 6 and 7 using the values obtained by Corbett[18] at 30 °C, as described in the figure caption. Comparison of k_{obs} with those in Figure 1 shows how similar the observed rate constants are, although individual species can have quite different values. The values of the individual constants are summarized in Table 1 for three compounds. The decrease in k_{obs} found electrochemically below pH 1 is probably due to additional complications in the mechanism[21,26].

To be complete, there probably should be another term in equation 6 to describe the specific buffer effect of the phosphate shown in Figure 2. Insufficient study precludes describing this as a general base or general acid catalysis. Tong's data above pH 11, not included in Figure 2, show a rate increase, indicating that direct attack by hydroxide is responsible[17].

B. *N,N*-Dialkylquinoneimines

The mechanism is somewhat different for *N,N*-dialkyl *p*-quinonediimines. In alkaline solution, the logarithm of k_{obs} for hydrolysis increases linearly with pH, i.e. the reaction is first order in $[OH^-]$[17]. The positively charged diimines are attacked by OH^- to cleave *N,N*-dialkylamine, forming *p*-quinoneimine (7).

The second-order rate constant, k_4, was measured by Tong and coworkers[28] for several compounds of photographic interest and the data are summarized in Table 2. Increasing

TABLE 2. Deamination rate constants of *p-N,N*-dialkylquinone diimines

R¹R²N⁺=C₆H₂(R³)(R⁴)=NH

R^1	R^2	R^3	R^4	pK_a	k_2 ($10^{-2}\ M^{-1}s^{-1}$)	k_3 ($10^{-5}\ M^{-1}s^{-1}$)	k_4 ($10^4\ M^{-1}s^{-1}$)	Reference
Me	Me	H	H	1.7 ± 0.1	1.1 ± 0.2	1.40	1.3	19[a]
				3.0[d]			1.5 ± 0.7	20
							2.5	17[b,c]
							1.1	23[e]
Me	Me	Me	H	1.74 ± 0.1	3.7 ± 0.4	—	—	19
Me	Me	Me	Me	1.48 ± 0.1	4.4 ± 0.4	—	—	19
Et	Et	H	H	1.60 ± 0.1	1.5 ± 0.2	0.5	0.9	19
							0.52	17
							1.0	28
Et	Et	Me	H	1.65 ± 0.1	3.1 ± 0.4	0.4	0.11	19
							0.16	17
Et	Et	Me	Me	—	—	0.10	0.02	19
							0.022	28
$-(CH_2)_4-$		H	H	—	—	—	0.60	28

[a] 25 °C, variable ionic strength; second-order rate constants k_2 and k_3 have been obtained by dividing the observed rate constants (see Figure 3) by the water activity (~ 55 M).

[b] 25 ± 0.1 °C, 0.2 M phosphate.

[c] 25 ± 0.1 °C, $\mu = 0.375$, phosphate buffers.

[d] Ref. 30.

[e] 25 °C, $\mu = 0.4$ M phosphate, based on published k_{obs} vs. pH above pH 10.

$+OH^- \xrightarrow{k_4}$ $+HNR^1R^2$

(7)

ionic strength slows the reaction[27] in agreement with the mechanism of direct reaction of two unlike charged species.

It was later shown by Nickel and Kemnitz[29], and Lelievre and coworkers[30] that, in neutral and acid solution, the mechanism and products change. Deamination now occurs by the attack of water on the primary imine, or its protonated form, to produce an *N,N*-dialkylquinoneimine (**9**). The observed first-order rate constant, followed by indirect

$+H_2O \xrightarrow[H^+]{k_2}$ $+NH_4^+$

(9)

$+H_2O \xrightarrow{k_3}$ $+NH_4^+$

methods[19], is small in neutral solution, increasing first order in $[H^+]$ to a constant value at a pH below the protonation equilibrium. The observed rate constant obeys an expression similar to equation 3, where the value of $k_1 = 0$ since no neutral diimine exists. The monoimine (**9**) is more stable than the diimine below pH 4 for R^1 and R^2 equal to alkyl, and can be observed spectrally[19]. The value of k_{obs} as a function of pH is plotted in Figure 3 for both *N,N*-dimethylquinonediimine and monoimine[19]. At high pH, the values reported by Tong[17] are in good agreement. The solid line, curve A, is given by equation 8

$$k_{obs} = \frac{k_3[H^+]/K_a[H_2O] + k_2[H_2O] + k_4[OH^-]}{1 + [H^+]/K_a} \tag{8}$$

with $pK_a = 1.7 \pm 0.1$. Notice that k_{obs} describe the formation of different products. The neutral monoimine (7) is produced in alkaline solution, represented by the term $k_4[OH^-]$. The *N,N*-dialkylmonoimine (**9**) is formed in acid solution, represented by the first two terms in equation 8. The product distribution will be 1:1 of **7** and **9** when the term $[H_2O](k_3[H^+]/K_a + k_2)$ equals $k_4[OH^-]$. This is near the pH of maximum stability of diimine, pH 6, represented by the minimum in curve A in Figure 3. The values of k_2 and k_3 in Table 2 are second-order rate constants since the water concentration is explicitly separated in equation 8[19].

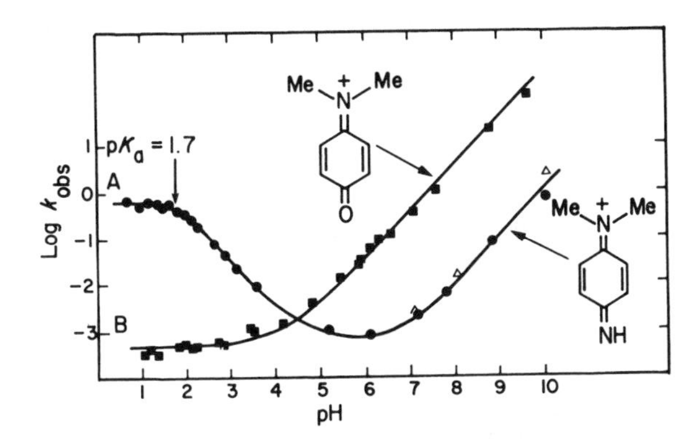

FIGURE 3. The effect of pH on the hydrolysis rate of N,N-dimethylquinonediimine (curve A) and monoimine (curve B)[19]. ● Diimine formed by Ce(IV) oxidation of diamine; ○ monoimine formed by I_2 oxidation of p-aminophenol; and Δ hydrolysis of diimine measured by Tong[17]. Curve A calculated from equation 8 using the values in Table 2. Curve B calculated from equation 9 using the values in Table 3

As methyl groups are added to the ring, the electron density on the charged nitrogen increases and k_4 decreases. Similarly, on going from Me to Et on this nitrogen, k_4 decreases.

The value of k_3 also decreases in the same manner, although the carbon to which the neutral amino group is attached is now the site of attack. Although the trend is small, k_2 actually increases slightly for deamination of the protonated amine as electron density increases. Given the variable ionic strength in the acid solutes (below pH 2) where these measurements are made[19], the differences may not be too meaningful.

Deamination of p-N,N-dialkylmonoimines has also been studied by Nickel and Jaenicke[20], and Tong and Glesmann[31]. The reaction product is a p-benzoquinone resulting from attack by OH^- in alkaline solution, or by H_2O in neutral and acid solution. The rate constant k_4 in alkaline solution is ~ 100 times larger than the value for the

$$\textbf{(9)}\ (R^1R^2\overset{+}{N}=\text{C}_6\text{H}_3(R^3)=O) + OH^- \xrightarrow{k_4} \textbf{(8)}\ (O=\text{C}_6\text{H}_3(R^3)=O) + HNR^1R^2$$

$$\textbf{(9)}\ (R^1R^2\overset{+}{N}=\text{C}_6\text{H}_3(R^3)=O) + H_2O \xrightarrow[H^+]{k_2} \textbf{(8)}\ (O=\text{C}_6\text{H}_3(R^3)=O) + H_2\overset{+}{N}R^1R^2$$

corresponding quinonediimine[31]. Nickel and Jaenicke[20] also studied acid solution and observed an additional influence due to buffer anions (general-base catalysis), particularly citrate. Their observed rate constants as a function of pH are given by equation 9, where

$$k_{obs} = k_4[OH^-] + k_2[H_2O] + k_5[B^-] \tag{9}$$

B^- represents a general base. At constant pH (pH = 4.5), the observed rate constant was proportional to the citrate concentration[20]. Above pH 6, the ionic strength dependence was consistent with the reaction of singly charged species of opposite sign; below pH 3, the dependence was very small[20]. A plot of log k_{obs} as a function of pH is shown in Figure 3 (curve B) for *N,N*-dimethylquinoneimine deamination. The rate is very slow in acid solution except when citrate ion is also present. Rate constants k_2 and k_4 are given in Table 3 for several compounds.

Comparing k_4 values in Tables 2 and 3, it is clear that the *N,N*-dialkylmonoimines are ~ 100 to 200 times more reactive toward hydroxide than are the corresponding diimines. In neutral-to-slightly-acid solution, they are slightly less reactive, although different products are formed. In a strongly acid solution, the diimine clearly is more reactive due to the formation of a protonated species.

TABLE 3. Deamination rate constants of *p-N,N*-dialkylquinonemonoimines

R^1	R^2	R^3	R^4	k_4 (10^4 M^{-1} s^{-1})	k_2 (10^{-6} M^{-1} s^{-1})	Reference
Me	Me	H	H	240 ± 10	7.4 ± 0.6	20[a]
				200		[b]
Me	CH_2CH_2OH	H	H	310	—	31[c]
Me	Me	Me	H	26 ± 2	0.96 ± 0.09	20
Me	Me	Me	Me	3.1 ± 0.03	0.16 ± 0.03	20
Et	Et	H	H	100 ± 5	1.4 ± 0.2	20
Et	Et	Me	H	11 ± 1	0.17 ± 0.03	20
				15.8	—	31
Et	Et	Me	Me	1.3 ± 0.1	0.04 ± 0.02	20
$-(CH_2)_4-$		H	H	61.7	—	[b]

[a] 25 °C, $\mu = 0.3$
[b] E. R. Brown, unpublished work.
[c] 25 °C, $\mu = 0.37$.

The preceding results have prompted Nickel and Jaenicke[20] to propose Scheme 2 as the common reaction mechanism for loss of *N,N*-dialkylamine from *N,N*-dialkylquinonediimines and monoimines.

A preceding solvation equilibrium ($K > 10^{-5}$ M^{-1}) is assumed to occur. This solvation complex reacts via k_4 in alkaline solution, or via k_2 or k_3 in neutral and acid solution. In pathway k_4 a proton is transferred to OH^-, or to a general base, from the adduct X^+, which rapidly loses water to form Y in an irreversible step. The dialkylamine is split off by

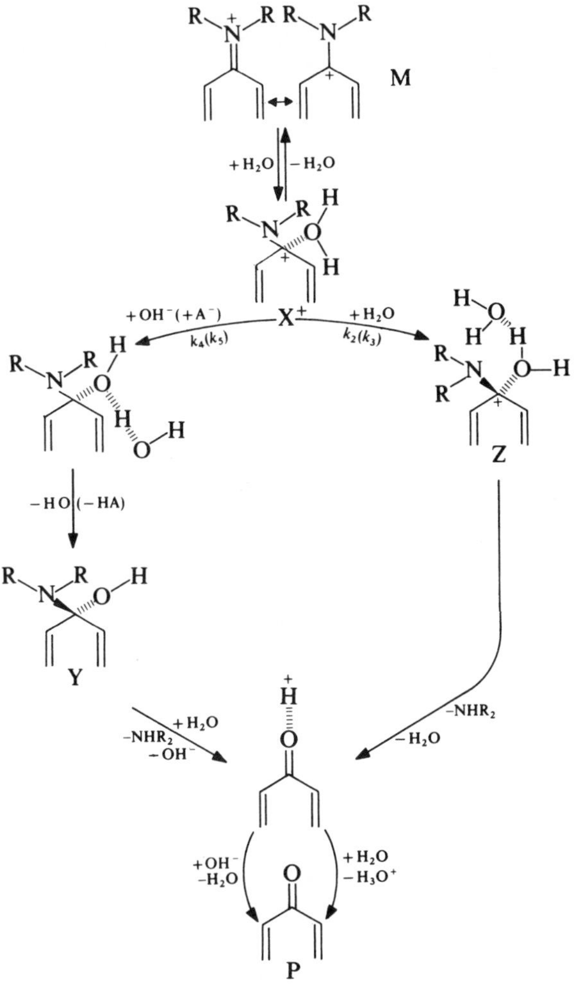

SCHEME 2. Common reaction mechanism for deamination of *N,N*-dialkylquinone-imines[20]

the attack of water and the loss of OH^-, forming a transient, positively charged intermediate which leads to product P. In acid solution, water attacks the adduct X^+, leading to the ring Z^+. This intermediate decomposes in a concerted reaction, losing both water and dialkylamine, to form the same positively charged intermediate which decomposes to P. In strongly alkaline solution, direct attack of OH^- (not shown in Scheme 2) upon the quinonediimine or monoimine is possible, without formation of the solvated species X^+ [17].

C. Pseudobase Formation

The positively charged quinonediimine, formed by the oxidation of *N,N*-dialkylphenylenediamine, can be stabilized toward deamination by internal cyclization

TABLE 4. Pseudobase equilibrium constants and deamination rate constants of *N*-β-hydroxyethyldiimines[28]

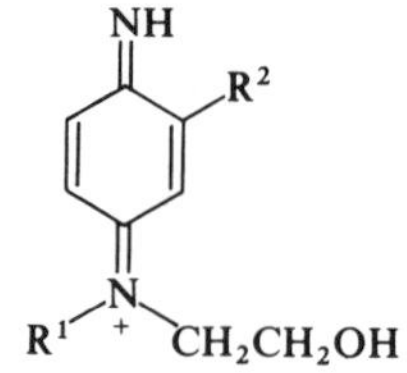

R^1	R^2	log K	k'_4 (M^{-1} s^{-1})
Et	H	5.10	1.4×10^{-1}
CH_2CH_2OH	H	5.88	0.8×10^{-1}
Et	Me	4.18	2.5×10^{-1}
CH_2CH_2OH	Me	5.05	1.4×10^{-1}

when the alkyl group is replaced by a β-hydroxyethyl group, or a β-sulfonamidoethyl group. Such quinonediimines form pseudobases (**10**), probably through intramolecular hydrogen bonding, after the addition of base[28]. Such adducts form reversibly and deaminate several orders of magnitude more slowly than their dialkyl counterparts. The equilibrium constant for pseudobase formation and the deamination rate for several adducts of diimines were measured by Tong and coworkers[28]. Values for K and k'_4 are given in Table 4 for several β-hydroxyethyl-substituted diimines. Comparison with values of k_4 in Table 2 shows that the rate constants are 10^4 to 10^5 times smaller for deamination of the adduct.

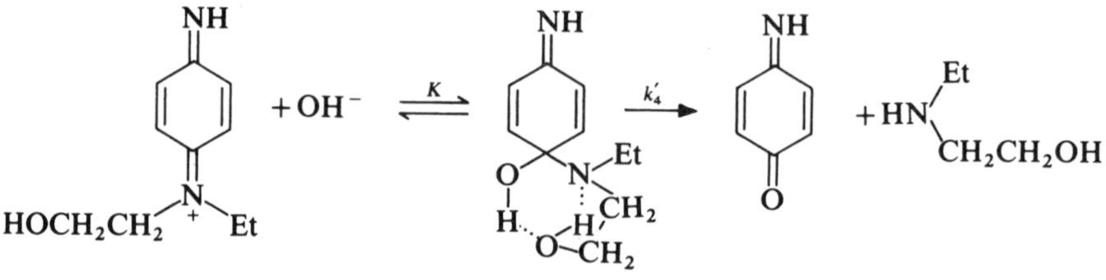

(**11**) (**10**) (7)

In the case of β-methyl sulfonamidoethyl substitution, the sulfonamido proton ionizes on the quinoneimine forming zwitterionic pseudobase (**13**)[17]. The rate constant for deamination in this case is only ~ 100 times smaller than the rate constant for deamination of **12**, due to the overall charge reduction on the diimine. The effect is much greater for monoimines, as shown in Table 5. The net effect of pseudobase formation is to yield observed kinetics that approach or become independent of pH as pH increases because the dominant species in solution, the pseudobase, is so stable.

D. *N*-Alkylquinonemonoimines

These compounds react in base, but do not undergo deamination. Instead, they undergo slow ionization to a tautomeric Schiff base, which then cleaves by hydrolysis. The slow tautomerization of several compounds has been studied in alkaline phosphate buffer[32].

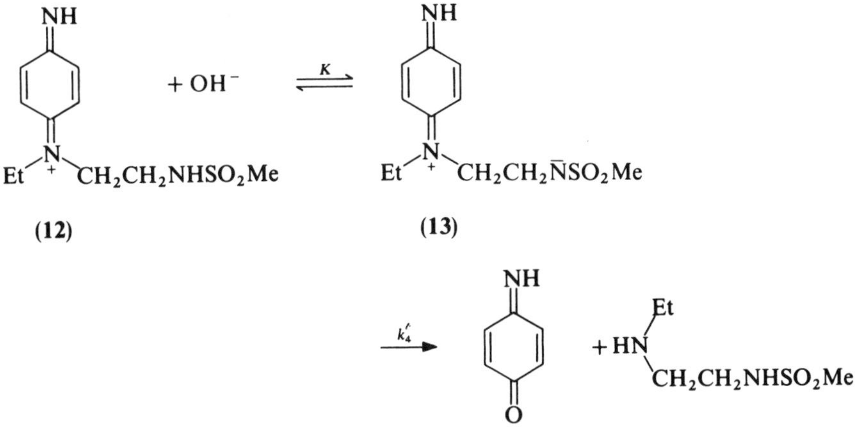

NH=C6H4=N+(Et)CH2CH2NHSO2Me (12) + OH⁻ ⇌(K) NH=C6H4=N+(Et)CH2CH2N⁻SO2Me (13)

→(k'_4) O=C6H4=NH + HN(Et)CH2CH2NHSO2Me

TABLE 5. Ionization equilibrium constants and deamination rate constants of *N*-β-methyl sulfonamidoethylimines

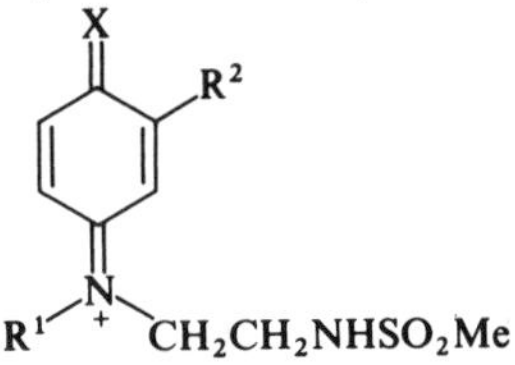

Structure R[1]	R[2]	X	log K	QDI[+] log k_4	QDI (ionized) log k'_4
Et	H	=NH[a]	5.13	4.30	2.34
Et	Me	=NH[a]	4.48	3.65	2.48
Et	H	=O[b]	8.15	6.69	1.75
Et	H	=O[b]	9.50	7.36	0.83

[a] Ref. 28.
[b] Ref. 31.

The observed rate constants for **14** and **15** are ~ 0.02 s^{-1} at pH 11, and the products are the corresponding Schiff bases (**16**), formed by general base-catalyzed ionization of the alkyl group.

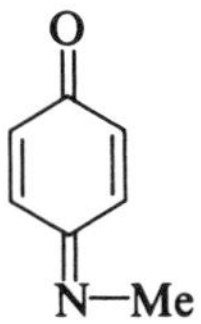

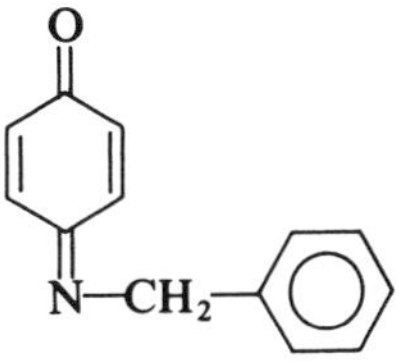

(**14**) (**15**)

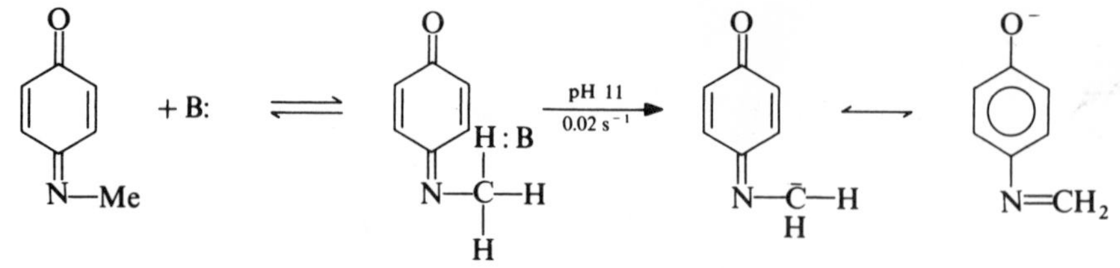

(14) + B: ⇌ [H:B complex] $\xrightarrow[0.02\ s^{-1}]{pH\ 11}$ + HB ⟷ (16)

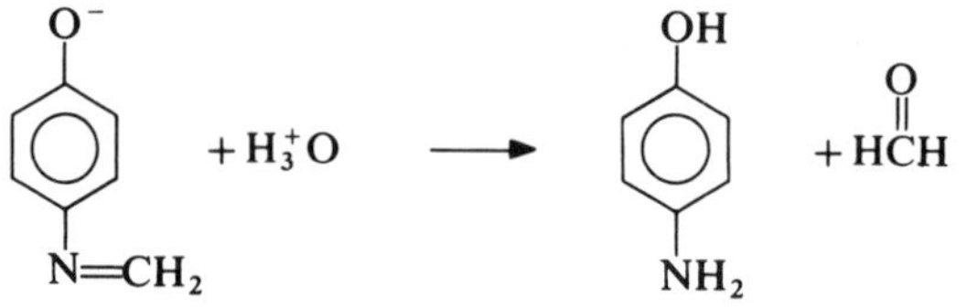

(16) + H_3^+O ⟶ p-aminophenol + HCHO

Upon acidification, rapid and quantitative formation of the corresponding aminophenol is observed along with the corresponding aldehyde, which in the case of **14** is formaldehyde, detected by chromotropic acid[32]. A detailed hydrolysis of the Schiff base of **17** has been published by Reeves[33].

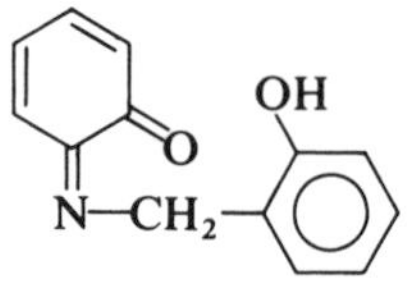

(17)

The pH dependence of tautomerization to the Schiff base is shown in Figure 4 for four monoimines in 0.375 ionic strength phosphate buffers. At high pH, log k vs. pH has a slope

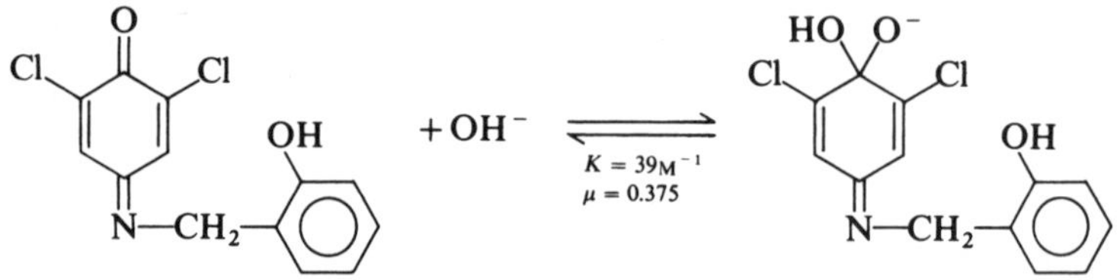

(18) + OH^- $\underset{\mu = 0.375}{\overset{K = 39M^{-1}}{\rightleftharpoons}}$

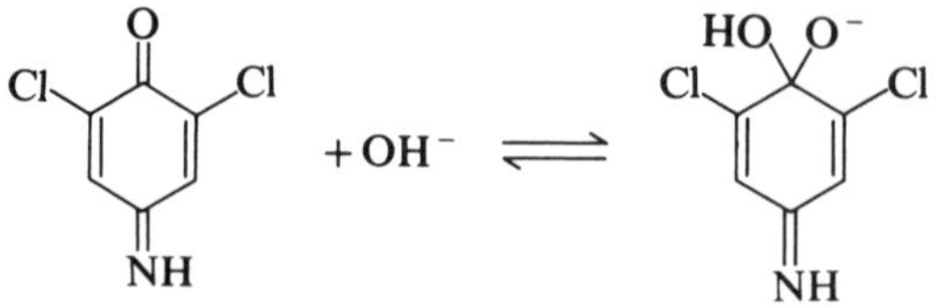

(19) + OH^- ⇌

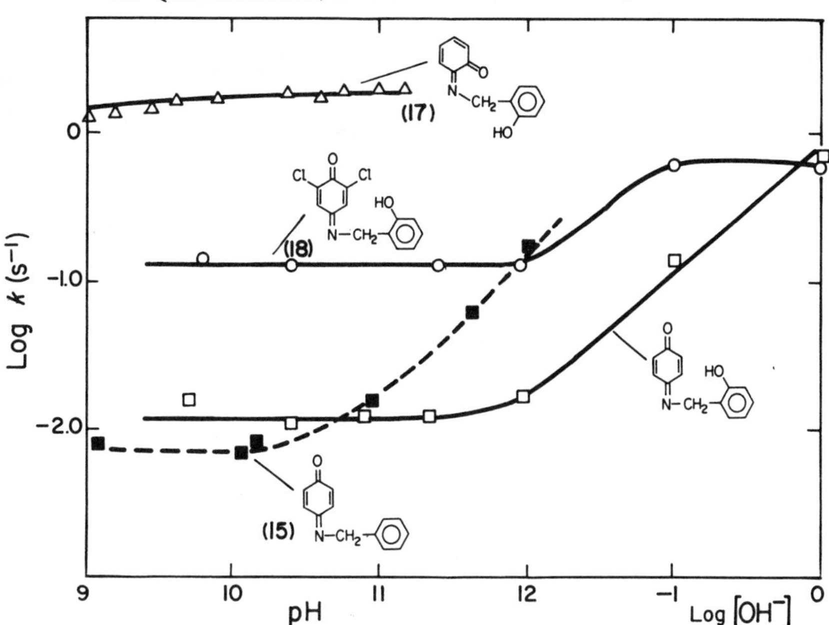

FIGURE 4. The effect of pH on the conversion of quinonemonoimines to the corresponding Schiff bases in phosphate buffer ($\mu = 0.375$)

of unity, indicating that OH^- is the most reactive base attacking the monoimine. The pH dependence of **18** is caused by formation of a reversible hydroxide adduct of the quinoneimine, observed spectrally[32]. The author has observed similar adduct formation with 2,6-dichloro-*p*-quinoneimine (**19**) at $\mu = 0.375$ with $K = 63\ \mathrm{M}^{-1}$.

A more detailed study of the tautomerization of monoimine (**20**) has yielded the separate rate constants for each of the four different bases present in phosphate buffer in alkaline solution.

(**20**)

The rate constants listed in Table 6 are used to calculate the solid line in Figure 5 for the observed tautomerization rate constants (points) of **20**. The hydroxide adduct equilibrium for **20** is 39 M^{-1}, causing the observed rate constant to level off at high pH, as shown in Figure 5. Tautomerization of the corresponding *N*-alkyldiimines has not been studied.

Hartke and Lohmann[34] studied the reaction of primary and secondary amines with 1,2-naphthoquinone-4-sulfonate in aqueous solution. In order to explain their products, they

TABLE 6. Rate constants for base-catalyzed tautomerization of **20**[32]

Base	k_{base} (M^{-1} s^{-1})
OH^-	159
PO_4^{-3}	0.75
HPO_4^{-2}	0.25
H_2O	0.02[a]

[a] Value of $k_{H_2O}[H_2O]$ in s^{-1}.

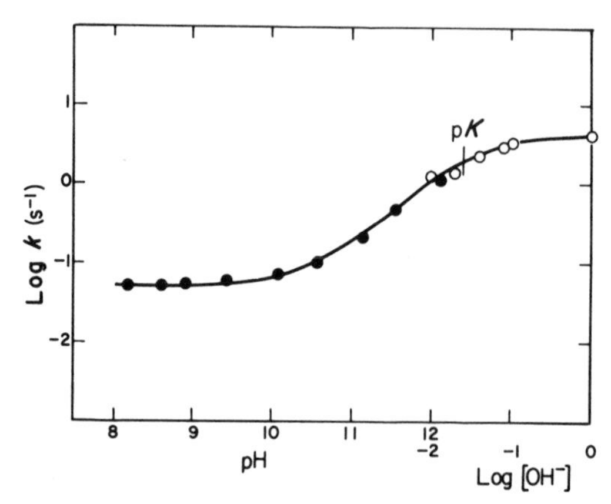

FIGURE 5. The effect of pH on conversion of monoimine **20** to Schiff base at 25 °C[32]. ● Phosphate buffer, $\mu = 0.375$; ○ NaOH solution, $\mu = 0.375$ with added NaCl. Solid line calculated using the rate constants in Table 6

postulate the formation of an *o*-quinoneimine (**21**) which they suggest can tautomerize to the Schiff base (**22**). This unisolated intermediate then undergoes hydrolysis to the 2-

$+ H_2NCHR^1R^2 \longrightarrow$ **(21)** $\rightleftharpoons$

(22) $\xrightarrow{+H_2O}$ **(23)** $+ R^1R^2C{=}O$

aminonaphthol-4-sulfonate (**23**). When the amine was benzylamine, benzaldehyde was detected in the reaction mixture. The corresponding carbonyl compound was also isolated when the amine was methylamine, ethylamine and *i*-propylamine.

E. Monoimides and Diimides

The hydrolysis of several monoimides and diimides has been studied in varying detail. The types of compounds discussed here are shown by structures **24**–**28** below. *N*-acetyl-*p*-quinoneimine (**24**) is the two-electron oxidation product of acetaminophen(*N*-acetyl-*p*-aminophenol), formed electrochemically[35, 36] on a carbon electrode in aqueous solution, or by oxidation with lead tetraacetate in benzene, or with freshly prepared silver oxide in either benzene or chloroform in almost quantitative yield[37].

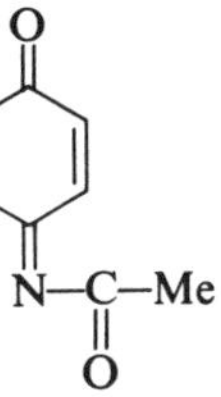

(**24**) 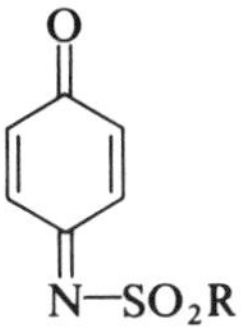 (**25a**) 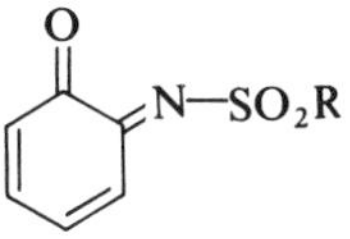 (**25b**) 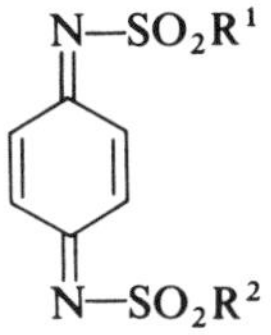 (**26**)

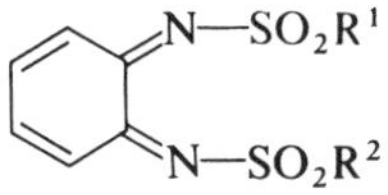

(**27**) 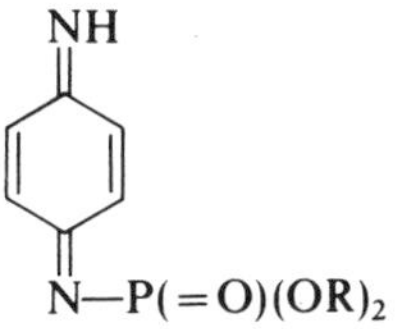 (**28**)

The hydrolysis mechanism of **24** in acid solution has been studied electrochemically in some detail by Kissinger and coworkers[35]. The initial attack by water on either **24** or its protonated form is the same as described for other diimines earlier (Scheme 1). The intermediate carbinolamide (**29**), however, is quite stable, decomposing to produce *p*-benzoquinone and acetamide on a much slower time scale than observed for the loss of **24**. The hydration kinetics to form **29** are shown in Figure 6 at $\mu = 0.5$ at 25 °C. The solid line through the points is given by equation 10, where the two terms represent attack by H_2O

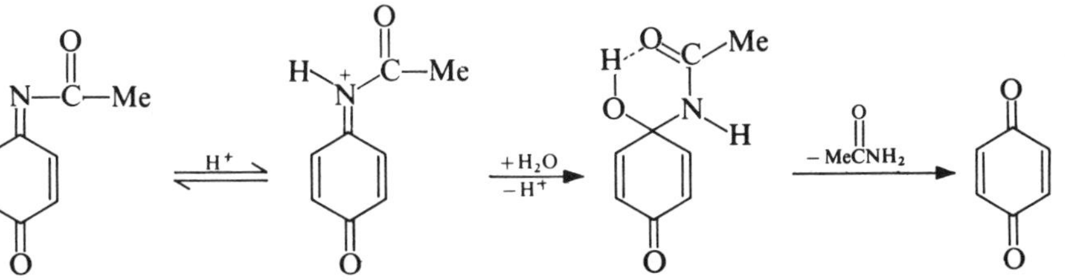

(**24**) $\xrightarrow{H^+}$ $\xrightarrow[-H^+]{+H_2O}$ (**29**) $\xrightarrow{-MeC(O)NH_2}$

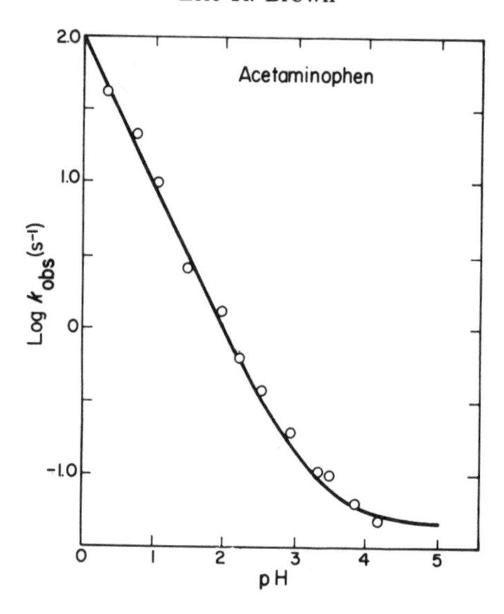

FIGURE 6. The pH dependence of the hydration rate constant of *N*-acetyl-*p*-quinoneimine (**24**) formed electrochemically[34]. Points experimentally determined at 25 °C, $\mu = 0.5$; line calculated using equation 10; $k_1 K_a = 98.2\ \text{M}^{-1}\text{s}^{-1}$; and $k_2 = 0.047\ \text{s}^{-1}$

on the protonated form of **24** (K_a is the protonation constant) and on **24** itself. The line is calculated using $k_1 K_a = 98.2\ \text{M}^{-1}\text{s}^{-1}$ and $k_2 = 0.047\ \text{s}^{-1}$. The protonation of **24** must

$$k_{obs} = k_1 K_a[\text{H}^+] + k_2 \tag{10}$$

occur well below pH 0, because the slope of the line in Figure 6 is still unity at 0.5 M $HClO_4$, the most acid solution studied.

Both HSO_4^- and phosphate ($H_2PO_4^-$) accelerate the rate in a linear fashion in this pH range. The effect is not general; citrate and SO_4^{2-} show no buffer concentration effect[35].

Deamidation of **29** to form quinone was studied only at 0.5 M and 0.1 M $HClO_4$. The rate constant increases with acid concentration, yielding a second-order rate constant, $k_{obs}/[\text{H}^+]$, of $0.25\ \text{M}^{-1}\text{s}^{-1}$. Details of the deamidation at higher pH are not available.

A brief study of the hydrolysis of a related compound, *N*-hydroxy-*N*-acetamido-*p*-quinoneimine (**30**), shows that acetic acid is rapidly released, forming nitrosophenol. The reaction is shown as an attack by hydroxide, because nitrosophenol appears in greater quantities at pH 6 than in 0.1 M $HClO_4$ in the electrochemical generation of **30**[36].

(**30**) $+ \text{OH}^- \longrightarrow$ [N—OH quinone oxime] $\rightleftharpoons$ [N=O, OH nitrosophenol] + MeCOOH

The hydrolysis of *N*-arylsulfonyl-*p*-quinoneimines (**25a**), R = Ar, and the related *p*-naphthoquinoneimines (**31**) with high molecular weight 'ballast' groups has been described by Hanson[38] for the use of these compounds in photographic imaging, where R is a dye species. Similar utility of the *ortho* analogs (**32**) has been described by Fujita[13, 39], along with synthesis and isolation of the appropriate quinoneimides[13]. The hydrolysis kinetics of the quinoneimides have been studied only in alkaline solution where hydroxide is the reactive species, attacking the sulfonimide group to release that group, and forming the corresponding quinone.

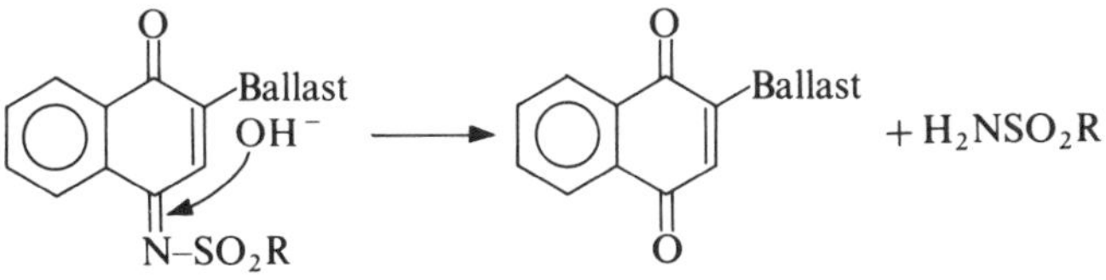

$+ H_2NSO_2R$

(**31**)

$+ H_2NSO_2R^1$

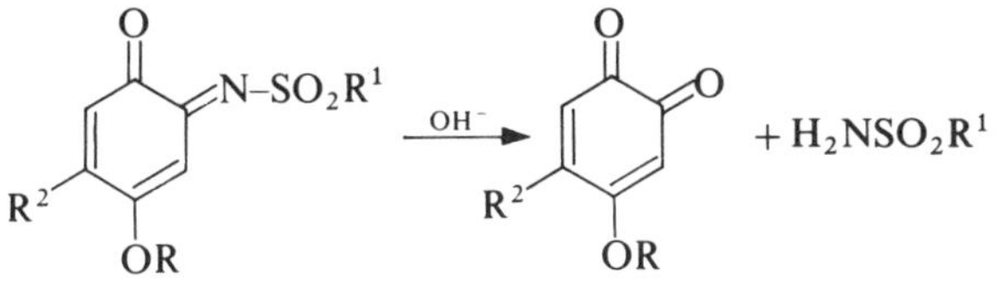

(**32**)

The reaction of *p*-quinonesulfonimides is first order in OH^- and goes to completion[40]. Electron-withdrawing groups on the ring, or on R when it is an aryl group, accelerate the reaction, whereas electron-donating groups decrease the rate constant. Table 7 shows the second-order rate constants ($k_{obs}/[OH^-]$) obtained from rate-constant measurements between pH 8 and 11 in phosphate buffer ($\mu = 0.375$). The rate constants vary over two orders of magnitude, depending on the substituent. Notice that *p*-naphthoquinonesulfonimide is about as reactive as 2-methyl-*p*-benzoquinonesulfonimide. Changing from phenylsulfonimide to methylsulfonimide has only a minor effect on the rate constant. 2,6-Dichloro-*p*-quinonesulfonimide (**33**) undergoes reversible adduct formation at pH 9.3 ($K = 50\,000$) and the adduct is only about 1/8 as reactive as the parent compound.

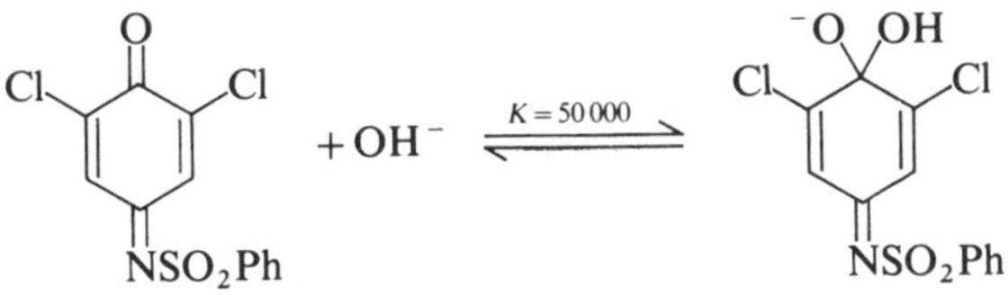

$+ OH^- \underset{}{\overset{K = 50\,000}{\rightleftharpoons}}$

(**33**)

The hydrolysis of *o*-quinonesulfonimide is more complicated, and several species are formed. In alkaline methanol, Fujita[13] found significant amounts of the *p*-quinone monoacetal (**34**) formed initially, slowing the release of sulfonamide. The substituent R

TABLE 7. Bimolecular rate constants for deamidation of *p*-quinonesulfonimides in phosphate buffer[a]

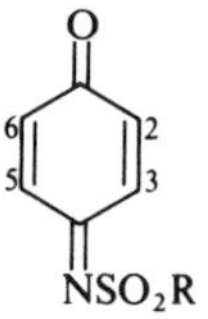
O
6
2
5
3
NSO2R

Substituent		R	k (10^4 M^{-1} s^{-1})
2-position	3-position		
H	H	Ph	2.6
H	H	Me	2.1
H	Me	Ph	2.6
Me	H	Ph	0.70
2,6-Cl_2	H	Ph	7.95 1.17 for OH^- adduct
H	H	4-O_2NPh	5.2
H	H	3-O_2NPh	3.7
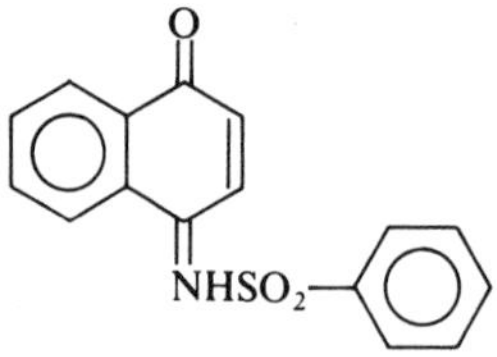			0.52

O
NHSO2

[a] Quinonesulfonimide is formed by oxidation of the corresponding *p*-sulfonamidophenol with two or four equivalents of ferricyanide in phosphate buffer ($\mu = 0.375$)[40].

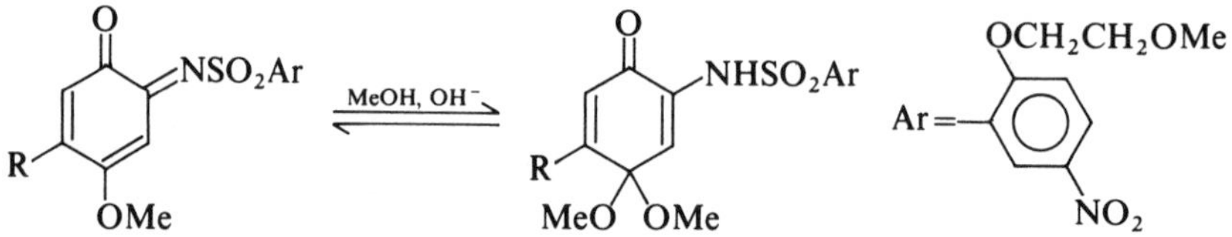
O
NSO2Ar
R
OMe
MeOH, OH−
O
NHSO2Ar
R
MeO
OMe
OCH2CH2OMe
Ar =
NO2

(34)

also affects the reaction. When R is methyl only ~ 60% of the sulfonamide is released, but when R is *t*-butyl > 80% is released. Two other products were identified, both a result of 1,4 addition to the ring. Compound **35** results from attack by hydroxide and **36** results

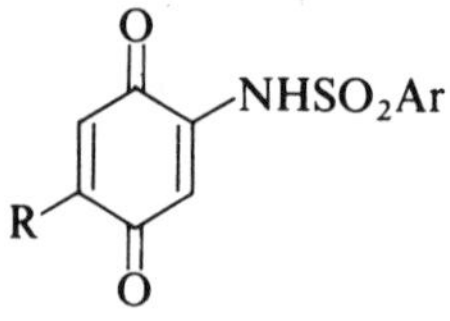
O
NHSO2Ar
R
O

(35)

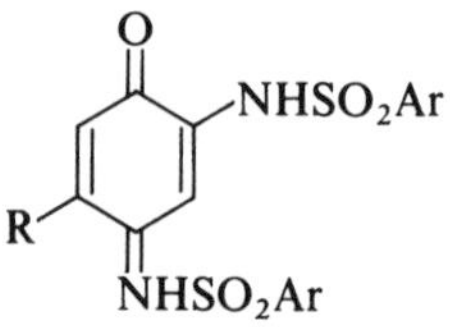
O
NHSO2Ar
R
NHSO2Ar

(36)

from attack by the sulfonamide released from the original *o*-quinonesulfonimide. As much as 12% **36** is produced when R is methyl. Rates of sulfonamide release in alkaline methanol were slow with a $t_{1/2} = 0.4$ h for R = methyl and $t_{1/2} = 1.3$ h for R = *t*-butyl[13]. No kinetic studies for the hydrolysis of bis-sulfonimido-*o*-quinone (**27**) have been reported.

Quinoneimine phosphonamides (**28**), where R = Ph, have been formed by oxidation of the corresponding *p*-aminophenol with ferricyanide at pH 11[41]. No detailed kinetic study was performed, but both the quinone and the phosphate fragment were isolated from the reaction mixture by thin-layer chromatography and comparison with authentic compounds. The half-life for deamidation at pH 11 is < 0.3 s in phosphate buffer.

IV. COUPLING CHEMISTRY

Coupling of quinoneimines to nucleophiles has generally been studied by chemical or electrochemical oxidation of the parent aminophenol or phenylenediamine to form the corresponding quinoneimine. This means that the quinoneimine often exists in the presence of starting material, which is, itself, a reasonable nucleophile and can react in competition with other nucleophiles in the system. These self-coupling reactions have been studied in some detail to identify products, which are often highly colored dyes, and to determine the exact coupling mechanism.

A. Self-coupling

The oxidative coupling of *p*-phenylenediamine (**37**) produces the trimeric species, Bandrowski's base (**41**), originally isolated from alkaline oxidation of the starting material by ferricyanide[42] or oxygen[43]. The structure was identified later by Corbett[44] and Dolinsky and coworkers[45] and the mechanism was studied by mixing less than two molar proportions of ferricyanide with *p*-phenylenediamine as a function of pH[46]. The reaction is first order in both diamine and protonated diimine (**38**), but the diamine concentration remains constant throughout the reaction, yielding an observed pseudo-first-order reaction. The pH dependence of the second-order rate constant is consistent with the attack of protonated diimine (**38**) on the neutral diamine, as described in Scheme 3. The reaction consumes three diimine species for each molecule of **41** formed, regenerating one molecule of diamine. The observed second-order rate constant increases linearly as the pH drops from pH 10 and pH 7.5, after correction for the deamination reaction described in Section III.

At low pH, the reaction is complicated by the formation of the intermediate radical cation, causing a decrease in the observed rate constant. Correcting for this effect, and for the amount of the reactive species in solution at each pH, Corbett[46] reports a pH-independent rate constant of 3.43×10^3 $\text{M}^{-1}\,\text{s}^{-1}$ for the slow step in Scheme 3 at 30 °C. This yields a maximum in the observed rate at about pH 6, due to the protonation equilibria of both diimine (pK = 5.75) and diamine (pK = 6.44). The observed half-life for the formation of **41** at pH 5.7 for a 10^{-4} M solution of *p*-phenylenediamine, mixed with 10^{-4} M diimine is 35 s. In alkaline solution, substantial amounts of 4,4′-diaminoazobenzene are also observed[44]. Sakata and coworkers[47] obtained **41** in ethanol solution by combining the intermediate radical cation with *p*-phenylenediamine. About 6.5 radical species were required to form one molecule of **41** and ~12% 4,4′-diaminoazobenzene, independent of the amount of diamine present initially. Oxidation of diamine by Br_2 also produced **41** in ethanol.

Similar studies by Sakata and coworkers[47] were made using *N,N*-dimethyl-*p*-phenylenediamine radical and substrate. After 4 days, the hexamethylated Bandrowski's

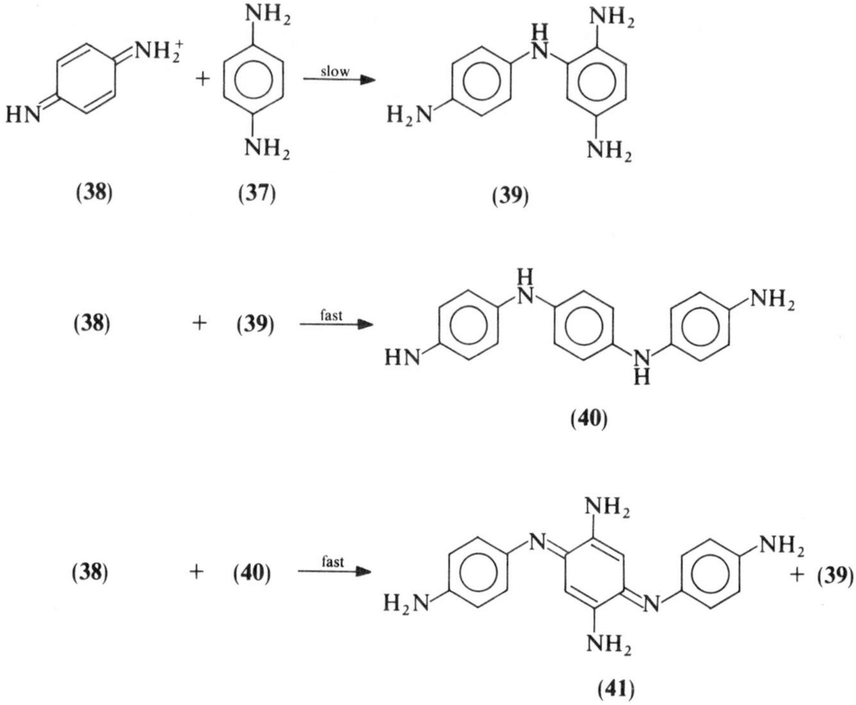

(38) (37) (39)

(38) + (39) → (40)

(38) + (40) → (41) + (39)

SCHEME 3. Mechanism for formation of Bandrowski's base[46]

base (**42**) was isolated, along with an equivalent amount of 4,4′-bis(dimethylamino)-azobenzene (**43**) and an unknown substance.

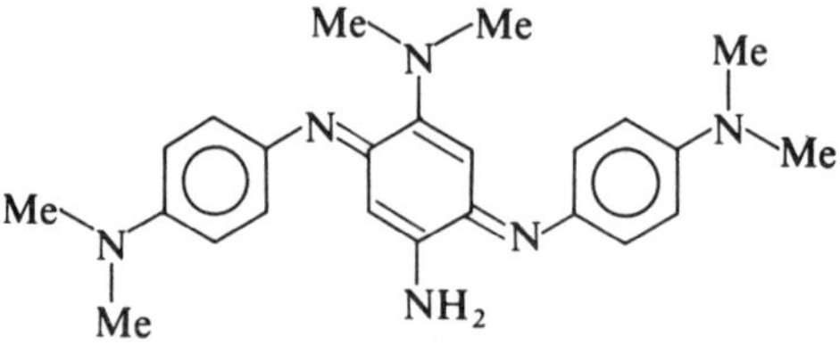

(42) 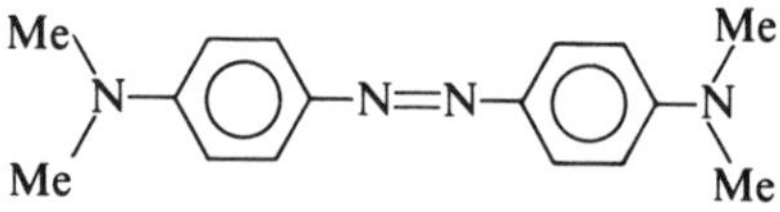 (43)

Electrochemical oxidation of **44** in aqueous solution at pH 4 produces a violet solution ($\lambda_{max} = 550$) with the consumption of three electrons per mole of **44**, enough required to form **42**. The explanation given (without product isolation) is that some deamination occurs and only a coupling with the quinone immonium salt, similar to the first coupling reaction of Scheme 3, occurs[48]. Species **45** thus formed is then oxidized by the loss of four electrons to form the colored species **46** in solution. In alkali, the solution is blue ($\lambda_{max} = 720$), which compares with the alkaline ethanol solution color of **42** mixed with **43**

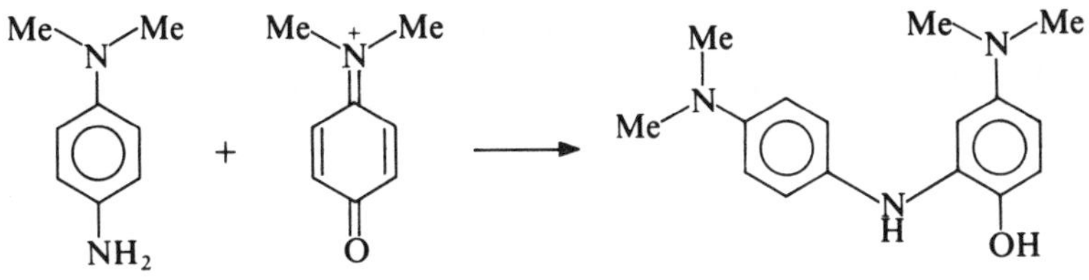

(44)

(45)

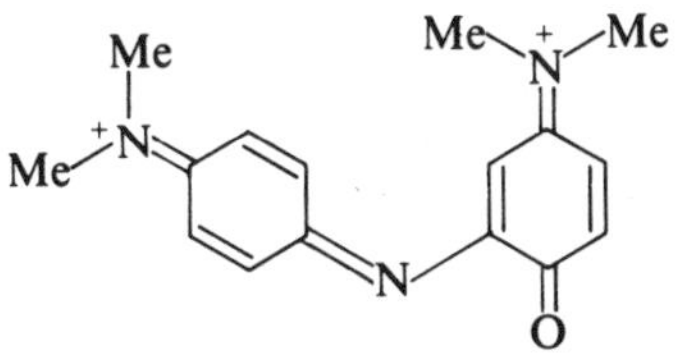

(46)

(λ_{max} = 660 nm)[47]. Both Sakata and coworkers[47] and Lelievre and coworkers[30, 48] refer to radical coupling as a possible mechanism to form these colored species.

Another electrochemical oxidation study of *N*-phenyl-*p*-phenylenediamine in aqueous acetate buffer also yielded three electrons per mole of substrate, which the authors[49] suggest is a result of the formation of the trimeric condensation product (**47**), shown in two tautomers. The relatively insoluble product is capable of undergoing a two-electron oxidation and a two-electron reduction, but no products were isolated. It is clear that N-substitution on *p*-phenylenediamines complicates the self-coupling chemistry of these

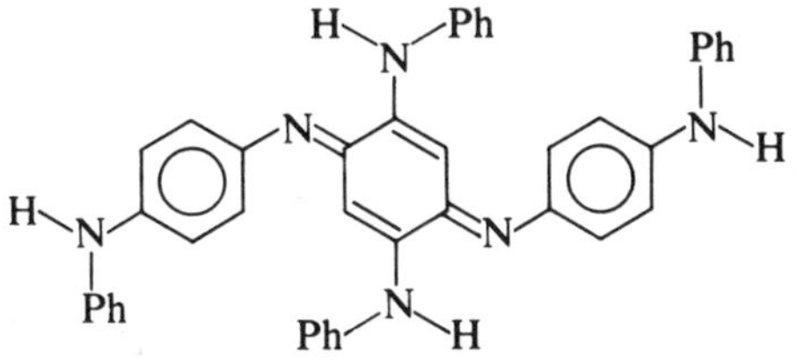

(**47**)

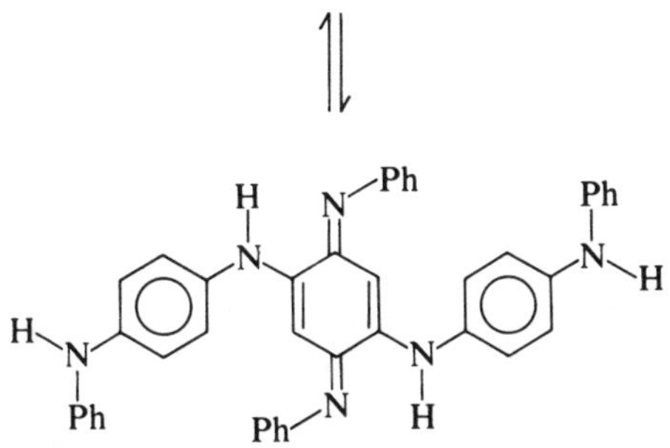

species, leading to a number of highly colored condensation products, which are difficult to isolate and characterize.

In an attempt to understand these self-coupling reactions in more detail, Bishop and Tong[50] studied the kinetics of azo dye formation from quinonediimines of *N*,*N*-diethyl-*p*-phenylenediamines. Only small amounts of dye are formed by coupling at the primary nitrogen group, in competition with deamination of the dialkyl amino group in the alkaline pH range studied. The amount of dye formed was independent of pH, and was at a maximum at half oxidation of the diamine. Thus, the mechanism has the same $[OH^-]$ dependence as deamination, and involves both diamine and diimine. It is not possible to distinguish a mechanism involving coupling of these two species from a mechanism involving radical–radical coupling, since they have the same dependence on the initial concentration of diamine and diimine. In addition, the radical would be the neutral radical, which accounts for the hydroxide dependence, and this species does not exist in measurable concentrations[50].

The amount of azo dye increases with the substitution of electron-withdrawing groups on the aromatic ring, changing from ~1.8% for no substitution to 24% for a single chlorine atom. Complete chloro substitution of simple *p*-phenylenediamine shows essentially complete azo dye formation (77–84%), when oxidized by chlorine or bromine in anhydrous methanol[51]. The reaction also requires the presence of acid, and presumably involves direct coupling of the diimine with its protonated form because only the diimine salts are initially present.

Brown and Corbett[52] extended earlier mechanistic studies of *p*-quinonediimine self-coupling to the same reaction by *p*-quinonemonoimine in the presence of its parent, *p*-aminophenol. The reaction mechanism is essentially the same as outlined in Scheme 3 to form the product trimer (**48**), but it is kinetically complicated due to the various ionic forms of monoimine and aminophenol which exist from pH 7 to 12. The rate-determining

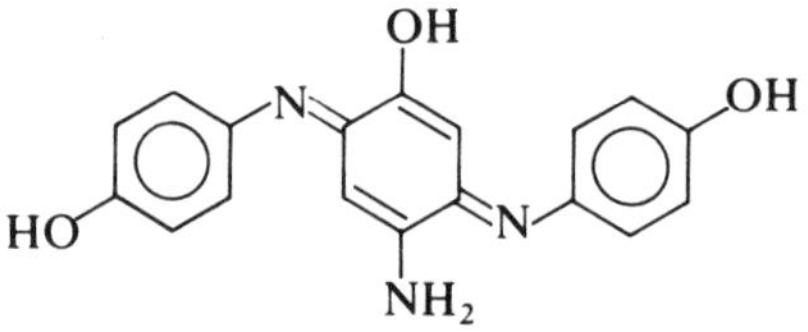

(**48**)

step is the first coupling reaction between monoimine and aminophenol. The rate constant between protonated monoimine and parent aminophenol is $1.9 \times 10^4\ M^{-1}\,s^{-1}$, which is about six times faster than for diimine coupling with its parent phenylenediamine. The rate constant for coupling of the protonated monoimine with the aminophenolate anion is 130 times faster than with the neutral aminophenol. Reaction between neutral monoimine and aminophenolate anion is very slow, and it is complicated by another reaction at high pH, as indicated by the lowered yield of **48** formed. Between pH 7 and 11.5 **48** forms in 100% yield due to slow deamination of the monoimine, which is the only other known competing reaction.

o-Quinonediimines and *o*-quinonemonoimines also undergo self-condensation reactions to form dimeric products. In the case of the diimine, the two major products are 2,3-diaminophenazine (**49**) and 2,2′-diaminoazobenzene (**50**)[15]. The ratio of these products depends on solvent and pH. In aqueous buffer the product is almost entirely **49**, and in diethyl ether, it is exclusively **50**[53, 54]. The products form by partial oxidation of the *o*-diamine to the *o*-diimine (**52**), which couples slowly with the parent in two parallel pathways[53, 54]. The coupling of diimine with diamine is assumed by analogy with

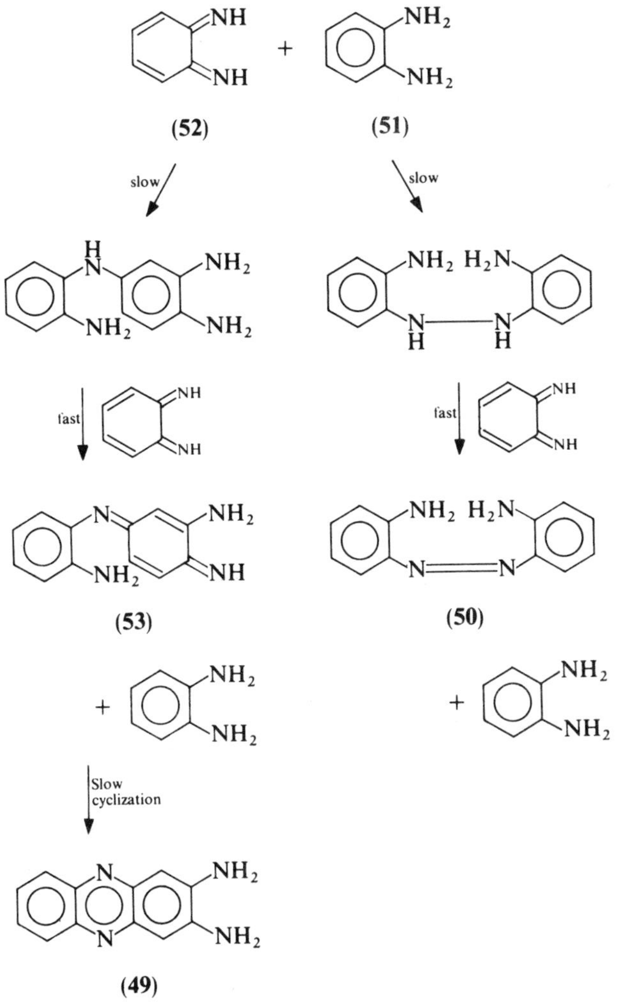

(52) (51)

(53) (50)

(49)

Corbett's work on *p*-diimines[44, 46] and by the observation that the formation of **50** is accelerated in diethyl ether by addition of diamine to a diimine solution[54].

o-Quinonemonoimine couples with its parent aminophenol slowly in aqueous solution at pH 4.5–10 to form 2-aminophenoxazin-3-one (**54**), along with some *o*-benzoquinone[53]. The reaction mechanism is probably similar to that proposed for the formation of **49** involving coupling of *o*-quinoneimine and subsequent cyclization[53].

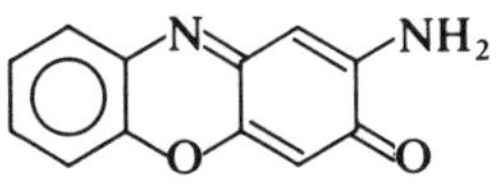

(54)

The cyclization reaction involving formation of phenazines from several *N*-(2′-aminophenyl)quinoneimines has been studied in aqueous solution as a function of pH[55]. In examples **55–58**, the cyclization reaction occurs only at pH 6–10. For compounds **56**, **57**

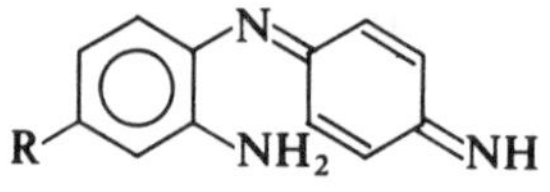

(**55**) **a**: R = H
b: R = Me
c: R = Cl

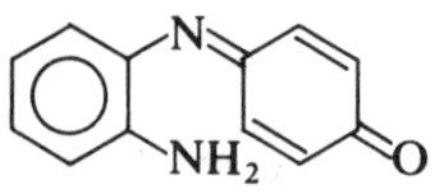

(**56**)

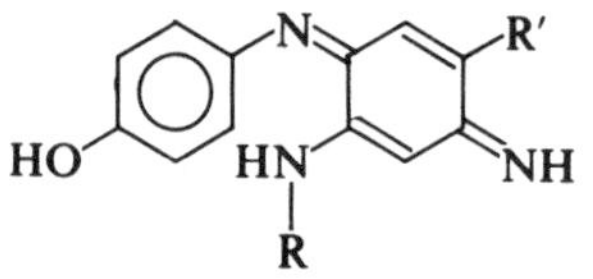

(**57**) **a**: R = R′ = H
b: R = H, R′ = Me
c: R = R′ = Me
d: R = Ph, R′ = Me

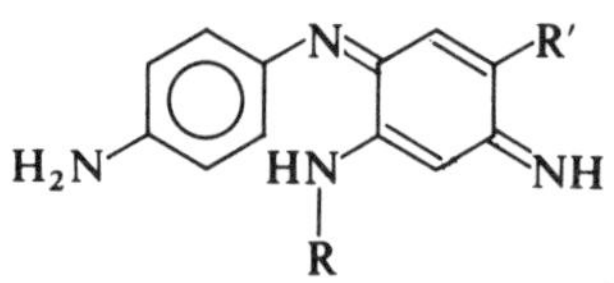

(**58**) **a**: R = R′ = H
b: R = H, R′ = Me
c: R = R′ = Me
d: R = Ph, R′ = Me

and **58**, the reaction clearly involves the protonated diimine, which, in the case of **57**, is a zwitterion[56]. The monoimine couples quantitatively from pH 5 to 12[55]. At low pH, hydrolysis at the azomethine bond competes with cyclization, forming *o*-phenylenediamine and quinoneimine from **55** and *p*-aminophenol or *p*-phenylenediamine

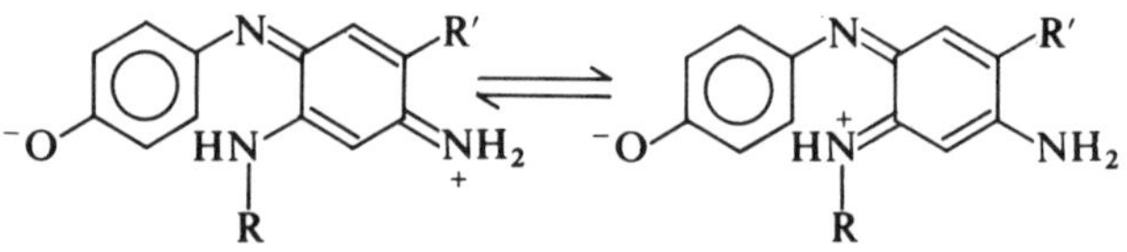

reactive form of **57**

and *p*-quinoneimine from **57** and **58**, respectively. At high pH, the competing reaction is the hydrolysis of the primary imine group. These reactions are in contrast to those of *N*,*N*-dialkyl-*p*-quinonediimines (described earlier), where the primary imine is preferentially cleaved in acid solution and the substituted imine is cleaved in alkaline solution.

The cyclization rate constants, expressed in terms of the reactive species, vary widely depending on substituents. For **55**, the observed rate constant decreases by a factor of 10 per pH unit from pH 6 to 9, which is above the p*K* for ionization of the protonated diimine. Rate constants and p*K* values are summarized in Table 8, which also includes data for **56** for comparison. Interestingly, **56** reacts slowly in acid to form 2,3-diaminophenazine **49** in high yield. The cleavage products, benzoquinone and *o*-phenylenediamine, apparently cross-oxidize to form **49** according to Nogami and coworkers[54].

TABLE 8. Rate constants for cyclization of **55** at 30 °C[a]

Structure	k_c (s^{-1})[b]	p$K^{a,c}$
55a	27.6	4.3
55b	13.1	4.2
55c	443.0	4.2
56	0.0013	—

[a] Ref. 55.
[b] Rate constant calculated for protonated form.
[c] Measured spectrally for protonated diimines.

Cyclization of **57** is quantitative between pH 7 and 10 to form **59b** or **60b**, but the rate constants are quite different[56]. As R goes from H, to Me, to phenyl, the rate constant, expressed in terms of the diimine cation concentration, increases significantly as shown in Table 9. The same rate constants for structure **58**, forming **59a** or **60a**, are 10–100 times

TABLE 9. Rate constants for cyclization to phenazine[a]

R	R′	Structure	k (s^{-1})[b]	Structure	k (s^{-1})[b]
H	H	**57a**	8.4×10^{-5}	**58a**	6.1×10^{-6}
H	Me	**57b**	7.0×10^{-5}	**58b**	2.7×10^{-6}
Me	Me	**57c**	1.36×10^{-3}	**58c**	1.4×10^{-5}
Ph	Me	**57d**	2.30×10^{-2}	**58d**	2.2×10^{-3}

[a] Measured at 30 °C; Ref. 56.
[b] Based on protonated form of reactant.

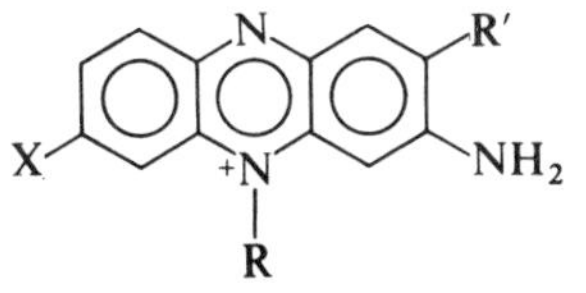

lower ($t_{1/2} \sim 5$ h at 30 °C), and may well be closer to the rate constant for formation of diaminophenazine (**49**), from *o*-phenylenediamine and *o*-quinonediimine.

X R′ NH$_2$ N +N R

(**59**) **a:** X=NH$_2$
b: X=OH

X R′ NH$_2$ N N

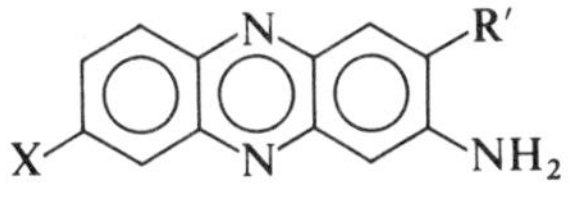

(**60**) **a:** X=NH$_2$
b: X=OH

In some compounds, such as **57d**, reactant loss rate is faster than the rate of appearance of the phenazine (**59b**), due to slow oxidation of the intermediate dehydrophenazine[56]. Most experimental studies[55–57] rely on oxygen dissolved in solution to bring about the oxidation to form **59** and **60**. Ferricyanide, added to the solution, rapidly increases the phenazine product formation, even when added after initial mixing of **57d** with pH 8 buffer.

The self-coupling of N-substituted *o*-quinoneimines has received little study. Berkenkotter and Neison[58] studied the oxidative coupling of *N,N′,N′*-triphenyl-*o*-phenylenediamine (**61**) in acetonitrile at a platinum electrode. The product, in 88% yield, is

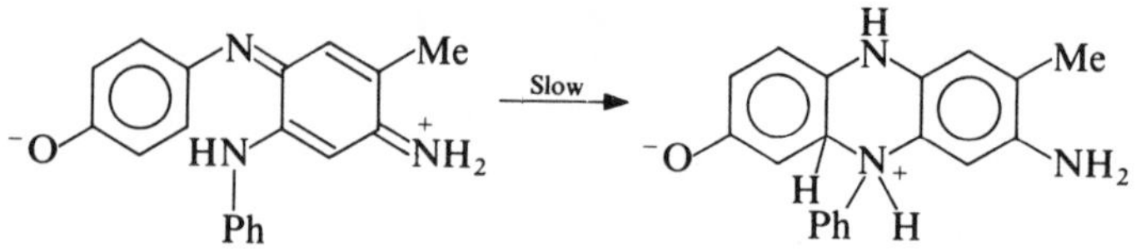

(57d)

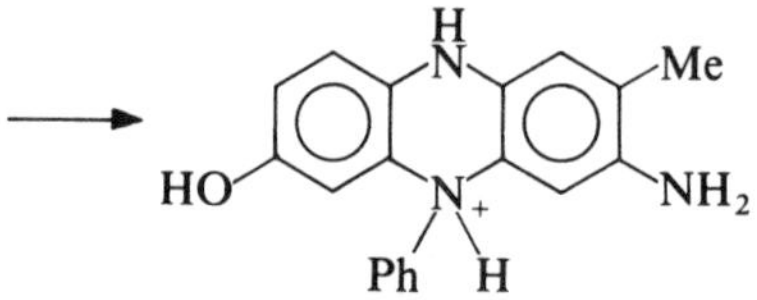

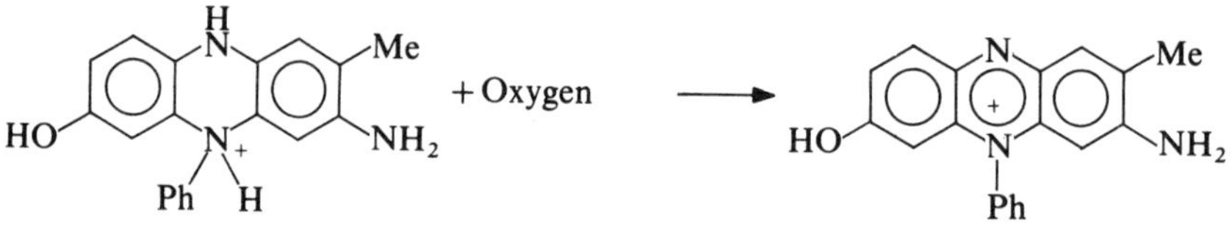

(59b)

the corresponding 5,10-diphenyldihydrophenazine (**62**). The mechanism is presumed to occur by cyclization of the two-electron oxidation product, but contribution by the radical intermediate could not be ruled out. Similar complex coupling schemes were postulated by

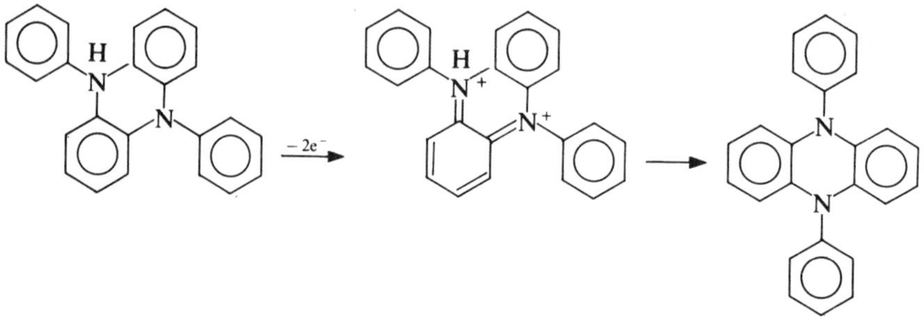

(61) (62)

Haynes and Hewgill[59] to account for formation of N-substituted phenazines isolated from the chemical oxidation of methoxy-substituted anilines.

B. Coupling with Amines

Although oxidative coupling of substituted *p*-phenylenediamines and *p*-aminophenols with the parent anilines yields highly colored solutions containing a mixture of products as discussed earlier, Corbett and coworkers[60–63], in a series of papers, studied the initial coupling step in aqueous solution.

In the case of *p*-quinonediimines, the initial coupling kinetics indicate that anilines couple with the protonated diimine (**63**) in the rate-limiting step. The fast-oxidation step,

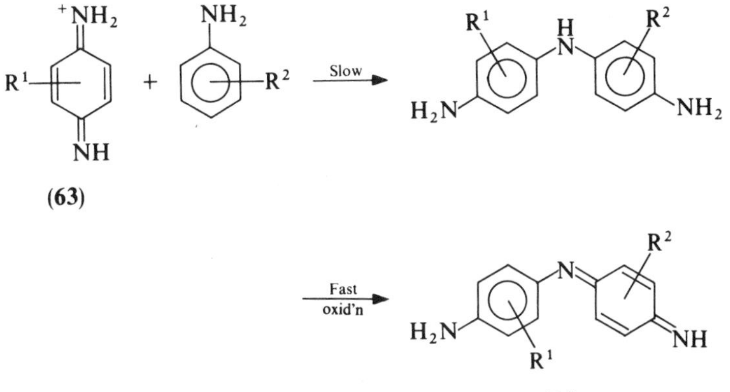

(**63**)

(**64**)

which produces indamine dye (**64**) occurs with excess diimine or ferricyanide in solution. In alkaline solution, above the pK_a of **63**, log k_{obs} decreases linearly as pH increases. The reaction is complicated by hydrolysis of the product dye, forming *p*-quinonemonoimine and *p*-phenylenediamine. Because the coupling reaction is slow, large excesses of aniline are necessary to form significant amout of dye.

A careful study[60] of substituent effects on both aniline and **63** with 21 compound combinations, involving three substituents, –Me, –OMe and –Cl, was carried out at 30 °C at pH 9. Electron-donating groups on aniline increase the rate constant, whereas the same group on **63** decreases the rate constant. The substituent effect on the diimine also includes the effect on pK_a of the protonated species, which was not measured separately. The results are summarized in Table 10. The second-order rate constant for coupling of the unsubstituted aniline with unsubstituted **63** (factor = 1.0) is 165 $M^{-1} s^{-1}$ at 30 °C[60]. The rate constant for substituted species is obtained by multiplying this value by the factors in Table 10.

An additional complication to the formation of indamine dye is the additional coupling of diimine and the cyclization of the product to form phenosafranine dyes (**65**)[61].

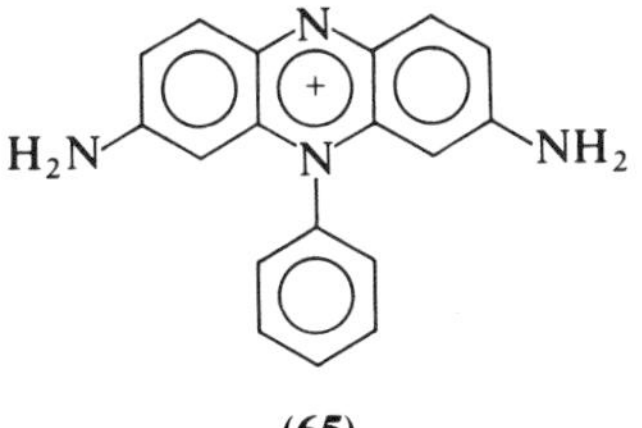

(**65**)

Substituents on aniline block this reaction, resulting in almost quantitative indamine dye formation.

Corbett[62] also studied the coupling kinetics of *p*-quinonediimines with *m*-aminoanilines. Again, the rate-limiting step in alkaline solution is the reaction of the

TABLE 10. Mean effect of substituents on coupling of **63** with aniline and *m*-aminoaniline at 30 °C

Diimine		Aniline[a]		*m*-Aminoaniline[b]	
Substituent	Factor	Substituent	Factor	Substituent	Factor
Me	0.234	2-Me	4.84	2-Me	5.33
Cl	1.98	3-Me	8.94	4-Me	3.03
		2-OMe	7.5	2,4-$(Me)_2$	14.3
		3-OMe	35	4-OMe	25
		3-Cl	0.37	4,6-$(OMe)_2$	800
		2-NH_2	1210		

[a] Ref. 60.
[b] Ref. 63.

protonated diimine with neutral *m*-aminoaniline. The second-order rate constant for coupling *m*-aminoaniline to *p*-quinonediimine is 2×10^5 M^{-1} s^{-1} at 30 °C[63], which is three orders of magnitude faster than the reaction of aniline with the same diimine. This increase is a result of the strong electron-donating effect of the second amino group on aniline.

The effect of methyl substituents in both reactants was also studied in alkaline solution. A methyl group on protonated diimine **63** yields a relative rate factor of 0.28 for coupling to *m*-aminoaniline, compared to a factor of 0.234 for coupling to aniline (Table 10). Surprisingly, addition of a second methyl group in either the 5- or 6-position of **63** had almost no additional effect on the rate constant. In the addition of methyl groups to *m*-aminoaniline, the effect is additive when two methyl groups are present (Table 10). A methoxy group accelerates the reaction even more and is additive, even when it is in the coupling position[63]. The rate-limiting step is still the initial coupling reaction; elimination of methanol is rapid.

The situation is more complex with 2,4-diamino-5-methoxytoluene (**66**), as shown in the scheme below. Initial coupling proceeds at two positions, but only the leuco-dye adduct

(68)

(66)

(67)

Slow
− MeOH

(**67**), formed by coupling at the *ipso* position to the methoxy group can eliminate methanol to form dye. The rate-limiting step now is the elimination reaction. The reaction proceeds in several steps. The fast process is through formation of adduct **67** and elimination. More slowly, adduct **68**, which is formed competitively with **67**, dissociates and dye is formed via **67**. Similar complicated coupling kinetics were also observed when diimine coupling with methyl-substituted aminophenols was studied, where methyl groups existed at the coupling position, *para* to OH or to NH_2[64].

Monoquinoneimines (**69**) also couple to aniline, although the observed reactions are slow. Competing reactions such as deamination and self-coupling can interfere. Corbett[63,65] studied monoimine coupling with the more reactive *m*-aminoanilines, using excess aminoaniline to assure that coupling is the predominant reaction. The pH dependence of the kinetics suggests that, at high pH, significant contribution to the reaction occurs through reaction of the neutral monoimine **69**.

The reaction is controlled by the coupling step; the intermediate leuco dye is oxidized rapidly to the indophenol (**70**) either by the monoimine or by other oxidants in the system.

NH O + H_2N NH_2 —Slow→ H H N ^-O H_2N NH

(**69**)

—Fast→ N $_-O$ H_2N NH

(**70**)

The rate constant for the conjugate acid of **69** is 3.53×10^4 M^{-1} s^{-1}, about six times lower than the rate constant for diimine (**63**; R^1 = H) conjugate acid. The rate constant for **69** coupling is only 11 M^{-1} s^{-1}, but the corresponding reaction of neutral diimine is not observed. Experimentally, the observed rate constants for monoimine coupling are almost 10^3 slower at all pH values than for diimine, because the conjugate acid pK is more than two pH units lower. This lowers the concentration of the active species by more than 100-fold, and it is six times less reactive anyway.

The effect of substituents on the reactivity of **69** are as expected. The contributions of methyl substitution on *m*-aminoaniline are additive, the rate constants being a factor of 4.8-fold higher for 2-Me, 3.7-fold higher for 4-Me, and 12-fold higher for 2,4-dimethyl substitution. Substitution of a methyl group on **69** decreases the rate by ~ 20 in either the 2- or 3-position. One methoxy substituent on *m*-aminoaniline increases the rate constant by a factor of 17.3, and two groups increase it by a factor of 300[63].

The reaction of 1,5-dimethoxy-2,4-diaminobenzene (**71**) with **69** is somewhat more complicated than observed for other substituted anilines. Dye formation occurs through methanol elimination, which is faster than the forward coupling step at millimolar concentrations, but not fast at all pH values compared with reversal of the leuco-dye adduct (**72**) back to reactants, k_{-1}. The result is a complex pH dependence of the apparent rate constants, but the rate constant k_1 can be obtained[63].

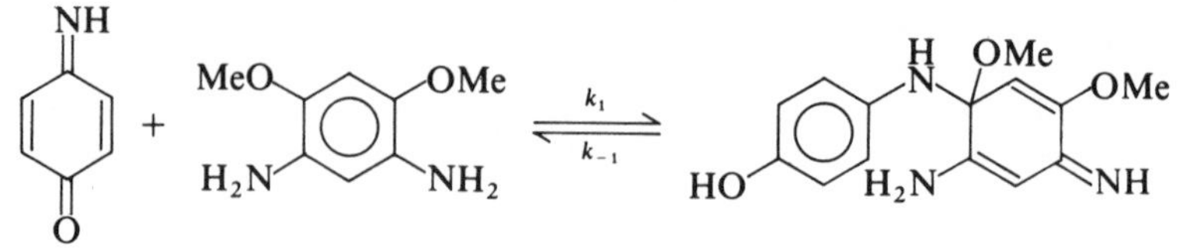

(69) (71) (72)

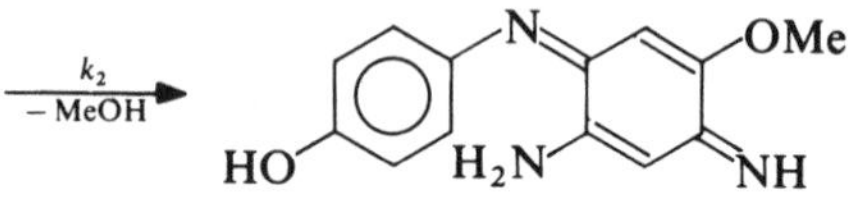

(73)

C. Coupling with Phenols and Naphthols

The interest in coupling reactions of diimines and monoimines with phenols has been extensive, due to the importance of the resulting dyes in color photography. Mechanistic studies of the reactions in alkaline and acid solution began with Tong and Glesmann[16, 66, 67] using ferricyanide to rapidly produce *N,N*-dialkylquinonediimines from the corresponding *p*-phenylenediamines. Later, work by Corbett[63, 68], using simple quinonediimine and phenols, confirmed the coupling mechanism postulated earlier. More recently, Pelizzetti and Saini[69] reproduced much of the earlier work and extended their studies to several other phenols and naphthols. Mann and coworkers[70, 71], and Baetzold and Tong[72] extended the experimental techniques of forming reactive diimine **75** by using flash photolysis of an azide precursor (**74**) in solution with the naphthol coupler. In this way, dye formation was studied in octanol, a solvent in which **75** is not very soluble[72].

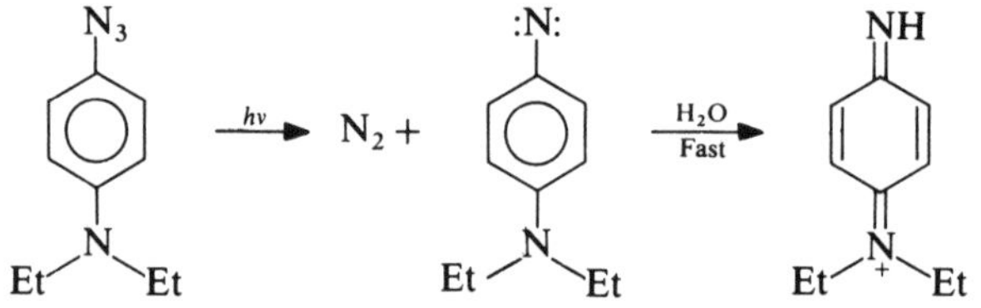

(74) (75)

In aqueous solution, the rate-limiting step for dye formation is the initial coupling reaction of the conjugate acid of quinonediimine (**76**) with either phenol or the phenolate anion, depending on the pH[63, 68]. An exception to this mechanism is when X is Me, and no elimination reaction occurs. If X is H, the dye forms by oxidation of the leuco-dye adduct **77** with either another molecule of **76**, or with ferricyanide. This reaction is not rate limiting.

Because **76** reacts with both ionic forms of the phenol, the pH dependence of log k_2 has the form shown in Figure 7[68]. The dotted lines show the contribution to the reaction by each ionic form. In the approximately pH-independent region, the concentration of **76** is decreasing at the same time that the concentration of the phenolate anion is increasing.

(76) + (OH / O⁻ phenol, R, X) $\xrightarrow[k_2]{\text{Slow}}$ (77) $\xrightarrow[\text{Fast}]{-\text{HX}}$

Above the phenolate pK_a, the anion concentration is constant, but **76** is still decreasing and the observed rate constant decreases as pH increases.

The effect of substituents on both **76** and phenol has been studied for several phenols and diimines. Electron-donating substituents on phenol increase the reactivity and the effect is additive, but the position is also important[68, 69]. For example, the second-order rate constant of **76** with phenolate ion is 8.43×10^4 M^{-1} s^{-1} at 30 °C. Substitution of a methoxy group in the 2-, 3-, or 4-position increases the rate constant by a factor of 37.0, 7.9 and 13.0, respectively[63, 68]. Similarly a methyl group in the 2- or 3-position increases the rate constant by a factor of 28.1 and 8.7, while 2,5-dimethyl substitution yields a factor of 243 ($28.1 \times 8.7 = 244$). The positional difference is not steric because the 2,3-dimethyl

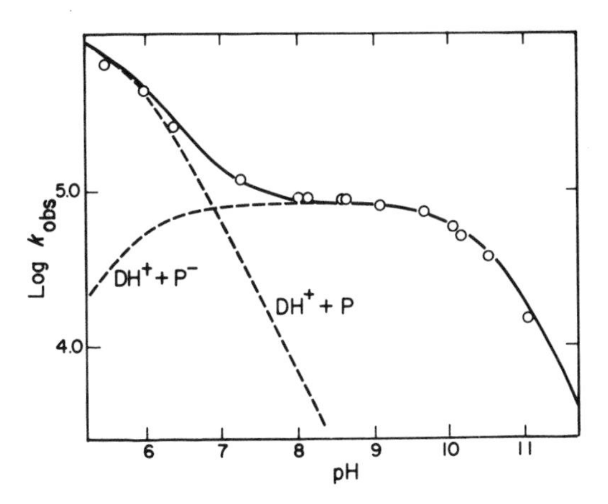

FIGURE 7. The effect of pH on coupling of *p*-benzoquinonediimine with 2,6-dimethylphenol (P) at 30 °C[68]. Full line calculated from equation below; dotted lines, contribution from each ionic form of phenol; $k_{obs} = k_1 f_{DH^+} f_p + k_2 f_{DH^+} f_{p^-}$; f, fraction of each species existing in solution; $k_1 = 1.56 \times 10^4 \text{M}^{-1}\text{s}^{-1}$; and $k_2 = 8.0 \times 10^4$ M^{-1} (Table 11)

substitution, in which a buttressing effect might be expected, yields a more reactive phenol than 2,5-dimethyl substitution. When naphthol replaces phenol, the coupling rate constant increases by a factor of 4500 with **76** and 1500 with **75**[69].

When extensive substitution occurs on either the phenol or the diimine adjacent to the coupling site, significant steric inhibition has been observed. For example, 3-*t*-butyl substitution lowers the reactivity by a factor of 6.6 compared with 3-methyl substitution[69]. Similarly, 2,3,5,6-tetramethylphenol couples seven times slower than predicted from the additivity of one methyl[68, 69]. Electron-withdrawing substituents on phenol decrease the reactivity, and this effect is additive also, although only chloro substitution has been studied in detail[68, 69].

Substitution on quinonediimine has the opposite effect. Electron-donating groups, such as methyl, decrease the rate constant by a factor of 4–9, depending on position while a chloro group increases the rate constant by a factor of 4[68]. These effects include the change in pK_a of the diimine conjugate acid **76** which was not factored out of the rate studies. A 3-Me group on *N,N*-diethylquinoneimine decreases the rate constant by a factor of 26 with three different phenols[69].

Tong and Glesmann[67] also observed significant steric effects due to ring substituents on *N,N*-diethylquinonediimines when a coupling-off group other than hydrogen is involved and substitution occurs adjacent to the imino coupling site. With 2,6-bishydroxymethyl-4-methoxyphenolate ion (**78**), where a methoxy group occupies the coupling site, a 3-chloro

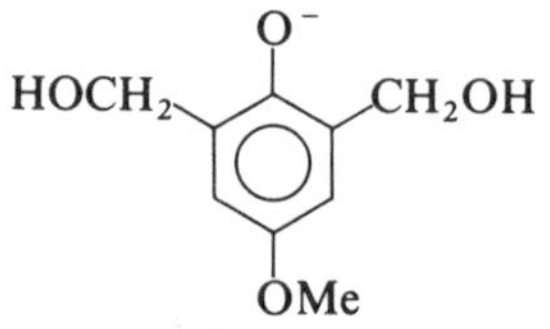

O^-

$HOCH_2$ CH_2OH

OMe

(78)

group on the diimine has no effect on the rate constant, even though it should accelerate the reaction. Substitution of 3-*i*-propyl or 3-*t*-butyl on the diimine decreases the rate constant by a factor of 3 and 3000 respectively, compared with 3-methyl substitution, even though there is no significant change in the inductive effect.

A summary of measured rate constants for the coupling of phenols and naphthols with several quinonediimines, the structures of the reactive species and the reaction conditions are given in Table 11. Of interest is how well the substituent effects of the phenols agree, independent of the diimine involved, unless steric effects are also present. Pelizzetti and Saini[69] showed that substituents on phenol had almost the same effect (rate factors compared with the parent) for the coupling of **81** as Corbett[68] observed for coupling with **79**.

In all reactions except one considered so far, the rate-limiting step has been the addition of the phenol anion to positively charged diimine. Tong and Glesmann[66, 67] examined several reactions in which the inetrmediate leuco dye accumulated in the reaction, because elimination of a coupling-off group from the reaction site was rate limiting. They studied *N,N*-dialkylquinonediimines (**82**) coupling with 4-substituted naphthols (**83**), where the leaving group (X) was methoxy or chloro.

At low naphtholate concentrations or excess **82**, k_1 can be the rate-limiting step, and the observed rate increases with anion concentration. At excess anion concentrations, the intermediate **84** rapidly accumulates and dye formation through k_2 is independent of anion concentration. In addition, it was possible to add another nucleophile to compete with diimine **82** through the reversible formation of **84**, thus obtaining a measure of k_{-1}.

TABLE 11. Quinonediimine coupling rate constants with phenolate anions

O^- (phenolate)	$\overset{+}{N}H_2$ / NH (quinonediimine)	$Et_2\overset{+}{N}$ / NH (**79**)	$Et_2\overset{+}{N}$ / NH, Me (**80**)	$Et(C_2H_4NHSO_2Me)\overset{+}{N}$ / NH, Me (**81**)
X	k_c^- ($M^{-1}s^{-1}$)[a]	k_c^- ($M^{-1}s^{-1}$)	k_c^- ($M^{-1}s^{-1}$)	k_c^- ($M^{-1}s^{-1}$)[b]
H	8.43×10^4	5.5×10^3	2.0×10^2	2.7×10^3
2-Me	2.37×10^6	1.8×10^5	—	1.0×10^5
2-Cl	No dye			7.94×10^2
2-Ph	—	—	—	2.2×10^4
2-CONHPh	—			4.0×10^2
2-OMe	3.12×10^6	—	—	—
2-*i*-Pr	2.71×10^6	—	—	—
2-*t*-Bu	3.15×10^6	—	—	—
3-Me	2.37×10^6			2.0×10^4
3-Cl	No dye			5.0×10^2
3-Et	—			1.3×10^4
3-*t*-Bu	—			3.0×10^3
3-NO_2	—			5.0×10^0
3-COMe	—			1.7×10^2
3-NH_2	1.73×10^7[c]			—
3-NEt_2	—			1.6×10^6
3-OMe	6.65×10^5	1.3×10^4	5.00×10^2	8.3×10^3
4-Cl	—			8.0×10^1
4-OMe	1.1×10^6[d]			1.0×10^4
2,3-Me_2	3.07×10^7			6.45×10^5
2,5-Me_2	2.05×10^7			—
2,6-Me_2	8.0×10^7[d]			4.0×10^6
2,5-Cl_2	—			2.45×10^2
2,6-Cl_2	—			1.7×10^2
2-Me, 4-Cl	—			5.5×10^3
2,5-Me_2, 4-Cl	—			6.9×10^5
2,6-Me_2, 4-Cl	2.2×10^6			—
Naphthol	3.8×10^8			4.46×10^6
2-COOH Naphthol	—	5.0×10^4	2.0×10^3	—

[a] 30 °C, phosphate buffer 0.1 M, Ref. 68.
[b] 25 °C, Borax buffer 0.125 M, Ref. 69.
[c] Ref. 64.
[d] Ref. 63.

The value of k_2 is ~ 100 times larger for the elimination of chloride ion than for the elimination of methoxide to form methanol.

When k_1 is the rate-limiting step, the effect of electron-donating substituents on naphthol increases the coupling rate constant, as described earlier. Such substituents also increase the pK_a for phenol ionization, so one might expect a good correlation between $\log k_1$ and pK_a. Such a correlation was found[67] for 4-substituted 2,6-dimethylphenolates (**86**) coupling with **79**, where the substituents (X) varied from SO_3^- to OMe. On the other

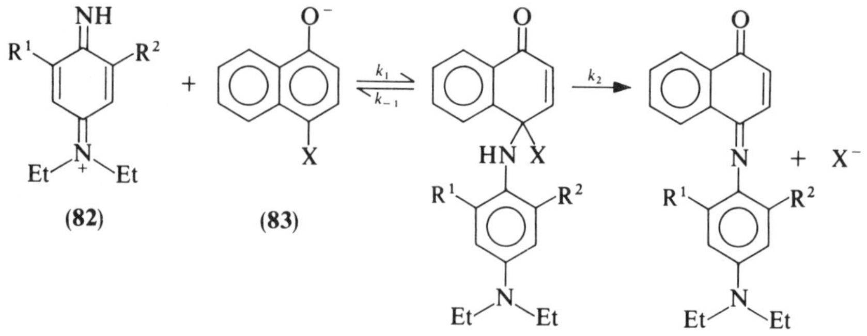

(82) (83) (84) (85)

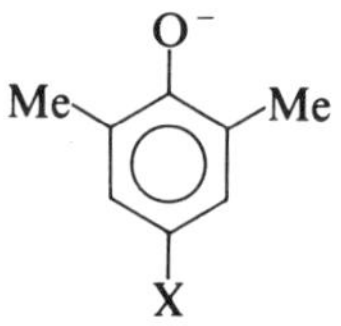

(86)

hand, Pelizzetti and Saini[69] said that no such correlation existed for the phenols they studied, many of which are shown in Table 11.

Corbett and coworkers[63, 73] studied the coupling kinetics of benzoquinonemonoimine with several substituted phenols to form indophenol dyes. The leuco-dye intermediate is rapidly oxidized by monoimine, making the initial step rate limiting. The reaction is fastest in alkaline solution, where the phenolate ion reacts with the neutral monoimine faster than with the conjugate acid. This is in contrast to the diimine reactivity, where there is no reported reactivity with the neutral diimine[68]. Effects of substituents on phenol are additive, although the observed rate constants are significantly lower. The neutral monoimine is $\sim 10^5$ less reactive than the corresponding diimine conjugate acid[73].

An interesting route to the formation of thiazine dyes (**87**) through coupling of *N*,*N*-diethylquinonediimine (**79**) with naphtho[2,1-*d*]-1,3-oxathiol-2-ones (**88**) has been reported by Mann and coworkers[74]. Essentially, the reaction involves coupling to a substituted naphthol, which rearranges to form the final phenothiazine dye (**87**), as outlined below.

Since the oxidative coupling products of *p*-phenylenediamines with phenols or anilines yield highly colored products, attempts have been made to use these reactions in the analytical methods of assay for low concentrations of such species. For example, Rao and Sastry[75] used metaperiodate, IO_4^-, to oxidize *N*,*N*-dimethyl-*p*-phenylenediamine at pH 7.8 in the presence of several phenols to measure the phenol concentration spectrophotometrically. Both *o*- and *p*-aminophenol can be determined by alkaline oxidation with hypochlorite in a self-coupling reaction[76]. Differential dye-forming kinetics have been used to analyze mixtures of anilines[77]. Additional studies have been evaluated by Corbett[78], who pointed out that some of these procedures do not adequately account for deamination reactions which compete with the dye formation step. He provided some guidelines for future attempts to use dye-forming reactions for quantitative analysis, which take into account the kinetics of both dye formation and deamination.

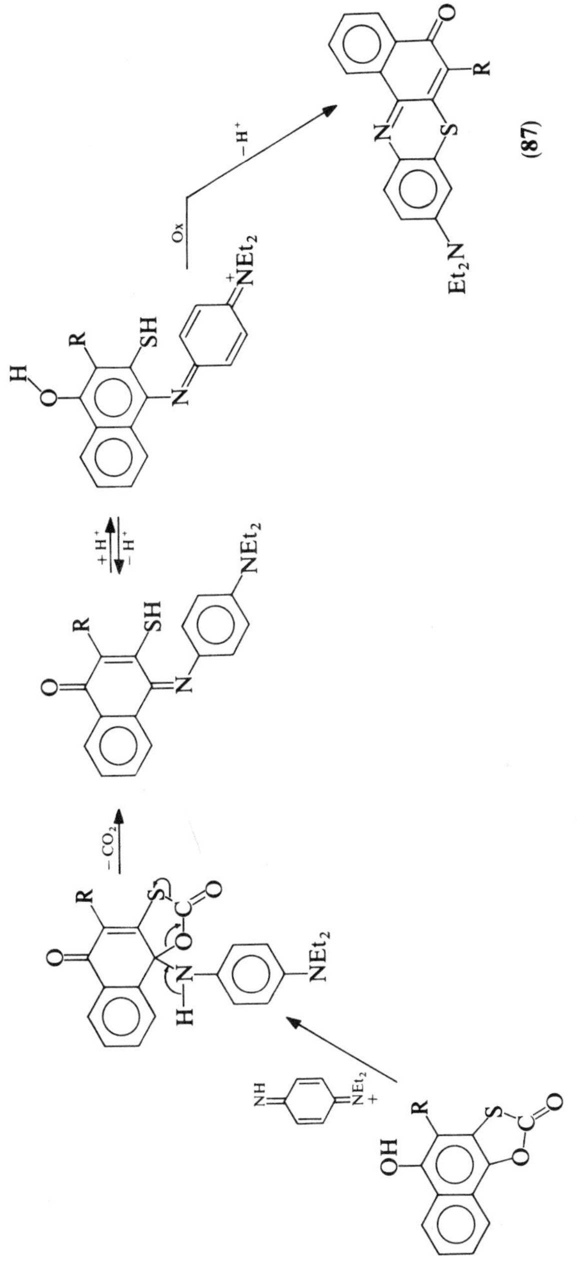

(88)

(87)

At least a 25% excess of coupling reagent over the species to be determined, and at least five equivalents of oxidant (preferably ferricyanide) should be used.

Some other interesting analytical coupling procedures have been developed to determine metal ions. For example, Cu(II) catalyzes the oxidation of *N,N*-dimethyl-*p*-phenylenediamine by H_2O_2 and the product quinonediimine forms a dye with *N,N*-dimethylaniline absorbing at 728 nm[79]. Concentrations of Cu(II) down to 10^{-9} M can be analyzed with few interfering metals. A similar method has been employed for Mn(II) and Ag(I) determinations, where the H_2O_2 catalyzed dye-forming reaction is used to detect the end point of a titration[80].

D. Other Coupling Reactions

Several other acidic methylene functions can ionize and couple with quinonediimines and monoimines to form photographically important dyes. Yellow dyes are formed by reaction with pivaloyl-[81] or benzoylacetanilides[82] (**89** and **90**). With 4-equivalent couplers

$$Me_3C{-}CO{-}CH_2{-}CO{-}NH{-}C_6H_5 \qquad C_6H_5{-}CO{-}CH_2{-}CO{-}NH{-}C_6H_5$$

(89) **(90)**

like **89** and **90**, the kinetics of dye formation are controlled by initial leuco-dye formation from the coupler anion and the diimine, which is followed by rapid oxidation to form the azomethine dye. The initial ionization of the coupler may be rate limiting, causing an induction period for dye formation[2], unless it is preionized[82].

$$-CO{-}CH_2{-}CO- \rightleftharpoons -CO{-}\bar{C}H{-}CO- + H^+$$

$$-CO{-}\bar{C}H{-}CO- + O{=}C_6H_4{=}\overset{+}{N}R^1R^2\ (\text{HN}{=}C_6H_4{=}\overset{+}{N}R^1R^2) \xrightarrow[k_c]{\text{Slow}} -CO{-}CH(NH{-}C_6H_4{-}NR^1R^2){-}CO- \xrightarrow[-2H^+]{-2e^-} -CO{-}C({=}N{-}C_6H_4{-}NR^1R^2){-}CO-$$

dye

Pelizzetti and Saini[82] studied the kinetics of dye formation as a function of substituent in both aromatic rings of **90**, using three different quinonediimines shown in Table 11 (**79**, **80** and **81**). Electron-donating substituents on the acetanilide (**90**), such as methyl and methoxy, increase the coupling rate constant, although they also increase the pK_a for coupler ionization. For a given substituent, the effect on $\log k_c$ is about twice as large as the effect on pK_a. The effect of a substituent in the benzoyl ring ($\rho = -1.3$) is ~ 9 times larger than the effect in the acetanilide ring ($\rho = -0.48$).

A methyl substituent on the quinonediimine decreases the rate constant by about a factor of six, compared with the factor of 25 decrease for coupling with phenolate ions[69]. The $C_2H_4NHSO_2Me$ group on **81** dissociates near pH 9.6[17], forming a zwitterion, which is also capable of coupling to form dye[66]. For each coupler, the rate constant for the zwitterion is slower than for the cationic diimine by a different amount ranging from 0.22 to 0.44. This compares with a rate of 0.05 for coupling with all substituted phenols[69].

The same yellow dyes can be formed by coupling oxidized *N,N*-diethyl-*p*-phenylenediamine with 2-oxo-4-oxazoline-5-carboxamide derivatives (**91**) of the benzoylacetanilides by ring-opening and CO_2 elimination, as shown below[83]. The oxazoline (**91**) ionizes in

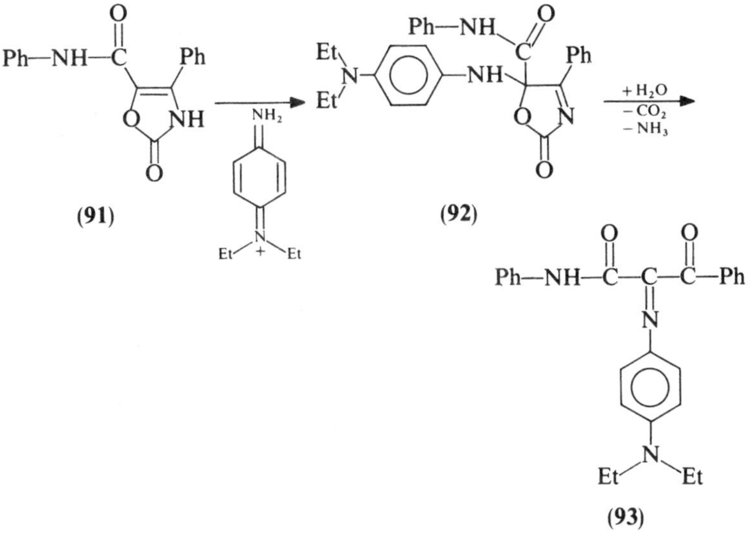

(**91**) (**92**) (**93**)

structures **95** and **96**. The leuco dye **92** formed in the initial coupling reaction undergoes acid hydrolysis with loss of CO_2 and NH_3 to form the stable azomethine dye **93**.

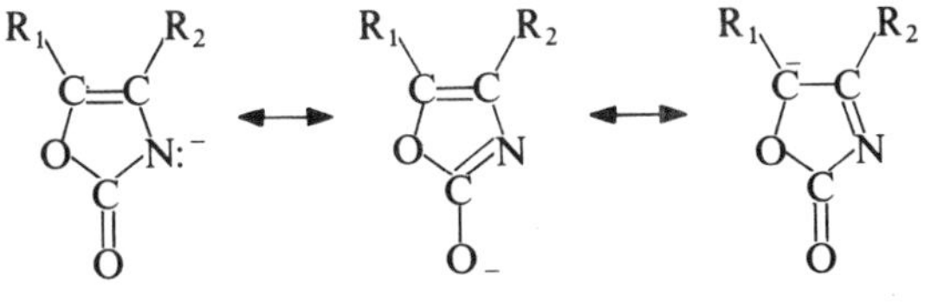

(**94**) (**95**) (**96**)

Magenta dyes (λ_{max} between 520 and 550 nm) are formed by coupling quinonediimines with nitrogen heterocycles containing an ionizable methylene group. Four classes of heterocycles studied are pyrazolin-5-ones (**97**)[70, 84], bis-pyrazolin-5-ones (**98**)[85], pyrazolo[1,5-*a*]imidazoles (**99**)[86] and pyrazolo[3,2-*c*]-*s*-triazoles[87] (**100**).

Pyrazolin-5-one coupling has received the most mechanistic study for both X = H (4-equivalent coupling) and for X equal to some other good anionic leaving group (2-equivalent coupling). When X = H, the coupling reaction proceeds to 100% dye yield with

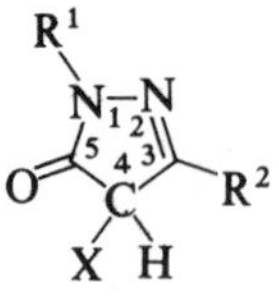

(97) 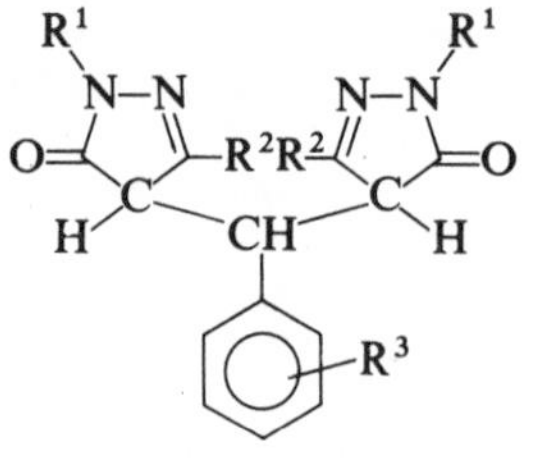 **(98)** 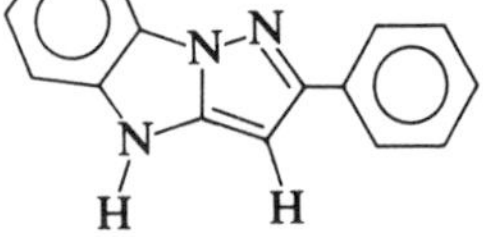 **(99)** 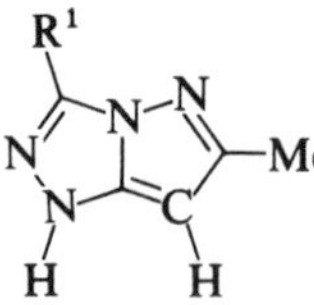 **(100)**

N,N-diethylquinonediimine (**79**). The same is true when X = Br or Cl. Most other groups produce only small quantities of dye, due to side reactions of the leuco dye or coupler[2]. With R^1 = phenyl, the rate constant for coupling with R^2 = Me and X = H is 1×10^5 $M^{-1}\,s^{-1}$. When R^2 = NHCOMe, the rate constant is somewhat lower[84], 1.6×10^4 $M^{-1}\,s^{-1}$ and the pK_a of the coupler is lower[70], 7.3 compared to 8.0. When X = Br, the rate constants are further decreased and < 100% dye is formed[84], particularly as the solution pH increases. No dye is formed when X = NHCOMe, COMe or Me. The reaction mechanism appears to be a rate-limiting formation of leuco dye, which can rapidly form dye or competitively react to form colorless products[2].

Bis-pyrazolin-5-ones (**98**) were reacted with quinonediimines **79** and **80** by flash photolysis of the azide precursor[70–72] in 30% *n*-propanol aqueous solution using phosphate buffers at pH 10 with an ionic strength of 0.18[86]. The reaction is monitored by following dye **101** formation and analyzed according to a second-order rate expression since the coupler is not in large excess. In studying the ionic strength dependence of the observed rate constant Fanghänel and coworkers[86] determined the rate-limiting step with the reaction of the quinonediimine cations **79** and **80**, with the dianion of the bis-pyrazoline-5-one. The intermediate (**102**) undergoes slow hydrolysis, forming a benzaldehyde and the simple pyrazolin-5-one (**97**; X = H) anion according to the following reactions.

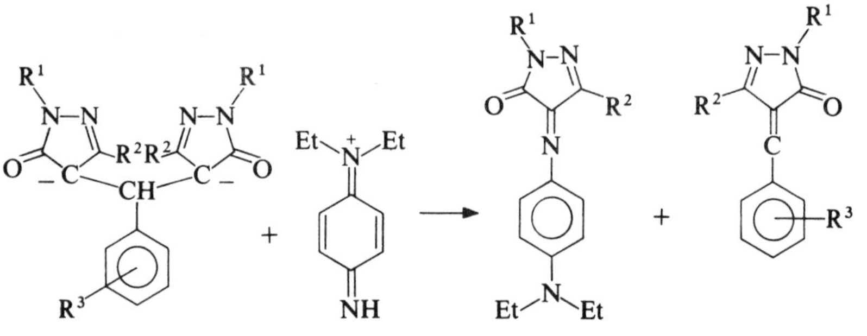

(98) dianion + **(79)** → **(101)** + **(102)**

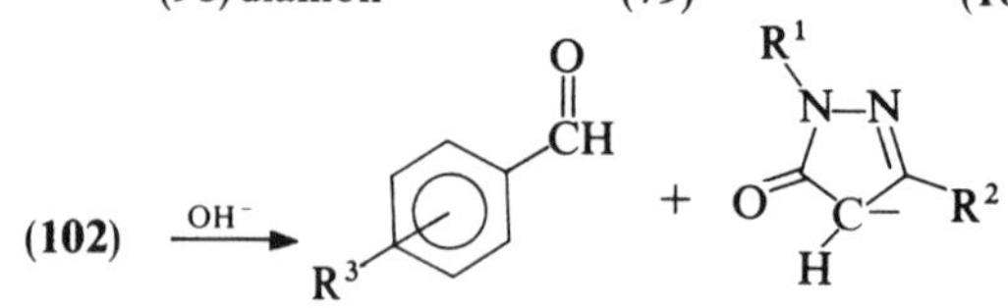

(102) $\xrightarrow{OH^-}$ benzaldehyde + **(97)** X = H anion

(97) + **(79)** → **(101)**

The hydrolysis rate of **102** is slow with second-order rate constants of 0.24 $M^{-1}\,s^{-1}$ for R^2 = Me and R^3 = 4-NMe_2 to 1400 $M^{-1}\,s^{-1}$ for R^2 = NHCOMe and R^3 = 4-OMe. Since these reactions are slower than hydrolysis of the quinonediimine (see k_4 in Table 2), only a small amount of dye forms in the second coupling step. For example, hydrolysis of **79** is 1×10^4 $M^{-1}\,s^{-1}$.

Couplers **99** and **100** react with **79** or **80**, formed by oxidation of the *p*-phenylenediamine with ferricyanide[86] or persulfate[87], to form the corresponding azomethine dyes. No mechanistic studies were made, but dyes **103** and **104** were isolated and characterized by analysis and by spectral comparison with dyes formed in an independent synthesis.

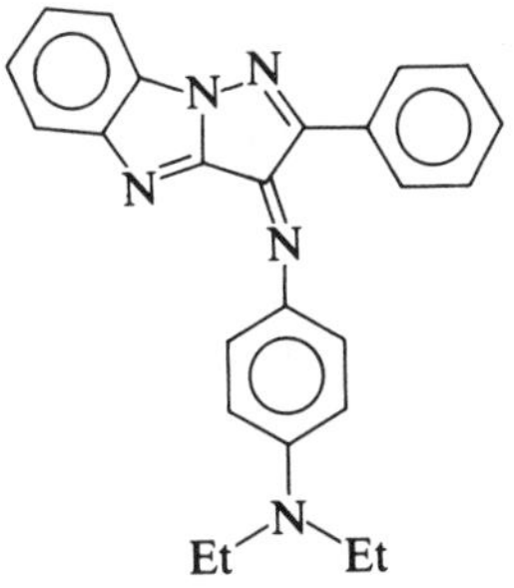

(103) 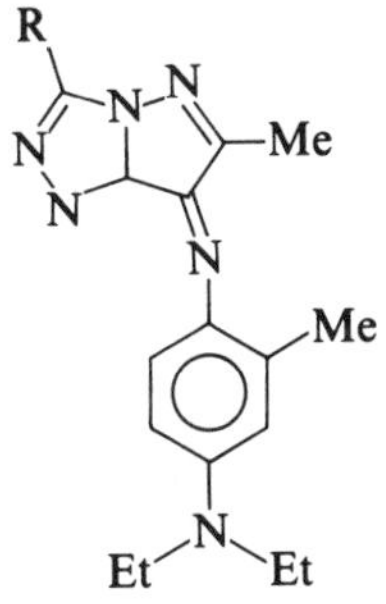 **(104)**

V. SULFONATION

A. Addition of Sulfites

Upon oxidation of *p*-phenylenediamine in the presence of sulfite at an electrode surface, the potential of oxidation shifts negative in proportion to the relative amount of sulfite added[88]. A second wave appears at a more positive potential. Similar studies with *N,N*-dimethyl-*p*-phenylenediamine[89] (**105**) and *N,N,N',N'*-tetramethyl-*p*-phenylenediamine[90] (**106**) in acid solution showed that even when only one electron is added to form a cation

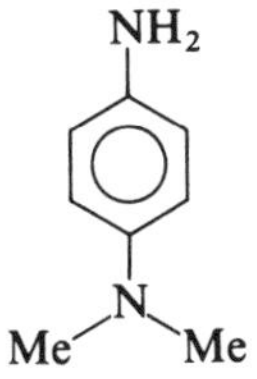

(105) 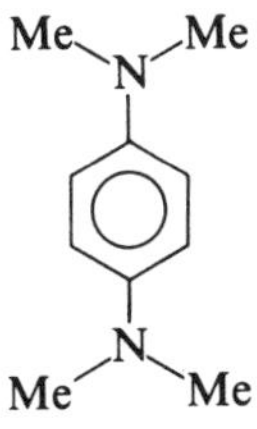 **(106)**

radical, rapid follow-up sulfonation occurs. Again, a second, more positive, oxidation wave occurs requiring two electrons. Both groups of workers explain their results by saying that sulfonation proceeds much more rapidly than oxidation, such that no intermediate sulfite–phenylenediamine complex can be observed. In excess sulfite, a second sulfite can add to the first product producing disulfonated *N,N*-dimethyl-*p*-phenylenediamine[89].

A careful kinetic study[91] of sulfonation using ferricyanide oxidation in phosphate buffer, showed definite accumulation of an intermediate, especially in an acid solution with a large excess of sulfite. For unsubstituted quinonediimine, the pH dependence of the rate indicates that the intermediate complex (**107**) must be protonated. This intermediate loses a proton in the rate-determining step. The reaction is independent of phosphate buffer

(**6**) $+ 2H^+ + SO_3^{2-} \underset{}{\overset{K_I}{\rightleftharpoons}}$ (**107**)

(**107**) $\xrightarrow[k_{H_2O}[H_2O]]{k_{OH}[OH^-] \text{ or}}$

concentration, indicating no general catalysis to form the product. The reaction rate is given by equation 11.

$$\text{rate} = (k_{OH}[OH^-] + k_{H_2O})K_I[H^+]^2[SO_3^{2-}][\mathbf{6}] \qquad (11)$$

Between pH 8 and 9.5, the observed second-order rate constant, $k_{SO_3^{2-}}$ in equation 12, is second order in $[H^+]$, but shifts to first order above pH 9.5. At lower pH, the reaction is complicated by incomplete oxidation and, above pH 11, deamination begins to compete. At pH 9, $k_{SO_3^{2-}} = 300\ \text{M}^{-1}\ \text{s}^{-1}$. A plot of log $k_{SO_3^{2-}}$ vs. pH is given in Figure 8.

$$k_{SO_3^{2-}} = K_I[H^+]^2(k_{OH}[OH^-] + k_{H_2O}) \qquad (12)$$

In the case of *N,N*-diethylquinonediimine (**79**), the reaction is complicated by the catalysis of phosphate ions in the buffer. Extrapolation to zero phosphate concentration at constant ionic strength for several pH values between 8 and 10 yields a constant value for the decay of the intermediate due to the water reaction. There is no pH dependence and $k_{SO_3^{2-}} = 3.7 \times 10^4\ \text{M}^{-1}\ \text{s}^{-1}$. The phosphate effect is suggestive of general base catalysis

(**79**) $+ SO_3^{2-} + H^+ \rightleftharpoons$ $\xrightarrow{:B}$ (**108**)

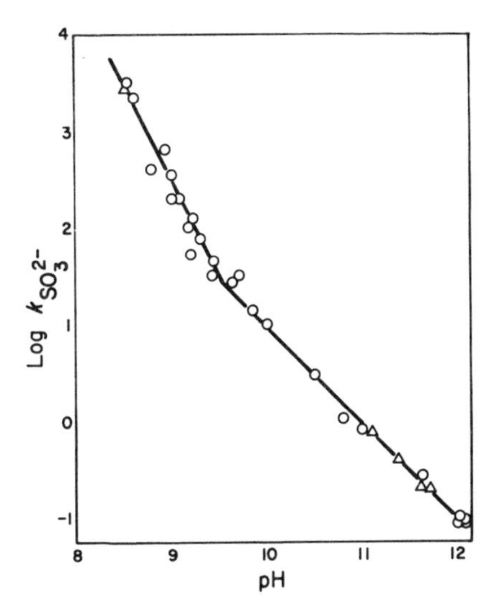

FIGURE 8. Effect of pH on the sulfonation of *p*-benzoquinonediimine in phosphate buffer (μ = 0.375)[91]. ○ Phosphate buffer only; Δ phosphate buffer diluted 1:4 with 0.375 M NaCl; and solid line calculated according to equation 12

assisting to break the C–H bond. The product of sulfonation (**108**) was confirmed by separate preparation[92].

Sulfonation of *N,N*-dimethylquinonediimine is more complicated, because extrapolation to zero phosphate concentrations as a function of pH yields different intercepts. The intercept value cannot be divided into a water rate and a hydroxide rate within experimental error, as done for unsubstituted quinonediimine. The diimine can also be sulfonated *ortho* to the alkylated imine due to reduced steric hindrance on going from ethyl to methyl. The observed rate constant in 0.375 M phosphate buffer is ~ 4 times larger than for *N,N*-diethylquinonediimine (**79**)[91].

A study[93] of the sulfonation of **81** was made by following the formation of the sulfonated product by stopped-flow fluorescence spectrometry. The observed pseudo-first-order observed rate constants were linear with sulfite only at low phosphate buffer

NH
Me
N
Et + $C_2H_4NHSO_2Me$

(**81**)

concentration and at high pH. The second-order rate constant, obtained when the observed rate constant is linear with sulfite, increases with phosphate concentration and with decreasing pH, but not in a linear way. The product mix varies with pH and includes at least three species, only two of which were identified. Unidentified product, based on the

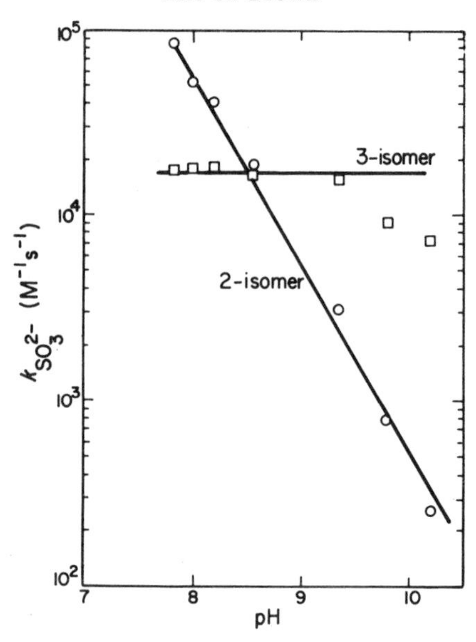

FIGURE 9. Effect of pH on sulfonation rate of **108** to form the 2-isomer (**109**) and the 3-isomer (**110**)[93] in 0.2 M phosphate buffer. ○ $k_{SO_3^{2-}}$ for 2-isomer; □ $k_{SO_3^{2-}}$ for 3-isomer

theoretical yield expected from **81**, amounted to 10–30% of the reaction depending on pH. Multiplying the product yield by k_{obs} for total product formation (or **81** loss) and dividing by total sulfite concentration yields $k_{SO_3^{2-}}$ as a function of pH for the formation of the 2-isomer (**109**) and the 3-isomer (**110**). The result at total phosphate = 0.2 M is given in Figure 9.

NH_2, Me, SO_3^-, N, Et, $C_2H_4NHSO_2Me$

(109)

2-isomer

NH_2, Me, SO_3^-, N, Et, $C_2H_4NHSO_2Me$

(110)

3-isomer

The results are consistent with a parallel reaction to form either the 2-isomer or the 3-isomer with no interconversion between the two paths, i.e. they are irreversible overall. The lack of sulfite dependence at low pH for formation of the 3-isomer is consistent with reversible formation of a sulfite complex, which rapidly builds up at high sulfite concentrations and decomposes in the rate-limiting, but sulfite-independent, step. The pH dependence for 2-isomer formation occurs because a proton must be added to the primary amine to prevent formation of a negative charge at that site, which would not be stabilized by resonance in the ring.

(81)

2-isomer

3-isomer

In addition to the above reaction, the drop in $k_{SO_3^{2-}}$ at high pH for formation of the 3-isomer in Figure 9 can be explained by ionization of the sulfamyl substituent on **81**, which apparently reacts more slowly, if at all[93].

Although phosphate increases the reaction rate, the effect is not linear. Possibly chloride ion, added to maintain $\mu = 0.75$, is acting as a weak base and has some effect on the irreversible, rate-limiting, base-catalyzed proton removal step. The overall expression for $k_{SO_3^{2-}}$ is given by equation 13, where k_2 and k_3 are dependent on phosphate and perhaps chloride. The value of pK_a for ionization of the sulfamyl group is 9.95 at $\mu = 0.75$[93], somewhat higher than the value of 9.5 obtained by Tong and coworkers[28] from deamination kinetics.

$$k_{SO_3^{2-}} = (k_2K_2[H^+] + k_3K_3)[H^+]/([H^+] + K_a) \qquad (13)$$

Attempts to identify other products in the reaction mixture were inconclusive. No disulfonates were obtained, even with large excesses of sulfite. Deamination products should not form competitively, and were not observed below pH 12. Sulfamate formation from addition to N-, rather than C-, may occur in small amounts, based on product analysis using high performance liquid chromatography (HPLC)[93]. No other species in the product mix were identified.

B. Addition of Sulfinates

The reaction of arylsulfonic acid with *N,N*-dialkylbenzoquinonediimines in alkaline solution should be similar to the reaction of sulfite ion. Finley and coworkers[94] studied the addition reaction for both *N,N*-dimethyl- and *N,N*-diethylquinonediimines formed oxidatively from ferricyanide and the corresponding phenylenediamine. They observed three addition products, a sulfonamide (**111**), a sulfone (**112**) and a disulfonamide (**113**) in oxygen-free phosphate buffer.

The product distribution was obtained from a characteristic vector analysis of the UV spectrum of chloroform extracts of the reaction mixture diluted with ethanol. The major product above pH 7 was the sulfonamide (**111**) (> 80%), independent of whether R was methyl or ethyl, and almost independent of X.

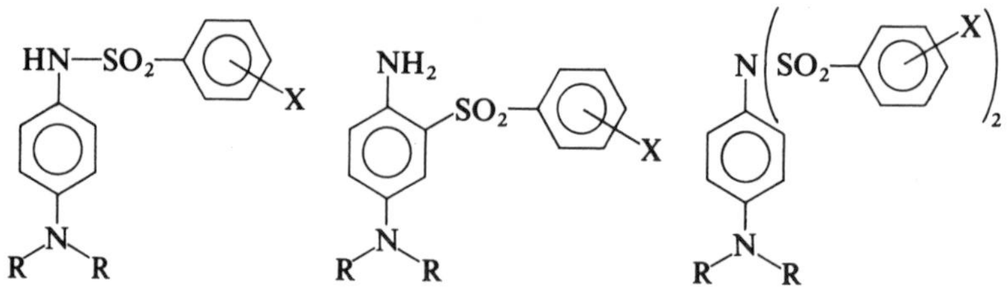

(111) (112) (113)

At pH 5, ~40% of the sulfone (**112**) was observed, while essentially none was formed above pH 7. The disulfonamide (**113**) occurred only at 5% or less, along with a similarly small amount of the parent *p*-phenylenediamine. The products are apparently formed

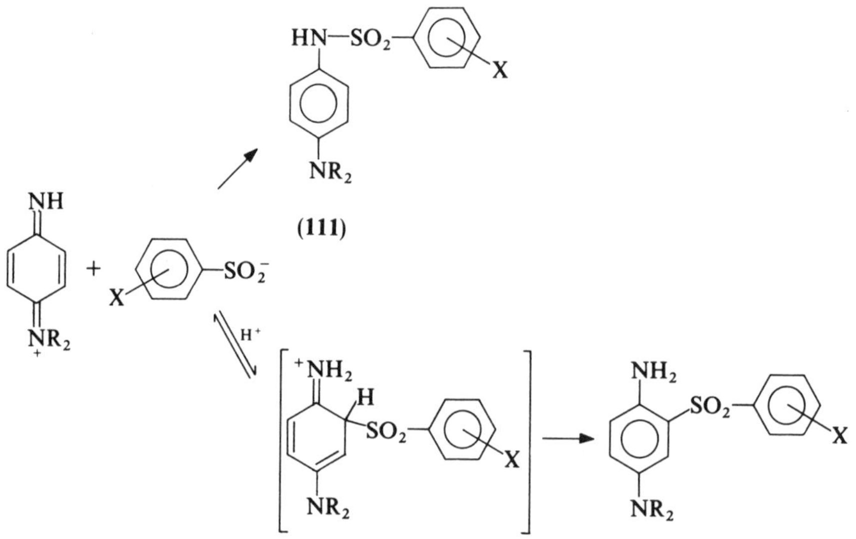

(111)

(112)

competitively, with the sulfone product requiring the addition of a proton to form the initial adduct as observed for sulfite. The disulfonamide apparently forms through further reaction of **111** anion with diimine[94].

Burmistrov and coworkers[95] studied the products of substituted benzenesulfinate addition to *N*-phenylquinonemonoimine (**114**). The single product identified (68–96% yield) was **115**, after oxidation with lead tetraacetate. When hydroxymethylsulfinic acid was used, a dimer was formed, which was oxidized to **116** with $Pb(OAc)_4$.

The reaction of benzenesulfinic acid with *N*-(thioaryl)quinonemonoimine (**117**) was studied by Kolesnikov and coworkers[96]. Two products were formed, the sulfone (**118**) and a thiosulfonate (**119**).

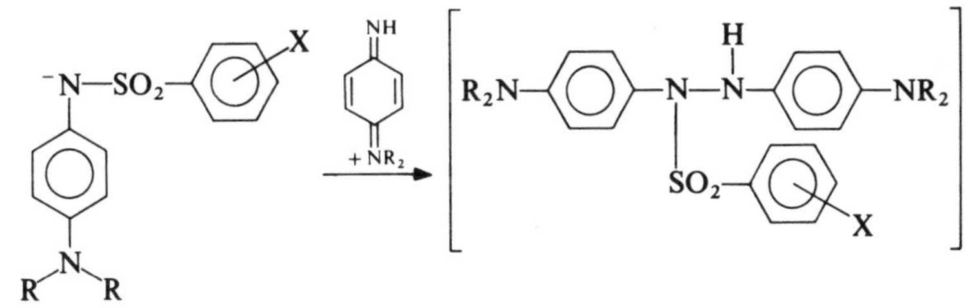

(111) anion

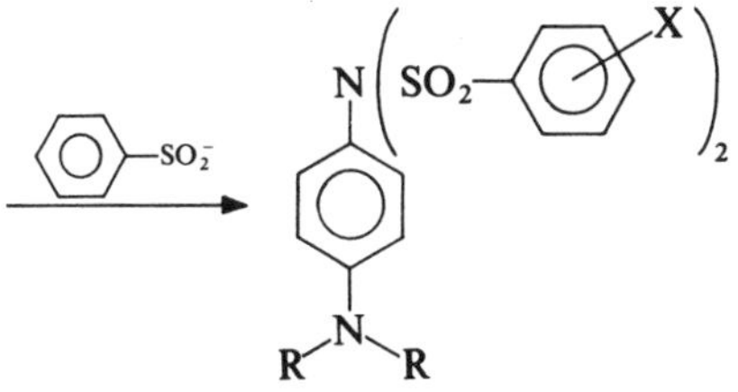

(113)

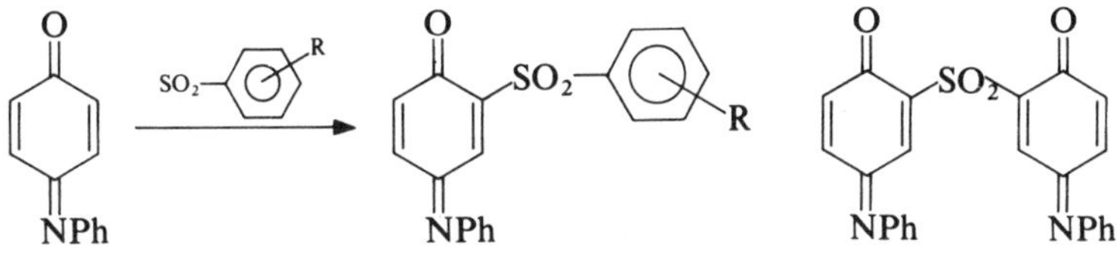

R = H
= 4-Me
= 4-Cl

(114) (115) (116)

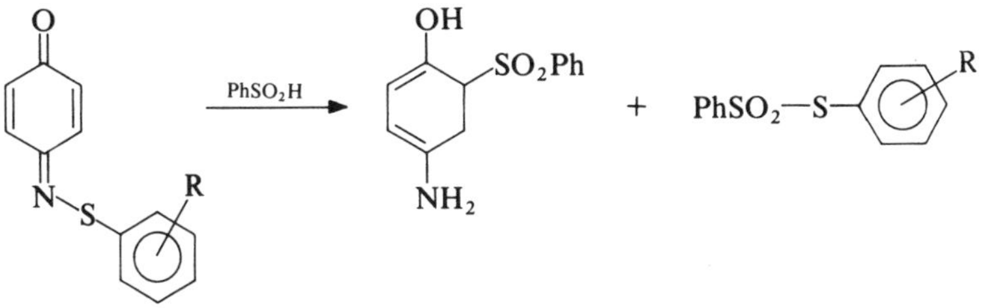

(117) (118) (119)

VI. OTHER QUINONEIMINE CHEMISTRY

A. *N*-(Arylthio)quinoneimines

Compounds **120–123**, formed by several different reactions, have been described in the literature. For compound **120**, with R equal to H, 4-OMe, 4-Cl, 4-NO_2, or 2-NHCOMe

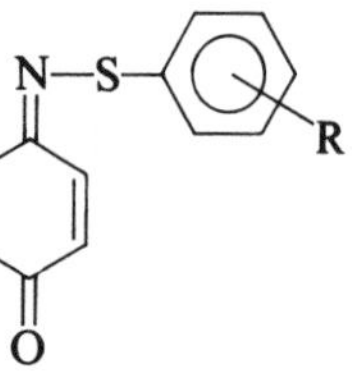

(120) 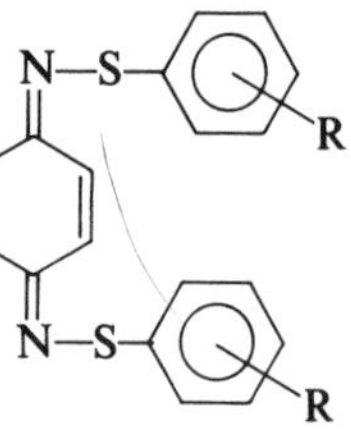 (121) 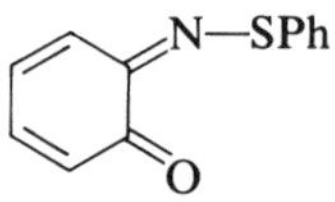 (122) 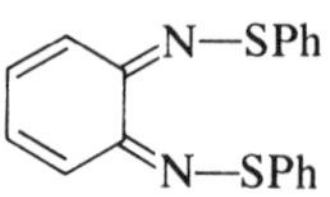 (123)

the following reaction was used[96, 97] in which the *N*-chloroquinoneimine in dioxane solution was added to the thiol in aqueous 10% sodium carbonate. The deep-red precipitate was filtered and identified by infrared and nuclear magnetic resonance spectroscopy.

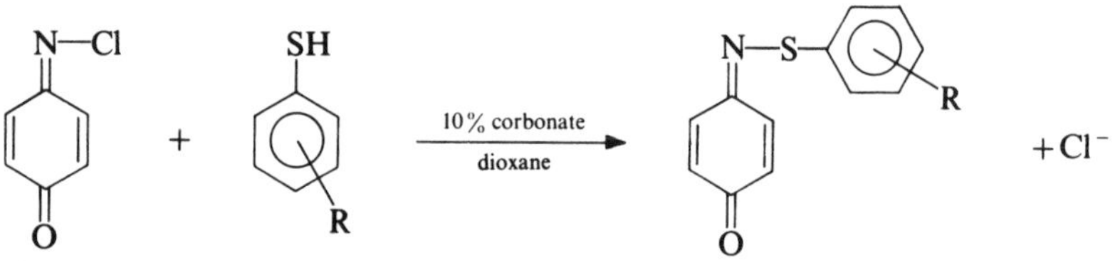

10% corbonate, dioxane; $+Cl^-$

(120)

Both *N*-(aryl)- and *N*-(alkyl)thioquinoneimines can be prepared by reacting the appropriate thiol with N_2O_4 to produce the unisolated thionitrate, which oxidizes and adds to *p*-aminophenols[98]. The compounds were yellow, and were identified by IR, NMR and mass spectroscopy.

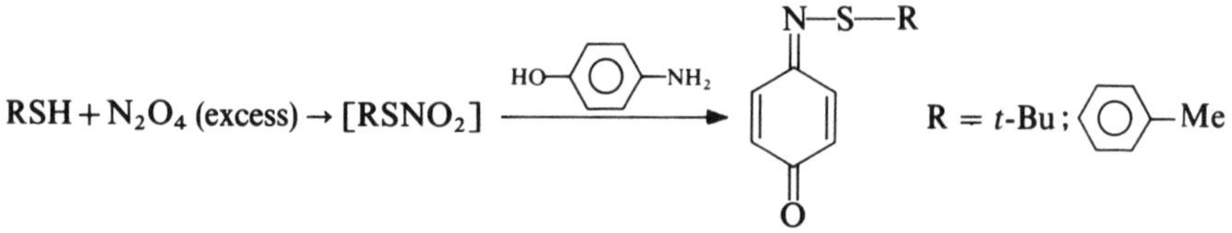

$$RSH + N_2O_4\ (excess) \rightarrow [RSNO_2] \xrightarrow{HO-C_6H_4-NH_2}$$ (124)   R = *t*-Bu; C_6H_4-Me

In a study of acylthiol addition to oxidized color developing agents, Wightman and colleagues[99] observed the formation of **120** with R = 4-NO_2. Their mechanistic work, using a competing dye-forming reaction with a substituted phenol (**125**), suggested the following scheme for the formation of **120**.

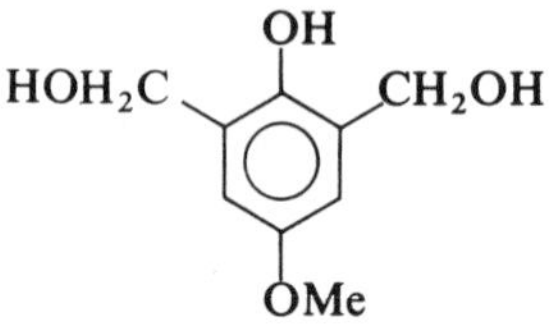

(125)

(80) (126)

(80) (126) (127) (128)

(127) (120) R = 4-O_2N

The initial product (**126**) is oxidized more easily than the starting *p*-phenylenediamine (**128**) and irreversible deamination of **127** drives the reversible redox reaction to form **120**. The oxidized intermediate (**127**) can also couple with **125** to form the same dye as formed from **125** and the diimine (**80**), but the reaction is much slower.

The initial rate of *p*-nitrothiophenolate addition to diimine **80** exceeds the rate of coupling with **125**. When R = 4-Me instead of 4-NO_2, the reaction is significantly faster, due to the enhanced nucleophilicity of the thiol. The reactions were studied at a pH well above the pK of the thiols.

Another method for the formation of **120** and **122** is the reaction of trisulphenamides (**129**) with phenols and naphthols[100]. Thermal decomposition of **129** yields the free radical (**130**), which reacts with phenols in two steps to form the *N*-(arylthio)quinoneimines **120** and **122** in dichloromethane. The predominant products with unhindered phenols are *o*-*N*-(arylthio)quinoneimines (**122**), because the 2- and 6-positions have the highest electron density in the radical formed in the first step. The radical mechanism outlined above is corroborated, in part, by kinetic isotope studies in dichloromethane in the presence of either H_2O or D_2O. The presence of D_2O decreases the initial atom abstraction rate constant, k_1, when the atom is deuterium instead of hydrogen.

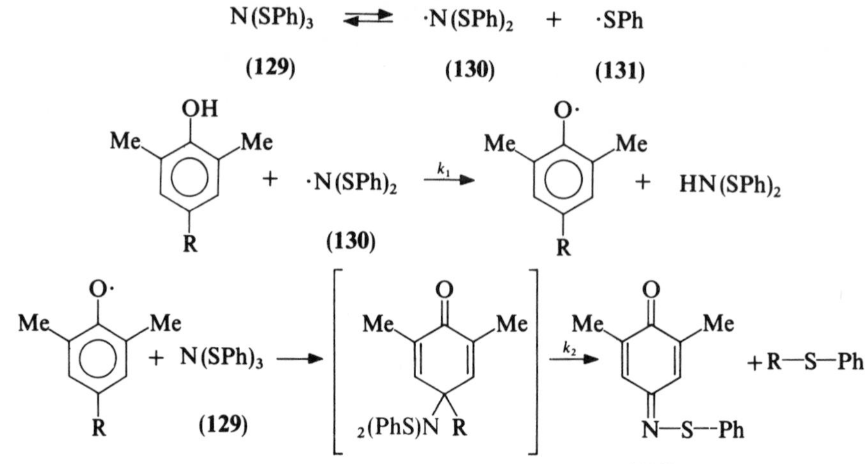

$$N(SPh)_3 \rightleftharpoons \cdot N(SPh)_2 + \cdot SPh$$

(129) (130) (131)

(130)

(129) (132)

When the substrate is β-naphthylamine, the disubstituted species (**133**) is formed in 33–38% yield in dichloromethane. 2,4-Dimethylaniline yielded modest formation of **134**, but aniline and *N*-methylaniline gave only intractable mixtures. Compound **135** was isolated when 2,6-dimethylaniline was the substrate.

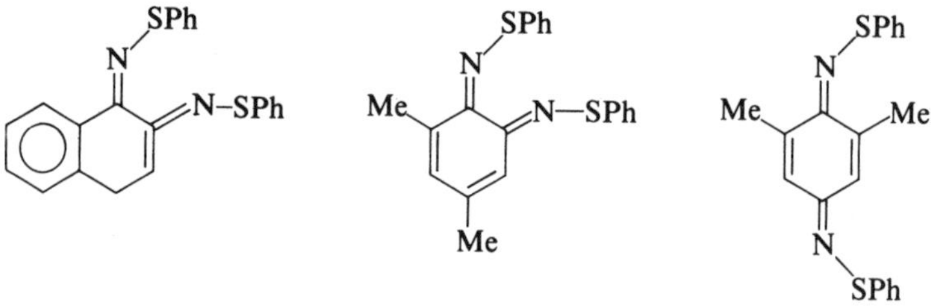

(133) (134) (135)

From liquid ammonia solutions of methyl phenyl sulfoxide **136**, Armitage and Clark[101] isolated **123**, which presumably forms by reaction of the phenylthiol radical **131** and

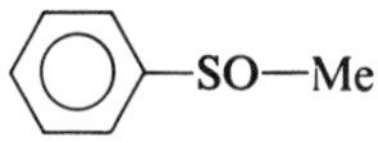

(136)

the dibenzenesulphenamide radical **130** in the reaction mixture. These species are also the same radicals suggested as forming from the thermal decomposition of tribenzenesulphenamide (**129**)[100].

The reactivity of *N*-(arylthio)quinonemonoimines has received very little study. The only study found in the literature[96] describes the reaction of benzenesulfinate (**137**) on **120** in acetic acid. The substrate (**137**) reacts at two sites of the molecule, both to cleave the

(120) R = 4-O_2N + $2PhSO_2^-$ (137) $\xrightarrow[H^+]{AcOH}$ $NH_2 \cdot HO_2S—Ph$ / $SO_2—Ph$ / OH + $Ph—SO_2—S—C_6H_4—R$

arylthio group from nitrogen and to add to the ring. The intermediate species (**138**), where only one molecule of **137** reacts, was isolated only in the case of dinitro substitution[96].

(**138**)

A series of *N*-(arylthio), *N'*-phenylbenzoquinonediimines (**139**) has been prepared by oxidation of benzenesulfenanilide (**140**) with PbO_2 in benzene[102]. The reaction apparently proceeds by a radical coupling mechanism involving **141.**

R—C_6H_4—NH—S—Ph (**140**) $\xrightarrow[C_6H_6]{PbO_2}$ R—C_6H_4—$\dot{N}$—S—Ph (**141**)

2 R—C_6H_4—$\dot{N}$—S—Ph (**141**) ⟶ Ph—S—N=C_6H_4(R)—N(S—Ph)—C_6H_4—R ⟶ PhSH

Ph—S—N=C_6H_4=N—C_6H_4—R

(**139**) 40% R = H

B. *N*-(Arylsulfonyl)quinoneimides

Many publications exist in the Russian literature describing the addition of many nucleophiles to both *o*- and *p*-*N,N'*-bisarylsulfonylquinonediimines and monoimides. Unfortunately, most of these papers are not readily available, so only a brief survey is possible from entries in *Chemical Abstracts*. For example, Kremlev and coworkers[103] studied the oxidative addition of *N*-chlorobenzenesulfonamide (**143**) to both *o*- and *p*-*N,N'*-bis(phenylsulfonyl)quinonediimide (**142**) to yield disubstituted oxidation products

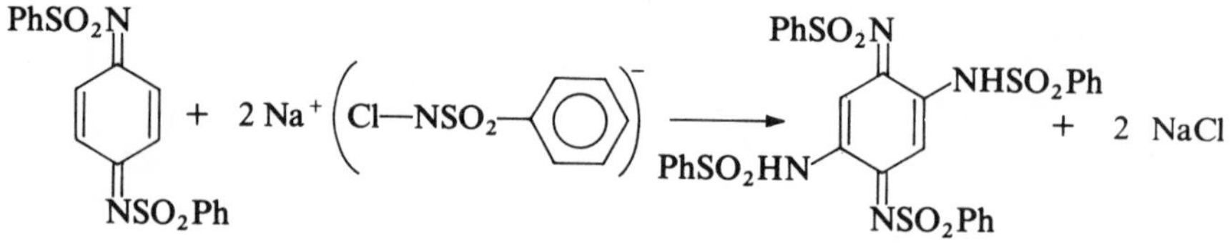

(142) (143) (144)

(**144**) in acetone. No monosubstituted product was isolated. The same product (**144**) was formed by addition of benzenesulfonyl chloride to **145**, followed by oxidation by lead

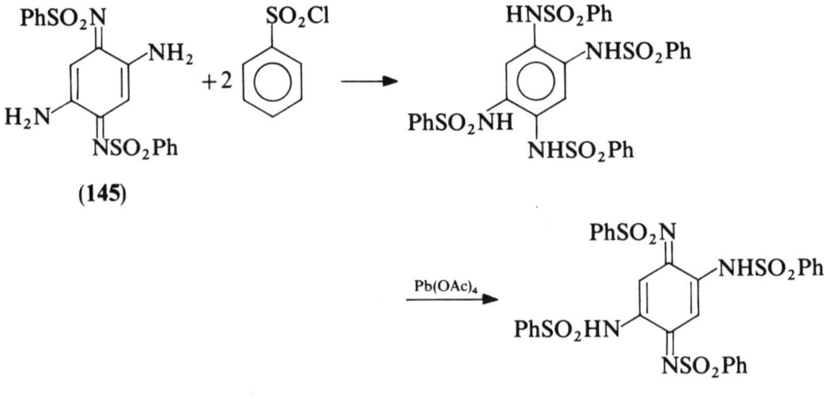

(145)

(144)

tetraacetate. Treatment of **142** or **146** with $BF_3 \cdot OEt_2$ and $Pb(OAc)_4$ in refluxing acetic acid yields significant amounts of **148** and **147**[104] respectively. Addition of acylchloramide

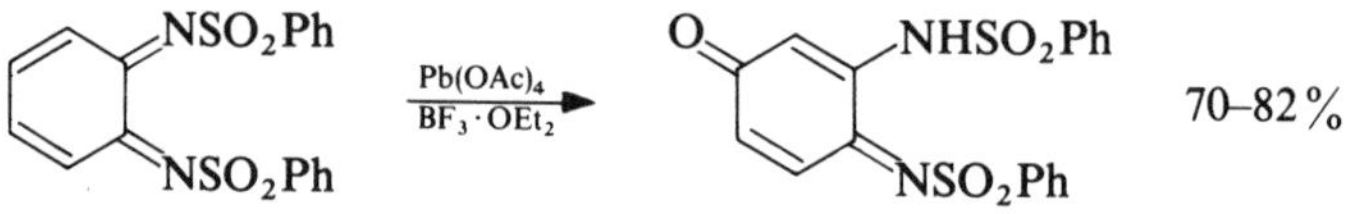

(146) (147) 70–82%

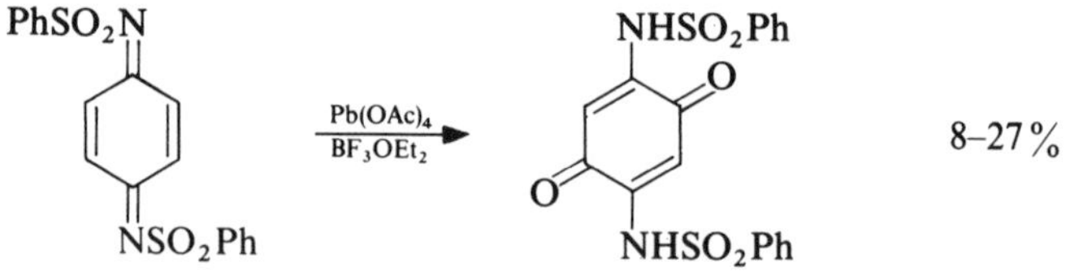

(142) (148) 8–27%

(**149**) to **142** in the presence of base yields disubstitution on the ring, unless one ring position is blocked by chloro, methyl, or benzylamino[105].

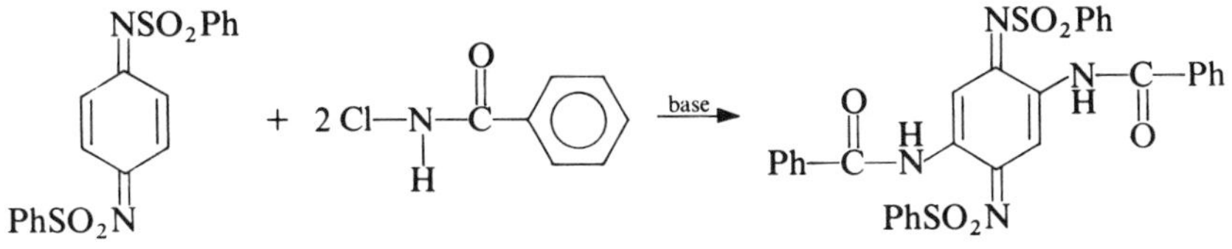

NSO$_2$Ph + 2 Cl—N(H)—C(=O)—C$_6$H$_5$ →(base)

(142) (149)

The addition of several nucleophiles to monosulfonamides has been studied by Titov and coworkers[106–110], primarily by product isolation. *N*,*N*-Bis-hydroxyethylamine (**150**) adds to **151** to form a novel ring compound (**152**), which forms **153** by treatment with zinc in acetic acid[106]. A mechanism was proposed. Simpler amines, such as phenylethylamine,

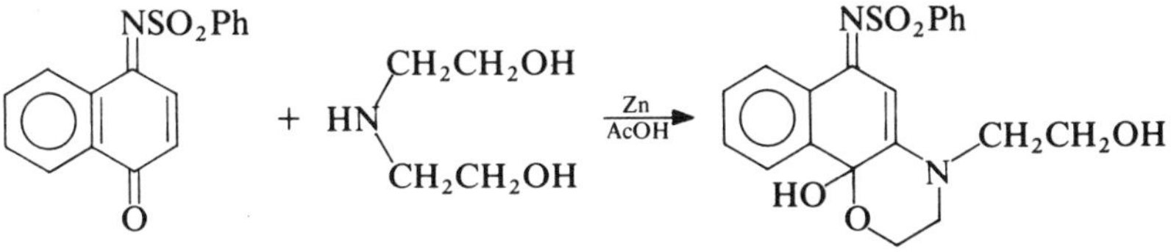

(151) (150) →(Zn, AcOH) (152) 91%

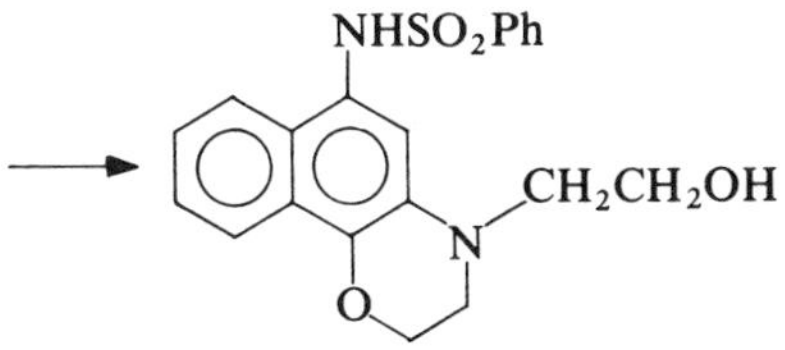

→ (153) 84%

add once to **151**, forming the 1,4-addition product, which is oxidized by the starting material to yield **154** and the *p*-sulfonamidonaphthol (**155**)[107]. A similar reductive

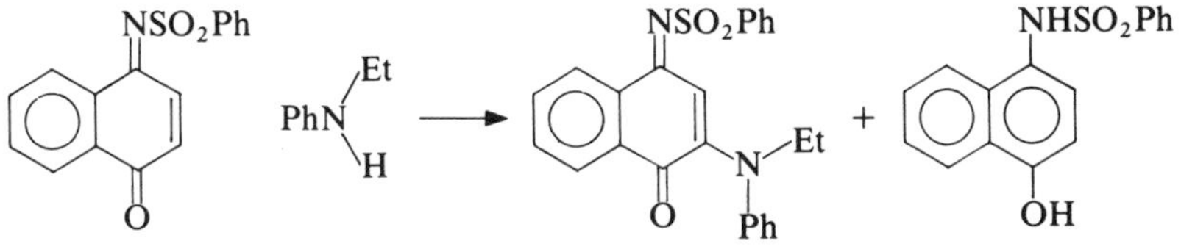

(151) + PhN(Et)H → (154) + (155)

addition reaction of bipyridine **156** to **157** was reported, forming the bipyridinium dication salt **158**[108].

In contrast to the examples of ring-addition cited above, dialkyl and trialkyl phosphites add to the sulfonamide group, forming a P–N bond[109–111]. Adveenko and Koshechko[111] suggest that the reaction proceeds through a free radical mechanism, based on ESR studies of the reaction mixture. The reactions occur on addition of solid quinoneimine (**159**) to

NSO_2Ph 2 + N N (157) $\xrightarrow{(156)}$ $[NHSO_2Ph \; NHSO_2Ph \; N \; N \; OH \; OH]^{2+} \; 2Cl^-$ (158)

NSO_2Ph $PhSO_2N$ (159) + $:P(OAlk)_3$ (160) $\xrightarrow{+H^+}$ $\cdot NSO_2Ph$ $PhSO_2NH$ + $P(OAlk)_3^{+}$

$\longrightarrow$ $(AlkO)_3\overset{+}{P}$ N SO_2Ph $PhSO_2NH$ $\longrightarrow$ O $(AlkO)_2P$ N SO_2Ph $PhSO_2NH$ (161)

liquid trialkyl phosphite (**160**) with heating. Alkyl groups ranged from methyl to ethyl, *i*-propyl and butyl. Both phenyl and *p*-tolyl aromatic sulfonamido groups react. When *N*-

O $PhSO_2N$ (157) + $HOP(OAlk)_2$ $\longrightarrow$ O $OP(OAlk)_2$ $PhSO_2NH$ (162)

(arylsulfonyl)-*p*-benzoquinonemonoimine (**157**) is the substrate, dialkyl phosphites form the corresponding phosphate ester (**162**)[109]. Alkyl groups studied were ethyl and *i*-propyl.

C. Quinone Oximes

The tautomeric equilibrium of quinone oximes (**163**) with nitrosophenol (**164**) is well established in the literature, and was reviewed in the earlier edition of this chapter[1]. This

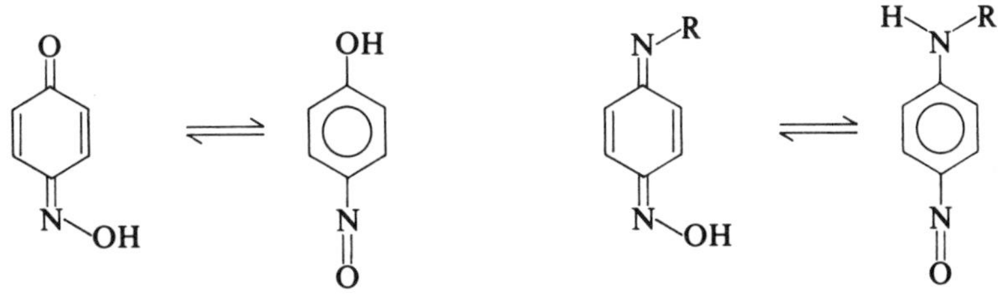

(**163**) (**164**) (**165**) (**166**)

equilibrium can complicate product identification when quinone oxime chemistry is studied. For example, addition of acid chlorides to quinoneimine oximes (**165**) could produce addition at either nitrogen, forming **167** or the oxime ester **168**, where X is the acid

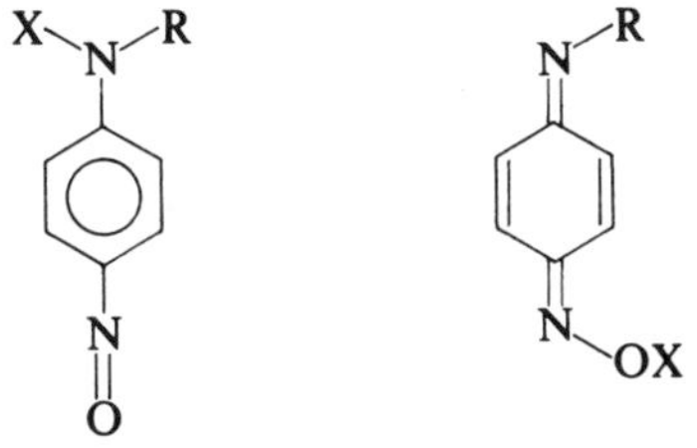

(**167**) (**168**)

group C(= O)R or SO_2R[112]. Titov and coworkers[112–114] have carried out extensive studies of oxime ester formation using acid chlorides in both aqueous base and organic solvents, such as acetone or ether. Products were identified by chromatography, and by hydrolysis and reaction with phenol or naphthol to form indophenol or indoaniline dyes.

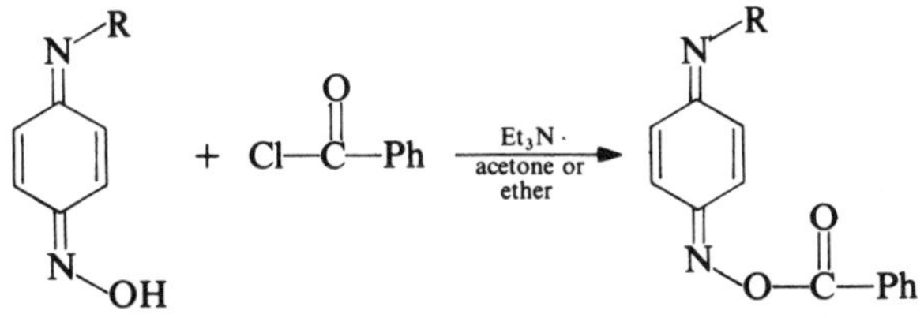

Et_3N · acetone or ether

(**165**) (**168**)

When R is alkyl, such as ethyl or methyl, ether is the preferred solvent with the use of a strong base such as triethylamine. When R is phenyl, or substituted phenyl, acetone is the solvent of choice and a weaker base, such as pyridine, can be used. The acid chloride of benzenesulfonic acid reacts cleanly in acetone at 0 °C to form the sulfonate ester, but the simple acetate ester required using excess acetic anhydride in benzene with heating. Esters have also been prepared when the R group is an aromatic acyl function, such as benzoyl or toluyl, using acetone–water mixtures and bicarbonate as the base[113].

When phenol or naphthol is added to an alkaline aqueous–alcohol solution of oxime ester (**168**), the blue indoaniline (**169**) is formed. If base is not added, the reaction proceeds

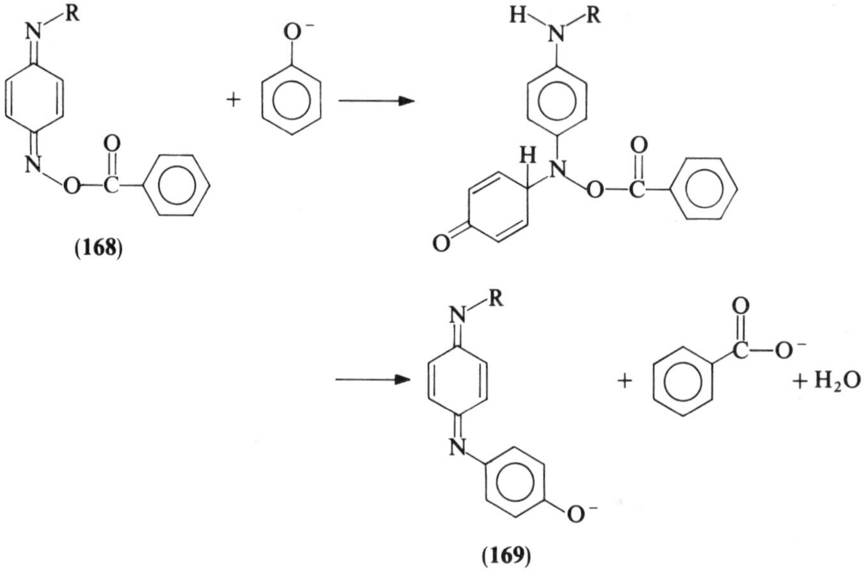

(**168**)

$+ H_2O$

(**169**)

even faster, suggesting that the mechanism involves the phenoxide ion and the protonated imine oxime ester[114]. This is consistent with the dye formation mechanism from quinonediimines described earlier. The reactivity of the *N*-alkyl (acyl)quinoneimine oxime esters (**168**) parallels the p*K* values for protonation of the imine; i.e. alkyl groups increase the reactivity because they are more basic and the substrate is protonated at higher pH.

The same dyes formed from the oxime ester above are also formed by oxidation of *N*-alkyl (aryl)phenylenediamines in the presence of phenol and by reaction of *p*-nitroso-*N*-alkyl(acyl)anilines with phenol. The latter reaction is a consequence of the tautomeric

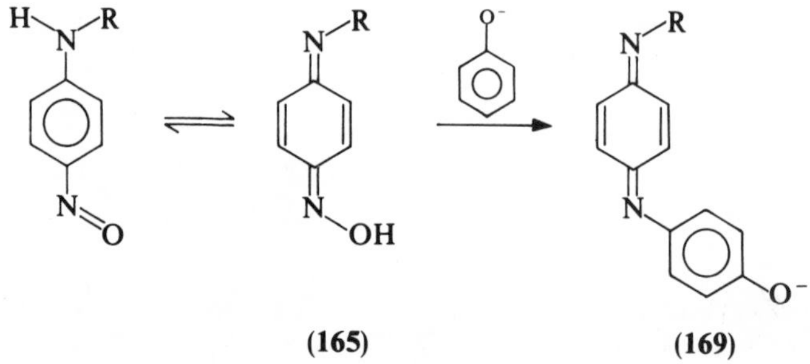

(**165**) (**169**)

equilibrium mentioned earlier. The same reaction with quinone oxime esters (**170**) produces an indophenol dye (**171**)[112].

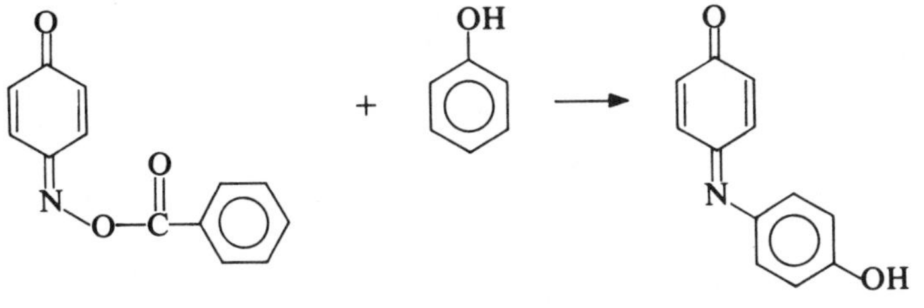

(**170**) (**171**)

The formation of quinone oxime ethers has also been described briefly[115].

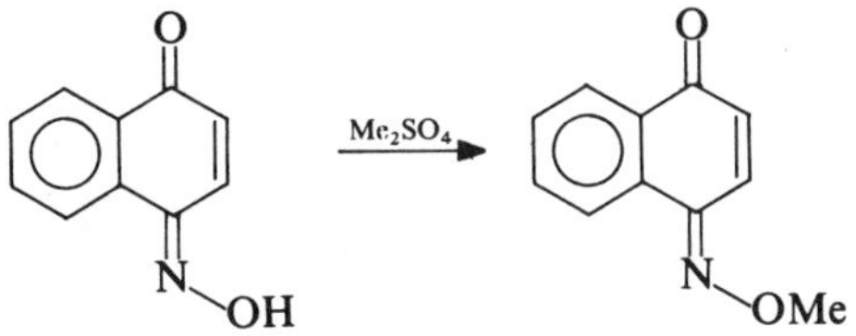

Me_2SO_4

VII. REFERENCES

1. K. T. Finley and L. K. J. Tong, in *The Chemistry of the Carbon–Nitrogen Double Bond* (Ed. S. Patai), Interscience (John Wiley and Sons), London, 1970, Chap. 14.
2. L. K. J. Tong, in *The Theory of the Photographic Process*, 4th ed. (Ed. T. H. James), Macmillan, New York, 1977, Chap. 12, Part II.
3. J. F. Corbett, *J. Soc. Cosmet. Chem.*, **24**, 103 (1973); **35**, 297 (1984); K. C. Brown and J. F. Corbett, *J. Soc. Cosmet. Chem.*, **30**, 191 (1979).
4. R. Willstätter and E. Mayer, *Chem. Ber.*, **37**, 1494 (1904); R. Willstätter and A. Pfannenstiehl, *Chem. Ber.*, **37**, 4605 (1904).
5. U. Nickel and W. Jaenicke, *Ber. Bunsenges. Phys. Chem.*, **86**, 695 (1982).
6. U. Nickel, M. Borchardt, M. R. Bapat and W. Jaenicke, *Ber. Bunsenges. Phys. Chem.*, **83**, 877 (1979); U. Nickel, E. Haase and W. Jaenicke, *Ber. Bunsenges. Phys. Chem.*, **81**, 849 (1977).
7. H. Firouzabadi and Z. Mostafavipoor, *Bull. Chem. Soc. Jpn*, **56**, 914 (1983); R. G. Srivastava, R. L. Pandey and P. Venkataramani, *Indian J. Chem.*, **20B**, 995 (1981).
8. M. S. Ramachandran, T. S. Vivekanandam, N. R. Subbaratnam and N. Rajaram, *Indian J. Chem.*, **22A**, 897 (1983).
9. H. E. Moore, M. J. Garmendia and W. J. Cooper, *Environ. Sci. Technol.*, **18**, 348 (1984).
10. R. N. Adams, *Electrochemistry at Solid Electrodes*, Marcel Dekker, New York, 1969.
11. E. R. Brown and J. R. Sandifer, in *Electrochemical Methods*, Vol. IIb, *Techniques in Chemistry* (Eds. B. W. Rossiter and J. Hamilton), John Wiley and Sons, New York, 1986, Chap. 4.
12. R. Adams and W. Reinfschneider, *Bull. Soc. Chim. Fr.*, 23 (1958), and references therein.
13. S. Fujita, *J. Chem. Soc., Chem. Commun.*, 425 (1981); *J. Org. Chem.*, **48**, 177 (1983).
14. H. W. Heine, B. J. Barchiesi and E. A. Williams, *J. Org. Chem.*, **49**, 2560 (1984); H. W. Heine, J. R. Empfield, T. D. Golobish, E. A. Williams and M. F. Garbauskas, *J. Org. Chem.*, **51**, 829 (1986).
15. W. Friedrichsen and A. Böttcher, *Heterocycles*, **16**, 1009 (1981).
16. L. K. J. Tong and M. C. Glesmann, *J. Am. Chem. Soc.*, **79**, 583 (1957).
17. L. K. J. Tong, *J. Phys. Chem.*, **58**, 1090 (1954).
18. J. F. Corbett, *J. Chem. Soc. B*, 213 (1969).

19. U. Nickel, K. Kemnitz and W. Jaenicke, *J. Chem. Soc., Perkin Trans. 2*, 1188 (1978).
20. U. Nickel and W. Jaenicke, *J. Chem. Soc., Perkin Trans. 2*, 1601 (1980).
21. A. C. Testa and W. H. Reinmuth, *Anal. Chem.*, **32**, 1512 (1960).
22. D. Hawley and R. N. Adams, *J. Electroanal. Chem.*, **10**, 376 (1965).
23. V. Plichon and G. Faure, *J. Electroanal. Chem.*, **44**, 275 (1973).
24. D. Lelievre and V. Plichon, *Electrochim. Acta*, **23**, 725 (1978).
25. U. Nickel and W. Jaenicke, *J. Photogr. Sci.*, **31**, 177 (1983).
26. W. J. Albery, R. G. Compton and I. S. Kerr, *J. Chem. Soc. Perkin Trans. 2*, 825 (1981).
27. L. K. J. Tong and M. C. Glesmann, *J. Am. Chem. Soc.*, **78**, 5827 (1956).
28. L. K. J. Tong, M. C. Glesmann and R. L. Bent, *J. Am. Chem. Soc.*, **82**, 1988 (1960).
29. U. Nickel and K. Kemnitz, *Angew. Chem.*, **89**, 273 (1977).
30. D. Lelievre, A. Henriet and V. Plichon, *J. Electroanal. Chem.*, **78**, 281 (1977).
31. L. K. J. Tong and M. C. Glesmann, unpublished results.
32. L. K. J. Tong and R. L. Reeves, unpublished results, 1963; M. C. Glesmann and J. O. Young, unpublished results, 1974.
33. R. L. Reeves, *J. Org. Chem.*, **30**, 3129 (1965).
34. K. Hartke and U. Lohmann, *Chem. Lett.*, 693 (1983).
35. C. R. Preddy, D. J. Miner, D. A. Meinsma and P. T. Kissinger, *Current Separations*, **6(4)**, 57 (1985), Bioanalytical Systems, Lafayette, Ind.
36. D. J. Miner, J. R. Rice, R. M. Riggin and P. T. Kissinger, *Anal. Chem.*, **53**, 2258 (1981).
37. I. A. Blair, A. R. Boobis and D. S. Davies, *Tetrahedron Lett.*, **21**, 4947 (1980).
38. W. T. Hanson, *Photogr. Sci. Eng.*, **20**, 155 (1976).
39. S. Fujita, *J. Syn. Org. Chem., Jpn*, **39**, 331 (1981); **40**, 176 (1982).
40. K. K. Lum and M. C. Glesmann, unpublished work, 1970–1973.
41. A. E. Anderson and M. C. Glesmann, unpublished work, 1974.
42. E. Bandrowski, *Chem. Ber.*, **27**, 480 (1894).
43. J. F. Corbett, *J. Soc. Cosmet. Chem.*, **23**, 683 (1972).
44. J. F. Corbett, *J. Soc. Dyers Colour.*, **85**, 71 (1969).
45. M. Dolinsky, C. H. Wilson, H. H. Wisneski and F. X. Demers, *J. Soc. Cosmet. Chem.*, **19**, 411 (1968).
46. J. F. Corbett, *J. Chem. Soc. B*, 818 (1969).
47. T. Sakata, M. Hiromoto, T. Yamagoshi and H. Tsubomura, *Bull. Chem. Soc. Jpn*, **50**, 43 (1977).
48. D. Lelievre and V. Plichon, *J. Electroanal. Chem.*, **78**, 301 (1977).
49. G. Durand, G. Morin and B. Trémillon, *Nouveau J. Chim.*, **3**, 463 (1979).
50. C. A. Bishop and L. K. J. Tong, *Photogr. Sci. Eng.*, **11**, 30 (1967).
51. M. Ballester, J. Riera, J. Castañer, A. Bandrés and S. Olivella, *Tetrahedron Lett.*, **21**, 4119 (1980).
52. K. C. Brown and J. F. Corbett, *J. Chem. Soc., Perkin Trans. 2*, 308 (1979).
53. T. Nogami, T. Hishida, M. Yamada, H. Mikawa and Y. Shirota, *Bull. Chem. Soc. Jpn*, **48**, 3709 (1975).
54. T. Nogami, T. Hishida, Y. Shirota and H. Mikawa, *Chem. Lett.*, 1019 (1973).
55. N. P. Loveless and K. C. Brown, *J. Org. Chem.*, **46**, 1182 (1981).
56. K. C. Brown and J. F. Corbett, *J. Org. Chem.*, **44**, 25 (1979); *J. Chem. Soc., Perkin Trans. 2*, 304 (1979).
57. K. C. Brown and J. F. Corbett, *J. Chem. Soc., Perkin Trans. 2*, 1125 (1977); J. F. Corbett, S. Pohl and I. Rodriguez, *J. Chem. Soc., Perkin Trans. 2*, 728 (1975).
58. P. Berkenkotter and R. F. Nelson, *J. Electrochem. Soc.*, **120**, 346 (1973).
59. K. K. Haynes and F. R. Hewgill, *J. Chem. Soc., Perkin Trans. 1*, 396, 408 (1972).
60. J. F. Corbett and E. P. Gamson, *J. Chem. Soc., Perkin Trans. 2*, 1531 (1972).
61. J. F. Corbett, *J. Soc. Dyers Colour.*, **88**, 438 (1972).
62. J. F. Corbett, *J. Chem. Soc. B*, 827 (1969).
63. J. F. Corbett, *J. Chem. Soc., Perkin Trans. 2*, 999 (1972).
64. J. F. Corbett, *J. Chem. Soc., Perkin Trans. 2*, 539 (1972).
65. J. F. Corbett, *J. Chem. Soc. B*, 823 (1969).
66. L. K. J. Tong and M. C. Glesmann, *J. Am. Chem. Soc.*, **79**, 592, 4305, 4310 (1957).
67. L. K. J. Tong and M. C. Glesmann, *J. Am. Chem. Soc.*, **90**, 5164 (1968).
68. J. F. Corbett, *J. Chem. Soc. B*, 1418 (1970).

69. E. Pelizzetti and G. Saini, *J. Photogr. Sci.*, **22**, 49 (1974).
70. G. Mann, H. Wilde, J. Lehmann, D. Labus, W. Schindler and U. Sydow, *Z. Chem.*, **15**, 141 (1975); G. Mann, H. Wilde, D. Labus, F. Hoeppner and W. Reissig, *J. Prakt. Chem.*, **320**, 705 (1978).
71. U. Sydow, H. Böttcher, D. Labus, G. Mann and P. Schumacher, *J. Signalaufzeichnungsmaterialien*, **6**, 299 (1978).
72. R. C. Baetzold and L. K. J. Tong, *J. Am. Chem. Soc.*, **93**, 1394 (1971).
73. J. F. Corbett, *J. Chem. Soc. B*, 1502 (1970); K. C. Brown, J. F. Corbett and R. Labinson, *J. Chem. Soc., Perkin Trans. 2*, 1292 (1978).
74. G. Mann, H. Wilde, S. Hauptmann, J. Lehmann, N. Naumann and P. Lepom, *J. Prakt. Chem.*, **323**, 776, 785 (1981).
75. K. E. Rao and C. S. P. Sastry, *J. Inst. Chem. (Calcutta)*, **55**, 161 (1983).
76. T. T. Ngo and C. F. Yam, *Anal. Lett.*, **71(A15)**, 1771 (1984).
77. R. Tawa and S. Hirose, *Chem. Pharm. Bull.*, **28**, 2136 (1980).
78. J. F. Corbett, *Anal. Chem.*, **47**, 308 (1975).
79. S. Nakano, M. Tanaka, N. Fushihara and T. Kawashima, *Mikrochim. Acta*, **403**, 457 (1983).
80. S. Abe, K. Watanabe and T. Sugai, *Bunseki Kagaku*, **32**, 398 (1983).
81. J. Korinek, J. Poskocil, J. Arient and J. Janosova, *Collect. Czech. Chem. Commun.*, **44**, 2101 (1979); J. Korinek, J. Poskocil and J. Arient, *Collect. Czech. Chem. Commun.*, **44**, 1460 (1979).
82. E. Pelizzetti and G. Saini, *J. Chem. Soc., Perkin Trans. 2*, 1766 (1973).
83. H. H. Credner, E. Maier and W. Lässig, *Chem. Ber.*, **112**, 3098 (1979).
84. G. Mann, L. Hennig, H. Wilde, D. Labus and U. Sydow, *Z. Chem.*, **19**, 293 (1979); H. Wilde, G. Mann, J. Lehman, D. Labus, W. Schindler and U. Sydow, *Z. Chem.*, **15**, 217 (1975).
85. H. Wilde, G. Mann, U. Burkhardt, G. Weber, D. Labus and W. Schindler, *J. Prakt. Chem.*, **321**, 495 (1979); H. Wilde, S. Hauptmann, G. Mann, G. Ostermann, D. Reifegerste and W. Schindler, *J. Signalaufzeichnungsmaterialien*, **4**, 285 (1981).
86. E. Fanghänel, M. S. Akhlaq and N. Grossmann, *J. Photogr. Sci.*, **34**, 15 (1986).
87. J. Bailey, *J. Chem. Soc., Perkin Trans. 2*, 2047 (1977).
88. K. Sasaki, *Ashahi Garasu Kogyo Gijutsu Shoreikai Kenkyu Kokoku*, **15**, 281 (1969); K. Sasaki, H. Imai, Y. Tanimizu and H. Shiba, *Nippon Kagaku Zasshi*, **91**, 1030 (1970).
89. T. Erabi, F. Arifuku and M. Tanaka, *Bull. Chem. Soc. Jpn*, **46**, 3582 (1973).
90. T. Erabi, Y. Shimotsu and M. Tanaka, *Denki Kagaku*, **41**, 32 (1973).
91. L. K. J. Tong, M. C. Glesmann and R. W. Andrus, unpublished work, 1967. These results were described in Ref. 1.
92. K. Meyer and H. Ulbricht, *Z. Wiss. Photogr.*, **45**, 222 (1950).
93. J. J. Lauff, K. G. Harbison and L. Weinstein, unpublished work, 1977.
94. K. T. Finley, R. S. Kaiser, R. L. Reeves and G. Wernimot, *J. Org. Chem.*, **34**, 2083 (1969).
95. S. I. Burmistrov, N. V. Toropin and K. S. Burmistrov, *Vopr. Khim. Khim. Tekhnol.*, **61**, 36 (1980).
96. V. T. Kolesnikov, L. V. Vid and L. O. Kuz'menko, *Zh. Org. Khim.*, **18**, 2163 (1982).
97. D. Kramer and R. N. Gamsonn, *J. Org. Chem.*, **24**, 1154 (1959).
98. S. Oae, K. Shinhama, K. Fujimori and Y. H. Kim, *Bull. Chem. Soc. Jpn*, **53**, 775 (1980).
99. P. J. Wightman, J. A. Kapecki and J. O. Young, unpublished work, 1982.
100. D. H. R. Barton, I. A. Blair, P. D. Magnus and R. K. Norris, *J. Chem. Soc., Perkin Trans. 1*, 1031, 1037 (1973).
101. D. A. Armitage and M. J. Clark, *Phosphorus and Sulfur*, **5**, 41 (1978).
102. C. Balboni, L. Benati, P. C. Montevecchi and P. Spagnolo, *J. Chem. Soc., Perkin Trans. 1*, 2111 (1983).
103. N. P. Bezverkhii and M. M. Kremlev, *Zh. Org. Khim.*, **14**, 2596 (1978); M. M. Kremlev, N. P. Bezverkhii and S. I. Burmistrov, *Zh. Org. Khim.*, **12**, 2479 (1976).
104. N. P. Bezverkhii, V. D. Zinukhov and M. M. Kremlev, *Zh. Org. Khim.*, **20**, 339 (1984); **18**, 2222 (1982); N. P. Bezverkhii, V. D. Zinukhov, G. V. Tikhonova and M. M. Kremlev, *Zh. Org. Khim.*, **19**, 1559 (1983); E. A. Titov, A. P. Avdeenko and V. F. Rudchenko, *Zh. Org. Khim.*, **8**, 2546 (1972).
105. N. P. Bezverkhii, V. D. Zinukhov, M. M. Kremlev, A. V. Kachanov and T. N. Litvincheva, *Zh. Org. Khim.*, **20**, 1040 (1984).
106. E. A. Titov, G. A. Podobuev and B. N. Nachul'skii, *Zh. Org. Khim.*, **8**, 2541 (1972).
107. E. A. Titov and G. A. Podobuev, *Vopr. Khim. Khim. Tekhnol.*, **27**, 22 (1972).

108. E. A. Titov and G. A. Podobuev, *Khim. Tekhnol. (Kharkov)*, **24**, 16 (1971).
109. E. A. Titov and A. P. Avdeeko, *Zh. Obshch. Khim.*, **41**, 797 (1971).
110. E. A. Titov and A. P. Avdeeko, *Zh. Obshch. Khim.*, **43**, 1686 (1973).
111. A. P. Avdeenko and V. G. Koshechko, *Zh. Obshch. Khim.*, **44**, 1459 (1974).
112. E. A. Titov, S. I. Burmistrov and T. A. Didyk, *Zh. Org. Khim.*, **1**, 1077 (1965); E. A. Titov, *Zh. Org. Khim.*, **4**, 882 (1968).
113. E. A. Titov and N. K. Sukhina, *Zh. Org. Khim.*, **6**, 1834 (1970).
114. E. A. Titov, S. I. Burmistrov, N. K. Sukhina, E. A. Ivanichenko and E. B. Gorbenko, *Zh. Org. Khim.*, **5**, 2210 (1969); E. A. Titov and N. K. Sukhina, *Khim. Tekhnol. (Kharkov)*, **23**, 34 (1971).
115. J. F. Chantot and A. Dargelos, *C. R. Acad. Sci., Ser. C*, **274**, 2001 (1972).

The Chemistry of Quinonoid Compounds, Vol. II
Edited by S. Patai and Z. Rappoport

CHAPTER **22**

Biochemistry of quinones

HIROYUKI INOUYE
Faculty of Pharmaceutical Sciences, Kyoto University, Sakyo-ku, Kyoto, Japan
ECKHARD LEISTNER
Institut für Pharmazeutische Biologie, Rheinische Friedrich-Wilhelms-Universität Bonn, Bonn, FRG

I. INTRODUCTION

In *The Chemistry of the Quinonoid Compounds*, Part 2 of the present series, Bentley and Campbell presented the monograph 'Biological reactions of quinones'[1]. It was concerned with the development achieved up to 1973 in the research of the biochemistry of quinones including biosynthesis. The present chapter mainly deals with the results obtained in the field of the biosynthesis of quinones since then. Recent studies have clarified more detailed processes for many of the quinones whose basic biosynthetic pathway had been outlined by Bentley and Campbell. There are still many other quinones of different types, which were found before or even after 1973, and their biosynthetic pathways have meanwhile been made clear. Many of these new quinones are antibiotics, such as polyketide derived quinones and ansamycins.

Compounds labelled with stable isotopes such as ^{13}C, ^{2}H and ^{18}O have frequently been used in studies of biosynthesis since the 1970s, because the sensitivity of detection by NMR spectroscopy has markedly improved. One of the groups of natural products which benefited greatly from this technical advance is the group of microbial metabolites including quinones. In particular, the NMR studies of [$^{13}C_2$]acetate enriched metabolites provided the decisive information about the polyketide folding mode which had not been disclosed with [^{14}C]acetate feeding. Spectral analysis of metabolites enriched by doubly labelled precursor (^{13}C and ^{2}H or ^{18}O) also provided a lot of evidence about biosynthetic intermediates as will be seen later.

The number of experiments in which plant cell cultures were used to elucidate a biosynthetic pathway leading to quinones and other natural products of higher plants have dramatically increased. This method enables the observation of biosynthetic processes under defined and reproducible conditions and facilitates the use of stable isotopes because incorporation of labelled compounds can be high when compared to experiments in which intact plants are employed. It is impossible to cover all the results of recent intensive studies on the biosynthesis of quinonoid compounds in a limited number of pages, and so attention was given to some representative compounds of each group. Some quinones are biosynthesized by a simple polyketide-, shikimate-, or isoprenoid pathway, but many others are formed by mixed pathways, as can be observed in other groups of natural products. These quinones formed by mixed pathways were classified by the most characteristic feature of their biosynthetic pathway.

II. BIOSYNTHESIS OF QUINONES

A. Polyketide-derived Quinones

Representatives of this group of quinones are believed to be formed from varying numbers of coenzyme A esters of short chain aliphatic acids (such as acetyl coenzyme A, malonyl coenzyme A, methylmalonyl coenzyme A, butyryl coenzyme A etc.) giving rise to a hypothetical ketide (e.g. heptaketide, octaketide, polyketide, etc.) which folds and aromatizes. The prefix (e.g. hepta-, octa- or poly- etc.) corresponds to the amount of keto functions in the hypothetical ketide and the amount of acids involved in the biosynthesis of the resulting quinone. This type of quinone is found mainly in microorganisms and only a few were detected in plants. Major advances in studies of quinone biosynthesis over the last decade have been made in this group of quinones. Most of these quinones are of the hepta- to decaketide origin.

1. Penta- and hexaketide quinones

Experiments with ^{14}C-labelled acetate suggested that flaviolin (**1**) produced by *Aspergillus niger* and 2,7-dimethoxynaphthazarin (**2**) produced by *Streptomyces* sp. are probably of pentaketide origin although they lack the methyl- or carboxyl group which is a polyketide diagnostic feature[2]. Flaviolin (**1**) is also found to be produced by some other fungi such as *Philaphora lagerbergii*, often occurring together with a trihydroxytetralone, viz. scytalone (**3**). Scytalone (**3**) was shown to incorporate [^{13}C]acetates and especially [$^{13}C_2$]acetate with accompanying randomization of ^{13}C–^{13}C coupling[3–5]. This is the evidence demonstrating the intermediacy of a symmetrical compound such as 1,3,6,8-tetrahydroxynaphthalene (**4**) in the biosynthesis of scytalone (**3**). Feeding [2-^{13}C, 2-$^{2}H_3$]acetate and [$^{2}H_3$]acetate to *P. lagerbergii* revealed that ^{2}H is retained at C(4) and C(5), but not at C(2) and C(7) of **3**[6], while no acetate starter effect was found by NMR studies of **3** enriched by [2-^{13}C]malonate[7]. It seems, therefore, likely that scytalone (**3**), and hence the cometabolite flaviolin (**1**) may not originally be a pentaketide, but may be formed from a hexaketide chain via 2-acetyl-1,3,6,8-tetrahydroxynaphthalene (**5**) and tetrahydroxynaphthalene (**4**) as shown in Figure 1. This pathway could also be valid for the above-mentioned 2,7-dimethoxynaphthazarin (**2**) and mompain (**6**)[8], a metabolite of *Helicobasidium mompa*, although experimental proof is still missing. Studies using [$^{13}C_2$]acetate showed that 6-ethyl-5-hydroxy-2,7-dimethoxy-1,4-naphthoquinone (**7**), a metabolite of *Hendersonula toluroideae*, was formed by the acetate assembly pattern as depicted in Figure 1. It was also shown by further feeding studies with [2-^{13}C, 2-$^{2}H_3$]acetate that only two ^{2}H were retained in the 12-position of this quinone, suggesting that at the ethyl side chain of this quinone sequential reduction, dehydration and reduction would occur after cyclization[9].

2. Heptaketide quinones

This group of quinones is represented by dihydrofusarubin (**8**), javanicin (**9**), norjavanicin (**10**), and fusarubin (**11**), metabolites of *Fusarium solani*. Feeding of ^{13}C-labelled acetates and [2-^{13}C, 2-$^{2}H_3$]acetate to *F. solani* demonstrated that dihydrofusarubin (**8**) was formed by folding of the heptaketide chain as shown in Figure 2[10]. Fusarubin (**11**) and norjavanicin (**10**) were shown to be produced from dihydrofusarubin (**8**)[11]. The terminal carboxyl group is reduced to alcohol during conversion to **8** and **11**, and to the methyl group during conversion to **9**. The terminal alcohol group of **8** has been lost in **10** through a retroaldol reaction. Marticin (**12**), a phytotoxic metabolite of *Fusarium martii*, also belongs to this group. The acetate assembly pattern of **12**[12] is that shown for **11** in Figure 2. Marticin (**12**) appears to be formed by addition of a C_3 unit originating from an

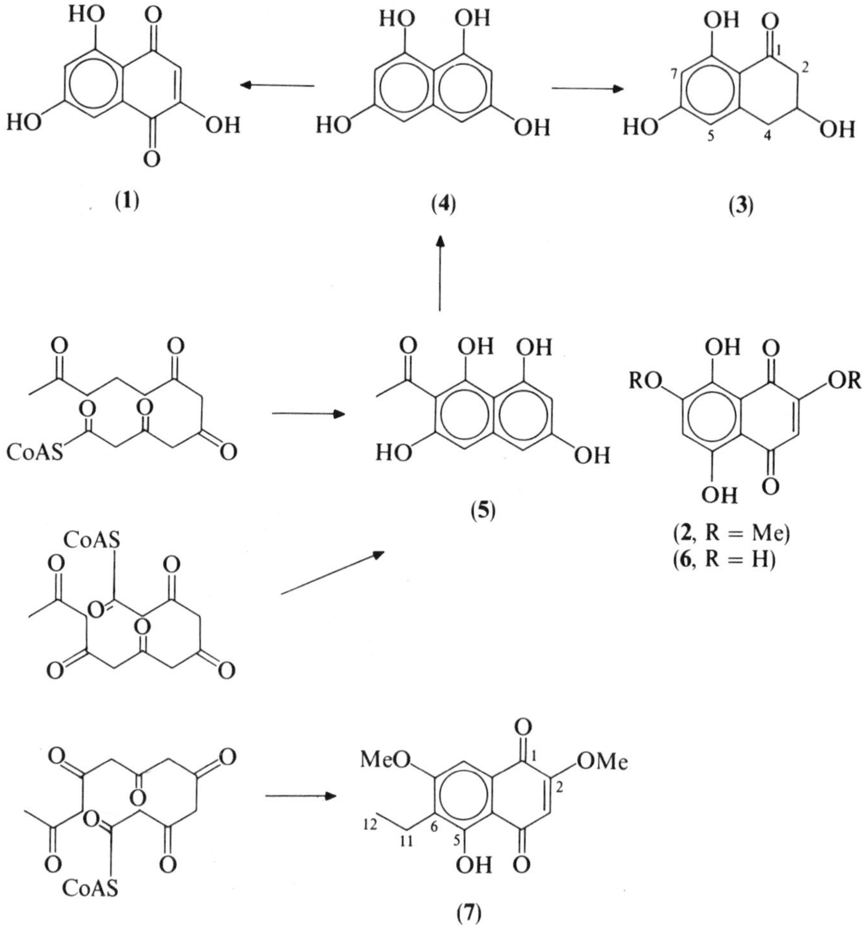

(1) (4) (3)

(5) (2, R = Me) (6, R = H)

(7)

FIGURE 1

intermediate of the Krebs cycle such as succinate or oxaloacetate to the heptaketide skeleton of fusarubin (**11**) or one of its precursors.

Xanthomegnin (**13**) and viomellein (**14**) isolated from *Aspergillus sulphureus* and *A. melleus* are dimers of heptaketide quinones formed by the same acetate assembly pattern[13, 14]. The terminal carboxyl group remains in the form of a lactone in these molecules. Cercosporin (**15**)[15] and elsinochromes C (**16**) and D (**17**)[16], respectively, elaborated by plant pathogenic fungi *Cercospora kikuchii* and *Pyrenochaeta terrestris* are quinones of the same origin as above. These are probably biosynthesized by oxidative coupling of substituted naphthalenes which are formed by cyclization of the heptaketide chain, either before or after decarboxylation.

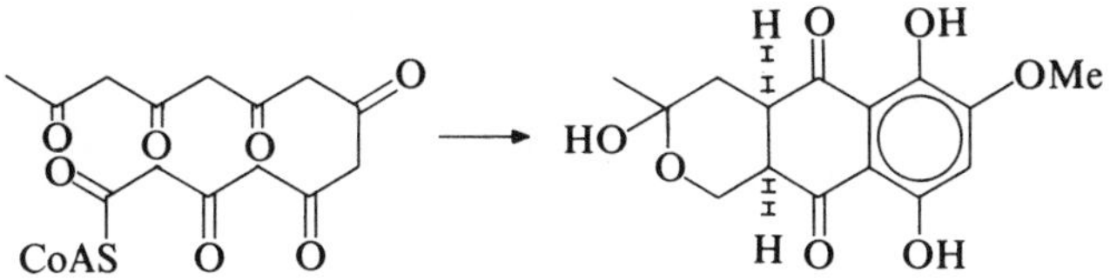

(8)

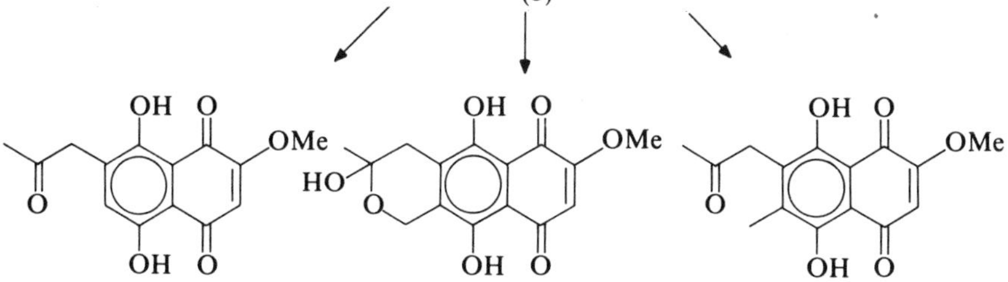

(10) (11) (9)

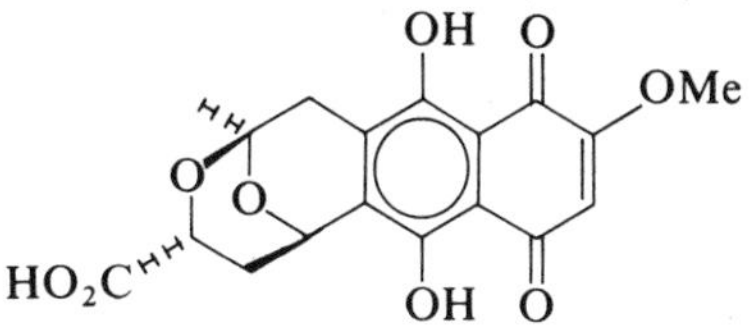

(12) 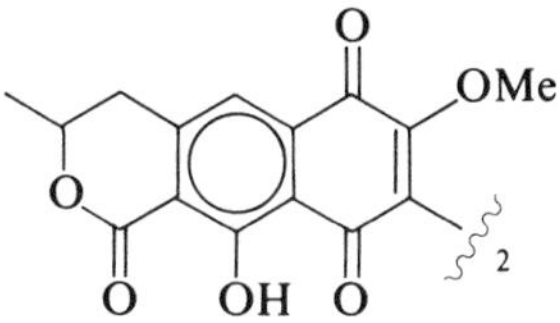 (13)

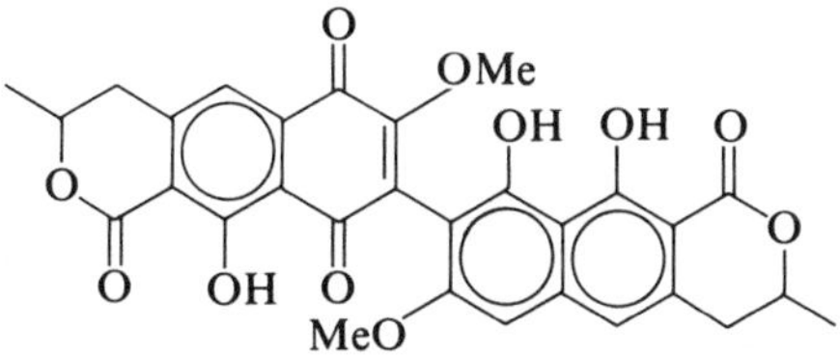

(14)

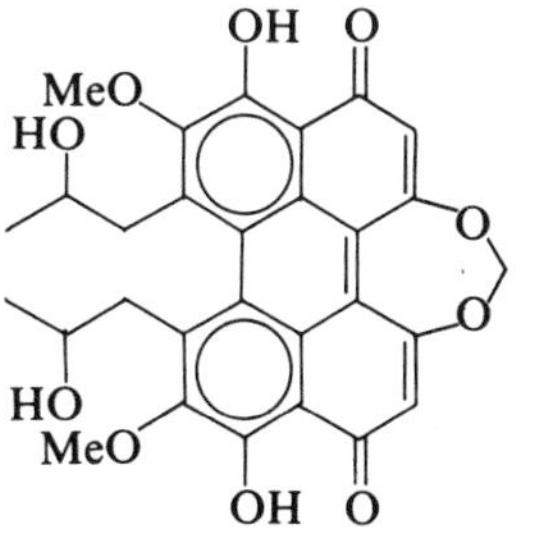

(15)

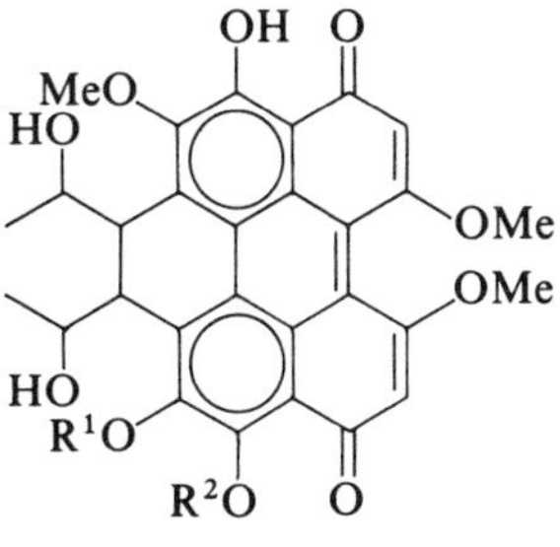

(16, R^1 = Me, R^2 = H)

(17, $R^1 + R^2$ = $>CH_2$

FIGURE 2

3. Octaketide quinones

a. Benzoisochromane quinones Antibiotics, such as nanaomycins A (**18**)–D (**19**) of *Streptomyces rosa*, granaticin of *S. olivaceous*, and *S. violaceoruber*, naphthocyclinones of *S. arenae*, and actinorhodin (**20**) of *S. coelicolor*, belong to this group (Figure 3).

Nanaomycins were shown to be formed by folding of the octaketide chain as depicted in Figure 3[17]. The biosynthetic relationship of the nanaomycins was studied by bioconversion experiments of **18**, **19**, etc., using *S. rosa* grown in the presence of cerulenin, a specific inhibitor of polyketide biosynthesis. Time-course studies revealed that the production of nanaomycin D (**19**), the first component formed from the hypothetical octaketide

SCoA

NADH NAD⁺

(**19**) (**23**)

(**22**) (**21**) (**18**)

(**25**) (**24**)

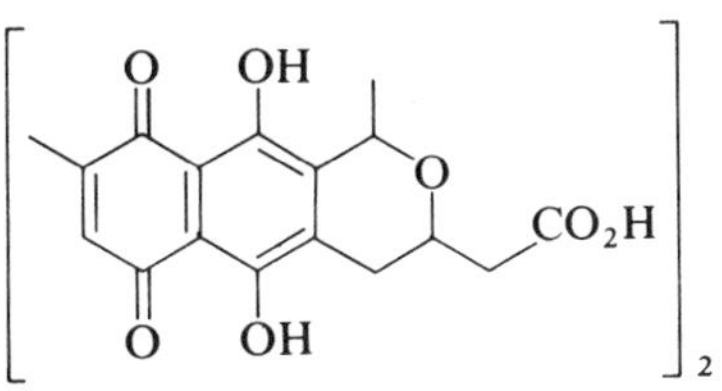

(20)

FIGURE 3

intermediate, was followed by sequential conversion to nanaomycins A (**18**), E (**21**) and B (**22**)[18]. Nanaomycin D reductase was isolated from *S. rosa*. This enzyme is a NADH dehydrogenase, which reduces **19** to a hydroquinone intermediate (**23**) under anaerobic conditions. The resulting hydroquinone intermediate is then non-enzymatically converted to **18**[19].

The ketide folding pattern of granaticin (**24**) is the same as that of nanaomycins[20,21]. Portions not derived from acetate originate from glucose, which was first converted to 2,6-dideoxyhexose, and then bound to the aromatic ring at the 1′,4′-positions of the sugar. During these processes, protons at the 3,5-positions of glucose were lost, while the hydroxy group at the 6-position was substituted by the proton at the 4-position with inversion of configuration. ^{13}C-NMR spectral studies of **24**, which was obtained by feeding [1-^{13}C, $^{18}O_2$]acetate to *S. violaceoruber*, demonstrated that acetate-derived oxygen was retained on C(1), C(3) and C(11) and probably on C(13)[20]. An isotope shift was found on C(3) and not on C(15), indicating the retention of a C–O bond on C(3) during formation of the dihydropyran ring. The cell-free extract of *S. violaceoruber* was found to catalyse the conversion of dihydrogranaticin (**25**) to granaticin (**24**) (Figure 3) without incorporation of ^{18}O from $^{18}O_2$. This suggests that the final step of granaticin (**24**) formation is not hydroxylation at C(4) and lactonization, but the direct cyclization of the carboxyl group onto the 4–5 bond. This reaction is the reverse of the conversion of nanaomycin D (**19**) to A (**18**).

There are seven homologues of naphthocyclinones including α-(**26**), β-(**27**), γ-(**28**), and δ-compounds (**29**) and all these are unsymmetrical dimers of isochromane quinones. Feeding experiments of biosynthetically prepared naphthocyclinones including the monomer (**30**) to *S. arenae* demonstrated the biosynthetic relationships shown in Figure 4[22], which parallels those of nanaomycins D (**19**), A (**18**) and E (**20**) (epoxide). β-Naphthocyclinone epoxide (**31**) corresponding to **19** was further converted to α-naphthocyclinone (**26**) through epoxide ring-opening followed by extrusion of a two-carbon unit.

Actinorhodin (**20**) is a symmetrical dimer of a benzoisochromane quinone[23] (Figure 3). The genes for actinorhodin biosynthesis have been mapped and since they are located in a short segment on the chromosome it was suggested that they form an uninterrupted cluster[24]. Indeed a large continuous segment of *Streptomyces coelicolor* DNA was isolated which contains the complete information required for the synthesis of **20** from simple primary metabolites. This DNA fragment was introduced into a vector which enabled expression of actinorhodin (**20**) biosynthesis after transformation into protoplasts of *Streptomyces parvulus*, an organism which usually does not produce **20**[25].

A series of 76 mutants of *Streptomyces coelicolor* unable to produce **20** were isolated[26]. These mutants were grouped together according to their ability to carry out cosynthesis. Cosynthesis is the ability of two mutant strains placed in close neighbourhood on a single agar plate to carry out antibiotic formation. This joint production of actinorhodin (**20**) is

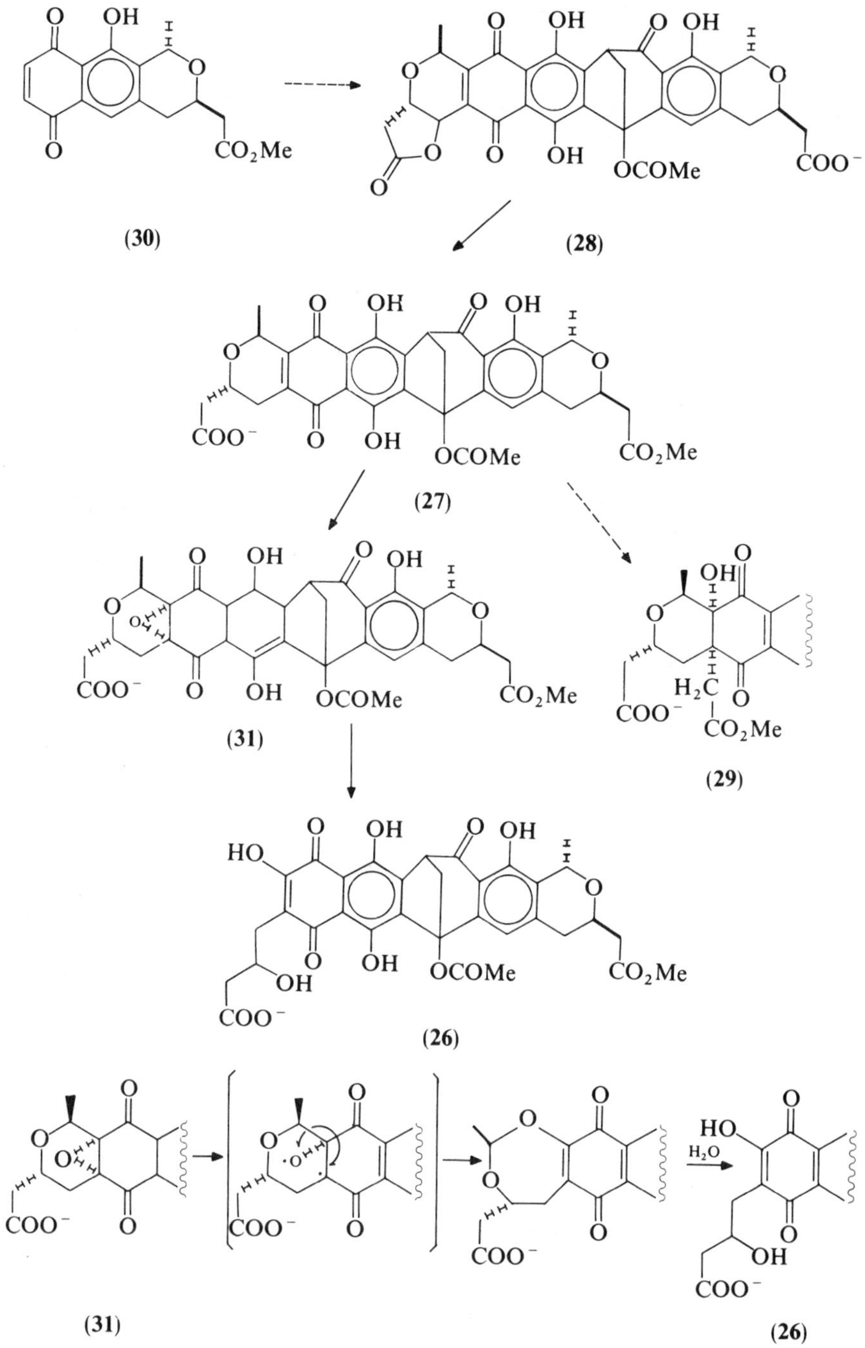

(30) (28)

(27)

(31) (29)

(26)

(31) (26)

FIGURE 4

possible because one mutant excretes an intermediate which is converted by the second mutant to the product (actinorhodin). Isolation and identification of the intermediates made it possible to elucidate late steps in the biosynthetic sequence leading to **20**[26,27].

The regulatory phenomena involved in actinorhodin (**20**) biosynthesis are also highly interesting because they reveal a close relationship between the biosynthesis of **20**, formation of a so-called autoregulator (also called 'A-factor', i.e. 25-isocapryloyl-3S-hydroxymethyl-γ-butyrolactone)[28] and morphological changes of the *Streptomyces* strain such as sporulation and formation of aerial mycelium[29].

Benzoisochromane quinones are found not only in microorganisms, but also in higher plants. Eleutherin (**32**) and isoeleutherin of *Eleutherine bulbosa* and ventiloquinones A (**33**)–K[30] and ventilatones A (**34**) and B[31] which have recently been isolated from *Ventilago* species are included in this group. Besides the acetate assembly mode (a) depicted in Figure 5, mode (b) is also feasible for these plant quinones. But, it is not possible to predict the mode of folding without biosynthetic experiments, because the terminal carboxyl group seems to be lost in these quinones.

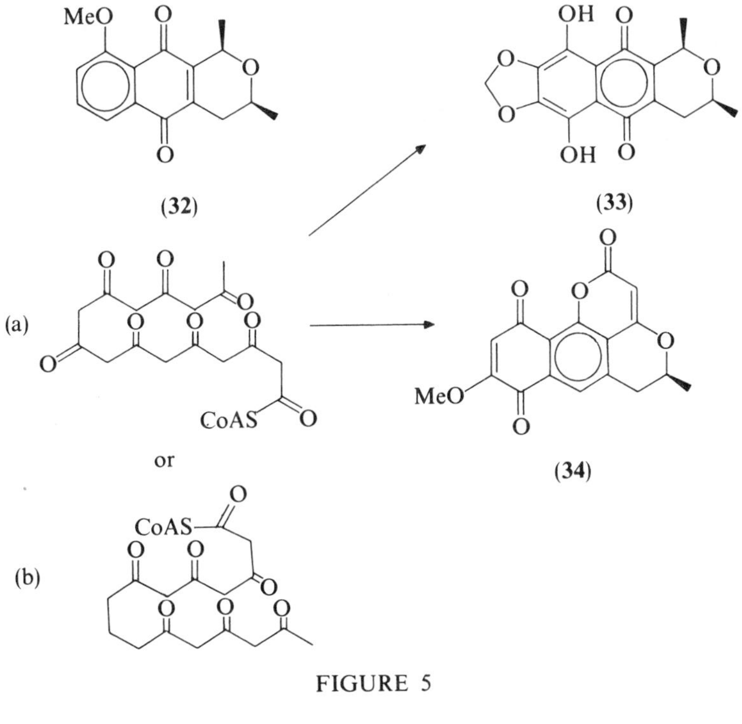

FIGURE 5

b. Anthraquinones It has been known for a long time that some anthraquinones are formed from an octaketide chain. It was believed that emodin (**35**), for instance, is biosynthesized by folding mode (a) shown in Figure 6. This assumption was based on results of [^{14}C]acetate feeding experiments. But, there was no definite evidence indicating that quinones of this type are not formed by folding pattern (b). However, [$^{13}C_2$]acetate feeding experiments demonstrated that islandicin (**36**)[32,33] produced by *Penicillium islandicum*, and altersolanol A (**37**), dactylariol (**38**) and macrosporin (**39**)[34,35] elaborated by *Alternaria solani*, a pathogenic fungus of tomatoes and potatoes, were all formed by route (a). Application of **37** to *A. solani* suggested that this metabolite would be converted to **39** via **38**. NMR studies of **37** and **39** enriched by [^{13}C, $^{2}H_3$]acetate showed that ^{2}H was retained in the C-methyl groups and at the 1-, 6- and 8-positions of both compounds, but did not remain at the 3-position of altersolanol A (**37**). The absence of ^{2}H at the 3-position of **37** seems to be of biosynthetic significance with respect to the introduction of an OH group in this position.

Aloesaponol (**40**), laccaic acid D methyl ester (**41**) and aloesaponarin (**42**) isolated from *Aloe saponaria* are also octaketide anthraquinones, but these compounds are evidently formed by a different folding mode of the ketide chain, which is delineated in Figure 6. The metabolic relationship[36] between these natural products is also depicted.

Interesting results on the biosynthesis of chrysophanol (**43**) (Figure 6) have been obtained recently. Structurally this metabolite differs from emodin (**35**) in that the OH group in position 6 of the anthraquinone skeleton is missing. The assumption that in the biosynthesis of chrysophanol removal of the oxygen function would occur at a non-aromatic stage seemed to be obvious. A cell-free system from *Pyrenochaeta terrestris* has been obtained, however, that dehydroxylates emodin (**35**)[37]. The dehydroxylation is stimulated in the presence of $NADPH_2$. Fe(II), ATP and under anaerobic conditions (N_2). The role of ATP in this reaction is unexplained. To the authors' knowledge this is the first case that enzymatic dehydroxylation of an aromatic compound has been demonstrated.

It is noteworthy that *P. islandicum* converted a ^{14}C-labelled naphthol derivative (**44**) to islandicin (**36**) and its dimer, viz. skyrin (**45**)[38].

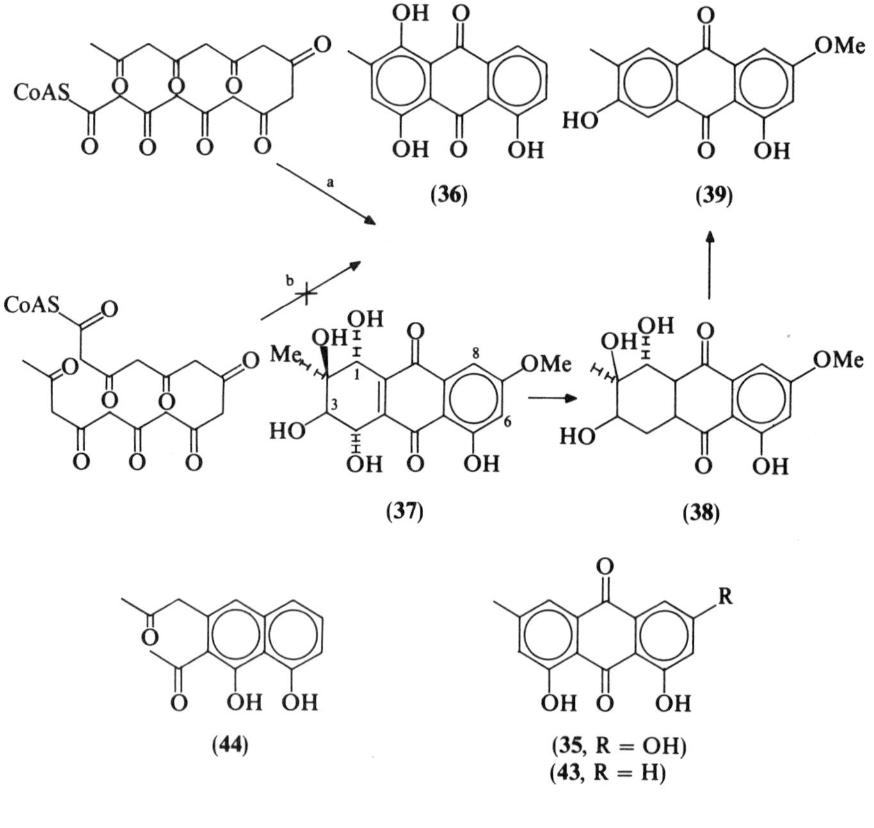

CoAS O O O O O O O O

(**36**)

(**39**)

CoAS O O O O O O O O

(**37**)

(**38**)

(**44**)

(**35**, R = OH)
(**43**, R = H)

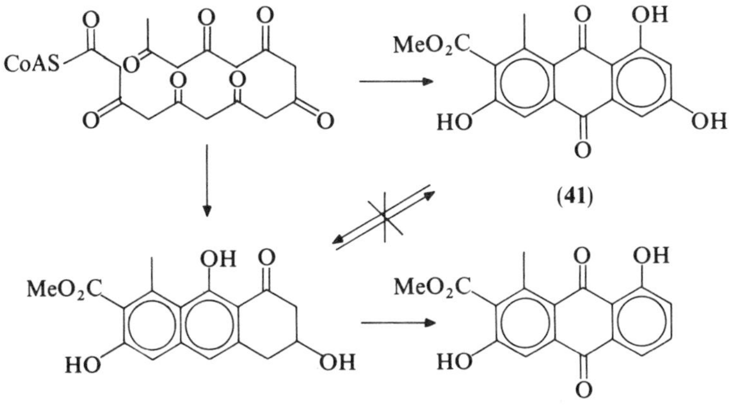

(40)

(41)

(42)

FIGURE 6

c. Dimerization of octaketide anthraquinones *P. islandicum* and *P. brunneum* also produce several anthraquinone dimers, such as skyrin (**45**), iridoskyrin (**46**), rubroskyrin (**47**), flavoskyrin (**48**) and rugulosin (**49**) (Figure 7). These dimers are certainly formed by dimerization of monomeric anthraquinones or monomeric anthrones. However, there is evidence that labelled emodinanthrone (**50**) was incorporated into skyrin (**45**) and iridoskyrin (**46**) in a higher ratio than labelled emodin (**35**)[39]. The formation of flavoskyrin produced by *P. islandicum* is rationally explained by Diels–Alder type 4S + 2S cycloaddition of an enolic form (**51**) of tetrahydroemodin[40]. (−)-Regulosin (**49**) is produced from flavoskyrin (**48**). Diels–Alder type cycloaddition is often found in natural products other than quinones[41,42].

4. Nonaketide quinones

There are only two nonaketide quinones, i.e. bikaverin (**52**)[43], an antiprotozoal metabolite of *Fusarium oxysporum* and phomazarin (**53**)[44], an azaanthraquinone of *Pyrenochaeta terrestris*. Several experiments using labelled acetates, particularly those using [$^{13}C_2$] acetate, demonstrated that these quinones were biosynthesized by the acetate assembly modes shown in Figure 8. But, as regards phomazarin, the assembly mode by two ketide chains such as mode (c) could not be ruled out by acetate feeding experiments only. This quinone enriched by ^{14}C- or ^{13}C-labelled malonate showed only a low level of labelling at C(15), but a normal level at C(11). These results disaffirm the possibility of the two chain assembly. It seems very likely that the ring system which was formed by mode (a) underwent oxidative fission at the stage of biquinone and then a nitrogen function was introduced to form phomazarin (**53**).

5. Decaketide quinones

a. Anthracyclins Anthracyclins are produced by species of the genus *Streptomyces*. Anthracyclins are glycosides with a 7,8,9,10-tetrahydro-5,12-naphthacenequinone skeleton (anthracyclinones). All anthracyclinones are decaketides. In some of them, the starter unit is the acetyl group as usual, while in others it is the propionyl-, isobutyryl-, butyryl or acetoacetyl group[45].

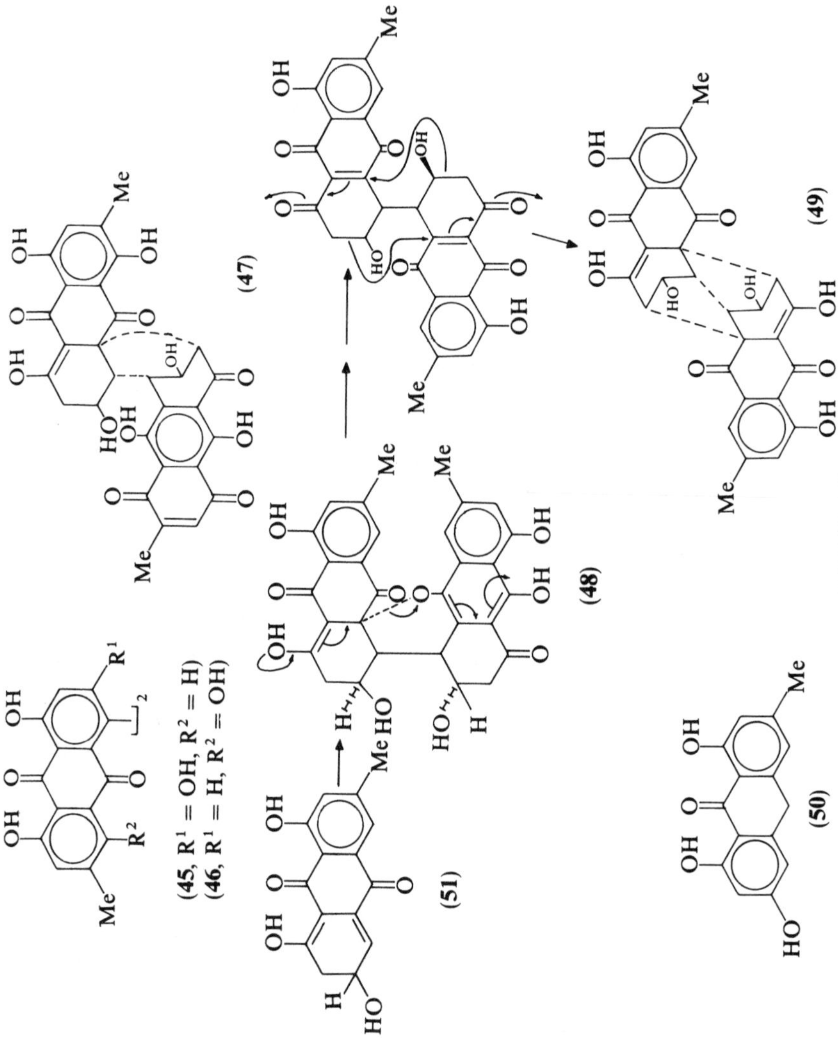

(45, R^1 = OH, R^2 = H)
(46, R^1 = H, R^2 = OH)

(47)

(48)

(49)

(50)

(51)

FIGURE 7

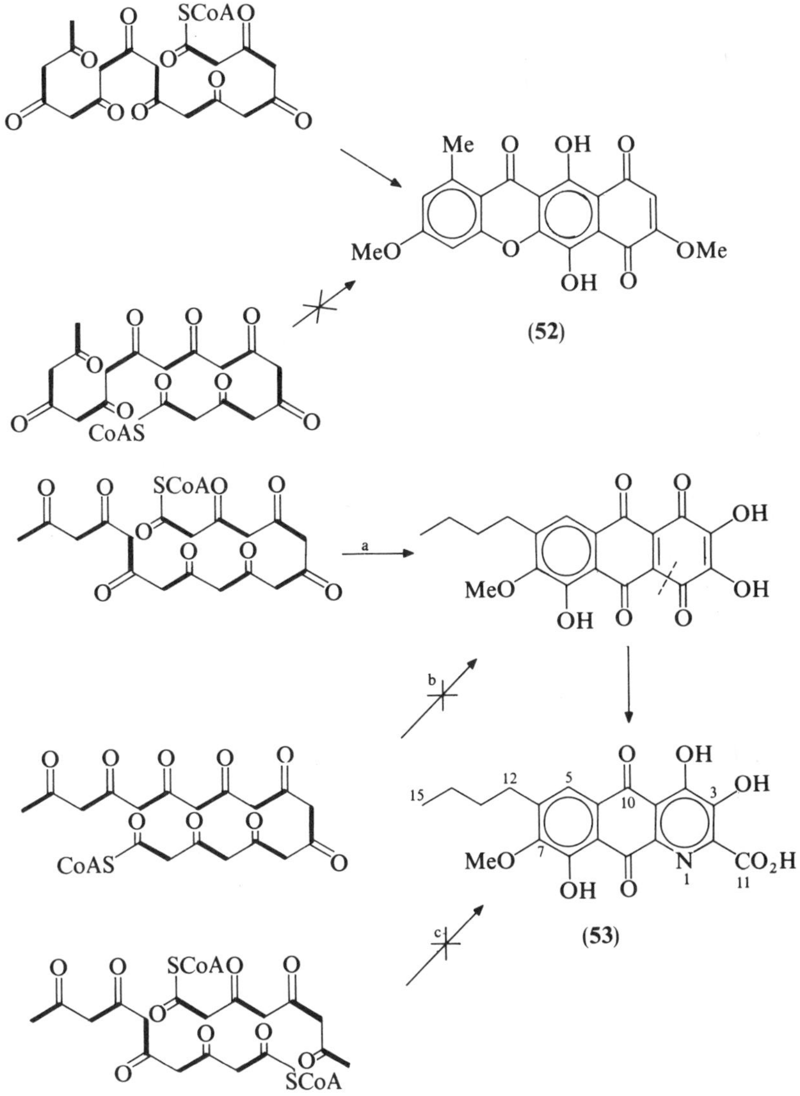

FIGURE 8

The biosynthesis of daunomycinone (**54**), the aglycone of a representative anthracyclin, daunomycin (**55**), was studied by feeding [^{13}C]acetates to *S. peucetius*. The results showed that the acetate assembly pattern of this compound was as shown in Figure 9[46]. It was recently demonstrated that alkanoate (**56**) isolated from *S.* sp. ZIMET or its methyl ester (**57**) was converted to ε-rhodomycinone (**58**), daunomycinone (**54**), or other anthracyclinones when fed to the fermentation broth of mutant strains of *S. griseus*[46,47]. This result suggests the likely intermediacy of alkanoate **56** or its methyl ester (**57**) in the biosynthesis

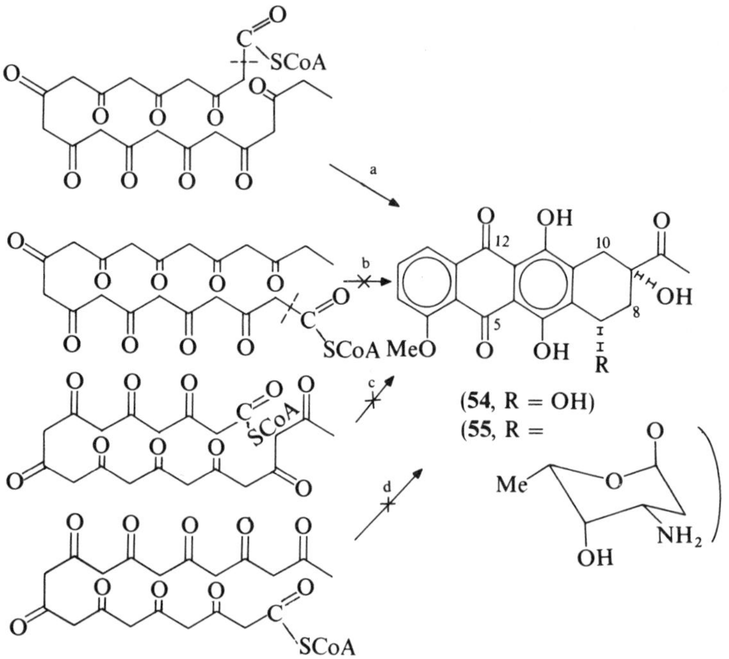

(**54**, R = OH)
(**55**, R =)

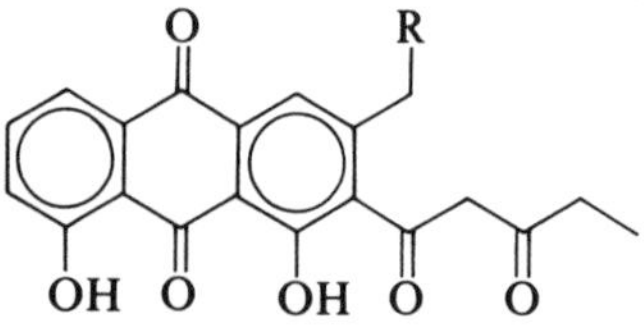

(**56**, R = COO^-)
(**57**, R = CO_2Me)
(**59**, R = H)

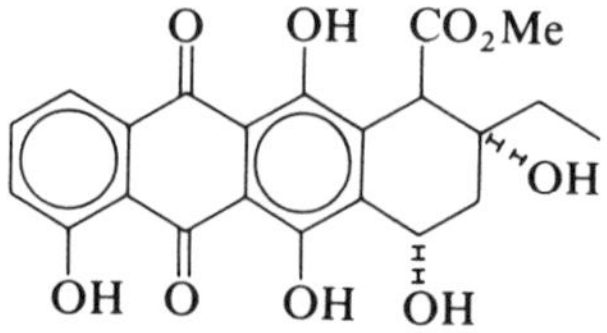

(**58**)

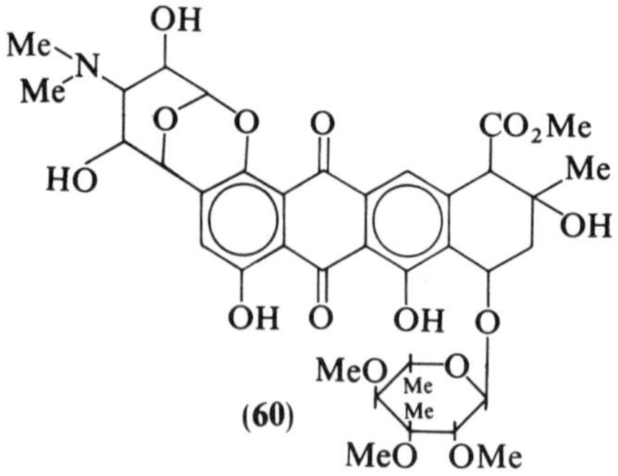

(**60**)

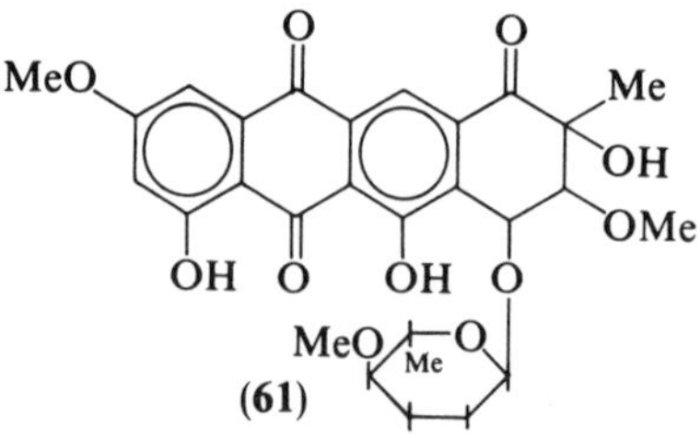

(**61**)

FIGURE 9

of daunomycinone (**54**) or other congeneric products formed from the decaketide chain. In accordance with the decaketide hypothesis alkanone (**59**), a decarboxylation product of alkanoate (**56**), does not undergo this conversion.

Application to *S. nogalata* and *S. elegreteus* of ^{13}C-labelled acetate disclosed decaketide assembly patterns for nogalamycin (**60**) and steffimycin B (**61**) similar to that of daunomycin (**55**). In the former compounds an acetyl group is the starter unit. The sugar moieties of both **60** and **61** are derived from D-glucose, and the methyl groups of MeO and MeN are derived from methionine[48].

The sequence of hydroxylation, methylation, decarboxylation and glycosidation in the anthracyclinone skeleton was also studied using various mutant strains of *Streptomyces* sp.[49–51]

b. Other decaketide quinones Vineomycin A_1 (**62**) and vineomycin B_2 (**63**), antibacterial and antitumour metabolites produced by *Streptomyces matensis*, have a unique benzanthraquinone skeleton. The chromophore portion of these compounds is biosynthesized by the acetate assembly mode shown in Figure 10[52]. Vineomycin B_2 is formed

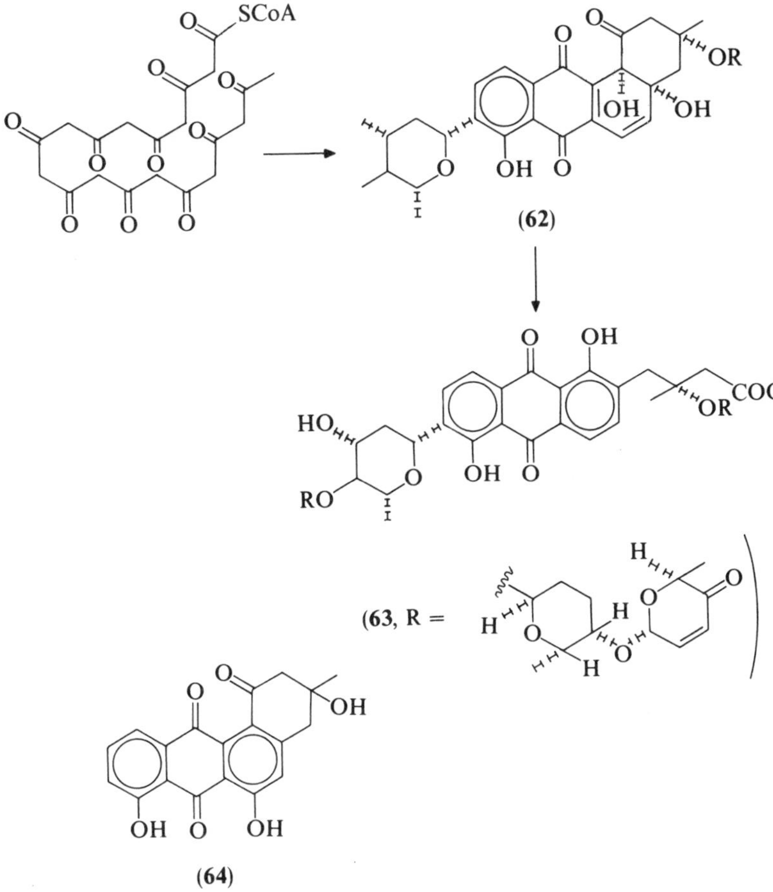

SCoA

OR

OH

OH

OH O

(**62**)

OH

COO⁻

OR

HO

RO

OH O

(**63**, R =

OH

OH O OH

(**64**)

FIGURE 10

by ring fission of vineomycin A_1 at the site shown by a dotted line. Along with these two compounds, a congeneric metabolite, rabelomycin (**64**), is also produced by the same fungus.

6. Other polyketide-derived quinones

a. Mollisin Mollisin (**65**), a naphthoquinone produced by *Mollisia caesia*, has a peculiar structure from the biosynthetic point of view, and seems to be formed from a triketide and a pentaketide chain. Although three modes of acetate assembly, a, b, and c, were presumed, the results of [$^{13}C_2$]acetate feeding experiments were consistent with mode a[53,54]. Another possibility is route d, in which the compound is formed by cleavage of a single octaketide chain[54].

b. Mavioquinone Mavioquinone (**66**) is a benzoquinone with a long alkyl side chain isolated from the lipid extract of *Mycobacterium avium*[55]. It is a dodecaketide biosynthesized from the acetate starter unit, four propionates and seven further acetates. The Me group attached to the oxygen is derived from methionine[56].

c. Cochlioquinones Cochlioquinones A (**67**) and B (**68**), metabolites of *Cochliobolus miyabeanus*, consist of a bis-C-methylated hexaketide and a sesquiterpene portion. The studies of the mass spectra of the degradation products of **67** and **68** obtained by incubation of the fungus in an atmosphere of $^{18}O_2$ and $^{16}O_2$ demonstrated that two oxygen atoms in this uncommon 2-(2-hydroxypropyl)tetrahydropyran structure were introduced at separate stages during the biosynthetic processes[57].

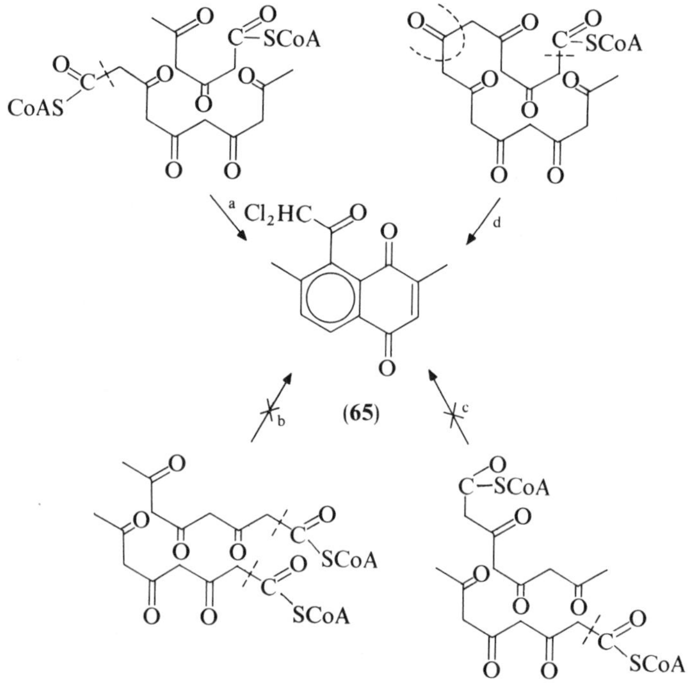

(**65**)

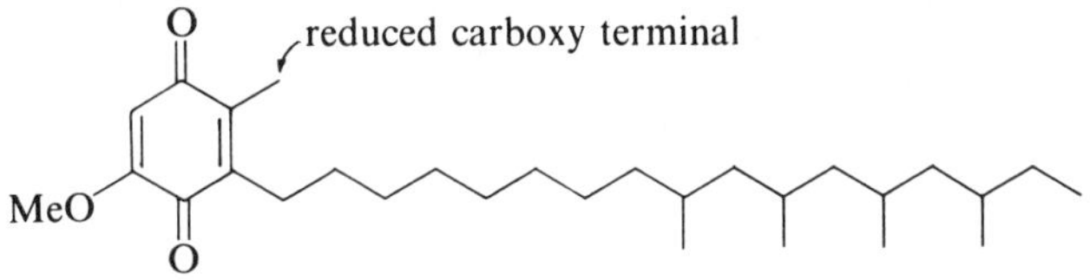

reduced carboxy terminal

(**66**)

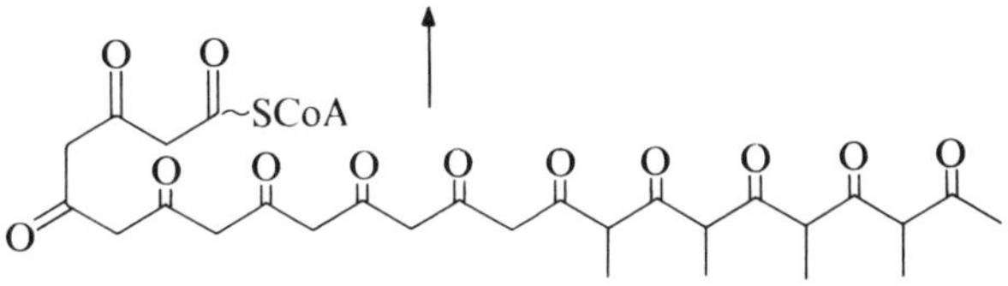

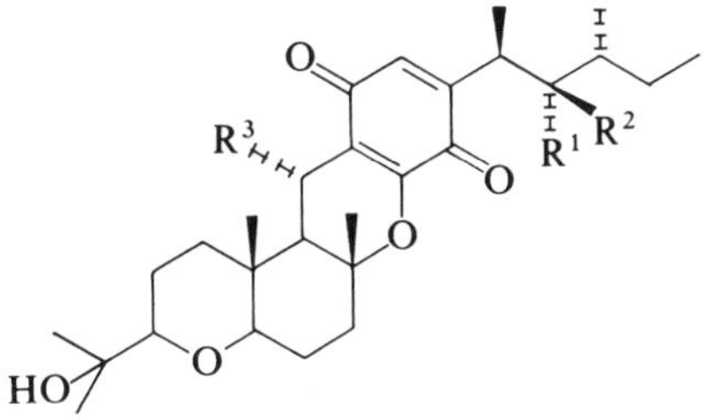

(**67**, R^1 = OAc, R^2 = H, R^3 = OH)
(**68**, R^1 R^2 = O, R^3 = H)

FIGURE 11

B. Shikimate-derived Quinones

There are as many quinones in this group as there are in the polyketide group. However, most of the quinones dealt with in this section are of mixed origin, and are biosynthesized, not only by the shikimate pathway, but also have a mevalonate and/or polyketide moiety.

1. p-Hydroxybenzoate-derived quinones

This group of compounds is represented by ubiquinones which are present in almost all organisms. Several quinones, which are biosynthesized via geranylhydroquinone (**69**) (Figure 12) and are found in boraginaceous plants, are also included in this group.

a. Ubiquinones Ubiquinones (**70**) play a role as lipid-soluble electron carriers in the membrane-bound respiratory chain. The structure of ubiquinones is shown in Figure 12. The number of isoprene units usually varies from 6 to 10. Biosynthetic studies on bacterial ubiquinones have already been outlined by Bentley and Campbell[1]. The present review includes the eucaryotic pathway. It has already been demonstrated in the 1960s that 4-hydroxybenzoate (**71**) formed by the shikimate pathway is the pivotal precursor in the biosynthesis of ubiquinones[58–61]. Up to the early 1970s evidence was obtained that in prokaryotes such as *Rhodospirillum rubrum*[62,63] and *E. coli*[64–68] ubiquinone-10 (or ubiquinone-8) (**70**) is biosynthesized from 4-hydroxy-3-deca-(or octa)prenylbenzoate

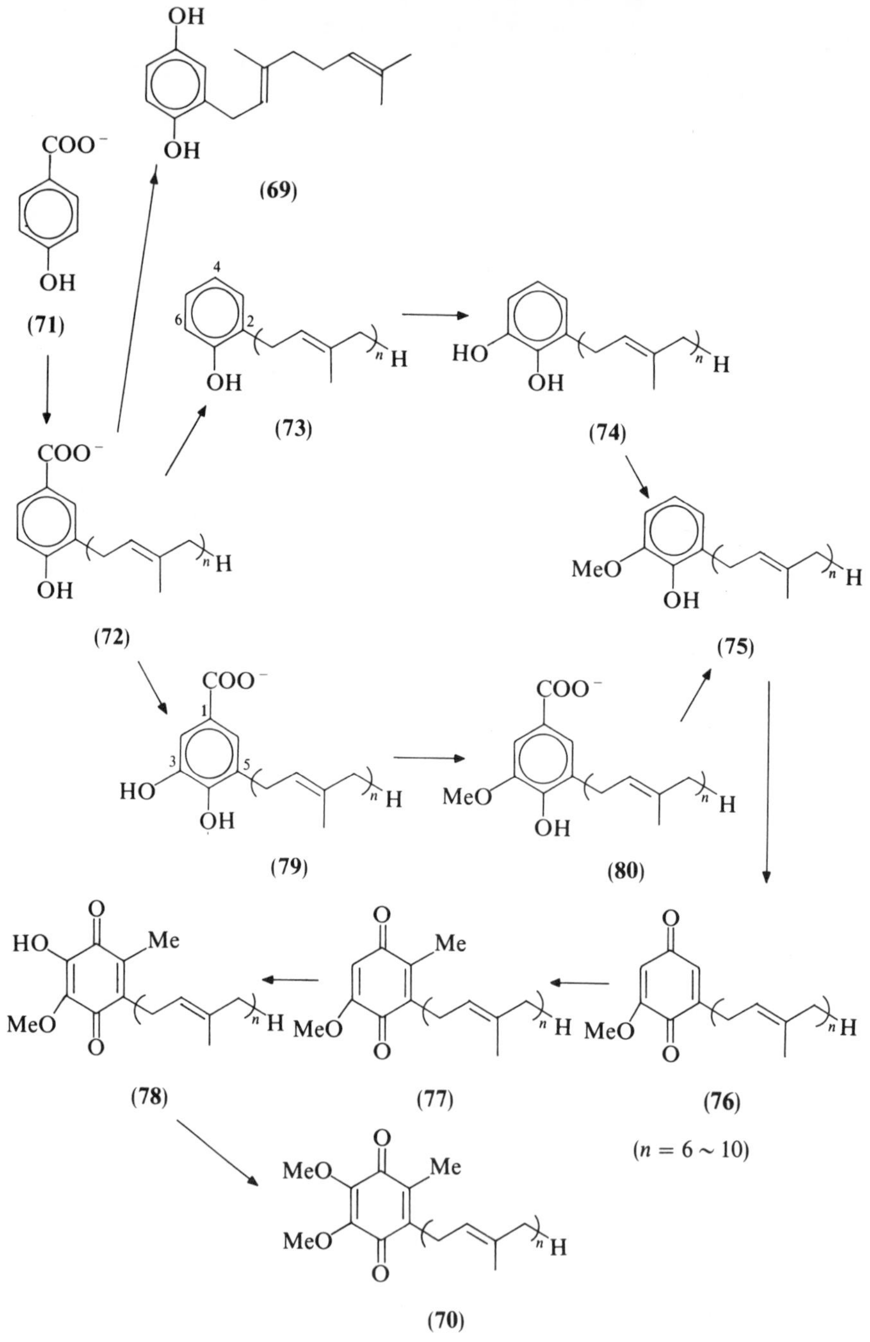

OH
OH
(69)
COO⁻
OH
(71)
4
6
2
OH
(73)
HO
OH
(74)
COO⁻
OH
(72)
MeO
OH
(75)
COO⁻
1
3
5
HO
OH
(79)
COO⁻
MeO
OH
(80)
HO
Me
MeO
(78)
Me
MeO
(77)
MeO
(76)
($n = 6 \sim 10$)
MeO
Me
MeO
(70)

FIGURE 12

(**72**) via 2-prenylphenol (**73**), 3-prenyl catechol (**74**), 6-methoxy-2-prenylphenol (**75**), 6-methoxy-2-prenyl-1,4-benzoquinone (**76**), 6-methoxy-3-methyl-2-prenyl-1,4-benzoquinone (**77**), and 5-hydroxy-6-methoxy-3-methyl-2-prenyl-1,4-benzoquinone (**78**) as shown in Figure 12. As regards the pathway in eukaryotes (**75**, $n=9$), 5-demethoxy-ubiquinone-9 (**77**, $n=9$), and 5-demethyl-ubiquinone-9 (**78**, $n=9$) were shown to be intermediates in ubiquinone-9 biosynthesis in rats[69-71]. The corresponding $n = 6$ compounds are involved in ubiquinone-6 (**70**, $n=6$) biosynthesis in yeast[72]. The later stages of ubiquinone biosynthesis in eukaryotes, therefore, are identical to those in prokaryotes.

It was also shown that nonaprenylphenol (**73**, $n=9$), an intermediate in the biosynthesis of ubiquinone-9 (**70**, $n = 9$) in prokaryotes, cannot be an intermediate in rats[73]. Thus, it was recognized that the biosynthesis of ubiquinones (**70**) in eukaryotes and prokaryotes differs significantly in the middle stage, though both groups of organisms share certain portions of the ubiquinone biosynthetic pathway. A new intermediate, **79** ($n=7$), was recently isolated from the ubiquinone-deficient strain E3-24 of *Saccharomyces cerevisiae*[74]. This new intermediate accumulates in a yeast mutant impaired in methionine biosynthesis. Radioactivity from this product was incorporated into ubiquinone-6 (70, $n=6$) when growth of the auxotrophic yeast mutant was supported by methionine[75]. In the ubiquinone-deficient strain 26*H*,3-methoxy-4-hydroxy-5-hexaprenylbenzoate (**80**, $n = 6$) accumulates. This compound was found to be converted to **70** by mitochondria isolated from wild-type yeast or rat liver[76,77]. All this evidence shows that 3,4-dihydroxy-5-polyprenyl benzoate (**79**) and 3-methoxy-4-hydroxy-5-polyprenylbenzoate (**80**) are intermediates between 4-hydroxy-3-polyprenylbenzoate (**72**) and 6-methoxy-2-polyprenylphenol (**75**) in eukaryotes.

Besides the biosynthetic pathways leading to ubiquinones as stated above, there seem to be alternate pathways in animal cells. For example, when norepinephrine (**81**), which is derivable from tyrosine, was incubated together with rat liver mitochondria, it was metabolized not only to vanillate (**82**) and protocatechuate (**83**), but also to their prenylated products[78]. If rat heart slices are used, 4-hydroxybenzoate (**71**) can be converted to **82** and **83**[79]. Thus in animals there may be variations in the metabolic route to 2-polyprenyl-6-methoxyphenol (**75**) depending on the substrates employed.

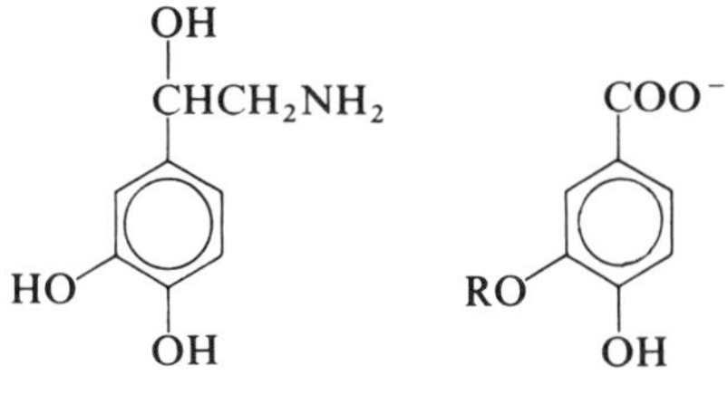

OH
$CHCH_2NH_2$
HO
OH
(**81**)

COO^-
RO
OH
(**82**, R = Me)
(**83**, R = H)

FIGURE 13

b. Geranylhydroquinone-derived quinones in boraginaceous plants It has already been shown in 1971 that the carbon skeleton of alkannin (**84**), a naphthoquinone occurring in some boraginaceous plants, is formed from *p*-hydroxybenzoate (**71**), and two molecules of mevalonate (MVA). In these experiments labelled precursors were fed to *Plagiobothrys arizonicus*[80]. Recent administration experiments of **71**, *m*-geranyl-*p*-hydroxybenzoate (**85**) and geranylhydroquinone (**69**) to shikonin producing tissue cultures of *Lithospermum* *erythrorhizon* demonstrated that shikonin (**86**), the enantiomer of **84**, is formed via **71**, **85** and **69**. Furthermore, in a quinone non-producing strain and in callus tissue which was impaired in pigment production by addition of 2,4-D to the medium or by blue light illumination, it was shown that the biosynthetic pathway was blocked at a step between **85** and **69**. These results were obtained by isotope dilution experiments in which [5-^{3}H]shikimate (**87**) was employed[81]. It is possible that blue light exerts its effect on pigment production via FMN[82].

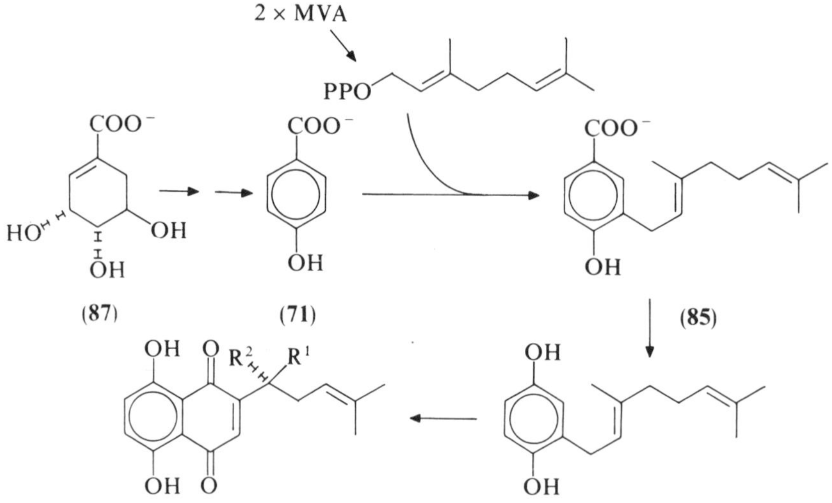

2 × MVA

PPO

(**87**) (**71**) (**85**)

(**84**, R^1 = OH, R^2 = H)
(**86**, R^1 = H, R^2 = OH)

(**69**)

FIGURE 14

It is evident from the above results that **69** is the pivotal intermediate in the biosynthesis of shikonin (**86**) and alkannin (**84**). **69** also seems to be a precursor of many phenols including quinones in boraginaceous plants. For example, echinone (**88**) and echinofuran (**89**)[83] produced by tissue cultures of *Echium lycopsis* and cordiachromes A (**90**), B (**91**) and C (**92**)[84] (Figure 15) isolated from *Cordia alliodora* are likely to be biogenetically related.

2. *Homogentisate-derived quinones*

a. Plastoquinones and tocopherols Plastoquinones (**93**) and tocopherols (**98**–**101**) belong to this group. They have the general structure shown in Figures 16 and 17. Phytylplastoquinone (**94**) (with a phytyl side chain) is structurally closely related to tocopherols. Bentley and Campbell[1] also outlined the work on the biosynthesis of these quinones. The pathway postulated up to 1973 was verified by recent studies.

The common precursor of the aromatic portion of these compounds is homogentisate (**95**) derived from shikimate (**87**)[85]. Experiments using labelled **95** and lettuce chloroplasts indicated that both plastoquinone-9 (**93**) and phytylplastoquinone (**94**) are biosynthesized from **95** via 2-demethylprenylplastoquinol (**96**) and prenylplastoquinol (**97**)

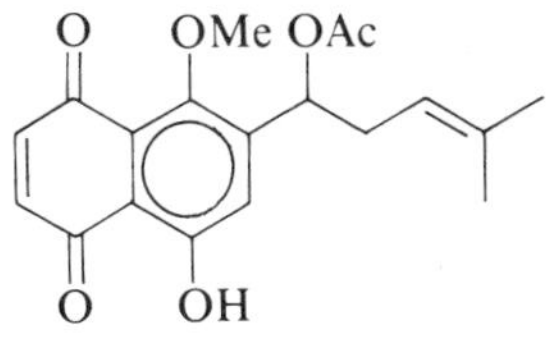

(88) 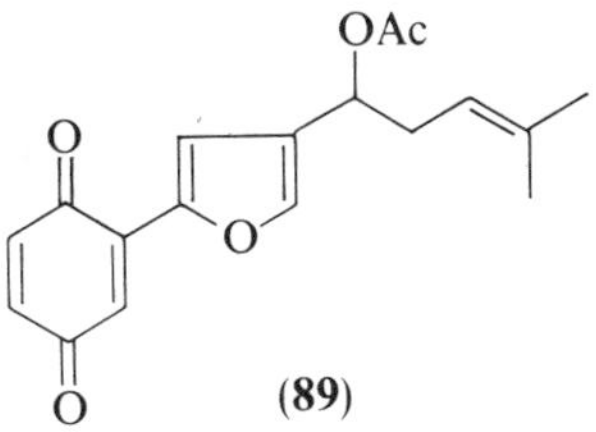 (89)

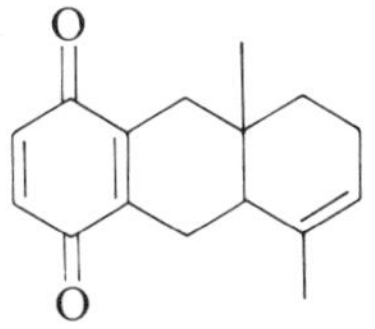

(90) 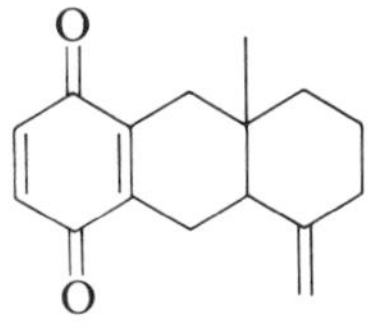 (91) 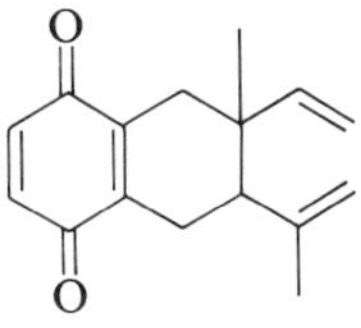 (92)

FIGURE 15

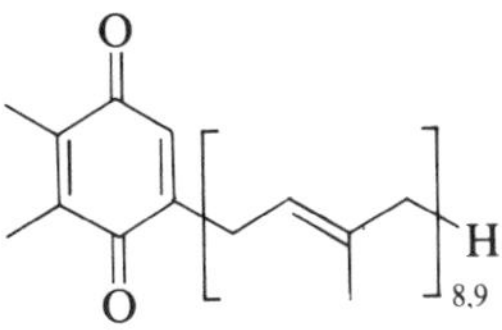

(93) 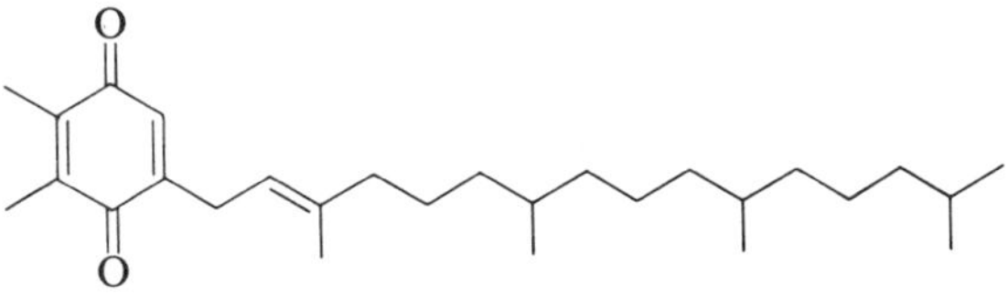 (94)

FIGURE 16

(prenyl = nonaprenyl or phytyl)[86] (Figure 17). While **95** is known as the precursor of α-(**98**), β-(**99**), γ-(**100**) and δ-tocopherols (**101**), recent experiments showed that **95** was also converted to α-(**98**) and δ-tocopherols (**101**) by lettuce chloroplasts[87]. The methyl group in the above series of compounds, which is not derived from *S*-adenosylmethionine, originates from **95** by decarboxylation.

The stereochemistry of this decarboxylation reaction was investigated after feeding homogentisate (**95**), chirally labelled in the methylene group, to *Raphanus sativus* seedling[88]. The chirally labelled homogentisate (**95**) was prepared as follows. Chemical and enzymatic exchange reactions were carried out with 4′-hydroxyphenyl pyruvate in the presence of base or 4′-hydroxyphenylpyruvate tautomerase, HTO or D_2O. The enantiotopically labelled products were converted to correspondingly labelled homogentisate (**95**) samples using 4′-hydroxyphenylpyruvate dioxygenase. After application to *Raphanus* of the homogentisate (**95**) so obtained, tocopherol and plastoquinone (**93**) were isolated and submitted to a mild Kuhn–Roth oxidation. The chirally labelled acetate samples carrying the desired methyl groups of the quinones were enzymically analyzed with known procedures[89]. The chirality of the acetate samples indicated stereochemical retention of the decarboxylation of homogentisate (**95**) during tocopherol and plastoquinone biosynthesis.

It is assumed that a decarboxylation reaction as outlined in Figure 18 takes place in which a quinone intermediate occurs and that the same group of the enzyme which deprotonates the carboxyl group also serves as a proton donor for an intermediate enolate.

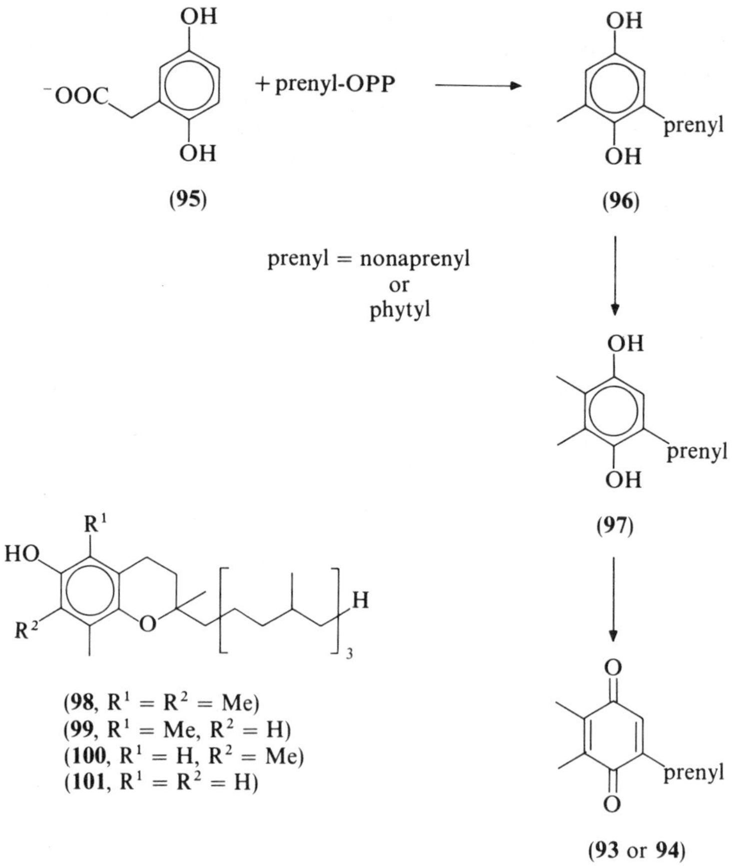

(95) + prenyl-OPP → (96) → (97) → (93 or 94)

prenyl = nonaprenyl or phytyl

(98, $R^1 = R^2 = Me$)
(99, $R^1 = Me, R^2 = H$)
(100, $R^1 = H, R^2 = Me$)
(101, $R^1 = R^2 = H$)

FIGURE 17

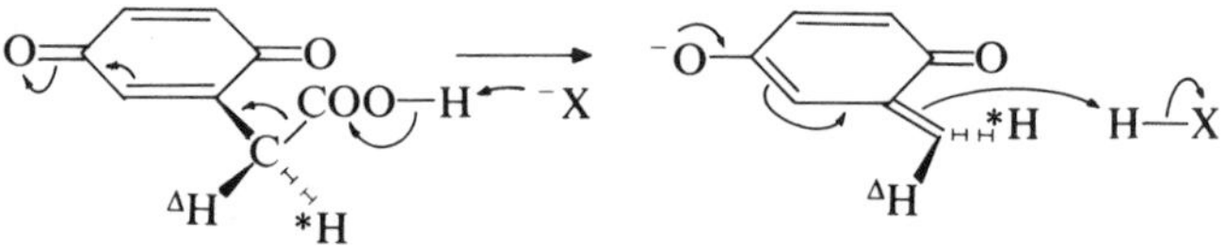

(95)

FIGURE 18

This would ensure removal of CO_2 and introduction of a proton from the same side and consequently stereochemical retention during decarboxylation.

Tocopherols (**98–101**) and plastoquinone (**93**) belong to the so-called lipoquinones. They are localized in chloroplasts and thus are assumed to be implicated in photosynthesis and the functioning of the chloroplasts. The steps of the biosynthesis of lipoquinones have been outlined (*vide supra*). The compartmentalization of the single steps has been extensively investigated by plant physiologists[90]. The distribution of lipoquinones within the chloroplast has also been investigated[91,92]. Thus plastoquinone (**93**) which is associated with the light reaction of photosynthesis is contained mainly in the thylakoids, whereas **98** is contained mainly in the envelope of the chloroplast.

The aromatic precursors of lipoquinones are supplied by the shikimate pathway which is mainly localized in chloroplasts (rather than the cytosol)[93]. Homogentisate (**95**), the precursor of tocopherols and plastoquinone (*vide supra*), is formed in the stroma of the chloroplasts from 4'-hydroxyphenylpyruvate. The enzyme involved is 4'-hydroxyphenylpyruvate dioxygenase. Part of the enzyme, however, seems to be associated with the envelope membrane facing the stroma. The prenyl side chain introduced into homogentisate (**95**) on the way to tocopherols is phytylpyrophosphate. This introduction is catalysed by homogentisate-phytylpyrophosphate prenyltransferase which is localized in the inner membrane of the envelope. Subsequent methylation steps also proceed in this membrane.

Since the inner membrane of the chloroplast is the main site of lipoquinone synthesis a transport of lipoquinones from the inner envelope membrane to the thylakoids must occur. Vesicles may be involved in such a transport.

3. *o-Succinylbenzoate-derived quinones*

o-Succinylbenzoate (OSB, **102**)-derived quinones are widely distributed in higher plants. The naphthoquinones of this group include phylloquinone (vitamin K_1, **103**), which is located in chloroplasts of green plants. Many other quinones such as prenylnaphthoquinones, lawsone (**104**) and juglone (**105**), as well as anthraquinones such as alizarin (**106**), tectoquinone (**107**) (Figure 19) and congeneric quinones contained in Rubiaceae, Bignoniaceae, Verbenaceae, etc. belong to this group of biogenetically related natural products. In microorganisms, menaquinones (vitamin K_2, **108**) are biosynthesized by the OSB pathway.

a. Pathway from shikimate to o-succinylbenzoate Details of the work in the early 1970s, which disclosed the precursorship of shikimate (**87**) and OSB (**102**) for the biosynthesis of some of these quinones, have already been reviewed by Bentley and Campbell[1]. These results are summarized as follows. It was first shown that shikimate (**87**) is the precursor of menaquinone (**108**), lawsone (**104**), juglone (**105**), etc. and that all seven carbons of shikimate (**87**) are involved in the biosynthesis of these quinones. The remaining three carbons in the naphthoquinone skeleton then proved to be derived from glutamate or 2-oxoglutarate (**109**). Therefore, the first aromatic ring was assumed to be formed by Michael-type addition of the succinylsemialdehyde thiamine pyrophosphate complex (**110**) to shikimate (**87**). The key intermediacy of OSB (**102**) was thus postulated and actually confirmed. It was also suggested that OSB (**102**) might be formed by condensation of chorismate (**111**) rather than shikimate (**87**) with **110**. Furthermore it was demonstrated that prenylation in menaquinone (**108**) biosynthesis and hydroxylation in lawsone (**104**) biosynthesis occur on an unsymmetrical intermediate. In contrast, in juglone (**105**) biosynthesis a symmetrical precursor is involved.

Until recently, it was believed that chorismate (**111**) links the shikimate pathway to OSB (**102**) because cell-free preparations of various strains of menaquinone-producing *Escherichia coli* were found to catalyse the conversion of chorismate and 2-oxoglutarate (**109**) to OSB (**102**) in the presence of thiamine pyrophosphate [94,95]. This is at variance with the observation that isochorismate (**112**) is converted to OSB (**102**) in a yield of about 90% by cell-free extracts of *E. coli* strains which are free of 2-oxoglutarate dehydrogenase[96]. (Both OSB synthase and 2-oxoglutarate dehydrogenase decarboxylate oxoglutarate in the presence of thiamine pyrophosphate, generating a carbanion which in

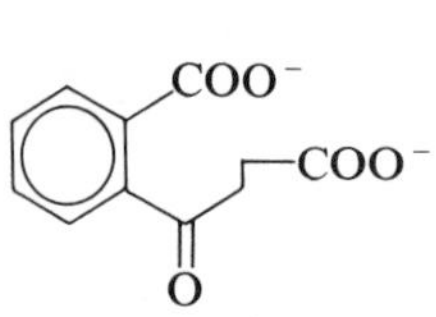

(102) 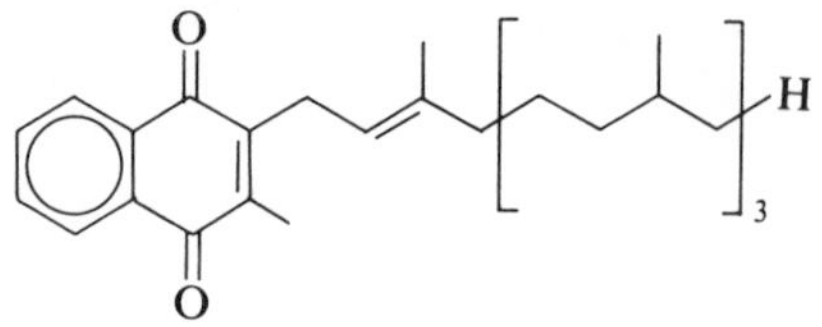 (103)

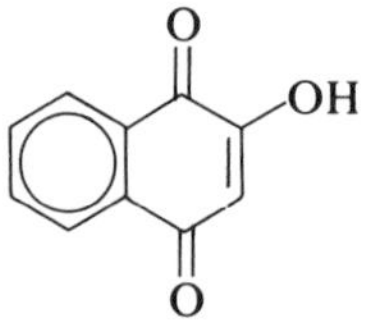

(104) 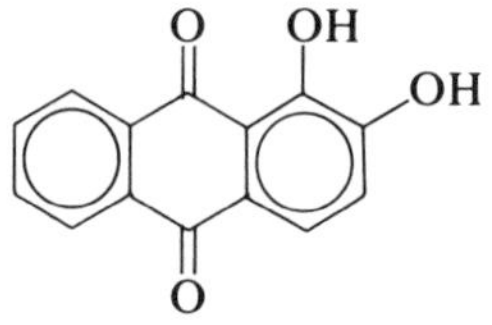 (106)

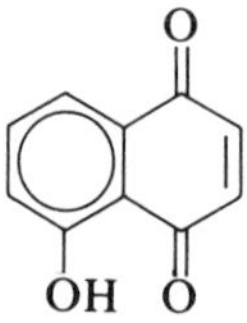

(105) 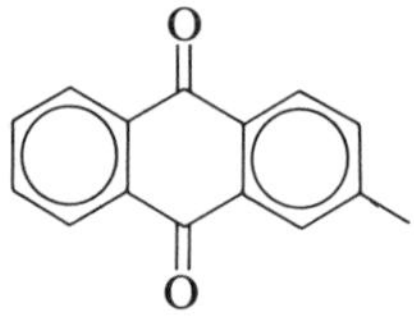 (107)

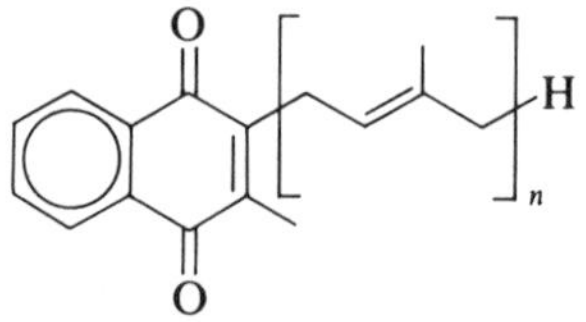

(108)

FIGURE 19

the case of the former enzyme reduces lipoic acid, whereas in the case of the latter attacks isochorismate.) Thus, it is concluded that the immediate precursor of OSB (**102**) is not chorismate, but rather isochorismate (**112**). The precursorship of **112** has also been previously suggested from a mechanistic point of view[97]. The reason for the mistaken belief that **111** was the immediate precursor of **102** could be accounted for by the fact that crude enzyme preparations which are able to convert **111** to **112** were employed. In addition the presence of a trace amount of **112** in the commercial preparations of **111** was overlooked[95].

The finding that **112** is the immediate precursor of OSB is at variance with genetic experiments[98] in which the formation of vitamin K_2 (**108**) was observed in *E. coli* mutants unable to convert **111** to **112**. Meanwhile it became evident, however, that the mutants used (*E. coli* AN 154[94,95] and *E. coli* AN 191[98]) are leaky[99].

An intermediate in the biosynthetic reaction from isochorismate (**112**) to OSB was recently detected[100]. The structure of this intermediate also strongly suggests isochorismate (**112**) as the starting material for OSB (**102**) biosynthesis. The intermediate is 2-succinyl-6-hydroxy-2,4-cyclohexadiene-1-carboxylate (**113**) (Figure 20). The structure of this intermediate shows that in the process of the conversion of **112** to **102** elimination of the pyruvate residue precedes removal of the hydroxyl function and aromatization.

Precisely speaking, the intermediacy of isochorismate (**112**) was proved only in the menaquinone biosynthesis of *E. coli*. However, it probably plays a similar role in the biosynthesis of other OSB-derived quinones.

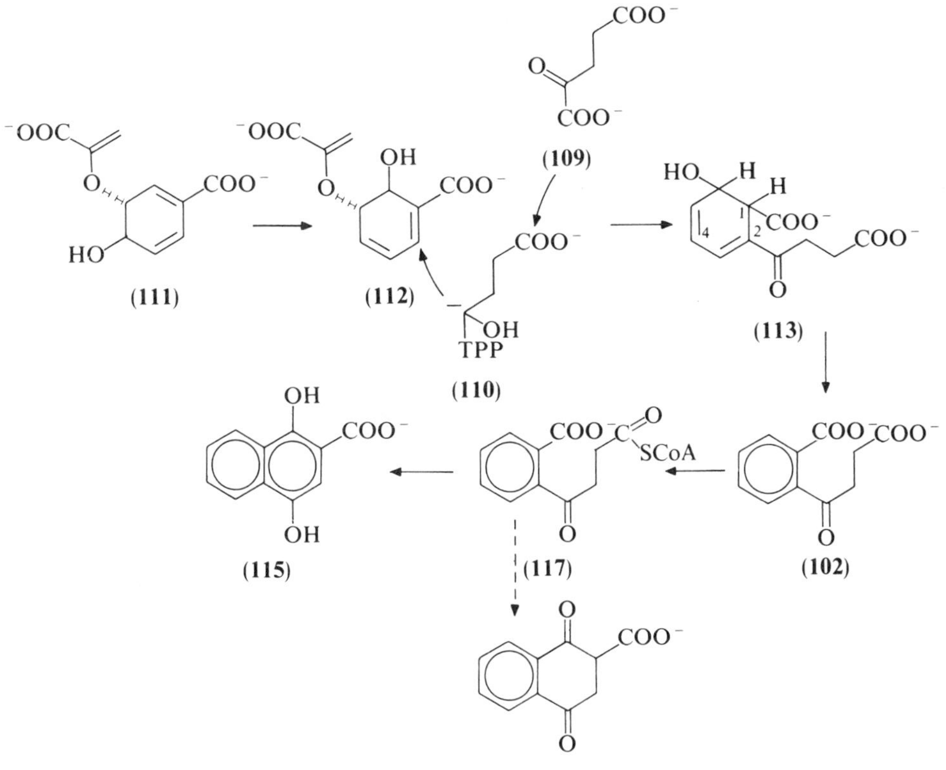

FIGURE 20

b. Formation of the naphthoquinone skeleton from o-succinylbenzoate One of the likely intermediates following OSB is 2-carboxy-4-oxotetralone (COT, **114**) formed by Dieckmann-type condensation and/or 1,4-dihydroxy-2-naphthoate (DHNA, **115**), an enol form of COT. As will be shown later, DHNA (**115**) proved to be a precursor of menaquinones (**108**) and phylloquinone (**103**). DHNA was efficiently and specifically incorporated into **108** in *E. coli* under anaerobic conditions[101]. It was also found that production of menaquinones (**108**) and demethylmenaquinones (**116**) increased by addition of **102** or **115** in some mutants of *E. coli*, while **115** accumulated after feeding **102** to other mutants[98]. Partially purified naphthoate synthase which converts OSB (**102**) to DHNA (**115**) was obtained from cell-free extracts of *E. coli*, *Mycobacterium phlei* and *Micrococcus luteus*. This enzyme system needs ATP, CoA and Mg^{2+} as cofactors, and consists of OSB CoA synthetase which converts OSB to OSB CoA ester (**117**) and DHNA synthase which catalyses cyclization of the OSB CoA ester (**117**) to DHNA (**115**)[102–105]. Conversion of **102** to **115** proceeds through a highly labile OSB monocoenzyme A ester **117**[105] in which the 'aliphatic' rather than 'aromatic' carboxyl group of **102** is activated[106].

In addition to menaquinones (**108**), lawsone (**104**) in *Impatiens balsamina* and juglone (**105**) in *Juglans regia* were shown to be formed via DHNA (**115**) (*vide infra*). However, it is not yet known if the quinones and quinone congeners of *Catalpa ovata* are formed via the OSB CoA ester (**117**).

c. Pathway after formation of 1,4-*dihydroxynaphthoate* (*DHNA*) *or* 2-*carboxy*-4-*oxo tetralone* (*COT*) DHNA (**115**) or COT (**114**) formed from OSB (**102**) undergoes decarboxylation, hydroxylation and prenylation. Naphthoquinones and anthraquinones formed via **102** are similar in structure, but their biosynthetic pathways subsequent to DHNA or COT are different. They will be explained for each group of quinones.

Precise degradation studies of menaquinone isolated after administration of [7-^{14}C]shikimate (**87**) to *Mycobacterium phlei* revealed that the radiolabel was specifically incorporated into C(4)[107] (Figure 21). This means that prenylation occurs on the carbon next to the carbonyl group which was derived from the carboxy group of shikimate (**87**). In the cell-free extract of *E. Coli*, DHNA-octaprenyltransferase, which catalyses the conversion of DHNA (**115**) to demethyl menaquinone (**116**), was detected[108]. It is thus established that menaquinones (**108**) are formed by *ipso* attack of the prenyl group at the C(2) position of DHNA, followed by decarboxylation and methylation.

Structurally phylloquinone (**103**) differs from menaquinone (**108**) in that the polyprenyl side chain is replaced by the phytyl residue. The quinonoid chromophore of both types of compounds, however, is identical. Therefore, the biosynthetic pathway of the chromophore is likely to be identical in both cases. It was demonstrated that OSB was incorporated into phylloquinone (**103**) in maize shoots and that the mode of incorporation was the same as that demonstrated for menaquinones (**108**)[109]. Prenylation of DHNA by phytol was also found to occur in the presence of ATP in spinach chloroplasts[110].

The resulting demethylphylloquinone is eventually methylated at the thylakoid membrane with *S*-adenosylmethionine being the methyl donor. *In vitro* this reaction requires the presence of a fraction of the stroma[111].

As stated above, lawsone (**104**) is formed via an unsymmetrical intermediate, while juglone (**105**) is derived from a symmetrical intermediate. The results of feeding experiments of [1,4-^{14}C]-DHNA (**115**) to *Impatiens balsamina* showed that **104** is biosynthesized through oxidation at C(2) of DHNA (**115**) with accompanying decarboxylation[112]. It was also proved that, in *Juglans regia*, juglone (**105**) is biosynthesized via **115** and its decarboxylation product, naphthohydroquinone (**118**)[112]. It seems likely that hydroxylation of the naphthalene nucleus occurs at the stage of 1,4-naphthoquinone (**119**) or 4-oxotetralone (**120**) (a decarboxylation product of **114**)[113]. These compounds have a symmetrical structure.

Catalpa ovata (Bignoniaceae) contains four α-lapachones (**121**), catalponol (**122**) and catalpalactone (**123**). In addition callus tissue of this plant produces four dehydro-iso-α-lapachones (**124**), menaquinone-1 (**125**) (with one prenyl unit), and 1-hydroxy-2-methylanthraquinone (**126**). When [1-carboxy-^{14}C] OSB (**102**) was administered to the plant, it was incorporated into **121**, **122**, and **123**. In particular, ^{14}C was specifically incorporated into the phthalide alcohol carbon of **123** (Figure 22). Examination of the $^{3}H/^{14}C$ ratio in **122** isolated after administration of [1-carboxy-^{14}C, 2′-$^{3}H_2$] OSB (**102**) to the plant revealed that the two protons at the 2′-position of **102** were both retained in the 3-

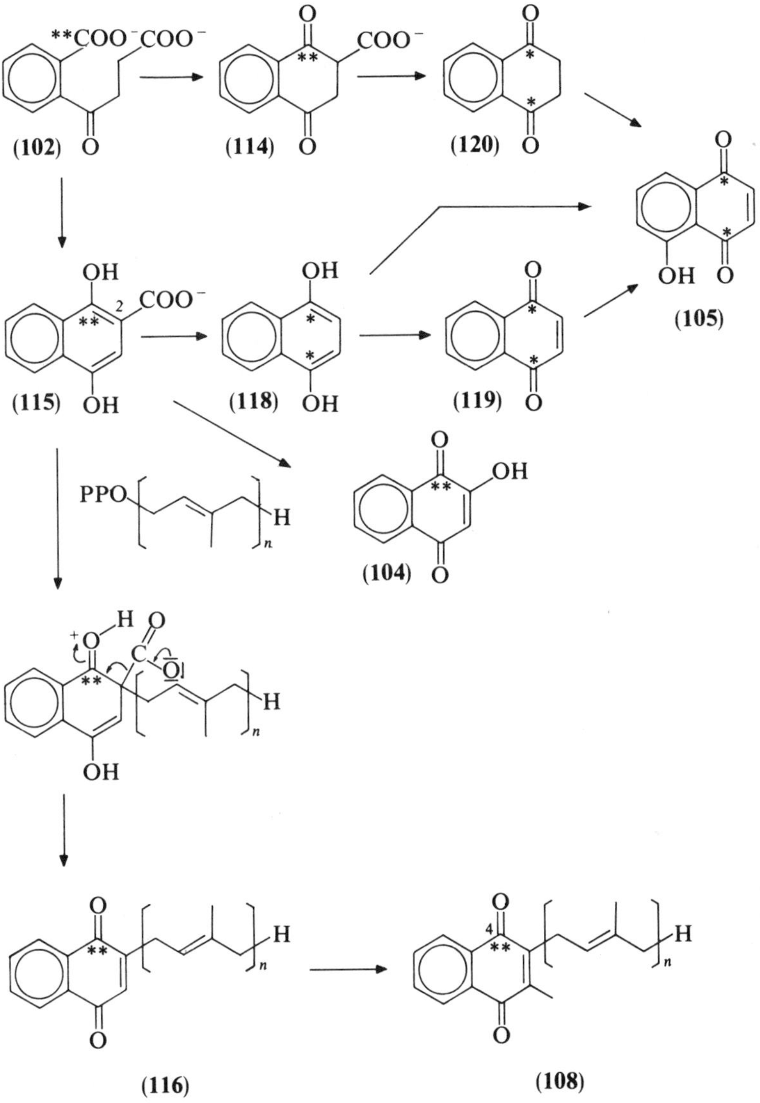

FIGURE 21

position of **122**. It is therefore concluded that, in the biosynthesis of these compounds, prenylation occurs at the same site as in microbial menaquinone (**108**) biosynthesis. Moreover, prenylation does not occur at an aromatic compound such as DHNA (**115**), but at a non-aromatic stage with COT (**114**) being a possible acceptor for the isoprene unit. Considering the structure of catalponol (**122**), it also seems possible that prenylation occurs at the state of 2-carboxy-4-hydroxy-1-tetralone (CHT) (**127**). Catalpalactone (**123**) is biosynthesized by cleavage of the quinone ring of a prenylnaphthoquinone congener which was formed in the above-mentioned way[114–116].

The intermediacy of COT (**114**) and CHT (**127**) was supported by administration experiments in which ^{3}H-labelled **127** and its methyl ester were applied to the *Catalpa* plant. **114** and its methyl ester are likely to be too unstable to be synthetically accessible[117]. Prenyl-COT (**128**) and prenyl-CHT (**129**) were then shown to be the next intermediates[118]. They were trapped after application to the callus tissue of [1-carboxy-^{14}C]OSB (**102**). The cold extract of the plant cells were treated with diazomethane, resulting in the isolation of methylated prenyl-CHT and prenyl-COT. Incorporation of [^{14}C]OSB (**102**) into prenyl-COT (**128**) was higher when compared to prenyl-CHT (**129**). All these results demonstrated that a series of quinone congeners of *C. ovata* are biosynthesized through OSB (**102**), COT (**114**), prenyl-COT (**128**) and catalponone (**130**). The chirality of **128** and **130** was examined by dilution analyses carried out with both enantiomers of both substances. The results showed that (2*S*)-prenyl-COT (**128**) and (2*R*)-catalponone (**130**) are the intermediates. It was also found that **102** was incorporated into dehydro-iso-α-lapachones (**124**), menaquinone-1 (**125**) and 1-hydroxy-2-methylanthraquinone (**126**) in tissue cultures, indicating that these compounds are also formed by the pathway shown in Figure 22. The most important feature of this pathway is that prenylation occurs at the 2-position of COT (**114**) or CHT (**127**).

Anthraquinones are often found to coexist with naphthol derivatives such as 2,2-dimethylnaphthochromane in rubiaceous plants, suggesting that these anthraquinones are biogenetically closely related to prenylnaphthoquinones. This relationship of both groups was actually verified by the observation that shikimate (**87**), OSB (**102**) and mevalonate (MVA) were incorporated into alizarin (**106**) of *Rubia tinctorum*[119] and shikimate (**87**) and OSB (**102**) into morindone (**131**) and alizarin (**106**) of intact plants or cell suspension cultures of *Morinda citrifolia*[120]. Furthermore, specific incorporation of ^{14}C from [7-^{14}C]shikimate (**87**) into the 9-position of anthraquinone provided evidence that prenylation occurs at the position corresponding to C(3′) of OSB (**102**) in the biosynthesis of these anthraquinones[119]. These results coupled with the co-occurrence of mollugin (**132**)[121] and 2-methoxycarbonyl-3-prenyl-1,4-naphthoquinone (**133**)[122] in *Galium mollugo* (this plant belongs also to the family Rubiaceae) led to the assumption that prenylation occurs at the 3-position of DHNA (**115**) in the biosynthesis of these quinones (Figure 23).

This was recently verified by administration of [1-carboxy-^{13}C]OSB (**102**) to cell cultures of *G. mollugo*[123, 124]. Incorporation of ^{13}C into the 9-position of lucidin (**134**) at a rate of about 80% and a significant increase in the formation of a diglucoside (**135**) from 3-prenyl-DHNA (**136**) was observed. ^{13}C was incorporated into **135** at a rate of higher than 90%. Thus, it is very likely that DHNA (**115**) also is an intermediate in the biosynthesis of anthraquinones in rubiaceous plants. Prenylation, however, takes place at the 3-position of **115** during the biosynthesis of these quinones, differing from what is known about menaquinone biosynthesis.

Intact plants and cell cultures of *S. dunnii* (Gesneriaceae) are found to contain several 1,2-naphthoquinones with a reversed prenyl side chain as seen in dunnione (**137**). 1-Hydroxy-2-methylanthraquinone (**126**) and 1-hydroxy-2-(hydroxymethyl) anthraquinone (**138**) were also isolated from this tissue. Administration of [1-carboxy-^{13}C]OSB (**102**) to these cultures revealed that ^{13}C was incorporated into the 1-position of dunnione (**137**) and the 10-position of anthraquinones. Administration of OSB triggered formation of tectoquinone (**107**). These results together with those of feeding experiments in which [7-^{2}H]lawsone (**104**) and its 2-prenyl ether (**139**) were applied suggested that **137** was formed by a Claisen-type rearrangement of lawsone 2-prenyl ether (**139**) whereas anthraquinones were formed by prenylation at the 2-position of COT (**114**) or DHNA (**115**)[125, 126]. Whether **114** or **115** is the intermediate in the biosynthesis of **104** and this type of anthraquinones is still not known. However, data available to date suggest that **114**

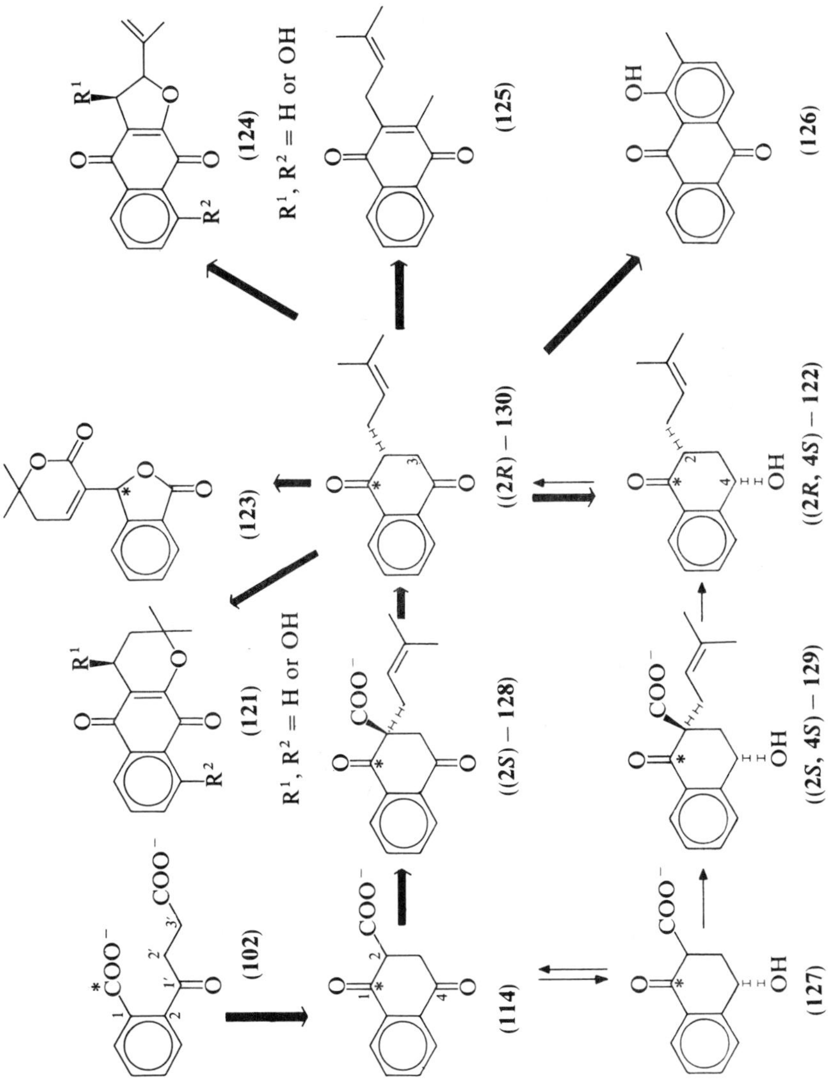

(102)

(114)

(127)

(121)

$R^1, R^2 = H$ or OH

((2*S*) − 128)

((2*S*, 4*S*) − 129)

(123)

((2*R*) − 130)

((2*R*, 4*S*) − 122)

(124)

$R^1, R^2 = H$ or OH

(125)

(126)

FIGURE 22

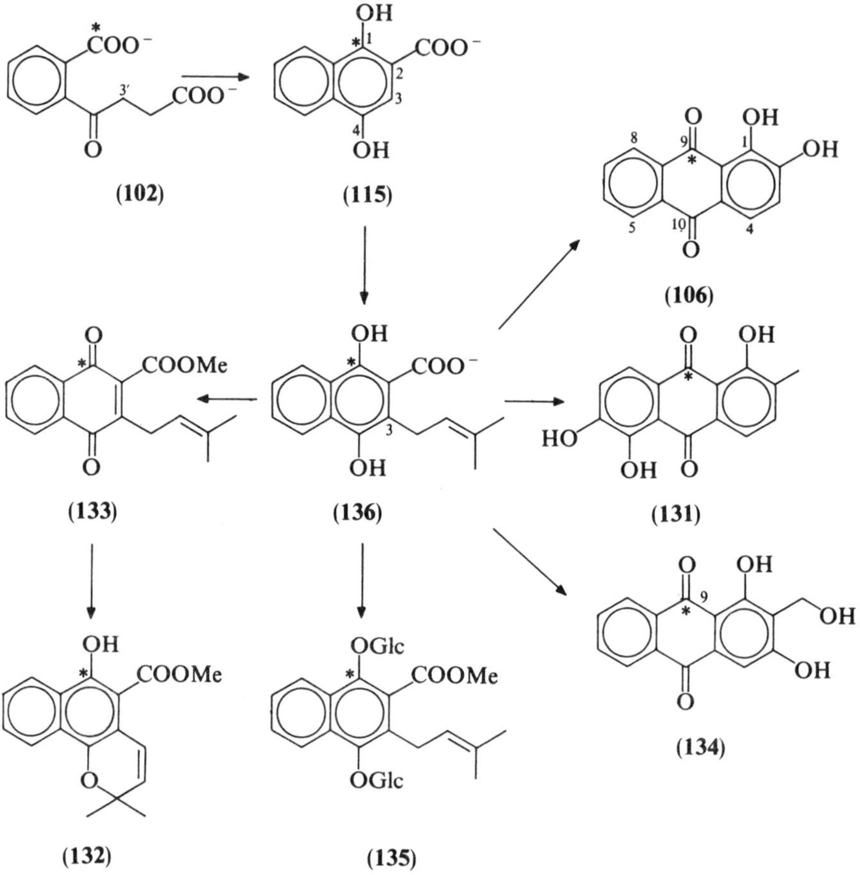

FIGURE 23

is the more likely candidate[127]. On the basis of the above results the biosynthetic pathway of quinones in *S. dunnii* is proposed as shown in Figure 24.

In spite of the fact that quinones are all formed from the same key intermediate, viz OSB (**102**), prenylation is different in the biosynthesis of menaquinones (**108**) produced by microorganisms, menaquinone-1 (**125**) produced by tissue culture of *C. ovata*, anthraquinones in rubiaceous plants or in tissue cultures of *C. ovata* as well as cell cultures of *S. dunnii*.

4. Quinones belonging to C_7N-antibiotics

Ansamycins are antibiotics with a so-called 'C_7N-ring system'. Many ansamycins are quinones. Ansamycins consist of an aromatic portion and a long aliphatic chain (*ansa* bridge) which connects non-adjacent sites of the aromatic portion. The aromatic portion is naphthalenoid or benzenoid. Quinones of the former type include rifamycins, damavaricins, streptovaricins and actamycins, while those of the latter type include geldanamycins,

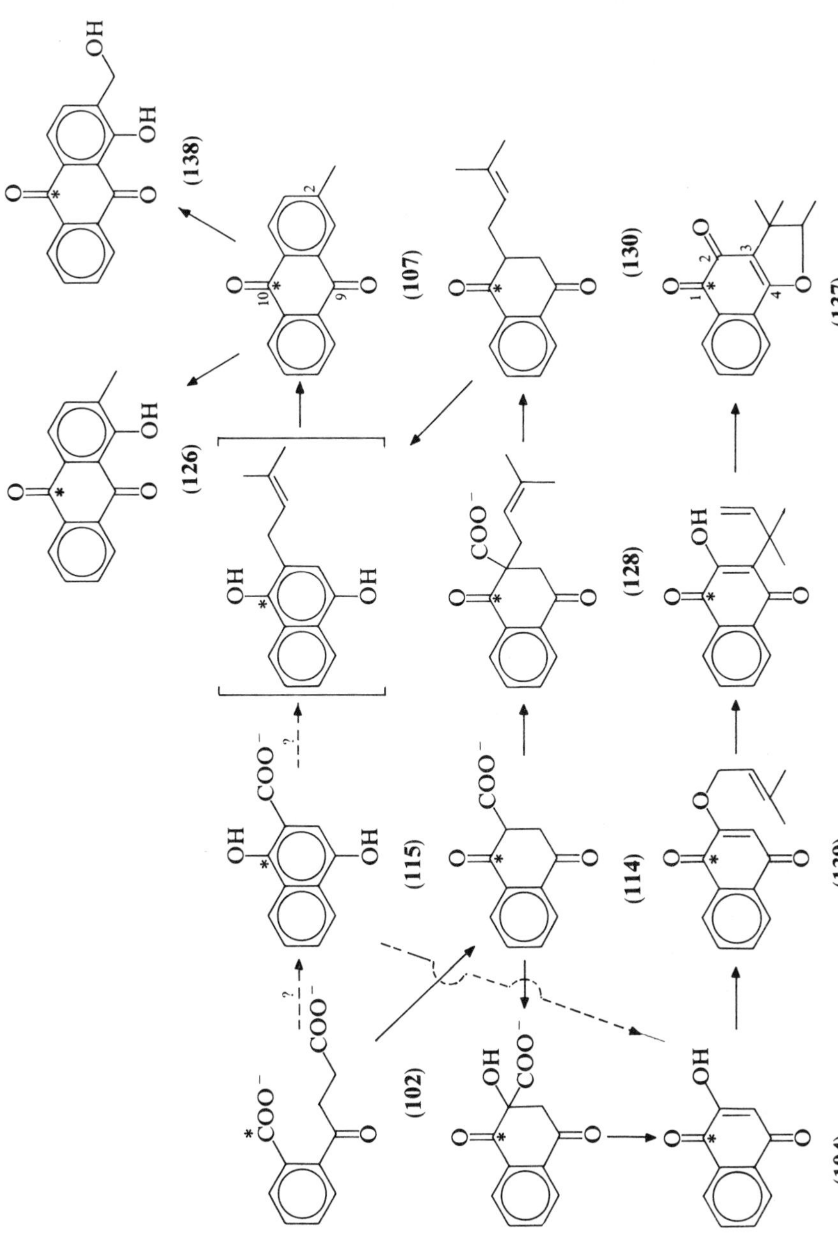

FIGURE 24

macbesin and so on. This section is mainly concerned with the biosynthesis of rifamycins including rifamycin S (**140**). Some related compounds are also mentioned.

Studies on the biosynthesis of rifamycins began with feeding ^{14}C- or ^{13}C-labelled precursors to *Nocardia mediterranei*[128–130]. The results showed that rifamycins are formed from a C_7N starter unit and a polyketide chain derived from acetate, propionate and methionine. There are various homologues of rifamycins. For example, rifamycin W (**141**)[131] produced by a mutant of *N. mediterranei*. Rifamycin W is known as an intermediate in the biosynthesis of rifamycins S (**140**) and B (**142**). C-atom 34a which originates from propionate is retained in rifamycin W (**141**), whereas it is lost in rifamycin S (**140**). Furthermore, C(12) and C(29) are directly linked together in rifamycin W (**141**), while they are separated by the introduction of an oxygen atom in rifamycin S (**140**)[132].

The C_7N-unit was presumed to be derived from an intermediate of the shikimate pathway because D-[1-^{13}C]glucose and D-[1-^{13}C]glycerate were incorporated into rifamycin[129, 133]. However, [U-^{14}C]shikimate (**87**) was not incorporated into rifamycin, and it was later found that shikimate does not permeate the cellular membrane of *N. mediterranei*[134]. Genetic experiments gave a clue as to the branch point of the C_7N-unit from the shikimate pathway. Two aromatic amino acid-deficient mutants of *N. mediterranei* were isolated and characterized. Of these two, the mutant strain A8 which was auxotrophic for aromatic amino acids was free from transketolase activity and produced much less rifamycin than the parent[135]. The other mutant strain A10 was devoid of shikimate kinase activity, but produced rifamycin[136]. 3-Deoxy-D-arabino-heptulosonic acid-7P(DAHP, **143**) cannot be formed without transketolase activity and products subsequent to shikimate-3P are formed without shikimate kinase. Therefore, the precursor of the C_7N-unit was assumed to be a compound of the shikimate pathway derived from metabolites between DAHP (**143**) and shikimate (**87**).

From this information and the structural features of ansamycins and their congeners such as maytansinoids it was assumed that the C_7N-unit would be 3-amino-5-hydroxybenzoate (AHBA) (**144**). This assumption was verified by the fact that AHBA (**144**) was incorporated into actamycin (**145**), an ansamycin produced by *Streptomyces* sp. E/784[137]. About at the same time additional evidence was obtained indicating that the C_7N-unit of rifamycins is actually AHBA (**144**). A number of mutants of *N. mediterranei* were found to accumulate a very early precursor of rifamycins, viz. product P8/1-OG (**146**) which consisted of **144** and the first propionate–acetate–propionate units of the *ansa* chain[138]. The isolation of this compound indicates that the starter for the biosynthesis of rifamycins certainly is 3-amino-5-hydroxybenzoyl CoA. Supplementation with AHBA (**144**) of the culture of the above-mentioned mutant A8 lacking transketolase activity strongly stimulated rifamycin production[139].

Supplementation of the culture of the mutant A8 with several 4-substituted AHBA derivatives demonstrated, however, that they were not able to substitute for the C_7N-unit, suggesting that C(3) substitutents are introduced in a later biosynthetic step leading to 3-substituted rifamycins[140]. The same results were also obtained in similar experiments using actamycin producing *Streptomyces* species[141]. The nitrogen function of **144** is derived from the amide nitrogen of glutamine[142].

The next problem to be solved was the mutual relationship among rifamycins. Protorifamycin I (**147**)[138] isolated from a mutant strain F1/24 of *N. mediterranei* is the earliest so far known precursor in the rifamycin group, while protostreptovaricin I (**148**)[143] isolated from *S. spectabilis* is the earliest occurring in the damavaricin and streptovaricin biosynthesis.

Two common hypothetical precursors, proansamycins A (**149**) and B (**150**), have been proposed for these compounds[144]. From proansamycin B (**150**) protorifamycin I (**147**) would be formed by oxidation at C(34a) while protostreptovaricin I (**148**) by methylation at C(3). From proansamycin A (**149**) protorifamycin I (**147**) and protostreptovaricin I

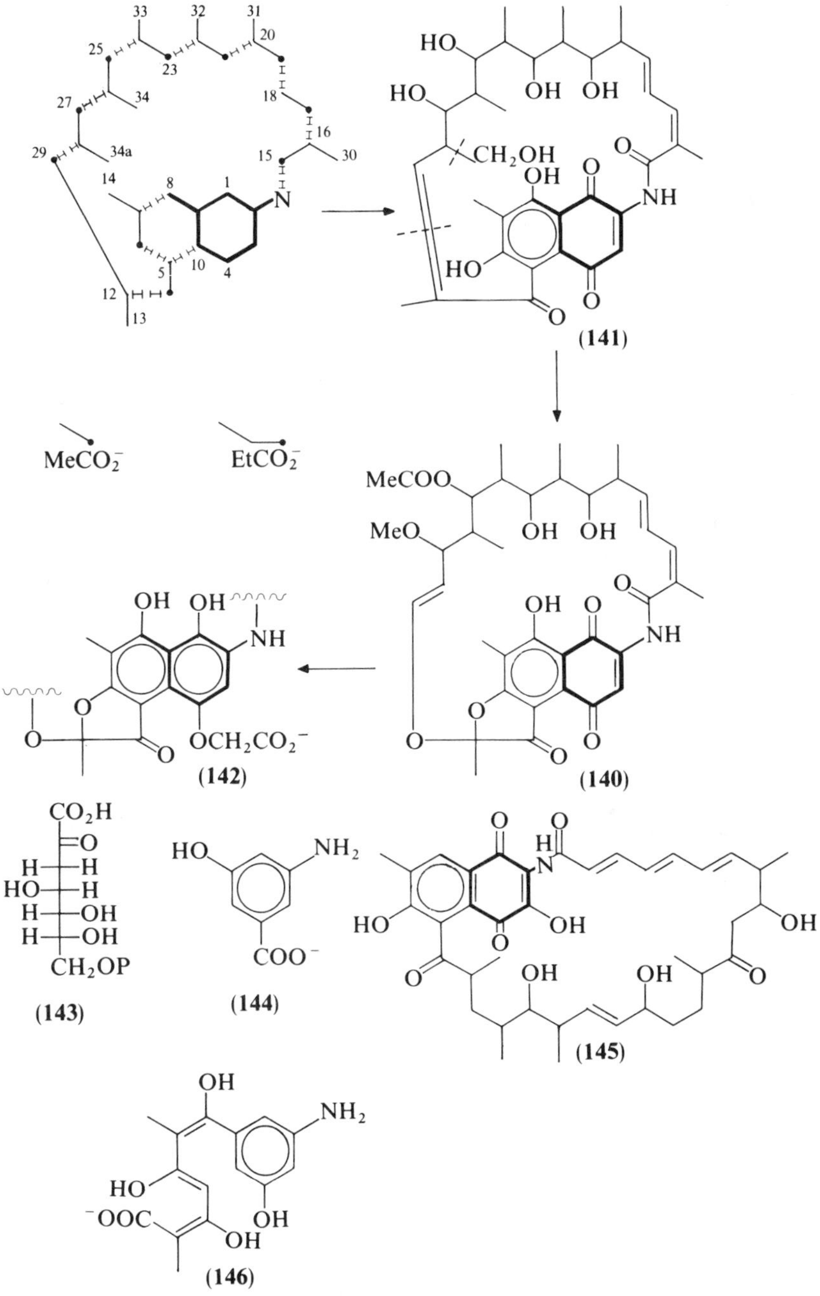

$MeCO_2^-$ $EtCO_2^-$

(141)

(140)

(142)

(143)

(144)

(145)

(146)

FIGURE 25

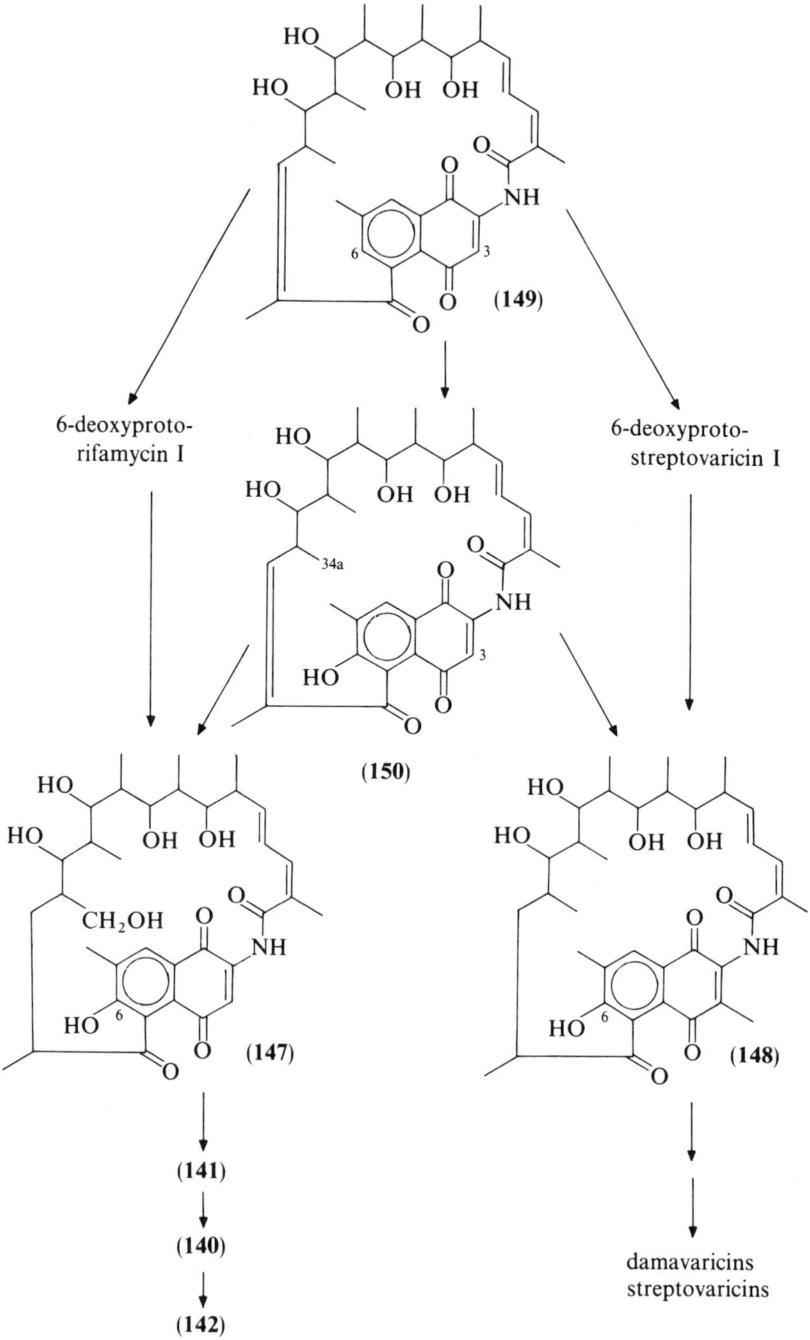

HO HO OH OH O O NH 6 3 O O (**149**)

6-deoxyproto-rifamycin I

6-deoxyproto-streptovaricin I

HO HO OH OH 34a O O NH HO 3 O O (**150**)

HO HO OH OH CH_2OH O O NH HO 6 O O (**147**)

HO HO OH OH O O NH HO 6 O O (**148**)

(**141**)

(**140**)

(**142**)

damavaricins
streptovaricins

FIGURE 26

(**148**) would be formed by two different steps. At present, it is not known which of these hypothetical proansamycins is more likely to be the common precursor for the above ansamycins. **147** gives rifamycin S (**140**) via rifamycin W (**141**), while **148** is converted to streptovaricins and damavaricins. Rifamycin S (**140**) is, on the other hand, the key intermediate for several other rifamycins (A, B, C, D, E and some others)[145].

The ansa chain in benzoquinone-type ansamycins such as geldanamycin (**151**)[146–148] and mycotrienin I (**152**)[149] also proved to be formed from acetate (or glycerate/glycolate in geldanamycin) and propionate units. The incorporation pattern of [^{13}C]glucose in the benzoquinone portion of geldanamycin (**151**) was in accordance with the presumed precursorship of AHBA (**144**)[150]. AHBA is also shown to be involved in the biosynthesis of profiromycin (**153**), a mitomycin antibiotic[151]. Regarding the biosynthesis of mitomycin, it was found that the chain of six carbons from C(3) to C(10) and the attached aziridine nitrogen are derived from D-glucosamine[152–155] and the carbamate function from L-citrulline[154, 155].

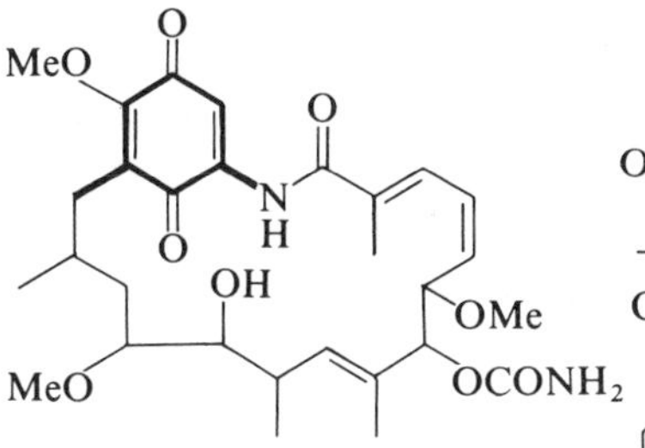

MeO O O N H O OH OMe MeO OCONH$_2$

(**151**)

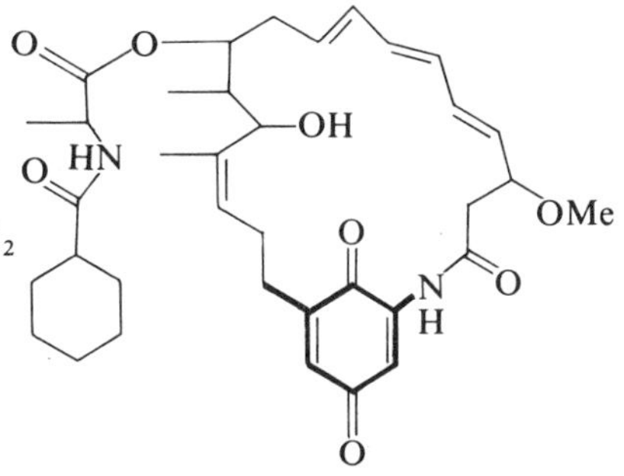

O O HN OH O OMe O N H O O

(**152**)

H$_2$N O ^{10}CH$_2$OCONH$_2$ OMe 1 N NMe 3 O

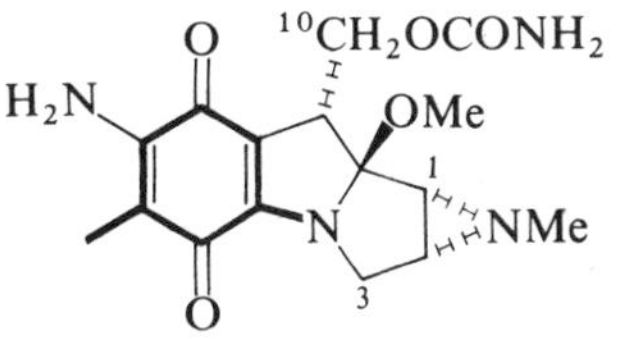

(**153**)

FIGURE 27

5. Others

a. Streptonigrin This quinone is a very potent anticancer antibiotic produced by *Streptomyces flocculus*. The 4-phenylpicolinate portion proved to be derived from tryptophan (**154**)[156]. NMR spectra of streptonigrin (**155**) which was isolated after feeding synthetic [2-^{13}C, ^{15}Nb]tryptophan (**154**) to *S. flocculus* showed a coupling of ^{13}C–^{15}N, suggesting that an unprecedented cleavage occurred in the hetero ring of tryptophan[157]. The tryptophan-derived portion underwent hydroxylation and methylation during the biosynthetic process. In spite of various feeding experiments, the origin of the quinoline quinone portion remained at first unknown. However, NMR studies of [U-^{13}C$_6$]-D-glucose (**156**)-enriched streptonigrin (**155**) revealed that D-glucose was incorporated into the quinone by the mode shown in Figure 28. This fact strongly suggests that the intermediate would be 4-aminoanthranilic acid (**157**) derived from shikimic acid[158]. Glucose was also incorporated into the phenylpicolinic acid portion. It was used as the internal standard in the NMR analysis because the biosynthetic pathway of tryptophan and its mode of incorporation into streptonigrin (**155**) were already known.

4-Aminoanthranilic acid (**157**) also seems to play an intermediary role in the biosynthesis of some antibiotics such as nybomycin (**158**)[159] and lavendamycin (**159**)[160]. The structure of the latter compound is particularly suggestive of the biosynthesis of streptonigrin (**155**). Streptonigrin would be biosynthesized by cleavage of the hetero ring of an intermediate which has a β-carboline structure like **159**.

b. Bis-indolylbenzoquinones Cochlidinol (**160**)[161], isocochlidinol (**161**) and neocochlidinol (**162**)[162] produced by *Chaetomium* spp., asterriquinone (**163**)[163], produced by *Aspergillus terreus*, and hinuliquinone (**164**)[164, 165], elaborated by *Nodulisporium hinuleum*, are bis-indolyl-2,5-dihydroxybenzoquinones prenylated at different positions (Figure 29). The biogenesis of these compounds is explained by the self-condensation of two indolylpyruvic acid (**165**) molecules which are formed by transamination of tryptophan in the same way as in the formation of bis-phenyl-benzoquinones from two phenylpyruvic acid molecules and subsequent prenylation. This interpretation was supported by feeding tryptophan and mevalonic acid to some fungi[161, 163–166].

c. Naphthyridinomycin Naphthyridinomycin (**166**) is an antibiotic which is a metabolite of *Streptomyces lusitanus*. Feeding experiments of labelled precursors demonstrated that its carbon skeleton is derived from the three amino acids tyrosine (**167**), serine (**168**) and ornithine (**169**)[167–169] (Figure 29). Three methyl groups were shown to be derived from methionine and the nitrogen atom from the amino acids mentioned above. However, the origin of C(9) and C(9′) is still unknown.

C. Pure Isoprenoid Quinones

In addition to the quinones described in the preceding section, there are many other quinones with complete isoprenoid carbon skeleton, particularly among the diterpenoids. However, there have been virtually no administration experiments on these quinones in the last decade. The natural products chemistry of this type of quinones has been treated by Eugster[170]. Hibiscoquinone A (**170**) is given in Figure 30 as an example of this type of quinone. The structural relationship between **170** and a related diterpenoid phenol, viz. gossypol (**171**), is obvious. Feeding experiments carried out on roots of *Gossypium herbaceum* showed that mevalonic acid is incorporated into gossypol (**171**) with a folding mechanism as shown in Figure 30[171] and as is known from sesquiterpenes of the cadalane type. The cation of the ten-membered ring system undergoes a 1,3-hydride shift. It is safe to assume that hibiscoquinone A (**170**) is derived in a very similar way.

III. METABOLISM OF QUINONES

A. Naphthoquinones

Today it is very well accepted that natural products are no metabolic end products but subject to a turnover[172]. This turnover is also experienced in lipids[173] of chloroplasts including lipoquinones[174]. A photoautotrophic cell culture of *Morinda lucida* has been described recently in which the degradation of lipoquinones can be induced when the cell

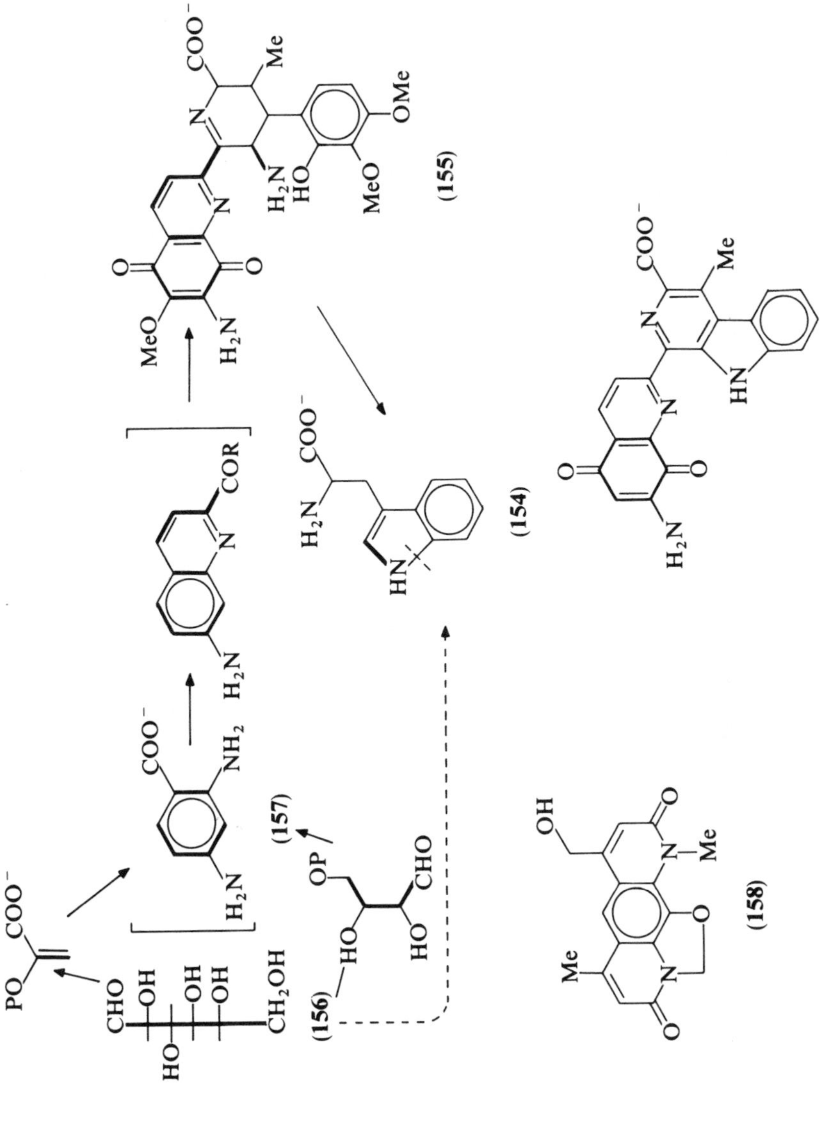

FIGURE 28

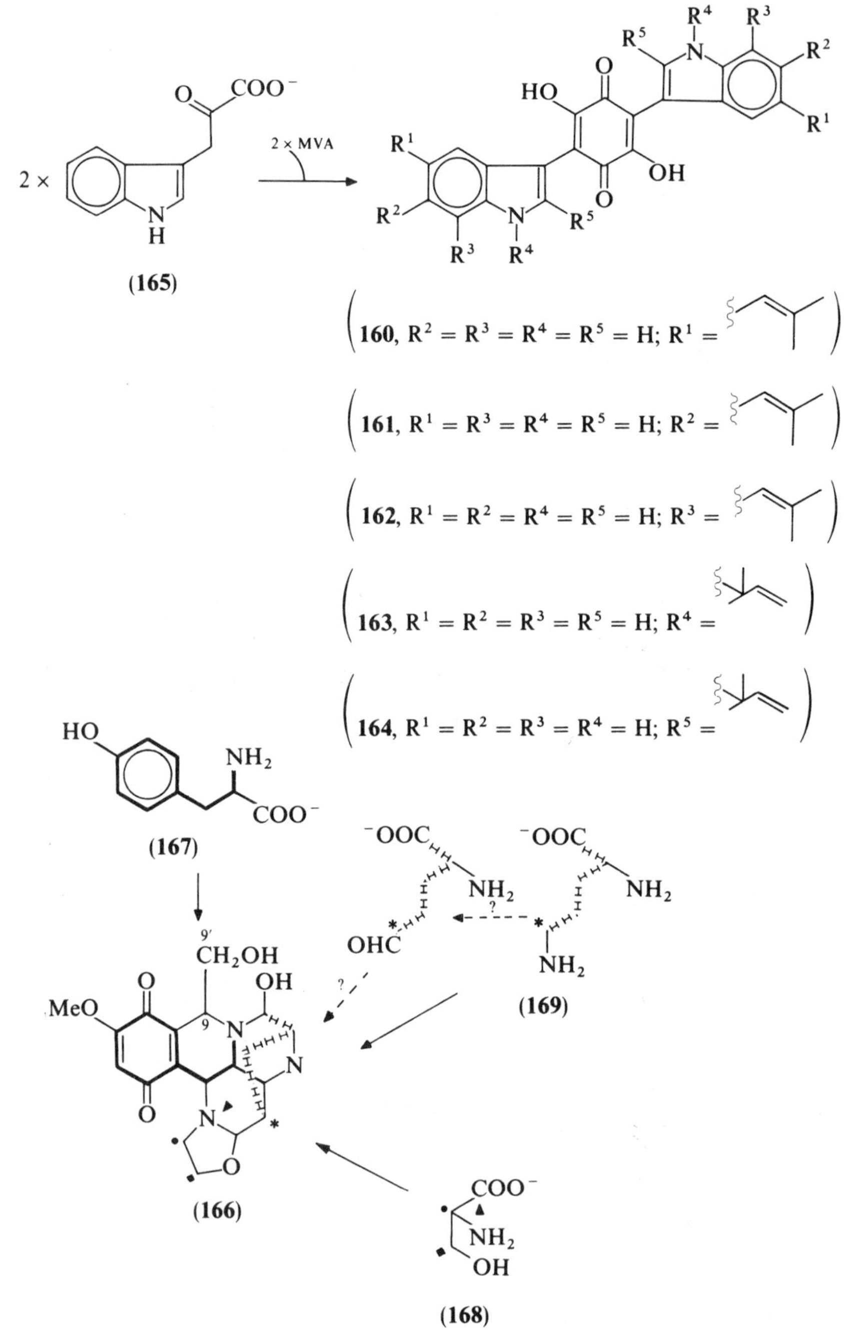

2 × (165) → (2 × MVA)

(**160**, $R^2 = R^3 = R^4 = R^5 = H$; R^1 =)

(**161**, $R^1 = R^3 = R^4 = R^5 = H$; R^2 =)

(**162**, $R^1 = R^2 = R^4 = R^5 = H$; R^3 =)

(**163**, $R^1 = R^2 = R^3 = R^5 = H$; R^4 =)

(**164**, $R^1 = R^2 = R^3 = R^4 = H$; R^5 =)

(**167**)

(**169**)

(**166**)

(**168**)

FIGURE 29

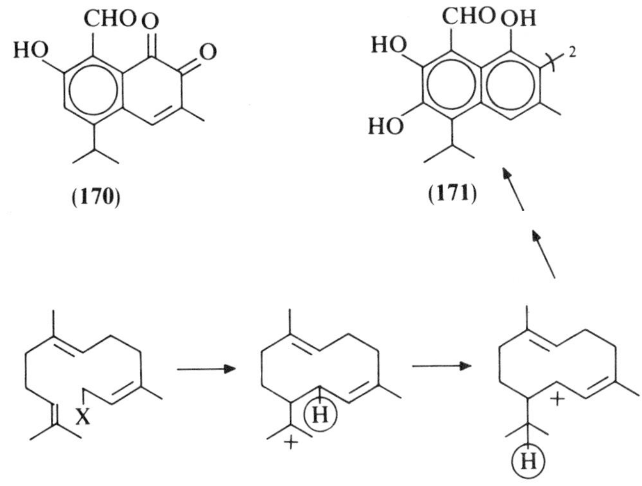

(170) (171)

FIGURE 30

culture is transferred to heterotrophic culture conditions[175]. Addition to the photoautotrophic culture of sucrose and cultivation in darkness results in concomitant degradation of chlorophyll and lipoquinones.

A typical secondary plant product, viz. lawsone (**104**), seems to be degraded in *Impatiens balsamina* to 3-(2′-carboxyphenyl)-3-oxopropionate (**172**) by ring fission of the 1,2-bond in the hydroxylated quinone[176].

Degradation of **104** was also observed in a bacterial culture isolated from soil[177]. The bacterium which was identified as a *Pseudomonas putida* strain grows on **104** as the only carbon and energy source. It is unknown whether this bacterium utilizes the same ring fission assumed to occur in *Impatiens balsamina*[176] because only late metabolites in the degradation pathway such as salicylate (**173**) and catechol (**174**) were detectable. This *Pseudomonas* strain is also capable of degrading other naphthoquinones including juglone (**105**). Another *Pseudomonas putida* strain was isolated from garden soil[178]. It grows slowly

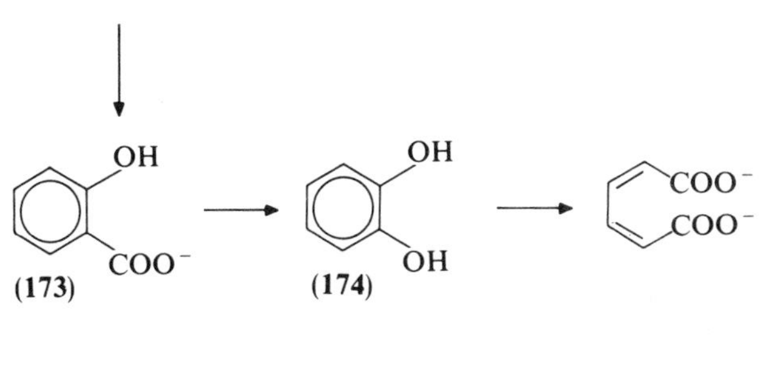

(104) (172)

(173) (174)

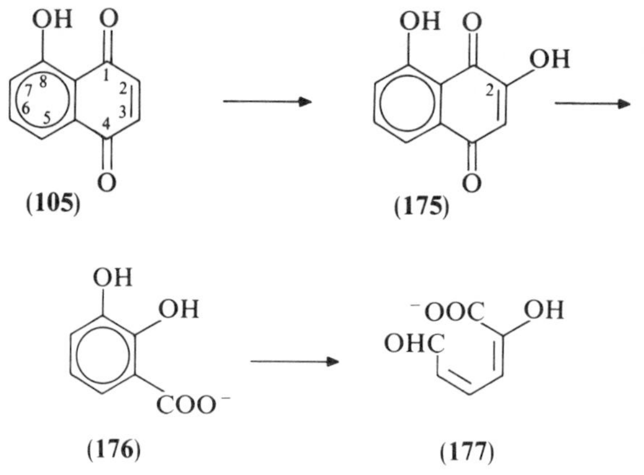

FIGURE 31

in the presence of **105** alone but growth is rapid in the presence of **105** and glucose. Degradation of juglone (**105**) in this strain proceeds via 2-hydroxyjuglone (**175**), 2,3-dihydroxybenzoate (**176**) and 2-hydroxymuconate semialdehyde (**177**) as depicted in Figure 31. It is unclear whether ring fission occurs between C atoms 1 and 2 or 2 and 3. The former mechanism would correspond to the assumed degradation of lawsone (**104**) in *Impatiens balsamina*[176] whereas the latter would correspond to what is known from degradation of lapachol (**178**) and dichloroallyllawsone by *Penicillium notatum* and other microorganisms, where the intermediacy of an epoxide (**179**) and ring fission between C atoms 2 and 3 in a monooxygenase reaction was postulated[179].

This epoxide (**179**) is assumed to be converted to the ketol **180** shown in Figure 32. Metabolism of lapachol (**178**) has also been observed in *Beauveria sulfurescens* and *Streptomyces albus*[180].

These organisms oxidize the prenyl side chain of lapachol (**178**) with excellent yield, a reaction which is not readily achieved chemically. The metabolites isolated are lomatiol (**181**), lomatate (**182**) and the acetate (**183**) of lomatiol. The side chain in these metabolites exhibits *E*-stereochemistry.

B. Anthraquinones

Anthraquinones may be stable in certain biological systems such as *Morinda citrifolia*[181] or *Galium mollugo*[182] cell cultures where they are deposited in the vacuole. In other biological systems, such as plants[183] or microorganisms (*vide infra*), however, they may be degraded.

1. Secoanthraquinones

Fungal metabolites, such as tajixanthone (**184**), ravenelin (**185**), secalonic acids, sulochrin (**186**) and geodin (**187**) belong to secoanthraquinones which are formed by oxidative cleavage of octaketide anthraquinones or anthrones. They are divided into two groups according to the site of cleavage of the central ring. One group is represented by tajixanthone (**184**), ravenelin (**185**), secalonic acids, etc., and the other group by sulochrin (**186**), geodin (**187**), etc. The ring fission is supposed to proceed by a Baeyer–Villiger-type oxidation of an anthraquinone[184] or by cleavage of a hydroperoxide derived from an anthrone or anthranol[185, 186].

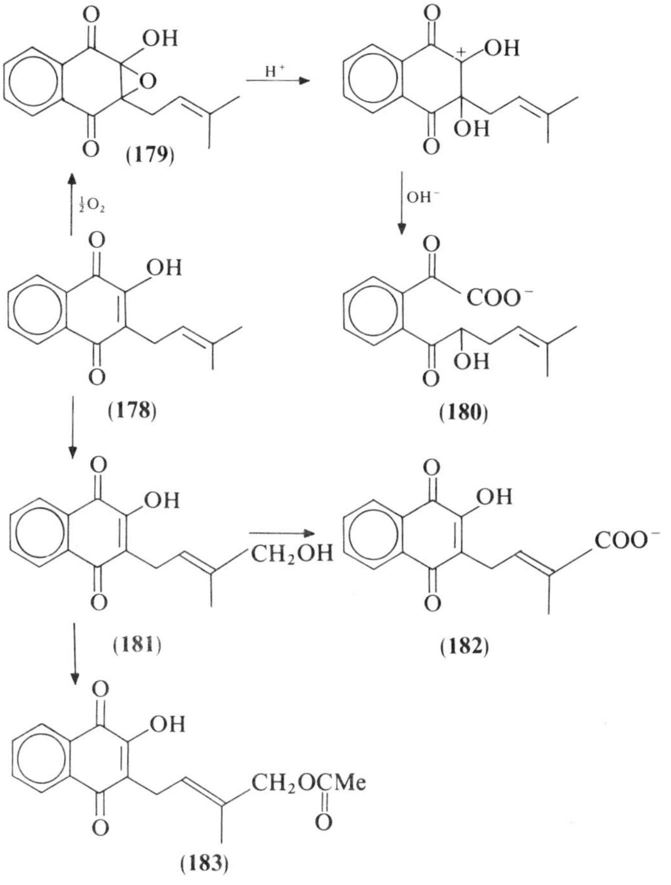

(179) (178) (180) (181) (182) (183)

FIGURE 32

The former mechanism seems to be involved in the ring fission of questin (**188**), an anthraquinone that is derived from emodin (**35**) by methylation[187]. The benzophenone sulochrin (**186**) formed in turn from questin (**188**) undergoes methylation and chlorination to give dihydrogeodin (**189**) (Figure 33). The latter compound is subject to an intramolecular phenol coupling reaction yielding (+)-geodin (**187**)[188]. The phenol oxidase (dihydrogeodin oxidase) involved in this reaction was purified to homogeneity. It is a blue copper protein with a molecular weight of 153000 consisting of two subunits. Phenol coupling reactions have been postulated many times but this work[188] represents the first example in which an enzyme catalysing a phenol oxidative coupling in natural product biosynthesis has been purified and characterized.

It has been reported that secalonic acid D (**190**) (Figure 34) incorporates anthrone at a rate 4.5 times higher than anthraquinone in *Penicillium oxalicum*[189]. However, there is other evidence suggesting that anthraquinone might be a precursor of tajixanthone (**184**) (*vide infra*). The mechanism of ring cleavage is still open for discussion.

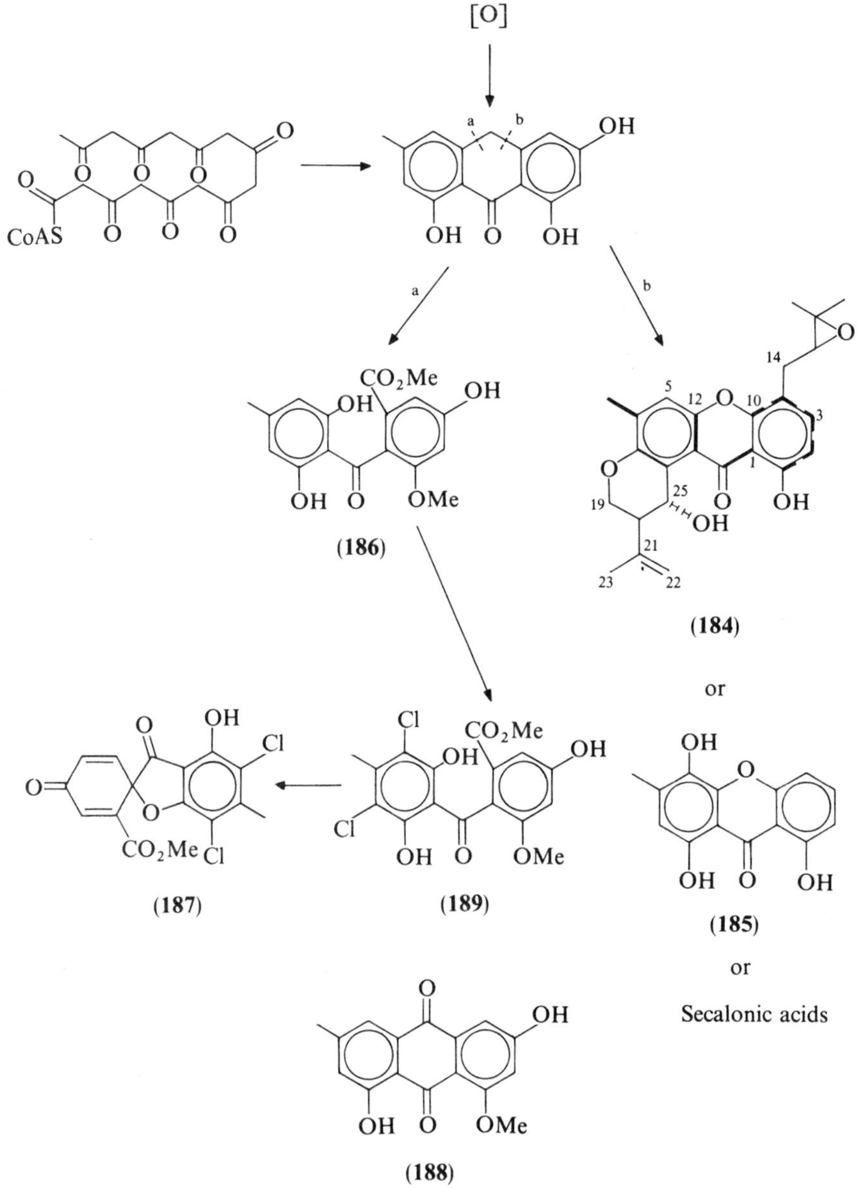

FIGURE 33

Tajixanthone (**184**) is a prenylated xanthone produced by *Aspergillus variecolor* as are shamixanthone (**191**) (Figure 35) and many other structurally related minor metabolites[190–192]. The following information on the biosynthetic processes was obtained by feeding experiments in which [^{13}C]acetates and [$^{2}H_3$]acetate[193, 194] were employed. The

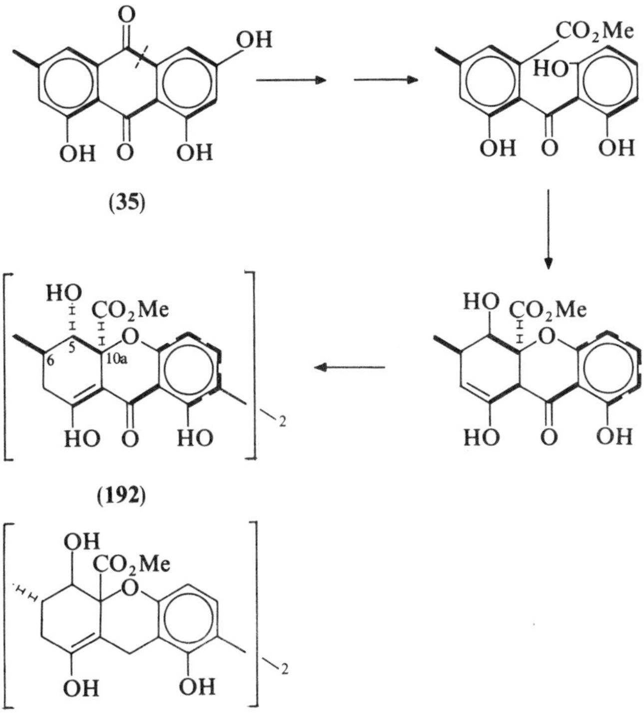

(35)

(192)

(190)

FIGURE 34

acetate assembly pattern was consistent with that of islandicin (**36**) and other similar compounds, but scrambling of ^{13}C–^{13}C coupling was observed in the C ring. This means that cycloaddition occurred prior to C-prenylation at two *ortho* positions of the symmetrical C ring of benzophenone which resulted from cleavage of the central ring. On the other hand, the sterospecificity of the dihydropyran ring indicates that this ring was formed by a concerted ene reaction at the *o*-prenylaldehyde portion, i.e. dihydropyran formation preceded cyclodehydration to the xanthone ring. If xanthone formation preceded, this reaction should be hindered by highly unfavourable interaction between the xanthone carbonyl and aldehyde groups. As mentioned before, the absence of the acetate-derived ^2H on the 25-position suggests that the ring was split at the stage of the anthraquinone, rather than an anthrone. Finally, non-retention of ^2H on C(5) suggests that decarboxylation occurred after cyclization and aromatization of the octaketide chain.

The acetate assembly pattern of ravenelin (**185**), elaborated by *Helminthosporium ravenelii* was consistent with that of tajixanthone (**184**). The central ring of islandicin (**36**), chrysophanol (**43**) or the corresponding anthrones is cleaved, followed by randomization in the C ring, and then cycloaddition[195, 196]. In **185** enriched by [1-^{13}C, $^{18}O_2$]acetate an ^{18}O-isotope shift was observed at C(1), C(8), C(9) and C(10a), but not at C(4a). This means that the ring-closure occurred by the nucleophilic attack of the hydroxy group of ring C on the *ortho* position of the ring A[197]. The substituent of the ring A, which should be eliminated, need not be hydroxyl. However, the benzophenone shown in Figure 35 is also a

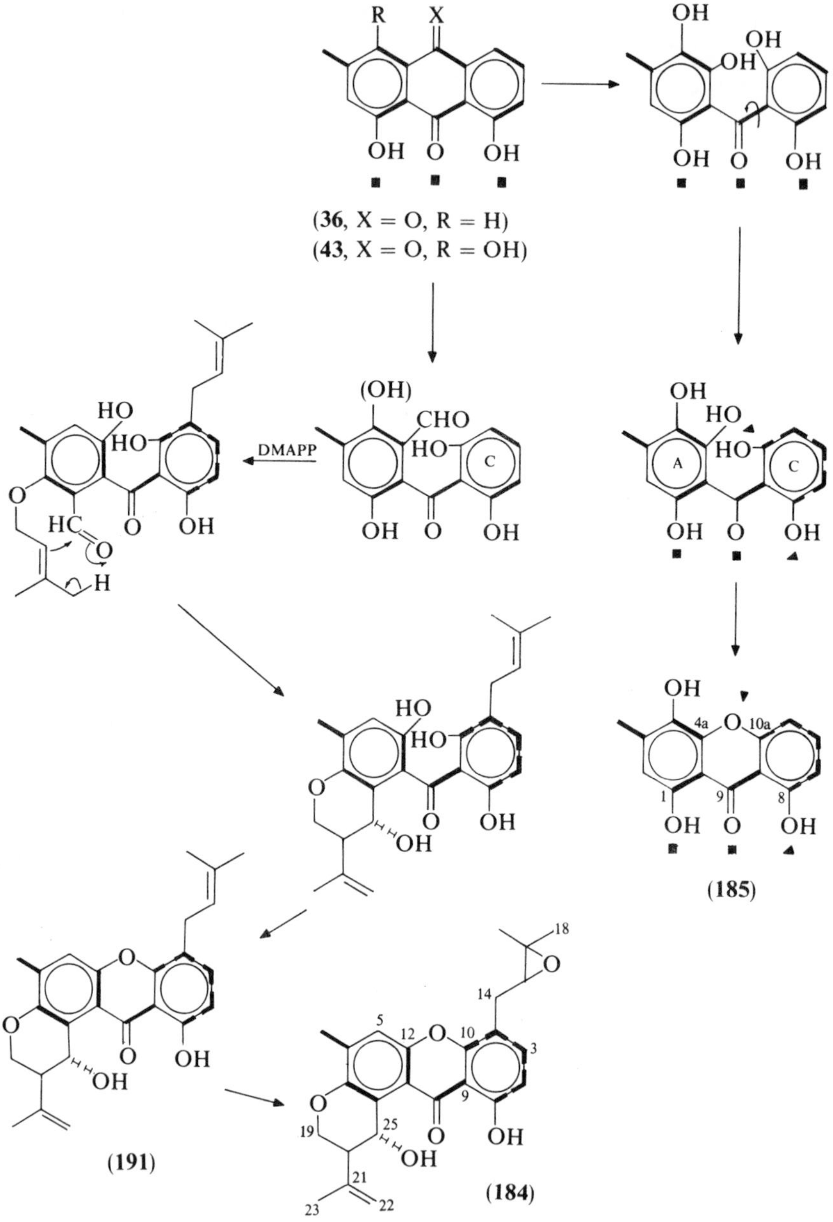

(36, X = O, R = H)
(43, X = O, R = OH)

DMAPP

(185)

(191)

(184)

FIGURE 35

chemically reasonable hypothetical intermediate. There is still no direct evidence that islandicin (**36**) or chrysophanol (**43**) is the genuine precursor of ravenelin (**185**), but trace amounts of both substances are contained in *H. ravenelii*.

Besides the above-mentioned experiments on secalonic acid D (**190**)[189], the biosynthesis of other secalonic acids was studied by feeding [$^{13}C_2$]acetate to *Pyrenochaeta terrestris*. Secalonic acid A (**192**) (Figure 34) is shown to be biosynthesized through cycloaddition of benzophenone, the ring cleavage product of emodin (**35**) or emodinanthrone (**50**), to tetrahydroxanthone under randomization of ^{13}C–^{13}C coupling in the ring and the subsequent dimerization of the resulting tetrahydroxanthone[198]. Monomeric tetrahydroxanthone has stereoisomers due to the configuration at positions 5, 6 and 10a. The dimerization of these stereoisomers in various combinations results in formation of various secalonic acids[199].

2. *Aflatoxins and congeners*

Aflatoxins represented by aflatoxin B_1 (**193**) are carcinogenic mycotoxins produced by some species of *Aspergillus* such as *A. flavus* and *A. parasiticus*. They themselves are not quinones, but coumarin derivatives, which are biosynthesized from a polyketide chain via various intermediates including several anthraquinones. These intermediates and their congeners are also produced by *A. versicolor* and other *Aspergillus* species. This group of substances has been one of the targets for very intensive studies during the last decade.

Studies on the biosynthesis of metabolites in the aflatoxin series began with feeding labelled acetates. Incorporation into the whole carbon skeleton of aflatoxin and an intermediate, sterigmatocystin (**194**) was demonstrated[200–202]. These studies were followed by feeding norsolorinic acid (**195**)[203], averufin (**196**)[204], versiconal acetate (**197**)[205], versicolorin A (**198**)[206] and sterigmatocystin (**194**)[207] to *A. parasiticus* and its mutants. Recently averantin (**199**) feeding was also carried out[208]. The observed significant incorporation of the fed susbstances into aflatoxin B_1 (**193**) suggested their likely precursorship.

For the construction of the carbon skeleton of these molecules two modes (a) and (b) of the acetate folding pattern were conceivable. All the demonstrated labelling pattern of averufin (**196**)[209–211], versiconal acetate (**197**)[212–214], versicolorin A (**198**)[215], sterigmatocystin (**194**)[216] and aflaxtoxin B_1 (**193**)[217] enriched by [^{13}C]acetates were consistent with mode (a) (Figure 36).

Very recently the side chain of these compounds proved to be formed from a starter hexanoate (**200**)[218, 219], i.e. when [1-^{13}C]hexanoate (**200**) was fed to an averufin (**196**)-accumulating mutant of *A. parasiticus*, the spectrum of isolated averufin (**196**) showed only the resonance of C(1′) to be strongly enhanced. Although the secondary incorporation of [1-^{13}C]acetate, which resulted from degradation of the original hexanoate, was also observed, resonances other than that of C(1′) were much weaker. Comparative studies using [1-^{13}C]butyrate, [1-^{13}C]-5-oxohexanoate, and [1-^{13}C]-3-oxooctanoate also showed only secondary incorporation.

The pivotal precursor in the biosynthesis of aflatoxins is averufin (**196**) which is formed from norsolorinic acid (**195**) via averantin (**199**). The biosynthetic relationship of averufin (**196**) and some subsequent intermediates were suggested by experiments using two mutants of *A. parasiticus*, deficient in aflatoxin production[220]. One accumulated averufin and the other versicolorin A (**198**). The averufin-accumulating mutant converted ^{14}C-labelled versiconal acetate (**197**), versicolorin A (**198**) and sterigmatocystin (**194**) to aflatoxin B_1 (**193**). In the presence of dichlorovos (dimethyl-2,2-dichlorovinyl phosphate), an inhibitor of aflatoxin biosynthesis, versicolorin A (**198**) and sterigmatocystin (**194**) were converted into aflatoxin by this mutant unaffectedly, but the conversion of versiconal acetate (**197**) into versicolorin A (**198**) was noticeably impeded. In contrast, the versicolorin A-accumulating mutant converted ^{14}C-labelled acetate, averufin (**196**) and versiconal acetate (**197**) up to versicolorin A (**198**) but in the presence of dichlorovos the conversion was impaired at the step of versiconal acetate. The intermediacy of averufin in

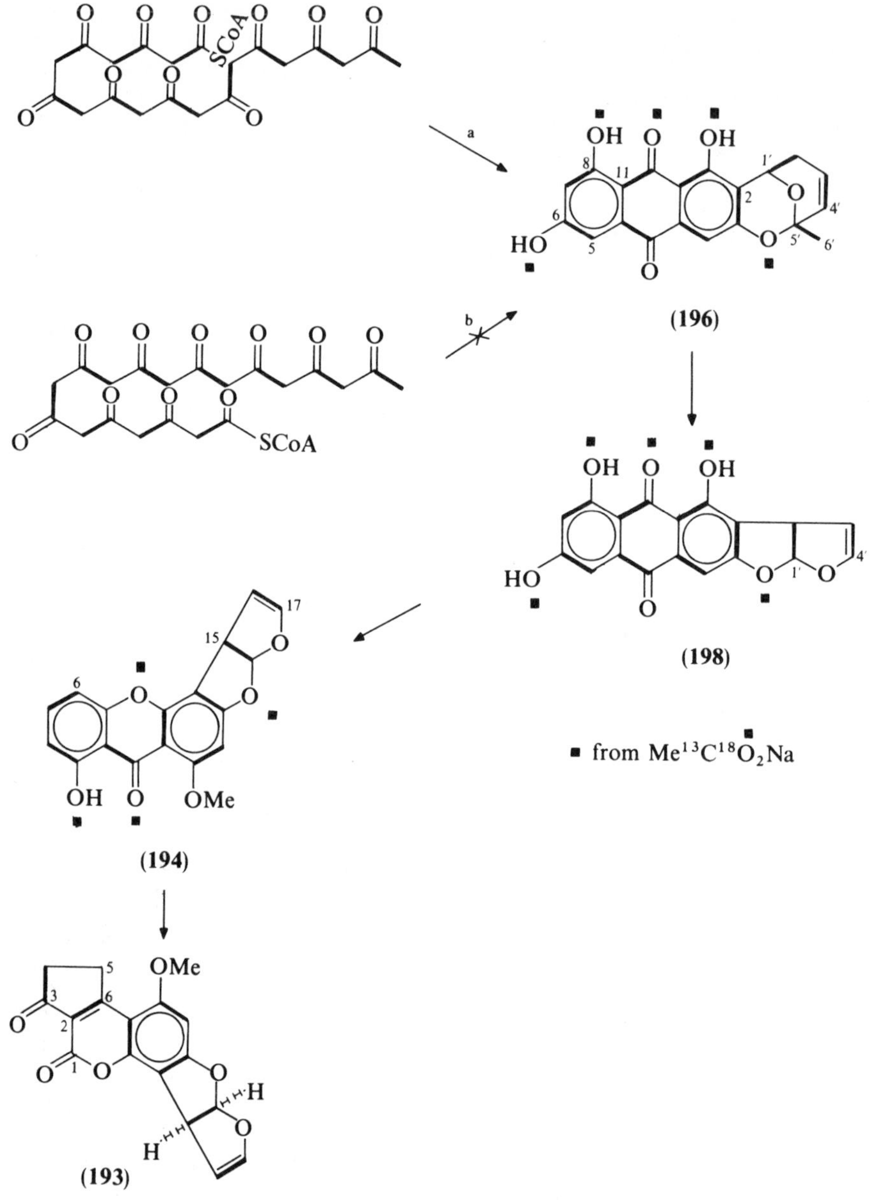

(196)

(198)

■ from $Me^{13}C^{18}O_2Na$

(194)

(193)

FIGURE 36

aflatoxin biosynthesis was finally unambiguously demonstrated by regiospecific incorporation of specifically labelled **196** into aflatoxin as follows. Feeding of [4′-^{2}H]averufin to *A. flavus* and of [4′-^{13}C]- and [1-^{13}C, 1′-^{2}H]averufin to *A. parasiticus* (ATCC 15517) indicated that 4-^{2}H and 4-^{13}C were incorporated into the 16-position of aflatoxin B_1 (**193**), while label from 1′-^{13}C and 1′-^{2}H were retained in the 13-position[221, 222]. Feeding of [5, 6-^{13}C]- and [8, 11-^{13}C]averufin to the same mutant of *A. parasiticus* further showed that C(8)–C(11) of averufin was intactly incorporated into C(2)–C(3) of **193**, whereas only C(6) of averufin was introduced into C(5) of **193** with loss of C(5) of averufin in the process, which is also in accordance with the assumed acetate assembly of **193**[223].

Averufin (**196**) was also shown to be regiospecifically incorporated into versicolorin A (**198**) and versiconal acetate (**197**). When [1′-^{13}C, 4′-$^{2}H_2$, 6′-$^{2}H_3$]averufin was fed to the dichlorovos-inhibited cultures of *A. parasiticus*, the labels were incorporated into the 1′,4′,6′-position of **197** without significant loss[224]. The retention of ^{2}H at the 6′-position indicates that **197** is formed by a Baeyer–Villiger-type oxidation. When the above-mentioned [4′-^{13}C]- and [1′-^{13}C, 1′-^{2}H]averufin was fed to a mutant (ATCC 36537) of *A. parasiticus*, 4′-^{13}C of averufin was incorporated into C(4′) of versicolorin A (**198**), whereas 1′-^{13}C of averufin was retained together with the bearing ^{2}H in the 1′-position of **198**[218]. The retention of the ^{2}H label at C(1′) of averufin throughout the process via **197**, **198** to aflatoxin excludes a Favorskii type reaction in the rearrangement of the side chain.

The above results clarified the pathway from averufin (**196**) to versicolorin A (**198**), which further leads to aflatoxin B_1 (**193**). For the explanation of the course of rearrangement in the side chain, nidurufin (**201**) (a known metabolite of *A. nidulans*) was proposed as an intermediate. The process was assumed to involve the rearrangement of **201** to an oxonium species (**202**) which would be converted to versiconal acetate (**197**) through hydrolysis followed by cyclization to a hemiacetal (**203**) and Baeyer–Villiger-type oxidation[218].

Application to *Aspergillus* species of acetates labelled with stable isotopes gave important suggestions on the process subsequent to versicolorin A (**198**). An example is the feeding of [^{13}C] acetate to a mutant (NRRL 5219) of *A. versicolor*[225]. The labelling pattern in the ring A of sterigmatocystin (**194**) enriched by [$^{13}C_2$] acetate in this way indicated that no randomization of ^{13}C–^{13}C coupling occurred, differing from what was seen in tajixanthone (**184**), ravenelin (**185**), etc. Therefore, a symmetrical structure would appear to be unlikely for the ring A of the intermediate which undergoes ring-closure to a xanthone structure. Sterigmatocystin (**194**) obtained after feeding [1–^{13}C, $^{2}H_3$] acetate to the same mutant incorporated ^{2}H only in positions 6, 15 and 17. The retention of ^{2}H at C(6) rules out any mechanism which proceeds through hydroxylation at this position in the course from versicolorin A (**198**) to sterigmatocystin (**194**). Feeding of [1-^{13}C, 1-$^{18}O_2$] acetate to mutants of *A. parasiticus* further indicated the distribution of the acetate-derived oxygen atoms in **196**, **198** and **194** as shown in Figure 36[226–229]. Coupled with the loss of the hydroxy group at C(6) of **198** in the process leading to **194**, the observed transformation of the hydroxy group at C(1) of **198** to the oxygen atom of the xanthone ring of **194** was also very suggestive of the process. All these findings appear most likely to be accounted for by assuming a process which involves the intramolecular oxidative coupling of benzophenone (**204**) to spirodienone (**205**) followed by reduction to spirodienol (**206**) and dienol–benzene rearrangement[229].

Finally, regarding the process from sterigmatocystin (**194**) to aflatoxin B_1 (**193**) an important suggestion came up by the finding that [$^{2}H_3$] acetate enriched aflatoxin B_1 (**193**) retained ^{2}H also at C(5)[230]. The presence of ^{2}H at C(5) which is originally derived from a carboxyl carbon indicates the migration of the ^{2}H label from the adjacent carbon in the biosynthetic process. Assuming 6-hydroxysterigmatocystin as an intermediate, it can be interpreted as the result of an NIH shift.

The complete pathway from the polyketide chain to aflatoxin B_1 (**193**) is shown in Figure 37. Aflatoxin B_1 is, on the other hand, a precursor of most other aflatoxins (B_2, **207**; B_2a, G, G_2 (**208**), G_2a)[231]. In addition to the above-mentioned products, many other biogenetically related metabolites are also isolated from *Dothistroma pini*[232], *Aspergillus*

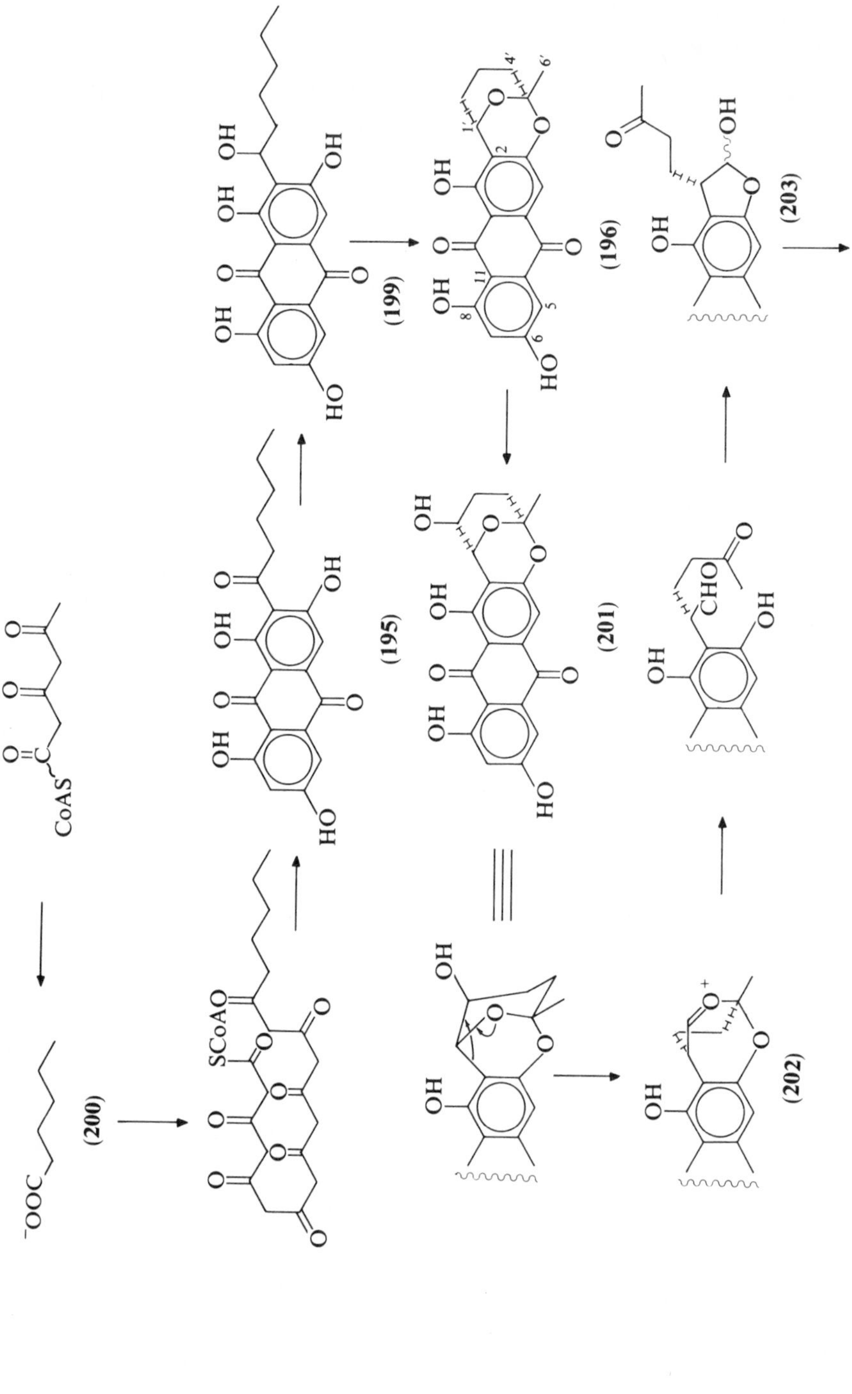
(200)
(195)
(199)
(196)
(201)
(202)
(203)

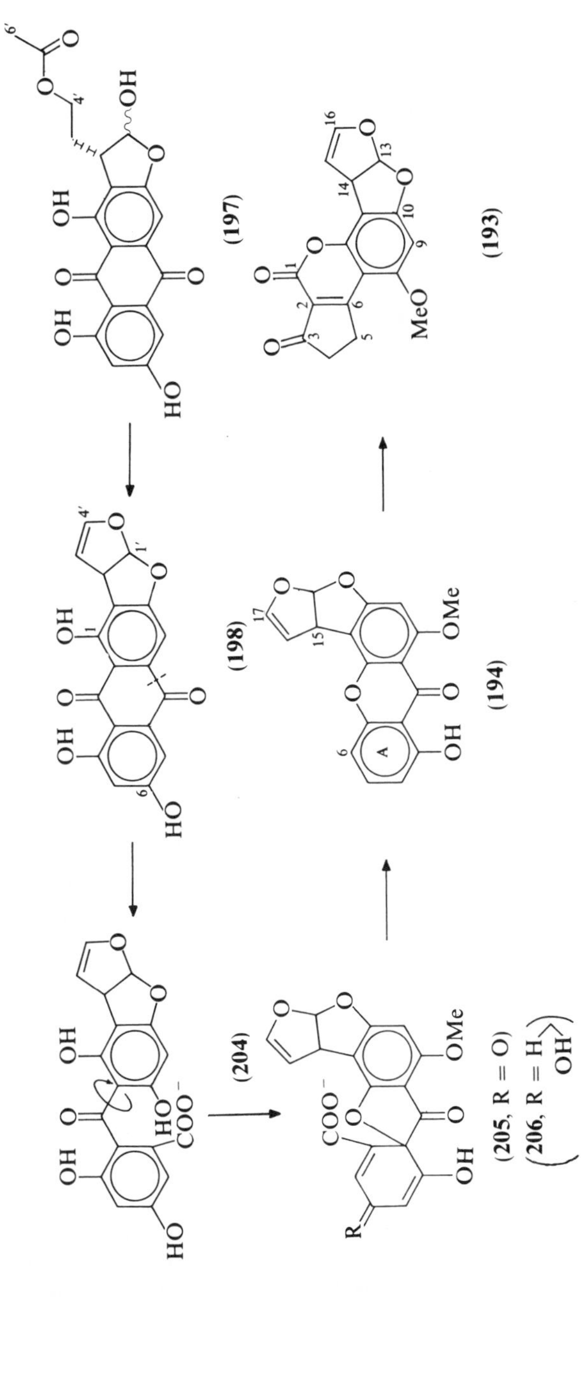

(197)

(198)

(204)

(193)

(194)

(205, R = O)

(206, R = H, OH)

FIGURE 37

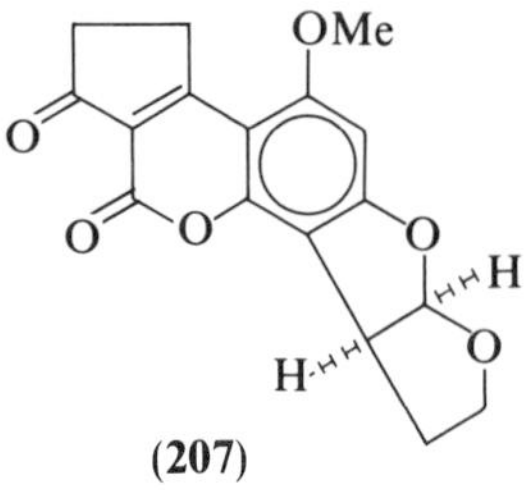

(207) 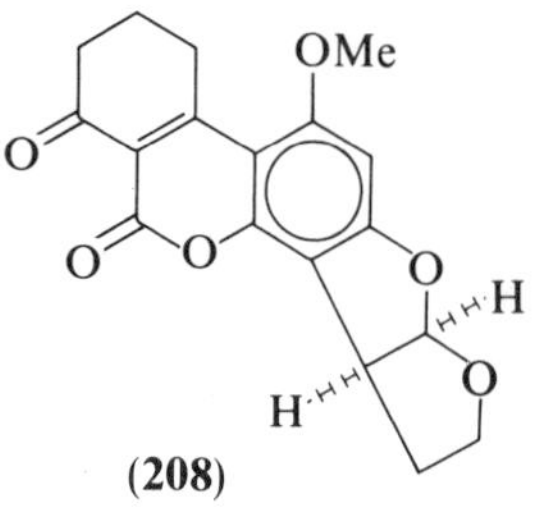 (208)

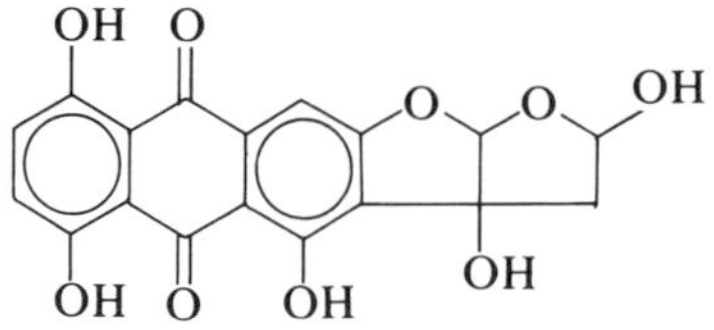

(209)

FIGURE 38

utsus[233], *Bipoloris sorokiniana*[234], etc. They include xanthones with linear fusion of the xanthone and bisdifuran portions such as dothiostromin (**209**)[232] (Figure 38).

IV. THE ROLE OF VITAMIN K IN BLOOD COAGULATION

Vitamin K_1 (i.e. phylloquinone, **103**) is a chloroplast-associated lipoquinone. Its possible role in photosynthesis is under discussion[174]. In contrast, a body of information has accumulated in the past ten years on the role of vitamin K in the blood coagulation process[235]. The vitamin K required in blood coagulation stems from two sources, viz. intestinal bacteria and green plants ingested with food. An important internal level of vitamin K_2 (**108**) is maintained by anaerobic intestinal bacteria like *Bacteroides fragilis*[236]. Vitamin K exhibits its function in the liver, where it is involved in the post-translational conversion of glutamyl residues (**210**) in precursor proteins to γ-carboxyglutamyl residues (**211**) in blood clotting factors (e.g. prothrombin) and other proteins. Mature carboxylated proteins are involved in the regulation of the activity of thrombin, a key event in the coagulation process[237]. The carboxylation leading to prothrombin takes place at the *N*-terminal region where approximately 10 glutamyl residues are carboxylated in the presence of a microsomal vitamin K-dependent carboxylase. The *in vitro* reaction requires CO_2, O_2, vitamin K hydroquinone (**212**) and a protein (e.g. 'preprothrombin') or a peptide, for example Phe-Leu-Glu-Glu-Val which is homologous to residues 5–9 of the bovine prothrombin precursor.

If the reaction is carried out in the absence of CO_2 the enzyme specifically exchanges the 4-pro S hydrogen of the glutamyl residue[238]. During this exchange a glutamyl carbanion is possibly formed which attacks CO_2. During the carboxylation vitamin K is converted to vitamin K epoxide (**213**) (Figure 39). The exact reaction mechanism for epoxide (**213**) formation and carboxylation is unknown but it has been observed that the reaction is inhibited in the presence of peroxidase. This and additional results[239] led to the

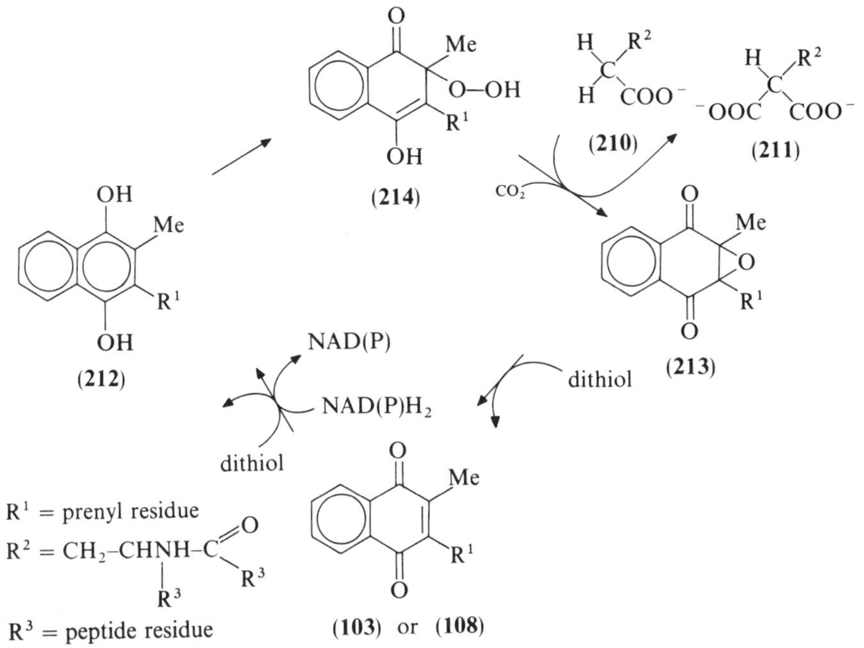

R^1 = prenyl residue

R^2 = CH_2–CHNH–C(=O)–R^3 (with R^3 on N)

R^3 = peptide residue

FIGURE 39

assumption that vitamin K hydroperoxide (**214**) (rather than vitamin K) is the true cofactor in the carboxylase reaction. Alternate reaction mechanisms have been discussed[235].

A prerequisite for the functioning of the carboxylation process is that the vitamin K hydroquinone (**212**) is regenerated from vitamin K epoxide (**213**)[240] (Figure 39). This reduction ensures that sufficient intracellular concentrations of vitamin K hydroquinone (**212**) are maintained to support normal rates of γ-carboxyglutamyl formation. The reduction is catalysed by microsomal fractions of the liver. The reduction is inhibited by anticoagulants like warfarin[240] resulting in a low vitamin K hydroquinone (**212**) level. This in turn leads to diminished carboxylation reactions and an impaired blood clotting process.

V. EPILOGUE

Quinones are very closely related to phenols which may easily be oxidized to give quinones provided an *ortho* or *para* diphenol grouping is present. Similarly, reduction of a quinone will easily give a phenol. Thus the quinonoid keto functions are rather artificial critera to classify the type of natural products treated in this review.

It follows that quinones are as heterogeneous as phenols when investigated under the aspects of structure, chemistry, biogenesis, physiology, chemosystematics or genetics.

It is the intent of this epilogue to call attention to this heterogeneity and to supply references for further readings. Where possible attention was drawn in this review to investigation on the genetics of quinone production; such data, however, are rare.

Physiological work on quinones such as localization of biosynthetic processes within chloroplasts or the participation of vitamin K in the blood clotting process has also been presented in this review. In this context attention should also be given to the preceding article in this series compiled by Campbell and Bentley[1].

Little information is available on the metabolism of quinones which to the authors' knowledge has been reviewed in this article for the first time. A relatively large body of information, however, is available on the biosynthesis of quinones. From the present and previous reviews it is evident that many different pathways lead to quinones[241–245]. Hence it is potentially dangerous to use quinones as chemotaxonomic markers because quinones of different biogenetic origin are taxonomically non-equivalent. The relation between distribution of quinones and biosynthesis has been outlined[242,245]. Occurrence of a quinone within a certain plant family and the structure of a quinone are usually sufficient to propose a reasonable hypothesis as to the origin of a quinone from primary precursors.

Thus acetate-derived quinones have a very characteristic substitution pattern. Enzyme systems, catalysing the synthesis of polyketides *in vitro*, however, have been detected in very few cases only[246]. Therefore NMR studies[247] mentioned in this review turned out to be particularly useful because they provided insight into reaction mechanisms involved in polyketide biosynthesis.

Among quinones enzymological work on plastoquinone, tocopherols (*vide supra*), ubiquinones[248] and vitamin K (see Ref. 249 and references therein) is most advanced.

Part of the work mentioned in this review deals with toxins[241] and antibiotics[250]. Quinones, however, may also exhibit other physiological properties. It is well known that they may be used as laxatives. Some are allergenes and others exhibit cytotoxic, mutagenic or even neurotoxic properties (see Ref. 241 and references therein).

A review on quinones would be incomplete if Thomson's work in this field was not mentioned[251].

VI. REFERENCES

1. R. Bentley and I. M. Campbell, in *The Chemistry of the Quinonoid Compounds*, Part 2 (Ed. S. Patai), John Wiley & Sons, New York, 1974, pp. 683–736.
2. E. P. McGovern and R. Bentley, *Biochemistry*, **14**, 3138 (1975).
3. D. C. Aldridge, A. B. Davis, M. R. Jackson and W. B. Turner, *J. Chem. Soc. Perkin Trans. 1*, 1540 (1974).
4. U. Sankawa, H. Shimada, T. Sato, T. Kinoshita and K. Yamasaki, *Tetrahedron Lett.*, 483 (1977).
5. H. Seto and H. Yonehara, *Tetrahedron Lett.*, 487 (1977).
6. U. Sankawa, H. Shimada and K. Yamasaki, *Tetrahedron Lett.*, 3375 (1978).
7. E. Bardshiri and T. J. Simpson, *Tetrahedron*, **39**, 3539 (1983).
8. S. Natori, Y. Inouye and H. Nishikawa, *Chem. Pharm. Bull.*, **15**, 380 (1967).
9. R. Bentley, W. J. Banach, A. G. McInnes and J. A. Walter, *Bioorg. Chem.*, **10**, 399 (1981).
10. I. Kurobane, L. C. Vining, A. G. McInnes and J. A. Walter, *Can. J. Chem.*, **58**, 1380 (1980).
11. A. W. McCulloch, A. G. McInnes, D. G. Smith, I. Kurobane and L. C. Vining, *Can. J. Chem.*, **60**, 2943 (1982).
12. J. E. Holenstein, H. Kern, A. Stoessl and J. B. Stothers, *Tetrahedron Lett.*, **24**, 4059 (1983).
13. T. J. Simpson, *J. Chem. Soc. Perkin Trans. 1*, 592 (1977).
14. G. Höfle and K. Röser, *J. Chem. Soc. Chem. Commun.*, 611 (1978).
15. A. Okubo, S. Yamazaki and K. Fuwa, *Agric. Biol. Chem.*, **39**, 1173 (1975).
16. I. Kurobane, L. C. Vining, A. G. McInnes, D. G. Smith and J. A. Walter, *Can. J. Chem.*, **59**, 422 (1981).
17. H. Tanaka, Y. Koyama, T. Nagai, H. Marumo and S. Omura, *J. Antibiot.*, **28**, 868 (1975).
18. C. Kitao, H. Tanaka, S. Minami and S. Omura, *J. Antibiot.*, **33**, 711 (1980).
19. H. Tanaka, S. Minami-Kakinuma and S. Omura, *J. Antibiot.*, **35**, 1565 (1982).
20. C. E. Snipes, C.-J. Chang and H. G. Floss, *J. Am. Chem. Soc.*, **101**, 701 (1979).
21. A. Arnone, L. Camarda, R. Cardillo, G. Fronza, L. Merlini, R. Mondelli and G. Nasini, *Helv. Chim. Acta*, **62**, 30 (1979).
22. B. Krone, A. Zeeck and H. G. Floss, *J. Org. Chem.*, **47**, 4721 (1982).
23. C. P. Gorst-Allman, B. A. M. Rudd, D.-J. Chang and H. G. Floss, *J. Org. Chem.*, **46**, 455 (1981).
24. B. A. M. Rudd and D. A. Hopwood, *J. Gen. Microbiol.*, **119**, 333 (1980).

25. F. Malpartida and D. A. Hopwood, *Nature*, **309**, 462 (1984).
26. B. A. M. Rudd and D. A. Hopwood, *J. Gen. Microbiol.*, **114**, 35 (1979).
27. H. G. Floss, *Biochemistry and Genetics of Antibiotic Biosynthesis.* Symposium report. Gemeinsame Frühjahrstagung der VAAM und der Sektion I der DGHM, Münster, F. R. G., March 3rd to 5th (1986).
28. S. Horinouchi, O. Hara and T. Beppu, *J. Bacteriol.*, **155**, 1238 (1983).
29. U. Gräfe, I. Eritt, G. Reinhardt, D. Krebs and W. F. Fleck, *Z. f. allgem. Mikrobiol.*, **24**, 515 (1984).
30. T. Hanumaiah, D. S. Marshall, B. K. Rao, G. S. R. Rao, J. U. M. Rao, K. V. J. Rao and R. H. Thomson, *Phytochemistry*, **24**, 2373 (1985).
31. T. Hanumaiah, D. S. Marshall, B. K. Rao, J. U. M. Rao, K. V. J. Rao and R. H. Thomson, *Phytochemistry*, **24**, 2669 (1985).
32. R. C. Paulick, M. L. Casey, D. F. Hillenbrand and H. W. Whitlock Jr, *J. Am. Chem. Soc.*, **97**, 5303 (1975).
33. M. L. Casey, R. C. Paulick and H. W. Whitlock, *J. Org. Chem.*, **43**, 1627 (1978).
34. A. Stoessl, C. H. Unwin and J. B. Stothers, *Tetrahedron Lett.*, 2481 (1979).
35. A. Stoessl, C. H. Unwin and J. B. Stothers, *Can. J. Chem.*, **61**, 372 (1983).
36. A. Yagi, M. Yamanouchi and I. Nishioka, *Phytochemistry*, **17**, 895 (1978).
37. J. A. Anderson, *Phytochemistry*, **25**, 103 (1986).
38. B. Franck and A. Stange, *Ann. Chem.*, 2106 (1981).
39. U. Sankawa, Y. Ebizuka and S. Shibata, *Tetrahedron Lett.*, 2125 (1973).
40. S. Seo, U. Sankawa, Y. Ogihara, Y. Iitaka and S. Shibata, *Tetrahedron*, **29**, 3721 (1973).
41. A. A. Bell, R. D. Stipanovic, D. H. O'Brien and P. A. Fryxell, *Phytochemistry*, **17**, 1297 (1978).
42. T. Nomura and T. Fukai, *Heterocycles*, **15**, 1531 (1981).
43. A. F. McInnes, D. G. Smith and J. A. Walter, *J. Chem. Soc. Chem. Commun.*, 66 (1975).
44. A. J. Birch and T. J. Simpson, *J. Chem. Soc. Perkin Trans. 1*, 816 (1979).
45. H. Brockmann, *Fortsch. Chem. Org. Naturstoffe*, **21**, 127 (1963).
46. K. Eckhardt, D. Tresselt, G. Schumann, W. Ihn and C. Wagner, *J. Antibiot.*, **38**, 1034 (1985).
47. C. Wagner, K. Eckhardt, G. Schumann, W. Ihn and D. Tresselt, *J. Antibiot.*, **37**, 691 (1984).
48. P. F. Wiley, D. W. Elrod and V. P. Marshall, *J. Org. Chem.*, **43**, 3457 (1978).
49. A. Yoshimoto, J. Ogasawara, I. Kitamura, T. Oki, T. Inui, T. Takeuchi and H. Umezawa, *J. Antibiot.*, **32**, 472 (1979).
50. T. Oki, A. Yoshimoto, Y. Matsuzawa, T. Takeuchi and H. Umezawa, *J. Antibiot.*, **33**, 1331 (1980).
51. T. Oki, Y. Takatsuki, H. Tobe, A. Yoshimoto, T. Takeuchi and H. Umezawa, *J. Antibiot.*, **34**, 1229 (1981).
52. N. Imamura, K. Kakinuma and N. Ikekawa, *J. Antibiot.*, **35**, 602 (1982).
53. H. Seto, L. W. Cary and M. Tanabe, *J. Chem. Soc. Chem. Commun.*, 867 (1973).
54. M. L. Casey, R. C. Paulick and H. W. Whitlock Jr, *J. Am. Chem. Soc.*, **98**, 2636 (1976).
55. F. Scherrer, H. A. Anderson and R. Azerad, *J. Chem. Soc. Chem. Commun.*, 127 (1976).
56. F. Scherrer and R. Azerad, *J. Chem. Soc. Chem. Commun.*, 128 (1976).
57. L. Canonica, L. Colombo, C. Gennari, B. M. Ranzi and C. Scolastico, *J. Chem. Soc. Chem. Commun.*, 679 (1978).
58. R. E. Olson, R. Bentley, G. H. Dialameh, P. H. Gold, V. G. Ramsay and C. M. Springer, *J. Biol. Chem.*, **238**, 3146 (1963).
59. H. Rudney and W. W. Parson, *J. Biol. Chem.*, **238**, 3137 (1963).
60. W. W. Parson and H. Rudney, *Proc. Natl. Acad. Sci. USA*, **51**, 444 (1964).
61. A. S. Aiyar and R. E. Olson, *Fed. Proc. Am. Soc. Exp. Biol.*, **23**, 425 (1964).
62. P. Friis, G. D. Daves Jr and K. Folkers, *J. Am. Chem. Soc.*, **88**, 4754 (1966).
63. J. L. G. Nilsson, T. M. Farley and K. Folkers, *Anal. Biochem.*, **23**, 422 (1968).
64. G. B. Cox, F. Gibson and J. Pittard, *J. Bacteriol.*, **95**, 1591 (1968).
65. G. B. Cox, I. G. Young, L. M. McCann and F. Gibson, *J. Bacteriol.*, **99**, 450 (1969).
66. I. G. Young, L. M. McCann, P. Stroobant and F. Gibson, *J. Bacteriol.*, **105**, 769 (1971).
67. P. Stroobant, I. G. Young and F. Gibson, *J. Bacteriol.*, **109**, 134 (1972).
68. F. Gibson, *Biochem. Soc. Trans.*, **1**, 317 (1973).
69. H. G. Nowicki, G. H. Dialameh and R. E. Olson, *Biochemistry*, **11**, 896 (1972).
70. B. L. Trumpower, A. S. Aiyar, C. E. Opliger and R. E. Olson, *J. Biol. Chem.*, **247**, 2499 (1972).
71. R. M. Houser and R. E. Olson, *Life Sci.*, **14**, 1211 (1974).

72. J. Casey and R. R. Threlfall, *FEBS Lett.*, **85**, 249 (1978).
73. B. L. Trumpower, R. M. Houser and R. E. Olson, *J. Biol. Chem.*, **249**, 3041 (1974).
74. R. R. Goewert, C. J. Sippel and R. E. Olson, *Biochem. Biophys. Res. Commun.*, **77**, 599 (1977).
75. R. R. Goewert, C. J. Sippel and R. E. Olson, *Biochemistry*, **20**, 217 (1981).
76. R. R. Goewert, C. J. Sippel, M. F. Grimm and R. E. Olson, *FEBS Lett.*, **87**, 219 (1978).
77. R. R. Goewert, C. J. Sippel, M. F. Grimm and R. E. Olson, *Biochemistry*, **20**, 5611 (1981).
78. H. Rudney, in *Biomedical and Clinical Aspects of Coenzyme Q* (Eds. K. Folkers and Y. Yamamura), Elsevier, Amsterdam, 1977, pp. 29–46.
79. A. M. P. Nambudiri, D. Brockmann, S. S. Alam and H. Rudney, *Biochem. Biophys. Res. Commun.*, **786**, 282 (1977).
80. H. Schmid and M. H. Zenk, *Tetrahedron Lett.*, 4151 (1971).
81. H. Inouye, S. Ueda, K. Inoue and H. Matsumura, *Phytochemistry*, **18**, 1301 (1979).
82. H. Mizukami, Ph.D. Thesis, Kyoto University, 1976, and unpublished results in Prof. Tabata's laboratories.
83. H. Inouye, H. Matsumura, M. Kawasaki, K. Inoue, M. Tsukada and M. Tabata, *Phytochemistry*, **20**, 1701 (1981).
84. M. Moir and R. H. Thomson, *J. Chem. Soc. Perkin Trans. 1*, 1352 (1973).
85. G. R. Whistance and D. R. Threlfall, *Biochem. J.*, **117**, 593 (1970).
86. K. G. Hutson and D. R. Threlfall, *Biochem. Biophys. Acta*, **632**, 630 (1980).
87. P. S. Marshall, S. R. Morris and D. R. Threlfall, *Phytochemistry*, **24**, 1705 (1985).
88. R. Krügel, K. H. Grumbach, H. Lichtenthaler and J. Rétey, *Bioorg. Chem.*, **13**, 187 (1985).
89. J. W. Cornforth, J. W. Redmond, H. Eggerer, W. Buckel and C. Gutschow. *Eur. J. Biochem.*, **14**, 1 (1970).
90. G. Schultz, J. Soll, E. Fiedler and D. Schulze-Siebert, *Physiol. Plant*, **64**, 123 (1985).
91. H. K. Lichtenthaler, U. Prenzel, R. Douce and J. Joyard, *Biochim. Biophys. Acta*, **641**, 99 (1981).
92. J. Soll, G. Schultz, J. Joyard, R. Douce and M. A. Block, *Arch. Biochem. Biophys.*, **238**, 290 (1985).
93. E. Fiedler, C. L. Schmidt, C. Gross and G. Schultz, *Proceedings of the 7th Hungarian Bioflavanoid Symposium, Szeged (Hungary)*, 1985.
94. R. Meganathan, *J. Biol. Chem.*, **256**, 9386 (1981).
95. R. Meganathan and R. Bentley, *J. Bacteriol.*, **153**, 739 (1983).
96. A. Weische and E. Leistner, *Tetrahedron Lett*, **26**, 1487 (1985).
97. P. Dansette and R. Azerad, *Biochem. Biophys. Res. Commun.*, **40**, 1090 (1970).
98. I. G. Young, *Biochemistry*, **14**, 399 (1975).
99. A. Weische, Ph.D. Thesis, Rheinische Friedrich-Wilhelms-Universität, Bonn, F.R.G, 1985.
100. G. T. Emmons, I. M. Campbell and R. Bentley, *Biochem. Biophys. Res. Commun.*, **131**, 956 (1985).
101. R. Bentley, *Pure Appl. Chem.*, **41**, 47 (1975).
102. R. Meganathan and R. Bentley, *J. Bacteriol.*, **140**, 92 (1979).
103. R. Meganathan, T. Folger and R. Bentley, *Biochemistry*, **19**, 785 (1980).
104. R. Bentley and R. Meganathan, *Microbiol. Rev.*, **46**, 241 (1982).
105. L. Heide, S. Arndt and E. Leistner, *J. Biol. Chem.*, **257**, 7396 (1982).
106. R. Kolkmann and E. Leistner, *Tetrahedron Lett.*, **26**, 1703 (1985).
107. R. M. Baldwin, C. D. Snyder and H. Rapoport, *J. Am. Chem. Soc.*, **95**, 276 (1973); *Biochemistry*, **13**, 1523 (1974).
108. B. Shineberg and I. G. Young, *Biochemistry*, **15**, 2754 (1976).
109. K. G. Hutson and D. R. Threlfall, *Phytochemistry*, **19**, 535 (1980).
110. G. Schultz, B. H. Ellerbrock and J. Soll, *Eur. J. Biochem.*, **117**, 329 (1981).
111. S. Kaiping, J. Soll and G. Schultz, *Phytochemistry*, **23**, 89 (1984).
112. W. U. Müller and E. Leistner, *Phytochemistry*, **15**, 407 (1976).
113. W. U. Müller and E. Leistner, *Phytochemistry*, **17**, 1735 (1978).
114. H. Inouye, S. Ueda, K. Inoue, T. Hayashi and T. Hibi, *Tetrahedron Lett.*, 2395 (1975).
115. S. Ueda, K. Inoue, T. Hayashi and H. Inouye, *Tetrahedron Lett.*, 2399 (1975).
116. H. Inouye, S. Ueda, K. Inoue, T. Hayashi and T. Hibi, *Chem. Pharm. Bull.*, **23**, 2523 (1975).
117. K. Inoue, S. Ueda, Y. Shiobara and H. Inouye, *Phytochemistry.*, **16**, 1689 (1977).
118. K. Inoue, S. Ueda, Y. Shiobara, I. Kimura and H. Inouye, *J. Chem. Soc. Perkin Trans. 1*, 1246 (1981).

119. E. Leistner, *Phytochemistry*, **12**, 337 (1973).
120. E. Leistner, *Phytochemistry*, **12**, 1669 (1973).
121. H. Schildknecht, F. Straub and V. Scheidel, *Liebigs Ann. Chem.*, 1295 (1976).
122. L. Heide and E. Leistner, *J. Chem. Soc. Chem. Commun.*, 334 (1981).
123. K. Inoue, Y. Shiobara, H. Nayeshiro, H. Inouye, G. Wilson and M. H. Zenk, *J. Chem. Soc. Chem. Commun.*, 957 (1979).
124. K. Inoue, Y. Shiobara, H. Nayeshiro, H. Inouye, G. Wilson and M. H. Zenk, *Phytochemistry*, **23**, 307 (1984).
125. K. Inoue, S. Ueda, H. Nayeshiro and H. Inouye, *J. Chem. Soc. Chem. Commun.*, 993 (1982).
126. K. Inouye, S. Ueda, H. Nayeshiro, N. Moritome and H. Inouye, *Phytochemistry.*, **23**, 313 (1984).
127. K. Inoue, T. Kido and H. Inouye, unpublished results.
128. M. Brufani, D. Kluepfel, G. C. Lancini, J. Leitich, A. S. Messentrev, V. Prelog, F. P. Schmook and P. Sensi, *Helv. Chim. Acta*, **56**, 2315 (1973).
129. A. Karlsson, G. Sartori and R. J. White, *Eur. J. Biochem.*, **47**, 251 (1974).
130. R. J. White, E. Martinelli, G. G. Gallo, G. C. Lancini and P. Beynon, *Nature*, **243**, 273 (1973).
131. E. Martinelli, G. C. Gallo, P. Antonini and R. J. White, *Tetrahedron*, **30**, 3087 (1974).
132. R. J. White, E. Martinelli and G. C. Lancini, *Proc. Natl. Acad. Sci. USA*, **71**, 3260 (1974).
133. R. J. White and E. Martinelli, *FEBS Lett.*, **49**, 233 (1974).
134. S. Bruggisser, Dissertation Nr. 5435 Federal School of Technology, Switzerland, 1975.
135. O. Ghisalba and J. Nüesch, *J. Antibiot.*, **31**, 202 (1978).
136. O. Ghisalba and J. Nüesch, *J. Antibiot.*, **31**, 215 (1978).
137. J. J. Kibby, I. A. McDonald and R. W. Rickards, *J. Chem. Soc. Chem. Commun.*, 768 (1980).
138. O. Ghisalba, H. Fuhrer, W. J. Richter and S. Moss, *J. Antibiot.*, **34**, 58 (1981).
139. O. Ghisalba and J. Nüesch, *J. Antibiot.*, **34**, 64 (1981).
140. P. Traxler and O. Ghisalba, *J. Antibiot.*, **35**, 1361 (1982).
141. A. M. Becker, A. J. Herlt, G. L. Hilton, J. J. Kibby and R. W. Rickards, *J. Antibiot.*, **36**, 1323 (1983).
142. J. Ruishen, L. Cijun, J. Zhikun, Z. Xuecong, N. Linying and L. Zhaomin, *Sci. Sin. Ser. B*, **XXVII**, 380 (1984).
143. P. V. Deshmukh, K. Kakinuma, J. J. Ameel and K. L. Rinehart Jr, *J. Am. Chem. Soc.*, **98**, 870 (1976).
144. O. Ghisalba, P. Traxler and J. Nüesch, *J. Antibiot.*, **31**, 1124 (1978).
145. O. Ghisalba, J. A. L. Auden, T. Schupp and J. Nüesch, *Biotechnology of Industrial Antibiotics*, Marcel Dekker, New York, 1984, p. 281.
146. R. D. Johnson, A. Haber and K. L. Rinehart Jr, *J. Am. Chem. Soc.*, **96**, 3316 (1974).
147. R. D. Johnson, A. Haber and K. L. Rinehart Jr, *Abstr. Pap. Am. Chem. Soc.*, **166**, 124 (1973).
148. A. Haber, R. D. Johnson and K. L. Rinehart Jr, *J. Am. Chem. Soc.*, **99**, 3541 (1977).
149. M. Sugita, T. Sasaki, K. Furihata, H. Seto and N. Otake, *J. Antibiot.*, **35**, 1467 (1982).
150. K. L. Rinehart Jr, M. Potgieter and D. A. Wright, *J. Am. Chem. Soc.*, **104**, 2649 (1982).
151. M. G. Anderson, J. J. Kibby, R. W. Rickards and J. M. Rothschild, *J. Chem. Soc. Chem. Commun.*, 1277 (1980).
152. U. Hornemann and J. C. Cloyd, *Chem. Commun.*, 301 (1971).
153. U. Hornemann and M. J. Aikman, *J. Chem. Soc. Chem. Commun.*, 88 (1973).
154. U. Hornemann, J. P. Kehrer, C. S. Nunez and R. L. Ranieri, *J. Am. Chem. Soc.*, **96**, 320 (1974).
155. U. Hornemann and J. H. Eggert, *J. Antibiot.*, **28**, 841 (1975).
156. S. J. Gould and C. C. Chang, *J. Am. Chem. Soc.*, **102**, 1702 (1980).
157. S. J. Gould, C. C. Chang, D. S. Darling, J. D. Roberts and M. Squillacote, *J. Am. Chem. Soc.*, **102**, 1707 (1980).
158. S. J. Gould and D. E. Cane, *J. Am. Chem. Soc.*, **104**, 343 (1982).
159. A. M. Nadzan and K. L. Rinehart Jr, *J. Am. Chem. Soc.*, **98**, 5012 (1976).
160. T. W. Doyle, D. M. Balitz, R. E. Grulich, D. E. Nettletone, S. J. Gould, C. H. Tann and A. E. Moews, *Tetrahedron Lett.*, **22**, 4595 (1981).
161. W. A. Jerram, A. G. McInnes, W. S. G. Maass, D. G. Smith, A. Taylor and J. A. Walter, *Can. J. Chem.*, **53**, 727 (1975).
162. S. Sekita, *Chem. Pharm. Bull.*, **31**, 2998 (1983).
163. Y. Yamamoto, K. Nishimura and N. Kiriyama, *Chem. Pharm. Bull.*, **24**, 1853 (1976).
164. M. A. O'Leary and J. R. Hanson, *Tetrahedron Lett.*, **23**, 1855 (1982).

165. M. A. O'Leary, J. R. Hanson and B. L. Yeoh, *J. Chem. Soc. Perkin Trans. 1*, 567 (1984).
166. A. Taylor and J. A. Walter, *Can. J. Chem.*, **53**, 727 (1975).
167. M. J. Zmijewski Jr, M. Mikolajczak, V. Viswanatha and V. J. Hruby, *J. Am. Chem. Soc.*, **104**, 4969 (1982).
168. M. J. Zmijewski Jr, *J. Antibiot.*, **38**, 819 (1985).
169. M. J. Zmijewski Jr, V. A. Plamiswamy and S. J. Gould, *J. Chem. Soc. Chem. Commun.*, 1261 (1985).
170. C. H. Eugster, in *Pigments in Plants*, 2nd edn (Ed. F. C. Czygan), G. Fischer, Stuttgart, New York, 1980, pp. 149–186.
171. R. Masciadri, W. Angst and D. Arigoni, *J. Chem. Soc. Chem. Commun.*, 1573 (1985).
172. W. Barz and J. Köster, in *The Biochemistry of Plants*, Vol. 7 (Ed. P. K. Stumpf and E. E. Conn), Academic Press, New York, 1981, pp. 35–84.
173. K. P. Heise and G. Harnischfeger, *Z. Naturforsch.*, **33c**, 537 (1978).
174. K. H. Grumbach and H. K. Lichtenthaler, *Planta*, **141**, 253 (1978).
175. U. Igbavboa, H. J. Sieweke, E. Leistner, I. Röwer, W. Hüsemann and W. Barz, *Planta*, **166**, 537 (1985).
176. E. Grotzinger and I. M. Campbell, *Phytochemistry*, **13**, 923 (1974).
177. J. Weisendorf, H. Rettenmaier and F. Lingens, *Biol. Chem. Hoppe-Seyler*, **366**, 945 (1985).
178. H. Rettenmaier, U. Kupas and F. Lingens, *FEMS Microbiology L.*, **19**, 193 (1983).
179. S. L. Otten and J. P. Rosazza, *J. Biol. Chem.*, **258**, 1610 (1983).
180. L. D. Gayet, J. C. Gayet and H. Veschambre, *Agric. Biol. Chem.*, **49**, 2693 (1985).
181. M. H. Zenk, H. El-Shagi and U. Schulte, *Planta medica, Suppl.*, 79 (1975).
182. H. J. Bauch and E. Leistner, *Planta medica*, **33**, 124 (1978).
183. J. W. Fairbairn and F. J. Muhtadi, *Phytochemistry*, **11**, 215 (1972).
184. J. M. Schwab, *J. Am. Chem. Soc.*, **103**, 1876 (1981).
185. B. Franck and B. Berger-Lohr, *Angew. Chem.*, **87**, 845 (1975).
186. T. Money, *Nature*, **199**, 592 (1963).
187. I. Fujii, Y. Ebizuka and U. Sankawa, *Chem. Pharm. Bull.*, **30**, 2283 (1982).
188. I. Fujii, H. Irijima, Y. Ebizuka and U. Sankawa, *Chem. Pharm. Bull.*, **31**, 337 (1983).
189. B. Franck, H. Backhaus and M. Rolf, *Tetrahedron Lett.*, 1185 (1980).
190. K. K. Chexal, C. Fouwether, J. S. E. Holker, T. J. Simpson and K. Young, *J. Chem. Soc. Perkin Trans. 1*, 1584 (1974).
191. K. K. Chexal, J. S. Holker and T. J. Simpson, *J. Chem. Soc. Perkin Trans. 1*, 543 (1975).
192. K. K. Chexal, J. S. E. Holker, T. J. Simpson and K. Young, *J. Chem. Soc. Perkin Trans. 1*, 549 (1975).
193. J. S. E. Holker, R. D. Lapper and T. J. Simpson, *J. Chem. Soc. Perkin Trans. 1*, 2135 (1974).
194. F. Bardshiri and T. J. Simpson, *J. Chem. Soc. Chem. Commun.*, 195 (1981).
195. A. J. Birch, T. J. Simpson and P. W. Westerman, *Tetrahedron Lett.*, 4173 (1975).
196. A. J. Birch, J. Baldas, J. R. Hlubucek, T. J. Simpson and P. W. Westerman, *J. Chem. Soc. Perkin Trans. 1*, 898 (1976).
197. J. G. Hill, T. T. Nakashima and J. C. Vederas, *J. Am. Chem. Soc.*, **104**, 1745 (1982).
198. I. Kurobane and L. C. Vining, *Tetrahedron Lett.*, 1379 (1978).
199. I. Kurobane and L. C. Vining, *J. Antibiot.*, **32**, 1256 (1979).
200. J. S. E. Holker and L. J. Mulheirn, *Chem. Commun.*, 1576 (1968).
201. M. Biollaz, G. Büchi and G. Milne, *J. Am. Chem. Soc.*, **92**, 1035 (1970).
202. M. Tanabe, T. Hamasaki and H. Seto, *Chem. Commun.*, 1539 (1970).
203. D. P. H. Hsieh, M. T. Lin, R. C. Yao and R. Singh, *J. Agric. Food Chem.*, **24**, 1170 (1976).
204. D. P. H. Hsieh, D. L. Fitzell and C. A. Reece, *J. Am. Chem. Soc.*, **98**, 1020 (1976).
205. R. C. Yao and D. P. H. Hsieh, *Appl. Microbiol.*, **28**, 52 (1974).
206. L. S. Lee, J. W. Bennett, A. F. Cucullu and R. L. Ory, *J. Agric. Food Chem.*, **24**, 1167 (1976).
207. D. P. H. Hsieh, M. T. Lin and R. C. Yao, *Biochem. Biophys. Res. Commun.*, **52**, 992 (1973).
208. J. W. Bennett, L. S. Lee, S. M. Shoss and G. H. Boudreaux, *Appl. Environ. Microbiol.*, **39**, 835 (1980).
209. D. L. Fitzell, D. P. H. Hsieh, R. C. Yao and G. N. La Mar, *J. Agric. Food Chem.*, **23**, 443 (1975).
210. C. P. Gorst-Allman, K. G. R. Pachler, P. S. Steyn, P. L. Wessels and D. B. Scott, *J. Chem. Soc. Chem. Commun.*, 916 (1976).
211. C. P. Gorst-Allman, K. G. R. Pachler, P. S. Steyn and P. L. Wessels, *J. Chem. Soc. Perkin Trans. 1*, 2181 (1977).

212. D. L. Fitzell, R. Singh, D. P. H. Hsieh and E. L. Motell, *J. Agric. Food Chem.*, **25**, 1193 (1977).
213. R. H. Cox, F. Churchill, R. J. Cole and J. W. Dorner, *J. Am. Chem. Soc.*, **99**, 3159 (1977).
214. P. S. Steyn, R. Vleggaar and P. L. Wessels, *J. Chem. Soc. Perkin Trans. 1*, 460 (1979).
215. C. P. Gorst-Allman, P. S. Steyn and P. L. Wessels, *J. Chem. Soc. Perkin Trans. 1*, 961 (1978).
216. K. G. R. Pachler, P. S. Steyn, R. Vleggaar and P. L. Wessels, *J. Chem. Soc. Chem. Commun.*, 355 (1975).
217. P. S. Steyn, R. Vleggaar, P. L. Wessels and D. B. Scott, *J. Chem. Soc. Chem. Commun.*, 193 (1975).
218. C. A. Townsend and S. B. Christensen, *Tetrahedron*, **39**, 3575 (1983).
219. C. A. Townsend, S. B. Christensen and K. Trautwein, *J. Am. Chem. Soc.*, **106**, 3868 (1984).
220. R. Singh and D. P. H. Hsieh, *Arch. Biochem. Biophys.*, **178**, 285 (1977).
221. T. J. Simpson, A. E. de Jesus, P. S. Steyn and R. Vleggaar, *J. Chem. Soc. Chem. Commun.*, 631 (1982).
222. C. A. Townsend, S. B. Christensen and S. G. Davis, *J. Am. Chem. Soc.*, **104**, 6152 (1982).
223. C. A. Townsend and S. G. Davis, *J. Chem. Soc. Chem. Commun.*, 1420 (1983).
224. C. A. Townsend, S. B. Christensen and S. G. Davis, *J. Am. Chem. Soc.*, **104**, 6154 (1982).
225. T. J. Simpson and D. J. Stenzel., *J. Chem. Soc. Chem. Commun.*, 890 (1982).
226. J. C. Vederas and T. T. Nakashima, *J. Chem. Soc. Chem. Commun.*, 183 (1980).
227. T. T. Nakashima and J. C. Vederas, *J. Chem. Soc. Chem. Commun.*, 206 (1982).
228. J. C. Vederas, *Can. J. Chem.*, **60**, 1637 (1982).
229. U. Sankawa, H. Shimada, T. Kobayashi, Y. Ebizuka, Y. Yamamoto, H. Noguchi and H. Seto, *Heterocycles*, **19**, 1053 (1982).
230. T. J. Simpson, A. E. de Jesus, P. S. Steyn and R. Vleggaar, *J. Chem. Soc. Chem. Commun.*, 338 (1983).
231. J. G. Heathcote, M. F. Dutton and J. R. Hibbert, *Chem. & Ind.*, 270 (1976).
232. G. J. Shaw, M. Chick and R. Hodges, *Phytochemistry*, **17**, 1743 (1978).
233. R. M. Horak, P. S. Steyn and R. Vleggaar, *J. Chem. Soc. Perkin Trans. 1*, 1745 (1983).
234. C. M. Maes and P. S. Steyn, *J. Chem. Soc. Perkin Trans. 1*, 1137 (1984).
235. J. W. Suttie, *Ann. Rev. Biochem.*, **54**, 459 (1985).
236. R. J. Gibbons and L. P. Engle, *Science*, **146**, 1307 (1964).
237. U. Delvos and G. Müller-Berghaus, *Naturwissenschaften*, **72**, 461 (1985).
238. P. Decottignies-LeMarechal, C. Ducrocq, A. Marquet and R. Azerad, *J. Biol. Chem.*, **259**, 15010 (1984).
239. M. de Metz, B. A. M. Soute, H. C. Hemker, R. Fokkens, J. Lugtenburg and C. Vermeer, *J. Biol. Chem.*, **257**, 5326 (1982).
240. J. J. Lee, L. M. Principe and M. J. Fasco, *Biochemistry*, **24**, 7063 (1985).
241. E. Leistner, in *Biosynthesis and Biodegradation of Wood Components* (Ed. T. Higuchi), Academic Press, New York, 1985, pp. 273–290.
242. M. H. Zenk and E. Leistner, *Lloydia*, **31**, 275 (1968).
243. E. Leistner, in *Pigments in Plants* (Ed. F. C. Czygan), Fischer, New York, 1980, pp. 352–369.
244. E. Leistner, in *Primary and Secondary Metabolism of Plant Cell Cultures* (Eds. K. H. Neumann, W. Barz and E. Reinhard), Springer, New York, 1985, pp. 215–224.
245. E. Leistner, in *The Biochemistry of Plants* (Ed. E. E. Conn), Vol. 7, Academic Press, New York, 1981, pp. 403–423.
246. N. M. Packter, in *The Biochemistry of Plants* (Ed. P. K. Stumpf), Vol. 4, Academic Press, New York, 1980, pp. 535–570.
247. T. J. Simpson, in *A Specialist Periodicals Report*, *Biosynthesis* (The Royal Society of Chemistry), London, 1983, pp. 1–44.
248. R. E. Olson and H. Rudney, in *Vitamins and Hormones* (Eds P. L. Munson, E. Diczfalusy, J. Glover and R. E. Olson), Vol. 39, Academic Press, New York, 1982, pp. 1–43.
249. E. Leistner, in *Recent Advances in Phytochemistry* (Ed. E. E. Conn), Plenum, New York, 1986, pp. 243–261.
250. O. Ghisalba, *Chimia*, **39**, 79 (1985).
251. R. H. Thomson, *Naturally Occurring Quinones*, Academic Press, London and New York, 1971.

The Chemistry of Quinonoid Compounds, Vol. II
Edited by S. Patai and Z. Rappoport

CHAPTER **23**

Quinones as oxidants and dehydrogenating agents*

HANS-DIETER BECKER and ALAN B. TURNER
Department of Organic Chemistry, Chalmers University of Technology and University of Gothenburg, S-412 96 Gothenburg, Sweden

* Professor Erich Adler (1905–1985) in memoriam.

I. INTRODUCTION

When the Chapter on 'Quinones as oxidants and dehydrogenating agents' was written for the first volume of *The Chemistry of the Quinonoid Compounds* which appeared in 1974, two decades of fascinating and fruitful exploratory research, originating with the investigations by Braude, Linstead, and their coworkers, were reviewed and evaluated[1]. A stepwise mechanism of dehydrogenation by quinones involving initial transfer of a hydride ion had been proposed, quinones were being applied as convenient oxidants in various branches of organic chemistry, and numerous types of compounds were found to be oxidizable by quinones. As a result, typical applications of quinones, such as in the aromatization of hydroaromatic compounds, benzylic oxidations, or the selective oxidation of allylic alcohols have become textbook examples in organic chemistry. When the present chapter was being written in the beginning of 1986, quinone dehydrogenation as a laboratory procedure had become so commonplace as to make selective rather then exhaustive coverage of the literature of the past 12 years mandatory.

Out of the large number of known quinones, the high potential quinones, namely, 2,3-dichloro-5,6-dicyano-1,4-benzoquinone (DDQ), tetrachloro-1,4-benzoquinone (*p*-chloranil), and tetrachloro-1,2-benzoquinone (*o*-chloranil) are most frequently applied as oxidants in synthetic organic chemistry. They have become 'reagents for organic synthesis', and examples of their application are found in every volume of the series of books by Fieser and Fieser[2]. For mechanism studies, however, numerous other quinones are also investigated, and less common quinones may find very specific applications. For example, 2,5-dibromo-6-isopropyl-3-methyl-1,4-benzoquinone is commercially available because of its use as an inhibitor of electron transfer in the photosynthesis system of chloroplasts[3]. On the other hand, anthraquinone and its derivatives are rarely used as dehydrogenating agents in synthetic organic chemistry, though they continue to be oxidants of outstanding industrial importance[4].

The examples of quinone oxidations discussed below were selected from the literature so as to supplement the chapter on 'Quinones as oxidants and dehydrogenating agents' published in 1974[1]. In particular, novel results and reactions, such as the oxidation of various silyl derivatives (cf. Sections II, III.B and VI.B), or the oxidative removal of the 4-methoxybenzyl group as a deprotection procedure (Section IV.D), have been considered. As for a comprehensive review of virtually all aspects of dehydrogenation of polycyclic hydroaromatic compounds, however, the reader is referred to an authoritative article published in 1978[5]. Also, DDQ and its applications as oxidant in organic chemistry, first comprehensively reviewed in 1967[6], were discussed in 1977[7], and in 1983[8].

By and large, the disposition of topics reviewed below is similar to that of the corresponding contribution to the first volume of *The Chemistry of the Quinonoid Compounds*[1]. Thus, the role of quinones in biological oxidations has been considered to be beyond the scope of the present review, though some recent results pertinent to the mechanism of quinone dehydrogenation are included in Section II.

The terms 'dehydrogenation' and 'oxidation' will be used indiscriminately throughout this chapter.

II. MECHANISM STUDIES

The seemingly simple two-step mechanism of quinone dehydrogenation of compounds AH_2, involving rate-determining hydride ion transfer to the quinone followed by fast loss of proton from AH^+ according to reactions 1 and 2 (cf. Ref. 1, p. 340), is a subject of recurring investigations[9–16]. However, neither the hydrogen group transfer by a concerted mechanism[10, 11], nor the stepwise ene-mechanism involving the formation of an intermediate[12], postulated in recent years, has been substantiated in subsequent investigations[13–16]. The results of a recent kinetic study dealing with the aromatization of numerous 1,4-dihydroarenes by DDQ convincingly support the stepwise ionic mechanism[14]. Likewise, in the dehydrogenation of 1,2-dihydronaphthalene by *o*-chloranil and *o*-bromanil, regioselective hydride ion transfer from the 2-position to the carbonyl oxygen was found to be the rate-determining step[15]. The reaction proceeds *cis*-stereoselectively, as is in agreement with the involvement of tight ion pair. The regioselectivity of dehydrogenation by DDQ is such as to be explicable by hydride ion transfer from the 1-position of 1,2-dihydronaphthalene, suggestive of a transition state geometry of parallel donor–acceptor arrangement.*

$$AH_2 + Q \rightarrow AH^+ + QH^- \qquad (1)$$

$$AH^+ + QH^- \rightarrow A + QH_2 \qquad (2)$$

Important experimental results of mechanistic significance which provide evidence against a concerted hydrogen transfer in quinone dehydrogenation were obtained in a study dealing with the aromatization of trimethylsilyl derivatives of isomeric methoxycarbonyl-substituted cyclohexa-1,4-dienes[16]. Dehydrogenation of cyclohexadiene **1** with DDQ in benzene at room temperature was found to proceed smoothly to give the *m*-trimethylsilyl derivative of methyl benzoate (**3**) in 89% yield. Unexpectedly, the *meta*-substituted compound **3** also was obtained (94% yield) by DDQ oxidation of the vicinal-substituted cyclohexa-1,4-diene **4**. As depicted in reactions 3 and 4, this latter result is explicable by a stepwise ionic mechanism in which the trimethylsilyl group in the intermediate cationic species **5** undergoes 1,2 migration.

(**1**) → (**2**) → (**3**) (3)

R = Me

(**4**) → (**5**) → (**6**) → (**3**) (4)

The formation of the rearranged product **3** from **4** is remarkable and mechanistically significant because the route of 'normal' dehydrogenation to give the *ortho*-trimethylsilyl methyl benzoate is available. In the dehydrogenation with palladium, which does not

* Added in proof: Recent results on the *cis*-selective hydrogen transfer from deuterium labelled 1,4-dihydroaromatic substrates are in agreement with the two-step mechanism: M. Brock, H. Hintze and A. Heesing, *Chem. Ber.*, **119**, 3727 (1986).

proceed according to reactions 1 and 2, cyclohexadiene **4** is indeed aromatized without concomitant rearrangement[17].

The syn-stereoselectivity of the dehydrogenations by DDQ of cyclohexa-1,4-dienes as outlined above is supported by the results of the DDQ oxidation of *cis*- and *trans*-3,6-dideuterio-cyclohexa-1,4-dienes. Also, the observed enchanced rate of aromatization by DDQ of *cis*-9,10-diisopropyl-9,10-dihydroanthracene, relative to that of the *trans* isomer, in refluxing benzene is explicable in terms of *syn*-stereoselectivity[18].

In the ionic mechanism of quinone dehydrogenations discussed above, the hydride ion is transferred to the carbonyl oxygen of the acceptor to give the hydroquinone anion. By contrast, dehydrogenation of hydroaromatic systems by concerted transfer of two hydrogens should proceed either by 1,4 addition to the enone system of the quinone (from 1,4-dihydroaromatic compounds), or by 1,2 addition (from 1,2-dihydroaromatic compounds) to the carbon–carbon double bond of the quinone. In both cases, the hydroquinone will be formed by subsequent tautomerization. Most recently, the kinetics and the isotope effects of the dehydrogenation of tetralin by DDQ, thymoquinone and anthraquinone were interpreted in these terms[19].

Oxidations by anthraquinone in alkaline solution are of considerable current interest in conjunction with recent progress in wood pulping [20,21]. Several mechanisms involving either the formation of addition products and/or electron transfer reactions from anthrahydroquinone anion have been advanced and may be operative. For the reduction of anthraquinone by hydroxide ion in aprotic media, experimental evidence has been obtained for a one-electron mechanism which involves nucleophilic addition of hydroxide to the carbonyl double bond[22].

The oxidation of coenzyme NADH or its model compounds by quinones continues to be of mechanistic interest within the general context of electron transfer phenomena[23–26]. Unequivocal evidence for one-electron transfer from *N*-methylacridan to 2,3-dicyano-benzoquinone was obtained by spin-trapping technique[27]. For the oxidation of 1-benzyl-1,4-dihydropyridine by a series of quinones of different redox potentials, the rate constants were found to span over a range of 11 orders of magnitude, so as to suggest a fleeting transition from one-electron transfer to hydride ion transfer mechanism, depending on the nature of the substrate and that of the quinone[28]. However, for numerous other model compounds and NADH itself, the kinetics of the oxidation by various *o*- and *p*-quinones were found to be in agreement with the hydride ion transfer mechanism[25].

In strongly acidic media, quinones undergo protonation (cf. Ref. 1, p. 413), and may then oxidize aromatic compounds by way of one-electron transfer. For example, 9,10-diphenylanthracene (DPA) upon oxidation with DDQ or *p*-chloranil in methylene chloride containing trifluoroacetic acid gives the DPA radical cation according to reactions 5 and 6[29]

$$\mathrm{DPA} + \mathrm{Q} + \mathrm{H}^+ \rightarrow \mathrm{DPA}^+ + \mathrm{QH}^{\cdot} \qquad (5)$$

$$\mathrm{DPA} + \mathrm{QH}^{\cdot} + \mathrm{H}^+ \rightarrow \mathrm{DPA}^{+\cdot} + \mathrm{QH}_2 \qquad (6)$$

$$2\,\mathrm{DPA} + \mathrm{Q} + 2\mathrm{H}^+ \rightarrow 2\,\mathrm{DPA}^{+\cdot} + \mathrm{QH}_2 \qquad (7)$$

The intermediate semiquinone radical is not detectable by spectroscopic means, and the kinetics of the overall reaction (7) indicate that QH in acidic media is more rapidly reduced than the quinone Q. (For a recent comparison of redox potentials of various quinonoid compounds as established by cyclic voltammetry, see Ref. 30).

III. DEHYDROGENATION OF HYDROAROMATIC COMPOUNDS

The dehydrogenation of hydroaromatic compounds was reviewed comprehensively in 1978[5]. Consequently, the discussion in this section will be limited to novel examples of quinone dehydrogenation.

A. Aromatization of Polycyclic Hydrocarbons

A survey of the literature readily reveals that high-potential quinones frequently are the reagents of choice for the aromatization of hydroaromatic compounds. Thus, in the synthesis of kekulene (**8**)[31], the final step involving dehydrogenation of its octahydro-derivative **7** was accomplished in 80% yield by oxidation with DDQ in 1,2,4-trichlorobenzene (reaction 8). The solvent used in this reaction was 'freshly purified by two distillations', and the dehydrogenation was carried out by keeping the reaction mixture at 100°C for three days under nitrogen.

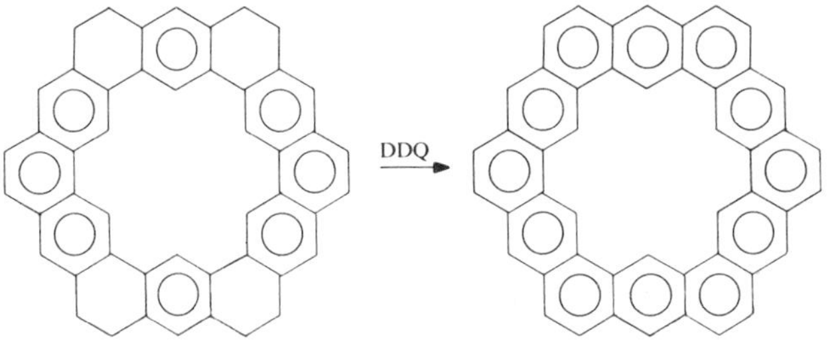

(**7**) —DDQ→ (**8**) (8)

More typical examples of dehydrogenation, generally by DDQ in benzene solution, include the aromatization of 1,4-dihydrophenanthrenes[32], various hydroanthracenes[33], photodimers of tetrahydronaphthacene[34], numerous hydropyrenes[35–37], 4,5-dihydro-1,12-methylenebenz[a]anthracene[38], 3,4-dihydrobenzo[g]chrysene[39], and various hydrobenzofluoranthenes[40].

Oxidation of hexahydrochrysene (**9**; R=H) with DDQ yields chrysene. However, with 5-alkylhexahydrochrysenes (**9**; R=Me or Et) the quinone oxidation surprisingly stops at the stage of dihydrochrysene **10** (reaction 9), although complete aromatization of **9** can be accomplished with palladium-charcoal[41].

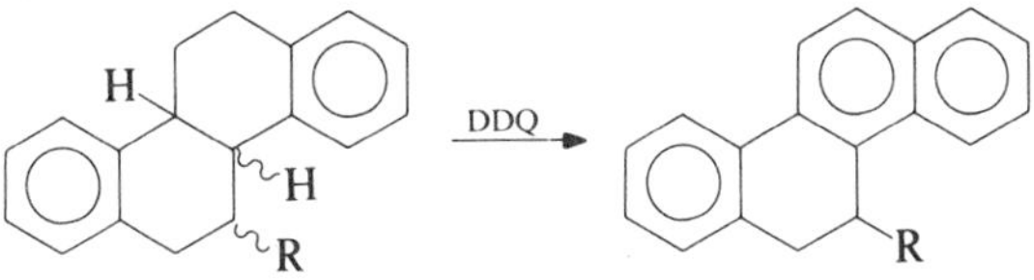

(**9**) —DDQ→ (**10**) (9)

Both aceanthrylene (**12**)[42, 43] and acephenanthrylene (**13**)[44] have been prepared recently from hydroaromatic precursors by dehydrogenation with DDQ (cf. reaction 10). The formation of acenaphthylene from acenaphthene by quinone dehydrogenation also proceeds smoothly (cf. Ref. 1, p. 365), but attempted dehydrogenation of the cycloheptenedione annelated acenaphthene **14** by DDQ failed[45] (cf. also Section VI.A).

Bridged bicyclic hydrocarbons such as 9,10-dihydro-9,10-ethenoanthracene have been prepared by dehydrogenation of hydroaromatic precursors, but the yields can be quite low[46]. Moderate yields (30%) were obtained in the synthesis of chiral triptycenes with an anthracene chromophore when the dehydrogenations were carried out in boiling benzene for prolonged periods of time[47]. It is not obvious why these dehydrogenations should require both elevated temperature and long reaction times. Similar aromatizations were accomplished with structurally related 'dihydrotriptycenes' under mild conditions in higher yields. For example, the formation of the novel triptycene **16** by DDQ oxidation of

DDQ

(11) (12)

(10)

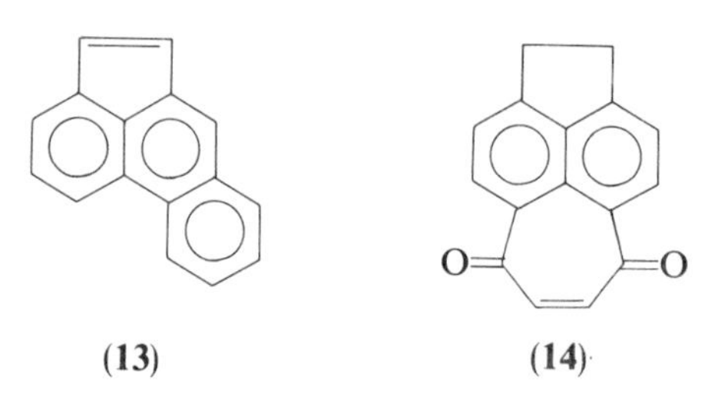

(13) (14)

15 for 1 h proceeded at room temperature in dichloromethane (reaction 11)[48]. (Under rather drastic conditions, i.e. in benzene solution at 240°C in a sealed tube, the hydrocarbon **16** can be obtained in 70% yield directly from 1,2-di(9-anthryl)ethane by oxidation with two molar equivalents of DDQ[49].) Dehydrogenations of tetrahydro or hexahydro derivatives of **16** indeed were found to require longer reaction times and elevated temperature than those of 1,2-dihydroaromatic systems[49].

DDQ

(11)

(15) (16)

An unexpected aromatization involving oxidative ring expansion was encountered in the dehydrogenation by DDQ of the difluoromethyl-substituted polycyclic hydrocarbon **17**[50]. Reaction 12 proceeds in 60% yield in refluxing benzene, but the mechanism of this intriguing transformation remains to be elucidated.

DDQ

CHF_2 F

(12)

(17) (18)

B. Dehydrogenation of Silyl-substituted Hydroaromatic Compounds

Bis-trimethylsilyl-substituted hydroaromatic compounds react with *p*-quinones predominantly by intermolecular transfer of one silyl group to give mono-silyl substituted aromatic hydrocarbons. Thus, oxidation of the *trans*-substituted 1,4-cyclohexadiene derivative **19** with chloranil in refluxing toluene gives mainly the mono-trimethylsilyl-substituted *m*-xylene **20** (80%), but only 10% of the bis-trimethylsilyl derivative **21**[51]. The oxidation of the 4,7-dihydroindole derivative **22** with *p*-benzoquinone under nitrogen at 0°C in methylene chloride affords 1,4-bis(trimethylsilyl)indole **23**[52]. The stereochemistry of compounds **19** and **22** has not been established, but the dehydrosilylations as shown in reactions 13 and 14 are in agreement with the *syn*-selectivity of quinone dehydrogenation discussed in Section II for trimethylsilyl-substituted 1,4-cyclohexadienes[16]. The dehydrosilylation of bis-trimethylsilyl acenaphthene by DDQ or chloranil to give the mono-silyl-substituted acenaphthylene[53] also is indicative of a *syn*-selective reaction. The observed regioselectivity of reaction 14 has been attributed to steric factors which, most likely, also govern the course of reaction 13. Although the fate of the silyl moiety in dehydrosilylations has not been investigated in the reactions presented above, the first step probably involves hydride ion transfer to the quinone. The subsequent step will be the transfer of the trimethylsilyl group to the hydroquinone mono-anion (cf. the analogous transfer of trimethylstannyl groups to *p*-benzoquinone, in Ref. 1, p. 403).

$SiMe_3$ ... *p*-chloranil → ... + ...

(**19**) (**20**; 80%) (**21**; 10%) (13)

benzoquinone →

(**22**) (**23**) (14)

C. Selective Dehydrogenation of Polycyclic Hydroaromatic Compounds

The application of quinones for the selective dehydrogenation, rather than straightforward aromatization, of hydroaromatic compounds has become of increasing importance in recent years. In connection with studies on carcinogenic metabolites of polyarenes, which are mainly dihydrodiols and diol epoxides, the selective dehydrogenation of various diol derivatives of hydroaromatic compounds frequently is part of synthetic schemes[54–59]. In general, dehydrogenations are accomplished with DDQ or *o*-chloranil. For example, in the chrysene series the hexahydrochrysene **24** undergoes rapid selective dehydrogenation with one molar equivalent of DDQ in refluxing benzene to give mainly the tetrahydrochrysene **25** (reaction 15)[54]. The observed regioselectivity of the oxidation correlates with the delocalization energies of the various possible benzylic carbocations, as calculated by perturbational MO methods[5, 54], unless the course of the oxidation is affected by steric factors. For example, oxidation of the dibenzoate **26** with two molar equivalents of DDQ in refluxing dioxane for three days gives the tetrahydrochrysene **27** in 91% yield (reaction 16)[54]. Differences in the rate of dehydrogenation of hydroaromatic diol esters are observed, but they are not easily rationalized. In certain instances, quinone dehydrogenation of hydroaromatic esters or carbinols has been found to lead to aromatized products by way of elimination reactions[56, 60].

DDQ

(24) (25) (15)

OBz OBz DDQ OBz OBz

(26) (27) (16)

D. Functionalization of Hydroaromatic Compounds

The reaction of DDQ with hydroaromatic compounds in the presence of water in either chloroform or dioxane results in benzylic oxidation to give ketones in fair to good yields (reaction 17)[61]. The formation of ketone **29** is explicable by the hydride ion transfer mechanism in which the carbocation formed from hydrocarbon **28** is trapped by water. The resulting benzylic alcohol then undergoes further oxidation. If there is more than one benzylic position, the functionalization takes place preferentially at the site for which the calculation of delocalization energies suggests the cationic intermediate to be the most stable. Side reactions competing with the functionalization may be the straightforward dehydrogenation of the hydroaromatic compound, or elimination of water from the intermediate benzylic carbinol. The choice of solvent appears to be of critical importance, most likely because charge-transfer complex formation between the substrate and the quinone is essential for the benzylic oxidation (cf. also Ref. 1, page 392.). Consequently, the functionalization is preferably carried out in solvents like chloroform or dioxane where donor:acceptor complex formation appears to be favored. Nevertheless, the yields of ketones greatly depend on the solvent. In the case of reaction 17, the yield of ketone **29** is 40% in chloroform, but 75% in dioxane.

DDQ dioxane H_2O O

(28) (29) (17)

IV. BENZYLIC AND ALLYLIC OXIDATIONS

A. Benzylic and Allylic Dehydrogenations

The effect of aryl substitution on the dehydrogenation by quinones has been discussed in detail in Ref. 1. Mechanistic features of benzylic and allylic dehydrogenations closely parallel those outlined for the dehydrogenation of hydroaromatic compounds in Section III. Therefore, the following discussion can be limited to selected novel examples of benzylic and allylic oxidations.

DDQ oxidation of the des-A-steroids **30** to the styrenes **31** (reaction 18) proceeds rapidly in dioxane at room temperature[62]. Interestingly, the rate of dehydrogenation is sensitive to changes in the substituent at the 17-position. Similar rate ratios for alcohol/acetate pairs are found for both free phenols and their methyl ethers (R = OH and OMe, respectively), suggesting direct attack by DDQ at the tertiary benzylic C–H bond. In aromatic steroids, seemingly subtle and remote structural changes may drastically affect the rate and course of benzylic dehydrogenation by high-potential quinones[63–65]. Differences in molecular geometry indeed affect the reactivity of benzylic or allylic carbon–hydrogen bonds, not only because attack by the hydride ion acceptor may be impaired for steric reasons (which may explain why only one of the stereoisomeric photodimers of 6,6-dimethyl-2,3-benzo-2,4-cycloheptadienone undergoes dehydrogenation by *o*-chloranil[66]), but also because the sterochemistry of conjugated systems has electronic consequences. For example, the *trans*-isomeric *p*-methoxystyrene derivatives **32** upon oxidation with DDQ give tetrahydrofurans **33** (reaction 19), but the *cis*-isomeric compounds are stable towards DDQ[67]. Obviously, only in the *trans* isomer **32** is the molecular geometry such as to permit coplanar arrangement of the carbon–carbon double bond with the aromatic π system. Reaction 19 also has been extended to the synthesis of tetrahydropyrans[67].

OR^1 OR^1 DDQ H R R **(30)** **(31)** (18)

R OH OMe DDQ R O H $\overset{+}{O}$Me **(32)** (19) R O (An)

An = 4-methoxyphenyl **(33)**

As for the formation of cyclic conjugated olefins, quinone dehydrogenation has been applied to the synthesis of azulenes (cf. Ref 1, p. 366), but the yields can vary greatly. For example, oxidation of 4,5-dihydro-4-phenylazulene with chloranil in boiling xylene gives 4-phenylazulene in only 11% yield[68], while azulene-5-carboxylic acid (**35**) was obtained in 65% yield from its hexahydro precursor **34** by DDQ oxidation in boiling benzene (reaction 20)[69].

COOEt —DDQ→ COOH (20)

(**34**) (**35**)

The dehydrogenation of aryl-substituted compounds for which double bond formation is structurally impossible may give rise to products whose formation is explicable by either ionic or radical intermediates. The oxidation of lignan **36** with two molar equivalents of DDQ in boiling benzene affords the naphthalide **37** (reaction 21), most likely by way of an electrophilic substitution following hydride ion abstraction from the benzylic position[70]. By contrast, benzylic oxidation of the fluorenylidene derivative **38** with *o*-chloranil in benzene (four weeks at room temperature) was found to give 1,4-di(fluorenylidene)-2,3-diphenylbutane (**39**), conceivably by radical dimerization (reaction 22)[71].

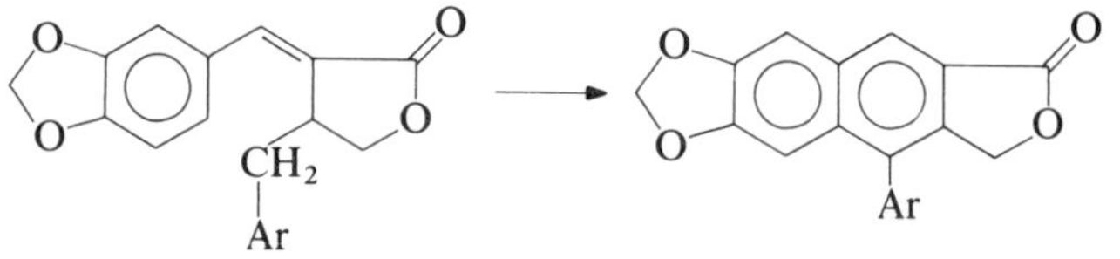

(21)

(**36**) (**37**)

Ar = 3,4-dimethoxyphenyl

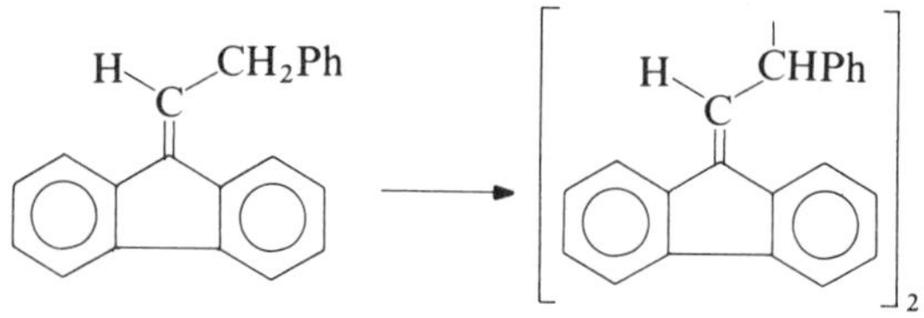

(22)

(**38**) (**39**)

Even if straightforward dehydrogenation is formally possible, quinone oxidation may result in the formation of thermodynamically more stable products by way of ionic fragmentation. In the recently reported DDQ oxidation of a *t*-butoxy-substituted methanoannulene in refluxing benzene, the cationic intermediate formed by hydride ion abstraction from a benzylic position apparently gives the final product by elimination of *t*-butyl cation[72].

B. Oxidation of Aryl Carbinols and Related Alcohols

The results of a kinetic study of the oxidation of benzylic alcohols by DDQ and *p*-chloranil are in agreement with the two-step ionic mechanism in which the transfer of a hydride ion from the benzylic C–H bond is rate determining[73]. The yields of propiophenone by oxidation of phenylpropanol with *p*-chloranil were found to be highest in solvents like chloroform and methylene chloride which favor the formation of an alcohol: quinone charge-transfer complex.

The experimental details of the oxidation of numerous primary and secondary benzylic alcohols with DDQ in dioxane, to give the corresponding carbonyl compounds (cf. Ref. 1, pp. 379–382), have now been published in full[74]. Significantly, the nature of the products obtained in the DDQ oxidation of phenolic benzyl alcohols is dependent on the solvent. The formation of oxidatively coupled products obtained in methanol solution is indicative of hydrogen transfer from the phenolic hydroxyl group to the quinone[74].

A stereoelectronic effect is apparent in the oxidation of benzocycloalkenols **40** by DDQ in refluxing benzene which gives the corresponding ketones **41** (reaction 23)[75]. The observed order of reactivity, i.e. α-tetralol > indan-1-ol ≫ 2,3-benzocyclohepten-1-ol parallels the decrease in the dihedral angle (82, 72, and 48 degrees) between the benzylic C–H bond and the plane of the benzene ring. Steric hindrance probably also accounts for the fact that the benzylic alcohol moiety in an azaphenanthrene alkaloid resisted oxidation by DDQ (and other oxidants as well)[76].

H OH
$(CH_2)_n$ → O $(CH_2)_n$ (23)

(**40**) (**41**)

As for the oxidation of allylic alcohols, the usefulness of DDQ is apparent from a recent investigation dealing with the selective oxidation of cyclohexene-3,4-diols[77]. Thus, oxidation of the *cis*-enediol **42** with 2.7 molar equivalents of DDQ in benzene at 60°C for 12 hours afforded ketol **43** in quantitative yield. Likewise, oxidation of shikimic acid with three molar equivalents of DDQ in THF at 65°C gave dehydroshikimic acid (60% yield). DDQ is superior to other oxidants such as manganese dioxide which lead to the corresponding catechols. Interestingly the selective oxidation by DDQ also was found to be applicable to the bis (*t*-butyldimethylsilyl) ether **44** which gave the ketol derivative **45** (reaction 24)[77].

OR OR → O OR (24)

(**42**; R = H) (**43**; R = H)
(**44**; R = TBDMS) (**45**; R = TBDMS = *t*-$BuMe_2Si$)

An apparent economical improvement for the oxidation of allylic alcohols involves catalytic amounts of DDQ in a two-phase system consisting of benzene, aqueous hydrochloric acid (0.1 normal), and periodic acid[78]. The yields of propenones **47** obtained in this fashion from propenols **46** range between 80 and 90% (reaction 25), and steroidal allylic alcohols can be oxidized selectively in the same manner. The role of the periodic acid in these reactions is that of an oxidant for $DDQH_2$ to regenerate DDQ. However, the mechanism of the catalytic oxidation is not readily understood: high-yield conversions are brought about with 10 mol% of DDQ in conjunction with a molar ratio of DDQ:periodic acid of 1:3. Consequently, also oxidation states of iodine lower than + 7 must be involved in the oxidation.

$$\mathrm{RCH=CHCHOHR'} \rightarrow \mathrm{RCH=CHCOR'} \quad (25)$$

e.g. R = phenyl, R′ = H

(46) **(47)**

Analogous to benzylic and allylic alcohols, cyclopropyl carbinols undergo oxidation by DDQ according to the hydride ion transfer mechanism. Oxidation of dicyclopropylmethanol with two molar equivalents of DDQ in anhydrous benzene gives dicyclopropyl ketone[79]. Likewise, DDQ oxidation of triasteranol **48** affords the triasteranone **50** via the intermediate triasteryl cation **49** (reaction 26). By contrast, carbinol **51**, in which hydride ion abstraction no longer gives rise to a cyclopropyl conjugated ionic intermediate, was found to be stable towards DDQ[79].

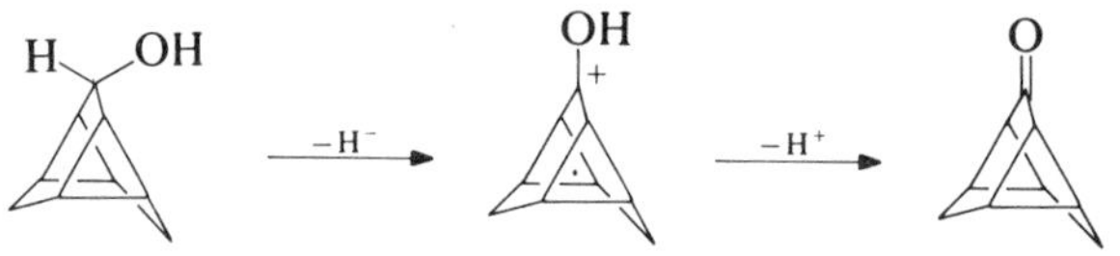

H OH —$-H^-$→ OH + —$-H^+$→ O (26)

(48) **(49)** **(50)**

OH

(51)

C. Functionalization of Benzylic and Allylic Positions*

Analogous to the functionalization of hydroaromatic compounds described in Section III. D, benzylic or allylic positions will be functionalized by oxidation with quinones, if the cationic intermediate formed by hydride ion abstraction can be trapped by nucleophiles such as water, alcohols, or acids (cf. Ref. 1, pp. 372–378). Thus, oxidation of methyl- and ethyl-substituted polycyclic aromatic hydrocarbons with DDQ in chloroform in the presence of water gives the corresponding carbonyl derivatives[61]. By and large, the yields of aldehydes or ketones decrease with decreasing number of aromatic rings. The oxidation of 1-methylpyrene and 9-methylphenanthrene gave the corresponding aldehydes in 33% and 20% yield, respectively, but 1-naphthaldehyde could not be obtained by oxidation of 1-methylnaphthalene[61]. Alkyl-substituted azulenes undergo regioselective oxidation by DDQ in aqueous acetone or dioxane to give acyl-substituted azulenes in good yields, as exemplified by the conversion of **52** into **53** (reaction 27)[80, 81].

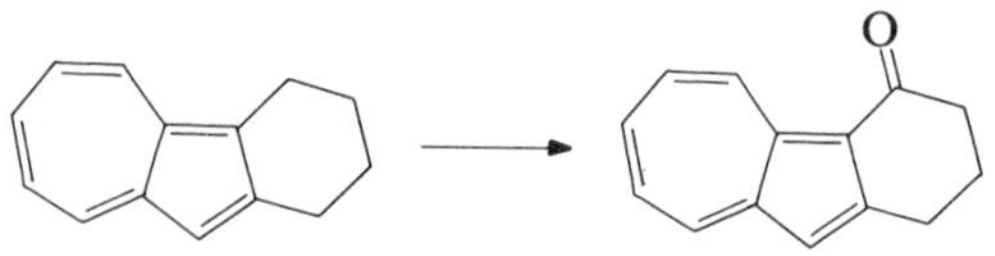

O (27)

(52) **(53)**

* Added in proof: Since the completion of the manuscript, the asymmetric functionalization by DDQ of the benzylic position has been reported. The asymmetric control of the oxidative acetoxylation is attributed to stereoselective donor:acceptor interaction between the substrate and DDQ: M. Lemaire, A. Guy, D. Imbert and J.-P. Guetté, *J. Chem. Soc., Chem. Commun.*, 741 (1986).

Selective oxidation of the side chain of indoles can be achieved by using DDQ in aqueous tetrahydrofuran[82]. Thus, tetrahydrocarbazole (**54**; $n = 2$) upon treatment with two molar equivalents of DDQ at 0°C gives tetrahydrocabazol-4-one (**55**; $n = 2$) in 83% yield (reaction 28). Analogous selective oxidations were accomplished with various cycloalkan-[b] indoles **54**, and their *N*-substituted derivatives[82]. However, 2,3,5-trimethylindole is oxidized less selectively giving the 3-formyl and 2-formyl derivatives in 30% and 8% yield, respectively[82].

$(CH_2)_n$ → $(CH_2)_n$ (28)

(**54**) (**55**)

The functionalization of indol derivatives has been extended to the synthesis of 3- and 4-acylindoles from 1,2,3,4-tetrahydro-β-carbolines by DDQ oxidation in aqueous THF[83]. The regioselectivity of the oxidation increases by lowering the reaction temperature.

DDQ has been shown to oxidize *p*-methyl groups of *o*-substituted aromatic amines[84]. For example, mesidine in dioxane is converted into 4-amino-3,5-dimethylbenzaldehyde in 49% yield. Likewise, *o*-chloro- and *o*-bromo-toluidines are oxidized to give the corresponding aldehydes in 46% and 64% yield, respectively. Steric hindrance of the amino group is a prerequisite for aldehyde formation, as *p*-toluidine gave no aldehyde at all, and 2,4-dimethylaniline gave only a 10% yield of aldehyde. (*N*-Acetyl derivatives are unreactive.) Obviously, the oxidation of *p*-methyl-substituted anilines resembles the benzylic oxidation of *p*-methyl-substituted phenols (Ref. 1, p. 392). Similar selective benzylic oxidations by DDQ in refluxing benzene to give aldehydes have been reported for 4-methoxytoluene and similarly substituted 1,2-dihydronaphthalenes and chromenes[85], and methyl-substituted tetralins[86]. More remarkable, however, is the one-pot conversion of the substituted tetralone **56** to the trimethoxy-substituted naphthaldehyde **57** (reaction 29)[87]. Three equivalents of DDQ are required, and the reaction is run for 24 hours in refluxing methanol in the presence of trimethyl orthoformate for *in situ* methylation of the intermediate enol.

O OMe Me OMe —DDQ, $HC(OMe)_3$, MeOH→ OMe OMe CHO OMe (29)

(**56**) (**57**)

Benzylic oxidation with DDQ in the presence of methanol under anhydrous conditions leads to benzyl methyl ethers or dimethyl ketals of arylcarbonyl compounds[88]. The oxidation of tetramethoxy-substituted flavan-3-ol derivatives **58** by DDQ in chloroform containing methanol has been found to result in stereospecific methoxylation of the benzylic position to give compounds **59** (reaction 30)[89]. A double molar excess of DDQ was employed in order to achieve short reaction times so as to avoid side reactions.

Examples of intramolecular benzyl ether formation by way of quinone oxidation have previously been encountered in the dehydrogenation of phenolic arylalkanes (cf. Ref. 1, pp. 386–387).

MeO, O, Ar, OR, OMe (58) —DDQ, $CHCl_3$, MeOH→ MeO, O, Ar, OR, OMe, OMe (59) (30)

R = H or Ac
Ar = 3,4-dimethoxyphenyl

The reaction of DDQ in anhydrous acetic acid at room temperature with electron-rich arylalkanes gives benzyl acetates in high yields. Thus, mesitol (**60**) is smoothly converted into 3,5-dimethyl-4-hydroxybenzyl acetate (**61**; reaction 31)[90]. Likewise, 4-methoxydiphenylmethane upon oxidation with DDQ in acetic acid gives the acetate of 4-methoxydiphenyl carbinol. Both 4-hydroxy- and 4-methoxy-substituted ethylbenzenes undergo benzylic acetoxylation in the same fashion[90].

OH, Me, Me, Me (60) —DDQ, AcOH→ OH, Me, Me, CH_2OAc (61) (31)

Toluene in refluxing acetic acid was found to be stable towards oxidation with DDQ, but a variety of other methyl-substituted benzenes reacted under the same conditions to give mono-acetoxylated products in drastically varying yields (see Table 1)[91]. The observed effect of substituents on the rate of reaction was found to be such as to support the hydride ion transfer mechanism. The absence of nuclear acetoxylation products precludes the involvement of radical ion intermediates in the oxidation.

$$\text{ArMe} \xrightarrow[\text{AcOH}]{\text{DDQ}} \text{ArCH}_2\text{OAc} \qquad (32)$$

TABLE 1. Acetoxylation of methyl-substituted aromatic compounds by DDQ in acetic acid (reaction 32)

Compound	Yield (%) of benzyl acetate
Hexamethylbenzene	80
P-Xylene	44
m-Xylene	10
Toluene	0

D. Oxidative Conversions of 4-Methoxybenzyl Derivatives by DDQ

1. O-Methoxybenzyl deprotection

The functionalization of benzylic positions is greatly facilitated by electron-donating substituents such as hydroxy or methoxy groups in the *p*-position, and the oxidative cleavage of 4-hydroxy and 4-methoxybenzyl ethers by DDQ in the presence of water has been discussed previously (Ref. 1, p. 392). However, the synthetic potential of this cleavage reaction has only recently been recognized, and its usefulness as a deprotection method has been demonstrated in the synthesis of numerous natural products[92]. The significance of the methoxybenzyl ether deprotection lies in the remarkable selectivity of the DDQ oxidation, as other common hydroxyl protecting groups, such as benzyl, methoxymethyl, *t*-butyl-dimethylsilyl, or acetyl, virtually remain unchanged. Moreover, the reaction is carried out in neutral solution, so that acid- or base-sensitive groups are left intact.

The principle of deprotection by oxidative cleavage of *p*-methoxybenzyl ethers **62** rests on the electron acceptor property of DDQ to form charge-transfer complexes with electron-rich aromatic compounds. Hydride ion transfer to the quinone from the benzylic position by way of heterolytic dissociation of the CT complex gives the semiquinone anion and a cationic species **63** which undergoes nucleophilic attack by water. The final products of the reaction, i.e. *p*-methoxybenzaldehyde (**65**) and the deprotected alcohol **66**, are obtained in high yields from the intermediate hemiacetal (reaction 33)[92, 93].

OMe
H—C—H
OR
(**62**)
DDQ
CT complex
+OMe
H C OR
(**63**)
$-DDQH_2$
$+H_2O$
OMe
H—COH
OR
(**64**)

OMe
CHO
(**65**)
+ ROH
(**66**)
(33)

e.g. R = phenethyl

The stability of the charge-transfer complex and, consequently, the rate of oxidation is greatly affected by solvent polarity (cf. Table 2). In 18:1 methylene chloride/water mixtures, the oxidative deprotection of *p*-methoxybenzyl ethers proceeds rapidly at room temperature.

TABLE 2. Effect of solvent composition on the oxidative cleavage of *p*-methoxybenzyl ether **62** at room temperature

Solvent	**62**: DDQ	Reaction time (h)	Yield of alcohol **66** (%)
MeOH	1:1	24	86
THF/H_2O (10:1)	1:1	24	85
$CH_2Cl_2/MeOH$ (4:1)	1:1	6	87
CH_2Cl_2/H_2O (18:1)	1:1	0.6	89
CH_2Cl_2/H_2O (18:1)	1:1.1	0.2	84

The oxidative 3,4-dimethoxybenzyl ether cleavage also has been used for the deprotection of hydroxyl groups[94]. Because of their lower oxidation potential, 3,4-dimethoxybenzyl ethers undergo DDQ oxidation even more rapidly than *p*-methoxybenzyl ethers. The selectivity of the deprotection reaction is apparent from the DDQ oxidation of the protected tetraol **67** which gives deprotected alcohols **68** and **69** in a ratio 92:8 (reaction 34)[94].

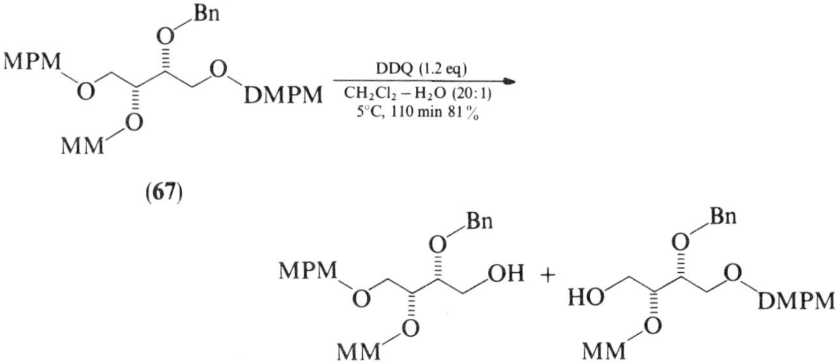

DDQ (1.2 eq)
$CH_2Cl_2 - H_2O$ (20:1)
5°C, 110 min 81%

(67)

(92:8)

(68) **(69)** (34)

Bn: benzyl; MM: methoxymethyl; MPM: 4-methoxybenzyl; DMPM: 3,4-dimethoxybenzyl

Both *p*-methoxybenzyl ether and 3,4-dimethoxybenzyl ether deprotection are finding extensive use in the synthesis of natural products[95-97].*

(The oxidation of 2,6-dimethoxybenzyl esters by DDQ also has been described as a deprotection method[98]. However, the formulae shown in Ref. 98 actually show 2,4-dimethoxybenzyl esters of carboxylic acids to undergo oxidative deprotection.)

2. *Oxidative acetalization of 1,2- and 1,3-diols*

In agreement with the ionic mechanism for the DDQ oxidation of methoxy-substituted benzyl ethers (cf. reaction 33), intramolecular nucleophilic attack by hydroxyl groups results in the formation of *p*-methoxybenzal acetals **71** (reaction 35)[99]. The DDQ oxidation of ethers **70** proceeds smoothly in anhydrous methylene chloride, and the reaction has also been extended to 3,4-dimethoxybenzyl ethers[99].

The absence of water is of critical importance for the formation of acetals **71** in high yields, as they may undergo further benzylic oxidation by DDQ to give 'deprotected' hydroxy-substituted esters **72** (reaction 36)[99].

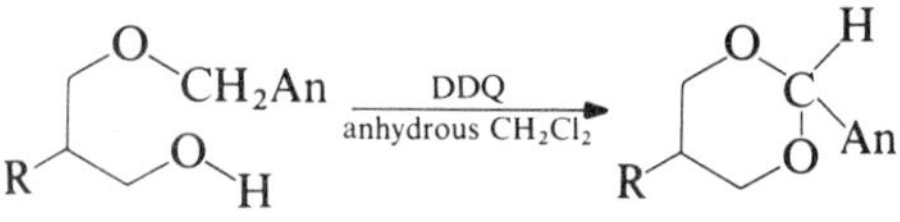

DDQ
anhydrous CH_2Cl_2

(70) **(71)** (35)

* Added in proof: Several interesting examples of selective deprotection by DDQ oxidation in the synthesis of macrolides have been reported since the completion of this manuscript: Y. Oikawa, T. Tanaka and O. Yonemitsu, *Tetrahedron Lett.*, **27**, 3647 (1986); T. Tanaka, Y. Oikawa, T. Hamada and O. Yonemitsu, *Tetrahedron Lett.*, **27**, 3651 (1986); N. Nakajima, T. Hamada, T. Tanaka, Y. Oikawa and O. Yonemitsu, *J. Am. Chem. Soc.*, **108**, 4645 (1986).

Interestingly, the protection of 1,2- and 1,3-diols by *p*-methoxybenzal acetalization can be accomplished by DDQ oxidation of *p*-methoxybenzyl methyl ether (**73**) in the presence of an appropriate 1,2- or 1,3-diol which reacts as a nucleophile (reaction 37)[100].

(**71**) $\xrightarrow[CH_2Cl_2,\ H_2O]{DDQ}$ (**72**) (36)

(**73**) + (**74**) $\xrightarrow[CH_2Cl_2]{DDQ}$ (**75**) + MeOH + $DDQH_2$ (37)

As the reaction does not require acid catalysis, the resulting acetals **75** do not undergo the stereochemical equilibration by which thermodynamically controlled mixtures of isomers are formed. This 'kinetic acetalization' of 1,2- and 1,3-diols by DDQ oxidation of *p*-methoxybenzyl methyl ether (**73**) is of considerable interest for the synthesis of chiral compounds[101] and may even be applied in those cases where conventional acid-catalyzed acetalization fails.

V. DEHYDROGENATION OF AROMATIC HYDROXY AND AMINO COMPOUNDS

A. Oxidation of Monohydric and Dihydric Phenols

High-potential quinones continue to find application as convenient oxidants for catechols and hydroquinones. A recent detailed study of the kinetics of oxidation of triazoliothiohydroquinones by 1,4-benzoquinones reveals deuterium isotope effects of ca. 3–6, and the results are in agreement with an apparent hydride transfer mechanism[102]. It is conceivable that the overall two-electron oxidation consists of two sequential one-electron transfers from the hydroquinone to the quinone. However, as the first step would lead to two semiquinones of drastically different redox potentials, the second step is expected to be so fast as to preclude kinetic detection of one-electron transfer products.

For preparative oxidations of catechols, hydroquinones, or related aromatic dihydroxy compounds, both DDQ and *o*-chloranil are the reagents of choice. Thus, DDQ in methylene chloride at 20°C has been used in the synthesis of benzoquinone-bridged porphyrins from the corresponding hydroquinones[103]. Likewise, oxidation of the di-*t*-butyl-substituted 1,5-dihydroxynaphthalene **76** with DDQ in methylene chloride under nitrogen gives the 1,5-naphthoquinone **77** in excellent yield (reaction 38)[104].

(**76**) $\xrightarrow[CH_2Cl_2]{DDQ}$ (**77**) (38)

As for the oxidation of 1,10-dihydroxy-substituted anthracenes (or their keto forms), appropriate substitution by alkyl groups is a prerequisite for the stability of the 1,10-anthraquinone system with respect to reaction with water and subsequent dehydrogenation, which ultimately leads to hydroxy-substituted 9,10-anthraquinones. Thus, 3-*t*-butyl-5,8-dimethyl-1,10-anthraquinone (**79**) was obtained by oxidation of the hydroxyanthrone **78** with DDQ in dry ether at $-5°C$ (reaction 39)[105]. The course of the dehydrogenation of 1,4,9,10-tetrahydroxy-substituted anthracenes can be governed by the choice of the quinone. For example, DDQ oxidation may give rise to the 9,10-anthraquinone, while oxidation by *o*-chloranil can lead to the 1,4-anthraquinone[106]. *o*-Chloranil also is used for the oxidation of catechols to give the corresponding *o*-quinones (cf. Ref. 1, p. 384) in remarkably high yields[107].

Me OH DDQ ether Me O Me Bu-*t* Me Bu-*t* O O (39)

(**78**) (**79**)

The selective oxidation of the hydroquinone system in compound **80**, and concomitant partial aromatization to give the naphthoquinone derivative **81** was accomplished by oxidation with two molar equivalents of DDQ in benzene at room temperature (reaction 40)[108].

OH R DDQ benzene O R OH O OH O OH O (40)

(**80**) (**81**)

R = CH=CHCH$_3$

The course of the oxidation of hydroxy-substituted anthracenes can be governed by the nature of the solvent which affects the position of the keto–enol equilibrium. Thus, DDQ oxidation of the dihydroxy-substituted dianthrylethane **82** in refluxing dioxane gives the stilbenequinone **83** (reaction 41), but the reaction of the keto-tautomer **84** with DDQ in refluxing chloroform containing ethanol affords the diethoxy derivative **85**[109]. The formation of **85** is indicative of a carbocationic intermediate which has been trapped by ethanol.

[OH CH$_2$]$_2$ DDQ dioxane [O C H]$_2$ (41)

(**82**) (**83**)

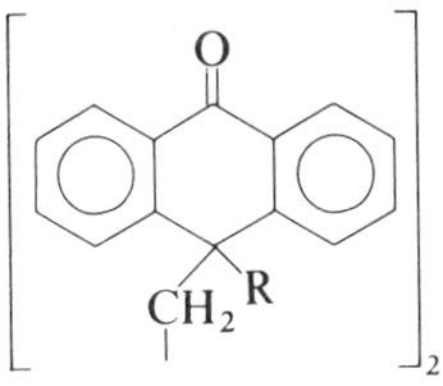

(41)

(**84**; R = H)
(**85**; R = ethoxy)

The reaction of 2-naphthol (**86**) with DDQ in solvents such as methanol, ethylene glycol, and 2,2-dimethylpropane-1,3-diol results in the formation of *o*-quinone ketals **87** by way of oxidative nucleophilic substitution (reaction 42)[110,111]. *p*-Quinone ketals **89** are formed in similar fashion by DDQ oxidation of hydroquinone monoalkyl ethers **88** in methanol solution containing catalytic amounts of *p*-nitrophenol (reaction 43)[112]. (The reaction is of synthetic interest insofar as it complements other means of oxidative ketalization, such as the oxidation of phenols with thallium(III) nitrate.) Analogous *p*-naphthoquinone dimethyl ketals have been obtained by DDQ oxidation in methanol in the absence of *p*-nitrophenol[113]. Presumably, DDQ itself may act as a catalyst by virtue of its Lewis acid properties.

The reaction of substituted 1- and 2-naphthols with either DDQ or *o*-chloranil gives rise to a variety of coupling products whose formation may be rationalized by one-electron transfer processes. Thus, oxidation of *o*-methyl-substituted naphthols leads to spiro-compounds via intermediate *o*-quinone methides[114,115]. The straightforward dehydrogenation by DDQ of bis(2-hydroxy-1-naphthyl)methanes results in the formation of spiro-quinol ethers by carbon–oxygen coupling[116,117]. In a recent detailed study, the oxidation of dinaphthol methane **90** was found to give *cis*- and *trans*-isomeric spiro-quinol ethers whose formation has been rationalized by the involvement of the 2,3-dichloro-5,6-dicyanohydroquinone ether **91** (reaction 44)[117].

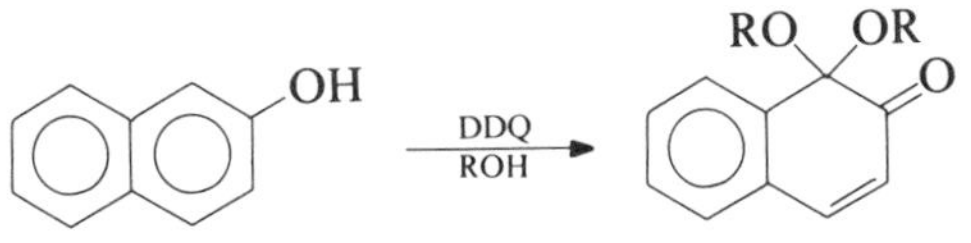

(**86**) —DDQ / ROH→ (**87**) (42)

(**88**) —DDQ / methanol→ (**89**) e.g. R = allyl (43)

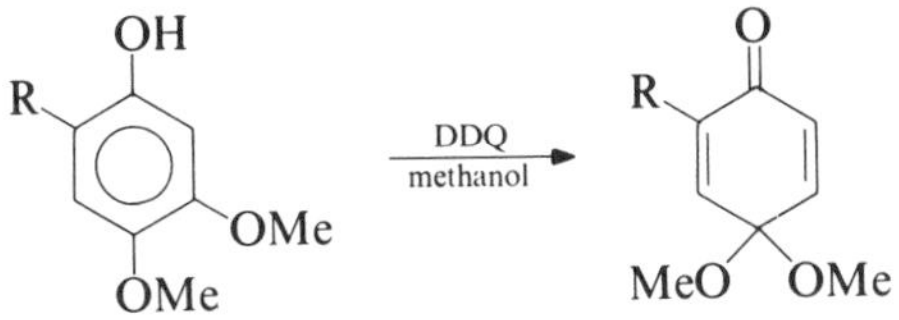

The oxidation of 2-hydroxystilbenes with DDQ may lead to either benzofuran structures by way of intramolecular coupling, or to benzopyrans derived from intermolecular carbon–oxygen coupling to give flavones[119].* DDQ oxidation of prenylated hydroxy-substituted isoflavones results in the formation of cyclodehydrogenation pro-

* Added in proof: 2′-Hydroxychalcones upon dehydrogenation with DDQ in dioxane have recently been reported to undergo cyclization to flavanones, flavones and aurones: K. Imafuku, M. Honda and J. F. W. McOmie, *Synthesis*, 199 (1987).

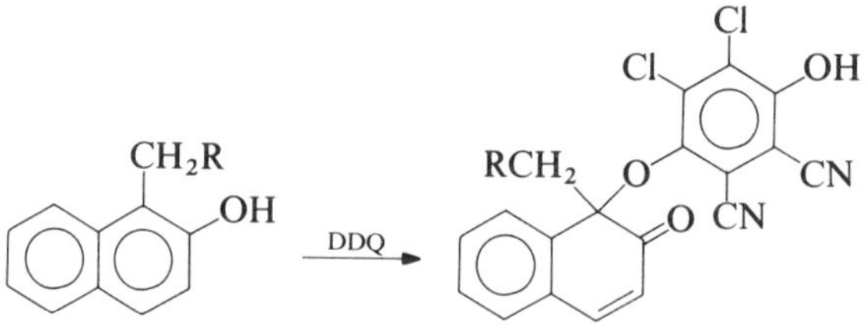

(44)

(90) (91)

R = 2-hydroxynaphthyl

molecular carbon–oxygen coupling to give flavones[119]. DDQ oxidation of prenylated hydroxy-substituted isoflavones results in the formation cyclodehydrogenation products[120]. Oxidation of 4-hydroxy-3-methoxyphenylpropan-2-one by DDQ in dioxane gives mainly a polymer linked through the benzylic position[88]. In the case of 2-cinnamyl-4,5-methylenedioxyphenol, DDQ oxidation in acetone or ether solution was found to give a crystalline *o*-quinone methide[121].

B. Dehydrogenation of Aromatic Amino Compounds

The oxidation of sterically hindered *p*-methyl-substituted anilines resulting in benzylic oxidation has been discussed in Section IV. C. Two novel quinone oxidations of aromatic amino compounds were reported recently. The oxidation of dicyano-substituted *p*-phenylenediamines **92** by DDQ in refluxing benzene gives *N,N'*-dicyanoquinonediimines **93** (reaction 45)[122]. DDQ oxidation failed in the case of the dichloro derivative (**92**, R = Cl) whose oxidation potential exceeds that of DDQ.

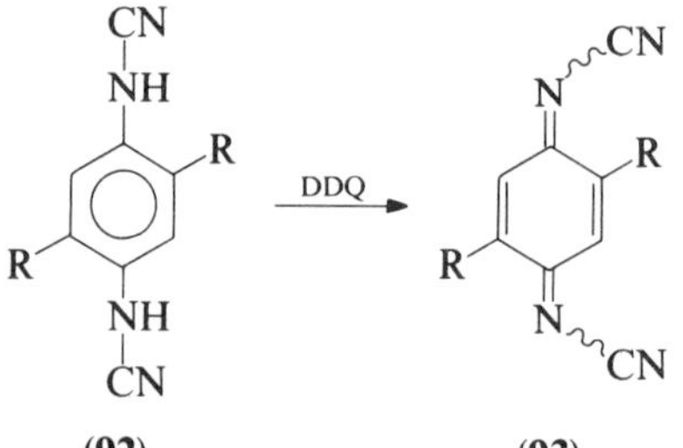

(45)

(92) (93)

An interesting oxidation by DDQ involving hydrogen abstraction from the nitrogen of stannyl-substituted hydrazone **94** was found to give azocyclopropane **95** in 88% yield (reaction 46). The reaction proceeds rapidly at −20°C in methylene chloride, and is suggested to occur by a radical mechanism[123].

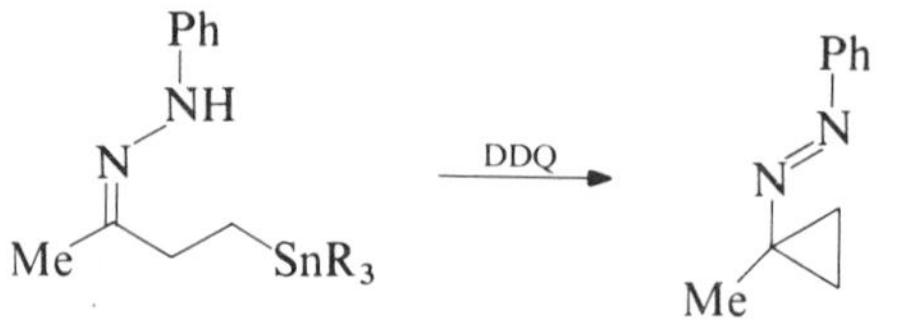

(46)

(94) (95)

R = *n*-butyl

VI. OXIDATION OF CYCLIC KETONES, ENOLS AND SILYL ENOL ETHERS

A. Dehydrogenation of Cyclic Ketones and Their Silyl Enol Ethers

Saturated cyclic ketones are difficult to dehydrogenate with quinones under normal conditions, but the reaction can be catalyzed by acids which enhance formation of the corresponding enols (cf. Ref. 1, pp. 352, 354). Certain derivatives of cyclohexanone have recently been found to give the corresponding cyclohexenones by oxidation with DDQ in dry HCl-saturated dioxane, though chloro-enones also are formed in substantial yields[124]. More conveniently, the dehydrogenation of cyclic ketones can be carried out with DDQ in benzene solution in the presence of a catalytic amount of *p*-toluenesulfonic acid[125].

Cyclohexenone **96** upon treatment with DDQ smoothly aromatizes in refluxing dioxane to give phenol **97** in 80% yield (reaction 47)[126]. Conceivably, the dehydrogenation of **96** is facilitated by naphthyl substitution. Attempts to convert 4-acetyl-4-methylcyclohexenone into the corresponding 2,5-cyclohexadienone by dehydrogenation with DDQ in refluxing benzene were unsuccessful. Instead, the reaction afforded (after 11 days!) a modest yield of 4-methylphenyl acetate[127]. Acyl migration was also observed in the DDQ oxidation of a bicyclic enone (Wieland–Miescher ketone)[127].

O R R DDQ OH R R (47)

(**96**) (**97**)

R = 1-naphthyl

The reaction of DDQ with the tricyclic compound **98** in refluxing benzene is interesting because formation of the tropolone system **99** (reaction 48) is favored over the structurally possible benzylic dehydrogenation[128].

R O R R R DDQ R O R R R (48)

(**98**) (**99**)

R = methoxy

If cyclic ketones are first converted into their enol acetates, dehydrogenation by high-potential quinones will then lead to the formation of the acetate of the corresponding linearly conjugated dienol. Numerous acetates of hydroxy-substituted aromatic hydrocarbons have been prepared in that fashion by oxidation with either *o*-chloranil or DDQ[129–131].

A significant improvement of the dehydrogenative conversion of cyclic ketones into the corresponding cyclic enones consists in the two-electron oxidation of silyl enol ethers[132,133]. These oxidations with molar equivalents of DDQ generally proceed smoothly in benzene at room temperature. In some cases, oxidations are carried out with small excess of DDQ in the presence of 2,4,6-collidine, which may suppress hydrolysis of the silyl enol ether. Cyclohexenone has been prepared in 60% yield from 1-trimethylsiloxycyclohexene by oxidation with DDQ (but not with chloranil)[133]. DDQ oxidation of 1-methoxycyclohexene, by contrast, gives anisole. The DDQ oxidation of silyl enol ethers has been utilized in key steps (reactions 49 and 50) of the synthesis of carvone (**103**)[134] and for the synthesis of heterocyclic enones[135,136]. As the rate of dehydrosilylation exceeds that of straightforward dehydrogenation, the oxidation of silyl enol ether moieties with high-potential quinones proceeds selectively[137]. The mechanism of dehydrosilylation has not been studied in detail, but the reaction most likely proceeds stepwise[132,133] and parallels the dehydrosilylation of hydroaromatic compounds discussed in Section III. B.

$OSiMe_3$ Me (**100**) —DDQ→ O Me (**101**) (49)

$OSiMe_3$ Me Me C CH_2 (**102**) —DDQ→ O Me Me C CH_2 (**103**) (50)

Trimethylsilyl enol ethers also may be dehydrosilylated to give enones by oxidation with Pd(II) acetate in acetonitrile in the presence of *p*-benzoquinone[138–140]. As these oxidations proceed smoothly in the absence of *p*-benzoquinone if equimolar amounts of Pd(II) acetate are used, *p*-benzoquinone obviously only serves to oxidize reduced Pd(II) acetate (cf. also Section VIII).

Interestingly, bis(trimethylsilyl) ethers of hydroquinones also undergo oxidative desilylation by high-potential quinones. This reaction has been applied in the synthesis of various quinones of azulenes, as exemplified by the conversion of silyl derivative **104** into azuloquinone **105** (reaction 51)[141,142].

$OSiMe_3$ $OSiMe_3$ (**104**) —DDQ or *p*-chloranil→ O O (**105**) (51)

B. Oxidation of Stable Enols and Enolized 1,3-Dicarbonyl Compounds

Enols and dienediols can be oxidized by quinones in a fashion which, in principle, is analogous to the oxidation of phenols and catechols or hydroquinones (cf. Ref. 1, p. 357).

For example, the dihydroxy tropolone **106** upon oxidation with DDQ in methanol at room temperature is converted into tropoquinone **107** (reaction 52)[143,144].

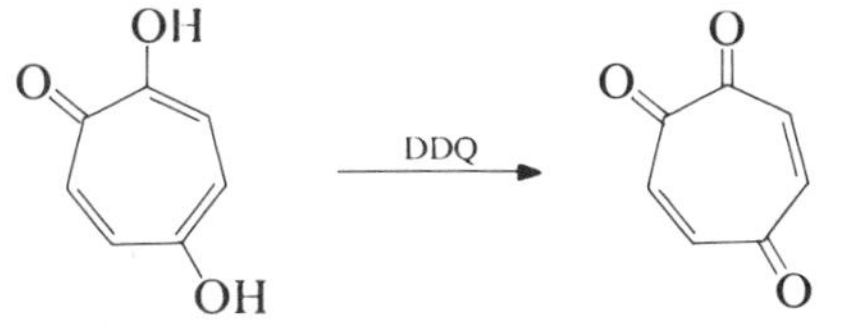

(52)

(106) **(107)**

The dehydrogenation of numerous enolized 2-acyl-substituted cyclohexanones by DDQ has recently been investigated[145]. Oxidation of enolized acylcyclohexanones **108** with one molar equivalent of DDQ in dioxane at room temperature was found to proceed rapidly and gave acyl ketones **109** in generally good yields (reaction 53). DDQ oxidation of enolized 2-acetyl-4,4-dimethylcyclohexanone (**108**, R = Me) does give the 2-acetylcyclohexenone **109** (R = Me) in 67% yield, but the reaction is very much slower than in the case of the formyl derivatives, and requires the presence of a catalytic amount of acetic acid[145].

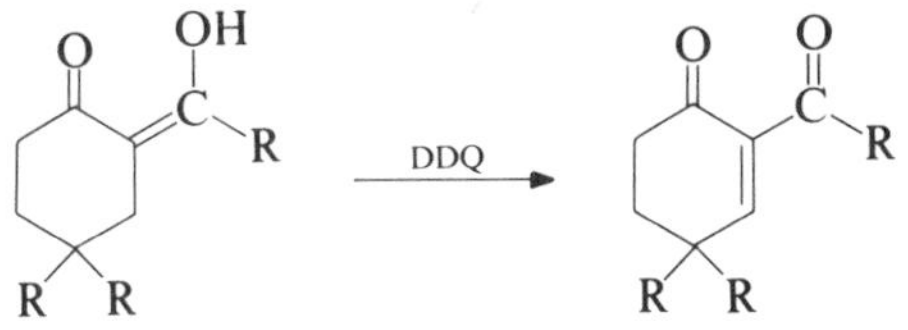

(53)

(108) **(109)**

e.g. R = H
R = Me

Quite recently, the enol of (cycloheptatrienyl)malonaldehyde (**110**) was found to undergo a remarkable skeletal rearrangement upon oxidation with DDQ (and also with silver oxide). Rather than giving (cycloheptatrienylidene)malonaldehyde, the reaction of DDQ with **110** in methylene chloride at −10°C afforded benzylidenemalonaldehyde (**111**) in 94% yield (reaction 54)[146]. (The mechanism of the oxidative ring contraction has not been investigated.)

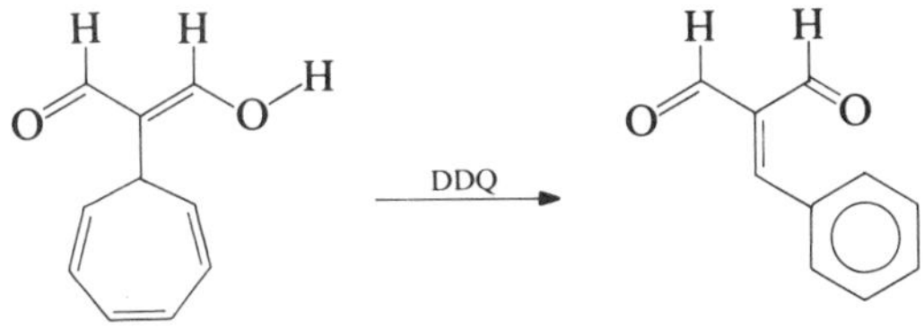

(54)

(110) **(111)**

VII. DEHYDROGENATION OF HETEROCYCLIC COMPOUNDS

A. Nitrogen Heterocycles

The DDQ oxidation of enantiomerically pure naphthyl-substituted 1,4-dihydroquinoline derivatives described recently[147] represents a significant example of quinone dehydrogenation insofar as it has provided unequivocal experimental evidence for the intramolecular transfer of a central chiral element to an axial chiral element. Dehydrogenation of the *R*-enantiomer **112** with DDQ in THF at $-78°C$ results in the formation of the *R*-enantiomeric naphthylquinoline **113** (reaction 55). The correspondingly substituted *S*-enantiomer was obtained by DDQ oxidation in THF at $-78°C$ of the *S*-enantiomeric dihydroquinoline[147].

THF, $-78°C$

(55)

(**112**) (**113**)

e.g. X = CHO

Less stringent temperature control is usually required in straightforward dehydrogenations of various nitrogen heterocycles. For example, the oxidation of the piperidine derivative **114** with DDQ in refluxing benzene proceeds selectively and gives the 1,2,3,4-tetrahydropyridine **115** (reaction 56)[148]. Similarly, phthalazines and quinazolines are obtained from their dihydro precursors by oxidation with DDQ in refluxing dioxane[149] and pyrrolines are converted into pyrroles by DDQ oxidation in benzene at 70°C[150].

DDQ

(56)

(**114**) (**115**)

Ar = *o*-tolyl
R = methoxycarbonyl
R^1 = Me

However, proper choice of the reaction temperature was recently found to be of critical importance in the quinone dehydrogenation of 3,4-dihydro-2*H*-pyrroles **116**[151]. Oxidation with DDQ in benzene at room temperature gives the expected 2*H*-pyrrole **117**, whereas oxidation in boiling benzene results in the loss of one carboethoxy group and the formation of the 1*H*-pyrrole **119**. Oxidation of **116** with chloranil in boiling xylene, by contrast, gives the 1*H*-pyrrole **118** whose formation involves a thermally induced acyl migration from carbon to nitrogen (cf. Scheme I; reactions 57–59)[151].

Quinone dehydrogenation of heterocyclic compounds containing more than one hetero atom usually proceeds in straightforward fashion.* 4-Aryl- and 4-phenacyl-substituted 3,4-dihydropyrimidinones are readily converted to the fully conjugated pyrimidinones by DDQ in benzene at room temperature[152,153]. The dehydrogenation of 4,5-dihydropyridazines with DDQ in boiling anhydrous benzene in the absence of oxygen affords pyrazines[154]. Δ^2-Isoxazolines are converted into isoxazoles by dehydrogenation with excess DDQ in refluxing benzene in excellent yield[155], but dehydrogenation of oxazolines under similar conditions appears to be more difficult[76].

* Added in proof: In a recent report on the synthesis of a heteroaromatic 14π system, both *p*-chloranil and DDQ have been used as dehydrogenation agents, but only the application of DDQ was found to lead to the aromatic ring system: R. Neidlein and L. Tadesse, *Chem. Ber.*, **119**, 3862 (1986).

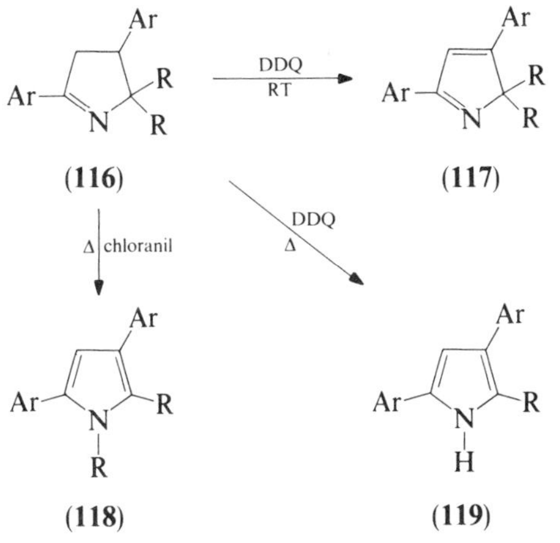

Ar = phenyl; R = ethoxycarbonyl

SCHEME I

(57–59)

During recent years, high-potential quinones such as *o*-chloranil and DDQ have found useful synthetic application in the dehydrogenation of dipeptide azlactones **120**[156]. These oxidations usually proceed at room temperature in dioxane or 1,2-dimethoxyethane in the presence of a base such as pyridine, imidazole, or collidine, but yields of dehydropeptides **121** are only close to 50% (reaction 60) (cf. also Ref. 136, dealing with the formation of dehydrolactones by way of dehydrosilylation as discussed in Section VI. A).

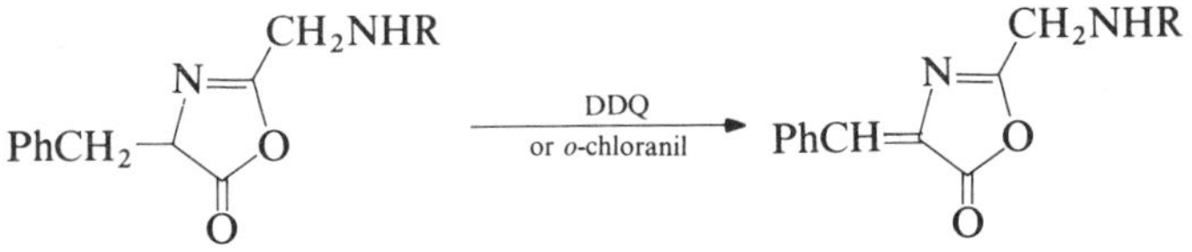

(60)

For the conversion of porphyrinogens into porphyrins (cf. Ref. 1, p. 408), high-potential quinones are most convenient oxidizing agents. For example, the dehydrogenation of the *o*-nitrophenyl-substituted compound **122** by *o*-chloranil in THF proceeds at room temperature, and subsequent reduction of the nitro group affords atropisomeric *meso*-diphenylporphyrins **123a** and **123b** (reaction 61)[157]. Similarly, the dehydrogenation of porphyrinogens has been carried out with DDQ at room temperature in degassed acetonitrile/ether[158].

The formation of biliverdin (**125**) by quinone dehydrogenation of bilirubin (**124**) has been studied in detail, and DDQ in DMSO at room temperature was found to be the reagent of choice (reaction 62)[159]. Low concentration of bilirubin and excess of DDQ are necessary in order to avoid concomitant formation of biliverdin isomers. (The order of mixing of the reagents is also of importance.) It has been suggested that the oxidation of bilirubin by DDQ proceeds by one-electron transfer steps involving resonance-stabilized tetrapyrrole radicals[159, 160].

In the dehydrogenation of nitrogen heterocycles by DDQ, the formation of colored charge-transfer complexes is frequently noticeable and, for that reason, DDQ may be used

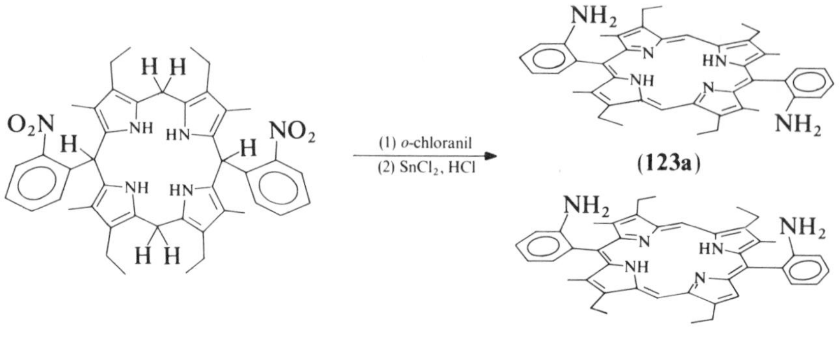

(1) *o*-chloranil
(2) $SnCl_2$, HCl

(122) **(123a)** **(123b)**

(61)

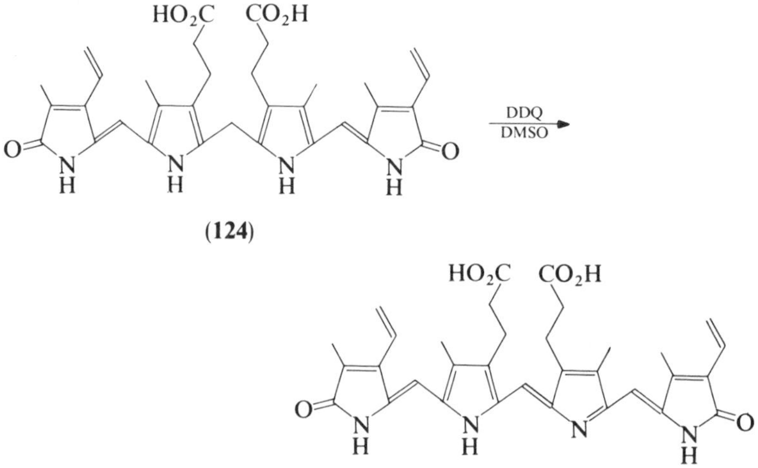

DDQ
DMSO

(124)

(125)

(62)

as a spray reagent for chromatographic detection[161]. The stability of these colored complexes depends largely on the nature of the solvent. For example, indole and DDQ form a crystalline charge-transfer complex which is stable in methylene chloride. In methanol or dioxane, by contrast, indole reacts with DDQ to give a substitution product whose formation can be rationalized by nucleophilic attack on the quinone[162]. Upon heating, the substitution product eliminates HCN and gives an indoloquinone. Similar substitution–elimination reactions have previously been noted in the reaction of *p*-chloranil with quinindine derivatives (cf. Ref. 1, pp. 408–409).

As for 3-alkyl-substituted indoles **126**, their reaction with DDQ results in straightforward dehydrogenation and gives 3-alkylidene-3*H*-indoles **127** (reaction 63)[162]. The oxidation of an *N*-acetyl-dihydropyridylindole with DDQ in ethyl acetate was found to result in an acyl group transfer[162].

Finally, the reaction of DDQ with the bis-trimethylsilyl-substituted nitrogen heterocycle **128** in dichlorobenzene at 180°C deserves mention because the product **129** seemingly is not formed by dehydrogenation but by twofold oxidative desilylation (reaction 64)[163].

(126) —DDQ→ (127) (63)

(128) —DDQ, 180°C→ (129) (64)

B. Oxygen and Sulfur Heterocycles

In most examples of oxygen heterocycles undergoing dehydrogenation by high-potential quinones, substrate activation is provided by the presence of aromatic substituents. However, reaction times may vary greatly, and yields of dehydrogenation products frequently are modest (cf. Ref. 1, p. 409). Oxidation of dihydrobenzofurans with DDQ in refluxing dioxane gives benzofurans in yields of 46–70%[164, 165]. Remarkably, in the case of dihydrofuropyridines **130**, their oxidation with two molar equivalents of DDQ in refluxing dioxane affords furopyridines **131** only when the pyridine ring is substituted (reaction 65)[166].

(130) —DDQ→ (131) (65)

As for six-membered oxygen heterocycles, DDQ in benzene has been used for the dehydrogenation of chroman moieties in alkaloids[167]; (cf. also Ref. 168). However, attempted dehydrogenation of a 4-chlorochroman with DDQ in benzene at room temperature was found to result instead in dehydrochlorination[169].

The dehydrogenation of heterocyclic ketones by quinones generally proceeds smoothly. For example, the conversion of flavanones into flavones by DDQ oxidation in refluxing dioxane proceeds in high yields in far shorter reaction times than in refluxing benzene[170]. Oxidation by DDQ in benzene was found to be the method of choice for the dehydrogenation of benzothienoannelated dihydrocoumarins (dihydrothiacoumestans)[171]. In methanol solution, however, the reaction of high-potential quinones with 3,4-dihydrocoumarins results mainly in the formation of ring-opened products by way of lactone solvolysis[172].

The oxidation of the tetrakis(t-butylthio)-substituted thiophene derivative **132** with chloranil in refluxing acetonitrile was recently reported to give the cyclic trithioanhydride **133** in 92% yield (reaction 66)[173].

(66)

(132) (133)

An interesting formation of the thiadiazole system **135** (90% yield) involves the oxidative cyclization of a semithiacarbazone **134** by DDQ in refluxing dioxane (reaction 67)[174].

$$RCH{=}NNHC(S)NMe_2 \xrightarrow{DDQ} \text{(135)} \qquad (67)$$

(134) (135)

VIII. OXIDATIONS INVOLVING ORGANOMETALLIC COMPOUNDS

The oxidation of various organometallic compounds by quinones has been discussed in Ref. 1, p. 411. The reaction of ferrocene with DDQ was then described to give rise to ferrocenium cation radical and the hydroquinone anion radical. In a subsequent study, the crystalline 1:1 charge-transfer complex between DDQ and decamethylferrocene has been investigated by X-ray diffraction, and was found to consist of the expected ferrocenium cation radical and the hydroquinone anion (rather than the anion radical)[175]. In the oxidation of nickel(0) complexes by certain *p*-quinones, hydroquinone anion radical formation has been established by ESR spectroscopy[176, 177].

The role of *p*-benzoquinone in the oxidative formation of enones from silyl enol ethers in the presence of palladium(II) chloride has been mentioned in Section VI. *p*-Benzoquinone is also used frequently as oxidant in synthetically interesting palladium-catalyzed reactions of olefins, such as the rearrangement of 1-vinyl-1-cyclobutanols[178], or the Wacker-type oxidative conversion of terminal olefins into methyl ketones[179] (cf. also Ref. 1, p. 345). Likewise, *p*-benzoquinone functions as oxidant in the conversion of cyclohexene into 2-cyclohexenyl acetate which is catalyzed by palladium(II) chloride in acetic acid[180]. Similarly, *p*-benzoquinone is used in the palladium(II) chloride-catalyzed 1,4-diacetoxylation of 1,3-dienes[181–183]. In these and related palladium(II)-catalyzed reactions, benzoquinone acts both as electron acceptor and as ligand for the metal complex.

The oxidation of arylmagnesium halides with quinones gives rise to diaryl compounds (cf. Ref. 1, p. 411). In a novel reaction of considerable synthetic potential, organomagnesium compounds like **136** and **137**, formed by conjugate addition of alkyl-

R = methyl; ethyl (136) (137) (138) (139)

(68)

Grignard reagents to nitroarenes, are smoothly oxidized by DDQ in THF at 0°C to give the corresponding alkyl-substituted nitroarenes **138** and **139** in quantitative yields (reaction 68)[184, 185].

IX. MISCELLANEOUS OXIDATIONS BY DDQ

Among the few quinones commonly used as oxidants and dehydrogenating agents in synthetic organic chemistry, DDQ is unique because of its high oxidation potential and its versatile reactivity. Being an exceptionally strong electron acceptor, DDQ has been found to induce and bring about some remarkable reactions. For example, oxepinobenzofurans **140** are valence isomers of 2,2′-diphenoquinones (previously believed to have a spiroquinol ether structure; cf. Ref. 1, p. 390) which in the presence of water are oxidized by DDQ to give benzofuranylidene derivatives such as **142**[186]. It has been suggested that DDQ catalyzes the addition of water to compound **140** (reaction 69) so as to give the intermediate **141** which is then dehydrogenated (reaction 70).

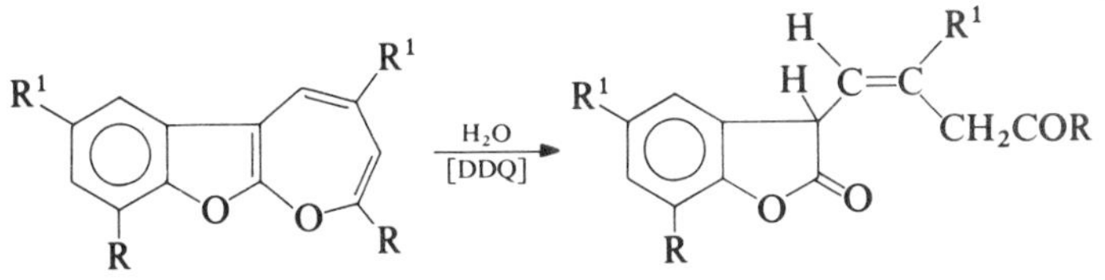

H_2O / [DDQ]; CH_2COR (69)

(140) **(141)**

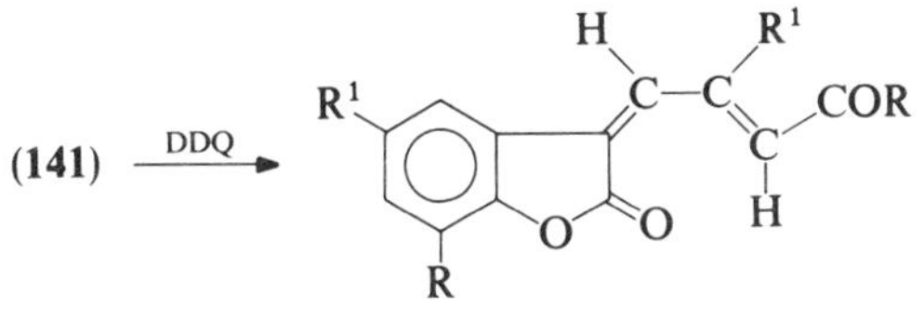

(141) DDQ → (70)

(142)

Upon oxidation with two molar equivalents of DDQ in the presence of water, benzofuranylidene derivatives **143** are converted into isoxindigos **144** in high yield[186]. The mechanism of this remarkable reaction is also believed to involve an initial benzylic oxidation of an intermediate which is formed by DDQ-catalyzed addition of water to the exocyclic carbon–carbon double bond of compound **143**[186].

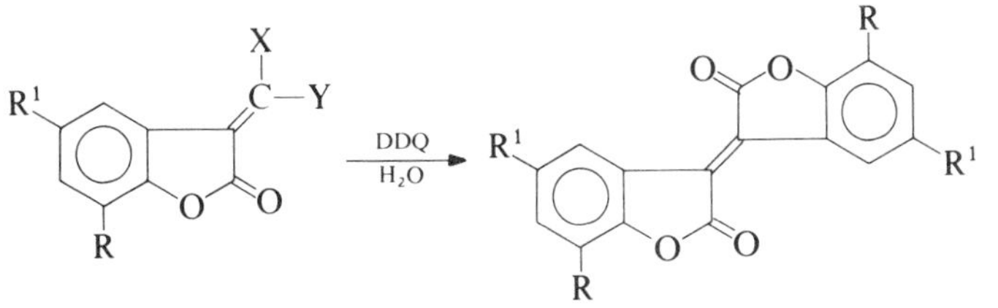

DDQ / H_2O (71)

(143) **(144)**

Methyleneanthrone (**145**) was found to react with DDQ in refluxing dioxane to give a tetramer whose formation conceivably involves the oxidative dimerization of the intermediate dimer **146**[187].

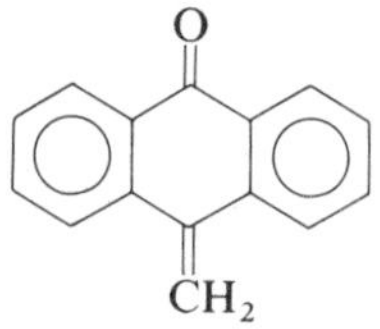

(**145**) 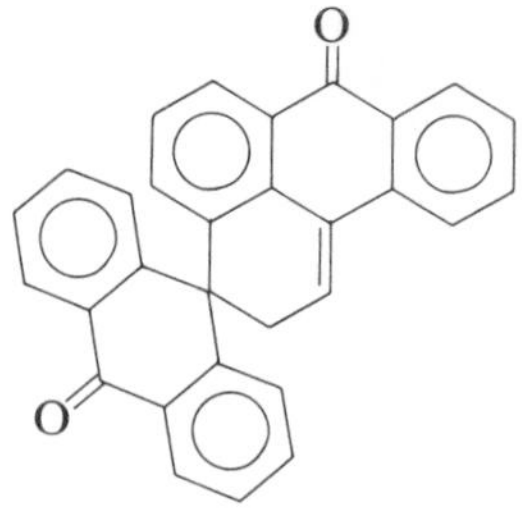 (**146**)

Diazomethane reacts with DDQ by elimination of nitrogen and addition of methylene to the cyano-substituted carbon–carbon double bond to give a bicycloheptene derivative[188]. The reaction of DDQ with diphenyldiazomethane (**147**) in the presence of alcohols results in the reduction of DDQ and gives acetals **148** in excellent yields (reaction 72)[188–190]. Macrocyclic acetals have been prepared by this method from α, ω-diols[191]. It has been suggested that acetals **148** are formed via an intermediate DDQ:diazonium betaine.

$$Ph_2C{=}N_2 + DDQ + ROH \longrightarrow Ph_2C(OR)_2 + DDQH_2 \qquad (72)$$

(**147**) (**148**)

X. REFERENCES

1. H.-D. Becker, in *The Chemistry of the Quinonoid Compounds* (Ed. S. Patai), Wiley-Interscience, New York, 1974, p. 335.
2. M. Fieser, *Reagents for Organic Synthesis*, Vol. 11, John Wiley & Sons, New York, 1984; and preceding volumes.
3. A. Trebst, E. Harth and W. Draber, *Z. Naturforsch.*, **25b**, 1157 (1970).
4. O. Bayer, in *Methoden der Organischen Chemie* (Houben-Weyl-Müller), 4th edn, Vol. 7/3c, Georg Thieme Verlag, Stuttgart, 1979, p. 271.
5. P. P. Fu and R. G. Harvey, *Chem. Rev.*, **78**, 317 (1978).
6. D. Walker and J. D. Hiebert, *Chem. Rev.*, **67**, 153 (1967).
7. A. B. Turner, in *Synthetic Reagents*, Vol. 3 (Ed. J. S. Pizey), Halsted Press, Ellis Harwood, Chichester, 1977, p. 193.
8. A. J. Fatiadi, in *The Chemistry of Functional Groups, Supplement C* (Ed. S. Patai and Z. Rappoport), John Wiley & Sons, New York, 1983, pp. 1057, 1209.
9. Z. M. Hashish and J. M. Hoodless, *Can. J. Chem.*, **54**, 2261 (1976).
10. F. Stoos and J. Rocec, *J. Am. Chem. Soc.*, **94**, 2719 (1972).
11. P. Müller, *Helv. Chim. Acta*, **56**, 1243 (1973).
12. B. M. Jacobson, *J. Am. Chem. Soc.*, **102**, 886 (1980).
13. P. Müller and D. Joly, *Helv. Chim. Acta*, **66**, 1110 (1983); P. Müller, D. Joly and F. Mermoud, *Helv. Chim. Acta*, **67**, 105 (1984).
14. R. P. Thummel, W. E. Cravey and D. B. Cantu, *J. Org. Chem.*, **45**, 1633 (1980).
15. R. Paukstat, M. Brock and A. Helsing, *Chem. Ber.*, **118**, 2579 (1985).
16. M. J. Carter, I. Fleming and A. Percival, *J. Chem. Soc., Perkin Trans. 1*, 2415 (1981).
17. M. E. Jung and B. Gaede, *Tetrahedron*, **35**, 621 (1979).
18. R. G. Harvey and P. P. Fu, unpublished results, cited in Ref. 5, p. 340.
19. J. Pajak and K. R. Brower, *J. Org. Chem.*, **50**, 2210 (1985).
20. D. R. Dimmel, *J. Wood Chem. Technol.*, **5**, 1 (1985); and references cited therein.
21. R. R. Dimmel, L. F. Perry, P. D. Palasz and H. L. Chum, *J. Wood Chem. Technol.*, **5**, 15 (1985).

22. J. L. Roberts, H. Sugimoto, W. C. Barrette and D. T. Sawyer, *J. Am. Chem. Soc.*, **107**, 4556 (1985).
23. F. M. Martens, J. W. Verhoeven, R. A. Gase, U. K. Pandit and Th. J. de Boer, *Tetrahedron*, **34**, 443 (1978).
24. A. K. Colter, A. G. Parsons and K. Foohey, *Can. J. Chem.*, **63**, 2237 (1985); and references cited therein.
25. B. W. Carlson and L. L. Miller, *J. Am. Chem. Soc.*, **107**, 479 (1985).
26. L. Eberson, *Adv. Phys. Org. Chem.*, **18**, 79 (1982).
27. C. C. Lai and A. K. Colter, *J. Chem. Soc. Chem. Commun.*, 1115 (1980).
28. S. Fukuzumi, N. Nishizawa and T. Tanaka, *J. Org. Chem.*, **49**, 3571 (1984).
29. W. J. Sep, J. W. Verhoeven and Th. J. de Boer, *Tetrahedron*, **35**, 2161 (1979).
30. A. Aumüller and S. Hünig, *Liebigs Ann. Chem.*, 165 (1986).
31. H. A. Staab and F. Diederich, *Chem. Ber.*, **116**, 3487 (1983).
32. R. Lapouyade, A. Veyres, N. Hanafi, A. Couture and A. Lablache-Combier, *J. Org. Chem.*, **47**, 1361 (1982).
33. R. Sangaiah, A. Gold and G. E. Toney, *J. Org. Chem.*, **48**, 1632 (1983).
34. R. Lapouyade, A. Nourmamode and H. Bouas-Laurent, *Tetrahedron*, **36**, 2311 (1980).
35. J. Pataki, M. Konieczny and R. G. Harvey, *J. Org. Chem.*, **47**, 1133 (1982).
36. M. Tashiro and T. Yamato, *J. Org. Chem.*, **50**, 2939 (1985).
37. R. G. Harvey, M. Konieczny and J. Pataki, *J. Org. Chem.*, **48**, 2930 (1983).
38. J. K. Ray and R. G. Harvey, *J. Org. Chem.*, **48**, 1352 (1983).
39. S. K. Agarwal, D. R. Boyd and W. B. Jennings, *J. Chem. Soc. Perkin Trans. 1*, 857 (1985).
40. S. Amin, K. Huie, N. Hussain, G. Balanikas and S. S. Hecht, *J. Org. Chem.*, **50**, 1948 (1985).
41. L. A. Levy and V. P. Sashikumar, *J. Org. Chem.*, **50**, 1760 (1985).
42. R. Sangaiah and A. Gold, *Org. Prep. Proc. Int.*, **17**, 53 (1985).
43. H.-D. Becker, L. Hansen and K. Andersson, *J. Org. Chem.*, **50**, 277 (1985).
44. L. T. Scott, G. Reinhardt and N. H. Roelofs, *J. Org. Chem.*, **50**, 5886 (1985).
45. J. Tsunetsugu, T. Ikeda, N. Suzuki, M. Yaguchi, M. Sato, S. Ebine and K. Morinaga, *J. Chem. Soc. Perkin Trans. 1*, 785 (1985).
46. M. Kimura, S. Sagara and S. Morasawa, *J. Org. Chem.*, **47**, 4344 (1982).
47. N. Harada, Y. Tamai and H. Uda, *J. Org. Chem.*, **49**, 4266 (1984).
48. H.-D. Becker, K. Sandros and K. Andersson, *Angew. Chem. Int. Ed. Engl.*, **22**, 495 (1983).
49. H.-D. Becker and K. Andersson, *Tetrahedron Lett.*, **24**, 3273 (1983).
50. M. S. Newman, V. K. Khanna and K. Kanakarajan, *J. Am. Chem. Soc.*, **101**, 6788 (1979).
51. G. Felix, M. Laguerre, J. Dunogues and R. Calas, *J. Chem. Res. (S)*, 236 (1980).
52. A. G. M. Barret, D. Dauzonne, I. A. O'Neil and A. Renaud, *J. Org. Chem.*, **49**, 4409 (1984).
53. M. Laguerre, G. Felix, J. Dunogues and R. Calas, *J. Org. Chem.*, **44**, 4275 (1979).
54. P. P. Fu and R. G. Harvey, *J. Org. Chem.*, **44**, 3778 (1979).
55. H. M. Lee and R. G. Harvey, *J. Org. Chem.*, **44**, 4948 (1979).
56. H. Lee, J. Sheth and R. G. Harvey, *Carcinogenesis*, **4**, 1297 (1983).
57. S. Amin, N. Hussain, H. Brielmann and S. S. Hecht, *J. Org. Chem.*, **49**, 1091 (1984).
58. D. M. Jerina, P. J. van Bladeren, H. Yagi, D. T. Gibson, V. Mahadevan, A. S. Neese, M. Koreeda, N. D. Sharma and D. R. Boyd, *J. Org. Chem.*, **49**, 3621 (1984).
59. S. Kumar, *Tetrahedron Lett.*, **26**, 6417 (1985).
60. H. Lee, N. Shyamasundar and R. G. Harvey, *Tetrahedron*, **37**, 2563 (1981).
61. H. Lee and R. G. Harvey, *J. Org. Chem.*, **48**, 749 (1983).
62. A. B. Turner and S. Kerr, *J. Chem. Soc. Perkin Trans. 1*, 1322 (1979).
63. D. R. Brown and A. B. Turner, *J. Chem. Soc. Perkin Trans. 1*, 165 (1978).
64. D. J. Collins and J. Sjövall, *Austr. J. Chem.*, **36**, 339 (1983).
65. K. Bischofberger and J. R. Bull, *Tetrahedron*, **41**, 365 (1985).
66. H. Hart, T. Miyashi, D. N. Buchanan and S. Sasson, *J. Am. Chem. Soc.*, **96**, 4857 (1974).
67. Y. Oikawa, K. Horita and O. Yonemitsu, *Heterocycles*, **23**, 553 (1985).
68. L. C. Dunn, Y.-M. Chang and K. N. Houk, *J. Am. Chem. Soc.*, **98**, 7095 (1976).
69. V. T. Ravi Kumar, S. Swaminathan and K. Rayagopalan, *J. Org. Chem.*, **50**, 5867 (1985).
70. S. Ghosal and S. Banerjee, *J. Chem. Soc. Chem. Commun.*, 165 (1979).
71. A. Schönberg, E. Singer and H. Schulze-Pannier, *Chem. Ber.*, **110**, 3714 (1977).
72. R. Neidlein, G. Hartz, A. Gieren, H. Betz and T. Hübner, *Chem. Ber.*, **118**, 1455 (1985).
73. A. Ohki, T. Nishiguchi and K. Fukuzumi, *Tetrahedron*, **35**, 1737 (1979).

74. H.-D. Becker, A. Björk and E. Adler, *J. Org. Chem.*, **45**, 1596 (1980).
75. D. R. Brown and A. B. Turner, *J. Chem. Soc. Perkin Trans. 1*, 1307 (1975).
76. J. I. Levin and S. M. Weinreb, *J. Org. Chem.*, **49**, 4325 (1984).
77. B. A. McKittrick and B. Ganem, *J. Org. Chem.*, **50**, 5897 (1985).
78. S. Cacchi, F. La Torre and G. Paolucci, *Synthesis*, 848 (1978).
79. U. Biethan, W. Fauth and H. Musso, *Chem. Ber.*, **110**, 3636 (1977).
80. T. Ameniga, M. Yasunami and K. Takase, *Chem. Lett.*, 587 (1977).
81. M. Yasunami, T. Amemiya and K. Takase, *Tetrahedron Lett.*, **24**, 69 (1983).
82. Y. Oikawa and O. Yonemitsu, *J. Org. Chem.*, **42**, 1213 (1977). Cf. also J. G. Rodriguez, F. Temprano, C. Esteba-Calderon, M. Martinez-Ripoll and S. Garcia-Blanco, *Tetrahedron*, **41**, 3813 (1985).
83. M. Cain, R. Mantei and J. M. Cook, *J. Org. Chem.*, **47**, 4933 (1982).
84. B. Lal, R. M. Gidwani, J. Reden and N. J. de Souza, *Tetrahedron Lett.*, **25**, 2901 (1984).
85. M. V. Naidu and G. S. K. Rao, *Synthesis*, 144 (1979).
86. R. S. Ward, P. Satyanarayana and B. V. G. Rao, *Tetrahedron Lett.*, **22**, 3021 (1981).
87. A. S. Kende, J.-P. Gesson and T. P. Demuth, *Tetrahedron Lett.*, **22**, 1667 (1981).
88. G. M. Buchan and A. B. Turner, *J. Chem. Soc. Perkin Trans. 1*, 1326 (1979).
89. J. A. Steenkamp, D. Ferreira and D. G. Roux, *Tetrahedron Lett.*, **26**, 3045 (1985).
90. M. Bouquet, A. Guy, M. Lemaire and J. P. Guette, *Synth. Commun.*, **15**, 1153 (1985).
91. L. Eberson, L. Jönsson and L.-G. Wistrand, *Acta Chem. Scand.*, **B 33**, 413 (1979).
92. Y. Oikawa, K. Horita, T. Yoshioka, T. Tanaka and O. Yonemitsu, *Tetrahedron*, **41**, 3021 (1986).
93. O. Yonemitsu, *J. Synth. Org. Chem., Japan*, **43**, 691 (1985).
94. Y. Oikawa, T. Tanaka, K. Horita, T. Yoshioka and O. Yonemitsu, *Tetrahedron Lett.*, **25**, 5393 (1984).
95. Y. Oikawa, T. Tanaka, K. Horita and O. Yonemitsu, *Tetrahedron Lett.*, **25**, 5397 (1984).
96. Y. Oikawa, T. Nishi and O. Yonemitsu, *J. Chem. Soc. Perkin Trans. 1*, 1 (1985).
97. Y. Oikawa, T. Nishi and O. Yonemitsu, *J. Chem. Soc. Perkin Trans. 1*, 19 (1985).
98. C. U. Kim and P. F. Misco, *Tetrahedron Lett.*, **26**, 2027 (1985).
99. Y. Oikawa, T. Yoshioka and O. Yonemitsu, *Tetrahedron Lett.*, **23**, 889 (1982).
100. Y. Oikawa, T. Nishi and O. Yonemitsu, *Tetrahedron Lett.*, **24**, 4037 (1983).
101. Y. Oikawa, T. Nishi and O. Yonemitsu, *J. Chem. Soc. Perkin Trans. 1*, 7 (1985).
102. M. P. Youngblood, *J. Am. Chem. Soc.*, **107**, 6987 (1985).
103. J. Weiser and H. A. Staab, *Tetrahedron Lett.*, **26**, 6059 (1985).
104. H. L. K. Schmand and P. Boldt, *J. Am. Chem. Soc.*, **97**, 447 (1975).
105. F. Setiabudi and P. Boldt, *Liebigs Ann. Chem.*, 1272 (1985).
106. D. W. Cameron, G. I. Feutrill, C. L. Gibson and R. W. Read, *Tetrahedron Lett.*, **26**, 3887 (1985).
107. W. Steglich, H.-T. Huppertz and B. Steffan, *Angew. Chem.*, **97**, 716 (1985).
108. G. A. Kraus and B. S. Fulton, *J. Org. Chem.*, **50**, 1782 (1985).
109. H.-D. Becker, D. Sanchez and A. Arvidsson, *J. Org. Chem.*, **44**, 4247 (1979).
110. H. Arzeno, D. H. R. Barton, R.-M. Berge-Lurion, X. Lusinchi and B. M. Pinto, *J. Chem. Soc. Perkin Trans. 1*, 2069 (1984).
111. D. H. R. Barton, R.-M. Berge-Lurion, X. Lusinchi and B. M. Pinto, *J. Chem. Soc. Perkin Trans. 1*, 2077 (1984).
112. G. Buchi, P.-S. Chu, A. Hoppman, C.-P. Mak and A. Pearce, *J. Org. Chem.*, **43**, 3983 (1978).
113. H. Laatsch, *Liebigs Ann. Chem.*, 1808 (1982).
114. L. Hageman and E. McNelis, *J. Org. Chem.*, **40**, 3300 (1975).
115. T. R. Kasturi and R. Sivaramakrishnan, *Proc. Ind. Acad. Sci.*, **86A**, 309 (1977).
116. D. J. Bennett, F. M. Dean, G. A. Herbin, D. A. Matkin, A. W. Price and M. L. Robinson, *J. Chem. Soc. Perkin Trans. 1*, 1978 (1980).
117. T. R. Kasturi, B. Rajasekhar, G. J. Raju, G. M. Reddy, R. Sivaramakrishnan, N. Ramasubbu and K. Ventakesan, *J. Chem. Soc. Perkin Trans. 1*, 2375 (1985).
118. B. Cardillo, M. Cornia and L. Merlini, *Gazzetta Chim. Ital.*, **105**, 1151 (1975).
119. A. C. Jain, R. C. Gupta and R. Khazanchi, *Tetrahedron*, **35**, 413 (1979).
120. A. C. Jain, A. Kumar and R. C. Gupta, *J. Chem. Soc. Perkin Trans. 1*, 279 (1979).
121. L. Jurd, *Tetrahedron*, **33**, 163 (1977).
122. A. Aumüller and S. Hünig, *Liebigs Ann. Chem.*, 142 (1986).

123. H. Nishiyama, H. Arai, Y. Kanai, H. Kawashima and K. Itoh, *Tetrahedron Lett.*, **27**, 361 (1986).
124. A. G. Schultz, R. D. Lucci, J. J. Napier, H. Kinoshita, R. Ravichandran, P. Shannon and Y. K. Lee, *J. Org. Chem.*, **50**, 217 (1985).
125. N. Harada, J. Kohori, H. Uda, K. Nakanishi and R. Takeda, *J. Am. Chem. Soc.*, **107**, 423 (1985).
126. H. E. Zimmerman and D. C. Lynch, *J. Am. Chem. Soc.*, **107**, 7745 (1985).
127. A. J. Waring and J. H. Zaidi, *J. Chem. Soc. Perkin Trans. 1*, 631 (1985).
128. D. A. Evans, S. P. Tanis and D. J. Hart, *J. Am. Chem. Soc.*, **103**, 5813 (1981).
129. P. P. Fu, C. Cortez, K. B. Sukamaran and R. G. Harvey, *J. Org. Chem.*, **44**, 4265 (1979).
130. J. Pataki and R. G. Harvey, *J. Org. Chem.*, **47**, 20 (1982).
131. R. G. Harvey and C. Cortez, *Carcinogenesis*, **4**, 941 (1983).
132. M. E. Jung, Y.-G. Pan, M. W. Rathke, D. F. Sullivan and R. P. Woodbury, *J. Org. Chem.*, **42**, 3961 (1977).
133. I. Ryu, S. Murai, Y. Hatayama and N. Sonoda, *Tetrahedron Lett.*, 3455 (1978).
134. I. Fleming and I. Paterson, *Synthesis*, 736 (1979).
135. H. Hofmann and H. Djafari, *Liebigs Ann. Chem.*, 599 (1985).
136. R. S. Lott, E. G. Breitholle and C. H. Stammer, *J. Org. Chem.*, **45**, 1151 (1980).
137. L. A. Paquette, D. T. Belmont and Y.-L. Hsu, *J. Org. Chem.*, **50**, 4667 (1985).
138. Y. Ito, T. Hirao and T. Saegusa, *J. Org. Chem.*, **43**, 1011 (1978).
139. M. T. Crimmins, S. W. Mascarella and L. D. Bredon, *Tetrahedron Lett.*, **26**, 997 (1985).
140. G. A. Kraus and Y.-S. Hon, *J. Org. Chem.*, **51**, 116 (1986).
141. L. T. Scott, P. Grütter and R. E. Chamberlain III, *J. Am. Chem. Soc.*, **106**, 4852 (1984).
142. L. T. Scott and C. M. Adams, *J. Am. Chem. Soc.*, **106**, 4857 (1984).
143. M. Hirayama and S. Ito, *Tetrahedron Lett.*, 1071 (1975).
144. S. Ito, Y. Shoji, H. Takeshita and K. Takahashi, *Tetrahedron Lett.*, 1075 (1975).
145. W. L. Meyer, M. J. Brannon, C. da G. Burgos, T. E. Goodwin and R. W. Howard, *J. Org. Chem.*, **50**, 438 (1985).
146. C. Reichardt, K.-Y. Yun, W. Massa and R. E. Schmidt, *Liebigs Ann. Chem.*, 1987 (1985).
147. A. I. Meyers and D. G. Wettlaufer, *J. Am. Chem. Soc.*, **106**, 1135 (1984).
148. D. Claremon and S. D. Young, *Tetrahedron Lett.*, **26**, 5420 (1985).
149. K. T. Potts, K. G. Bordeaux, W. R. Kuehnling and R. L. Salsbury, *J. Org. Chem.*, **50**, 1666, 1677 (1985).
150. A. Padwa, Y.-Y. Chen, W. Dent and H. Nimmesgern, *J. Org. Chem.*, **50**, 4006 (1985).
151. M. P. Sammes, M. W. L. Chung and A. R. Katritzky, *J. Chem. Soc. Perkin Trans. 1*, 1773 (1985).
152. F. Rise and K. Undheim, *Acta Chem. Scand.*, **B39**, 195 (1985).
153. F. Rise and K. Undheim, *J. Chem. Soc. Perkin Trans. 1*, 1997 (1985).
154. M. Christl and S. Freund, *Chem. Ber.*, **118**, 979 (1985).
155. G. Bianchi and M. De Amici, *J. Chem. Res. (S)*, 311 (1979).
156. S. Konno and C. H. Stammer, *Synthesis*, 598 (1978).
157. R. Young and C. K. Chang, *J. Am. Chem. Soc.*, **107**, 898 (1985).
158. U. Kämpfen and A. Eschenmoser, *Tetrahedron Lett.*, **26**, 5899 (1985).
159. A. F. McDonagh and L. A. Palma, *Biochem. J.*, **189**, 193 (1980).
160. D. A. Lightner and A. F. McDonagh, *Accts Chem. Res.*, **17**, 417 (1984).
161. S. Roy and Chakraborty, *J. Chromatogr.*, **96**, 266 (1974).
162. J. Bergman, R. Carlsson and S. Misztal, *Acta Chem. Scand.*, **B30**, 853 (1976).
163. T. Kumagai, S. Tanaka and T. Mukai, *Tetrahedron Lett.*, **25**, 5669 (1984).
164. Y. Y. Lin, E. Thom and A. A. Libman, *J. Heterocycl. Chem.*, 799 (1979).
165. R. B. Gammill, *Tetrahedron Lett.*, **26**, 1385 (1985).
166. E. C. Taylor and J. E. Macor, *Tetrahedron Lett.*, **27**, 431 (1986).
167. M. F. Grundon and M. J. Rutherford, *J. Chem. Soc. Perkin Trans. 1*, 197 (1985).
168. M. F. Grundon, V. N. Ramachandran and B. M. Sloan, *Tetrahedron Lett.*, **22**, 3105 (1981).
169. P. E. Brown, W. Y. Marcus and P. Anastasis, *J. Chem. Soc. Perkin Trans. 1*, 1127 (1985).
170. S. Matsuura, M. Iinuma, K. Ishikawa and K. Kagei, *Chem. Pharm. Bull. Japan*, **26**, 305 (1978).
171. R. A. Conley and N. D. Heindel, *J. Org. Chem.*, **40**, 3169 (1975).
172. S. M. Ali, J. W. Findlay and A. B. Turner, *J. Chem. Soc. Perkin Trans. 1*, 407 (1976).
173. S. Yoneda, K. Ozaki, K. Yanagi and M. Minobe, *J. Chem. Soc., Chem. Commun.*, 19 (1986).

174. T. Sugawara, H. Masuya, T. Matsuo and T. Miki, *Chem. Pharm. Bull. Japan*, **27**, 2544 (1979).
175. E. Gebert, A. H. Reis Jr, J. S. Miller, H. Rommelmann and A. J. Epstein, *J. Am. Chem. Soc.*, **104**, 4403 (1982).
176. E. Uhlig and R. Fischer, *J. Organomet. Chem.*, **239**, 385 (1982).
177. E. Uhlig, R. Fischer and W. Ludwig, *Z. Chem.*, **24**, 386 (1984).
178. G. R. Clark and S. Thiensathit, *Tetrahedron Lett.*, **26**, 2503 (1985).
179. T. Antonsson, S. Hansson and C. Moberg, *Acta Chem. Scand.*, **B 39**, 593 (1985).
180. A. Heumann and B. Åkermark, *Angew. Chem. Int. Ed. Engl.*, **23**, 453 (1984).
181. J.-E. Bäckvall, S. E. Byström and R. E. Nordberg, *J. Org. Chem.*, **49**, 4619 (1984).
182. J.-E. Bäckvall, J.-E. Nyström and R. E. Nordberg, *J. Am. Chem. Soc.*, **107**, 3676 (1985).
183. J.-E. Bäckvall, R. E. Nordberg and D. Wilhelm, *J. Am. Chem. Soc.*, **107**, 6892 (1985).
184. G. Bartoli, *Accts Chem. Res.*, **17**, 109 (1984).
185. G. Bartoli, M. Bosco and R. Dalpozzo, *Tetrahedron Lett.*, **26**, 115 (1985). Cf. also F. Kienzle, *Helv. Chim. Acta*, **61**, 449 (1978).
186. H.-D. Becker and H. Lingnert, *J. Org. Chem.*, **47**, 1095 (1982).
187. H.-D. Becker and D. Sanchez, *J. Org. Chem.*, **44**, 1787 (1979).
188. H.-D. Becker, B. W. Skelton and A. H. White, *Aust. J. Chem.*, **34**, 2189 (1981).
189. T. Oshima, R. Nishioka and T. Nagai, *Tetrahedron Lett.*, **21**, 3919 (1980).
190. T. Oshima, R. Nishioka, S. Ueno and T. Nagai, *J. Org. Chem.*, **47**, 2114 (1982).
191. S. Ueno, T. Oshima and T. Nagai, *J. Org. Chem.*, **49**, 4060 (1984).

The Chemistry of Quinonoid Compounds, Vol. II
Edited by S. Patai and Z. Rappoport

CHAPTER **24**

Azulene quinones

LAWRENCE T. SCOTT
Department of Chemistry and Center for Advanced Study, College of Arts and Science, University of Nevada-Reno, Reno, Nevada 89557, USA

I. INTRODUCTION

From the earliest days of organic chemistry, quinones have figured prominently in the chemistry of aromatic compounds[1]. Most quinones can be viewed as derivatives of benzene, naphthalene, or some higher benzenoid aromatic hydrocarbon, and collectively these constitute the class of benzenoid quinones. The family of aromatic hydrocarbons, however, is no longer confined solely to benzenoid molecules; the last three decades have witnessed an almost explosive growth in the field of non-benzenoid aromatics, i.e. compounds that display aromatic character despite the absence of benzene rings[2–5]. Quinones of these hydrocarbons constitute the class of non-benzenoid quinones. This volume on *The Chemistry of the Quinonoid Compounds* and its predecessor[6] both contain chapters devoted to non-benzenoid quinones. The present chapter focuses on an important subset of this family, the azulene quinones. In keeping with the traditional naming of benzoquinones and naphthoquinones, the quinones of azulene have sometimes been referred to as azuloquinones.

Foremost among the non-benzenoid aromatic hydrocarbons, azulene has attracted considerable attention for many years[7,8]. Numerous pathways to azulenes have been reported since the first synthesis of the parent hydrocarbon in 1937 by Plattner and Pfau[9]; however, work on the corresponding quinones did not begin for almost 40 years. As of 1974, when the first volume of this book was published, no quinones of azulene had yet been reported.

Figure 1 shows the numbering system for azulene and the eleven possible azulene quinones that can be constructed with two carbonyl groups and four double bonds (Kekulé quinones). Azulene quinones related to *meta*-benzoquinone (non-Kekulé quinones) are discussed briefly in Section V. For ease of reference, abbreviated names rather than arbitrary numbers will be used for the azulene quinones throughout this chapter. Thus, for example, 1,2-azuloquinone will be referred to simply as 1,2-AQ (see Figure 1).

In the present context, azulene quinones are defined as the fully unsaturated derivatives of the various isomeric bicyclo[5.3.0]decadiones. Benzenoid quinones bearing an aromatic azulene nucleus, either fused to one side or attached as a pendant group, fall outside the scope of this chapter. Polycyclic molecules with an embedded azuloquinone and quinomethides have also been excluded.

II. THEORY

It has long been recognized that the special effects of cyclic conjugation normally associated with a ring of p orbitals are virtually absent in quinones such as *para*-benzoquinone (**1**) as a result of the cross-conjugating interruptions introduced by two carbonyl groups in the cycle[1,10]. Odd-membered ring annulenones with only a single carbonyl group, such as tropone (**2**) and cyclopentadienone (**3**), on the other hand, retain a measure of aromatic or antiaromatic character that depends on the number of π electrons in the olefinic bonds[11]. The presence of these two annulenones in many of the azulene quinones provides a basis for preliminary predictions about the properties of these novel compounds.

Tropone (**2**) is a stable, planar molecule that exhibits little tendency to dimerize, polymerize, or decompose under ordinary laboratory conditions[12,13]. By contrast, **3** is

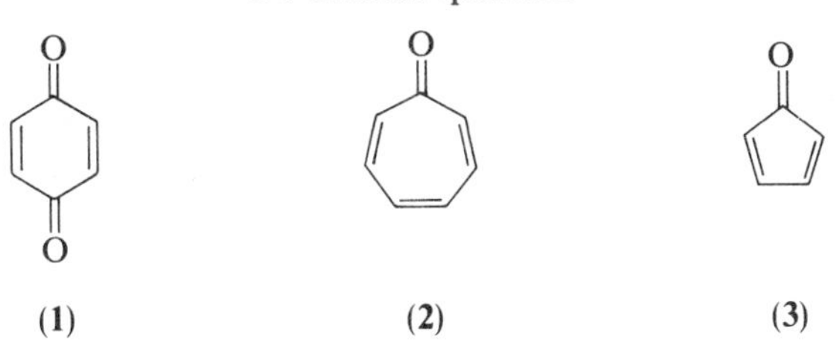

exceedingly unstable and dimerizes rapidly even at very low temperatures[14, 15]. Accordingly, one might expect those azulene quinones that contain a tropone but no cyclopentadienone (1,5-AQ and 1,7-AQ) to be more stable than those that contain a cyclopentadienone but no tropone (2,4-AQ and 2,6-AQ) and those that contain both subunits (1,4-AQ, 1,6-AQ and 1,8-AQ) to be intermediate in stability. The term 'stability',

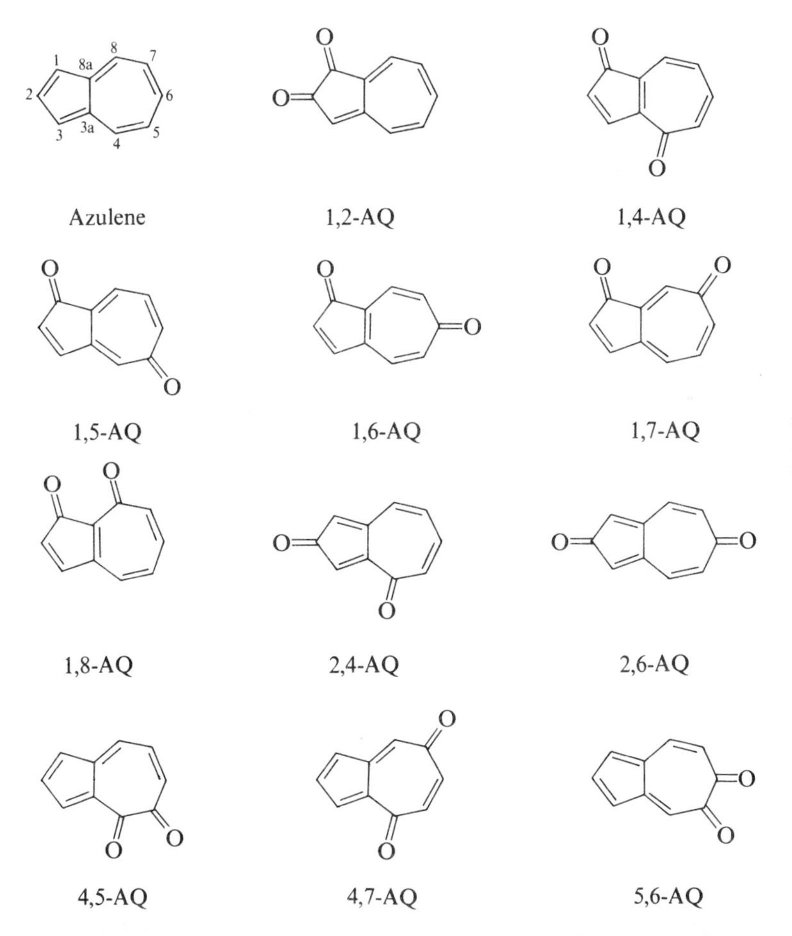

FIGURE 1. Azulene and the eleven possible Kekulé quinones thereof

as used here, refers to kinetic stability toward bimolecular destruction, i.e. isolability, which may or may not correlate with thermodynamic stability or heats of formation.

This rudimentary analysis suggests which azulene quinones might be better than others as targets for synthesis (1,5-AQ and 1,7-AQ), but it leaves many questions unanswered. The relative ordering of quinones *within* each set above, e.g. 1,4-AQ vs. 1,6-AQ vs. 1,8-AQ, cannot be predicted easily 'by inspection', nor can this treatment handle the isomers in which both carbonyl groups reside in the same ring (1,2-AQ, 4,5-AQ, 4,7-AQ and 5,6-AQ).

Unsatisfied with such a crude and incomplete theoretical prognosis, the author of this chapter persuaded chemists from several other laboratories in the late 1970s to carry out detailed molecular orbital calculations on the azulene quinones in a systematic manner. The resulting international collaborative effort produced comprehensive sets of data from both Hückel and MINDO/3 calculations[16], selected portions of which are collected in Table 1. In several instances, predictions based on the Hückel and semi-empirical MINDO/3 results were reinforced by single point *ab initio* STO-3G calculations on the MINDO/3-optimized geometries of representative compounds. PPP π-electron calculations of electronic transitions were employed for predictions on the UV–VIS spectra and colors of the azulene quinones. Calibration of the various theoretical methods was achieved by extending them to include calculations on the isomeric naphthoquinones for which experimental data could be obtained from the literature.

TABLE 1. Selected calculated energies of azulene quinones[16]

	ΔH_f	π-DE	E-LUMO	
Compound	MINDO/3 kcal mol^{-1}	MINDO/3 kcal mol^{-1}	MINDO/3 eV	Hückel β
1,2-AQ	−0.4	+12.7	−0.70	−0.24
1,4-AQ	−0.8	+9.1	−1.03	−0.11
1,5-AQ	−6.4	+14.7	−0.63	−0.24
1,6-AQ	−0.6	+8.9	−1.17	−0.11
1,7-AQ	−5.3	+13.6	−0.66	−0.23
1,8-AQ	−0.6	+8.9	−0.94	−0.11
2,4-AQ	+4.7	+3.6	−1.16	−0.05
2,6-AQ	+4.7	+3.6	−1.28	−0.05
4,5-AQ	+7.0	+5.3	−0.95	−0.03
4,7-AQ	+7.2	+1.1	−1.16	−0.04
5,6-AQ	+9.5	+2.8	−0.96	−0.04

When these calculations and predictions were first published in 1980[16], very little experimental work on the azulene quinones had been reported. Since that time, however, certain indices of reactivity have emerged as more reliable than others[17]. The following discussion covers only those aspects of the theoretical treatment that have withstood preliminary testing and still appear valid.

First it should be noted that the reactions most common in quinone chemistry, e.g. Michael additions, Diels–Alder cycloadditions, charge-transfer complex formation, reduction, etc., all involve the interaction of external electrons with the lowest unoccupied molecular orbital (LUMO) of the quinone. As a general rule, the lower the energy of the LUMO (E-LUMO), the greater the susceptibility of the quinone to all of these reactions[18,19]. Thus, the E-LUMO should represent a fairly universal index of reactivity for quinones[20].

Within the family of azulene quinones, the order of chemical stability (isolability) of the various isomers can be expected to follow the E-LUMOs, those with the lowest-lying

LUMOs being the most easily destroyed by bimolecular processes and, therefore, the most difficult to isolate. Those isomers with high-lying LUMOs, on the other hand, ought to enjoy somewhat greater kinetic stability. In accord with this theoretical viewpoint, the three azulene quinones with relatively high E-LUMOs (1,2-AQ, 1,5-AQ and 1,7-AQ) have now all been synthesized and isolated in crystalline form (*vida infra*). By contrast, all attempts to prepare isomers with lower E-LUMOs or derivatives thereof under ordinary laboratory conditions have failed to yield monomeric quinones, although their existence as fleeting intermediates has been established in several cases. The E-LUMO, therefore, does indeed seem to provide a good index for isolability, at least among the azulene quinones. It was gratifying that the calculations affirmed the qualitative theoretical predictions, which took into account only the presence or absence of certain annulenone subunits.

The first three isomers to be isolated (1,2-AQ, 1,5-AQ and 1,7-AQ) not only have the highest E-LUMOs, they also have the largest π-delocalization energies. The π-DEs listed in Table 1 were obtained simply by taking the difference between the MINDO/3 calculated heats of formation (ΔH_f) and the ΔH_f^{ref} values obtained from the additivity scheme of Dewar and coworkers[21, 22], as extended by Gleicher and coworkers[10], and correcting for ring strain[16]. The calculated heats of formation clearly reveal the thermodynamic stabilizing effect of a tropone moiety (cf. 1,4-AQ vs. 2,4-AQ: $\Delta\Delta H_f = 5.5$ kcal mol^{-1}) and the destabilizing effect of a cyclopentadienone moiety (cf. 1,4-AQ vs 1,5-AQ: $\Delta\Delta H_f = 5.6$ kcal mol^{-1}).

Electronic $\pi\pi^*$ transition energies for some of the azulene quinones were calculated by the PPP π-electron method (see Table 2)[16]. MINDO/3 geometries served as initial input, and a π-bond order/length criterion was used to reoptimize geometries. Confidence in the results derives from the reasonable agreement observed between the experimental absorption spectra of 1,2- and 1,4-naphthoquinone and those calculated by this method. Although these calculations do not include the long-wavelength $n\pi^*$ transitions, it was predicted[16] that the colors of the azulene quinones should range from yellow (1,5-AQ and 1,7-AQ) to purple (e.g., 4,7-AQ) to blue-green (e.g., 2,6-AQ). As discussed in Section IV.B, these predictions have proven surprisingly accurate.

TABLE 2. Calculated electronic absorption spectra (PPP)[16]

Compound	$\lambda_{max}(f)$
1,2-AQ	527 (0.04), 397 (0.40), 309 (0.13)
1,5-AQ	371 (0.38), 347 (0.00), 341 (0.14)
1,6-AQ	549 (0.05), 332 (0.07), 321 (0.23)
1,7-AQ	394 (0.31), 349 (0.19), 314 (0.03)
2,4-AQ	596 (0.07), 367 (0.22), 316 (0.08)
2,6-AQ	597 (0.02), 350 (0.60), 304 (0.03)
4,7-AQ	529 (0.05), 344 (0.13), 304 (0.62)

In the chemical reactions of any single azulene quinone, competing pathways involving attack at different sites must be considered. These will be important, for example, in all Michael additions and cycloaddition reactions. Houk and coworkers have shown[23] that LUMO coefficients can be used as reliable indicators for the site of nucleophilic attack (Michael addition) on quinones, and electron-rich cycloaddends will also attack a quinone where the LUMO coefficients are large. Table 3 gives the calculated LUMO coefficients for all atoms in all eleven of the Kekulé azulene quinones[16].

Incorporation of donor substituents on an electron-deficient azulene quinone framework should raise the LUMO energy, with the greatest effect resulting from substitution on the position with the largest π-LUMO coefficient[19]. Since those same positions are

TABLE 3. Calculated LUMO coefficients of azulene quinones (MINDO/3)[16]

Compound	C(1)	C(2)	C(3)	C(3a)	C(4)	C(5)	C(6)	C(7)	C(8)	C(8a)	O(1)	O(2)
1,2-AQ	0.28	0.22	0.15	−0.25	−0.38	0.33	0.34	−0.36	−0.22	0.33	−0.27	−0.21
1,4-AQ	0.29	0.29	−0.28	−0.48	−0.12	−0.25	−0.02	−0.36	0.17	0.44	−0.28	−0.11
1,5-AQ	0.33	0.25	−0.33	−0.23	0.08	0.28	0.34	−0.35	−0.28	0.36	−0.29	−0.24
1,6-AQ	−0.29	−0.31	0.28	0.47	0.13	−0.34	−0.01	0.32	−0.11	−0.44	0.29	0.01
1,7-AQ	0.25	0.37	−0.34	−0.41	0.34	0.34	−0.37	−0.22	0.03	0.12	−0.22	0.19
1,8-AQ	0.28	0.31	−0.25	−0.45	−0.19	0.37	0.38	−0.23	0.08	0.46	−0.28	−0.08
2,4-AQ	0.37	0.30	0.37	−0.27	−0.23	−0.24	0.24	0.28	−0.26	−0.32	−0.30	−0.24
2,6-AQ	0.38	0.30	0.38	−0.31	−0.23	0.25	0.23	0.25	−0.23	−0.31	−0.30	−0.23
4,5-AQ	−0.33	0.27	0.39	−0.27	−0.16	0.12	−0.33	0.24	0.43	−0.33	0.11	0.05
4,7-AQ	0.33	−0.25	−0.42	0.25	0.22	0.23	−0.20	−0.29	−0.38	0.32	−0.22	−0.28
5,6-AQ	−0.42	−0.28	0.34	0.31	−0.43	−0.19	−0.12	−0.24	0.21	0.31	0.13	0.05

predicted to be the most likely sites for nucleophilic attack, a judicious placement of alkyl groups could impede Michael additions to the quinones not only by raising the LUMO energy but also by simple steric hindrance. The positions at which alkyl substitution should have the most 'stabilizing' effect can be read directly from Table 3. On 2,4-AQ and 2,6-AQ, for example, bulky alkyl groups at the 1- and 3-positions should have the most stabilizing effect.

Throughout the foregoing theoretical treatment, the dominance of the LUMO has been apparent. Other molecular orbitals will also contribute to the overall reactivity of azulene quinones; however, it is not unreasonable to expect the effects of the LUMO to prevail in most circumstances[18, 19].

III. SYNTHESIS

A. 1,2-Azulene Quinone

Morita, Karasawa and Takase reported the first unsubstituted quinone of azulene, the parent 1,2-AQ, in 1980[24]. Their synthesis (equation 1) begins with the preparation of diethyl 2-hydroxyazulene-1,3-dicarboxylate (**5**) from 2-chlorotropone (**4**) according to a procedure developed much earlier under the direction of Professor Nozoe in the same laboratories at Tohoku University. Partial deethoxycarbonylation of **5** and acetylation gives the disubstituted azulene **6**, which, on direct oxygenation with benzoyl peroxide, yields the hydroquinone derivative **7**. Hydrolysis and deethoxycarbonylation then gives the very unstable 1,2-dihydroxyazulene (**8**), which can be oxidized with DDQ to the green

O, Cl (**4**) —$CH_2(COOEt)_2$, NaOEt→ HO, COOEt, COOEt (**5**) —(1) H_3PO_4, (2) Ac_2O, pyr→

AcO, COOEt (**6**) —$(PhCOO)_2$→ PhCOO, AcO, COOEt (**7**) —H_3PO_4→

HO, HO (unstable) (**8**) —DDQ→ O, O 1,2-AQ (1)

1,2-quinone. Variations on this synthetic scheme have provided also the 3-ethoxycarbonyl and the 3-cyano derivatives (**9**)[24].

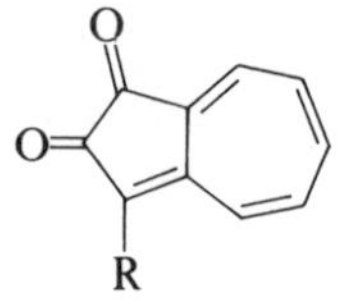

(9, R = COOEt, CN)

B. 1,4-Azulene Quinone

In Reno, a general approach to the synthesis of 1,n-AQs has been developed[25, 26]. The first step involves intramolecular addition to a benzene ring by the carbene derived from diazoketone **10** (equation 2)[27]. Rhodium acetate is superior to the copper salts formerly used to catalyze the loss of nitrogen in this reaction[28]. Norcaradiene (**11**), the initial product of this cyclization reaction, opens spontaneously to the acid-sensitive bicyclic trienone **12**, which isomerizes to the more stable trienone **13** on exposure to basic alumina. The readily available **13** serves as a common precursor to several of the azulene quinones (*vida infra*).

For the synthesis of 1,4-AQ, **13** was oxidized with chromium trioxide pyridine[25]. This reaction gives the two isomeric tropones **14** and **15** in roughly equal amounts. It should be

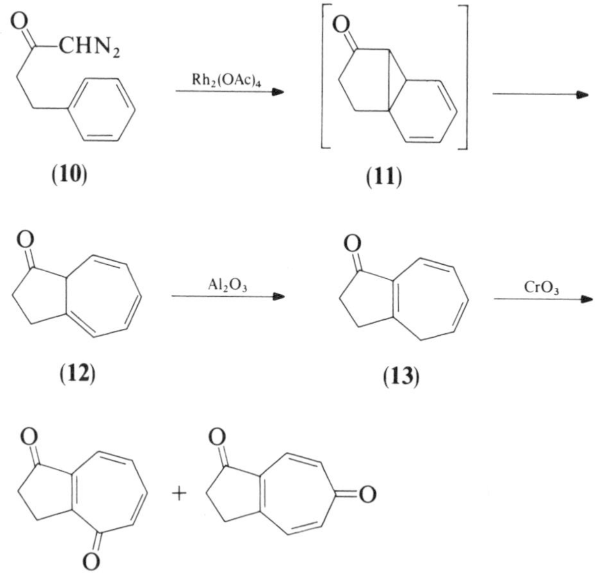

CHN_2 (**10**) $\xrightarrow{Rh_2(OAc)_4}$ [(**11**)] $\longrightarrow$ (**12**) $\xrightarrow{Al_2O_3}$ (**13**) $\xrightarrow{CrO_3}$ (**14**) + (**15**) (2)

noted that neither of these tropones exhibits any tendency to tautomerize to the corresponding dihydroxyazulene. Apparently, the aromaticity of azulene does not suffice to overcome the strength of two carbonyl π bonds and any special thermodynamic stability associated with the tropone ring. Such behavior contrasts with that of 1,2-dihydroxyazulene (**8**)[24], which would gain no tropone moiety on tautomerization to a diketone.

Unlike compound **8**, neither **14** nor **15** could be oxidized to the corresponding azulene quinone with DDQ[28]. Conventional methods for introducing a double bond directly into the five-membered ring of these 2,3-dihydroquinones also proved unsatisfactory[25]. Fortunately, the beautiful blue azulene diacetate **16**, a hydroquinone derivative of **14**, could be obtained simply by stirring the tropone with acetic anhydride and pyridine in hot ethyl acetate (equation 3).

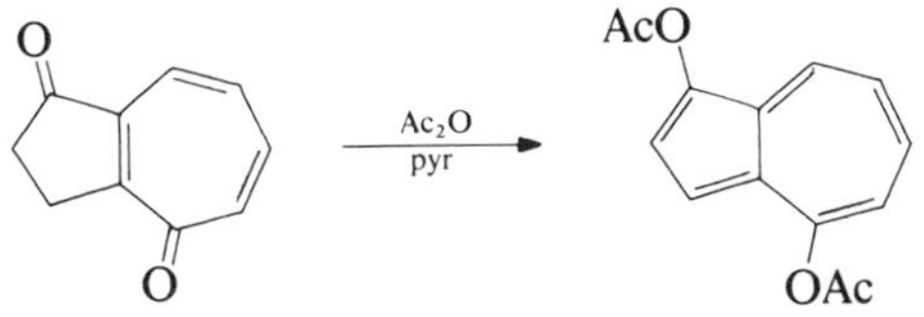

Ac_2O, pyr (3)

(**14**) (**16**)

On treatment with an excess of methyllithium in tetrahydrofuran, 1,4-diacetoxyazulene (**16**) yields a blue-green solution of the azulene-1,4-hydroquinone dianion by nucleophilic cleavage of the ester groups. Quenching the reaction mixture with chlorotrimethylsilane then gives the labile bis-trimethylsilyl ether **17**. Subsequent oxidation of **17** in the presence of cyclopentadiene produces 1,4-AQ, which is immediately trapped in a Diels–Alder reaction (equation 4). An *endo* stereochemistry best accounts for the observed coupling

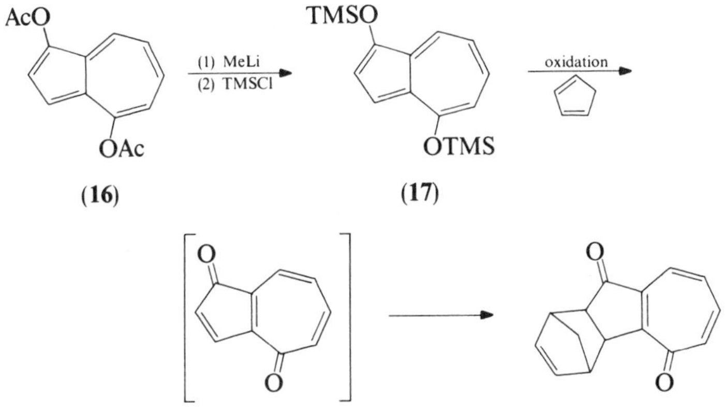

(1) MeLi (2) TMSCl; oxidation (4)

(**16**) (**17**) 1,4-AQ (**18**)

constants in the ^{1}H-NMR spectrum of **18**. Omission of the cyclopentadiene from this reaction has not yet permitted isolation of the monomeric quinone; under the conditions explored to date, only higher molecular weight materials are obtained. Useful oxidizing agents for the reaction in equation 4 include pyridinium chlorochromate (PCC) and tetrachloro-*p*-benzoquinone (*p*-chloranil)[25].

An alternative route to 1,4-AQ[25] takes advantage of the well-known susceptibility of azulenes to electrophilic substitution at the α-postition in the smaller ring. Thus, bromination of 1,4-diacetoxyazulene (**16**) with *N*-bromosuccinimide yields the 3-bromo derivative **19**. Cleavage of the ester groups at this stage followed by protonation with acetic acid gives back the tropone ring system, substituted now with a good leaving group β to the carbonyl in the five-membered ring (**20**). Addition of **20** to a solution of pyridine and cyclopentadiene smoothly generates 1,4-AQ, which once again is trapped to produce the Diels–Alder adduct **18** (equation 5).

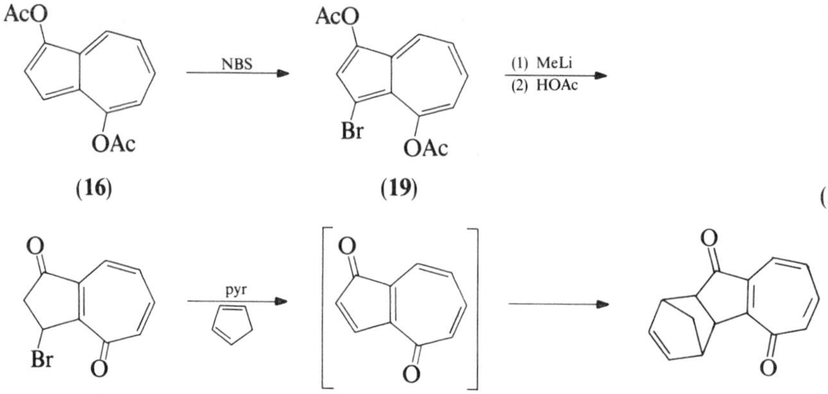

(5)

As anticipated from the theoretical considerations presented above, this quinone is too reactive under ordinary laboratory conditions to be isolated.

C. 1,5-Azulene Quinone

In sharp contrast to the 1,4-quinone of azulene, the 1,5-quinone is quite stable, just as predicted.

The first example of a 1,5-AQ to be prepared was the 3-methoxycarbonyl derivative reported by Morita and coworkers in 1982[29]. Using again the Nozoe method for constructing substituted azulenes from tropolone derivatives, they began by preparing the 5-chloroazulene **21**. Deamination of **21** to **22** occurs smoothly on diazotization with isopentyl nitrite and sulfuric acid in the presence of hydroquinone (equation 6). Heating of **22** with NaOMe–MeOH in anhydrous benzene gives the 5-methoxyazulene **23**. Partial demethoxycarbonylation of diester **23** then gives a mixture (1 : 1) of the monoesters **24** and **25**, which can be separated by column chromatography.

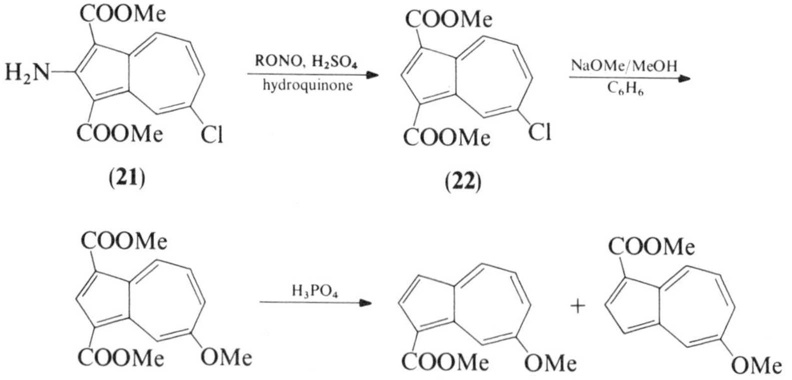

(6)

Direct oxygenation of **24** is effected with 3 equivalents of lead tetraacetate in benzene–pyridine–DMSO (equation 7). Subsequent oxidation of the resulting hydroquinone derivative (**26**) with ceric ammonium nitrate gives the corresponding yellow 1,5-quinone **27**.

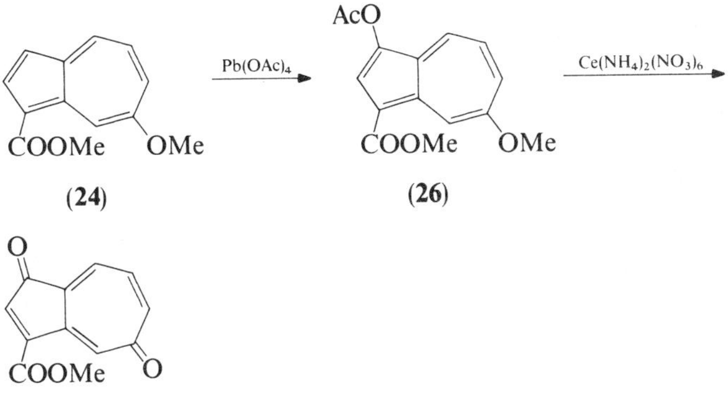

Pb(OAc)$_4$ → Ce(NH$_4$)$_2$(NO$_3$)$_6$ →

AcO

COOMe OMe COOMe OMe

(**24**) (**26**)

O

COOMe O

(**27**)

(7)

Attempts to extend this route to a synthesis of the parent 1,5-AQ have been thwarted by oxidative coupling at the (no longer blocked) 3-position of the azulene hydroquinone derivatives[30].

The first synthesis of unsubstituted 1,5-AQ was published by Scott and Adams in 1984[26]. They began with the versatile bicyclic trienone **13**, prepared as in equation 2. Photooxygenation of **13** gives the two endoperoxides **28** and **29** in high overall yield, with the former predominating (equation 8). Separation of **28** from **29** can be achieved by

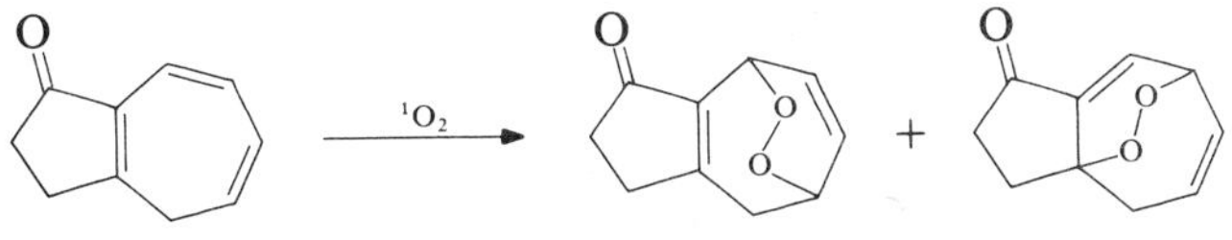

$^1O_2$ →

(**13**) (**28**) + (**29**)

(8)

chromatography, and each endoperoxide can be carried on to a single azulene quinone (**28** to 1,5-AQ; **29** to 1,7-AQ); however, it is far more efficient to carry both isomers through

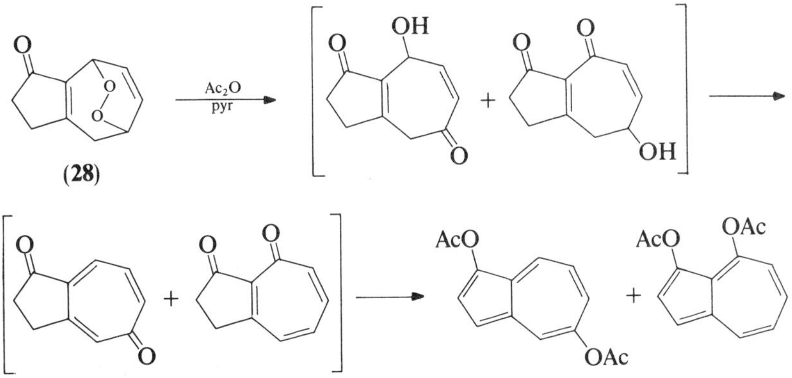

Ac$_2$O, pyr →

(**28**)

OH O O O

O OH

AcO AcO OAc

OAc

(**30**) (**31**)

(9)

together and separate the two quinones at the end. For clarity, only the synthesis of 1,5-AQ will be described in this section.

Treatment of endoperoxide **28** with pyridine and acetic anhydride triggers a marvelous cascade of events which continues all the way to the new azulene diacetates **30** and **31**, presumably via the pathway depicted in equation 9. The base-catalyzed isomerization of such singlet oxygen adducts to γ-hydroxy enones has ample precedent[31], and the final conversion of tropone intermediates to diacetoxyazulenes parallels the reaction in equation 3. The 1,8-isomer (**31**) is easily removed by chromatography and has been examined as a potential precursor to 1,8-AQ (*vida infra*).

Cleavage of the two acetoxy groups in **30** with methyllithium, quenching with chlorotrimethylsilane, and oxidizing as before (equation 10, cf. equation 4) gives the parent pale yellow 1,5-AQ; PCC, DDQ and *p*-chloranil have all proven effective for the final oxidation.

AcO, OAc (**30**) —(1) MeLi, (2) TMSCl→ TMSO, OTMS (**32**) —oxidation→ O, O 1,5-AQ (10)

Still other routes to the parent 1,5-AQ have also been explored. Equation 11 illustrates two variations on the intramolecular carbene addition reaction described above (equation 2). Both the *meta*-substituted anisole (**33**)[32, 33] and the acetanilide (**34**)[34] cyclize to mixtures of 1,5- and 1,7-difunctionalized hydroazulenes; however, difficulties in the subsequent transformations of these bicyclic trienones caused both routes to be abandoned.

O, CHN_2, X —$-N_2$→ O, X + O, X (11)

(**33**, X = OMe)
(**34**, X = NHAc)

A remarkably short synthesis of 3-*t*-butyl-1,5-AQ, starting from azulene itself, has been developed by Scott and Gingerich (equation 12)[35]. Direct oxygenation of azulene with benzoyl peroxide followed by Friedel–Crafts alkylation with *t*-butyl bromide yields the 1,3-disubstituted azulene **35**. Chromium trioxide oxidation in wet acetic acid then gives the 3-*t*-butyl derivatives of 1,5-AQ and 1,7-AQ (**36** and **37**, respectively), each in about 15% yield.

An independent synthesis (equation 13) was carried out[35] to confirm the structural assignment of **36**. Analogous confirmation was obtained[35] for the structure of **37**.

Extension of the route depicted in equation 12 to guaiazulene (**39**) gives guaiazulquinone (**41**), albeit in low yield (equation 14)[35]. This synthesis was inspired by the discovery of guaiazuloquinone among marine natural products collected deep in the Pacific Ocean (– 350 meters) by Scheuer's group in Hawaii[36]. More recently, Nozoe and coworkers have found that autoxidation of guaiazulene (**39**) at 100 °C in *N,N*-

PhCOO

(1) $(PhCOO)_2$
(2) t-BuBr
$ZnCl_2$

t-Bu

(35)

CrO_3
H_2O–HOAc

O O O O

t-Bu t-Bu

(36) + (37) (12)

AcO AcO

OAc OAc

t-BuBr
$ZnCl_2$

t-Bu

as in eqn 10 → **36** (13)

(30) (38)

PhCOO

$(PhCOO)_2$

CrO_3
H_2O–HOAc

(39) (40)

O O

(41) (14)

dimethylformamide produces guaiazuloquinone (**41**) directly in 1% yield, along with a plethora of other interesting products[37]. Under similar conditions, 4,6,8-trimethylazulene (**42**) gives the corresponding 1,5- and 1,7-AQs (**43** and **44**, respectively), each in 1.5% yield

O_2/DMF
120 °C

O O O

O

(42) (43) + (44) (15)

(equation 15)[38]. Somewhat higher yields can be achieved under less drastic conditions of autoxidation, but anodic oxidation shows the greatest promise (60% yield of **41** from **39**)[39].

D. 1,6-Azulene Quinone

The first two syntheses of unsubstituted 1,6-AQ were reported by Scott and coworkers in 1984[25]. Starting from tropone **15**, these routes (equations 16 and 17) employ the same chemistry as that used above to prepare the parent 1,4-AQ. Like the 1,4-AQ, this quinone also is too reactive to be isolated under ordinary laboratory conditions and must be trapped as a fleeting intermediate.

O O (15) $\xrightarrow[\text{pyr}]{Ac_2O}$ OAc OAc (45) $\xrightarrow[\text{(2) TMSCl}]{\text{(1) MeLi}}$

TMSO OTMS (46) $\xrightarrow{\text{oxidation}}$ [O O 1,6-AQ] ⟶

O O (47) (16)

OAc OAc (45) $\xrightarrow{\text{NBS}}$ OAc OAc Br (48) $\xrightarrow[\text{(2) HOAc}]{\text{(1) MeLi}}$

O O Br (49) $\xrightarrow{\text{pyr}}$ [O O 1,6-AQ] ⟶ O O (47) (17)

The development of high-yield routes to tropone **15** (equation 18)[25] have made this precursor to the 1,6-AQ ring system available for extensive additional experimentation. In this connection, it has been found that direct bromination of **15** with NBS and benzoyl

(50) (1) $Rh_2(OAc)_4$ (2) Al_2O_3 → (51) NBS/MeOH →

(52) (1) Et_3N (2) H_2O → (15)

(13) NBS, H_2O, DMSO → (53) CrO_3, pyr → (15) (18)

peroxide in carbon tetrachloride followed by treatment of the crude reaction mixture with excess pyridine and cyclopotentadiene also gives the Diels–Alder adduct **47**, albeit in poor yield (equation 19)[40].

(15) NBS, $(PhCOO)_2$ → pyr, cyclopentadiene →

[1,6-AQ] → (47) (19)

Two sterically hindered derivatives of 1,6-AQ were prepared in the hope that bulky substituents might render the quinones stable enough to survive in monomeric form. The 3-*t*-butyl derivative (**56**), synthesized as in equation 20, does in fact show greatly diminished

OAc, OAc (45) $\xrightarrow[ZnCl_2]{t\text{-BuBr}}$ OAc, OAc, t-Bu (54) $\xrightarrow[(2)\ TMSCl]{(1)\ MeLi}$

TMSO, OTMS, t-Bu (55) $\xrightarrow{\text{oxidation}}$ [O, O, t-Bu (56)] ⟶

O, CO, t-Bu (57) (20)

reactivity at the double bond of the cyclopentadienone ring; however, other regions of the molecule remain quite reactive, as evidenced by the efficient trapping of **56** with cyclopentadiene in a [6 + 4] cycloaddition reaction[40]. Unfortunately, all attempts to isolate the substituted 1,6-AQ **56** have failed.

The 2,3-disubstituted derivative **59**, prepared as in equation 21, likewise proved too reactive to be isolated[40]. In this case, however, a dimer could be obtained when the quinone was generated in the absence of trapping agents. From spectroscopic data, the gross structure of the dimer was determined to be that of a [4 + 2] cycloadduct formed by addition of the norbornene double bond of one molecule across a diene moiety in the tropone ring of the second molecule[40]. Although some aspects of the structure remain uncertain, it should be noted that this reaction mode simultaneously disrupts the cyclopentadienone rings in both quinones.

O, O (47) $\xrightarrow{HN=NH}$ O, O (58) $\xrightarrow[(PhCOO)_2]{NBS}$

$\xrightarrow[Et_3N]{DBN}$ [O, O (59)] ⟶ dimer (21)

Unpublished experiments from the laboratory of Professor Hafner in Darmstadt[41] have established that compound **60**, the oxime of a 1,6-AQ obtained by nitrosation of 4,8-dimethyl-6-hydroxyazulene, behaves very much like the parent 1,6-AQ (equation 22).

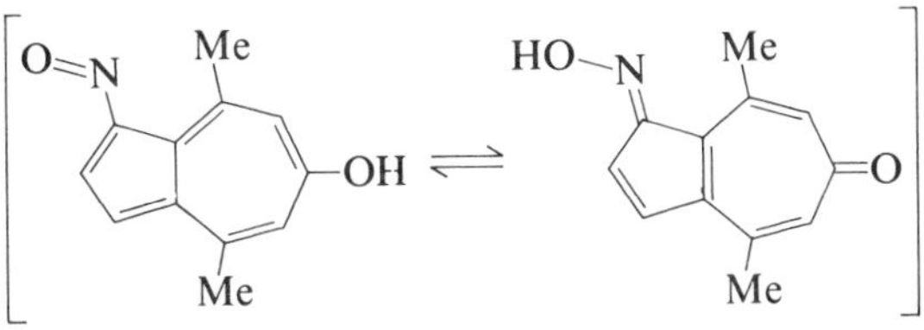

(**60**) (22)

E. 1,7-Azulene Quinone

To date, all syntheses of 1,7-AQs, except that of guaiazuloquinone (equation 14), have been carried out in conjunction with syntheses of the isomeric 1,5-AQs. The 3-methoxycarbonyl derivative **61** was the first 1,7-AQ to be reported[29]. Precursor **25**, prepared as in equation 6, is converted into the quinone by the reactions indicated in equation 23 (cf. equation 7). Attempts to extend this route to a synthesis of the parent 1,7-AQ have not been successful[30].

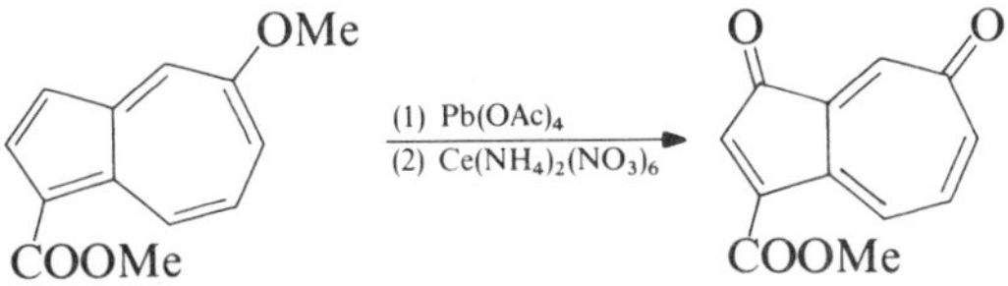

(1) $Pb(OAc)_4$ (2) $Ce(NH_4)_2(NO_3)_6$ (23)

(**25**) (**61**)

The first synthesis of the unsubstituted lemon yellow 1,7-AQ is shown in equation 24 (cf. equations 9 and 10)[26]. Preparation of endoperoxide **29** was described in equation 8.

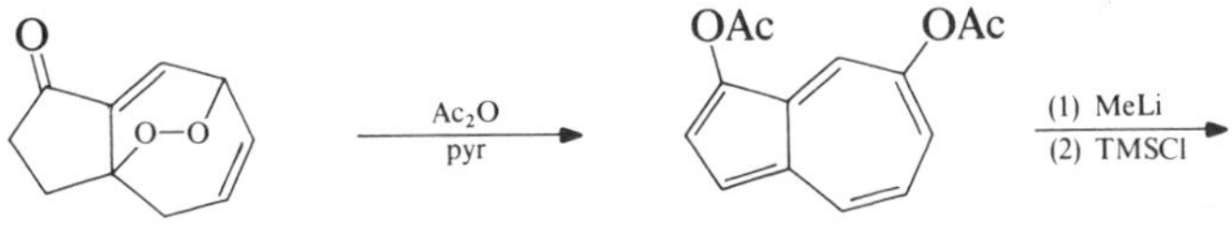

Ac_2O, pyr; (1) MeLi (2) TMSCl

(**29**) (**62**)

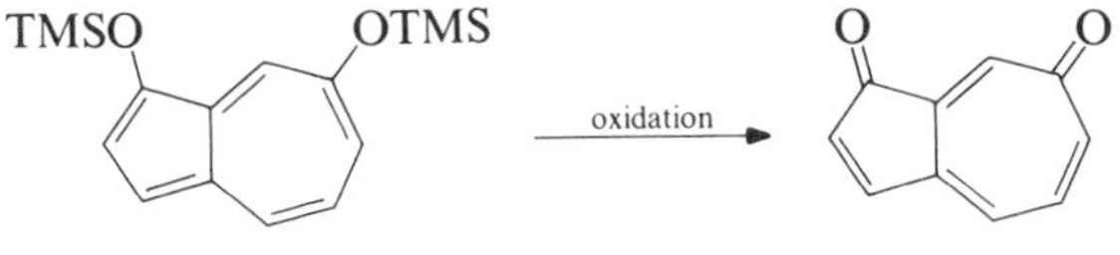

oxidation (24)

(**63**) 1,7-AQ

Alternative routes to the parent 1,7-AQ have been explored, e.g. equation 11, but none has yet been carried to completion[32-34]. The 3-*t*-butyl derivative (**37**)[35] and the 4,6,8-trimethyl derivative (**44**)[38], on the other hand, have both been obtained from readily available azulenes by oxidative methods (equations 12 and 15, respectively).

F. 1,8-Azulene Quinone

This quinone is expected to exhibit reactivity comparable to that of the unstable 1,4-AQ and 1,6-AQ. Since these two quinones can be generated from diacetoxy azulenes via the corresponding bis-trimethylsilyl ethers and trapped by cyclopentadiene (equations 4 and 16), analogous reactions have been attempted[42] starting with the 1,8-diacetoxy azulene **31** (from equation 9). Cleavage of the two acetoxy groups with methyllithium and silylation of the resulting dianion with chlorotrimethylsilane proceed smoothly without any complication from *peri*-interactions of the 1,8-difunctionality (equation 25). Unfortunately, no Diels–Alder adduct could be isolated when bis-trimethylsilyl ether **64** was oxidized under a variety of conditions in the presence of cyclopentadiene; only polymers were obtained[42].

AcO OAc → (1) MeLi (2) TMSCl → TMSO OTMS (25)

(**31**) (**64**)

Other trapping agents, such as 1-hexyne, likewise failed to give any products arising from reactions with 1,8-AQ[43]. Thus, there is still no evidence that 1,8-AQ has ever been generated even as a transient intermediate.

G. 2,4-Azulene Quinone

No work on the synthesis of 2,4-AQ has been reported.

H. 2,6-Azulene Quinone

The very first (bicyclic) azulene quinone ever prepared, the 1,3-diethoxycarbonyl derivative of 2,6-AQ (**66**), was reported by Morita and Takase in 1977[44]. Oxidation of the corresponding hydroquinone (**65**) with DDQ gives the substituted 2,6-AQ, which dimerizes in a [4+4] manner under the reaction conditions (equation 26). A *syn* stereochemistry has been assigned to **67** on the basis of a dipole moment measurement.

COOEt, HO, OH, COOEt → DDQ → [COOEt, O, O, COOEt] →

(**65**) (**66**)

COOEt COOEt O O O O COOEt COOEt (26)

(**67**)

Regeneration of the monomeric quinone and interception by an external trapping agent has been achieved by heating a xylene solution of dimer **67** under reflux with dimethyl acetylenedicarboxylate[44].

The synthesis of unsubstituted 2,6-dihydroxyazulene was also published by Morita and coworkers in 1977, but attempted oxidations of this hydroquinone to the parent 2,6-AQ have not been reported[45a].

I. 4,5-Azulene Quinone

No work on the synthesis of 4,5-AQ has been reported.

J. 4,7-Azulene Quinone

No work on the synthesis of 4,7-AQ has been reported.

K. 5,6-Azulene Quinone

Preliminary work on a potential synthesis of 5,6-AQ (equation 27) has appeared in an MS thesis from Reno[32].

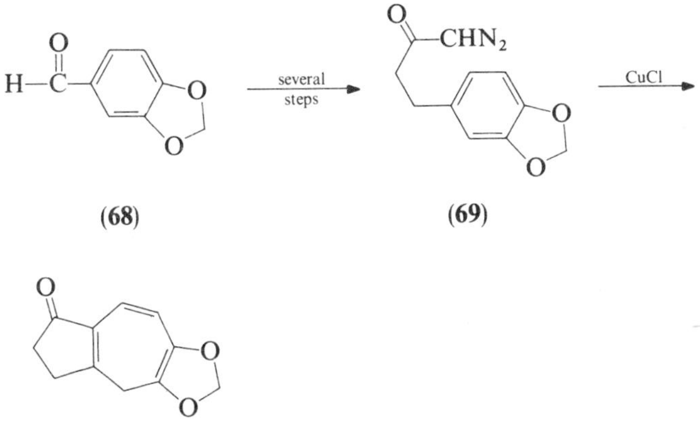

several steps

CuCl

(68) **(69)**

(27)

(70)

IV. PROPERTIES

A. Chemical Properties

1. Isolability

As of mid-1987, only three of the eleven possible quinones of azulene had been isolated in monomeric form, viz. 1,2-AQ, 1,5-AQ and 1,7-AQ (see Sections III.A, III.C and III.E). Several derivatives of each of these were also parpared and characterized in the early 1980s. The parent 1,4-AQ and 1,6-AQ have been trapped as fleeting intermediates, but both have proven too reactive to be isolated under ordinary laboratory conditions (see Sections III.B and III.D). Substituted derivatives of 1,6-AQ and 2,6-AQ have likewise been generated and trapped, although none has been stable enough to be isolated (see Sections III.D and III.H). The chemical stabilities (isolability) of 1,8-AQ, 2,4-AQ, 4,5-AQ, 4,7-AQ and 5,6-AQ remain unknown.

This property of the azulene quinones correlates with both the MINDO/3 π-delocalization energies and the calculated E-LUMOs (Table 1). The three isolable quinones have significantly more positive π-DEs and significantly higher-lying LUMOs than the other eight isomers, and it would be reasonable to expect difficulty in isolating any of the remaining unsubstituted quinones of azulene.

2. Cycloadditions

Both 1,4-AQ and 1,6-AQ exhibit high reactivity as dienophiles in the Diels–Alder reaction[25]. From the chemistry described in Sections III.B and III.D, it is evident that the cycloadditions of these quinones with cyclopentadiene at 0 °C must occur quite rapidly in order to compete so successfully with the alternative biomolecular processes that preclude their isolation. A [4 + 2] cycloaddition on the 2,3-double bond in the cyclopentadienone ring is the preferred mode of reaction, unless that site is blocked, in which case a [6 + 4] cycloaddition on the tropone ring occurs (equation 20).

In sharp contrast to these elusive quinones, both 1,5-AQ and 1,7-AQ can be recovered unchanged after mixing with cyclopentadiene[26]. These results are in complete accord with the theoretical calculations[16], which indicate that the LUMOs of 1,5-AQ and 1,7-AQ lie considerably higher in energy than those of 1,4-AQ and 1,6-AQ (Table 1); dienophilicity toward cyclopentadiene should be greatest for those quinones with the lowest-lying LUMOs[18, 19].

When a 1:1 mixture of 1,5-AQ and *p*-benzoquinone is treated with 1 equivalent of cyclopentadiene, only the *p*-benzoquinone reacts[45b]. This direct competition experiment demonstrates unequivocally the low dienophilicity of 1,5-AQ relative to that of *p*-

(71) + [6+4] → (72) [4+2] → (74)

(71) [4+2] → (73) [6+4] → (74) (28)

benzoquinone. The analogous experiment with 1,7-AQ and *p*-benzoquinone gave the same result[45b]. The outcome of these experiments was anticipated on the basis of the theoretical calculations, which predict that the LUMOs of 1,5-AQ and 1,7-AQ should lie even higher in energy than that of *p*-benzoquinone[16].

With diphenylisobenzofuran (DPIBF, **71**), a more reactive diene, 1,5-AQ combines to give a 2:1 adduct (equation 28)[26]. Whether the [6 + 4] cycloaddition preceeds or follows the [4 + 2] addition has not been established, since the 1:1 adduct (either **72** or **73**) reacts more rapidly with DPIBF than does the original quinone. An equimolar mixture of the two cycloaddends gives only the 2:1 adduct **74** and recovered quinone. The 1,7-quinone reacts similarly (equation 29)[26].

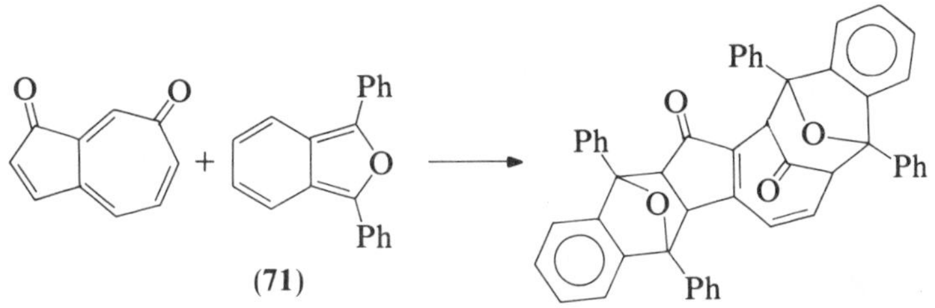

(**71**) (**75**) (29)

The substituted 2,6-AQ **66** (generated by cracking of the dimer, **67**, in refluxing xylene) combines with dimethyl acetylenedicarboxylate in a Diels–Alder fashion to give intermediate **76**, which spontaneously decarbonylates to yield the benzotropone **77**[44] (equation 30).

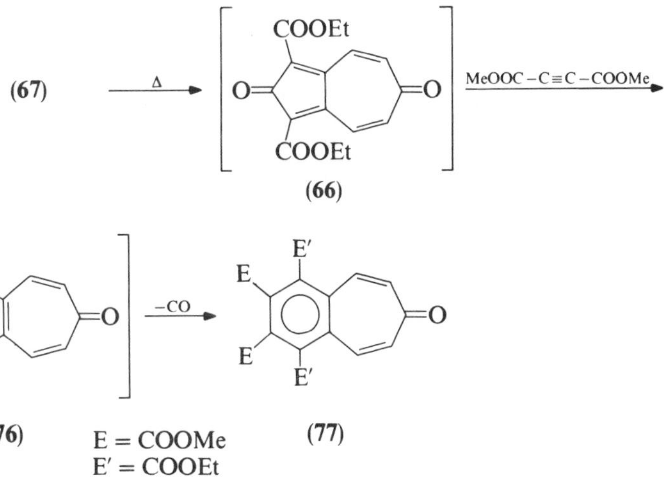

(**67**) $\xrightarrow{\Delta}$ [(**66**)] $\xrightarrow{\text{MeOOC}-\text{C}\equiv\text{C}-\text{COOMe}}$ [(**76**)] $\xrightarrow{-\text{CO}}$ (**77**) (30)

E = COOMe
E′ = COOEt

No cycloaddition chemistry has been reported for 1,2-AQ.

3. Reduction

Polarographic half-wave potentials for derivatives of 1,2-AQ, 1,5-AQ and 1,7-AQ as well as those for the parent 1,2-AQ are summarized in Figure 2[24, 29]. On an absolute scale,

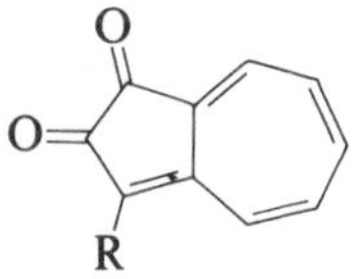

R	E_1	E_2
H	−0.55	−1.17
COOEt	−0.40	−1.00
CN	−0.34	−0.88

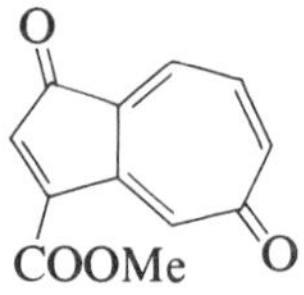

COOMe

$E_1 = -0.54$

COOMe

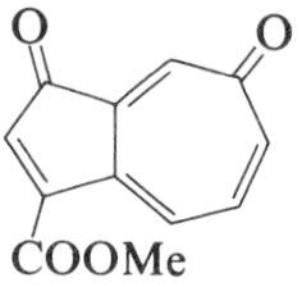

$E_1 = -0.50$

FIGURE 2. Polarographic half-wave potentials (V vs. SCE)[24, 29] in anhydrous MeCN at 25 °C, dropping-mercury electrode, supporting electrolyte 0.1 M Et_4NClO_4

these data should be compared with the half-wave potentials for *p*-benzoquinone ($E_{1/2}$ = −0.51, −1.14 V)[46]. As predicted by theory, the parent 1,2-AQ is slightly less easily reduced than *p*-benzoquinone. A comparison of the data for the three alkoxycarbonyl derivatives further reveals that the ease of reduction of the isolable azulene quinone ring systems (1,5-AQ < 1,7-AQ < 1,2-AQ) correlates well with the calculated LUMO energies (Table 1).

As expected, electron-withdrawing groups increase the potential while donor substituents lower the potential. Nozoe and coworkers have measured the potentials for several trialkylazulene quinones[39] (V vs. SCE in anhydrous MeCN at 25°C, platinum electrode, 0.1 M Et_4NClO_4): 4,6,8-trimethylazulene-1,5-quinone **43** ($E_{1/2}$ = −1.05 and −1.46); 4,6,8-trimethylazulene-1,7-quinone **44** ($E_{1/2}$ = −1.05 and −1.5); guaiazulene-1,7-quinone **41** ($E_{1/2}$ = −1.13 and −1.52).

Although the potentials of unsubstituted 1,4-AQ, 1,5-AQ, 1,6-AQ and 1,7-AQ have not been reported, it is clear that they cannot exceed the potential of *p*-chloranil ($E_{1/2}$ = +0.01 V)[46], an oxidizing agent used to synthesize these quinones from their hydroquinone derivatives[25, 26]. By the same reasoning, the potential of 1,3-diethoxycarbonyl-2,6-azuloquinone (**66**)[44] cannot exceed that of DDQ ($E_{1/2}$ = +0.51 V)[46].

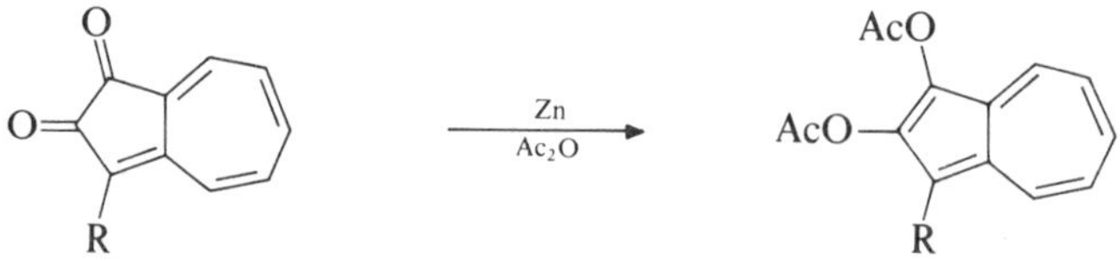

R = H, COOEt, CN —(Zn, Ac_2O)→ (**78**) (31)

R = H, COOMe —(Zn, Ac_2O)→ (**79**) (32)

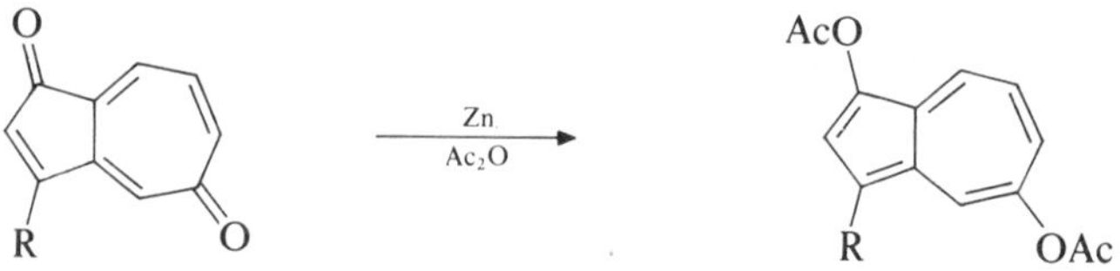

Just as benzenoid quinones can be reduced with zinc and acetic anhydride back to the corresponding hydroquinone diacetates, so too have many of the stable azulene quinones been reduced to their corresponding diacetoxyazulenes (equations 31–33)[24, 26, 29].

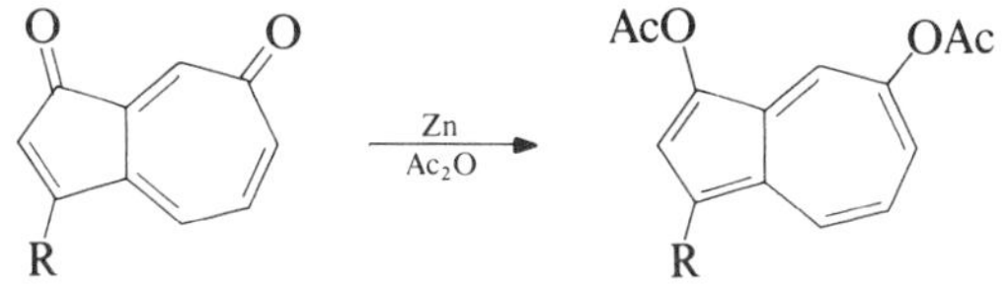

(33)

R = H, COOMe **(80)**

4. Other chemical reactions

Condensation of 1,2-AQ and its derivatives with *o*-phenylenediamine gives the expected azuleno[1,2]quinoxalines **81** (equation 34)[24].

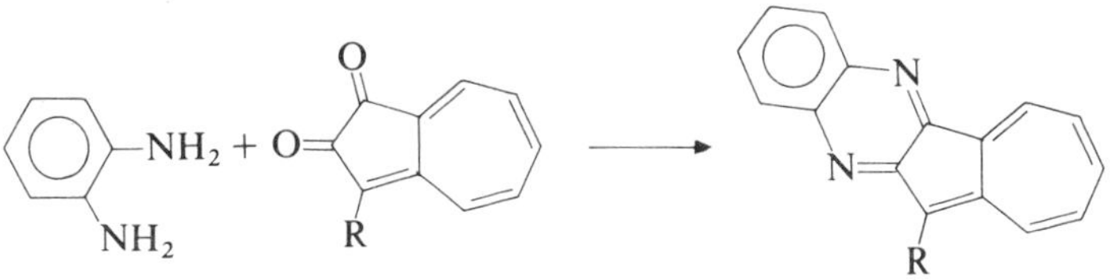

(34)

R = H, COOEt, CN **(81)**

In methanol, the parent 1,2-AQ exists in equilibrium with the hemiketal (**82**) formed by addition of solvent to the carbonyl group at the 1-position; electron-withdrawing groups at the 3-position drive this equilibrium completely over to the hemiketal (equation 35)[24].

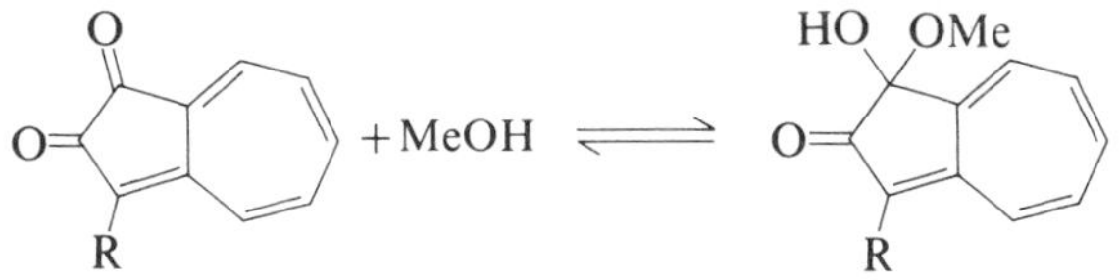

(35)

R = H, COOEt, CN **(82)**

The stability of the residual heptafulvene moiety (a vinylogous tropone) presumably accounts for the site selectivity in this reaction[24].

B. Spectroscopic Properties

1. UV–VIS absorption spectra and color

The UV–VIS absorption spectral data available for 1,2-AQ, 1,5-AQ, 1,7-AQ and several derivatives thereof are summarized in Table 4.

The 1,2-AQs each exhibit a maximum in the long wavelength portion of the spectrum between 565 and 595 nm, with lower intensity absorptions extending all the way out to 800 nm. From PPP calculation (Table 2, Section II), the strongest long wavelength $\pi\pi^*$

TABLE 4. UV–VIS absorption spectra of azulene quinones

Compound	R^2	R^3	R^4	R^5	R^6	R^7	R^8	Solvent	λ_{max} nm (log ε)	Reference
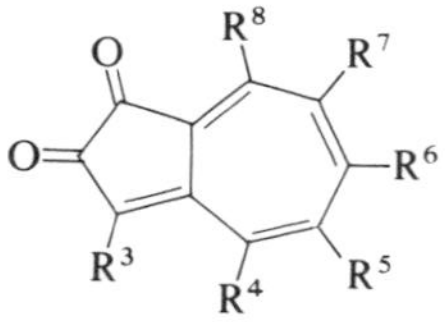	—	H	H	H	H	H	H	$CHCl_3$	800 (2.05), 705 (2.63), 655 (2.74), 594 (2.79), 553 (2.74), 516 (2.71), 479 (2.68), 411sh (3.39), 386 (3.95), 361sh (3.84), 261 (4.42)[a]	24
	—	CO_2Et	H	H	H	H	H	$CHCl_3$	750 (2.21), 679 (2.53), 616 (2.76), 566 (2.82), 521 (2.76), 424sh (4.00), 397 (4.13), 371sh (4.02), 331 (3.68), 268 (4.29)[a]	24
	—	CN	H	H	H	H	H	$CHCl_3$	632 (2.55), 566 (2.66), 526 (2.63), 484 (2.63), 431 (3.84), 418 (3.84), 405 (4.00), 360sh (3.68), 305 (3.53), 279 (4.01), 263 (4.12)[a]	24
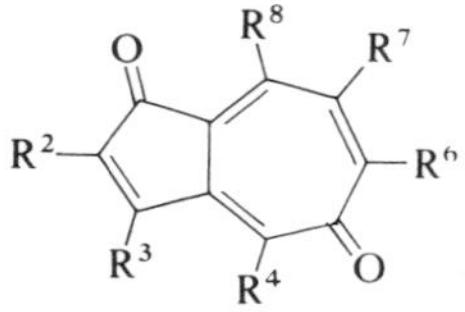	H	H	H	—	H	H	H	MeCN	389sh (3.29), 373 (3.53), 350 (3.60), 338 (3.59), 324 (3.57), 307sh (3.54), 264 (4.27), 254 (4.36), 216 (3.89)	26
	H	H	H	—	H	H	H	EtOH	389 sh, 373, 350, 337, 322[b]	26
	H	t-Bu	H	—	H	H	H	MeCN	386 (3.43), 369 (3.71), 350 (3.76), 335 (3.77), 324 (3.75), 270 (4.35), 261 (4.43), 254 (4.31), 222 (4.07)	35

Structure	R^2	R^3	R^4	R^5	R^6	R^7	R^8	Solvent	λ_{max} (log ε)	Ref.
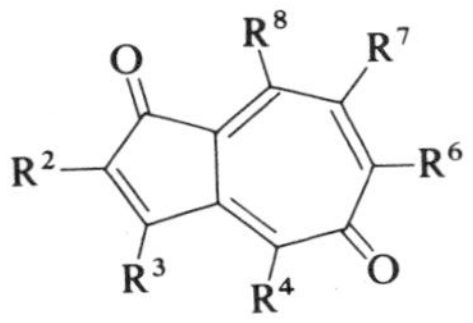 azulene-1,6-dione (R^2, R^3, R^4, R^6, R^7, R^8)	H	CO_2Me	H	—	H	H	H	$CHCl_3$	414sh (3.65), 390 (3.94), 372 (3.82)[a]	29
	H	H	Me	—	Me	H	Me	MeOH	400 (3.65)[c]	38
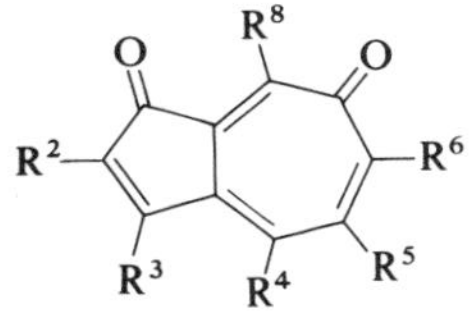 azulene-1,8-dione (R^2, R^3, R^4, R^5, R^6, R^8)	H	H	H	H	H	—	H	MeCN	425sh (3.21), 401 (3.51), 384 (3.54), 345sh (3.59), 333sh (3.75), 319 (3.84), 236 (4.38)	26
	H	H	H	H	H	—	H	EtOH	430sh, 399, 384, 345sh, 327sh, 319[b]	26
	H	*t*-Bu	H	H	H	—	H	MeCN	391 (3.47), 375 (3.50), 328 (3.79), 317 (3.81), 270 (3.88), 235 (4.34), 226 (4.37)	35
	H	CO_2Me	H	H	H	—	H	$CHCl_3$	423 (3.50), 403 (3.52), 312 (3.80)[a]	29
	H	H	Me	H	Me	—	Me	MeOH	386 (3.95)[c]	38
	H	Me	H	*i*-Pr	H	—	Me	MeOH	398 (3.95)[c]	37

[a] Spectrum was not reported below 250 nm.
[b] Spectrum was not reported below 300 nm, sample decomposes on standing in EtOH.
[c] Only the longest wavelength absorption maximum was reported.

electronic transition for 1,2-AQ was predicted to occur at 572 nm. A second, stronger absorption in the spectrum of 1,2-AQ and its derivatives appears in the 385–405 nm region (calculated: 397 nm). The green color reported for 1,2-AQ was accurately predicted[16] on the basis of the calculations.

Also as predicted[16], the 1,5-AQs and 1,7-AQs absorb at much shorter wavelength than the 1,2-AQs and appear yellow in color. The 1,7 isomer was predicted[16] to absorb at slightly longer wavelength than the 1,5 isomer, and this too was borne out by experiment.

Especially noteworthy is the observation that these latter two quinones of azulene absorb light at significantly *shorter* wavelength than does azulene itself (λ_{max} = 579 nm). Such behavior is precisely the reverse of that which is observed for alternant systems; the quinones of benzene and naphthalene, for example, are yellow, orange and red, whereas the parent hydrocarbons are colorless. The peculiarity of the azulenic compounds is a direct consequence of the non-alternant nature of the π system and is well accounted for by the theoretical calculations[16].

In the homoazulenic series, this same peculiarity is also seen (Figure 3)[47]. Thus, quinones **84** and **85** absorb light at significantly *shorter* wavelength than does

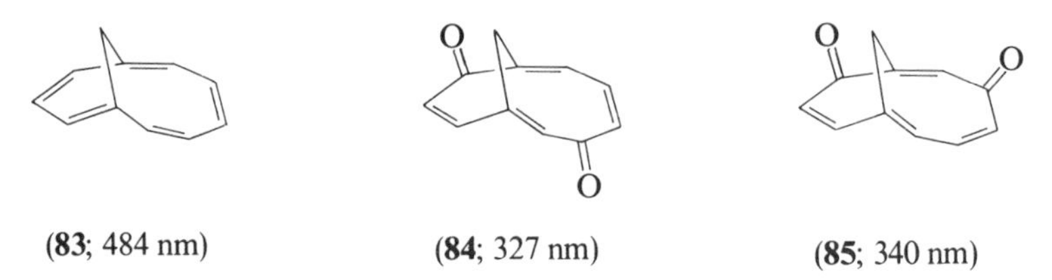

FIGURE 3. Long wavelength maxima in the UV–VIS absorption spectra of homoazulene (**83**), homoazulene-1,5-quinone (**84**), and homoazulene-1,7-quinone (**85**)[47].

homoazulene itself (**83**), and in this respect they behave more like the nonalternant quinones of azulene than like quinones of a purely alternant [10]annulene. Clearly, the non-alternant homoconjugative perturbation enforced by the homoazulene skeleton[48] has a dramatic effect on the electronic properties of these bridged relatives of azulene quinones.

The success of the PPP π-electron method in predicting the long wavelength absorption maxima of 1,2-AQ, 1,5-AQ and 1,7-AQ, which range over more than 200 nm, engenders confidence in the predictions of colors for the other quinones of azulene.

2. *Infrared spectra*

The IR spectral data available for 1,2-AQ, 1,5-AQ, 1,7-AQ and several derivatives thereof are summarized in Table 5.

The five-membered ring α-diketone and the C=C bonds of 1,2-AQ give rise to three prominent bands in the 1800–1600 cm^{-1} region of the spectrum, including a relatively high-frequency band at 1751 cm^{-1}. Derivatives of 1,2-AQ give similar IR spectra.

The IR spectra of 1,5-AQ and 1,7-AQ are very similar to one another and look much like what one would expect from the individual component rings. Tropone gives rise to two strong bands at 1643 and 1594 cm^{-1} intermingled with several weaker bands in the IR spectrum; assignment of the 1594 cm^{-1} band to the C=O stretching mode has been confirmed by an elegant ^{18}O-labeling experiment[49]. In the spectrum of 1,5-AQ, two strong bands appear at 1650 and 1590 cm^{-1}, adorned with several weak shoulders; 1,7-AQ absorbs at 1649 and 1586 cm^{-1}. Such similarities in the characteristic vibrational frequencies of these molecules indicate that the geometries and bond orders in the seven-

TABLE 5. IR spectra of azulene quinones (C=O and C=C region)

Compound	R^2	R^3	R^4	R^5	R^6	R^7	R^8	Medium	ν_{max} (cm^{-1})	Reference
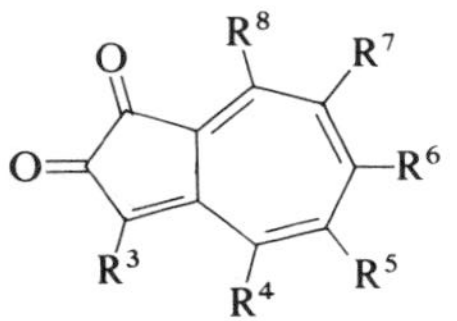	—	H	H	H	H	H	H	$CHCl_3$	1751 (m), 1687 (vs), 1643 (m)	24
	—	CO_2Et	H	H	H	H	H	$CHCl_3$	1752 (m), 1699 (vs), 1688[a], 1638 (m)	24
	—	CN	H	H	H	H	H	$CHCl_3$	1757 (m), 1707 (vs), 1693 (m)	24
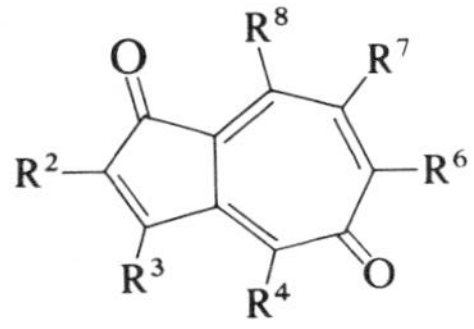	H	H	H	—	H	H	H	KBr	1706 (s), 1650 (s), 1590 (s)	26
	H	*t*-Bu	H	—	H	H	H	CCl_4	1720 (s), 1650 (s), 1605 (s)	35
	H	CO_2Me	H	—	H	H	H	KBr	1725sh, 1711, 1648, 1590	29
	H	H	Me	—	Me	H	Me	KBr	1695, 1575[b]	38
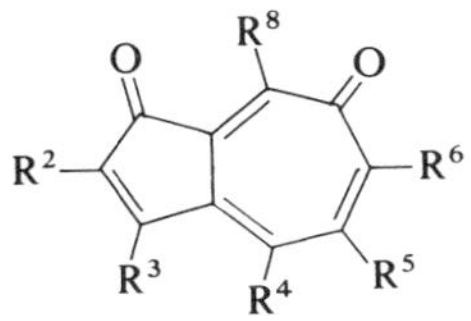	H	H	H	H	H	—	H	KBr	1709 (s), 1649 (s), 1586 (s)	26
	H	*t*-Bu	H	H	H	—	H	CCl_4	1720 (s), 1650 (w), 1635 (w), 1605 (s)	35
	H	CO_2Me	H	H	H	—	H	KBr	1724sh, 1711, 1641, 1590	29
	H	H	Me	H	Me	—	Me	KBr	1695, 1580[b]	38
	H	Me	H	*i*-Pr	H	—	Me	KBr	1680, 1590[b]	37

[a] Assigned as the ethoxycarbonyl group by Morita and coworkers.
[b] Only the two C=O bands reported.

membered rings of 1,5-AQ and 1,7-AQ must differ very little from those in tropone. The five-membered ring C=O stretching band is seen at 1706 cm^{-1} for 1,5-AQ and 1709 cm^{-1} for 1,7-AQ.

3. ^{1}H-NMR spectra

The ^{1}H-NMR spectral data available for 1,2-AQ, 1,5-AQ, 1,7-AQ and several derivatives thereof are summarized in Table 6.

The signals for the seven-membered ring protons in 1,5-AQ and 1,7-AQ appear at slightly lower field than those in tropone (broad singlet at $\delta 6.8$), while those in 1,2-AQ appear at slightly higher field, as expected for a vinylogous tropone. The only unusual feature is the exceptionally low-field signal for the proton at position-3 in 1,5-AQ and 1,7-AQ. Substituents perturb these spectra in the anticipated manner.

The coupling constants for vicinal hydrogens in the seven-membered ring of 1,5-AQ, 1,7-AQ and their derivatives are completely consistent with a strong alternation of single and double bonds, as in tropone[50]. Unfortunately, the spectra for 1,2-AQ and its derivatives were insufficiently resolved at 100 MHz to permit the measurement of coupling constants for most of the protons in these molecules.

4. ^{13}C-NMR spectra

The ^{13}C-NMR spectral data available for azulene quinones are limited to 1,5-AQ, 1,7-AQ[26], the 3-*t*-butyl derivatives thereof[35] and the 3-methoxycarbonyl derivatives thereof[29]. For all six compounds, the resonance of C(1) appears in the region $\delta 192$–195, and the seven-membered ring carbonyl carbon absorbs at $\delta 187$–188. Morita and coworkers have pointed out[29] that the order of average chemical shift of the ring carbons in the ^{13}C-NMR spectra for the 3-methoxycarbonyl derivatives of 1,5-AQ and 1,7-AQ ($\delta 149.1$ and 148.6, respectively) agrees with the order of the polarographic half-wave potentials.

5. Mass spectra

The mass spectra of 1,2-AQ, 1,5-AQ and 1,7-AQ all show prominent molecular ions and sequential loss of two carbonyl groups to give a base peak at m/z 102:

1,2-AQ[24] (25 eV) m/z (relative abundance) 158 (M^+, 30), 130 (M–CO, 11), 102 (M–2CO, 100).
1,5-AQ[26] (70 eV) m/z (relative abundance) 158 (M^+, 20), 130 (M–CO, 85), 102 (M–2CO, 100), 76 (24).
1,7-AQ[26] (70 eV) m/z (relative abundance) 158 (M^+, 38), 130 (M–CO, 71), 102 (M–2CO, 100), 76 (31).

Similar fragmentation patterns are observed for the 3-alkoxycarbonyl derivatives of these quinones and for the 3-cyano derivative of 1,2-AQ, although the substituents do complicate the picture to some extent[24,29]. Mass spectra of the trialkylazulene 1,5- and 1,7-quinones **41**, **43** and **44** show base peaks for the molecular ions but otherwise fragment normally[37–39].

C. Biological Properties

A number of azuloquinone and hydroquinone derivatives have been found to exhibit significant cytotoxic activity. In the standard KB cell culture screen, compounds with an $ED_{50} < 4\ \mu g\ ml^{-1}$ are considered significantly cytotoxic by the US National Cancer

TABLE 6. ^{1}H-NMR spectra of azulene quinones in $CDCl_3$

Compound	R^2	R^3	R^4	R^5	R^6	R^7	R^8	MHz	δ (ppm from $SiMe_4$)	Reference
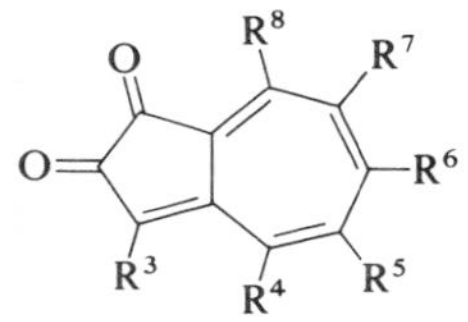	—	H	H	H	H	H	H	100	6.24–6.04 (m, H^4–H^8), 5.78 (s, H^3)	24
	—	CO_2Me	H	H	H	H	H	100	8.29 (complex d, $J = 12$ Hz, H^4), 7.12–6.72 (m, H^5–H^8), 4.36 (q, $J = 7.0$ Hz, CH_2), 1.40 (t, $J = 7.0$ Hz, Me)	24
	—	CN	H	H	H	H	H	100	7.20 (dd, $J = 10$, 1.4 Hz, H^4), 7.08–6.42 (m, H^5–H^8)	24
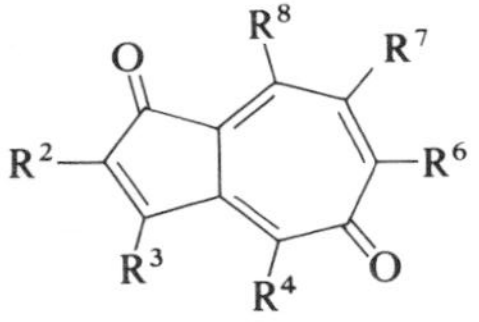	H	H	H	—	H	H	H	360	7.78 (d, $J = 5.9$ Hz, H^3), 7.31 (dd, $J = 7.8$, 1.1 Hz, H^8), 7.11 (dd, $J = 12.2$, 7.8 Hz, H^7), 6.94 (ddd, $J = 12.2$, 2.6, 1.1 Hz, H^6), 6.82 (d, $J = 2.6$ Hz, H^4), 6.56 (d, $J = 5.9$ Hz, H^2)	26
	H	*t*-Bu	H	—	H	H	H	100	7.35–6.80 (m, H^4–H^8), 6.36 (s, H^2), 1.40 (s, *t*-Bu)	35
	H	CO_2Me	H	—	H	H	H	200	7.81 (H^4), 7.43 (H^8), 7.19 (H^7), 7.09 (H^2), 7.03 (H^6), 3.98 (COOMe)	29
	H	H	Me	—	Me	H	Me	200	7.94 (d, $J = 6.0$ Hz, H^3), 7.04 (s, H^7), 6.25 (d, $J = 6.0$ Hz, H^2), 2.60 (s, Me^8), 2.31 (s, Me^4), 2.24 (s, Me^6)	38
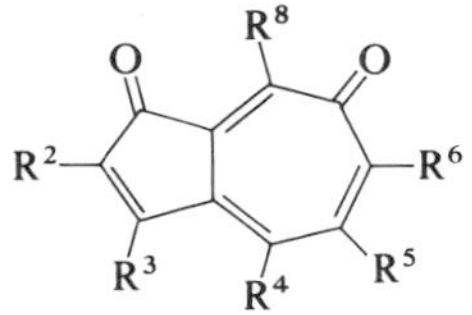	H	H	H	H	H	—	H	360	7.84 (d, $J = 5.8$ Hz, H^3), 7.24 (d, $J = 2.8$ Hz, H^8), 7.04 (dd, $J = 12.4$, 8.0 Hz, H^5), 6.82 (dd, $J = 12.4$, 2,8 Hz, H^6), 6.76 (d, $J = 8.0$ Hz, H^4), 6.50 (d, $J = 5.8$ Hz, H^2)	26
	H	*t*-Bu	H	H	H	—	H	100	7.30–6.70 (m, H^4–H^8), 6.33 (s, H^2), 1.42 (s, *t*-Bu)	35
	H	CO_2Me	H	H	H	—	H	200	7.83 (H^4), 7.35 (H^8), 7.20 (H^5), 7.02 (H^2), 6.95 (H^6) 3.99 (COOMe)	29
	H	H	Me	H	Me	—	Me	200	8.06 (d, $J = 6.0$ Hz, H^3), 7.09 (q, $J = 1.5$ Hz, H^5), 6.38 (d, $J = 6.0$ Hz, H^2), 2.65 (s, Me^8), 2.33 (s, Me^4), 2.27 (d, $J = 1.5$ Hz, Me^6)	38
	H	Me	H	*i*-Pr	H	—	Me	200	6.76 (d, $J = 2.0$ Hz, H^6), 6.63 (dd, $J = 2.0$, 0.5 Hz, H^4), 6.23 (qd, $J = 1.5$, 0.5 Hz, H^2), 2.76 (sept, $J = 7.0$ Hz, $CHMe_2$), 2.64 (s, Me^8), 2.29 (d, $J = 1.5$ Hz, Me^3), 1.26 (d, $J = 7.0$ Hz, $CHMe_2$)	37

Institute. More than half of the compounds submitted for testing by Scott and coworkers have proven active at this level or below (Figure 4), and the parent 1,5-AQ and 1,7-AQ exhibit 100% activity at the lowest dosage measured (1.0 μg ml^{-1})[51].

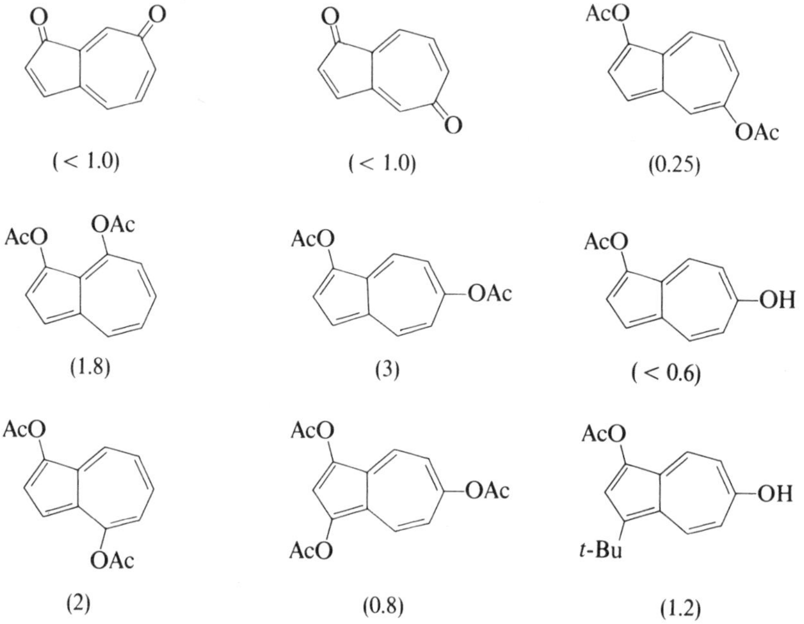

FIGURE 4. KB cell cytotoxicity (ED$_{50}$ μg ml^{-1})[51]

The hydroquinone derivatives presumably owe their activity to an *in vivo* oxidation which generates reactive quinones within the cell. In agreement with this hypothesis, several of the most active hydroquinone derivatives were found to produce a dramatic stimulation of O_2 uptake in Sarcoma 180 whole cell respiration[51]. Also, those hydroquinone derivatives with a free OH group show greater activity than the corresponding doubly acetylated hydroquinones, while the methyl ethers exhibit little activity[51].

Seven of the most cytotoxic compounds have been tested against P-388 leukemia in mice[51]. All seven compounds exhibit pronounced biological activity at relatively low dosages; however, no dosages have yet been found which effect remission (T/C > 125) for any of the compounds. The parent 1,5-AQ and 1,7-AQ are toxic in mice (T/C < 100) at a level of 1.0 mg ml^{-1}.

V. NON-KEKULÉ QUINONES

In addition to the eleven possible Kekulé quinones of azulene illustrated in Figure 1, five non-Kekulé quinones can be derived from the azulene framework (Figure 5, cf. *meta*-benzoquinone). None of these has ever been isolated or even trapped as a reactive intermediate, but the entire family warrants attention in light of the growing interest in non-Kekulé benzenoid quinones[52] (cf. Chapter 10 of this volume).

As expected, MINDO/3 calculations[16] predict the non-Kekulé azulene quinones all to have less favorable heats of formation, lower-lying LUMOs, larger LUMO coefficients,

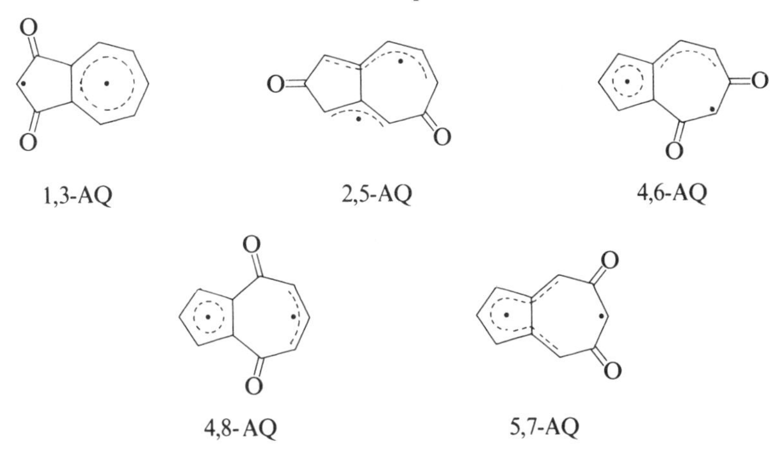

FIGURE 5. The five possible non-Kekulé quinones of azulene

and smaller HOMO–LUMO energy gaps than the isomers shown in Figure 1. Thus, all the non-Kekulé quinones of azulene are predicted to suffer easy dimerization, polymerization, nucleophilic addition and reduction. The 2,5-AQ is actually not even a minimum on the MINDO/3 energy surface but collapses to a cyclopropanone by formation of a bond between C(1) and C(3).

Of these hypothetical quinones, however, 1,3-AQ stands out as a particularly intriguing compound. Inspection of the calculated geometry and charge density pattern[16] suggests that this isomer can best be represented as

1,3- AQ

It is difficult to conceive of *any* non-Kekulé quinone (non-benzenoid or benzenoid) with a more stabilized zwitterionic form than this one. Compared to the other quinones in Figure 5, 1,3-AQ is predicted to have, by far, the most favorable heat of formation, the highest-lying LUMO, the smallest LUMO coefficients, and the largest HOMO–LUMO energy gap. Of the five possible non-Kekulé azulene quinones, this one should have the best chance for survival.

VI. FUTURE PROSPECTS

It is inevitable that all of the remaining Kekulé quinones of azulene will sooner or later become known, but it now seems unlikely that any of the parent quinones other than 1,2-AQ, 1,5-AQ and 1,7-AQ will be isolable under ordinary laboratory conditions. Direct spectroscopic study of the more reactive quinones in solid matrices should be possible, even for the non-Kekulé quinones, and some of the Kekulé isomers may even survive in dilute solution at low temperatures. Judiciously positioned bulky alkyl substituents could stabilize some of the more reactive quinones sufficiently to permit studies of their chemical and spectroscopic properties. This ploy might even render the non-Kekulé 1,3-AQ observable in solution. Clearly there is still much work to be done on azulene quinones.

VII. ACKNOWLEDGEMENTS

The author is deeply indebted to all of the principal investigators in this field during the last decade for their free communication of new results and information prior to publication. The warm hospitality of K. Takase and T. Morita (Sendai), T. Nozoe (Tokyo) and K. Hafner (Darmstadt) during visits by the author was especially appreciated. Most of the research conducted on azulene quinones in Reno was supported by the US National Institutes of Health and the National Science Foundation.

VIII. REFERENCES

1. L. F. Fieser and M. Fieser, *Advanced Organic Chemistry*, Reinhold Publishing Corp., New York, 1961, Chapt. 26.
2. *Nonbenzenoid Aromatic Hydrocarbons* (Ed. D. Ginsburg), Interscience, New York, 1959.
3. G. M. Badger, *Aromatic Character and Aromaticity*, Cambridge University Press, Cambridge, 1969.
4. D. Lewis and D. Peters, *Facts and Theories of Aromaticity*, Macmillan, London, 1975.
5. D. Lloyd, *Nonbenzenoid Conjugated Carbocyclic Compounds*, Elsevier, Amsterdam, 1984.
6. *The Chemistry of the Quinonoid Compounds*, Vol. 1, Parts 1 and 2 (Ed. S. Patai), Wiley, New York, 1974.
7. E. Heilbronner, in *Nonbenzenoid Aromatic Hydrocarbons* (Ed. D. Ginsburg), Interscience, New York, 1959, Chapt. 5.
8. W. Keller-Schierlein and E. Heilbronner, in *Nonbenzenoid Aromatic Hydrocarbons* (Ed. D. Ginsburg), Interscience, New York, 1959, Chapt. 6.
9. P. A. Plattner and A. S. Pfau, *Helv. Chim. Acta*, **20**, 224 (1937).
10. G. J. Gleicher, D. F. Church and J. C. Arnold, *J. Am. Chem. Soc.*, **96**, 2403 (1974) and references cited therein.
11. B. A. Hess Jr, L. J. Schaad and C. W. Holyoke Jr, *Tetrahedron*, **28**, 3229 (1972).
12. F. Pietra, *Chem. Rev.*, **73**, 293 (1973).
13. D. Lloyd, *Nonbenzenoid Conjugated Carbocyclic Compounds*, Elsevier, Amsterdam, 1984, pp. 89–106.
14. M. A. Ogliaruso, M. G. Romanelli and E. I. Becker, *Chem. Rev.*, **65**, 261 (1965).
15. O. L. Chapman and C. L. McIntosh, *Chem. Commun.*, 770 (1971).
16. L. T. Scott, M. D. Rozeboom, K. N. Houk, T. Fukunaga, H. J. Lindner and K. Hafner, *J. Am. Chem. Soc.*, **102**, 5169 (1980).
17. L. T. Scott, *Pure Appl. Chem.*, **55**, 363 (1983).
18. K. N. Houk, *Acc. Chem. Res.*, **8**, 361 (1975).
19. I. Fleming, *Frontier Orbitals and Organic Chemical Reactions*, Wiley, New York, 1976.
20. H. L. K. Schmand and P. Boldt, *J. Am. Chem. Soc.*, **97**, 447 (1975) and references cited therein.
21. M. J. S. Dewar and C. deLlano, *J. Am. Chem. Soc.*, **91**, 789 (1969).
22. M. J. S. Dewar and T. Morita, *J. Am. Chem. Soc.*, **91**, 796 (1969).
23. M. D. Rozeboom, I.-M. Tegmo-Larsson and K. N. Houk, *J. Org. Chem.*, **46**, 2338 (1981).
24. T. Morita, M. Karasawa and K. Takase, *Chem. Lett.*, 197 (1980).
25. L. T. Scott, P. Grütter and R. E. Chamberlain III, *J. Am. Chem. Soc.*, **106**, 4852 (1984).
26. L. T. Scott and C. M. Adams, *J. Am. Chem. Soc.*, **106**, 4857 (1984).
27. L. T. Scott, M. A. Minton and M. A. Kirms, *J. Am. Chem. Soc.*, **102**, 6311 (1980).
28. L. T. Scott, P. Grütter and R. E. Chamberlain III, unpublished results.
29. T. Morita, F. Ise and K. Takase, *Chem. Lett.*, 1303 (1982).
30. T. Morita and K. Takase, personal communication.
31. M. Balchi, *Chem. Rev.*, **81**, 91 (1981).
32. D. G. Michels, MS Thesis, University of Nevada-Reno, 1980; cf. reference 33.
33. M. A. McKervey, S. M. Tuladhar and M. F. Twohig, *J. Chem. Soc., Chem. Commun.*, 129 (1984).
34. J. L. Donovan, *Gov. Rep. Announce. Index (U.S.)*, **83(12)**, 2605 (1983); *Chem. Abstr.*, **99**, 104468u (1983); C. F. Rowell, U.S. Naval Academy, personal communication.

35. L. T. Scott and S. Gingerich, unpublished results.
36. P. J. Scheuer, personal communication; M. K. W. Li, PhD Dissertation, University of Hawaii, 1985.
37. T. Nozoe, S. Takekuma, M. Doi, Y. Matsubara and H. Yamamoto, *Chem. Lett.*, 627 (1984).
38. Y. Matsubara, S. Takekuma, K. Yokoi, H. Yamamoto and T. Nozoe, *Chem. Lett.*, 631 (1984).
39. T. Nozoe, personal communication; Y. Matsubara, S. Takekuma, K. Yokoi, H. Yamamoto and T. Nozoe, *J. Org. Chem.*, submitted for publication.
40. R. E. Chamberlain III, PhD Dissertation, University of Nevada-Reno, 1986.
41. A. Grund, Dr.-Ing. Dissertation, Technischen Hochschule Darmstadt, West Germany, 1980.
42. C. M. Adams, PhD Dissertation, University of Nevada-Reno, 1983.
43. L. T. Scott and J. P. DeLuca, unpublished results.
44. T. Morita and K. Takase, *Chem. Lett.*, 513 (1977).
45. (a) T. Morita, H. Kanzawa and K. Takase, *Chem. Lett.*, 753 (1977); (b) L. T. Scott and J. Solbach, unpublished results.
46. M. E. Peover, *J. Chem. Soc.*, 4540 (1962).
47. L. T. Scott and M. Oda, *Tetrahedron Lett.*, 779 (1986).
48. L. T. Scott, M. Oda and I. Erden, *J. Am. Chem. Soc.*, **107**, 7213 (1985).
49. A. Krebs and B. Schrader, *Ann. Chem.*, **709**, 46 (1967).
50. D. J. Bertelli, in *Topics in Nonbenzenoid Aromatic Chemistry*, Vol. 1 (Eds. T. Nozoe, R. Breslow, K. Hafner, S. Ito and I. Murata), Wiley, New York, 1973, pp. 29–46.
51. L. T. Scott and R. S. Pardini, unpublished results.
52. P. M. Lahti, A. R. Rossi and J. A. Berson, *J. Am. Chem. Soc.*, **107**, 2273 (1985).

The Chemistry of Quinonoid Compounds, Vol. II
Edited by S. Patai and Z. Rappoport

CHAPTER **25**

Extended quinones

PETER BOLDT
Institut für Organische Chemie der Technischen Universität Braunschweig, D-3300 Braunschweig, FRG

I. INTRODUCTION

Extended quinones are those quinones which bear the quinonoid carbonyl groups in different rings. Three major classes may be envisaged: first compounds which are treated in Chapter II and in which the two carbonyl groups are present in rings or ring systems connected by a double bond, cumulene bonds, or conjugated double bonds. The parent quinones are called diphenoquinones (**1**), cumulenoquinones or diquinoethylenes (**2**) and (with two conjugated double bonds) stilbenoquinones (**3**), respectively.

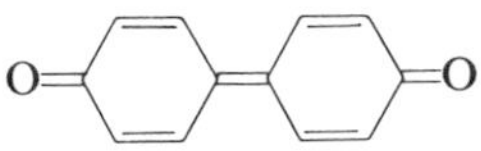

(**1**)  (**2**)

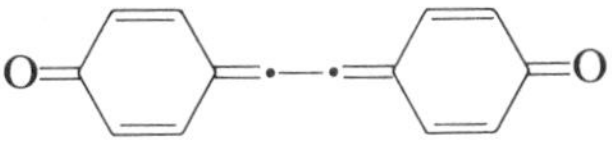

(**3**)

Many members of this group of compounds are known. Publications until 1978 have been reviewed in detail, including some benzo-annelated homologues[1]. No important developments in this field have since been published with the exception perhaps of the oligoquinocycloalkanes (which may be considered as substituted stilbenoquinones) and the cumulenoquinones. Therefore only these two classes are reviewed in Section II.

Another large class of extended quinones are derived from polycyclic aromatic compounds. Some of them have been known for about 70 years and many have attained technical importance as vat dyes. To my knowledge no modern, comprehensive review of this interesting class of quinones has been published. They are treated in Section III.

Section IV is dedicated to perhaps the most exciting class of extended quinones. They may be regarded as non-classical quinones, because they are related to non-classical aromatic systems such as azulene or the annulenes. They have been synthesized and studied within the last two decades, many of them within the last few years.

II. POLYQUINOCYCLOALKANES AND DIQUINOETHYLENES

A. Polyquinocycloalkanes

This class of quinones has been explored by West and coworkers[2-7]. The strongly coloured, dye-like polyquinocycloalkanes of type **4** or **5** may be regarded as substituted

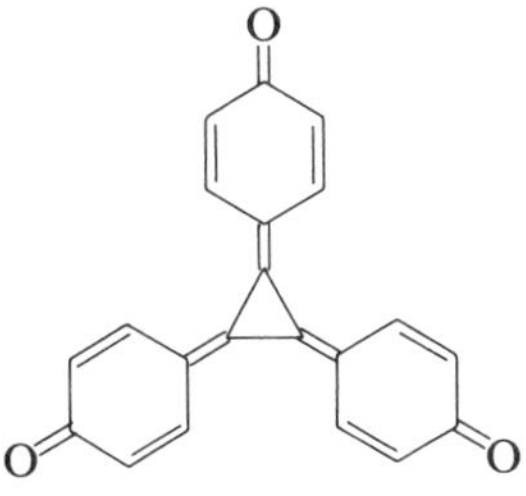

(**4**) 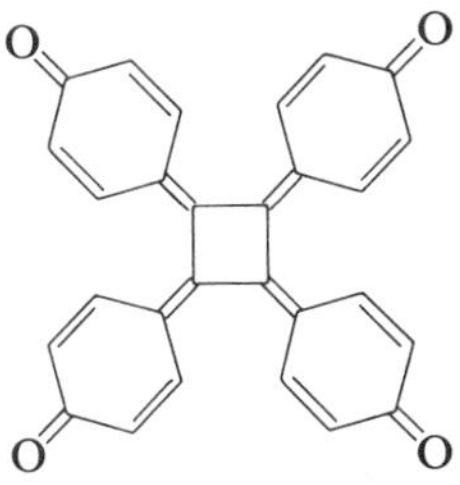 (**5**)

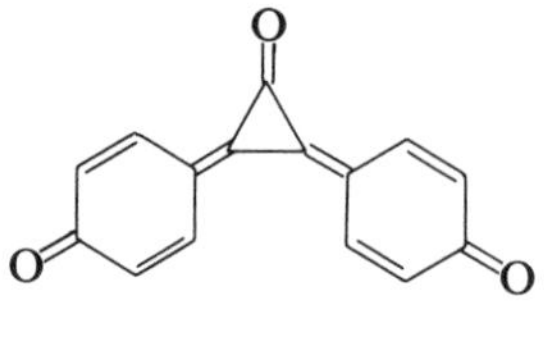

(6) 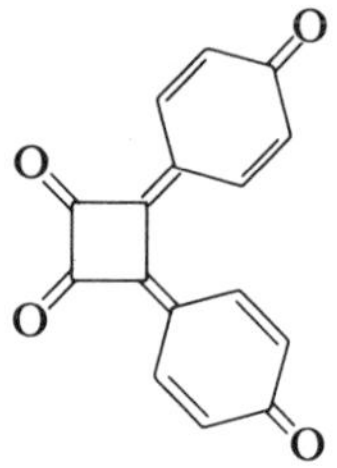 (7)

stilbenoquinones. The same is true for the isolable but unstable diquinocyclopropanone **6**, and the rather labile 1,2-diquinocyclobutanedione **7**. The last step of the preparation is usually the oxidation of the conjugate hydroquinone. Thus for the preparation of tris(9-anthron-10-ylidene)cyclopropane (**12**), the hydroquinone is typically prepared by Friedel–Crafts reaction of 9-methoxyanthracene (**9**) with trichloropropenylium tetrachloroaluminate (**8**), followed by demethylation of the resulting **10**. The bis(9-hydroxy-10-anthryl)cyclopropylideneanthrone (**11**) thus formed, a stilbene derivative, is finally oxidized to **12**.

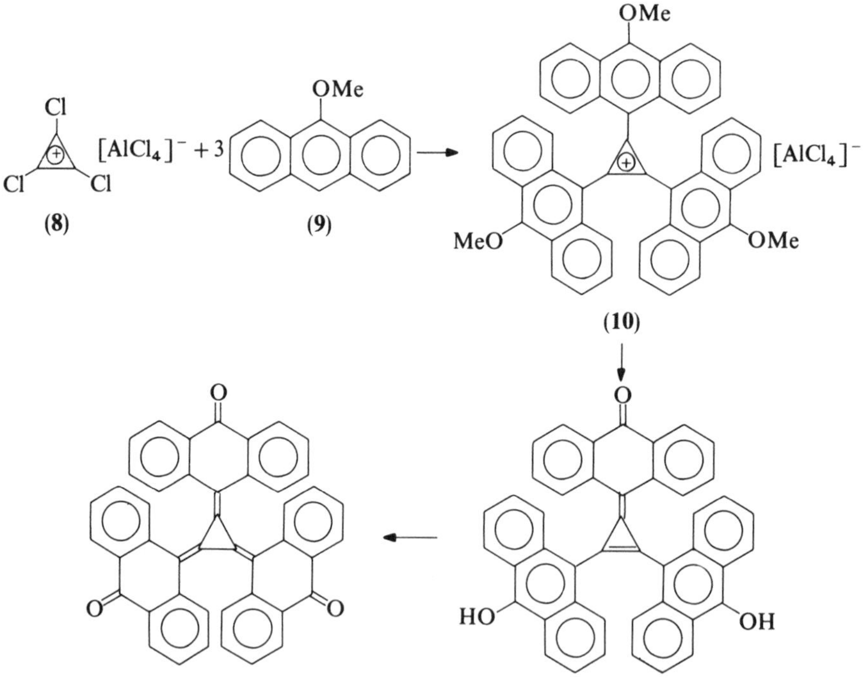

Cl Cl Cl $[AlCl_4]^-$ + 3 OMe → OMe MeO OMe $[AlCl_4]^-$

(8) (9) (10)

HO OH

(12) (11)

Reaction of two equivalents of 9-methoxyanthracene results after ether cleavage, hydrolysis and oxidation in the formation of 2,3-bis(9-anthron-10-ylidene)cyclopropanone (**13**)[2]. An analogous procedure has been used for the preparation

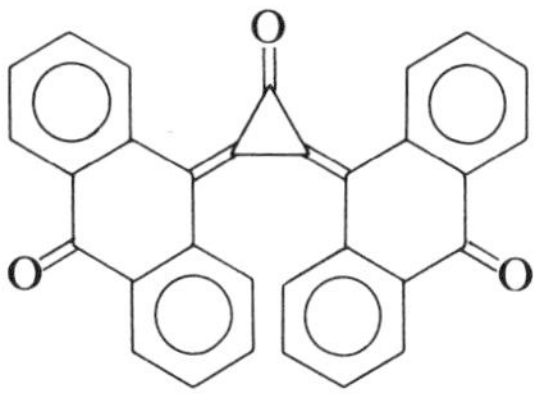

(13)

of 1,2-diquinocyclobutanediones, as shown in the reaction scheme for the synthesis of 1,2-bis(3,5-di-*t*-butyl-4-oxo-2,5-cyclohexadien-1-ylidene)cyclobutanedione (**14**)[3]:

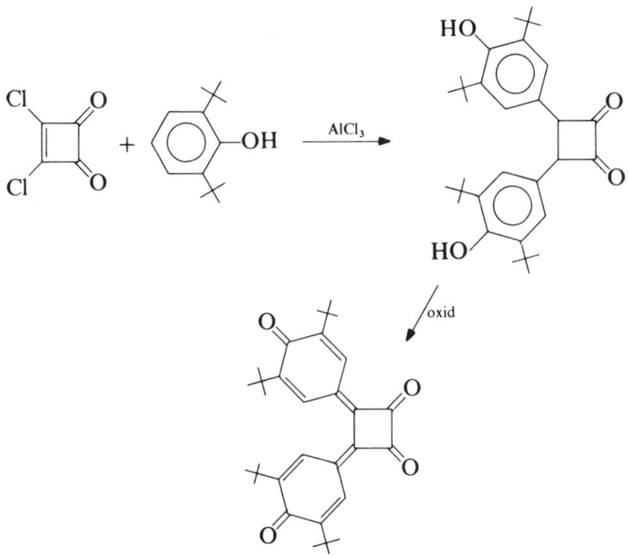

$AlCl_3$

oxid

(14)

A variety of diquinocyclopropanones and triquinocyclopropanes has been synthesized by this method[2-4]. An outstanding example is **15**[4] with a remarkable electronic excitation absorption at 1300 nm[5]. This absorption is well into the near-infrared region, an area approaching molecular bond vibrational energy and not often seen for electronic excitation. The ESR spectrum suggests that **15** exists predominantly in the diradical form **16**, which dimerizes to give the diradical dimer **17**. **18** was found to be a little more stable than **15**. It could be isolated in crystalline form whereas **15** has a half-life time of 92 min in solution at room temperature. **18** exhibits very strong absorption at 672 nm and no other bands are observed up to 2000 nm.

Another synthetic method to a quinocycloalkane has been used in the case of the bright purple tetraquinocyclobutane **20**[6], which was obtained by heating the cumulenoquinone **19** in cyclooctane for 3–4 hours. X-ray analysis showed that **20** is not planar and exists in a propeller-like conformation with an average twist angle of 36°. The central four-membered ring is also distorted from planarity. The reduction of **20** to the hydroquinone **21** is not accomplished easily. Although **20** is stable indefinitely in hydrocarbon

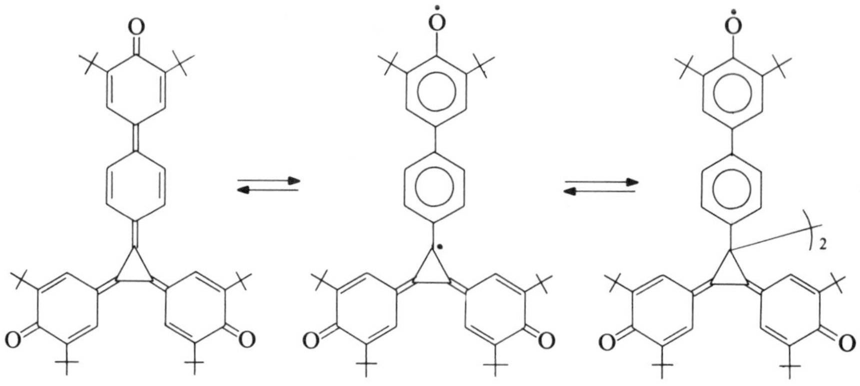

(15) (16) (17)

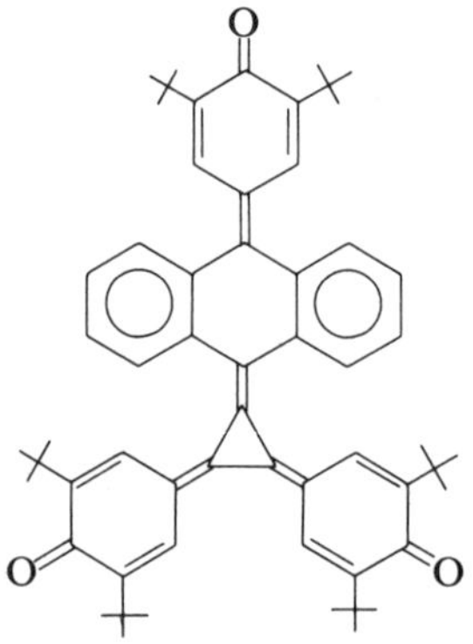

(18)

solutions, it reacts with nucleophiles. Solution in methanol results in the addition of one mole of methanol to give **22**, and conventional reduction techniques fail because of this reactivity with nucleophiles. However, on refluxing with benzopinacol in cyclooctane the diaryldiquinocyclobutene **21** is formed and is readily reoxidized to **20** by atmospheric oxygen.

The redox behaviour of the triquinocyclopropanes **23a** and **b**, **12**, and of **13** has been examined using cyclic voltammetry[7]. Two-wave redox cycles were observed for each compound corresponding to the formation of the semiquinone radical and the dianion. Compounds **23a** and **b**, **12**, and **13** are powerful oxidizing agents with first reduction waves at $+0.02$–$+0.05$ and second waves at about -0.27 V (vs. saturated calomel). The ESR spectra of the semiquinone anions has been measured. The hyperfine splitting constants are in accoordance with that predicted by MO calculations.

B. Diquinoethylenes

West[8] proposed the name diquinoethylenes for compounds containing quinone nuclei connected by cumulated double bonds. They are genuine quinones since they are reduced

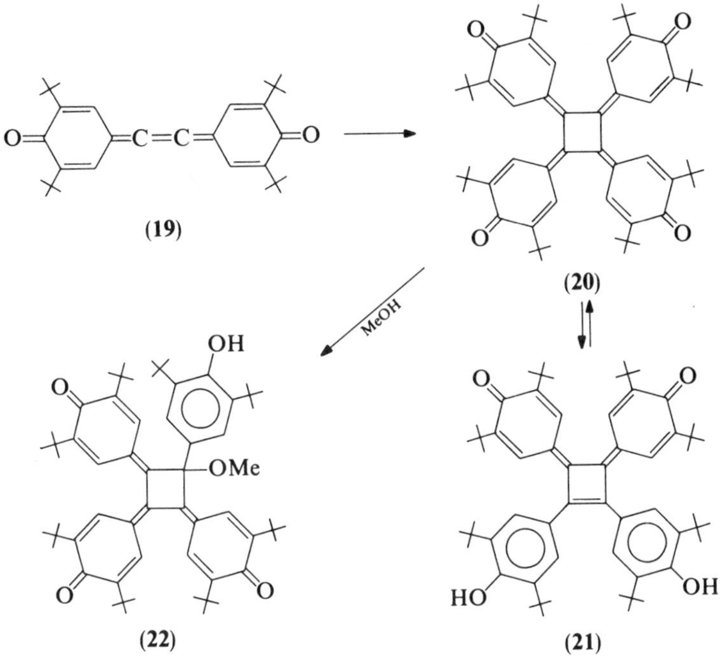

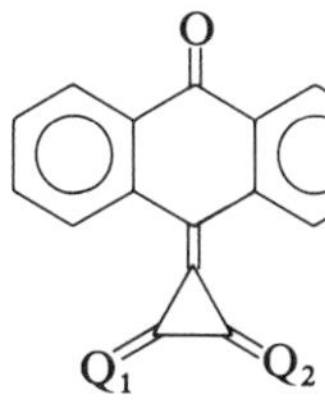

(19)

(20)

MeOH

(22)

(21)

(23)

a: $Q_1 = Q_2$ = 4-oxo-3,5-di-*t*-butyl-2,5-cyclohexadien-1-ylidene
b: Q_1 = 4-oxo-3,5-di-*t*-butyl-2,5-cyclohexadien-1-ylidene;
Q_2 = 9-anthron-10-ylidene

reversibly to the conjugate hydroquinones with two aromatic nuclei connected by an acetylene moiety (see e.g. **26** and **27**).

Diquinoethylenes are formed by the spontaneously proceeding decarbonylation of diquinocyclopropanones **25** at room temperature or by photochemical-induced decarbonylation of their conjugate hydroquinones, the bis(*p*-hydroxyaryl)cyclopropenones (**24**, R = H) to the alkynes **26**, with subsequent oxidation[8].

The diquinoethylenes are magenta-coloured solids. Their stability depends on the alkyl groups: **27c** (R = *t*-butyl) is stable and unreactive, **27b** (R = isopropyl) is isolable but reacts with water, and **27a** (R = methyl) was so reactive that it could not be isolated and was detected only by its UV–VIS spectrum in solution.

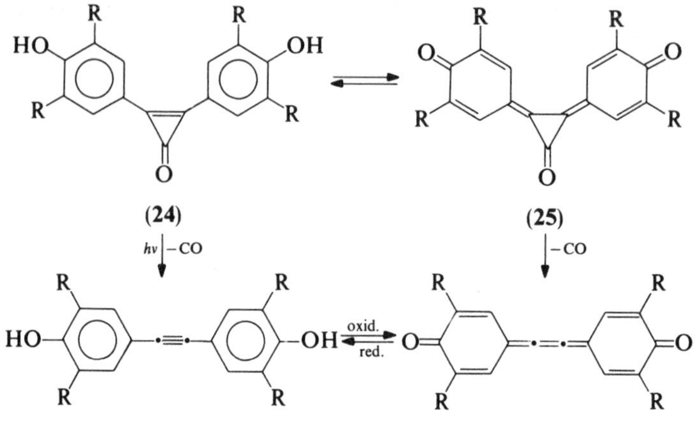

(24) (25)

(26) (27)

a: R = Me **b**: R = *i*-Pr **c**: R = *t*-Bu

In the same way dianthraquinoethylene (**29**)[2] and compound **30**[4] have been synthesized. the latter could not be isolated, but its existence was indicated by the UV–VIS spectrum of the blue-green solution.

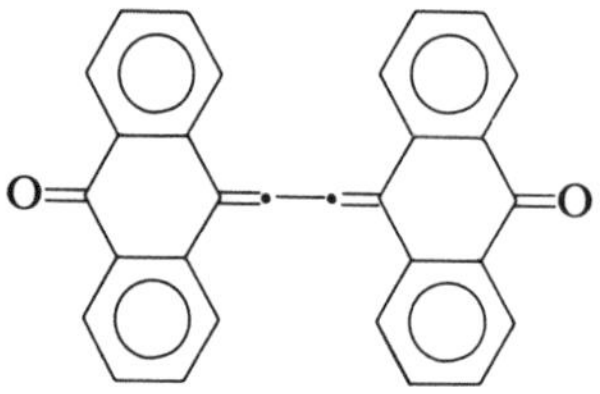

(28) 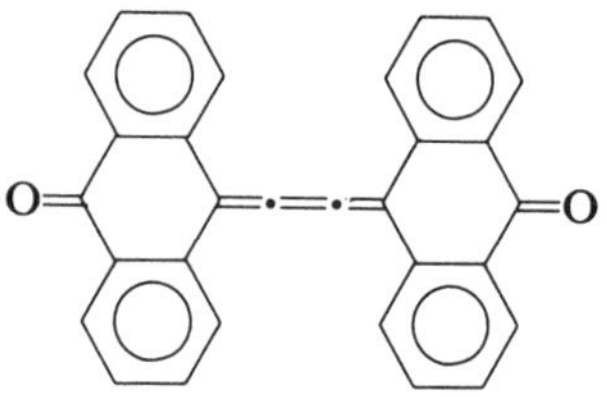 (29)

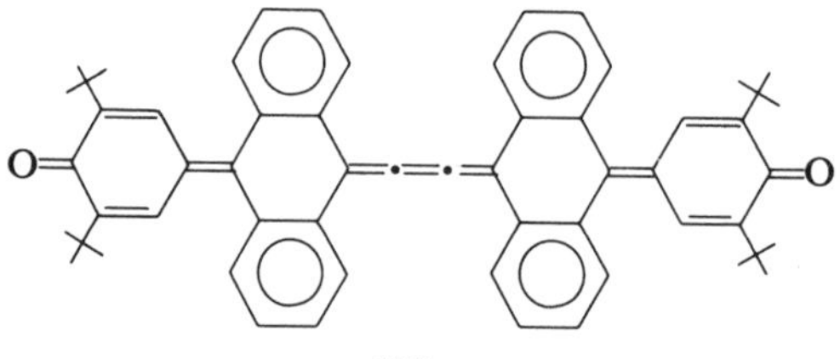

(30)

In contrast **29** is a stable and long-known bordeaux-red dye, which has also been prepared from anthrone. By treatment with glyoxal sulphate, the stilbenoquinone **28** is easily formed and can be oxidized by heating with ethanolic potassium hydroxide to **29**, forming a cherry-red vat dye[9]. The reversible reduction of **29** has been investigated by cyclic voltammetry[8]. Two wave cycles were observed in alkaline solution, corresponding to the semiquinone anion and to the dianion with $E_{(1/2)1} = -0.42$ V and $E_{(1/2)2} = -0.61$ V. The ESR spectrum of the semiquinone anion has been measured.

III. QUINONES OF POLYCYCLIC AROMATIC HYDROCARBONS

A. General Aspects

Extended quinones of condensed aromatics have some synthetic importance as precursors for the conjugate aromatic hydrocarbon itself[10], since the quinones having carbonyl groups are more readily available by syntheses.

Commercially, some extended quinones have gained importance as dyes[11]. They have been used as vat dyes for many decades to dye cotton and other cellulose fibres. Despite their high cost and not so brilliant colours these dyes are extremely important because of their superior fastness. Because of their low solubility they can be used also as pigments, but in this case they have to possess a high degree of purity and certain physical properties of the particles. Their relatively high cost restricts their use to special applications, e.g. with high requirements for fastness.

The reactivities of extended quinones with nucleophiles, electrophiles and free radicals[12], their ground-state properties and electronic structure[13] as well as heats of atomization, dipole moments, carbonyl stretching frequencies, and reduction potentials[14] have been calculated.

With some quinones of the highly condensed aromatic hydrocarbons the question arises, whether the compound is a quinone or a simple aromatic diketone. 'Violongthrone'[15] (**31**), e.g. may be viewed as quinone insofar as on reduction a 54 π-electron system with cyclic conjugated double bonds is formed, and is formally a Hückel aromatic with $4n + 2$ electrons, $n = 13$. However, the quinone itself represents a very stable aromatic ketone, which forms no vat with alkaline dithionite. In contrast, iso-violongthrone[16] (**32**) forms an (unstable) dark-blue vat dye on treatment with pyridine and alkaline dithionite.

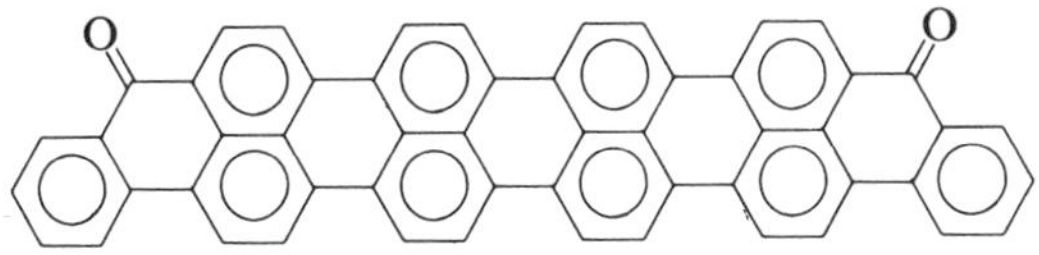

(**31**)

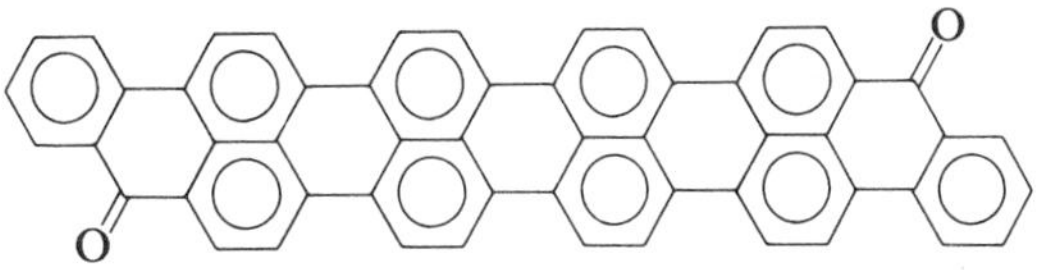

(**32**)

The formation of a vat with alkaline dithionite is no unambiguous proof of a quinonoid character. It may be that the redox potential in the case of **31** was insufficient, or that in the case of **32** not the alkali salt of the conjugate hydroquinone was formed but overreduction took place. This is observed in some cases[10a]. The investigation of the electrochemical behaviour, if possible with cyclic voltammetry, seems to be a more reliable method to answer the question, whether a certain compound is a quinone, i.e. whether it possesses reversible redox properties or not. Little is known about the photochemistry of the extended quinones with the exception of their photosensitizing ability, which has practical importance in the use of some vat dyes[17].

The most important method for the synthesis of extended quinones of condensed aromatics seems to be the 'Scholl cyclization' of benzoyl or naphthoyl derivatives of condensed aromatics which in an aluminium chloride melt yield quinones[18]. Later this method was improved by introducing dry oxygen into the well-stirred melt[19] when, e.g. 1,6-dibenzoylpyrene (**33**) gives pyranthrone (**35**) in 80% yield. Other oxidants such as nitrates have also been used.

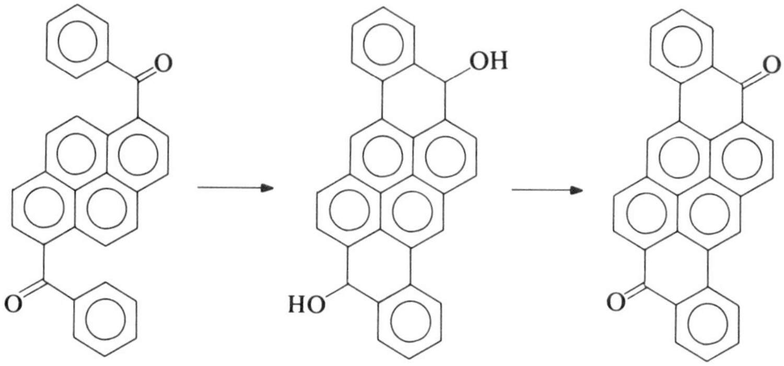

(**33**) (**34**) (**35**)

Attention should be given to the fact that under the conditions of Scholl cyclization migrations of the aroyl substitutents are sometimes observed.

Another route to extended quinones is oxidation of the condensed aromatic hydrocarbon with chromic acid or selenium dioxide, for example, of **36** to **37**[20].

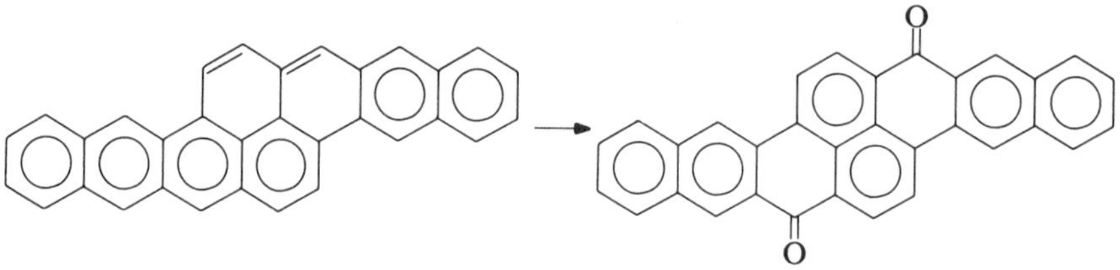

(**36**) (**37**)

The products may be predicted by the simple rule of thumb that a maximum number of intact benzene and/or naphthalene rings should be retained in the process. The ease of oxidation increases with increasing annelation.

The synthetic value of this method is limited because, as mentioned, the aromatic hydrocarbons themselves are synthesized very often by reduction of the conjugate quinones[10].

Benzanthrone, its derivatives and homologues with free 6- and 7-positions can be dimerized by melting with potassium hydroxide, with or without melting-point reducing additives. Thus naphthanthrone (**38**) gives, in a potassium hydroxide/potassium acetate melt at 240°C, the quinone **39**[21]. Benzanthrone and its derivatives are therefore important initial products for technical syntheses of vat dyes.

In the following sections the quinones are classified according to number of rings they contain.

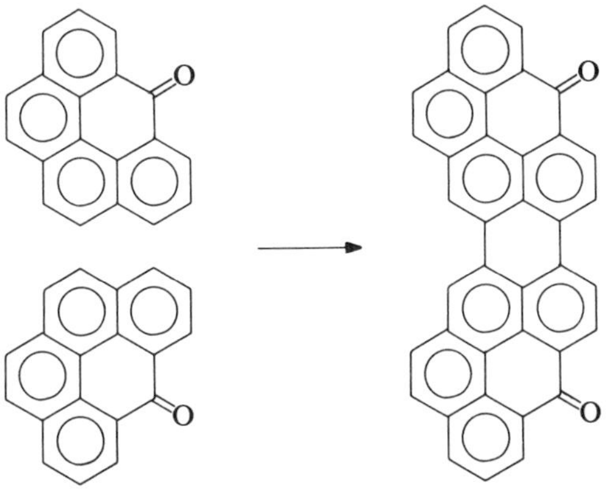

(38) (39)

B. Two-ring Quinones

The extended naphthoquinones **40–42** are unstable and the only known one is the 2,6-quinone **40**, which was prepared by oxidation of the conjugate hydroquinone with lead dioxide[22]. The yield was improved later by the use of active lead dioxide[23]. The 2,6-quinone is stable in absence of water but reacts quickly with traces of moisture.

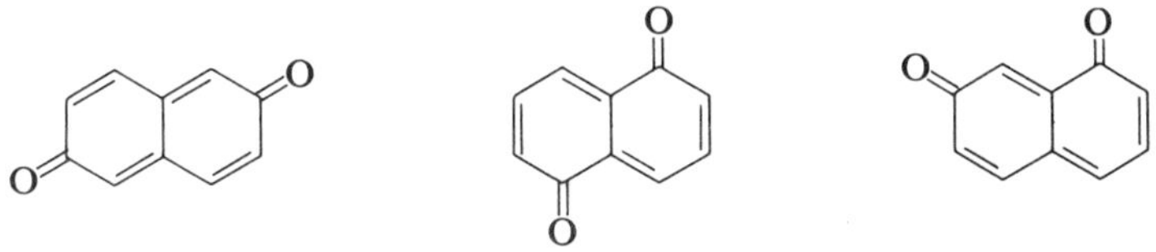

(40) (41) (42)

1,5-Disubstituted derivates of **40**, such as 1,5-dichloro-2,6-naphthoquinone (**43**), are more stable. **43** can be prepared by oxidation of the conjugate hydroquinone with chromic acid[1a, 22]. Prolonged treatment with an excess of sodium bichromate yields **44**.

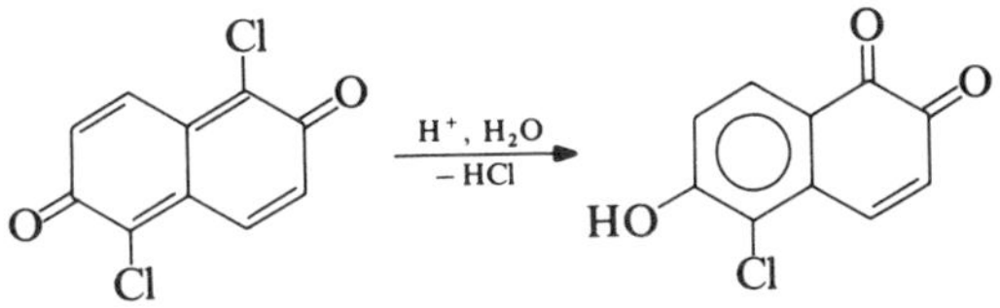

H^+, H_2O
$-HCl$

(43) (44)

For the oxidation product of 1,2,5,6-tetrahydroxynaphthalene structure **45** was claimed[24], but the prototropic structure **46** seems to be more stable. In contrast the structure of 1,5-diamino-2,6-naphthoquinone seems to be well established[25].

Attempts to prepare 1,5-naphthoquinone (**41**) in the same manner as **40** failed, even in the absence of water[26]. The reason is probably that **41** possesses conjugated double bonds

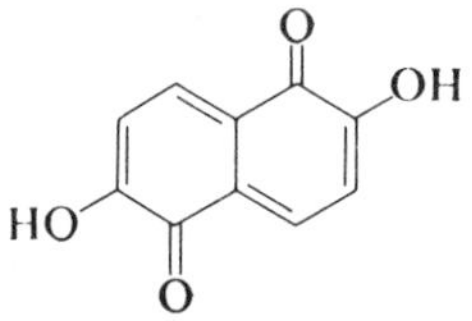

(45) 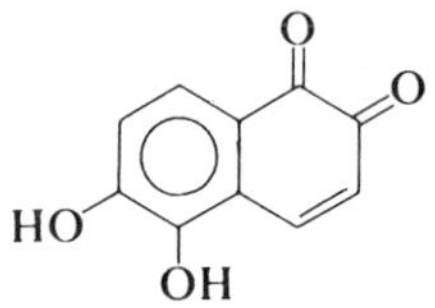 (46)

with a *s-cis* partial structure and, unlike **40**, it can undergo Diels–Alder addition reactions with itself. Only by shielding the molecule with two *t*-butyl groups in the 3- and 7-positions is the sufficiently stable 1,5-quinone **47** obtainable[27]. In the presence of water it is converted slowly to **48**.

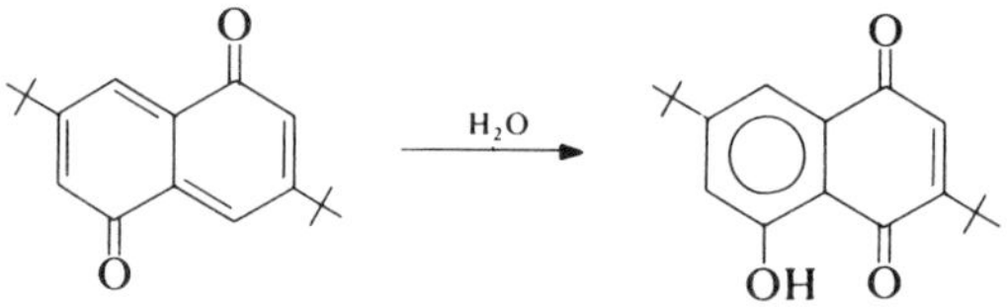

H_2O

(47) (48)

The stable 4,8-diamino-1,5-naphthoquinone (**50**) is an intermediate in the conversion of 1,5-dinitronaphthalene (**49**) to naphthazarin (**51**) with fuming sulphuric acid and sulphur[28].

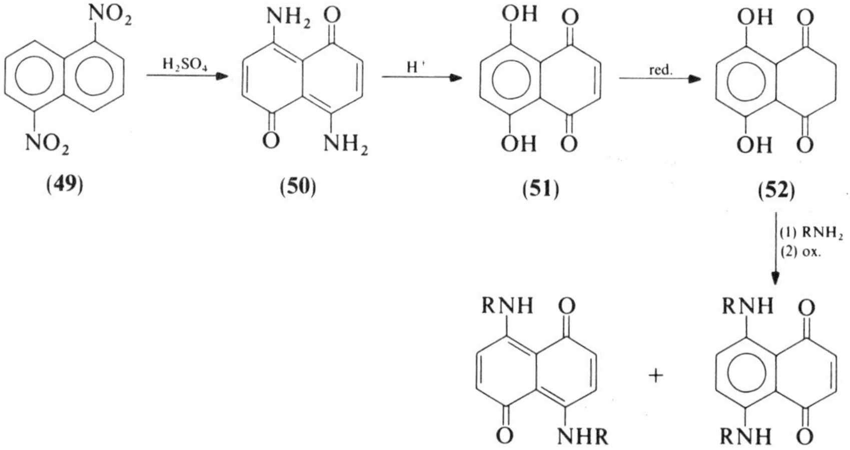

H_2SO_4 H^+ red. (1) RNH_2 (2) ox.

(49) (50) (51) (52)

(53) + (54)

For naphthazarin (**51**) a 1,5-quinonoid structure was proposed, but the 1,4-quinonoid structure is now well established[27]. However, compound **50** clearly has the 1,5-quinonoid structure, as shown[29].

50 or *N,N'*-dialkyl derivates (**53**) can generally be prepared starting with **51**, which is reduced to the leuco form **52** and then treated with an amine. The substances are then reoxidized with atmospheric oxygen. The 5,8-dialkylamino-1,4-naphthoquinones (**54**) are sometimes formed as by-product[29].

PMO/MNDO calculations suggested that 1,7-naphthoquinones should undergo Diels–Alder reactions with themselves as easily as 1,5-naphthoquinones[30]. On the basis of PMO calculations and by comparison with analogous cases, *t*-butyl groups in the 3,6-positions should ensure sufficient kinetic stability with respect to reactions with nucleophiles, such as water or Diels–Alder self-condensations. Nevertheless, 3,6-di-*t*-butyl-8-methyl-1,7-naphthoquinone (**55**) showed an unexpected low stability. According to MNDO calculations the alkyl groups, especially the methyl group, cause a high steric strain in the molecule that is possibly responsible for the low stability of **55**. But the 8-methyl group proved to be indispensible, since in its absence the corresponding hydroquinone underwent an oxidative coupling reaction in the 8-position[30].

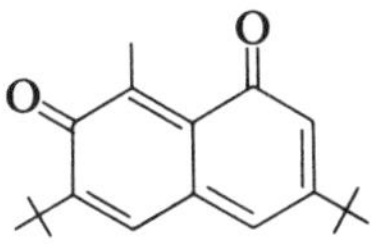

(**55**)

C. Three-ring Quinones

1. Anthraquinones

Of the possible nine anthraquinones only those with the quinone carbonyl groups in one ring are known, i.e. the 1,2-, 1,4- and 9,10-quinones. The reason is the inherent instability of the extended quinones. On the basis of PMO/MNDO calculations it has been predicted that the reactivity of the extended anthraquinones toward water[27], and in some cases the dimerization tendency[30], should be very high. 1,10-Anthraquinones are known with chlorine, and/or amino groups, as well as alkyl groups as stabilizing substituents.

Boiling of 1,4-dihydroxyanthraquinone with thionyl chloride in the presence of bases yielded 2,4,9-trichloro-1,10-anthraquinone (**56**) and 2,3,4,9-tetrachloro-1,10-anthraquinone (**57**). Both compounds can also be obtained from other 1-hydroxy-anthraquinones, such as 1-hydroxyanthraquinone, 4-chloro-1-hydroxyanthraquinone,

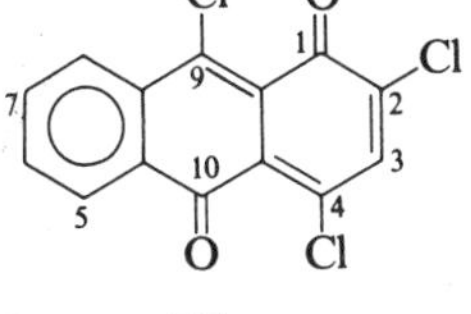

(**56**) 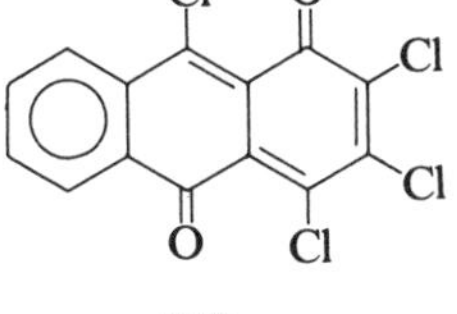 (**57**) 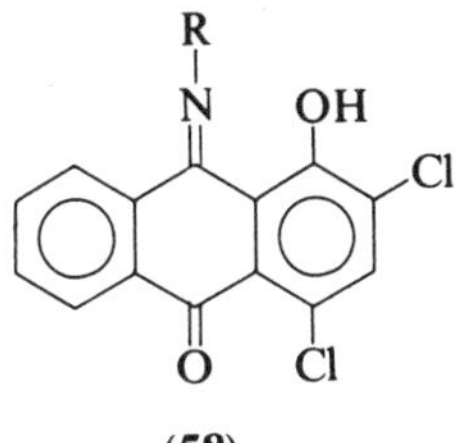 (**58**)

2,4-dichloro-1-hydroxyanthraquinone and 2-chloro-1,4-dihydroxyanthraquinone. This reaction is evidently common to compounds containing a hydroxy group in the 1-position[31].

With ammonia, aliphatic and aromatic primary amines **56** gives 2,4-dichloro-1-hydroxyanthraquinone-9-imines (**58**)[32]. Derivatives of 1-phenoxyanthraquinone containing amino-, methylamino-, or benzoylamino groups at positions 2, 4 and 5 show photochromism, which involves the reversible photoisomerization of the 9,10-quinonoid structure (**59**) to the 1,10-quinonoid structure (**60**). The ability of the compounds **59** to undergo photoisomerization depends on the position and electronic nature of the substituents. 4-Amino-9-phenoxy-1,10-anthraquinone (**60a**) was obtained in pure crystalline form by irradiation of **59a**[33].

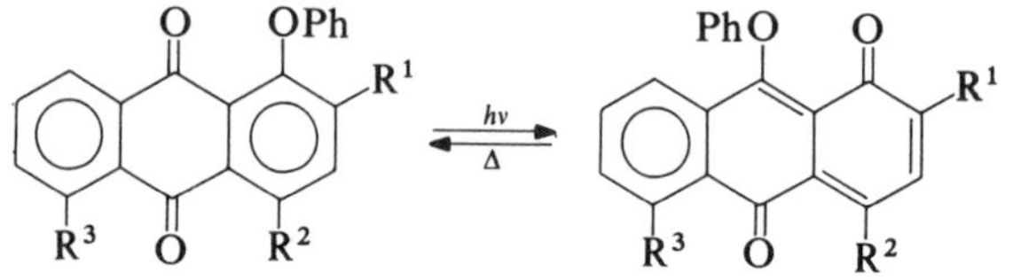

(59) **(60)**

a: $R^1 = R^3 = H$; $R^2 = NH_2$

Attempts were made to synthesize the 1,10-anthraquinone (**61**) shielded with methyl groups, but only the hemiketal **62** could be isolated in pure form[34], probably due to overcrowding at the 10-carbonyl group.

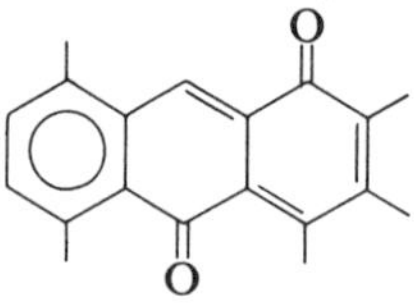

(61) 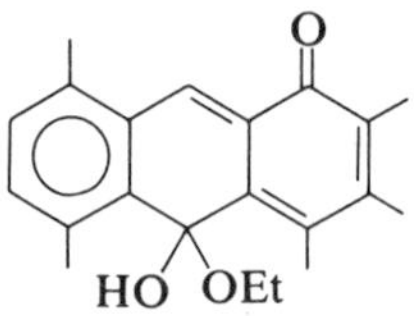 **(62)**

The 3-*t*-butyl-5,8-dimethyl-1,10-anthraquinone (**63**) could be prepared in a five-step synthesis and was stable enough to be isolated. Despite the shielding of the *meso*- position by the 1-carbonyl and 8-methyl groups **63** reacts rapidly with water and oxygen to give 3-*t*-butyl-1-hydroxy-5,8-dimethyl-9,10-anthraquinone (**64**)[35]. Stable derivatives of 2,6-anthraquinone, 3,7-dihydroxy-9,10-dimethylanthraquinone (**65**)[36] and 3,7-di-*t*-butyl-9,10-dimethyl-2,6-anthraquinone (**66**)[37] are also described. The latter has been prepared by oxidation of the conjugate hydroquinone with a mixture of nitrogen oxides.

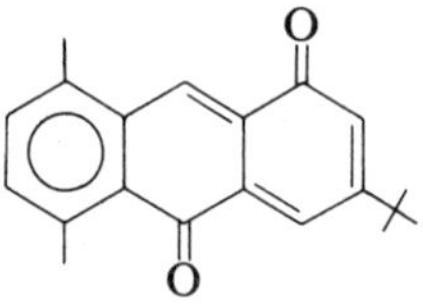

(63) 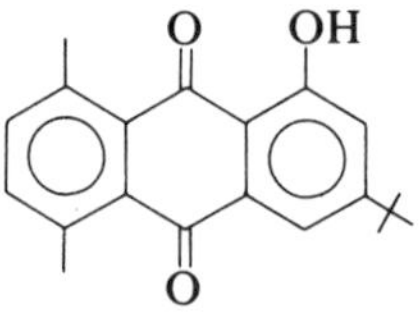 **(64)** 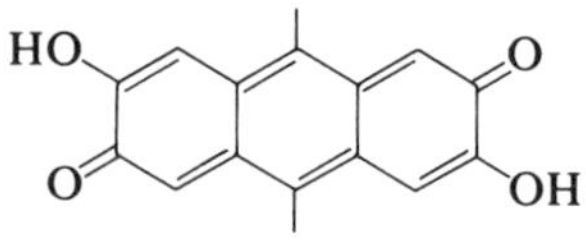 **(65)**

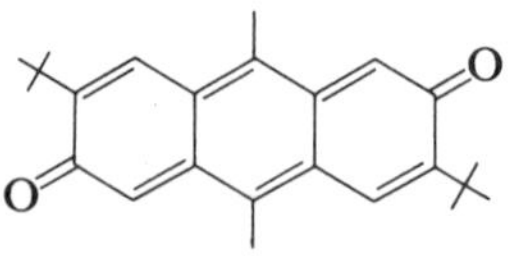

(66)

2. *Phenanthrenequinones*

As with the anthracene system only phenanthrenequinones with the carbonyl groups in the same ring are known. Newman and Childers[38] have synthesized and oxidized several 4,5-phenanthrenediols, most giving 1,4-phenanthrenequinone derivatives. 4,5-Dihydroxy-1,3,6,8-tetramethylphenanthrene gave an apparently polymeric product[39]. Only 1,3,6,8-tetra-*t*-butyl-4,5-phenanthrenequinone (**68**) has been detected as a short-lived species in solution: 1,3,6,8-tetra-*t*-butyl-4,5-dihydroxyphenanthrene (**67**), prepared in a five-step synthesis starting with 2,4-di-*t*-butyl-5-methylphenol, gave on oxidation with lead dioxide in benzene a fleeting green solution containing **68**, which rapidly rearranges to the dienone **69**. The corresponding 9,10-dihydro compound **70**, prepared in a similar way, crystallizes as its oxepine valence isomer **71**[40].

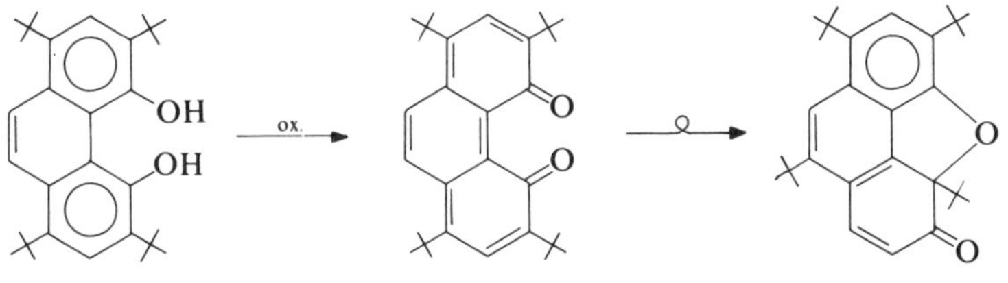

OH OH ox. O O O O

(**67**) (**68**) (**69**)

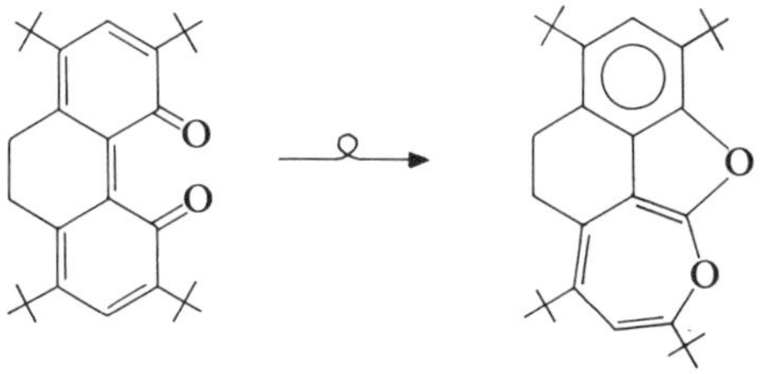

O O O O

(**70**) (**71**)

D. Four-ring Quinones

1. *Pyrenediones*

A mixture of the 1,6-pyrenequinone (**72**) and 1,8-pyrenequinone (**73**) was obtained more than one hundred years ago by chromic-acid oxidation of pyrene[41]. The ratio 1,6/1,8-quinone was determined to be 1:2[19]. Since then several methods for the preparation of pyrenequinones by oxidation of the parent hydrocarbon have been described, especially

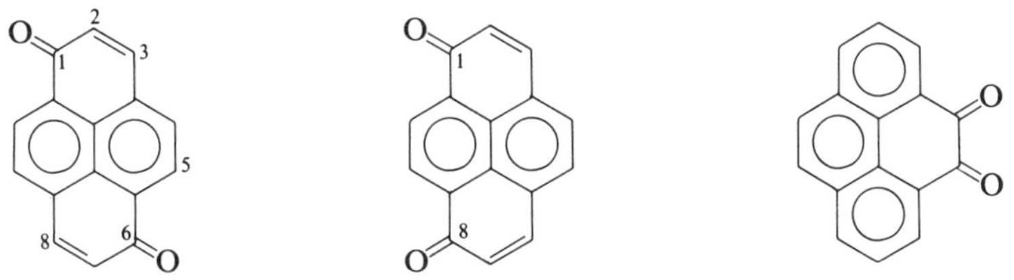

O 2 1 3 5 8 6 O O 1 8 O O O

(**72**) (**73**) (**74**)

vapour-phase oxidation with oxygen over vanadium and/or titanium oxide catalysts[42]. The 4,5-pyrenequinone (**74**) was obtained as by-product. The addition of ammonia in this oxidation increases the yields of the 1,6- and 1,8-quinones while the yields of the 4,5-quinone **74** and of higher oxidation products were unaffected. This effect is probably caused by shielding those active centres of the catalyst that were responsible for the oxidation of the 4–5 and 9–10 bonds of pyrene[43].

Especially under basic conditions, the quinones are oxidized further to yield mainly aromatic carboxylic acids. Thus a mixture of 1,6- and 1,8-pyrenequinone with 1.5 mole of aqueous potassium hydroxide at 140°C was completely oxidized by air within 1 hour. The main product (37%) proved to be naphthalene-1,4,5,8-tetracarboxylic acid[44]. 1,6-Dibenzoylpyrene gives 3,8-dibenzoyl-1,6-pyrenequinone on cautious oxidation with chromic acid in acetic acid[19].

Chloro derivatives of pyrenequinones can be obtained by oxidation of chloropyrenes: a mixture of 3,8-dichloro-1,6-pyrenequinone and 3,6-dichloro-1,8-pyrenequinone was obtained by treatment of 1,3,6,8-tetrachloropyrene with 20% oleum at 85°C[19] and 3,5,8,10-tetrachloro-1,6-pyrenequinone (**75**) could be synthesized by treatment of 1,3,5,6,8,10-hexachloropyrene with nitric acid at room temperature. In the same way 3,4,5,8,9,10-hexachloro-1,6-pyrenequinone (**76**) and 2,3,4,5,7,8,9,10-octachloro-1,6-pyrenequinone (**77**) are formed from the corresponding octa- and deca-chloropyrenes[19].

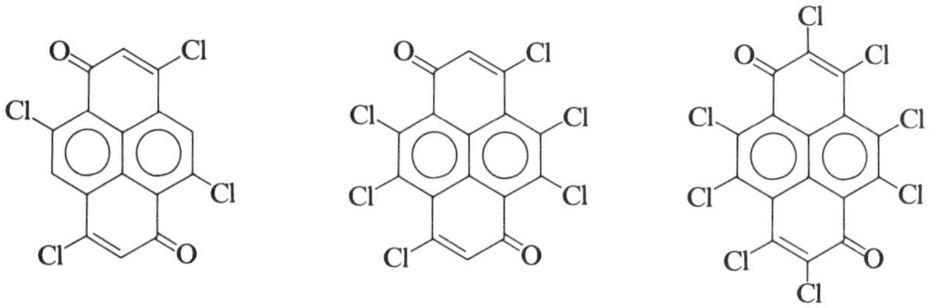

(**75**) (**76**) (**77**)

Chloro derivatives of 1,6-pyrenequinone can also be prepared by chlorination of 1,6-pyrenequinone with chlorine in trichlorobenzene at 100°C. 2,7-Dichloro-1,6-pyrenequinone, and under more vigorous conditions 2,3,7,8-tetrachloro-1,6-pyrenequinone, is formed[19].

The reactivity for the substitution of chlorine atoms by amines[19] depends strongly on the position of the chlorine in 1,6-pyrenequinone. Chlorine in the *peri*-position (5,10-position) is exchanged very easily, followed by chlorine in the 3,8-position. But it was not possible, even under drastic conditions, to substitute chlorines in the 2,7-position. Thus 2,3,7,8-tetrachloro-1,6-pyrenequinone is converted by heating with arylamines to 3,8-diarylamino-2,7-dichloroquinone and in 3,5,8,10-tetrachloro-1,6-pyrenequinone (**75**) the chlorine can be substituted stepwise by aniline, at first in the 5-position (60°C) and at higher temperature and in the presence of bases also in the 10-position. Boiling with an excess of aniline under copper salt catalysis leads in high yield, via the trianilino derivative, to 3,5,8,10-tetraanilino-1,6-quinone. Other amines react similarly, and several unsymmetrically substituted 1,6-pyrenequinones have been synthesized in this way.

3,5,8,10-Tetrachloro-1,6-pyrenequinone (**75**) is converted to the 10-hydroxy derivate by heating with potassium acetate in nitrobenzene. Other compounds with active hydrogens, such as phenols and thiophenols, also substitute easily the reactive chlorines in 1,6-pyrenequinone[19,45].

Characteristic of the 5,10-diarylamino-1,6-pyrenequinones is the cyclization reaction with sulphuric acid or aluminium chloride/pyridine or in an aluminium chloride/sodium chloride melt. The cyclization product is oxidized by air. Thus 5,10-dianilino-3,8-dichloro-1,6-pyrenequinone (**78**) forms the dicarbazole (**79**)[19].

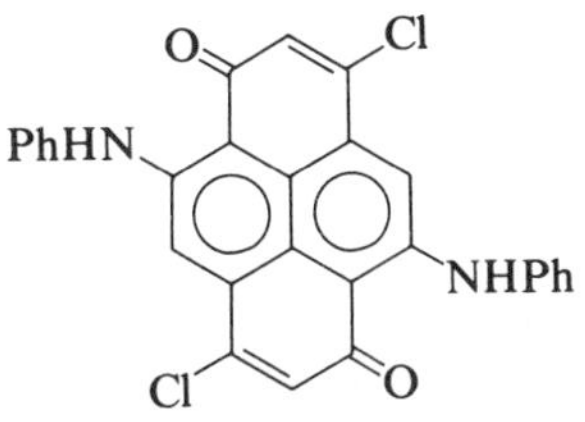

(**78**) 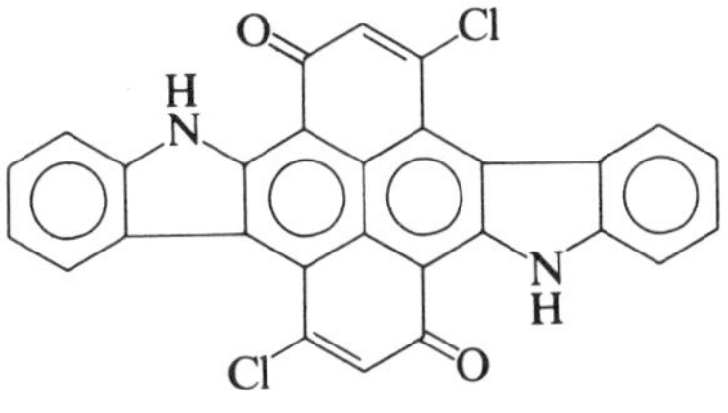 (**79**)

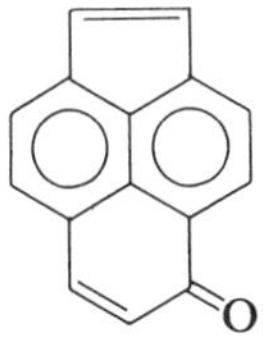

(**80**) 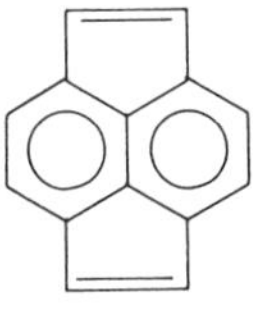 (**81**)

Vacuum flow pyrolysis of 1,6- and 1,8-pyrenedione (**72** and **73**) at 1100°C leads to stepwise decarbonylation under formation of 5*H*-cyclopent[*cd*]phenalen-5-one (**80**) and, finally, cyclopent[*fg*]acenaphthylene (**81**, pyracyclene)[46]. The use of pyrenequinones as sensitizers in photoresist compositions for reproduction techniques has been proposed[47].

The remarkable conductive properties of the tetrathiofulvalene/7,7,8,8-tetracyanoquinodimethane complex led to a search for further 'organic metals'. In this connection INDO and π-SCF calculations were performed on the electronic properties of the unknown 2,7-pyrenequinone (**82**), its quinodimethane, and the 13,13,14,14-tetracyano-2,7-pyrenoquinodimethane (**83**)[13]. The synthesis of **83** and its electrical properties have been described by Cowan and coworkers[48].

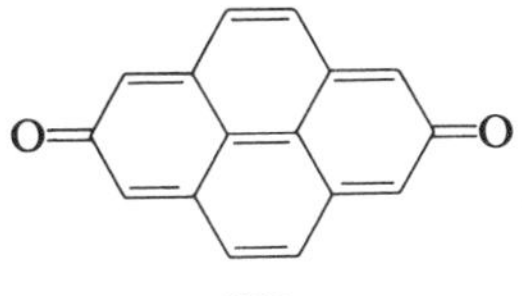

(**82**) 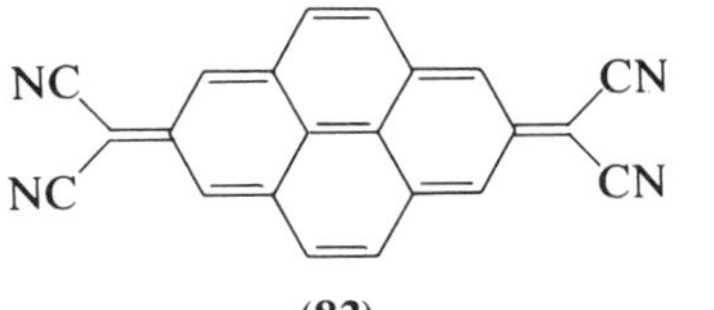 (**83**)

2. *Chrysene-6,12-dione*

The only known extended quinone of chrysene is the 6,12-dione (**86**). An interesting synthesis was performed via the dicarboxylic acid **84**, which can be synthesized by Reformatzky reaction of benzil with bromoacetic acid. Cyclization gave the hydroquinone **85**, which gave **86** on treatment with lead dioxide[49, 50]. **86** is converted in alkaline solution in the presence of oxygen to the 12-hydroxy-5,6-chrysenequinone[49].

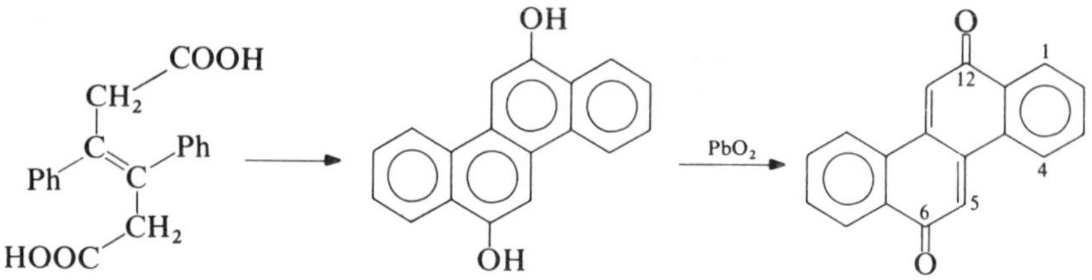

(84) (85) (86)

Pyrolysis of **86** at 900°C leades to decarbonylation and formation (15%) of indeno [2, 1-*a*] inden (**87**), which was not easily accessible previously. This and other aromatic carbonyl compounds were chosen, because they readily lose carbon monoxide in the mass spectrometer[51].

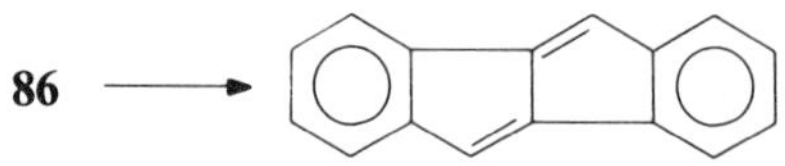

(87)

3. *Naphthacene-5,11-dione*

The 5,11-naphthacenequinone (**88**) is the only known extended naphthacenequinone. It is prepared by heating 5,11-dibromonaphthacene with 88% sulphuric acid[52].

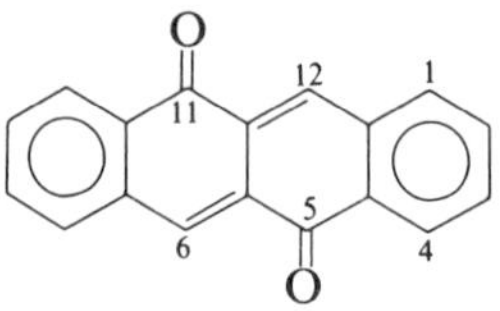

(88)

The photochromism of 6-phenoxy-5,12-naphthacenequinone (**89**) is due to the formation of 12-phenoxy-5,11-naphthacenequinone (**90**)[53]. Orange crystals of **90** are converted to the original quinone merely on melting. In benzene, **90** was readily converted under the influence of ammonia or aniline into the 12-amino-5,11-naphthacenequinones (**91a, b**), earlier synthesized by an alternative method[54–56].

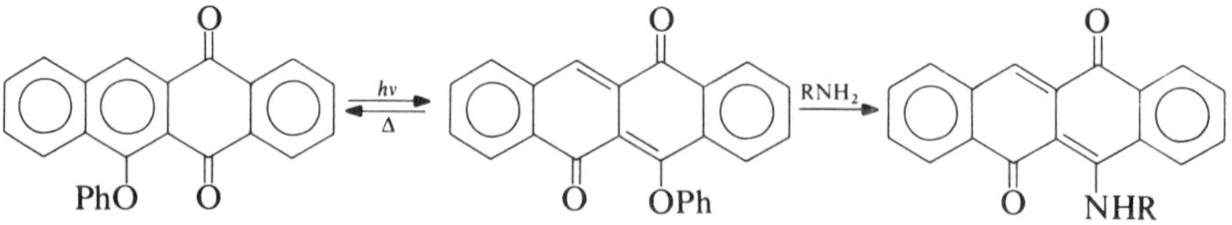

(89) (90) (91)

a: R = Ph
b: R = H

The 5,12-naphthacenequinones **92a** and **b** are converted under the influence of concentrated sulphuric acid or aluminium chloride (in benzene) into mixtures with the tautomeric 5,11-naphthacenequinone derivatives **93a** and **b**. With substituents **a** the 5,11-quinonoid form **93** dominates, and with substituents **b** the 5,12-quinonoid form (**92**)[57].

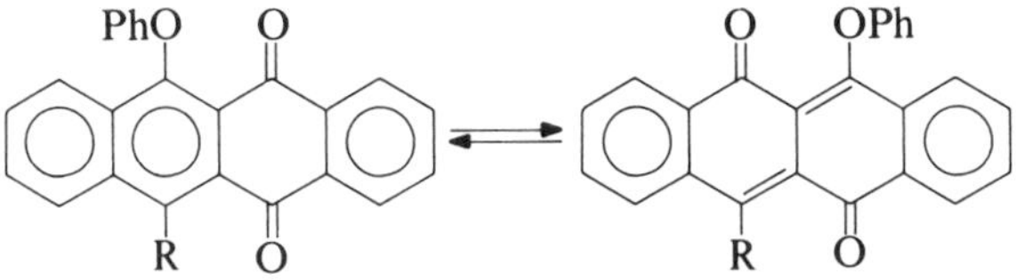

(**92**) (**93**)

a: R = NH_2, NHMe, NHPh, NHAc
b: R = H, OPh, OMe, Cl

With bromine 1,4-addition to positions 6 and 12 takes place giving **94**, and no substitution reaction is observed[54].

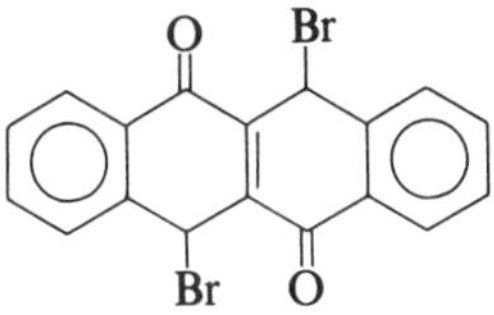

(**94**)

5,11- and 5,12-naphthacenequinone and 5,6,11,12-diquinone have been proposed for use as cathode material for a battery with alkali metals or alkali earth metals as anode[58a].

E. Five-ring Quinones

Of the common five-ring aromatics no extended quinones are known of picene (**95**) and pentaphene (**96**).

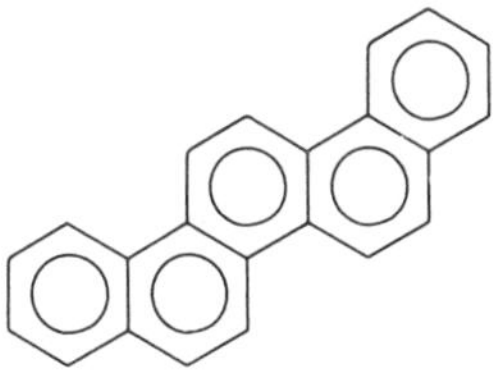

(**95**) 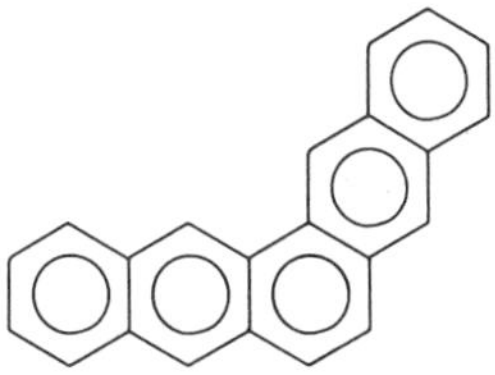 (**96**)

Extended quinones of perylene, pentacene, benzopyrene and benzochrysene are treated below.

1. Perylenediones

At the present time four quinones of perylene are known, the perylene-3,10-, 3,9-, 1,12-diones and perylene-3,4,9,10-tetraone (**98**, **99**, **100** and **102**, respectively). Most important is 3,10-perylenequinone (**98**), which can be obtained by oxidation of perylene with aqueous chromic acid[58b] or by heating 3,10-dihalogenoperylene with sulphuric acid[59, 60].

4,4′-Dihydroxy-1,1′-binaphthyl (**97**) is also converted to 3,10-perylenequinone by heating with aluminium chloride/manganese dioxide and subsequent oxidation[61] or by heating with concentrated sulphuric acid containing a little nitric acid and iron(II) salt[62]. 3,10-Perylenequinone forms a red vat.

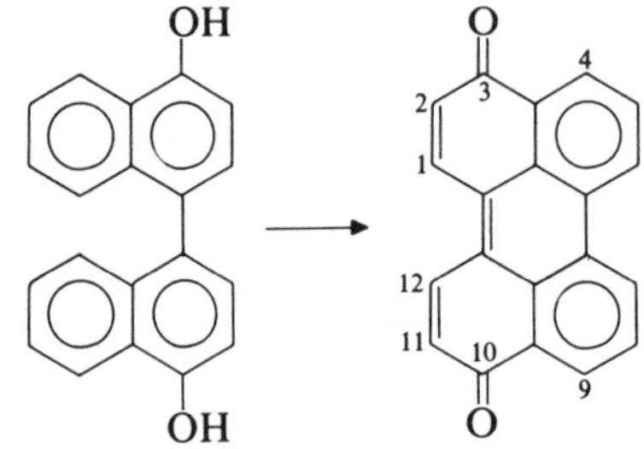

(**97**) (**98**) 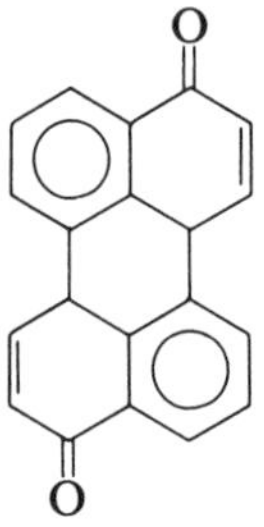 (**99**) 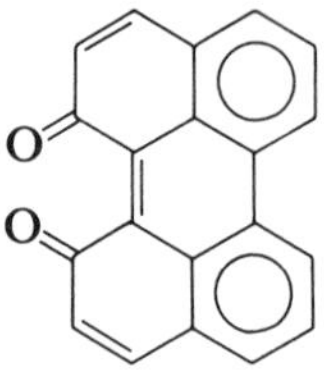 (**100**)

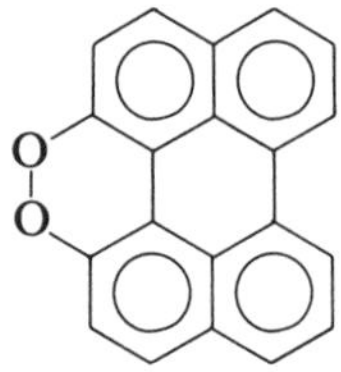

(**101**) 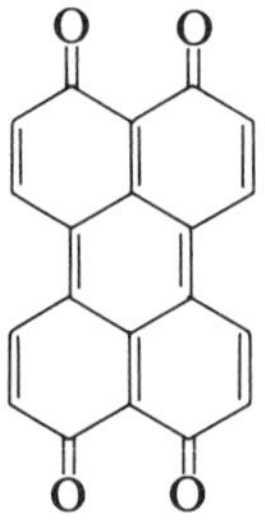 (**102**)

2,11-Dihydroxy-3,10-perylenequinone (**104**) has been synthesized in a similar way by treating 1,1′-binaphthyl-3,4,3′,4′-diquinone (**103**) with aluminium chloride[61]. **104** cannot be oxidized to the diquinone.

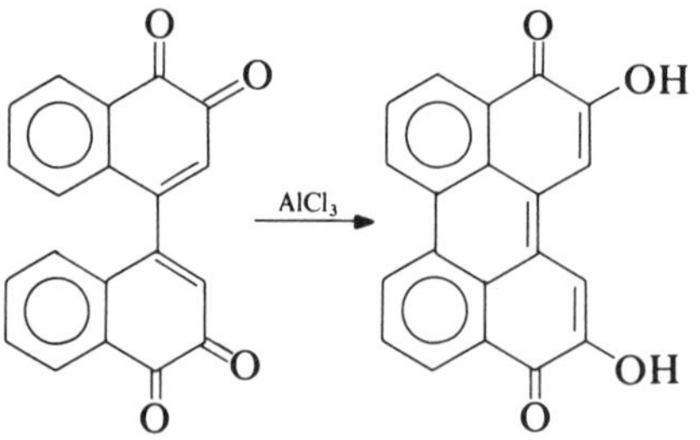

(**103**) (**104**) 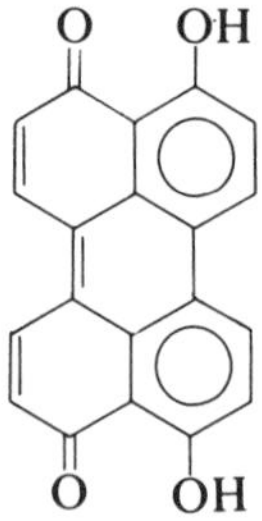 (**105**) 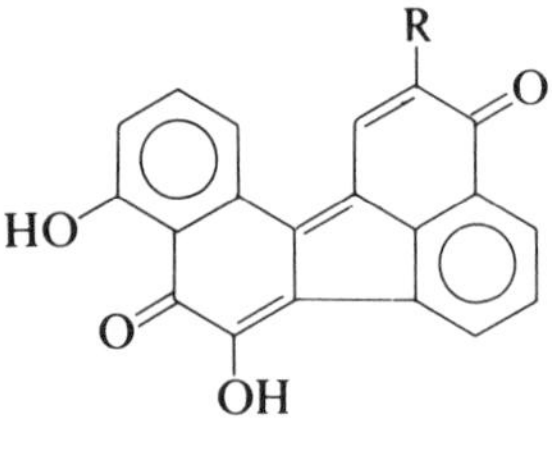 (**106**, R = OH) (**107**, R = H)

4,9-Dihydroxy-3,10-perylenequinone (**105**) was found as a component of the fruiting bodies of the fungus *Daldinia concentrica*[63]. It has been established that perylene derivates can also be produced in nature by oxidative coupling of naphthalene derivatives[64] and 4,5,4′,5′-tetrahydroxy-1,1′-binaphthyl is considered to be the precursor of **105** in *Daldinia concentrica*[63]. It seems reasonable to assume that similar precursors are involved in the formation of bulgarhodin (**106**) and bulgarein (**107**), two other extended quinones found together with **105** in the fruiting bodies of *Bulgaria inquinans*[65]. This fungus grows, e.g. on the bark of freshly felled oaks. In order to produce the benzo[*j*]fluoranthene nucleus of **106** and **107**, a *para,meta* coupling of a binaphthyl precursor would be required. This is unlikely in 4,5,4′,5′-tetrahydroxy-1,1′-binaphthyl itself but should be possible if further hydroxy groups are introduced into the 3- and 3′-positions.

3,9-Perylenequinone (**99**) can be obtained by heating 3,9-dibromo- or 3,9-dichloroperylene with sulphuric acid[66]. Dehydrogenation of 1,2,7,8-tetrahydroperylene-3,9-quinone also yields (**99**)[67]. (**99**) is very easily oxidized to 3,4,9,10-diquinone **102**. It forms a red vat.

1,12-Perylenequinone (**100**) is obtained by oxidation of the conjugate hydroquinone in alkaline solution with air. The hydroquinone is formed by heating 2,2′-dihydroxy-1,1′-binaphthyl[68]. **100** is an isomer of perylene-1,12-peroxide (**101**), which is formed from 1,12-dihydroxyperylene with zinc chloride[69]. In contrast to **100**, **101** is not reduced with dithionite or hydrogen iodide.

3,4,9,10-Perylene-diquinone (**102**) is obtained by heating 3,9-dichloro-4,10-dinitro-, 3,4,9,10-tetranitro-, or 3,10-dinitroperylene with sulphuric acid[70]. It gives a dark-red vat dye with alkaline dithionite.

2. Pentacenediones

5,12-Pentacenequinone (**110**) has been synthesized starting with 5,7,12,14-tetrahydroxy-6,13-pentacenequinone (**108**)[71], which was reduced with zinc powder in acid or alkaline medium to 5,7,12,14-tetrahydroxy-6,13-dihydropentacene (**109**), which in turn loses water to form **110**. Oxidation of **110** with chromic acid yields the diquinone **111**.

OH O OH / OH O OH — Zn → OH OH / OH OH

(108) **(109)**

$-H_2O$

O O / O O ← CrO_3 — O, 1, 2, 3, 4, 5, 12, O

(111) **(110)**

The phenoxy derivative of 5,13-pentacenequinone **113** can be prepared from 5-phenoxy-6,13-pentacenequinone (**112**) by UV irradiation[72]. On exposure of a solution of **113** to visible light or on keeping it in the dark, the reverse isomerization takes place. These reactions are paralleled by those of *peri*-aryloxyanthraquinone (see Section III.C.1) and naphthacenequinone (see Section III.D.3).

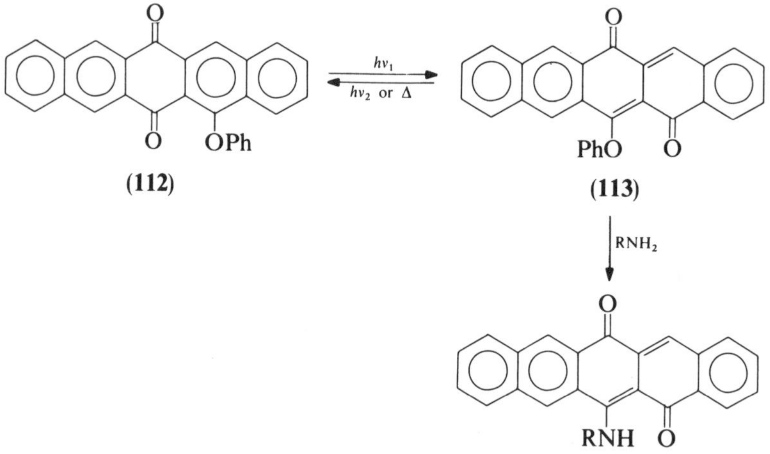

(112) (113) (114, R = H, ph)

On account of the slow thermal isomerization of **113** it was possible to isolate it in the crystalline state. With amines, nucleophilic substitution of the phenoxy group by an amino group occurs at room temperature with formation of **114**.

3. *Benzo*[def]*chrysenediones (benzo*[a]*pyrenediones)*

The IUPAC nomenclature prescribes benzo[*def*]chrysene instead of benzo[*a*]pyrene for the conjugate hydrocarbon. Since the name benzo[*a*]pyrene (or formerly 3,4-benzopyrene) has always been used in the literature this name is retained in this chapter.

3,6-benzo[*a*]pyrenequinone (**120**) has been synthesized starting with the 9-anthracene aldehyde which, on Knoevenagel condensation with malonic ester, yielded **115**. Reduction and cyclization with hydrogen fluoride gave **116**. Knoevenagel condensation of **116** with ethyl succinate led to **117** which, after saponification, was decarboxylated, reduced, and cyclisized to **118**. Dehydrogenation by heating with palladium gave 3-hydroxybenzo[*a*]pyrene (**119**), which was easily oxidized to 3,6-benzo[*a*]pyrenequinone (**120**)[73].

The synthesis of 6,12-benzo[*a*]pyrenequinone (**123**) was effected by condensation of the phthalidene acid **120a** with naphthalene in anhydrous hydrogen fluoride, probably with **121** as an intermediate[74]. A similar synthesis has been described by Norman and Waters[75]. Attempts to use the equivalent synthon **122** or its lactone for the synthesis of **123** have also been successful (77% yield)[76]. The first synthesis gave appreciable yields only with naphthalene, while the latter was more versatile also in the synthesis of derivatives.

Oxidation of benzo[*a*]pyrene always gave mixtures of benzo[*a*]pyrenequinones. As with pyrene itself chromic acid attacks the 3,6- and 1,6-positions and 3,6-benzo[*a*]pyrenequinone (**120**) and 1,6-benzo[*a*]pyrenequinone (**124**)[19] are formed. The same is probably true for 10-azabenzo[*a*]pyrene, but only the 1,6-quinone **125** has been isolated[19].

Pure 3,6-benzo[*a*]pyrenequinone (**120**) is also obtained by oxidation of benzo[*a*]pyrene-1-carboxylic acid followed by decarboxylation[76].

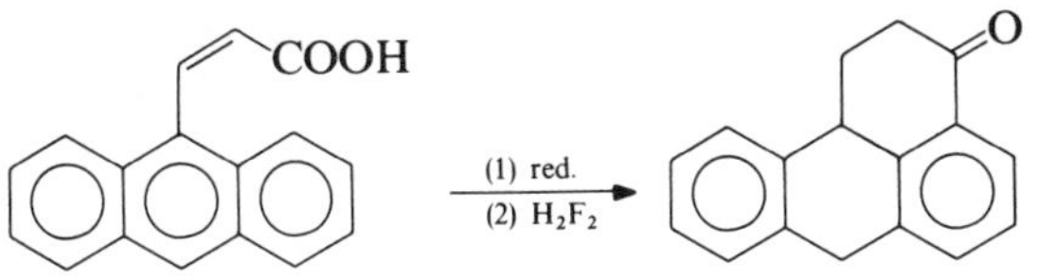

COOH

(1) red.
(2) H_2F_2

(115) **(116)**

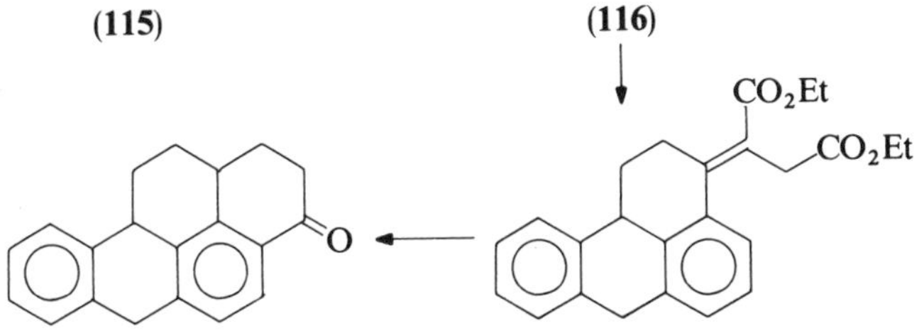

CO_2Et
CO_2Et

(118) **(117)**

Pd | $-H_2$

OH ox.

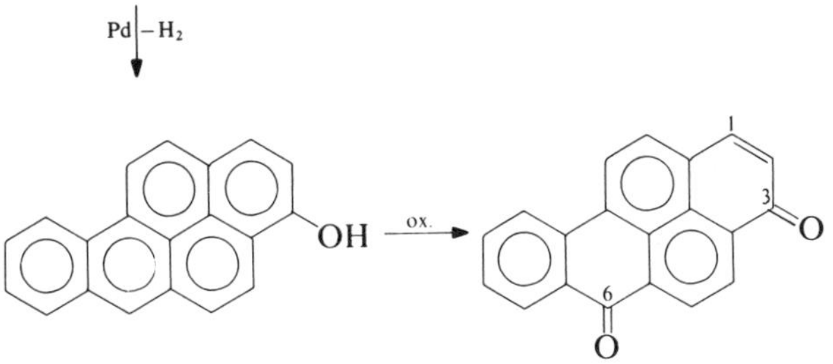

(119) **(120)**

An important aspect of the oxidation of polycyclic aromatic compounds is the removal of these potential carcinogens from tap water or other sources. The $NaClO_2$ oxidation products of benzo[*a*]pyrene were separated by thin-layer chromatography and identified

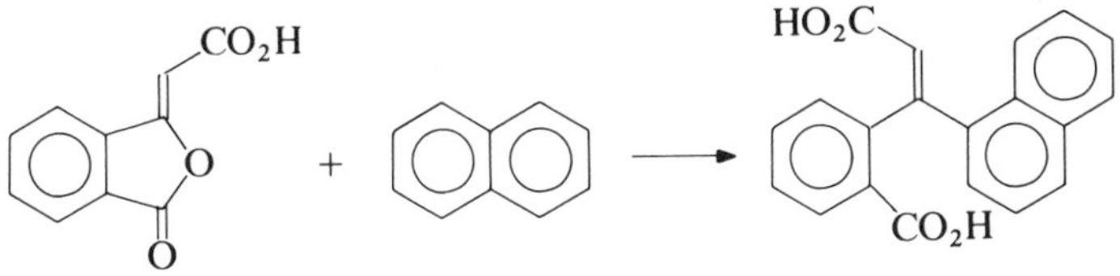

CO_2H + HO_2C CO_2H

(120a) **(121)**

EtO_2C OH CO_2H

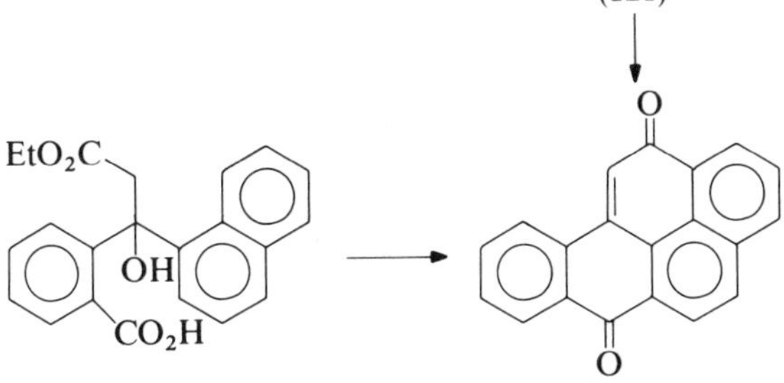

(122) **(123)**

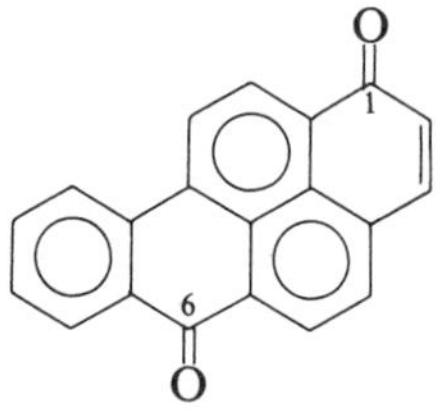

(124)

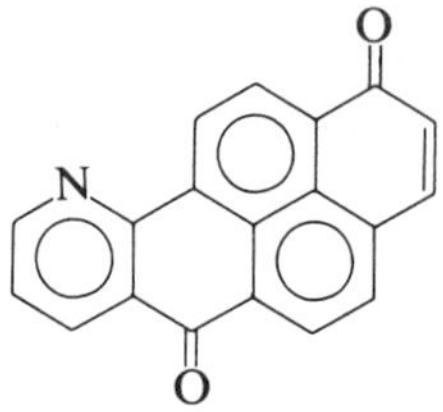

(125)

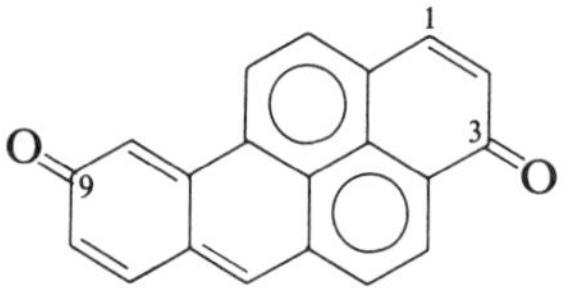

(126)

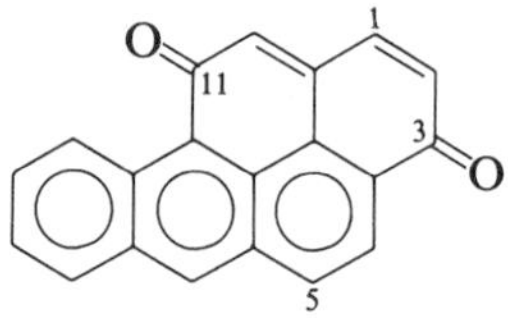

(127)

as 3,6-benzo[*a*]pyrenequinone (**120**), 3,9-benzo[*a*]pyrenequinone (**126**), and 3,11-benzo[*a*]pyrenequinone (**127**)[77]. The products of the chlorine dioxide treatment of benzo[*a*]pyrene have been investigated. Three of the eight isolated derivatives were again identified as **120**, **126** and **127**, which represent about 90% of the products and are considered inactive with respect to carcinogenesis. The other products are chloro derivates of benzo[*a*]pyrene. Accordingly, the treatment of drinking water with chlorine dioxide seems to be a method to reduce the possible carcinogenic danger[78]. For the same reason, the oxidation of benzo[*a*]pyrene with iron(III) chloride/hydrogen peroxide has been investigated. Oxidation in nitromethane and acetone as solvent yielded, among other products, 1.8% of **126**, 0.7% of **120**, and 1.4% of **127**[79].

The destruction of benzo[*a*]pyrene by light and oxygen or ozone has been studied[80], using the hydrocarbon in low concentrations and an excess of ozone. From about eight products detected in these experiments three have been identified as 1,6-, 3,6- and 6,12-quinones (**124**, **120** and **123** respectively). Considerable evidence has been accumulated in support of the hypothesis that cellular metabolism is a prerequisite for the carcinogenic activity of the polycylic aromatic hydrocarbons. Hence much effort has been directed to the elucidation of the metabolism of the carcinogenic hydrocarbons. Often found metabolic products of benzo[*a*]pyrene are **123**, **124** and **120**. A common precursor seems to be 6-hydroxybenzo[*a*]pyrene (**128**), which is indeed oxidized in rat liver homogenate and is autoxidized in aqueous buffer–ethanol solution to produce the three quinones in yields of 36%, 27% and 29%, respectively. Thus the carcinogenic activity of benzo[*a*]pyrene seems to be connected with 6-hydroxybenzo[*a*]-pyrene (**128**) and its cellular oxidation[81]. The mechanism of the formation of **128** and two alternative pathways for the oxidation to the three quinones are discussed[82].

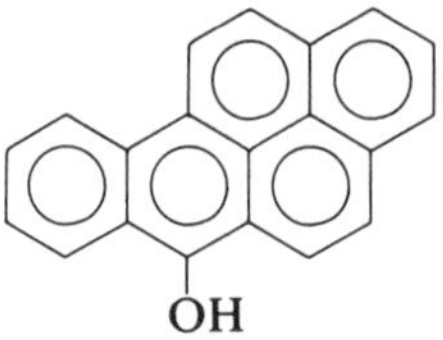

(128)

F. Six-ring Quinones

No extended quinones of hexaphene (**128a**) and hexacene (**129**) seem to have been described.

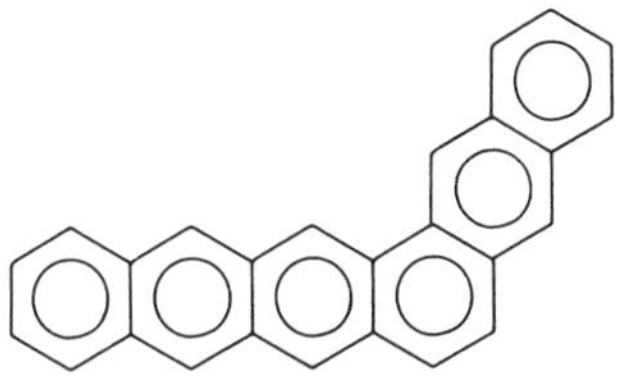

(**128a**) 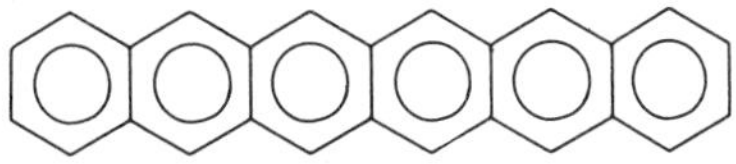 (**129**)

The known extended quinones with six rings are derived from aceanthreno[1,2:2′,1′]aceanthrene, anthanthrene, zethrene, benzoperylene, dibenzochrysene, dibenzopyrene and dibenzonaphthacene, which are reviewed below.

1. Aceanthrono[1,2:2′,1′]aceanthrone (acedianthrone)

Aceanthrono[1,2:2′,1′]aceanthrone, usually called acedianthrone (**133**), can be synthesized easily starting with anthrone (**130**). Condensation with glyoxal sulphate[82a] generates the stilbene-quinone **131**, which yields on alkaline oxidation[83] or on heating with nitrobenzene and organic acid chlorides **132**[84, 85].

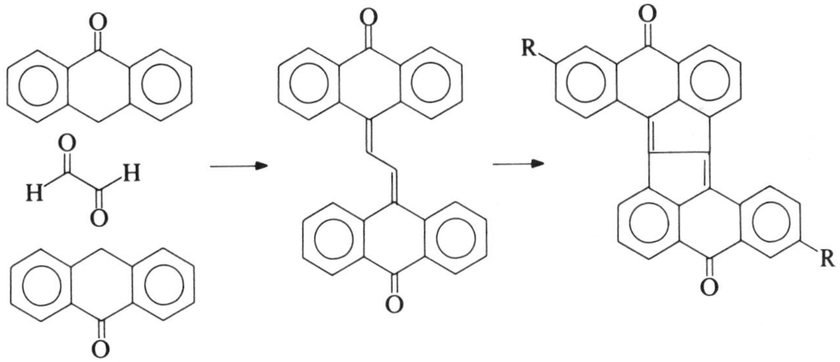

(**130**) (**131**) (**132**, R = H)
(**133**, R = Cl)

This process is used in industry for the synthesis of the vat dye Indanthrene Red Brown RR (**133**) from 2-chloroanthrone[11b]. In a similar manner many other derivatives of **132** have been synthesized[86].

2. Anthanthrone

Quinones may undergo Diels–Alder reactions acting either as dienophiles or as dienes. Therefore an interesting path to polycyclic quinones could be a Diels–Alder reaction of angular anellated quinones with dienophiles. Indeed 6,12-chrysenedione (**134**) reacts in boiling maleic anhydride in the presence of chloroanil as dehydrogenating agent to give anthanthronetetracarboxylic acid-(5,6,12,13)-dianhydride (**135**), which in turn was decarboxylated to anthanthrone (**136**) (overall yield about 35%).

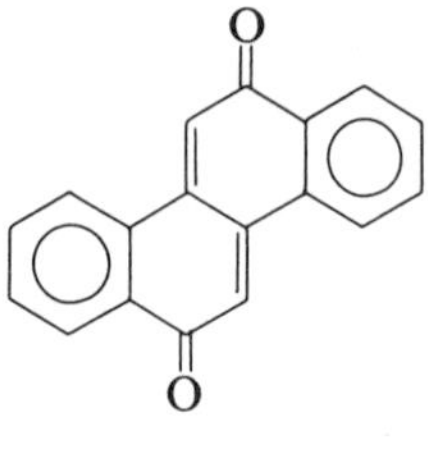

(**134**) 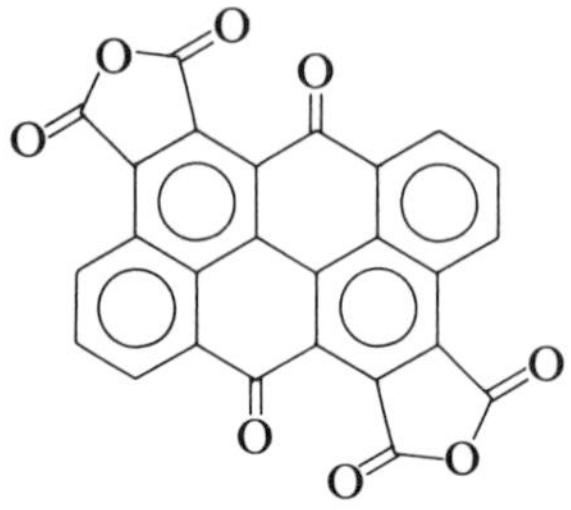 (**135**) 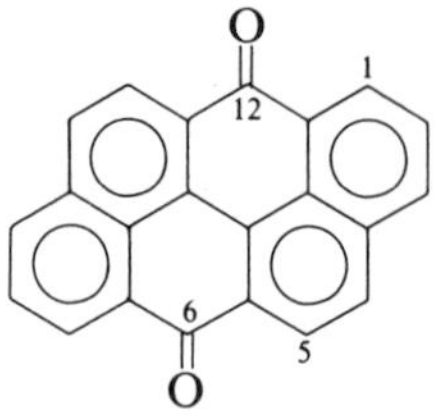 (**136**)

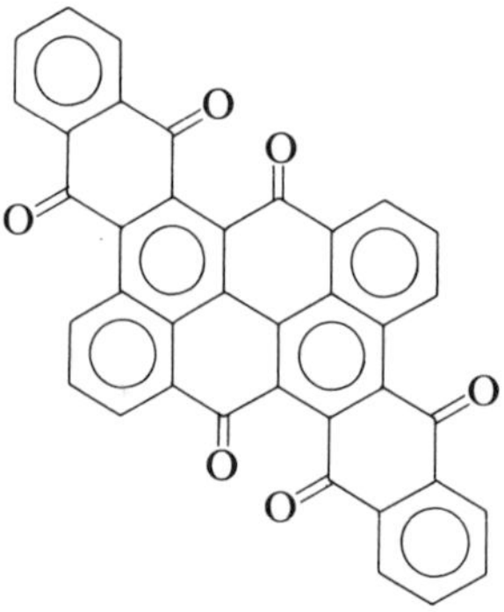

(**137**) 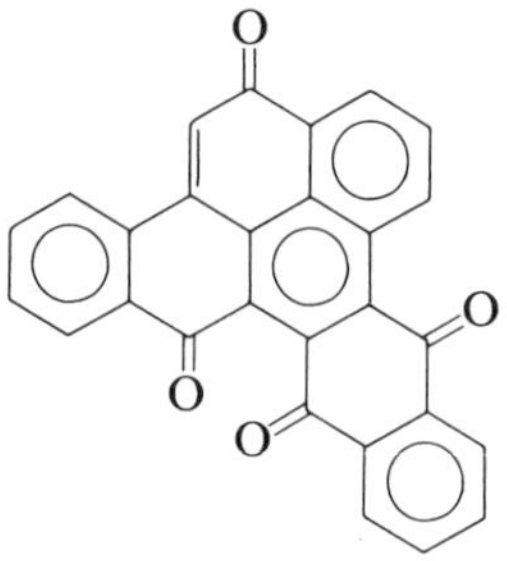 (**138**)

With nitrobenzene as solvent and dehydrogenating agent the addition reaction of maleic anhydride probably gave only the mono-addition product (as the hydroquinone) whereas naphthoquinone yielded both the mono- and di-adducts (**137**, **138**)[87].

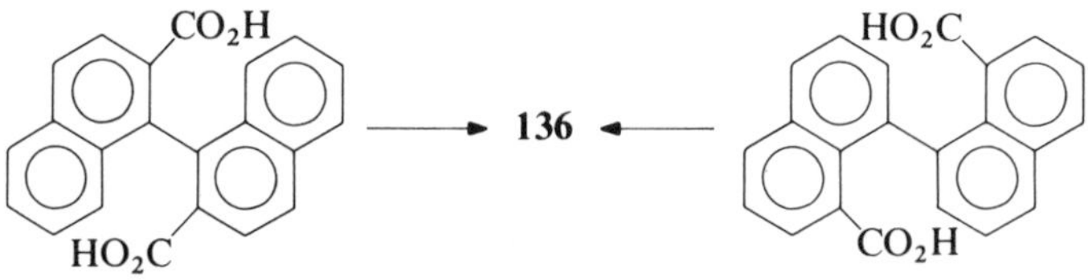

(**139**) → **136** ← (**140**)

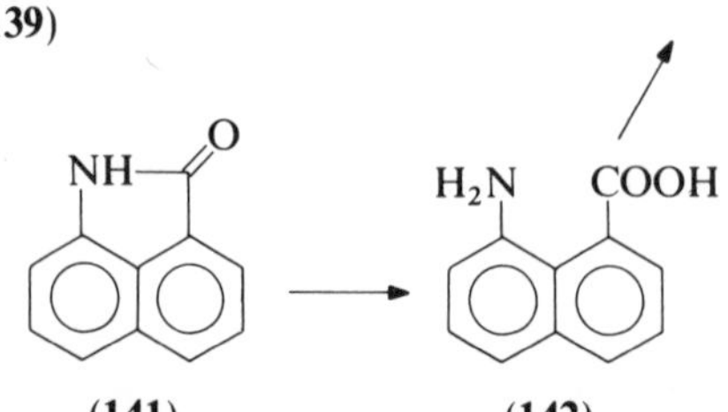

(**141**) → (**142**)

A synthesis of **136** starting with acenaphthene and the preparation of some derivatives has been described[88]. **136** was first synthesized by a double ring-closure of either 1,1′-binaphthyl-2,2′-dicarboxylic acid (**139**), or of the 8,8′-dicarboxylic acid **140** with sulphuric acid, or of the dichlorides with aluminium chloride[89]. This procedure is used for commercial syntheses. The starting material is naphthostyril (**141**), which is hydrolyzed to the amino acid **142**, diazotized, and dimerized, losing nitrogen[90]. The ring-closure is effected by sulphuric acid.

In the same way the novel vat dyes, such as 6,12-anthanthrenedione-3,4,9,10-tetracarboxylic diimides (**144a–c**), have been prepared from **143**[91].

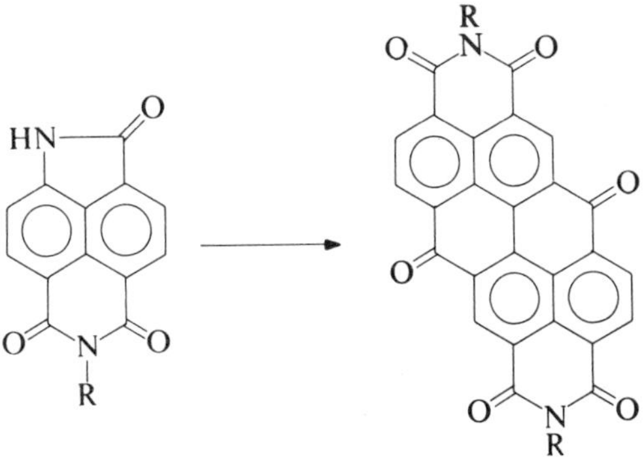

(**143**) (**144**)

a: R = H
b: R = Et
c: R = Ph

Anthanthrone is an intensely coloured orange vat dye, but has only little fibre affinity. By halogenation, 4,10-dichloroanthanthrone (**145a**) and 4,10-dibromo-anthanthrone (**145b**) are formed in high yield[92] and show greater affinity to fibres and possess brighter, more intense red shades.

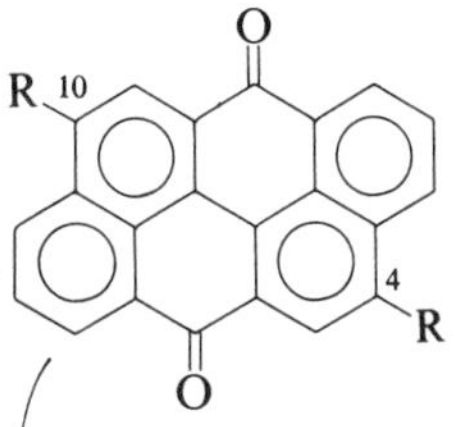

145, **a**: R = Cl; **b**: R = Br; **c**: R = (1-benzoylamino-anthraquinon-4-yl)-NH–

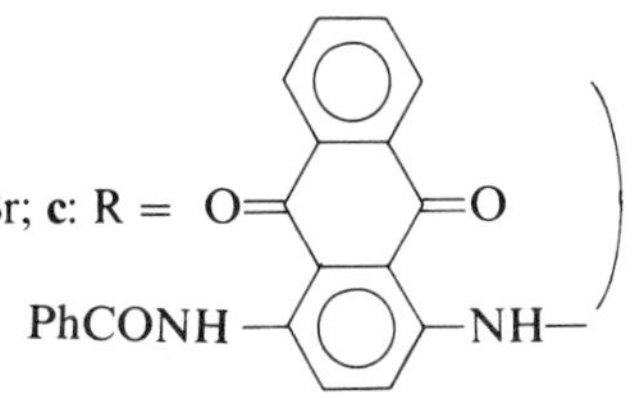

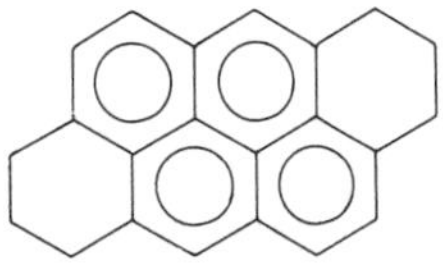

(**146**)

The bromine atoms in the 4- and 10-positions undergo nucleophilic substitution: heating of **145b** with 1-amino-4-benzoylaminoanthraquinone gives, e.g. another valuable vat dye, **145c** (Indanthren Grey BG)[93]. For the reduction of anthanthrone to anthanthrene the zinc powder melt is, as usual, the simplest and best procedure[94].

Hydroiodic acid/red phosphorous reduces anthanthrone **136** to 1,2,3,7,8,9-hexahydrodibenzo[*def, mno*]chrysene (**146**), with replacement of the oxygen atoms and partial hydrogenation of the aromatic rings[95]. The electrochemical behaviour of anthanthrone has been examined and it was found that ion pairs with divalent metal ions were absorbed at the surface of mercury electrodes. This phenomenon is not observed with gold electrodes. The possible structure and orientation of the ion pairs on the surface has been discussed[96].

3. *Dibenzo*[hi:qr]*naphthacene-7,14-dione (7,14-zethrenequinone)*

Of the derivatives of zethrene, the 7,14-quinone **150** has been described[10b].

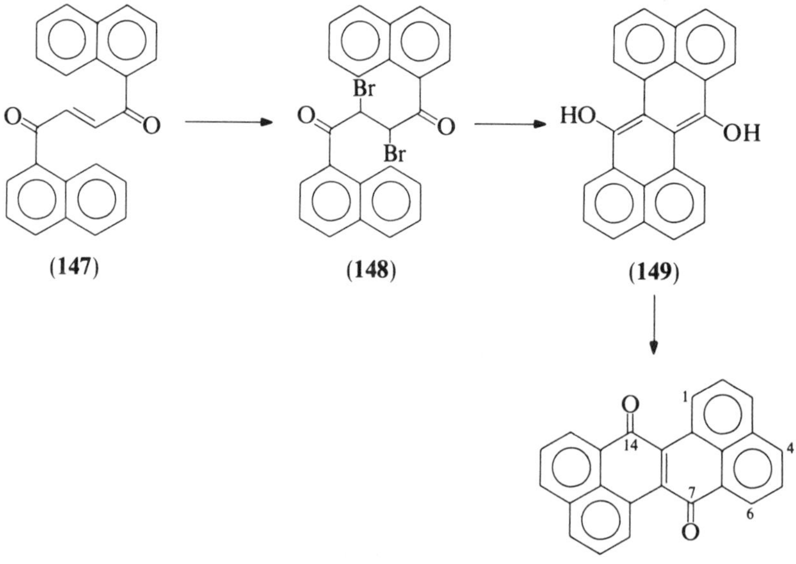

Br
Br
HO
OH
O
O
O
O
(147) (148) (149)

O
1
14
4
7
6
O
(150)

The diketone **147** is obtained by Friedel–Crafts condensation of fumaryl chloride with naphthalene. **147** adds bromine to yield **148**, which cyclizes to **149** in an aluminium chloride/sodium chloride melt. When oxygen is passed in during melting, **150** is formed immediately. 7,14-Zethrenequinone gives a blue vat on short reduction with alkaline sodium dithionite.

4. *Benzo*[a]*perylene-7,14-dione*

Benzo[*a*]perylene-7,14-quinone (**152**) has been synthesized in two steps, starting with an aluminium chloride catalysed condensation of 10,10-dichloroanthrone with napthalene to 7-hydroxybenzo[*a*]perylene (**151**), which gives the quinone **152** on oxidation with chromic acid[97]. **152** forms a greenish-blue vat.

The quinone with an oxygen bridge between the 11- and 12-positions, **155**, is obtained by oxidation of the conjugate hydrocarbon **153** or by heating 1-(β-

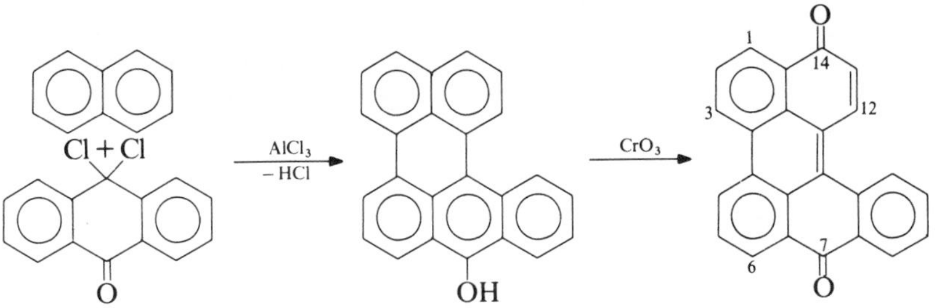

(151) (152)

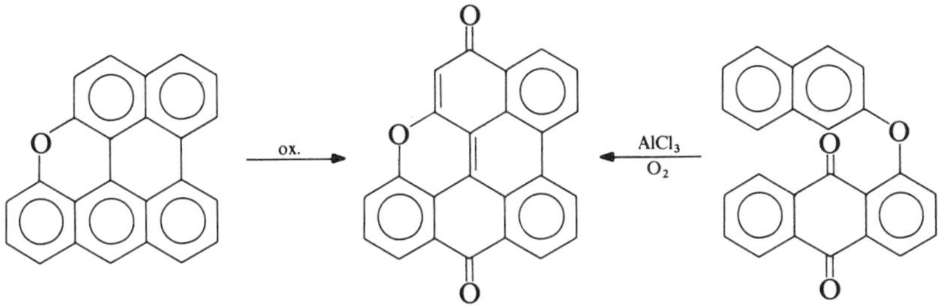

(153) (155) (154)

naphthoxy)anthraquinone (**154**) in an aluminium chloride/sodium chloride melt passing in oxygen[98].

5. *Benzo*[rst]*pentaphene-5,8-dione (dibenzo*[a,i]*pyrene-5,8-quinone)*

The correct IUPAC name is the first one, but the second is mostly used in the literature and is therefore retained here.

Dibenzo[*a, i*]pyrene-5,8-quinone (**157**) can be easily synthesized by Scholl cyclization of 1,4-dibenzoylnapthalene (**156**)[99] or 4-benzoylbenzanthrone (**158**)[100]. Also chromium trioxide[99] and selenium dioxide oxidation of dibenzo[*a, i*]pyrene leads to **157**[101]. Ozonolysis of dibenzo[*a, i*]pyrene yielded in low yield (17%) the 5,8-dione. 56% of the starting material has been recovered[102].

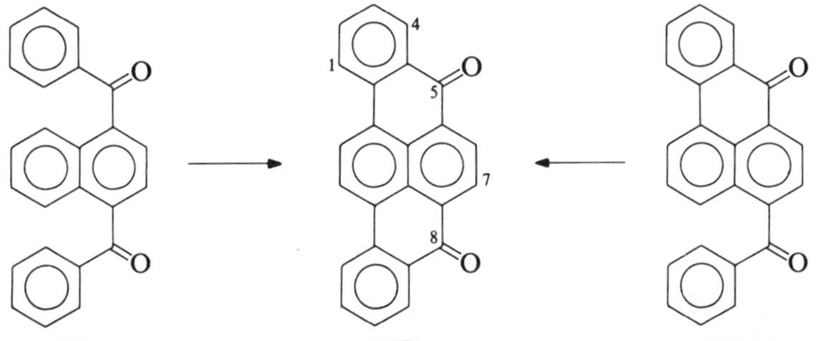

(156) (157) (158)

Oxidation of certain condensed aromatic carcinogenic hydrocarbons with chlorine dioxide to quinones has been recommended as a possible procedure to purify tap water (see also Section III.E.3). The quinones are considered as not dangerous. It has even been claimed that **157**, which was found in cigarette-smoke condensates, stops the progress of already established cancer in mice[103].

157 was formed in an attempt to perform a Diels–Alder condensation between *trans*-1,2-dibenzoylethylene and diethyl muconate. The product was not the expected one but **157**. A mechanistic explanation for this surprising result has been given[104]. **157** gives a yellow-red vat dye. In contrast to the dibenzo[*b*, *i*]pyrene-7,14-quinone (Section III.F.6) **157** is not used commercially as a vat dye.

The synthesis of 1,2-diazadibenzo[*a*, *i*]pyrene-5,8-quinone has also been described[105].

157 is attacked by bromine to yield a dibromo compound. Though this compound was used as vat dye, the constitution has not been published[11c].

6. *Dibenzo*[b,def]*chrysene-7,14-dione (dibenzo*[b,i]*pyrene-7,14-quinone)*

The IUPAC name is the first one. However dibenzo[*b*, *i*]pyrene-7,14-quinone (**161**) and some of its derivatives are of technical interest as vat dyes, and the conventional name will be retained here. The quinone (**161**) is used as golden-yellow dye (Indanthren Goldgelb-GK or Cibanongelb-GK). It was first synthesized by heating benzanthrone (**159**), benzoyl chloride, and aluminium chloride in the presence of oxidants or oxygen. **160** is an intermediate[106]. This synthesis is now obsolete. The commercially used synthesis is a Scholl cyclization of 1,5-dibenzoylnaphthalene (**162**)[107] which can be performed also without oxygen or other oxidants[108], although an improvement was achieved by addition

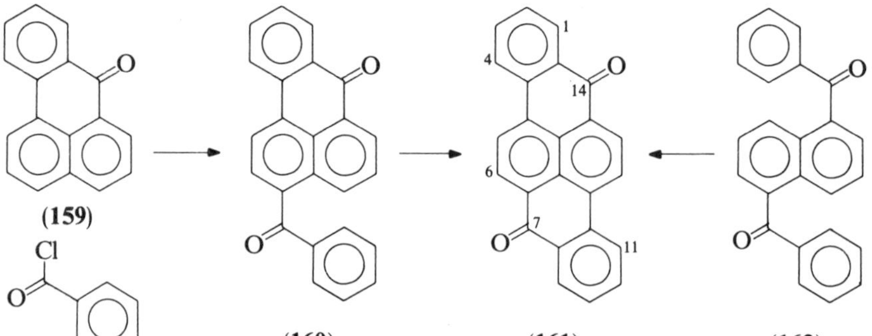

(**159**) (**160**) (**161**) (**162**)

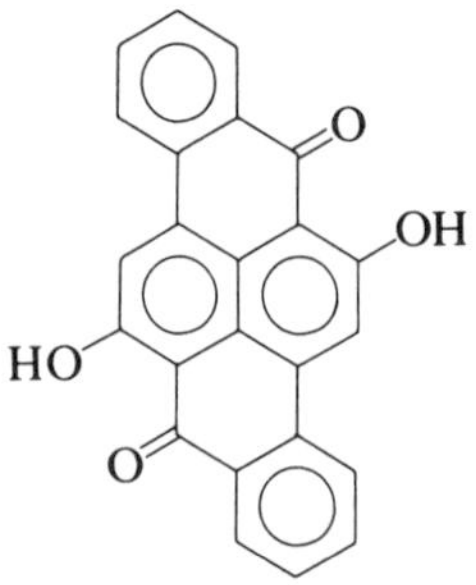

(**163**)

of m-$O_2NC_6H_4SO_3K$ as hydrogen acceptor to the aluminium chloride/sodium chloride melt[109]. The reaction was also performed with aluminium chloride in chlorobenzene, in the presence of 2,4-dinitrobenzene and melting-point reducing agents such as alkali chlorides, organic amines or amides (urea)[110].

Many derivatives of **161** have been synthesized by Scholl cyclization or by direct substitution reactions. 8-Bromodibenzo[*b*, *i*]pyrene-7,14-quinone, another vat dye (Indanthren Goldgelb-RK or Cibanongoldgelb-RK), is prepared by bromination of **161** in the melt[111]. 6,13-Dihydroxydibenzo[*b*, *i*]pyrene-7,14-quinone (**163**) is formed (70%) from 2,6-dihydroxy-1,5-dibenzoylnaphthalene by Scholl cyclization[19]. It is converted by phosphorus pentachloride to 6,13-dichlorodibenzo[*b*, *i*]pyrene-7,14-quinone, which in turn reacts with aniline to give 6-anilino-13-chloroquinone and, on further heating, yields 6,13-dianilinoquinone.

The same product can be obtained by methylating the hydroxyquinone and heating the dimethyl ether with aniline. The methoxy groups are substituted as easily as the chlorine atoms[19]. **161** was also used for the synthesis of dibenzo[*b*, *def*]chrysene-7,14-^{14}C (**164**). The radioactive carbon was introduced via Grignard reaction of 1,5-dibromonaphthalene with [^{14}C]carbon dioxide. The dicarboxylic acid was converted to 1,5-dibenzoylnaphthalene by Friedel–Crafts reaction with benzene. For the Scholl cyclization of the dibenzoylnaphthalene in an aluminium chloride/sodium choride melt *m*-dinitrobenzene was used as oxidant. Reduction of the quinone to the dibenzochrysene was effected as usual by zinc dust in a sodium chloride/zinc chloride melt at 210° C with about 70% yield[112].

161 on vacuum flow pyrolysis successively eliminates two molecules of carbon monoxide and forms 13*H*-naphtho[3,2,1-*cd*]fluoranthene-13-one (**165**) and indeno[1,2,3-*cd*]fluoranthene (**166**)[113].

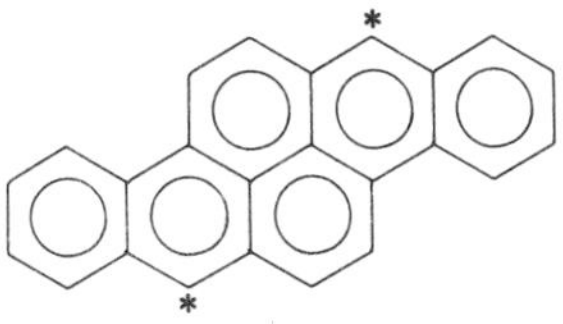

(**164**)

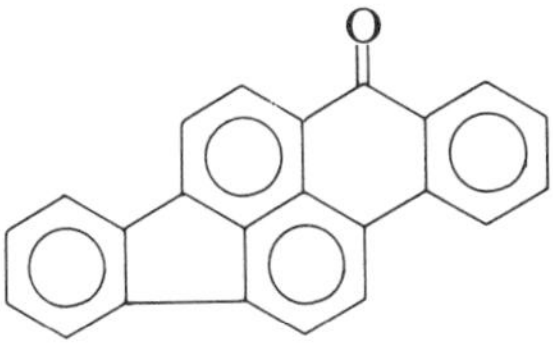

(**165**) 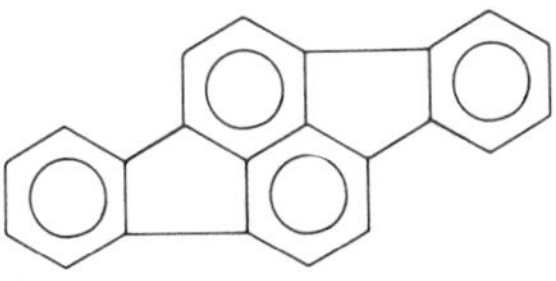 (**166**)

7. *Dibenzo*[fg,op]*naphthacenediones*

A surprising, elegant and effective synthesis of dibenzopyrene quinones was possible by oxidation of 3,3′,4,4′-tetraalkoxydiphenyl (**167**) and of 1,2-dialkoxybenzenes (**169**), respectively, with chloranil in aqueous sulphuric acid. 2,5,6,9,12,13-Hexamethoxydibenzo[*fg*, *op*]naphthacene-1,8-dione (**168a**) was formed from the tetramethoxydiphenyl (**167**) in 76% and the isomeric 1,10-dione (**170a**) from 1,2-dimethoxybenzene in 72% yield[115]. Though **168** and **170** may be regarded as derived from dibenzopyrene, the IUPAC notation as dibenzonaphthacene derivatives is used here. Ethers with other alkyl groups (**168b, c** and **170b**) have also been synthesized.

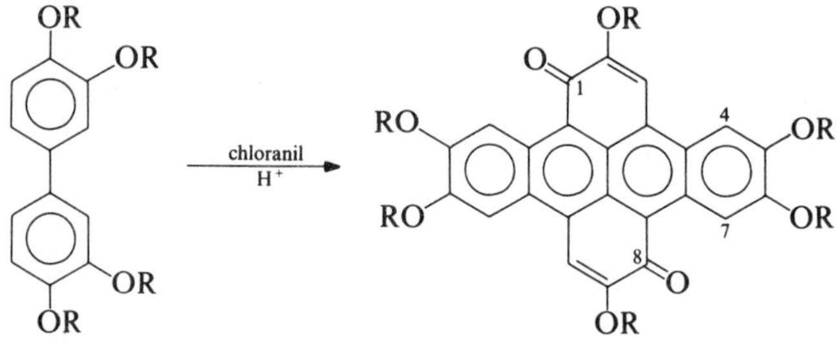

(167) (168)

a: R = Me; b: R = CD_3; c: R = Et

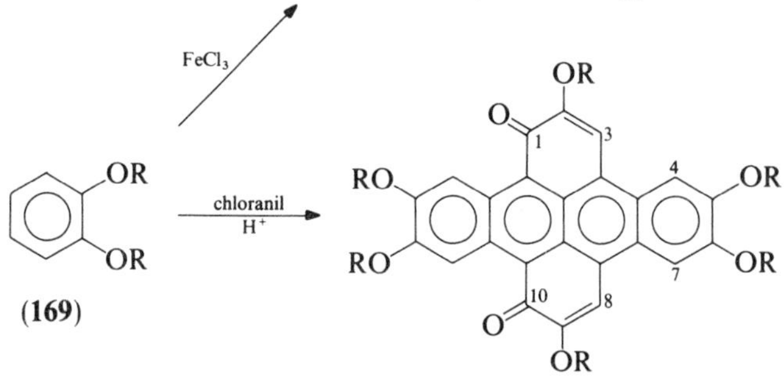

(169) (170)

a: R = Me; b: R = Et

The mechanism of the reactions has been discussed[114, 115]. The ability of chloranil to abstract hydride ions, i.e. the oxidative power, seems to be enhanced considerably by protonation. Moist iron(III) chloride transforms 1,2-dimethoxybenzene to **168a**. The 1,8-dione **168** dyes cotton to a dull violet which lacks fastness.

8. *Naphtho[1,2,3,4,-*def*]chrysene-8,14-dione*

The 5-methyl derivative of naphtho[1,2,3,4-*def*]chrysene-8,14-dione (**171**) has been obtained by chromic acid oxidation of the corresponding hydrocarbon[116]. This reaction seems to be remarkable because, e.g. the formation of the *o*-quinone **172** would retain a larger aromatic system. Indeed Zincke and coworkers supposed falsely that **172** was formed[117].

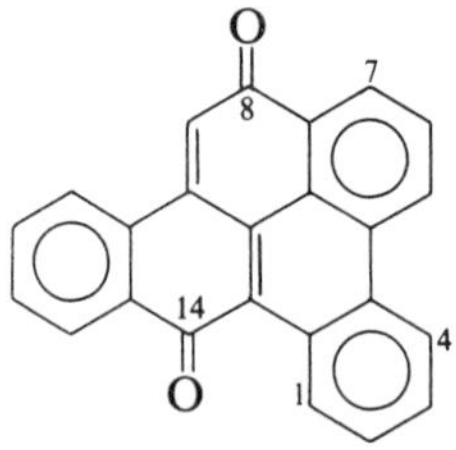

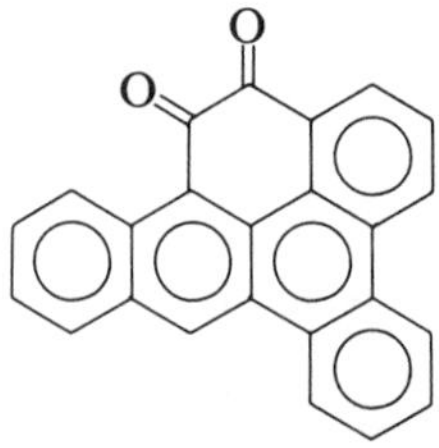

(171) (172)

G. Seven-ring Quinones

Extended quinones of the common seven-ring aromatics coronene (**173**), trinaphthylene (**174**), heptaphene (**175**) and heptacene (**176**) have not yet been described. The quinones of the other seven-ring aromatics are treated in alphabetical order.

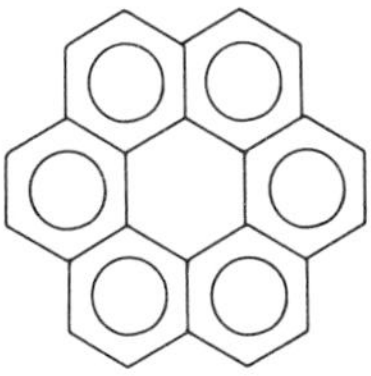

(**173**) 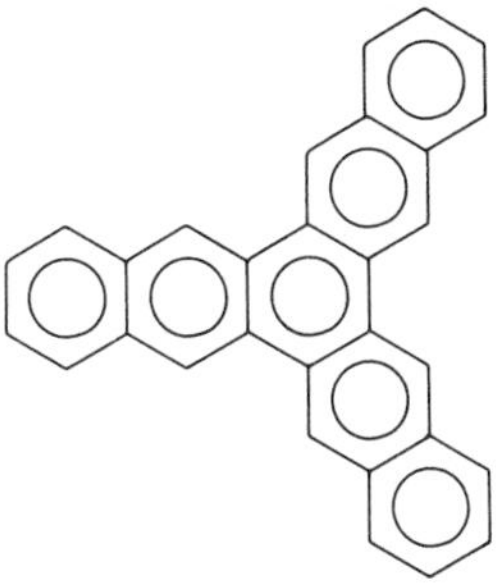 (**174**) 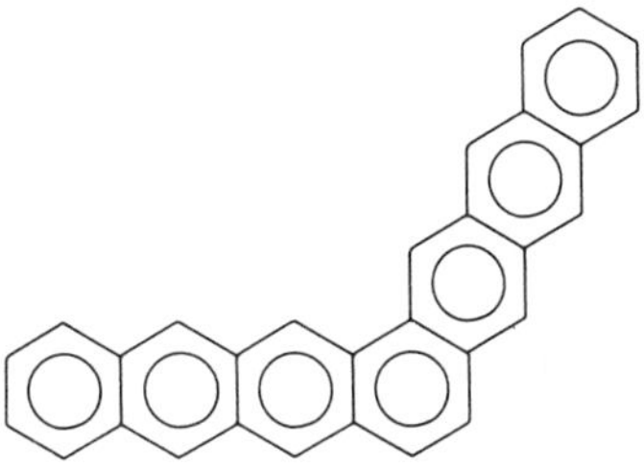 (**175**)

(**176**)

1. *Dibenzo*[hi:st]*pentacenedione*

Oxidation of dibenzo[*hi*:*st*]pentacene (**177**) yields a quinone of uncertain structure which contains three oxygen atoms[114]. On reduction with alkaline dithionite and reoxidation with air a compound with two oxygens is formed[115], very probably **178**. Structure **179**[115] should be excluded[10c]. **178** gives with alkaline dithionite an orange-red vat.

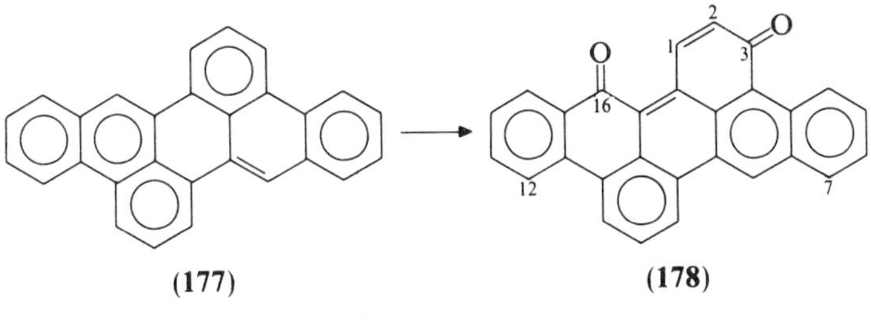

(**177**) (**178**)

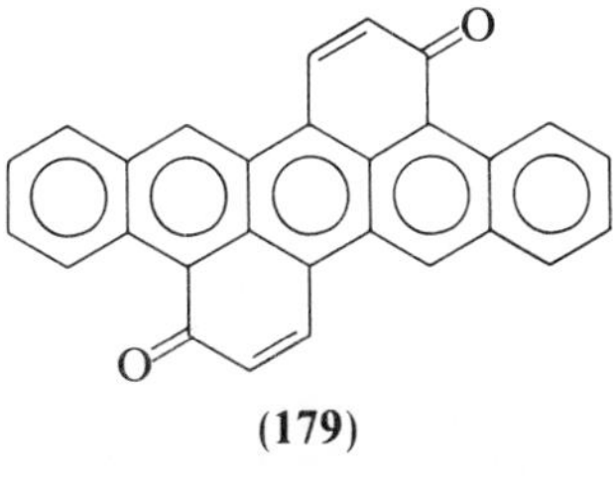

(**179**)

At fusion of 9,10-dibenzoylanthracene with aluminium chloride a red violet or bluish condensation product is formed[116]. Its constitution can be the dibenzo [*h*:*rst*]pentaphene-5,10-dione (**180**) or the dibenzo[*hi*:*st*]pentacene-8,16-dione (**181**) or a mixture of both, and since migrations of benzoyl groups are sometimes observed in the aluminium chloride melt, other structures cannot be excluded.

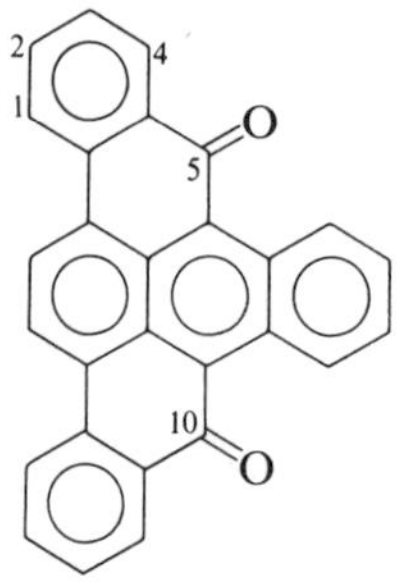

(**180**) 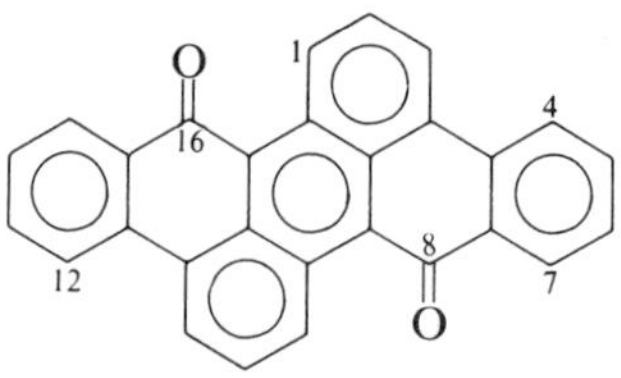 (**181**)

2. *Dibenzo*[jk:uv]*pentacene-7,15-dione*

Dibenzo[*jk*:*uv*]pentacene-7,15-dione (**183**) has been synthesized simply by heating 5,7,12,14-tetrahydroxy-6,13-dihydropentacene (**182**) with glycerol and sulphuric acid[117]. In a zinc dust melt **183** yields a hexahydro derivative of the conjugate hydrocarbon. With alkaline dithionite **183** gives no vat.

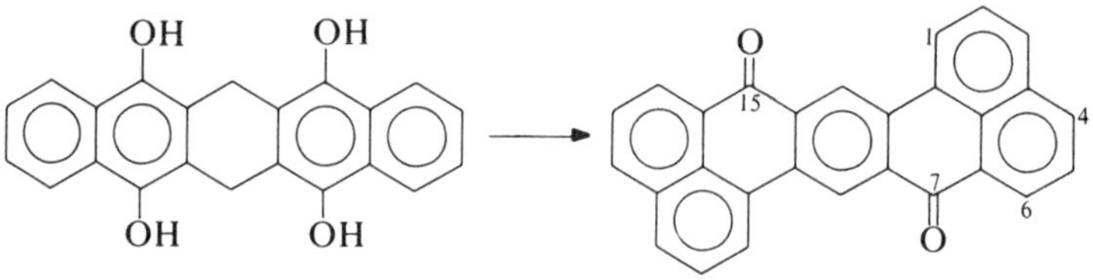

(**182**) (**183**)

3. *Dibenzo*[fg:ij]*pentaphene-15,16-dione*

On oxidation of dibenzo[*fg*:*ij*]pentaphene a quinone is formed which probably possesses the constitution **184**. With hydrazine it forms an azine[118].

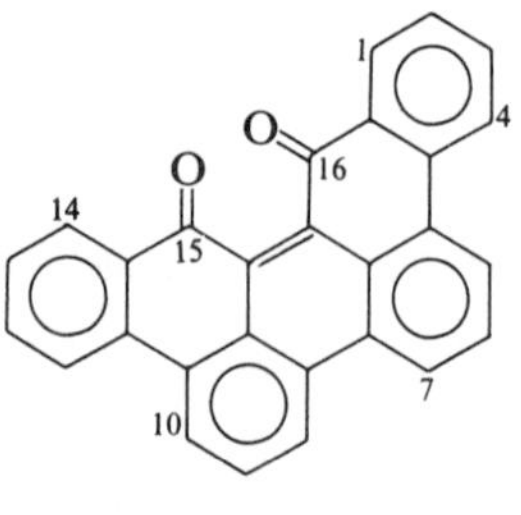

(**184**)

4. *Dibenzo*[h:rst]*pentaphene-5,10-dione*

On fusion of 9,10-dibenzoylanthracene (**186**) with aluminium chloride a red violet or bluish condensation product is formed[119]. The constitution is not clear. It may be the dibenzo[*h*:*rst*]pentaphene-5,10-dione (**187**) or the dibenzo [*hi*:*st*]pentacene-8,16-dione (**185**) or a mixture of both and yet other structures.

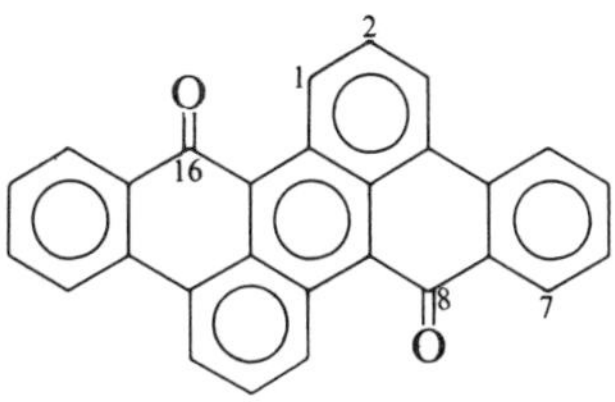

(**185**) 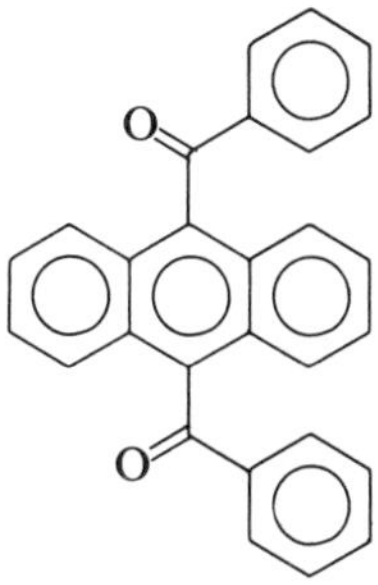 (**186**) 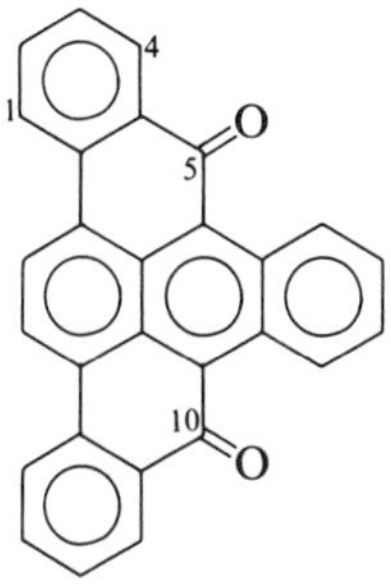 (**187**)

5. *Dibenzo*[b:tuv]*picene-9,16-dione*

The dibenzo[*b*:*tuv*]picene-9,16-dione (**188**) is formed by Scholl cyclization of 3-(1-naphthoyl)-benzanthrone in an aluminium chloride melt[120]. The position of the ring-closure is marked by a dotted line. **188** dyes cotton in brown-orange hues.

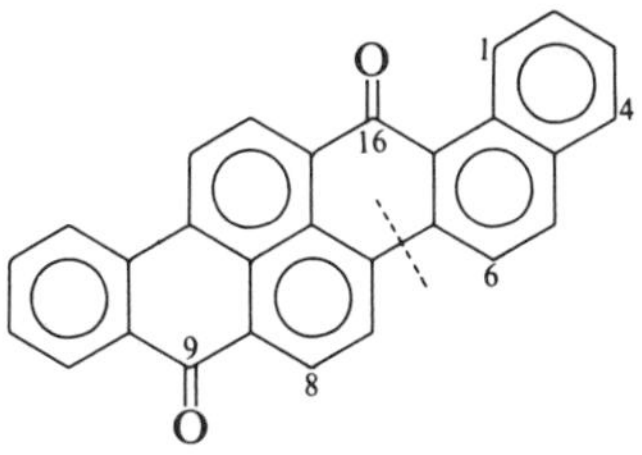

(**188**)

6. *Dibenzo*[a,n]*perylene-5,11-dione*

A compound with the tentative structure of a dibenzo[*a*, *n*] perylene-5,11-dione (**190**) is formed at oxidation of dibenzo[*a*, *n*]perylene (**189**) with selenium dioxide in acetic acid. The quinone forms a blue-green vat with alkaline dithionite[121,122].

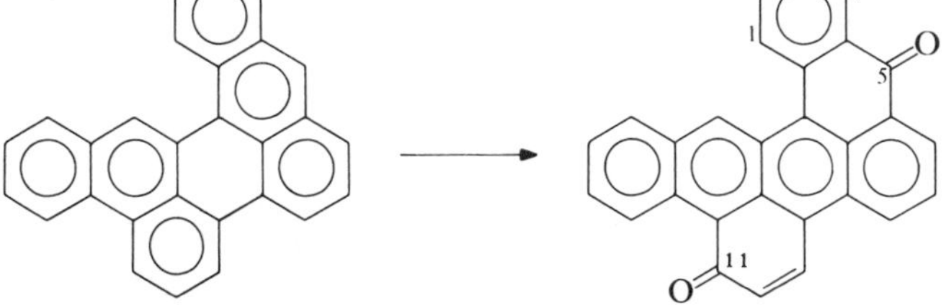

(**189**) (**190**)

7. *Dibenzo*[a,j]*perylene-8,16-dione (hetero-coerdianthrone)* (**193**)

In the older literature the name hetero-coerdianthrone is used for structure **193**. Several syntheses have been described. The simplest seems to be heating of 1-chloroanthrone-(10) (**191**) with zinc chloride and pyridine at 245°C. The first-formed hydroquinone **192** is oxidized to **193** during working up[123].

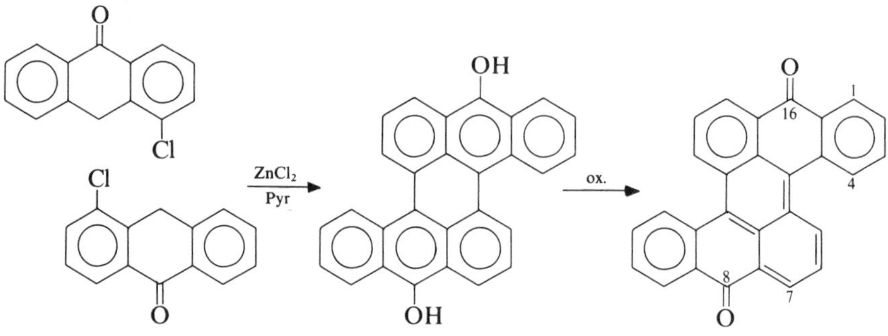

(**191**) (**192**) (**193**)

An interesting synthesis has been described by Scholl and coworkers starting with anthraquinone-1,5-dicarboxylic acid. The dichloride reacts with aluminium chloride in benzene to the dilactone **194**, which can be reduced with hydroiodic acid and phosphorous to 9,10-diphenylanthracene-1,5-dicarboxylic acid (**195**). On treatment with sulphuric acid **195** gives **193**[124].

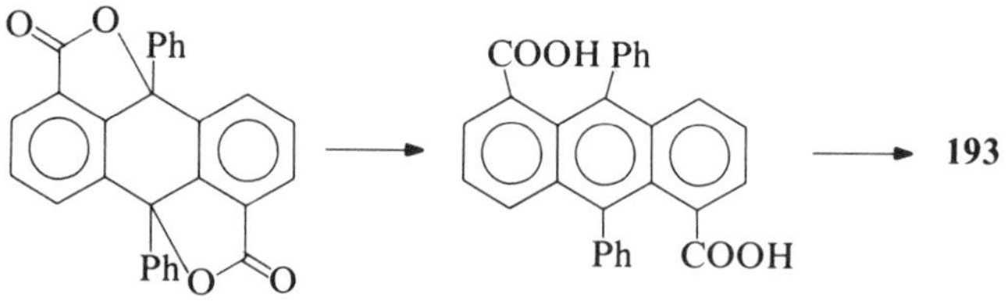

(**194**) (**195**)

Another very simple synthesis is achieved by melting methyleneanthrone (**196**) with aluminium chloride. In the first step the endocyclic ring system **197** is formed which, on heating with aluminium chloride or alone, splits off ethylene in a retro-Diels–Alder reaction and forms **193**[124].

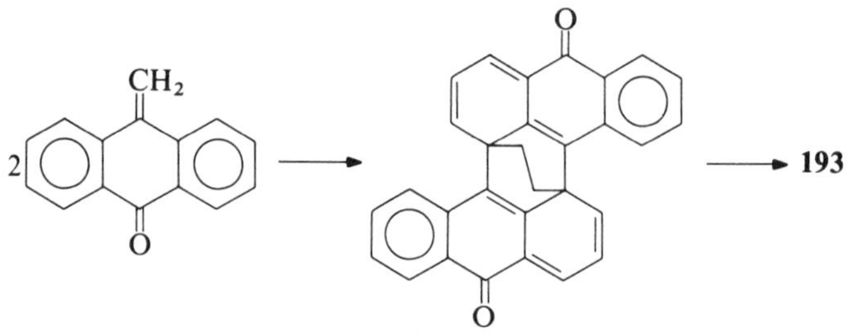

(**196**) (**197**)

The Scholl cyclization was also used for the synthesis of **193**, starting with 1,5-dibenzoylanthracene. As oxidant in the aluminium chloride melt manganese dioxide was used[125].

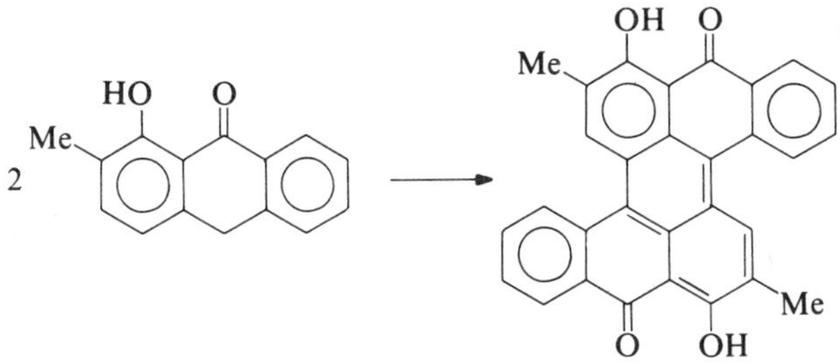

(198) **(199)**

The described syntheses have also been used to prepare some derivatives of **193**[126]. **199** was synthesized by mild oxidation of 1-hydroxy-2-methyl-9-anthrone (**198**) with iodine in pyridine[127]. **193** is easily reduced to the corresponding hydrocarbon with zinc in pyridine/acetic acid[123], when it forms in organic solvents a red-violet solution with strong red fluorescence. On irradiation with visible light in the presence of oxygen the extraordinary thermally stable 4*b*,12*b*-endoperoxide **200** is formed[128, 129]. This is split

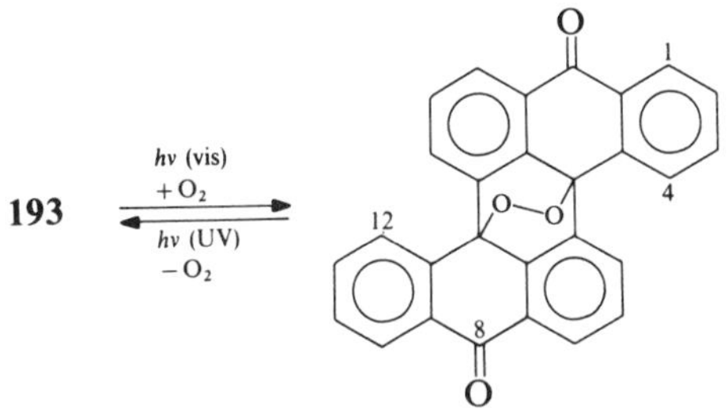

(200)

with light of wavelength between 248 and 334 nm to the parent compounds[129]. These high-quantum-yield processes are very selective with only one subordinate side-reaction (quantum yield 0.005)[130]. The system is proposed as a new reusable liquid chemical actinometer in the UV region[129–131]. The quantum yield for the splitting reaction in the mentioned wavelength range is nearly wavelength-independent. The system has high reproducibility and accuracy and can be used without loss of accuracy in more than 100 repeated actinometric cycles[129]. The kinetics of the self-sensitized photo-oxidation with visible light[131] and the photolysis of the *endo* peroxide[132] have been investigated.

A further derivative of **193**, 5,8,13,15-tetrahydroxy-7,16-dibenzo[*f*, *n*]perylene-7,16-dione (**201**), is formed (29%) at prolonged heating of an alkaline solution of 1,3-dihydroxy-9,10-anthraquinone under nitrogen in the presence of hydroquinone[133].

8. *Dibenzo*[a, o]*perylene-7,16-dione (helianthrone, ms-benzodianthrone)* (**203**)

203 is formed easily by reduction of 1,1′-dianthraquinonyl (**202**) with copper powder in concentrated sulphuric acid. Other reducing agents, such as zinc/acetic acid, tin(II)

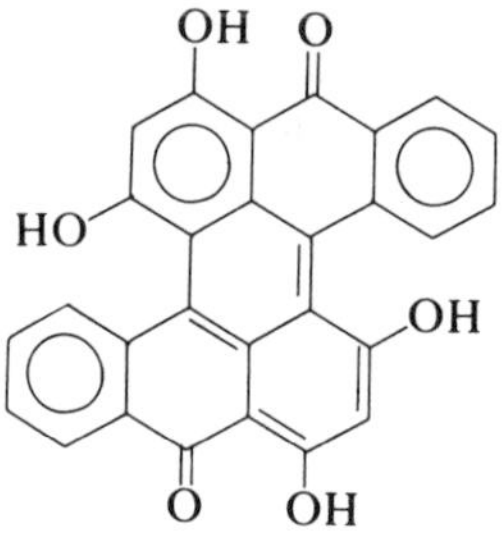

(**201**)

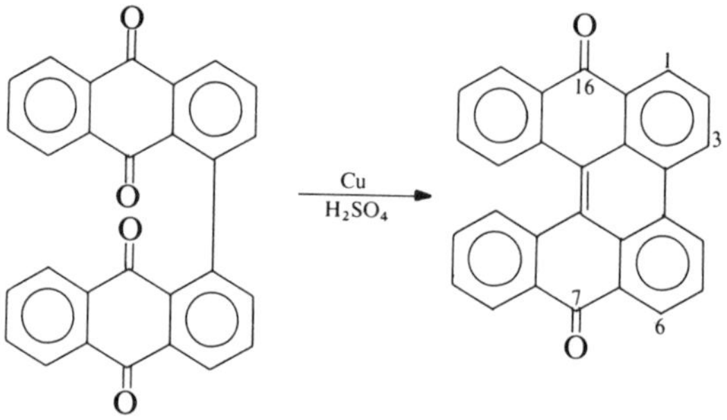

(**202**) (**203**)

chloride/alcoholic hydrochloric acid, and zinc in a zinc chloride melt or with alcoholic potassium hydroxide, can be used[134]. The formation of **203** is obviously facilitated by the proximity of the anthraquinone nuclei in **202** and its derivatives. It seems especially remarkable that the two helianthrone homologues **203a** and **203b** are not stable. They revert to **202a** and **202b**, respectively, even in the solid state[135].

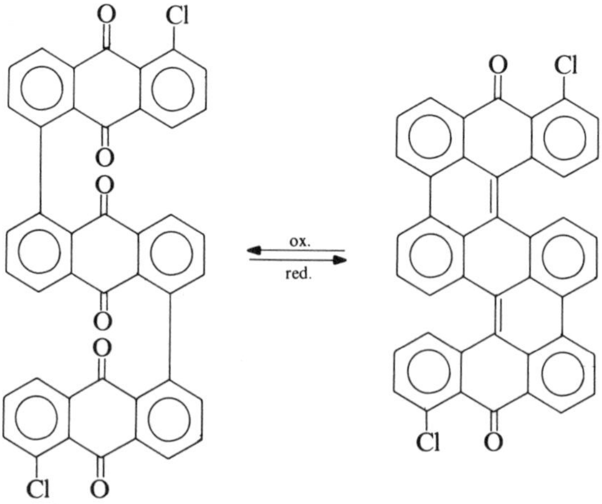

(**202a**) (**203a**)

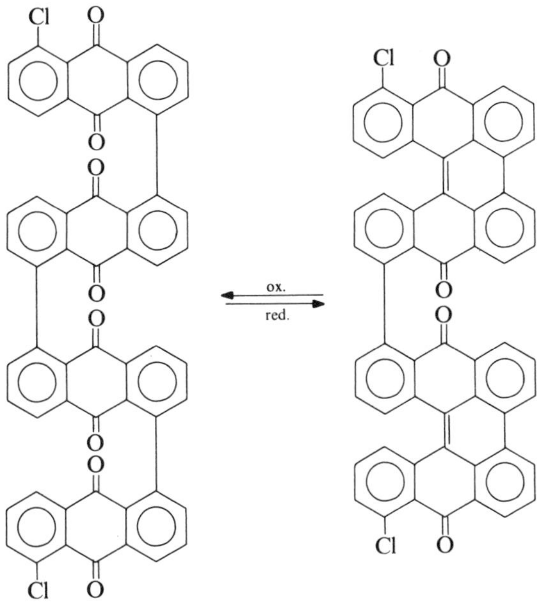

(202b) ox. / red. **(203b)**

Alkyl, carboxy[136], chloro[137], and hydroxy derivatives[138] have been synthesized by reductive cyclization. Hydroxy derivatives of **203**, e.g., **205**, can also be obtained by reduction of hydroxyanthraquinones, directly[139] or by oxidation of hydroxyanthrones[140].

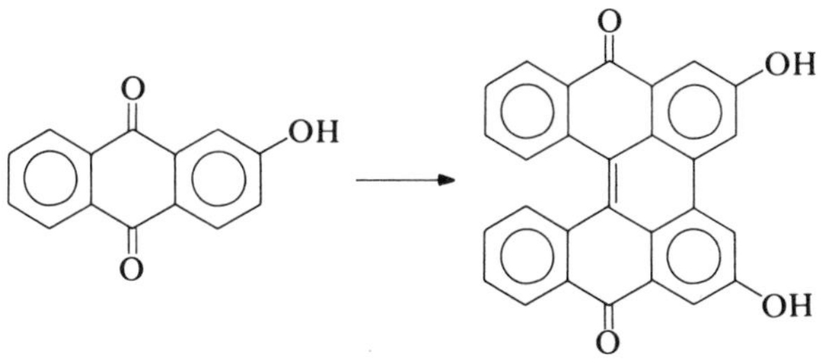

(204) **(205)**

An elegant synthetic access to helianthrones is the irradiation of dehydrodianthrones. On irradiation in the presence of oxygen **206** forms the hydroquinone **208**. This is oxidized *in situ* to the quinone **203**[141]. The irreversible photochemical primary process is a ring connection to **207** in the 4,4′-position starting from the S_1 state. The aromatization under formation of the hydroquinone is a solvent-dependent secondary reaction. Below -130°C the triplet state seems to be involved in the primary reaction[142].

The results of Pariser–Parr–Pople calculations suggest that the intermediate **207** is the most probable structure of the green photochromic form of dehydrodianthrone (**206**)[143]. Finely grounded **203** yields with alkaline dithionite a green vat.

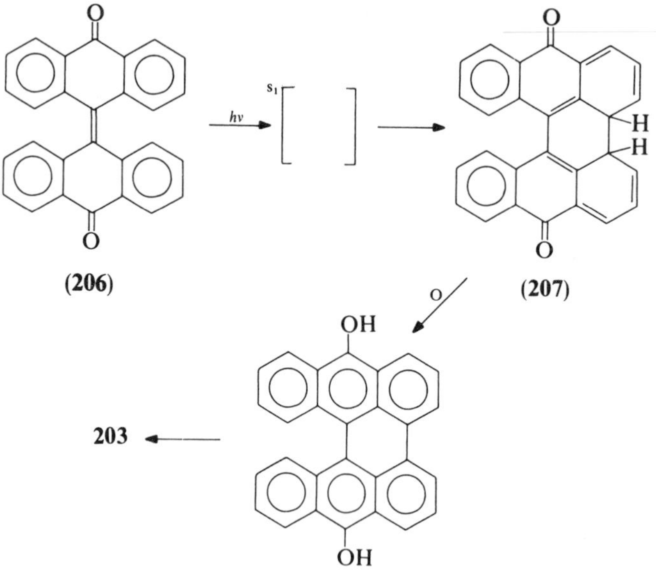

9. *Naphtho[1,2,3-*rst*]pentaphene-5,8-dione*

The 3,10-dimethyl- and 3,10-dimethoxy-naphtho[1,2,3-*rst*]pentaphene-5,8-dione (**211a** and **b**) are accessible by Friedel–Crafts reaction of anthraquinone-1,4-dicarboxylic acid with toluene or anisole to the dilactone **209**. **209** can be reduced by hydroiodic acid to **210**, which gives with concentrated sulphuric acid a double ring-closure to **211**[144]. **211a** is a blue and **211b** a green vat.

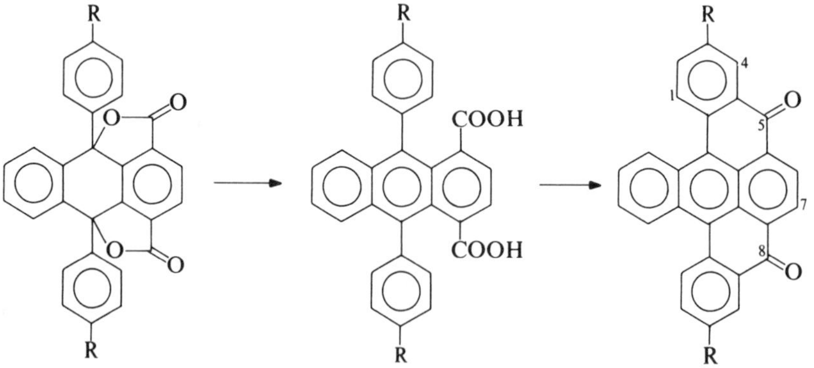

a: R = Me; **b**: R = OMe

10. Dibenzo[cd,lm]perylenediones (peropyrenequinones)

Dibenzo[*cd, lm*]perylene-3,8-dione (**213**) is formed by melting 1-phenalenone (**212**) with methanolic potassium hydroxide[145]. It is possible that the 1,8-dione **214** is also formed. **213** (or **214**) is a purple-red vat dye. Several derivatives have been prepared by

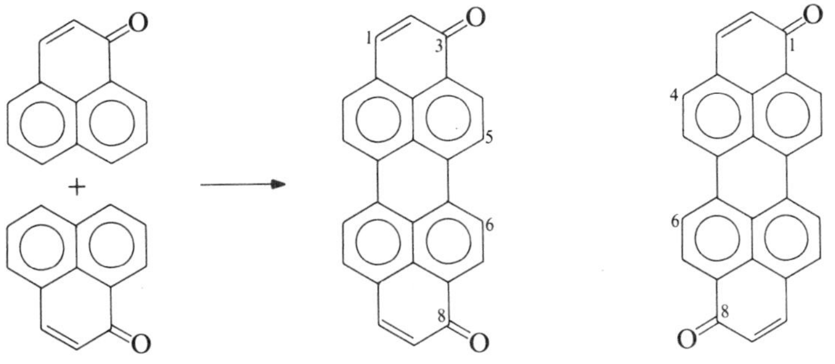

(**212**) (**213**) (**214**)

using substituted 1-phenalenones[146]. The corresponding hydrocarbon is sometimes also named peropyrene, but the IUPAC nomenclature is preferred.

An unequivocal two-step synthesis of the dibenzo[*cd, lm*]perylene-1,8-dione (**214**) started with 1-phenalenone (**212**) which, on heating with benzoyl chloride, gave 3,10-dibenzoyloxydibenzo[*cd, lm*]perylene (**215**). Saponification of **215** with concentrated sulphuric acid yielded the quinone **214**[147].

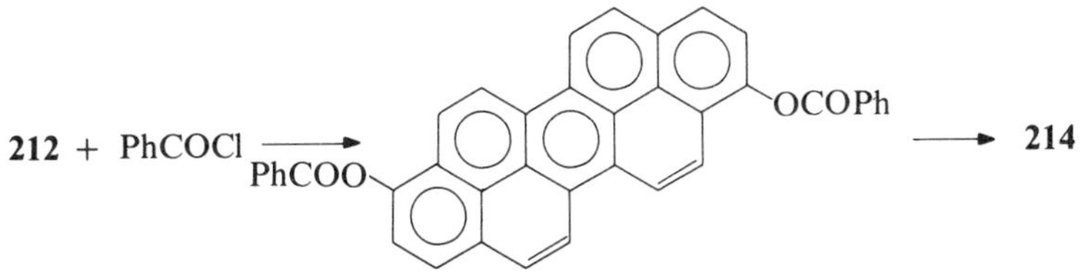

212 + PhCOCl → (**215**) → **214**

The 3,10-diamino derivative (**218**) of the 1,8-dione **214** has been obtained by heating 3,9-diacetyl-4,10-dichloroperylene (**216**) with copper(I) cyanide in quinoline. The dinitrile **217**

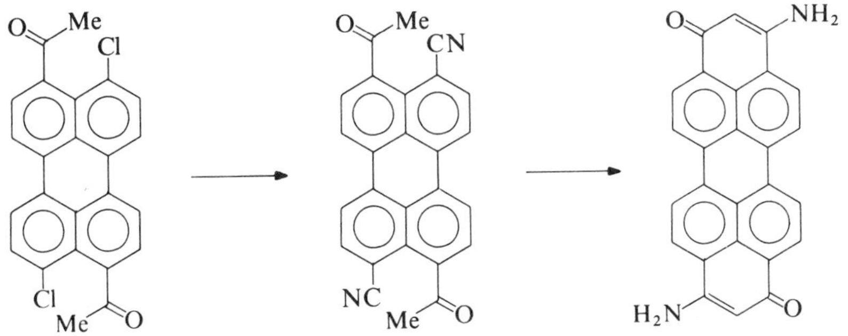

(**216**) (**217**) (**218**)

should be an intermediate. The reaction is also possible with other acyl groups in 3,9-positions[148].

3,3′-Biphenalene-1,1′-dione (**219**) gives, on Scholl cyclization, the dibenzo [*cd*, *lm*]perylene-4,7-dione (**220**). The hydroquinone of **220** is formed readily on reduction. Several ethers of the hydroquinone and the diacetate have been described[149].

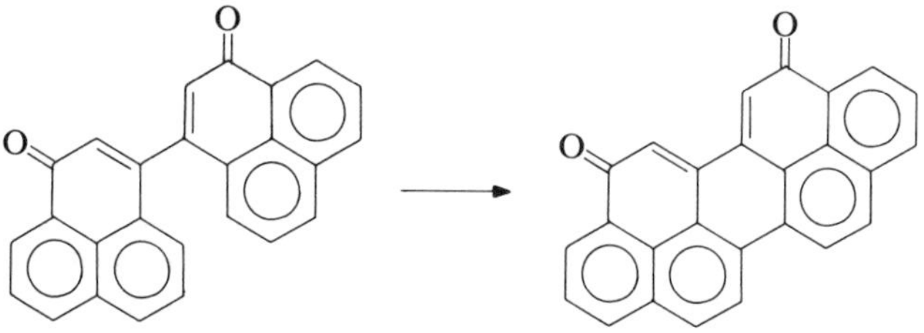

(**219**) (**220**)

11. *Tribenzo*[a:de:mn]*naphthacene-5,9-dione (***221***)*

The synthesis of **221** involved a Diels–Alder condensation between 1-phenalenon and methyleneanthrone in boiling nitrobenzene[10d]. Reduction of **221** in a zinc powder melt leads to the corresponding aromatic hydrocarbon, showing the instability of the central double bonds in naphthacene. Nevertheless it is possible to generate an unstable green vat from **221** with alkaline dithionite.

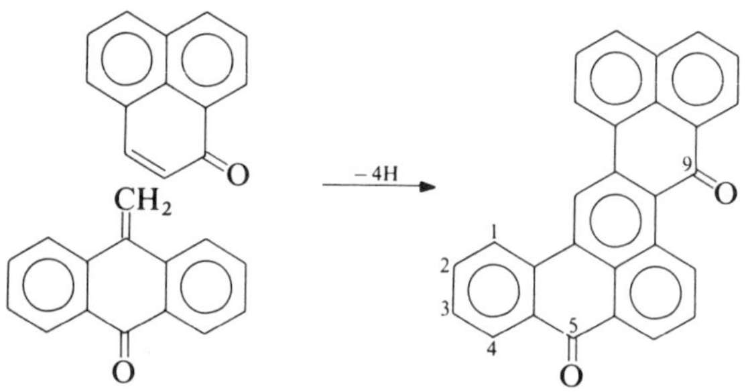

CH_2 — 4H

(**221**)

H. Eight-ring Quinones

The most important eight-ring quinones which are derived from pyranthrene and *meso*-naphthodianthrene are treated first. The others follow in alphabetical order.

1. *Pyranthrenediones (pyranthrones)*

Two pyranthrenediones, the 8,16- and 5,13-isomers, are known. Only the 8,16-dione, also named pyranthrone, and some of its derivatives are of technical importance as fast vat dyes.

Pyranthrone (**223**) is prepared commercially by a Knoevenagel-type double ring-closure of 2,2′-dimethyl-1-1′-bianthraquinonyl (**222**), which is produced by Ullmann reaction of 1-chloro-2-methylanthraquinone[150]. The ring-closure is effected by alkali, e.g. by heating

222 at 105°C with potassium hydroxide/isobutanol, but it is also possible by heating **222** alone, with zinc chloride or with potassium hydroxide/sodium acetate. Some process improvements have been described in patents[151].

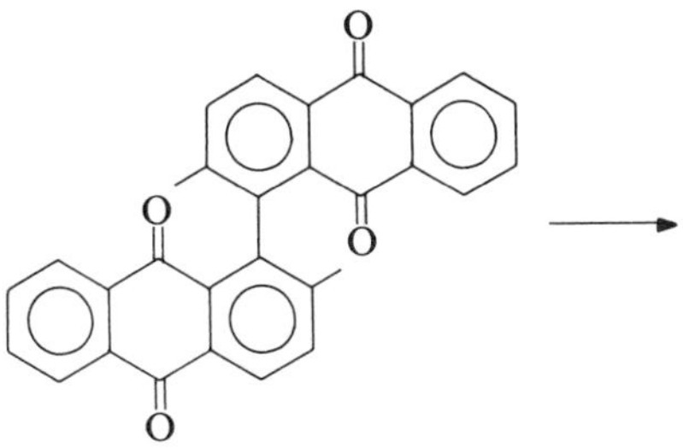

(222) 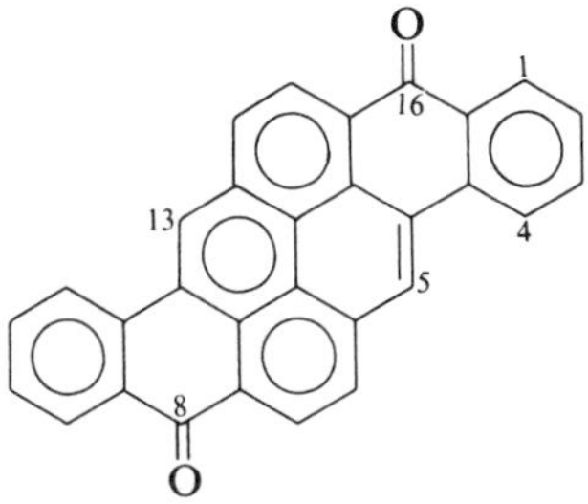 **(223)**

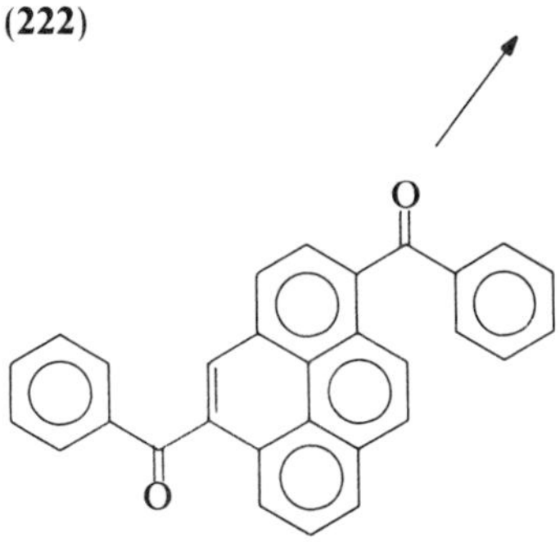

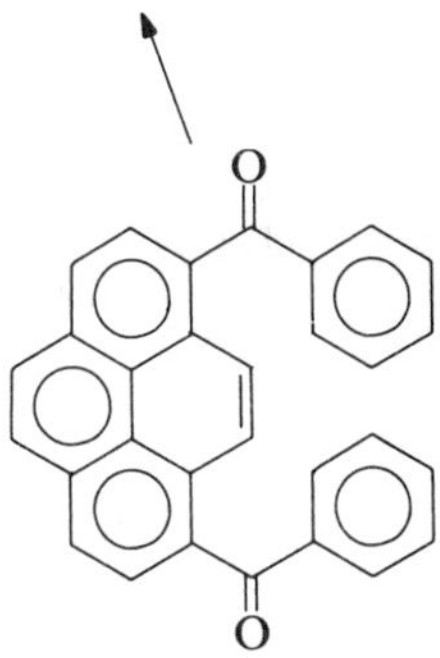

(224) **(225)**

Scholl cyclization of 1,6-dibenzoylpyrene (**224**) also yields pyranthrone (**223**) (85%)[19,152] which is formed in addition from 3,6-dibenzoylpyrene (**225**), but the yield is lower (30%) and decomposition is observed[19]. This result shows that rearrangements of the aryl ketones can occur under the reaction conditions and that the course of Scholl cyclization is not always unambiguous.

It is possible to synthesize pyranthrone also in a one-step reaction of pyrene, aluminium chloride and benzoyl chloride[153]. 1,6-Dibenzoyl-3,8-pyrenequinone (**226**) on melting with aluminium chloride/sodium chloride forms 6,14-dihydroxypyranthrone (**227**)[19].

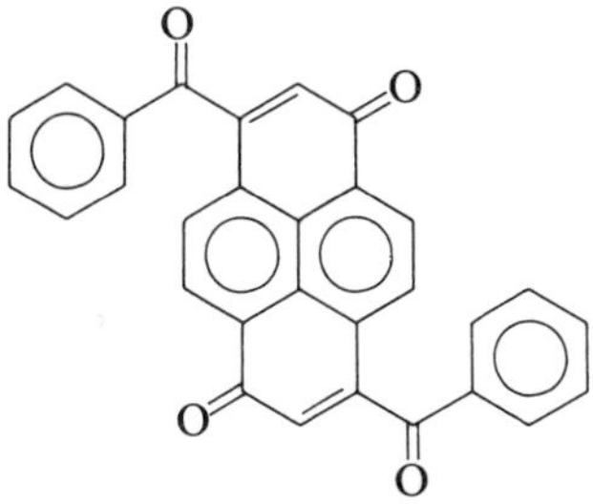

(226) 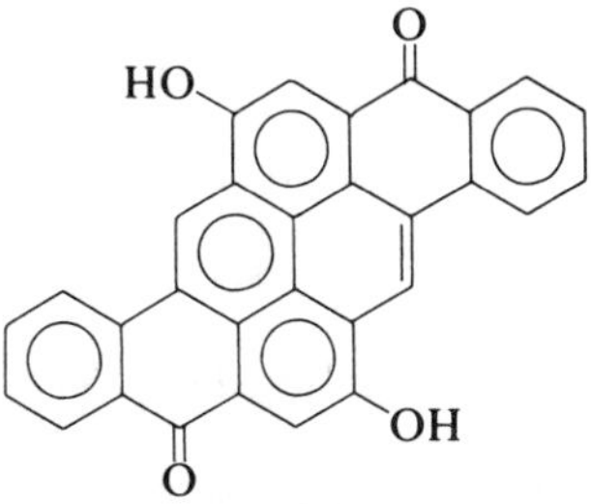 **(227)**

Pyranthrone forms a purple red vat and the salt of tetrahydropyranthrone is formed on reduction of **223** with alkaline dithionite[154].

Since pyranthrone is a valuable vat dye (Indanthren Goldorange G, Caledon Goldorange-G), there were many attempts to synthesize its derivatives. Alkyl homologues have been synthesized starting from the corresponding 1,1′-bianthraquinonyls[155 156] and 1,3-diarylpyranthrones are obtained by reduction of 2,2′-diaroyl-1,1′-bianthraquinonyls. Pyranthrone itself is formed from the 1,1′-bianthraquinonyl-2,2′-aldehyde already at vatting[157]. Halogen derivatives of **223** can be obtained by synthesis[158] or direct halogenation[159]. It is also possible to introduce nitro groups by mixed acid or with nitric acid in nitrobenzene[160]. A dibenzoylpyranthrone is obtained by heating tetrachlorotetrabenzoylpyrene with potassium hydroxide in quinoline[161].

4-Bromopyranthrone (Caledon Goldorange-2RTS), 4,7-dibromopyranthrone (Indanthren Orange RRTS, Caledon brilliantorange-4R), and a mixture of a di- and tribromo derivatives (Indanthren Orange 4R) are of technical interest as vat dyes. They are prepared by bromination of pyranthrone in chlorosulphonic acid in the presence of iodine or sulphur[11b]. For the reduction potentials of the pyranthrone dyes see Marshall and Peters[162] and Gupta[163].

On oxidation with chromic acid pyranthrone gives the dicarboxylic acid **228**, which can be reduced with ammonia/zinc to the bianthracene **229**. Heating **229** with zinc chloride or phosphorous pentachloride yields pyranthrene-5,13-dione (**230**)[164]. **230** forms with alkaline dithionite[230] a blue-green vat at room temperature the blue salt of the conjugate hydroquinone, at elevated temperatures the green salt of the tetrahydro derivative **231**[154].

230 can be dihydroxylated directly in the 8,16-position to **232**. The tautomeric 5,13-dihydroxypyranthrone **233** is obtained by hydroxylation of pyranthrone[165].

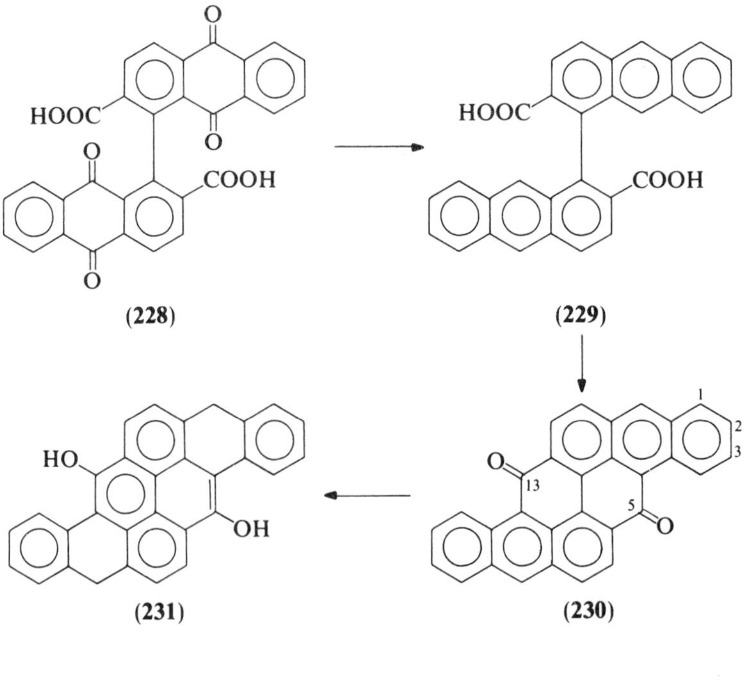

HOOC O O COOH O O

(228)

HOOC COOH

(229)

HO OH

(231)

1 2 3 O 13 5 O

(230)

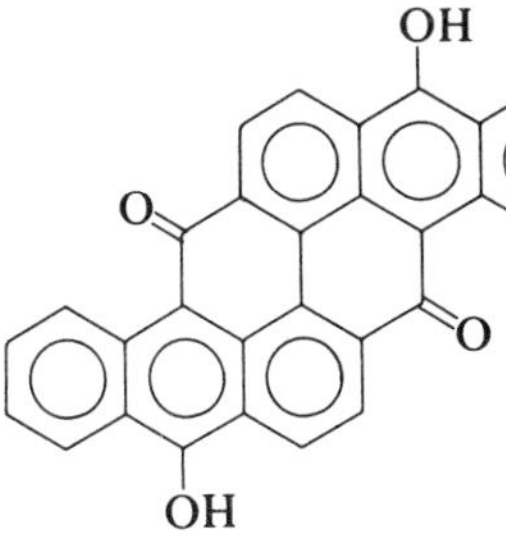

(232) 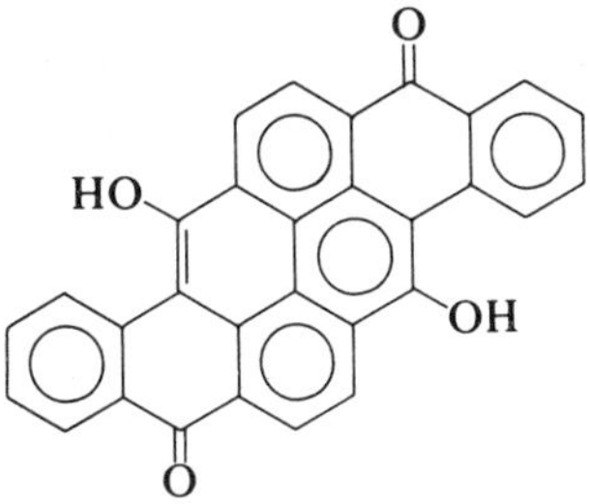 (233)

2. *Phenanthro[1,10,9,8-*opqra*]perylene-7,14-dione(*meso*-naphthodianthrone) (235)*

235 and its derivatives are synthesized mostly from helianthrones (**236**) or dehydrodianthrone (**234**) and the corresponding derivatives, respectively. The ease of formation of **235** from **234** and **236** seems to be influenced by the close proximity of the two anthraquinone ring systems. **236** is converted to **235** by Scholl cyclization or by oxidation with chromic acid in sulphuric acid[166]. Irradiation in the presence of oxygen also converts **236** and even **234** to **235**[141, 167].

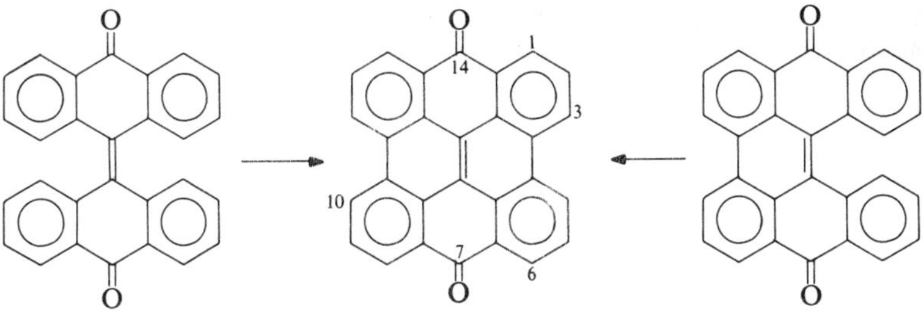

(234) (235) (236)

235 is also formed in a one-step synthesis by irradiation of 9-bromo-, 9,9-dichloro-, or 9,9-dibromo-anthrone (**237**). On short exposure to light the 9,9′-bianthrones are formed, in the case of **237** 9,9′-dibromodianthrone (**238**) via dimerization of 10-anthryloxy radicals involving triplet states, followed by the conversion of the diantbrone **238** to *meso*-naphthodianthrone (**235**) at prolonged irradiation[168, 169].

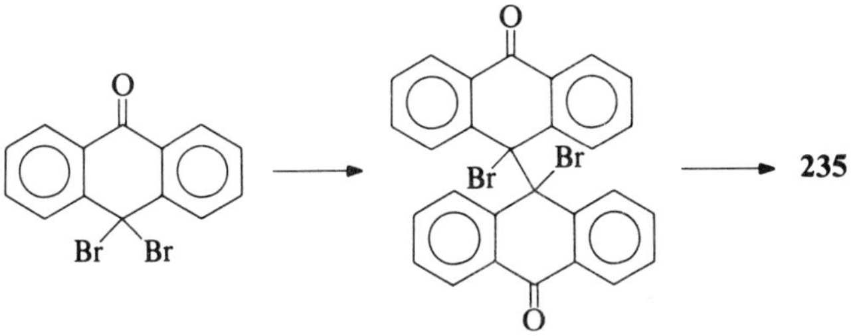

(237) (238) → 235

Formation of **235** from 1,1′-dianthraquinonyl on zinc powder distillation should also be mentioned[170]. **235** yields *meso*-naphthodianthrene on reduction with zinc in pyridine/acetic acid[171]. With alkaline dithionite **235** forms only a vat dye if zinc powder is added.

Finally, it is of interest that thin films of **235** exhibit semiconductivity[172] with two activation energies at 0.74 and 0.43 eV. In the presence of oxygen the conductivity associated with the lower activation energy is decreased[173].

Halogen derivatives of **235** have been synthesized mostly via Scholl cyclization of halogenated helianthrones[137, 174]. Another route includes photochemical reactions; however, if the positions in bianthrone (**234**) or helianthrone (**236**) where the photochemical cyclization has to take place are occupied by chlorine atoms, no reaction occurs on irradiation in organic solvents. On irradiation in concentrated sulphuric acid, even in these cases ring-closure to *meso*-naphthodianthrones is observed under elimination of hydrochloric acid[135]. Thus, 1,4,5,8,10,15-hexachlorohelianthrone (**239**) is unchanged, e.g.

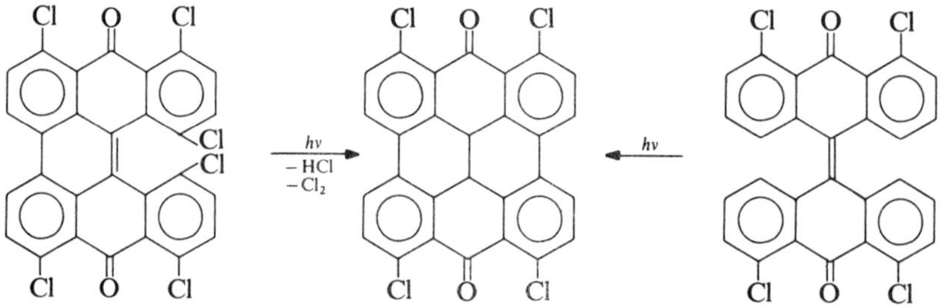

(**239**) $\xrightarrow[-HCl,\ -Cl_2]{h\nu}$ (**240**) $\xleftarrow{h\nu}$ (**241**)

(**242**) $\xrightarrow[H_2SO_4]{Cu}$ (**243**) $\xrightarrow{h\nu}$ (**244**)

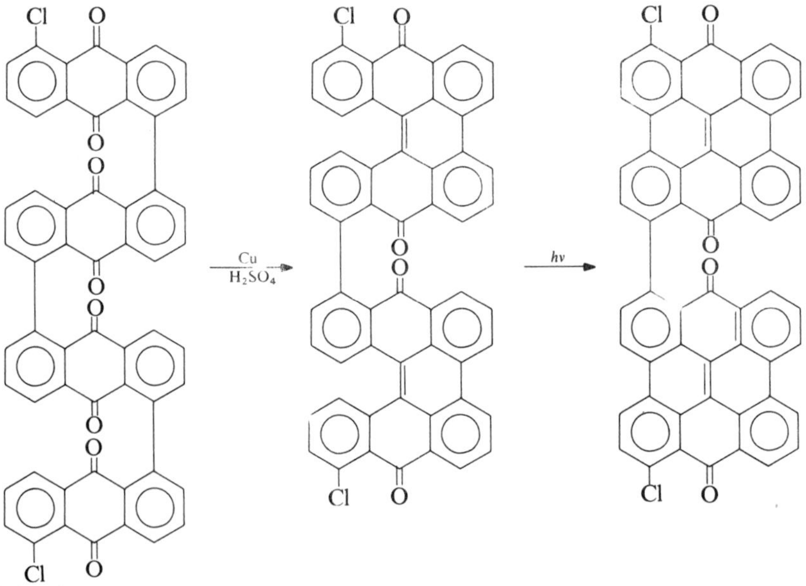

by irradiation in nitrobenzene, but gives 1,6,8,13-tetrachloro-*meso*-naphthodianthrone (**240**), eliminating hydrochloric acid and subsequently chlorine on irradiation in sulphuric acid.

Irradiation of the tetrachlorobianthrone **241** also leads to **240**. Many other derivatives of **235** are known, e.g. methyl[136,175,176] and hydroxy derivatives[177] as well as carboxylic acids[136].

The dimeric *meso*-naphthodianthrone **244** has been synthesized by reduction of the tetra-anthraquinonyl **242** with powdered copper in sulphuric acid to the dimeric helianthrone **243**, followed by irradiation of **243** to give by further cyclization **244**[135].

Derivatives of special interest are hypericine (**248**) and pseudohypericine (**248a**), which occur in nature[178]. A mixture of both are found in *Hypericum perforatum* (St. John's-wort),

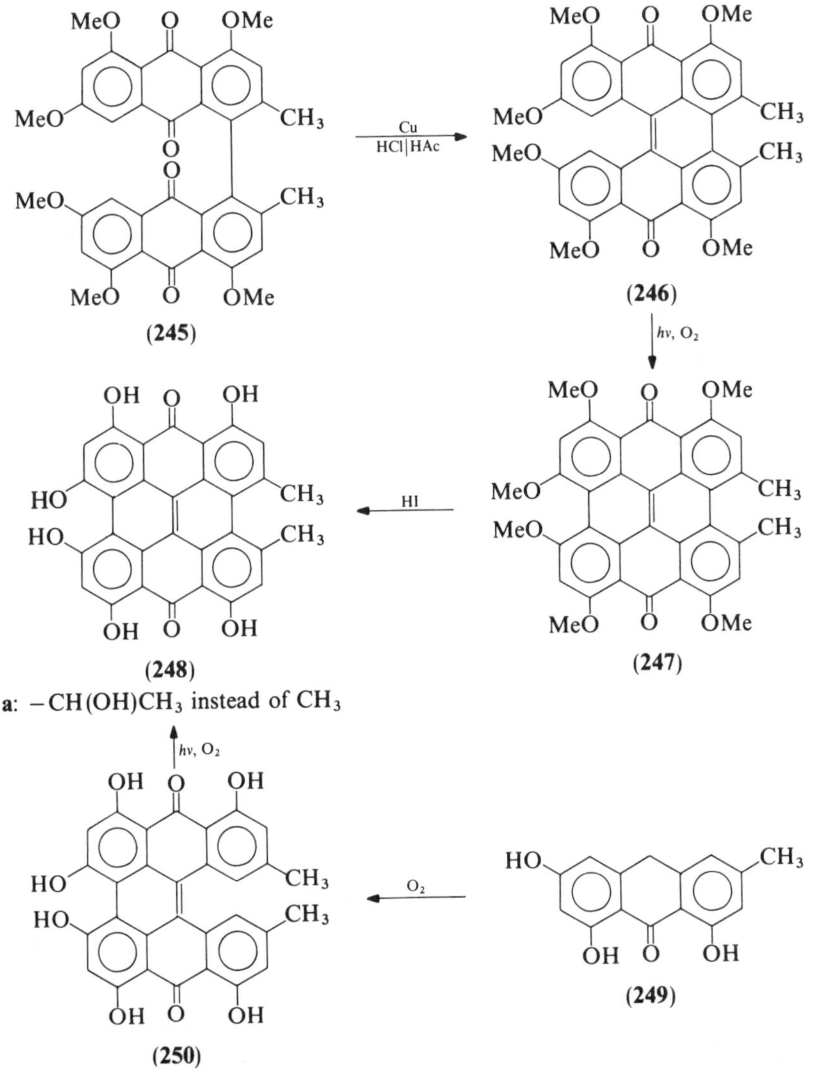

(**245**)

Cu, HCl|HAc

(**246**)

hν, O_2

(**247**)

HI

(**248**)

a: $-CH(OH)CH_3$ instead of CH_3

hν, O_2

(**250**)

O_2

(**249**)

(251)

250

(248)

b: H instead of CH_3
c: COOH instead of CH_3

the former in *Hypericum hirsutum*. From St. John's-wort the red quinones may be set free by pressing and rubbing the buds or yellow blossoms between two fingers. The hypericines are responsible for a light illness of sheep, cows, goats and horses with fair coat that have eaten St. John's-wort (hypericism).

Two main approaches have been used for the synthesis of hypericine and related compounds, the 1,1′-dianthraquinonyl route and the dianthrone route, respectively. In the first case 4,5,7,4′,5′,7′-tetramethoxy-2,2′-dimethyldianthraquinonyl-(1,1′) (**245**) was heated with copper powder in acetic acid/hydrochloric acid to yield the corresponding helianthrone derivative **246**. The third ring connection to yield **247** could be effected by irradiation with visible light in the presence of oxygen. In the absence of oxygen a second helianthrone molecule serves as an oxidant and is reduced to the conjugate hydroquinone. The last step was ether cleavage by heating with potassium iodide in phosphoric acid. A similar synthesis of 2,2′-dimethyl-*meso*-naphthodianthrone was possible. The second, biomimetic route starts from emodine-9-anthrone (**249**), which is oxidized by air to give the helianthrone **250** directly via the dianthrone. The latter is also a precursor of hypericine in the plant and was named protohypericine. The ring-closure to hypericine was effected again by photochemical dehydrogenation.

In further biomimetic synthesis[179], hypericine (**248**) is formed (29%) on prolonged heating of an alkaline solution of emodine (**251**) under nitrogen and in the presence of hydroquinone. Similar yields are obtained for the syntheses of bisdesmethylhypericine (**248b**) and hypericine dicarboxylic acid (**248c**).

It was shown[180] that protohypericine (**250**) is formed under these conditions. Ring-closure to **248** occurs on working up by light.

3. *Aceanthryleno[2,1-a]aceanthrylene-5,13-dione (acedianthrone)* (253)

For the sake of brevity, we will use the name acedianthrone. Anthrone can be condensed with glyoxal sulphate in acetic acid or with chloral and tin(II) chloride to 1,2-bis(10-oxo-10*H*-[9]anthrylidene)ethane (**252**)[181,182]. With aluminium chloride in the presence of diluents and oxidants[182,183], or on heating a nitrobenzene solution with benzoyl chloride or other acid chlorides[184], **252** gives cyclization to acedianthrone (**253**).

Several derivatives of **253** have been synthesized[183,184]. Despite two five-membered rings in the molecule, **253** has the properties of a real quinone. With alkaline dithionite it is reduced to a yellow-brown vat, which dyes cotton to a red-brown. **253** can be regarded as stilbenoquinone. The stilbenoquinone **252** and its 4,4′-dichloro derivative (**252a**) are also vat dyes[185].

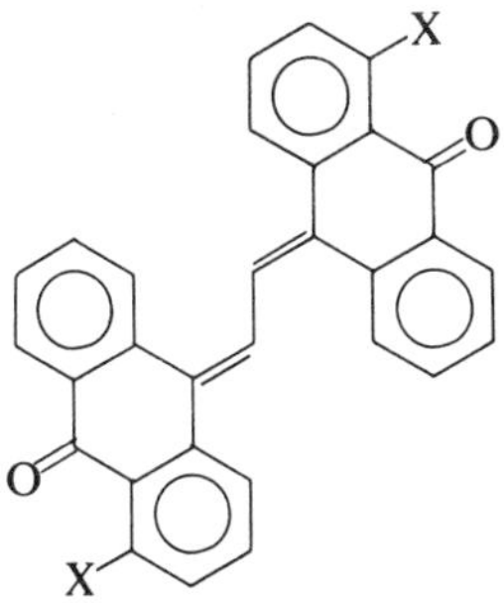

(**252**, X = H)
a: X = Cl

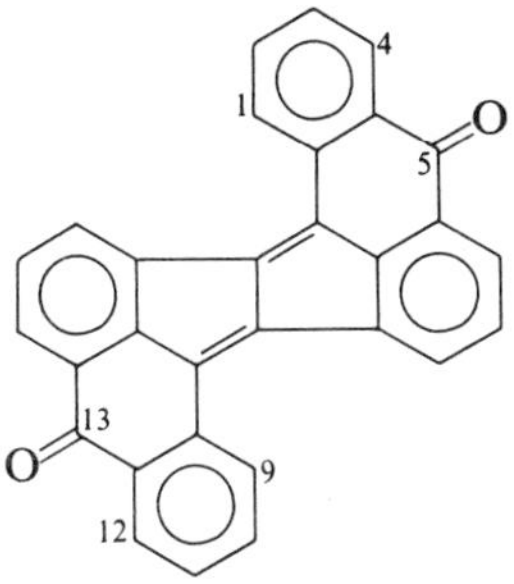

(**253**)

*4. Benzo[3,4]anthraceno[2,1,9,8-*aopqr*]naphthacene-5,11-dione* (**254**)

254 has been prepared by selenium dioxide oxidation of benzo[3,4]anthraceno[2,1,9,8-*aopqr*]naphthacene[186].

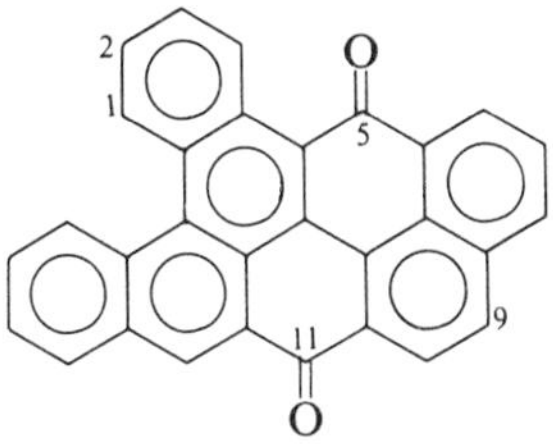

(**254**)

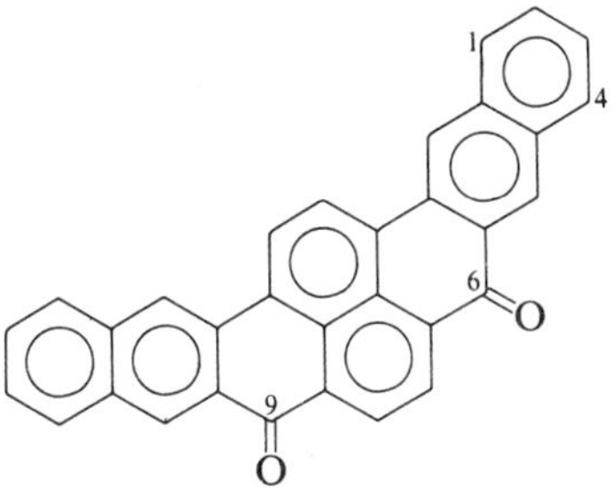

(**255**)

*5. Dibenzo[*b:vwx*]hexaphene-6,9-dione* (**255**)

255 has been prepared by selenium dioxide oxidation of dibenzo[*b*:*vwx*]hexaphene in refluxing nitrobenzene. It gives a violet vat with alkaline dithionite[187].

*6. Dibenzo[*jk:wx*]hexacene-8,17-dione* (**256**)

256 is obtained by Scholl cyclization of 5,11-dibenzoylchrysene. The newly formed bonds are marked with a dotted line in the structure. The blue-green colour of the seems to justify structure **256**[188].

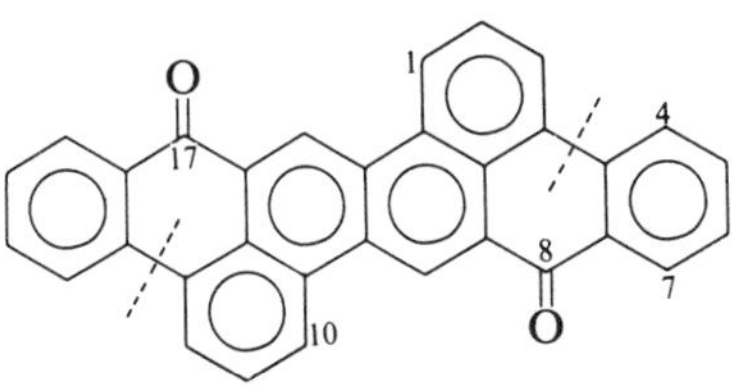

(**256**)

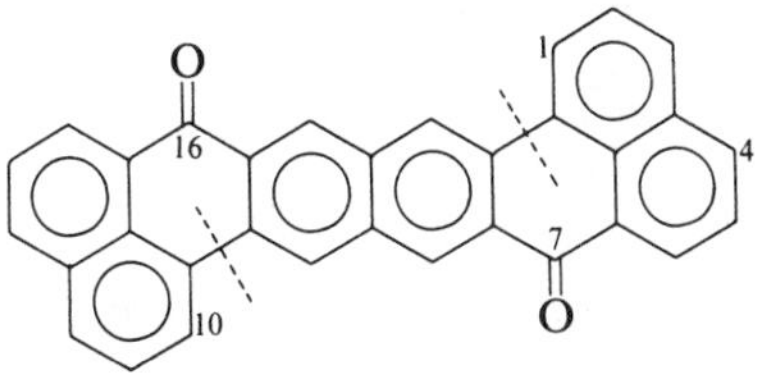

(**257**)

7. *Dibenzo*[lm:yz]*hexacene-7,16-dione (octacethrene-7,16-dione)* (**257**)

257 has been prepared by heating the easily accessible 2,6-bis(1-naphthoyl) naphthalene in a sodium chloride–aluminium chloride melt at 140°C under passing in oxygen (40%)[189]. The bonds formed in this Scholl cyclization are marked with a dotted line in structure **257**. Other quinones could possibly be formed, but the structure **257** has been proved by oxidative degradation.

Zinc powder melt of **257** yielded not the corresponding aromatic hydrocarbon but the 7,16-dihydro derivative.

8. *Dinaphtho*[*1,2,3*-fg:*3,2,1*-op]*naphthacene-9,18-dione* (**259**)

259 could be synthesized by heating the ester of 6,12-diphenylnaphthacene-5,11-dicarboxylic acid (**258**) with concentrated sulphuric acid[190].

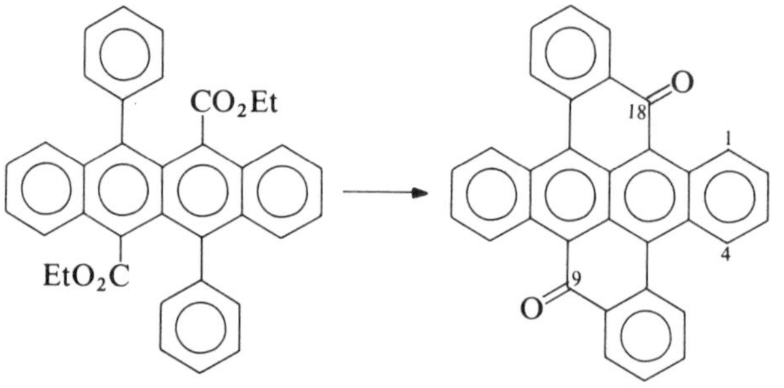

CO_2Et

EtO_2C

(**258**)

(**259**)

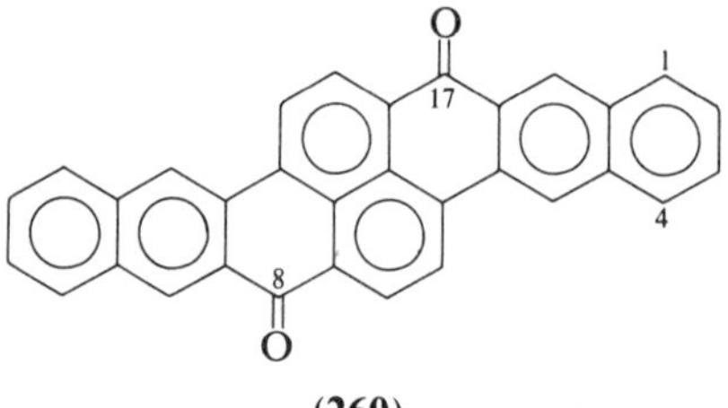

(**260**)

9. *Naphthaceno*[*2,1,12*-aqr]*naphthacene-8,17-dione* (**260**)

260 has been synthesized by selenium dioxide oxidation of naphthaceno(2,1,12-*aqr*)naphthacene[87] and it forms a brownish olive-red vat.

10. *Naphthaceno*[*2,1,12,11*-aopqr]*naphthacene-8,16-dione* (**261**)

Scholl cyclization of 4,9-dibenzoylpyrene gives quickly and almost quantitatively **261**[19, 191]. The formed bonds are marked with a dotted line in the structure. **261** is a bluish-red vat dye.

11. *Tetrabenzo*[a:de:j:mn]*naphthacene-5,14-dione* (**264**)

Condensation of benzanthrone-3-aldehyde (**262**), obtainable from methyleneanthrone and acrolein, with anthrone in pyridine/piperidine gives **263**, which on Scholl cyclization

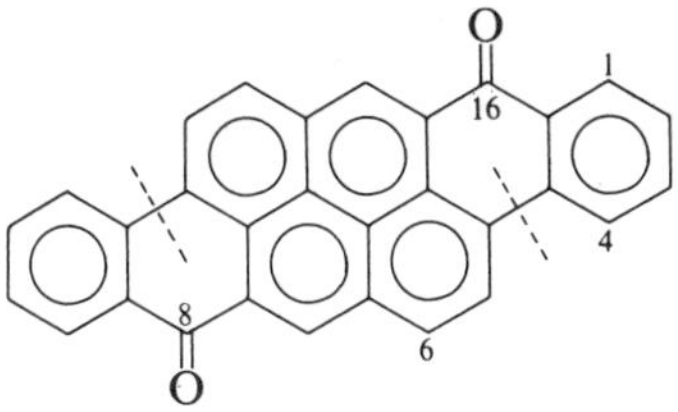

(**261**)

forms **264**[192]. With alkaline dithionite a ruby vat is formed, obviously derived from a tetrahydro derivative of **264**, since judging from similar compounds the vat of **264** should be coloured deep green.

Halogen derivatives of **264** can be prepared starting with halogenated methyleneanthrones or by direct halogenation[192].

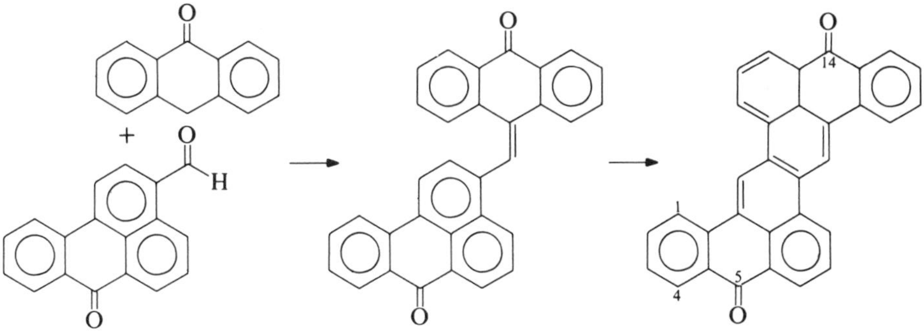

(**262**) (**263**) (**264**)

12. Tribenzo[a:ghi:o]*perylene-7,16-dione* (**266**)

266 is formed by boiling 3,4-dimethylhelianthrone (**265**) with barium hydroxide in nitrobenzene or on melting with a mixture of potassium hydroxide and aniline[193]. **266** forms a vat with alkaline dithionite.

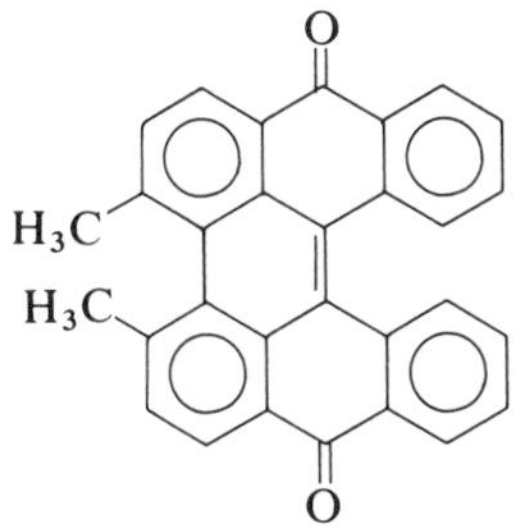

(**265**) 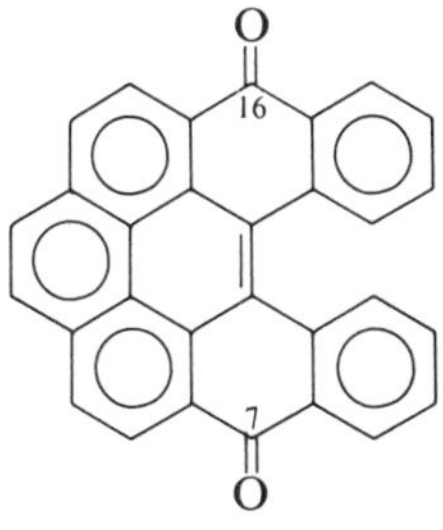 (**266**) 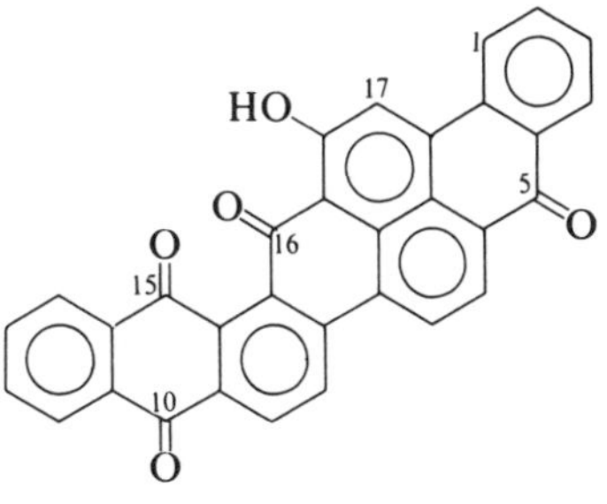 (**267**)

13. Tribenzo[b:n:tu]*picene-5,10,15,16-tetraone*

The 17-hydroxy derivative of tribenzo[*b*:*n*:*tu*]picene-5,10,15,16-tetraone (**267**) is obtained by oxidation of violanthrone. Alkylation of **267** leads to orange-red vat dyes[194].

I. Nine-ring Quinones

The two most important compounds of this group are violanthrone and isoviolanthrone. These are treated first, followed by the others in alphabetical order.

1. Violanthrone (268) and isoviolanthrone (269)

The correct IUPAC notation for **268** is anthra[9,1,2-*cde*]benzo[*rst*]pentaphene-5,10-dione and for **269** benzo[*rst*]phenanthro[10,1,2-*cde*]pentaphene-9,18-dione. In the literature the names violanthrone and isoviolanthrone are used almost exclusively.

268 and **269** are formed formally by symmetrical or unsymmetrical condensations of two benzanthrone molecules. The dotted lines in the structures show the place of connection of the two benzanthrone halves.

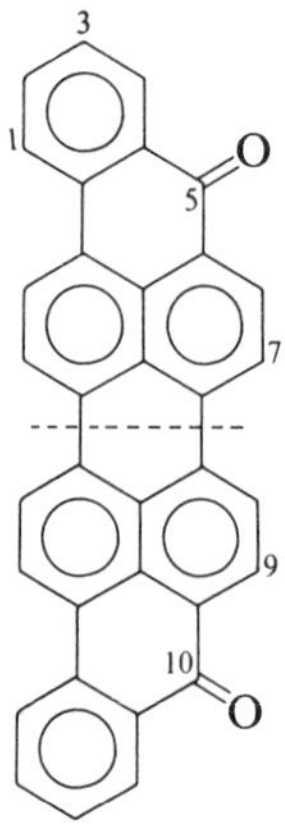

(**268**) 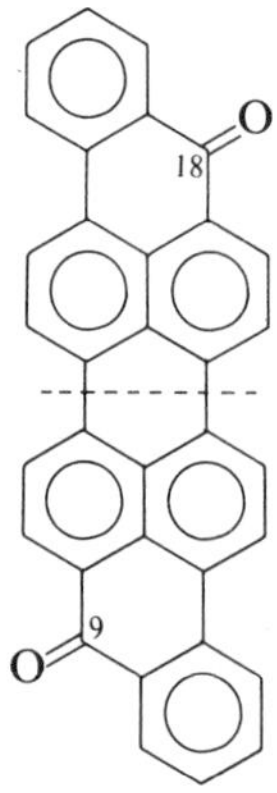 (**269**)

a. Violanthrone

Symmetrical condensation with generation of **268** takes place in potassium hydroxide or alcohol/potassium hydroxide melt of benzanthrone at 230–240°C[195]. 3,3′-Dibenzanthrone (**270**) is an intermediate[196] and is the main product at lower temperatures. Indeed **268** can be synthesized in especially pure form and in 96% yield by heating **270** at 430°C for 15 min[127]. Only small amounts of **268** are formed by Ullmann reaction of 3-halobenzanthrones[198].

The amounts of side products, such as 4-hydroxybenzanthrone or isoviolanthrone, can sometimes be considerable in the potash melt of benzanthrone. Several processes have been developed to decrease the amounts of undesirable side products or to purify the intermediate **270** or the crude violanthrone[199,200]. Effective and high-yield syntheses from benzanthrone are possible by adding to the melt, chlorates or nitrites, glycol- or polyglycol-ether and surfactants[201,202].

Another interesting but technically unimportant synthesis of **268** is by heating 4,4′-dibenzoyl-1,1′-binaphthyl (**271**) with aluminium chloride[152]. **271** is easily obtained by Friedel–Crafts reaction of benzoyl chloride with 1,1′-binaphthyl.

Violanthrone is a blue compound which dyes cotton to a fast violet. It is an important vat dye and is marketed as Indanthrene Dark Blue BOA. For the redox potential of violanthrone see Gupta[163]. Violanthrone as well as its corresponding aromatic hydrocarbon are semiconductors[172,203].

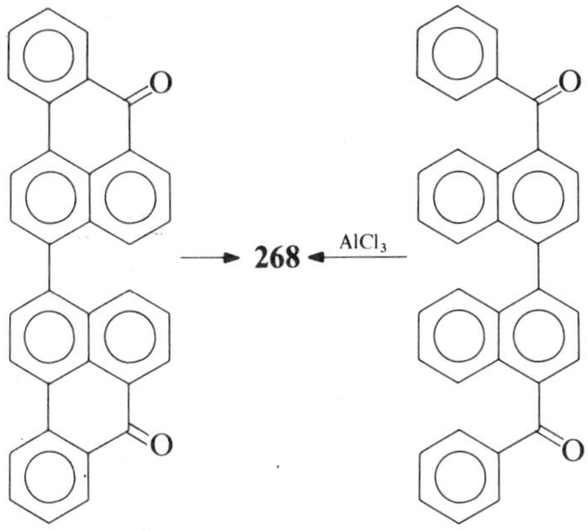

→ 268 ← $AlCl_3$

(270) (271)

Many derivatives of **268** have been synthesized in order to obtain other hues such as 3,12,16,17-tetrachloroviolanthrone (Indanthren Navy Blue RB), a monobromo (Indanthren Navy Blue BRF) and a dibromo derivative (Indanthrene Navy Blue BF). The halogen derivatives are prepared by direct halogenation of violanthrone in chlorosulphonic acid or in organic solvents, sometimes in the presence of sulphur or antimony[204]. 15,16,17-Trichloroviolanthrone is obtained, besides other products, by treating 16,17-dihydroxyviolanthrone with phosphorous pentachloride[205]. Nitro derivatives can be prepared by direct nitration of violanthrone[206]. The dinitro compound can be reduced to a green diamino compound, which in turn yields on oxidation with hypochlorite on the fibre a valuable black[207].

Of technical importance is the oxidation of violanthrone with manganese dioxide in sulphuric acid in the presence of boric acid[208]. Violanthrone-16,17-dione (**272**) is formed and, on reduction, gives 16,17-dihydroxyviolanthrone and, after methylation, 16,17-dimethoxyviolanthrone (**273**). **273** is a valuable green vat dye (Indanthren Brilliant Green FFB, Caledon Yade Green).

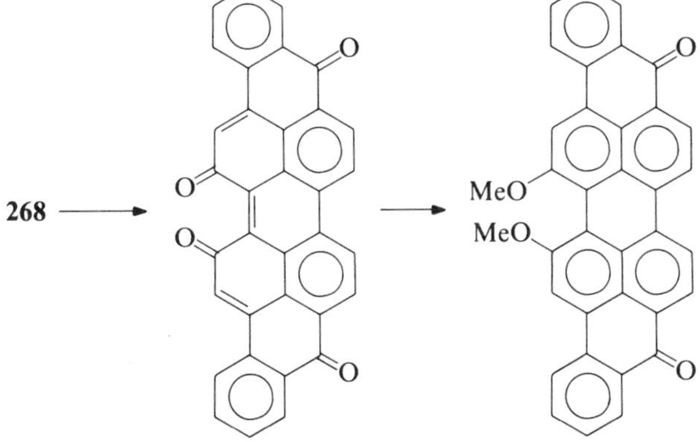

268 → →

MeO MeO

(272) (273)

Syntheses of homologues[209], of 6,9-diphenylviolanthrone[210], and of hydroxy and alkoxy derivatives[211] have been described. 3-(3-Benzanthranyl)violanthrone (**277**) was synthesized by Ullmann condensation of 3,9-diiodobenzanthrone (**274**) with 3-iodobenzanthrone (**275**). The Ullmann product (**276**, 16%) was quantitatively converted to **277** by alkaline condensation[212].

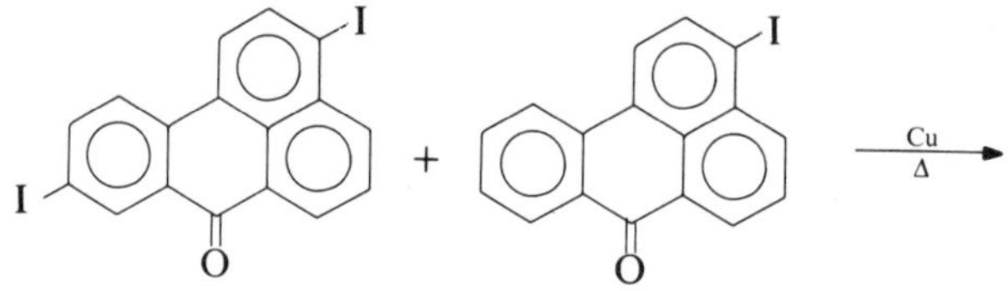

(**274**) (**275**)

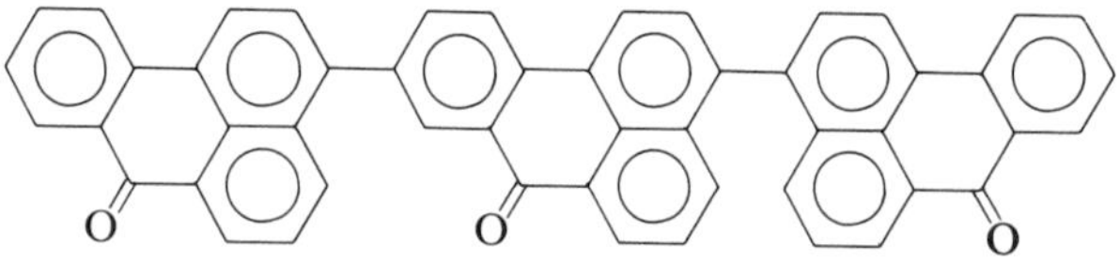

(**276**)

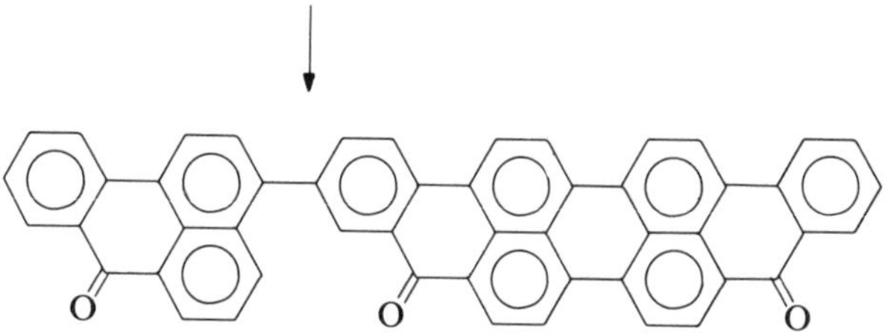

(**277**)

Violanthrone is oxidized by chromic acid in sulphuric acid to 2,2′-dianthraquinonyl-1,1′-dicarboxylic acid (**278**)[213]. Reduction of violanthrone in a zinc dust melt leads in

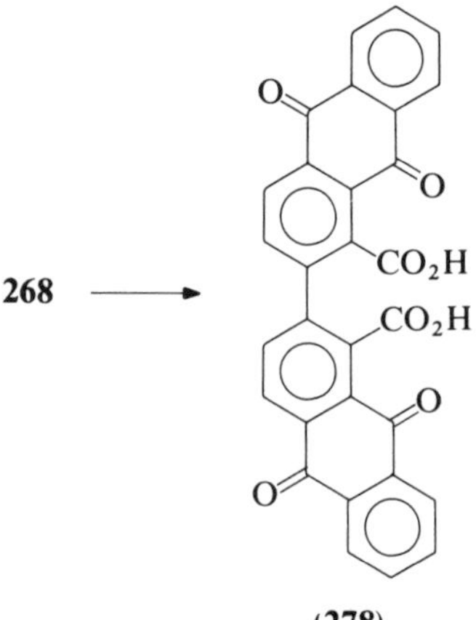

268 →

(**278**)

high yield to anthra[9,1,2-*cde*]benzo[*rst*]pentaphene, the corresponding aromatic hydrocarbon[214].

b. Isoviolanthrone

Isoviolanthrone (**269**) and some of its derivatives are of importance as vat dyes. When melting benzanthrone (**279**) with alcoholic potassium hydroxide[195] at 170–175°C the portion of **268** is at maximum, but both below and above this temperature range more **269** is formed[196]. **269** is also formed predominantly at low temperatures with potassium hydroxide in the presence of solvents as benzene or trichlorobenzene[215] or with metal anilides as condensating agents[216].

The separation of violanthrone and isoviolanthrone in the raw product is possible since the reduced form of violanthrone is insoluble in 4% sodium hydroxide[196].

Pure **269** can be synthesized by heating 4-chloro- or 4-bromo-benzanthrone (**280**) with alcoholic potassium hydroxide to 120–140°C[217]. Also, mixtures of 4-halogenobenzanthrones with benzanthrone[218] and the thioether **281**[219, 220] give **269** under alkaline conditions. The thioether route is the nowadays applied technical synthesis of isomer-free isoviolanthrone[221]. A synthesis starting from perylene has been described by Zincke and coworkers[59]. 4,10-Dibenzoylperylene (**282a**) is converted in low yields to **269** by heating with aluminium chloride. The yield can be raised (45%) by adding manganese dioxide[222]. Very pure isoviolanthrone has been obtained by heating 4,10-dibenzoyl-3,9-dibromoperylene (**282c**) with aluminium chloride[59]. The same condensation is also possible under alkaline conditions. In the case of the **282c** boiling with powdered potassium hydroxide is sufficient; with **282b**, boiling quinoline is needed[222]. Benzo[*rst*]phenanthro[10,1,2-*cde*]pentaphene, the aromatic hydrocarbon corresponding to **269**, can be prepared in high yield (85%) from isoviolanthrone in a zinc dust melt[214]. Isoviolanthrone and the hydrocarbon are both semiconductors[172,203].

Isoviolanthrone is marketed as Indanthren violet R Extra. Other important vat dyes are 6,15-dichloroisoviolanthrone (Indanthren Brilliant Violet RR and 4R) and tribromoisoviolanthrone (Indanthren brilliantviolet 3B and F3B). Chlorination of isoviolanthrone is effected with sulphuryl chloride[223], bromination with bromine in chlorosulphonic acid in the presence of sulphur[209]. 3,12-Dichloroisoviolanthrone has been obtained by Scholl cyclization of 3,9-dichloro-4,10-di-(4-chlorobenzoyl)perylene[222]. Nitro derivatives are obtained by direct nitration[206], and the corresponding amino derivatives by reduction of the nitro compounds either separately or during the vatting process[224]. Methyl homologues can be synthesized starting from the appropriate aroylperylenes or methylbenzanthrones[195]. 9,17-Diphenylisoviolanthrone has been prepared by alkali treatment of 3,6-diphenylbenzanthrone[225].

2. meso-*Anthrodianthrone*

The IUPAC notation for *ms*-anthrodianthrone (**284**) is dibenzo[*kl,no*]coronene-7,14-dione, but the first name is generally used. Syntheses of **284** had been performed already in 1926 by heating 12,13-dimethyl-*ms*-naphthodianthrone (**283**) with alcoholic alkali[226] or Scholl cyclization of tribenzoperylenequinone (**285**)[227]. The latter reaction can also be effected by exposure of **285** to light in the presence of oxygen. **284** has also been synthesized starting with a quick and quantitative Diels–Alder reaction between the blue, reactive *ms*-naphthodianthrene **286** and maleic anhydride in boiling nitrobenzene. The resulting anhydride **287** was oxidized with chromic acid and decarboxylated with soda lime to give **284**[228], which is a yellow vat dye (Indanthrenbraun).

1,6,8,10,11,13-Hexahydroxy-3,4-dimethyldibenzo[*kl, no*]coronene-7,14-dione (**289**) has been obtained by sulphuric acid treatment of the naturally occurring quinone pseudo-

O

\+

O

(**279**)

O

Hal + Hal

O

(**280**)

O

O

(**269**)

O

S

O

(**281**)

O

R

R

O

(**282**; **a**: R = H; **b**; R = Cl; **c**: R = Br)

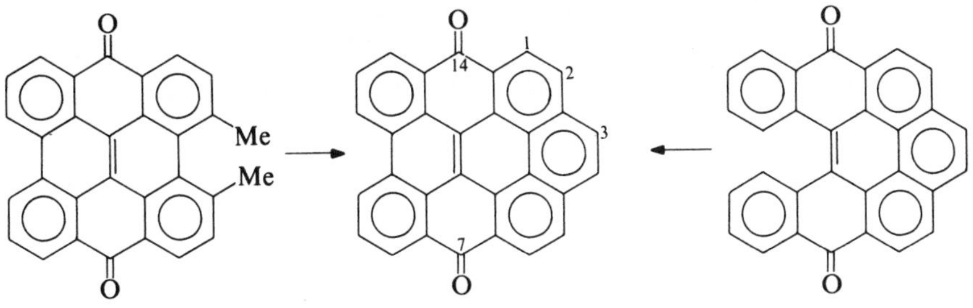

(283) (284) (285)

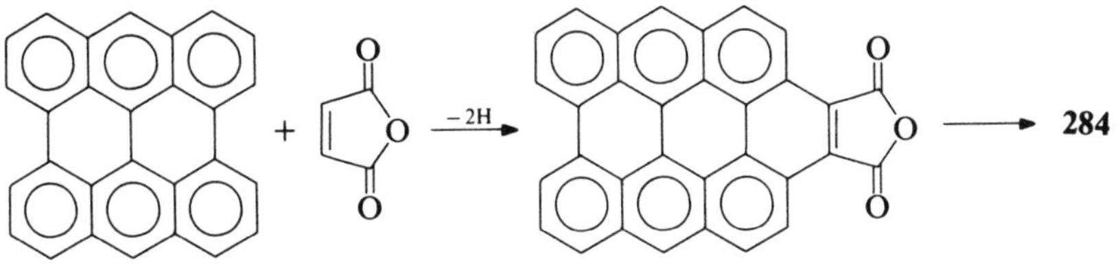

(286) (287)

hypericine (**288**)[178]. This reaction could constitute a principal way for the anellation of a benzene ring in analogous positions.

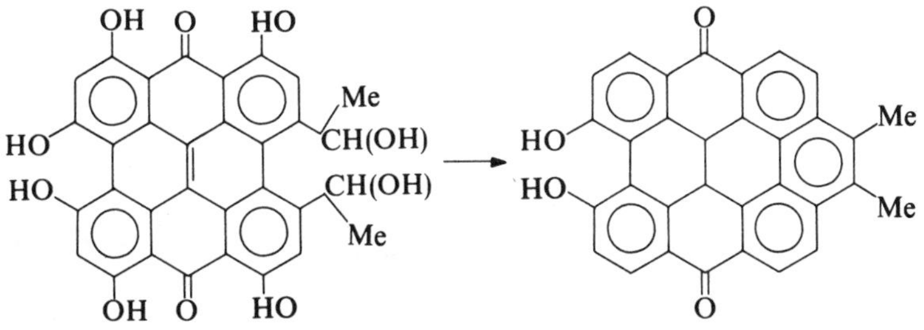

(288) (289)

3. *Benzo*[a]*naptho*[2,1-j]*anthanthrene-5,15-dione* (**290**)

290 has been obtained by Scholl cyclization of 3,8-dibenzoylpyrene (see dotted lines), which is accessible from pyrene by successive Friedel–Crafts acylation with benzoyl chloride and 1-napthoyl chloride[161]. **290** gives a blue vat.

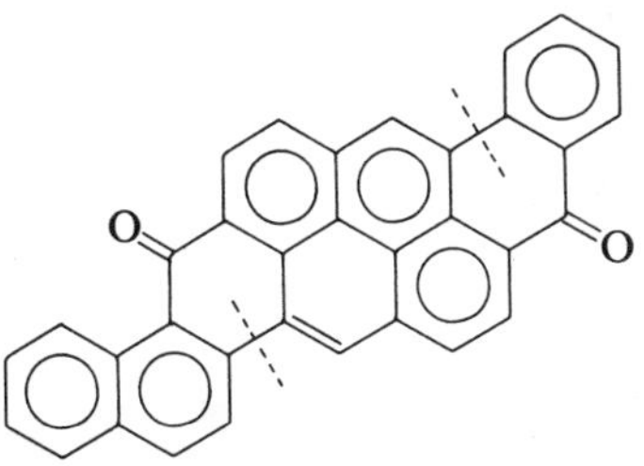

(290)

4. *Tetrabenzo*[a:de:l:op]*naphthacene-5,15-dione* (**294**)

Tetrabenzo[*a:de:l:op*]naphthacene-5,9,15,19-tetraone (**291**) is formed by boiling methyleneanthrone with chloranil or benzoquinone in nitrobenzene or acetic acid[229]. A *cis*-biangular connection instead of the *trans*-biangular, as in **291**, cannot be excluded with certainty, but seems to be not very probable. On reduction with zinc dust in pyridine/acetic acid **291** gives the dihydro derivative of the corresponding hydrocarbon, obviously an equilibrium mixture of **292** and **293**, which on oxidation with selenium dioxide yields a dione, probably **294**. No vat dye of **294** is obtainable with alkaline dithionite.

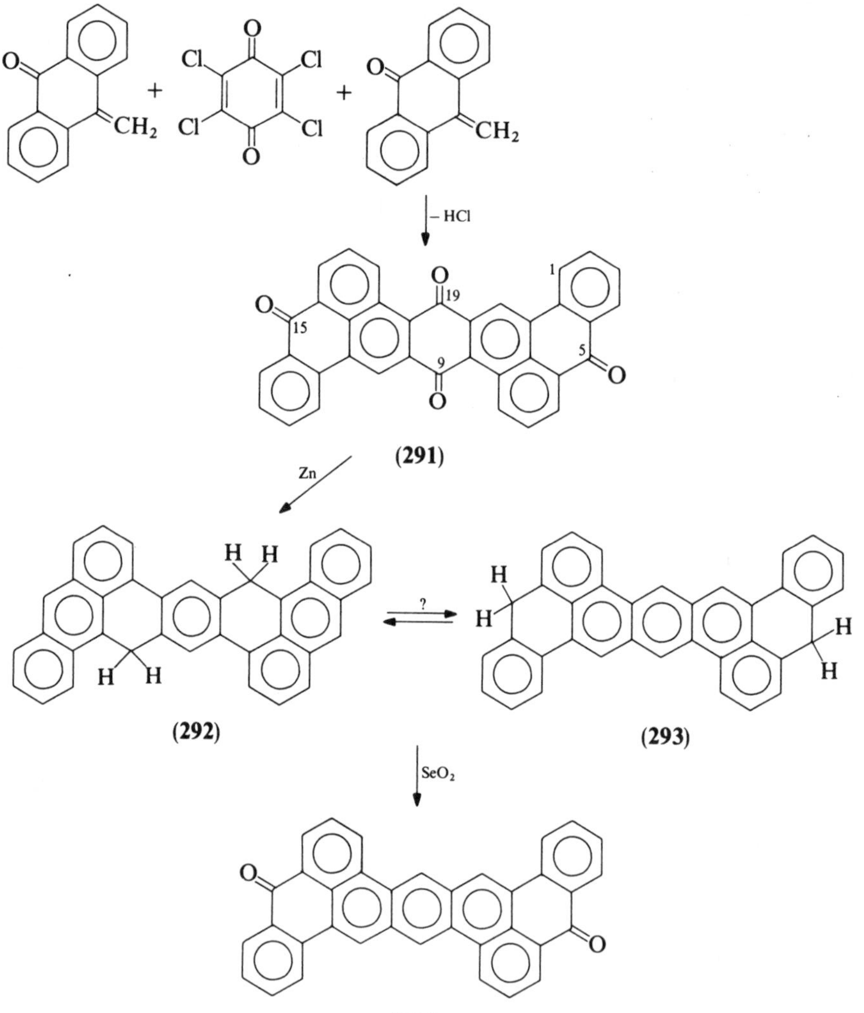

(**291**)

(**292**)

(**293**)

(**294**)

5. *Tetrabenzo*[a,f,j,o]*perylene-9,10-dione* (**298**)

298 has been obtained from **297** by reduction with copper powder in sulphuric acid. **297**

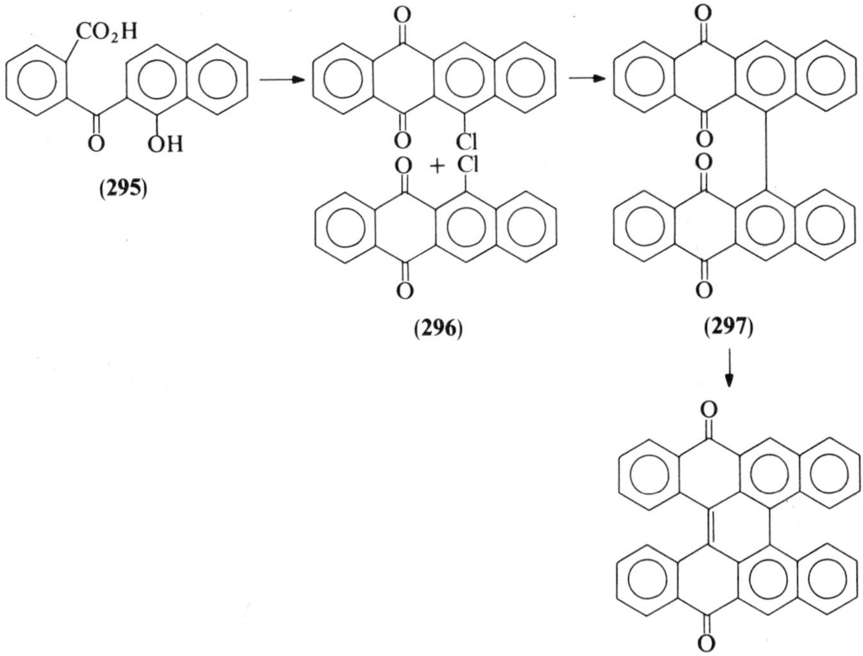

(295) (296) (297) (298)

was the product of the Ullmann reaction of 11-chloronaphthacene-5,12-dione (**296**), which is formed on treatment of **295** with phosphorous pentachloride[230].

J. Ten-ring Quinones

1. *Anthraceno*[*9,1,2*-klm]*dibenzo*[a,ghi]*perylene-5,14-dione* (**300**)

300 has been synthesized by copper reduction of the 1,1′-dibenzo[*a*]anthracene-7,12-dione (**299**) in sulphuric acid. **299** was obtained by Ullmann reaction of 1-chlorobenzo[*a*]anthracene-7,12-dione (**301**)[231].

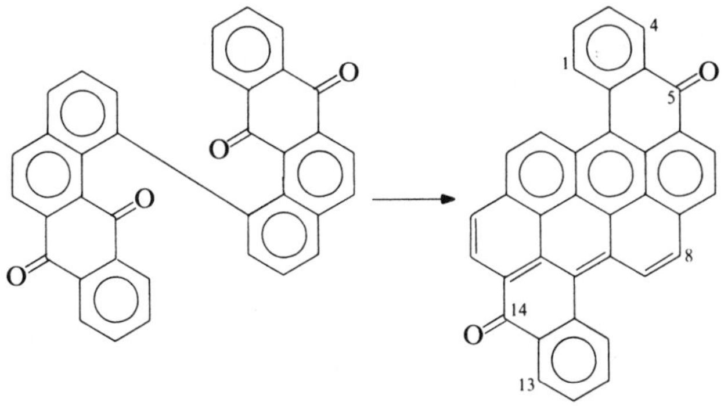

(299) (300)

300 is a violet vat dye. Halogeno and hydroxy derivatives of **300** with other shades, some with uncertain structures, have been described in the patent literature[231, 232].

2. *Dibenzopyranthrenediones*

Dibenzo[*a,def*]pyranthrene-5,10-dione (**303**) has been obtained in a simple three-step synthesis. On fusion of 2-methylbenzanthrone (**301**) in potassium hydroxide with glucose or naphthalene and manganese dioxide, the dimethylviolanthrone **302** is formed. **302** is transformed to **303** by boiling in nitrobenzene, especially in the presence of barium hydroxide[233].

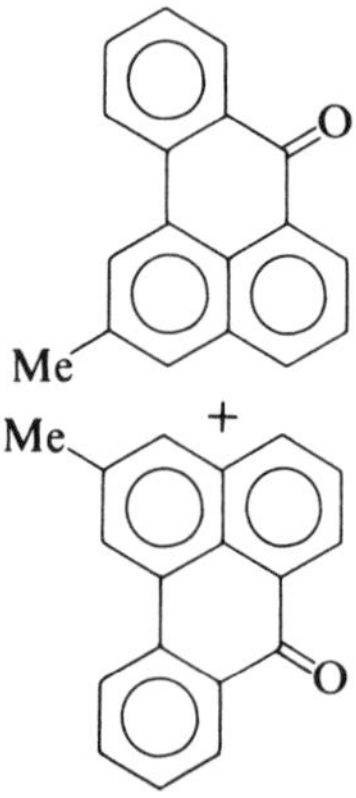

Me Me + O O

(**301**) 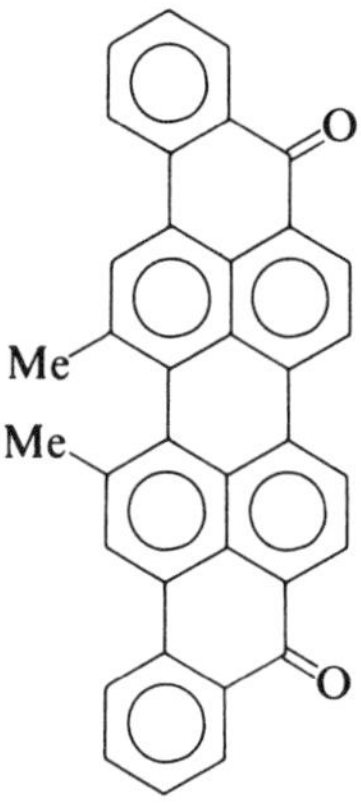 (**302**) 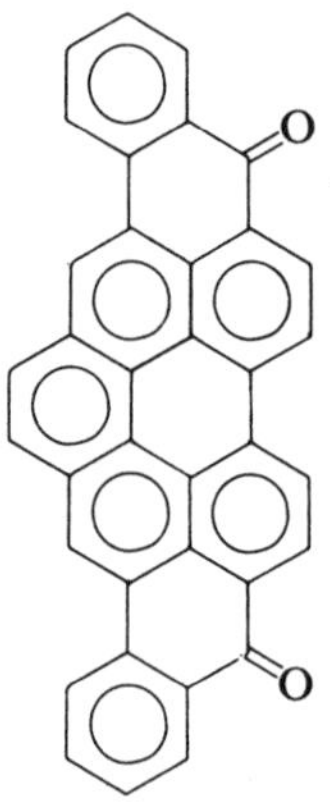 (**303**)

Another synthesis of **303** starts from 18,19-dihydrotetrabenzo[*c, m, pq, uv*]pentaphene-5,12-dione (**304**), which gives, on melting with alcoholic potassium hydroxide, 16,17-dihydrodibenzo[*a,def*]pyranthrene-5,10-dione (**305**). The dehydrogenation of **305** was effected with sodium nitrite[234]. **303** forms a blue vat with alkaline dithionite.

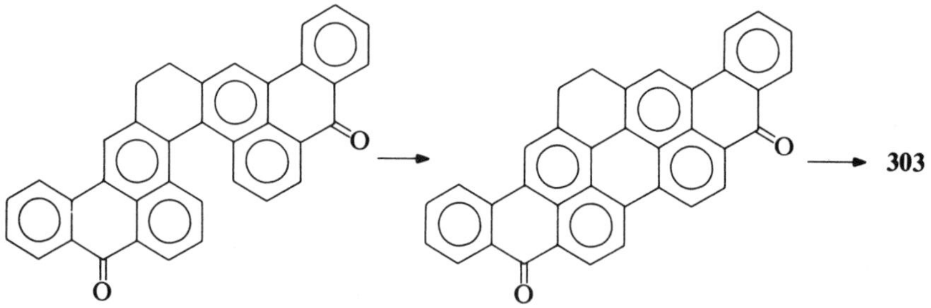

→ → **303**

(**304**) (**305**)

The 16,17-diphenyldibenzo[*a,def*]pyranthrene-5,10-dione (**307**) was obtained directly by fusion of 2-benzoylbenzanthrone (**306**) with alcoholic alkali[235].

Dibenzo[*a,n*]pyranthrene-10,20-dione (**309**) or dibenzo[*c,p*]pyranthrene-10,20-dione (**311**) is formed on heating 1,6-di-1-naphthoyl pyrene (**308**) and 1,6-di-2-naphthoylpyrene

(**310**), respectively, with aluminium chloride[152]. Both quinones have properties similar to pyranthrone, but cotton is dyed from the blue vats in redder shades. **309** and **311** are not used commercially, because there is no convenient technical source for pyrene, the precursor of **308** and **310**.

(**306**) (**307**)

(**308**) (**309**)

(**310**) (**311**)

3. *Dibenzo*[fgh:f′g′h′]*naphthaceno*[2,1-a]*naphthacene-9,20-dione* (**314**)

The synthesis of **314** has been accomplished by selenium dioxide oxidation of 9,20-dihydrodibenzo[*fgh*:*f*′*g*′*h*′]naphthaceno[2,1-*a*]naphthacene (**313**). **313** is readily available by heating the diketone **312** to 400°C in the presence of copper powder[236].

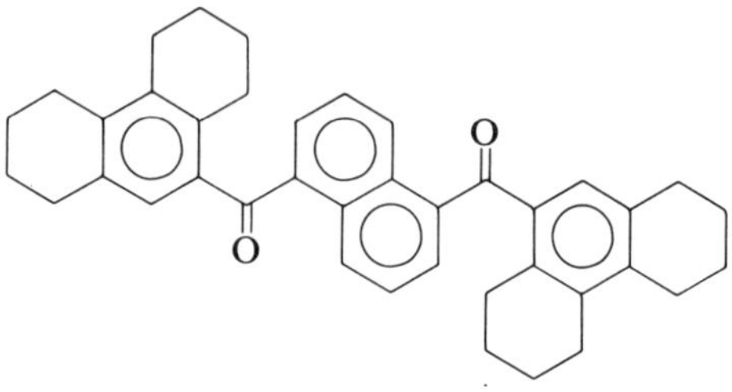

(**312**)

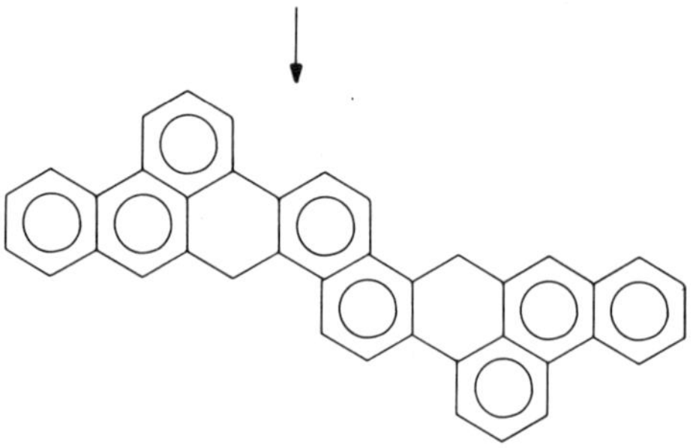

(**313**)

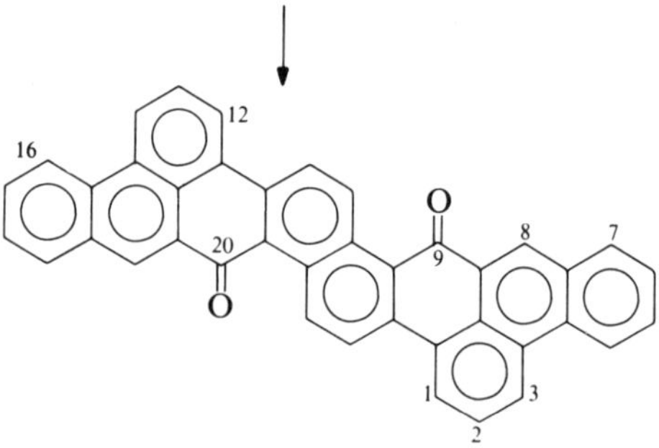

(**314**)

4. *Tetrabenzo*[a,de,kl,o]*pentaphene-5,14-dione* (**319**)

Naphthalene codenses twice with dichloroanthrone (**315**) with aluminium chloride as catalyst. **316** is formed as an intermediate. If the second dichloroanthrone attacks **316** at the 12-position, the hydroquinone of **319** is formed, while attack at the 11-position leads to the hydroquinone of tribenzo[*a,de,rst*]naphtho[4,3,2-*kl*]pentaphene-5,15-dione (**318**). The hydroquinone **317** is oxidized to **319** by boiling with nitrobenzene[237].

On reduction of **319** with zinc dust in pyridine-acetic acid the corresponding aromatic hydrocarbon is formed. With alkaline dithionite no vat is generated, probably because the (green) salt of the hydroquinone **317** is insoluble.

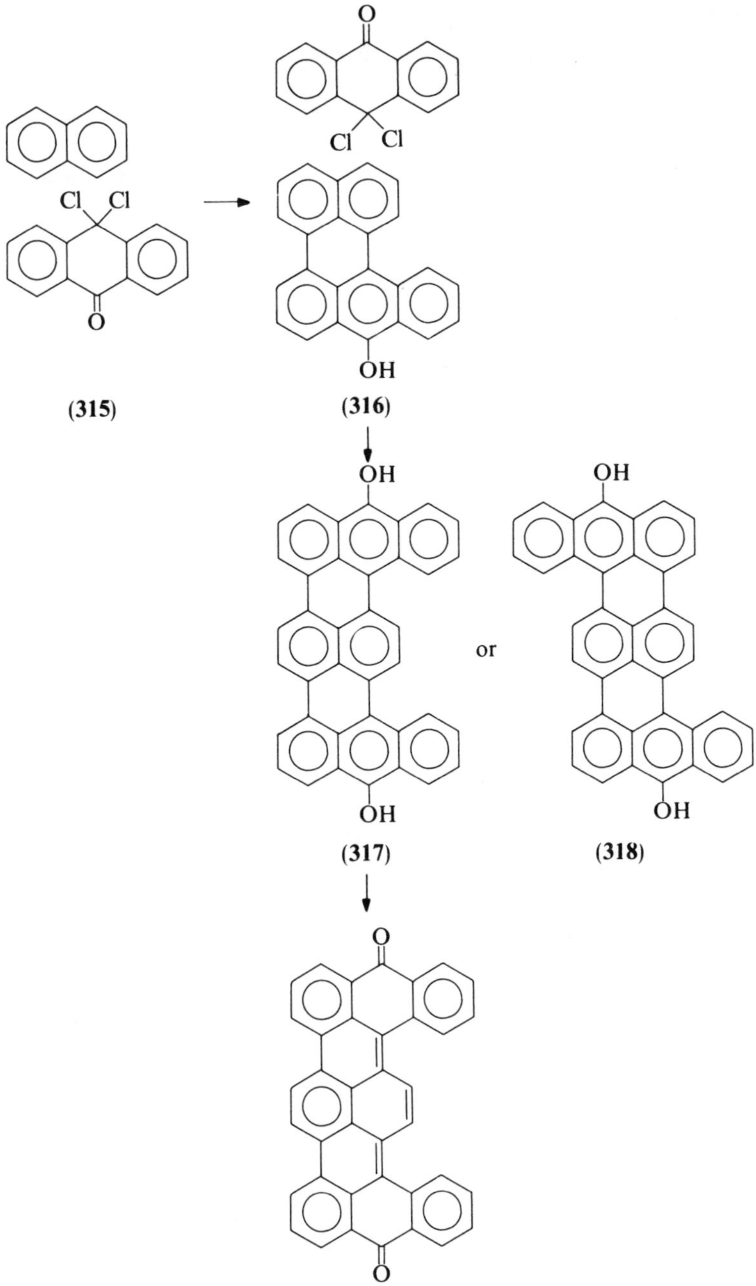

(315)

(316)

(317) or (318)

(319)

K. Eleven-ring Quinones

1. *Anthraceno*[*2,1,9,8*-klmno]*naphtho(3,2,1,8,7*-vwxyz)*hexaphene-4,9-dione* (**321**)

Fusion of benzo[*cd*]pyrene-6-one (**320**) resulted in the formation of a blue colouring matter, which contains 2-hydroxybenzo[*cd*]pyrene-6-one, and possibly a tetrahydroxy derivative of **321**[238]. **321** dissolves sparingly in alkaline dithionite, forming a bluish-green vat.

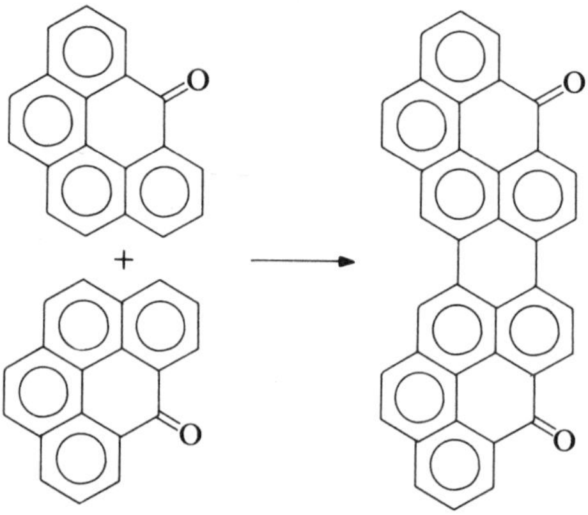

(**320**) (**321**)

2. *Dianthraceno*[2,1,9,8-stuva:2,1,9,8-hijkl]*pentacene-9,18-dione* (**323**)

323 has been synthesized by melting 2-bromobenzo[*cd*]pyrene-6-one (**322**) with alcoholic potassium hydroxide at 120°C[238]. **323** gives, with alkaline dithionite, a blue-green vat.

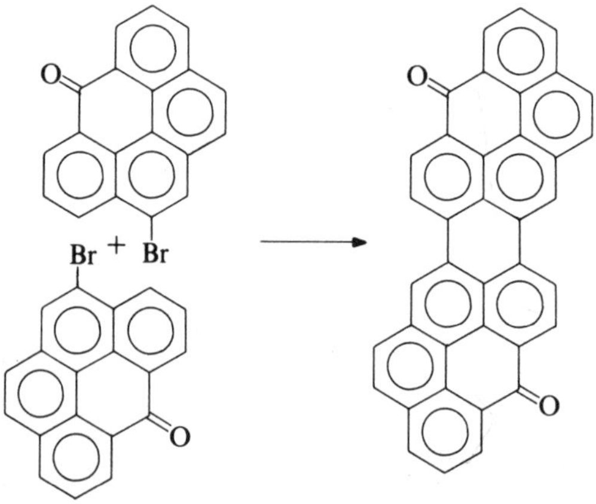

(**322**) (**323**)

3. *Dibenzo*[a, o]*dinaphtho*[3,2,1-cd:1,2,3-lm]*perylene*-5,14-*dione* (**324**)

The green **324** has been synthesized from the two halves (dibenzo[*a,de*]anthracene-3-one, see dotted line) by melting with potassium hydroxide/potassium acetate at 225°C[239]. **324** may be considered to be a dibenzo derivative of violanthrone. Indeed it resembles violanthrone in its spectral properties and forms with alkaline dithionite a purple solution.

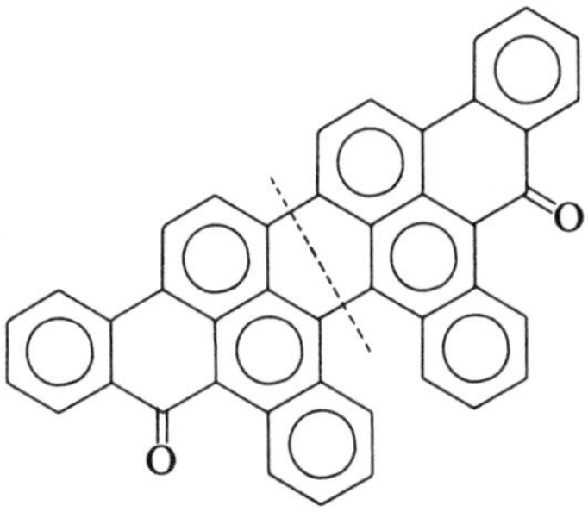

(**324**)

4. *Diphenanthrenoperylenediones*

Two blue diphenanthrenoperylenediones have been described in the literature, diphenanthreno[4,3,2-*cd*:5,6,7-*lm*]perylene-5,10-dione (**325**) and diphenanthreno[4,3,2-

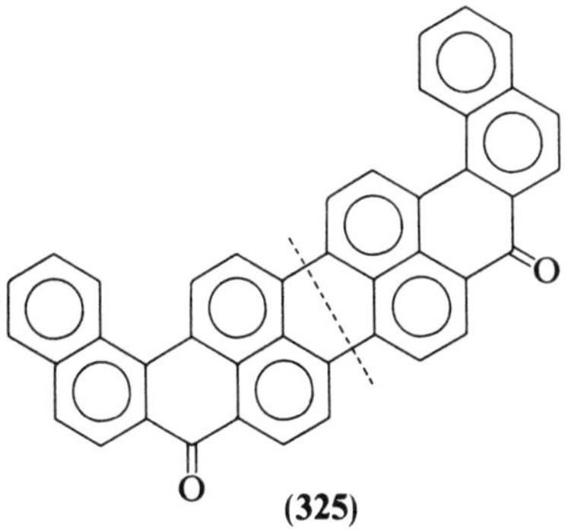

(**325**) 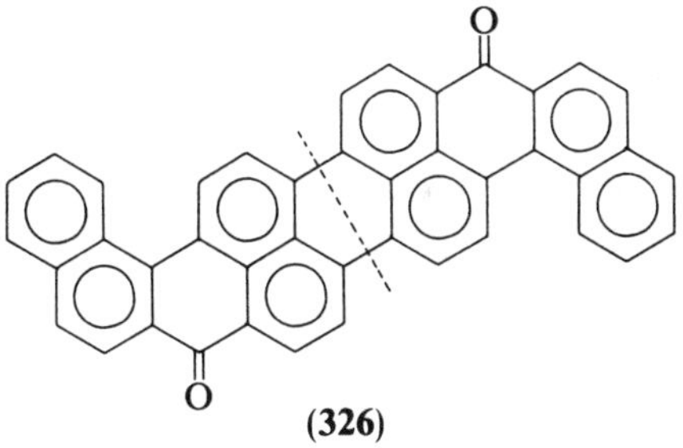 (**326**)

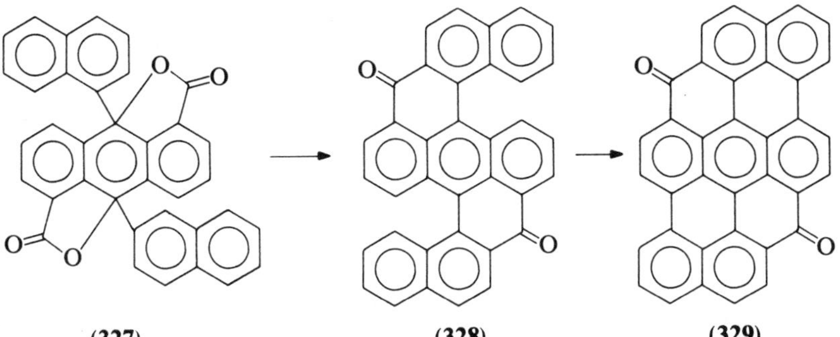

(**327**) (**328**) (**329**)

cd:4,3,2-*lm*]perylene-7,12-dione (**326**). Both quinones were obtained by alkaline treatment of the two halves of the molecule (see dotted lines), i.e., of benzo[*hi*]chrysene-7-one[240]. With sodium anilide **326** is formed, whereas potassium hydroxide, sodamide and sodium piperidide afford predominantly **325**.

5. *Dinaphtho[2,1,8-*apq*:2,1,8-*ghi*]coronene-8,16-dione* (329)

Anthraquinone-1,5-dicarboxylic acid reacts with naphthalene and aluminium chloride in nitrobenzene to yield the dilactone **327**. Further condensation yields the dinaphthoperylenequinone **328**[124]. At elevated temperatures **329** is eventually formed[228].

L. Thirteen-ring Quinones

1. *Benzo[*j*]dinaphtho[3,2,1-*cd*:4,3,2-*pq*]terrylene-5,12-dione* (331)

331 has been synthesized by Ullmann reaction of 3-chloro- or 3-bromo-benzanthrone with 9,10-dibromo- or 9,10-dichloro-anthracene. The resulting 9,10-bis(3-benzanthronyl)anthracene (**330**) gave **331** on heating in a potassium hydroxide/sodium acetate melt at 230°C. With alkaline dithionite it forms a dark violet vat which dyes cotton grey[241].

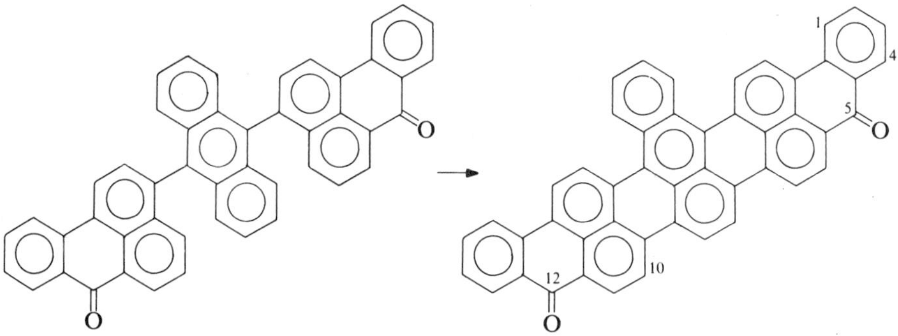

(**330**) (**331**)

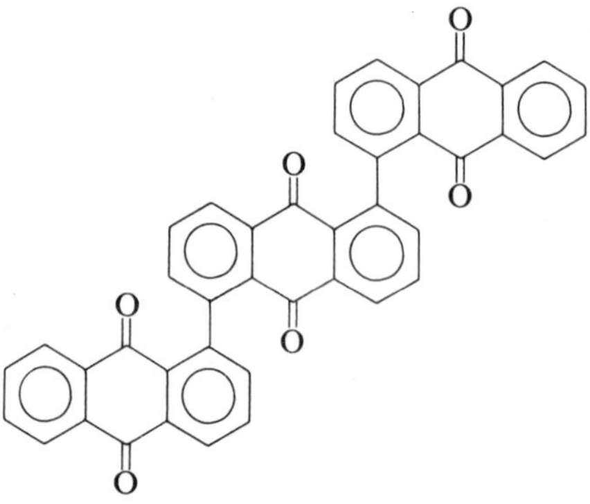

(**332**) 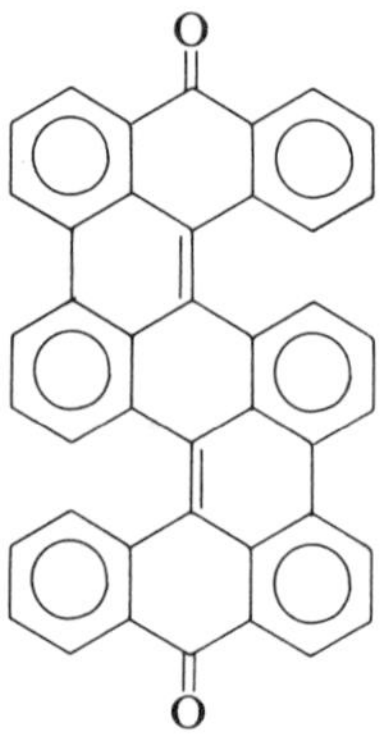 (**333**) 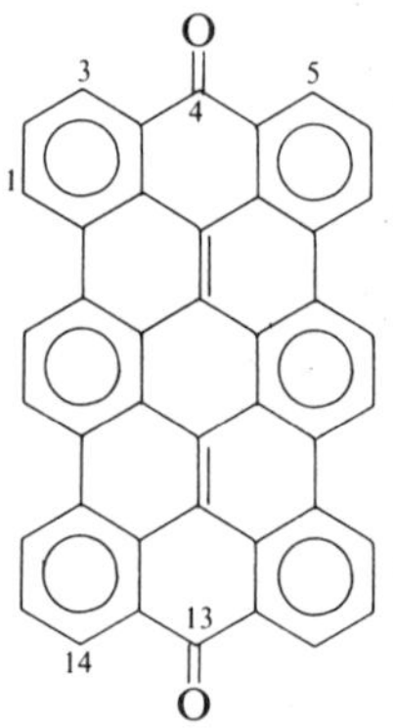 (**334**)

2. *Dianthraceno(1,9,8-*apqr*:1,9,8-*ghij*)-coronene-4,13-dione* (**334**)

The synthesis of the highly condensed **334** has been effected by irradiation of **333**. This quinone is easily accessible by treatment of the trianthraquinonyl **332** with powdered copper in sulphuric acid[135]. In contrast to simple helianthrones, **333** is unstable and reverts to **332**, even in the solid state.

The synthesis of the 5,14-dichloro derivative of **334** has also been described[135].

IV. ANNULENEDIONES

One of the most important properties of quinones is the reversible formation of dihydroxy aromatics or their dianions, on two-electron reduction. Therefore, non-benzenoid diones, which yield on reduction compounds containing cyclic conjugated double bonds with $4n$ π electrons, i.e., non-aromatic systems, cannot be regarded as quinones[242]. In contrast, dibenzo[*cd,gh*]pentalene-4,8-dione (**335**) produces on electrolytic reduction the radical anion (**336**). In **336**, 14 π electrons are contained in cyclic conjugated double bonds, hence it may be considered a Hückel aromatic and **335** a quinone.

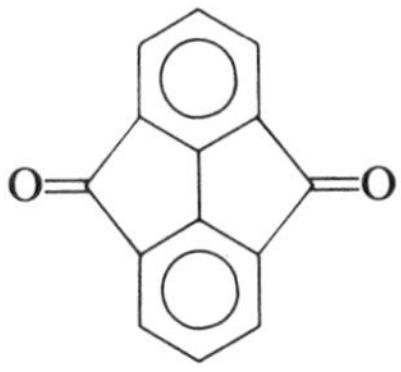

(**335**) 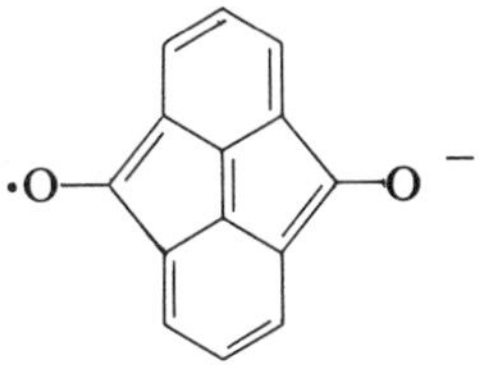 (**336**)

Another example has recently been described by Kuroda and coworkers[243]. They synthesized cyclohepta[*a*]phenalene-6,12-dione (**337a**) and the 5-bromo derivate (**337b**).

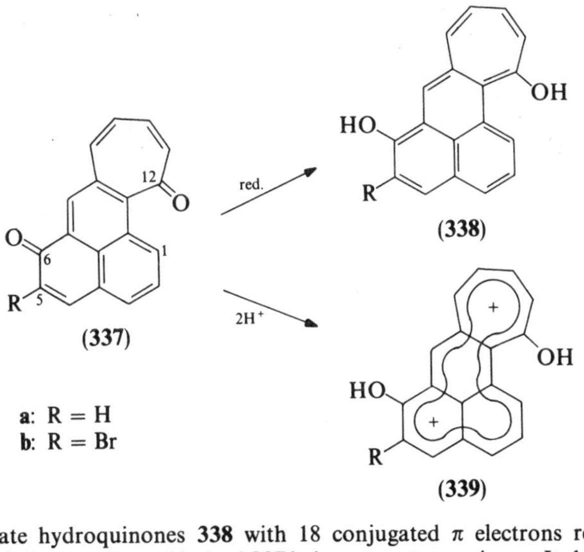

(**337**)

a: R = H
b: R = Br

(**338**)

(**339**)

The conjugate hydroquinones **338** with 18 conjugated π electrons represent non-benzenoid Hückel aromatics and indeed **337** behaves as a true quinone. It shows reversible redox properties, with reduction potentials at $E(\frac{1}{2})_1 = -1.05$, $E(\frac{1}{2})_2 = -1.44$ and $E(\frac{1}{2})_1 = -0.90$, $E(\frac{1}{2})_2 = -1.37$ V, respectively. These potentials resemble those of anthraquinone.

The dications (**339**) are formed in concentrated sulphuric acid. The shift of the proton signals in the ^{1}H-NMR spectrum suggests that the dications are surprisingly diatropic. Also quinones of azulene are known. They are reviewed in Chapter 27.

The first syntheses of annulenes constitute a milestone in the search for non-classical ($4n + 2$) Hückel aromatics. Soon after the question arose as to whether the annulenes parallel the classical aromatics also in their ability to form quinones. Indeed several annulenoquinones have meanwhile been described. The first was synthesized in 1967 by Boekelheide and Phillips[244]. In most cases the quinonoid character of the annulenediones was established carefully by means of their chemical and electrochemical properties.

A. [10]Annulenediones

1. Homoazulenequinones

Outstanding syntheses of the 1,5-, 1,7- and 4,7-homoazulenequinones (**340**–**342**) have recently been described by Scott and Oda[245]. The synthesis of the 1,5- and 1,7-quinones

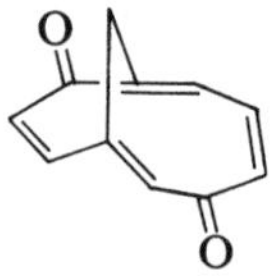

(**340**) 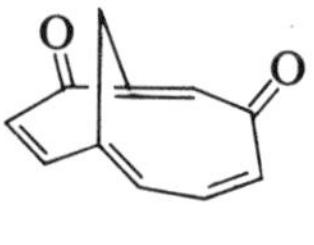 (**341**) 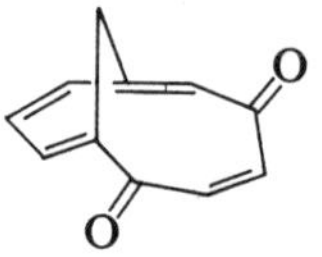 (**342**)

starts with the propellane **343**. Dehydrogenation via the alpha-phenyl selenide gave **344**, which was photo-oxidized to the endoperoxide **345**. A mixture of **346** and **347** was obtained by isomerization of **345** with Hünig's base. **346** yielded in a Grob fragmentation reaction (trifluoroacetic anhydride/triethylamine) the 1,5-quinone **340**–**348**. This last amazing step has been performed with 77% yield. A closely related synthesis was performed for the preparation of the 1,7-quinone **341**, starting from **349**, which could be obtained by isomerization of **344**.

The 4,7-quinone **342** could be obtained by NBS bromination of the known dione **350** followed by triethylamine treatment. Though prepared in pure form in solution, the 4,7-quinone could not be isolated. The stabilities of the homoazulenequinones **340**–**342**

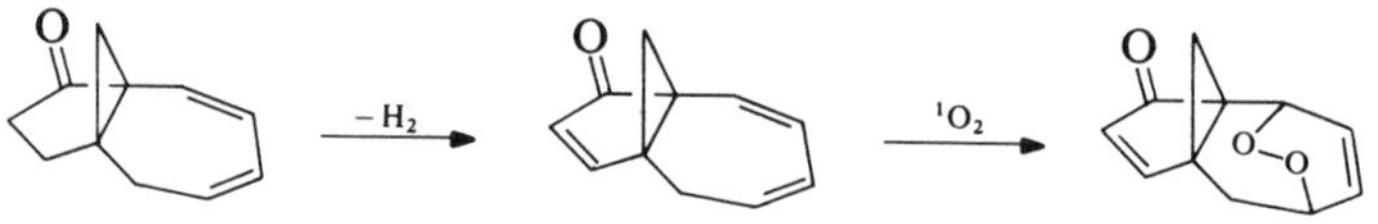

(**343**) $\xrightarrow{-H_2}$ (**344**) $\xrightarrow{^1O_2}$ (**345**)

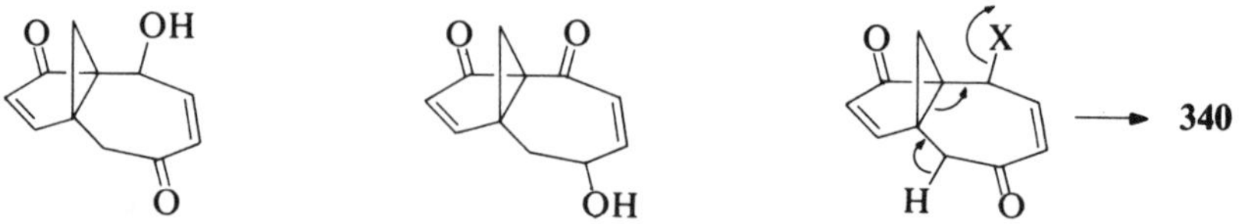

(**346**) (**347**) (**348**) → **340**

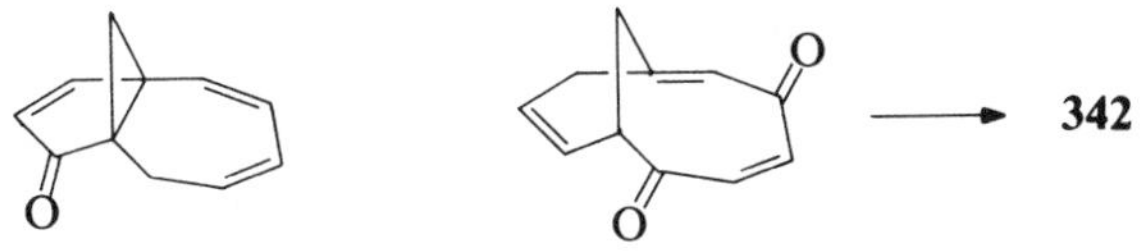

(349) (350) → 342

parallel those predicted for the corresponding azulenequinones[246]. Up to now only the 1,5- and 1,7-azulenequinones could be isolated in substance[247] (see also Chapter 24).

2. *Homonaphthoquinones*

Attempts have been made to synthesize bicyclo[4.4.1]undeca-3,6,8,10-tetraene-2,5-dione or 2,5-homonaphthoquinone (**351**). Surprisingly, the dynamic isomer with the norcaradiene structure **352** proved to possess the lower energy[248]. Treatment of **352** with a mild reducing agent, such as the enolate ion of propiophenone, yields a free radical anion wherein, judging from the ESR spectrum, the unpaired electron is extensively delocalized into the entire π system. Hence the semiquinone structure **352a** has been proposed for the radical anion[249].

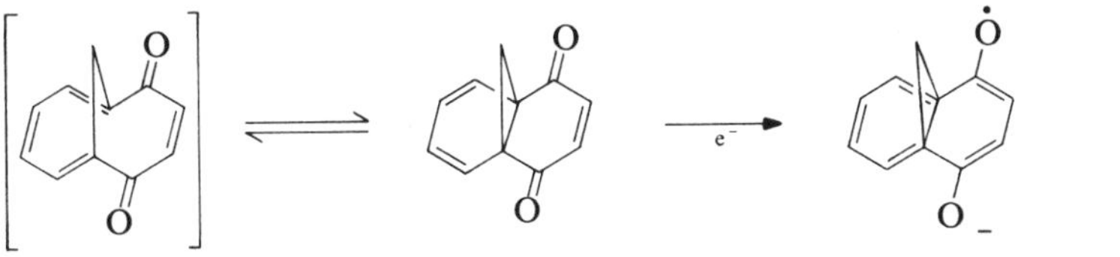

(351) ⇌ (352) —e^-→ (352a)

(353) —$Pb(OAc)_4$→ (354) —MeLi→ (355) —MnO_2→ (356) ⇌ (357)

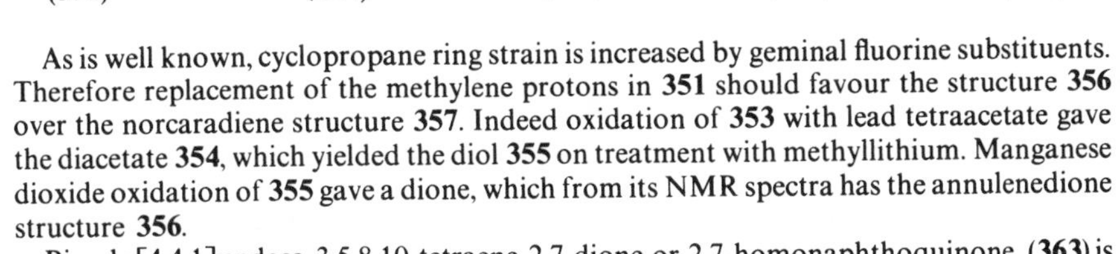

As is well known, cyclopropane ring strain is increased by geminal fluorine substituents. Therefore replacement of the methylene protons in **351** should favour the structure **356** over the norcaradiene structure **357**. Indeed oxidation of **353** with lead tetraacetate gave the diacetate **354**, which yielded the diol **355** on treatment with methyllithium. Manganese dioxide oxidation of **355** gave a dione, which from its NMR spectra has the annulenedione structure **356**.

Bicyclo[4.4.1]undeca-3,5,8,10-tetraene-2,7-dione or 2,7-homonaphthoquinone (**363**) is the annulene pendant of the unstable 1,5-naphthoquinone. **363** has been synthesized[250] starting with the 2,7-dibromo-1,6-methano-[10]annulene (**358**). Its Grignard compound reacted with perbenzoic acid *t*-butyl ester to give the ether (**360**). Cleavage of **360** with catalytic amounts of *p*-toluenesulphonic acid in benzene at 80° C (10 min) yielded the diketone **361**. It was not possible to dehydrogenate **361** directly with 2,3-dichloro-5,6-dicyanobenzoquinone, but with *N*-bromosuccinimide the dibromide **362** was formed and gave the quinone **363** on treatment with potassium iodide. **363** possesses at least one

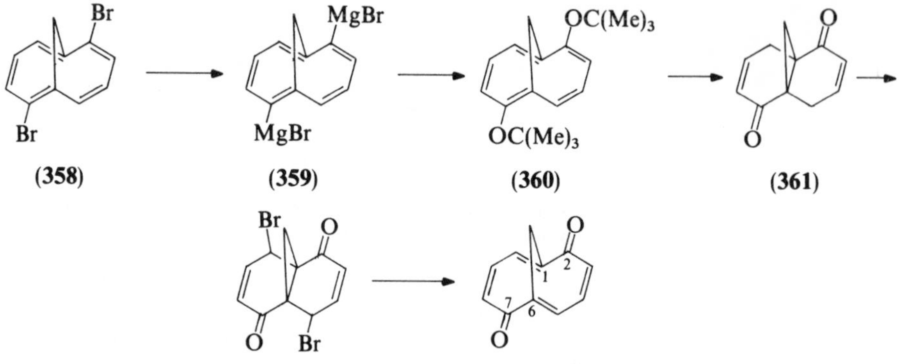

Br Br (358) → MgBr MgBr (359) → OC(Me)3 OC(Me)3 (360) → O O (361) →

Br O O Br (362) → O O 1 2 6 7 (363)

important property of quinones: it is transformed by reductive acetylation into the 2,7-diacetoxy-1,6-methano[10]annulene, the diacetate of the conjugate (quasi-)aromatic hydroquinone. **363** is a stable compound, which crystallizes from acetone in orange-yellow crystals. The stability of **363** is surprising because it was not possible to isolate the analogous 1,5-naphthoquinone[27]. However, the very low tendency of **363** to undergo Michael additions or Diels–Alder condensations and, on the other hand, the high reactivity of 1,5-napthoquinone are in accord with the results of PMO calculations on both compounds[27, 30].

B. [14]Annulenediones

Several [14]annulenediones with and without inner alkano bridges have been described. They are treated in alphabetical order.

1. *Bisdehydro[14]annulenediones*

The synthesis of bisdehydro[14]annulenediones seemed of interest considering the high aromatic nature of the bisdehydro[14]annulenes[251]. A very elegant synthesis of the di-*t*-butyl- and diphenyl-bisdehydro[14]annulenediones **365a** and **365b**, respectively, was possible by cyclodimerization of the acid chlorides **364a** and **b**, respectively, in the presence of a palladium/copper catalyst[252].

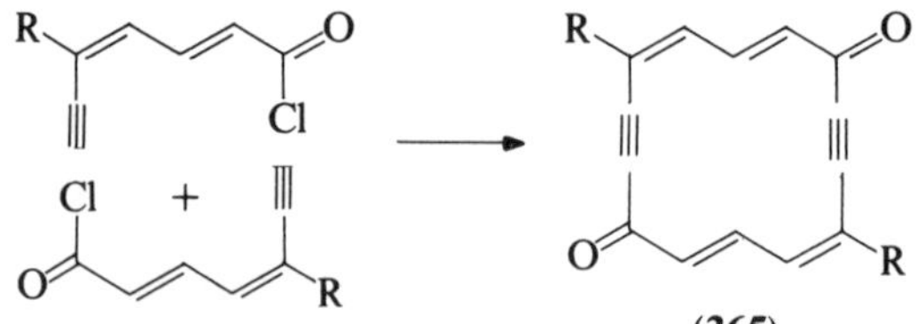

R O Cl Cl + O R (364) → R O O R (365)

a: R = *t*-Bu
b: R = Ph

The dibenzo derivative **367** has been obtained by the catalytic dimerization of the acid chloride **366**[252], or from the copper salt of *o*-iodocinnamoylacetylene[253].

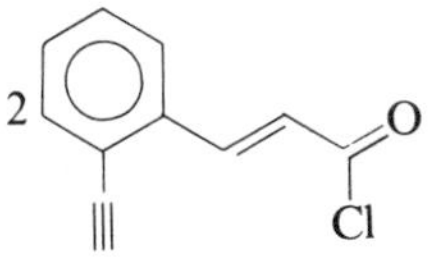

(366)

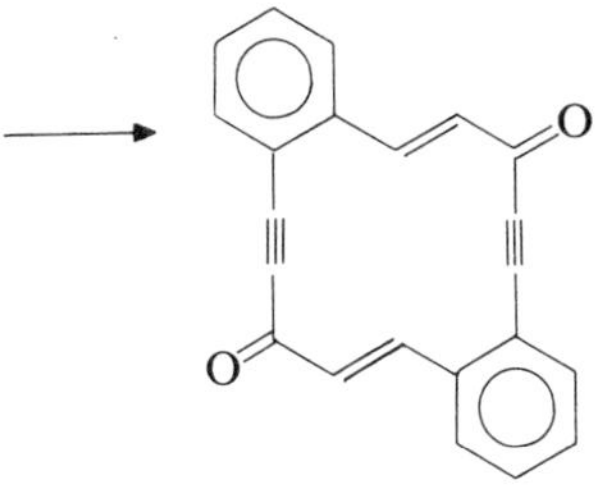

(367)

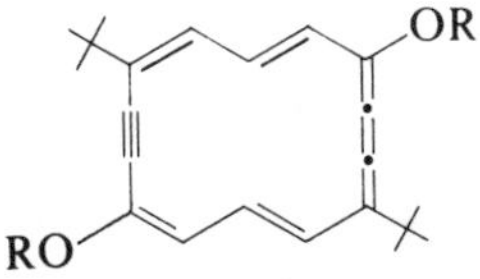

(368)

a: R = Ac
b: R = Me

The diones **365a, b** and **367** are stable compounds. They crystallize in the form of yellow to orange needles. The chemical and electrochemical reduction of **365a** has been investigated in order to obtain evidence for its quinonoid nature[254].

Reductive acetylation with zinc powder/acetic anhydride in the presence of pyridine yielded the diacetoxy-bisdehydro[14]annulene **368a** and reductive methylation with dimethyl sulphate/sulphuric acid and zinc powder the dimethyl ether **368b**.

On cyclic voltammetry **365a** exhibits electrochemical reversibility at both the first and second waves even at low scan rates (16 mV s^{-1}). The well-defined two-wave pattern corresponds to the two discrete one-electron transfer processes forming radical anions initially and then dianions. This behaviour is consistent with that observed for quinones. The reduction potential of **365a** ($E(\frac{1}{2})_1)_1$: -0.63, $E(\frac{1}{2})_1)_2$: -1.02 V) is similar to that of 1,4-naphthoquinone ($E(\frac{1}{2})_1)_1$: -0.59, $E(\frac{1}{2})_1)_2$: -1.40 V).

2. *1,6;8,13-Bismethano-*[14]*annulene-7,14-diones (bishomoanthraquinones)*

Two stereoisomers of bishomoanthraquinones are possible, the *syn* and *anti* form (**372** and **374**, respectively). Both have been synthesized. *syn*-Bishomoanthraquinone (**372**) was accessible from the *syn*-bishomoanthranthracene (**369**)[255]. Bromination yielded the 7,14-addition product **370**, which could be transformed to the diol **371** by reaction with silver nitrate in wet acetone. **371** yielded **372** on oxidation. The best results were obtained by oxidation with dimethyl sulphoxide/trifluoroacetic anhydride/triethylamine, a reagent especially suited for the oxidation of sterically hindered alcohols.

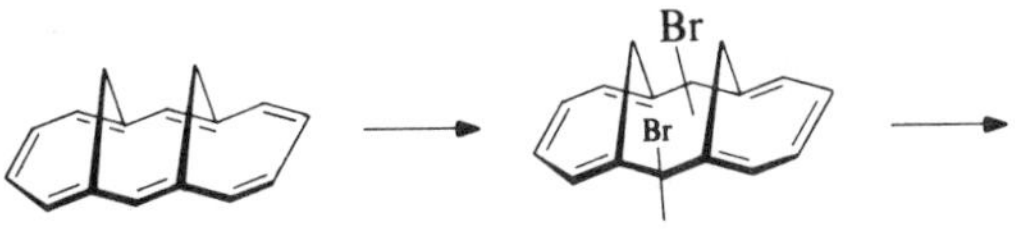

(369)

(370)

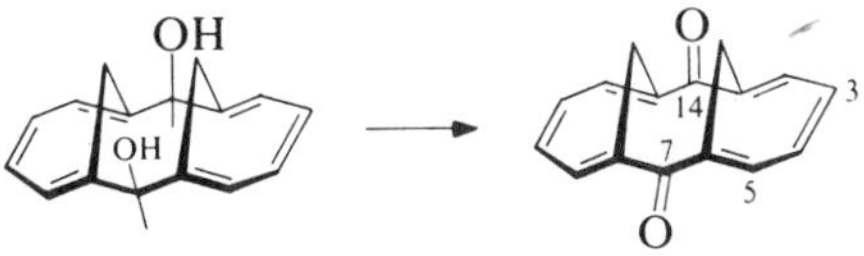

(371)

(372)

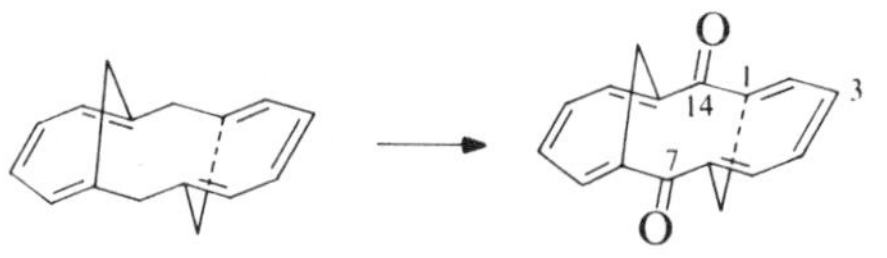

(373) (374)

374 has been obtained by successive oxidation of the hydrocarbon **373** with selenium dioxide in dioxane/water and chromium(III) oxide in pyridine[256]. For steric reasons and in contrast to the *syn*- isomer (**369**) the *anti*-1,6;8,13-bismethano-[14]annulene (**373**) exhibits no aromatic character but behaves as a readily polymerizable olefin. Nevertheless, with zinc/acetic anhydride in pyridine both the *syn* and *anti* diones (**372** and **374**) undergo facile reductive acetylation to the corresponding 7,14-diacetoxy-1,6;8,13-bismethano-[14]annulenes.

At cyclic voltammetry both isomers exhibit typical behaviour of quinones in that they show two reversible one-electron transitions to the corresponding radical anions and dianions[255]. The reduction potential of **374** (−2.29 V) is lower than that of **372** (−1.79 V), reflecting the differences in the aromaticity of the conjugate hydrocarbons.

3. trans-*15,16-Dimethyldihydropyrene-2,7-dione* (**381**)

An elegant synthesis of a fascinating quinonoid system (**381**) has been described by Boekelheide and Phillips[244]. The 5,13-dimethoxy-8,16-dimethyl[2.2]metacyclophane (**375**) undergoes a smooth reaction with chromic acid in acetone or with iron(III) chloride in dry chloroform to give the bisdienone **376** in nearly quantitative yield. The fact that a coupling of two phenol ethers was accomplished under the conditions of the oxidative-radicalic phenolic coupling is quite surprising.

The bisdienone **376** might be expected to undergo a dienone–phenol rearrangement, but it is recovered unchanged from boiling hydrochloric acid. In contrast, it is readily soluble in aqueous alkali. This surprising property can be readily explained by a double enolization followed by valence tautomerism to give the dianion of the metacyclophane **377**. This in turn should undergo readily phenolic oxidation. In fact further oxidation by bubbling air through a basic solution of **376** or with *N*-bromosuccinimide occcurs smoothly to give **381**, again in nearly quantitative yields. The transient green-violet colour that appears during this oxidation suggests a pathway involving the semiquinone anion **380**, which is formed by enolization/valence tautomerism of **378** to **379** and further radicalic coupling.

The oxidation steps leading from **376** to **381** are reversible. This has been shown in a detailed study of the ESR spectrum of the violet semiquinone **380**[257], which is obtained by treatment of the yellow **381** with glucose in the presence of alkali. Further addition of glucose leads to the formation of the colourless hydroquinone-dianion **379**. This sequence of colour changes is exactly reversed by introduction of oxygen. Undoubtedly the hydroquinone dianion **379** is in equilibrium with the corresponding dihydropyrene valence tautomer, the proper hydroquinone dianion of **381**, but, in this case, relief of charge repulsion shifts the equilibrium in favour of **379**. As by-product of the alkaline oxidation, minor amounts of 4-hydroxy-*trans*-15,16-dimethyldihydropyrene-2,7-dione could be isolated. When the quinone **381** was subjected to reductive acetylation with zinc/acetic anhydride it was converted in 90% yield to the hydroquinone diacetate **382**, a further proof of the quinonoid character of **381**.

The conversion of **381** to the parent hydrocarbon, the *trans*-15,16-dimethyldihydropyrene (**383**), was effected by treatment with a lithium aluminium hydride/aluminium chloride mixture at room temperature.

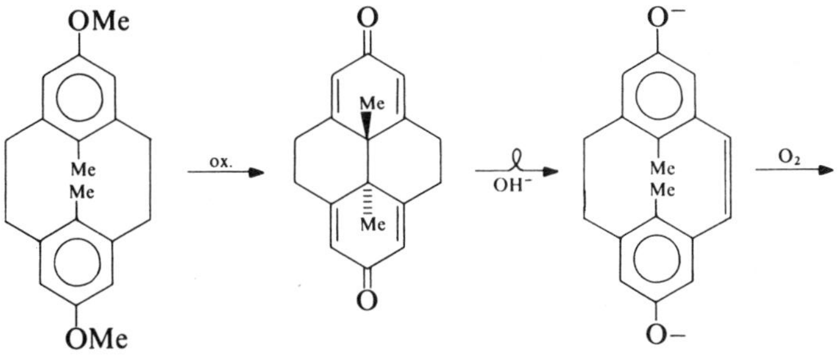

(375) (376) (377)

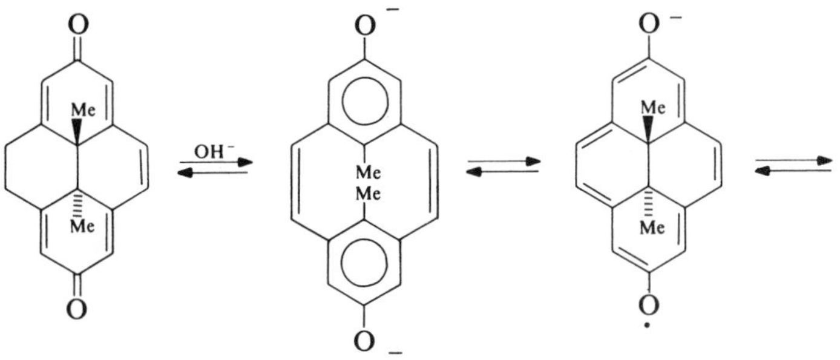

(378) (379) (380) (381)

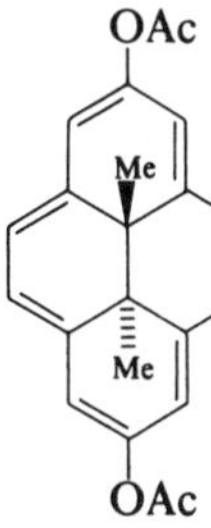

(382) 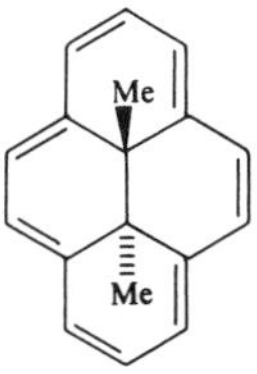 (383)

C. [18]Annulenediones

1. *Cyclooctadecahexaenediynedione (bisdehydro[18]annulenedione)*

The 4,13-di-*t*-butylcyclooctadeca-4,6,8,13,14,15-hexaene-2,11-diyne-1,10-dione (**385**) has been prepared in the same way as the bisdehydro[14]annulene homologue (see Section IV.B.1) by cyclodimerization of the acid chloride **384** under the influence of a palladium/copper(I) catalyst[252]. Reductive acetylation of **385** with zinc/acetic anhydride in the presence of pyridine gave the diacetoxybisdehydro[18]annulene **386a** as a dark red solution. **386a** was found to be unstable even at $-10°C$, but the structure is obvious from the spectrum of the solution. The product of reductive methylation of **385** with dimethyl

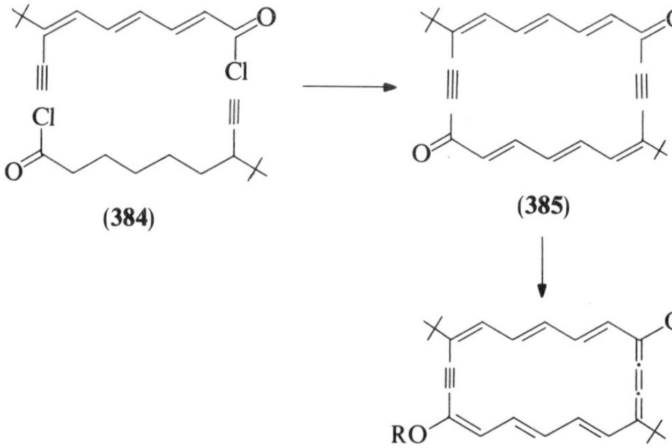

(384) (385)

(386)
a: R = Ac
b: R = Me

sulphate/sulphuric acid and zinc, the dimethoxybisdehydro[18]annulene **386b**, could be isolated in the form of dark reddish violet crystals.

Further proof for the quinonoid character of **385** is given on electrochemical reduction by cyclic voltammetry[254], which showed electrochemical reversibility at both the first and second waves, corresponding to the formation of the anion radical and hydroquinone dianion. The redox potential of **385** (−0.92 V) is more positive than that of benzoquinone (−1.18 V) and by 0.1 V even more positive than that of the [14]annulenequinone homologue. This is by no means a measure for the growing aromaticity of the corresponding hydrocarbons, but may be connected with the increasing charge separation in the larger ring systems. Indeed the first half-wave potentials become more *negative* going from benzoquinone, over [14]annulenedione to **385** ($E(\frac{1}{2})_1$ = −0.42, −0.63 and −0.72 V, respectively).

2. *Cyclooctadecatetraenetetraynediones (tetradehydro[18]annulenediones)*

A mixture of 10,15-dimethylcyclooctadeca-7,9,15,17-tetraene-2,4,11,13-tetrayne-1,6-dione (**388**) and 6,15-dimethylcyclooctadeca-6,8,15,17-tetraene-2,4,11,13-tetrayne-1,6-dione (**389**) has been obtained by oxidative coupling of **387** with oxygen, copper(I) chloride, ammonium chloride and concentrated hydrochloric acid in aqueous ethanol and benzene[258]. **388** is bright red and relatively soluble, while **389** is bright yellow and very insoluble.

The dicyclohexenotetradehydro[18]annulenediones **390** and **391** have been synthesized in a similar way[258]. **390** has also been obtained by another unambiguous route[259].

The electrochemical reduction of the diones **388**–**390** was examined by cyclic voltammetry[260]. All compounds exhibited chemical and electrochemical reversibility at both the first and second waves. The first wave corresponds to the addition of one electron to produce the radical anion and the second wave corresponds to the addition of a second electron to produce the dianion. The reduction potentials of **388**–**390** lie at more positive values (−1.60 to −1.74 V) than that of benzoquinone (−1.92 V). This is by no means a measure for an increased aromatic character of the conjugate hydroquinone anions of **388**–**390** but may be partially due to the greater charge separation in the larger rings.

2 (387) → (388), (389)

(387)

(388)

(389)

(390)

(391)

Indeed the potentials of the first wave ($E(\frac{1}{2})_1$: -0.66 to -0.70 V) are more negative than that of benzoquinone ($E(\frac{1}{2})_1$: -0.52 V). At any rate **388–390** are very easily and reversibly reduced at both waves and by this experimental criterion it is reasonable to regard these annulenediones as quinones of an aromatic system.

V. REFERENCES

1. P. Boldt, in Houben-Weyl, *Methoden der Organischen Chemie*, Vol. 7/3b (Ed. Ch. Grundmann), G. Thieme Verlag, Stuttgart, 1979, pp. 187–232.
2. J. L. Benham, R. West and J. A. T. Norman, *J. Am. Chem. Soc.*, **102**, 5047 (1980).
3. L. Wendling, S. K. Koster and R. West, *J. Org. Chem.*, **42**, 1126 (1977).
4. D. E. Wellman, K. R. Lassila and R. West, *J. Org. Chem.*, **49**, 965 (1984).
5. L. A. Wendling and R. West, *J. Org. Chem.*, **43**, 1573 (1978).
6. S. K. Koster and R. West, *J. Org. Chem.*, **40**, 2300 (1975).
7. J. L. Benham and R. West, *J. Am. Chem. Soc.*, **102**, 5054 (1980).
8. R. West, D. C. Zecher, S. K. Koster and D. Eggerling, *J. Org. Chem.*, **40**, 2295 (1975).
9. IG-Farbenind., *DRP*, 470501 (1926).
10. E. Clar, *Aromatische Kohlenwasserstoffe*, Springer-Verlag, Berlin, 1952; (a) p. 390; (b) p. 387; (c) p. 310; (d) p. 390.

11. H. S. Bien, J. Stawitz and K. Wunderlich, in *Ullmann's Encyclopedia of Industrial Chemistry*, Vol. A2, VCH Verlagsgesellschaft, Weinheim, 1985, pp. 355–417; (a) p. 397; (b) p. 395; (c) p. 394.
12. H. Hopff and H. R. Schweizer, *Helv. Chim. Acta*, **45**, 1044 (1962).
13. H. Vogler and M. C. Böhm, *Theor. Chim. Acta*, **66**, 51 (1984).
14. G. J. Gleicher, D. F. Church and J. C. Arnold, *J. Am. Chem. Soc.*, **96**, 2403 (1974).
15. Y. Nagai and K. Yamamoto, *Kogyo Kagaku Zasshi*, **68**, 2257 (1965); *Chem. Abstr.*, **66**, 47291y (1967).
16. Y. Nagai and K. Nagasawa, *Nippon Kagaku Zasshi*, **87**, 284 (1966); *Chem. Abstr.*, **65**, 15293 (1966).
17. J. Griffiths and C. Hawkins, *J. Soc. Dyers Colour.*, **89**, 173 (1973); *Chem. Abstr.*, **79**, 93284q.
18. R. Scholl and C. Seer, *Ann. Chem.*, **394**, 111 (1912).
19. H. Vollmann, H. Becker, M. Corell and H. Streeck, *Ann. Chem.*, **531**, 38 (1937).
20. E. Clar, *J. Chem. Soc.*, 2013 (1949).
21. W. Bradley and F. Sutcliffe, *J. Chem. Soc.*, 2118 (1951).
22. R. Willstätter and J. Parnas, *Chem. Ber.*, **40**, 1406 (1907).
23. R. Kuhn and J. Hammer, *Chem. Ber.*, **83**, 413 (1950).
24. L. Horner and K. H. Weber, *Chem. Ber.*, **98**, 1246 (1965).
25. H. Paul and G. Zinner, *J. Prakt. Chem.* (4), **18**, 219 (1962).
26. R. Willstätter and A. S. Wheeler, *Chem. Ber.*, **47**, 2798 (1914).
27. H. L. K. Schmand, H. Kratzin and P. Boldt, *Ann. Chem.*, 1560 (1976).
28. F. Farina, R. Martinez-Utrilla, M. C. Paredes and V. Stefani, *Synthesis*, 781 (1985).
29. M. S. Bloom and G. O. Dudek, *Tetrahedron*, **26**, 1267 (1970).
30. K. H. Menting, W. Eichel, K. Riemenschneider, H. L. K. Schmand and P. Boldt, *J. Org. Chem.*, **48**, 2814 (1983).
31. M. V. Gorelik, S. P. Titova and V. A. Trdatyan, *Zh. Org. Khim.*, **13**, 424 (1977); *J. Org. Chem (USSR)*, **15**, 147 (1979).
32. M. V. Gorelik, S. P. Titova and V. A. Trdatyan, *Zh. Org. Khim.*, **15**, 157 (1979).
33. Yu. E. Gerasimenko, N. T. Poteleshchenko and V. V. Romanov, *Zh. Org. Khim.*, **14**, 2387 (1978).
34. A. Topp, P. Boldt and H. Schmand, *Ann. Chem.*, 1167 (1974).
35. F. Setiabudi and P. Boldt, *Ann. Chem.*, 1272 (1985).
36. P. Boldt, *Chem. Ber.*, **100**, 1270 (1967).
37. P. Boldt, P. Hilmert-Schimmel, R. Müller and D. Heuer, *Chem. Ber.*, **120**, 797 (1987).
38. M. S. Newman and R. L. Childers, *J. Org. Chem.*, **32**, 62 (1967).
39. M. S. Newman and H. M. Chung, *J. Org. Chem.*, **39**, 1036 (1974); M. S. Newman and H. M. Dale, *J. Org. Chem.*, **42**, 734 (1977).
40. F. R. Hewgill and J. M. Stewart, *J. Chem. Soc., Chem. Commun.*, 1419 (1984).
41. C. Graebe and C. Liebermann, *Chem. Ber.*, **3**, 742 (1870); *Ann. Chem.*, **158**, 285 (1871).
42. V. K. Duplyakin, D. Kh. Sembaev and B. V. Suvurov, *Izv. Akad. Nauk Kaz. SSR, Ser. Khim.*, **19**, 65 (1969); *Chem. Abstr.*, **71**, 70381q (1969); Goftman and Goheb, *Z. Prikl. Chim.*, **29**, 1256 (1956); Engl. ed., p. 1355.
43. V. K. Duplyakin, D. Kh. Sembaev and B. V. Suvurov, *Izv. Akad. Nauk Kaz. SSR, Ser. Khim.*, **22**, 62 (1972); *Chem. Abstr.*, **77**, 101282p (1972).
44. V. A. Proskurjakov and A. N. Cistjakov, Obislenie Uglevodorodov, *ich proiz. i. bitumov. Sbornik statij*, **9**, 24 (1971); *Chem. Abstr.*, **77**, 4896c (1972).
45. Yu. E. Gerasimenko and V. A. Shigalevski, *Probl. Poluch. Poluprod. Prom. Org. Sin., Akad. Nauk SSSR,Otd. Obshch. Tekh. Khim.*, 217 (1967) and 220 (1967); *Chem. Abstr.*, **68**, 114288j and 114289k (1968).
46. G. Schaden, *J. Org. Chem.*, **48**, 5385 (1983).
47. M. Hirohashi, T. Tsumoda, A. Yamao-ka and G. Nagamatsu, Japan. Kokai, *Jp*, 7347, 826 (1973); *Chem. Abstr.*, **79**, 127037d (1973).
48. Mc. R. Maxfield, A. N. Bloch and D. O. Cowan, *J. Org. Chem.*, **50**, 1789 (1985).
49. E. Beschke, *Ann. Chem.*, **384**, 143 (1911).
50. G. M. Badger, *J. Chem. Soc.*, 999 (1948).
51. G. Schaden, *Angew. Chem.*, **89**, 50 (1977).
52. C. Marshalk and C. Stamm, *Bull. Soc. Chim. Fr.*, 418 (1948).

53. Yu. E. Gerasimenko and N. T. Poteleshenko, *Zh. Org. Khim.*, **7**, 2413 (1971).
54. C. Marshalk, *Bull. Soc. Chim. Fr.*, 155 (1952).
55. C. Dufraisse and J. Houpillart, *C. R. Hebd. Séances Acad. Sci., Ser. C*, **206**, 756 (1938).
56. H. Stobbe and P. Naoum, *Chem. Ber.*, **37**, 2240 (1904).
57. Yu. E. Gerasimenko, N. T. Poteleshenko and V. V. Romanov, *Zh. Org. Khim.*, **16**, 2014 (1980).
58. (a) Toshiba Corp., Jpn Kokai Tokyo Koho, *Jp*, 57, 187, 870 (1982); *Chem. Abstr.*, **98**, 188035r (1983); (b) A. Zincke and E. Unterkreuter, *Monatsh. Chem.*, **40**, 405 (1919); K. Brass and E. Tengler, *Chem. Ber.*, **64**, 1646 (1931).
59. A. Zincke, F. Limmer and O. Wolfbauer, *Chem. Ber.*, **58**, 323 (1925).
60. A. Zincke, A. Pongratz and K. Funke, *Chem. Ber.*, **58**, 330 (1925).
61. A. Zincke, F. Stimmler and E. Reuss, *Monatsh. Chem.*, **64**, 415 (1934).
62. A. Zincke, *Monatsh. Chem.*, **61**, 1 (1932).
63. D. C. Allport and J. D. Bu'Lock, *J. Chem. Soc.*, 4090 (1958).
64. D. W. Cameron and H. W. S. Chan, *J. Chem.Soc.*, 1825 (1966).
65. R. L. Edwards and H. J. Lockett, *J. Chem. Soc., Perkin Trans. 1*, 2149 (1976).
66. A. Zincke, R. Springer and A. Schmidt, *Chem. Ber.*, **58**, 2386 (1925).
67. E. Haslam, *Tetrahedron*, **5**, 99 (1959).
68. A. Zincke and R. Dengg, *Monatsh. Chem.*, **43**, 125 (1922).
69. A. Zincke and K. J. v. Schieszl, *Monatsh. Chem.*, **67**, 196 (1936).
70. A. Zincke, W. Hirsch and E. Brozek, *Monatsh. Chem.*, **51**, 205 (1929).
71. E. Clar, *Chem. Ber.*, **73**, 409 (1940).
72. Yu. E. Gerasimenko and N. T. Poteleshenko, *Zh. Org. Khim.*, **15**, 393 (1979).
73. J. W. Cook, R. S. Ludwiczak and R. Schoental, *J. Chem. Soc.*, 1112 (1950).
74. H. E. Schroeder, F. B. Stilman and F. S. Palmer, *J. Am. Chem. Soc.*, **78**, 446 (1956).
75. R. O. C. Norman and W. A. Waters, *J. Chem. Soc.*, 2379 (1956).
76. A. Windaus and K. Raichle, *Ann. Chem.*, **537**, 157 (1939).
77. H. Thielemann, *Microchim. Acta*, **1**, 145 (1973).
78. J. K. Reichert, *Arch. Hyg. Bakteriol.*, **152**, 265 (1968); *Chem. Abstr.*, **69**, 89651h (1968).
79. E. Cavalieri and R. Auerbach, *J. Natl Cancer Inst.*, **53**, 393 (1974); *Chem. Abstr.*, **82**, 3427p (1975).
80. R. Rajagopalan, K. G. Vohra and A. M. Mohan Rao, *The Science Total Environment*, **27**, 33 (1983).
81. C. Nagata, M. Kodama and Y. Joki, in *Polycyclic Hydrocarbons and Cancer*, Vol. 1, Academic Press, New York, 1978, pp. 247–260.
82. R. J. Lorentzen, W. J. Caspari, S. A. Lesko and P. O. P. Tso, *Biochemistry*, **14**, 3970 (1975).
82a. E. Clar, *Chem. Ber.*, **72**, 2134 (1939).
83. IG-Farbenind., *DRP*, 453 768 (1925); 470 501 (1926).
84. IG-Farbenind., *DRP*, 550 712 (1930); *Chem. Zentralbl.*, 783 (1932, II).
85. IG-Farbenind., *DRP*, 576 466 (1931); *Chem. Zentralbl.*, 791 (1933, II).
86. IG-Farbenind., *DRP*, 589 639 (1932); *Chem. Zentralbl.*, 1890 (1934, I); *DRP*. 589 639 (1932), 611 512 (1933); *Chem. Zentralbl.*, 1890 (1934, I), 282 (1935, II).
87. M. Zander and W. H. Franke, *Chem. Ber.*, **101**, 2404 (1968).
88. A. Corbellini, *Gazz. Chim. Ind. Appl.*, **13**, 109 (1931); *R. Ist. Lombardo Sci. Lettere Rend.* (2), 69, 287, 429, 580 (1936).
89. L. Kalb, *Chem. Ber.*, **47**, 1724 (1914).
90. Casella, *DRP*, 445 390 (1913).
91. S. S. Dalvi and S. Seshadri, *Indian J. Chem., Sect. B.*, **24**, 377 (1985).
92. Casella, *DRP*, 458 598 (1925).
93. IG-Farbenind., *DRP*, 485 961 (1927).
94. E. Clar, *Chem. Ber.*, **72**, 1645 (1939).
95. L. L. Ansell, T. Rangarajan, W. M. Burgess, E. J. Eisenbraun, G. W. Keen and M. C. Hamming, *Org. Prep. Proc. Intern.*, **8**, 133 (1976); *Chem. Abstr.*, **85**, 177123y (1976).
96. T. Nagaoka and S. Okazaki, *J. Electroanal. Chem.*, **138**, 139 (1983).
97. E. Clar and H. Frömmel, *Chem. Ber.*, **82**, 52 (1949).
98. E. Clar, *Chem. Ber.*, **73**, 351 (1940).
99. A. Scholl and H. Neumann, *Chem. Ber.*, **55**, 118 (1922).
100. IG-Farbenind., *DRP*, 518 316 (1927), 555 180 (1929); *Chem. Zentralbl.*, 3627 (1932, II), 3000 (1928, I).

101. E. Unseren and L. F. Fieser, *J. Org. Chem.*, **27**, 1386 (1962).
102. E. J. Moriconi and L. Salce, *J. Org. Chem.*, **32**, 2829 (1967).
103. F. Homburger, A. Treger and E. Boger, *Natl Cancer Inst., Monogr.*, **28**, 259 (1967); *Chem. Abstr.*, **69**, 85184k (1968).
104. G. Freslon and Y. Lepage, *C. R. Hebd. Séances Acad. Sci., Ser. C*, **280**, 961 (1975).
105. E. Clar and F. John, *Chem. Ber.*, **64**, 986 (1931).
106. IG-Farbenind., *DRP*, 430 558 (1924), 440 890 (1924).
107. IG-Farbenind., *DRP*, 426 711 (1924), 483 229 (1927), 518 316, 555 180 (1931); *Chem. Zentralbl.*, 3627 (1932, II).
108. A. V. Artynkhov, V. V. Varifat'ev, J. J. Krasynk, O. A. Chussak, K. J. Evstraf'eva and V. T. Kens, *Otkrytiya, Izobret., Prom. Obratztsy Tovarnye Znaki*, **46**, 54 (1969); *Chem. Abstr.*, **71**, 112692z (1969).
109. J. P. Savoca, Am. Cyanamid, US 37 967 33 (1974); *Chem. Abstr.*, **81**, 65209s (1974).
110. A. B. Tarasenko, V. P. Esip, N. N. Kuz'menko, V. J. Dyachenko and V. M. Il'mushkin, *USSR* 810 662 (1981); *Chem. Abstr.*, **95**, 61858u (1981).
111. Hoechst, *DRP*, 423 720 (1924).
112. G. Chatelain, *Comm. Energie At. (France) Rappt.* CEA-R-2858 (1965); *Chem. Abstr.*, **65**, 13626d (1966).
113. G. Schaden, *Z. Naturforsch.*, **B35**, 1328 (1980).
114. E. Clar, *Chem. Ber.*, **65**, 846 (1932).
115. A. Zincke, E. Ziegler and H. Gottschall, *Chem. Ber.*, **75**, 148 (1942).
116. IG-Farbenind., *DRP*, 430 557 (1924).
117. E. Clar, *Chem. Ber.*, **73**, 409 (1940).
118. A. Zincke, *Monatsh. Chem.*, **80**, 202 (1949).
119. IG-Farbenind., *DRP*, 430 557 (1924).
120. IG-Farbenind., *DRP*, 446 187 (1925); *Chem. Zentralbl.*, 1096 (1927, II); 1228 (1927, I).
121. IG-Farbenind., *DRP*, 430 557 (1924).
122. E. Clar and H. Frömmel, *Chem. Ber.*, **82**, 50 (1949).
123. E. Clar, *Chem. Ber.*, **82**, 46 (1949).
124. R. Scholl, H. K. Meyer and W. Winkler, *Ann. Chem.*, **494**, 201 (1932).
125. E. I. du Pont de Nemours, *A.P.* 1991 687 (1933); *Chem. Zentralbl.*, 2454 (1935, II).
126. R. Scholl, O. Böttger and L. Wanka, *Chem. Ber.*, **67**, 599 (1934).
127. A. G. Perkin and N. H. Haddock, *J. Chem. Soc.*, 1512 (1933).
128. C. Dufraisse and M. T. Mellier, *C. R. Hebd. Séances Acad. Sci., Ser. C*, **215**, 541 (1942).
129. H.-D. Brauer and R. Schmidt, *Photochem. Photobiol.*, **37**, 587 (1983).
130. H.-D. Brauer, W. Drews and R. Schmidt, *J. Photochem.*, **12**, 293 (1980).
131. W. Drews, R. Schmidt and H.-D. Brauer, *J. Photochem.*, **6**, 391 (1976/77).
132. R. Schmidt, W. Drews and H.-D. Brauer, *J. Am. Chem. Soc.*, **102**, 2791 (1980).
133. G. Rodewald, R. Arnold, J. Griesler and W. Steglich, *Angew. Chem.*, **89**, 56 (1977); *Angew. Chem. Int. Ed.*, **16**, 46 (1977).
134. R. Scholl and J. Mansfeld, *Chem. Ber.*, **43**, 1734 (1910).
135. S. Tokita, S. Mijazaki and I. Iwamoto, *Nippon Kagaku Zasshipon Kagaku Kaishi*, 440 (1982); *Chem. Abstr.*, **97**, 676 (1982).
136. F. Ullmann and M. Minajew, *Chem. Ber.*, **55**, 687 (1912); R. Scoll and C. Tänzer, *Ann. Chem.* **433**, 172 (1923).
137. A. Eckert and R. Tomascheck, *Monatsh. Chem.*, **39**, 839 (1918).
138. R. Scholl andd C. Seer, *Chem. Ber.*, **44**, 1091 (1911).
139. A. G. Perkin and J. W. Haller, *J. Chem. Soc.*, 125, 231 (1924).
140. A. G. Perkin and N. H. Haddock, *J. Chem. Soc.*, 1512 (1933).
141. H. Brockmann and R. Mühlmann, *Chem. Ber.*, **82**, 348 (1949).
142. R. Korenstein, K. Muszkat and E. Fischer, *Helv. Chim. Acta*, **59**, 1826 (1976).
143. U. P. Wild, *Chimia*, **22**, 473 (1968).
144. R. Scholl and H. K. Meyer, *Ann. Chem.*, **512**, 112 (1934).
145. BASF, *DRP*, 283 066 (1913).
146. I. G. Farbenind., *F.P.* 823 261 (1937); *Chem. Zentralbl.*, 3539 (1938, I).
147. N. S. Dokuniklin and L. Solodar, *Zh. Org. Khim.*, **4**, 1857 (1968); *Chem. Abstr.*, **70**, 28704b (1969).

148. A. Pongratz, *Monatsh. Chem.*, **50**, 87 (1928).
149. N. S. Dokuniklin, S. L. Solodar and A. V. Reznichenko, *Zh. Org. Khim.*, **12**, 1064 (1976); *Chem. Abstr.*, **85**, 108449x (1976).
150. R. Scholl, *Chem. Ber.*, **43**, 346 (1910).
151. BASF, *DE-OS* 1951 708 (1969); 2115 093 (1971); 2115 131 (1972).
152. R. Scholl and C. Seer, *Ann. Chem.*, **394**, 111 (1912).
153. IG-Farbenind., *E.P.* 382 877 (1932); *Chem. Zentralbl.*, 1525 (1933, I).
154. J. Potschiwanschek, *Chem. Ber.*, **43**, 1748 (1910).
155. R. Scholl, *Chem. Ber.*, **43**, 353 (1910).
156. R. Scholl, J. Potschiwanschek and J. Lenko, *Monatsh. Chem.*, **32**, 687 (1911).
157. BASF, *DRP*, 238 980 (1910); 278 424 (1913).
158. BASF, *DRP*, 211 927 (1908).
159. BASF, *DRP*, 186 596 (1906).
160. BASF, *DRP*, 268 504 (1912).
161. R. Scholl, K. Meyer and J. Donat, *Chem. Ber.*, **70**, 2180 (1937).
162. Marshall and Peters, *J. Soc. Dyers Colour.*, **68**, 289, 292 (1952).
163. A. K. Gupta, *J. Chem. Soc.*, 3479 (1952).
164. R. Scholl and C. Tänzer, *Ann. Chem.*, **433**, 163 (1923).
165. H. Hopff and H. R. Schweizer, *Helv. Chim. Acta*, **45**, 1045 (1962).
166. R. Scholl and J. Mansfeld, *Chem. Ber.*, **43**, 1734 (1910).
167. H. Meyer, R. Bondy and A. Eckert, *Monatsh. Chem.*, **33**, 1451 (1912).
168. W. Koch, T. Saito and Z. Yoshida, *Nippon Kagaku Zasshi*, **88**, 684 (1967); *Chem. Abstr.*, **68**, 59365 (1968).
169. W. Koch, T. Saito and Z. Yoshida, *Tetrahedron*, **28**, 319 (1972).
170. R. Scholl, *Chem. Ber.*, **52**, 1829 (1919).
171. E. Clar, *Chem. Ber.*, **81**, 62 (1948); **82**, 54 (1949).
172. H. Inokuchi, *Bull. Chem. Soc. Japan*, **24**, 222 (1951).
173. H. Kuroda and E. A. Flood, *Can. J. Chem.*, **39**, 1475 (1961).
174. A. Eckert, *Chem. Ber.*, **58**, 322 (1925).
175. G. Kortüm, W. Theilacker, H. Zeissinger and H. Elliehausen, *Chem. Ber.*, **86**, 294 (1953).
176. H. Brockmann and R. Randebrock, *Chem. Ber.*, **84**, 533, 541, 543 (1951).
177. H. Brockmann, R. Neef and E. Mühlmann, *Chem. Ber.*, **83**, 467 (1950) and literature cited there.
178. H. Brockmann, in *Fortschritte der Chemie Organischer Naturstotte.* (Ed. L. Zechmeister), Springer-Verlag, Wien, 1957, pp. 141–182.
179. G. Rodenwald, R. Arnold, J. Griesler and W. Steglich, *Angew. Chem.*, **89**, 56 (1977); *Angew. Chem. Int. Ed.*, **16**, 46 (1977).
180. D. Spitzner, *Angew. Chem.*, **89**, 55 (1977); *Angew. Chem. (Int. Ed.)*, **16**, 46 (1977).
181. E. Clar, *Chem. Ber.*, **72**, 2134 (1939).
182. K. Inukai and A. Ueda, *J. Chem. Soc. Japan, Ind. Chem. Sect.*, **53**, 175 (1950); *Chem. Abstr.*, **46**, 9847 (1952).
183. IG-Farbenind., *DRP*, 550 712 (1930); 576 466 (1931); *Chem. Zentralbl.*, 783 (1932, II); 791 (1933, II).
184. IG-Farbenind., *F.P.* 644 782 (1927); *Chem. Zentralbl.*, 580 (1929, I); *DRP*, 589 639 (1932); *Chem. Zentralbl.*, 1890 (1934, I).
185. K. Inukai and T. Ueda, *Rep. Gov. Ind. Res. Inst. Tokyo*, **47**, 21, 26 (1952); *Chem. Abstr.*, Vol. **47**, 2989 (1953).
186. E. Clar and W. Willicks, *Chem. Ber.*, **89**, 743, 748 (1956).
187. E. Clar, *J. Chem. Soc.*, 2013 (1949).
188. IG-Farbenind., *DRP*, 691 644 (1934).
189. R. K. Erünlü, *Ann. Chem.*, **721**, 43 (1969).
190. C. Dufraisse and M. Loury, *C. R. Hebd. Séances Acad. Sci., Ser. C.*, **213**, 689 (1941).
191. E. Clar, *Chem. Ber.*, **76**, 331 (1943).
192. IG-Farbenind., *DRP*, 724 833 (1938); *Chem. Zentralbl.*, 98 (1943, I).
193. IG-Farbenind., *DRP*, 456 583 (1926); *Chem. Zentralbl.*, 2011 (1928, I).
194. IG-Farbenind., *E.P.* 480 882 (1936); *Chem. Zentralbl.*, 1134 (1938, II); *DRP*, 695 031 (1936); *Chem. Zentralbl.*, 582 (1941, I).
195. R. Bohn, *Chem. Ber.*, **28**, 195 (1905).

196. A. Lüttringhaus and H. Neresheimer, *Ann. Chem.*, **473**, 259 (1929).
197. K. Nagasawa and Y. Nagai, Jpn. Tokyo Koho, *Jp.* 42/21032 (1967); *Chem. Abstr.*, **68**, 87070y (1968).
198. Y. Nagai and K. Nagasawa, *Kogyo Kagaku Zasshi*, **69**, 666 (1966); *Chem. Abstr.*, **66**, 37696v (1967).
199. BASF, *DRP*, 416 028 (1924).
200. Du Pont, *US* 2872 459 (1965); *DE-AS* 1068 687 (1956).
201. Czech., *CS* 202 946 (1982); *Chem. Abstr.*, **98**, 199838f (1983).
202. Allied Chem. Corp., *US* 3446 810 (1967); *Chem. Abstr.*, **71**, 62282t (1969).
203. H. Akamaku and H. Inokuchi, *J. Chem. Phys.*, **18**, 810 (1950).
204. BASF, *DRP*, 217 570 (1909); 402 640 (1922); IG-Farbenind., *DRP*, 436 828 (1922); 595 461 (1929); 608 442 (1933).
205. Y. Maezawa, *Bull. Chem. Soc. Japan*, **28**, 77 (1955).
206. BASF, *DRP*, 185 222 (1904).
207. BASF, *DRP*, 226 215 (1909).
208. BASF, *DRP*, 259 370, 260 020 (1912); 280 710 (1913); 395 691; 411 013 (1922); Scottish Dyes Ltd., *DRP*, 416 208; 418 639 (1921); Höchst. Farb., *DRP*, 420 146 (1923); IG-Farbenind., *DRP*, 436 829; 438 478 (1922).
209. BASF, *DRP*, 188 193 (1905); IG-Farbenind., *DRP*, 435 533 (1925).
210. A. J. Backhouse, *J. Chem. Soc.*, 93 (1955).
211. Höchst. Farb., *DRP*, 413 738; 414 203; 414 924 (1923); 442 511 (1924); IG-Farbenind., *DRP*, 436 887 (1924).
212. J. H. Li and N. Gotoh, *Yuki Gosei Kagaku Kyokai Shi*, **33**, 274 (1975); *Chem. Abstr.*, **83**, 78947x (1975).
213. R. Scholl, E. J. Müller and O. Böttger, *Chem. Ber.*, **68**, 45 (1935).
214. E. Clar, *Chem. Ber.*, **72**, 1684 (1939); **76**, 458 (1943).
215. IG-Farbenind., *DRP*, 431 775 (1924).
216. IG-Farbenind., *DRP*, 436 533 (1925).
217. BASF, *DRP*, 194 252 (1906).
218. IG-Farbenind., *DRP*, 436 888 (1924).
219. E. Schwenk, *Chem.-Ztg.*, 62 (1928).
220. *Ger.* 2 704 964 (1978); *Chem. Abstr.*, **89**, 131063h (1978); *USSR* 598 866 (1978); *Chem. Abstr.*, **89**, 108839g (1978).
221. BASF, *Ger*, 2 704 964 (1977); for a comprehensive treatment of this reaction see Reference 196.
222. A. Zincke, K. Funcke and A. Pongratz, *Chem. Ber.*, **58**, 799, 2222 (1925).
223. BASF, *DRP*, 217 570 (1909).
224. BASF, *DRP*, 267 418 (1912).
225. A. J. Backhouse, W. Bradley and F. K. Sutcliffe, *J. Chem. Soc.*, 91 (1955).
226. IG-Farbenind., *DRP*, 458 710 (1926); *Chem. Zentralbl.*, 398 (1928, II).
227. IG-Farbenind., *DRP*, 457 494 (1926); *Chem. Zentralbl.*, 2544 (1928, I).
228. R. Scholl and K. Meyer, *Chem. Ber.*, **67**, 1229 (1934).
229. E. Clar, *Chem. Ber.*, **69**, 1686 (1936).
230. H. Waldmann and G. Pollack, *J. Prakt. Chem.(NF)*, **150**, 113 (1938).
231. IG-Farbenind., *DRP*, 576 131 (1931); 553 000 (1930); *Chem. Zentralbl.*, 2245 (1932, II); 288 (1933, II).
232. IG-Farbenind., *DRP*, 551 447 (1930); *E. P.* 362 965 (1930); *Chem. Zentralbl.*, 2245 (1932, II); 288 (1933, II).
233. D. H. Hey, R. J. Nicholls and C. W. Pritchett, *J. Chem. Soc.*, 97 (1944).
234. Eastman Kodak, *US.P.* 2 637 733 (1951).
235. IG-Farbenind., *DRP*, 718 704 (1939); *Chem. Zentralbl.*, 101 (1942, II).
236. R. K. Erünlü, *Chem. Ber.*, **100**, 533 (1967).
237. E. Clar, *Chem. Ber.*, **82**, 52 (1949).
238. W. Bradley and F. Sutcliffe, *J. Chem. Soc.*, 2118 (1951).
239. W. Bradley and F. Sutcliffe, *J. Chem. Soc.*, 1247 (1952).
240. M. Stephenson and F. K. Sutcliffe, *J. Chem. Soc.*, 3516 (1962).
241. Y. Nagai and K. Nagasawa, *Nippon Kagaku Zasshi*, **87**, 281 (1966); *Chem. Abstr.*, **65**, 15293c (1966).

242. T. A. Turney, in *The Chemistry of the Quinonoid Compounds*, pt. 2 (Ed. S. Patai), John Wiley, and Sons, New York, 1974, pp. 857–876.
243. S. Kuroda, Y. Fukuyawa, T. Tsuchida, E. Tanaka and S. Hirooka, *Angew. Chem.*, **97**, 770 (1985).
244. V. Boekelheide and J. B. Phillips, *J. Am. Chem. Soc.*, **89**, 1695 (1967).
245. L. T. Scott and M. Oda, *Tetrahedron Lett.*, **27**, 799 (1986).
246. L. T. Scott, M. D. Roozeboom, K. N. Houk, T. Fukunaga, H. J. Lindner and K. Hafner, *J. Am. Chem. Soc.*, **102**, 5169 (1980).
247. L. T. Scott and C. M. Adams, *J. Am. Chem. Soc.*, **106**, 4857 (1984).
248. E. Vogel, E. Lohmar, W. A. Böll, B. Söhngen, K. Müller and H. Günther, *Angew. Chem*, **83**, 401 (1971).
249. G. A. Russell, T. Ku and J. Lockensgard, *J. Am. Chem. Soc.*, **92**, 3833 (1970).
250. E. Vogel, W. A. Böll and E. Lohmar, *Angew. Chem.*, **83**, 403 (1971).
251. M. Nakagawa, *Pure Appl. Chem.*, **44**, 885 (1975).
252. Y. Onishi, M. Iyoda and M. Nakagawa, *Tetrahedron Lett.*, **22**, 3641 (1981).
253. T. Kojima, Y. Sakato and S. Misumi, *J. Chem. Soc. Japan*, **45**, 2834 (1972).
254. M. Iyoda, Y. Onishi and M. Nakagawa, *Tetrahedron Lett.*, **22**, 3645 (1981).
255. E. Vogel, S. Böhm, A. Hedwig, B. O. Hergarten, J. Lex, J. Uschmann and R. Gleiter, *Angew. Chem.*, to be published.
256. E. Vogel, M. Biskup, A. Vogel, U. Haberland and J. Eimer, *Angew. Chem.*, **78**, 642 (1966).
257. F. Gerson, E. Heilbronner and V. Boekelheide, *Helv. Chim. Acta*, **47**, 1123 (1964).
258. N. Darby, K. Yamamoto and F. Sondheimer, *J. Am. Chem. Soc.*, **96**, 249 (1974).
259. K. Yamamoto and F. Sondheimer, *Angew. Chem. Int. Ed.*, **12**, 411 (1973).
260. R. Breslow, D. Murajama and R. Drury, *J. Am. Chem. Soc.*, **96**, 249 (1974).

The Chemistry of Quinonoid Compounds, Vol. II
Edited by S. Patai and Z. Rappoport

CHAPTER **26**

Non-benzenoid quinones

HENRY N. C. WONG, TZE-LOCK CHAN and TIEN-YAU LUH
Department of Chemistry, The Chinese University of Hong Kong, Shatin, New Territories, Hong Kong

I. INTRODUCTION

The format of this report on non-benzenoid quinones follows the one so neatly set by our predecessor[1]. Although the present authors seek to cover material published from 1974 to 1985, a few developments in this area which were omitted previously have also been included. As no encyclopaedic coverage is intended, the progress on the chemistry of non-benzenoid quinones is presented here mainly in conformity with the interests of the authors. Nevertheless, we hope there is sufficient material in this chapter to give the reader a bird-eye's view of the subject matter.

II. EVEN-MEMBERED RINGS

A. General Formula

In accord to the previous review[1], non-benzenoid quinones under the titled classification may be regarded as being generated from the following general formula:

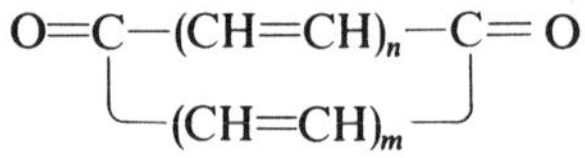

$$O{=}C{-}(CH{=}CH)_n{-}C{=}O \quad \text{(ring closed by } (CH{=}CH)_m\text{)}$$

in which either *m* or *n*, but not both, may be zero, and *m* and *n* are positive integers.

Cyclobutenequinone (**1**) is the case where $m = 1$, $n = 0$. The cases $m = 0$, $n = 2$ (*o*-benzoquinone) and $m = 1, n = 1$ (*p*-benzoquinone) are of course outside the scope of this chapter.

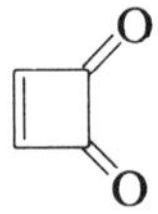

(**1**)

When $m = 2$, $n = 1$ or $m = 3$, $n = 0$, cycloocta-2,5,7-triene-1,4-dione (**2**) and cycloocta-3,5,7-triene-1,2-dione (**3**) are generated respectively.

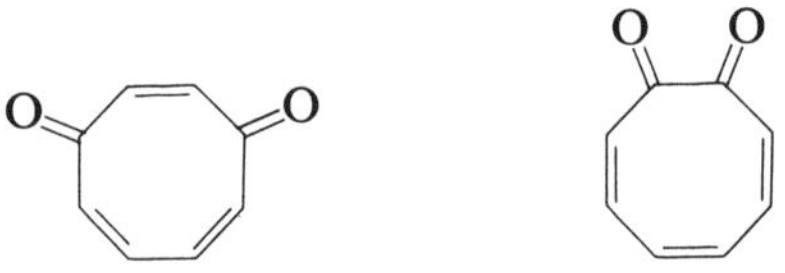

(**2**) (**3**)

Annulenequinones, which can be generated by the general formula where $m+n \geqslant 6$ (where some of the double bonds may be replaced by triple bonds) will also be discussed here.

B. Four-membered Ring Systems

The chemistry of cyclobutenequinone[2] and derivatives[3-5], benzocyclobutenequinone (**4**) and derivatives[6,7], squaric acid (**5**) and derivatives[6,8-17] has been extensively

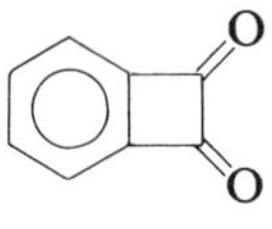

(**4**) 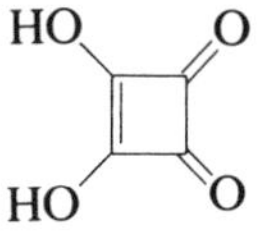 (**5**)

reviewed[2-17]. Hence we will restrict ourselves in this chapter to the discussion of the latest progress in this field.

1. Cyclobutenequinone, squaric acid and their derivatives

The chemistry of four-membered ring quinones was highlighted by the isolation of a new microbial toxin moniliformin (**6**)[18], the structure of which was confirmed by X-ray crystallography[19].

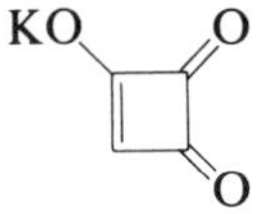

(**6**) 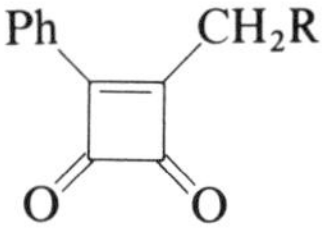 (**7**)

New synthetic routes to 3-alkyl-4-phenylcyclobutene-1,2-dione (**7**) have been reported[20]. In one of them, quinone **8** reacts with the sodium salt **9** to produce diester **10**,

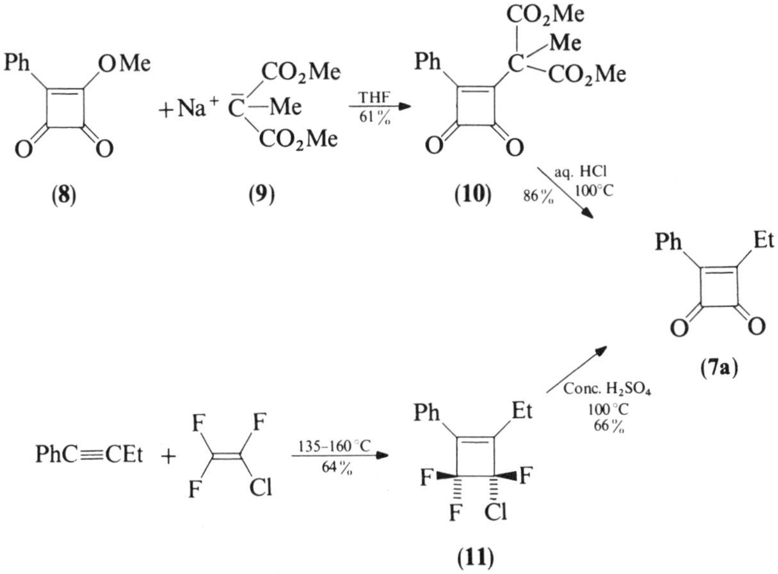

(**8**) (**9**) (**10**) (**7a**) (**11**)

which can be hydrolysed to **7a**[20]. Alternatively, **7a** can also be prepared by the thermal [2 + 2] cycloaddition of ethyl phenyl acetylene and chlorotrifluoroethylene which yields cyclobutene **11**. Hydrolysis of **11** furnishes **7a**[20].

The methylene group of **7** is sufficiently activated to undergo condensation with various aldehydes[21]. For example, **7a** reacts with aldehyde **12** to provide the dehydration product **13**[21]. Furthermore, the methylene group of **7** can also be brominated[22] to **14**, which upon treatment with silver acetate gives **15**[22].

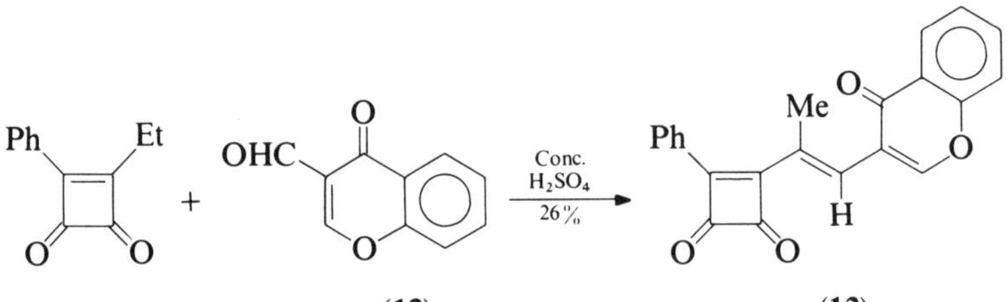

(**7a**) (**12**) (**13**)

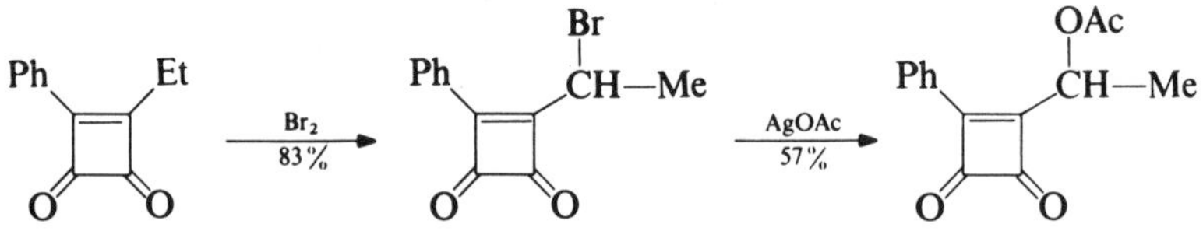

(**7a**) (**14**) (**15**)

It has been reported that 3,4-dichloro-3-cyclobutene-1,2-dione (**16**) readily reacts with dithiol or dithiolate to give 1,4-dithiin- as well as 1,3-dithio-derivatives[23]. At $-30°C$, **16** reacts with **17** and **18**, to give **19a** and **19b** in 50% and 30% yields, respectively[23]. Moreover, Lawesson reagent has been used to convert **16** to dicyclobuta[1,4]dithiin-1,2,4,5-tetraone (**19c**) in quantitative yields[23].

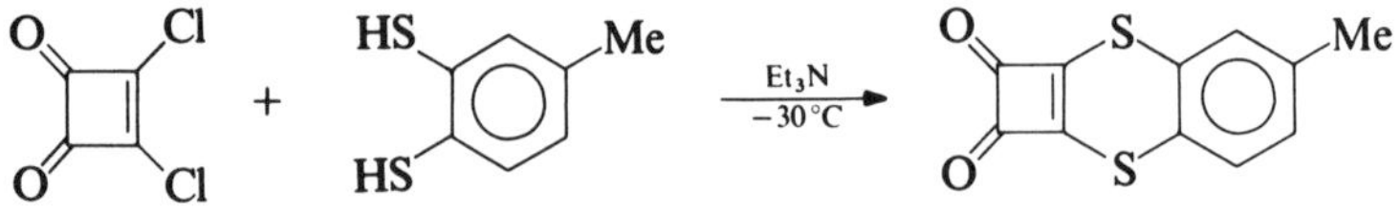

(**16**) (**17**) (**19a**)

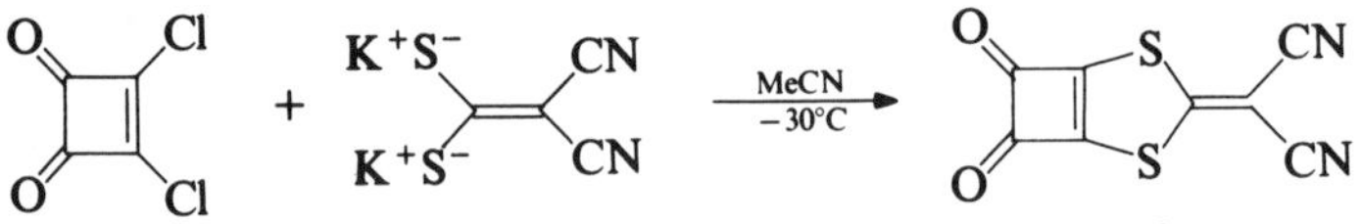

(**16**) (**18**) (**19b**)

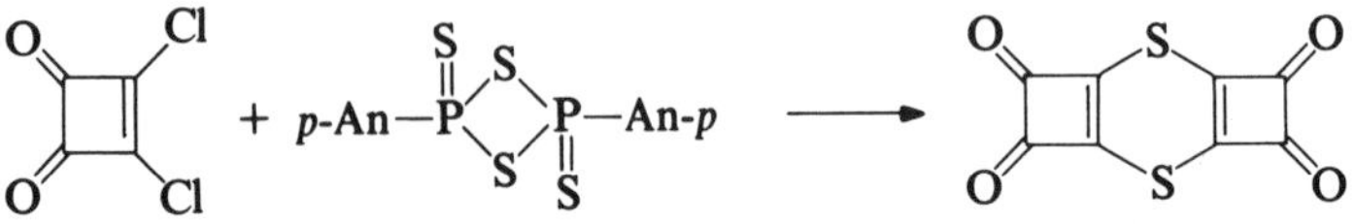

(**16**) (**19c**)

Slow diffusion of atmospheric moisture to an acetonitrile solution of **20** affords the colourless, water-soluble 4-diethylaminocyclobutenedione-3-*N*, *N*-diethylcarboxamide (**21**)[24], whose structure has been confirmed by X-ray diffraction study[24].

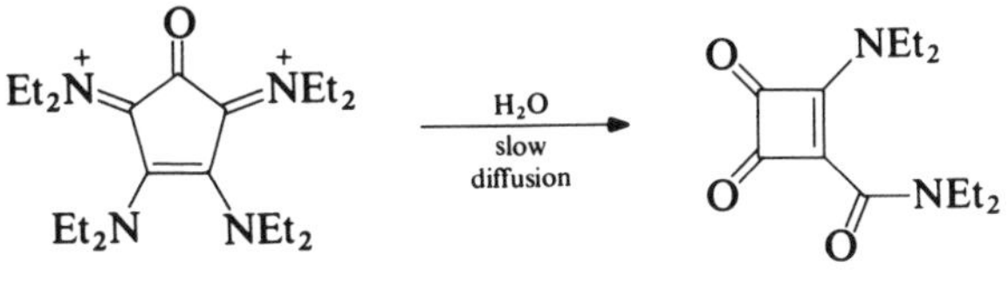

(**20**) (**21**)

2. *Betaines of squaric acid derivatives*

Compounds **22** and **23** are of interest because their resonance forms **22b** and **23b** are examples of 'push–pull' cyclobutadienes. The synthesis of **22** has been realized by addition

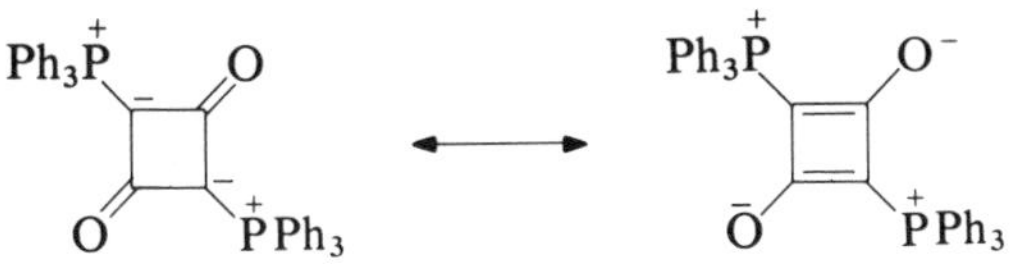

(**22a**) (**22b**)

(**23a**) (**23b**)

of HCl to a solution of the heterocumulene ylide **24**. Presumably the ketenic phosphonium salt **25** is generated first, which undergoes [2 + 2]cycloaddition with excess **24** to give **26**.

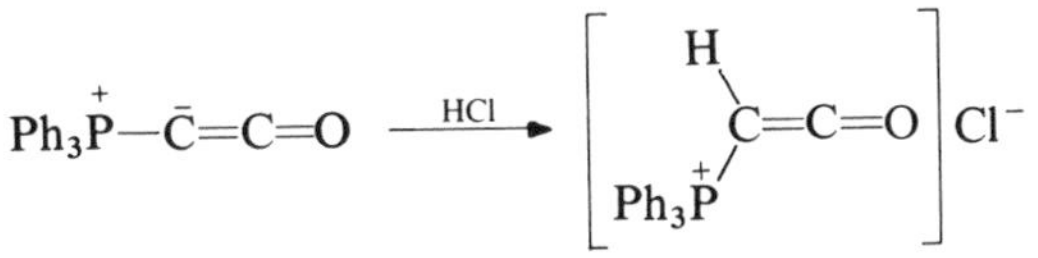

(**24**) (**25**)

Subsequent treatment of **26** with sodium bis-(trimethylsilyl)amide affords the bis-ylide **22**[25]. It is interesting to note that, oxidation of **22** with ozone–triphenylphosphite adduct affords the cyclobutanetrione derivative **27**[25]. Similarly, **28** gives the ketenimine **29** which leads to **23**. The CO stretching frequencies which have been located at 1650 and 1610 cm^{-1} and at 1563 and 1527 cm^{-1} respectively for **22** and **23** lend support to the notion that **22b** and **23b** contribute significantly to the overall structures[25].

24 + **25** → (**26**) $\xrightarrow{NaN[SiMe_3]_2}$ (**22**) → (**27**)

$Ph_3\overset{+}{P}-\overset{-}{C}=C=NPh$ $\xrightarrow{HCl}$ $[H(Ph_3\overset{+}{P})C=C=NPh]\ Cl^-$

(**28**) (**29**)

Compound **27** has also been prepared by addition of triphenylphosphine to perchlorocyclobutenone (**30**), which leads to **31**. Upon hydrolysis, **31** provides **27**[26].

(**30**) + PPh_3 → [(**31**) Cl^-] → **27**

In principle, the cyclobutanetrione **27** can possess four resonance structures **27a**, **27b**, **27c** and **27d**[26]. However, their relative importance has not been settled.

(**27a**) ↔ (**27b**) ↔ (**27c**) ↔ (**27d**)

The chemistry of the *N*-betaines, namely, squaraines (**32**) has been comprehensively reviewed[27]. A new class of *N*-betaines of squaric acid (**5**), namely **33**, have been prepared by reaction between an amine and 3,4-dichloro-3-cyclobutene-1,2-dione (**16**) in aqueous THF[28]. In the IR spectra of betaines **33**, the existence of strong bands at 1650–1570 cm^{-1} indicates that they should have the cyclobutenequinone structure[28].

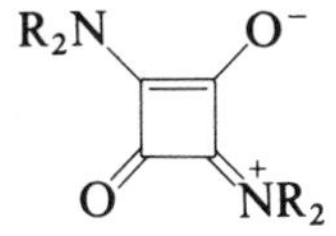

(32)

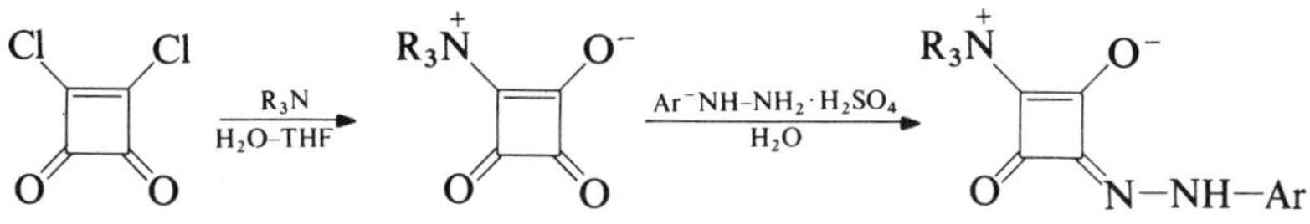

(16) (33) (34)

Condensation of **33** with arylhydrazines yields arylhydrazones **34**[28]. By the same strategy, **16** can be converted to **35**, **36**, **37** and **38**[29].

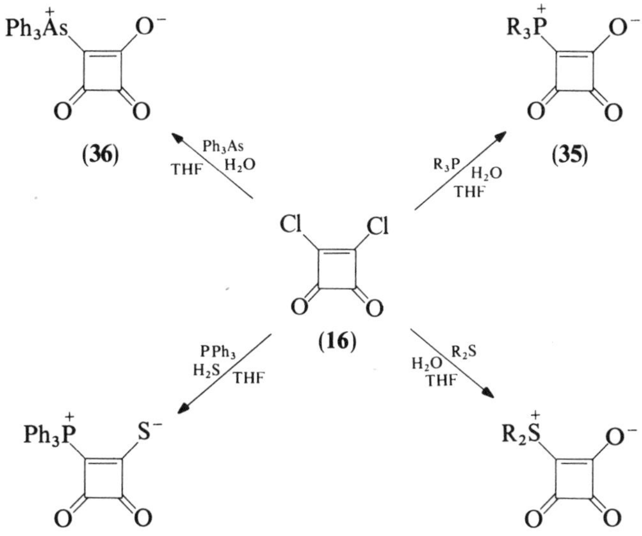

(36) (35) (16) (38) (37)

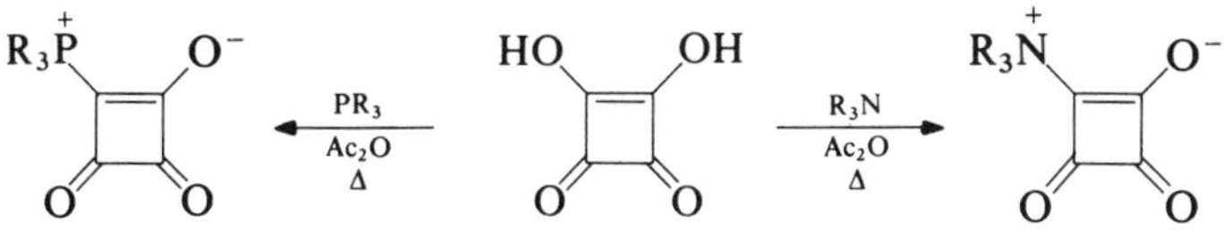

(35) (5) (33)

Alternatively, **33** and **35** can also be obtained from squaric acid (**5**)[30].

3. Pseudooxocarbon dianions

The chemistry of thioxocarbon dianions and their derivatives has been reviewed[31]. Novel pseudooxocarbon dianions of the C_4 series such as **39** and **40** can be prepared from

41 and **42** respectively[32, 33]. Similarly, **42** can also be converted to dianion **43**[32, 33]. The highly symmetrical structures of **39**, **40** and **43** have been revealed by ^{13}C-NMR spectroscopy[32, 33]. The X-ray structural analysis of **39**, however, indicates that the dianion is twisted and the central four-membered ring is non-planar[32, 33].

Me$_2$N, NTos, Me$_2$N, NTos — $\xrightarrow[\text{pyridine}]{4CH_2(CN)_2}$ — NC—C, CN, CN, C—CN, NC—C, C—CN, CN, CN, $2H_2\overset{+}{N}Me_2$

(41) **(39)**

BuO, S, BuO, S — $\xrightarrow[\text{NaOBu}]{4CH_2(CN)_2}$ — CN, CN, NC—C, C—CN, NC—C, C—CN, CN, CN, 2Na$^+$ **(40)**

(42) — $\xrightarrow[\text{NaOBu}]{4H_2NCN}$ — NC—N, N—CN, NC—N, N—CN, 2Na$^+$ **(43)**

4. Benzocyclobutenequinone and derivatives

Making use of two known procedures, a large scale synthesis of benzocyclobutenequinone (**4**) has recently been described[34–36]. Thus, conversion of anthranilic acid to benzyne and the subsequent [2 + 2]cycloaddition with vinylidene chloride yields benzocyclobutenone (**44**)[34, 35]. Bromination and hydrolysis finally transform **44** to **4**[34, 36].

CO$_2$H, NH$_2$ — $\xrightarrow[\text{HCl}]{\text{isoamyl nitrite}}$ — (1) $H_2C{=}CCl_2$ (2) dil. H_2SO_4 — O — $\xleftarrow[(2)\ H_2SO_4]{(1)\ NBS}$ — O, O

(4) **(44)**

Alternatively, reaction of cyclobutene **45** with 1-trimethylsiloxydiene **46a** gives **47a** after desilylation[34]. Aromatization and hydrolysis affords **4** in 63% overall yield[34].

(**46a**, R=H)
(**46b**, R=Me)

(**45**)

(**47a**, R=H)
(**47b**, R=Me)

NaOMe
MeOH
0°C

70% H_2SO_4
100°C

(**4**, R=H)
(**49**, R=Me)

(**48a**, R=H)
(**48b**, R=Me)

Similarly, 4-methylbenzocyclobutenequinone (**49**) can also be obtained in 72% overall yield from **46b**[34]. Furthermore, **47a** can be oxidized to **50a**, which is aromatized and

PCC
CH_2Cl_2

(**47a**, R = H)
(**47b**, R = Me)

(**50a**, R = H)
(**50b**, R = Me)

(1) NaOMe
MeOH 0°C
(2) dil·HCl
MeOH

conc. H_2SO_4
HOAc

(**52a**, R = H)
(**52b**, R = Me)

(**51a**, R = H)
(**51b**, R = Me)

hydrolysed to **51a**. More vigorous hydrolysis converts **51a** to 3-hydroxybenzocyclobutenequinone (**52a**) in 36% overall yield[34]. Similarly, 3-hydroxy-5-methylbenzocyclobutenequinone (**52b**) can be prepared in 63% overall yield[34].

The Danishefsky's diene **53** reacts smoothly with **45** to give cycloadduct **54** after hydrolysis. Adduct **54** can be converted to enone **55**, which undergoes aromatization as well as mild acid hydrolysis to afford the gem-difluoride **56**. More vigorous acid hydrolysis of **56** yields 4-hydroxybenzocyclobutenequinone (**57**) in 43% overall yield[34].

OMe Cl Cl F F Me$_3$SiO F + 120°C OMe Cl Cl F F O H F

(**53**) (**45**) (**54**)

pTsOH C_6H_6 Δ

O conc. H_2SO_4 HOAc O HO O F F HO Cl Cl F F O H F

(**57**) (**56**) (**55**)

The McOmie–Rees procedure[37, 38] has been applied extensively to prepare substituted benzocyclobutenequinones[39]. As an example, the protected Diels–Alder adduct **58** can be pyrolysed in the vapour phase to afford 3-alkoxybenzocyclobutenequinones (**59**)[39], which

O NH NH HO O + Pb(OAc)$_4$ CH_2Cl_2 O N N HO O

O N N RO O 500 °C O O RO

(**58a**, R = Me)
(**58b**, R = CH_2OMe)

(**59a**, R = Me)
(**59b**, R = CH_2OMe)

serve as starting material for the total synthesis of islandicin and digitopurpone[39]. Similarly, 4-chlorobenzocyclobutenequinone (**60**)[40], 3,6-dichlorobenzocyclobutenequinone (**61**)[40], 4,5-dichlorobenzocyclobutenequinone (**62**)[40], 4,5-dibromobenzocyclobutenequinone (**63**)[40], 4,5-dimethylbenzocyclobutenequinone (**64**)[40],

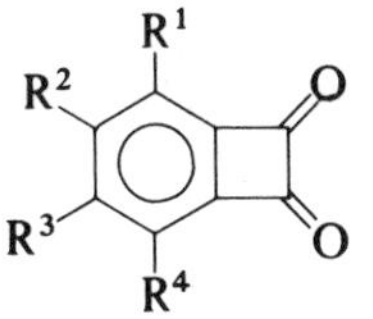

(60–64, 69, 70)

(60, R^3 = Cl; R^1 = R^2 = R^4 = H)
(61, R^1 = R^4 = Cl; R^2 = R^3 = H)
(62, R^2 = R^3 = Cl; R^1 = R^4 = H)
(63, R^2 = R^3 = Br; R^1 = R^4 = H)
(64, R^2 = R^3 = Me; R^1 = R^4 = H)
(69, R^3 = OMe; R^1 = R^2 = R^4 = H)
(70, R^2 = R^3 = OMe; R^1 = R^4 = H)

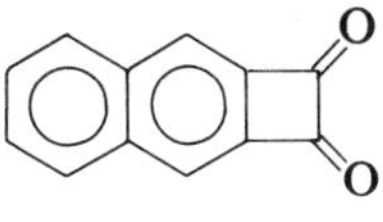

(65)

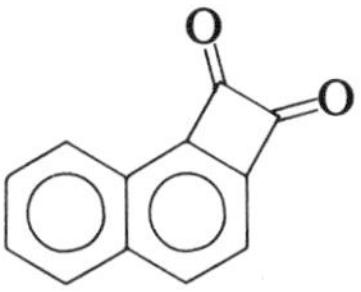

(66) 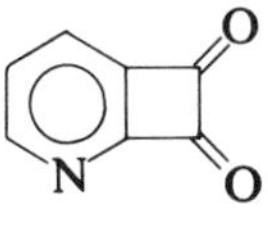 (67) 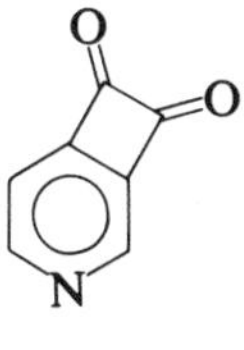 (68)

cyclobuta[*b*]naphthalene-1,2-dione (**65**)[40], cyclobuta[*a*]naphthalene-1,2-dione (**66**)[40], cyclobuta[*a*]pyridine-1,2-dione (**67**)[40], cyclobuta[*b*]pyridine-1,2-dione (**68**)[40], 4-methoxybenzocyclobutenequinone (**69**)[41] and 4,5-dimethoxybenzocyclobutenequinone (**70**)[41] have been prepared by applying the same methodology[37, 38]. Compounds **66**, **67** and **68** are reported to be very unstable[40].

The amino acid **71** can be converted to the benzyne **72**, which is trapped with vinylidene chloride. Subsequent hydrolysis of the resulting dichloride intermediate affords the ketone

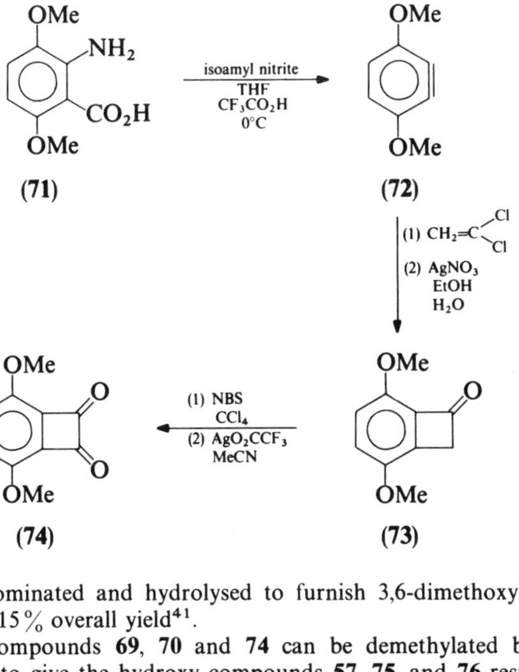

isoamyl nitrite, THF, CF_3CO_2H, 0°C

(71) (72)

(1) $CH_2{=}CCl_2$ (2) $AgNO_3$, EtOH, H_2O

(1) NBS, CCl_4 (2) AgO_2CCF_3, MeCN

(74) (73)

73[42], which is brominated and hydrolysed to furnish 3,6-dimethoxybenzocyclobutenequinone (**74**) in 15% overall yield[41].

The methoxy compounds **69**, **70** and **74** can be demethylated by heating with hydrobromic acid to give the hydroxy compounds **57**, **75**, and **76** respectively[41]. It is interesting to point out that these products are all relatively strong acids, and the acidities of **57** and **75** are stronger than those of **52a** and **76** respectively[41]. Furthermore, the acidities of **75** and **76** are weaker than that of squaric acid (**5**)[41].

69 —HBr→ **57**

70 —HBr→ (**75**)

74 —HBr→ (**76**)

Thermal dimerization of **77** leads to a radical intermediate which rearranges to benzocyclobutene **78**[43]. Acid hydrolysis converts **78** to the benzocyclobutenequinone **79**[43].

$Cl_2C{=}C(Me){-}C{\equiv}C{-}Cl$ (**77**) —CCl_4, 80 °C→ [intermediates] —MeCN, Δ, 50%→ (**78**) —H_2SO_4, 100°C, 56%→ (**79**)

Photo[2 + 2]cycloaddition of tetrachloroethene to phenanthrene gives the adduct **80**, which can be aromatized to **81** by consecutive treatment of NBS and alumina. Treatment of **81** with silver trifluoroacetate, followed by hydrolysis of the resulting tetrakis (trifluoroacetate) with water finally leads to cyclobuta[l]phenanthrene-1,2-dione (**82**)[44]. The molecular structure of **82** has been determined by X-ray crystallography[44].

hv, 17%

(80)

(1) NBS
(2) Al_2O_3
71%

(1) AgO_2CCF_3
(2) H_2O

(82) **(81)**

5. *Eight-membered ring-fused cyclobutenequinones*

The monoacetylene **83** undergoes smooth [2 + 2]cycloaddition with dichloroketene, whereby the dichloroketone **84** is obtained[45,46]. Subsequent treatment of **84** with excess silver trifluoroacetate and hydrolysis leads to the eight-membered ring-fused cyclobutenequinone **85**[45,46], which is a relatively stable crystalline compound. Similar treatment of

$Cl_2CHCOCl$, Et_3N, 17%

(1) AgO_2CCF_3, (2) H_2O, 6.7%

(83) **(84)** **(85)**

$Cl_2CHCOCl$, Et_3N, 34%

(1) AgO_2CCF_3, (2) H_2O

(86) **(87)** **(88)**

the diacetylene **86** leads to the presumably coplanar cyclobutenequinone **88** which is obtained as an unstable red crystalline solid[45,46]. The electrochemical reduction of **88** to its radical anion and then to the dianion **89** has been studied[46] and was found to be a facile process[46]. This observation lends some support to the general idea that the fusion of two planar antiaromatic $4n\pi$ systems as is the case in **89** would constitute an overall $(4n+2)\pi$ aromatic periphery, and hence would give rise to certain degree of aromatic stabilization[46].

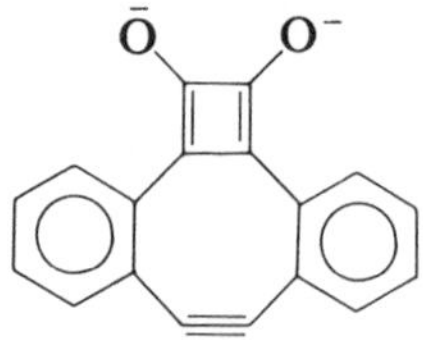

(89)

C. Eight-membered Ring Systems

Cycloocta-2,5,7-triene-1,4-dione (**2**) and cycloocta-3,5,7-triene-1,2-dione (**3**) have attracted wide interest due primarily to the possible existence of their corresponding eight-

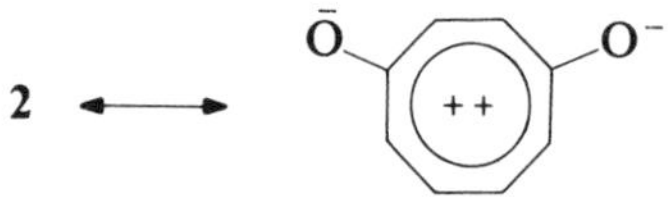

(2a)

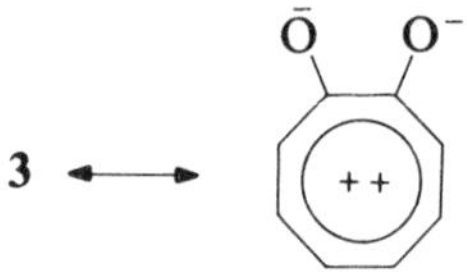

(3a)

carbon 6π canonical forms **2a** and **3a**. If **2a** and **3a** indeed make appreciable contribution, **2** and **3** should be planar and exhibit diatropic properties.

1. Cycloocta-2,5,7-triene-1,4-dione and derivatives

The first known derivatives of **2** were reported by Cava in 1962[47]. Compounds **91** were obtained in good yields by oxidation of the corresponding bromide **90a** and iodide **90b**[47].

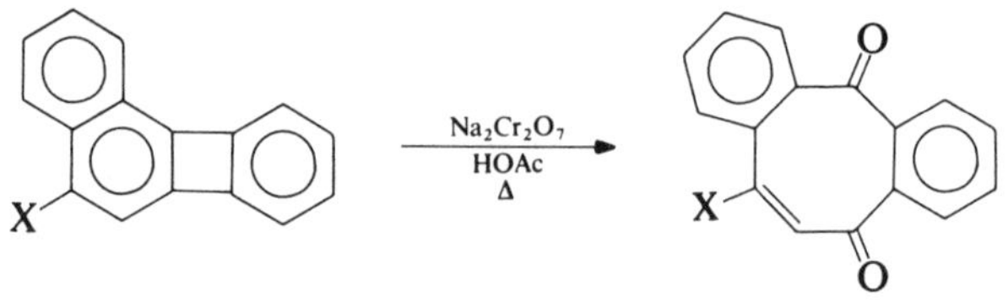

$Na_2Cr_2O_7$, HOAc, Δ

(**90a**, X = Br)
(**90b**, X = I)

(**91a**, X = Br, 76.6%)
(**91b**, X = I, 81.3%)

Attempts to convert the diketone **92** to the tetraketone **93**, were in vain[48]. Compound **93** is of interest because its enol form **93a** may yet be another derivative of **2**[48]. An attempt to

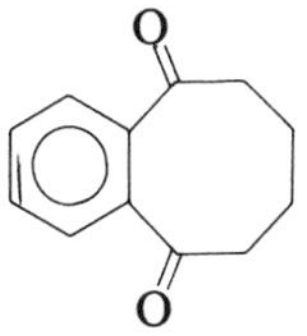

(**92**) 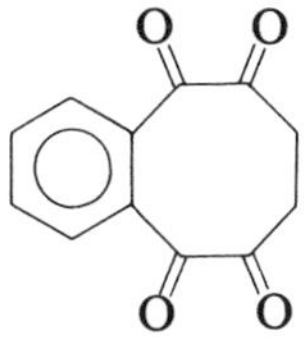 (**93**) 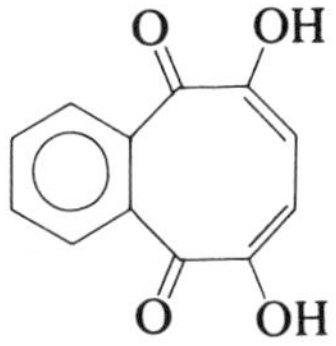 (**93a**)

generate **95** from **94** by thermolysis resulted in the formation of **96**. However, compound **95** has been proposed as an intermediate in association with the rearrangement[49].

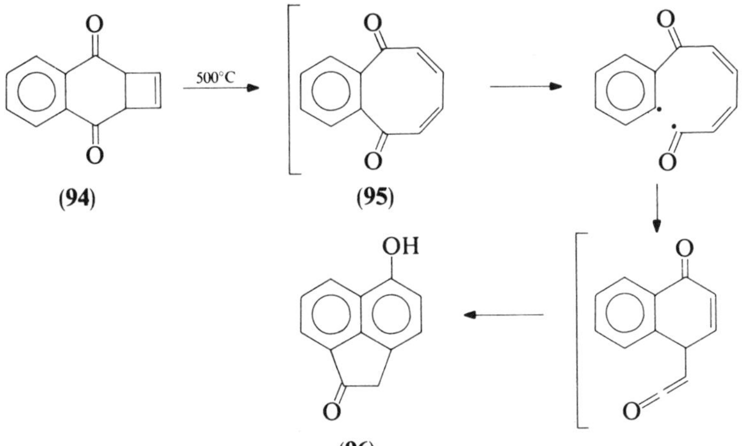

500°C

(**94**) (**95**)

(**96**)

In the presence of air, 2,2′-bis(phenylacetyl)biphenyl reacts with sodium methoxide in methanol to give **97**, which is a dibenzo derivative of **2**[50].

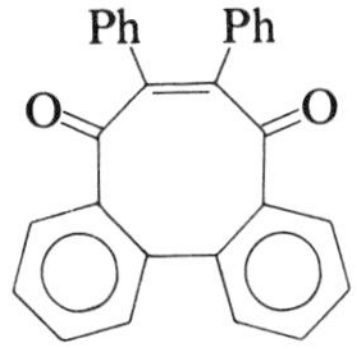

(**97**)

The conversion of diketone **98** to **99** has been achieved by bromination and dehydrobromination. Similarly, **100** has been converted to **101**[51]. However, no extended conjugation is observed for **99a**, **99b** and **101** in their corresponding UV spectra[51]. Moreover, the fact that the vinylic protons of **99a** and **101** do not exhibit detectable diamagnetic shift leads to the conclusion that these compounds do not behave as eight-carbon 6π electron aromatic species[51]. Nevertheless, downfield shift is observed when the

R = H (1) PTAB THF (2) SiO_2 60%

R = Me (1) Br_2/dioxane (2) $NaHCO_3$ 82%

(**98a**, R = H)
(**98b**, R = Me)

(**99a**, R = H)
(**99b**, R = Me)

PTAB: phenyltrimethylammonium perbromide

(1) PTAB THF (2) $CaCO_3$ HMPA 30%

(**100**) (**101**)

NMR spectra are recorded in deuteriotrifluoroacetic acid, suggesting that some mono- or diprotonated diatropic species might have been formed[51].

Addition of large excess of dichlorocarbene to **102** results in the isolation of **103**, whose eight-membered moiety does not indicate any aromatic property[52, 53].

$:CCl_2$ 12.5%

(**102**) (**103**)

In view of the lack of evidence for possible contribution of their charged distropic canonical forms in **91**, **97**, **99**, **101** and **103**, (viz. **2a**), the attention of the organic chemists was then drawn to the realization of their parent compound **2**. Starting from cyclooctane-1,4-dione (**104**), cyclooctatriene-1,4-dione bisethylene ketal (**105**) was prepared, but all attempts have been unsuccessful in the hydrolysis of **105** to **2**[54, 55].

The synthesis of some valence tautomers of **2** have been recorded. Thus, photolysis of a mixture of methoxyl-*p*-benzoquinone and various acetylenes leads to isolation of **106**[56].

The direct synthesis of the parent valence tautomer **107** of **2**, however, poses considerable difficulty because *p*-benzoquinone would react photochemically with acetylenes at the carbonyl bond. To avoid this problem, Yates has converted *p*-benzoquinone to **108** by reaction with anthracene. Photoaddition of the adduct **108** to acetylenes gives compounds **109** which undergo retro Diels–Alder reaction to yield **110**[57].

OH OH
$(CO_2H)_2$
$HC(OMe)_3$
MeCN
50%

(**104**)

Br_2 30%
Et_2O

Br Br Br

t-BuOK
t-BuOH
50%

(**105**)

OMe + R^1 ≡ R^2 *hν* MeCN

(**106a**, $R^1 = R^2 = Ph$)
(**106b**, $R^1 = R^2 = Me$)
(**106c**, $R^1 = Ph$, $R^2 = H$)
(**106d**, $R^1 = H$, $R^2 = Ph$)

(**107**)

R^1–≡–R^2
hν

220°C
20 mm Hg

(**108**)

(**109a**, $R^1 = R^2 = Me$)
(**109b**, $R^1 = H$, $R^2 = Ph$)

(**110a**, $R^1 = R^2 = Me$)
(**110b**, $R^1 = H$, $R^2 = Ph$)

The synthesis of the parent compound **107** was finally realized by Kitahara using the scheme shown below[58].

Br_2

O_2
$h\nu$
sensitizer
78%

Br
Br

$LiAlH_4$
THF 90%

OH
OH

Ac_2O
pyridine
90%

OAc
OAc

Zn
DMSO
90°C
100%

$LiAlH_4$
THF
100%

CrO_3
66%

107

All attempts to convert **107** into its tautomer **2** have failed. Thermolysis of **107** at 500°C affords tropone (**111**) in 49% yield together with trace amounts of an unstable and unidentified compound[59]. It has been assumed that either **2** or its *trans* isomer may be the possible intermediate in the thermal reaction[59].

107 500°C, 0.2 mmHg → or

(2)

(111)

The parent compound **2** was eventually isolated in 1975[60]. Prior reduction of **107** to **112**, followed by thermolysis of the latter gives **113** in excellent yield. Bromination of **113** can be achieved by using NBS in CF_3CO_2H–CH_2Cl_2 at room temperature.

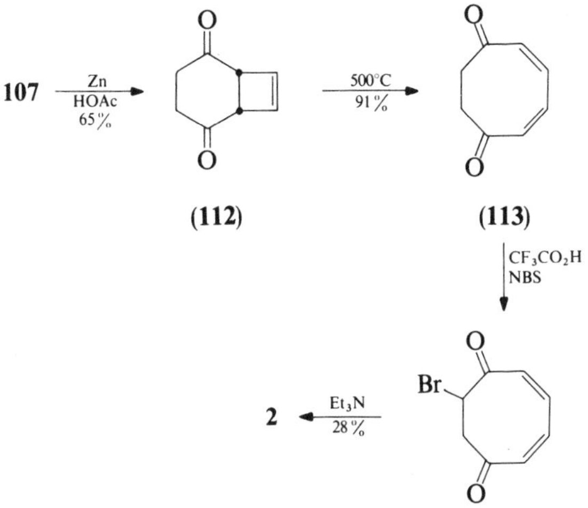

107 → (Zn, HOAc, 65%) → **(112)** → (500°C, 91%) → **(113)** → (CF_3CO_2H, NBS) → **(114)** → (Et_3N, 28%) → 2

Dehydrobromination of the unstable bromide **114** furnishes **2** as a yellow liquid. The UV and NMR spectra unequivocally show that **2** is an ordinary olefinic ketone. Therefore the contribution of **2a** can be neglected. Compound **2** is stable in aprotic solvents, but decomposes readily in acidic solvents.

2. Cycloocta-3,5,7-triene-1,2-dione and derivatives

The synthesis and structure of dibenzo[*a*, *e*]cyclooctene-5,6-dione (**115**) has been recorded[61,62] and reviewed[1]. It is not surprising that, like other benzannelated derivatives of **2**, **115** is non-planar and non-aromatic. This situation is also true for compound **117**

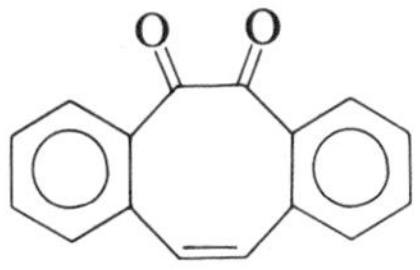

(115)

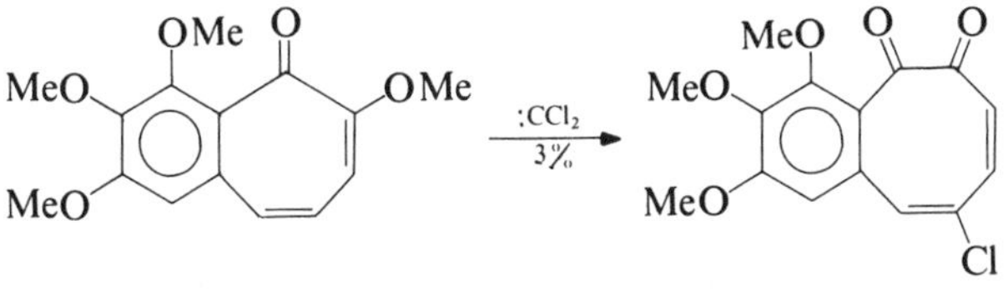

(116) → (:CCl_2, 3%) → **(117)**

which can be conveniently synthesized by subjecting **116** to a large excess of dichlorocarbene[52, 53].

The parent compound **3** itself and its valence tautomers **118** as well as **119** have become target molecules of various research groups. Carpino has reported the synthesis of **120**, which resisted transformation into **118**[63]. The synthesis of the other valence tautomer **119**

(118) (119) (120)

has however been successful[64]. Starting from cyclooctatetraene, **119** can be obtained from the series of reactions[64] outlined below.

Br_2; O_2, $h\nu$, sensitizer; PPh_3 71%; HOAc, $BF_3 \cdot Et_2O$ 88%; isomer +; Zn DMSO; isomer +; Jones' oxidation 42%; (1) HO OH H^+ (2) $LiAlH_4$ 61%; CrO_3 pyridine HOAc 48%; 12N H_2SO_4 81%

(119)

The thermolysis of **119** has been examined with the aim to effect its valence tautomerization into **3**. However, **119** undergoes an intriguing rearrangement to give bicyclo[3.2.1]octa-3,6-diene-2,8-dione (**121**), which is decarbonylated rapidly at 200°C to tropone (**111**)[65].

160°C

200°C
−CO
111

(119)

(121)

Dehydrobromination of **122** fails to give **3**[66]. On the other hand, reaction of **122** with *o*-phenylenediamine furnishes the quinoxaline **123**, which is dehydrobrominated to **124**, a quinoxaline derivative of **3**[66].

Br
O
O
Br
NH_2
NH_2
HOAc
72%
Br
N
N
Br

(122)

(123)

DBN
DMSO
48.5%

N
N

(124)

Dehydrobromination of the dibromide **126**[67,68] with triethylamine at −50°C leads successfully to **3** which can be detected by NMR spectroscopy[69]. However, if the NMR solution of **3** in $CDCl_3$ is brought up to room temperature, the relevant NMR signals vanish and from this solution, compounds **4** and **125** can be isolated. This result indicates that at room temperature, **3** equilibrates with its valence tautomer **118**, and the mixture undergoes oxidation–reduction reaction to afford **4** and **125**[69]. Furthermore, a crude yellow solution of **3** can be obtained by low temperature chromatography at −78°C. When this yellow solution is allowed to react with *N*-phenyltriazolinedione at room temperature, the Diels–Alder adduct **127** can be isolated. The existence of **3** is further confirmed by action of bromine at −30°C which results in the formation of the dibromide **128**[69].

3-Bromo-cycloocta-3,5,7-triene-1,2-dione (**132**) was first reported to be a transient intermediate in the dehydrobromination of **122**[66]. Its existence has subsequently been established spectroscopically at low temperature[69]. Thus, treatment of **129**[68] with NBS leads to the dibromide **130** which rearranges smoothly to **131**. Dehydrobromination of

SeO_2, THF, 41%

Et_3N, C_6H_6, 80°C, 65%

(125)

Br_2, room temp, 1 day, 85%

(118) ⇌ (3)

Et_3N, −50°C

(126)

oxidation–reduction

N—Ph

Br_2

4 + (125)

(127)

(128)

NBS

Florisil, 60%

(125)

(129)

NBS

room temp. in solution

Et_3N, −50°C

(132)

(131)

(130)

−20°C, −HBr → 4

(132a)

131 at $-50°C$ with Et_3N gives **132**, whose structure is shown to be non-planar and non-aromatic by NMR spectroscopy at $-50°C$. When the $CDCl_3$ solution of **132** is allowed to warm up to $-20°C$, the signals of **132** vanish and benzocyclobutenequinone (**4**) and an unidentified compound can be isolated. This observation indicates that **132** is even more unstable than **3**[69].

In several related studies, the preparations of a number of potential precursors of **3** such as **133**[70], **134**[71], **135**[72] and **136**[73] have also been described.

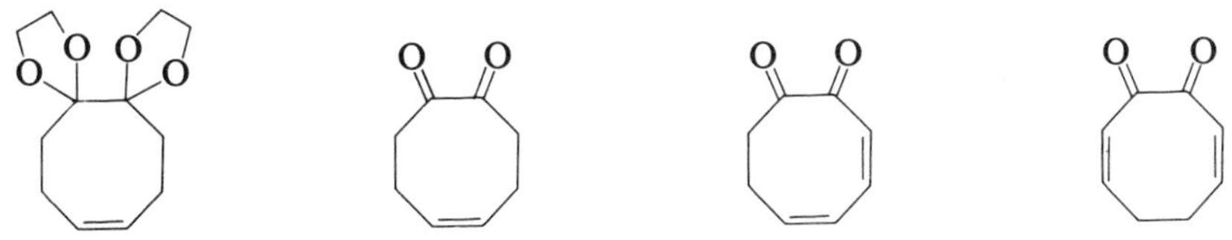

(**133**) (**134**) (**135**) (**136**)

D. Annulenequinones

The synthesis of [14]annulenequinone **137** has been reported[74]. Protonation of **137** leads to the 12π paratropic dication **138**, whose inner protons exhibit low field shift[74]. The [18]annulenequinones **139**[75], **140**[76], **141**[76] and **142**[76] have been synthesized. The proton NMR spectra of these compounds in $CDCl_3$ with added CF_3CO_2D are consistent with the ketonic structures and as expected no detectable ring current can be observed[76].

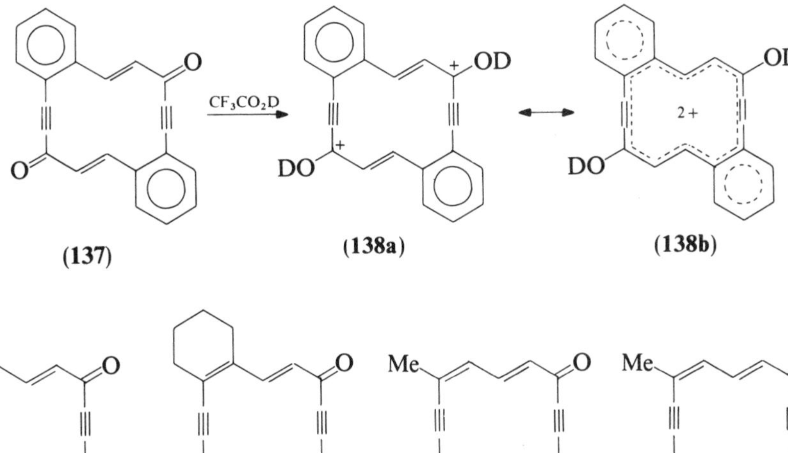

CF_3CO_2D

(**137**) (**138a**) (**138b**)

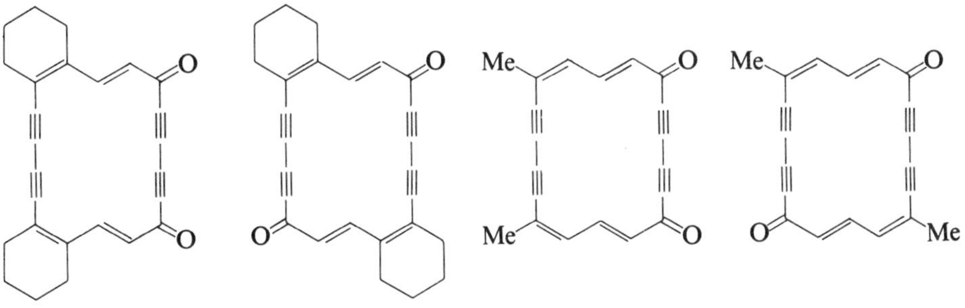

(**139**) (**140**) (**141**) (**142**)

However, results obtained from the electrochemical reduction of **139**, **141** and **142** to their respective 18π aromatic dianions[77] are in accord with the reduction behaviour of quinones. It has been found that **139**, **141** and **142** are more easily reduced than is *p*-benzoquinone, possibly because the resulting negatively charged oxygen atoms are further apart than in a six-membered ring, so that electrostatic repulsion is greatly diminished.

The [16]annulenequinone **143** has also been prepared[78]. Surprisingly, the proton NMR spectrum shows that **143** is rather diatropic, probably arising from a significant contribution from its 14π aromatic canonical form **143a**[78].

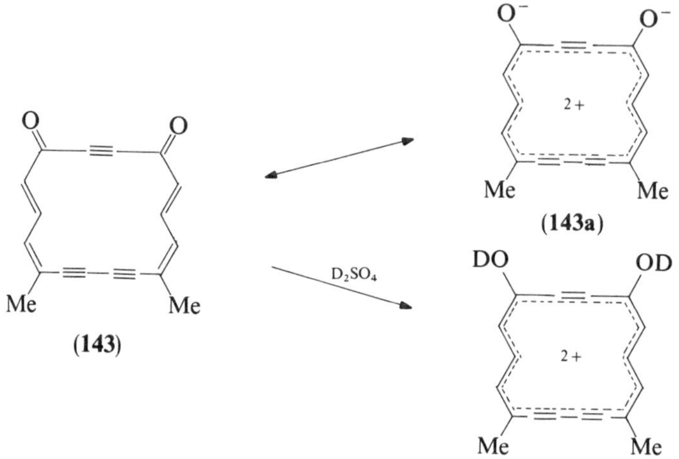

O$^-$ O$^-$ 2+ Me Me **(143a)**

O O Me Me **(143)**

D_2SO_4

DO OD 2+ Me Me **(144)**

Treatment of **143** with deuteriated sulphuric acid gives **144** which is strongly diatropic[78]. Moreover, electrochemical reduction of **143** to its corresponding dianion has been found to be more difficult than is **141**. This observation can be explained in terms of an antiaromatic 16π dianion[78].

Radical anions of **141**, **142** and **143** have been investigated by ESR spectroscopy[79]. The results demonstrate that **141**$^{-\cdot}$ and **142**$^{-\cdot}$ are aromatic whereas **143**$^{-\cdot}$ is antiaromatic[79].

The [26]annulenequinone **145** has been isolated and used as a precursor towards the synthesis of an [18]annuleno[18]annulene. The quinonoid properties of **145** have not, however, been examined[80].

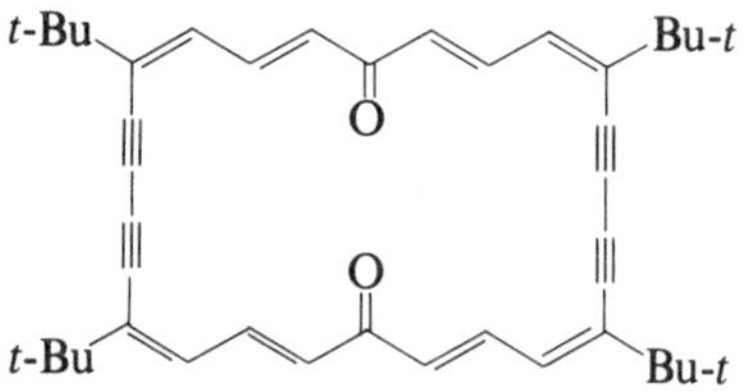

t-Bu Bu-*t* O O *t*-Bu Bu-*t*

(145)

The synthesis of [14]annulenequinones **146**, **147** and [18]annulenequinone **148** together with a modified method for the preparation of **137** have been described[81]. The proton NMR spectra of these annulenequinones, exhibit no sign of any appreciable ring current[81]. The electrochemical reduction of **146** and **148** to their 14π and 18π aromatic dianions respectively is relatively easy and reversible, which indicates that **146** and **148** indeed possess characteristic quinonoid properties[82].

(146) (147) (148)

III. ODD-MEMBERED RINGS

A. General Formula

In agreement with the previous review[1], these compounds may be regarded as being generated from the formula in which n can be 0, 1, 2, 3, etc., and X must not produce an immobile electron system on the carbon attached to the two carbonyls. Hence, X can be oxygen, methylene or an imine moiety, etc.

B. Three-membered Ring Systems

1. Cyclopropanetrione derivatives

The simplest member of this class is cyclopropanetriquinone (**149**) which remains an unknown compound. However, the deltate dianion (**150**) is well known and its chemistry has been reviewed[12, 13, 16, 83]. The application of the graph theory of aromaticity has shown that **149** has a very small resonance energy[84]. On the other hand, the deltate dianion (**150**) has been predicted to be both highly aromatic and highly diatropic[84].

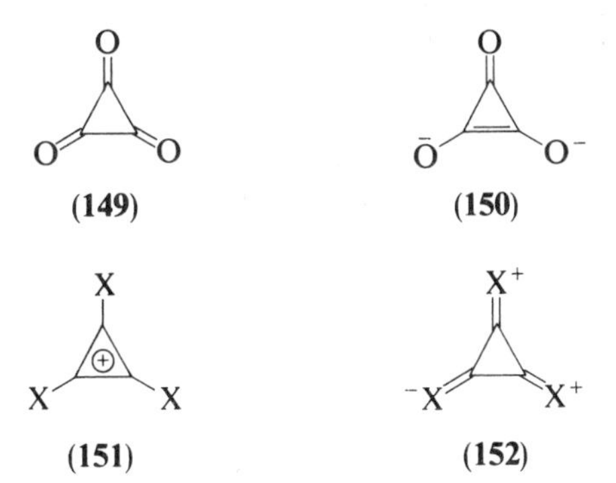

(149) (150)

(151) (152)

The heteroatom-substituted cyclopropenyl cation system **151** has received widespread attention[85]. One of its resonance forms may have the quinonoid or hetero-[3]-radialene structure **152**. Treatment of tetrachlorocyclopropene (**153**) with dimethylamine yields trisdimethylaminocyclopropenyl cation (**154**) which has been isolated as the perchlorate salt[86]. Interestingly, diethylamine affords not the tris, but only the 1,2-bisdiethylaminocyclopropenyl perchlorate (**155**)[86].

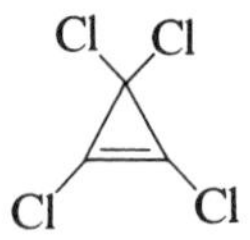

(**153**) 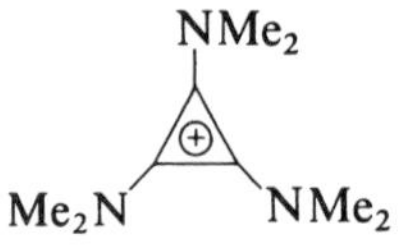 (**154**) 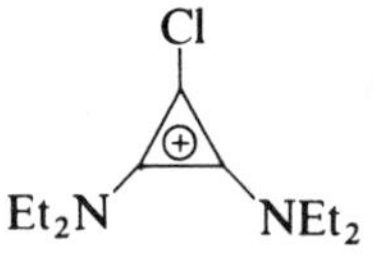 (**155**)

It has been noted that the C–N bond in **155** has partial double bond character. As a result, the rotational barrier about the C–N bond is increased[86]. The chlorine atom in the related diisopropyl derivative **156** is readily substituted by redox hydrolysis with Ph_3P/H_2O to give **157**[87]. The synthesis of cyclopropeniumyldiazonium salts **158** has recently been achieved in excellent yields from diazotization of **159** with $NO^+BF_4^-$, or from the oxidation of **160** with SO_2Cl_2 and ICl[88]. MNDO calculations on the parent

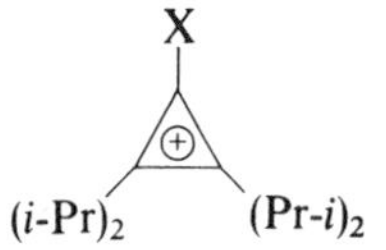

(**156**, X = Cl)
(**157**, X = H)

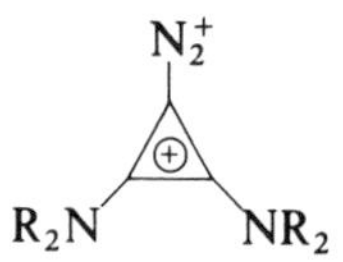

(**158a**, R = Me)
(**158b**, R = *i*-Pr)
(**158c**, R = H)

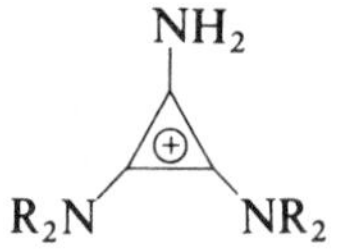

(**159a**, R = Me)
(**159b**, R = *i*-Pr)

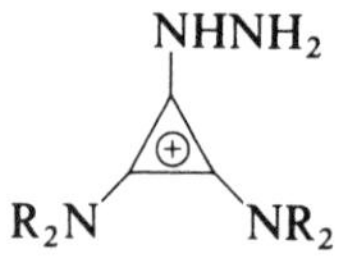

(**160a**, R = Me)
(**160b**, R = *i*-Pr)

system suggests that **158c** can be conceived of as a resonance hybrid made up from an acceptor-stabilized/cycloaliphatic diazo system **161**, **162** and a donor-stabilized aromatic diazonium system **163**[88]. The cations **158** undergo hydrolysis to give, after ring-opening,

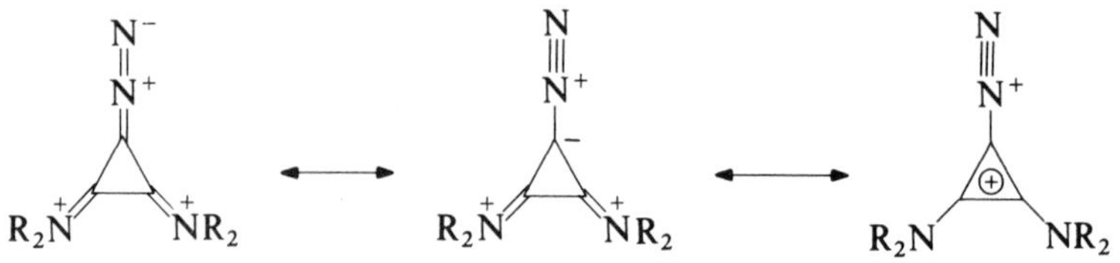

(**161**) (**162**) (**163**)

vinyldiazonium salts **164**[88]. Trialkylthio- and triphenylthiocyclopropenyl cations **165** have also been prepared[89]. However, contribution of the hetero-[3]radialene structure

152 (X = SR) might be neglected based on the NMR data in comparison with those of related compounds[89].

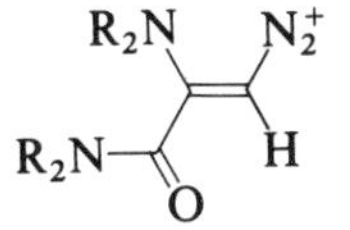

(**164a**, R = Me)
(**164b**, R = *i*-Pr)

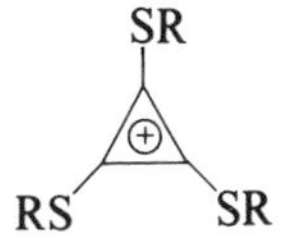

(**165a**, R = alkyl)
(**165b**, R = Ph)

Tetraaminotrifulvalene dications **166**[90, 91] and the 1,3-bis(diamino-cyclopropenylio)cyclopenta-dienide system **167**[92] have been prepared. Tris(cycloprop-enylio)cyclopropenylium salt (**168**) can be prepared from the reaction of **169** with KF followed by addition of **153** and perchloric acid[93]. Compound **168** is the first isolated tetravalent cation[93]. Radialenoid, fulvenoid and aromatic resonance structures can be formulated for each of the four rings of **168**[93].

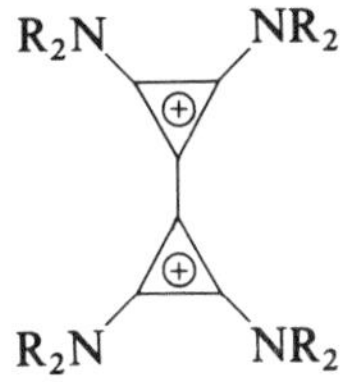

(**166**, R = Me or *i*-Pr)

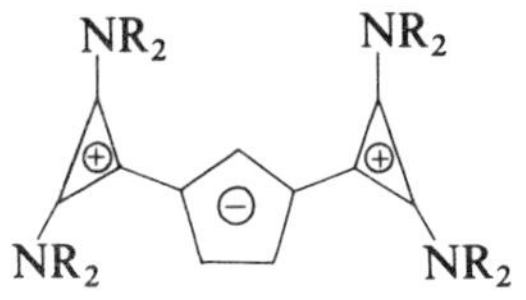

(**167**, R = *i*-Pr)

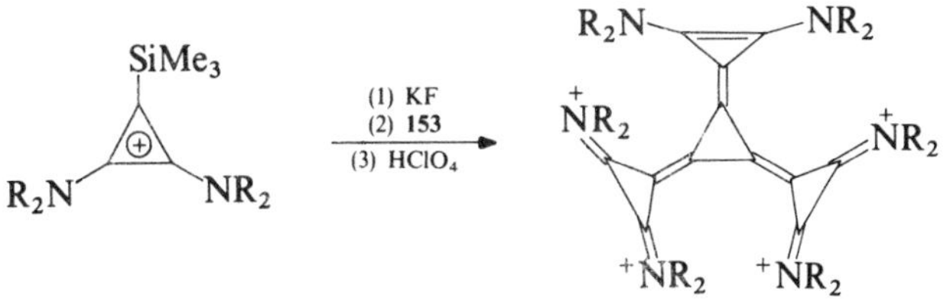

(1) KF
(2) **153**
(3) $HClO_4$

(**169**, R = *i*-Pr)

(**168**, R = *i*-Pr)

Treatment of tetrachlorocyclopropene (**153**) with malononitrile and sodium hydride affords the dianion **170**[94]. The tetrabutylammonium salt of **170** has been isolated as crystalline solid[94]. Upon oxidation with potassium persulphate, the dianion **170** gives an

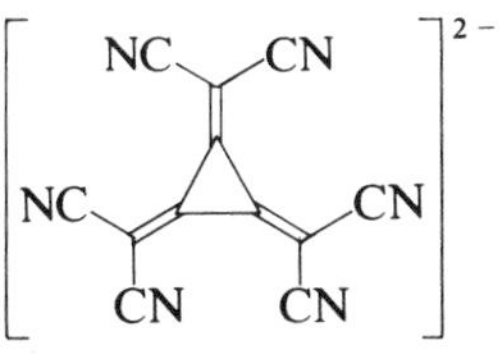

(**170**)

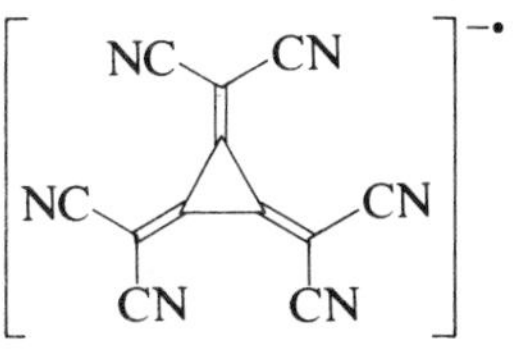

(**171**)

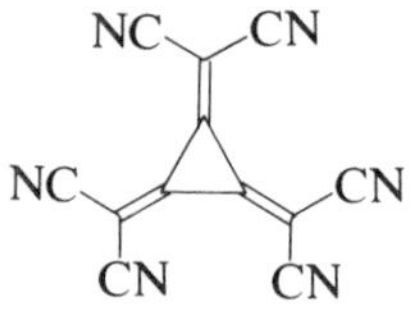

(172)

equilibrium mixture containing **170**, **171** and **172**[94]. The ESR spectrum of **171** exhibits thirteen lines due to the C≡N groups[94].

2. *Quinocyclopropane derivatives*

The chemistry of triquinocyclopropanes **173**[95] has been discussed in the previous review[1]. Syntheses and properties of quinocyclopropanes **174**[96] and quinoiminocyclopropanes **175**[97] have been reported.

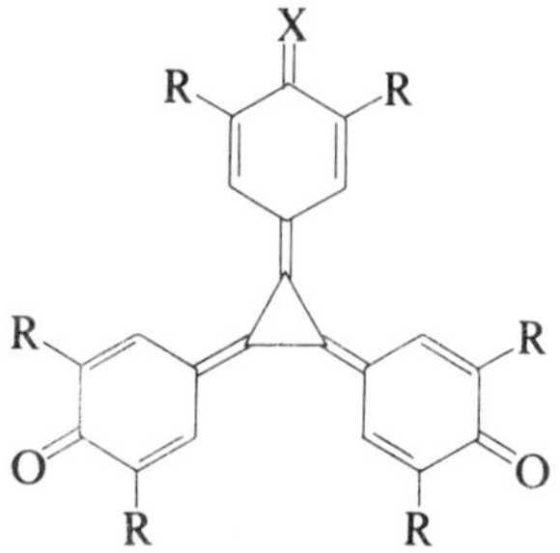

(**173**, X = O)
(**174**, X = O; R = *t*-Bu)
(**175**, X = NH; R = *t*-Bu)

Bis(*p*-hydroxyaryl)cyclopropenones (**176**) can be oxidized to the bright purple bis-quinonoid derivatives **177**[98]. Compounds **177** are highly unstable and readily

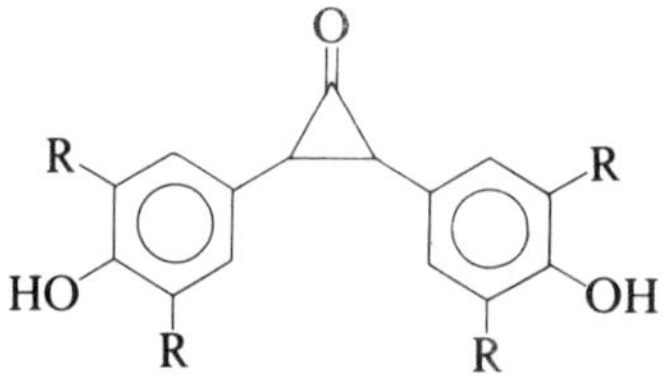

(**176**) 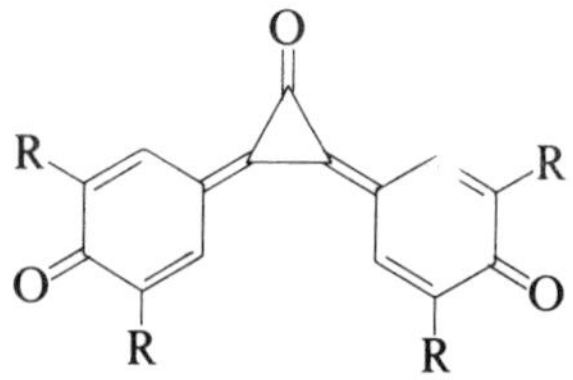 (**177**)

extrude carbon monoxide. However, immediate reduction of **177** with hydroquinone reverts them back to **176**[98]. Upon treatment with sodium–potassium alloy, compounds **177** are converted into the corresponding anion radicals[98]. Tris(9-anthron-10-ylidene)cyclopropane (**178**) has been prepared by oxidation with ferricyanide or PbO_2[99]. The related compounds **179**, **180** and **181** can also be prepared by similar

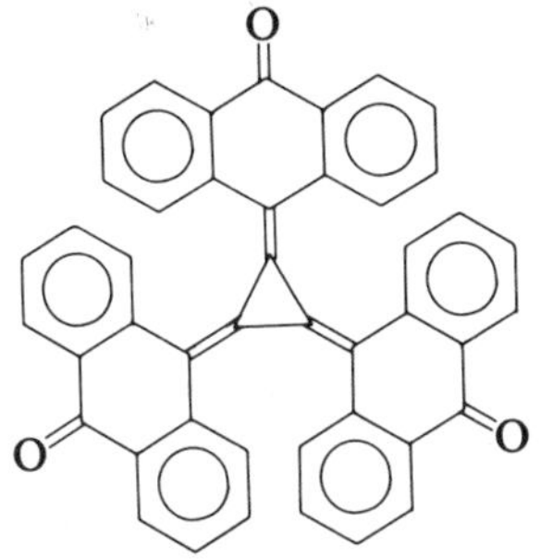

(178)

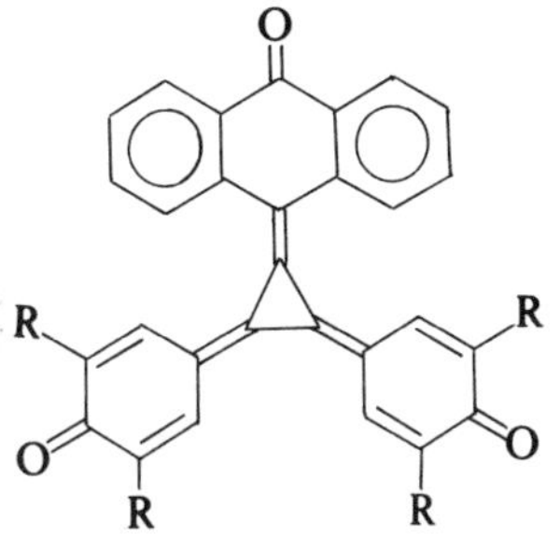

(180)

procedures[99]. It is noteworthy that the electronic absorptions for these intensely coloured materials appear in the near infrared region[99]. Furthermore, the anthraquinone derivatives **178–181**, in general, are more stable than the corresponding benzoquinone analogues **177**. For example, **180** is stable indefinitely in benzene, although it is photolytically decomposed to dianthraquinoethylene (**182**)[99]. The redox properties of anthraquinocyclopropanes **178–181** have been studied in detail[100]. Compounds **183** have been synthesized from **176** by condensation with appropriate substrates followed by oxidation[101]. These compounds are luminously coloured dichroic solids which are potentially useful dyes and photographic agents[102]. Furthermore, they are powerful oxidants[101].

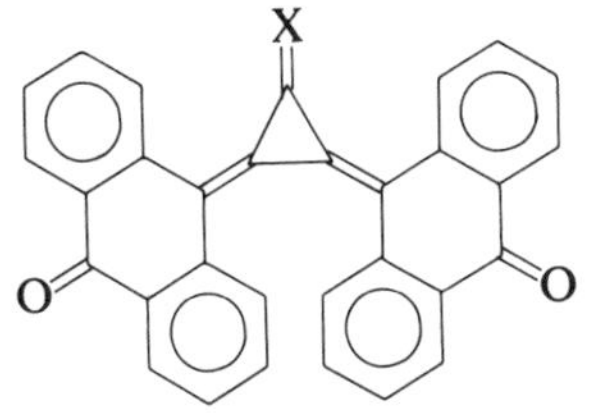

(**181**, X = O)

(**179**, X =)

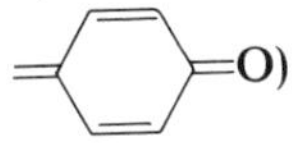

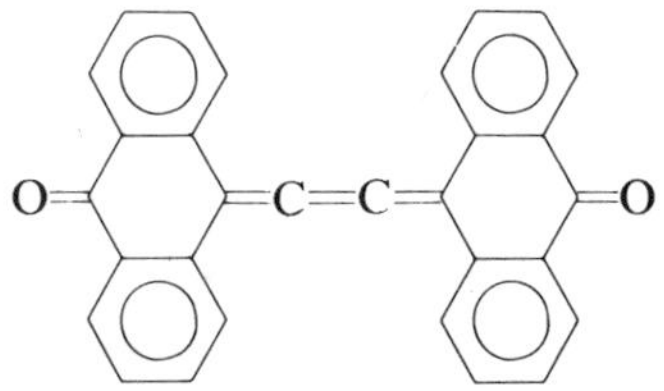

(182)

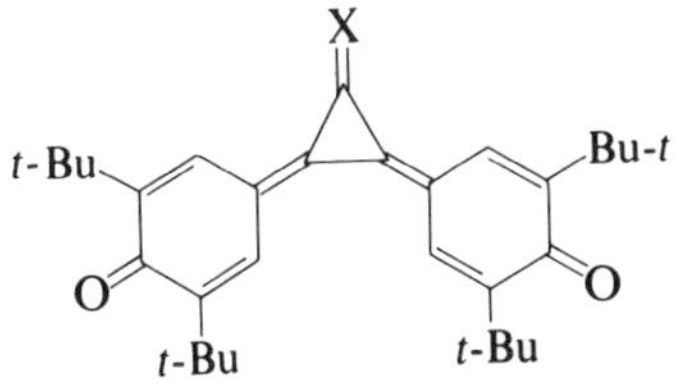

(183)

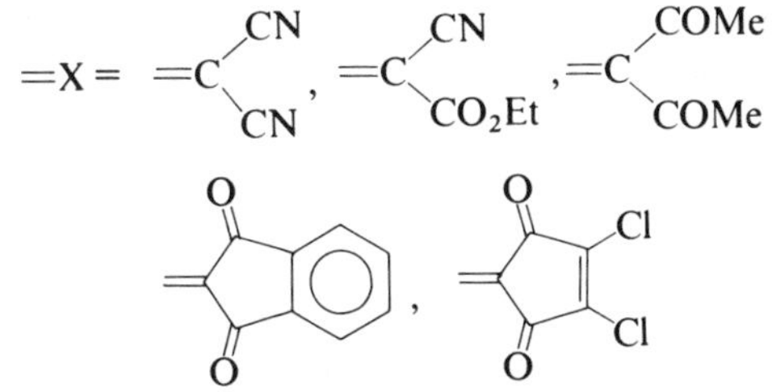

C. Five-membered Ring Systems

1. *Cyclopentenequinone and derivatives*

The parent compound in this class may be regarded as **184** where quinonoid properties are achieved by the attachment of a mobile electron system X. The simplest compound in

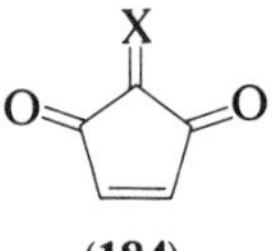

(**184**)

this family is **184a**[103,104] whose structure has been determined by X-ray crystallography[105]. Pyrolysis of **184a** affords the decarbonylated intermediate **185** which cyclizes to **1** or can be trapped with methanol to yield dimethyl succinate[104]. Oxidation of **7** (R = H) with selenium dioxide furnishes the cyclopentenetrione **186** which can also be obtained from the reaction of **187** with bromine or selenium dioxide[106]. The He(I) photoelectron spectra of **184a** have been measured[103]. The results show that the lone pair electrons on the oxygen atoms interact strongly with the σ framework[103].

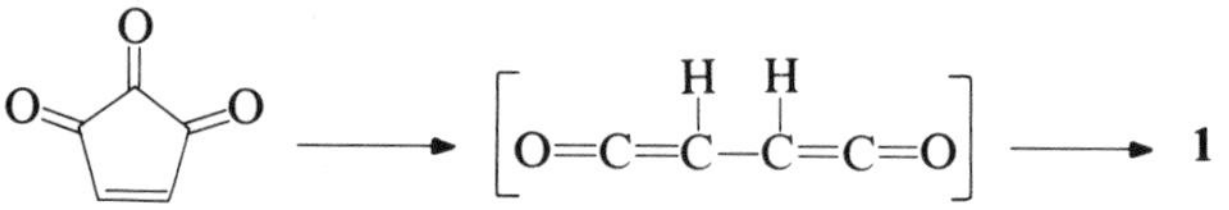

$$\textbf{(184a)} \longrightarrow \left[\mathrm{O{=}C{=}CH{-}CH{=}C{=}O}\right] \textbf{(185)} \longrightarrow \textbf{1}$$

$$\textbf{7} \xrightarrow[\mathrm{R\,=\,H}]{\mathrm{SeO_2}} \textbf{(186)} \xleftarrow[\mathrm{SeO_2}]{\mathrm{Br_2\ or}} \textbf{(187)}$$

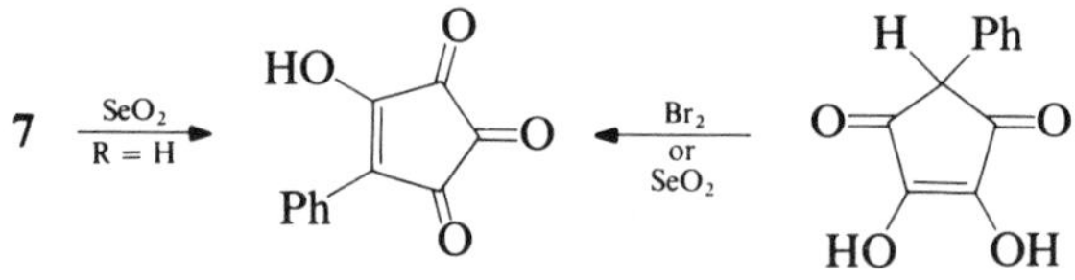

(**186**) (**187**)

The chemistry of croconic acid (**188**) is well documented[12,13,16,107]. A mass spectral study of **188** has been carried out[108]. The ^{13}C-NMR spectra of **188** are solvent dependent[109]. In DMSO solution, **188** exhibits signal averaging. On the other hand, **188** can be observed as an non-dynamic species in anhydrous THF[109]. Compound **188** readily forms hydrate or hemiketal with water or methanol[109].

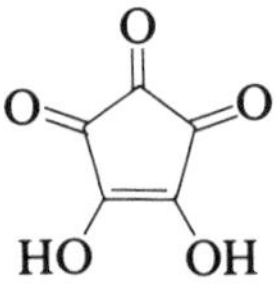

(**188**)

The chemistry of indane-1,2,3-trione (**189**) has been extensively reviewed[110]. The mass spectral and pyrolytic fragmentation of **189** has been studied in detail[111]. The structure of **189** has been determined by X-ray crystallography[112]. The reaction of its 'monohydrate' (ninhydrin) with α-amino acids gives the well-known purple coloured product **190**[113].

The synthesis of 5,8-dithiafulvalene-1,4-diquinone (**191**) has been described[114]. Alkylidene-1,3-indanedione (**192**) can easily be prepared by the condensation of 1,3-indanedione and the corresponding aldehyde or ketone[115,116]. A complete kinetic analysis of the four-step hydrolysis of **192** (R = Ph) to benzaldehyde and 1,3-indanedione

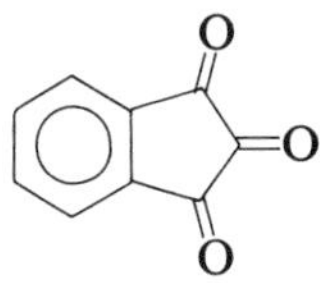

(189) 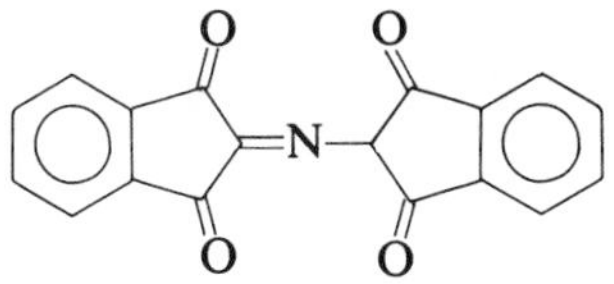 (190)

in aqueous DMSO has recently been recorded[117]. The rate and equilibrium constants of the reversible addition of 1,3-indanedione anion to **192** (R = Ph) in 50% aqueous DMSO at 20 °C have been determined[118].

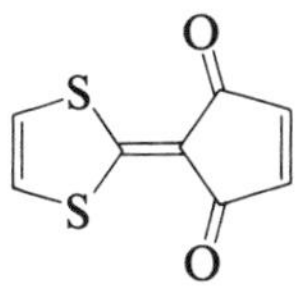

(191) 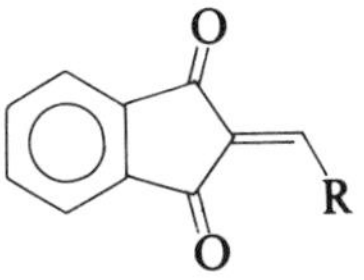 (192)

Compound **193** has been shown to be an electrophotographic photoconductive material, sensitive to an (AlGa)As semiconductor laser[119].

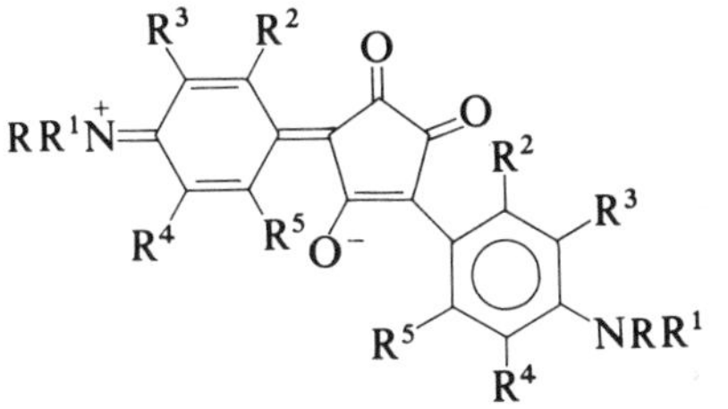

(193)

2. *Cyclopentanepentaone*

The parent compound **194** exists as its hydrate, or leuconic acid. It can be synthesized by nitric acid oxidation of croconic acid (**188**)[109, 120]. The mass spectral[108] and ^{13}C-NMR data[109] of **194** have been recorded.

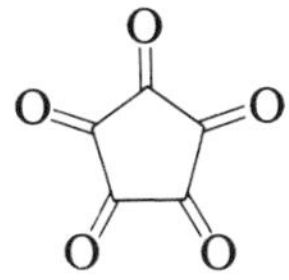

(194)

D. Seven Membered Ring Systems

1. *Diketones*

No monocyclic seven-membered diones can be formulated according to the general formula described earlier in this section. When tropolone **195** is treated with 2,3-dichloro-

5,6-dicyano-1,4-benzoquinone (DDQ) in methylene chloride, a bright red coloration appears which may be caused by the formation of heptalene-2,3-dione (**196**)[121]. However, the solution turns quickly to brown and no clear product can be isolated[121].

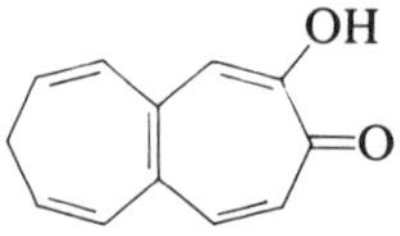

(**195**) 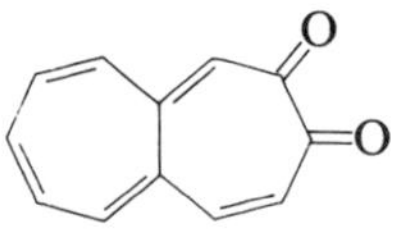 (**196**)

2. Triketones

a. o-*Tropoquinone and related compounds* The title compound **197** was first synthesized in solution in 1975 by the oxidation of the 3-hydroxytropolone (**198**) with DDQ[122]. It was subsequently isolated in crystal form in 1978 and an X-ray analysis was also reported[123]. Compound **197** instantaneously reacts with water or methanol to give

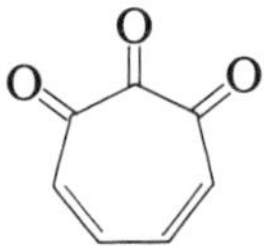

(**197**) 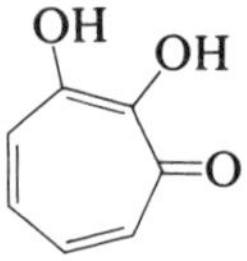 (**198**)

the hydrate **199** or hemiacetal **200**, respectively[122]. The hydrate **199** can be converted to the corresponding oxime **201** as reddish purple crystals which are stable in anhydrous form[122]. The half-wave potential of **197** is more positive than that of *o*-benzoquinone, and its electronic spectrum exhibits maxima, at 334.5 and 574 nm[123]. The latter absorption maximum coincides with that of *o*-benzoquinone[123]. The He(I) photoelectron spectra of **197** and *o*-benzoquinone also reveal such a resemblance[103, 124]. Most of the reactions of **197** are similar to those of *o*-benzoquinone and other cyclic vicinal triketones[125]. In line with expectation, compound **197** can readily be reduced to **198**[125].

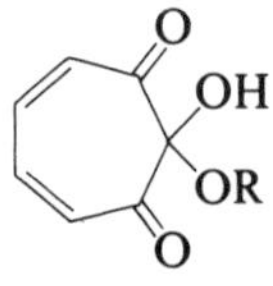

(**199**, R = H)
(**200**, R = Me)

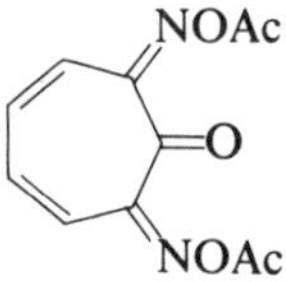

(**201**)

5-Substituted tropolones **202** couple at the 3-position with aryldiazonium ion containing *para*-electron-withdrawing groups. Mixtures of azo compounds **203** and hydrazones **204** are formed[126, 127]. The preparations of benzo-*o*-tropoquinone (**205**)[128] and diazo compound **206**[129] have been described.

3,5-Dibromo-7,8-diphenylheptatriafulvalene-1,2-quinone (**207**) has been prepared by reaction of 3,5-dibromotropolone (**208**) and 1,2-diphenyl-3-ethoxycyclopropenium ion (**209**) in the presence of Et_3N in MeCN[130]. The product **207** recrystallizes as reddish

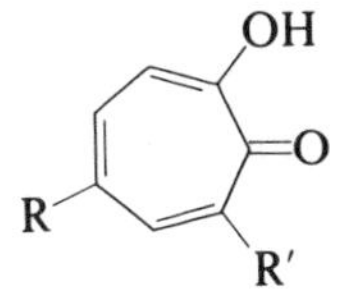

(**202**, R′ = H)
(**203**, R′ = N=NAr)

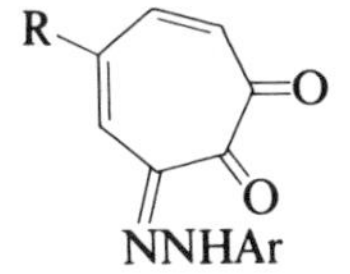

(**204**)

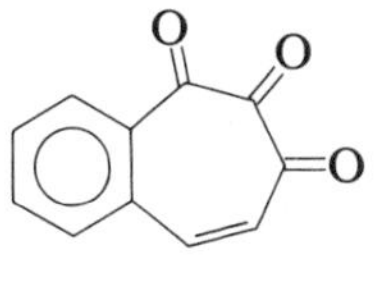

(**205**) 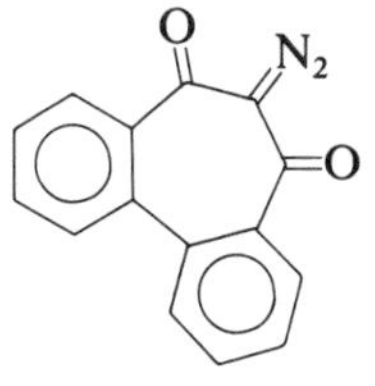 (**206**)

violet needles[130]. The electronic spectrum of **207** indicates extended conjugation and semiempirical calculations likewise predict a nearly coplanar structure with strong interaction between the three- and seven-membered rings[130]. 7,10-Dithiasesquifulvalene-1,6-quinone (**210**) has also been synthesized[131].

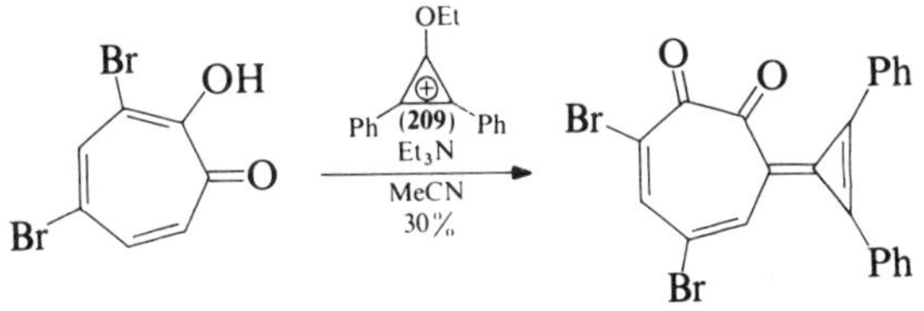

(**209**), Et_3N, MeCN, 30%

(**208**) (**207**)

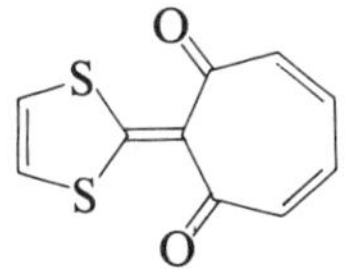

(**210**)

b. p-*Tropoquinone and related compounds* The parent quinone **211** has been obtained as pale yellow needles from the haematoporphyrin (**212**)-sensitized photooxidation of tropolone or 5-hydroxytropolone (**213**) followed by treatment in each case with dimethyl sulphide[132]. A similar result has been observed for 2-chloro-5-ethoxytropone (**214**). Alternatively, **211** can also be prepared from the oxidation of **213** with chloranil or DDQ[132]. Compound **211** is stable in non-polar solvents, but gradually decomposes in acids and in DMSO. In contrast to **197**, the quinone **211** forms hydrate **215** and hemiacetal **216** reversibly at room temperature[132]. In other words, **216** easily reverts to **211** by evaporation of methanol or on addition of non-polar solvents[132].

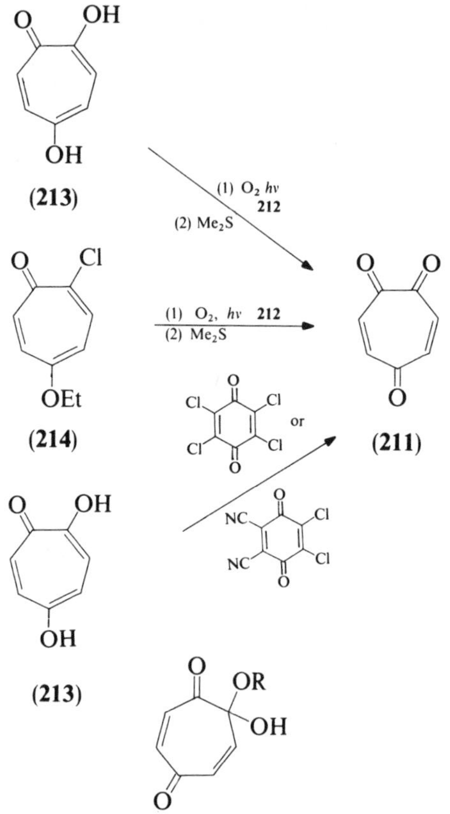

(213)

(1) O_2 $h\nu$ 212
(2) Me_2S

(214)

(1) O_2, $h\nu$ 212
(2) Me_2S

or

(211)

(213)

(215, R = H)
(216, R = Me)

Reaction of **211** with *o*-phenylenediamine gives quinoxalotropone (**217**) quantitatively[132, 133]. Oxidation of **217** with *m*-chloroperbenzoic acid or with anhydrous acetaldehyde and aerial oxygen affords the eight-membered acid anhydride **218**[133].

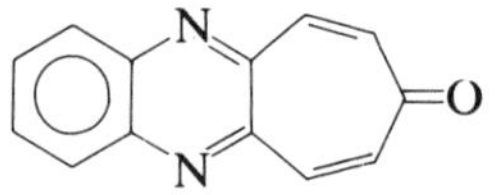

(217) 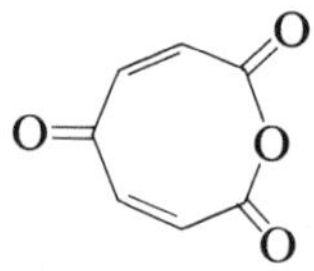 (218)

The *p*-tosylhydrazone **219** is converted into the diazoketone **220** which upon photolysis or thermolysis in methanol gives methyl 4-hydroxybenzoate in quantitative yield[133]. In agreement with theoretical prediction, most of the nucleophiles are found to attack at the C(4) position of **211** under standard conditions. As an example, hydrogen chloride reacts with **211** to yield 4-chloro-5-hydroxytropolone (**221**). Sodium benzenesulphinate in acetic

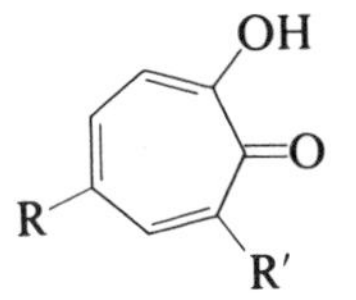

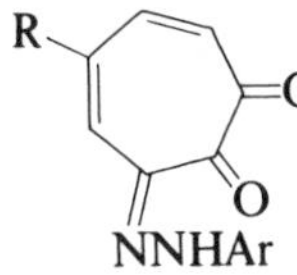

(**204**)

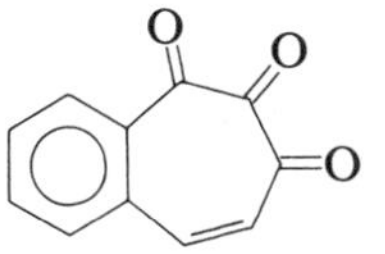

(**205**)

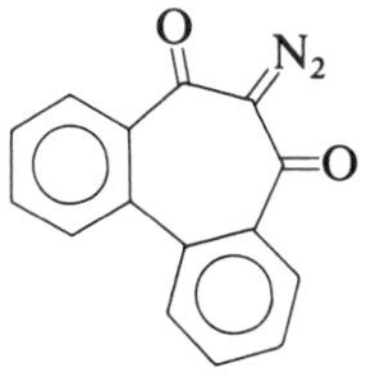

(**206**)

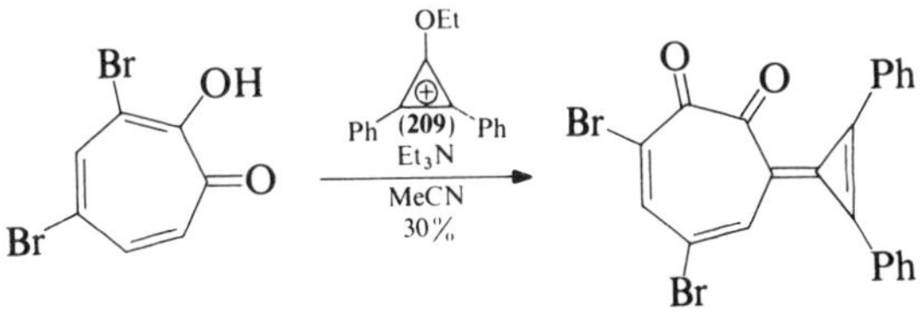

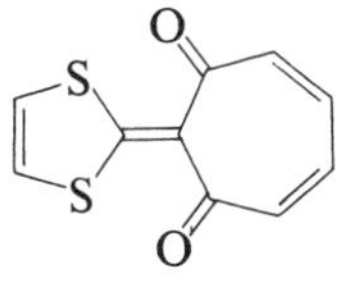

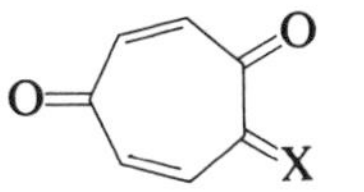

(**219**, =X = =NNHTs)
(**220**, =X = $=N_2$)

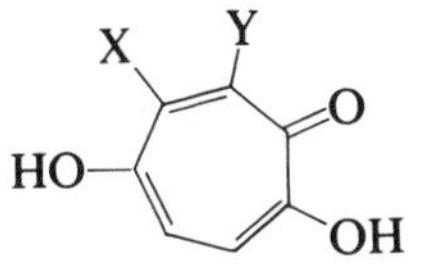

(**221**, X = Cl; Y = H)
(**222**, X = SO_2Ph; Y = H)
(**223**, X = H; Y = SO_2Ph)

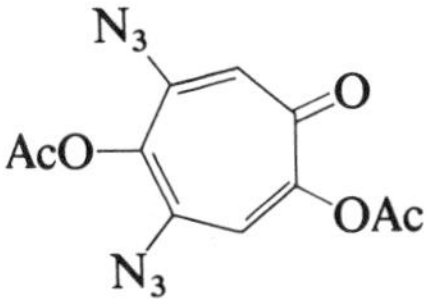

(**224**)

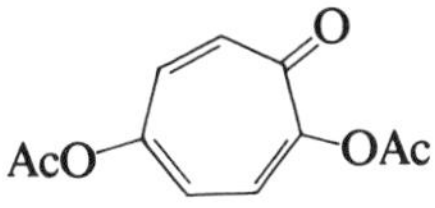

(**225**) 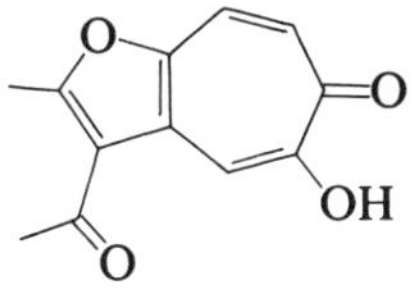 (**226**)

acid solution has been found to add in the same manner to give **222**[133]. Interestingly, when the reaction is carried out in aqueous solution, the isomeric product **223** can also be obtained[133]. Sodium azide in acetic acid also adds to **211**, and after acetylation, the 4,6-diazido compound **224** along with 5-hydroxytropolone diacetate (**225**) are produced[133]. Reaction of acetylacetonate with **211** in the presence of sodium acetate gives the adduct **226**[133]. Thiele type acetylation of **211** in the presence of sulphuric acid yields **227** and

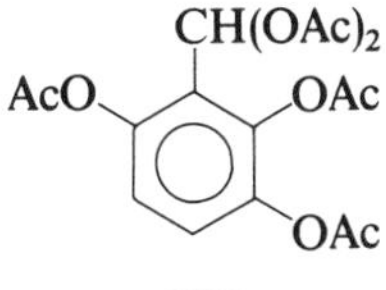

(**227**) 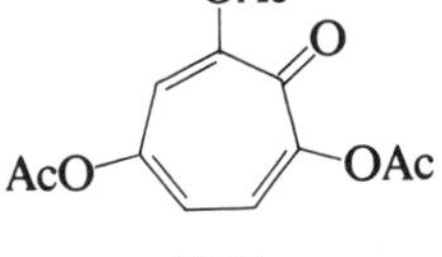 (**228**)

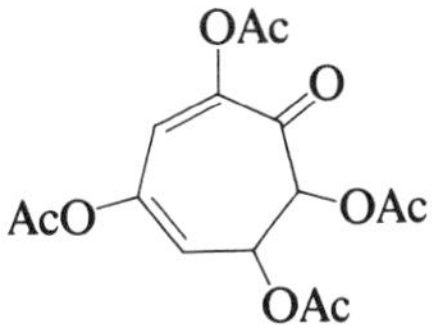

(**229**)

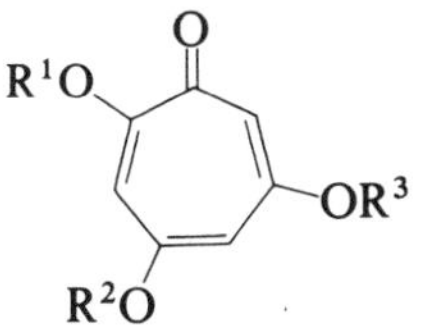

(**230a**, $R^1 = R^2$ = OAc; $R^3 = BF_2$)
(**230b**, $R^1 = BF_2$; $R^2 = R^3$ = OAc)

228[134]. The acetoxy group attacks the C(3) position exclusively and as a result C(3) becomes the aldehydic carbon in 227[134]. When BF_3 is employed, the regioselectivity of the nucleophilic attack diminishes and a mixture of 229 and 230 is thus obtained[134]. Based on He(I) photoelectron spectral data, the HOMO of *p*-tropoquinone (211) resembles that of *p*-benzoquinone[103, 124].

Treatment of dibromo-5-hydroxytropolone (231) with silver acetate in acetone affords the corresponding *p*-tropoquinone 232 in good yield[135]. When 231 is oxidized in acetic acid, the Michael-type adduct 233 is formed, while oxidation in ethanol gives rise to the ring contraction product 234[135]. It is worthwhile to point out that, in the course of the oxidative rearrangment of 5-hydroxytropolone to hydroquinone with cerium(IV) salts in a faintly alkaline solution, 211 has been proposed as an intermediate[136].

AgOAc, acetone

(231) **(232)**

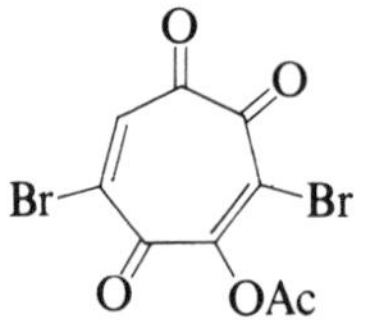

(233) 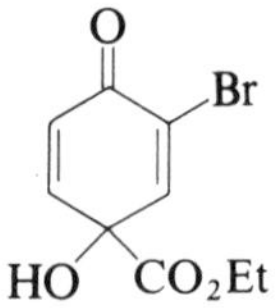 **(234)**

Benzo-*p*-tropoquinone (**235**)[137] and dibenzo-*p*-tropoquinone (**236**)[138] have been synthesized. 4-Hydroxy-3,6,7-triphenyl-*p*-tropoquinone (**237**)[139] and some of the 5-

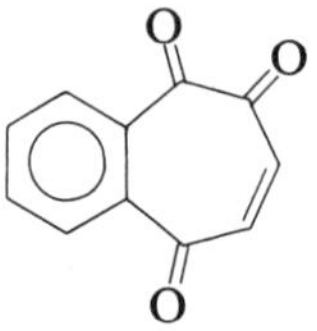

(235) 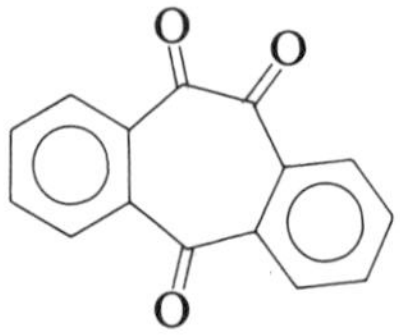 **(236)**

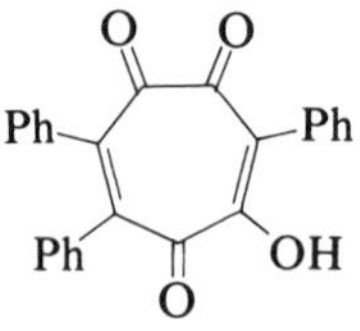

(237) 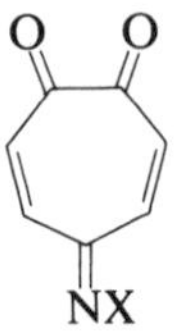 (**238**, X = OH)
(**239**, X = NHAr)

substituted tropolones such as **238**[140] and **239**[141] are also known to behave as quinonoid derivatives.

More recently, a novel cyclophane which contains a *p*-tropoquinone **240** has been synthesized[142] by the scheme outlined below. Preliminary X-ray crystallographic analysis

(240)

has shown that the four carbons and the two oxygens in the α-diketone moiety form a plane parallel to the mean plane of the benzene ring, and the remaining three carbons and oxygen in the dienone part are away from the mean plane[142]. Electronic spectral data and polarographic measurements further disclose the presence of sizeable intramolecular charge transfer interaction and the deformation of tropoquinone ring in **240**[142].

The 7,10-dithiasesquifulvalene-3,4-quinone **241** has been prepared. A noticeable dipolar character has been observed in this compound[143]. Cyclopropenylation of dibromo- or diiodo-tropolones (**242**) with **209** in the presence of triethylamine affords the diphenylheptatriafulvalene-3,4-diones **243** as orange needles[144]. These compounds are stable to light at room temperature and to heat up to about 150° C in solid states. However, they slowly decompose in solution[144].

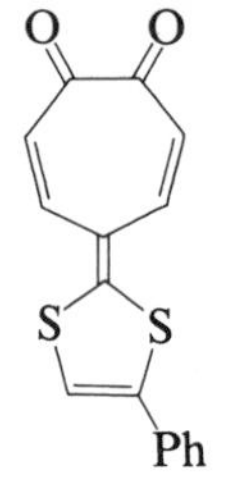

(**241**)

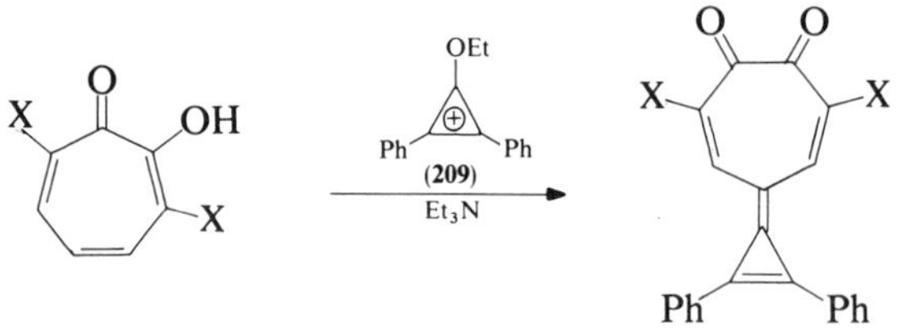

(**242**, X = Br or I) (**243**, X = Br or I)

In a similar manner, **244** and **245** are obtained when tropolone is treated with an equimolar amount of **209** in acetonitrile in the presence of triethylamine. It is interesting to note that the addition of excess triethylamine to **244** in acetonitrile promotes its immediate

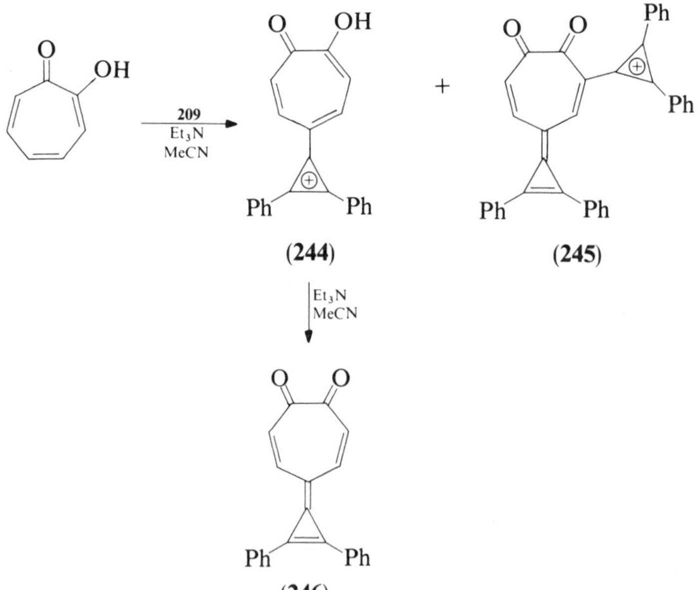

(**244**) (**245**)

(**246**)

and complete conversion to the corresponding dione **246** as orange needles[145] Compound **244** can be regenerated upon treatment with acid[145].

IV. OTHER DICARBONYL SYSTEMS

A. 1,2-Dicarbonyl Systems

1. *1,2-Acenaphthylenedione*

1,2-Acenaphthylenedione (**247**) is a well-known compound and a standard procedure for its preparation has been described[146]. Much of the chemistry investigated on this dione

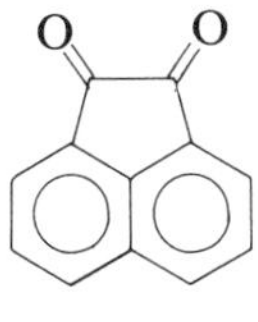

(**247**)

in recent years has centred on its application as a starting material for the synthesis of new heterocyclic and carbocyclic systems, and we believe it is instructive and worthwhile to discuss several of these syntheses.

A majority of the heterocyclic compounds prepared from **247** have been intended as potential dyes or pharmacological agents. In a large number of cases, the reactions

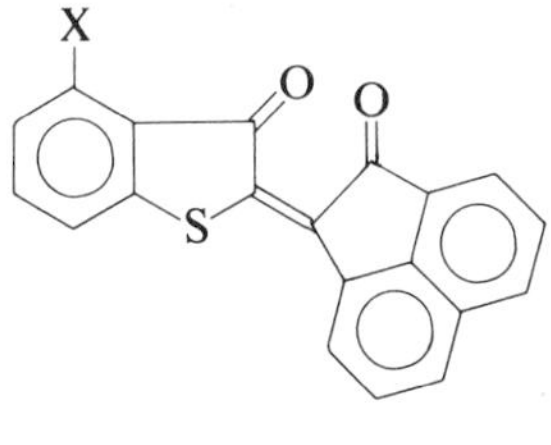

(**248**, X = Br, I)

involved are rather standard. Among the recent examples are the indigoid dyes **248**[147, 148] which can be prepared by condensing **247** with 4-halo-3-hydroxythionaphthenes. The

Ar = 2-furyl, phenyl,
p-chlorophenyl,
p-*N*,*N*-dimethylaminophenyl.

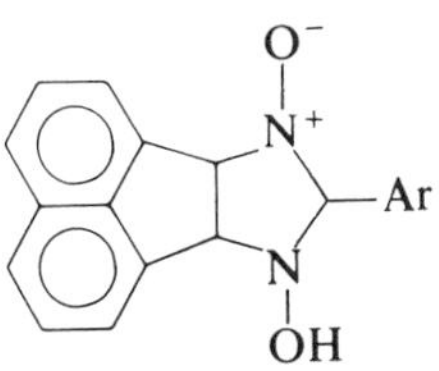

(**249**)

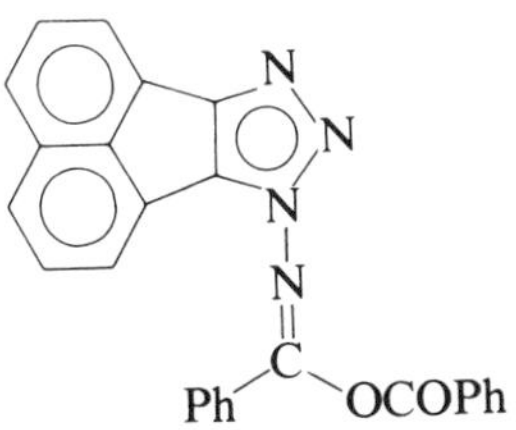

(**250**)

acenaphthylene-fused 1-hydroxy-2-arylimidazole-3-oxides **249** and related compounds[149] have been prepared by reacting **247** with hydroxylamine hydrochloride in the presence of an aromatic aldehyde. Oxidative cyclization of the bisbenzoylhydrazone of **247** has been found to give the acenaphtho[1,2-*d*]triazole **250**[150]. The 9-(3-substituted

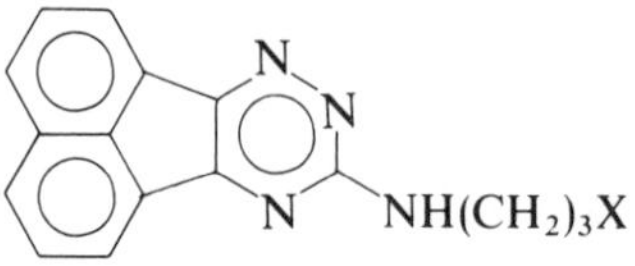

(**251**, X = OH, NMe_2, NEt_2, NBu_2)

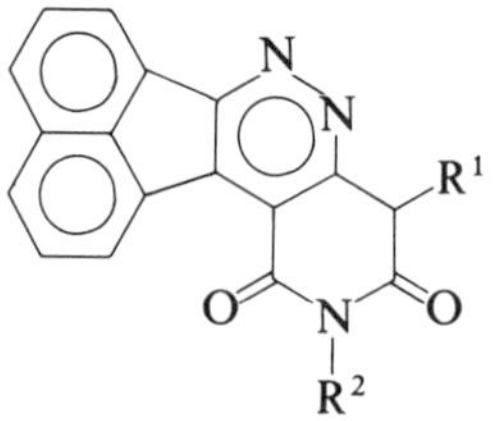

(**252**)

propylamino)-acenaphtho[1,2-*e*]-*as*-triazines **251**[151], of which some are active against vesicular stomatitis virus, can be synthesized by condensation of **247** with thiosemicarbazide followed by treatment of the resulting acenaphtho[1,2-*e*]-*as*-triazine-9(8*H*)-thione with 3-substituted propylamines. The 3,4-acenaphtho-5,7-dioxotetrahydropyrimido[4,5-*c*]pyridazine systems **252**[152] are conveniently constructed by reacting **247** with 6-hydrazinouracils. Along the same line, the synthesis of a series of 2,4-substituted 6,7-acenaphthopteridines **253**[153] is effected by condensing **247** with various 5,6-diaminopyrimidines. The list of the phenazine group of dyes has been expanded to include the

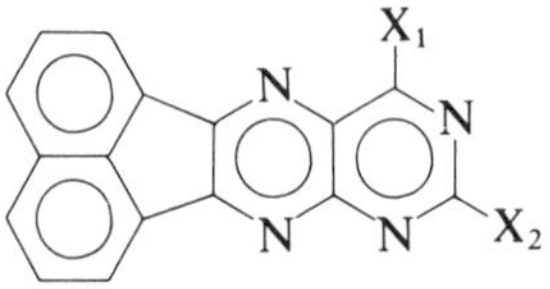

(**253**)

fluoroacenaphthoquinoxalines **254**[154], which are obtainable by treatment of substituted **247** with 4-fluoro-*o*-phenylenediamine. It has also been found that the methyl groups of

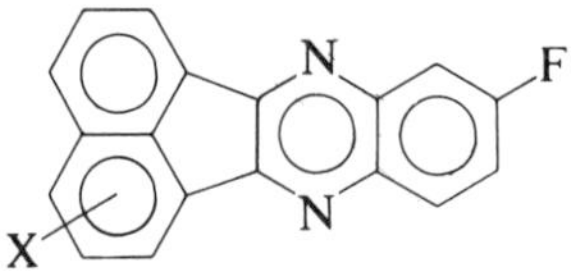

(**254**)

2,3-dimethylquinoxaline-1,4-dioxide are sufficiently activated to undergo facile cyclocondensation with **247** to give acenaphtho[1,2-*b*]phenazine (**255**)[155]. The hexahydro-1,2,4,5-tetrazine derivative **256** is obtained when equimolar quantities of **247** and 2,4-dimethylcarbonohydrazide are heated in boiling methanol. Upon heating in glacial acetic acid, **256** is converted to the cyclic dihydrazone **257**[156].

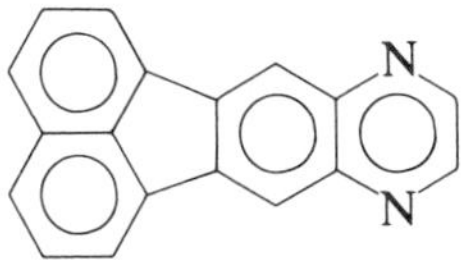

(**255**)

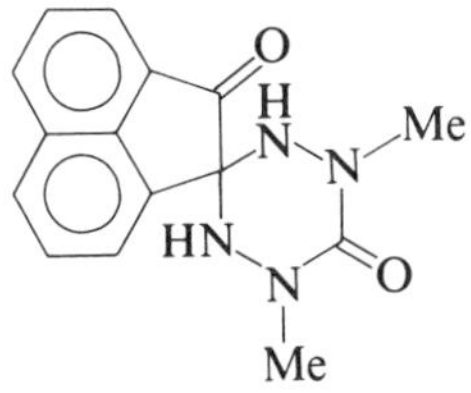

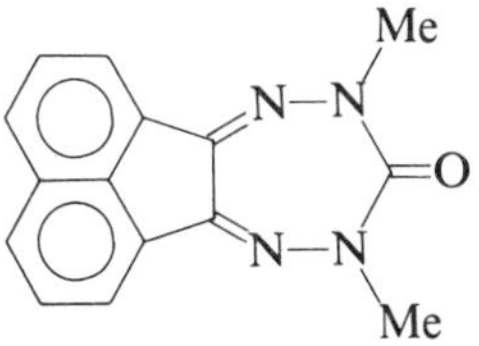

(**256**) (**257**)

In the past decade, a number of interesting carbocyclic systems have been prepared from (**247**). An improved procedure for the preparation of benzo[*k*]fluoranthene (**259**) has been described in which **247** is first condensed with *o*-phenylenediacetonitrile to 7,12-dicyanobenzo[*k*]fluoranthene (**258**). Heating of **258** in phosphoric acid affords **259**[157] in 60% overall yield. The related naphthofluoranthene **260** has been synthesized by a bis-Wittig reaction of the bis(triphenylphosphonium) salt of 2,3-bis(bromomethyl)naphthalene and **247**[158]. It is noteworthy that the reaction can be carried out in a heterogeneous mixture of dichloromethane and an alkaline aqueous solution, in which the phosphonium salt serves dually as a phase-transfer catalyst.

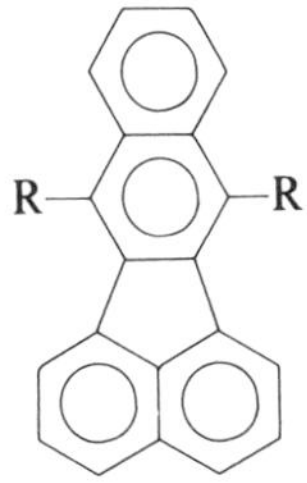

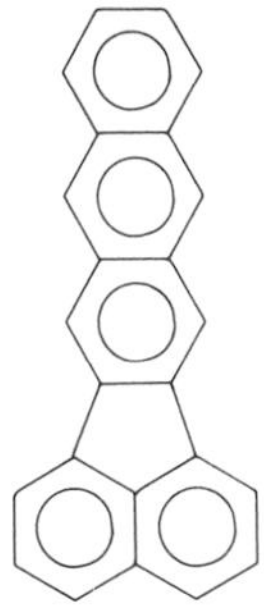

(**258**, R = CN)
(**259**, R = H)

(**260**)

The synthesis of 1,2-acenaphtho-3,8-disubstituted derivatives of cyclooctatetraene **261**[159] from **247** has been accomplished in several steps. Base-catalysed condensation of

247 with ketones **262** readily gives cyclopentadienones **263** which upon heating with the cyclooctatetraene dimethyl acetylenedicarboxylate adduct **264** affords **266**, presumably via

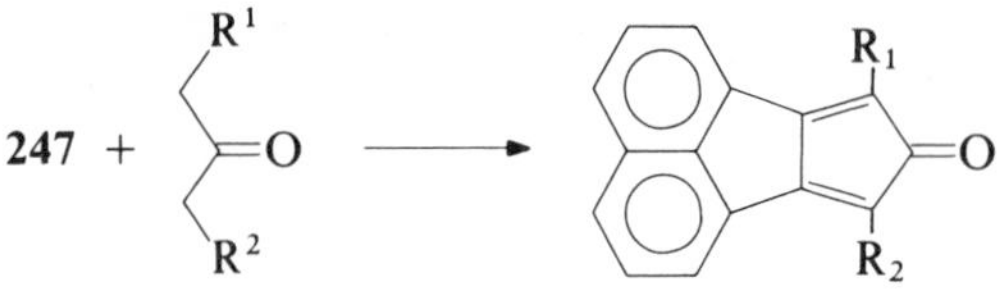

(262) **(263)**

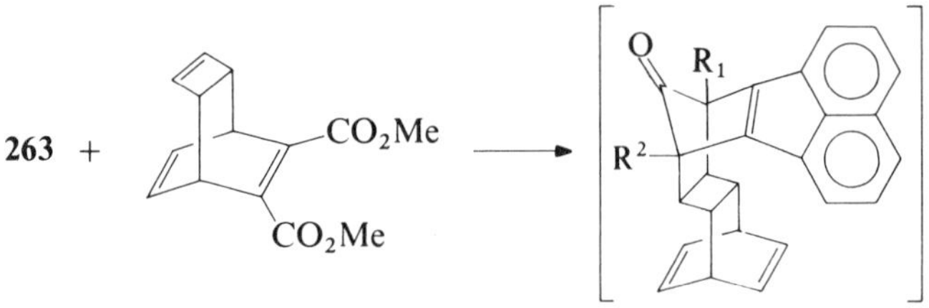

(264) **(265)**

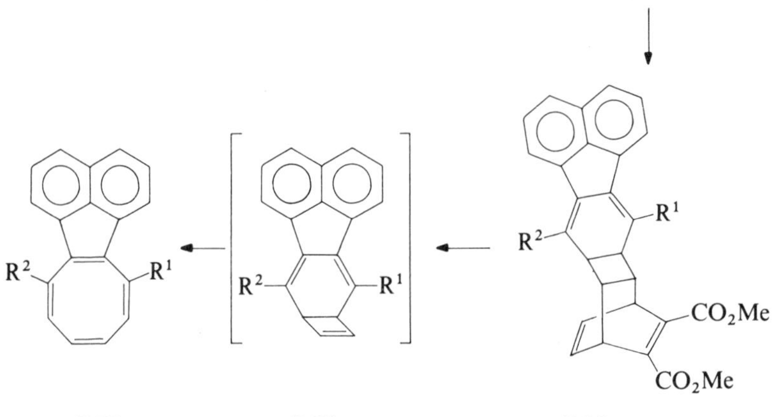

(261) **(267)** **(266)**

cheletropic loss of carbon monoxide from the initial [4 + 2] adduct **265**. Further heating of **265** in boiling xylenes leads to extrusion of dimethyl phthalate by thermal [4 + 2] cycloreversion to give **261** by way of its valence tautomer **267**. NMR evidence[159] indicates

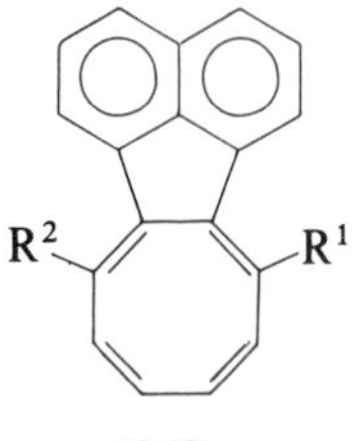

(268)

that compounds **261** contain conformationally rigid, bond-fixed, non-planar cycloocta-tetraene rings, and rules out the possibility of the bond-shift isomers **268**.

Three reports[160–162] of the synthesis of acenaphth[1,2-*a*]acenaphthylene (**269**) from **247** have appeared in the literature. The best yield[162] can be obtained by treatment with anhydrous HF of diol **270** which is derived from reaction of 1,8-dilithionaphthalene with **247**.

HO OH (1) HF (2) H_2O

(270) **(269)**

As expected, **247** reacts readily with Grignard reagents to give 2-hydroxyacenaphthe-nones **271** which undergo a facile base-catalysed carbon-to-oxygen acyl rearrangment[163]

247 (1) RMgBr (2) H_2O O OH R base O O R H

(271) **(272)**

to peri ring-expanded naphthalides **272**. The synthesis of the torsionally rigid *cis*-1-phenyl-2-mesitylacenaphthylene **273** involves a first step of addition of mesitylmagnesium bromide to **247**. Reduction of the hydroxy ketone **274** to the corresponding diol followed by a pinacol rearrangment yields ketone **275**. Addition of phenyllithium to **275** gives

247 (1) Mes Mg Br (2) H_2O O Mes OH (1) $LiAlH_4$ (2) HCO_2H Δ O Mes

(274) **(275)**

PhLi

Ph Mes Li, NH_3 HO Mes Ph H

(273) **(276)**

benzyl alcohol **276** which is converted to **273** via a Li/Na alloy variation of the Birch reduction. Hydrocarbon **273** together with *cis*-1-phenyl-2-(2,4,6-triisopropyl)acenaphthylene has been used for the investigation of internal rotation of the face-to-face aromatic rings[164]. The carbonyl functions in **247** are also susceptible to attack by other organometallic reagents. It has been found that π-2-methylallylnickel bromide selectively attacks one of the carbonyl groups to give the homoallylic alcohol **277**[165], even

247 + (π-2-methylallyl)Ni(μ-Br)₂Ni(π-2-methylallyl) —DMF→

(277)

if an excess of reagent is used. In addition, BF_3-catalysed allylation of **247** with allyltributyltin gives high yield of 2-allyl-2-hydroxyacenaphthenone (**278**)[166].

247 + $CH_2{=}CHCH_2SnBu_3$ —(1) $BF_3\cdot Et_2O$, CH_2Cl_2; (2) H_3O^+→

(278)

Several enediol derivatives of 1,2-acenaphthylenedione (**247**) have been reported. Electroreductive methylation of **247** gives 1,2-dimethoxyacenaphthylene (**279**) among other products, the formation of which are accountable by the coupling of a radical anion derived from **247** with a methyl radical[167]. As expected, similar cathodic reduction of **247**

(**279**, R = Me)
(**280**, R = COR′)

in the presence of acylating reagents yields the corresponding enediol diesters **280**[168]. Extension of the electroreductive alkylation methodology has led to the synthesis of acenaphtho crown ethers **281**[169]. More recently, it has been found that reductive cyclization of **247** can also be effected by treatment with sodium followed by addition of 1,4-dichlorobutane, as demonstrated by the preparation of the 1,4-dioxacine **282**[170]. In a related study, the synthesis of the bisacenaphtho-[18]crown-6, **283**, has been described[171].

The photochemistry of **247** has been investigated by several groups[172–176]. In the presence of oxygen photolysis of **247** leads to naphthalene-1,8-dioic acid anhydride (**284**) in good yield[172, 173]. On the other hand, irradiation of **247** in degassed tetrahydrofuran

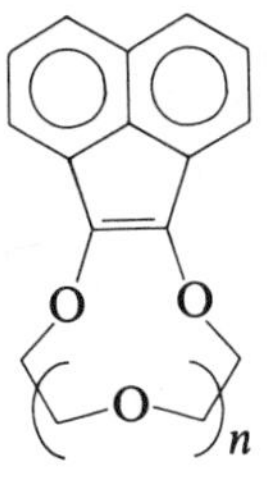

(**281**, $n = 1, 2$) 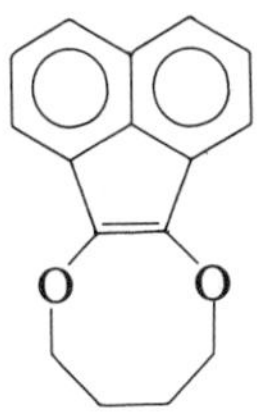 (**282**) 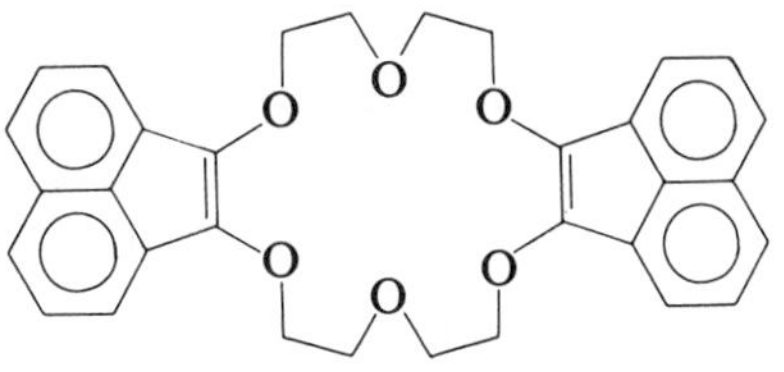 (**283**)

gives mainly the hydroxyacenaphthenone **285**[172]. Interestingly, when an olefin, e.g. cyclohexene is photoxidized with **247** as the sensitizer, the formation of cyclohexene oxide,

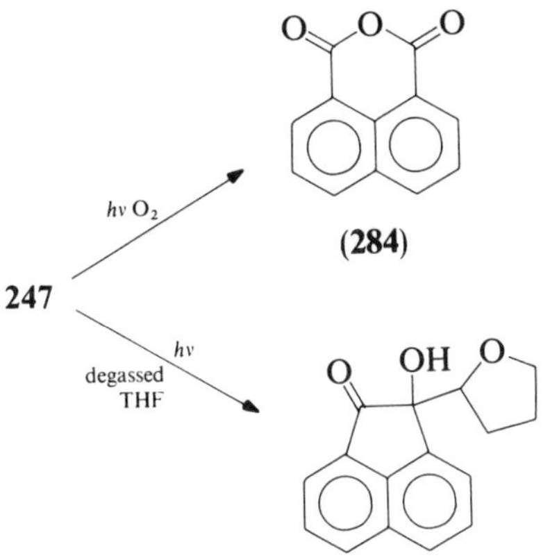

247 → ($h\nu$ O_2) (**284**)

247 → ($h\nu$, degassed THF) (**285**)

3-hydroperoxycyclohexene and adipaldehyde is accompanied by the oxidation of **247** to **284**[173].

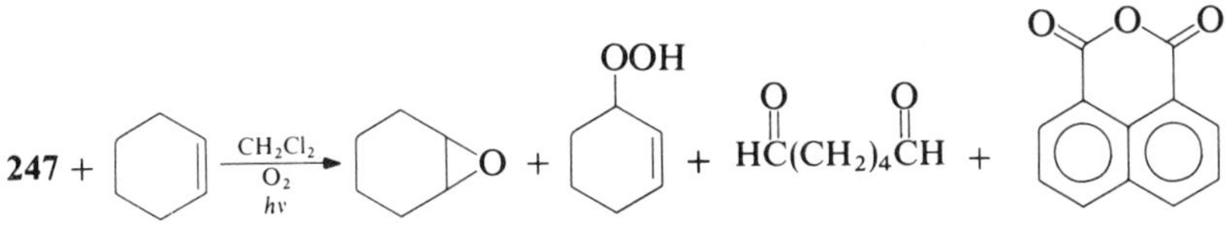

247 + cyclohexene → (CH_2Cl_2, O_2, $h\nu$) cyclohexene oxide (33%) + 3-hydroperoxycyclohexene (44%) + $HC(O)(CH_2)_4C(O)H$ (trace) + (**284**) 65% with respect to **247**

The photochemical cycloadditions of **247** to several olefinic systems have been reported. With cycloheptatriene[174], both the [2+2] and [2+6] cycloadducts, **286** and **287**, respectively, are obtained together with the ene product **288**. With norbornadiene the sole product is the [2+2] keto oxetane **289**, but with quadricyclane, the photorearrangement product **290** is also formed[175]. In the photochemical addition of **247** to ketene acetals, 1,1-dimethoxypropene gives mainly the [2+2] cycloadducts oxetane and bisoxetane.

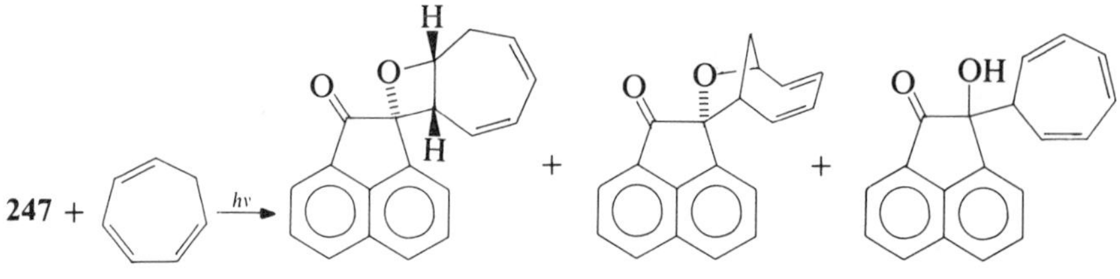

(286) (287) (288)

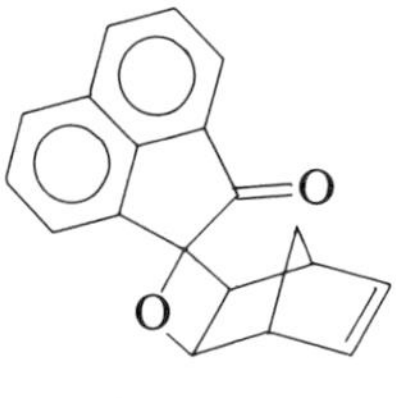

(289) 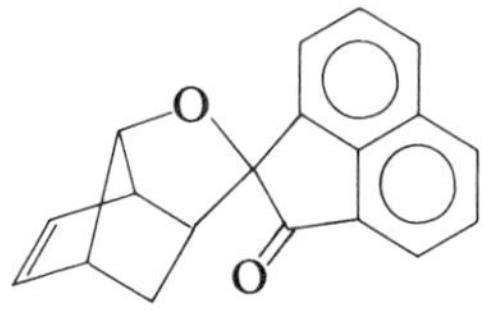 (290)

However, in a similar reaction with tetramethoxyethene, the [4+2] product dihydrodioxin is obtained[176].

The chemistry of the diazo compounds derived from **247** has attracted some attention recently. It has been found that the thermolysis of the dilithium salt of 1,2-acenaphthylenedione bistosylhydrazone (**291**) gives 1,8-dicyanonaphthalene (**292**)[177]. The reaction occurs presumably by way of the intermediacy of the bisdiazo derivative **293** which cyclizes to 1,2,3,4-tetrazine **294** and loses a molecule of nitrogen to produce **292**.

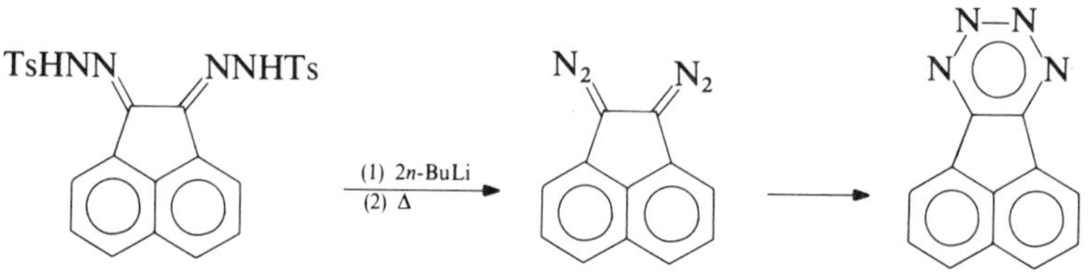

(291) (293) (294) 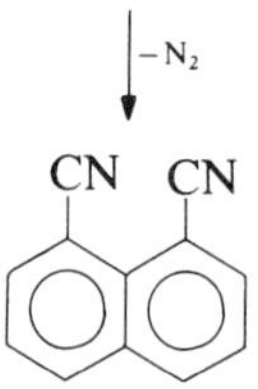 (292)

The thermolysis of 2-diazo-1(2*H*)-acenaphthylenone (**295**)[178] has been reported[179, 180] to yield biacenedione (**296**) together with a small amount of the azine **297**. On the other hand, the photosensitized (by *meso*-tetraphenylporphine or methylene blue) oxidation of **295** has been found to give **247** and the anhydride **284**[181].

(295) (296) (297)

The thermolysis and photolysis of **295** in various environments have been investigated again recently [182–186] in relation to the mechanism of the Wolff rearrangement. Under standard conditions, neither thermolysis nor photolysis of **295** gives rise to any product attributable to the Wolff rearrangement[184]. Thus, thermal and photochemical decompositions of **295** in cyclooctane result in loss of nitrogen and formation of 2-cyclooctylacenaphthenone (**298**). In 2-propanol solution containing oxygen, photolysis

cyclooctanol $h\nu$ or Δ

(298)

295 i-PrOH O_2 $h\nu$ + MeCOMe + **284** + **296**

(299)

$OCMe_3$

t-BuOH O_2 $h\nu$

(300)

leads to a mixture of acenaphthenone (**299**), acetone, the anhydride **284** and **296**. Irradiation of **295** in oxygen-saturated *t*-butanol gives rise to the solvent-captured product **300**. Furthermore, thermolysis of **295** in benzonitrile with added cupric sulphate results in the formation of 8-phenylacenaphth[1,2-*d*]oxazole (**301**), and heating **295** in acrylonitrile containing palladium acetate leads to a mixture of the isomeric spirocyclopropanes **302**

295 → (PhCN, $CuSO_4$, Δ) → (301)

295 → (CH_2=CHCN) → (302) (*Z*)– + (303) (*E*)–

and **303**. These and other related results can be accounted for by the intermediacy of 2-oxoacenaphthylidene (**304**) either in its singlet or triplet state[184]. Photolysis of **295** in an

(304)

argon matrix at 10–15 K produces the triplet ground state ketocarbene **305** as the primary photoproduct. The identity of ketocarbene **305** has been established by UV–VIS, IR and

295 → (*hν*, Ar, 15 K) → (305) → (*hν*) → (306)

ESR spectroscopy[182, 183, 185, 186]. Subsequent excitation (T_0–T_1) of **305** results in ring contraction to the strained ketene **306**. It has been suggested that the overall non-concerted Wolff rearrangement proceeds via the S‴ state of ketocarbene **304** which is reached from T_1 by internal conversion[186].

2. *Cyclopent*[fg]*acenaphthylene-1,2-dione (pyracycloquinone) and 5,6-dihydrocyclopent*[fg]*acenaphthylene-1,2-dione (pyracenequinone)*

The work of Trost in the late 1960s on the preparation[187,188], ESR studies[189–191] and some aspects of the chemistry[188] of cyclopent[*fg*]acenaphthylene-1,2-dione (pyracycloquinone, **307**) and the related 5,6-dihydro derivative (pyracenequinone, **308**) has been presented in the previous review[1].

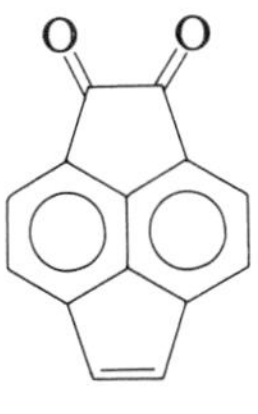

(307) 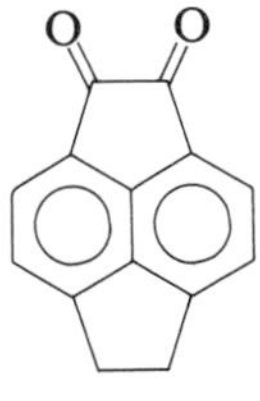 **(308)**

The photochemical behaviour of **307** was subsequently reported by Castellano and coworkers[192,193]. Irradiation of **307** under argon in protic solvents such as methanol, ethanol and 2-propanol gives the corresponding 5,6-acenaphthenedicarboxylic acid diester **309**, while under identical conditions **308** fails to give a product. A detailed

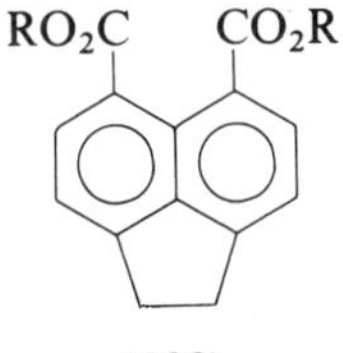

RO_2C CO_2R

(309)

investigation of the luminescence spectra of **307** coupled with kinetic data have led to the conclusion that the photolysis of **307** in protic solvents proceeds through a singlet encounter complex represented as structure **310**[193].

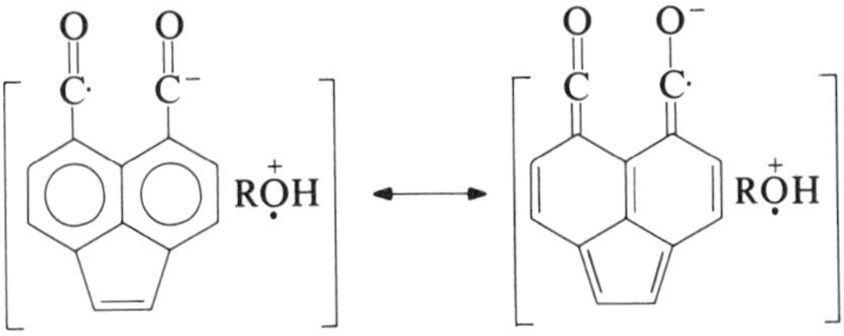

(310)

Reaction of **308** with dialkyl phosphites yields the adducts **311**[194]. With trialkyl phosphites, the products are the cyclic phosphates **312** which can be hydrolysed to **313**[194].

In conjunction with the study of the strain in acenaphth[1,2-*a*]acenaphthylene (**269**), the structurally similar alkene **314** has been synthesized from **308**[162]. Reaction of **308** with 5,6-dilithioacenaphthene (**315**) gives the diol **316** which on treatment with

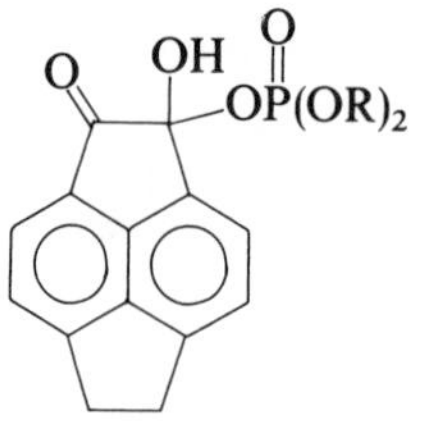

(**311**) 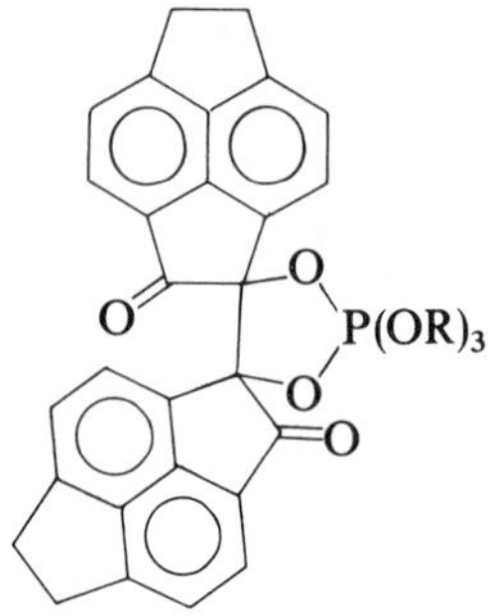 (**312**) 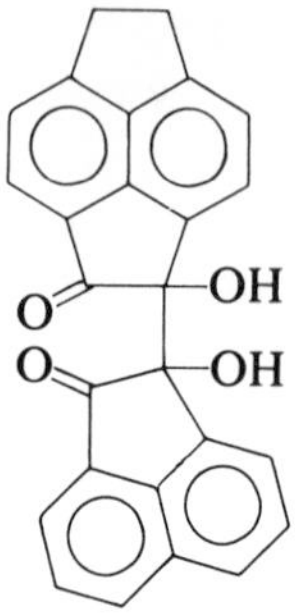 (**313**)

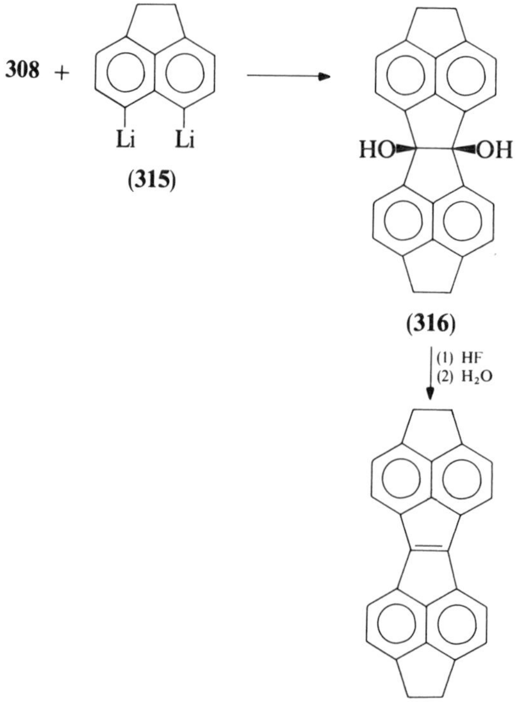

308 + (**315**) → (**316**) → (1) HF (2) H_2O → (**314**)

HF followed by H_2O yields **314**. Alternatively, **314** may also be prepared from **316** by the Corey–Winter procedure[162, 195]. Attempts to convert **314** into the [4n + 4n] fused system **317** were unsuccessful.

The additions of organomagnesium and organolithium reagents to pyracenequinone (**308**) have been examined by Tanaka[196]. The reaction of **308** with methylmagnesium bromide gives a mixture of *cis* and *trans* diols **318**, whereas the reaction of **308** with 5,6-dilithioacenaphthene–*N*,*N*,*N'*,*N'*-tetramethylethylenediamine complex yields the *cis* diol **316**, as previously noted by Mitchell and coworkers[162].

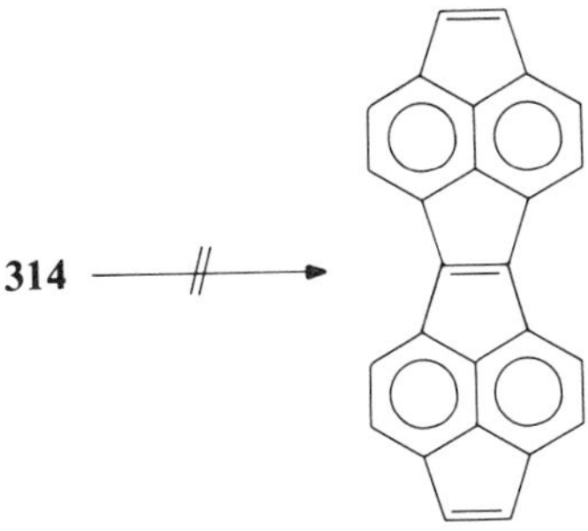

314 —//→

(317)

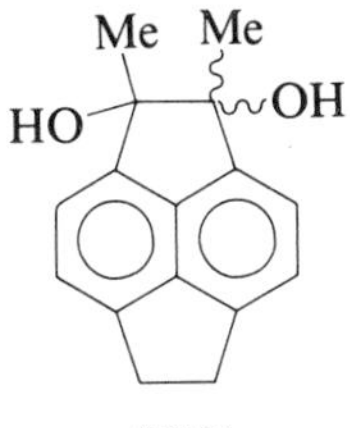

Me Me
HO OH

(318)

The photolysis of diazoketones **319** and **320** have also been included in the recent studies by Chapman and coworkers[182, 183, 185, 186] on the mechanism of the Wolff rearrangment. As with the results observed for 2-diazo-1(2*H*)-acenaphthylenone (**295**) described above, photochemical extrusion of nitrogen from **319** in an argon matrix at 15 K produces the triplet ground state (T_0) ketocarbene **321**, which is characterized by ESR and IR spectroscopy. Subsequent excitation (T_0–T_1) of **321** leads to slow ring contraction to **322**.

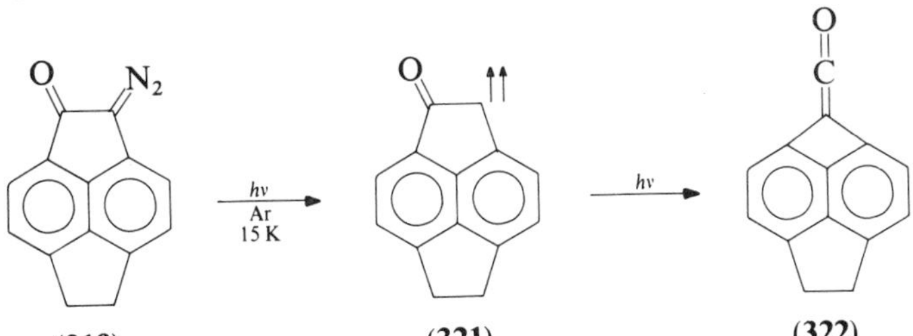

$\xrightarrow[\text{Ar, 15 K}]{h\nu}$ $\xrightarrow{h\nu}$

(319) (321) (322)

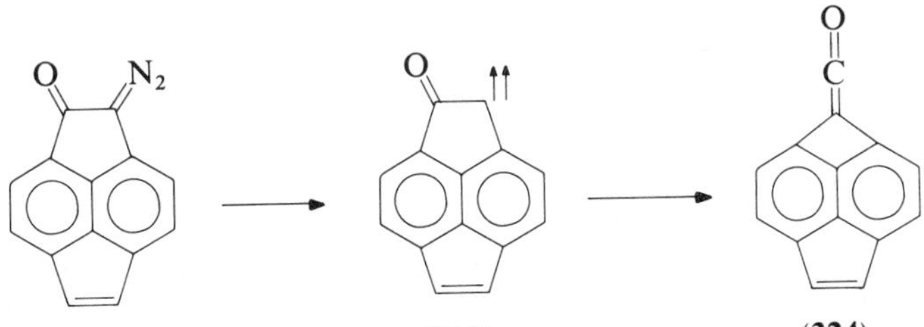

(320) (323) (324)

However, due to experimental complications the transformation sequence **320** → **323** → **324** has not been rigorously established[186].

3. *Cyclohepta*[de]*naphthalene-7,8-dione (*o*-pleiadienequinone) and acepleiadylene-5,6-dione*

o-Pleiadienequinone (**325**), which may be regarded as a higher analogue of 1,2-acenaphthylenedione (**247**), has been synthesized by Tsunetsugu and coworkers[197, 198] by cycloaddition of acenaphthylene **326** with dichloroketene followed by hydrolysis of the

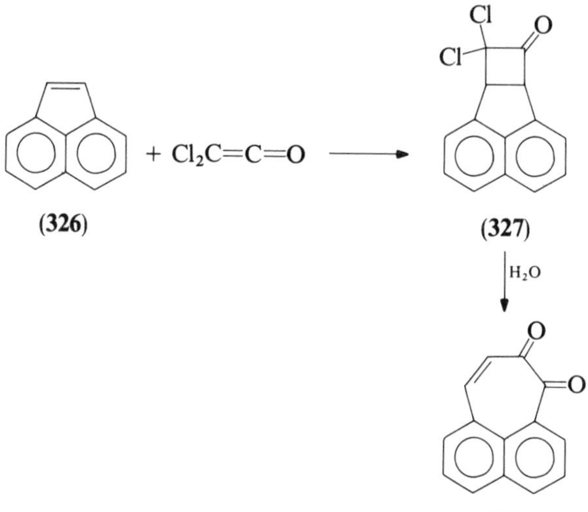

+ $Cl_2C{=}C{=}O$

(**326**) (**327**)

H_2O

(**325**)

resulting adduct **327**. NMR data suggest that **325** has some contribution from such canonical forms as 2,3-(**325a**) and/or 4,5-benzotropolonate (**325b**) ions. The half-wave reduction potentials for **325** in aqueous ethanol have been determined[198].

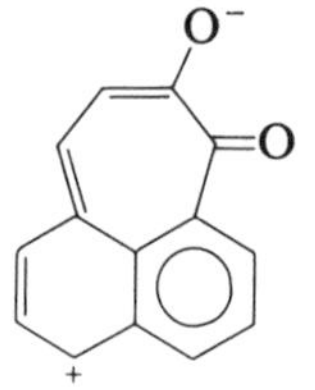

(**325a**) 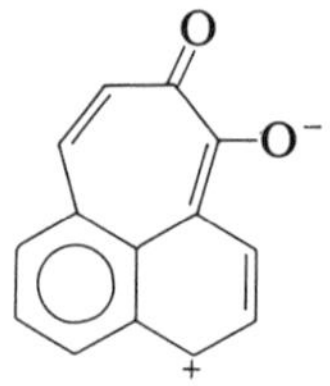 (**325b**)

As expected, dione **325** reacts with *o*-phenylenediamine to give the quinoxaline **328**. Reaction of **325** with acetic anhydride in the presence of a catalytic amount of sulphuric acid yields the triacetate **329**. Treatment of **325** with ethanol and acid gives the diethyl acetal **330**, and with 1 M sodium hydroxide, a rearranged product, 1-hydroxyphenalene-1-carboxylic acid (**331**) is formed. Attempts to effect the Diels–Alder reaction of **325** with cyclopentadiene, cycloheptatriene, furan or anthracene have not been successful[198].

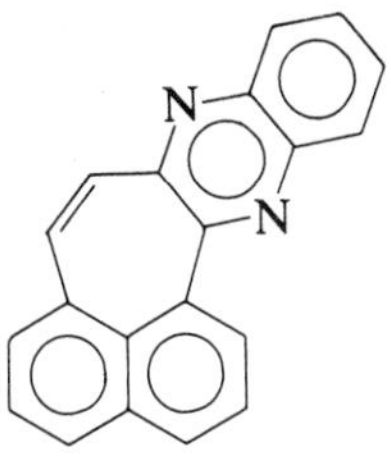

N
N

(328)

OAc
AcO
OAc
Ac_2O
H_2SO_4

(329)

CO_2H
OH
325
(1) OH^-
(2) H_3O^+

(331)

OEt
OEt
O
EtOH
H^+

(330)

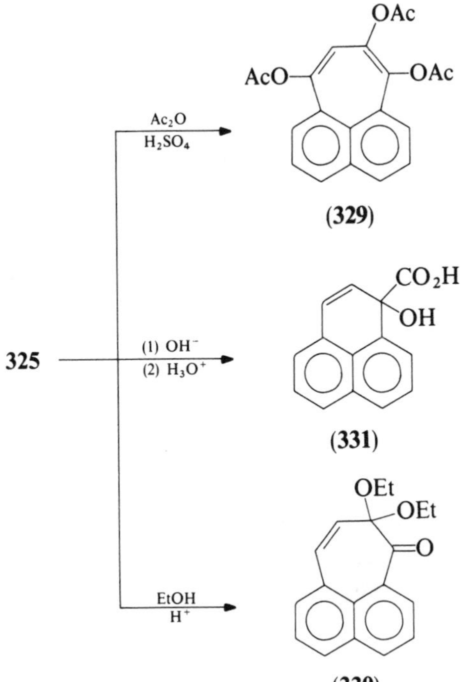

It has been found subsequently[199] that the reaction of **325** with acetic anhydride in the presence of perchloric acid yields keto triacetate **332** instead of **329**. The conversion of **325** into the corresponding epoxide **333** has also been described[199].

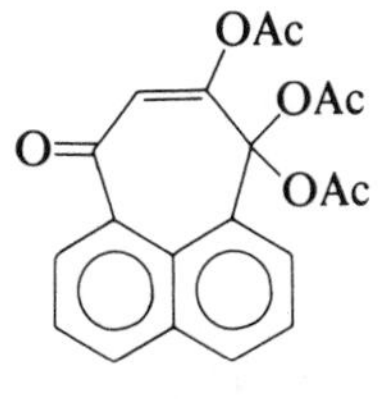

OAc
OAc
O
OAc

(332)

325 —NaOCl, dioxane, H_2O, pH 8→ (333)

Acepleiadylene-5,6-dione (**334**), a higher analogue of pyracycloquinone (**307**), has been synthesized by Tsunetsugu and coworkers[200] from the diketone **335** in seven steps. The

(335) —(1) $HOCH_2CH_2OH$, TsOH (2) NH_2NH_2, OH^-, $HOCH_2CH_2OH$ (3) H_3O^+ (4) $MeSSC_6H_4NO_2$-*o*, NaH→ (SMe, SMe) —(1) $Tl(NO_3)_3 \cdot 3H_2O$, MeOH (2) H_3O^+→ —DDQ→ (334)

values of E_1, E_2 and $(E_1 + E_2)$ for **334** are appreciably higher than those of pyracycloquinone (**307**). These data, as well as the low-field carbonyl carbon resonances (190.6 and 192.3 ppm for C(5) and C(6), respectively) in **334** are taken as evidence for the high quinonoid character of this compound[200].

B. Other Dicarbonyl Systems

1. *Cyclopenta*[def]*fluorene-4,8-dione (dibenzo*[cd,gh]*pentaleno-4,8-quinone)*

The synthesis of cyclopenta[*def*]fluorene-4,8-dione (**336**) was first reported by Kinson and Trost[201] in 1971. Since this compound was last reviewed[1], a full paper[202] detailing its preparation as well as its spectroscopic properties has appeared. Both UV and IR data reflect decreased conjugation between the carbonyl groups and the benzene rings, which has been ascribed to the unusually long bonds α to the carbonyl functions. This notion is further supported by the NMR spectrum of the bisprotonated quinone **337**, which indicates poor delocalization of excess positive charge into the benzene rings. The

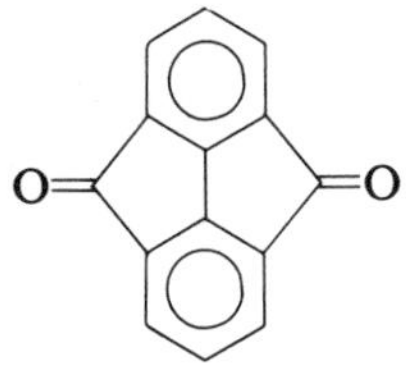

(336)

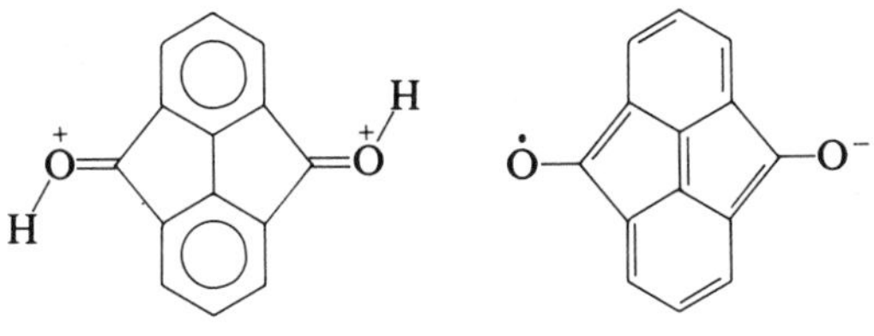

(337) **(338)**

polarographic characteristics of **336** and the ESR spectral properties of the semiquinone radical anion **338** strongly support the view that electron delocalization in **338**, which possesses a 4n-π periphery, leads to destabilization[202].

2. *Cyclohepta*[de]*naphthalene-7,10-dione (1,4-pleiadienequinone) and cyclohept*[fg]*acenaphthylene-5,8-dione (acepleiadylene-5,8-dione)*

Cyclohepta[*de*]naphthalene-7,10-dione (**339**) has been prepared by dehydrogenation of dione **340** with selenium dioxide[203]. On treatment with sulphuric acid in methanol, **339**

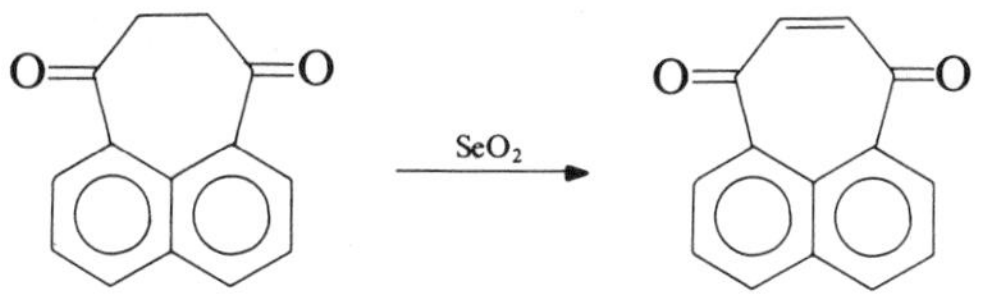

SeO_2

(340) **(339)**

undergoes ring contraction to give 3-(dimethoxymethyl)phenalenone (**341**). The hydroxy derivative **343**[199] has been prepared subsequently by treatment of **332** above with methanolic potassium carbonate followed by acid hydrolysis of the acetal **342**.

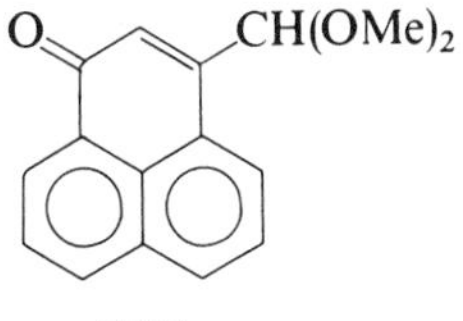

$CH(OMe)_2$

(341)

332 $\xrightarrow[\text{MeOH}]{K_2CO_3}$ (342) $\xrightarrow{H_3O^+}$ (343)

Cyclohept[*fg*]acenaphthylene-5,8-dione(acepleiadylene-5,8-dione, **344**) has been synthesized from diketone **335** in four steps[200]. Comparison of the reduction potentials E_1 and E_2 of **344** with those of pyracycloquinone (**307**) lends support to the notion that the former may be regarded as [14]annulenequinone with a vinyl cross-link[200].

335 (1) Br_2, $CHCl_3$, −10°C; (2) DBU; (3) NBS → ; DBU → (344)

3. Bridged Annulenediones

An initial attempt to synthesize bicyclo[4.4.1]undeca-3,6,8,10-tetraene-2,5-dione (**345**) from 1,6-methano[10]annulene (**346**) by the steps outlined below failed to materialize because the anticipated valence tautomerization rests on the side of the norcaradiene-enedione structure **347**[204]. However, an analogous reaction sequence starting from 11,11-fluoro-1,6-methano[10]annulene (**348**) yields the bridged annulenedione **349**[204]. The ESR spectrum[205] of the radical anion of **347** can be interpreted in terms of a structure intermediate to **347** and **345**, but is probably closer to the ketonic structure **347**. On the

(346) $\xrightarrow{Pb(OAc)_4}$ $\xrightarrow{MeLi}$ $\xrightarrow{MnO_2}$ (347) ⇸ (345)

other hand, the ESR spectrum of the radical anion of **349** reflects some degree of quinonoid character in **349**.

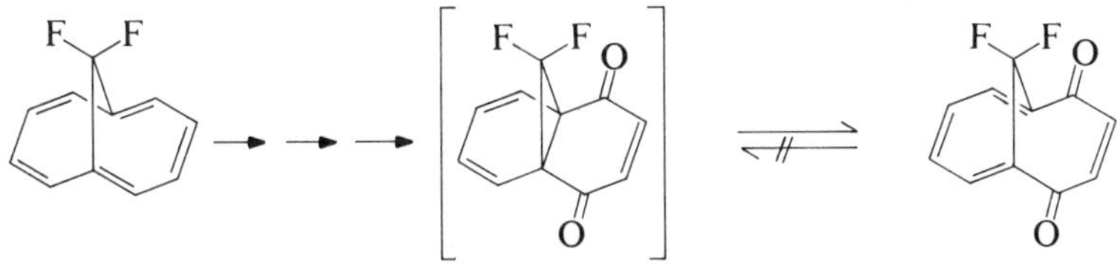

(**348**) (**349**)

The synthesis of bicyclo[4.4.1]undeca-3,5,8,10-tetraene-2,5-dione (**350**) from 2,7-dibromo-1,6-methano[10]annulene (**351**) has also been reported[206]. Whether dione **350**

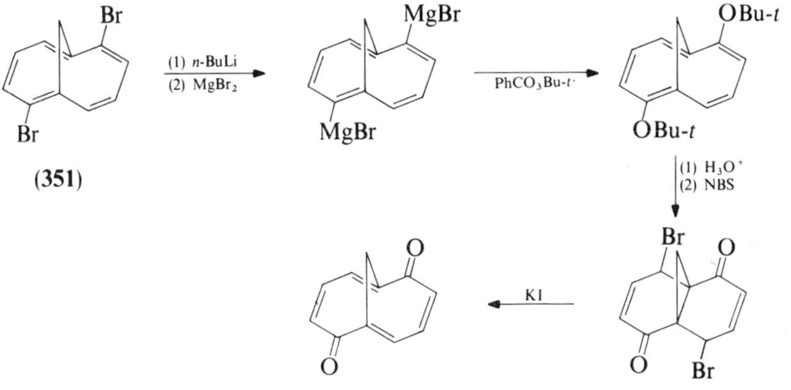

(**351**) (**350**)

can be regarded as a quinone has not been settled on the basis of its spectral properties[206]. However, that **350** undergoes reductive acetylation to 2,7-diacetoxy-1,6-methano [10]annulene **352** gives an indication of its quinonoid property.

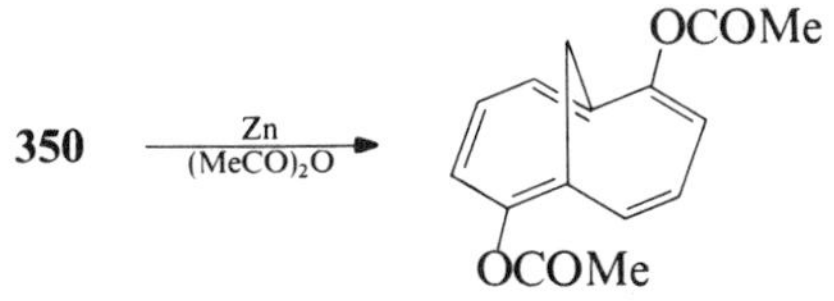

(**352**)

The monohydrazones of **345**, namely, compounds **353**, have been prepared[207–210] by coupling 2-methoxy-1,6-methano[10]annulene (**354**) with aryldiazonium salts. A combination of cycloheptatriene-norcaradiene valence tautomerization and hydrazone keto–azo enol tautomerization may lead to a total of four isomers in equilibrium. Both ^{13}C-NMR and IR data indicate that these compounds exist predominantly as quinone hydrazones[207]. However, if the aryl group in **353** carries an electron-withdrawing group at the 4-position, a slow isomerization can be observed by NMR spectroscopy[210]. For the

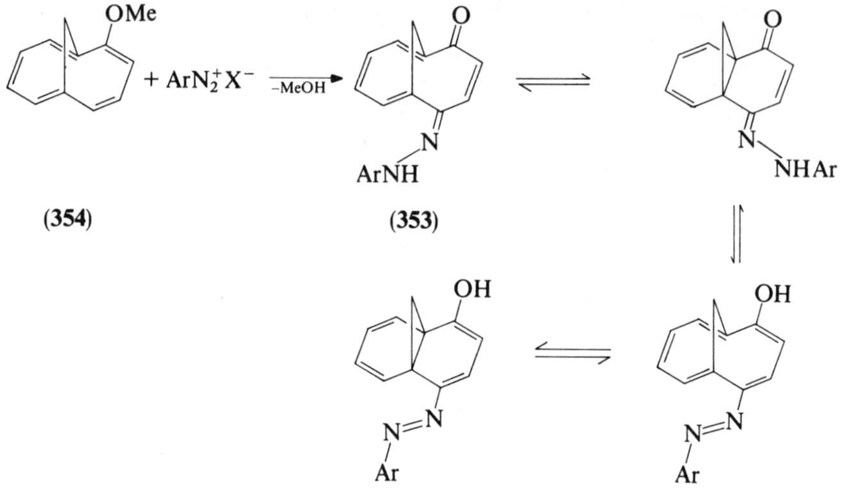

OMe + $ArN_2^+X^-$ $\xrightarrow{-MeOH}$ (353) ⇌ ⇌ ⇌

(354) (353)

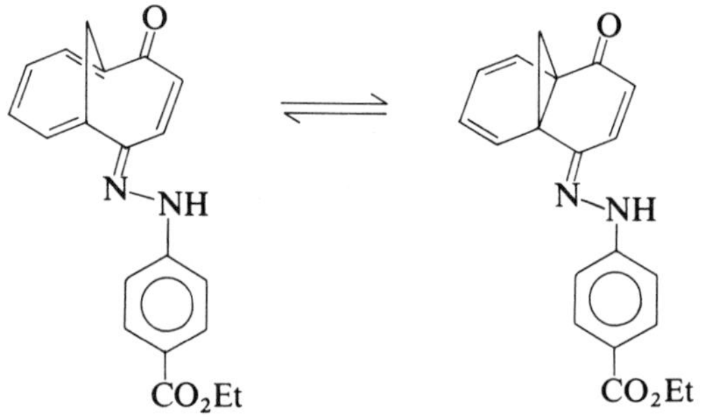

(355) ⇌ (356)

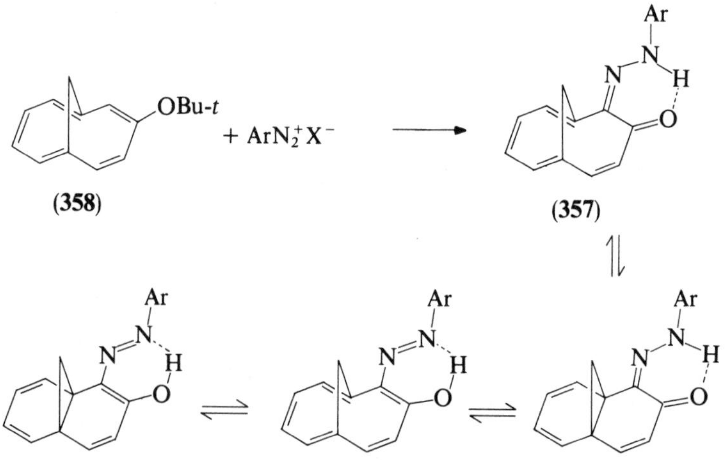

CO_2Et CO_2Et

(358) OBu-*t* + $ArN_2^+X^-$ → (357)

⇌ ⇌ (359)

interconversion of **355** and **356**, the activation energy in either direction has been estimated to be $\sim 60\,\mathrm{KJ\,mol^{-1}}$.

In a similar manner, hydrazones **357** have been prepared[207] from 3-*t*-butoxy-1,6-methano[10]annulene (**358**), although the parent bridged[10]annulene-2,3-dione, i.e. bicyclo[4.4.1]undeca-4,6,8,10-tetraene-2,3-dione, remains unknown. It has been suggested that a fast equilibrium exists between **357** and **359**.

V. REFERENCES

1. T. A. Turney, in *The Chemistry of Quinonoid Compounds*, Part 2 (Ed. S. Patai), John Wiley and Sons, New York, 1974, Chapter 16.
2. M. P. Cava and M. J. Mitchell, *Cyclobutadienes and Related Compounds*, Academic Press, New York, 1967, p. 128.
3. A. H. Schmidt and W. Ried, *Synthesis*, 1 (1978).
4. H. Knorr and W. Ried, *Synthesis*, 649 (1978).
5. W. Ried and A. H. Schmidt, *Angew. Chem., Int. Ed. Engl.*, **11**, 997 (1972).
6. A. H. Schmidt and W. Ried, *Synthesis*, 869 (1978).
7. Ref. 2, p. 219.
8. G. Maahs and P. Hegenberg, *Angew. Chem. Int. Ed. Engl.*, **5**, 888 (1966).
9. R. West and J. Niu, *Nonbenzenoid Aromatics*, Vol. 1 (Ed. J. Snyder), Academic Press, New York, 1969, p. 311.
10. R. West and H. Y. Niu, in *The Chemistry of the Carbonyl Group*, Vol. 2 (Ed. J. Zabicky), John Wiley and Sons, New York, 1970, p. 241.
11. A. H. Schmidt, *Synthesis*, 961 (1980).
12. R. West, *Israel J. Chem.*, **20**, 300 (1980).
13. G. Seitz, *Nachr. Chem. Tech. Lab.*, **28**, 804 (1980).
14. A. H. Schmidt, *Chem. Unserer Zeit*, **16**, 57 (1982).
15. G. Oremek and H. Kozlowski, *Chem. Ztg.*, **107**, 295 (1983).
16. F. Serratosa, *Acc. Chem. Res.*, **16**, 170 (1983).
17. R. West (ed.), *Oxocarbons*, Academic Press, New York, 1980.
18. For reviews, see D. Bellus and H. P. Fisher, *Advances in Pesticide Science*, Part 2 (Ed. H. Geissbühler), Pergamon Press, Oxford, 1979, p. 373; Ref. 17, p. 101.
19. J. P. Springer, J. Clardy, R. J. Cole, J. W. Kirksey, R. K. Hill, R. M. Carlson and J. L. Isidor, *J. Am. Chem. Soc.*, **96**, 2267 (1974).
20. W. Ried and M. Vogl, *Justus Liebigs Ann. Chem.*, 355 (1982).
21. W. Ried and M. Vogl, *Chem. Ber.*, **115**, 403 (1982).
22. W. Ried, M. Vogl and H. Knorr, *Justus Liebigs Ann. Chem.*, 396 (1982).
23. G. Seitz and B. Gerecht, *Chem. Ztg.*, **107**, 105 (1983).
24. T. LePage, K. Nakasuji and R. Breslow, *Tetrahedron Lett.*, **26**, 5919 (1985).
25. H. J. Bestmann, G. Schmid, D. Sandmeier and L. Kisielowski, *Angew. Chem., Int. Ed. Engl.*, **16**, 268 (1977).
26. A. H. Schmidt and A. Aimène, *Chem. Ztg.*, **107**, 299 (1983).
27. For a review on the chemistry of squaraines, see Ref. 17, p. 185.
28. A. H. Schmidt, A. Aimène and M. Schneider, *Synthesis*, 436 (1984).
29. A. H. Schmidt, A. Aimène and M. Hoch, *Synthesis*, 754 (1984).
30. A. H. Schmidt, U. Becker and A. Aimène, *Tetrahedron Lett.*, **25**, 4475 (1984).
31. See T. Kämpchen, G. Seitz and R. Sutrisno, *Chem. Ber.*, **114**, 3448 (1981) and Ref. 17, p. 15.
32. G. Seitz, R. Sutrisno, B. Gerecht, G. Offermann, R. Schmidt and W. Massa, *Angew. Chem., Int. Ed. Engl.*, **21**, 283 (1982).
33. B. Gerecht, T. Kämpchen, K. Köhler, W. Massa, G. Offermann, R. E. Schmidt, G. Seitz and R. Sutrisno, *Chem. Ber.*, **117**, 2714 (1984).
34. M. S. South and L. S. Liebeskind, *J. Org. Chem.*, **47**, 3815 (1982).
35. H. Dürr, H. Nickels, L. A., Pacala and M. Jones, Jr, *J. Org. Chem.*, **45**, 973 (1980).
36. M. P. Cava, D. Mangold and K. Muth, *J. Org. Chem.*, **29**, 2947 (1964).
37. J. F. W. McOmie and D. H. Perry, *J. Chem. Soc., Chem. Commun.*, 248 (1973).
38. D. L. Forster, T. L. Gilchrist, C. W. Rees and E. Stanton, *J. Chem. Soc., Chem. Commun.*, 695 (1971).

39. M. E. Jung and J. A. Lowe, *J. Org. Chem.*, **42**, 2371 (1977).
40. K. J. Gould, N. P. Hacker, J. F. W. McOmie and D. H. Perry, *J. Chem. Soc., Perkin Trans. 1*, 1834 (1980).
41. O. Abou-Teim, R. E. Jansen, J. F. W. McOmie and D. H. Perry, *J. Chem. Soc., Perkin Trans.. 1*, 1841 (1980).
42. M. Azadi-Ardakani and T. W. Wallace, *Tetrahedron Lett.*, **24**, 1829 (1983).
43. A. Roedig, B. Heinrich and V. Kimmel, *Justus Liebigs Ann. Chem.*, 1195 (1975).
44. N. P. Hacker, J. F. W. McOmie, J. Meunier-Piret and M. Van Meerssche, *J. Chem. Soc., Perkin Trans. 1*, 19 (1982).
45. H. N. C. Wong and F. Sondheimer, *Tetrahedron*, **37**(S1), 99 (1981).
46. H. N. C. Wong, F. Sondheimer, R. Goodin and R. Breslow, *Tetrahedron Lett.*, 2715 (1976).
47. M. P. Cava and K. W. Ratts, *J. Org. Chem.*, **27**, 752 (1962).
48. D. McIntyre, G. R. Proctor and L. Rees, *J. Chem. Soc. (C)*, 985 (1966).
49. M. Oda, H. Miyazaki, Y. Kayama and Y. Kitahara, *Chem. Lett.*, 627 (1975).
50. M. Nôgrádi, W. D. Ollis and I. O. Sutherland, *Acta Chim. Acad. Sci. Hung.*, **78**, 311 (1973).
51. E. Ghera, Y. Gaoni and S. Shoua, *J. Am. Chem. Soc.*, **98**, 3627 (1976).
52. J. Tsunetsugu, M. Sato and S. Ebine, *J. Chem. Soc., Chem. Commun.*, 363 (1973).
53. J. Tsunetsugu, M. Sugahara, K. Heima, Y. Ogawa, M. Kosugi, M. Sato and S. Ebine, *J. Chem. Soc., Perkin Trans. 1*, 1983 (1983).
54. P. A. Chaloner, A. B. Holmes, M. A. McKervey and R. A. Raphael, *Tetrahedron Lett.*, 265 (1975).
55. P. A. Chaloner, A. B. Holmes, M. A. McKervey and R. A. Raphael, *J. Chem. Soc., Perkin Trans. 1*, 2524 (1977).
56. S. P. Pappas and B. C. Pappas, *Tetrahedron Lett.*, 1597 (1967).
57. P. Yates and G. V. Nair, *Syn. Commun.*, **3**, 337 (1973).
58. M. Oda, Y. Kayama and Y. Kitahara, *Tetrahedron Lett.*, 2019 (1974).
59. Y. Kayama, M. Oda and Y. Kitahara, *Tetrahedron Lett.*, 3293 (1974).
60. M. Oda, Y. Kayama, H. Miyazaki and Y. Kitahara, *Angew. Chem., Int. Ed. Engl.*, **14**, 418 (1975).
61. V. I. Bendall, *Dissertation Abstr.*, **25**, 4400 (1965).
62. P. Yates, E. G. Lewars and P. H. McCabe, *Can. J. Chem.*, **48**, 788 (1970).
63. L. A. Carpino and J.-H. Tsao, *J. Org. Chem.*, **44**, 2387 (1979).
64. M. Oda, M. Oda and Y. Kitahara, *Tetrahedron Lett.*, 839 (1976).
65. Y. Kitahara, M. Oda and M. Oda, *J. Chem. Soc., Chem. Commun.*, 446 (1976).
66. T. R. Kowar and E. LeGoff, *J. Org. Chem.*, **41**, 3760 (1976).
67. Y. Kitahara, M. Oda, S. Miyakoshi and S. Nakanishi, *Tetrahedron Lett.*, 2149 (1976).
68. Y. Kitahara, M. Oda and S. Miyakoshi, *Tetrahedron Lett.*, 4141 (1975).
69. M. Oda, M. Oda, S. Miyakoshi and Y. Kitahara, *Chem. Lett.*, 293 (1977).
70. P. Yates and E. G. Lewars, *J. Chem. Soc., Chem. Commun.*, 1537 (1971).
71. P. Yates, E. G. Lewars and P. H. McCabe, *Can. J. Chem.*, **50**, 1548 (1972).
72. Y. Itô, M. Oda and Y. Kitahara, *Tetrahedron Lett.*, 239 (1975).
73. T. Echter and H. Meier, *Chem. Ber.*, **118**, 182 (1985).
74. T. Kojima, Y. Sakata and S. Misumi, *Bull. Chem. Soc. Japan*, **45**, 2834 (1972).
75. K. Yamamoto and F. Sondheimer, *Angew. Chem., Int. Ed. Engl.*, **12**, 68 (1973).
76. N. Darby, K. Yamamoto and F. Sondheimer, *J. Am. Chem. Soc.*, **96**, 248 (1974).
77. R. Breslow, D. Murayama, R. Drury and F. Sondheimer, *J. Am. Chem. Soc.*, **96**, 249 (1974).
78. L. Lombardo and F. Sondheimer, *Tetrahedron Lett.*, 3841 (1976).
79. F. Gerson, H.-D. Beckhaus and F. Sondheimer, *Israel J. Chem.*, **20**, 240 (1980).
80. T. Kashitani, S. Akiyama, M. Iyoda and M. Nakagawa, *J. Am. Chem. Soc.*, **97**, 4424 (1975).
81. Y. Onishi, M. Iyoda and M. Nakagawa, *Tetrahedron Lett.*, **22**, 3641 (1981).
82. M. Iyoda, Y. Onishi and M. Nakagawa, *Tetrahedron Lett.*, **22**, 3645 (1981).
83. D. Eggerding and R. West, *J. Am. Chem. Soc.*, **97**, 207 (1975).
84. J. Aihara, *J. Am. Chem. Soc.*, **103**, 1633 (1981).
85. Z. Yoshida, H. Konishi, and H. Ogoshi, *Israel J. Chem.*, **21**, 139 (1981); Z. Yoshida, *Top. Curr. Chem.*, **40**, 47 (1973).
86. Z. Yoshida and Y. Tawara, *J. Am. Chem. Soc.*, **93**, 2573 (1971).
87. R. Weiss and C. Priesner, *Angew. Chem. Int. Ed. Engl.*, **17**, 445 (1978).
88. R. Weiss, K.-G. Wagner, G. Priesner and J. Macheleid, *J. Am. Chem. Soc.*, **107**, 4491 (1985).

89. Z. Yoshida, S. Mike and S. Yoneda, *Tetrahedron Lett.*, 4731 (1973).
90. Z. Yoshida, H. Konishi, S. Wawada and H. Ogoshi, *J. Chem. Soc., Chem. Commun.*, 850 (1977).
91. R. Weiss, C. Priesner and H. Wolf, *Angew. Chem. Int. Ed. Engl.*, **17**, 446 (1978).
92. Z. Yoshida, S. Arakai and H. Ogoshi, *Tetrahedron Lett.*, 19 (1975).
93. R. Weiss, M. Hertel and H. Wolf, *Angew. Chem. Int. Ed. Engl.*, **18**, 473 (1979).
94. T. Fukunaga, *J. Am. Chem. Soc.*, **98**, 610 (1976) T. Fukunaga, M. D. Gordon and P. J. Krusic, *J. Am. Chem. Soc.*, **98**, 611 (1976).
95. R. West and D. C. Zecher, *J. Am. Chem. Soc.*, **89**, 152, 153 (1967); **92**, 155, 161 (1970).
96. L. A. Wendling and R. West, *J. Org. Chem.*, **43**, 1573 (1978).
97. L. A. Wendling and R. West, *J. Org. Chem.*, **43**, 1577 (1978).
98. R. West, D. C. Zecher, C. K. Koster and D. Eggerding, *J. Org. Chem.*, **40**, 2295 (1975).
99. J. L. Benham, R. West and J. A. T. Norman, *J. Am. Chem. Soc.*, **102**, 5047 (1980).
100. J. L. Benham and R. West, *J. Am. Chem. Soc.*, **102**, 5054 (1980).
101. K. Komatsu, R. West and D. Beyer, *J. Am. Chem. Soc.*, **99**, 6290 (1977).
102. W. A. Huffman, S. P. Birkeland and K. P. O'Leary, U. S. Patent 4,052,209 (October, 1977).
103. R. Gleiter, W. Dobler and M. Eckert-Maksic, *Nouv. J. Chim.*, **6**, 123 (1982).
104. M. Kasai, M. Oda and Y. Kitahara, *Chem. Lett.*, 217 (1978).
105. H. Irngartinger, M. Nixdorf, W. Dobler and R. Gleiter, *Acta Cryst.*, **C40**, 1481 (1984).
106. W. Ried and M. Vogl, *Chem. Ber.*, **115**, 783 (1982).
107. A. J. Fatiadi, in Ref. 17, p. 59.
108. S. Skujins, J. Delderfield, and G. A. Webb, *Tetrahedron*, **24**, 4805 (1968).
109. W. Städeli, R. Hollenstein, and W. V. Philipsborn, *Helv. Chim. Acta*, **60**, 948 (1977).
110. T. Laird, in *Comprehensive Organic Chemistry* (Eds. D. H. R. Barton and W. D. Ollis), Pergamon, Oxford, 1979, Vol. 1, p. 1205.
111. R. F. C. Brown and R. K. Solly, *Aust. J. Chem.*, **19**, 1045 (1966).
112. W. Bolton, *Acta Cryst.*, **18**, 5 (1965).
113. A. Schoenberg and E. Singer, *Tetrahedron*, **34**, 1285 (1978).
114. J. Nakayama, J. Miyoko and H. Masamatsu, *Chem. Lett.*, 77 (1977).
115. R. K. Behera and A. Nayak, *Indian J. Chem.*, **14B**, 223 (1976).
116. R. B. Prichard, C. E. Lough, D. J. Currie and H. L. Holmes, *Can. J. Chem.*, **46**, 775 (1968).
117. C. F. Bernasconi, A. Laibelman and J. L. Zitomer, *J. Am. Chem. Soc.*, **107**, 6563 (1985).
118. C. F. Bernasconi, A. Laibelman and J. L. Zitomer, *J. Am. Chem. Soc.*, **107**, 6573 (1985).
119. K. K. Canon, Japan Kokai Tokkyo Koho JP Patent, 59 14, 150, 59, 60, 443 April 1984.
120. A. J. Fatiadi, H. S. Isbell and W. F. Sager, *J. Res. Natl. Bureau of Standards*, **67A**, 153 (1963).
121. S. Kuroda and T. Asao, *Tetrahedron Lett.*, 285 (1977).
122. M. Hirama and S. Itô, *Tetrahedron Lett.*, 1071 (1975).
123. M. Hirama, Y. Fukazawa, and S. Itô, *Tetrahedron Lett.*, 1299 (1978).
124. R. Gleiter, W. Dobler, and M. Eckert-Maksic, *Angew. Chem.*, **94**, 62 (1982).
125. M. Hirama, A. Kawamata and S. Itô, *Chem. Lett.*, 855 (1979).
126. T. Toda, H. Horino, T. Mukai and T. Nozoe, *Tetrahedron Lett.*, 1387 (1968).
127. T. Ide, K. Imafuku and H. Matsumura, *Chem. Lett.*, 717 (1977).
128. A. Kawamata, E. Kikuchi, M. Hirama, Y. Fujise and S. Itô, *Chem. Lett.*, 859 (1979).
129. W. Ried and R. Conte, *Chem. Ber.*, **105**, 799 (1972).
130. K. Takahashi, K. Morita and K. Takase, *Tetrahedron Lett.*, 1511 (1977).
131. J. Nakayama, I. Miyoko and H. Masamatsu, *Chem. Lett.*, 287 (1977).
132. S. Itô, Y. Shoji, H. Takeshita, M. Hirama and K. Takahashi, *Tetrahedron Lett.*, 1075 (1975).
133. M. Hirama and S. Itô, *Tetrahedron Lett.*, 2339 (1976).
134. M. Hirama and S. Itô, *Chem. Lett.*, 627 (1977).
135. H. Takeshita, T. Kusaba and A. Mori, *Bull. Chem. Soc. Japan*, **55**, 1659 (1982).
136. W. T. Dixon and D. Murphy, *J. Chem. Soc., Perkin Trans. 2*, 1430 (1964).
137. M. Hirama, Y. Koyama, Y. Shoji and S. Itô, *Tetrahedron Lett.*, 2289 (1978).
138. J. Rigaudy and L. Nedelec, *Bull. Soc. Chim. France*, 655 (1955).
139. H. Wittmann, V. Illi, H. Rathmayer, H. Sterk and E. Ziegler, *Naturforsch.*, **27B**, 524, 528 (1972).
140. T. Nozoe, M. Sato and T. Matsuda, *Sci. Repts. Tohoku Univ., Ser. 1*, **37**, 407 (1953); S. Itô, *Sci. Repts. Tohoku Univ., Ser. 1*, **42**, 236, 247 (1959).
141. T. Nozoe, S. Itô, S. Suzuki and K. Hiraga, *Proc. Japan Acad.*, **32**, 344 (1956).
142. A. Kawamoto, Y. Fukazawa, Y. Fujise and S. Itô, *Tetrahedron Lett.*, **23**, 1083 (1982).

143. K. Takahashi, K. Morita and K. Takase, *Chem. Lett.*, 1505 (1977).
144. K. Takahashi, T. Fujita and K. Takase, *Tetrahedron Lett.*, 4507 (1971).
145. K. Takahashi and K. Takase, *Tetrahedron Lett.*, 2227 (1972).
146. C. S. Maxwell and C. F. H. Allen, *Org. Syn. Coll. Vol. 3*, 1 (1955).
147. A. K. Das and A. K. Sinha, *J. Indian Chem. Soc.*, **51**, 569 (1974).
148. K. D. Banerji, A. K. D. Mazumdar and S. K. Guha, *J. Indian Chem. Soc.*, **54**, 969 (1977).
149. E. K. Dora, B. Dash and C. S. Panda, *J. Indian Chem. Soc.*, **56**, 620 (1979).
150. E. S. H. El Ashry, M. A. M. Nassr, A. A. Abdallah and M. Shoukry, *Indian J. Chem.*, **19B**, 612 (1980).
151. M. W. Davidson and D. W. Boykin, *J. Pharm. Sci.*, **67**, 737 (1978).
152. V. J. Ram, H. K. Pandey and A. J. Vlietinck, *J. Heterocyclic Chem.*, **17**, 1305 (1980).
153. V. J. Ram, H. K. Pandey and A. J. Vlietinck, *J. Heterocyclic Chem.*, **18**, 15 (1981).
154. K. D. Banerji, A. K. D. Mazumda, K. Kuma and S. K. Guha, *J. Indian Chem. Soc.*, **56**, 396 (1979).
155. C. H. Issidorides, M. A. Atfah, J. J. Sabounji, A. R. Sidani and M. J. Haddadin, *Tetrahedron*, **34**, 217 (1978).
156. F. A. Neugebauer, H. Fischer, R. Siegel and C. Krieger, *Chem. Ber.*, **116**, 3461 (1983).
157. E. H. Vickery and E. J. Eisenbrau, *Org. Prep. Proc. Int.*, **11**, 259 (1979).
158. A. Minsky and M. Rabinovitz, *Synthesis*, 487 (1983).
159. G. I. Fray, G. R. Geen, K. Mackenzie and D. L. William-Smith, *Tetrahedron*, **35**, 1173 (1979).
160. R. L. Letsinger and J. A. Gilpin, *J. Org. Chem.*, **29**, 243 (1964).
161. W. C. Agosta, *J. Am. Chem. Soc.*, **89**, 3505 (1967); *Tetrahedron Lett.*, 3635 (1966).
162. R. H. Mitchell, T. Fyles and L. M. Ralph, *Can. J. Chem.*, **55**, 1480 (1977).
163. A. R. Miller, *J. Org. Chem.*, **44**, 1931 (1979).
164. A. R. Miller and D. Y. Curtin, *J. Am. Chem. Soc.*, **98**, 1860 (1976).
165. L. S. Hegedus, S. D. Wagner, E. L. Waterman and K. Siirala-Hansen, *J. Org. Chem.*, **40**, 593 (1975).
166. Y. Naruta, *J. Am. Chem. Soc.*, **102**, 3774 (1980).
167. J. Simonet and H. Lund, *Bull. Soc. Chim. France*, 2547 (1975).
168. M. Rabinovitz and D. Tamarkin, *Syn. Commun.*, **14**, 333 (1984).
169. K. Boujlel and J. Simonet, *Tetrahedron Lett.*, **20**, 1497 (1979).
170. M. S. Singh and K. N. Mehrota, *Indian J. Chem.*, **23B**, 1289 (1984).
171. A. Merz, F. Dietl, R. Tomahogh, G. Weber and G. M. Sheldrick, *Tetrahedron*, **40**, 665 (1984).
172. K. Maruyama, K. Ono and J. Osugi, *Bull. Chem. Soc. Japan*, **45**, 847 (1972).
173. J.-Y. Koo and G. B. Schuster, *J. Org. Chem.*, **44**, 847 (1979).
174. H. Takeshita, A. Mori, M. Funakura and H. Mametsuka, *Bull. Chem. Soc. Japan*, **50**, 315 (1977).
175. T. Sasaki, K. Kanematsu, I. Ando and O. Yamashita, *J. Am. Chem. Soc.*, **99**, 871 (1977).
176. C. G. Bakker, J. W. Scheeren and R. J. F. Nivard, *Recl. Trav. Chim. Pays-Bas*, **102**, 96 (1983).
177. S.-J. Chang, *Synth. Commun.*, **12**, 673 (1982).
178. M. P. Cava, R. L. Little and D. R. Napier, *J. Am. Chem. Soc.*, **80**, 2257 (1958).
179. W. Ried and H. Lohwasser, *Justus Liebigs Ann. Chem.*, **683**, 118 (1965).
180. O. Tsuge, I. Shinkai and M. Koga, *J. Org. Chem.*, **36**, 745 (1971).
181. K. Okada and K. Mukai, *Tetrahedron Lett.*, **21**, 359 (1980).
182. O. L. Chapman, *Chem. Eng. News*, **56** (38), 17 (1978).
183. O. L. Chapman, *Pure Appl. Chem.*, **51**, 331 (1979).
184. S.-J. Chang, B. K. Ravi Shankar and H. Shechter, *J. Org. Chem.*, **47**, 4226 (1982).
185. R. A. Hayes, T. C. Hess, R. J. McMahon and O. L. Chapman, *J. Am. Chem. Soc.*, **105**, 7786 (1983).
186. R. J. McMahon, O. L. Chapman, R. A. Hayes, T. C. Hess and H.-P. Krimmer, *J. Am. Chem. Soc.*, **107**, 7597 (1985).
187. B. M. Trost, *J. Am. Chem. Soc.*, **88**, 853 (1966).
188. B. M. Trost, *J. Am. Chem. Soc.*, **91**, 918 (1969).
189. B. M. Trost and S. F. Nelsen, *J. Am. Chem. Soc.*, **88**, 2876 (1966).
190. S. F. Nelsen, B. M. Trost and D. H. Evans, *J. Am. Chem. Soc.*, **89**, 3034 (1967).
191. S. F. Nelsen and B. M. Trost, *Tetrahedron Lett.*, 5737 (1966).
192. F. M. Beringer, R. E. K. Winter and J. A. Castellano, *Tetrahedron Lett.*, 6183 (1968).
193. J. A. Castellano, F. M. Beringer and R. E. K. Winter, *J. Org. Chem.*, **37**, 3151 (1972).

194. M. M. Sidky, F. M. Soliman and A. A. El-Kateb, *Indian J. Chem.*, **14B**, 961 (1976).
195. E. J. Corey and R. A. Winter, *J. Am. Chem. Soc.*, **85**, 2677 (1963).
196. N. Tanaka and T. Kasai, *Bull. Chem. Soc. Japan*, **54**, 3020 (1981).
197. J. Tsunetsugu, M. Sato, M. Kanda, M. Takahashi and S. Ebine, *Chem. Lett.*, 885 (1977).
198. J. Tsunetsugu, M. Kanda, M. Takahashi, K. Yoshida, H. Koyama, K. Shiraishi, Y. Tankano, M. Sato and S. Ebine, *J. Chem. Soc., Perkin Trans. 1*, 1465 (1984).
199. R. Gleiter and W. Dobler, *Chem. Ber.*, **118**, 4725 (1985).
200. J. Tsunetsugu, T. Ikeda, N. Suzuki, M. Yaguchi, M. Sato, S. Ebine and K. Morinaga, *J. Chem. Soc., Chem. Commun.*, 28 (1983).
201. P. S. Kinson and B. M. Trost, *J. Am. Chem. Soc.*, **93**, 3823 (1971).
202. B. M. Trost and P. L. Kinson, *J. Am. Chem. Soc.*, **97**, 2438 (1975).
203. G. Ashworth, D. Berry and D. C. C. Smith, *J. Chem. Soc., Perkin Trans. 1*, 2995 (1977).
204. E. Vogel, E. Lohmar, W. A. Böll, B. Sohngen, K. Müllen and H. Günther, *Angew. Chem. Int. Ed. Engl.*, **10**, 398 (1971).
205. F. Gerson, K. Müllen and E. Vogel, *Helv. Chim. Acta*, **54**, 1046 (1971).
206. E. Vogel, W. A. Böll and E. Lohmar , *Angew. Chem. Int. Ed. Engl.*, **10**, 399 (1971).
207. R. Neidlein, C. M. Radke and R. Gottfried, *Chem. Lett.*, 653 (1983).
208. R. Neidlein and C. M. Radke, *Helv. Chim. Acta*, **66**, 2369 (1983).
209. R. Neidlein and C. M. Radke, *Helv. Chim. Acta*, **66**, 2621 (1983).
210. R. Neidlein and C. M. Radke, *Helv. Chim. Acta*, **66**, 2626 (1983).

Author Index

Note – This author index is designed to enable the reader to locate an author's name and work with the aid of the reference numbers appearing in the text. The page numbers are printed in normal type in ascending numerical order, followed by the reference number in parentheses. The numbers in *italics* refer to the pages on which the references are actually listed.

Subject Index